The IMA Volumes in Mathematics and its Applications

Volume 159

More information about this series at http://www.springer.com/series/811

Institute for Mathematics and its Applications (IMA)

The Institute for Mathematics and its Applications (IMA) was established in 1982 as a result of a National Science Foundation competition. The mission of the IMA is to connect scientists, engineers, and mathematicians in order to address scientific and technological challenges in a collaborative, engaging environment, developing transformative, new mathematics and exploring its applications, while training the next generation of researchers and educators. To this end the IMA organizes a wide variety of programs, ranging from short intense workshops in areas of exceptional interest and opportunity to extensive thematic programs lasting nine months. The IMA Volumes are used to disseminate results of these programs to the broader scientific community.

The full list of IMA books can be found at the Web site of the Institute for Mathematics and its Applications:

http://www.ima.umn.edu/springer/volumes.html.

Presentation materials from the IMA talks are available at

http://www.ima.umn.edu/talks/.

Video library is at

http://www.ima.umn.edu/videos/.

Fadil Santosa, Director of the IMA

Andrew Beveridge • Jerrold R. Griggs
Leslie Hogben • Gregg Musiker • Prasad Tetali
Editors

Recent Trends in Combinatorics

Springer

Editors
Andrew Beveridge
Department of Mathematics, Statistics
and Computer Science
Macalester College
St. Paul, MN, USA

Leslie Hogben
Department of Mathematics
Iowa State University
Ames, IA, USA

Prasad Tetali
School of Mathematics
and Computer Science
Georgia Institute of Technology
Atlanta, GA, USA

Jerrold R. Griggs
Department of Mathematics
University of South Carolina
Columbia, SC, USA

Gregg Musiker
School of Mathematics
University of Minnesota
Minneapolis, MN, USA

ISSN 0940-6573 ISSN 2198-3224 (electronic)
The IMA Volumes in Mathematics and its Applications
ISBN 978-3-319-24296-5 ISBN 978-3-319-24298-9 (eBook)
DOI 10.1007/978-3-319-24298-9

Library of Congress Control Number: 2016932007

Mathematics Subject Classification (2010): 05

Printed on acid-free paper

This Springer imprint is published by Springer Nature
The registered company is Springer International Publishing AG Switzerland

Foreword

This volume is based on the research focus at the IMA during the Fall semester of 2014. The Annual Thematic Program covering this period was "Discrete Structures: Analysis and Applications". The program was organized by Sergey Bobkov, Jerrold Griggs, Penny Haxell, Michel Ledoux, Benny Sudakov, and Prasad Tetali. Many of the topics presented in this volume were discussed in the first three workshops that took place during the year. We thank the organizers of the workshops, the speakers, workshop participants, and visitors to the IMA who contributed to the scientific life at the institute and to the successful program. In particular, we thank Andrew Beveridge, Jerrold Griggs, Leslie Hogben, Gregg Musiker, and Prasad Tetali for taking the lead to edit this volume. We also thank the National Science Foundation for its support of the IMA.

Minneapolis, MN, USA Fadil Santosa

Preface

Combinatorics is a research field driven by collaboration, with a large number of applications to different areas of pure and applied mathematics. The Institute for Mathematics and Its Applications (IMA) is an ideal setting for such collaborations and applications to develop.

The 2014–2015 Annual Thematic Program at the IMA was *Discrete Structures: Analysis and Applications*. The program was organized by Sergey Bobkov (University of Minnesota), Jerrold Griggs (University of South Carolina), Penny Haxell (University of Waterloo), Michel Ledoux (Paul Sabatier University of Toulouse), Benny Sudakov (University of California, Los Angeles), and Prasad Tetali (Georgia Institute of Technology).

Combinatorics was the focus during Fall 2014, and this volume presents some of the research topics discussed during this intense semester. We have particularly encouraged authors to write surveys of research problems, thus making state-of-the-art results more conveniently and widely available.

This volume is organized into parts, following the themes of the three workshops held during Fall 2014:

- *Probabilistic and Extremal Combinatorics*, held September 8–12, 2014, at IMA and organized by Penny Haxell (University of Waterloo), Eyal Lubetzky (Microsoft Research), Dhruv Mubayi (University of Illinois, Chicago), and Benny Sudakov (Eidgenössische TH Zürich-Zentrum).
- *Additive and Analytic Combinatorics*, held September 29–October 3, 2014, at IMA and organized by David Conlon (University of Oxford), Ernie Croot (Georgia Institute of Technology), Van Vu (Yale University), and Tamar Ziegler (Hebrew University).
- *Geometric and Enumerative Combinatorics*, held November 10–14, 2014, at IMA and organized by Zoltan Furedi (Hungarian Academy of Sciences MTA), Jerrold Griggs (University of South Carolina), Victor Reiner (University of Minnesota, Twin Cities), and Carla Savage (North Carolina State University).

Part 1: Extremal and probabilistic combinatorics. Extremal and probabilistic combinatorics are central to modern combinatorial theory, and both have developed dramatically over the last few decades. Extremal combinatorics studies problems of finding the maximum or minimum possible cardinality of a set of finite objects satisfying certain requirements. Frequently such problems originate in other areas, such as computer science, information theory, analysis, number theory, and geometry. Probabilistic combinatorics, as the name suggests, blends combinatorics and probability. It is the foundation of the study of random graphs and other random discrete structures, and probabilistic arguments have been very powerfully applied to problems in other areas of combinatorics and in theoretical computer science. Major research topics in extremal and probabilistic combinatorics include extremal problems for graphs and set systems, Ramsey theory, random graphs, and application of probabilistic methods.

Part 2: Additive and analytic combinatorics. Additive combinatorics counts additive structures in sets; there have been exciting developments in recent years. Tools from Fourier and harmonic analysis have expanded the realm of additive combinatorics into the analytic while contributing to more effective applications. Many combinatorial ideas known to the combinatorics community can be used effectively to attack difficult problems in other areas of mathematics. For example, a famous theorem of Szemerédi on arithmetic progressions in dense sets is a key tool for the proof of the Green-Tao theorem on the existence of long arithmetic progressions in primes. The work of Breuillard, Green, and Tao, which established an analogue of the Freiman inverse theorem for noncommutative groups, is another example. This theorem was first stated and proved for integers by Freiman. Ruzsa's subsequent different proof was extended to abelian groups by Green and Ruzsa a few years ago. The extension to noncommutative groups is much more difficult. Research in additive and analytic combinatorics is also of interest to computer scientists; techniques and results have been applied to communication complexity, property testing, and the design of randomness extractors.

Part 3: Enumerative and geometric combinatorics. Geometric combinatorics studies discrete objects with geometric or topological structure, such as convex polytopes, arrangements of vectors, points, subspaces, triangulations, tilings, and partially ordered sets. Enumerative combinatorics, often called the mathematics of counting, has broad applications to probability, statistical physics, optimization, and computer science. Problems in geometric combinatorics give rise to counting problems that are sometimes difficult even to estimate and sometimes involve objects with interesting symmetry groups. Such problems often dovetail nicely with topics from enumerative combinatorics via calculations of partition functions, f-vectors, Ehrhart polynomials, and other quantities. Enumerative combinatorics also includes the study of permutation patterns, the complexity of tilings, and bijections between families of objects counted by the same numerical sequences or with related generating functions. In recent years, problems in both of these areas have stimulated the development of many new results and tools and enhanced connections with other areas of mathematics.

Discrete Structures: Analysis and Applications attracted intense interest from the mathematical sciences community, with each of the three workshops drawing more than 100 visitors and often filling Keller 3-180 to capacity. There are many other aspects to an annual thematic year at the IMA besides workshops, with the relaxed but stimulating environment of the IMA fostering new collaborations and approaches to solving problems old and new. This program drew an eclectic mix of experts and junior researchers in various aspects of combinatorics together with numerous people who apply combinatorics to other fields. This volume reflects many of the aspects of the semester, with chapters drawn from workshop talks, annual program seminars, and research interests of the many visitors.

No single volume could possibly cover all the active and important areas of combinatorics research that were presented at the IMA, and we make no claim of comprehensiveness. But we think this volume presents a reasonable selection of interesting areas, written by leading experts who have surveyed the current state of knowledge and posed conjectures and open questions to stimulate further research. We thank the authors for their generous donations of time and expertise; needless to say, without them this volume would not have been possible.

We thank the IMA for wonderfully stimulating and productive long-term visits. We believe that the IMA is a critical national resource for mathematics. The *Discrete Structures: Analysis and Applications* program will have a lasting impact on research in combinatorics and related fields, and we hope this volume will enhance that impact. We are grateful for the opportunity to be part of it.

St. Paul, MN, USA — Andrew Beveridge
Columbia, SC, USA — Jerrold R. Griggs
Ames, IA, USA — Leslie Hogben
Minneapolis, MN, USA — Gregg Musiker
Atlanta, GA, USA — Prasad Tetali

Contents

Part I
Extremal and probabilistic combinatorics

Problems related to graph indices in trees

László Székely, Stephan Wagner, and Hua Wang

Abstract In this chapter we explore recent development on various problems related to graph indices in trees. We focus on indices based on distances between vertices, vertex degrees, or on counting vertex or edge subsets of different kinds. Some of the indices arise naturally in applications, e.g., in chemistry, statistical physics, bioinformatics, and other fields, and connections are also made to other branches of graph theory, such as spectral graph theory. We will be particularly interested in the extremal values (maxima and minima) for different families of trees and the corresponding extremal trees. Moreover, we review results for random trees, consider localized versions of different graph indices and the associated notions of centrality, and finally discuss inverse problems, where one wants to find trees for which a specific graph index has a prescribed value.

Keywords Wiener index • Randić index • Merrifield-Simmons index • Hosoya index • Random tree • Inverse problems • Dominating set • Number of subtrees

Mathematics Subject Classification (2000): Primary: 05C05, Secondary 05C12, 05C31, 05C35, 05C69, 05C70, 05C80, 92E10

1 Introduction

Enumeration of trees [27, 116] and of spanning trees [80] goes back to the XIXth century. The concepts of the center and centroid are also that old [77]. However, the topic of this survey really starts with the papers of Wiener [153, 154]. He noted

L. Székely (✉)
Department of Mathematics, University of South Carolina, Columbia, SC, USA
e-mail: laszlo@mailbox.sc.edu; szekely@math.sc.edu

S. Wagner
Department of Mathematical Sciences, Stellenbosch University, Stellenbosch, South Africa
e-mail: swagner@sun.ac.za

H. Wang
Department of Mathematical Sciences, Georgia Southern University, Statesboro, GA, USA
e-mail: hwang@georgiasouthern.edu

A. Beveridge et al. (eds.), *Recent Trends in Combinatorics*, The IMA Volumes in Mathematics and its Applications 159, DOI 10.1007/978-3-319-24298-9_1

that the boiling temperatures of alkanes correlate with the sum of the lengths of the shortest paths between all pairs of vertices in the chemical graph representing the non-hydrogen atoms in the molecule. This quantity (now called the Wiener index), and its variants will be discussed in Section 2.

The Wiener index started chemical graph theory. Graphs arise as representation of atoms with vertices, bonds with edges, usually suppressing hydrogen atoms. (Such representation appeared already in [26], as pointed out in [12].) The point of chemical graph theory is to come up with graph invariants, which have predictive power for chemical properties of the molecule, if computed for the molecular graph. In chemical graph theory such invariants are called (topological) indices, as the expectation is that the shape of the molecule is the ultimate source of information. The discriminating power of an index is high, if different tree shapes tend to show distinct indices. The range of an index limits its discriminating power. A chemical graph (resp. tree) is understood to have maximum degree four, reflecting the valence of a carbon atom. Several books are fully or partly devoted to chemical graph theory [8, 132, 133]. Our survey is focused on indices of trees although many of these indices have been investigated on graphs as well. This survey does not even try to be complete, as the number of indices is exploding: one of the main organizing principles of selection is the interest of the authors.

Randić [111] introduced a very influential index, which is now named after him. Earlier it was called the branching index or connectivity index. The Randić index, which is the prototype of degree-based indices, and its variants will be reviewed in Section 4. The survey papers [58] and [94] are particularly informative about degree-based indices. There are even three books devoted to the Randić index [78, 79, 95]. Unexpectedly, the generalized Randić index turned useful in an entirely different context: to find an analogue of the Crossing Lemma to set lower bounds for the minor crossing number of graphs [14].

The Merrifield-Simmons index of a graph is the number of its independent vertex sets [103] and the Hosoya index (also called topological Z index) of a graph is the number of its independent edge sets, i.e., matchings without size restriction [66]. These quantities have been relevant in statistical physics in the hard square model [10] and the monomer-dimer model [64]. The number of subtrees of a tree came from bioinformatics [85], where the number of subtrees of a binary tree containing at least one leaf was involved in the complexity of an algorithm. The maximum of the number of subtrees among binary trees was determined in [125], and the systematic study of this quantity determined these maximum values exactly using a novel number representation [124]. The original problem of determining the maximum number of subtrees of a binary tree containing at least one leaf was also solved [123]. It turned out that for several classes of trees, those trees that minimize the Wiener index happen to maximize the number of subtrees and vice versa. A similar phenomenon is present for the Merrifield-Simmons and Hosoya indices, see Section 3. The paper [136] contains an analysis of the correlation between pairs of tree indices, and found the highest (negative) correlation between the Wiener index and the number of subtrees. An intriguing problem would be to give an "explanation" for the negative correlation between the number of subtrees and the

Wiener index—although we admit that we cannot tell criteria for a "satisfactory" explanation.

From a practical point of view, identifying the extremal structures alone is far from being sufficient for understanding the behavior of an index. A natural and important question appears to be the following.

For a given index, what is the distribution of its value over all possible trees under given restriction (i.e. of given order, with given degree sequence, etc..)?

The answer to this question more specifically presents the behavior of a graph index and hence can be used to further examine the similarities or differences between different indices.

How similar or different are two indices according to the distributions of their values over different categories of trees?

Random trees are a huge topic of their own right, there is a whole book [43] devoted to them. In Section 5 we focus on only a few questions about expected values and distributions of tree indices that we consider in this survey.

There are several centrality concepts for trees. Two of them, the center and the centroid, are age-old [77]. Section 6 investigates the behavior of some local versions of the indices in central positions compared to the behavior elsewhere.

Lepović and Gutman [91] made the beautiful conjecture that almost all positive integers are the Wiener indices of some trees. This conjecture was verified about a decade later, in [135] and [152]. The conference paper [55] proposed a more general problem, the inverse problem of indices in combinatorial chemistry: when we have to synthesize a new molecule with expected properties, we may want to create first molecular graphs with prescribed indices to narrow the search. Inverse problems will be discussed in Section 7.

2 Distance-Based Graph Indices

The most classic and widely used distance-based index is the *Wiener index*, named after the chemist Harry Wiener, who proposed this concept in 1947 [153, 154]. For a graph G, the Wiener index of G is

$$W(G) = \sum_{u,v \in V(G)} d(u,v)$$

where $d(u, v)$ is the distance between vertices u and v in G. The mathematical examination of this concept frequently happened independently of applications in chemistry. Because of the many acyclic molecular structures in applications of the Wiener index, the study of its behavior in trees has been of particular interest. Over the past several decades, many interesting results have been obtained regarding the Wiener index of trees under various restrictions as well as variations of the Wiener index. An early informative survey on the Wiener index is [42].

2.1 The Wiener index

Among general trees of given order, the trees with maximum and minimum Wiener index have been characterized:

Theorem 1 ([44], [99, Ex. 6.23]). *Among all trees of the same order, the star minimizes the Wiener index and the path maximizes the Wiener index.*

As we will see, the star and the path are the extremal trees for many other indices as well. In fact, [30] argues that any acceptable branching index should attain the unique minimum for a star and the unique maximum for the path (or vice versa) among trees on the same number of vertices. If the degrees of vertices in the graph correspond to valences of atoms in a molecule, then they are severely restricted. Hence it is relevant to consider the extremal values of the Wiener index when the tree has a bounded maximum degree. Fischermann, Hoffmann, Rautenbach, Székely and Volkmann [52], and independently Jelen and Triesch [76] identified the trees with minimum Wiener index among all trees (of given order) with a bounded maximum degree, and the trees with maximum Wiener index among all trees (of given order) whose vertex degrees are 1 or k.

The work in [52] was further generalized to trees with a given degree sequence. For a tree T, the *degree sequence* is simply the non-increasing sequence of vertex degrees. In [149] and [164], respectively, it is shown that the minimum Wiener index is attained by the so-called *greedy tree*.

Definition 1 (Greedy trees). Given a tree degree sequence d, the greedy tree is achieved through the following "greedy" algorithm:

i) Start with a single vertex v as the root and give v the appropriate number of children so that it has the largest degree;
ii) Label the neighbors of v as $v_1, v_2, \ldots$, and assign to them the largest available degrees such that $d(v_1) \geq d(v_2) \geq \cdots$;
iii) Label the neighbors of v_1 (except v) as $v_{11}, v_{12}, \ldots$ such that they take all the largest degrees available and that $d(v_{11}) \geq d(v_{12}) \geq \cdots$, then do the same for $v_2, v_3, \ldots$;
iv) Repeat (iii) for all the newly labeled vertices, always start with the neighbors of the labeled vertex with largest degree whose neighbors are not labeled yet.

For example, Fig. 1 shows a greedy tree with degree sequence

$$\{4, 4, 4, 3, 3, 3, 3, 3, 3, 3, 2, 2, 1, \dots, 1\}.$$

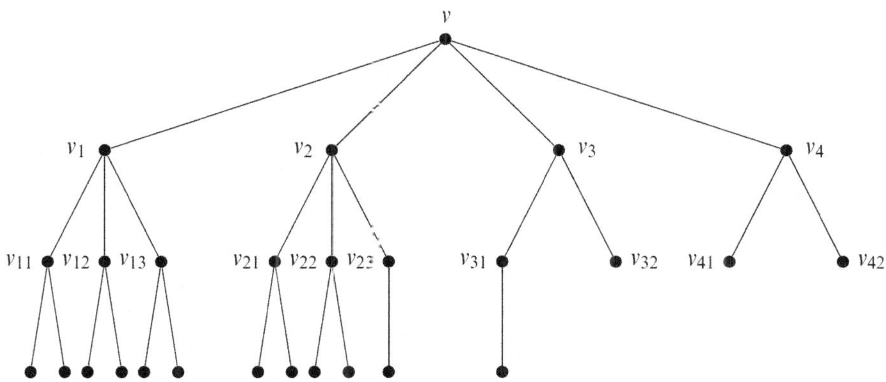

Fig. 1 A greedy tree.

Theorem 2 ([149, 164]). *Among all trees with a given degree sequence (and hence given order), the greedy tree minimizes the Wiener index.*

To maximize the Wiener index among trees with a given degree sequence turned out to be a much more difficult question. Such extremal trees were examined in detail in both [118] and [163]. While the extremal structure was already shown by Shi [117] to be a *caterpillar* (a tree for which the removal of leaves yields a path) with certain properties, the specific characteristics of such extremal trees depend on the particular degree sequence. This question was also examined as a quadratic assignment problem in [28], where an efficient algorithm was provided.

2.2 Variations of the Wiener index

In the past three decades many variations of the Wiener index have been introduced, including, but certainly not limited to the following:

- In 1993, the *hyper-Wiener index* [112] was introduced for trees and later generalized to cyclic graphs [84]:

$$WW(T) = \frac{1}{2} \sum_{u,v \in V(T)} \left(d(u,v) + d(u,v)^2 \right),$$

- the *Harary index* was defined in [71, 109]:

$$H(T) = \sum_{u,v \in V(T)} \frac{1}{d(u,v)},$$

- and the *terminal Wiener index* was proposed in [61]:

$$TW(T) = \sum_{u,v \in L(T)} d(u,v),$$

  where $L(T)$ stands for the set of leaves of T. In addition to its application in chemistry, the terminal Wiener index, being simply the sum of distances between leaves, is also found to be of importance in the study of phylogenetic trees and is known there as the *gamma index* [128].

Much work has been done on these indices. Among general trees of given order, it has been shown that the star minimizes $WW(T)$ [57, 59] and maximizes $H(T)$ [57, 158], while the path maximizes $WW(T)$ [57, 59], minimizes $TW(T)$ [61], and minimizes $H(T)$ [57, 158].

Besides finding extremal trees with the maximum (or minimum) value of these indices, a significant amount of work has also been done on characterizing trees with the first few largest or smallest values of a certain index. For a summary of such studies on the aforementioned indices we suggest Section 3 of [159].

2.3 General results and unified approaches

Given all these different graph indices defined in terms of distances with identical extremal structures, it is natural to explore unified approaches that provide more general results on such extremal questions. While studying $TW(T)$ of trees with given degree sequence, a "semi-regular" property was introduced that is satisfied by the extremal trees with a given degree sequence with respect to many indices [128]. Furthermore, satisfying this semi-regular property forces the tree to be a greedy tree.

Theorem 3 ([128]). *Among all trees of given degree sequence, $TW(T)$ is minimized by the greedy tree.*

Shortly thereafter, it was shown in [115] that if a tree is a *level-greedy tree* (defined below) with respect to any choice of root, then it satisfies the semi-regular property.

Definition 2. [Level-greedy trees] For $i = 0, 1, \ldots, H$, partition a degree sequence into multisets $\{a_{i1}, a_{i2}, \ldots, a_{i\ell_i}\}$ such that $\ell_0 = 1$, $\ell_1 = a_{01}$, and for $1 \le i < H$

$$\ell_{i+1} = \sum_{j=1}^{\ell_i} (a_{ij} - 1).$$

Assume that the elements of each multiset are sorted, i.e., $a_{i1} \geq a_{i2} \geq \cdots \geq a_{i\ell_i}$. The level-greedy tree corresponding to this sequence of multisets is the rooted tree whose j-th vertex at level i has degree a_{ij}.

Likewise, for a degree sequence that is given as sorted multisets $\{a_{i1}, a_{i2}, \ldots, a_{i\ell_i}\}$ for $i = 0, 1, \ldots, H$ such that $\ell_0 = 2$ and for $0 \leq i < H$

$$\ell_{i+1} = \sum_{j=1}^{\ell_i}(a_{ij} - 1),$$

the level-greedy tree corresponding to this sequence is the edge-rooted tree (i.e., there are two vertices at level 0, connected by an edge) whose j-th vertex at level i has degree a_{ij}.

It is obvious that every greedy tree is also level greedy, but not necessarily vice versa. Fig. 2 shows a level-greedy tree with the level degree sequences: $\{a_{01} = 3\}$, $\{a_{11} = 5, a_{12} = 3, a_{13} = 2\}$, $\{3, 3, 3, 2, 2, 1, 1\}$, $\{2, 2, 1, 1, 1, 1, 1, 1\}$, and $\{1, 1\}$. This level-greedy tree is not greedy.

However, it was shown in [115] that a tree that is level-greedy with respect to any choice of vertex root or edge root must indeed be greedy. This led in particular to the following stability results for the extremal trees:

Theorem 4 ([115, 147]). *Let $f(x)$ be any non-negative, non-decreasing function of x. Then the graph invariant*

$$W_f(T) = \sum_{u,v \in V(T)} f(d(u,v))$$

is minimized by the greedy tree among all trees with given degree sequence.

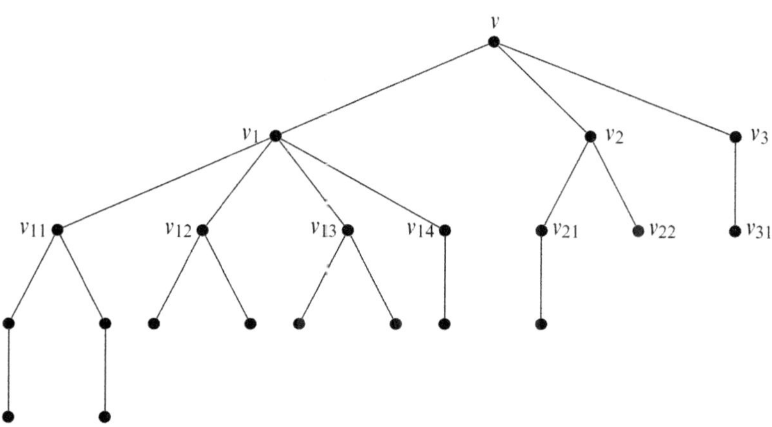

Fig. 2 A level-greedy tree.

Likewise, if $f(x)$ is any non-negative, non-increasing function of x, then the graph invariant W_f is maximized by the greedy tree among all trees with given degree sequence.

With different choices of the function f, the result above provides a general statement for various indices, including the Wiener index, the hyper-Wiener index, and the Harary index.

An important partial order of degree sequences is defined by a relation known as *majorization*:

Definition 3. For non-increasing sequences $\pi = (d_0, \ldots, d_{n-1})$ and $\pi' = (d'_0, \ldots, d'_{n-1})$, π' is said to *majorize* π if $\sum_{i=0}^{n-1} d_i = \sum_{i=0}^{n-1} d'_i$, and for $k = 0, \ldots, n-2$

$$\sum_{i=0}^{k} d_i \leq \sum_{i=0}^{k} d'_i.$$

The notion of majorization provides a means of comparing the extremal trees for different degree sequences, which yields a number of corollaries. Specifically, we have

Theorem 5 ([147]). *Let $f(x)$ be any non-negative, non-decreasing function of x, and let π and π' be two degree sequences of trees of the same length such that π' majorizes π. If $G(\pi)$ and $G(\pi')$ are the greedy trees associated with π and π', respectively, we have*

$$W_f(G(\pi)) \geq W_f(G(\pi')).$$

Likewise, if f is a non-negative, non-increasing function, then

$$W_f(G(\pi)) \leq W_f(G(\pi')).$$

Among other things, this shows that the star is always extremal, that the greedy tree associated with the sequence $(\Delta, \Delta, \ldots, \Delta, r, 1, 1, \ldots, 1)$ is extremal among trees of given order whose maximum degree is Δ (here, $r \in \{1, 2, \ldots, \Delta\}$ is chosen in such a way that the correct number of vertices is obtained), and that the greedy tree associated with the sequence $(\ell, 2, 2, \ldots, 2, 1, 1, \ldots, 1)$ is extremal among trees with exactly ℓ leaves. A similar comparison works for the Merrifield-Simmons and Hosoya indices, although the extremal trees in those cases are not greedy trees, see Section 3.

2.4 Other distance-based indices

Besides variations of the Wiener index, some other distance-based indices were also of interest. For example, the sum of distances between internal vertices and leaves is studied in [151], the sum of the *eccentricity* (the largest distance from any vertex to a given vertex) and equivalently the average eccentricity in trees are considered in [70, 121, 129]. The star, path, greedy trees, and caterpillars continue to be extremal with respect to these indices. Some lesser known distance-based indices are mentioned in [159].

3 Graph Indices Based on Counting

3.1 Independent sets and matchings

Several important graph invariants are based on counting particular sets of vertices or edges. Two prominent examples in chemical graph theory are the *Merrifield-Simmons index* and the *Hosoya index*, defined as the number of independent sets and matchings, respectively. There is a vast amount of literature devoted to the extremal values of these invariants in various families of graphs, enough to fill a survey of its own (see [143]). This includes in particular trees with various restrictions on the maximum degree, diameter, number of leaves, etc. Remarkably, one observes the general phenomenon that the graphs that maximize the Merrifield-Simmons index in some given class of graphs also minimize the Hosoya index, and vice versa. While there are many examples (and a few counterexamples) of this connection, it is still poorly understood. The paper [136] makes an attempt by studying the correlation of the two for random trees, and [53] gives a number of inequalities between the number of independent or 2-independent sets and matchings. It would be highly desirable to have a better understanding of the relation between the two indices.

The connection applies in particular to trees: unsurprisingly, if we do not impose any further restrictions, the extremal trees are the star and the path. The star S_n has the greatest number of independent sets (namely $2^{n-1}+1$) and the smallest number of matchings (namely n), while the path P_n has the greatest number of matchings and the smallest number of independent sets. Both are Fibonacci numbers: the number of independent sets of a path with n vertices is the Fibonacci number F_{n+2} (where $F_1 = F_2 = 1$, $F_{n+1} = F_n + F_{n-1}$), while the number of matchings is F_{n+1}. This illustrates the general fact that the Hosoya index of a graph is the Merrifield-Simmons index of the line graph, since matchings of the original graph correspond exactly to independent sets of the line graph. The occurrence of Fibonacci numbers is also the reason why the number of independent sets was called the *Fibonacci number* of a graph in what is probably its earliest occurrence in the mathematical literature [110].

Let us remark that the Merrifield-Simmons index and the Hosoya index are intimately tied to two important graph polynomials. If $m(G,k)$ denotes the number of matchings of cardinality k in a graph G, the *matching polynomial* of G is defined as

$$\sum_{k\geq 0}(-1)^k m(G,k)x^{n-2k}.$$

Remarkably, the matching polynomial coincides with the characteristic polynomial (of the adjacency matrix) for trees (see, e.g., Section 5 in [47]). This is also one of the reasons for the way the matching polynomial is defined, which is perhaps somewhat less intuitive than the definition of the *matching-generating polynomial*

$$\sum_{k\geq 0} m(G,k)x^k.$$

Note that the Hosoya index, henceforth denoted by $Z(G)$, is simply the value of this polynomial at $x = 1$, i.e., $Z(G) = \sum_{k\geq 0} m(G,k)$. Since the Hosoya index is so closely related to the characteristic polynomial, it is also unsurprising that there are connections to graph invariants based on the spectrum, such as the *graph energy* (the sum of the absolute values of all eigenvalues, see the recent book [97]). Likewise, the Merrifield-Simmons index is the value of the *independence polynomial* at $x = 1$: if $i(G,k)$ is the number of independent sets of cardinality k in G, this polynomial is defined by

$$\sum_{k\geq 0} i(G,k)x^k,$$

see [92] for a survey on this polynomial. The Merrifield-Simmons index, in the following denoted by $\sigma(G)$, is of course given by $\sigma(G) = \sum_{k\geq 0} i(G,k)$.

While it is impossible to give a complete account of extremal results on the Merrifield-Simmons index and the Hosoya index of trees (the reader is again referred to [143] for a more comprehensive survey), let us state a fairly general theorem due to Andriantiana [4] that implies many other results as corollaries. Since the path and the star are extremal trees, as they are for the Wiener index, one might assume that the greedy trees (see Theorem 2 in Section 2) are also extremal again, but this is not the case.

Definition 4. Let $(d_1, d_2, \ldots, d_k, 1, 1, \ldots, 1)$ be a degree sequence of a tree, where $d_k \geq 2$, in non-increasing order. Let the associated tree $\mathcal{M}(d_1, d_2, \ldots, d_k, 1, 1, \ldots, 1)$ be defined recursively as follows: if $k \leq d_k + 1$, then $\mathcal{M}(d_1, d_2, \ldots, d_k, 1, 1, \ldots, 1)$ is the tree obtained from a star S_{d_k+1} with d_k leaves by identifying $k-1$ of its leaves with the centers of stars $S_{d_1}, S_{d_2}, \ldots, S_{d_{k-1}}$, respectively.

The internal vertices (non-leaves) are labeled $v_1, \ldots, v_k$ in such a way that their degrees are exactly $d_1, d_2, \ldots, d_k$ (in particular, v_1 is the center of the star S_{d_k+1} that started the construction).

If $k \geq d_k + 2$, we define $\mathcal{M}(d_1, d_2, \ldots, d_k, 1, 1, \ldots, 1)$ as follows: let l be the greatest integer such that v_l is a label in $\mathcal{M}(d_{d_k}, \ldots, d_{k-1}, 1, 1, \ldots, 1)$, and let s be the smallest integer such that v_s is adjacent to a leaf in $\mathcal{M}(d_{d_k}, \ldots, d_{k-1}, 1, 1, \ldots, 1)$. Now $\mathcal{M}(d_1, d_2, \ldots, d_k, 1, 1, \ldots, 1)$ is obtained from $\mathcal{M}(d_{d_k}, \ldots, d_{k-1}, 1, 1, \ldots, 1)$ by connecting a leaf that is adjacent to v_s to the centers of $d_k - 1$ disjoint stars $S_{d_1}, S_{d_2}, \ldots, S_{d_{d_k-1}}$. The centers of these stars receive the labels $v_{l+1}, \ldots, v_{l+d_k-1}$, in increasing order of degree.

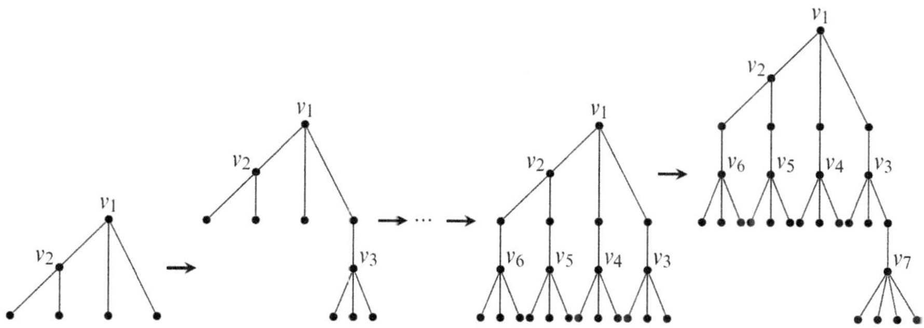

Fig. 3 Construction of the tree $\mathscr{M}(5,4,4,4,4,3,3,2,2,2,2,2,1,1,\dots,1)$.

See Figure 3 for an example of this step-by-step construction. Note that unlike greedy trees, large and small degrees alternate in $\mathscr{M}(d_1,d_2,\dots,d_n)$.

Theorem 6. *Among all trees with degree sequence $(d_1,d_2,\dots,d_n)$, the tree $\mathscr{M}(d_1, d_2,\dots,d_n)$ has the greatest number of independent sets and the least number of matchings (and it is unique with either of these two properties).*

In spite of the fact that the extremal trees in this scenario are no longer the greedy trees of Section 2, a majorization result (see Definition 3) akin to greedy trees holds.

Theorem 7. *If a degree sequence $(d_1',d_2',\dots,d_n')$ majorizes another degree sequence $(d_1,d_2,\dots,d_n)$, then we have*

$$\sigma(\mathscr{M}(d_1,d_2,\dots,d_n)) \le \sigma(\mathscr{M}(d_1',d_2',\dots,d_n'))$$

and

$$Z(\mathscr{M}(d_1,d_2,\dots,d_n)) \ge Z(\mathscr{M}(d_1',d_2',\dots,d_n')).$$

As an immediate application, one finds (in a similar way as for distance-based indices, see Theorem 5) that the tree $\mathscr{M}(\Delta,\Delta,\dots,\Delta,r,1,1,\dots,1)$ is extremal among trees whose maximum degree is Δ, $\mathscr{M}(\ell,2,2,\dots,2,1,1,\dots,1)$ is extremal among trees with exactly ℓ leaves, and $\mathscr{M}(n-D+1,2,2,\dots,2,1,1,\dots,1)$ is extremal among trees with diameter D (and fixed number n of vertices in all cases).

The analogous problems for the minimum of the Merrifield-Simmons index and the maximum of the Hosoya index generally appear to be harder. A complete characterization is known for trees with given maximum degree [139], and partial results are available for trees with given number of leaves [40, 160].

If one is interested in the maximum or minimum number of independent sets or matchings, it makes sense to include a restriction on the independence number or matching number. Note that in view of the König-Egerváry theorem, these two quantities are related: the sum of the matching number and the independence number in trees (and, more generally, in bipartite graphs) equals the number of vertices.

The trees of given order and independence number that maximize the number of independent sets were determined by Bruyère and Mélot [21] as part of a more general result, and a partial characterization for the minimum was given by Bruyère, Joret, and Mélot in [22]. The number of matchings in trees with given matching number, on the other hand, was studied in [67].

Instead of imposing additional restrictions on the trees, it is also natural and interesting to restrict the sets that are counted. In particular, independent sets and matchings that are *maximal* (with respect to set inclusion, i.e., not contained in any larger independent set or matching) or *maximum* (i.e., of greatest possible cardinality) have been studied. Things change considerably in that the extremal trees are no longer simply the star and the path.

While the star has the minimum number of maximal independent sets (namely 2) and the minimum number of maximum independent sets (namely 1, but it is not the only tree with this property), the path is not extremal for these two quantities. Wilf [155] showed that a tree of order n has at most $2^{(n-1)/2}$ maximal independent sets if n is odd, and $2^{(n-2)/2} + 1$ if n is even. Sagan [113] gave an alternative proof and also characterized the extremal graphs. For odd order, they are simply subdivided stars, for even order they are *batons*, obtained by attaching one or more paths of length 2 to the two ends of a path of length 1 or 3. The maximum of the number of maximum independent sets is also $2^{(n-2)/2} + 1$ if n is even, but only $2^{(n-3)/2}$ for odd n, as proven by Zito in [168]. These results have been extended further to graphs with at most r cycles [114].

For maximal and maximum matchings, the situation is even more complicated. The number of maximum matchings is trivially bounded below by 1, and this value is attained for any tree with a perfect matching. On the other hand, the trees with the greatest number of maximum matchings have one of seven different shapes (depending on the number of vertices modulo 7) composed of a repeated pattern of seven vertices, see [65]. There are also exceptions to the general pattern up to order 34 and two different extremal trees if the number of vertices is either 6 or 34. The maximum is of order $\Theta(1.391664^n)$, the exact constant being $(\frac{1}{2}(11 + \sqrt{85}))^{1/7}$. This also gives a lower bound for the maximum number of maximal matchings in a tree (since every maximum matching is automatically maximal), and an upper bound of $O(1.395337^n)$ was given for this maximum number by Górska and Skupień [56]. The exact shape of the trees with greatest number of maximal matchings and the asymptotic behavior of this greatest number remain an open problem.

Finally, we mention one further variant, namely the number of independent or maximal independent sets containing the leaves of a tree, which was studied in [156] and [157].

3.2 Subtrees and related invariants

Research on the number of *subtrees* (connected induced subgraphs of a tree) and a related quantity, the number of subtrees containing at least one leaf, started with [123–125], motivated by the study of the complexity of an algorithm in bioinformatics [85]. Remarkably, trees that maximize the number of subtrees are typically also those that minimize the Wiener index and vice versa, a similar connection as mentioned earlier for the Merrifield-Simmons index and the Hosoya index. In particular, a tree with n vertices has at least $\binom{n+1}{2}$ and at most $2^{n-1}+n-1$ subtrees (the minimum occurs for the path, the maximum for the star).

A lot of further work was done in particular on trees with given degree sequence [81, 166, 167], where the greedy trees of Theorem 2 occur again as extremal structures. In fact, as shown in [7], they have the greatest number of subtrees of any given cardinality among trees with a specific degree sequence. The situation for the minimum is more complicated, but it is known that it always occurs for a caterpillar, and partial results on the precise structure have been obtained as well [119, 165]. Other families of trees (with given number of leaves, bipartition, etc.) were considered in [93].

Some closely related concepts can be found in two much earlier papers of Jamison [72, 73], who studied the *average subtree order* (average number of vertices in a subtree). Among other things, he showed that the path has the smallest average subtree order among trees with n vertices, namely $\frac{n+2}{3}$. He posed many open questions, some of which were only resolved recently [63, 134, 144, 145] or are still open. In particular, it is not known which trees maximize the average subtree order. The star is only extremal if the number of vertices is very small, while the extremal trees for larger orders are quite "path-like" (in that they must have many vertices of degree 2). Jamison conjectured that the maximum might always be attained for a caterpillar. Since the average subtree order is also a relatively complicated parameter to calculate, not a lot of evidence for or against this conjecture is available.

As a final remark, let us mention an interesting variant obtained by counting subtrees only up to isomorphism. In this context, the path and star of order n are "equally bad," in that they only have n non-isomorphic subtrees, which is clearly also the minimum. As it turns out, the maximum is of order $\Theta(5^{n/4})$, see [37]. A simple construction yields trees that reach this bound: take a path of length $\frac{n}{4}$ and attach a leaf and a path of length 2 to each vertex.

3.3 Dominating sets

The number of *dominating sets* is again very different from the other quantities mentioned in this section so far. An important feature is the fact that leaves can only be dominated by their neighbors or themselves, which often severely limits the possibilities. Interestingly, the number of dominating sets of any graph is always odd, see [20].

The maximum number of dominating sets among trees of order n is obtained for a star (and also for the path if the number of vertices is at most 5, see [18]), but the lower bound of $5^{n/3}$, $9 \cdot 5^{(n-4)/3}$, or $3 \cdot 5^{(n-2)/3}$ (depending on whether n is congruent to 0, 1, or 2 modulo 3) is attained for a rich class of trees: the condition is simply that each internal vertex must be adjacent to exactly two leaves, with one exception (adjacent to only one leaf) if $n \equiv 2 \bmod 3$ and one exception (adjacent to three leaves) or two exceptions (each adjacent to one leaf) if $n \equiv 1 \bmod 3$. However, there is nothing particularly special about trees in this context, as was pointed out in [141] (see also [120]): the same values are extremal for connected graphs (and even for graphs without isolated vertices), and the characterization of extremal connected graphs is also essentially the same.

Similar observations can be made for the number of *efficient dominating sets* (i.e., dominating sets with the property that no vertex is dominated by more than one of the vertices in the dominating set) and *total dominating sets*. Similar to the 3-periodicity for the total number of dominating sets, a period of 7 can be observed in the structure of the extremal trees, see [19] and [89].

Krzywkowski [86–88] studied several other types of domination and provided algorithms for listing minimal dominating sets, minimal 2-dominating sets, or minimal double dominating sets. The running time analysis of these algorithms provides upper bounds for the respective numbers.

3.4 Walks

Let $A(G)$ denote the adjacency matrix of a graph. It is well known that the entries of the k-th power $A(G)^k$ count the number of walks of length k in G starting and ending at specified vertices. In particular, the total number of closed walks of length k (each with a fixed starting vertex), which we denote by $\overline{w}_k(G)$, equals the trace of $A(G)^k$ (which in turn is equal to the k-th spectral moment), and the total number of all walks of length k, denoted by $w_k(G)$, is the sum of all entries in $A(G)^k$, which is $\mathbf{1}^T A(G)^k \mathbf{1}$.

The number of walks is closely connected to the spectrum, in particular the spectral radius, which is equal to the limits $\lim_{k\to\infty} w_k(G)^{1/k}$ and $\lim_{k\to\infty} \overline{w}_k(G)^{1/k}$; in the latter case, the limit needs to be taken only over even numbers k if G is bipartite (in particular if it is a tree). There are also explicit inequalities relating the spectral radius and the number of walks of different lengths, see [107] and the references therein. Spectral graph theory is of course a vast area of its own right, and a lot more could be said about spectral parameters of trees, but since our space is limited, let us refer to the books [29, 34, 35] and the surveys in [33].

We specifically mention two recent papers [130, 131] by Täubig et al. that also deal with inequalities for walks in graphs, especially because of an interesting conjecture on trees. It is shown in [130] that

$$w_{2a+c}(G)w_{2a+2b+c}(G) \leq w_{2a}(G)w_{2(a+b+c)}(G)$$

for all graphs G and non-negative integers a, b, c, which generalizes several inequalities proved earlier by other authors. Since $w_0(G) = |V(G)|$ and $w_1(G) = 2|E(G)|$, this implies in particular that

$$\frac{2|E(G)|}{|V(G)|} \cdot w_{k-1}(G) \leq w_k(G)$$

for all graphs G if k is even. Täubig et al. [130] conjecture that this inequality holds for *all* positive integers k if G is a tree. It is noteworthy that it is not valid for arbitrary graphs or even for the narrower class of bipartite graphs, and counterexamples are given in [130]. On the other hand, it is proven for trees in the special cases $k = 3$ and $k = 5$ in [130].

The problem of maximizing the number of closed walks in trees was motivated in particular by the aforementioned connection to spectral moments and thus also the so-called *Estrada index* [62]. Csikvári [31] proved that the path has the minimum number of closed walks of any fixed length among trees of given order, and the star has the maximum number of such walks. This was generalized to arbitrary walks by Bollobás and Tyomkyn [17], who also gave an alternative proof. Csikvári's proof is based on defining a poset structure on trees of given order, and he showed [32] that this idea could also be applied to many other tree invariants (specifically, coefficients and roots of different graph polynomials).

Trees with given degree sequence were studied in [5], and it turns out that greedy trees as defined in Section 2 are extremal once again, which has several implications on spectral invariants: they maximize all spectral moments, thus also the spectral radius (which was proven earlier in [13]) and also any invariant of the form

$$E_f(G) = \sum_i f(\lambda_i(G)),$$

where $\lambda_1(G), \lambda_2(G), \ldots$ are the eigenvalues of G and f is any entire function whose Taylor series at 0 has only non-negative coefficients. This includes the aforementioned Estrada index (corresponding to $f(x) = e^x$) as a particular instance. Again, a majorization theorem holds as well, which implies one half of Csikvári's result (extremality of the star).

It is very likely that the greedy trees also have the greatest possible total number of walks (not necessarily closed), but it seems that this is more difficult to prove (the proof for closed walks is already quite long and technical).

3.5 Rooted trees regarded as posets

A rooted tree in its standard drawing can be interpreted in a natural way as the Hasse diagram of a partially ordered set, so it makes sense to study concepts stemming from the theory of posets, such as *chains* and *antichains*.

There is a natural bijection between antichains in a rooted tree and subtrees containing the root (the leaves of such a subtree forming the corresponding antichain), so the results of [7] also show that greedy trees (with their natural root) also maximize the number of antichains, even antichains of any fixed cardinality.

Another invariant of a similar nature that was studied recently is the number of *transversals*, i.e., subsets of vertices with the property that every path from the root to a leaf contains at least one of the elements (this can be the root or leaf itself). Thus removal of a transversal "destroys" the connection between the root and the leaves, and in fact one of the motivations for the authors of [23] to study transversals was a mathematical model in counterterrorism. Among other results, the d-ary trees with minimum and maximum number of transversals are characterized in [23]: they are caterpillars and full d-ary trees, respectively.

Finally, let us mention a problem due to Klazar [83] that was only solved very recently [6]. An *infima closed set* in a poset is a set with the property that the infimum (greatest common lower bound) of any subset exists and is contained in the set as well. In the context of rooted trees, this means that the closest common ancestor of any two vertices in such a set must be contained in the set as well. Both the star and the path have "many" infima closed sets: in fact, all subsets of a path, rooted at one of its ends, are infima closed, and all sets containing the root of a star, rooted at its center, are infima closed. Klazar's question was therefore to characterize the trees with the least number of infima closed sets, and it turns out that these trees are essentially complete binary trees (all internal vertices have two children), except for the vertices that are directly adjacent to leaves: these vertices must have three leaves as children, with a bounded number of exceptions (that depend on the precise order of the tree).

4 Degree-Based Graph Indices

Randić [111] studied a new index, earlier called branching index or connectivity index,

$$R_\alpha(G) = \sum_{\{u,v\}\in E(G)} (d_G(u)d_G(v))^\alpha,$$

where the summation is over all edges of the graph, $d_G(u)$ denotes the degree of vertex u in graph G, and $\alpha = -1/2$. This is known today as Randić index, and for other values of α it is known as the generalized Randić index. (When $\alpha < 0$, the index is undefined for graphs with isolated vertices, but molecular graphs are connected.) $R_1(G)$ is also known as the second Zagreb index. (The first Zagreb index is just the sum of degree squares; more general versions of it allow any exponent instead of 2.)

There could be many graphs with the same degree sequence, and they could look strikingly different, depending on whether vertices of similar or very different degrees tend to be connected. In network science the concepts of assortativity and disassortativity describe this phenomenon, where assortativity is usually measured by the Pearson correlation coefficient of degree pairs of adjacent vertices. Assortativity is determined by the joint degree matrix [38], whose entries give the number m_{ij} of edges between degree i and degree j vertices. The Randić index can be written alternatively as

$$R_\alpha(G) = \frac{1}{2} \sum_{u \in V(G)} d_G(u)^{2\alpha+1} - \frac{1}{2} \sum_{i<j} m_{ij}(i^\alpha - j^\alpha)^2, \tag{1}$$

in which the first sum turns just n for $\alpha = -1/2$. We suspect that the power of the Randić index comes from using information from the whole joint degree matrix.

The Randić index became better known among graph theorists after the work of Bollobás and Erdős [15, 16]. They [15] proved that among graphs on n vertices with minimum degree at least 1, the star minimizes $R_{-1/2}$. [94] discusses the development of minimizing $R_{-1/2}$ among graphs on n vertices with minimum degree at least k, as this problem is not yet completely solved. On the other hand, formula (1) shows that among graphs on n vertices, the regular graphs maximize $R_{-1/2}$. Further restricting graphs to trees, the path maximizes $R_{-1/2}$ [25, 161].

Adding an edge to the graph, $R_{-1/2}$ can go up or down. This is a delicate quantity, and it is no surprise that a few published conjectures about it were refuted. Conjecture making and proving software like Graffiti [46], and AutographiX [3, 24] included $R_{-1/2}$ among the graph parameters they compared. Aouchiche et al. [3] made the conjecture that for any connected graph of order $n \geq 3$ with diameter $D(G)$,

$$R_{-1/2}(G) - D(G) \geq \sqrt{2} - \frac{n+1}{2} \quad \text{and} \quad \frac{R_{-1/2}(G)}{D(G)} \geq \frac{n-3+2\sqrt{2}}{2n-2},$$

and equality holds in any of them iff G is a path. [100] proved a stronger result that easily implies this conjecture: for any connected graph of order $n \geq 3$ with diameter $D(G)$, $R_{-1/2}(G) - \frac{1}{2}D(G) \geq \sqrt{2} - 1$, and equality holds iff G is a path.

Fajtlowicz [46] conjectured for connected graphs that $R_{-1/2}(G) \geq \frac{W(G)}{\binom{n}{2}}$ where the average distance on the RHS is written with the Wiener index in the numerator. Caporossi and Hansen [24] strengthened this conjecture to $R_{-1/2} \geq \frac{W(G)}{\binom{n}{2}} + \sqrt{n-1} + \frac{2}{n} - 2$, which, if true, is tight. They [24] also conjectured that $r(G) \leq R_{-1/2}(G)$, where $r(G)$ is the radius of the graph G, unless G is an even path and verified this conjecture if G is a tree.

The survey [94] further discusses extremal results for $R_{-1/2}(T)$, where (a) T is a tree with given number of vertices and leaves; (b) T is a tree with given diameter and number of vertices; (c) T is a chemical tree with given number of vertices; (d) T is a chemical tree with given number of vertices and leaves. As chemical trees are the relevant trees for chemical graph theory, [60] worked out the three largest possible and the three smallest possible values of $R_{-1/2}(T)$, where T is a chemical tree with given number of vertices, and the realizing trees.

For the generalized Randić index R_α, the minimizing tree is a path for positive α and $n \geq 5$, and is a star for negative α [68]. Randić himself considered R_{-1} [111]. Building on considerable earlier work, [108] and [69] showed that the maximum of R_{-1} on trees with n vertices is $\frac{15}{56}n+o(n)$. The survey [94] further discusses extremal results for $R_{-1}(T)$, where (a) T is a tree with given number of vertices and leaves; (b) T is a tree with given number of vertices and has maximum degree 3; (c) T is a chemical tree with given number of vertices.

Gutman [58] lists a number of variants of the Randić index and notes "we have far too many descriptors, and there seems to lack a firm criterion to stop or slow down their proliferation." We include here one more, the higher order Randić index proposed in [78]:

$$ {}^{i}R(G) = \sum_{v_0 v_1 \ldots v_i} \frac{1}{\sqrt{d_G(v_0) d_G(v_1) \cdots d_G(v_i)}}, $$

where the sum is taken over paths $v_0 v_1 \ldots v_i$ of length i in the graph G. $i = 1$ gives back $R_{-1/2}(G)$, while $i = 0$ gives $\sum_v d_G(v)^{-1/2}$, the first Zagreb index with exponent $-1/2$. We note that there is some ambiguity in the definition of higher order Randić indices, as some papers allow any $v_0 v_1 \ldots v_i$ walks that do not repeat edges—for $i \geq 3$, if the graph has cycles, this may bring extra terms. Regarding the second order Randić index, [100] showed that among trees on n vertices, the star maximizes the second order Randić index, and among trees on n vertices with maximum degree 3, the path is the unique minimizer of the second order Randić index.

As for many other graph indices discussed in this survey, trees with a given degree sequence have been considered as well. The first article of this kind is [39], where the trees with given degree sequence that maximize R_1 were found. This was further generalized to arbitrary α in [148]. It is noteworthy that the extremal trees are generally not unique, and it would be interesting to find necessary and sufficient conditions on the degree sequence for the extremal tree to be unique. Finally, an even more general type of graph invariant, where $(d_G(u)d_G(v))^\alpha$ is replaced by a symmetric function $f(d_G(u), d_G(v))$ (so that the Randić index occurs as a special case) is studied in [150].

5 Random Trees

To have an idea whether the Wiener index of a tree is "small" or "large," and to facilitate statistical analysis, it is desirable to know the expectation and the variance of the Wiener index of a randomly selected tree.

Entringer et al. [45] gave a method to compute the asymptotics for the expected Wiener index of a random tree selected uniformly from a simply generated family of trees and found that the asymptotics is constant times $n^{5/2}$. Simply generated families include several important families, for example, binary trees, ordered trees, and unordered labeled trees. Asymptotic results for the average Wiener index of star-like trees were obtained in [135], and results relevant for chemical trees can be found in [41] and [137].

Janson [74] obtained asymptotics for the mean, variance, and higher moments as well as for the distribution of the Wiener index of a random tree from a simply generated family. This distribution can be expressed in terms of a Brownian excursion. He also determined the joint asymptotic distribution of the Wiener index and the internal path length, as well as asymptotics for the covariance and other mixed moments.

The study of the expected Wiener index extended to classes of random trees relevant in computer science. For them, the motivation is no longer chemical graph theory. Neininger [106] investigated the Wiener index of uniformly selected random recursive trees and random binary search trees, obtained asymptotics for the expectation, the variance, and also for the correlation and covariance with the path length. Munsonius [104] investigated the Wiener index and path length of random split trees asymptotically; Fuchs and Lee [54] obtained asymptotic expansions of moments of the Wiener index and showed a central limit law for the Wiener index of digital search trees, tries and PATRICIA tries.

Upper bounds for the tail distribution of the Wiener index in several models were obtained by [1, 51, 75], and [105].

Comparatively little work has been done on other graph invariants mentioned in this survey: [48] studies the distribution of the Randić index, proving a normal limit law for different random binary tree models. Central limit theorems for the (first) Zagreb index (i.e., the sum of the squared degrees) are proven in [49, 50].

Averages of various enumerative invariants (independent sets, matchings, subtrees, etc.) have been determined for different types of trees. A whole collection of results of this type can be found in a paper of Klazar [83]; see also [82, 101, 102, 138] for other instances. For the number of subtrees, the distribution was shown to be log-normal in a recent paper [142]. Although the results in this paper are quite general, they do not seem to apply directly to similar invariants such as the number of independent sets.

6 Localized Tree Indices and Concepts of Centrality

Recall that the function $ecc_T(.)$ tells the largest distance from a vertex in the tree T. The *center* $C(T)$ of the tree T is the set of vertices where $ecc_T(.)$ is minimized.

The *distance* $\sigma_T(.)$ *of a vertex* (unrelated to the Merrifield-Simmons index $\sigma(T)$ that was defined in Section 3) in a tree T is the sum of distances from the variable vertex to all other vertices. Clearly, $W(T) = \frac{1}{2}\sum_v \sigma_T(v)$. According to Zelinka [162], the *centroid* $CT(T)$ of a tree coincides with the set of vertices minimizing the distance function. Let $F(T)$ denote the number of subtrees of the tree T, and $F_T(v)$ denote the number of subtrees of the tree T that contain vertex v.

Let the function $F_T(.)$ assign to vertex v of the tree T the number of subtrees of T containing the vertex v. [124] defined the *subtree core* $Core(T)$ of the tree T, a new concept analogous to, but different from the concepts of center $C(T)$ and centroid $CT(T)$, where $F_T(v)$ is maximized. We see here an interaction of "local" and "global" tree indices: $\sigma_T(.)$ is the local version of the Wiener index, $F_T(.)$ is the local version of the number of subtrees; while $ecc_T(.)$ is inherently local, one can define the global version by $Ecc(T) = \sum_v ecc_T(v)$.

The center, the centroid, and the subtree core contain either a single vertex or two adjacent vertices of the tree. The reason is the following. Given the vertices along any path of a tree, the sequence of the values of $F_T(.)$ is strictly concave down ([124, proof to Theorem 9.1]), the sequence of the values of $\sigma_T(.)$ is strictly concave up ([99, Ex. 6.22]), and the sequence of the values of $ecc_T(.)$ is concave up ([99, Ex. 6.21]). Strict concavity immediately implies that $Core(T)$ and $CT(T)$ consist of one or two adjacent vertices. The fact that $C(T)$ also consists of one or two adjacent vertices was already known to Jordan [77] (see also [99, Ex. 6.21a]). It is well known that the center and centroid are generally distinct, and moreover, they can be arbitrary far from each other ([99, Ex. 6.22c]). The paper [122] investigates how far these different middle parts can be in a tree.

Behavior of ratios of local and global indices can be subtle. The investigation of extreme values of ratios started in [9], which determined extremal values of $W(T)/\sigma_T(v)$, $W(T)/\sigma_T(w)$, $\sigma_T(w)/\sigma_T(v)$, and $\sigma_T(w)/\sigma_T(u)$, where T is a tree on n vertices, v is in the centroid of the tree T, and u, w are leaves in T. The two papers [126] and [127] went further to see how far the negative correlation between distances and subtrees, discovered in [136], goes. They characterized the extremal values of $F(T)/F_T(v)$, $F(T)/F_T(w)$, $F_T(w)/F_T(v)$, and $F_T(w)/F_T(u)$, where T is a tree on n vertices, v is in the subtree core of the tree T, and u, w are leaves in T—the complete analogue of [9], changing distances to subtrees. It is instructive to look at the table in [126] to see the extremal trees corresponding to "dual" problems—there is a striking similarity.

The most recent investigation in this area [121] characterized the extremal values of $Ecc(T)/ecc_T(v)$, $Ecc(T)/ecc_T(w)$, $ecc_T(w)/ecc_T(v)$, and $ecc_T(w)/ecc_T(u)$, where T is a tree on n vertices, v is in the center of the tree T, and u, w are leaves in T—another complete analogue of [9], changing distances to eccentricities.

7 Inverse Problems

Perhaps [55] was the first paper that called for the systematic study of inverse problems in combinatorial chemistry: create molecular graphs with prescribed topological index. [91], however, asked for trees with a given Wiener index, and made the conjecture that with a finite number of exceptions all positive integers are Wiener indices of some trees.

Li et al. [96] study four indices and show that all positive integers are the number of cliques in some graphs. It discusses the first Zagreb index (sum of degree squares) and observes that it is even for any graph, and shows that all even integers $\neq 4, 8$ are the sum of squares of the degrees of some tree. An early but relevant paper [98] shows that any positive integer is the Merrifield-Simmons index of a bipartite graph, and there is a conjecture that any sufficiently large positive integer is the Merrifield-Simmons index of a tree (clearly every positive integer is the Hosoya index of some star). An interesting feature of this conjecture (which is also one reason why the problem might be very difficult) is the distribution of the Merrifield-Simmons index in residue classes: it was shown in [140] (and by a slightly different method in [2]) that for every modulus m, almost all trees (in the sense that the proportion converges to 1 as the order tends to infinity) have a number of independent sets that is divisible by m. However, every residue class modulo m occurs for some tree, as was shown by Law [90].

The Lepović-Gutman conjecture that all sufficiently large numbers are Wiener indices of some trees was proven independently in [152] and [135]. In fact, 49 exceptional numbers were found. Part of the difficulty of the problem is the large number of tree shapes—the first paper only used short caterpillars, while the second used star-like trees, which are obtained from a star by changing its leaves to some stars. Both papers approached the problem through representation of integers as values of certain quadratic forms. This approach had to use arbitrarily large degree vertices, which is not an attractive approach for chemical graph theorists. Combining their efforts the authors of these two papers [146] showed that every sufficiently large integer n is the Wiener index of some caterpillar tree with degrees at most 3, and every sufficiently large even integer is the Wiener index of some hexagon type graph. [11] conjectures that every sufficiently large integer is the Wiener index of a binary tree, a stronger form of the Lepović-Gutman conjecture.

In analogy to the Lepović-Gutman conjecture for the Wiener index, it was shown in [36] that all but 34 positive integers are the number of subtrees of some trees. In fact, the proof shows that all sufficiently large positive integers are equal to the number of subtrees of a caterpillar tree, and then bridges the gap by explicit computation. Here the quadratic form representation appears as a representation of a number as a certain kind of sum of pentagonal numbers.

The paper [36] poses a metaconjecture: if a tree index on n vertices takes sufficiently many values, and for rooted trees the index can be computed from the indices of the subtrees rooted at the children of the root with a reasonable polynomial formula (the paper gives a number of examples of such indices), and if the values of the tree index are not constrained to some residue class, then every sufficiently large positive integer is a value of this tree index.

A result of [135] is instructive to accept this conjecture: let

$$W_\lambda(T) = \sum_{u,v \in V(T)} d_T(u, v)^\lambda,$$

where λ is a positive integer, a variant of the Wiener index. If there is a star-like tree T such that $W_\lambda(T) \equiv r \bmod 2^\lambda(2^\lambda - 1)$, then all members of the residue class $r \bmod 2^\lambda(2^\lambda - 1)$— with only finitely many exceptions—are Wiener indices of trees. For $\lambda = 2, 3, 5, 6, 7, 9, 10$, this implies that all integers, with finitely many exceptions, can be written as $W_\lambda(T)$ for some star-like tree T, as all residue classes mod $2^\lambda(2^\lambda - 1)$ are covered. (For $\lambda = 4$ and all other multiples of 4, this is not the case any more.)

Acknowledgements This research was supported in part by the NSF DMS contract 1300547 and by the DARPA and AFOSR under the contract FA9550-12-1-0405 (László Székely), National Research Foundation of South Africa, grant number 96236 (Stephan Wagner), and Simons Foundation grant 245307 (Hua Wang).

References

1. T. Ali Khan, R. Neininger, Tail bounds for the Wiener index of random trees, in *Discrete Math. Theor. Comput. Sci. Proc., Proceedings of the 2007 Conference on the Analysis of Algorithms* (2007), pp. 279–289
2. N. Alon, S. Haber, M. Krivelevich, The number of F-matchings in almost every tree is a zero residue. Electron. J. Comb. **18**(1), paper no. #P30 (2011)
3. M. Aouchiche, P. Hansen, M. Zheng, Variable neighborhood search for extremal graphs. 19. Further conjectures and results about the Randić index. MATCH Commun. Math. Comput. Chem. **58**, 83–102 (2007)
4. E.O.D. Andriantiana, Energy, Hosoya index and Merrifield-Simmons index of trees with prescribed degree sequence. Discrete Appl. Math. **161**(6), 724–741 (2013)
5. E.O.D. Andriantiana, S. Wagner, Spectral moments of trees with given degree sequence. Linear Algebra Appl. **439**(12), 3980–4002 (2013)
6. E.O.D. Andriantiana, S. Wagner, Trees with minimum number of infima closed sets. In preparation (2015)
7. E.O.D. Andriantiana, S. Wagner, H. Wang, Greedy trees, subtrees and antichains. Electron. J. Comb. **20**(3), paper no. #P28 (2013)
8. A.T. Balaban, *Chemical Applications of Graph Theory* (Academic Press, New York, 1976)
9. C.A. Barefoot, R.C. Entringer, L.A. Székely, Extremal values for ratios of distances in trees. Discrete Appl. Math. **80**, 37–56 (1997)
10. R.J. Baxter, I.G. Enting, S.K. Tsang, Hard-square lattice gas. J. Stat. Phys. **22**(4), 465–489 (1980)
11. S. Bereg, H. Wang, Wiener indices of balanced binary trees. Discrete Appl. Math. **155**, 457–467 (2007)
12. N.L. Biggs, E.K. Lloyd, R.J. Wilson, *Graph Theory 1736–1936* (Clarendon Press, Oxford, 1976; Oxford University Press, Oxford, 1986)
13. T. Bıyıkoğlu, J. Leydold, Graphs with given degree sequence and maximal spectral radius. Electron. J. Comb. **15**(1), paper no. #P119 (2008)
14. D. Bokal, É. Czabarka, L.A. Székely, I. Vrt'o, General lower bounds for the minor crossing numbers of graphs. Discrete Comput. Geom. **44**, 463–483 (2010)
15. B. Bollobás, P. Erdős, Graphs of extremal weights. Ars Comb. **50**, 225–233 (1998)
16. B. Bollobás, P. Erdős, A. Sarkar, Extremal graphs for weights. Discrete Math. **200**, 5–19 (1999)
17. B. Bollobás, M. Tyomkyn, Walks and paths in trees. J. Graph Theory **70**(1), 54–66 (2012)

18. D. Bród, Z. Skupień, Trees with extremal numbers of dominating sets. Australas. J. Comb. **35**, 273–290 (2006)
19. D. Bród, Z. Skupień, Recurrence among trees with most numerous efficient dominating sets. Discrete Math. Theor. Comput. Sci. **10**(1), 43–55 (2008)
20. A.E. Brouwer, The number of dominating sets of a finite graph is odd. Preprint (2009). http://www.win.tue.nl/~aeb/preprints/domin2.pdf
21. V. Bruyère, H. Mélot, Fibonacci index and stability number of graphs: a polyhedral study. J. Comb. Optim. **18**(3), 207–228 (2009)
22. V. Bruyère, G. Joret, H. Mélot, Trees with given stability number and minimum number of stable sets. Graphs Comb. **28**(2), 167–187 (2012)
23. V. Campos, V. Chvátal, L. Devroye, P. Taslakian, Transversals in trees. J. Graph Theory **73**(1), 32–43 (2013)
24. G. Caporossi, P. Hansen, Variable neighborhood search for extremal graphs 1: the AutographiX system. Discrete Math. **212**, 29–44 (2000)
25. G. Caporossi, I. Gutman, P. Hansen, L. Pavlović, Graphs with maximum connectivity index. Comput. Biol. Chem. **27**, 85–90 (2000)
26. A. Cayley, On the mathematical theory of isomers. Philos. Mag. **47**, 444–446 (1874)
27. A. Cayley, A theorem on trees. Quart. J. Math. **23**, 376–378 (1889)
28. E. Çela, N. Schmuck, S. Wimer, G.J. Woeginger, The Wiener maximum quadratic assignment problem. Discrete Optim. **8**, 411–416 (2011)
29. F.R.K. Chung, Spectral graph theory, in *CBMS Regional Conference Series in Mathematics*, vol. 92 (American Mathematical Society, Providence, 1997)
30. L. Clark, I. Gutman, The exponent in the general Randić index. J. Math. Chem. **43** (1), 32–44 (2006)
31. P. Csikvári, On a poset of trees. Combinatorica **30**(2), 125–137 (2010)
32. P. Csikvári, On a poset of trees II. J. Graph Theory **74**(1), 81–103 (2013)
33. D.M. Cvetković, I. Gutman, (eds.), Selected topics on applications of graph spectra. Zb. Rad. (Beogr.) **14**(22), 155–174 (2011)
34. D.M. Cvetković, M. Doob, H. Sachs, *Spectra of Graphs. Theory and Applications*, 3rd edn. (Johann Ambrosius Barth, Heidelberg, 1995)
35. D.M. Cvetković, P. Rowlinson, S. Simić, *An Introduction to the Theory of Graph Spectra*. London Mathematical Society Student Texts, vol. 75 (Cambridge University Press, Cambridge, 2010)
36. É. Czabarka, L.A. Székely, S. Wagner, The inverse problem for certain tree parameters. Discrete Appl. Math. **157**(15), 3314–3319 (2009)
37. É. Czabarka, L.A. Székely, S. Wagner, On the number of nonisomorphic subtrees of a tree. (2016, submitted, arXiv:1601.00944)
38. É. Czabarka, A. Dutle, P.L. Erdős, I. Miklós, On realizations of a Joint Degree Matrix. Discrete Appl. Math. **181**, 283–288 (2015)
39. C. Delorme, O. Favaron, D. Rautenbach, Closed formulas for the numbers of small independent sets and matchings and an extremal problem for trees. Discrete Appl. Math. **130**, 503–512 (2003)
40. H. Deng, Q. Guo, On the minimal Merrifield-Simmons index of trees of order n with at least $[n/2]+1$ pendent vertices. MATCH Commun. Math. Comput. Chem. **60**(2), 601–608 (2001)
41. A.A. Dobrynin, I. Gutman, The average Wiener index of trees and chemical trees. J. Chem. Inf. Comput. Sci. **39**, 679–683 (1999)
42. A.A. Dobrynin, R. Entringer, I. Gutman, Wiener index of trees. Theory and applications. Acta Appl. Math. **66**, 211–249 (2001)
43. M. Drmota, *Random Trees. An Interplay Between Combinatorics and Probability* (Springer, Wien/New York, 2009)
44. R.C. Entringer, D.E. Jackson, D.A. Snyder, Distance in graphs. Czech. Math. J. **26**, 283–296 (1976)
45. R.C. Entringer, A. Meir, J.W. Moon, L.A. Székely, On the Wiener index of trees from certain families. Australas. J. Comb. **10**, 211–224 (1994)

46. S. Fajtlowicz, On conjectures of Graffiti. Discrete Math. **72**, 113–118 (1988)
47. E.J. Farrell, An introduction to matching polynomials. J. Comb. Theory Ser. B **27**(1), 75–86 (1979)
48. Q. Feng, H.M. Mahmoud, A. Panholzer, Limit laws for the Randić index of random binary tree models. Ann. Inst. Stat. Math. **60**(2), 319–343 (2008)
49. Q. Feng, Z. Hu, On the Zagreb index of random recursive trees. J. Appl. Probab. **48**(4), 1189–1196 (2011)
50. Q. Feng, Z. Hu, Phase changes in the topological indices of scale-free trees. J. Appl. Probab. **50**(2), 516–532 (2013)
51. J.A. Fill, S. Janson, Precise logarithmic asymptotics of the right tails of some limit random variables for random trees. Ann. Comb. **12**, 403–416 (2009)
52. M. Fischermann, A. Hoffmann, D. Rautenbach, L. Szekely, L. Volkmann, Wiener index versus maximum degree in trees. Discrete Appl. Math. **122**, 127–137 (2002)
53. M. Fischermann, L. Volkmann, D. Rautenbach, A note on the number of matchings and independent sets in trees. Discrete Appl. Math. **145**(3), 483–489 (2005)
54. M. Fuchs, C.-K. Lee, The Wiener index of random digital trees. SIAM J. Discrete Math. **29**(1), 586–614 (2015)
55. D. Goldman, S. Istrail, G. Lancia, A. Piccolboni, B. Walenz, Algorithmic strategies in combinatorial chemistry, in *SODA '00 Proceedings of the eleventh annual ACM-SIAM Symposium on Discrete Algorithms* (2000), pp. 275–284
56. J. Górska, Z. Skupień, Trees with maximum number of maximal matchings. Discrete Math. **307**(11–12), 1367–1377 (2007)
57. I. Gutman, A property of the Wiener number and its modifications. Indian J. Chem. **36A**, 128–132 (1997)
58. I. Gutman, Degree-based topological indices. Croat Chem. Acta **86**(4), 351–361 (2013)
59. I. Gutman, W. Linert, I. Lukovits, A.A. Dobrynin, Trees with extremal hyper-Wiener index: Mathematical basis and chemical applications. J. Chem. Inf. Comput. Sci. **37**, 349–354 (1997)
60. I. Gutman, O. Miljković, G. Caporossi, P. Hansen, Alkanes with small and large Randić connectivity indices. Chem. Phys. Lett. **306**(5), 366–372 (1999)
61. I. Gutman, B. Furtula, M. Petrović, Terminal Wiener index. J. Math. Chem. **46**, 522–531 (2009)
62. I. Gutman, H. Deng, S. Radenković, The Estrada index: an updated survey. Zb. Rad. (Beogr.) **14**(22), 155–174 (2011)
63. J. Haslegrave, Extremal results on average subtree density of series-reduced trees. J. Comb. Theory Ser. B **107**, 26–41 (2014)
64. O.J. Heilmann, E.H. Lieb, Theory of monomer-dimer systems. Commun. Math. Phys. **25**(3), 190–232 (1972)
65. C. Heuberger, S. Wagner, The number of maximum matchings in a tree. Discrete Math. **311**(21), 2512–2542 (2011)
66. H. Hosoya, Topological index. A newly proposed quantity characterizing the topological nature of structural isomers of saturated hydrocarbons. Bull. Chem. Soc. Jpn. **44**(9), 2332–2339 (1971)
67. Y. Hou, On acyclic systems with minimal Hosoya index. Discrete Appl. Math. **119**(3), 251–257 (2002)
68. Y. Hu, X. Li, Y. Yuan, Trees with minimum general Randic index. MATCH Commun. Math. Comput. Chem. **52**, 119–128 (2004)
69. Y. Hu, X. Li, Y. Yuan, Solutions to two unsolved questions on the best upper bound for the Randic index R_{-1} of trees. MATCH Commun. Math. Comput. Chem. **54**, 441–454 (2005)
70. A. Ilić, On the extremal properties of the average eccentricity. Comput. Math. Appl. **64**(9), 2877–2885 (2012)
71. O. Ivanciuc, T.S. Balaban, A.T. Balaban, Reciprocal distance matrix, related local vertex invariants and topological indices. J. Math. Chem. **12**, 309–318 (1993)
72. R.E. Jamison, On the average number of nodes in a subtree of a tree. J. Comb. Theory Ser. B **35**(3), 207–223 (1983)

73. R.E. Jamison, Monotonicity of the mean order of subtrees. J. Comb. Theory Ser. B **37**(1), 70–78 (1984)
74. S. Janson, The Wiener index of simply generated random trees. Random Struct. Algoritm. **22**(4), 337–358 (2003)
75. S. Janson, P. Chassaing, The center of mass of the ISE and the Wiener index of trees. Electron. Commun. Probab. **9**, 178–187 (2004)
76. F. Jelen, E. Triesch, Superdominance order and distance of trees with bounded maximum degree. Discrete Appl. Math. **125** (2–3), 225–233 (2003)
77. C. Jordan, Sur les assemblages de lignes. J. Reine Angew. Math. **70**, 185–190 (1869)
78. L.B. Kier, L.H. Hall, *Molecular Connectivity in Chemistry and Drug Research* (Academic Press, New York, 1976)
79. L.B. Kier, L.H. Hall, *Molecular Connectivity in Structure-Activity Analysis* (Wiley, New York, 1986)
80. G. Kirchoff, Über die Auflösung der Gleichungen, auf welche man bei der Untersuchung der linearen Vertheilung galvanischer Strōme gefürt wird. Ann. Phys. Chem. **72**, 497–508 (1847)
81. R. Kirk, H. Wang, Largest number of subtrees of trees with a given maximum degree. SIAM J. Discrete Math. **22**(3), 985–995 (2008)
82. P. Kirschenhofer, H. Prodinger, R.F. Tichy, Fibonacci numbers of graphs. III. Planted plane trees, in *Fibonacci Numbers and Their Applications (Patras, 1984)* (Reidel, Dordrecht, 1986). Math. Appl. **28**, 105–120
83. M. Klazar, Twelve countings with rooted plane trees. Eur. J. Comb. **18**(2), 195–210 (1997)
84. D.J. Klein, I. Lukovits, I. Gutman, On the definition of the hyper-Wiener index for cycle-containing structures. J. Chem. Inf. Comput. Sci. **35**, 50–52 (1995)
85. B. Knudsen, Optimal Multiple Parsimony Alignment with Affine Gap Cost Using a Phylogenetic Tree. Lecture Notes in Bioinformatics, vol. 2812 (Springer, New York, 2003), pp. 433–446
86. M. Krzywkowski, Trees having many minimal dominating sets. Inform. Process. Lett. **113**(8), 276–279 (2013)
87. M. Krzywkowski, Minimal 2-dominating sets in trees. RAIRO Theor. Inform. Appl. **47**(3), 235–240 (2013)
88. M. Krzywkowski, An algorithm for listing all minimal double dominating sets of a tree. Fund. Inform. **130**(4), 415–421 (2014)
89. M. Krzywkowski, S. Wagner, Trees having few total dominating sets. In preparation (2015)
90. H.-F. Law, On the number of independent sets in a tree. Electron. J. Comb. **17**(1), paper no. #P18 (2010)
91. M. Lepović, I. Gutman, A collective property of trees and chemical trees. J. Chem. Inf. Comput. Sci. **38**, 823–826 (1998)
92. V.E. Levit, E. Mandrescu, The independence polynomial of a graph—a survey, in *Proceedings of the 1st International Conference on Algebraic Informatics* (Aristotle University of Thessaloniki, Thessaloniki, 2005), pp. 233–254
93. S. Li, S. Wang, Further analysis on the total number of subtrees of trees. Electron. J. Comb. **19**(4), paper no. #P48 (2012)
94. X. Li, Y. Shi, A survey on the Randić index. MATCH Commun. Math. Comput. Chem. **59**(1), 127–156 (2008)
95. X. Li, I. Gutman, *Mathematical Aspects of Randić-type Molecular Structure Descriptors*. Mathematical Chemistry Monographs, vol. 1 (University of Kragujevac, Kragujevac, 2006)
96. X. Li, Z. Li, L. Wang, The inverse problems for some topological indices in combinatorial chemistry. J. Comput. Biol. **10** (1), 47–55 (2003)
97. X. Li, Y. Shi, I. Gutman, *Graph Energy* (Springer, New York, 2012)
98. V. Linek, Bipartite graphs can have any number of independent sets. Discrete Math. **76** (2), 131–136 (1989)
99. L. Lovász, *Combinatorial Problems and Exercises* (North-Holland, Amsterdam, 1979)
100. L. Lu, Y. Yang, A theorem on Randić index and the diameter of a graph. Discrete Math. **311**(14), 1333–1343 (2011)

101. A. Meir, J.W. Moon, On subtrees of certain families of rooted trees. Ars Comb. **16B**, 305–318 (1983)
102. A. Meir, J.W. Moon, On maximal independent sets of nodes in trees. J. Graph Theory **12**(2), 265–283 (1988)
103. R.E. Merrifield, H.E. Simmons, *Topological Methods in Chemistry* (Wiley, New York, 1989)
104. G.O. Munsonius, On the asymptotic internal path length and the asymptotic Wiener index of random split trees. Electron. J. Probab. **16** paper no. 35, 1020–1047 (2011)
105. G.O. Munsonius, On tail bounds for random recursive trees. J. Appl. Probab. **49**, 566–581 (2012)
106. R. Neininger, The Wiener index of random trees. Comb. Probab. Comput. **11**(6), 587–597 (2002)
107. V. Nikiforov, Walks and the spectral radius of graphs. Linear Algebra Appl. **418**(1), 257–268 (2006)
108. L. Pavlovic, M. Stojanovic, X. Li, More on "Solutions to two unsolved questions on the best upper bound for the Randić Index R_{-1} of trees". MATCH Commun. Math. Comput. Chem. **58**(1), 167–182 (2007)
109. D. Plavsić, S. Nikolić, N. Trinajstić, Z. Mihalić, On the Harary index for the characterization of chemical graphs. J. Math. Chem. **12**, 235–250 (1993)
110. H. Prodinger, R.F. Tichy, Fibonacci numbers of graphs. Fibonacci Quart. **20**(1), 16–21 (1982)
111. M. Randić, Characterization of molecular branching. J. Am. Chem. Soc. **97** (23), 6609–6615 (1975)
112. M. Randić, Novel molecular descriptor for structure-property studies. Chem. Phys. Lett. **211**, 478–483 (1993)
113. B.E. Sagan, A note on independent sets in trees. SIAM J. Discrete Math. **1**(1), 105–108 (1988)
114. B.E. Sagan, V.R. Vatter, Maximal and maximum independent sets in graphs with at most r cycles. J. Graph Theory **53**(4), 283–314 (2006)
115. N.S. Schmuck, S. Wagner, H. Wang, Greedy trees, caterpillars, and Wiener-type graph invariants. MATCH Commun. Math. Comput. Chem. **68**, 273–292 (2012)
116. E. Schroeder, Vier combinatorische Probleme. Z. f. Math. Phys. **15**, 361–376 (1870)
117. R. Shi, The average distance of trees. Syst. Sci. Math. Sci. **6**(1), 18–24 (1993)
118. A.V. Sills, H. Wang, On the maximal Wiener index and related questions. Discrete Appl. Math. **160**, 1615–1623 (2012)
119. A.V. Sills, H. Wang, The minimal number of subtrees of a tree. Graphs Comb. **31**(1), 255–264 (2015)
120. Z. Skupień, Majorization and the minimum number of dominating sets. Discrete Appl. Math. **165**, 295–302 (2014)
121. H. Smith, L.A. Székely, H. Wang, Eccentricity in trees (2014, submitted, arXiv:1408.5865)
122. H. Smith, L.A. Székely, H. Wang, S. Yuan, On different "middle parts" of a tree. In preparation (2015)
123. L.A. Székely, H. Wang, Binary trees with the largest number of subtrees with at least one leaf. Congr. Numer. **177**, 147–169 (2005),
124. L.A. Székely, H. Wang, On subtrees of trees. Adv. Appl. Math. **34**, 138–155 (2005)
125. L.A. Székely, H. Wang, Binary trees with the largest number of subtrees. Discrete Appl. Math. **155**(3), 374–385 (2006)
126. L.A. Székely, H. Wang, Extremal values of ratios: distance problems vs. subtree problems in trees. Electron. J. Comb. **20**(1), #P67, 1–20 (2013)
127. L.A. Székely, H. Wang, Extremal values of ratios: distance problems vs. subtree problems in trees II. Discrete Math. **322**, 36–47 (2014)
128. L.A. Székely, H. Wang, T. Wu, The sum of distances between the leaves of a tree and the 'semi-regular' property. Discrete Math. **311**, 1197–1203 (2011)
129. Y. Tang, B. Zhou, On Average Eccentricity. MATCH Commun. Math. Comput. Chem. **67**, 405–423 (2012)
130. H. Täubig, J. Weihmann, S. Kosub, R. Hemmecke, E.W. Mayr, Inequalities for the number of walks in graphs. Algorithmica **66**(4), 804–828 (2013)

131. H. Täubig, J. Weihmann, Matrix power inequalities and the number of walks in graphs. Discrete Appl. Math. **176**, 122–129 (2014)
132. R. Todeschini, V. Consonni, *Handbook of Molecular Descriptors* (Wiley-VCH, Weinheim, 2000)
133. N. Trinajstić, *Chemical Graph Theory* (CRC Press, Boca Raton, 1992)
134. A. Vince, H. Wang, The average order of a subtree of a tree. J. Comb. Theory Ser. B **100**(2), 161–170 (2010)
135. S. Wagner, A class of trees and its Wiener index. Acta Appl. Math. **91**(2), 119–132 (2006)
136. S. Wagner, Correlation of graph-theoretical indices. SIAM J. Discrete Math. **21**(1), 33–46 (2007)
137. S. Wagner, On the average Wiener index of degree-restricted trees. Australas. J. Comb. **37**, 187–203 (2007)
138. S. Wagner, On the number of matchings of a tree. Eur. J. Comb. **28**(4), 1322–1330 (2007)
139. S. Wagner, Extremal trees with respect to Hosoya index and Merrifield-Simmons index. MATCH Commun. Math. Comput. Chem. **57**(1), 221–233 (2007)
140. S. Wagner, Almost all trees have an even number of independent sets. Electron. J. Comb. **16**(1) paper no. #93 (2009)
141. S. Wagner, A note on the number of dominating sets of a graph. Util. Math. **92**, 25–31 (2013)
142. S. Wagner, Central limit theorems for additive tree parameters with small toll functions. Comb. Probab. Comput. **24**(1), 329–353 (2015)
143. S. Wagner, I. Gutman, Maxima and minima of the Hosoya index and the Merrifield-Simmons index: a survey of results and techniques. Acta Appl. Math. **112**(3), 323–346 (2010)
144. S. Wagner, H. Wang, Indistinguishable trees and graphs. Graphs Comb. **30**(6), 1593–1605 (2014)
145. S. Wagner, H. Wang, On the local and global means of subtree orders. J. Graph Theory **85**(2), 154–166 (2016)
146. S. Wagner, H. Wang, G. Yu, Molecular graphs and the inverse Wiener index problem, Discrete Appl. Math. **157**(7), 1544–1554 (2009)
147. S. Wagner, H. Wang, X.D. Zhang, Distance-based graph invariants of trees and the Harary index. Filomat. **27**, 39–48 (2013)
148. H. Wang, Extremal trees with given degree sequence for the Randić index. Discrete Math. **308**, 3407–3411 (2008)
149. H. Wang, The extremal values of the Wiener index of a tree with given degree sequence. Discrete Appl. Math. **156**, 2647–2654 (2009)
150. H. Wang, Functions on adjacent vertex degrees of trees with given degree sequence. Cent. Eur. J. Math. **12**, 1656–1663 (2014)
151. H. Wang, The distances between internal vertices and leaves of a tree. Eur. J. Comb. **41**, 79–99 (2014)
152. H. Wang, G. Yu, All but 49 numbers are Wiener indices of trees. Acta Appl. Math. **92**(1) 15–20 (2006)
153. H. Wiener, Structural determination of paraffin boiling points. J. Am. Chem. Soc. **69**, 17–20 (1947)
154. H. Wiener, Correlation of heats of isomerization, and differences in heats of vaporization of isomers, among the paraffin hydrocarbons. J. Am. Chem. Soc. **69**, 2636–2638 (1947)
155. H.S. Wilf, The number of maximal independent sets in a tree. SIAM J. Algebraic Discrete Methods **7**(1), 125–130 (1986)
156. I. Włoch, Trees with extremal numbers of maximal independent sets including the set of leaves. Discrete Math. **308**(20), 4768–4772 (2008)
157. I. Włoch, A. Włoch, The number of independent sets intersecting the set of leaves in trees. Ars Comb. **85**, 225–231 (2007)
158. K. Xu, Trees with the seven smallest and eight greatest Harary indices. Discrete Appl. Math. **160**, 321–331 (2012)
159. K. Xu, M. Liu, K. Das, I. Gutman, B. Furtula, A survey on graphs extremal with respect to distance-based topological indices. MATCH Commun. Math. Comput. Chem. **71**, 461–508 (2014)

160. W. Yan, L. Ye, On the maximal energy and the Hosoya index of a type of trees with many pendant vertices. MATCH Commun. Math. Comput. Chem. **53**(2), 449–459 (2005)
161. P. Yu, An upper bound on the Randić index of trees, J. Math. Study **31**, 225–230 (1998) (in Chinese)
162. B. Zelinka, Medians and peripherans of trees. Arch. Math. (Brno) **4**, 87–95 (1968)
163. X.D. Zhang, Y. Liu, M. Han, The maximum Wiener index of trees with given degree sequences. MATCH Commun. Math. Comput. Chem. **64**, 661–682 (2010)
164. X.D. Zhang, Q. Xiang, L. Xu, R. Pan, The Wiener index of trees with given degree sequences. MATCH Commun. Math. Comput. Chem. **60**, 623–644 (2008)
165. X.M. Zhang, X.D. Zhang, The minimal number of subtrees with a given degree sequence. Graphs Comb. **31**(1), 309–318 (2015)
166. X.M. Zhang, X.D. Zhang, D. Gray, H. Wang, Trees with the most subtrees—an algorithmic approach. J. Comb. **3**(2), 207–223 (2012)
167. X.M. Zhang, X.D. Zhang, D. Gray, H. Wang, The number of subtrees of trees with given degree sequence. J. Graph Theory **73**(3), 280–295 (2013)
168. J. Zito, The structure and maximum number of maximum independent sets in trees. J. Graph Theory **15**(2), 207–221 (1991)

The edit distance in graphs: Methods, results, and generalizations

Ryan R. Martin

Abstract The edit distance is a very simple and natural metric on the space of graphs. In the edit distance problem, we fix a hereditary property of graphs and compute the asymptotically largest edit distance of a graph from the property. This quantity is very difficult to compute directly but in many cases, it can be derived as the maximum of the edit distance function. Szemerédi's regularity lemma, strongly regular graphs, constructions related to the Zarankiewicz problem – all these play a role in the computing of edit distance functions. The most powerful tool is derived from symmetrization, which we use to optimize quadratic programs that define the edit distance function. In this paper, we describe some of the most common tools used for computing the edit distance function, summarize the major current results, outline generalizations to other combinatorial structures, and pose some open problems.

Keywords Edit distance • Colored regularity graphs • Quadratic programming

Mathematics Subject Classification (2010): 05C35, 05C80

1 Introduction

The edit distance in graphs was originally studied to answer two different and independent problems: one to answer questions on property testing [5] and the other, to answer a question regarding consensus trees from evolutionary biology [10]. In metabolic networks, the presence or absence of edges in a certain graph corresponds to pairs of genes which activate or deactivate one another. In evolutionary theory, avoiding forbidden induced subgraphs [23] is studied, which is equivalent to a similar edit problem of bipartite graphs or matrices. Edit distance problems with respect to more general classes of graphs are important in the algorithmic aspects of property testing [3, 5–7] and in the techniques involved in computing the speed of dense graph properties [19, 46].

R.R. Martin (✉)
Iowa State University, Ames, IA 50011-2064, USA
e-mail: rymartin@iastate.edu

A. Beveridge et al. (eds.), *Recent Trends in Combinatorics*, The IMA Volumes in Mathematics and its Applications 159, DOI 10.1007/978-3-319-24298-9_2

The (normalized) edit metric is a metric on the set of simple, labeled n-vertex graphs. The distance between two graphs is the symmetric difference of the edge sets divided by the total number of possible edges. If $\mathrm{dist}(G,G')$ denotes the edit distance between G and G' on the same labeled vertex set, then

$$\mathrm{dist}(G,G') = |E(G)\triangle E(G')|/\tbinom{n}{2}.$$

As with any metric, we may take a *property* of graphs $\mathscr{H}$ (that is, a set of graphs), and compute the distance of a graph from that property:

$$\mathrm{dist}(G,\mathscr{H}) = \min\left\{\mathrm{dist}(G,G') : V(G') = V(G)\right\}. \tag{1}$$

The properties that we study in this paper are *hereditary properties.* A property of graphs is hereditary if is closed under isomorphism and deletion of vertices. Alon and Stav [5] suggest that "In fact, almost all interesting graph properties are hereditary." Planarity having chromatic number at most k or not having a given H as an induced subgraph all are commonly studied hereditary properties. The property of having no graph H as an induced subgraph is called a *principal hereditary property* and we denote it by $\mathrm{Forb}(H)$. For every hereditary property $\mathscr{H}$ there exists a family of graphs $\mathscr{F}(\mathscr{H})$ ("forbidden graphs") such that $\mathscr{H} = \bigcap_{H\in\mathscr{F}(H)} \mathrm{Forb}(H)$. A hereditary property is said to be *nontrivial* if there is an infinite sequence of graphs that is in the property.

In the seminal papers by Alon and Stav [4, 5] and by Axenovich, Kézdy, and Martin [10], the fundamental question was the maximum distance of a graph G on n vertices from hereditary property $\mathscr{H}$. In fact, the maximum distance is asymptotically the same as that of the Erdős-Rényi random graph $G(n,p)$, for some value of p.

Theorem 1 (Alon-Stav [5]) *Let $\mathscr{H}$ be an arbitrary graph property. There exists $p^* = p^*_{\mathscr{H}} \in [0,1]$ such that*

$$\max\{\mathrm{dist}(G,\mathscr{H}) : |V(G)| = n\} = \mathbb{E}[\mathrm{dist}(G(n,p^*),\mathscr{H})] + o(1). \tag{2}$$

We denote the limit of the quantity in (2) by $d^*_{\mathscr{H}}$. This is, asymptotically, the maximum distance of a graph from $\mathscr{H}$. Although $d^*_{\mathscr{H}}$ is the quantity in which we are most interested, determining its value is most often done by generalizing the result in Theorem 1. We do so by instead finding the maximum edit distance of a density-p graph from $\mathscr{H}$, for **all** values of p.

Balogh and Martin [11] introduced the *edit distance function* of a hereditary property.

Definition 2 *Let $\mathscr{H}$ be a nontrivial hereditary property of graphs. The* edit distance function *of $\mathscr{H}$ is*

$$\mathrm{ed}_{\mathscr{H}}(p) := \lim_{n\to\infty} \max\left\{\mathrm{dist}(G,\mathscr{H}) : |V(G)| = n, |E(G)| = \left\lfloor p\tbinom{n}{2}\right\rfloor\right\}. \tag{3}$$

The existence of the limit in (3) was proven in [11].[1]

Theorem 3 (Balogh-Martin [11]) *Let $\mathcal{H}$ be an arbitrary nontrivial graph property. Then*

$$\mathrm{ed}_{\mathcal{H}}(p) = \lim_{n\to\infty} \mathbb{E}[\mathrm{dist}(G(n,p),\mathcal{H})].$$

Theorems 1 and 3 make use of Szemerédi's regularity lemma [48] but in a way that detects induced subgraphs. The idea of applying Szemerédi's regularity lemma to hereditary properties has been studied in a number of contexts, including pioneering work by Prömel and Steger [40–42], Scheinerman and Zito [46], and Bollobás and Thomason [18–20]. The essential technique is to apply the regularity lemma twice – once to the graph itself and a second time to each of the graphs induced by the non-exceptional clusters. More directly, one can use a variant of Szemerédi's regularity lemma due to Alon et al. [7] that has been used in a number of papers, including the edit distance papers [5, 11].

The edit distance function is symmetric with respect to complementation. It is easy to see that $\mathrm{ed}_{\mathrm{Forb}(H)}(p) = \mathrm{ed}_{\mathrm{Forb}(\overline{H})}(1-p)$ and, in fact, a general case is true.

Proposition 4 *Let $\mathcal{H} = \bigcap_{H\in\mathcal{F}(\mathcal{H})} \mathrm{Forb}(H)$ be a nontrivial hereditary property and let $\mathcal{H}^* = \bigcap_{H\in\mathcal{F}(\mathcal{H})} \mathrm{Forb}(\overline{H})$. Then $\mathrm{ed}_{\mathcal{H}}(p) = \mathrm{ed}_{\mathcal{H}^*}(1-p)$.*

A very similar setting to the edit distance problem was studied by Richer [43] and as further investigated by Marchant and Thomason [31, 32] regarding the two-coloring of the edges of the complete graph. Many of the most vital results for solving the edit distance problem come from this setting. In solving the problems they pose on a hereditary property $\mathcal{H}$, they obtain the function $1 - \mathrm{ed}_{\mathcal{H}}(p)$. The connection between the two settings is addressed in [32] as well as by Thomason [49] in a survey.

2 Colored regularity graphs

The key observation in computing the edit distance is that a graph can be approximated by a graph-like structure in which the clusters either behave like cliques or independent sets and the ϵ-regular pairs either behave like complete bipartite graphs, empty bipartite graphs or random graphs with density bounded away from both 0 and 1.

Alon and Stav [5] defined a *colored regularity graph (CRG) K* to be a simple complete graph, together with a partition of the vertices into white and black,

[1] It should be noted that early papers on the edit distance do not normalize the distance. That is, the distance is merely $|E(G)\Delta E(G')|$. Normalization, however, is required in order to define the edit distance function and it seems most natural to put the normalization in the metric itself, rather than doing so in order to define $\mathrm{ed}_{\mathcal{H}}$.

$V(K) = \mathrm{VW}(K) \cup \mathrm{VB}(K)$ and a partition of the edges into white, gray, and black, $E(K) = \mathrm{EW}(K) \cup \mathrm{EG}(K) \cup \mathrm{EB}(K)$.[2]

We say that a graph H embeds in K, writing $H \mapsto K$, if there is a function $\varphi : V(H) \to V(K)$ so that the following occur:

- If $h_1h_2 \in E(H)$, then either $\varphi(h_1) = \varphi(h_2) \in \mathrm{VB}(K)$ or $\varphi(h_1)\varphi(h_2) \in \mathrm{EB}(K) \cup \mathrm{EG}(K)$.
- If $h_1h_2 \notin E(H)$, then either $\varphi(h_1) = \varphi(h_2) \in \mathrm{VW}(K)$ or $\varphi(h_1)\varphi(h_2) \in \mathrm{EW}(K) \cup \mathrm{EG}(K)$.

A CRG K' is said to be a *sub-CRG* of K if K' can be obtained by deleting vertices of K and is a *proper sub-CRG* if $K' \neq K$.

If a graph H embeds in CRG K, then a large enough graph that is approximated by K will have an induced copy of H. This is stated and proven more formally in Section 4 of [11]. However, the main idea is that for any large graph in a hereditary property $\mathscr{H} = \bigcap_{H\in\mathscr{F}(\mathscr{H})} \mathrm{Forb}(H)$, the CRG K that approximates the graph satisfies the property that $H \not\mapsto K$ for all $H \in \mathscr{F}(\mathscr{H})$. We denote $\mathscr{K}(\mathscr{H})$ to be the subset of CRGs K such that no forbidden graph maps into K. Formally, $\mathscr{K}(\mathscr{H}) = \{K : H \not\mapsto K, \forall H \in \mathscr{F}(\mathscr{H})\}$.

2.1 The *f* and *g* functions

There is a matrix associated with a CRG called $\mathbf{M}_K(p)$ that plays a role similar to the role the adjacency matrix does for graphs. We can use this matrix to help define the functions f_K and g_K, that are essential for understanding edit distance.

Definition 5 *Let K be a CRG on vertex set $\{v_1, \ldots, v_k\}$ with* VW *and* VB *denoting the white and black vertices, respectively, and* EW, EG, *and* EB *denoting the white, gray, and black edges, respectively. Let* $\mathbf{M}_K(p)$ *denote the matrix with entries defined as follows:*

$$m_K(p)_{ij} = \begin{cases} p, & \text{if } i \neq j \text{ and } v_iv_j \in \mathrm{EW} \text{ or } i = j \text{ and } v_i \in \mathrm{VW}; \\ 0, & \text{if } i \neq j \text{ and } v_iv_j \in \mathrm{EG}; \\ 1-p, & \text{if } i \neq j \text{ and } v_iv_j \in \mathrm{EB} \text{ or } i = j \text{ and } v_i \in \mathrm{VB}. \end{cases} \tag{4}$$

The functions f_K and g_K are defined as follows:

$$f_K(p) = \frac{1}{k^2}\left[p(|\mathrm{VW}| + 2|\mathrm{EW}|) + (1-p)(|\mathrm{VB}| + 2|\mathrm{EB}|)\right] = \frac{1}{k^2}\mathbf{1}^T\mathbf{M}_K(p)\mathbf{1} \tag{5}$$

[2] Papers by Bollobás and Thomason [18–20] and others such as [49] use the term "type" rather than CRG.

$$g_K(p) = \min \left\{ \mathbf{x}^T \mathbf{M}_K(p) \mathbf{x} : \mathbf{x}^T \mathbf{1} = 1, \mathbf{x} \geq \mathbf{0} \right\}. \tag{6}$$

The vector $\mathbf{0}$ *is the all-zeroes vector,* $\mathbf{1}$ *is the all-ones vector, and vector inequalities are coordinatewise.*

Clearly, for any CRG K and any $p \in [0, 1]$, we have $g_K(p) \leq f_K(p)$. Although the linearity of f_K makes proving general results about $\mathrm{ed}_{\mathscr{H}}$ possible, the g function is more useful in computing the edit distance function. In fact, if an optimal vector of (6) has a zero entry, we may obtain K' by deleting the corresponding entry and achieve $g_{K'}(p) = g_K(p)$. We say that a CRG K is *p-core* if, for any proper sub-CRG K' of K, we have $g_{K'}(p) > g_K(p)$.

The edit distance function can be defined in terms of the f and g functions:

Theorem 6 *Let* $\mathscr{H}$ *be a nontrivial hereditary property. For any* $p \in [0, 1]$,

$$\begin{aligned} \mathrm{ed}_{\mathscr{H}}(p) &= \inf \{ f_K(p) : K \in \mathscr{K}(\mathscr{H}) \} = \inf \{ g_K(p) : K \in \mathscr{K}(\mathscr{H}) \} \qquad (7) \\ &= \min \{ g_K(p) : K \in \mathscr{K}(\mathscr{H}) \}. \qquad (8) \end{aligned}$$

Equation (7) is due to Balogh and Martin [11]. Equation (8) is from the results of Marchant and Thomason [32] and gives rise to the question as to whether only a finite set of p-core CRGs is sufficient to define the edit distance function for any nontrivial hereditary property and all $p \in [0, 1]$.

There is some evidence (see Theorem 31(d) and Theorem 30(b) below) that for some hereditary properties, determining the edit distance function requires knowledge of an infinite sequence of CRGs. Nonetheless, we believe that the bulk of the edit distance function can be determined from a finite number of CRGs. That is, for any $\epsilon > 0$, we believe a finite set of CRGs can simultaneously define $\mathrm{ed}_{\mathscr{H}}$ for all $p \in [\epsilon, 1 - \epsilon]$. This is Conjecture 1 in Section 8.2.

2.2 Clique spectrum

Certain colored regularity graphs play a key role in the computation of the edit distance. A *gray-edge CRG* is the CRG K with all $\binom{|V(K)|}{2}$ edges gray. The gray-edge CRG with r white vertices and s black vertices is denoted $K(r, s)$. The *clique spectrum* of $\mathscr{H}$ is the set

$$\Gamma(\mathscr{H}) \stackrel{\text{def}}{=} \{ (r, s) : H \not\mapsto K(r, s), \forall H \in \mathscr{F}(\mathscr{H}) \}.$$

For example, if $\mathscr{H} = \mathrm{Forb}(H)$ is a hereditary property, then pairs (r, s) are in the clique spectrum of $\mathrm{Forb}(H)$ if and only if H **cannot** be partitioned into r independent sets and s cliques.

The clique spectrum has a number of useful properties. For example, it is monotone in the sense that if $(r, s) \in \Gamma(\mathscr{H})$ and $0 \leq r' \leq r$ and $0 \leq s' \leq s$, then $(r', s') \in \Gamma(\mathscr{H})$. As a result, the clique spectrum of a hereditary property can

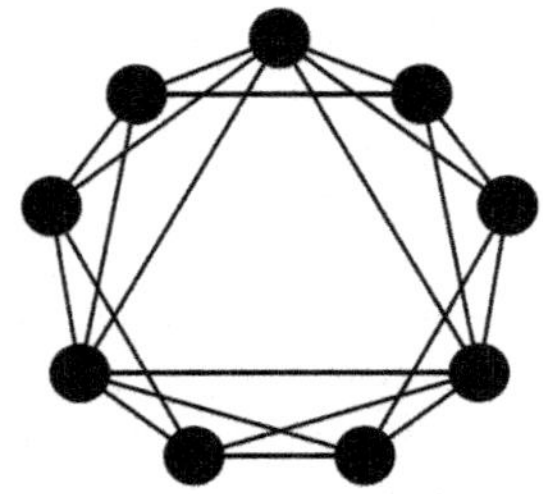

Fig. 1 A graph H_9 on 9 vertices.

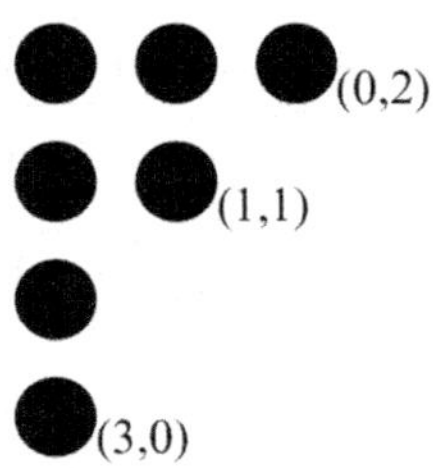

Fig. 2 The Ferrers diagram of the clique spectrum of H_9 with the extreme points labeled.

be expressed as a Ferrers diagram. An *extreme point* of the clique spectrum Γ is a pair $(r, s) \in \Gamma$ for which both $(r+1, s) \notin \Gamma$ and $(r, s+1) \notin \Gamma$. Figure 1 shows the graph H_9, and Figure 2 shows its clique spectrum, expressed as a Ferrers diagram.

Since the matrix $\mathbf{M}_{K(r,s)}(p)$ is a diagonal matrix with r entries of value p and s entries with value $1-p$, it is easy to compute that for all $p \in (0, 1)$

$$g_{K(r,s)}(p) = \left(\frac{r}{p} + \frac{s}{1-p}\right)^{-1} = \frac{p(1-p)}{r(1-p)+sp}. \tag{9}$$

We have the natural convention that if $r = 0$ then $g_{K(r,s)}(0) = 1$ and if $s = 0$ then $g_{K(r,s)}(1) = 1$.

In fact, we have a more general way of computing the edit distance function if the matrix $\mathbf{M}_K(p)$ is block diagonal matrix, where the blocks correspond to a CRG notion of components.

Definition 7 *A sub-CRG, K', of a CRG K is a* component *if it is maximal with respect to the property that, for all $v, w \in V(K')$, there exists a path consisting of white and black edges entirely within K'.*

More simply, components of K are the components of the graph G with vertex set $V(K)$ and the nonedges of G the gray edges of K. This leads to the generalization of (9).

Proposition 8 (Martin [33]) *Let K be a CRG with components $K_{(1)}, \ldots, K_{(\ell)}$. Then*

$$(g_K(p))^{-1} = \sum_{i=1}^{\ell} (g_{K_{(i)}}(p))^{-1}.$$

2.3 Characterization of p-core CRGs

Marchant and Thomason [32] gave a characterization of all p-core CRGs.

Theorem 9 (Marchant-Thomason [32]) *Let K be a p-core CRG.*

(a) If $p \leq 1/2$, then there are no black edges and the white edges are only incident to black vertices.
(b) If $p \geq 1/2$, then there are no white edges and the black edges are only incident to white vertices.

Consequently, if $p = 1/2$, then all edges are gray.

Theorem 9 is an essential tool and is used in most results on the edit distance function as we shall see below.

3 Estimating the edit distance function

Although it is difficult to compute the edit distance function for general hereditary properties, we can estimate the function through a variety of techniques. First, we can use the clique spectrum and (9) to construct an upper bound.

3.1 Upper bound via the clique spectrum

We begin with a trio of results with elementary proofs, followed by a result concerning the nature of the edit distance function.

Theorem 10 *Let $\mathcal{H}$ be a nontrivial hereditary property and let $\Gamma(\mathcal{H})$ denote the clique spectrum of $\mathcal{H}$. If we define*

$$\gamma_{\mathcal{H}}(p) := \min_{(r,s)\in\Gamma(\mathcal{H})} g_{K(r,s)}(p) = \min_{(r,s)\in\Gamma(\mathcal{H})} \frac{p(1-p)}{r(1-p)+sp}, \tag{10}$$

then $\mathrm{ed}_{\mathcal{H}}(p) \leq \gamma_{\mathcal{H}}(p)$.

There are three (not necessarily distinct) extreme points of a clique spectrum that are of particular interest. First, if $(r,0) \in \Gamma(\mathcal{H})$ but $(r+1,0) \notin \Gamma(\mathcal{H})$, then $r+1$ is the *chromatic number* of $\mathcal{H}$, denoted $\chi(\mathcal{H})$ or just χ, when the hereditary property is understood. Second, if $(0,s) \in \Gamma(\mathcal{H})$ but $(0,s+1) \notin \Gamma(\mathcal{H})$, then $s+1$ is the *complementary chromatic number* of $\mathcal{H}$, denoted $\overline{\chi}(\mathcal{H})$ or just $\overline{\chi}$. Note that if $\mathcal{H} = \mathrm{Forb}(H)$ for some graph H, then $\chi(\mathcal{H}) = \chi(H)$ and $\overline{\chi}(\mathcal{H}) = \chi(\overline{H})$.

We observe that if $\chi(\mathcal{H}) \geq 2$ then $(\chi-1,0) \in \Gamma(\mathcal{H})$ and if $\overline{\chi}(\mathcal{H}) \geq 2$ then $(0,\overline{\chi}-1) \in \Gamma(\mathcal{H})$. Therefore, we have the following corollary.

Corollary 11 *Let $\mathcal{H}$ be a nontrivial hereditary property with chromatic number χ and complementary chromatic number $\overline{\chi}$.*

(a) If $\chi \geq 2$, then $\mathrm{ed}_{\mathcal{H}}(p) \leq p/(\chi - 1)$.
(b) If $\overline{\chi} \geq 2$, then $\mathrm{ed}_{\mathcal{H}}(p) \leq (1-p)/(\overline{\chi} - 1)$.

The chromatic and complementary chromatic numbers of a hereditary property $\mathcal{H}$ can be defined in terms of $\mathcal{F}(\mathcal{H})$ as in Proposition 12.

Proposition 12 *Let $\mathcal{H} = \bigcap_{H \in \mathcal{F}} \mathrm{Forb}(H)$ be a nontrivial hereditary property. Then,*

(a) $\chi(\mathcal{H}) = \min\{\chi(H) : H \in \mathcal{F}\}$ and
(b) $\overline{\chi}(\mathcal{H}) = \min\{\chi(\overline{H}) : H \in \mathcal{F}\}$.

The third extreme point we address is evaluated as follows: the largest value of $r + s + 1$ such that $(r, s) \in \Gamma(\mathcal{H})$ is called the *binary chromatic number* of $\mathcal{H}$ and is denoted $\chi_B(\mathcal{H})$ or just χ_B. This quantity has appeared in the literature previously. Prömel and Steger [40–42] called $\chi_B - 1$ simply τ. Bollobás and Thomason [18, 20] called $\chi_B - 1$ the *coloring number*.

Since Theorem 9 establishes that every 1/2-core CRG is a gray-edge CRG (that is, of the form $K(r, s)$) we can compute $\mathrm{ed}_{\mathcal{H}}(1/2)$ in terms of $\chi_B(\mathcal{H})$. Combining this with other basic facts, we obtain Theorem 13.

Theorem 13 *Let $\mathcal{H}$ be a nontrivial hereditary property.*

(a) $\mathrm{ed}_{\mathcal{H}}(p)$ is continuous.
(b) $\mathrm{ed}_{\mathcal{H}}(p)$ is concave down.
(c) $\mathrm{ed}_{\mathcal{H}}(1/2) = \frac{1}{2(\chi_B(\mathcal{H})-1)}$.

Theorem 13(a) was established by Marchant and Thomason [32]. We note that a different, analysis-based proof of this is in [11]. Theorem 13(b) was proven in [11]. Theorem 13(c) was proven in [10] in the case where $\mathcal{H}$ is a principal hereditary property. More sophisticated knowledge of the edit distance function has made it a simple corollary.

Using only Corollary 11 and Theorem 13 we can already find edit distance functions for some important hereditary properties. If P_4 denotes the path on 4 vertices, then $\mathrm{ed}_{\mathrm{Forb}(P_4)}(p) = \min\{p, 1-p\}$. If C_5 denotes the cycle on 5 vertices, then $\mathrm{ed}_{\mathrm{Forb}(C_5)}(p) = \frac{1}{2}\min\{p, 1-p\}$. More about hereditary properties forbidding self-complementary graphs is below in Corollary 15.

Because $\mathrm{ed}_{\mathcal{H}}(p)$ is continuous and concave down, the function achieves its maximum on the interval [0, 1]. Thus, both $d^*_{\mathcal{H}}$ and $p^*_{\mathcal{H}}$ are well-defined and the coordinate $(p^*_{\mathcal{H}}, d^*_{\mathcal{H}})$ is the point at which $\mathrm{ed}_{\mathcal{H}}$ achieves its maximum value.

Note: Although $p^*_{\mathcal{H}}$ is formally defined to be a closed interval, in all but a few (very) interesting cases[3] the interval is degenerate. That is, $p^*_{\mathcal{H}}$ is usually a single value. We will often abuse notation and terminology by referring to $p^*_{\mathcal{H}} = p$, rather than $p^*_{\mathcal{H}} = [p, p]$ and instead indicate explicitly where the interval is not degenerate.

---

[3]See, e.g., Section 5.5.2.

3.2 Upper bound using $\chi_B(\mathscr{H})$

If $\mathscr{H} = \bigcap_{H\in\mathscr{F}} \mathrm{Forb}(H)$ is a hereditary property such that $\mathscr{F}$ contains no complete graph and no empty graph, then it is trivial that $\mathrm{ed}_{\mathscr{H}}(0) = \mathrm{ed}_{\mathscr{H}}(1) = 0$. Indeed, Proposition 12 gives that $\chi(\mathscr{H}) \geq 2$ and $\overline{\chi}(\mathscr{H}) \geq 2$. The statement then follows from the simple bounds given by Corollary 11.

Using only the $\gamma_{\mathscr{H}}$ function, we may narrow down the possible values for $p^*_{\mathscr{H}}$ and for $d^*_{\mathscr{H}}$.

Theorem 14 *Let* $\mathscr{H} = \bigcap_{H\in\mathscr{F}} \mathrm{Forb}(H)$ *with* $(r, s) \in \Gamma(\mathscr{H})$ *such that* $r + s = \chi_B(\mathscr{H}) - 1$.

(a) $\mathrm{ed}_{\mathscr{H}}(p) \leq \gamma_{\mathscr{H}}(p) \leq \frac{p(1-p)}{r(1-p)+sp}$ *for all* $p \in [0, 1]$.
(b) $d^*_{\mathscr{H}} \leq \frac{1}{r+s+2\sqrt{rs}}$.
(c) $\mathrm{ed}_{\mathscr{H}}(p) \geq \min\left\{\frac{p}{\chi_B(\mathscr{H})-1}, \frac{1-p}{\chi_B(\mathscr{H})-1}\right\}$.
(d) *If* $r \leq s$, *then* $p^*_{\mathscr{H}} \in \left[\frac{r}{r+s}, \frac{1}{2}\right]$.
(e) *If* $s \leq r$, *then* $p^*_{\mathscr{H}} \in \left[\frac{1}{2}, \frac{r}{r+s}\right]$.

Theorem 14(a) comes from Theorem 10. Theorem 14(b) is simply the maximum value of $g_{K(r,s)}(p)$. Theorem 14(c) follows from Theorem 13 – continuity, concavity, and the value of $\mathrm{ed}_{\mathscr{H}}(1/2)$ – and the fact that $\mathrm{ed}_{\mathscr{H}}(0) = \mathrm{ed}_{\mathscr{H}}(1) = 0$. Theorem 14(d) and (e) follow from the fact that these are the intervals over which $g_{K(r,s)}(p) \geq 1/(2(\chi_B(\mathscr{H}) - 1))$.

Corollary 15 gives the values of $p^*_{\mathscr{H}}$ and of $d^*_{\mathscr{H}}$ if $\mathscr{H} = \mathrm{Forb}(H)$ for a self-complementary graph H.

Corollary 15 *Let* $\mathscr{H} = \bigcap_{H\in\mathscr{F}} \mathrm{Forb}(H)$ *with* $(r_1, s_1) \in \Gamma(\mathscr{H})$ *and* $(r_2, s_2) \in \Gamma(\mathscr{H})$ *(not necessarily distinct) such that* $r_1 + s_1 = r_2 + s_2 = \chi_B(\mathscr{H}) - 1$, $r_1 \leq s_1$, *and* $r_2 \geq s_2$. *Then*

$$p^*_{\mathscr{H}} = 1/2 \qquad \text{and} \qquad d^*_{\mathscr{H}} = 1/(2(\chi_B(\mathscr{H}) - 1)).$$

In particular, if $\mathscr{H} = \mathrm{Forb}(H)$, *where* H *is a self-complementary graph, then* $p^*_{\mathscr{H}} = 1/2$ *and* $d^*_{\mathscr{H}} = 1/(2(\chi_B(H) - 1))$.

4 Symmetrization

The most powerful tool for determining the edit distance function of a hereditary property is called symmetrization. This is a term Pikhurko [38] used for a method due to Sidorenko [47]. In fact, symmetrization can be traced back to Zykov [50] and his proof of Turán's theorem. Our version of symmetrization comes directly from the matrix defined by a CRG.

Theorem 16 (Martin [33]) *Let $p \in [0,1]$ and let K be a p-core CRG with associated matrix $\mathbf{M}_K(p)$, as defined in (4). If $\mathbf{x}^*$ is an optimal solution of the quadratic program from (6), namely that $\mathbf{x}^* \geq \mathbf{0}$, $\mathbf{x}^*\mathbf{1} = 1$, and $g_K(p) = (\mathbf{x}^*)^T \mathbf{M}_K(p)\mathbf{x}^*$, then*

$$\mathbf{M}_K(p) \cdot \mathbf{x}^* = g_K(p)\mathbf{1}.$$

In addition, by virtue of K being p-core, the vector $\mathbf{x}^*$ has no zero entries and $\mathbf{x}^*$ is unique for any fixed labeling of the vertices of K.

4.1 The weighted gray degree of a vertex

In order to interpret Theorem 16, we define the *white neighborhood* of vertex v in CRG K to be $N_W(v) := \{v' \in V(K) : vv' \in \mathrm{EW}(K)\} \cup \{v : \text{if } v \in \mathrm{VW}(K)\}$. The *black neighborhood* of v is $N_B(v) := \{v' \in V(K) : vv' \in \mathrm{EB}(K)\} \cup \{v : \text{if } v \in \mathrm{VB}(K)\}$. The *gray neighborhood* of v is $N_G(v) := \{v' \in V(K) : vv' \in \mathrm{EG}(K)\}$.

If $\mathbf{x}$ is the optimum weight vector in the quadratic program (6) that defines $g_K(p)$, then the *weighted white degree* of vertex $v \in V(K)$ is $\mathrm{d}_W(v) := \sum_{v' \in N_W(v)} \mathbf{x}(v')$. The *weighted black degree* of vertex $v \in V(K)$ is $\mathrm{d}_B(v) := \sum_{v' \in N_B(v)} \mathbf{x}(v')$. The *weighted gray degree* of vertex $v \in V(K)$ is $\mathrm{d}_G(v) := \sum_{v' \in N_G(v)} \mathbf{x}(v')$.

Theorem 16 gives that, for any $v \in \mathrm{VW}(K)$,

$$p\mathrm{d}_W(v) + (1-p)\mathrm{d}_B(v) = g_K(p). \tag{11}$$

Using the characterization of p-core CRGs from Theorem 9, we can apply (11) to compute the gray degree of each vertex.

Theorem 17 (Martin [33]) *Let $p \in (0,1)$ and K be a p-core CRG with optimum weight vector $\mathbf{x}$.*

(a) If $p \leq 1/2$, then $\mathbf{x}(v) = g_K(p)/p$ for all $v \in \mathrm{VW}(K)$ and

$$\mathrm{d}_G(v) = \frac{p - g_K(p)}{p} + \frac{1-2p}{p}\mathbf{x}(v), \qquad \text{for all } v \in \mathrm{VB}(K).$$

(b) If $p \geq 1/2$, then $\mathbf{x}(v) = g_K(p)/(1-p)$ for all $v \in \mathrm{VB}(K)$ and

$$\mathrm{d}_G(v) = \frac{1-p-g_K(p)}{1-p} + \frac{2p-1}{1-p}\mathbf{x}(v), \qquad \text{for all } v \in \mathrm{VW}(K).$$

Most of the results below use Theorem 17 as a primary tool. Intuitively, if $g_K(p)$ is small, then $\mathrm{d}_G(v)$ is large for each vertex and so K has a large amount of gray.

However, if K has too much gray, then some $H \in \mathcal{F}(\mathcal{H})$ would map to K, which contradicts the choice of $K \in \mathcal{K}(\mathcal{H})$. This general paradigm is made more precise by knowing more about the structure of the CRGs $K \in \mathcal{K}(\mathcal{H})$.

4.2 Basic structural facts of p-core CRGs

We can use Theorem 17 to obtain some basic helpful results on certain types of CRGs:

Corollary 18 *Let $t \geq 2$ and $k \geq 2$ be integers.*

(a) Let $p \leq 1/2$ and let K be a p-core CRG on k black vertices.

(i) If K has no gray edges, then $g_K(p) = \frac{1}{k}[1 + (k-2)p]$.
(ii) If K has no gray clique of order t, then $g_K(p) > p/(t-1)$.

(b) Let $p \geq 1/2$ and let K be a p-core CRG on k white vertices.

(i) If K has no gray edges, then $g_K(p) = \frac{1}{k}[1 + (k-2)(1-p)]$.
(ii) If K has no gray clique of order t, then $g_K(p) > (1-p)/(t-1)$.

Proof By symmetry, it is sufficient to prove (a). For (a)(i) we observe that, by Theorem 9(a), all edges are white and it is easy to see that the optimum weight vector in equation (6) is constant. Thus, all vertices have the same weight and the result follows.

For (a)(ii), we use a well-worn trick, used, e.g., in [35]. Let the maximum-sized clique of K (in terms of the number of vertices) be on vertex set $\{v_1, \dots, v_c\}$ where $c \geq 2$. For every $w \notin \{v_1, \dots, v_c\}$ we know that wv_i is a gray edge for at most $c-1$ values of i. Using Theorem 17(a), we have

$$\sum_{i=1}^{c} \left(\frac{p - g_K(p)}{p} + \frac{1-2p}{p} \mathbf{x}(v_i) \right) \leq (c-1) \left(1 - \sum_{i=1}^{c} \mathbf{x}(v_i) \right)$$

$$(c - 3 + 1/p) \sum_{i=1}^{c} \mathbf{x}(v_i) + 1 \leq \frac{c}{p} g_K(p).$$

Since $c - 3 + 1/p \geq c - 1 > 0$, we can conclude that $g_K(p) > p/c \geq p/(t-1)$, which concludes the proof. □

Remark 19 *The bound in Corollary 18(a)(ii) can be approached by a CRG on black vertices where the gray edges induce a blow-up of K_{t-1}.*

5 Known results

5.1 *Hereditary properties that forbid either a complete or an empty graph*

If $\mathcal{H} \subseteq \mathrm{Forb}(K_h)$, then $\mathrm{ed}_{\mathcal{H}}(1) > 0$ and, by Proposition 4, if $\mathcal{H} \subseteq \mathrm{Forb}(\overline{K_h})$, then $\mathrm{ed}_{\mathcal{H}}(0) > 0$. We can produce bounds on the edit distance function for such properties.

Theorem 20 (Martin [33]) *Let* $\mathcal{H} = \bigcap_{H\in\mathcal{F}(\mathcal{H})} \mathrm{Forb}(H)$ *be a nontrivial hereditary property with* $\mathcal{H} \subseteq \mathrm{Forb}(K_h)$ *for some* $h \geq 2$ *such that*

- *m is the least positive integer such that $\mathcal{F}(\mathcal{H})$ contains a complete multipartite graph with m parts, and*
- *χ is the chromatic number of $\mathcal{H}$.*

Note $\chi \geq 2$ because $\mathcal{H}$ is nontrivial and $\chi \leq m \leq h$. Then

(a) $\mathrm{ed}_{\mathcal{H}}(p) = \frac{p}{\chi-1}$, *for all* $p \in [0, 1/2]$, *and*
(b) $\frac{1-p}{\chi-1} + \frac{2p-1}{m-1} \leq \mathrm{ed}_{\mathcal{H}}(p) \leq \min\left\{1-p+\frac{2p-1}{m-1}, \frac{p}{\chi-1}\right\}$, *for all* $p \in [1/2, 1]$.

In particular, if $\mathcal{H} = \mathrm{Forb}(K_h)$, *then* $\mathrm{ed}_{\mathcal{H}}(p) = \frac{p}{h-1}$.

Remark 21 *The bound* $\mathrm{ed}_{\mathcal{H}}(p) \leq p/(\chi-1)$ *in Theorem 20 was not expressed explicitly in [33] but follows directly from the concavity of the edit distance function. By Proposition 4, there are similar bounds for* $\mathcal{H}$ *where* $\mathcal{H} \subseteq \mathrm{Forb}(\overline{K_h})$. *Consequently,* $\mathrm{ed}_{\mathrm{Forb}(\overline{K_h})}(p) = (1-p)/(h-1)$.

5.2 C_6^* *and* H_9

5.2.1 Forb(C_6^*)

In [32], Marchant and Thomason address the graph C_6^*, which is the 6-cycle with a diagonal. The extreme points of the clique spectrum of $\mathrm{Forb}(C_6^*)$ are $(1,1)$ and $(0,2)$. Thus, if $\mathcal{H} = \mathrm{Forb}(C_6^*)$, then $\gamma_{\mathcal{H}}(p) = \min\{p(1-p), (1-p)/2\}$.

In fact, the edit distance function has a smaller value for $p \in (0,1)$.

Theorem 22 (Marchant-Thomason [32]) *Let* $\mathcal{H} = \mathrm{Forb}(C_6^*)$, *where* C_6^* *is the 6-cycle with a diagonal.*

(a) $\mathrm{ed}_{\mathcal{H}}(p) = \min\left\{\frac{p}{1+2p}, \frac{1-p}{2}\right\}$, *for* $p \in [0, 1]$.
(b) $p_{\mathcal{H}}^* = 1/2$ *and* $d_{\mathcal{H}}^* = 1/4$.

The CRG that corresponds to the $p/(1+2p)$ part of the function has 1 white vertex, 2 black vertices, one white edge between the black vertices, and two gray edges incident to the white vertex. See Figure 3. Although the edit distance function cannot

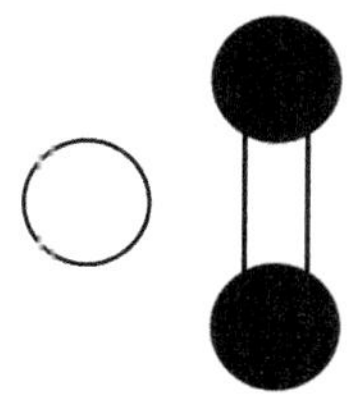

Fig. 3 The 3-vertex CRG that gives $p/(1+2p)$ in Theorem 22. The white edge is indicated, the two gray edges are not.

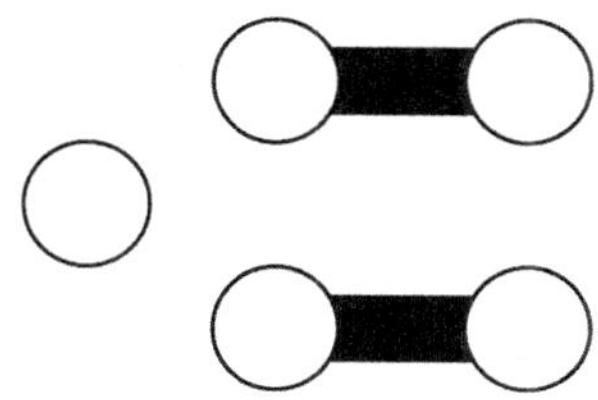

Fig. 4 The 5-vertex CRG that gives $p/(1+4p)$ in Theorem 23. The two black edges are indicated, the eight gray edges are not.

be determined by the clique spectrum, the values of $p^*_{\mathscr{H}} = 1/2$ and $d^*_{\mathscr{H}} = 1/4$ can be computed by knowing only the clique spectrum.

Using the tools in Section 4, the proof of Theorem 22 is much easier than the original proof.

Proof By Theorem 13, we may use continuity, concavity, and the knowledge of the value at $p = 1/2$ to conclude that $\mathrm{ed}_{\mathscr{H}}(p) = (1-p)/2$ for all $p \in [1/2, 1]$. Let $p \in [0, 1/2)$ and K be a p-core CRG. If $C_6^* \not\mapsto K$ and K has no white vertices, then it has no gray triangle and Corollary 18(a)(ii) gives that $g_K(p) > p/2$. If $C_6^* \not\mapsto K$ and K has one white vertex, then there are at most two black vertices, which cannot have a gray edge. Corollary 18(a)(ii) and Proposition 8 give that $g_K(p) \geq \left(p^{-1} + (1/2)^{-1}\right)^{-1} = p/(1+2p)$, as required. □

5.2.2 Forb(H_9)

Balogh and Martin [11] introduced the graph H_9, which is drawn in Figure 1. For $\mathscr{H} = \mathrm{Forb}(H_9)$, the values of $p^*_{\mathscr{H}}$ and $d^*_{\mathscr{H}}$ cannot be determined by the clique spectrum and this was established in [11]. Later, the author determined the edit distance function completely.

Theorem 23 (Martin [34]) *Let H_9 be the graph drawn in Figure 1 and let $\mathscr{H} =$* Forb(H_9).

(a) $\mathrm{ed}_{\mathscr{H}}(p) = \min\left\{\frac{p}{3}, \frac{p}{1+4p}, \frac{1-p}{2}\right\}$ *for* $p \in [0,1]$.
(b) $p^*_{\mathscr{H}} = \frac{1}{8}(1+\sqrt{17})$ *and* $d^*_{\mathscr{H}} = \frac{1}{8}(7-\sqrt{17})$.

The CRG that corresponds to the $p/(1+4p)$ part of the function has 5 white vertices, 2 nonadjacent white edges, and the remaining 8 edges gray. See Figure 4.

5.3 Cycles

The case of $\mathrm{Forb}(C_h)$, where C_h is a cycle on $h \geq 3$ vertices, has been widely investigated. Theorem 20 gives immediately that $\mathrm{ed}_{\mathrm{Forb}(C_3)}(p) = \mathrm{ed}_{\mathrm{Forb}(K_3)}(p) = p/2$.

In her Master's thesis, Peck almost completely settled the edit distance function for hereditary properties that forbid a cycle. Utilizing techniques inspired by the cycle arguments of Pósa [39], she determined the edit distance function for $\mathrm{Forb}(C_h)$ for odd $h \geq 5$. For even $h \geq 4$, she was able to determine enough of the function to find the maximum.

Theorem 24 (Peck [37]) *Let* $\mathscr{H} = \mathrm{Forb}(C_h)$ *where* C_h *is the cycle on* $h \geq 4$ *vertices.*

(a) *If* h *is odd, then* $\mathrm{ed}_{\mathscr{H}}(p) = \min\left\{\frac{p}{2}, \frac{p(1-p)}{1-p+(\lceil h/3\rceil-1)p}, \frac{1-p}{\lceil h/2\rceil-1}\right\}$ *for all* $p \in [0,1]$.
(b) *If* h *is even, then* $\mathrm{ed}_{\mathscr{H}}(p) = \min\left\{\frac{p(1-p)}{1-p+(\lceil h/3\rceil-1)p}, \frac{1-p}{\lceil h/2\rceil-1}\right\}$ *for all* $p \in [\lceil h/3\rceil^{-1}, 1]$.

Marchant and Thomason [32] first proved the case of $h = 4$ and, in fact, proved

Theorem 25 (Marchant-Thomason [32]) $\mathrm{ed}_{\mathrm{Forb}(C_4)}(p) = p(1-p)$ *for all* $p \in [0,1]$.

Marchant [31] proved the case for $h = 5, 7$. The cases of $h = 6, 8, 9, 10$ were first proven in [33] and, in fact, a larger range of p was proven in [33] for small even h.

Theorem 26 (Martin [33]) *Let* $\mathscr{H} = \mathrm{Forb}(C_h)$, *where* C_h *is the cycle on* $h \geq 4$ *vertices.*

(a) *If* $h = 6$, *then* $\mathrm{ed}_{\mathscr{H}}(p) = \min\left\{p(1-p), \frac{1-p}{2}\right\}$ *for all* $p \in [0,1]$.
(b) *If* $h = 8$, *then* $\mathrm{ed}_{\mathscr{H}}(p) = \min\left\{\frac{p(1-p)}{1+p}, \frac{1-p}{3}\right\}$ *for all* $p \in [0,1]$.
(c) *If* $h = 10$, *then* $\mathrm{ed}_{\mathscr{H}}(p) = \min\left\{\frac{p(1-p)}{1+2p}, \frac{1-p}{4}\right\}$ *for all* $p \in [1/7,1]$.

Corollary 27 *Let* $\mathscr{H} = \mathrm{Forb}(C_h)$ *where* C_h *is the cycle on* $h \geq 4$ *vertices.*

(a) *If* $h \notin \{4,7,8,10,16\}$, *then*

$$p^*_{\mathscr{H}} = \frac{1}{\lceil h/2\rceil - \lceil h/3\rceil + 1} \qquad \text{and} \qquad d^*_{\mathscr{H}} = \frac{\lceil h/2\rceil - \lceil h/3\rceil}{(\lceil h/2\rceil - 1)(\lceil h/2\rceil - \lceil h/3\rceil + 1)}.$$

(b) If $h \in \{4, 7, 8, 10, 16\}$*, then*

$$p^*_{\mathcal{H}} = \frac{1}{1+\sqrt{\lceil h/3\rceil - 1}} \qquad \text{and} \qquad d^*_{\mathcal{H}} = \frac{1}{\lceil h/3\rceil + 2\sqrt{\lceil h/3\rceil - 1}}.$$

It is interesting that $p^*_{\mathrm{Forb}(C_h)}$ and $d^*_{\mathrm{Forb}(C_h)}$ are both rational and result from the intersection of the g functions of two p-core CRGs, except in the cases $h \in \{7, 8, 10, 16\}$.

5.4 Powers of cycles

A natural extension of hereditary properties defined by forbidding certain cycles are hereditary properties defined by forbidding certain powers of cycles. For $h \geq 2t+1$, we define C_h^t to be the graph with vertex set $\{1, \dots, h\}$ and $ij \in E(C_h^t)$ if and only if $|i-j| \leq t \pmod h$. We consider the case for $t = 2$, that is, the case of the squared cycle.

For $h = 5$, C_5^2 is complete and from Theorem 20 we see that

$$\mathrm{ed}_{\mathrm{Forb}(C_5^2)}(p) = \mathrm{ed}_{\mathrm{Forb}(K_5)}(p) = p/4.$$

In the case of C_6^2, the complement is a perfect matching and we can use Proposition 4 if we know the edit distance function for $\mathrm{Forb}(M_6)$, where M_6 is the perfect matching on 6 vertices. It is easy to see that the CRGs $K \in \mathcal{F}(\mathrm{Forb}(M_6))$ have at most 2 black vertices and no pair of white vertices can have a gray edge between them, otherwise $M_6 \mapsto K$. By Theorem 9(a), we can conclude that $\mathrm{ed}_{\mathrm{Forb}(M_6)} = \frac{p(1-p)}{1+p}$ for $p \in [0, 1/2]$. Some more work verifies that $\mathrm{ed}_{\mathrm{Forb}(M_6)} = \frac{p(1-p)}{1+p}$ for $p \in [1/2, 1]$ also. Hence,

$$\mathrm{ed}_{\mathrm{Forb}(C_6^2)}(p) = \mathrm{ed}_{\mathrm{Forb}(M_6)}(1-p) = \frac{p(1-p)}{2-p}.$$

The complement of C_7^2 is simply C_7 and so Proposition 4 gives

$$\mathrm{ed}_{\mathrm{Forb}(C_7^2)}(p) = \mathrm{ed}_{\mathrm{Forb}(C_7)}(1-p) = \min\left\{\frac{p}{3}, \frac{p(1-p)}{2-p}, \frac{1-p}{2}\right\}.$$

Peck established some more values of $\mathrm{ed}_{\mathrm{Forb}(C_h^2)}(p)$ for $h \in \{8, 9, 10\}$, which we give in Theorem 28.

Theorem 28 (Peck [37]) *Let* $\mathcal{H} = \mathrm{Forb}(C_h^2)$ *where* C_h^2 *is the square of the cycle on h vertices.*

(a) If $\mathcal{H} = \mathrm{Forb}(C_8^2)$*, then* $\mathrm{ed}_{\mathcal{H}}(p) = \min\left\{\frac{p}{3}, \frac{p(1-p)}{2-p}, \frac{1-p}{2}\right\}$ *for all* $p \in [0, 1]$.

(b) If $\mathcal{H} = \mathrm{Forb}(C_9^2)$*, then* $\mathrm{ed}_{\mathcal{H}}(p) = \min\left\{\frac{p(1-p)}{2-p}, \frac{p(1-p)}{1+p}\right\}$ *for all* $p \in [0, 1]$.

(c) *If* $\mathscr{H} = \mathrm{Forb}(C_{10}^2)$, *then* $\mathrm{ed}_{\mathscr{H}}(p) = \min\left\{\frac{p}{3}, \frac{1-p}{3}\right\}$ *for all* $p \in [0,1]$.

(d) *If* $\mathscr{H} = \mathrm{Forb}(C_{11}^2)$, *then* $\mathrm{ed}_{\mathscr{H}}(p) = \min\left\{\frac{p}{3}, \frac{p(1-p)}{2}\right\}$ *for all* $p \in [0,1/2]$ *and* $\mathrm{ed}_{\mathscr{H}}(p) \leq \min\left\{\frac{p(1-p)}{2}, \frac{1-p}{3}\right\}$ *for all* $p \in [1/2,1]$.

(e) *If* $\mathscr{H} = \mathrm{Forb}(C_{12}^2)$, *then* $\mathrm{ed}_{\mathscr{H}}(p) = \frac{p(1-p)}{2}$ *for all* $p \in [0,1/2]$ *and* $\mathrm{ed}_{\mathscr{H}}(p) \leq \min\left\{\frac{p(1-p)}{2}, \frac{1-p}{3}\right\}$ *for all* $p \in [1/2,1]$.

Theorem 28, together with Theorem 13, is enough to determine the value of $p^*_{\mathrm{Forb}(C_h^2)}$ and of $d^*_{\mathrm{Forb}(C_h^2)}$ for $h \in \{5,\ldots,12\}$.

In work in progress, Berikkyzy, Peck, and Martin have extended the results from [37] to apply to powers of cycles, provided the number of vertices is large enough.

Theorem 29 (Berikkyzy-Martin-Peck [17]) *Let* $\mathscr{H} = \mathrm{Forb}(C_h^t)$ *where* C_h^t *is the* t^{th} *power of the cycle on* h *vertices. For* $t \geq 1$ *and* h *sufficiently large, let* $\ell_0 = \lceil h/(t+1)\rceil$, $\ell_t = \lceil h/(2t+1)\rceil$, *and* $p_0 = \ell_t^{-1}$.

(a) *If* $(t+1) \nmid h$, *then* $\mathrm{ed}_{\mathscr{H}}(p) = \min\left\{\frac{p}{t+1}, \frac{p(1-p)}{t(1-p)+(\ell_t-1)p}, \frac{1-p}{\ell_0-1}\right\}$ *for* $p \in [0,1]$.

(b) *If* $(t+1) \mid h$, *then* $\mathrm{ed}_{\mathscr{H}}(p) = \min\left\{\frac{p(1-p)}{t(1-p)+(\ell_t-1)p}, \frac{1-p}{\ell_0-1}\right\}$ *for* $p \in [p_0,1]$.

If $t = 2$, then $h \geq 13$ suffices for Theorem 29. In general, the bound that is proven to suffice is $h \geq 4t^2 + \Omega(t)$ although this is likely not best possible.

5.5 Complete bipartite graphs

5.5.1 Forb($K_{s,s}$)

The case of $\mathscr{H} = \mathrm{Forb}(K_{2,2})$ was established by Marchant and Thomason [32] where it was shown that $\mathrm{ed}_{\mathrm{Forb}(K_{2,2})}(p) = p(1-p)$ for all $p \in [0,1]$.

In the case of $\mathscr{H} = \mathrm{Forb}(K_{3,3})$, the values of $p^*_{\mathrm{Forb}(K_{3,3})} = \sqrt{2}-1$ and $d^*_{\mathrm{Forb}(K_{3,3})} = 3-2\sqrt{2}$ were established by Balogh and Martin [11].

For p not too small, the edit distance function for $\mathrm{Forb}(K_{3,3})$ coincides with $\gamma_{\mathrm{Forb}(K_{3,3})}(p) = \frac{p(1-p)}{1+p}$, but for p very small, the edit distance function is strictly smaller.

Theorem 30 (Marchant-Thomason [32]) *Let* $\mathscr{H} = \mathrm{Forb}(K_{3,3})$ *where* $K_{3,3}$ *is the complete bipartite graph with 3 vertices in each part.*

(a) $\mathrm{ed}_{\mathscr{H}}(p) = \frac{p(1-p)}{1+p}$, *for* $p \in [1/9,1]$.

(b) $\mathrm{ed}_{\mathscr{H}}(p) < \frac{p(1-p)}{1+p}$, *for* $p \in (0,1/124]$.

The CRGs used to establish Theorem 30(b) are defined by constructions due to Brown [22] to address a related Zarankiewicz problem. Specifically, for a prime power r, the constructions are $(r^2 - r)$-regular bipartite graphs on $2r^3$ vertices. Such graphs have no copy of $K_{3,3}$ and, of course, no copy of K_3. For such a graph G, we construct the CRG K for which the vertices of G are (black) vertices of K, the edges of G are gray edges of K, and the nonedges of G are white edges of K. By (5), this construction gives

$$f_K(p) = \frac{1}{2r^3}\left[1 + p\left(2r^3 - r^2 + r - 2\right)\right].$$

With $r = 19$, we obtain strict inequality for $p = 1/124$ and the continuity and concavity of the edit distance function give Theorem 30(b) for all $p \leq 1/124$.

It is also established in [32] that the value of $d^*_{\mathrm{Forb}(K_{s,s})}$ cannot be determined by the clique spectrum. The only extreme point of the clique spectrum is $(1, s-1)$ and the resulting CRG has $g_{K(1,s-1)} = \frac{p(1-p)}{1+(s-2)p} = \gamma_{\mathrm{Forb}(K_{s,s})}(p)$. The construction is a CRG $K^{(s-1)}$ on $2s-2$ black vertices consisting of $s-1$ disjoint white edges. It is easy to show that $K_{s,s} \not\mapsto K^{(s-1)}$ and since the g function of each component is $1/2$, Proposition 8 gives $g_{K^{(2s-2)}}(p) = 1/(2s-2)$.

So, $g_{K^{(2s-2)}}(p)$ is less than the maximum value of $\gamma_{\mathrm{Forb}(K_{s,s})}(p)$ for $s \geq 7$. We ask in Problem 2 if $1/(2s-2)$ is, indeed, the maximum value of $\mathrm{ed}_{\mathrm{Forb}(K_{s,s})}(p)$ and if that value is achieved for a positive length interval.

5.5.2 Forb($K_{2,t}$)

McKay and Martin [35] establish some surprising results for the hereditary property Forb($K_{2,t}$).

Theorem 31 (Martin-McKay [35]) *Let* $\mathcal{H} = \mathrm{Forb}(K_{2,t})$ *where* $K_{2,t}$ *is the complete bipartite graph with 2 vertices in one part and t vertices in the other part. For all* $t \geq 2$, $\gamma_{\mathcal{H}}(p) = \min\{p(1-p), (1-p)/(t-1)\}$.

(a) If $t = 3$, *then* $\mathrm{ed}_{\mathcal{H}}(p) = \min\left\{p(1-p), \frac{1-p}{2}\right\}$ *for all* $p \in [0, 1]$.

(b) If $t = 4$, *then* $\mathrm{ed}_{\mathcal{H}}(p) = \min\left\{p(1-p), \frac{7p+1}{15}, \frac{1-p}{3}\right\}$ *for all* $p \in [0, 1]$.

(c) If $t \geq 5$ *and is odd, then* $d^*_{\mathcal{H}} = \frac{1}{t+1}$ *and* $p^*_{\mathcal{H}} \supseteq \left[\frac{2t-1}{t(t+1)}, \frac{2}{t+1}\right]$.

(d) If $t \geq 9$, *there exists a* $p_0(t) < 1/2$ *such that* $\mathrm{ed}_{\mathcal{H}}(p) < p(1-p)$.

There are a number of interesting consequences arising from the study of these hereditary properties. First, we note that the CRG that gives the portion of the function in Theorem 31(b) corresponding to $(1 + 7p)/15$ results from a strongly regular graph construction.

Definition 32 *A* (n, k, λ, μ)-strongly regular graph *or* (n, k, λ, μ)-SRG *is a k-regular graph on n vertices for which each pair of adjacent vertices has exactly* λ *common neighbors and for which each pair of nonadjacent vertices has exactly* μ *common neighbors.*

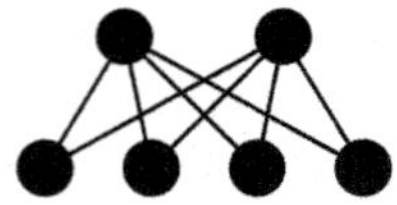

Fig. 5 The complete bipartite graph $K_{2,4}$.

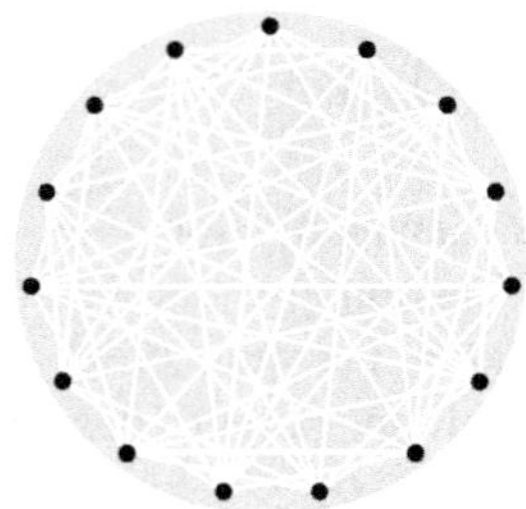

Fig. 6 The 15-vertex CRG that gives $(1 + 7p)/15$ in Theorem 31(b). The white edges are shown. The remaining edges are gray, and form a graph isomorphic to GQ(2, 2).

The CRG we use for Theorem 31(b) is constructed from a (15, 6, 1, 3)-SRG, commonly called GQ(2, 2). It is a member of the family of so-called generalized quadrangles. Given a GQ(2, 2) G', the CRG K' has 15 white vertices that correspond to the vertices of the graph. An edge of K' is gray if and only if the corresponding pairs of vertices are adjacent in G'. See Figure 5 for the graph $K_{2,4}$ and Figure 6 for the 15-vertex CRG mentioned above. In [35], it is shown that $K_{2,4} \not\mapsto K'$ and $g_{K'}(p) = (1 + 7p)/15$.

Similar constructions from strongly regular graphs are in $\mathscr{K}(\mathrm{Forb}(K_{2,t}))$ and have a smaller g function than $\gamma_{\mathscr{H}}(p) = \min\{p(1-p), (1-p)/(t-1)\}$ for certain values of t and p.

For Theorem 31(c), the corresponding CRG $K^{(t)}$ has $t + 1$ black vertices and a perfect matching of $(t + 1)/2$ white edges. The remaining edges are gray. It is easy to show that $K_{2,t} \not\mapsto K^{(t+1)}$ and $g_{K^{(t+1)}}(p) = 1/(t + 1)$.

For Theorem 31(d), the constructions are due to Füredi [28] to address a related Zarankiewicz problem. If q is a prime power such that $t-1$ divides $q-1$, then there exists a graph on $2(q^2-1)/(t-1)$ vertices that is q-regular with no copy of $K_{2,t}$ and no triangle. This is enough to ensure that $K_{2,t}$ does not map to the corresponding CRG. By (5), this construction gives

$$f_K(p) = \frac{t-1}{2(q^2-1)}\left[1 + p\left(\frac{2(q^2-1)}{t-1} - q - 2\right)\right].$$

With $t \geq 9$, we can find a sufficiently large prime power q so that $f_k(p) < p(1-p)$.

5.6 *Split graphs*

A graph H on at least two vertices is a *split graph* if there is a partition of $V(H)$ into one independent set and one clique. For a split graph with independence number $\alpha \geq 2$ and clique number $\omega \geq 2$, either $\alpha + \omega = h$ or $\alpha + \omega = h + 1$. We can compute the edit distance function of hereditary properties defined by such graphs.

Theorem 33 (Martin [34]) *Let H be a split graph which has independence number $\alpha = \alpha(H) \geq 2$ and clique number $\omega = \omega(H) \geq 2$. If $\mathcal{H} = \mathrm{Forb}(H)$, then*

$$\mathrm{ed}_{\mathcal{H}}(p) = \min\left\{\frac{p}{\omega-1}, \frac{1-p}{\alpha-1}\right\}.$$

Hence $p^*_{\mathcal{H}} = (\omega-1)/(\alpha+\omega-2)$ and $d^*_{\mathcal{H}} = 1/(\alpha+\omega-2)$.

6 Quantities related to the edit distance function

We get the following notation from Balogh et al. [14]. For a graph property[4] $\mathcal{H}$, the *labeled slice* of $\mathcal{H}$ is the set $\mathcal{H}^n$ of graphs in $\mathcal{H}$ with vertex set $\{1,\ldots,n\}$. The *labeled speed* of $\mathcal{H}$ is the function $n \mapsto |\mathcal{H}^n|$.

Theorem 34 ([12, 13, 16, 19]) *If $\mathcal{H}$ is a hereditary property of graphs, then one of the following holds:*

(i) There exist $N, k \in \mathbb{N}$ and polynomials $\{p_i(n)\}_{i=0}^k$ such that, for all $n > N$, $|\mathcal{H}^n| = \sum_{i=0}^k p_i(n)i^n$.
(ii) For some $t \in \mathbb{N}$, $t > 1$, we have $|\mathcal{H}^n| = n^{(1-1/t+o(1))n}$.
(iii) For n sufficiently large, $n^{(1+o(1))n} \leq |\mathcal{H}^n| \leq 2^{o(n^2)}$.
(iv) For some $k \in \mathbb{N}$, $k > 1$, we have $|\mathcal{H}^n| = 2^{(1-1/k+o(1))n^2/2}$.

Here $k = \chi_B(\mathcal{H}) - 1$.

This partition of hereditary properties was first discovered by Scheinerman and Zito [46]. As for the precise results, parts (i) and (ii) were established by Balogh, Bollobás, and Weinreich [16], part (iii) was also established by Balogh, Bollobás, and Weinreich [12, 13], and part (iv) was established by Bollobás and Thomason [19].

Part (iv) has been well-studied, including the $o(1)$ error term [15]. It has also been generalized. Bollobás and Thomason [20] define $c_p(\mathcal{H})$ as follows:

$$c_{\mathcal{H}}(p) = \lim_{n\to\infty} -\log_2 \Pr(G(n,p) \in \mathcal{H})/\binom{n}{2}. \tag{12}$$

[4] For us, the property will be hereditary, although that is not necessary in order to define the speed.

They showed that the limit exists, based on work by Alekseev [1] (see also [2]). Thomason [49] compiled these results to show the relationship to the edit distance function:

Theorem 35 (Thomason [49]) *Let $\mathcal{H}$ be a nontrivial hereditary property and let $c_{\mathcal{H}}(p)$ be defined as in (12). Then*

$$c_{\mathcal{H}}(p) = (-\log_2 p(1-p))\,\mathrm{ed}_{\mathcal{H}}\left(\frac{\log_2(1-p)}{\log_2 p(1-p)}\right).$$

As we see, $c_{\mathcal{H}}(p)$ can be derived directly from the edit distance function.

Remark 36 *The function $c_{\mathcal{H}}(p)$ is not necessarily concave down, however $\mathrm{ed}_{\mathcal{H}}(p)$ is. Concavity is a key tool in finding the elusive lower bounds on the edit distance function which can then be used to compute lower bounds for $c_{\mathcal{H}}(p)$.*

Perhaps other functions of hereditary properties can be defined from the edit distance function. We are particularly interested in other metrics. For each positive integer n, let d be a metric on the space of graphs with vertex set $\{1,\ldots,n\}$. For any hereditary property $\mathcal{H}$, define

$$\mathrm{d}(G,\mathcal{H}) = \min\left\{\mathrm{d}(G,G') : V(G') = V(G), G' \in \mathcal{H}\right\},$$

and define the function

$$\phi_{\mathcal{H}}(p) := \limsup_{n\to\infty}\ \max\left\{\mathrm{d}(G,\mathcal{H}) : |V(G)| = n, |E(G)| = \left\lfloor p\tbinom{n}{2}\right\rfloor\right\}. \tag{13}$$

In Question 3 from Section 8.2, we ask whether $\phi_{\mathcal{H}}(p)$ can be expressed as a function of p and $\mathrm{ed}_{\mathcal{H}}(\cdot)$ if d satisfies a natural property. Recall that dist represents the edit metric. It is clear that in order for any such result to exist, the metric d should be *continuous with respect to the edit metric*. That is, for every $\epsilon > 0$, there exists a δ such that $\mathrm{d}(G,G') < \epsilon$ whenever $\mathrm{dist}(G,G') < \delta$. This is a natural restriction because, for example, the trivial metric where $\mathrm{d}(G,G) = 0$ but $\mathrm{d}(G,G') = 1$ when $G \neq G'$ produces no useful results.

A well-studied metric is the so-called *cut metric* (or *cut norm*). Frieze and Kannan [27] introduced the cut norm and it is used extensively in the theorey of graph limits (see, e.g., Borgs et al. [21]). The cut norm is defined as follows for graphs on the same labeled vertex set $V = \{1,\ldots,n\}$:

$$\mathrm{d}_{\square}(G,G') = \max_{S,T\subset V}\ \frac{1}{n^2}\left|e_G(S,T) - e_{G'}(S,T)\right|,$$

where $e_G(S,T)$ is the number of ordered pairs (i,j) with $i \in S$ and $j \in T$ and $ij \in E(G)$. If S and T are disjoint, it counts the number of edges between S and T in G. The cut metric is useful for comparing random graphs. Although two typical graphs selected according to $G(n,p)$ have edit distance close to $2p(1-p)$, their $\mathrm{d}_{\square}$ distance is $O(1/n)$.

7 Generalizations of edit distance

Axenovich and Martin have investigated natural generalizations of the edit distance problem. The paper [8] addressed editing matrices (Section 7.1 below). The paper [9] addressed both editing the edges of multicolorings of a complete graph (Section 7.2) and editing the edges of a directed graph (Section 7.3).

7.1 Matrices

Let $\mathscr{A} = \{A_1, \ldots, A_r\}$ be a partition of pairs from $[m] \times [n]$ into r nonemtpy classes. An $m \times n$ matrix $A = (a_{ij})$ is said to have a *pattern* $\mathscr{A}$ provided that $a_{ij} = a_{i'j'}$ if and only if $(i,j), (i',j') \in A_t$ for some $t \in \{1, \ldots, r\}$. A pattern is *nontrivial* if $r \geq 2$. For a matrix M, if there is a submatrix M' with pattern $\mathscr{A}$, then we say that M has a *subpattern* $\mathscr{A}$.

For a pattern $\mathscr{A}$ and positive integers m, n, s, we define $\mathrm{Forb}(m, n; s, \mathscr{A})$ to be the set of all $m \times n$ matrices with at most s distinct entries and not containing subpattern $\mathscr{A}$.

For two matrices A and B of the same dimensions, we say that $\mathrm{dist}(A, B)$ is the number of positions in which A and B differ; i.e., it is the matrix Hamming distance. For a class of matrices $\mathscr{F}$ and a matrix A, all of the same dimensions, we denote $\mathrm{dist}(A, \mathscr{F}) = \min\{\mathrm{dist}(A, \mathscr{F}) : F \in \mathscr{F}\}$. Finally,

$$f(m, n; s, \mathscr{A}) := \max\left\{\mathrm{dist}(A, \mathscr{F}) : A \in \mathscr{M}(m, n; s), \mathscr{F} = \mathrm{Forb}(m, n; s, \mathscr{A})\right\}/mn. \tag{14}$$

The function f in (14) counts the maximum proportion of edits required to remove a pattern with r places from an $m \times n$ matrix with s distinct entries.[5]

Theorem 37 (Axenovich-Martin [8]) *Let s, r be positive integers, $s \geq r$. Let b_1, b_2 be positive constants such that $b_1 \leq m/n \leq b_2$. Let $\mathscr{A}$ be a nontrivial pattern with r distinct entries. Then*

$$f(m, n; s, \mathscr{A}) = (1 + o(1))\left(\frac{s - r + 1}{s}\right).$$

Without loss of generality, the case of $s = 2$ corresponds to a $\{0, 1\}$-matrix. If the pattern has both zeros and ones, then $r = 2$ and the edit distance is $1/2$; i.e., an asymptotically most efficient editing algorithm is to make all entries zero or all entries one, whichever is most prevalent and the worst case is that there is the same

[5] In [8], f counts the number of edits, but we normalize by dividing by mn to make it consistent with the rest of this paper.

number of each. If the pattern has, say only zeroes, then the edit distance is 1 because the worst case is that the original matrix is all zeros and almost all of them must be changed to one.

The setting for matrices is identical to the case of editing the $m \times n$ complete bipartite graph in which the edges are colored with s distinct colors.

7.2 Multicolor edit distance

We will use slightly different terminology from [9] so as not to confuse it with similar notation for hypergraphs in Section 8.1. For any integer $r \geq 2$, an *r-colored graph* is pair (V, c) such that V is a finite labeled set and $c : E \to \{1, \ldots, r\}$. For r-colored graph G and $\rho \in \{1, \ldots, r\}$, we denote $E_\rho(G)$ to be the set of edges colored ρ.

If G and G' are two r-colored graphs on the same labeled vertex set, then the edit distance between them, $\mathrm{dist}(G, G')$ is the proportion of edges that receives a different color. For example, if $r = 2$, then graphs correspond to black edges and the complement corresponds to white edges. The following definitions for $r = 2$ are consistent with the graph case.

Further, we may define $\mathrm{dist}(G, \mathscr{H})$ for any hereditary property of r-colored graphs as in (1). In this setting, a property is still hereditary if it is closed under isomorphism and the deletion of vertices. For any r-colored graph H, we write $\mathrm{Forb}(H)$ to be the set of all r-colored graphs that have no copy of H. Note that "induced" is not necessary here because all edges receive a color. For an integer $r \geq 2$, a *density vector* $\mathbf{p} = (p_1, \ldots, p_r)$ is a nonnegative real vector with the property that $\sum_{\rho=1}^{r} p_i = 1$. The domain of r dimensional density vectors is the *(standard)* $(r-1)$*-simplex.*

If $\mathscr{H}$ is a hereditary property of r-colored graphs, then we may define the edit distance function parallel to (3) as follows.

$$\mathrm{ed}_{\mathscr{H}}(\mathbf{p}) := \lim_{n \to \infty} \max \left\{ \mathrm{dist}(G, \mathscr{H}) : |V(G)| = n, |E_\rho(G)| = p_\rho \tbinom{n}{2}, \rho = 1, \ldots, r \right\}.$$

The limit was proven to exist in [9]. We omit floors and ceilings in defining $|E_\rho(G)|$ because they play no role in the limit.

We can also define the equivalent of CRGs in this setting. In [9], the term *type* is used, though for consistency of this paper, we will just call them r-CRGs.[6]

Definition 38 *An* r-CRG *K is a pair (U, ϕ) where U is a finite set of vertices and $\phi : U \times U \to 2^{\{1,\ldots,r\}} - \emptyset$ such that $\phi(x, y) = \phi(y, x)$ and $\phi(x, x) \neq \{1, \ldots, r\}$. The* sub-$r$-CRG induced by *$W \subseteq U$ is the r-CRG that results from deleting $U - W$.*

[6]In Section 8.1, we refer to r-CRHs when discussing the edit distance on r-uniform hypergraphs.

We say that an r-colored-graph $H = (V, c)$ embeds in r-CRG K, *and write* $H \mapsto K$, *if there is a map* $\gamma : V \to U$ *such that* $c(vv') = c_0$ *implies* $c_0 \in \phi(\gamma(v)\gamma(v'))$. *For any hereditary property* $\mathscr{H} = \bigcap_{H \in \mathscr{F}(\mathscr{H})} \mathrm{Forb}(H)$, *let* $\mathscr{K}(H)$ *be the set of r-CRGs for which none of* $\mathscr{F}(\mathscr{H})$ *embeds in that* r-CRG.

The notion of the binary chromatic number is more complicated in the r-colored graph case when $r > 2$. There are weak and strong colorings.

Definition 39 *Let* $\mathscr{H} = \bigcap_{H \in \mathscr{F}(\mathscr{H})} \mathrm{Forb}(H)$ *be a hereditary property of r-colored-graphs.*

- *An r-tuple* $(a_1, \ldots, a_r)$ *of nonnegative integers is* weakly good *if for some* $H \in \mathscr{F}(\mathscr{H})$ *the vertex set* $V(H)$ *can be partitioned into sets* $S_1, \ldots, S_r$ *such that for each* $\rho \in \{1, \ldots, r\}$ with $a_\rho \neq 0$, *the partition can be further refined* $S_\rho = V_{\rho,1} \cup \cdots \cup V_{\rho,a_\rho}$ *where each edge in* $V_{\rho,j}$ **does not** *have color* ρ.
- *An r-tuple* $(a_1, \ldots, a_r)$ *of nonnegative integers is* strongly good *if for some* $H \in \mathscr{F}(\mathscr{H})$ *the vertex set* $V(H)$ *can be partitioned into sets* $S_1, \ldots, S_r$ *such that for each* $\rho \in \{1, \ldots, r\}$ *with* $a_\rho \neq 0$, *the partitioned can be further refined* $S_\rho = V_{\rho,1} \cup \cdots \cup V_{\rho,a_\rho}$ *where each edge in* $V_{\rho,j}$ **must have** *color* ρ.

We can then define spectra and r-ary chromatic numbers based on weak and strong colorings.

Definition 40 *Let* $\mathscr{H} = \bigcap_{H \in \mathscr{F}(\mathscr{H})} \mathrm{Forb}(H)$ *be a hereditary property of r-colored-graphs.*

- *The* weak clique spectrum *of* $\mathscr{H}$ *is the set of all tuples* $(a_1, \ldots, a_r)$ *that are* **not** *weakly good. The* weak r-ary chromatic number *of* $\mathscr{H}$, *denoted* $\chi_r^{\mathrm{wk}}(\mathscr{H})$, *is the largest* $a_1 + \cdots + a_r + 1$ *such that* $(a_1, \ldots, a_r)$ *is in the weak clique spectrum of* $\mathscr{H}$.
- *The* strong clique spectrum *of* $\mathscr{H}$ *is the set of all tuples* $(a_1, \ldots, a_r)$ *that are* **not** *strongly good. The* strong r-ary chromatic number *of* $\mathscr{H}$, *denoted* $\chi_r^{\mathrm{st}}(\mathscr{H})$, *is the largest* $a_1 + \cdots + a_r + 1$ *such that* $(a_1, \ldots, a_r)$ *is in the strong clique spectrum of* $\mathscr{H}$.

The f and g functions are defined similar to the graph case.

Definition 41 *Let* $K = (\{u_1, \ldots, u_k\}, \phi)$ *be an r-CRG and for* $\mathbf{p} = (p_1, \ldots, p_r)$, *let* $\mathbf{M}_K(\mathbf{p})$ *denote the matrix with entries defined as follows:*

$$m_K(\mathbf{p})_{ij} = 1 - \sum_{\rho \in \phi(u_i, u_j)} p_\rho.$$

The functions f_K *and* g_K *are defined as follows:*

$$f_K(\mathbf{p}) = \frac{1}{k^2} \mathbf{1}^T \mathbf{M}_K(\mathbf{p}) \mathbf{1} \tag{15}$$

$$g_K(\mathbf{p}) = \min \left\{ \mathbf{x}^T \mathbf{M}_K(\mathbf{p}) \mathbf{x} : \mathbf{x}^T \mathbf{1} = 1, \mathbf{x} \geq \mathbf{0} \right\}. \tag{16}$$

We say that a CRG K is $\mathbf{p}$-*core* if, for any proper sub-r-CRG K' of K, $g_{K'}(\mathbf{p}) > g_K(\mathbf{p})$.

We summarize the basic properties of this version of the edit distance function that generalize Theorems 13, 3, and 6. For $\mathbf{p} = (p_1, \ldots, p_r)$, the random r-colored graph $G(n, \mathbf{p})$ is the complete graph on n vertices in which each edge independently receives color ρ with probability p_ρ.

Theorem 42 (Axenovich-Martin [9]) *Let $\mathscr{H}$ be a hereditary property of r-colored graphs.*

(a) $\mathrm{ed}_{\mathscr{H}}(\mathbf{p})$ *is continuous over the* $(r-1)$*-simplex.*
(b) $\mathrm{ed}_{\mathscr{H}}(\mathbf{p})$ *is concave down over the* $(r-1)$*-simplex.*
(c) $\mathrm{ed}_{\mathscr{H}}(r^{-1}\mathbf{1}) \geq 1/(r(\chi_r^{\mathrm{st}}(\mathscr{H})-1))$.
(d) $\mathrm{ed}_{\mathscr{H}}(\mathbf{p}) \leq 1/(\chi_r^{\mathrm{wk}}(\mathscr{H})-1)$ *for all* $\mathbf{p}$ *in the* $(r-1)$*-simplex.*
(e) $\mathrm{ed}_{\mathscr{H}}(\mathbf{p}) = \lim_{n\to\infty} \mathbb{E}[\mathrm{dist}(G(n,\mathbf{p}), \mathscr{H})]$ *for all* $\mathbf{p}$ *in the* $(r-1)$*-simplex.*
(f) $\mathrm{ed}_{\mathscr{H}}(\mathbf{p}) = \inf\{f_K(\mathbf{p}) : K \in \mathscr{K}(\mathscr{H})\} = \inf\{g_K(\mathbf{p}) : K \in \mathscr{K}(\mathscr{H})\}$ *for all* $\mathbf{p}$ *in the* $(r-1)$*-simplex.*

Finally, we give some examples of results in the case $r = 3$.

Theorem 43 (Axenovich-Martin [9]) *Let $r = 3$ and let $\mathscr{H} = \bigcap_{H\in\mathscr{F}} \mathrm{Forb}(H)$ be a hereditary property of r-colored graphs. Let* $d^*_{\mathscr{H}} := \max\{\mathrm{ed}_{\mathscr{H}}(\mathbf{p}) : \mathbf{p}^T\mathbf{1} = 1, \mathbf{p} \geq \mathbf{0}\}$.

(a) If $\mathscr{F}$ is a family that consists of a single monochromatic triangle, then $d^*_{\mathscr{H}} = 1/2$.
(b) If $\mathscr{F}$ is a family that consists of a single triangle with two edges colored 1 and the other edge colored 2, then $d^*_{\mathscr{H}} = 1/2$.
(c) If $\mathscr{F}$ is a family that consists of two monochromatic triangles of different colors, then $d^*_{\mathscr{H}} = 1/2$.
(d) If $\mathscr{F}$ is a family that consists of all six bi-chromatic triangles, then $d^*_{\mathscr{H}} = 2/3$.
(e) If $\mathscr{F}$ is a family that consists of a single rainbow triangle, then $d^*_{\mathscr{H}} = 1/3$.

7.3 Directed edit distance

A *simple directed graph* or *digraph* G is a pair (V, c) such that V is a finite labeled set and, if $(V)_2 = V \times V - \{(v, v) : v \in V\}$ then $c : (V)_2 \to \{\bigcirc, -, \leftarrow, \rightarrow\}$ where

- $c(v, w) = c(w, v)$ if and only if $c(v, w) \in \{\bigcirc, -\}$ and
- $c(v, w) = \rightarrow$ if and only if $c(w, v) = \leftarrow$.

In the standard representation of digraphs as a pair (V, E) where $E \subseteq (V)_2$, we interpret $c(v, w) = \bigcirc$ to mean that neither (v, w) nor (w, v) is in E, $c(v, w) = -$ to mean that both (v, w) and (w, v) are in E, and $c(v, w) = \rightarrow$ to mean that $(v, w) \in E$ but $(w, v) \notin E$. We also define the following for any digraph G:

- $E_{\bigcirc}(G)$ is the set of all **unordered** pairs $\{v, w\}$ such that $c(v, w) = \bigcirc$.
- $E_{\leftarrow}(G)$ is the set of all **ordered** pairs $\{v, w\}$ such that $c(v, w) = \leftarrow$.
- $E_{\rightarrow}(G)$ is the set of all **ordered** pairs $\{v, w\}$ such that $c(v, w) = \rightarrow$.
- $E_{-}(G)$ is the set of all **unordered** pairs $\{v, w\}$ such that $c(v, w) = -$.

If $G = (V, c)$ and $G' = (V, c)$ are two digraphs on the same labeled vertex set with the same fixed palette, then the edit distance between them, $\mathrm{dist}(G, G')$ is the proportion of ordered pairs on which G and G' differ. We define $\mathrm{dist}(G, \mathscr{H})$ for any hereditary property of digraphs on any palette as in (1). A property is, of course, hereditary if it is closed under isomorphism and the deletion of vertices. For any digraph, we write $\mathrm{Forb}(H)$ to be the set of all digraphs that have no induced copy of H.

The digraph case encompasses several well-studied subclasses of digraphs. Just as the number of colors must be specified in Section 7.2, the palette must be specified for the digraph case.

Definition 44 *We say that* $\mathscr{P} \subseteq \{\bigcirc, -, \leftarrow, \rightarrow\}$ *is a* palette *if either none or both of "$\leftarrow$" and "$\rightarrow$" are in $\mathscr{P}$. There are 5 possible nontrivial palettes:*

(0) $\mathscr{P}_0 = \{\bigcirc, -, \leftarrow, \rightarrow\}$ *is the general case.*
(1) $\mathscr{P}_{\mathrm{compl}} = \{-, \leftarrow, \rightarrow\}$ *is the case of simple digraphs such that every pair of vertices has at least one arc between them.*
(2) $\mathscr{P}_{\mathrm{orien}} = \{\bigcirc, \leftarrow, \rightarrow\}$ *is the case of oriented graphs.*
(3) $\mathscr{P}_{\mathrm{undir}} = \{\bigcirc, -\}$ *is the usual case of simple, undirected graphs.*
(4) $\mathscr{P}_{\mathrm{tourn}} = \{\leftarrow, \rightarrow\}$ *is the case of tournaments.*

Definition 45 *A* directed density vector (p, q) *is a pair such that* $p \geq 0$, $q \geq 0$, *and* $p + 2q \leq 1$. *For different palettes, there are further restrictions.*

(0) If $\mathscr{P} = \mathscr{P}_{\mathrm{compl}}$, *then* $p + 2q = 1$.
(1) If $\mathscr{P} = \mathscr{P}_{\mathrm{orien}}$, *then* $p = 0$ *and* $q \leq 1/2$.
(2) If $\mathscr{P} = \mathscr{P}_{\mathrm{undir}}$, *then* $q = 0$ *and* $p \leq 1$; *i.e., the usual graph case.*
(3) If $\mathscr{P} = \mathscr{P}_{\mathrm{tourn}}$, *then* $p = 0$ *and* $q = 1/2$.

If $\mathscr{H}$ is a hereditary property of digraphs with palette $\mathscr{P}$, then for all directed density vectors $\mathbf{p} = (p, q)$, we define the edit distance function for hereditary property $\mathscr{H}$ as follows:

$$\mathrm{ed}_{\mathscr{H}}(\mathbf{p}) := \lim_{n\to\infty} \max\left\{\mathrm{dist}(G, \mathscr{H}) : \begin{array}{l} |V(G)| = n, |E_{-}(G)| = \lfloor p\binom{n}{2}\rfloor, \\ |E_{\leftarrow}(G)| = |E_{\rightarrow}(G)| = \lfloor q\binom{n}{2}\rfloor \end{array}\right\}.$$

The limit was proven to exist in [9]

In [9], the equivalent of CRGs (called *dir-types* in [9], but it would be natural to call them $\mathscr{P}$-dir-CRGs for palette $\mathscr{P}$) are defined as well as the notion of $H \mapsto K$ for any digraph H and any K a $\mathscr{P}$-dir-CRG. The matrix $\mathbf{M}_K(\mathbf{p})$ and functions $f_K(\mathbf{p})$ and $g_K(\mathbf{p})$ are defined analogously. In addition, the *strong directed clique spectrum, strong directed chromatic number* $\chi_{\mathscr{P}}^{\mathrm{st,dir}}(\mathscr{H})$, *weak directed clique spectrum*, and *weak directed chromatic number* $\chi_{\mathscr{P}}^{\mathrm{wk,dir}}(\mathscr{H})$ are defined for each palette, although for $\mathscr{P}_{\mathrm{undir}}$ and $\mathscr{P}_{\mathrm{tourn}}$ "strong" and "weak" are the same, where we use the notation $\chi_{\mathscr{P}}^{\mathrm{dir}}(\mathscr{H})$.

We will not give the detailed definitions of these quantities or of the random digraph $G(n, \mathbf{p})$. The natural notions are defined precisely in [9]. We have similar basic results for the directed case as for the multicolored case in Theorem 42.

Theorem 46 (Axenovich-Martin [9]) *Let $\mathscr{P}$ be a palette and let $\mathscr{H}$ be a hereditary property of digraphs with palette $\mathscr{P}$. Let the domain be defined as in Definition 45.*

(a) $\mathrm{ed}_{\mathscr{H}}(\mathbf{p})$ *is continuous over the domain.*
(b) $\mathrm{ed}_{\mathscr{H}}(\mathbf{p})$ *is concave down over the domain.*
(c) $\mathrm{ed}_{\mathscr{H}}(r^{-1}\mathbf{1}) \geq 1/(r(\chi_{\mathscr{P}}^{\mathrm{st,dir}}(\mathscr{H}) - 1))$.
(d) $\mathrm{ed}_{\mathscr{H}}(\mathbf{p}) \leq 1/(\chi_{\mathscr{P}}^{\mathrm{wk,dir}}(\mathscr{H}) - 1)$ *for all* $\mathbf{p}$ *in the domain.*
(e) $\mathrm{ed}_{\mathscr{H}}(\mathbf{p}) = \lim_{n\to\infty} \mathbb{E}[\mathrm{dist}(G(n, \mathbf{p}), \mathscr{H})]$ *for all* $\mathbf{p}$ *in the domain.*
(f) $\mathrm{ed}_{\mathscr{H}}(\mathbf{p}) = \inf\{f_K(\mathbf{p}) : K \in \mathscr{K}(\mathscr{H})\} = \inf\{g_K(\mathbf{p}) : K \in \mathscr{K}(\mathscr{H})\}$ *for all* $\mathbf{p}$ *in the domain.*

We give some examples involving triangles.

Theorem 47 (Axenovich-Martin [9]) *Let $\mathscr{H}$ be a hereditary property of digraphs with palette $\mathscr{P}$.*

(a) If $\mathscr{H} = \mathrm{Forb}(H_{\mathrm{dir}})$ *where* H_{dir} *is a directed triangle, then* $d^*_{\mathscr{H}} = 1/2$ *regardless of* $\mathscr{P}$.
(b) If $\mathscr{H} = \mathrm{Forb}(H_{\mathrm{tra}})$ *where* H_{tra} *is a transitive triangle, then* $\mathscr{H}$ *is a trivial hereditary property as long as* $\mathscr{P} = \mathscr{P}_{\mathrm{tourn}}$.
(c) If $\mathscr{H} = \mathrm{Forb}(H_{\mathrm{tra}})$ *where* H_{tra} *is a transitive triangle, then* $d^*_{\mathscr{H}} = 1/2$, *as long as* $\mathscr{P} \neq \mathscr{P}_{\mathrm{tourn}}$.
(d) If $\mathscr{H} = \mathrm{Forb}(H_{\mathrm{dir}}) \cap \mathrm{Forb}(H_{\mathrm{tra}})$ *where* H_{dir} *is a directed triangle and* H_{tra} *is a transitive triangle, then* $d^*_{\mathscr{H}} = 1/2$, *as long as* $\mathscr{P} \neq \mathscr{P}_{\mathrm{tourn}}$.

The case of tournaments turns out to be trivial. Theorem 47(b) is a simple consequence of Ramsey theory, a hereditary property $\mathscr{H} = \bigcap_{H\in\mathscr{F}(\mathscr{H})} \mathrm{Forb}(H)$ is nontrivial if and only if no member of $\mathscr{F}(\mathscr{H})$ is transitive. In the case of tournaments, the density vector must be $\mathbf{p} = (0, 1/2)$. The edit distance function is, therefore, a constant.

Theorem 48 *Let $\mathscr{H}$ be a nontrivial hereditary property of tournaments and $\mathscr{P} = \mathscr{P}_{\mathrm{tourn}}$. Then*

$$\mathrm{ed}_{\mathscr{H}}(0, 1/2) = \frac{1}{2(\chi_{\mathscr{P}}^{\mathrm{dir}}(\mathscr{H}) - 1)}.$$

8 Future directions

8.1 Hypergraph edit distance

Berikkyzy and the author have been investigating the extension of the edit distance problem to r-uniform hypergraphs (r-graphs). A *colored regularity hypergraph of order r (r-CRH)* is a triple (V, E, ϕ) in which V is a vertex set, E is the collection of all r-multisets on V, and $\phi : E \to \{\mathbf{W}, \mathbf{G}, \mathbf{B}\}$ with the restrictions that, (a) for any $v \in V$, $\phi(\{v, \ldots, v\}) \in \{\mathbf{W}, \mathbf{B}\}$ and (b) for any permutation $\sigma \in \Sigma_r$, $\phi(\{v_1, \ldots, v_r\}) = \phi(\{v_{\sigma(1)}, \ldots, v_{\sigma(r)}\})$. Therefore, a 2-CRH is just a CRG.

In parallel to the graph case, we can define colored homomorphisms from r-uniform hypergraphs to r-CRHs so that if r-graph H does not map to a r-CRH K, then an r-graph G which is edited according to the "recipe" defined by K will have no induced copy of H.

We can then define, for each r-CRH K, an r-linear form which we can also call $g_K(p)$. It is easy to prove, for a hereditary property $\mathcal{H}$ of r-graphs, that there is a family $\mathcal{F}(\mathcal{H})$ of r-CRHs such that

$$\mathrm{ed}_{\mathcal{H}}(p) \leq \inf_{K \in \mathcal{K}(\mathcal{H})} g_K(p). \tag{17}$$

The difficulty in extending the theory of the edit distance in graphs to hypergraphs is in proving that (17) is, in fact, an equality.

The above definition of the r-CRH would not seem to be adequate to capture the subtleties of hypergraphs. Consider the common example of a hypergraph whose 3-edges are cyclic triangles in an underlying random tournament. See, e.g., the survey of hypergraph Turán theory by Keevash [30]. This hypergraph has no copy of the tetrahedron K_4^3 but crossing triples would be gray in any 3-CRH that models it.

Strong hypergraph regularity was developed in the 3-uniform case by Frankl and Rödl [26] and then for the general r-uniform case by Gowers [29], Rödl and Skokan [44, 45], and Nagle, Rödl, and Schacht [36]. In these formulations, the notion of how overlapping hyperedges interact is captured by structures known as *complexes*. The structure of complexes inherent in strong hypergraph regularity would seem to be necessary in order to define r-CRHs and colored homomorphisms in order for the existence of a particular induced hypergraph to be determined.

The edit distance problem is, asymptotically, a general case of the Turán problem. In the context of Turán-type problems, a hypergraph property is *monotone* if it is closed under the taking of (not necessarily induced) subgraphs. Therefore, a monotone property is also hereditary. For a monotone property $\mathcal{M}$, the Turán density is $\pi(\mathcal{M}) = \limsup_{n\to\infty} \max\{|E(G)|/\binom{n}{2} : |V(G)| = n, G \in \mathcal{M}\}$. It is easy to see that

$$\pi(\mathcal{M}) = 1 - \mathrm{ed}_{\mathcal{M}}(1).$$

The Turán density for most monotone properties is not currently known, even though a great deal of work has been done on the subject.

In the graph case, it is trivial to derive $\mathrm{ed}_{\mathscr{M}}(1)$ using symmetrization. In addition, the classification of p-core CRGs established by Marchant and Thomason [32] allows for a trivial proof of the asymptotic Erdős-Stone-Simonovits result [24, 25].

In Question 4 from Section 8.2, we ask several questions that are related to a general theory of edit distance in r-uniform hypergraphs.

8.2 Open Problems

We first ask about powers of cycles and the questions left open in Section 5.4.

Question 1 *Let* $\mathscr{H} = \mathrm{Forb}(C_h^t)$. *What is* $\mathrm{ed}_{\mathscr{H}}(p)$ *for small values of* p, *where* $t+1$ *divides* h? *What is* $\mathrm{ed}_{\mathscr{H}}(p)$ *for small values of* h? *In particular:*

- *For* $\mathscr{H} = \mathrm{Forb}(C_h)$, *what is* $\mathrm{ed}_{\mathscr{H}}(p)$ *for even values of* h *and all values of* p?
- *For* $\mathscr{H} = \mathrm{Forb}(C_h)$, *what is* $d^*_{\mathscr{H}}$ *for all* $t \geq 2$ *and* $h \geq 2t+1$?

Next we consider complete bipartite graphs and some interesting questions from Section 5.5

Question 2 *What is* $\mathrm{ed}_{\mathscr{H}}(p)$ *for* $\mathscr{H} = \mathrm{Forb}(K_{s,t})$? *In particular:*

- *For* $\mathscr{H} = \mathrm{Forb}(K_{s,s})$, *is* $d^*_{\mathscr{H}} = 1/(2s-2)$ *if* $s \geq 7$?
- *For* $\mathscr{H} = \mathrm{Forb}(K_{s,s})$, *is* $p^*_{\mathscr{H}}$ *is an interval of positive length if* $s \geq 7$?
- *For* $\mathscr{H} = \mathrm{Forb}(K_{s,t})$, *which values of* s *and* t *give that* $p^*_{\mathscr{H}}$ *is an interval of positive length?*

Other metrics on the space of graphs are of interest, as we discussed in Section 6.

Question 3 *Let* $\mathscr{H}$ *be a nontrivial hereditary property of graphs.*

- *For the cut metric* $\mathrm{d}_\square$, *is it the case that the function* $\phi_{\mathscr{H}}$, *as defined in (13), can be expressed only in terms of* p *and of* $\mathrm{ed}_{\mathscr{H}}(\cdot)$?
- *For any metric* d *that is continuous with respect to the edit metric, is it the case that the function* $\phi_{\mathscr{H}}$, *as defined in (13), can be expressed only in terms of* p *and of* $\mathrm{ed}_{\mathscr{H}}(\cdot)$?

The question of the edit distance in hypergraphs is wide open, as we discussed in Section 8.1.

Question 4 *Let* $\mathscr{H}$ *be a nontrivial hereditary property of* r*-uniform hypergraphs.*

- *Is it the case that* $\mathrm{ed}_{\mathscr{H}}(p) = \inf_{K\in\mathscr{K}(\mathscr{H})} g_K(p)$ *for all* $p \in [0,1]$?
- *If* $\mathscr{H}$ *is monotone, is it the case that* $\mathrm{ed}_{\mathscr{H}}(1) = \inf_{K\in\mathscr{K}(\mathscr{H})} g_K(1)$?
- *Is there a useful form of generalizations properties do* r*-linear forms have?*
- *Can we provide a structural characterization for* r*-CRHs that are* p*-core, à la Theorem 9?*

The cases of $\mathcal{H} = \text{Forb}(K_{3,3})$ and $\mathcal{H} = \text{Forb}(K_{2,t})$ for $t \geq 9$ suggest an infinite number of CRGs are necessary to define an edit distance function, but only if one wants to compute it for p arbitrarily close to 0 (or by considering the property $\overline{\mathcal{H}}$, arbitrarily close to 1).

Conjecture 1 *Let $\mathcal{H}$ be a nontrivial hereditary property. For every $\epsilon > 0$ there exists a $\mathcal{K}' = \mathcal{K}'(\epsilon, \mathcal{H})$ such that*

$$\text{ed}_{\mathcal{H}}(p) = \min\{g_K(p) : K \in \mathcal{K}'\} \qquad \textit{for all } p \in (\epsilon, 1-\epsilon).$$

We ask if the behavior we seem to observe for $\text{Forb}(K_{3,3})$ and $\text{Forb}(K_{2,t})$ for $t \geq 9$ – that is, that an infinite sequence of CRGs are required to compute the edit distance function for all values of p – does, in fact, occur.

Question 5 *Are there hereditary properties of graphs for which the edit distance function cannot be determined from the G functions of a finite number of CRGs?*

Finally, we conclude with an open problem for the random graph. Recall that $G(n,p)$ denotes the Erdős-Rényi random graph on n vertices with probability p.

Conjecture 2 (Martin [33]) *Fix $p_0 \in (0,1)$ and let $\mathcal{H} = \text{Forb}(G(n_0, p_0))$. Then*

$$\text{ed}_{\mathcal{H}}(p) = (1+o(1))\frac{2\log_2 n_0}{n_0} \min\left\{\frac{p}{-\log_2(1-p_0)}, \frac{1-p}{-\log_2 p_0}\right\}$$

with probability approaching 1 as $n_0 \to \infty$.

The functions that define this bound are of the form $p/(\chi - 1)$ and $(1-p)/(\overline{\chi} - 1)$. The case of $p_0 = 1/2$ was proved to be true by Alon and Stav [4].

If Conjecture 2 is true, then it implies that $p^*_{\mathcal{H}} = \frac{\log_2(1-p_0)}{\log_2 p_0(1-p_0)}$, which is only equal to p_0 itself when $p_0 \in \{0, 1/2, 1\}$. Informally, this implies the counterintuitive notion that it is harder to remove induced copies of $G(n_0, p_0)$ from $G(n, p^*_{\mathcal{H}})$ than it is to remove them from $G(n, p_0)$.

If Conjecture 2 is false, then it implies that the structure of random graphs and the behavior of editing induced graphs are quite complex and very unexpected.

Acknowledgements The author would like to extend a heartfelt appreciation many people for valuable conversations on the topic of edit distance, including Noga Alon, Ed Marchant, Richard Mycroft, Jozef Skokan, and Andrew Thomason. Thanks also to his collaborators József Balogh, Zhanar Berikkyzy, André Kézdy, Tracy McKay, and Chelsea Peck. Special appreciation to Maria Axenovich, who suggested the problem in the first place and without whom this work would not be possible.

The author would like to profoundly thank the National Security Agency (NSA) which partially supported his research via grant H98230-13-1-0226 as well as the Institute for Mathematics and its Applications (IMA). This paper was written while the author was a long-term visitor at the IMA and would not have been possible without their resources and generous support.

References

1. V.E. Alekseev, Hereditary classes and coding of graphs. Problemy Kibernet. **39**, 151–164 (1982). MR 694829 (85a:05071)
2. V.E. Alekseev, Range of values of entropy of hereditary classes of graphs. Diskret. Mat. **4**(2), 148–157 (1992). MR 1181539 (93k:05140)
3. N. Alon, Testing subgraphs in large graphs. Random Struct. Algorithm **21**(3–4), 359–370 (2002). Random structures and algorithms (Poznan, 2001). MR 1945375 (2003k:05129)
4. N. Alon, U. Stav, The maximum edit distance from hereditary graph properties. J. Comb. Theory Ser. B **98**(4), 672–697 (2008). MR 2418765 (2009d:05111)
5. N. Alon, U. Stav, What is the furthest graph from a hereditary property? Random Struct. Algorithm **33**(1), 87–104 (2008). MR 2428979 (2009c:05219)
6. N. Alon, R.A. Duke, H. Lefmann, V. Rödl, R. Yuster, The algorithmic aspects of the regularity lemma. J. Algorithm **16**(1), 80–109 (1994). MR 1251840 (94j:05112)
7. N. Alon, E. Fischer, M. Krivelevich, M. Szegedy, Efficient testing of large graphs. Combinatorica **20**(4), 451–476 (2000). MR 1804820 (2002f:05143)
8. M. Axenovich, R.R. Martin, Avoiding patterns in matrices via a small number of changes. SIAM J. Discrete Math. **20**(1), 49–54 (electronic) (2006). MR 2257244 (2007j:05015)
9. M. Axenovich, R.R. Martin, Multicolor and directed edit distance. J. Comb. **2**(4), 525–556 (2011). MR 2911190
10. M. Axenovich, A. Kézdy, R.R. Martin, On the editing distance of graphs. J. Graph Theory **58**(2), 123–138 (2008). MR 2407000 (2009c:05107)
11. J. Balogh, R.R. Martin, Edit distance and its computation. Electron. J. Comb. **15**(1), Research Paper 20, 27 (2008). MR 2383440 (2008j:05175)
12. J. Balogh, B. Bollobás, D. Weinreich, The penultimate rate of growth for graph properties. Eur. J. Comb. **22**(3), 277–289 (2001). MR 1822715 (2002b:05079)
13. J. Balogh, B. Bollobás, D. Weinreich, Measures on monotone properties of graphs. Discrete Appl. Math. **116**(1–2), 17–36 (2002). MR 1877113 (2002i:05066)
14. J. Balogh, B. Bollobás, M. Saks, V.T. Sós, The unlabelled speed of a hereditary graph property. J. Comb. Theory Ser. B **99**(1), 9–19 (2009). MR 2467814 (2009k:05105)
15. J. Balogh, B. Bollobás, M. Simonovits, The fine structure of octahedron-free graphs. J. Comb. Theory Ser. B **101**(2), 67–84 (2011). MR 2763070 (2012a:05157)
16. J. Balogh, B. Bollobás, D. Weinreich, The speed of hereditary properties of graphs. J. Comb. Theory Ser. B **79**(2), 131–156 (2000). MR 1769217 (2001d:05092)
17. Z. Berikkyzy, R.R. Martin, C. Peck, On the edit distance of powers of cycles. Submitted. arXiv: 1509.07438
18. B. Bollobás, A. Thomason, Projections of bodies and hereditary properties of hypergraphs. Bull. Lond. Math. Soc. **27**(5), 417–424 (1995). MR 1338683 (96e:52006)
19. B. Bollobás, A. Thomason, Hereditary and monotone properties of graphs, in *The mathematics of Paul Erdős, II*, Algorithms Combination, vol. 14 (Springer, Berlin, 1997), pp. 70–78. MR 1425205 (98a:05085)
20. B. Bollobás, A. Thomason, The structure of hereditary properties and colourings of random graphs. Combinatorica **20**(2), 173–202 (2000). MR 1767020 (2001e:05114)
21. C. Borgs, J.T. Chayes, L. Lovász, V.T. Sós, K. Vesztergombi, Convergent sequences of dense graphs. I. Subgraph frequencies, metric properties and testing. Adv. Math. **219**(6), 1801–1851 (2008). MR 2455626 (2009m:05161)
22. W.G. Brown, On graphs that do not contain a Thomsen graph. Canad. Math. Bull. **9**, 281–285 (1966). MR 0200182 (34 #81)
23. D. Chen, O. Eulenstein, D. Fernández-Baca, M. Sanderson, Supertrees by flipping, in *Computing and Combinatorics*. Lecture Notes in Computer Science, vol. 2387 (Springer, Berlin, 2002), pp. 391–400. MR 2064534 (2005b:92030)
24. P. Erdős, M. Simonovits, A limit theorem in graph theory. Studia Sci. Math. Hungar. **1**, 51–57 (1966). MR 0205876 (34 #5702)

25. P. Erdös, A.H. Stone, On the structure of linear graphs. Bull. Am. Math. Soc. **52**, 1087–1091 (1946). MR 0018807 (8,333b)
26. P. Frankl, V. Rödl, Extremal problems on set systems. Random Struct. Algorithm **20**(2), 131–164 (2002). MR 1884430 (2002m:05192)
27. A. Frieze, R. Kannan, Quick approximation to matrices and applications. Combinatorica **19**(2), 175–220 (1999). MR 1723039 (2001i:68066)
28. Z. Füredi, New asymptotics for bipartite Turán numbers, J. Comb. Theory Ser. A **75**(1), 141–144 (1996). MR 1395763 (97f:05096)
29. W.T. Gowers, Hypergraph regularity and the multidimensional Szemerédi theorem. Ann. Math. (2) **166**(3), 897–946 (2007). MR 2373376 (2009d:05250)
30. P. Keevash, Hypergraph Turán problems, in *Surveys in Combinatorics 2011*, The London Mathematical Society, Lecture Notes Series, vol. 392 (Cambridge University Press, Cambridge, 2011), pp. 83–139. MR 2866732
31. E.J. Marchant, *Graphs with Weighted Colours and Hypergraphs*. Ph.D. thesis, University of Cambridge, 2011
32. E. Marchant, A. Thomason, Extremal graphs and multigraphs with two weighted colours, in *Fete of Combinatorics and Computer Science*, Bolyai Society Mathematical Studies, vol. 20 (János Bolyai Mathematical Society, Budapest, 2010), pp. 239–286. MR 2797967 (2012f:05149)
33. R. Martin, The edit distance function and symmetrization Electron. J. Comb. **20**(3), Paper 26, 25 (2013). MR 3104524
34. R.R. Martin, On the computation of edit distance functions. Discrete Math. **338**(2), 291–305 (2015). MR 3279281
35. R.R. Martin, T. McKay, On the edit distance from $K_{2,t}$-free graphs. J. Graph Theory **77**(2), 117–143 (2014). MR 3246171
36. B. Nagle, V. Rödl, M. Schacht, The counting lemma for regular k-uniform hypergraphs. Random Struct. Algorithm **28**(2), 113–179 (2006). MR 2198495 (2007d:05084)
37. C. Peck, *On the Edit Distance from a Cycle- and Squared Cycle-Free Graph*. Master's thesis, Iowa State University, 2013
38. O. Pikhurko, An exact Turán result for the generalized triangle. Combinatorica **28**(2), 187–208 (2008). MR 2399018 (2009c:05161)
39. L. Pósa, Hamiltonian circuits in random graphs. Discrete Math. **14**(4), 359–364 (1976). MR 0389666 (52 #10497)
40. H.J. Prömel, A. Steger, Excluding induced subgraphs: quadrilaterals. Random Struct. Algorithm **2**(1), 55–71 (1991). MR 1099580 (92j:05157)
41. H.J. Prömel, A. Steger, Excluding induced subgraphs. III. A general asymptotic. Random Struct. Algorithm **3**(1), 19–31 (1992). MR 1139486 (93d:05065)
42. H.J. Prömel, A. Steger, Excluding induced subgraphs. II. Extremal graphs. Discrete Appl. Math. **44**(1–3), 283–294 (1993). MR 1227710 (94f:05121)
43. D.C. Richer, *Graph Theory and Combinatorial Games*. Ph.D. thesis, University of Cambridge, 2000
44. V. Rödl, J. Skokan, Regularity lemma for k-uniform hypergraphs. Random Struct. Algorithm **25**(1), 1–42 (2004). MR 2069663 (2005d:05144)
45. V. Rödl, J. Skokan, Applications of the regularity lemma for uniform hypergraphs. Random Struct. Algorithm **28**(2), 180–194 (2006). MR 2198496 (2006j:05099)
46. E.R. Scheinerman, J. Zito, On the size of hereditary classes of graphs. J. Comb. Theory Ser. B **61**(1), 16–39 (1994). MR 1275261 (95e:05061)
47. A. Sidorenko, Boundedness of optimal matrices in extremal multigraph and digraph problems. Combinatorica **13**(1), 109–120 (1993). MR 1221180 (94c:05043)
48. E. Szemerédi, Regular partitions of graphs, in *Problèmes combinatoires et théorie des graphes (Colloq. Internat. CNRS, Univ. Orsay, Orsay, 1976)*. Colloq. Internat. CNRS, vol. 260 (CNRS, Paris, 1978), pp. 399–401. MR 540024 (81i:05095)

49. A. Thomason, *Graphs, Colours, Weights and Hereditary Properties*. Surveys in combinatorics 2011, The London Mathematical Society, Lecture Notes Series, vol. 392 (Cambridge University Press, Cambridge, 2011), pp. 333–364. MR 2866736
50. A.A. Zykov, On some properties of linear complexes. Am. Math. Soc. Translat. **1952**(79), 33 (1952). MR 0051516 (14,493a)

Repetitions in graphs and sequences

Maria Axenovich

Abstract The existence of unavoidable repeated substructures is a known phenomenon implied by the pigeonhole principle and its generalizations. A fundamental problem is to determine the largest size of a repeated substructure in any combinatorial structure from a given class. The strongest notion of repetition is a pair of isomorphic substructures, such as a pair of vertex-disjoint or edge-disjoint isomorphic subgraphs or a pair of disjoint identical subsequences of a sequence. A weaker notion of repetition is a pair of substructures that have the same value on a certain set of parameters. This includes vertex-disjoint induced subgraphs of the same order and size, disjoint vertex sets with the same multiset of pairwise distances, subgraphs with the same maximum degree. This paper surveys results on unavoidable repetitions, also referred to as twins, with a focus on three asymptotically tight results obtained over the past 5 years.

Keywords Sequence • Subword • Identical subwords • Twins in sequences • Twins in graphs • Isomorphic subgraphs • Repeated sequences • Self-similarity • Divisibility of graphs • Twins in hypergraphs.

Mathematics Subject Classification (2010): 05D05, 05C35, 05C60, 05C70

1 Introduction

There are many repetitions that one can observe in nature: identical twins, two leaves on a tree that look alike, repetitive motifs in a bird's song, segments of the DNA that are identical, etc.

Discrete mathematical structures possess repetitions as well, as is implied by the pigeonhole principle and its generalizations. A fundamental problem is to determine the largest size of a repeated substructure in any combinatorial structure from a given class. Here, we shall often refer to repeated structures as twins even if these twins are not identical.

M. Axenovich (✉)
Karlsruher Institut für Technologie, Karlsruhe, Germany
e-mail: maria.aksenovich@kit.edu

A. Beveridge et al. (eds.), *Recent Trends in Combinatorics*, The IMA Volumes in Mathematics and its Applications 159, DOI 10.1007/978-3-319-24298-9_3

The strongest notion of repetition deals with "identical" twins, i.e., with a pair of isomorphic substructures. A weaker notion of repetition is a pair of substructures with the same value on a certain set of parameters.

This survey focuses on three types of "identical" twins, which have corresponding size functions $f_v(n), f_e(n), f(n)$ and a few types of weaker twins, which have size functions such as $t(n), h(n)$.

Let $\mathcal{G}_n$ be the class of all n-vertex graphs, and $\mathcal{G}^m$ be the class of all graphs with m edges, let k be an integer. For all standard graph theoretic notions we refer the reader to the book of West [54]. All graphs here are simple graphs with no repeated edges and no loops. We define the size functions:

$$
\begin{aligned}
f_v(n) &= \max\{\, k : \text{ any } G \in \mathcal{G}_n \text{ has two isomorphic vertex-disjoint induced} \\
&\qquad \text{subgraphs on } k \text{ vertices each}\},\\
f_e(m) &= \max\{\, k : \text{ any } G \in \mathcal{G}^m \text{ has two isomorphic edge-disjoint subgraphs} \\
&\qquad \text{on } k \text{ edges each}\},\\
f(n) &= \max\{\, k : \text{ any binary sequence with } n \text{ elements contains} \\
&\qquad \text{two disjoint identical subsequences of } k \text{ elements each}\},\\
t(n) &= \max\{\, k : \text{ any } G \in \mathcal{G}_n \text{ has two vertex-disjoint induced subgraphs} \\
&\qquad \text{on } k \text{ vertices and with the same number of edges}\},\\
h(n) &= \max\{\, k : \text{ any } G \in \mathcal{G}_n \text{ has two disjoint subsets of vertices} \\
&\qquad \text{whose multisets of pairwise distances are identical}\}.
\end{aligned}
$$

For all of these functions, except for the last one, we now know exact asymptotic behavior. The following theorem is an easy consequence of Ramsey theorem [22, 49] and a property of random graphs, see Section 2.

Theorem 1.1.

$$f_v(n) = \Theta(\log n).$$

The next three theorems proved in 2012, 2014, and 2012, respectively, involve more sophisticated proof techniques such as random methods on graphs, regular partitions of sequences, and balanced partitions of integers.

Theorem 1.2. (Lee, Loh, and Sudakov [41]). *There are absolute positive constants c and C for which*

$$c(m \log m)^{2/3} \leq f_e(m) \leq C(m \log m)^{2/3}.$$

Theorem 1.3. (Axenovich, Person, and Puzynina [6]). *There exists an absolute constant C such that*

$$\left(1 - C\left(\frac{\log n}{\log\log n}\right)^{-1/4}\right)\frac{n}{2} \leq f(n) \leq \frac{n}{2} - \frac{1}{2}\log n.$$

Theorem 1.4. (Bollobás, Kittipassorn, Narayanan, and Scott [14]). *For every $\epsilon > 0$, there is a natural number $N = N(\epsilon)$ such that for all $n > N$*

$$\frac{n}{2} - \epsilon n \leq t(n) \leq \frac{n}{2} - \log\log n.$$

Compared to the above theorems giving tight bounds, it is still not known what is the correct asymptotic behavior of the function $h(n)$. The gap between the upper and the lower bound is large. Here, the slight $\log\log n$ improvement of the upper bound is due to a result from [5].

Theorem 1.5. (Albertson, Pach, Young [1, 5]). *There are positive constants c, c', such that for any $n > 3$,*

$$\frac{c\log n}{\log\log n} < h(n) \leq \frac{n}{4} - c'\log\log n.$$

We address functions $f_v(n), f_e(m), f(n), t(n)$, and $h(n)$ in Sections 2, 3, 4, 5, and 6, respectively. We provide some insight into proof techniques, state generalizations and open problems. At the end of the survey, we mention some other weak twin problems.

2 Vertex-disjoint isomorphic subgraphs

Among all twin problems we consider here, the problem of finding the largest order of two vertex-disjoint isomorphic subgraphs is relatively easy and Theorem 1.1 answers it asymptotically. We prove it here for completeness.

Proof of Theorem 1.1. For the lower bound, consider an n-vertex graph G. By Ramsey theorem [22, 49] there is a complete subgraph or an independent set on $c\log n$ vertices. Thus G has two vertex-disjoint induced subgraphs on $\frac{c}{2}\log n$ vertices both isomorphic to either a complete graph or an empty graph.

For the upper bound, we shall consider Erdős-Renyi random graph $G=G(n, 1/2)$ and follow the simple union bound approach of Lee, Loh, and Sudakov [41]. The fact that G has two vertex-disjoint isomorphic twins on t vertices each is equivalent to the existence of a $2t$-vertex subgraph H' that can be partitioned into a graph H and a vertex-disjoint isomorphic copy of H. The expected number of such graphs H' is at most $\binom{n}{2t}\binom{2t}{t}t!2^{-\binom{t}{2}}$, where the first binomial coefficient gives the number of ways to choose $2t$ vertices in G, the second counts the number of ways to split $2t$ vertices in two equal parts, $t!$ is the number of ways to permute the vertices in one of the parts, and finally $2^{-\binom{t}{2}}$ is the probability that the second part with ordered vertices is isomorphic to the first part. If $t > 12\log n$, then $\binom{n}{2t}\binom{2t}{t}t!2^{-\binom{t}{2}} \leq (en/2t)^{2t}2^{2t}t^t2^{-t^2/4} \leq n^{2t}2^{-t^2/4} = 2^{t(2\log n - t/4)} \leq 2^{t(2\log n - 3\log n)} < 1$.

Thus there is a graph G with no two vertex-disjoint isomorphic subgraphs on t vertices each. Note that this graph has no such subgraphs on $t' > t$ vertices each since otherwise one can delete the corresponding $t' - t$ vertices from each of these subgraphs and obtain isomorphic vertex-disjoint subgraphs on t vertices each. So, any two induced vertex-disjoint subgraphs on at least t vertices in G are not isomorphic. □

Algorithmic problems of finding vertex-disjoint isomorphic subgraphs of a given graph and finding a largest induced subgraph common to two given graphs were studied extensively [8, 39, 43]. It was observed that one can map this twin problem to the problem of finding a largest clique in an auxiliary graph. Let G be a graph with $A, B \subseteq V(G), A \cap B = \emptyset$. Let $G' = (V', E')$, where $V' = \{(u, v) : u, v \in V(G),\ u \neq v\}$, the edge set E' consists of pairs $((u, v), (u', v'))$, where u, v, u', v' are pairwise distinct and either $\{u, u'\}, \{v, v'\} \in E(G)$ or $\{u, u'\}, \{v, v'\} \notin E(G)$. Then it is clear that there is an isomorphism ϕ between $G[A]$ and $G[B]$, graphs induced by A and B in G, if and only if $\{(v, \phi(v)) : v \in A\}$ induces a clique in G'. Similarly, a clique in G' gives an isomorphism. Dzido and Krzywdziński [26] studied so-called local similarity of graphs. Specifically, for two parameters k, ℓ, they defined a function $g(k, \ell)$ that is the smallest n such that any ℓ graphs on n vertices have a common induced subgraph on k vertices. They proved that $g(k, 3) = R(k, k)$, where $R(k, k)$ is the classical diagonal Ramsey number, and determined $g(3, \ell)$ and $g(4, \ell)$ for all $\ell > 1$. In general, a question about repeated and not necessarily disjoint induced subgraphs was addressed by Alon and Bollobás in [2].

3 Isomorphic edge-disjoint subgraphs

Let's consider a more complex situation with edge-disjoint twins in graphs. Let $f_e(G)$ be the largest k such that a graph or hypergraph G has two edge-disjoint isomorphic subgraphs on k edges. It is clear that $f_e(G) \leq |E(G)|/2$, thus the optimization problem is set over graphs of the same size, m. In a general setting of hypergraphs, let

$$f_e(m, r) = \min\{f_e(G) : \ |E(G)| = m,\ G \text{ is an } r - \text{uniform hypergraph}\}.$$

The notion of $f_e(G)$ was introduced by Jacobson and Schönheim, where they called this parameter the self-similarity of G. This name may seem confusing, since self-similarity is most commonly used for the fractal structure of an object, i.e., for example, we would want a graph to be isomorphic to its subgraph.

Erdős, Pach, and Pyber [27] proved that there are positive absolute constants c and C, such that

$$cm^{2/(2r-1)} \leq f_e(m, r) \leq Cm^{2/(r+1)} \frac{\log m}{\log \log m}.$$

Lee, Loh, and Sudakov [41] improved the upper and the lower bounds to obtain a tight asymptotic result for graphs. Let $f_e(m) = f_e(m, 2)$.

Theorem 3.1. ([41]). *There are absolute constants c and C for which*

$$c(m \log m)^{2/3} \leq f_e(m) \leq C(m \log m)^{2/3}.$$

The upper bound is based on a random graph with modified edge-probability. For the lower bound, the authors observe that a graph in which the vertex-disjoint induced twins are small behaves in a sense like a random graph.

When $r = 3$, Gould and Rödl [29] proved that $f_e(m, 3) \geq c_3 m^{1/2}$ for a constant c_3. Horn, Koubek, and Rödl [36] studied the hypergraph case further, proving the tightness, up to a logarithmic factor, of the upper bound in the result of Erdős, Pach, and Pyber [27] in case when $r = 4, 5, 6$:

Theorem 3.2. ([36]). *There are constants c_4, c_5, c_6, such that $f_e(m, 4) \geq c_4 m^{2/5}$, $f_e(m, 5) \geq c_5 \frac{m^{1/3}}{\log m}$, and $f_e(m, 6) \geq c_6 \frac{m^{2/7}}{\log^{35} m}$. For any $r \geq 7$,*

$$f_e(m, r) \geq \Omega(m^{\frac{2}{2r - \log_2 r}}).$$

Lee, Loh, and Sudakov [41] also generalized this problem to multiple twins. Let $f_e^s(m)$ be the largest k such that any graph on m edges has s pairwise edge-disjoint isomorphic subgraphs. Then, they showed in [41], that

$$f_e^s(m) = \Theta\left(m^{\frac{s}{2s-1}} (\log m)^{\frac{2s-2}{2s-1}}\right).$$

The earlier research was done on *perfect* edge-disjoint isomorphic twins. Here, we say that G has t perfect edge-disjoint isomorphic twins if the edge set of G is partitioned into t isomorphic subgraphs. Graham, Harary, Robinson, Wallis, and Wormald considered this notion, also called t-divisibility of graphs, or isomorphic factorization of graphs.

Harary, Robinson, Wallis, and Wormald [33, 34], see also Guidotti [31], proved that K_n is t-divisible if t divides the total number of edges, $\binom{n}{2}$, in K_n. Alon, Caro, and Krasikov [3] showed that any m-edge tree could be made 2-divisible by deleting $O(m/\log \log m)$ edges. The notion of 2-divisibility was addressed also under the names *bisectable* and 2-*splittable*, see [32, 52]. There is an extensive research done on isomorphic factorizations, with a lot of papers devoted to factorizations of complete or complete multipartite graphs into specific trees or cycles. See also a book on decompositions by Bosák [15].

4 Repeated Subsequences

An *alphabet* Σ is a finite non-empty set of symbols called *letters*. Let $S = s_1 \dots s_n$ be a sequence $s_1, s_2, \dots, s_n$, where $s_i \in \Sigma$, $i = 1, \dots, n$. Sometimes, one refers to S as a word over an alphabet Σ. A *subsequence* of S is a sequence $S' = s_{i_1} s_{i_2} \dots s_{i_s}$, where $i_1 < i_2 < \dots < i_s$. A *factor* of S is a subsequence obtained by taking consecutive elements from S, i.e., $s_i s_{i+1} \dots s_j$, for some $1 \le i \le j \le n$.

The study of repeated sequences is largely motivated by problems investigated in combinatorics on words and in the theory of formal languages with special attention given to counting subword occurrences, different complexity questions, and the word reconstruction problem (see, e.g., [25, 46, 51]). We say that two subsequences $s_{i_1} s_{i_2} \dots s_{i_s}$ and $s_{j_1} s_{j_2} \dots s_{j_t}$ of S are disjoint if the index sets are disjoint, i.e., $\{i_1, \dots, i_s\} \cap \{j_1, \dots, j_t\} = \emptyset$. For a sequence S, let $f(S)$ be the largest integer m such that there are two disjoint identical subsequences of S, each of length m. We call such subsequences *twins*. For example, if $S = s_1 s_2 s_3 s_4 s_5 s_6 s_7 = 0001010$, then $S_1 = s_1 s_2 s_4$ and $S_2 = s_3 s_5 s_6$ are both equal to 001, in addition $S'_1 = s_1 s_4 s_5$ and $S'_2 = s_2 s_6 s_7$ are both equal to 010. So, both pairs, S_1, S_2 and S'_1, S'_2 are pairs of twins. In this example $f(S) = 3$.

The twin problem in sequences is closely related to the classical problem on common subsequences, see, for example, [23, 35, 55]. Indeed, if we consider two disjoint subsequences S_1 and S_2 of a given sequence S and find subsequences S'_1 of S_1 and S'_2 of S_2 such that $S'_1 = S'_2$, then these subsequences correspond to twins in S. Optimizing over all such S_1 and S_2 gives largest twins. Let Σ^n be the set of sequences of length n over the alphabet Σ, let

$$f(n, \Sigma) = \min\{f(S) : S \in \Sigma^n\}.$$

Observe first that $f(n, \{0, 1\}) \ge \lfloor (1/3)n \rfloor$. Indeed, consider any $S \in \Sigma^n$ and split it into consecutive triples. Each triple has either two zeros or two ones, so we can build a subsequence S_1 by choosing a repeated element from each triple, and similarly build a subsequence S_2 by choosing the second repeated element from each triple. For example, if $S = 001\ 101\ 111\ 010$, then there are twins S_1, S_2, each equal to 0 1 1 0. If we mark one twin red and another bold, then $S =$ **0**01 **1**01 **1**11 **0**10.

We shall sometimes refer to two disjoint identical subsequences as *twins*, three disjoint identical subsequences as 3-*twins*, k disjoint identical subsequences as k-*twins*.

4.1 Two twins in a binary alphabet

The main result of Axenovich, Person, and Puzynina [6] shows that any sufficiently large binary sequence can be split into two identical subsequences almost perfectly. This is one of the main theorems stated in the introduction.

Theorem 4.1. ([6]). *There exists an absolute constant C such that*

$$\left(1 - C\left(\frac{\log n}{\log\log n}\right)^{-1/4}\right) n \leq 2f(n, \{0,1\}) \leq n - \log n.$$

The proof employs a classical density increment argument successfully applied in combinatorics and number theory, see, e.g., the survey of Komlós and Simonovits [40] and some important applications [30] and [53]. One can partition any sequence S into factors that look as if they were random in a certain weak sense (call them ϵ-regular). These ϵ-regular sequences can be partitioned (with the exception of ϵ-proportion of letters) into two identical subsequences. By appending these together for every ϵ-regular sequence, we eventually obtain identical subsequences of S whose lengths satisfy the claimed inequality.

4.2 More than two twins or larger alphabets

For a given sequence S, let $f(S,k)$ be the largest m so that S contains k pairwise disjoint identical subsequences of length m each. Finally, let $f(n,k,\Sigma) = \min\{f(S,k) : S \in \Sigma^n\}$ and $f(n,\Sigma) = f(n,2,\Sigma)$.

Axenovich, Person, and Puzynina [6] showed that the situation differs significantly depending on whether the number of twins is larger or smaller than the size of the alphabet. If the size of the alphabet is at most the number of twins, the given sequences can be almost partitioned by these twins. Otherwise, in most cases the largest twins cover at most $(1-c)n$ elements of a given sequence of length n, for some constant $c > 0$, depending on the number of twins.

Theorem 4.2. ([6]). *For any integer $k \geq 2$, and alphabet Σ, $|\Sigma| \leq k$,*

$$\left(1 - C|\Sigma|\left(\frac{\log n}{\log\log n}\right)^{-1/4}\right) n \leq kf(n,k,\Sigma).$$

Theorem 4.3. ([6]). *For any integer $k \geq 2$, and alphabet Σ, $|\Sigma| > k$,*

$$\left(\frac{k}{|\Sigma|} - C|\Sigma|\left(\frac{\log n}{\log\log n}\right)^{-1/4}\right) n \leq kf(n,k,\Sigma) \leq n - \max\{\alpha n, \log n\},$$

where $\alpha \in [0, 1/k]$ is the solution of the equation $|\Sigma|^{-(k-1)\alpha}\alpha^{-k\alpha}(1-k\alpha)^{k\alpha-1} = 1$, whenever such solution exists and 0 otherwise.

Bukh and Zhou [17] improved the above bounds in case when $k = 2$.

Theorem 4.4. ([17]). *Let Σ be an alphabet, $|\Sigma| \geq 3$. Then*

$$\max\left\{\frac{2.04}{|\Sigma|}n - o(n),\ 2\cdot 3^{-4/3}|\Sigma|^{-2/3}n - 3^{-1/3}|\Sigma|^{1/3}\right\} \leq 2\cdot f(n,2,\Sigma) \leq 2\alpha n,$$

where α is a real number in the interval $[1/|\Sigma|, 1/2]$ satisfying inequality

$$(1-2\alpha)\log\frac{1}{1-2\alpha} - \alpha\log(\alpha^2|\Sigma|) - 2\alpha\log\left(\frac{2}{1+\sqrt{1-1/|\Sigma|}}\right)$$
$$+ (1-2\alpha)\log(1-1/|\Sigma|) < 0.$$

The upper bound implies, in particular, that $2f(n,2,\Sigma) \leq 0.9864\,n$, for $|\Sigma| = 4$; $2f(n,2,\Sigma) \leq 0.96\,n$, for $|\Sigma| = 5$;

$$2f(n,2,\Sigma) \leq \left(\frac{2e}{\sqrt{|\Sigma|}} - (2e^2+1)/|\Sigma| + O(|\Sigma|^{-3/2})\right)n + o(n),$$

for all large $|\Sigma|$.

That is, for alphabets on at least four letters, there are sequences that could not be almost split into two twins. In comparison, any sufficiently large binary sequence has such almost perfect splitting. It remains unclear what happens for alphabets on 3 letters. From the above theorems it follows that $\frac{2.04}{3}n - o(n) \leq 2\cdot f(n,2,\Sigma) \leq n - \log n$.

Open Problem 4.5. *Find $f(n,2,\Sigma)$ for $|\Sigma| = 3$.*

4.3 *Tools*

There were two main tools used in the twin problem in sequences so far: a version of the Regularity Lemma for sequences and common subsequences.

The first, introduced in [6], employs the Regularity Lemma variant also developed in that paper. The idea is to split the given sequence into a constant number of factors such that almost all of them are regular, i.e., have about the same density of each letter in sufficiently large subfactors, and then to find twins in each of these factors. To do so, say in case of the alphabet $\{0,1\}$, delete at most an ϵ-proportion of elements, and split a factor further into subfactors each containing exactly a zeros and exactly b ones, for some a and b. Build one twin by taking all a zeros from the 1st subfactor, all b ones from the 2^{d} subfactor, all a zeros from the 3^{d}, all b ones from the 4th, etc. The second twin is build by taking all a zeros from the 2^{d} subfactor, all b ones from the 3^{d}, all a zeros from the 4th, etc.

The second tool, used for small cases in [6] and explored much more widely in [17], is based on common subsequences. For two sequences, a largest subsequence contained in both of these sequences is called a largest common subsequence. In order to find twins in a given sequence, one can split it into consecutive factors $S_1, \ldots, S_m$, split each S_i in two subsequences A_i, B_i, find a largest common subsequence C_i of A_i and B_i, and obtain twins each equal to $C_1C_2\cdots C_m$. A refinement of this natural approach together with Regularity Lemma provided the best lower bounds on f in case of large alphabets, see [16, 17]. Next, we describe these approaches a bit more formally.

4.3.1 Regularity Lemma for sequences

First, we shall introduce some notations (for more detail, see, for instance, [20, 44]). Let $S = s_1 \ldots s_n$. We define a *support* of a subsequence to be its index set, i.e., for a subsequence $S' = s_{i_1}s_{i_2}\ldots s_{i_s}$ of S, $\{i_1, i_2, \ldots, i_s\}$ a support of S' in S. We write $\text{supp}(S')$ to denote support. The *length* of a sequence S is the number of elements in S, denoted $|S|$. For example, $|(0010)| = 4$. If $S' = s_{i_1}s_{i_2}\ldots s_{i_s}$, we write $S' = S[I]$, for an index set $I = \{i_1, \ldots, i_s\}$. We use notation $S[i, i+m]$ for $s_is_{i+1}\ldots s_{i+m}$. If S is a sequence over alphabet Σ and $q \in \Sigma$, then we denote $|S|_q$ the number of elements of S equal to q. The *density* $d_q(S)$ is defined to be $|S|_q/|S|$. For two subsequences S' and S'' of S, we say that S' is contained in S'' if $\text{supp}(S') \subseteq \text{supp}(S'')$, we also denote by $S' \cap S''$ a subsequence of S, $S[\text{supp}(S') \cap \text{supp}(S'')]$. If $S = s_1 \ldots s_n$ and $S[1, i] = A$, $S[i+1, n] = B$, for some i, then we write $S = AB$ and call S a concatenation of A and B. In all of the calculations below we omit floors and ceilings to avoid making the presentation too cluttered.

Next, we define *ϵ-regular sequence*. For a positive ϵ, $\epsilon < 1/3$, call a sequence S of length n over an alphabet Σ *ϵ-regular* if for every i, $\epsilon n + 1 \leq i \leq n - 2\epsilon n + 1$ and every $q \in \Sigma$,

$$|d_q(S) - d_q(S[i, i + \epsilon n - 1])| < \epsilon.$$

Note that in the case of the binary alphabet $\Sigma = \{0, 1\}$, $d_0(S) = 1 - d_1(S)$ and thus $|d_0(S) - d_0(S[i, i + \epsilon n - 1])| < \epsilon \iff |d_1(S) - d_1(S[i, i + \epsilon n - 1])| < \epsilon$. In this case, we shall denote $d(S) = d_1(S)$. For example, the sequence

$$S = 011101010101000101001100110101010100110101010101111000010100$$

of length $n = 60$ and density $1/2$ is ϵ-regular for $\epsilon = 1/5$. This could be verified directly by considering all $60-12$ consecutive segments of length $\epsilon n = (1/5)\cdot 60 = 12$ and comparing their densities with the density $1/2$ of S. Such a 12-letter sequence starting at first position is $S' = 011101010101$, $d(S') = 7/12$, $|7/12 - 1/2| < \epsilon = 1/5$. Such a 12-letter sequence starting at position 8 is $S'' = 101010001010$, $d(S'') = 5/12$, $|5/12 - 1/2| < 1/5$.

The notion of an ϵ-regular sequence resembles the notion of pseudorandom (quasirandom) sequence, see [21]. However, these two notions are quite different. A sequence that consists of alternating 0s and 1s is ϵ-regular but not pseudorandom. Also, unlike the case of stronger notions of pseudorandomness, one can check in a linear time, by observing at most n sequences $S[i, i+\epsilon n-1]$, whether a sequence is ϵ-regular, cf. [4], in the graph case.

Definition. We call $\mathcal{S} := (S_1, \ldots, S_t)$ a partition of S if $S = S_1S_2\ldots S_t$, (S is concatenation of consecutive S_is). A partition $\mathcal{S}$ is an **ϵ-regular partition** of a sequence $S \in \Sigma^n$ if

$$\sum_{\substack{i\in[t] \\ S_i \text{ is not } \epsilon-\text{regular}}} |S_i| \leq \epsilon n,$$

i.e., the total length of ϵ-irregular subsequences is at most ϵn.

Lemma 4.6. (Regularity Lemma for sequences, [6]). *For every ϵ, t_0, and n such that $0 < \epsilon < 1/3$, $t_0 > 0$, and $n > n_0 = t_0\epsilon^{-\epsilon^{-4}}$, any sequence $S \in \Sigma^n$ admits an ϵ-regular partition into t parts with $t_0 \leq t \leq T_0 = t_0 3^{1/\epsilon^4}$.*

Using the Regularity Lemma, one can find large twins in sequences.

Outline of the proof of Theorem 4.1. We claim that if S is an ϵ-regular sequence, then $2f(S) \geq |S| - 5\epsilon|S|$. Let $|S| = m$. We partition S into $t = 1/\epsilon$ consecutive factors $S_1, \ldots, S_{1/\epsilon}$, each of length ϵm. Since S is ϵ-regular, $|d(S_i) - d(S)| < \epsilon$, for every $i \in \{2, \ldots, 1/\epsilon - 1\}$. Thus each S_i has at least $(d(S) - \epsilon)\epsilon m$ occurrences of 1's and at least $(1 - d(S) - \epsilon)\epsilon m$ occurrences of 0's. Let $S_i(1)$ be a subsequence of S_i consisting of exactly $(d(S) - \epsilon)\epsilon m$ letters 1 and $S_i(0)$ be a subsequence of S_i consisting of exactly $(1 - d(S) - \epsilon)\epsilon m$ letters 0. When t is even, consider the following two disjoint subsequences of S: $A = S_2(1)S_3(0)S_4(1)\cdots S_{t-2}(1)$ and $B = S_3(1)S_4(0)S_5(1)\cdots S_{t-2}(0)S_{t-1}(1)$. When t is odd, A and B are constructed similarly. We see that A and B together have at least $m - 2\epsilon^2m(1/\epsilon - 3) - 3\epsilon m$ elements, where $2\epsilon^2m(1/\epsilon - 3)$ is an upper bound on the number of 0's and 1's which we had to "throw away" to obtain *exactly* $(d(S) - \epsilon)\epsilon m$ letters 1 and $(1 - d(S) - \epsilon)\epsilon m$ letters 0 in each S_i, $2\epsilon m$ is the number of elements in S_1 and S_t, and ϵm is the upper bound on $|S_2(0)| + |S_{t-1}(1)|$. Thus, $2f(S) \geq m - 5\epsilon m$. This concludes the proof of the claim. Now, consider an ϵ-regular partition of a given sequence of length n. Ignore the parts that are not ϵ-regular. Note, that there are at most ϵn letters in such parts. In each of the regular parts, find large twins using the claim. Finally, build large twins of the original sequence by concatenating the corresponding twins of each regular part. This gives the lower bound on $f(n)$. For the upper bound, consider a binary sequence 0100011100000000011111111…., where consecutive blocks have lengths 3^i. One can show that the union of any two twins in such a sequence omits at least one letter from each block. □

4.3.2 Common Subsequences

The length of a longest common subsequence of sequences W_1 and W_2 is denoted by $LCS(W_1, W_2)$. For example, $LCS(001000, 110010) = 4$ with a common subsequence 1000. Here it is clear that there is no common subsequence on 5 letters because any 5-letter subsequence of the second sequence contains at least two 1's, and the first sequence has only one 1. The problem of finding common subsequences in sequences is of great importance in bioinformatics [55]. There is an extensive literature devoted to computational and algorithmic aspects of the problem, see, for example, [35]. Kiwi, Loebl, and Matoušek [38] considered random binary sequences. If $L'(n,k)$ is the length of the longest common subsequence of two n-element sequences chosen uniformly and independently at random over an alphabet of size k, then it was shown that

$$\lim_{k\to\infty} \sqrt{k} \lim_{n\to\infty} n^{-1}|\mathrm{Exp}(L'(n,k)| = 2.$$

Roughly speaking, the expected length of a common subsequence of two such random sequences is $\frac{2n}{\sqrt{k}}$, for sufficiently large n and k.

One of the very few extremal results regarding the common subsequences in general is done in case of permutations by Beane and Huynh-Ngoc [9], where it was shown that given three permutations W_1, W_2, W_3, the length of a longest common subsequence $LCS(W_i, W_j) \geq n^{1/3}$ for some $1 \leq i < j \leq 3$. An immediate corollary of this result implies that if W_1, W_2, W_3 are n-letter sequences over some alphabet such that each letter appears exactly the same number of times in each of W_i's, then for some $1 \leq i < j \leq 3$, $LCS(W_i, W_j) \geq n^{1/3}$. Bukh and Ma [16] built on this to provide asymptotically tight results concerning $L(n,k)$, the largest n' such that any two sequences over an alphabet of size k have a common subsequence of length n'. It was shown in [16], that for a fixed k, $L(n,k) = \frac{n}{k}(1 + o(1))$.

The idea of Bukh and Zhou [17] was to apply this result, the Regularity Lemma and generalizations, to the problem of twins in sequences. In order to prove the second lower bound in Theorem 4.4, an ϵ-regular sequence is split into consecutive factors $F_1, \ldots, F_{n/k}$ of length k. In each of the factors F_i, three largest subsequences are chosen such that each letter of the alphabet is repeated the same number of times in each of these three subsequences. This could be done by taking, for a fixed letter, a largest subsequence consisting only of this letter and splitting it between the three equally with perhaps one or two elements leftover. Finally, the above results on common subsequences, applied to the three subsequences, gives twins of size roughly $k^{1/3}$ in F_i for each i. Concatenating the corresponding twins from all F_i's gives long twins from the original sequence.

Bukh and Zhou [17] and Bukh and Ma [16] built on the idea of common subsequences much more and provided many interesting generalizations.

4.4 Small values of $f(n, k, \Sigma)$

We will slightly abuse notation and denote by $f(n,k,\ell)$ the value of $f(n,k,\Sigma)$ with $|\Sigma| = \ell$. In the introductory section it was observed that $f(3,2,2) = 1$ yielding immediately a weak lower bound on $f(n,2,2)$ to be $\lfloor n/3 \rfloor$. In general, it is clear that

$$f(n, k, \ell) \geq \lfloor \tfrac{n}{m} \rfloor f(m, k, \ell).$$

For example, Theorem 4.3 gives a lower bound on $f(n, 2, 3)$ to be $\frac{1}{3}n - o(n)$, Theorem 4.4 provides the better lower bound of $\frac{1.02}{3}n - o(n)$, and gives another bound $f(n, 2, 3) \geq n/9 - 1$. There are no good lower bounds without $o(n)$ term, i.e., there is no known (reasonable) number t such that $f(t, 2, 3) \geq \frac{t}{3}$. If one could find such t, this would immediately give another proof of $f(n, 2, 3) \geq \frac{1}{3}n - t$. However, the smallest value for such a possible t could be 21, which already presents a computationally challenging task. In Table 1 we summarize estimates on the values on $f(n, k, \ell)$, which were determined with the help of a computer. We see that for $n = 21, f(n, 2, 3) \leq 7 = n/3$, for $n = 22, f(n, 2, 3) \leq 7 < n/3$. Thus, the first "open" case which might improve lower bound $n/3$ on $f(n, 2, 3)$ is $f(22, 2, 3)$.

$\Sigma \backslash n$	6	7	8	9	10	11	12	13	14	15	16	17
$\{0, 1\}$	2	2	2	3	3	4	4	5	5	5	6	6
$\{0, 1, 2\}$	1	1	2	2	2	3	3	3	4	4	4	4

Table 1 Values for small t of $f(t, 2, 2)$ and $f(t, 2, 3)$.

$\Sigma \backslash n$	18	19	20	21	22	23	24
$\{0, 1\}$	7	7	8				
$\{0, 1, 2\}$	≤ 5	≤ 6	≤ 6	≤ 7	≤ 7	≤ 8	≤ 8

Further we remark, that a more careful analysis below of the increment argument in the proof of Proposition 4.6 leads to the bound $T_0 \leq t_0 3^{(-2\log \epsilon)/\epsilon^3}$, which in turn improves the bounds in Theorems 4.1 and 4.2 to

$$\left(1 - C|\Sigma| \left(\frac{(\log\log n)^2}{\log n}\right)^{1/3}\right) n \leq kf(n, k, \Sigma).$$

5 Subgraphs with the same number of edges and vertices

The following twin problem in graphs was introduced by Caro and Yuster in [18]. Let $t(G)$ be the largest k such that there are two induced vertex-disjoint subgraphs of G on k vertices each and having the same number of edges. Note that the function $t(n)$ introduced in the beginning is equal to $\min\{t(G) : |V(G)| = n\}$.

Caro and Yuster [18] provided some bounds first, see Theorem 8. We include the proof of this theorem for completeness. The key idea is to use the chromatic number of the Kneser graph. For positive integers k, n ($k \leq n/2$), the *Kneser graph* $\mathcal{K} = KG(n, k)$ is the graph on the vertex set $\binom{[n]}{k}$ whose edge set consists of pairs of disjoint k-sets. Lovász [45] proved that the chromatic number of the Kneser graph $KG(n,k)$ is $n-2k+2$.

Theorem 5.1. ([18]). *There exists a positive constant c such that $\sqrt{n} \leq t(n) \leq n/2 - c\log\log n$.*

Proof. For a graph G with vertex set $[n]$, we define a vertex coloring of $\mathcal{K}$ by letting the number of edges induced by each k-subset of $V(G)$ be the color of the corresponding vertex in $\mathcal{K}$. The number of colors is at most $\binom{k}{2}+1$. For $k=\sqrt{n}$, we have $\binom{k}{2}+1 < n-2k+2$. Thus the number of colors is less than the chromatic number of $\mathcal{K}$ and there are two adjacent vertices in $\mathcal{K}$ of the same color. Therefore there are two disjoint k-sets of G that induce the same number of edges.

To verify the upper bound, consider a disjoint union of cliques of odd orders $a_1,\ldots,a_m$, where $a_1 = 1$, $a_j > 2(a_1^2+a_2^2+\cdots+a_{j-1}^2)$, and a_j is the smallest possible such number; so $m = c\log\log n$. One can show that any pair of twins in this graph either has at most $n - c\log\log n$ vertices together or omits at least one vertex from each of the cliques. □

Ben-Eliezer and Krivelevich [11] proved that $t(G(n,p)) = \lfloor n/2\rfloor$ with high probability, where $G(n,p)$ is the Erdős-Renyi random graph. In addition, Caro and Yuster proved that in a sparse graph there are twins of size almost $n/2$:

Theorem 5.2. (Caro and Yuster [18]). *For every fixed $\alpha > 0$ and for every $\epsilon > 0$ there exists $N = N(\alpha,\epsilon)$ so that for all $n > N$, if G is a graph on n vertices and at most $n^{2-\alpha}$ edges then $t(G) \geq (1-\epsilon)n/2$.*

Axenovich, Martin, and Ueckerdt [7] made further progress on a twin problem in terms of the number of edges in a graph. The main approach there was to split a large subset of vertices into sets A', B' of equal size, inducing graphs on almost the same number of edges, say with A' inducing smaller number of edges than B'. Then, using carefully chosen ahead of time extra vertices, that induce a matching and send no edges to A' and B', add a matching to A' and an independent set to B'.

Theorem 5.3. ([7]). *If G is a graph on n vertices and e edges, then*

$$t(G) \geq \frac{n}{2}\left(1 - \frac{20\sqrt{e}\log n}{n}\right).$$

This theorem implies, in particular, that any n-vertex graph on $o\left(n^2/\log^2 n\right)$ edges has twins of size $n/2 - o(n)$ each and that planar graphs have twins of size at least $n/2 - c\sqrt{n}\log n$ each.

We say that G has *perfect twins* if the vertex set can be split in two sets of equal size inducing subgraphs with the same number of edges. Finding large twins is equivalent to finding a small subset of vertices whose deletion allows to find perfect twins in the remaining graph. Some criteria for the existence of perfect twins were provided in [7, 18]. The main observation here is that the graph has perfect twins if and only if the degree sequence could be split into two subsequences of the same length and having the same sum of the elements. We call this a *degree splitting property*. Indeed, sets A and B partitioning the vertex set of a graph G induce perfect twins iff $|A| = |B|$ and $|E(G[A])| = |E(G[B])|$. Since $2|E(A)| = \sum_{a\in A} deg(a) - |E(A,B)|$ and $2|E(B)| = \sum_{b\in B} deg(b) - |E(A,B)|$, where deg is the degree and $E(A,B)$ is the set of edges between A and B, the statement follows. Therefore the following theorem is basically number theoretic as it deals with degree sequence and not with the structure of the graph. Here Δ and δ are maximum and minimum degrees of a graph, respectively, V_i is the set of vertices of degree i.

Theorem 5.4. *Let G be a graph on n vertices, where n is even. If one of the following conditions holds, then $t(G) = n/2$.*

- *The degree sequence of G forms a set of consecutive integers.*
- *$|V_i|$ is even for each i.*
- *$n \geq 90$ and $|\{i : |V_i| \text{ is odd}\}| > n/2$.*
- *There are at least $\Delta(G) - \delta(G)$ disjoint consecutive pairs of vertices.*

In Theorem 5.4, the result of Karolyi [37] on partition of integers was used. We say that two integers are *almost equal* if they differ by $1, 0$, or -1.

Theorem 5.5. ([37]). *Let X be a set of m integers, each between 1 and $2m-2$. If $m \geq 89$, then one can partition X into two sets, X_1 and X_2 of almost equal sizes such that the sum of elements in X_1 is almost equal to the sum of elements in X_2.*

One of the few exact results in this area concerns trees and forests.

Theorem 5.6. ([7]). *If G is a forest, then $t(G) \geq \lceil n/2 \rceil - 1$.*

If n is odd, this is clearly best possible. For even n this is attained, for example, by a star. Such a graph has no perfect twins.

Finally, Bollobás, Kittipassorn, Narayanan, and Scott [14] proved the strongest result in the area showing that almost perfect twins exist in any sufficiently large graph. This is one of four main theorems we mention in the beginning of the survey.

Theorem 5.7. ([14]). *For every $\epsilon > 0$, there is a natural number $N = N(\epsilon)$ such that for any graph G on $n > N$ vertices, $t(G) \geq n/2 - \epsilon n$.*

The idea of the proof is to delete at most $2\epsilon n$ vertices of a given n-vertex graph such that the rest has perfect twins, or, as observed above, has the degree splitting property. This is done using probabilistic techniques. Specifically, the authors show that there is a probability p at most $\epsilon = \epsilon(G)$ such that removing vertices of G with probability p results in a graph that has degree splitting property with positive probability. In order to do so, many pairs of vertices with degree difference that is controlled are chosen and then the vertices of each pair are assigned to the first or the second part according to the following lemma.

Lemma 5.8. *Let H be a graph on an even number of vertices and $V(H)$ be partitioned into disjoint collections of pairs $\mathcal{P}_1, \ldots, \mathcal{P}_k$ such that the degree difference for vertices in each pair in $\mathcal{P}_i$ is between a_i and b_i, $0 \leq a_1 \leq b_1$, $0 < a_i \leq b_i$, $2 \leq i \leq k$. Then H can be partitioned into two sets A and B of* *the same size such that the sum of the degrees of vertices from A differs from the sum of the degrees of vertices from B by at most b_k. In particular, if $b_k = 1$, then H has perfect twins.*

Bollobás, Kittipassorn, Narayanan, and Scott [14] note that their proof leads to a more precise lower bound $t(n) \geq n/2 - n/(\log\log n)^c$, for a positive constant c. They suspect that $n/2 - t(n) = o(n^\epsilon)$ for each positive ϵ and even state that $n/2 - t(n)$ could be $\Theta(\log n)$.

The twin problem in graphs could be interpreted as a problem in complete graphs edge colored in 2 colors, say red and blue, where the twins are two vertex-disjoint complete subgraphs with the same number of red edges and the same number of blue edges. The authors of [14] define a natural generalization of the twin problem for multicolored complete graphs with multiple twins. Let G be an edge-colored complete graph on n vertices. Then $t(\ell, G)$ is the largest k such that there are ℓ pairwise disjoint sets of vertices, inducing the same number of edges of color c, for each color c. Finally, let

$$t(n, s, \ell) = \min\{t(\ell, G) : \quad G \text{ is an } s-\text{edge-colored complete graph on } n \text{ vertices}\}.$$

So, $t(n, 2, 2) = t(n)$. The authors conjecture that $f(n, s, 2) = n/2 - o(n)$ for any number of colors s.

6 Subgraphs with the same multiset of pairwise distances

In this section we consider one of the "weak" twins. For a vertex set $S \subseteq V(G)$ in a graph G, the *distance multiset*, $D(S)$, is the multiset of pairwise distances between vertices of S in G. Two vertex sets are called *homometric*, or homometric twins, if their distance multisets are identical. For a graph G, the largest integer h, such that there are two disjoint homometric sets of order h in G, is denoted by $h(G)$. This parameter was introduced by Albertson, Pach and Young [1].

Theorem 6.1. ([1]). $\frac{c\log n}{\log\log n} < h(n) \leq \frac{n}{4}$ *for $n > 3$, and a constant c.*

It is an easy observation that $h(G) = \lfloor |V(G)|/2 \rfloor$ when G is a path or G is a cycle. However, more is known. Note that the multisets of distances for a vertex subset of a path correspond to a multiset of pairwise differences between elements of a subset of positive integers. We shall say that two subsets of integers are homometric if their multisets of pairwise differences coincide. Piccard [48] claimed to prove that any two homometric sets of positive integers, whose multisets of distances contain no repeated elements are congruent to each other. Later, Bloom [12] found a counterexample to Piccard's claim. Recently, there has been more progress on this problem [10]. Among others, Rosenblatt and Seymour [50] proved that two multisets A and B of integers are homometric if and only if there are two multisets U, V of integers such that $A = U + V$ and $B = U - V$, where $U + V$ and $U - V$ are multisets, $U + V = \{u + v : u \in U, v \in V\}$, $U - V = \{u - v : u \in U, v \in V\}$. Lemke, Skiena and Smith [42] showed that if G is a cycle of length $2n$, then any subset of $V(G)$ with n vertices and its complement are homometric sets. Surprisingly, when the class $\mathcal{G}$ of graphs under consideration is not a path or a cycle, the problem of finding $h(\mathcal{G})$ becomes nontrivial. For a class of graphs of diameter 2, the function h is equal to the function t introduced in the beginning, i.e., the largest integer k such that any graph on n vertices has two vertex-disjoint induced subgraphs on k vertices and with the same number of edges.

The homometric set problem we consider here has its origins in Euclidean geometry, with applications in X-ray crystallography introduced in the 1930's with later applications in restriction site mapping of DNA. In particular, the fundamental problem that was considered is whether one could identify a given set of points from its multiset of distances. There are several related directions of research in the area, for example, the question of recognizing the multisets corresponding to a multiset of distances realized by a set of points in the Euclidean space of given dimension, see [47].

Theorem 6.2. *Let m_{ij}, $0 \le i,j \le n$ be nonnegative real numbers with $m_{ij} = m_{ji}$ for all i,j and $m_{ii} = 0$ for all i. There are points $P_0, \ldots, P_n \in \mathbb{R}^n$ with $||P_i - P_j|| = m_{ij}$ for all i,j if and only if the matrix Q with entries $q_{ij} = \frac{1}{2}(m_{0i}^2 + m_{0j}^2 - m_{ij}^2)$ is positive semidefinite.*

Here, a matrix Q is positive semidefinite if it is symmetric and $\mathbf{x}^T Q \mathbf{x} \ge 0$ for all $\mathbf{x} \in \mathbb{R}^n$.

Let us come back to the function $h(n)$. As we see from Theorem 6.1, the gap between the upper and the lower bound is significant. There were some improvements on the lower bound for graphs of given size or diameter.

Theorem 6.3. ([5]). *Let G be a graph on n vertices with e edges, diameter d, $n \ge 5$, $d \ge 2$. If for an integer k, $\binom{\binom{k}{2}+d-1}{d-1} < n - 2k + 2$, then*

$$h(G) \ge \max\{k,\ d/2,\ \sqrt{2e/n}\}.$$

In particular,

$$h(G) \ge \max\{0.5n^{1/(2d-2)},\ d/2,\ \sqrt{2e/n}\}$$

for $n \ge c(d) = d^{2d-2}$.

Since it is known that almost all graphs $G(n,p)$ have diameter 2, see [13], Theorem 5.7 implies that for any $\epsilon > 0$ and almost every graph $G = G(n,p)$, for $n > n(\epsilon)$, $h(G_{n,p}) = n/2 - \epsilon n$. As the problem of finding large homometric sets is quite difficult in general, there was an effort devoted to determining $h(G)$ for graphs from some specific classes, such as specific tree classes.

A *spider* is a tree that is a union of vertex-disjoint paths and a vertex that is adjacent to one of the endpoints of each path. A spider is k-legged, $k \ge 3$ if its maximum degree is k. A *caterpillar* is a tree whose vertex set consists of two parts, one inducing a path and another inducing an independent set of leaves adjacent to the path. A *haircomb* is a tree, that is a union of a path called the *spine* and a collection of vertex-disjoint paths, called legs, that have an endpoint on the spine. In the following theorems, ceilings and floors are omitted for a more clear presentation. Axenovich and Ozkahya [5] determined several bounds for these classes of trees.

Theorem 6.4. ([5]). *Let T be a tree on n vertices.*

- $h(T) \geq n^{1/3} - 1$.
- *If T is a caterpillar, then* $h(T) \geq n/6$.
- *If T is a haircomb, then* $h(T) \geq \sqrt{n}/2$.
- *If T is a k-legged spider, then* $h(T) \geq \frac{5}{12}n$ *for* $k = 3$; $h(T) \geq \frac{1}{3}n$ *for* $k = 4$; $h(T) \geq \left(\frac{1}{4} + \frac{3}{8k-12}\right) n$ *for* $k \geq 5$; $h(T) = (n+2)/4$ *for* $k = n/2$.

Fulek and Mitrović [28] improved the general result for trees and for haircombs:

Theorem 6.5. ([28]). *If T is an n-vertex tree, then* $h(T) \geq \sqrt{n/2} - 1/2$. *If T is an n-vertex haircomb tree, then* $h(H) \geq c \cdot n^{2/3}$, *for a constant c.*

Dubickas [24] generalized the problem of homometric sets to multiple copies. Let $h(n, r)$ be the largest integer h such that in any n-vertex graph there are r pairwise disjoint sets of vertices, each of size h which are pairwise homometric. In [24] it is proven that for any positive ϵ, there is n_ϵ such that for any $n > n_\epsilon$, $h(n, r) \geq (1 - \epsilon)\frac{1}{r}\frac{\log n}{\log \log n}$.

7 Subgraphs with the same maximum degree

One can set up a weak twin problem for any graph parameter or a set of parameters. When this parameter is the maximum degree, the twin problem has been addressed. Caro and Yuster [19] introduced a function $s_k(G)$ for a positive integer k and a graph G to be the largest integer m such that there are k induced pairwise vertex-disjoint subgraphs of G, each with m vertices, and with the same maximum degree.

Theorem 7.1. ([19]). *For any fixed k and for any graph G on n vertices* $s_k(G) \geq n/k - o(n)$.

The proof of this theorem implies that k could be taken to be n^ϵ with $o(n)$ replaced with $o(n^{1-\epsilon})$, for sufficiently small ϵ.

Acknowledgements The author thanks collaborators with whom she worked on various twin problems: Ryan Martin, Lale Özkahya, Yury Person, Svetlana Puzynina, and Torsten Ueckerdt, as well as Boris Bukh, Radoslav Fulek, Alex Scott, and Benny Sudakov for interesting discussions. In addition, the author acknowledges generous support and effort of IMA during the thematic year of 2014–2015 and a partial support of NSF grant DMS-0901008 of the twin problem project.

References

1. M. Albertson, J. Pach, M. Young, Disjoint homometric sets in graphs. Ars Math. Contemp. **4**, 1–4 (2011)
2. N. Alon, B. Bollobás, Graphs with a small number of distinct induced subgraphs. Discret. Math. **75**, 23–30 (1989)
3. N. Alon, Y. Caro, I. Krasikov, Bisection of trees and sequences, in *Combinatorics and Algorithms* (Jerusalem, 1988). Discrete Mathematics, vol. 114, no. 1–3 (Amsterdam, North-Holland, 1993), pp. 3–7
4. N. Alon, R.A. Duke, H. Lefmann, V. Rödl, R. Yuster, The algorithmic aspects of the regularity lemma. J. Algorithms **16**(1), 80–109 (1994)
5. M. Axenovich, L. Ozkahya, On homometric sets in graphs. Aust. J. Comb. **55**, 175–187 (2013)
6. M. Axenovich, Y. Person, S. Puzynina, Regularity lemma and twins in sequences. J. Comb. Theory Ser. A **120**, 733–743 (2013)
7. M. Axenovich, R. Martin, T. Ueckerdt, Twins in graphs. Eur. J. Comb. **39**, 188–197 (2014)
8. S. Bachl, F.-J. Brandenburg, D. Gmach, Computing and drawing isomorphic subgraphs. J. Graph Algorithm. Appl. **8**(2), 215–238 (2004)
9. P. Beame, D.T. Huynh-Ngoc, On the value of multiple read/write streams for approximating frequency moments. Electronic Colloquium on Computational Complexity (ECCC), May 2008. Technical Report TR08-024
10. A. Bekir, S.W. Golomb, There are no further counterexamples to S. Piccard's theorem. IEEE Trans. Inform. Theory **53**, 2864–2867 (2007)
11. I. Ben-Eliezer, M. Krivelevich, Perfectly balanced partitions of smoothed graphs. Electron. J. Comb. **16**(1), note N14 (2009)
12. G.S. Bloom, A counterexample to a theorem of S. Piccard. J. Comb. Theory. Ser. A **22**(3), 378–379 (1977)
13. B. Bollobás, *Modern Graph Theory*. Graduate Texts in Mathematics, vol. 184 (Springer, New York, 1998), xiv+394 pp.
14. B. Bollobás, T. Kittipassorn, B. Narayanan, A. Scott, Disjoint induced subgraphs of the same order and size. Eur. J. Comb. **49**, 153–166 (2015)
15. J. Bosák, *Decompositions of Graphs*. Translated from the Slovak. With a preface by Štefan Znám. Mathematics and Its Applications (East European Series), vol. 47 (Kluwer Academic, Dordrecht, 1990), xviii+248 pp.
16. B. Bukh, J. Ma, Longest common subsequences in sets of words. arXiv:1406.7017
17. B. Bukh, L. Zhou, Twins in words and long common subsequences in permutations. arXiv:1307.0088
18. Y. Caro, R. Yuster, Large disjoint subgraphs with the same order and size. Eur. J. Comb. **30**(4), 813–821 (2009)
19. Y. Caro, R. Yuster, Large induced subgraphs with equated maximum degree. Discret. Math. **310**(4), 742–747 (2010)
20. C. Choffrut, J. Karhumäki, Combinatorics of words, in *Handbook of Formal Languages* (Springer, Berlin, 1997)
21. F.R.K. Chung, R.L. Graham, Quasi-random subsets of $\mathbb{Z}_n$. J. Comb. Theory Ser. A **61**(1), 64–86 (1992)
22. D. Conlon, A new upper bound for diagonal Ramsey numbers. Ann. Math. **170**, 941–960 (2009)
23. T.H. Cormen, C.E. Leiserson, R.L. Rivest, C. Stein, *Introduction to Algorithms* (MIT Press/McGraw-Hill, Cambridge/New York, 2001), pp. 350–355
24. A. Dubickas, Equal distances between equal sets of vertices in graphs. Šiauliai Math. Semin. **8**(16), 37–47 (2013)
25. M. Dudík, L.J. Schulman, Reconstruction from subsequences. J. Comb. Theory Ser. A **103**(2), 337–348 (2003)
26. T. Dzido, K. Krzywdziński, On a local similarity of graphs. Discret. Math. **338**(6), 983–989 (2015)

27. P. Erdős, J. Pach, J. Pyber, Isomorphic subgraphs in a graph, in *Combinatorics* (Eger, 1987). Colloquia Mathematica Societatis János Bolyai, vol. 52 (North-Holland, Amsterdam, 1988), pp. 553–556
28. R. Fulek, S. Mitrović, Homometric sets in trees. Eur. J. Comb. **35**, 256–263 (2014)
29. R. Gould, V. Rödl, On isomorphic subgraphs. Discret. Math. **118**(1–3), 259–262 (1993)
30. W.T. Gowers, A new proof of Szemerédi's theorem. Geom. Funct. Anal. **11**(3), 465–588 (2001)
31. L. Guidotti, Sulla divisibilita dei grafi completi. Riv. Mat. Univ. Parma **1**, 231–237 (1972)
32. F. Harary, R. Robinson, Isomorphic factorizations. VIII: Bisectable trees. Combinatorica **4**(2–3), 169–179 (1984)
33. F. Harary, W.D. Wallis, Isomorphic factorizations II: Combinatorial designs. Congr. Numer. **19**, 13–28 (1978)
34. F. Harary, R.W. Robinson, N.C. Wormald, Isomorphic factorisations. I: Complete graphs. Trans. Am. Math. Soc. **242**, 243–260 (1978)
35. D.S. Hirschberg, A linear space algorithm for computing maximal common subsequences. Commun. ACM **18**(6), 341–343 (1975)
36. P. Horn, V. Koubek, V. Rödl, Isomorphic edge disjoint subgraphs of hypergraphs (2012, preprint)
37. G. Károlyi, Balanced subset sums in dense sets of integers. Integers **9**(A45), 591–603 (2009)
38. M. Kiwi, M. Loebl, J. Matoušek, Expected length of the longest common subsequence for large alphabets. Adv. Math. **197**(2), 480–498 (2005)
39. I. Koch, Enumerating all connected maximal common subgraphs in two graphs. Theor. Comput. Sci. **250**, 1–30 (2001)
40. J. Komlós, M. Simonovits, Szemerédi's regularity lemma and its applications in graph theory, in *Combinatorics, Paul Erdős is Eighty* (Keszthely, 1993). Bolyai Society Mathematical Studies, vol. 2 (János Bolyai Mathematical Society, Budapest, 1996), pp. 295–352
41. C. Lee, P. Loh, B. Sudakov, Self-similarity of graphs. SIAM J. Discret. Math. **27**(2), 959–972 (2013)
42. P. Lemke, S.S. Skiena, W.D. Smith, Reconstructing sets from interpoint distances. Discret. Comput. Geom. Algorithm. Comb. **25**, 507–631 (2003)
43. G. Levi, A note on the derivation of maximal common subgraphs of two directed or undirected graphs. Calcolo **9**, 341–352 (1972/1973)
44. M. Lothaire, *Algebraic Combinatorics on Words* (Cambridge University Press, Cambridge, 2002)
45. L. Lovász, Kneser's conjecture, chromatic number, and homotopy. J. Comb. Theory Ser. A **25**(3), 319–324 (1978)
46. A. Mateescu, A. Salomaa, S. Yu. Subword histories and Parikh matrices. J. Comput. Syst. Sci. **68**(1), 1–21 (2004)
47. J. Matoušek, Thirty-three miniatures, in *Student Mathematical Library*. Mathematical and Algorithmic Applications of Linear Algebra, vol. 53 (American Mathematical Society, Providence, 2010), x+182 pp.
48. S. Piccard, Sur les ensembles de distances des ensembles de points d'un espace Euclidien. Mém. Univ. Neuchâtel, vol. 13 (Secrétariat de l'Université, Neuchâtel, 1939), p. 212
49. F.P. Ramsey, On a problem of formal logic. Proc. Lond. Math. Soc. **S2–30**(1), 264 (1930)
50. J. Rosenblatt, P.D. Seymour, The structure of homometric sets. SIAM J. Algebraic Discret. Methods **3**, 343–350 (1982)
51. A. Salomaa, Counting (scattered) subwords. Bull. EATCS **81**, 165–179 (2003)
52. G. Stevens, Some caterpillars are 2-splittable, in Proceedings of the Twenty-Fifth Southeastern International Conference on Combinatorics, Graph Theory and Computing, Boca Raton, FL, 1994. Congr. Numer. vol. 104 (1994). pp. 181–186
53. E. Szemerédi, Regular Partitions of Graphs. Colloq. Internat. CNRS, Univ. Orsay, Orsay, 1976, CNRS, vol. 260 (CNRS, Paris, 1978). pp. 399–401
54. D.B. West, *Introduction to Graph Theory* (Prentice Hall Inc., Upper Saddle River, 1996)
55. X. Xia, *Bioinformatics and the Cell: Modern Computational Approaches in Genomics, Proteomics and Transcriptomics* (Springer, New York, 2007)

Extremal problems for cycles in graphs

Jacques Verstraëte

Abstract In this survey we consider extremal problems for cycles of prescribed lengths in graphs. The general extremal problem is cast as follows: if $\mathscr{C}$ is a set of cycles, determine the largest number of edges $\mathrm{ex}(n, \mathscr{C})$ in an n-vertex graph containing no cycle from $\mathscr{C}$. The survey contains short proofs of various known theorems, including the even cycle theorem of Erdős and Bondy and Simonovits. We also give proofs of new results and conjectures of Erdős on cycles, for instance, we find new sufficient conditions for cycles of length ℓ modulo k and for long cycles in triangle-free graphs of large chromatic number. We also review proofs of some conjectures of Erdős on the distribution of the lengths of cycles in graphs, as well as related problems on chromatic number and girth, counting graphs without short cycles, and extensions to cycles in uniform hypergraphs. Throughout the survey, we include a number of conjectures and open problems.

Keywords Turán numbers • Zarankiewicz numbers • Even cycle theorem • Chromatic number • Hypergraph cycles

Mathematics Subject Classification (2010): 05C35, 05C38, 05C15

1 Introduction

If $\mathscr{F}$ is any family of graphs, then $\mathrm{ex}(n, \mathscr{F})$ denotes the maximum number of edges in an n-vertex $\mathscr{F}$-free graph. These quantities are collectively referred to as the *Turán numbers* for $\mathscr{F}$. Let $z(m, n, \mathscr{F})$ denote the maximum number of edges in an $m \times n$ bipartite $\mathscr{F}$-free graph; we refer to these as the *Zarankiewicz numbers* for $\mathscr{F}$. If the smallest chromatic number of any graph in $\mathscr{F}$ is $r \geq 2$, then the Erdős-Stone Theorem [42] implies $\mathrm{ex}(n, \mathscr{F}) \sim (1 - \frac{1}{r-1})\binom{n}{2}$. In the case that $\mathscr{F}$ contains bipartite graphs, $\mathrm{ex}(n, \mathscr{F}) = o(n^2)$ and the notorious problem of determining the order of

J. Verstraëte (✉)
Department of Mathematics, University of California, San Diego, CA, USA
e-mail: jacques@ucsd.edu

A. Beveridge et al. (eds.), *Recent Trends in Combinatorics*, The IMA Volumes in Mathematics and its Applications 159, DOI 10.1007/978-3-319-24298-9_4

magnitude of $\mathrm{ex}(n, \mathscr{F})$ is known as the bipartite Turán problem. For an account of Turán-type problems for bipartite graphs, the reader is referred to the comprehensive survey of Füredi and Simonovits [57], and to the book of Bollobás [9]. The focus of this survey is the case that $\mathscr{F}$ contains even cycles. In particular, if $\mathscr{C}$ is a finite set of even cycles, then the determination of $\mathrm{ex}(n, \mathscr{C})$ is one of the notable open problems in combinatorics. We also consider the case that $\mathscr{C}$ is an infinite set of even cycles, for example, the set of cycles of length congruent to ℓ modulo k, a class of problems which has also been extensively studied in the literature.

1.1 Organization

In the rest of this section, we discuss the main problems and results in the area, on the extremal problems for excluding finite sets and infinite sets of cycles, as well as cycles in graphs of large chromatic number, the distribution of the set of lengths of cycles in graphs, and cycles in hypergraphs. This section presents a number of new results, including Theorem 11 on cycles of length ℓ mod k, Theorem 21 on cycles in triangle-free graphs of large chromatic number, proving a conjecture of Erdős. In addition, we cover a number of well-known conjectures, and introduce some new conjectures. In Section 3, we give a few graph-theoretic preliminaries to be used repeatedly in the material to follow. In Section 4, we briefly discuss various constructions of extremal graphs without short cycles, and in Section 5 we give a short proof of the even cycle theorem of Erdős and Bondy and Simonovits [12]. This method is then adapted in Section 6 to give various proofs of theorems on excluding infinite lists of cycles, including the well-researched case of the set of cycles of length ℓ modulo k. In Section 7, we discuss the distribution of cycle lengths in graphs relative to the density of the graph.

1.2 Excluding finitely many cycles

Let C_k denote the cycle of length k, and define $\mathscr{C}_k = \{C_3, C_4, \ldots, C_k\}$. In this notation, a graph has girth at least g if and only if it is $\mathscr{C}_{g-1}$-free. The following conjecture of Erdős and Simonovits [41] can be identified as one of the main open problems in the area:

Conjecture I. *For all* $k \geq 2$, $\mathrm{ex}(n, \mathscr{C}_{2k}) = \Theta(n^{1+\frac{1}{k}})$.

Upper bounds on $\mathrm{ex}(n, \mathscr{C}_{2k})$ and $\mathrm{ex}(n, \mathscr{C}_{2k+1})$ follow from the following result of Alon, Hoory, and Linial [3]: every n-vertex graph of average degree at least $d \geq 2$ contains at least $nd(d-1)^k$ non-backtracking walks of length k. In particular, if G is an n-vertex graph of girth at least g, then $n \geq \nu(d, g)$ where

$$\nu(d,g) = \begin{cases} 1 + \sum_{i=1}^{\frac{1}{2}(g-1)} d(d-1)^{i-1} & \text{if } g \text{ is odd} \\ 2\sum_{i=0}^{\frac{1}{2}(g-2)} (d-1)^i & \text{if } g \text{ is even} \end{cases}$$

This is known as the *Moore bound*, and a d-regular graph G with girth g and $\nu(d,g)$ vertices will be called a *Moore graph*.

Moore graphs of girths $g \in \{6,8,12\}$ exist, arising as bipartite incidence graphs of structures from projective geometry known as *generalized polygons*, and these are extremal $\mathscr{C}_{2k+1}$-free graphs when $k \in \{2,3,5\}$ – see Theorems 4.1 and 4.2 in Section 4.1. The existence of generalized polygons was first established by Tits [128], and translated into the language of extremal graph theory by Benson [5]. We discuss Moore graphs at greater length in Section 4. These constructions together with the Moore bound give the following theorem:

Theorem 1. *For any* $k \geq 2$, $\mathrm{ex}(n,\mathscr{C}_{2k}) \lesssim \frac{1}{2}n^{1+\frac{1}{k}}$ *and* $\mathrm{ex}(n,\mathscr{C}_{2k+1}) \lesssim (\frac{n}{2})^{1+\frac{1}{k}}$. *If* $k \in \{2,3,5\}$, *then* $\mathrm{ex}(n,\mathscr{C}_{2k+1}) \sim (\frac{n}{2})^{1+\frac{1}{k}}$.

While Moore graphs of even girth $g \in \{6,8,12\}$ exist and give the asymptotic value of $\mathrm{ex}(n,\mathscr{C}_{2k+1})$ for $k \in \{2,3,5\}$, it appears to be much more difficult to determine the asymptotic value of $\mathrm{ex}(n,\mathscr{C}_{2k})$, even for $k = 2$. A long-standing conjecture of Erdős [31, 34] states $\mathrm{ex}(n,\mathscr{C}_4) \sim (\frac{n}{2})^{\frac{3}{2}}$, while the best bounds available are $(\frac{n}{2})^{\frac{3}{2}} \lesssim \mathrm{ex}(n,\mathscr{C}_4) \lesssim \frac{1}{2}n^{\frac{3}{2}}$. We make the following conjecture:

Conjecture II. *There exists* $\delta > 0$ *such that*

$$(1+\delta)(\tfrac{n}{2})^{\frac{3}{2}} \lesssim \mathrm{ex}(n,\mathscr{C}_{2k}) \lesssim (1-\delta)\tfrac{1}{2}n^{\frac{3}{2}}.$$

This runs counter to the afore-mentioned conjecture of Erdős. We note that a construction of Parsons [104] shows that extremal $\mathscr{C}_4$-free graphs are not bipartite, and values of $\mathrm{ex}(n,\mathscr{C}_4)$ for $n \leq 30$ were computed by Garnick, Kwong, Lazebnik, and Nieuwejaar [59].

Next we consider $\tilde{\mathscr{C}}_{2k} = \{C_4, C_6, \ldots, C_{2k}\}$. The following theorem was proved by Lam and the author [81]:

Theorem 2. *For any* $k \geq 2$, $\mathrm{ex}(n,\tilde{\mathscr{C}}_{2k}) \lesssim \frac{1}{2}n^{1+\frac{1}{k}}$, *with equality for* $k \in \{2,3,5\}$.

The constructions for $k \in \{2,3,5\}$ are so-called *polarity graphs of generalized polygons*, described in detail in Lazebnik, Ustimenko, and Woldar [90]. These are n-vertex graphs G with $e(G) \sim \frac{1}{2}n^{1+\frac{1}{k}}$ and G does not contain $\tilde{\mathscr{C}}_{2k}$ and also does not contain any odd cycles of length at most k. However, these graphs contain all odd cycles of lengths between k and $2k$, and so are not $\mathscr{C}_{2k}$-free.

For $k \notin \{2,3,5\}$, the densest known constructions of $\mathscr{C}_{2k}$-free graphs are the algebraic constructions of Lazebnik, Ustimenko, and Woldar [88], which slightly improve the density of earlier constructions of Margulis [93] and Lubotzky, Phillips, and Sarnak [86]. The following theorem of Lazebnik, Ustimenko, and Woldar [88] summarizes these results:

Theorem 3. *For all $k \geq 2$,*

$$\mathrm{ex}(n, \mathscr{C}_{2k}) \gtrsim \begin{cases} (\frac{n}{2})^{1+\frac{2}{3k-2}} & \text{if } k \text{ is even} \\ (\frac{n}{2})^{1+\frac{2}{3k-3}} & \text{if } k \text{ is odd} \end{cases}$$

1.2.1 The Even Cycle Theorem

The problem of determining $\mathrm{ex}(n, C_{2k})$ is interesting in its own right, and Erdős and Simonovits [41] conjectured $\mathrm{ex}(n, C_{2k}) = \Theta(n^{1+\frac{1}{k}})$. The conjecture remains open for all $k \not\in \{2, 3, 5\}$, and in particular in the case $k = 4$. The even cycle theorem of Bondy and Simonovits [12], also attributed to Erdős in an unpublished form, gives an upper bound on $\mathrm{ex}(n, C_{2k})$:

Theorem 4. *For each $k \geq 2$, there exists $c_k > 0$ such that* $\mathrm{ex}(n, C_{2k}) \leq c_k n^{1+\frac{1}{k}}$.

We will give a short proof of this theorem in Section 5. A number of improvements of the value of c_k were obtained, including by the author [132], Pikhurko [105], and Bukh and Jiang [15], who showed $c_k = O(\sqrt{k \log k})$, answering a question of Bondy who asked if $c_k = o(k)$. It remains plausible that c_k is a bounded function of k:

Conjecture III. *There exists $c > 0$ such that for all $k \geq 2$,* $\mathrm{ex}(n, C_{2k}) \leq cn^{1+\frac{1}{k}}$.

In fact, Bukh believes [15] it is likely that if k is large enough, then $\mathrm{ex}(n, C_{2k}) = o(n^{1+\frac{1}{k}})$ as $n \to \infty$. We suggest the following conjecture:

Conjecture IV. *For all $k \geq 3$, there exists $\delta > 1$ such that* $\mathrm{ex}(n, C_{2k}) \gtrsim \delta \mathrm{ex}(n, \mathscr{C}_{2k})$.

This was proved for $k = 5$ by Lazebnik, Ustimenko, and Woldar [90], and by Füredi, Naor, and the author for $k = 3$ and $k = 5$, disproving a conjecture of Erdős and Simonovits [41]. In fact, it seems plausible that δ is not bounded as a function of k.

1.2.2 Quadrilaterals and hexagons

Upper bounds for $\mathrm{ex}(n, C_4)$ and $z(m, n, C_4)$ have quite a long history, going back to the works of Kövari, Sós, and Turán [77], Reiman [107], Mörs [98], and Hyltén-Cavallius [69]. Reiman [107] showed $z(m, n, C_4) \leq \frac{1}{2}(m + \sqrt{m^2 + 4mn(n-1)})$, and that equality holds when $m = n(n-1)/k(k-1)$ only if there exists a Steiner k-tuple system on n vertices. Györi [64] determined bounds on $z(m, n, C_4)$ and $z(m, n, C_6)$ when m is very small relative to n, and the case $z(m, n, C_6)$ was studied by de Caen and Székely [27] in the context of incidence graphs of points and lines in the plane. The Kövari-Sós-Turán Theorem [77] gives $\mathrm{ex}(n, C_4) \leq \frac{1}{2}n^{\frac{3}{2}} + \frac{n}{2}$, and this was shown to be asymptotically sharp by Erdős, Rényi, and Sós [44] via the construction of *polarity graphs* of projective planes. Finally, for infinite many n, Füredi [54] established the exactly value of $\mathrm{ex}(n, C_4)$:

Theorem 5. *For prime powers $q > 13$, $\mathrm{ex}(q^2 + q + 1, C_4) = \frac{1}{2}q(q + 1)^2$, and the extremal graphs are polarity graphs of a projective plane of order q.*

A brief description of constructions for Theorem 5 is given in Section 4.2. Using the fact that if p_n is the nth prime, then $p_n -_{n-1} = o(p_n)$ as $n \to \infty$, one deduces $\mathrm{ex}(n, C_4) \sim \frac{1}{2}n^{\frac{3}{2}}$ as $n \to \infty$. It is important to note that the inequality $\mathrm{ex}(q^2 + q + 1, C_4) \leq \frac{1}{2}q(q + 1)^2$ in Theorem 5 is non-trivial, and slightly better than the bound from the Kövari-Sós-Turán Theorem [77]. Further bounds on $\mathrm{ex}(n, C_4)$ and some exact values of $\mathrm{ex}(n, C_4)$ have been determined by Firke, Kosek, Nash, and Williford [50] and Tait and Timmons [121], and for small values of n in [19, 109].

In the case of hexagons, Füredi, Naor, and the author [58] proved the following theorem:

Theorem 6.

$$\frac{3(\sqrt{5}-2)}{(\sqrt{5}-1)^{4/3}}n^{4/3} \lesssim \mathrm{ex}(n, C_6) \lesssim \frac{1}{12}\left(\sqrt[3]{\frac{317+9\sqrt{1241}}{2}} - \sqrt[3]{\frac{8}{317+9\sqrt{1241}}} + 1\right)n^{4/3}.$$

The lower bound is roughly $0.5338n^{\frac{4}{3}}$ and the upper bound is roughly $0.6272n^{\frac{4}{3}}$. It seems to be a challenging problem to prove the existence and determine the limiting value of $\mathrm{ex}(n, C_6)n^{-4/3}$.

1.2.3 Zarankiewicz numbers

Extending the results of Alon, Hoory, and Linial [3] to bipartite graphs, Hoory [68] proved the following analog of the Moore bound:

Theorem 7. *For any $k \geq 2$,*

$$z(m, n, \mathscr{C}_{2k}) \leq \begin{cases} (mn)^{\frac{1}{2}+\frac{1}{2k}} + m + n & \text{if } k \text{ is odd} \\ m^{\frac{k+2}{2k}} n^{\frac{1}{2}} + m + n & \text{if } k \text{ is even.} \end{cases}$$

For $k = 2$, the bound in Theorem 7 is asymptotically tight, by taking $m \times n$ bipartite subgraphs of the $n \times n$ bipartite Moore graphs of girth six (see Section 4.1), and for $m = n$ and $k \in \{2, 3, 5\}$, these bounds are tight due to the Moore graphs of girth $2k$. Using the method for proving Theorem 4, the following was established by Naor and the author [101]:

Theorem 8. *For positive integers $m \leq n$ and $k \geq 2$,*

$$z(m, n, C_{2k}) \leq \begin{cases} (2k-3)[(mn)^{\frac{1}{2}+\frac{1}{2k}} + m + n] & \text{if } k \text{ is odd} \\ (2k-3)[m^{\frac{k+2}{2k}} n^{\frac{1}{2}} + m + n] & \text{if } k \text{ is even.} \end{cases}$$

For $k > 2$ it is not known if the bounds above describe the order of magnitude of $z(m,n,C_{2k})$. In [101], it is shown that for any positive integers k,m,n, $z((k-1)m,n,C_{2k}) \geq (k-1)z(m,n,\mathscr{C}_{2k})$. In particular, this shows $z(n,n,C_{10}) \gtrsim \frac{8}{5^{6/5}} z(n,n,\mathscr{C}_{10})$, and suggests following general conjecture, which is a bipartite analogue of Conjecture IV:

Conjecture V. *For all $k \geq 3$, there exists $\delta > 1$ such that $z(n,n,C_{2k}) \gtrsim \delta z(n,n,\mathscr{C}_{2k})$.*

The first open case is the case $k = 3$, despite the fact that we know $z(n,n,\mathscr{C}_6) \sim n^{\frac{4}{3}}$. For $k = 3$, the bounds in Theorem 8 are not tight: for instance, the Ruzsa-Szemerédi Theorem [110] shows that $z(m,n,\mathscr{C}_6) = o((mn)^{\frac{2}{3}})$ for certain values of $m = n^{2-o(1)}$. In general, the asymptotic behavior of $z(n,n,C_6)$ is not known, despite $z(n,n,\mathscr{C}_6) \sim n^{\frac{4}{3}}$. Erdős [35] conjectured that $z(m,n,\mathscr{C}_6) = O(n)$ when $m = O(n^{\frac{2}{3}})$, however de Caen and Székely [27] disproved this conjecture, and Lazebnik, Ustimenko, and Woldar [89] gave even denser constructions, showing $z(m,n,\mathscr{C}_6) = \Omega(n^{\frac{16}{15}})$ for $m = \Theta(n^{\frac{2}{3}})$. In fact, we observe that their constructions are matched by considering a subgraph of the incidence graph of an elliptic quadric with $(1+q)(1+q^3)$ points and $n = (1+q)(1+q^4)$ lines: the points from a typical set of $m \sim n^{\frac{2}{3}}$ points together with the set of all n lines in the quadric generate a $\mathscr{C}_6$-free incidence graph with roughly $(mn)^{\frac{2}{3}} \sim n^{\frac{16}{15}}$ edges. It would be interesting to determine for which m,n one has $z(m,n,\mathscr{C}_6) = o((mn)^{\frac{2}{3}})$ or $z(m,n,C_6) = o((mn)^{\frac{2}{3}})$. The generalized quadrangles, hexagons, and octagons based on groups of Lie type of rank two [129] also give constructions of $\mathscr{C}_{2k}$-free $m \times n$ bipartite graphs for certain values of m and n. These constructions agree with the bounds in Theorem 7, and give $z(m,n,\mathscr{C}_6) \sim (mn)^{\frac{2}{3}}$ when $m \sim n^{\frac{4}{5}}$ or $m \sim n^{\frac{7}{8}}$, and $z(m,n,\mathscr{C}_{10}) \sim (mn)^{\frac{3}{5}}$ if $m \sim n^{\frac{10}{11}}$ and $z(m,n,\mathscr{C}_{14}) \sim (mn)^{\frac{4}{7}}$ when $m = (1+q)(1+q^3+q^6+q^9)$ and $n = (1+q^2)(1+q^3+q^6+q^9)$ and $q = 2^{2t+1}$ for some positive integer t.

1.2.4 An application of sparse regularity

In the papers of Keevash, Sudakov, and the author [71] and Allen, Keevash, Sudakov, and the author [1], an approach to extremal problems for cycles using Scott's sparse regularity lemma [113] is developed. We briefly discuss the general framework. A family $\mathscr{F}$ of graphs is called *smooth* if there exist $\alpha, \beta, \gamma \in \mathbb{R}^+$ where $\alpha < 1$ and $\beta < \alpha$ such that for all $m \leq n$ with $m = \Theta(n)$, $z(m,n,\mathscr{F}) \leq \gamma mn^{\alpha} + O(n^{\beta})$ and $z(n,n,\mathscr{F}) = \gamma n^{1+\alpha} + O(n^{\beta})$. We have seen, for instance, that this is the case when $\mathscr{F} = \{C_4, C_6, \ldots, C_{2k}\}$ where $k \in \{2,3,5\}$ by Theorem 2 and Hoory's bounds on Zarankiewicz numbers $z(m,n,\mathscr{C}_{2k})$. Further examples of smooth families are given in [1]. It is not known whether every finite family of bipartite graphs is smooth, and in particular, whether $\{C_6\}$ is smooth. The main result in [1] is the following, which under the condition of smoothness answers one of the main conjectures of Erdős and Simonovits [41] on bipartite Turán numbers:

Theorem 9. *If $\mathscr{F}$ is any smooth family, then there exists ℓ such that for any odd $L \geq \ell$, every extremal $\mathscr{F} \cup \{C_L\}$-free n-vertex graph G can be made bipartite by the deletion of $o(e(G))$ edges. In particular, for $k \in \{2, 3, 5\}$ and $L \geq k$,*

$$\mathrm{ex}(n, \mathscr{C}_{2k} \cup \{C_{2L+1}\}) \sim (\tfrac{n}{2})^{1+\frac{1}{k}}.$$

The last statement was proved for $k = L = 2$ by Erdős and Simonovits [41]. Since it is not known whether $\{C_6\}$ is smooth, we make the following conjecture:

Conjecture VI. *For any odd $L \geq 3$, $\mathrm{ex}(n, \{C_6, C_{2L+1}\}) \sim (\frac{n}{2})^{\frac{4}{3}}$.*

We believe this conjecture to be tractable; however, it appears substantially harder to predict the asymptotic behavior of $\mathrm{ex}(n, \{C_5, C_6\})$, which we leave as an open problem. It would also be interesting to determine for which 3-colorable graphs F other than odd cycles one has $\mathrm{ex}(n, \{C_4, F\}) \sim (\frac{n}{2})^{\frac{3}{2}}$, and to determine if there exists a 4-colorable graph F for which every extremal $\{C_4, F\}$-free graph G can be made tripartite by deleting $o(e(G))$ edges. We mention that while $z(n, n, C_4)$ is known for infinitely many n, it appears to be more challenging to determine the maximum number of edges in a C_4-free tripartite graph with n vertices in each part (see [87] for some details).

1.2.5 Dense subgraphs with large girth

Generalizing the problem of determining $\mathrm{ex}(n, \mathscr{C}_k)$, one may ask for the largest $\mathscr{C}_k$-free subgraph of a graph of given average degree (the former problem is when the host graph is K_n). The following conjecture is due to Thomassen [122]:

Conjecture VII. *For every k, g there exists $h(k, g)$ such that any graph G with average degree at least $h(k, g)$ contains a subgraph of girth at least $2g + 1$ and average degree at least k.*

This conjecture, which is easy if $g = 1$, was verified by Kühn and Osthus [80] for $g = 2$ with $h(k, 2) = \Theta((\log\log d)^{\frac{1}{3}})$, but remains open for $g \geq 3$. A number of interesting new results are given in Foucaud, Krivelevich, and Perarnau [53], who showed, for instance, that a d-regular graph has a subgraph of girth at least six and minimum degree $\Omega(\frac{\sqrt{d}}{\log d})$, which is best possible up to the logarithmic factor in d. We propose the following conjecture:

Conjecture VIII. *For $k \geq g \geq 2$, there exists $c(k, g) > 0$ such that every C_{2k}-free graph G has a subgraph of girth at least $2g + 1$ with at least $c(k, g) \cdot e(G)$ edges.*

Kühn and Osthus [79] proved that the conjecture is true for $g = 2$ with $c(k, 2) = \frac{1}{k-1}$, which is best possible, and extends an earlier result of Györi [64] showing $c_{3,2} = \frac{1}{2}$. They also proved the conjecture is true when $g = O(\frac{\log k}{\log\log k})$. It is already an open question to determine the existence of $c(4, 3)$ or $c(5, 3)$.

1.3 Counting graphs without short cycles

Erdős conjectured that for every graph F, the number of F-free n-vertex graphs is $\exp(O(\mathrm{ex}(n,F)))$, and perhaps even $2^{\mathrm{ex}(n,F)+o(\mathrm{ex}(n,F))}$ – we refer to the latter as the strong form of the conjecture. The strong form is known to hold for non-bipartite F, but is false for $F = C_6$: Morris and Saxton [97] used the construction in [58] for Theorem 6 to show that the number of n-vertex C_6-free graphs is at least $2^{(1+c)\mathrm{ex}(n,C_6)}$ for some $c > 0$. The case $F = C_4$ received some attention, where Kleitman and Winston [73] showed that the number of C_4-free n-vertex graphs is at most $2^{a\mathrm{ex}(n,C_4)}$ where $a \approx 1.17$. It may be that the strong form of the conjecture is true for $F = C_4$, and perhaps the strong form of Erdős' conjecture holds when $F = \mathscr{C}_g$ for all odd $g \geq 5$. The development of new techniques of Balogh, Morris, and Samotij [4] and Saxton and Thomason [112], sometimes referred to as the method of containers, has led to the following theorem of Morris and Saxton [97]:

Theorem 10. *There exists a family $\mathcal{G}$ of graphs such that each C_{2k}-free graph is a subgraph of some graph in $\mathcal{G}$, the number of n-vertex graphs in $\mathcal{G}$ is $\exp(O(n^{1+\frac{1}{k}}))$, and every n-vertex graph in $\mathcal{G}$ has $O(n^{1+\frac{1}{k}})$ edges.*

An immediate consequence is that the number of C_{2k}-free n-vertex graphs is $\exp(O(n^{1+\frac{1}{k}}))$, which proves the conjecture of Erdős in the cases $k = 3$ and $k = 5$ using Theorem 1. The details of the proof of the above theorem are beyond the scope of this survey. For a history of the problem, the reader is referred to [97].

1.4 Excluding infinitely many cycles

For $k \geq 2$ and $\ell \geq 0$, let $\mathscr{C}_{\ell \bmod k} = \{C_\ell, C_{k+\ell}, C_{2k+\ell}, \ldots\}$ denote the family of all cycles of length ℓ modulo k. Bollobás [8] was the first to show that $\mathrm{ex}(n, \mathscr{C}_{\ell \bmod k})$ is linear in n whenever $\mathscr{C}_{\ell \bmod k}$ contains even cycles, and showed $\mathrm{ex}(n, \mathscr{C}_{\ell \bmod k}) \leq \frac{1}{k}[(k+1)^k - 1]n$. This upper bound was reduced by a number of authors [13, 17, 25, 26, 28, 48, 124]. The first linear bound in k was given by the author [132]. Diwan [28] slightly improved this using methods of Mader [92], by showing that a graph of minimum degree at least $2k-1$ contains cycles of all even lengths modulo k, from which one deduces $\mathrm{ex}(n, \mathscr{C}_{\ell \bmod k}) \leq (2k-2)n$ for all ℓ. The following general conjecture is due to Thomassen [124]:

Conjecture IX. *Let $k \geq 2$, and let G be a graph of minimum degree at least $k+1$. Then G contains cycles of all even lengths modulo k.*

Diwan [28] showed that a graph of minimum degree at least $k+1$ contains a cycle of length four modulo k. We shall see next that this can be greatly improved:

Theorem 11. *Let $k \geq 2\ell \geq 4$, and let G be a graph of minimum degree at least $48\ell k^{\frac{1}{\ell}}$. Then G contains a cycle of length 2ℓ modulo k.*

In particular this shows, perhaps surprisingly, that $\text{ex}(n, \mathscr{C}_{2\ell \bmod k}) = \Theta(\ell)n$ when $\ell \geq \log k$ and $\text{ex}(n, \mathscr{C}_{4 \bmod k}) = \Theta(k^{\frac{1}{2}})n$. In fact the bound $48\ell k^{\frac{1}{\ell}}$ in Theorem 11 can be replaced by $O(\frac{\text{ex}(k,2\ell)}{k})$. By taking an extremal graph with k vertices and no cycle of length 2ℓ, we obtain a graph average degree $\frac{2\text{ex}(k,2\ell)}{k}$ with no cycle of length 2ℓ modulo k, showing that this is tight up to an absolute constant factor.

1.4.1 Consecutive even cycle lengths

Using a structural argument based on a lemma on induced cycles of Thomassen and Toft [122], Bondy and Vince [13] showed that a graph with at most two vertices of degree less than three and at least four vertices contains two cycles of consecutive lengths or two cycles of consecutive even lengths. Häggkvist and Scott [70] asked whether there exists a constant c such that if G is a graph of minimum degree at least ck, then G contains cycles of k consecutive even lengths, and showed that every graph of minimum degree at least $100k^2$ contains cycles of k consecutive even lengths. The next theorem from [132] answers the question of Häggkvist and Scott, and was the first result showing $\text{ex}(n, \mathscr{C}_{\ell \bmod k}) = O(k) \cdot n$ for all even ℓ.

Theorem 12. *Let k be a positive integer. If G is an n-vertex graph with at least $3kn$ edges, then for some integer r, G contains cycles $C_{2r}, C_{2r+2}, \dots, C_{2r+2k-2}$.*

Theorem 12 will be proved in Section 6.1. An n-vertex graph whose blocks are all cliques of order $2k+1$ contains cycles $C_4, C_6, \dots, C_{2k}$ and has $\frac{1}{2}(2k+1)(n-1)$ edges, and therefore Theorem 12 is best possible up to a factor roughly three. We mention a related result of Fan [48]: every graph with minimum degree at least $3k-2$ contains k-cycles of consecutive even lengths or consecutive odd lengths. We make the following conjecture:

Conjecture X. *If G is an n-vertex graph not containing k-cycles of consecutive even lengths, then $e(G) \leq \frac{1}{2}(2k+1)(n-1)$, with equality only if every block of G is a clique of order $2k+1$.*

Very recently, a proof of Thommasen's conjecture (conjecture X) for even values of k has been claimed by Liu and Ma [82]. They combine the arguments of Bondy and Vince with additional structural arguments to prove the following:

Theorem 13. *If G is a graph of minimum degree at least $k+1$, then G contains $\lfloor \frac{k}{2} \rfloor$ cycles with consecutive even lengths. Furthermore, if G is 2-connected and non-bipartite, then G contains $\lfloor \frac{k}{2} \rfloor$ cycles with consecutive odd lengths.*

In their paper, Liu and Ma [82] establish a number of other interesting related results, in particular, they verify that any graph of minimum degree at least $k+4$ contains cycles of all possible lengths modulo k.

1.4.2 Distinct cycle lengths

Given a graph G, let $C(G) = \{\ell \in \mathbb{Z}^+ : C_\ell \subset G\}$ denote the set of lengths of cycles in G – this is sometimes called the *cycle spectrum* of G. It is well known that any graph of minimum degree $d \geq 2$ contains a cycle of length at least $d + 1$, as well as cycles of at least $d - 1$ distinct lengths. The length of a longest cycle $\max C(G)$ in a graph G is called the *circumference* of the graph, and is an extensively researched topic [11, 29, 125, 135]. Erdős [38] conjectured that if G is a graph of minimum degree at least d and girth at least $2k + 1$, then $|C(G)| = \Omega(d^k)$ as $d \to \infty$. The conjecture was proved by Erdős, Faudree, Rousseau, and Schelp [47] in the case $k = 5$. The conjecture was proved in full by Sudakov and the author [119]:

Theorem 14. *Let G be a graph of average degree d and girth at least $2k + 1$. Then G contains cycles of $\Omega(d^k)$ consecutive even lengths, and in particular $|C(G)| = \Omega(d^k)$.*

This theorem is best possible up to a constant factor for $k \in \{2, 3, 5\}$, due to the existence of Moore graphs of even girth (see Theorem 1). We give a short proof of this theorem in Section 6.3, and we make the following conjecture:

Conjecture XI. *Let $d \geq 2$, and let $n(d, g)$ denote the smallest number of vertices in a graph of girth g and minimum degree d. Then every graph G of minimum degree at least d contains cycles of at least $\frac{1}{2}n(d, g) - O(g)$ even lengths.*

This conjecture suggests that the minimal graphs of girth g and minimum degree d contain cycles of essentially all possible even lengths. It can be verified using spectral techniques as in Krivelevich and Sudakov [78] that for the $n \times n$ bipartite Moore graph G of girth $2k$ where $k \in \{3, 4, 6\}$, $C(G) = \{2k, 2k+2, \dots, 2n\}$. In fact, the results in [119] generalize to monotone properties in the following way: if $\mathcal{P}$ is a monotone property of graphs and $n(d, \mathcal{P})$ is the minimum number of vertices in a graph in $\mathcal{P}$ of minimum degree d, then for some constant $\delta > 0$, every graph $G \in \mathcal{P}$ of minimum degree at least $d \geq 2$ contains cycles of at least $n(\delta d, \mathcal{P})$ consecutive even lengths. This is applied to obtain sharp results when $\mathcal{P}$ is the family of $\mathcal{F}$-free graphs for various $\mathcal{F}$, and Theorem 14 is the case $\mathcal{P}$ consists of all graphs of girth at least $2k+1$. For instance, one deduces that every $K_{3,3}$-free graph of minimum degree d contains cycles of $\Omega(d^{5/3})$ consecutive lengths. We also propose the following conjecture:

Conjecture XII. *Let G be a graph of minimum degree at least three and circumference t. Then $|C(G)| = \Omega(t)$.*

In particular, the conjecture implies that a Hamiltonian n-vertex graph of minimum degree at least three has linearly many cycle lengths. It is not hard to see that this is true for Hamiltonian planar graphs, and Milans, Rautenbach, Pfender, Regen, and West [96] showed that $|C(G)| \geq \sqrt{m - n} + \frac{1}{2}\log(m - n) - 1$ when G is an n-vertex Hamiltonian graph with m edges.

1.4.3 Distribution of cycle lengths

Throughout this section, $\sigma = (\sigma_1, \sigma_2, \ldots)$ is an increasing sequence of positive integers. The *density* of σ, when it exists, is

$$\lim_{N\to\infty} \frac{|\{i : \sigma_i \leq N\}|}{N}.$$

Let $\mathscr{C}_\sigma = \{C_{\sigma_i} : i \in \mathbb{Z}^+\}$. The general problem of determining $\mathrm{ex}(n, \mathscr{C}_\sigma)$ when σ is a finite sequence has been discussed in the preceding sections. By Theorem 12, if σ is a sequence of even integers of positive density, then $\mathrm{ex}(n, \mathscr{C}_\sigma) = O(n)$, so the focus is on sequences of zero density, such as $\sigma_i = 2^i$ or σ_i equal to twice the ith prime. Erdős conjectured that for both of these sequences one has $\mathrm{ex}(n, \mathscr{C}_\sigma) = O(n)$ (see page 228 of [38]), and Erdős and Gyárfás [37] made the following conjecture:

Conjecture XIII. *Any graph of minimum degree at least three has a cycle of length a power of two.*

This conjecture remains open. Erdős later conjectured [18, 38] that there exists a sequence σ of density zero such that $\mathrm{ex}(n, \mathscr{C}_\sigma) = O(n)$. This conjecture was proved in [131]:

Theorem 15. *There exists a sequence σ of density zero such that* $\mathrm{ex}(n, \mathscr{C}_\sigma) \leq 5n$.

In fact, the proof of Theorem 15 shows that for a proportion $1 - o(1)$ of the increasing sequences σ of even integers containing at least $n^{0.99}$ elements from $\{1, 2, \ldots, n\}$, one has $\mathrm{ex}(n, \mathscr{C}_\sigma) \leq 5n$. The proof of Theorem 15 is beyond the scope of this survey. A general theorem for these problems was established by Sudakov and the author [119]. Let $\log^* n$ denote the smallest non-negative integer m such that $\log\log\log \ldots \log n \leq 1$, where the logarithm is iterated m times.

Theorem 16. *If σ is an increasing sequence of even integers such that $\sigma_1 = 1$ and $\sigma_i \leq 2^{\sigma_{i-1}}$ for all $i \geq 2$, then* $\mathrm{ex}(n, \mathscr{C}_\sigma) \leq \exp(8 \log^* n)$.

We prove this in Section 7. This shows that an n-vertex graph of average degree at least $\exp(8 \log^* n)$ contains a cycle of length a tower of twos. It is shown in [119] that Theorem 16 is best possible up to the factor 8 in the exponent (see Section 7.4). Furthermore, it is shown in [119] that the sequence σ defined by $\sigma_1 = 1$ and $\sigma_i = 2i^{\sigma_{i-1}}$ for $i \geq 2$ has $\mathrm{ex}(n, \mathscr{C}_\sigma) = (\frac{\log n}{\log\log n})^{\Omega(1)}$, and so the condition $\sigma_i \leq 2^{\sigma_{i-1}}$ in Theorem 16 is in some sense necessary.

1.4.4 Reciprocals of cycle lengths

Erdős proposed $\phi(G) = \sum_{\ell \in C(G)} \frac{1}{\ell}$ as a measure of the distribution of cycle lengths in a graph G. Gyárfás, Komlós, and Szemerédi [62] proved the following theorem:

Theorem 17. *There exists $\delta > 0$ such that any graph G of average degree $d \geq 2$ has $\phi(G) \geq \delta \log d$.*

In particular, this shows that if G is a graph with infinite chromatic number, then the sum of reciprocals of cycle lengths in G diverges, proving a conjecture of Erdős [62]. More precise results were given for graphs G with average degree close to two and large girth in Gyárfás, Prömel, Szemerédi, and Voigt [63]. The following is a consequence of the results in [119]:

Theorem 18. *If G is an n-vertex graph of average degree $d \geq 3$, then $\phi(G) \geq \frac{1}{2}\log d - 8\log^* n$.*

We prove this theorem in Section 7. This lends support to the following conjecture:

Conjecture XIV. *If G is a graph of average degree d, then $\phi(G) \gtrsim \frac{1}{2}\log d$ as $d \to \infty$.*

Erdős and Hajnal [40] conjectured that if G is a graph of infinite chromatic number, then the sum of reciprocals of lengths of odd cycles in G, denoted $\phi_\circ(G)$, is infinite. If a graph has uncountable chromatic number, Erdős, Hajnal, and Shelah [45] proved that it contains all sufficiently long odd cycles (see also Thomassen [123] and Erdős and Hajnal [40] for more on infinite graphs, cycles, and chromatic number). Define the *independence ratio* of a graph G by

$$\iota(G) = \max_{X \subset V(G)} \frac{|X|}{\alpha(G[X])},$$

where $\alpha(G)$ is the independence number of G. The independence ratio is a relaxation of the chromatic number $\chi(G)$ in the sense that $\chi(G) \geq \iota(G)$, while for many natural classes of graphs these quantities are almost equal, although $\chi(G)$ and $\iota(G)$ can differ substantially, for example, for Kneser graphs (see Godsil and Royle [60]). The following theorem is proved in [120]:

Theorem 19. *For any n-vertex graph G with $\phi_\circ(G) \geq 3$, $\phi_\circ(G) \geq \frac{1}{2}\log\iota(G) - 8\log^* n$.*

The sum of reciprocals of odd cycle lengths in a graph G whose components are complete graphs K_t is at most $\frac{1}{2}\log\iota(G)+1$, so the above result is best possible apart from the additive $8\log^* n$ term. It seems likely in Theorem 19 that this term can be replaced by an absolute constant and $\iota(G)$ can be replaced by $\chi(G)$. We therefore make the following conjecture:

Conjecture XV. *For any graph G, $\phi_\circ(G) \gtrsim \frac{1}{2}\log\chi(G)$ as $\chi(G) \to \infty$.*

1.5 Odd cycles, chromatic number, and girth

Gyárfás [61] showed that a graph of chromatic number $k \geq 3$ contains cycles of $\lfloor \frac{k}{2} \rfloor$ distinct odd lengths, which is tight according to any graph whose blocks are all cliques of order k, answering a question of Bollobás and Erdős [36]. Mihók and

Schiermeyer [95] proved a similar result for even cycles: every graph G of chromatic number k contains cycles of $\lfloor \frac{k}{2} \rfloor - 1$ even lengths. We make the following conjecture, which proposes a common generalization of the above results:

Conjecture XVI. *If G is a graph of chromatic number $k \geq 3$, then for some m, G contains $C_{2m+1}, C_{2m+2}, \ldots, C_{2m+k-2}$.*

We give a proof of the following theorem in Section 6.2, which is within a factor roughly two of the conjecture:

Theorem 20. *Let G be a graph of chromatic number at least $2k-1$ where $k \geq 4$. Then for some m, G contains cycles $C_{2m+1}, C_{2m+2}, \ldots, C_{2m+k-1}$.*

1.5.1 Chromatic number and girth

A well-known theorem of Erdős [32] using random graphs shows that there exist graphs of arbitrarily large chromatic number and girth. The first explicit examples for this result were given by Lovász [83] (see also Sachs and Stiebitz for a survey [111]). Erdős conjectured [38] that for any $\varepsilon > 0$, any triangle-free graph G of chromatic number k contains cycles of $\Omega(k^{2-\varepsilon})$ different lengths as $k \to \infty$. The following stronger theorem was recently proved by Kostochka, Sudakov, and the author [76], which in the case $g = 2$ verifies Erdős' conjecture:

Theorem 21. *Every graph G of odd girth at least $2g+1$ and chromatic number k contains cycles of $\Omega(k^g \log k)$ consecutive lengths.*

As there exist triangle-free graphs of chromatic number k with at most roughly $4k^2 \log k$ vertices for large k (see Kim [72], Bohman and Keevash [7], and Fiz Pontiveros, Griffiths, and Morris [52]), this result is tight up to a constant factor for $g = 2$. A related theorem of Ma [91] states that every non-bipartite 2-connected graph G with average degree at least k and girth at least $2g+1$ contains cycles of $\Omega(k^g)$ consecutive odd lengths, which is optimal up to a constant factor by taking a bipartite Moore graph of even girth $2g+2 \in \{6, 8, 10\}$ and adding one edge in one of the parts. Whether Theorem 21 is tight for $g > 2$ is likely to rely on cycle-complete graph Ramsey numbers, which are notoriously difficult to determine (see Caro [16], Shearer [114, 115], Erdős, Faudree, Rousseau, and Schelp [46], and Sudakov [118]). For quadrilaterals, the following conjecture of Erdős [18] remains wide open:

Conjecture XVII. *For some $\varepsilon > 0$, every n-vertex C_4-free graph G has $\alpha(G) = \Omega(n^{\frac{1}{2}+\varepsilon})$.*

The current record is the unpublished result of Szemerédi showing $\alpha(G) = \Omega(n^{\frac{1}{2}} \log n)$. In general, the Erdős-Rényi-Sós graphs (see Section 4) have much larger independence numbers – see Mubayi and Williford [100]. In fact, in [76], a stronger theorem is proved, namely if $\mathcal{P}$ is a monotone property of graphs and the smallest number of vertices in a graph of chromatic number at least k in $\mathcal{P}$ is n_k, then for some constant $\delta > 0$, every graph of chromatic number k in $\mathcal{P}$

contains cycles of $\Omega(n_{8k})$ consecutive lengths. For instance, this shows that a K_4-free graph of chromatic number k contains cycles of $\Omega(k^{\frac{3}{2}}\sqrt{\log k})$ consecutive lengths, due to known bounds on the Ramsey number of K_4 versus K_n, for instance, see Shearer [115].

One of the salient open questions on chromatic number and girth is the following conjecture of Erdős and Hajnal [33]:

Conjecture XVIII. *For every k, g there exists $f(k, g)$ such that any graph G with $\chi(G) \geq f(k, g)$ contains a subgraph of girth at least g and chromatic number at least k.*

This conjecture was verified for $g = 4$ by Rödl [108], whose proof shows $f(k, 4) \leq F(8k^2 \log k)$ where $F(k) = (k-1)^{F(k-1)} + 1$. This conjecture is akin to the conjecture of Thomassen on dense subgraphs of large girth (Conjecture VII).

1.5.2 Sequences of odd cycles

This section is motivated by many conjectures of Erdős [38] stating that a graph of large enough chromatic number should have cycles of lengths in certain infinite sequences of odd integers, such as odd primes. The following result in [120] extends Theorem 16 to accommodate odd cycles in graphs of large independence ratio:

Theorem 22. *Let σ be any increasing sequence of integers such that $\sigma_1 = 1$ and $\sigma_i \leq 2^{\sigma_{i-1}}$ for all $i \geq 2$. If G is $\mathscr{C}_\sigma$-free, then $\iota(G) \leq \exp(8 \log^* n)$.*

Theorem 22 shows that if σ is the sequence of primes, or powers of two plus one, then $\iota(G) = O(\log^* n)$. Theorem 22 distinguishes between the independence ratio and the chromatic number: generalizations of Mycielski's well-known construction (see Sachs and Stiebitz [111]) of graphs of large girth and chromatic number provide constructions of an n-vertex graph G_n of chromatic number $\chi(G_n) = \Omega(\frac{\log n}{\log \log n})$ with no cycle of length in a carefully constructed sequence σ satisfying the conditions of Theorem 22. The conclusion of Theorem 16 therefore does not hold if we replace $\iota(G)$ with $\chi(G)$ in the theorem.

1.6 Hypergraph cycles

Recent attention has been given to problems involving cycles in r-uniform hypergraphs (or simply, r-graphs). If $\mathscr{F}$ is a family of r-graphs, let $\mathrm{ex}_r(n, \mathscr{F})$ denote the maximum number of edges in an r-graph on n vertices that does not contain $\mathscr{F}$. There are many different notions of cycles in hypergraphs. We consider three different types of cycles in r-graphs, all of which coincide with the usual definition of cycles in graphs when $r = 2$. If C is a graph cycle of length k with vertices $v_0, v_1, \ldots, v_{k-1}$ in order around the cycle, then a *tight k-cycle*, denoted TC_k, is the

r-graph whose vertex set is $\{v_0, v_1, \ldots, v_{k-1}\}$ and whose edge set is the set of paths of r vertices in C, in other words, the edges are sets $\{v_i, v_{i+1}, \ldots, v_{i+r-1}\}$ with subscripts modulo k. A *Berge k-cycle* is any r-graph in the family BC_k consisting of all r-graphs whose edge set is of the form $\{v_i, v_{i+1}\} \cup f_i$ where $f_0, f_1, \ldots, f_{k-1}$ are arbitrary sets of $r-2$ vertices with $f_i \cap \{v_i, v_{i+1}\} = \emptyset$ and subscripts modulo k. In other words, we take a graph k-cycle and expand every edge to a set of size r to get a Berge cycle. If the sets f_i are pairwise disjoint and disjoint from $\{v_0, v_1, \ldots, v_{k-1}\}$, then we obtain a *loose k-cycle*, denoted LC_k. In this section, we discuss the Turán numbers for LC_k, BC_k, and TC_k. The methods used to prove upper bounds for $\mathrm{ex}_r(n, LC_k)$ and $\mathrm{ex}_r(n, BC_k)$ have substantial similarity with those used to prove Theorem 4, by careful analysis of an analog of breadth-first search trees in hypergraphs and local expansion, however the proofs are markedly more complicated.

1.6.1 Berge cycles

The problem of determining $\mathrm{ex}_r(n, BC_k)$ is related to the problem of determining Zarankiewicz numbers $z(m, n, C_{2k})$ as follows: if H is an n-vertex r-uniform hypergraph with m edges and no BC_k, then the bipartite incidence graph of H is an $m \times n$ bipartite C_{2k}-free graph with rm edges, and so $r|H| \leq z(m, n, C_{2k})$. If $\mathscr{B}_k = \{BC_2, BC_3, \ldots, BC_k\}$, then one can obtain Moore bounds for $\mathscr{B}_k$-free r-uniform hypergraphs (hypergraphs of girth at least $k+1$) directly from Theorem 7, improving bounds given by Ellis and Linial [30]. Furthermore, Theorem 8 immediately shows that both $\mathrm{ex}_r(BC_{2k+1})$ and $\mathrm{ex}_r(BC_{2k})$ are $O(n^{1+\frac{1}{k}})$ if $r \geq 4k$. However, Theorem 8 does not apply for k large relative to r. Györi [65] determined $\mathrm{ex}_3(n, BC_3)$ for all n, and Bollobás and Györi [10] showed $\mathrm{ex}_3(n, BC_5) = O(n^{\frac{3}{2}})$, which is matched by a construction of Mubayi [99], showing $\mathrm{ex}_3(n, BC_5) = \Theta(n^{\frac{3}{2}})$. Györi and Lemons [66] were the first to determine a general analog of Theorem 4 for all ranges of parameters, and an improvement of their result for $r = 3$ was given by Füredi and Özkahya [56] and Alon and Shikhelman [2]:

Theorem 23. *For all $k \geq 2$ and $r \geq 3$, $\mathrm{ex}_r(n, BC_{2k+1}) = O(n^{1+\frac{1}{k}})$ and $\mathrm{ex}_r(n, BC_{2k}) = O(n^{1+\frac{1}{k}})$.*

For $r = 3$, Alon and Shikhelman [2] showed $\mathrm{ex}_3(n, BC_{2k+1}) \leq \frac{16}{3}\mathrm{ex}(\lceil \frac{n}{2} \rceil, C_{2k})$ and Füredi and Özkahya [56] showed $\mathrm{ex}_3(n, BC_{2k}) \leq \frac{1}{3}(2k-3)\mathrm{ex}(n, C_{2k})$. Note that unlike the case of graphs, there is less of a difference between hypergraph Turán numbers of odd Berge cycles and even Berge cycles. It is not known if the upper bounds in this theorem are tight for all $r, k \geq 3$. Constructions of $\mathscr{B}_k$-free extremal r-uniform hypergraphs for any $r \geq 3$ and $k \geq 5$ appear to be difficult to find. For $r = 3$, Lazebnik and the author [87] determined the exact value of $\mathrm{ex}_3(n, \mathscr{B}_4)$ for infinitely many n, and in particular, $\mathrm{ex}_3(n, \mathscr{B}_4) \sim \frac{1}{6}n^{\frac{3}{2}}$. Timmons and the author [127] proved the following result:

Theorem 24. *For all $r \geq 2$, $\mathrm{ex}_r(n, \mathscr{B}_4) = n^{\frac{3}{2}-o(1)}$.*

It appears to be difficult to determine the order of magnitude of $\mathrm{ex}_4(n, \mathscr{B}_4)$ when $r \geq 4$. We make the following conjecture:

Conjecture XIX. *For all $r \geq 4$, $\mathrm{ex}_r(n, \mathscr{B}_4) = \Theta(n^{\frac{3}{2}})$.*

If this conjecture is true for $r = 4$, then it would disprove a conjecture of Solymosi [117], stating that an n-vertex graph in which every edge is in exactly one cycle of length four has $o(n^{\frac{3}{2}})$ edges: we would place inside each hyperedge a copy of C_4 to obtain a graph with $\Theta(n^{\frac{3}{2}})$ edges and every edge in exactly one quadrilateral. We also mention a problem of Fischer and Matoušek [51] related to *VC-dimension* (see Vapnik and Chervonenkis [130]): suppose we have a tripartite graph with n vertices in each part, such that the bipartite graph of edges between any two of the parts is C_4-free. Fischer and Matoušek asked for the maximum number of triangles in any such graph, observing that the maximum is between $\Theta(n^{\frac{3}{2}})$ and $\Theta(n^{\frac{7}{4}})$ – this upper bound comes from a careful application of Theorem 8. Finally, we also ask for an analogue of Theorem 12 for hypergraphs:

Conjecture XX. *Let $r \geq 2$. If H is an r-graph which does not contain Berge cycles of k consecutive lengths, then H has average degree $O(k^{r-1})$ as $k \to \infty$.*

The complete r-graph on k vertices shows that this conjecture, if true, is best possible up to a constant factor, and Theorem 12 verifies it for $r = 2$.

1.6.2 Loose cycles

The Turán problem for loose k-cycles was solved by Füredi and Jiang [55] for $r \geq 5$ and by Kostochka, Mubayi, and the author [75] for all $r \geq 3$: the problem is qualitatively different to the case $r = 2$, and in these cases $\mathrm{ex}_r(n, LC_k) \sim \lfloor \frac{k-1}{2} \rfloor \binom{n}{r-1}$ for $k, r \geq 3$. Let $\mathrm{ex}_r^*(n, LC_k)$ denote the maximum number of edges in a linear r-uniform hypergraph on n vertices which does not contain LC_k. The problem of determining $\mathrm{ex}_r^*(n, LC_k)$ seems to be a more faithful generalization of the even cycle problem in graphs. In this case, Collier, Graber, and Jiang [20], and later Füredi and Özkahya [56] proved the following result:

Theorem 25. *For all $k, r \geq 3$, $\mathrm{ex}_r^*(n, LC_{2k+1}) = O(n^{1+\frac{1}{k}})$ and $\mathrm{ex}_r^*(n, LC_{2k}) = O(n^{1+\frac{1}{k}})$.*

One of the interesting related problems is the extremal problem for rainbow cycles in properly colored graphs. Specifically, one may ask for the maximum number of edges $q(n, k)$ in a properly edge-colored n-vertex graph that does not contain a $2k$-cycle whose edges all have different colors. It turns out that $q(n, k) = \Omega(n^{1+\frac{1}{k}})$, but matching upper bounds are harder to come by. In fact, a simple reduction shows $\mathrm{ex}_r^*(n, LC_{2k}) = O(q(n, k))$ for all $r \geq 3$ – see [71]. In [71] it is shown that $q(n, k) = O(n^{1+\frac{1}{k}})$ for $k \in \{2, 3\}$, and Das, Lee, and Sudakov [24] showed that $q(n,k) = O(n^{1+\varepsilon_k})$ where $\varepsilon_k \to 0$ as $k \to \infty$. This leaves the open problem of determining whether $q(n,k) = \Theta(n^{1+\frac{1}{k}})$ for all $k \geq 2$. In [71] the question of determining the limiting value of $q(n,2)n^{-\frac{3}{2}}$ is also raised.

1.6.3 Tight cycles

Finally, we mention the problem of tight cycles. If k is not divisible by r, then TC_k is not r-partite, and $\mathrm{ex}(n, TC_k) = \Theta(n^r)$. If k is divisible by r, then it is likely that for some $\alpha \in (r-1, r)$, $\mathrm{ex}(n, TC_k) = \Theta(n^\alpha)$, however there is no single pair of values of k and r with $k \geq 3$ and $r \geq 3$ for which this is known. We propose the following conjecture:

Conjecture XXI. *For all $r \geq 2$ and $k \geq 2r$ such that $k|r$, $\mathrm{ex}_r(n, TC_k) = O(n^{r-1+\frac{r-1}{k}})$.*

Perhaps it is true that $\mathrm{ex}_3(n, TC_6) = \Theta(n^{\frac{8}{3}})$, however there is insufficient evidence to conjecture this. The best upper and lower bounds in this case are $\mathrm{ex}_3(n, TC_6) = O(n^{\frac{11}{4}})$ and $\mathrm{ex}_3(n, TC_6) = \Omega(n^{\frac{12}{5}})$ coming from random hypergraphs. This problem is interesting in the context of an application to an extremal problem in the hypercube due to Conlon [22]. Let TC denote the family of all tight cycles. Sós and the author also independently raised the problem of determining whether $\mathrm{ex}_r(n, TC) = \binom{n-1}{r-1}$ for $r \geq 3$: a construction is simply to take all edges containing a fixed vertex.

2 Notation and terminology

Throughout this survey, graphs are considered to be simple (without multiple edges). For standard graph-theoretic terminology pertinent to this survey, see Bondy [11]. If G is a graph, then $e(G)$ denotes the number of edges in G, $N_G(v)$ is the neighborhood of a vertex $v \in V(G)$, and $d_G(v) = |N_G(v)|$ is the degree of v. We suppress subscripts where the graph G is clear from the context. If $X \subset V(G)$, then the neighborhood of X is $N(X) = X \cup \bigcup_{x \in X} N(x)$. We write $G[X]$ for the subgraph induced by X, $e(X)$ for $e(G[X])$, and for disjoint $X, Y \subseteq V(G)$, $G(X, Y)$ denotes the bipartite graph consisting of all edges of G with one end in X and one end in Y, and $e(X, Y) = e(G(X, Y))$. If T is a breadth-first search tree in a graph G, let $L_i(T)$ denote the set of vertices at distance exactly i from the root of the tree. For $X \subseteq V(G)$ and a graph invariant f, let $f(X) := f(G[X])$. Let $\iota(G)$ denote the *independence ratio* of G, namely $\max\{\frac{|X|}{\alpha(X)} : X \subseteq V(G)\}$.

If $\mathscr{F}$ is a family of graphs, then a graph G is $\mathscr{F}$-free if no graph in $\mathscr{F}$ is a subgraph of G, and $\mathrm{ex}(n, \mathscr{F})$ is the Turán number for $\mathscr{F}$. For a graph G, $\chi(G)$ is the chromatic number of G and $\alpha(G)$ is the independence number of G. Denote by P_k and C_k the path and cycle of length k, respectively. We write $\mathscr{C}_k$ instead of

$\{C_3, C_4, \ldots, C_k\}$ and if L is a set of positive integers, then $\mathscr{C}_L = \{C_\ell : \ell \in L\}$. The *ends* of a path P are the vertices of degree 1 in P. A *chord* of a cycle C in a graph G is an edge $e \in E(G)\backslash E(C)$ with both ends in C.

For functions $f, g : \mathbb{Z}^+ \to \mathbb{R}^+$, we write $f = O(g)$ if there exists $c > 0$ such that $f(n) \leq cg(n)$ for all $n \in \mathbb{Z}^+$, or equivalently $g = \Omega(f)$, and $f = o(g)$ if $f(n)/g(n) \to 0$ as $n \to \infty$. We write $f \sim g$ if $f(n) = g(n) + o(g(n))$ and $f \lesssim g$ if $f(n) \leq g(n) + o(n)$. If $f = O(g)$ and $g = O(f)$, we write $f = \Theta(g)$. In words, we say that f and g have the same order of magnitude.

3 Graph-theoretic preliminaries

3.1 Subgraphs of large minimum degree

The *k-core* of a graph G is the largest subgraph of minimum degree at least k. The k-core of a graph is found by repeatedly deleting vertices of degree less than k in the graph – note the k-core may be empty. The extremal problem for k-cores is addressed by the following proposition. We write $H_{k,n}$ for the graph consisting of a complete bipartite graph $K_{k,n-k}$ plus a clique of size k in the part of size k – in other words, $H_{k,n}$ is the n-vertex complement of a clique of order $n-k$. By inspection, $H_{k,n}$ has no non-empty $(k+1)$-core, and the following shows that $H_{k,n}$ is extremal with this property.

Proposition 3.1. *Let $k \geq 2$ be an integer, and let G be an n-vertex graph with at least $k(n-k) + \binom{k}{2}$ edges. Then G has a non-empty $(k+1)$-core unless $G = H_{k,n}$.*

3.2 The Erdős-Gallai Theorem

A standard argument for the length of a longest path and a longest cycle in a graph of minimum degree k is to take a longest path, and observe that both ends of the path have all their neighbors on the path. Since these ends have at least k neighbors, the path has length at least k, and considering each neighbor on the path we find cycles of $k-1$ distinct lengths, including a cycle of length at least $k+1$. The complete graph K_{k+1} shows this is best possible. The Erdős-Gallai Theorem [39] (see also Kopylov [74]) transfers these statements to graphs of average degree at least k: the theorem states that if G is a graph of average degree at least k, then G contains a cycle of length at least $k+1$. The following is a consequence of the proof of this theorem:

Proposition 3.2. *Let $k \geq 3$ be an integer. If G is a graph of average degree at least k, then G contains a cycle of length at least $k+1$ with a chord.*

3.3 Pósa's lemma

If P is a path with ends u and v and we orient P from u to v, then P is called a *uv-path*. If $x \in V(P)$, let x^+ and x^- denote the immediate successor and immediately predecessor of x on P, respectively. For a set $S \subset V(P)$, let $S^+ = \{x^+ : x \in S\}$ Define S^- similarly. If $e \in E(G)$ and $H \subset G$, we write $H+e$ for the graph with vertex set $V(H) \cup e$ and edge set $E(H) \cup \{e\}$. If P is a longest path in a graph G with ends u and v, then for any $w \in N_G(v)$, we have $w \in V(P)$. Then $P' = P - \{w, w^+\} + \{v, w\}$ is a uw^+-path of the same length as P. The operation of passing from P to P' is sometimes called a *rotation*. Note that a rotation preserves the first vertex u of P. Let S be the set of ends of longest paths obtained by rotations of P. Then $N(S) \subset S \cup S^+ \cup S^-$ and therefore $|N(S)| \leq 3|S|$. This leads to Pósa's Lemma [106] (see also [85], Exercise 10.20):

Proposition 3.3. *Let $t \in \mathbb{Z}^+$. If G is a graph such that $|N(X)| > 3|X|$ for every set $X \subset V(G)$ of size at most t, then G contains a cycle of length at least $3t$ with a chord.*

This proposition is very useful in showing that locally expanding graphs have long paths and cycles.

3.4 Coloring lemma

A key ingredient of the proofs of the theorems is the following coloring lemma, which is implicit in the paper of Bondy and Simonovits [12]. The proof of this lemma can be found in [132]:

Lemma 3.4. *Let H be a graph comprising a cycle C plus a chord, and suppose that (A, B) is a non-trivial partition of $V(H)$. Then for every $\ell < |V(H)|$ there exists a path $P \subset H$ of length ℓ such that one end of P is in A and the other end of P is in B, unless (A, B) is a bipartition of H.*

The purpose of the chord of the cycle C is to break the possibility that the vertices of A are every mth vertex along C, for in that case, there is no path of length zero mod m with one end in A and one end in B. The key use of Lemma 3.4 is in the following:

Lemma 3.5. *Let T be a breadth-first search tree in a graph G, and suppose G contains a cycle C of length k plus a chord contained in $G[L_i(T)]$ or contained in $G(L_i(T), L_{i+1}(T))$, respectively. Then for some $m \leq i$, G, respectively, contains cycles $C_{2m+1}, C_{2m+2}, \ldots, C_{2m+k-1}$ or cycles $C_{2m+2}, C_{2m+4}, \ldots, C_{2m+\ell}$ where ℓ is the largest even integer less than k.*

Proof. Let $L_i = L_i(T)$ and suppose $G[L_i]$ contains a cycle C of length k plus a chord. Let U be a minimal subtree of T whose set of leaves is $L = V(C) \cap L_i$, so that U

branches at its root u. Let $r \leq i$ be the height of U. If A is the set of vertices of L in one branch, and B is the set of vertices of L in the remaining branches, then there are paths from A to B of all possible lengths in $G[C]$, unless (A, B) is the bipartition of $G[C]$, by Lemma 3.4. In the former case, there exists a path of length $2m$ from A to B in U through u, which combines with all the paths from A to B in $G[C]$ to obtain cycles $C_{2m+1}, C_{2m+2}, \ldots, C_{2m+k-1} \subset G$. If (A, B) is the bipartite of $G[C]$, we instead pick a descendant v of u in U, and let A' be the set of all leaves which are descendants of v, and $B' = L \backslash A$. Since (A, B) is the bipartition of $G[C]$, (A', B') is not, and using Lemma 3.4, we again find $C_{2m+1}, C_{2m+2}, \ldots, C_{2m+k-1}$. A similar argument applies if $G(L_i, L_{i+1})$ contains a cycle C of length k plus a chord. □

4 Moore graphs and extremal graphs

We recall that if G is an n-vertex graph of girth at least g, then $n \geq \nu(d, g)$ where

$$\nu(d, g) = \begin{cases} 1 + \sum_{i=1}^{\frac{1}{2}(g-1)} d(d-1)^{i-1} & \text{if } g \text{ is odd} \\ 2\sum_{i=0}^{\frac{1}{2}(g-2)} (d-1)^i & \text{if } g \text{ is even} \end{cases}$$

This is the Moore bound, and a d-regular graph G with girth g and $\nu(d, g)$ vertices will be called a Moore graph. For $g = 2k + 1$, equality holds in the Moore bound only if G is a d-regular graph of diameter k, and for $g = 2k$, equality holds only if G is a d-regular bipartite graph of diameter $k + 1$. Moore graphs of girth $g \in \{3, 4\}$ are clearly complete graphs and complete bipartite graphs. The interesting cases are the existence of d-regular Moore graphs of girth g when $d \geq 3$ and $g \geq 5$. Hoffman and Singleton [67] used linear algebra to show that if a d-regular Moore graph of girth five exists, then $d \in \{2, 3, 7, 57\}$. The unique d-regular Moore graphs of girth five for $d = 3$ and $d = 7$ are the Petersen graph and Hoffman-Singleton graph. It is not known if there exists a 57-regular Moore graph of girth five; such a graph would have 3250 vertices. Damerell [23] showed that for $d \geq 3$, no d-regular Moore graphs of odd girth $g \geq 7$ exist. However, it appears difficult to rule out the possibility that the Moore bound is asymptotically tight: in other words, for all $k \geq 2$, $\mathrm{ex}(n, \mathscr{C}_{2k}) \sim \frac{1}{2}n^{1+\frac{1}{k}}$.

4.1 Moore graphs of even girth

In the case of even girth $g \geq 6$, the Feit-Higman Theorem [129] shows that for $d \geq 3$, d-regular Moore graphs of even girth g do not exist unless $g \in \{6, 8, 12\}$. The existence of Moore graphs in these cases is due to the existence of *generalized polygons*, established by Tits [128]. From a combinatorial point of view, a *generalized k-gon of order q* is a $(q+1)$-regular bipartite graph with $q^{k-1}+q^{k-2}+\cdots+q+1$ vertices in each part and no cycles of length at most $k-1$. We give brief combinatorial descriptions of the constructions of Moore graphs of girth six and eight, which require some projective geometry. We refer the reader to Beutelspacher and Rosenbaum [6] for an elegant account of projective planes, and van Maldeghem [129] for a complete account of generalized polygons. To describe Moore graphs of even girth, we need some elementary linear algebra.

For a vector space V, let $\left[{V \atop k}\right]$ denote the set of all k-dimensional subspaces of V. For any $n \geq k \geq \ell$ and an n-dimensional vector space V over the finite field $\mathbb{F}_q$ of order q, one may form a bipartite graph $G_q[n,k,\ell]$ with parts $\binom{V}{\ell}$ and $\binom{V}{k}$, and where an ℓ-space and a k-space form an edge if the ℓ-space is a subspace of the k-space. These graphs are rich sources of extremal graphs, and in particular, it is easy to see that $G_q[3,2,1]$ is a Moore graph of girth six.

Theorem 4.1. *For any prime power q, $G_q[3,2,1]$ is a $(q+1)$-regular Moore graph of girth six. In particular, if $n=q^2+q+1$, then $z(n,n,\mathscr{C}_4)=(q+1)n=\mathrm{ex}(2n,\mathscr{C}_5)$ and for all n,*

$$z(n,n,\mathscr{C}_4) \sim n^{\frac{3}{2}} \quad \textit{and} \quad \mathrm{ex}(n,\mathscr{C}_5) \sim (\tfrac{n}{2})^{\frac{3}{2}}.$$

The lack of quadrilaterals comes simply from the fact that any pair of two-dimensional subspaces of V intersects in a unique one-dimensional space (any two lines share exactly one point). Next we come to Moore graphs of girth eight. The following was proved by Benson [5]:

Theorem 4.2. *Let V be a 5-dimensional vector space over $\mathbb{F}_q$, and let $\mathscr{P}=\{P\in \left[{V \atop 1}\right] : P \in P^{\perp}\}$ and let $\mathscr{L}=\{L \in \left[{V \atop 2}\right] : L \subset U\}$. Let G be the bipartite graph with parts $\mathscr{P}$ and $\mathscr{L}$ such that $P\in\mathscr{P}$ is joined to $L\in\mathscr{L}$ if $P\subset L$. Then G is a $(q+1)$-regular Moore graph of girth eight. In particular, if $n=q^3+q^2+q+1$, then $z(n,n,\mathscr{C}_6)=(q+1)n=\mathrm{ex}(2n,\mathscr{C}_7)$ and for all n,*

$$z(n,n,\mathscr{C}_6) \sim n^{\frac{4}{3}} \quad \textit{and} \quad \mathrm{ex}(n,\mathscr{C}_7) \sim (\tfrac{n}{2})^{\frac{4}{3}}.$$

The lack of hexagons in the construction comes simply from the fact that a hexagon corresponds to three self-orthogonal and pairwise orthogonal one-dimensional subspaces P_1,P_2,P_3, so if W is the subspace of V generated by P_1,P_2,P_3, then $W\subseteq W^{\perp}$. Since $\dim(W)+\dim(W^{\perp})\leq \dim(V)$, we conclude $\dim(W)\leq 2$ which implies $P_i=P_j$ for some $i\neq j$ (see [84] for details). We also point out that a construction exists for Moore graphs of girth twelve; the details are supplied in [5]. These Moore graphs give $\mathrm{ex}(n,\mathscr{C}_{2k+1}) \sim (\frac{n}{2})^{1+\frac{1}{k}}$ for $k\in\{2,3,5\}$, as claimed by Theorem 1. For the case of extremal $\tilde{\mathscr{C}}_{2k}$-free graphs, one already needs a more sophisticated construction.

4.2 Erdős-Rényi-Sós graphs

In this section, we briefly discuss constructions of dense C_{2k}-free graphs. The constructions all have a strong algebraic flavor, and for a large part the adjacency of vertices can be described by low-degree polynomial equations over finite fields. We begin with the extensively studied case of quadrilateral-free graphs [44, 54, 77, 107]. In this case, the theorem of Füredi [54] determines $\mathrm{ex}(n, C_4)$ whenever $n = q^2 + q + 1$ and $q > 13$ is a prime power. We give a brief description of an extremal construction. Let V be a three-dimensional vector space over $\mathbb{F}_q$. Let G_q be the graph with $V(G_q) = \left[{V \atop 1}\right]$ and the edge set of G_q is the set of pairs $\{U, W\}$ such that U and W are distinct orthogonal subspaces. These graphs are called *Erdős-Rényi-Sós polarity graphs*. It is straightforward to check that G_q does not have any quadrilaterals (although it contains triangles). Also, G_q has exactly q^2+q+1 vertices and every one-dimensional subspace in $\binom{V}{1}$ that is not orthogonal to itself has degree $q + 1$ in the graph. A counting argument shows that there are exactly $q + 1$ self-orthogonal subspaces, and as vertices these have degree q, and so

$$e(G_q) = \tfrac{1}{2}(q + 1)q + \tfrac{1}{2}q^2(q + 1) = \tfrac{1}{2}q(q + 1)^2.$$

Once more the distribution of primes shows $\mathrm{ex}(n, C_4) \gtrsim \frac{1}{2}n^{3/2}$. The graphs G_q provide extremal C_4-free graphs with $q^2 + q + 1$ vertices for $q > 13$, according to Theorem 5. The main difficulty is in improving the upper bound $d(d - 1) \leq n - 1$ for a C_4-free n-vertex graph of average degree d, since when $n = q^2 + q + 1$ this only gives $\mathrm{ex}(n, C_4) \leq \frac{1}{2}(q + 1)n$. We remark that the graphs G_q may be described geometrically by *polarities of projective planes* – see Lazebnik, Ustimenko, and Woldar [90] for details on various polarity graphs.

4.3 Algebraic equations

Let $f_1, f_2, \ldots, f_{k-1}$ be polynomials in $2k$ variables, which we denote by $x_1, x_2, \ldots, x_k, y_1, y_2, \ldots, y_k$, over $\mathbb{F}_q$. Then the graph $\mathscr{G}(f_1, f_2, \ldots, f_{k-1})$ is the bipartite graph with parts $X = \mathbb{F}_q^k$ and $Y = \mathbb{F}_q^k$ such that $x \in X$ is adjacent to $y \in Y$ if $f_1(x, y) = f_2(x, y) = \cdots = f_{k-1}(x, y) = 0$. A simple construction of C_{2k}-free graphs $W_{k,q}$ for $k \in \{2, 3, 5\}$ is due to Wenger [134]: $W_{k,q}$ is the instance of $\mathscr{G}(f_1, f_2, \ldots, f_{k-1})$ with $f_i(x, y) = x_{i+1} + y_{i+1} - x_1 y_1^i$ for $i = 1, 2, \ldots, k - 1$. The elegant proof [134] that $W_{k,q}$ is C_{2k}-free only uses basic properties of Vandermonde determinants. The graph $W_{2,q}$ corresponds to the incidence graph of points and lines in the affine plane. Since the definition is symmetric, in fact one gets an asymptotically extremal C_4-free graph with vertex set $\mathbb{F}_q^2$ and where x is joined to y if $x_2 + y_2 = x_1 y_1$. The constructions of Lazebnik, Ustimenko, and Woldar [90] are of this type, however their equations actually give disconnected graphs $D_{k,q}$ whose components $CD_{k,q}$ are the graphs which give the lower bounds in Theorem 3. We do not investigate further properties of these graphs and the remarkable Lubotzky-Phillips-Sarnak constructions of Ramanujan graphs [86] here.

It is natural to ask, however, whether there exist $f_1, f_2, \dots, f_{k-1}$ such that $\mathscr{G}(f_1, f_2, \dots, f_{k-1})$ is C_{2k}-free, or even perhaps $\mathscr{C}_{2k}$-free (the Wenger graph $W_{q,5}$ is C_{10}-free but contains C_8). This is open even for $k = 4$. Using the Lang-Weil Bound from algebraic geometry, Conlon [21] proved that if the polynomials $f_1, f_2, \dots, f_{k-1}$ are chosen randomly and independently with appropriate degree, then there exists t_k such that with positive probability $\mathscr{G}(f_1, f_2, \dots, f_k)$ has at least q^{k+1} edges and no two vertices are the ends of at least t_k paths of length k. Let $\theta_{k,t}$ denote a *theta graph*, consisting of t internally disjoint paths of length k between two vertices. Together with a result of Faudree and Simonovits [49], Conlon obtains the following:

Theorem 4.3. *For every $k \geq 2$, there exists t_k such that* $\mathrm{ex}(n, \theta_{k,t}) = \Theta(n^{1+\frac{1}{k}})$ *for $t \geq t_k$.*

Unfortunately, the constant t_k is rather large [21]. For $k = 4$, the author and Williford [133] showed that $\mathrm{ex}(n, \theta_{4,3}) = \Theta(n^{\frac{5}{4}})$ so we can take $t_4 = 3$, and Mellinger and Mubayi [94] showed $\mathrm{ex}(n, \theta_{7,3}) = \Theta(n^{\frac{8}{7}})$, so we can also take $t_7 = 3$.

4.4 Combinatorial number theory

A *B_k-set* in an abelian group Γ a set $A \subset \Gamma$ such that if $\{a_1, a_2, \dots, a_k\} \subset A$ and $\{b_1, b_2, \dots, b_k\}$ have the same sum, then $\{a_1, a_2, \dots, a_k\} = \{b_1, b_2, \dots, b_k\}$. In the case $k = 2$, A is called a *Sidon set* (see O'Bryant [102] for a survey of Sidon sets). A well-known conjecture of Erdős and Turán [43] states that there exists a constant $c > 0$ such that if $A \subset \{1, 2, \dots, N\}$ is a Sidon set, then $|A| \leq N^{\frac{1}{2}} + c$. This is certainly true if $A \subset \mathbb{Z}_N$, for in that case $|A|(|A| - 1) \leq \frac{1}{2}(\sqrt{4N-3} + 1)$. Sidon sets provide a way of describing certain projective planes: Singer [116] gave a construction for each odd prime q of a Sidon subset A of optimal size $q + 1$ in the cyclic group $\mathbb{Z}_N$ when $N = q^2 + q + 1$ – a Sidon set A of size $q + 1$ called a *perfect difference set*. Then the additive translates $A + \lambda : \lambda \in \mathbb{Z}_N$ of A form the set of lines of a projective plane of order q. Moore graphs of girth six can then be described by letting the parts of the bipartite graph be copies X and Y of $\mathbb{Z}_N$, and where $x \in X$ is joined to $y \in Y$ if $x + a = y$ for some $a \in A$. This is called the *bipartite Cayley graph* of A.

If Γ is an abelian group and $A \subseteq \Gamma$ is a B_k-set, then one can consider the bipartite Cayley graph $G(\Gamma, A)$ of A whose parts are copies of Γ. It turns out that for $k = 3$, this graph contains no $\theta_{3,3}$ but, however, does contain a cycle of length six. For $k > 3$, the graph contains $\theta_{k,t}$ where $t = \Omega(|\Gamma|)$, regardless of the structure of the ambient abelian group Γ. Bose and Chowla [14] were the first to give for each k a construction of a B_k-set A in $\mathbb{Z}_n$ with $|A| \sim n^{\frac{1}{k}}$, for infinitely many n. It is not hard to see that a B_{2k}-set $A \subset \mathbb{Z}_n$ has $|A| \lesssim (k!)^{\frac{1}{k}} n^{\frac{1}{2k}}$ and a B_{2k-1}-set $A \subset \mathbb{Z}_n$ has $|A| \lesssim (k!(k-1)!)^{\frac{1}{2k-1}} n^{\frac{1}{2k-1}}$. One may ask whether every B_k-set in $\mathbb{Z}_n$ has size at most $cn^{\frac{1}{k}}$ for some constant $c > 0$ independent of k, which is perhaps reminiscent of Conjecture III. It is however even an open question to show that every B_k-set in $\mathbb{Z}_n$ has size $o(k) \cdot n^{\frac{1}{k}}$ as $k \to \infty$.

In order to avoid $2k$-cycles in the bipartite Cayley graph $G(\Gamma, A)$ for a group Γ and a set $A \subset \Gamma$, one requires a version of B_k-sets in non-abelian groups: if Γ is a non-abelian group written multiplicatively, then a B_k-set in Γ is a set A such that whenever $a_1, a_2, \ldots, a_{2k}$ are elements of A such that $a_1 a_2^{-1} a_3 a_4^{-1} \ldots a_{2k-1} a_{2k}^{-1}$ is the identity, then $a_i = a_{i+1}$ for some $i < 2k$. Then the bipartite Cayley graph of a B_k-set in a non-abelian group is C_{2k}-free. This leads to the problem of determining whether there exists a group Γ_n and a B_k-set $A \subset \Gamma_n$ of size $\Theta(n^{\frac{1}{k}})$. A promising approach was given by Odlyzko and Smith [103], who define an *S_k-set* in a group Γ to be a set $A \subset \Gamma$ such that whenever $a_1 a_2 \ldots a_k = b_1 b_2 \ldots b_k$ with $a_i, b_i \in A$, then $(a_1, a_2, \ldots, a_k) = (b_1, b_2, \ldots, b_k)$. Odlyzo and Smith [103] ingeniously translate the Bose-Chowla B_k-sets in $\mathbb{Z}_n$ for certain n to S_k-sets A of asymptotically optimal size in some symmetric group.

Finally, we raise the following question from [87, 126], which is related to Conjecture XIX: for each $k \geq 1$, determine the maximum size $s_k(n)$ of a Sidon set $A \subset \mathbb{Z}_n$ such that if $a - b = i(c - d)$ for some $i \in \{2, \ldots, k\}$, then $a = b$ and $c = d$. In particular, for each $k \geq 1$, we believe that $s_k(n) = \Theta(n^{\frac{1}{2}})$, and this is open even for $k = 3$. In [127], it is shown that $s_k(n) = n^{\frac{1}{2} - o(1)}$ for all $k \geq 1$, and in [87], it is shown that $\frac{1}{2} n^{\frac{1}{2}} \lesssim s_2(n) \lesssim (\frac{n}{2})^{\frac{1}{2}}$.

5 Proof of Theorem 4

The aim of this short section is to prove Theorem 4. We show specifically that $\mathrm{ex}(n, C_{2k}) < (k-1)n^{1+\frac{1}{k}} + 4(k-1)$. This follows the proofs given in [105, 132]. Let G be an n-vertex C_{2k}-free graph with at least $(k-1)n^{1+\frac{1}{k}} + 4(k-1)n$ edges. By Proposition 3.1, G has a connected subgraph H of minimum degree at least $d > (k-1)n^{\frac{1}{k}} + 4(k-1)$. Let T be a breadth-first search tree in this subgraph with $L_i = L_i(T)$. We claim that for all $i < k$,

$$e(L_i) \leq (k-1)|L_i| \quad \text{and} \quad e(L_i, L_{i+1}) \leq (k-1)(|L_{i+1}| + |L_i|)$$

Suppose, for a contradiction, that $e(L_i) > (k-1)|L_i|$. By the Erdős-Gallai Theorem (Proposition 3.2), $G[L_i]$ contains a cycle C of length at least $2k-1$ with a chord. By Lemma 3.5, G contains cycles $C_{2m+1}, C_{2m+2}, \ldots, C_{2m+2k-2}$. Since $m \leq k-1$, one of these cycles has length exactly $2k$, a contradiction. Similarly, $e(L_i, L_{i+1}) \leq (k-1)(|L_{i+1}| + |L_i|)$. We now claim that $|L_i| \geq n^{\frac{1}{k}} |L_{i-1}|$ for all $i \in \{1, 2, \ldots, k\}$. This is a contradiction, since then $|L_k| \geq n$. The claim is true for $i = 1$ since H has minimum degree at least d. Having proved the claim for all $j < i$, we have

$$e(L_{i-1}, L_i) \geq d|L_{i-1}| - 2e(L_{i-1}) - e(L_{i-2}, L_{i-1}) \geq d|L_{i-1}| - 4(k-1)|L_{i-1}|.$$

Since $e(L_{i-1}, L_i) \leq (k-1)(|L_i|+|L_{i-1}|)$, the definition of d shows $|L_{i-1}| > n^{\frac{1}{k}}|L_{i-1}|$, as required. □

Some minor improvements are possible here. However, the substantial new idea in Bukh and Jiang [15] to get

$$\mathrm{ex}(n, C_{2k}) \leq 80\sqrt{k \log k} \cdot n^{1+\frac{1}{k}} + 10k^2 n$$

for $n \geq (2k)^{8k^2}$ is to consider subgraphs which span three consecutive layers of a breadth-first search tree instead of at most two layers as in the proof above. We remark that the proof above in fact shows that any n-vertex graph with minimum degree at least $cn^{\frac{1}{k}}$ has a subgraph of average degree at least c and radius at most k.

6 Consecutive cycle lengths

The aim of this section is to prove Theorems 11–14 and Theorem 20.

6.1 *Proof of Theorem 12*

Given the n-vertex graph G with at least $3kn$ edges, we aim to find cycles of k consecutive even lengths in G. We may assume G is connected, and let T be a breadth-first search tree in G, $L_i := L_i(T)$. Then

$$e(G) = \sum_{i=0}^{\infty}(e(L_i) + e(L_i, L_{i+1})) \quad \text{and} \quad \sum_{i=0}^{\infty} e(L_i, L_{i+1}) \leq 2n.$$

If $e(L_i) \geq k|L_i|$ for some i, then by Proposition 3.2, we find a cycle C of length at least $2k + 1$ with a chord, and Lemma 3.5 shows that G contains cycles $C_{2m}, C_{2m+2}, \ldots, C_{2m+2k-2}$, as required. Otherwise we conclude $\sum_{i=0}^{\infty} e(L_i) < kn$ which implies $e(L_i, L_{i+1}) > k(|L_i| + |L_{i+1}|)$ for some i, since $e(G) \geq 3kn$. Once again, Proposition 3.2 and then Lemma 3.5 apply to complete the proof. □

6.2 *Proof of Theorem 20*

Let G be a graph of chromatic number at least $2k - 1$. Let H be a component of G of chromatic number at least $2k - 1$. If T is a breadth-first search tree in H, then there exists i such that $H[L_i(T)]$ has chromatic number at least k. Now any graph of chromatic number $k \geq 4$ contains a cycle C inducing a non-bipartite graph with at least k vertices, such that C has at least one chord (see, for instance, [120]). Now as in the proof of Theorem 12, G contains cycles $C_{2m+1}, C_{2m+2}, \ldots, C_{2m+k-1}$ for some integer m. □

6.3 Proof of Theorem 14

Our starting point for the proof of Theorem 14 is the Moore bound (see Section 1.2). The following straightforward lemma states that graphs of large minimum degree and girth "expand" on small sets, due to the Moore bound.

Lemma 6.1. *Let G be a graph of minimum degree at least $6(d+1)$. If G has girth at least $2g+1$, then for every $X \subset V(G)$ of size at most $\frac{1}{3}d^g$, $|N(X)| > 3|X|$. If G is $C_{2\ell}$-free, then for every $X \subset V(G)$ of size at most $(\frac{d}{3\ell})^\ell$, $|N(X)| > 3|X|$.*

Proof. Suppose $|N(X)| \leq 3|X|$ for some $X \subset V(G)$. Let H be the subgraph of G spanned by $Y = N(X)$. Then $|Y| \leq 3|X|$ and

$$e(H) \geq \frac{1}{2}\sum_{x\in X} d(x) \geq 3(d+1)|X| \geq (d+1)|Y|.$$

By Proposition 3.1, H contains a subgraph Γ with minimum degree $d+1$. Applying the Moore bound to Γ, we obtain:

$$3|X| \;\geq\; |Y| \;\geq\; |V(\Gamma)| \;\geq\; 1 \,+\, (d+1)\sum_{i<g} d^i \;>\; d^g$$

and therefore $|X| > \frac{1}{3}d^g$, as required. For the second statement of the lemma, we use Theorem 4 instead of the Moore Bound. □

We now prove Theorem 14. Let G be a graph of average degree at least $48(d+1)$ and girth at least $2g+1$. Pass to a bipartite subgraph H of G with average degree at least $24(d+1)$. In a component of H of average degree least $24(d+1)$, there exists a breadth-first search tree T such that for some i, $G(L_i(T), L_{i+1}(T))$ is a bipartite subgraph F of average degree at least $12(d+1)$. Then F contains a $6(d+1)$-core Γ by Proposition 3.1. By Lemma 6.1 we have that $|N(X)| > 3|X|$ for every $X \subset V(\Gamma)$ of size at most $\frac{1}{3}d^g$. Hence Γ contains a cycle of length d^g with a chord, by Pósa's Lemma, Proposition 3.3. Now Lemma 3.5 gives the required cycle lengths. □

6.4 Proof of Theorem 11

In this section, we prove Theorem 11. Let G be an n-vertex graph containing no cycle of length 2ℓ modulo k. We show $e(G) \leq 48\ell k^{\frac{1}{\ell}} n$. If not, then there is a breadth-first search tree T in a subgraph of G such that for i, $G(L_i(T), L_{i+1}(T))$ has average degree at least $12\ell k^{\frac{1}{\ell}}$. This bipartite graph contains a subgraph F of minimum degree at least $6(\ell k^{\frac{1}{\ell}} + 1) := 6(d+1)$ by Proposition 3.1. By Lemma 6.1, for every subset X of F of size at least k, $|N(X)| > 3k$. By Pósa's Lemma (Proposition 3.3), F contains a cycle of length at least $3k$. By Lemma 3.5,

G contains $C_{2m}, C_{2m+2}, \ldots, C_{2m+2k-2}$ for some m, and therefore G contains a cycle of every even length modulo k, a contradiction. □

7 Distribution of cycle lengths

We write $\pi < \sigma$ to denote that π is a subsequence of σ. The following theorem from [119] is the main technical result which gives Theorem 16, and a similar result in [120] gives Theorem 22.

Theorem 7.1. *For any infinite increasing sequence σ of positive even integers and for any n-vertex graph G, if G is $\mathscr{C}_\sigma$-free, then G has average degree at most*

$$d(\sigma) := \inf_{\substack{\pi<\sigma \\ r\geq 1}} \exp\left(6r + \sum_{i=1}^{r} \frac{2\log\Delta_i}{\pi_{i-1}} + \frac{2\log n}{\pi_r}\right),$$

where $\pi_0 := 1$, $\Delta_1 := \pi_1$, and $\Delta_i = \max\{\sigma_j - \sigma_{j-1} : \sigma_j \leq \pi_i\}$ for $i \geq 2$.

The same theorem holds for any increasing sequence σ (which includes odd numbers) with average degree replaced by independence ratio. This then gives Theorem 22 and Theorem 19. For the sake of completeness, we derive Theorem 16 from Theorem 7.1.

7.1 Proof of Theorem 16

Let G be an n-vertex $\mathscr{C}_\sigma$-free graph, where $\sigma_i \leq 2^{\sigma_{i-1}}$ for $i \geq 1$ and $\sigma_1 = 1$. Let $r = \log^* n$ and let $\pi < \sigma$ be chosen so that $2^{\pi_{i-1}} \leq \pi_i \leq (2c)^{\pi_{i-1}}$ for all $i \geq 2$. Note that $\pi_r \geq n$. Then, since $\Delta_i \leq \pi_i$, $d(\sigma) \leq \exp(6r + 2r\log(2) + 2) \leq \exp(8\log^* n)$. This proves Theorem 16. □

7.2 Proof of Theorem 7.1

Fix $\pi < \sigma$, and let $\mathcal{P}_i, i \geq 1$ denote the monotone property of graphs containing no cycle of length σ_j for all $\sigma_j \leq \pi_i$, and recall $\Delta_i = \max\{\sigma_j - \sigma_{j-1} : \sigma_j \leq \pi_i\}$. To prove Theorem 7.1, we first prove the following claim.

Claim. *Let $(a_i)_{i\geq 1}$ be positive real numbers such that $a_1 = 4\pi_1$ and, for all $i \geq 2$,*

$$\pi_{i-1}\log\frac{a_i}{288a_{i-1}} \geq 2\log\Delta_i.$$

Then, for every n-vertex graph $G \in \mathcal{P}_i$, $e(G) \leq a_i n^{1+\frac{2}{\pi_i}} := h_i(n)$

We proceed by induction on i. Theorem 12 implies the claim for $i = 1$. Suppose we have proved the claim for $j < i$, and let $G \in \mathcal{P}_i$ be an n-vertex graph with $e(G) > h_i(n)$. By the induction hypothesis we have that every m-vertex graph in $\mathcal{P}_{i-1}$ has at most $h_{i-1}(m)$ edges. By an appropriate analogue of Lemma 6.1, we have that for any graph $F \in \mathcal{P}_{i-1}$ with minimum degree d every subset $X \subset V(F)$ of size at most

$$f(d) = \Big(\frac{d}{18a_{i-1}}\Big)^{\frac{1}{2}\pi_{i-1}}$$

has $|N(X)| > 3|X|$. Since G has n vertices and $e(G) \geq h_i(n)$, the proof of Theorem 4 gives a subgraph Γ of G of average degree at least a_i and radius at most $\frac{1}{2}\pi_i$. Note that Γ has property $\mathcal{P}_i$ and thus has also property $\mathcal{P}_{i-1}$. Applying Pósa's Lemma as in the proof of Theorem 14, we find a cycle of length at least $3f(\frac{a_i}{16}) + 1$ between two layers of some breadth-first search tree in Γ. By Lemma 3.5, Γ contains cycles of at least $3f(\frac{a_i}{16})$ consecutive even lengths, the shortest of which has length at most π_i. Since $\Gamma \in \mathcal{P}_i$, there must be less than Δ_i of these consecutive even lengths, otherwise Γ contains a cycle of length σ_j for some $j \leq \pi_i$. Therefore $f(\frac{a_i}{16}) < 3f(\frac{a_i}{16}) < \Delta_i$ which contradicts the lower bound on a_i in the claim.

Now we finish the proof of prove Theorem 7.1. Recall that $\pi_0 = 1$, $\Delta_1 = \pi_1$ and let

$$a_r = 4\pi_1(288)^{r-1}\prod_{i=2}^{r}\exp\left(\frac{2\log\Delta_i}{\pi_{i-1}}\right) < \frac{1}{2}\exp\left(6r + \sum_{i=1}^{r}\frac{2\log\Delta_i}{\pi_{i-1}}\right).$$

Since a_r satisfies the condition of the claim, we have that the estimate on the number of edges of G from this claim is valid for any r. Therefore the average degree of G is at most $\inf_{r\geq 1}\frac{2}{n}h_r(n)$, which gives the bound in Theorem 7.1 by minimizing over all $\pi < \sigma$. □

7.3 Proof of Theorem 18

We are given an n-vertex graph G with average degree at least d and we have to show $\phi(G) \geq \frac{1}{2}\log d - 8\log^* n$. This is clearly true if $d \leq \exp(8\log^* n)$ so we assume $d > \exp(8\log^* n)$. We let $s = \lceil d\exp(-8\log^* n)\rceil$, and consider the disjoint sets S_i of even numbers in $\{s^i, s^i+1, \ldots, s^{i+1}-1\}$ for $i \geq 0$. If for some i we have $S_i \subset C(G)$, then $\phi(G) \geq \frac{1}{2}\log s \geq \frac{1}{2}\log d - 8\log^* n$, and the proof is complete. Otherwise, for each i we pick $\sigma_i \in S_i\backslash C(G)$ and $\sigma_1 = 1$. We have defined a sequence σ with $\sigma_i \leq s\sigma_{i+1}$. Let π be a subsequence of σ such that $\log_2 \pi_r \leq \pi_{r-1}$ and such that π_r is contained in the interval S_i for which $T(r) \geq s^{i+1} - 1$ but $T(r) < s^{i+2} - 1$, where $T(r)$ is a tower of twos of height r. We also let $\pi_1 = 1$. Applying Theorem 7.1 with this sequence π and choosing r such that $\pi_r > 2\log n$, we have $r \leq \log^* n$

and the average degree of G is at most $\exp(8\log^* n) < s\exp(8\log^* n) < d$. This contradiction completes the proof. □

7.4 Constructions

The purpose of this section is to show that Theorem 7.1 is best possible up to a factor roughly $\frac{3}{2}$ in the exponent. Similar constructions starting with graphs of large chromatic number and girth [83, 111, 120] show that Theorem 22 is tight up to a constant factor in the exponent – see [119, 120] for details.

Let $n_0 = 2$ and let $G_1 = C_4$. We construct a sequence of graphs $(G_i)_{i\geq 1}$ such that $|V(G_i)| = n_i - 2$, G_i is $(d_i - 1)$-regular, and G_i has girth larger than n_{i-1}, where n_i is even for all i. By known explicit constructions of dense graphs of large girth (see, for instance, Theorem 3 and [88]), we may take $\log n_i \sim \frac{3}{4}n_{i-1}\log d_i$. Now let σ be defined by $\sigma_i = n_i$. Since G_i has girth larger than n_{i-1} and G_i has order less than n_i, each G_i is $\mathscr{C}_\sigma$-free and has average degree d_i. We choose d_i so that $\log_2\log_2\log_2 d_i \geq i$. Now we compute the bound in Theorem 7.1. If we take $\pi = \sigma$ and $r = i$, then we have $\Delta_i \sim n_i$ as $i \to \infty$, and so as $i \to \infty$, the bound in Theorem 7.1 is:

$$\begin{aligned}\exp\Big(6i + \sum_{j=1}^{i}\frac{2\log\Delta_j}{\pi_{j-1}} + \frac{2\log n_i}{n_i}\Big) &= \exp\Big(6i + (1+o(1))\sum_{j=1}^{i}\frac{2\log n_j}{n_{j-1}}\Big)\\ &= \exp\Big(6i + (\tfrac{3}{2}+o(1))\sum_{j=1}^{i}\log d_j\Big) = d_i^{\frac{3}{2}+o(1)}.\end{aligned}$$

Note that we chose d_i to be at least triple exponential in i, so that the $\log d_i$ term is the leading term in the exponent in the above computation. Since G_i has average degree $d_i + 1$, the bound given by Theorem 7.1 is tight up to a factor $\frac{3}{2} + o(1)$ in the exponent.

Acknowledgements Research of the author Jacques Verstraëte was supported by NSF Grant DMS-1362650.

References

1. P. Allen, P. Keevash, B. Sudakov, J. Verstraëte, Turán numbers of bipartite graphs plus an odd cycle. J. Comb. Theory Ser. B **106**, 134–162 (2014)
2. N. Alon, C. Shikhelman, Triangles in H-free graphs. arxiv.org/abs/1409.4192
3. N. Alon, S. Hoory, N. Linial, The Moore bound for irregular graphs. Graphs Comb. **18**(1), 53–57 (2002)

4. J. Balogh, R. Morris, W. Samotij, Independent sets in hypergraphs. J. Am. Math. Soc. **28**(3), 669–709 (2015)
5. C.T. Benson, Minimal regular graphs of girths eight and twelve. Can. J. Math. **18**, 1091–1094 (1966)
6. A. Beutelspacher, U. Rosenbaum, *Projective Geometry: From Foundations to Applications* (Cambridge University Press, Cambridge, 1998)
7. T. Bohman, P. Keevash, Dynamic concentration of the triangle-free process. arxiv.org/abs/1302.5963
8. B. Bollobás, Cycles modulo k. Bull. Lond. Math. Soc. **9**, 97–98 (1977)
9. B. Bollobás, *Extremal Graph Theory* (Academic Press, New York, 1978)
10. B. Bollobás, E. Györi, Pentagons vs. triangles. Discret. Math. **308**(19), 4332–4336 (2008)
11. J.A. Bondy, Basic graph theory: paths and circuits, in *Handbook of Combinatorics*, vol. I (Elsevier, Amsterdam, 1995), pp. 3–110
12. J.A. Bondy, M. Simonovits, Cycles of even length in graphs. J. Comb. Theory Ser. B **16**, 97–105 (1974)
13. J.A. Bondy, A. Vince, Cycles in a graph whose lengths differ by one or two. J. Graph Theory **27**(1), 11–15 (1998)
14. R.C. Bose, S. Chowla, Theorems in the additive theory of numbers. Math. Helvet. **37**, 141–147 (1962/1963)
15. B. Bukh, Z. Jiang, A bound on the number of edges in graphs without an even cycle. arxiv.org/pdf/1403.1601v1.pdf
16. Y. Caro, Y. Li, C. Rousseau, Y. Zhang, Asymptotic bounds for some bipartite graph-complete graph Ramsey numbers. Discret. Math. **220**, 51–56 (2000)
17. G.T. Chen, A. Saito, Graphs with a cycle of length divisible by three. J. Comb. Theory Ser. B **60**, 277–292 (1994)
18. F.R.K. Chung, R.L. Graham, *Erdős on Graphs: His Legacy of Unsolved Problems* (A K Peters, Wellesley, 1998)
19. C.R.J. Clapham, A. Flockart, J. Sheehan, Graphs without four-cycles. J. Graph Theory **13**, 29–47 (1989)
20. C. Collier, C.N. Graber, T. Jiang, Linear Turán numbers of r-uniform linear cycles and related Ramsey numbers. arxiv.org/abs/1404.5015
21. D. Conlon, Graphs with few paths of prescribed length between any two vertices. arxiv.org/pdf/1411.0856v1.pdf
22. D. Conlon, An extremal theorem in the hypercube. Electron. J. Comb. **17**(1), 7 (2010). Research Paper 111
23. R. Damerell, On Moore geometries. II. Math. Proc. Camb. Philos. Soc. **90**(1), 33–40 (1981)
24. S. Das, C. Lee, B. Sudakov, Rainbow Turán problem for even cycles. Eur. J. Comb. **34**, 905–915 (2013)
25. N. Dean, A. Kaneko, K. Ota, B. Toft, Cycles modulo 3. DIMACS Technical Report 91–32 (1991)
26. N. Dean, L. Lesniak, A. Saito, Cycles of length 0 modulo 4 in graphs. Discret. Math. **121**, 37–49 (1993)
27. D. de Caen, L. Székely, The maximum size of 4- and 6-cycle free bipartite graphs on m, n vertices, in *Sets, Graphs and Numbers*, Budapest, 1991. Colloquia Mathematica Societatis János Bolyai, vol. 60 (North-Holland, Amsterdam, 1992), pp. 135–142
28. A. Diwan, Cycles of even lengths modulo k. J. Graph Theory **65**(3), 246–252 (2010)
29. M.N. Ellingham, D.K. Menser, Girth, minimum degree, and circumference. J. Graph Theory **34**(3), 221–233 (2000)
30. D. Ellis, N. Linial, On regular hypergraphs of high girth. Electron. J. Comb. **21**(1), 17 (2014). Paper 1.54
31. P. Erdős, On sequences of integers no one of which divides the product of two others and on some related problems. Mitt. Forsch.-Inst. Math. Mech. Univ. Tomsk **2**, 74–82 (1938)
32. P. Erdős, Graph theory and probability. Can. J. Math. **11**, 34–38 (1959)

33. P. Erdős, Problems and results in combinatorial analysis and graph theory, in *Proof Techniques in Graph Theory* (Academic Press, New York, 1969), pp. 27–35
34. P. Erdős, Some recent progress on extremal problems in graph theory. Congr. Numer. **14** 3–14 (1975)
35. P. Erdős, Some old and new problems in various branches of combinatorics, in *Proceedings of the Tenth Southeastern Conference on Combinatorics, Graph Theory and Computing*, Boca Raton, 1979, vol. 1. Congressus Numerantium, vol. 23 (1979), pp. 19–38
36. P. Erdős, Some of my favourite unsolved problems, in *A Tribute to Paul Erdős*, ed. by A. Baker, B. Bollobás, A. Hajnal (Cambridge University Press, Cambridge, 1990), pp. 467–468
37. P. Erdős, Some of my favorite solved and unsolved problems in graph theory. Quaest. Math. **16**, 333–350 (1993)
38. P. Erdős, Some old and new problems in various branches of combinatorics, in *Graphs and Combinatorics* (Marseille, 1995). Discrete Mathematics, vol. 165/166 (1997), pp. 227–231
39. P. Erdős, T. Gallai, On maximal paths and circuits of graphs. Acta Math. Acad. Sci. Hung. **10**, 337–356 (1959)
40. P. Erdős, A. Hajnal, On chromatic number of graphs and set-systems. Acta Math. Acad. Sci. Hung. **17**, 61–99 (1966)
41. P. Erdős, M. Simonovits, Compactness results in extremal graph theory. Combinatorica **2**(3), 275–288 (1982)
42. P. Erdős, A.H. Stone, On the structure of linear graphs. Bull. Am. Math. Soc. **52**, 1087–1091 (1946)
43. P. Erdős, P. Turán, On a problem of Sidon in additive number theory, and on some related problems. J. Lond. Math. Soc. **16**, 212–216 (1941)
44. P. Erdős, A. Rényi, V.T. Sós, On a problem of graph theory. Stud. Sci. Math. Hung. **1**, 215–235 (1966)
45. P. Erdős, A. Hajnal, S. Shelah, Topics in Topology (Proc. Colloq. Keszthely, 1972). Colloquia Mathematica Societatis János Bolyai, vol. 8 (North-Holland, Amsterdam, 1974), pp. 243–245
46. P. Erdős, R. Faudree, C. Rousseau, R. Schelp, On cycle-complete graph Ramsey numbers. J. Graph Theory **2**, 53–64 (1978)
47. P. Erdős, R.J. Faudree, C. Rousseau, R. Schelp, The number of cycle lengths in graphs of given minimum degree and girth. Discret. Math. **200**, 55–60 (1999)
48. G. Fan, Distribution of cycle lengths in graphs. J. Comb. Theory Ser. B **84**, 187–202 (2002)
49. R.J. Faudree, M. Simonovits, On a class of degenerate extremal graph problems. Combinatorica **3**, 83–93 (1983)
50. F.A. Firke, P.M. Kosek, E.D. Nash, J. Williford, Extremal graphs without 4-cycles. arxiv.org/pdf/1201.4912v1.pdf
51. P. Fischer, J. Matoušek, A lower bound for families of Natarajan dimension d. J. Comb Theory Ser. A **95**(1), 189–195 (2001)
52. G. Fiz Pontiveros, S. Griffiths, R. Morris, The triangle-free process and $R(3,k)$ arxiv.org/abs/1302.6279
53. F. Foucaud, M. Krivelevich, G. Perarnau, Large subgraphs without short cycles. SIAM J. Discret. Math. **29**(1), 65–78 (2015)
54. Z. Füredi, Quadrilateral-free graphs with maximum number of edges. www.math.uiuc.edu/~z-furedi/PUBS/furediC4from1988.pdf (1988, preprint)
55. Z. Füredi, T. Jiang, Hypergraph Turán numbers of linear cycles. arXiv:1302.2387
56. Z. Füredi, L. Özkahya, On 3-uniform hypergraphs without a cycle of given length. arxiv.org/pdf/1412.8083.pdf
57. Z. Füredi, M. Simonovits, The history of degenerate (bipartite) extremal graph problems. arxiv.org/pdf/1306.5167.pdf
58. Z. Füredi, A. Naor, J. Verstraëte, On the Turán number for the hexagon. Adv. Math. **203**(2), 476–496 (2006)
59. D.K. Garnick, Y.H.H. Kwong, F. Lazebnik, Extremal graphs without three-cycles or four-cycles. J. Graph Theory **17**(5), 633–645 (1993)
60. C. Godsil, G. Royle, *Algebraic Graph Theory* (Springer, New York, 2001)

61. A. Gyárfás, Graphs with k odd cycle lengths. Discret. Math. **103**, 41–48 (1992)
62. A. Gyárfás, J. Komlós, E. Szemerédi, On the distribution of cycle lengths in graphs. J. Graph Theory **8**(4), 441–462 (1984)
63. A. Gyárfás, H.J. Prömel, E. Szemerédi, M. Voigt, On the sum of the reciprocals of cycle lengths in sparse graphs. Combinatorica **5**(1), 41–52 (1986)
64. E. Györi, C_6-free bipartite graphs and product representation of squares, in *Graphs and Combinatorics* (Marseille, 1995). Discrete Mathematics, vol. 165/166 (1997), pp. 371–375
65. E. Györi, Triangle-free hypergraphs. Comb. Probab. Comput. **15**(1–2), 185–191 (2006)
66. E. Györi, N. Lemons, Hypergraphs with no cycle of a given length. Comb. Probab. Comput. **21**(1–2), 193–201 (2012)
67. A. Hoffman, R. Singleton, On Moore graphs with diameters 2 and 3. IBM J. Res. Develop. **4**, 497–504 (1960)
68. S. Hoory, The size of bipartite graphs with a given girth. J. Comb. Theory Ser. B **86**(2), 215–220 (2002)
69. C. Hyltén-Cavallius, On a combinatorial problem. Colloq. Math. **6**, 59–65 (1958)
70. R. Häggkvist, A. Scott, Arithmetic progressions of cycles (Preprint)
71. P. Keevash, D. Mubayi, B. Sudakov, J. Verstraëte, On a conjecture of Erdős and Simonovits: even cycles. Combinatorica **33**, 699–732 (2013)
72. J. Kim, The Ramsey number $R(3;t)$ has order of magnitude $t^2/\log t$. Random Struct. Algorithm. **7**, 173–207 (1995)
73. D. Kleitman, K. Winston, On the number of graphs without 4-cycles. Discret. Math. **41**, 167–172 (1982)
74. G.N. Kopylov, On maximal path and cycles in a graph. Dokl. Akad. Nauk SSSR **234**, 19–21 (1977) [Soviet Math. Dokl. **18**, 593–596, 1977]
75. A.V. Kostochka, D. Mubayi, J. Verstraëte, Turán problems and shadows I: paths and cycles. J. Comb. Theory Ser. A **129**, 57–79 (2015)
76. A.V. Kostochka, B. Sudakov, J. Verstraëte, Cycles in triangle-free graphs of large chromatic number (to appear)
77. T. Kövari, V.T. Sós, P. Turán, On a problem of K. Zarankiewicz. Colloq. Math. **3**, 50–57 (1954)
78. M. Krivelevich, B. Sudakov, Sparse pseudo-random graphs are Hamiltonian. J. Graph Theory **42**, 17–33 (2003)
79. D. Kühn, D. Osthus, Every graph of sufficiently large average degree contains a C_4-free subgraph of large average degree. Combinatorica **24**, 155–162 (2004)
80. D. Kühn, D. Osthus, Four-cycles in graphs without a given even cycle. J. Graph Theory **48**, 147–156 (2005)
81. T. Lam, J. Verstraëte, A note on graphs without short even cycles. Electron. J. Comb. **12**, Note 5, 6 (2005)
82. C.-H. Liu, J. Ma, Cycle lengths and minimum degree of graphs, arXiv:1508.07912
83. L. Lovász, On chromatic number of finite set-systems. Acta Math. Acad. Sci. Hung. **19**, 59–67 (1968)
84. L. Lovász, Kneser's conjecture, chromatic number, and homotopy. J. Comb. Theory Ser. A **25**(3), 319–324 (1978)
85. L. Lovász, *Combinatorial Problems and Exercises*, 2nd edn. (North-Holland, Amsterdam, 1993)
86. A. Lubotsky, M. Phillips, P. Sarnak, Ramanujan graphs. Combinatorica **9**, 261–277 (1988)
87. F. Lazebnik, J. Verstraëte, On hypergraphs of girth five. Electron. J. Comb. **10**, Research Paper 25, 15 (2003)
88. F. Lazebnik, V.A. Ustimenko, A.J. Woldar, New constructions of Bipartite graphs on m, n vertices, with many edges, and without small cycles. J. Comb. Theory Ser. B **61**(1), 111–117 (1994)
89. F. Lazebnik, V.A. Ustimenko, A.J. Woldar, New upper bounds on the order of cages. Electron. J. Comb. **14**(R13), 1–11 (1997)

90. F. Lazebnik, V.A. Ustimenko, A.J. Woldar, Polarities and 2k-cycle-free graphs. Discret. Math. **197/198**, 503–513 (1999)
91. J. Ma, Cycles with consecutive odd lengths. arxiv.org/pdf/1410.0430v1.pdf
92. W. Mader, Existenz gewisser Konfigurationen in n-gesättigten Graphen und in Graphen genügend großer Kantendichte. Math. Ann. **194**, 295–312 (1971)
93. G. Margulis, Explicit group-theoretic constructions of combinatorial schemes and their applications in the construction of expanders and concentrators. Probl. Inform. Transm. **24**, 39–46 (1988)
94. K.E. Mellinger, D. Mubayi, Constructions of bipartite graphs from finite geometries. J. Graph Theory **49**(1) 1–10 (2005)
95. P. Mihók, I. Schiermeyer, Cycle lengths and chromatic number of graphs. Discret. Math. **286**, 147–149 (2004)
96. K. Milans, D. Rautenbach, F. Pfender, F. Regen, D. West, The cycle spectra of Hamiltonian graphs. J. Comb. Theory Ser. B **102**(4), 869–874 (2012)
97. R. Morris, D. Saxton, The number of $C_{2\ell}$-free graphs. arxiv.org/pdf/1309.2927v2.pdf
98. M. Mörs, A new result on the problem of Zarankiewicz. J. Comb. Theory Ser. A **31**(2), 126–130 (1981)
99. D. Mubayi, Personal communication
100. D. Mubayi, J. Williford, On the independence number of the Erdős-Rényi and projective Norm graphs and a related hypergraph. J. Graph Theory **56**(2), 113–127 (2007)
101. A. Naor, J. Verstraëte, A note on bipartite graphs without $2k$-cycles. Comb. Probab. Comput. **14**(5–6), 845–849 (2005)
102. K. O'Bryant, A complete annotated bibliography of work related to Sidon sequences. Electron. J. Comb. **11**, Dynamic Survey 39 (2004)
103. A. Odlyzko, W. D. Smith, Nonabelian sets with distinct k-sums. Discret. Math. **146**(1–3), 169–177 (1995)
104. T.D. Parsons, Graphs from projective planes. Aequationes Math. **14**, 167–189 (1976)
105. O. Pikhurko, A note on the Turán function of even cycles. Proc. Am. Math. Soc. **140**(11), 3687–3692 (2012)
106. L. Pósa, Hamiltonian circuits in random graphs. Discret. Math. **14**, 359–364 (1976)
107. I. Reiman, Über ein Problem von K. Zarankiewicz. Acta Math. Acad. Sci. Hung. **9**, 269–273 (1958)
108. V. Rödl, On the chromatic number of subgraphs of a given graph. Proc. Am. Math. Soc. **64**, 370–371 (1977)
109. P. Rowlinson, Y. Yuansheng, On extremal graphs without four-cycles. Utilitas Math. **41**, 204–210 (1992)
110. I. Ruzsa, E. Szemerédi, Triple systems with no six points carrying three triangles, in *Combinatorics*. Proceedings of the Fifth Hungarian Colloquium (Keszthely, 1976), vol. II, pp. 939–945. Colloquia Mathematica Societatis János Bolyai, vol. 18 (North-Holland, Amsterdam/New York, 1978)
111. H. Sachs, M. Stiebitz, On constructive methods in the theory of colour-critical graphs. Discret. Math. **74**(1–2), 201–226 (1989)
112. D. Saxton, A. Thomason, Hypergraph containers. arxiv.org/pdf/1204.6595.pdf
113. A. Scott, Szemerédi's Regularity Lemma for matrices and sparse graphs. Comb. Probab. Comput. **20**, 455–466 (2011)
114. J. Shearer, A note on the independence number of triangle-free graphs. Discret. Math. **46**(1), 83–87 (1983)
115. J. Shearer, A note on the independence number of triangle-free graphs, II. J. Comb. Theory Ser. B **53**(2), 300–307 (1991)
116. J. Singer, A theorem in finite projective geometry and some applications to number theory. Trans. Am. Math. Soc. **43**, 377–385 (1938)
117. J. Solymosi, C_4 Removal Lemma for sparse graphs, in *Open Problem Session*. Mathematisches Forschungsinstitut Oberwolfach, Report No. 01/2011 Combinatorics (2011). doi:10.4171/OWR/2011/01

118. B. Sudakov, A note on odd cycle-complete graph Ramsey numbers. Electron. J. Comb. **9**(1), Note 1, 4 (2002)
119. B. Sudakov, J. Verstraëte, Cycle lengths in sparse graphs. Combinatorica **28**(3), 357–372 (2008)
120. B. Sudakov, J. Verstraëte, Cycles in graphs with large independence ratio. J. Comb. **2**, 82–102 (2011)
121. M. Tait, C. Timmons, Small dense subgraphs of polarity graphs and the extremal number for the 4-cycle. arxiv.org/pdf/1502.02722.pdf
122. C. Thomassen, Cycles in graphs of uncountable chromatic number. Combinatorica **3**, 133–134 (1983)
123. C. Thomassen, Girth in graphs. J. Comb. Theory Ser. B **35**, 129–141 (1983)
124. C. Thomassen, Paths, circuits and subdivisions, in *Selected Topics in Graph Theory*, vol. 3, ed. by L. Beineke, R. Wilson (Academic Press, New York, 1988), pp. 97–133
125. C. Thomassen, B. Toft, Non-separating induced cycles in graphs. J. Comb. Theory Ser. B **31**, 199–224 (1981)
126. C. Timmons, J. Cilleruelo, k-fold Sidon sets. Electron. J. Comb. **21**(4), 12 (2014)
127. C. Timmons, J. Verstraëte, A counterexample to sparse removal. arxiv.org/pdf/1312.2994.pdf
128. J. Tits, Sur la trialité et certains groupes qui s'en déduisent. Inst. Hautes Études Sci. Publ. Math. **2**, 13–60 (1959)
129. H. van Maldeghem, *Generalized Polygons*. Monographs in Mathematics (Birkhäuser, Basel, 1998). http://cage.ugent.be/~fdc/contactforum/vanmaldeghem.pdf
130. V. Vapnik, A. Chervonenkis, On the uniform convergence of relative frequencies of events to their probabilities. Theory Probab. Appl. **16**(2), 264–280 (1971)
131. J. Verstraëte, On arithmetic progressions of cycle lengths in graphs. Comb. Probab. Comput. **9**, 369–373 (2000)
132. J. Verstraëte, Unavoidable cycle lengths in graphs. J. Graph Theory **49**(2), 151–167 (2005)
133. J. Verstraëte, J. Williford, Unpublished manuscript
134. R. Wenger, Extremal graphs with no C_4's, C_6's, or C_{10}'s. J. Comb. Theory Ser. B **52**(1), 113–116 (1991)
135. C.Q. Zhang, Circumference and girth. J. Graph Theory **13**(4), 485–490 (1989)

A survey of Turán problems for expansions

Dhruv Mubayi and Jacques Verstraëte

Abstract The r-expansion G^+ of a graph G is the r-uniform hypergraph obtained from G by enlarging each edge of G with a vertex subset of size $r-2$ disjoint from $V(G)$ such that distinct edges are enlarged by disjoint subsets. Let $\mathrm{ex}_r(n, F)$ denote the maximum number of edges in an r-uniform hypergraph with n vertices not containing any copy of the r-uniform hypergraph F. Many problems in extremal set theory ask for the determination of $\mathrm{ex}_r(n, G^+)$ for various graphs G. We survey these Turán-type problems, focusing on recent developments.

Keywords Expansions • Turan numbers • Shadows • Delta systems • Erdos-Ko-Rado theorem • Canonical Ramsey theorem

Mathematics Subject Classification (2010): 05C35, 05C65, 05D05, 05D40

1 Introduction

An r-uniform hypergraph F, or simply *r-graph*, is a pair (V, E) where V is a finite set and E is a family of r-element subsets of V. As is customary in extremal set theory, we identify F with its edge set E when the vertex set of F is clear from context. We will always write $|F|$ for the number of edges in F. Given an r-graph F, let $\mathrm{ex}_r(n, F)$ denote the maximum number of edges in an r-graph on n vertices that does not contain a copy of F (if the uniformity is obvious from context, we may omit the subscript r).

Definition 1.1. Fix $r \geq 2$. The *expansion* of a graph $G = (V, E)$ is the r-graph $G^+ = (V^+, E^+)$ with $E^+ = \{e \cup S_e : e \in E\}$, where $|S_e| = r-2$, $S_e \cap V = \emptyset$ for all $e \in G$, and $S_e \cap S_{e'} = \emptyset$ for $e \neq e'$, and $V^+ = V \cup \bigcup_{e \in E} S_e$.

D. Mubayi (✉)
Department of Mathematics, Statistics, and Computer Science,
University of Illinois, Chicago, IL 60607, USA
e-mail: mubayi@uic.edu

J. Verstraëte
Department of Mathematics, University of California, San Diego, CA 92093, USA
e-mail: jacques@ucsd.edu

A. Beveridge et al. (eds.), *Recent Trends in Combinatorics*, The IMA Volumes in Mathematics and its Applications 159, DOI 10.1007/978-3-319-24298-9_5

By definition, $|E^+| = |E|$ and $|V^+| = |V| + (r-2)|E|$ and, when $r = 2, G^+ = G$. Expansions include many important hypergraphs whose extremal functions have been investigated, for instance, the celebrated Erdős-Ko-Rado theorem [25] is the case of expansions of a matching of size two. In this article we survey these results focusing on recent developments and proof techniques.

A crucial parameter that determines the growth rate of $\mathrm{ex}_r(n, G^+)$ is the chromatic number of G. Indeed, an old result of Erdős [16] implies that for fixed $r \geq 2$, the function $\mathrm{ex}_r(n, G^+)$ has order of magnitude n^r if and only if $\chi(G) > r$. As we will see in the next section, the asymptotics in this case have been determined and we say that these problems fall under the nondegenerate case. The majority of open problems in the area arise when $\chi(G) \leq r$ and we will refer to this as the degenerate case.

Along with obtaining exact extremal results, one can ask about the structure of nearly extremal structures. The seminal result in this direction is the stability theorem for graphs, proved independently by Erdős and Simonovits (see [64]). Similar theorems for hypergraphs have been proven more recently. For example, papers [37–39, 44, 47, 48] each prove stability theorems for hypergraphs, [37, 47] for the Fano plane, [39] for $\{123, 124, 125, 345\}$, [38, 48] for certain types of hypergraph triangles, [44] for cancellative triple systems, and also for $\{123, 124, 345\}$. By now there are many other stability results and the list above is far from exhaustive. Stability theorems have also proved useful to obtain exact results. This approach, first used by Simonovits [64] to determine the exact Turán number of color critical graphs, was developed and applied to hypergraph problems in the above results. The method proceeds by first proving an approximate result, then a stability statement, and finally uses the stability result to guarantee an exact extremal result. All of these results concern the nondegenerate case, i.e., when the order of magnitude of $\mathrm{ex}_r(n, F)$ is n^r.

It was shown in [56] that this method can be used in the degenerate case as well by solving a generalization of an old question of Katona. Since then, this approach has been used many times, for example, to solve the extremal problem for k-regular subsystems [49, 62], k-uniform clusters and simplices [10, 43, 45, 57, 58], and for expansions of acyclic graphs and cycles [50, 51].

This survey is organized as follows. In Section 2 we state results for the nondegenerate case, and in Section 3 we survey the results in the degenerate case. Section 4 describes the important proof techniques, focusing on the classical Delta-system method and a new approach that uses shadows, random sampling, and canonical Ramsey theory. In Section 5 we will use these techniques to give short proofs of a variety of recent results in the degenerate case. Specifically, we will give proofs of the asymptotic results for $\mathrm{ex}_3(n, K_{s,t}^+)$ when $t > (s-1)!$, for $\mathrm{ex}_r(n, G^+)$ when $r \geq 4$ and G is a forest, and for $\mathrm{ex}_3(n, C_4^+)$. Our hope is that these will illustrate many of the important proof ideas used in this area. Section 6 lists some open problems.

1.1 Notation and Terminology

For a set V, let 2^V be the set of subsets V, and let $\binom{V}{r}$ be the set of r-element subsets of V. For a positive integer n, we write $[n] = \{1, 2, \ldots, n\}$. A hypergraph with vertex set V is a family of sets $H \subset 2^V$. If $H \subset \binom{V}{r}$ then H is an *r-graph*. If $r = 2$, H is a graph and if $r = 3$, H is a *triple system*. If $S \subset V$, then the subhypergraph of H induced by S is $H[S] = \{e \in H : e \subset S\}$. The *degree of* S in H is $d_H(S) = |\{e \in H : S \subset e\}|$, and we write $d_H(v)$ if $S = \{v\}$. The *neighborhood* of S is $N_H(S) = \{e \backslash S : S \subset e \in H\}$, and we write $N_H(v)$ if $S = \{v\}$. The subscript H is suppressed when H is clear from context. For $s \in \mathbb{N}$, define the *shadow s-graph*

$$\partial_s H = \left\{ S \in \binom{V}{s} : d_H(S) \geq 1 \right\}.$$

We write ∂H instead of $\partial_2 H$. For a set $S \subset V$, the *trace* of H on S is

$$H|_S = \{e \cap S : e \in H\}.$$

For a positive integer n, the *Turán Number* of an r-graph F is

$$\mathrm{ex}_r(n, F) = \max \left\{ |H| : H \subset \binom{[n]}{r}, F \not\subset H \right\}.$$

If $F \not\subset H$ we say that H is F-free. An n-vertex F-free r-graph H is *extremal for* F if $|H| = \mathrm{ex}_r(n, F)$. For standard graph-theoretic terminology, see Bondy and Murty [6].

2 The nondegenerate case

We begin by observing that when $l \geq r = 2$ and $G = K_{l+1}$, the complete graph with $l + 1$ vertices,

$$\mathrm{ex}_2(n, G^+) = \mathrm{ex}_2(n, G) = \mathrm{ex}_2(n, K_{l+1}).$$

Turán's theorem [66] states that $\mathrm{ex}(n, K_{l+1})$ is uniquely achieved (for all $n > l$) by the Turán graph, which is the complete l-partite graph on n vertices with part sizes that differ by at most one. More generally, for any graph G with $\chi(G) = l + 1 > 2$, the Erdős-Stone-Simonovits theorem [21, 24] determines the asymptotics of $\mathrm{ex}_2(n, G^+) = \mathrm{ex}_2(n, G)$. In general exact results are known for very few cases; one such case is when G is color critical [64].

In order to state the results for $r \geq 3$, we must generalize the definition of the Turán graph to hypergraphs. An r-graph is l-partite if its vertex set can be partitioned into l classes, such that every edge has at most one vertex from each class. Thus in particular, there are no edges if $l < r$. A complete l-partite r-graph is one where all of the allowable edges (given a vertex l-partition) are present. For $n, l, r \geq 1$, let $T_r(n, l)$ be the complete l-partite r-graph on n vertices with no two part sizes differing by more than one. Thus the part sizes are $n_i = \lfloor (n + i - 1)/l \rfloor$ for $i \in [l]$. Among all l-partite r-graphs on n vertices, $T_r(n, l)$ has the most edges. The number of edges in $T_r(n, l)$ is

$$t_r(n,l) = \sum_{S \in \binom{[l]}{r}} \prod_{i \in S} n_i \qquad \text{and} \qquad t_r(n,l) \sim \frac{l(l-1)\cdots(l-r+1)}{l^r}\binom{n}{r}$$

for fixed $l \geq r$. Clearly $K_{l+1}^+ \not\subset T_r(n, l)$ so $\mathrm{ex}_r(n, K_{l+1}^+) \geq t_r(n, l)$. The first author [55] proved that if $l \geq r$ is fixed, then $\mathrm{ex}_r(n, K_{l+1}^+) < t_r(n, l) + o(n^r)$ and conjectured that this could be further improved to an exact result for n sufficiently large. This was subsequently proved by Pikhurko [63].

Theorem 2.1. **([55, 63])** *Fix $l \geq r \geq 2$ and let n be sufficiently large. Then*

$$\mathrm{ex}_r(n, K_{l+1}^+) = t_r(n, l).$$

Moreover equality is achieved only by the Turán r-graph $T_r(n, l)$.

It was observed by Alon and Pikhurko [2] that the approach applied to prove Theorem 2.1 can be extended to any color critical graph G with $\chi(G) > r$. More generally, a result of Erdős [16], the supersaturation technique (see Erdős-Simonovits [22]), and the asymptotic result of the first author [55] give the following.

Theorem 2.2. *Fix integers $l \geq r \geq 2$ and a graph G with $\chi(G) = l + 1$. Then*

$$\mathrm{ex}_r(n, G^+) \sim t_r(n, l).$$

Theorem 2.2 determines the asymptotics of $\mathrm{ex}_r(n, G^+)$ for all G with $\chi(G) > r$.

The Erdős-Simonovits stability theorem [64] states that if an n-vertex K_{l+1}-free graph (n large) has almost as many edges as $T_2(n, l)$, then its structure is very similar to that of $T_2(n, l)$. The analogous result for the r-expansion of K_{l+1} was proved in [55].

Theorem 2.3. **([55])** *Fix $l \geq r \geq 2$, and $\delta > 0$. Then there exists an $\epsilon > 0$ and an n_0 such that the following holds for all $n > n_0$: If G is an n-vertex K_{l+1}^+-free r-graph with at least $t_r(n, l) - \epsilon n^r$ edges, then G can be transformed to $T_r(n, l)$ by adding and deleting at most δn^r edges.*

3 The degenerate case

As mentioned earlier, many results in extremal set theory can be phrased as results concerning $\mathrm{ex}_r(n, G^+)$ in the degenerate case, i.e., when $\chi(G) \le r$. In fact, many famous open problems in extremal set theory deal with very simple graphs G. In this section we survey these results. Section 3.1 deals with matchings, section 3.2 with stars, section 3.3 with paths and cycles, section 3.4 with trees, and sections 3.5–3.7 with a variety of graphs containing cycles.

3.1 Matchings

Let us assume that $G = M_t = tK_2$ is a matching with t edges. The simplest case is $t = 2$, where G^+ comprises two disjoint edges. Here the Erdős-Ko-Rado theorem determines $\mathrm{ex}_r(n, M_2^+)$ for all r, n. Say that a hypergraph H is a *star* if there is a vertex of H that lies in all edges of H.

Theorem 3.1. (Erdős-Ko-Rado [25]) *Let $n, r \ge 2$. If $n < 2r$, then $\mathrm{ex}_r(n, M_2^+) = \binom{n}{r}$ while if $n \ge 2r$, then $\mathrm{ex}_r(n, M_2^+) = \binom{n-1}{r-1}$. For $n > 2r$, equality holds above only for a star.*

The general case is an old conjecture of Erdős.

Conjecture 3.2. (Matching Conjecture, Erdős [18]) *Let $r \ge 2$, $t \ge 1$ and $n \ge r(t+1) - 1$. Then*

$$\mathrm{ex}_r(n, M_{t+1}^+) = \max\left\{\binom{r(t+1)-1}{r}, \binom{n}{r} - \binom{n-t}{r}\right\}.$$

For $r = 2$ it was proved by Erdős and Gallai [19]. For $n > n_0(r, t)$ it was proved by Erdős [18]. Bollobás, Daykin, and Erdős [5] lowered the value of $n_0(r, t)$ to $2tr^3$. More recently, Huang, Loh, and Sudakov [42] lowered it further to $3tr^2$, which was further improved slightly in [31] (see also [54]). Frankl and Füredi (unpublished) gave an improvement for large r by showing that $n_0(r, t) < crt^2$ for some positive c. Finally, Frankl [29] gave the following improvement that contains all of the previous results on this problem. Let

$$\mathcal{A}(n, r, t) := \left\{e \in \binom{[n]}{r} : e \cap [t] \ne \emptyset\right\}.$$

Theorem 3.3. **([29])** *Let $r, t \geq 1$ and $n \geq (2t+1)r - t$. Then*

$$\mathrm{ex}_r(n, M_{t+1}^+) = \binom{n}{r} - \binom{n-t}{r}.$$

Equality holds only for families isomorphic to $\mathcal{A}(n, r, t)$.

3.2 Stars

For $t \geq 2$, let $S_t = K_{1,t}$ be the star with t edges. The problem of determining $\mathrm{ex}_r(n, S_t^+)$ was first considered in 1976 by Erdős and Sós [65]. They determined $\mathrm{ex}_3(n, S_2^+)$ precisely for all n.

Theorem 3.4. **([65])**

$$\mathrm{ex}_3(n, S_2^+) = \begin{cases} n & \text{if } n \equiv 0 \text{ (mod 4)} \\ n-1 & \text{if } n \equiv 1 \text{ (mod 4)} \\ n-2 & \text{if } n \equiv 2, 3 \text{ (mod 4)}. \end{cases}$$

The extremal families above are basically the disjoint unions of triple systems consisting of all 3-subsets of a 4-subset (in the last case there are some more possibilities, namely to take some 3-sets containing two fixed points). The case $r \geq 4$ is more complicated. Erdős and Sós conjectured that

$$\mathrm{ex}_r(n, S_2^+) = \binom{n-2}{r-2} \qquad \text{for} \quad n > n_0(r) \tag{1}$$

and this was proved soon after by Frankl [27] (the lower bound is trivial for all $n \geq r \geq 4$). Determining the correct value of $n_0(r)$ above seems very difficult. The following bound proved by Keevash-Mubayi-Wilson [46] is slightly sub-optimal, but holds for all $n \geq r \geq 4$:

$$\mathrm{ex}_r(n, S_2^+) \leq \binom{n}{r-2}.$$

For fixed $r, t \geq 2$, Duke and Erdős [15] showed that $\mathrm{ex}_r(n, S_t^+) = \Theta(n^{r-2})$ and conjectured that $\mathrm{ex}_3(n, S_3^+) = 6(n-3) + 2$ (presumably for $n \geq 7$ though this is not explicitly mentioned). Frankl [28] proved that

$$\mathrm{ex}_3(n, S_3^+) = 6(n-3) + 2 \qquad \text{for} \quad n \geq 54$$

and $\mathrm{ex}_3(n, S_t^+) < (5/3)t(t-1)n$. Chung [11] substantially improved these bounds and finally, Frankl and Chung [12] improved them further as follows.

Theorem 3.5. **([12])**

$$\mathrm{ex}_3(n, S_t^+) = \begin{cases} nt(t-1) + 2\binom{t}{3} & \text{if} > t(t-1)(5t+2)/2 \text{ and } t \geq 3 \text{ is odd} \\ \frac{nt(2t-3)}{2} - \frac{2t^3-9t+6}{2} & \text{if } n \geq 2t^3 - 9t + 7 \text{ and } t \geq 4 \text{ is even.} \end{cases}$$

Frankl and Chung characterized the extremal example as well. The only other case where exact results are known for all n is $r = 4$ and $t = 2$. Before stating this result we should recall that full version of the Erdős-Ko-Rado [25] theorem. Say that an r-graph H is *t-intersecting* if for every $e, f \in H$ we have $|e \cap f| \geq t$. They showed that, if H is a t-intersecting r-graph on $[n]$ with n sufficiently large, then $|H| \leq \binom{n-t}{r-t}$. (The case $t = 2$ is pertinent to our current discussion.) Confirming a conjecture of Erdős, Wilson [67] showed that this bound in fact holds for $n \geq (t+1)(r-t+1)$ (which is the best possible strengthening), and furthermore that the unique maximum system consists of all r-sets containing some fixed t-set. To describe the complete solution for all n we need to define the t-intersecting systems

$$\mathcal{F}(n, r, t, i) = \left\{ e \in \binom{[n]}{r} : |e \cap [t + 2i]| \geq t + i \right\}$$

for $0 \leq i \leq r - t$. The complete intersection theorem, conjectured by Frankl, and proved by Ahlswede and Khachatrian [1], is that a maximum size t-intersecting r-graph on $[n]$ is isomorphic to $\mathcal{F}(n, r, t, i)$ for some i which can easily be computed given n. Note that $\mathcal{F}(n, r, t, 0)$ is isomorphic to the r-graph of all r-sets containing some fixed t-set.

Theorem 3.6. **([46])**

$$\mathrm{ex}_4(n, S_2^+) \leq \begin{cases} \binom{n}{4} & n = 4, 5, 6 \\ 15 & n = 7 \\ 17 & n = 8 \\ \binom{n-2}{2} & n \geq 9 \end{cases}$$

Furthermore, the only cases of equality are $\binom{[n]}{4}$ *for* $n = 4, 5$, $\mathcal{F}(n, 4, 2, 2) = \binom{[6]}{4}$ *for* $n = 6, 7$, $\mathcal{F}(8, 4, 2, 1)$ *for* $n = 8$ *and* $\mathcal{F}(n, 4, 2, 0)$ *for* $n \geq 9$.

The following stability version of the above result was also shown in [46]

Theorem 3.7. *For any* $\epsilon > 0$ *there is* $\delta > 0$ *such that if* H *is an* r*-graph on* $[n]$ *with no singleton intersection and* $|H| \geq (1 - \delta)\binom{n-2}{r-2}$, *then there are two points* x, y *so that all but at most* ϵn^{r-2} *sets of* H *contain both* x *and* y.

3.3 Paths and Cycles

In this section we consider the case when $G = P_t$ or $G = C_t$, where P_t (C_t) is the path (cycle) with t edges. Erdős and Gallai [19] proved that $\mathrm{ex}_2(n, P_t) \le \frac{t-1}{2}n$ and this is tight whenever $t|n$. Since $P_2 = S_2$, we have already seen in (1) that Frankl [27] determined $\mathrm{ex}_r(n, P_2^+)$ for large n. Füredi, Jiang, and Seiver [40] determined $\mathrm{ex}_r(n, P_t^+)$ precisely for all $r \ge 4$, $t \ge 3$, and n large while also characterizing the extremal examples. This improved results of the authors [61] and also settled a conjecture from [61]. They conjectured a similar result for $r = 3$.

The case $G = K_3 = C_3$ is also well-researched [13, 30, 60], indeed, when $r = 2$ this is precisely Mantel's theorem from 1907. Frankl and Füredi [30] showed that the unique extremal r-graph on $[n]$ not containing C_3^+ is a star, for large enough n. For $r = 3$, Csákány and Kahn [13] proved the same result for all $n \ge 6$. More recently, Füredi and Jiang [35] determined the extremal function for C_t^+ for all $t \ge 3$, $r \ge 5$, and large n; their results substantially extend earlier results of Erdős and settled a conjecture of the current authors [61] for $r \ge 5$. Recently, Kostochka and the current authors [50] further extended the results in [35] to the case of $r = 3, 4$. To state the result for all r, t, we need some notation. For $L \subset [n]$ let $S_L^r(n)$ denote the r-graph on $[n]$ consisting of all r-element subsets of $[n]$ intersecting L.

Theorem 3.8. **([35, 40, 50])**
Let $r \ge 3$, $t \ge 4$, and $\ell = \lfloor \frac{t-1}{2} \rfloor$. For sufficiently large n,

$$\mathrm{ex}_r(n, P_t^+) = \binom{n}{r} - \binom{n-\ell}{r} + \begin{cases} 0 & \text{if } t \text{ is odd} \\ \binom{n-\ell-2}{r-2} & \text{if } t \text{ is even} \end{cases}$$

with equality only for $S_L^r(n)$ if t is odd and $S_L^r(n) \cup F$ where F is extremal for $\{P_2^+, 2P_1^+\}$ on $n - \ell$ vertices if t is even. The same result holds for C_t^+ except the case $(t, r) = (4, 3)$, in which case

$$\mathrm{ex}_3(n, C_4^+) = \binom{n}{r} - \binom{n-1}{r} + \max\{n-3, 4\lfloor \tfrac{n-1}{4} \rfloor\}$$

with equality only for triple systems of the form $S_L^3(n) \cup F$ where F is extremal for P_2^+ on $n-1$ vertices.

It was recently shown by Bushaw and Kettle [9] that the Turán problem for disjoint t-paths can be easily solved once we know the extremal function for a single t-path. As Theorem 3.8 solves the t-paths problem for all $r \ge 3$, the corresponding extremal questions for disjoint t-paths are also completely solved (for large n). A similar situation likely holds for disjoint t-cycles, as recently observed by Gu, Li, and Shi [41].

Finally, we remark that the results of [35, 40, 50] also give stability versions of Theorem 3.8. Here is a sample result.

Theorem 3.9. **([50])** *Fix $r \geq 3, t \geq 4$ $\ell = \lfloor \frac{t-1}{2} \rfloor$ and let H be an n-vertex r-graph with $|H| \sim \ell\binom{n}{r-1}$ containing no P_t^+ or no C_t^+. Then there exists $H' \subset \partial_{r-1}H$ with $|H'| \sim \binom{n}{r-1}$ and a set L of ℓ vertices of H such that $N_{H'}(e) = L$ for every $e \in H'$. In particular, $|H - L| = o(n^{r-1})$.*

3.4 Trees

In this section we consider the case when G is acyclic. What makes this problem more complicated than the previous ones where G is a star or path, is that we can produce constructions that exploit the structure of G. Let us make this precise. A set of vertices in a hypergraph F containing exactly one vertex from every edge of F is called a *crosscut*, following Frankl and Füredi [30]. Let $\sigma(F)$ be the minimum size of a crosscut of F if it exists, i.e.,

$$\sigma(F) := \min\{|X| : X \subset V(F), \forall e \in F, |e \cap X| = 1\}$$

if such an X exists.

Since the r-graph on n vertices consisting of all edges containing exactly one vertex from a fixed subset of size $\sigma(F) - 1$ does not contain F, we have

$$\mathrm{ex}_r(n, F) \geq (\sigma(F) - 1)\binom{n - \sigma(F) + 1}{r - 1} \sim (\sigma(F) - 1)\binom{n}{r-1}. \tag{2}$$

An intriguing open question is when asymptotic equality holds above and this is one of the main open problems in this area. Indeed, it appears that the parameter $\sigma(F)$ often plays a crucial role in determining the extremal function for F. Füredi [34] determined the asymptotics of $\mathrm{ex}_r(n, G^+)$ when G is a forest and $r \geq 4$ and conjectured a similar result for $r = 3$. This conjecture was recently proved by Kostochka and the current authors [52].

Theorem 3.10. **([34] for $r \geq 4$, [52] for $r = 3$)** *Fix $r \geq 3$ and a forest G. Then*

$$\mathrm{ex}_r(n, G^+) \sim (\sigma(G^+) - 1)\binom{n}{r-1}.$$

Stability and exact results for Theorem 3.10 for $r = 3$ do not follow using current proof methods. It would be interesting to obtain such results. However, for $r \geq 4$, Füredi and Jiang [36] recently proved stability results for expansions of forests. Their methods also enabled them to obtain exact results in many cases.

3.5 Graphs with crosscuts of size two

If $\sigma(G^+) = 1$, then clearly $G = S_t$ and this case has already been discussed. Here we focus on the next case, $\sigma(G^+) = 2$. It is straightforward to see that any graph G with $\sigma(G^+) = 2$ is contained in a star with an edge added, or a complete bipartite graph with one part of size two. The following result of Kostochka and the current authors [51] gives an asymptotically exact result in these cases:

Theorem 3.11. ([51]) *For every fixed graph G with $\sigma(G^+) = 2$,*

$$\mathrm{ex}_3(n, G^+) \sim \binom{n}{2}.$$

In particular, for any graph G, $\mathrm{ex}_3(n, G^+) \leq \left(\frac{1}{2} + o(1)\right) n^2$ or $\mathrm{ex}_3(n, G^+) \geq (1 + o(1))n^2$.

The last statement in Theorem 3.11 follows easily from the first, since if $\sigma(G^+) = 3$ then $\mathrm{ex}_3(n, G^+) \geq (1 + o(1))n^2$. Theorem 3.11 determines the asymptotic behavior of $\mathrm{ex}_3(n, G^+)$ when $\sigma(G^+) = 2$; the next case $\sigma(G^+) = 3$ is an avenue of further research. It does appear, however, that determining the order of magnitude of $\mathrm{ex}_3(n, G^+)$ is more difficult when $\sigma(G^+) = 3$; it is shown in [52] that if $G = K_{3,t}$ then $\mathrm{ex}_3(n, G^+)/n^2 \to \infty$ as $t \to \infty$.

3.6 Complete bipartite graphs

The general problem of determining $\mathrm{ex}_3(n, G^+)$ when G is bipartite was studied in [52]. In this case, the problem is related to maximizing the number of triangles in a G-free graph: if one takes as the hypergraph the triangles in the graph, then one obtains a triple system not containing G^+. This framework was studied by Alon and Shikhelman [3] in the context of $K_{s,t}$-free graphs who considered the following question: what is the maximum number of triangles in a $K_{s,t}$-free graph on n vertices? Kostochka and the current authors [52] give a short proof of one of their results, and in particular, prove the following more general theorem.

Theorem 3.12. ([52]) *For $t \geq s \geq 3$ fixed, $\mathrm{ex}_3(n, K_{s,t}^+) = O(n^{3-\frac{3}{s}})$ and $\mathrm{ex}_3(n, K_{s,t}^+) = \Theta(n^{3-\frac{3}{s}})$ for $t > (s-1)!$.*

We give the proof of this theorem in Section 5.1. Theorem 3.12 does not extend to the case $s = 2$, and in that case $\mathrm{ex}_3(n, K_{2,t}^+) \sim \binom{n}{2}$ as shown by Theorem 3.11. For a general bipartite graph G, the following is proved in [52].

Theorem 3.13. *Let G be a bipartite graph such that $\mathrm{ex}(n, G) = O(n^{1+\alpha})$ and $\mathrm{ex}(n, G - \{v\}) = O(n^{1+\beta})$ for every $v \in V(G)$. Then $\mathrm{ex}_3(n, G^+) = O(n^{1+\alpha+\alpha\beta}) + O(n^2)$.*

Let Q denote the 3-dimensional cube graph. Erdős and Simonovits [23] showed $\mathrm{ex}(n, Q) = O(n^{8/5})$ and $\mathrm{ex}(n, Q - \{v\}) = O(n^{3/2})$ for all $v \in Q$. Applying Theorem 3.13, we find $\mathrm{ex}_3(n, Q^+) = O(n^{1.9}) + O(n^2) = O(n^2)$ and therefore $\mathrm{ex}_3(n, Q^+) = \Theta(n^2)$. In fact for any bipartite graph G (except a star) such that $\mathrm{ex}(n, G) = O(n^{1+\alpha})$ where $1 + \alpha$ is less than the golden ratio $(1 + \sqrt{5})/2$, one has $\mathrm{ex}_3(n, G^+) = \Theta(n^2)$. This includes the family of all bipartite graph in which the vertices in one part have degree two, since for any such graph G, $\mathrm{ex}_2(n, G) = O(n^{3/2})$ using the method of dependent random choice [26]:

Corollary 3.14. *Let G be a bipartite graph such that all the vertices in one part have degree at most two. Then* $\mathrm{ex}_2(n, G^+) = O(n^2)$.

3.7 General expansions

If G is a 3-colorable graph, then $\chi(G^+) \leq 3$ and therefore $\mathrm{ex}_3(n, G^+) = o(n^3)$. In the last section, we saw examples of bipartite graphs for which the order of magnitude of $\mathrm{ex}_3(n, G^+)$ can be determined. One may ask for which 3-colorable graphs G one has $\mathrm{ex}_3(n, G^+) = O(n^2)$. Theorem 3.8 shows that for any odd cycle C, $\mathrm{ex}_3(n, C^+) \sim (\sigma(C) - 1)\binom{n}{2}$. The following result gives a superquadratic lower bound in many cases.

Theorem 3.15. **([52])** *Let G be a graph such that in every proper coloring of G, every pair of color classes induces a graph containing a cycle. Then for some $c > 2$,* $\mathrm{ex}_3(n, G^+) = \Omega(n^c)$.

A construction proving Theorem 3.15 is simple. Suppose $|V(G)| = k$. Consider the 3-partite triple system H with parts U, V, and W where $|U| \leq |V| \leq |W| \leq |U| + 1$, and let J be an extremal bipartite graph containing no cycle of length less than $k + 1$ and with parts U and V. Then the edges of H consist of all triples $\{u, v, w\}$ such that $\{u, v\} \in J$ and $w \in W$. The number of edges in H is $|J||W|$. It is well-known that an extremal bipartite graph with n vertices and no cycles of length up to k has $\Omega(n^{c_k})$ edges, for some $c_k > 1$, so if H has n vertices, then $|H| = |J||W| = \Omega(n \cdot n^{c_k})$. This establishes Theorem 3.15 with $c = 1 + c_k$.

A second construction using random graphs works when G is a graph with many edges. Specifically, let G be a graph containing a cycle and let $m(G) = \max\{\frac{|H|-1}{|V(H)|-2} : H \subseteq G, |V(H)| \geq 3\}$. It is known that there exists $\delta > 0$ such that with positive probability, the random graph $G_{n,p}$ has a G-free subgraph with at least $\delta p^3 n^3$ triangles when $p = \delta n^{-1/m}$. If H is the triple system comprising the triangles in the subgraph of $G_{n,p}$, then H does not contain G^+. This gives the following theorem:

Theorem 3.16. *For any graph G, with $m = m(G)$,*

$$\mathrm{ex}_3(n, G^+) = \Omega(n^{3-\frac{3}{m}}).$$

In particular, for any graph G with more than $3|V(G)| - 5$ *edges, there exists* $c > 2$ *such that*

$$\mathrm{ex}_3(n, G^+) = \Omega(n^c).$$

An *acyclic coloring* of a graph is a proper vertex-coloring of the graph such that the union of any two color classes induces a forest. If G is a graph with an acyclic 3-coloring, then $|G| \leq 2|V(G)| - 3$ and Theorems 3.16 and 3.15 do not apply. It seems natural therefore to ask whether $\mathrm{ex}_3(n, G^+) = O(n^2)$ when G has an acyclic 3-coloring.

4 Proof Methods

In this section, we describe some of the important techniques that have been successful in proving results about extremal numbers of expansions.

4.1 Full subgraphs

It is a well-known fact that any graph G has a subgraph of minimum degree at least $d+1$ with at least $|G|-d|V(G)|$ edges. In this section, we state the simple extension of this fact to hypergraphs.

Definition 4.1. For $r > s \geq 1$ and $d \geq 1$, an r-graph H is (d, s)-*full* if $d(e) \geq d$ for all $e \in \partial_s H$.

Thus H is (d, s)-full is equivalent to saying that the minimum non-zero degree of $(s-1)$-sets in $V(H)$ is at least d. When $s = 2$ we will use the simpler notation d-full.

Lemma 4.2. *For* $d \geq 1$ *and* $r > s \geq 1$*, every n-vertex r-graph H has a* $(d+1, s)$*-full subgraph F with*

$$|F| \geq |H| - d|\partial_s H|.$$

Proof. A *d-sparse sequence* is a maximal sequence $e_1, e_2, \ldots, e_m \in \partial_s H$ such that $d_H(e_1) \leq d$, and for all $i > 1$, e_i is contained in at most d edges of H which contain none of $e_1, e_2, \ldots, e_{i-1}$. The r-graph F obtained by deleting all edges of H containing at least one of the e_i is $(d+1)$-full. Since a d-sparse sequence has length at most $|\partial_s H|$, we have $|F| \geq |H| - d|\partial_s H|$. □

Although Lemma 4.2 and its proof is simple, it is a very important tool and is used frequently in the proofs of the theorems in this survey.

4.2 Kövari-Sós-Turán Theorem

The problem of determining the order of magnitude of $ex_2(n, G)$ is generally notoriously difficult when G is a bipartite graph. Even for such small graphs as $G = C_8$, $G = Q_3$, and $G = K_{4,4}$, this order of magnitude is not known. Zarankiewicz [68] conjectured that $ex_2(n, K_{s,t}) = \Theta(n^{2-1/s})$ whenever $t \geq s \geq 2$, and this conjecture remains open in general. Kövari, Sós, and Turán [53] gave the following general upper bound for $ex_2(n, K_{s,t})$.

Theorem 4.3. *For all $t \geq s \geq 2$,*

$$ex_2(n, K_{s,t}) \leq \tfrac{1}{2}[(t-1)^{1/s}n^{2-1/s} + (s-1)n].$$

When $t > (s-1)! \geq 1$, the norm graph constructions of Alon, Rónyai, and Szabó [4] show that Theorem 3.12 is tight up to constants, and so this verifies the Zarankiewicz [68] conjecture in those cases. In addition, Füredi [32] determined the asymptotic behavior of $ex_2(n, K_{2,t})$ and also $ex_2(n, K_{3,3})$ – a construction of Brown [7] gives the tightness of Füredi's upper bound for $ex_2(n, K_{3,3})$.

The short proof of Theorem 3.12 exclusively uses Theorem 4.3 in combination with Lemma 4.2, and we will present this proof in Section 5.1.

4.3 Delta-systems

In this section we describe the Delta-system method initiated by Deza, Erdős, and Frankl [14]. The proof of Theorem 3.10 for $r \geq 4$ due to Füredi [34] uses this approach. That proof does not extend to $r = 3$, and different techniques are needed for this case (these will be discussed in detail in sections 4.4 and 5.1). We begin with the basic definitions.

A *Delta-system* is a hypergraph Δ such that for any distinct edges $e, f \in \Delta$, we have $e \cap f = \bigcap_{g \in \Delta} g$. We denote by $\Delta_{r,s}$ the r-uniform Delta-system with s edges and the *core* of a Delta-system Δ is $\text{core}(\Delta) = \bigcap_{g \in \Delta} g$. If H is any r-graph and $f \subset V(H)$, then the *core degree of f* is

$$d^*_H(f) = \max\{s : \exists \Delta_{r,s} \subset H, \text{core}(\Delta_{r,s}) = f\}.$$

The following basic lemma is observed in [40]:

Lemma 4.4. *Let $r \geq 3$, H be an r-graph and G be a graph with $V(G) \subset V(H)$. Suppose that $d^*_H(e) \geq r|G|$ for all $e \in G$. Then $G^+ \subset H$.*

Proof. Let $F = \{e_1, e_2, \dots, e_k\} \subset \binom{V(H)}{2}$ with $F \cong G$ and $F_i = \{e_1, \dots, e_i\}$ for $i \in [k]$. We will show by induction on i that $F_i^+ \subset H$. The case $i = 0$ is trivial, as we may assume that $F_0 = \emptyset$. For the induction step, assume that $F_i^+ \subset H$, so we have edges $f_1, f_2, \ldots, f_i \in H$ such that $e_j \subset f_j$ for $j \leq i$ and the sets $f_j \backslash e_j$ are pairwise disjoint. Since $d^*_H(e_{i+1}) \geq rk$, e_{i+1} is contained in edges $g_1, g_2, \ldots, g_{rk} \in H$ forming a Delta-system Δ with $\text{core}(\Delta) = e_{i+1}$. Since $\sum_{j=1}^{i} |f_i| = r(k-1)$, and the sets $g_j \backslash e_{i+1}$ are pairwise disjoint, there exists $f_{i+1} \in \{g_1, g_2, \ldots, g_{rk}\}$ such that $f_{i+1} \cap f_j = \emptyset$ for all $j \leq i$. Then $F^+_{i+1} = F^+_i \cup \{f_{i+1}\} \subset H$. □

4.3.1 Intersection semilattice lemma

A basic result about finite sets is the Erdős-Rado Sunflower Lemma [20]:

Lemma 4.5. (Erdős-Rado Sunflower Lemma [20]) *If F is a collection of sets of size at most k and $|F| \geq k!(s-1)^k$, then F contains a Delta-system with s sets.*

The driving force behind the Delta-system method is the *intersection semilattice lemma* of Füredi [33], which is a far reaching generalization of the Erdős-Rado sunflower lemma. If H is a hypergraph and $e \subseteq V(H)$, define

$$H|_e = \{e \cap f : f \in H, e \neq f\}.$$

If H is an r-graph with r-partition $(X_1, X_2, \ldots, X_r)$, define the projection of e to be $\text{proj}(e) = \{i : e \cap X_i \neq \emptyset\}$. Then the *intersection pattern of e* is the hypergraph

$$I_H(e) = \{\text{proj}(g) : g \in H|_e\} \subset 2^{[r]}.$$

Lemma 4.6. (The intersection semilattice lemma) *For any positive integers r, s there exists a positive real $c := c(r, s)$ such that if H is an r-graph, then there is an r-partite r-graph $H^* \subset H$ with parts $(X_1, X_2, \ldots, X_r)$ and $|H^*| > c|H|$ and a hypergraph $J \subset 2^{[r]}$ such that*

(1) *J is intersection-closed.*
(2) *For all $e \in H^*$, $I_{H^*}(e) = J$.*
(3) *For all $e \in H^*$ and $g \in H^*|_e$, $d^*_{H^*}(g) \geq s$.*

An r-partite r-graph H^* satisfying (1) – (3) is referred to as *(s, J)-homogeneous*. An r-partite r-graph is *J-homogeneous* if it is (s, J)-homogeneous for some positive integer s. For example, the complete r-partite r-graph with parts of size $t \geq 2$ is (t, J)-homogeneous with $J = 2^{[r]}$.

4.3.2 Homogeneous hypergraphs and rank

The goal of this section is to prove Lemma 4.11 below which produces a partition of an r-graph which does not contain the expansion of a forest into two parts. The first part has few edges, while the second part has a rich structure. The machinery used to prove this decomposition lemma heavily employs Lemma 4.6 and the entire approach constitutes the Delta-system method.

The *rank* of a hypergraph $J \subset 2^{[r]}$ is

$$\rho(J) = \min\{|e| : e \subset V(J), d_J(e) = 0\}.$$

For n-vertex J-homogeneous r-graphs H^* with $|H^*| > \binom{n}{k}$ in Lemma 4.6, we see that $\rho(J) > k$.

Lemma 4.7. *Let r be a positive integer and H^* an n-vertex J-homogeneous r-graph, such that $\rho(J) = k$. Then $|H^*| \leq \binom{n}{k}$.*

Proof. Since $\rho(J) = k$, every edge $e \in H^*$ contains a k-set S_e such that for every $f \in H^* \backslash \{e\}$, $S_e \not\subset f \cap e$. Then the map $\phi : H^* \to \binom{[n]}{k}$ defined by $\phi(e) = S_e$ is injective, so $|H^*| \leq \binom{n}{k}$. □

Let us now prove a very simple consequence of the semilattice lemma.

Corollary 4.8. *Let $r > s$, $q \geq 1$, let G be a q-edge forest, and let H^* be a non-empty (rq, J)-homogeneous r-graph such that $\rho(J) = r$. Then $G^+ \subset H^*$.*

Proof. Since $\rho(J) = r$, every edge $e \in H^*$ has the property that whenever $x \in e$, $d^*_{H^*}(e \backslash \{x\}) \geq rq$. Since J is intersection-closed, $d^*_{H^*}(f) \geq rq$ for every non-empty $g \subset V(J)$ with $|g| \leq r - 1$ and $\emptyset \neq f \subset \prod_{i \in g} X_i$. Therefore every edge f of ∂H^* has $d^*_{H^*}(f) \geq rq$, and ∂H^* is a graph of minimum degree at least rq. In particular, $G \subset \partial H^*$. Applying Lemma 4.4 completes the proof. □

The heart of the matter for embedding expansions in hypergraphs with roughly $\binom{n}{r-1}$ edges is therefore the case $\rho(J) = r-1$ in the semilattice lemma. If $\rho(J)=r-1$ in the semilattice lemma, then every $(r-2)$-set in $V(J)$ is contained in some edge of J (but is not necessarily an edge of J). The following lemma shows why $r \geq 4$ is needed to apply the Delta-system method to the current problem. A hypergraph $J \subset 2^{[r]}$ is *central* if there exists $x \in [r]$ such that J contains all $(r-1)$-sets containing x but not $[r]\backslash\{x\}$.

Lemma 4.9. *Let $J \subset 2^{[r]} \setminus [r]$ be an intersection-closed hypergraph with $\rho(J) = r-1$ such that J is not central. If $r \geq 4$, then J contains at least two singleton edges.*

Proof. Let $e_i = [r]\backslash\{i\}$. If $e_i \in J$ for $1 \leq i \leq r$, then since J is intersection-closed, every element of $[r]$ is a singleton edge in J. Suppose $e_1, e_2, \ldots, e_t \notin J$ and $e_{t+1}, e_{t+2}, \ldots, e_r \in J$. Since J is not central, $t \geq 2$. Since $\rho(J) = r - 1$ and $|e_i \cap e_j| = r - 2$, $e_i \cap e_j \in J$ for $1 \leq i < j \leq t$. Since J is intersection-closed, $e_i \cap e_j \in J$ for $t < i < j \leq r$. Since J is intersection-closed, $\{i\} \in J$ for each $i \in [t]$ if $t \geq 3$, and $\{i\} \in J$ for $3 \leq i \leq r$ if $t = 2$. Since $r \geq 4$, these are the required singletons. □

The purpose of the next lemma is to show that if H^* is an r-graph that is (s, J)-homogeneous and J is not central, then we can expand any forest in H^* unless $|H^*| = o(n^{r-1})$.

Lemma 4.10. *Let $q \geq 1$, $r \geq 4$. Let G be a q-edge tree and H^* an (rq, J)-homogeneous r-graph with $\rho(J) \geq r-1$ and J not central. Then $G^+ \subset H^*$.*

Proof. If $\rho(J) = r$, then apply Corollary 4.8. For $\rho(J) = r-1$, the proof is a standard greedy embedding algorithm, and we proceed by induction on q, the case $q = 1$ being trivial. By Lemma 4.9 and (2) and (3) in Lemma 4.6, for each $e \in H^*$ there exist vertices $u_e, v_e \in e$ such that $d^*_{H^*}(u_e) \geq rq$ and $d^*_{H^*}(v_e) \geq rq$. Let $C = \{u_e, v_e : e \in H^*\}$. The key observation is that in order to do the induction step, we will map $V(G)$ to C. For $q \geq 2$, let v be a leaf of G, u is its unique neighbor in G and $G_1 = G - \{v\}$. By induction, there exists a graph isomorphism $\phi : G_1 \to L$ such that $L \subset \partial H^*$ and $V(L) \subset C$ and $L^+ \subset H^*$. Then $d^*_{H^*}(\phi(u)) \geq rq$, so there is a Delta-system $\Delta = \Delta_{r,rq} \subset H^*$ with core$(\Delta) = \{\phi(u)\}$. Therefore there is an edge $f \in \Delta$ such that $f \setminus \{\phi(u)\}$ is disjoint from $V(L)$. Moreover, by Lemma 4.9 we can find another vertex $w \in f$ with $d^*_{H^*}(w) \geq rq$. Extending ϕ by defining $\phi(v) = w$, we have an isomorphism $\phi : G \to L \cup \{v, w\}$ such that $L \subset \partial H^*$ and $L^+ \cup \{f\}$ are an embedding of G^+ in H^*. □

The intersection semilattice lemma was tailored to the problem of embedding expansions of forests in r-graphs for $r \geq 4$ by Füredi [34, 40]. The following lemma is the main consequence of the Delta-system method that is used to embed expansions of forests in r-graphs. We will use it in Section 5.2 to give a proof of Theorem 3.10 for $r \geq 4$.

Lemma 4.11. *For $v \geq 1$ and $r \geq 4$, there exists a constant $c = c(v, r)$ such that if H is an n-vertex r-graph and G a v-vertex forest, and $G^+ \not\subset H$, then there is a partition $H = H_1 \sqcup H_2$ of H such that $|H_1| \leq c\binom{n-2}{r-2}$ and for all $f \in H_2$, there exists $x \in f$ such that $d^*_{H_2}(\{x, y\}) \geq rv$ for all $y \in f \backslash \{x\}$.*

Proof. Apply Lemma 4.6 to H to obtain an (r, s)-homogeneous family F_1 with intersection pattern J_1 and $|F_1| > c'|H|$ (where $c' := c(r, s)$). Now apply Lemma 4.6 again to $H - F_1$ to obtain F_2 and J_2, where $|F_2| > c'(|H| - |F_1|)$. Continue this process and let m be the smallest nonnegative integer such that $\rho(J_{m+1}) \leq r-2$. Put $H_2 = \bigcup_{i=1}^m F_i$ and $H_1 = H - H_2$. Then $|F_{m+1}| > c'|H_1|$ and, by Lemma 4.7, we have $|F_{m+1}| \leq \binom{n}{r-2}$. Setting $c = 2/c'$ we have $|H_1| \leq c\binom{n}{r-2}$ as desired. An edge not in H_1 is in some F_i for $i \leq m$, and $\rho(J_i) \geq r-1$. If $r(J_i) = r-1$ or $r(J_i) = r-2$ and J_i is not central, then $G^+ \subset H$ by Lemma 4.10, so if $f \in F_i \setminus H_1$, then J_i is central with rank $r-1$. Let x be a vertex of $[r]$ such that J_i contains all $(r-1)$-sets containing x but not $[r] \setminus \{x\}$. Since J_i is intersection-closed, all proper subsets of f containing x are cores of some Delta-system $\Delta_{r,s}$ in $F_i \subset H_2$, including all the subsets of the form $\{x, y\}$ where $y \in f \setminus \{x\}$. Consequently, $d^*_{H_2}(\{x, y\}) \geq rv$ for all $y \in f \backslash \{x\}$. □

4.4 *Ramsey theory, shadows, and random sampling*

In this section we introduce another method for proving results on expansions. This approach is more recent, and was developed by Kostochka and the current authors in [50–52] to prove Theorems 3.8, 3.10, and 3.11. The method applies to r-graphs for $r \geq 3$, but here we restrict attention to the case of triple systems.

4.4.1 Bipartite canonical Ramsey theorem

One of the main new ideas is to use the canonical Ramsey theorem of Erdős and Rado [20]. We need a bipartite version of this classical result, for which the following definitions are used.

Definition 4.12. Let F be a bipartite graph with parts X and Y and an edge-coloring χ. Then

- χ is *Z-canonical* for $Z \in \{X, Y\}$ if for each $z \in Z$, all edges of F on z have the same color and edges on different vertices in Z have different colors
- χ is *canonical* if χ is X-canonical or Y-canonical
- χ is *rainbow* if the colors of all the edges of F are different and
- χ is *monochromatic* if the colors of all the edges of F are the same.

If χ is an edge-coloring of a graph F and $G \subset F$, then $\chi|_G$ denotes the edge-coloring of G obtained by restricting χ to the edge set of G. The following bipartite version of the canonical Ramsey theorem was proved in [51].

Theorem 4.13. *For each $s > 0$ there exists $t > 0$ such that for any edge-coloring χ of $G = K_{t,t}$, there exists $K_{s,s} \subset G$ such that $\chi|_{K_{s,s}}$ is monochromatic or rainbow or canonical.*

Proof. Let X and Y be the parts of G and let $S = \{y_1, y_2, \dots, y_{2s^2}\} \subset Y$. Let W be the set of vertices $x \in X$ contained in at least s edges of the same color connecting x with S. If $|W| > m := s^2\binom{2s^2}{s}$, then there is a set $Y' \subset S$ of size s and a set $X' \subset W$ of size s^2 such that for every $x \in X'$, the edges xy with $y \in Y'$ all have the same color. In this case we recover either a monochromatic $K_{s,s}$ or an X'-canonical $K_{s,s}$. Now suppose $|W| \leq m$. For $x \in X_0 := X \backslash W$, let $C(x)$ be a set of $2s$ distinct colors on edges between x and S. By Lemma 4.5, if $|X_0| > (2s)!(s!m)^{2s}$, then there exists $X_1 \subset X_0$ such that $\{C(x) : x \in X_1\}$ is a Delta-system $\triangle$ of size $s!m$. Let $C = \text{core}(\triangle)$. First suppose $|C| \geq s$. Then we have a set X_2 of at least $|X_1|/\binom{2s^2}{s} \geq s!m\binom{2s^2}{s} > s \cdot s!$ vertices in X_1 which each send s edges with colors from C into a fixed subset Y_3 of S of size s. Since there are $s!$ orderings of the s edges with colors from C on each $x \in X_2$, this implies that for some set $X_3 \subset X_2$ of size s, the $K_{s,s}$ between X_3 and Y_3 is Y_3-canonical. Finally, suppose $|C| < s$. Pick $C'(x) \subset C(x) \backslash C$ of size s for $x \in X_0$. Since $|X_0| = s!m > s\binom{2s^2}{s}$, we find a set $Y^* \subset S$ of s vertices as well as a set $X^* \subset X$ of s vertices $x \in X_0$ such that the edges between x and Y^* have colors from $C'(x)$. Since the sets $C'(x)$ are disjoint, this is a rainbow copy of $K_{s,s}$. □

4.4.2 List colorings in shadows

In this section, we relate Theorem 4.13 to hypergraphs via the following definition. Recall that if H is a 3-graph and $e \in \partial H$, then the neighborhood $N_H(e)$ is the set of $v \in V(H) \setminus e$ such that $e \cup \{v\} \in H$.

Definition 4.14. Let H be a 3-graph. For $G \subset \partial H$ and $e \in G$, let

$$L_G(e) = N_H(e) \setminus V(G).$$

The set $L_G(e)$ is called the *list* of e and the elements of $L_G(e)$ are called *colors*.

Let $L_G = \bigcup_{e \in G} L_G(e)$ – this is the set of colors in the lists of edges of G.

Definition 4.15. A *list-edge-coloring of* G is a map $\chi : G \to L_G$ with $\chi(e) \in L_G(e)$ for all $e \in G$. List-edge-colorings $\chi_1, \chi_2 : G \to L_G$ are *disjoint* if $\chi_1(e) \neq \chi_2(f)$ for all $e, f \in G$.

If χ is an injection – the coloring is rainbow – then clearly $G^+ \subset H$. We require one more definition.

Definition 4.16. Let H be a 3-graph and $m \in \mathbb{N}$. An *m-multicoloring* of $G \subset \partial H$ is a family of list-edge-colorings $\chi_1, \chi_2, \ldots, \chi_m : G \to L_G$ such that $\chi_i(e) \neq \chi_j(e)$ for every $e \in G$ and $i \neq j$.

A necessary and sufficient condition for the existence of an m-multicoloring of G is that all edges of G have codegree at least m in H. We stress here that the definitions are all with respect to the fixed host 3-graph H. The following result will be key to the proofs of Theorem 3.8 and Theorem 3.10.

Theorem 4.17. *Let $m, s \in \mathbb{N}$, let H be a 3-graph, and let $G = K_{t,t} \subset \partial H$. Suppose G has an m-multicoloring. If t is large enough, then there exists $F = K_{s,s} \subset G$ such that F has either a rainbow list-edge-coloring or an m-multicoloring such that the colorings are pairwise disjoint, and each coloring is monochromatic or canonical.*

Proof. Set $s = t_m/m^2$ and $t_m < t_{m-1} < \cdots < t_1 < t_0 = t$ where Theorem 4.13 with input t_i has output t_{i-1}. Pick a color $c_1(e)$ on each edge $e \in G$ and apply Theorem 4.13 to G. We obtain a rainbow, monochromatic or canonical subgraph G_1 of G where $G_1 = K_{t_1,t_1}$. If it is rainbow, then we are done, so assume it is monochromatic or canonical. For every $e \in G_1$, remove $c_1(e)$ from its list. Now pick another color on each edge of G_1 and repeat. We obtain subgraphs $G_m \subset G_{m-1} \subset \cdots \subset G_1$ such that each G_i is monochromatic or canonical where $G_i = K_{t_i,t_i}$ has parts X_i, Y_i. In particular, each coloring of the m-multicoloring of G restricted to G_m is monochromatic or canonical.

Let us assume that we have a monochromatic colorings, b X_m-canonical colorings, and c Y_m-canonical colorings of G_m where $a + b + c = m$. It suffices to ensure that these colorings are pairwise disjoint. A color $\chi(xy)$ in an X_i-canonical coloring of G_i cannot appear in a $Y_{i'}$-canonical coloring of $G_{i'}$ for $i' > i$ as $\chi(xy)$ was deleted from all edges incident to x when forming G_{i+1}. A similar statement holds with X and Y interchanged, so every X_m-canonical coloring of G_m is disjoint from every Y_m-canonical coloring of G_m. The same argument shows that no color in a monochromatic coloring appears in a canonical coloring. It suffices to show that colors on different X_m-canonical colorings are disjoint (and the same for Y_m-canonical).

Let the b X_m-canonical colorings be $\chi_1, \ldots, \chi_b$. Construct an auxiliary graph K with $V(K) = X_m$ where $xx' \in K$ if there exist $i \neq i'$ and a color α that is canonical for x in χ_i and canonical for x' in $\chi_{i'}$. Then the maximum degree of K is at most $b(b-1)$, so K has an independent set of size $s = t_m/m^2 \geq |X_m|/b^2$. Let us restrict X_m to this independent set. We repeat this procedure for Y_m and finally obtain a subgraph $F = K_{s,s}$ with an m-multicoloring that satisfies the requirement of the theorem. □

4.4.3 Random sampling

Let H be a triple system and $K \subset \partial H$. We say that K has *out-codegree at least d* if for every edge $e \in K$, there exist vertices $v_1, v_2, \ldots, v_d \in V(H) \setminus V(K)$ such that $e \cup \{v_i\} \in H$ for $i \leq d$.

Lemma 4.18. *Let $m, t \in \mathbb{N}$, $\delta \in \mathbb{R}_+$ and H be an n-vertex triple system. Suppose that $F \subset \partial H$ and each $f \in F$ has $d_H(f) \geq m$. If $|F| \geq \delta n^2$ and n is large enough, then there is a complete bipartite graph $K_{t,t} \subset F$ with out-codegree at least m.*

Proof. Let T be a random subset of $V(H)$ obtained by picking each vertex independently with probability $p = 1/2$. Let G be the graph of all edges $f \in F$ with $f \subset T$ such that there exist $v_1, v_2, \ldots, v_m \in V(H) \setminus T$ such that $f \cup \{v_i\} \in H$ for $1 \leq i \leq m$. Then

$$\mathbb{E}(|G|) \geq |F| p^2 (1-p)^m \geq \frac{\delta}{2^{m+2}} n^2.$$

So there is a $T \subset V(H)$ with $|G|$ at least this large. If n is large enough, then Theorem 4.3 implies that there exists a complete bipartite graph $K \subset G \subset F$ with parts of size t. Due to the definition of G, the subgraph K satisfies the requirements of the lemma. □

By Lemma 4.18, for a bipartite graph G, the problem of embedding G^+ in a triple system H whose shadow is dense and d-full is reduced to the problem of embedding G^+ in a subgraph of H consisting of a $K_{t,t} \subset \partial H$ with out-codegree at least m.

Corollary 4.19. *Let $m \geq 2$ and $\delta > 0$, and let H be an m-full triple system with n vertices such that $|\partial H| \geq \delta n^2$. If n is large enough, then $G^+ \subset H$ for any bipartite graph G with $|G| \leq m$.*

Proof. Suppose $|V(G)| = t$. By Lemma 4.18, there exists a complete bipartite graph $K = K_{t,t} \subset \partial H$ such that K has out-codegree at least m. Since $G \subset K$ and $|G| \leq m$, we may greedily embed the m vertices of $V(G^+)\backslash V(G)$ in $V(H)$ to obtain $G^+ \subset H$. □

The corollary handles the case $m \geq |G|$, however, we will be applying the lemma with $m \ll |G|$, and in particular with $m = \sigma(G^+)$ in Theorems 3.8 and 3.10 for $r = 3$, and for this we require Theorem 4.17.

5 Proof of Theorems

In the next sections, we will show how to apply all the tools we have developed in section 4 to extremal problems for expansions.

5.1 *Proof of Theorem 3.12*

In this section we discuss how to apply Theorem 4.3 and Lemma 4.2 to obtain a short proof of Theorem 3.12. Since the proof of Theorem 3.13 is very similar, we omit that proof.

Proof of Theorem 3.12. The lower bound in Theorem 3.12 comes from a calculating the number of triangles in the norm graph constructions in [4]. In particular, a $K_{s,t}$-free graph G is constructed in [4] which has $\Theta(n^{3-3/s})$ triangles. Taking H to be the hypergraph whose vertex set is $V(G)$ and whose edge set is the set of triangles in G, we obtain a $K_{s,t}^+$-free triple system with $\Theta(n^{3-3/s})$ edges. Precise details are presented in [52] or [3].

Next we show $\mathrm{ex}_3(n, K_{s,t}^+) = O(n^{3-3/s})$ for $t \geq s \geq 3$. For convenience, set $f(n) = \mathrm{ex}_2(n, K_{s,t})$ and $g(n) = \mathrm{ex}_2(n, K_{s-1,t})$. Theorem 4.3 gives bounds on these quantities. By Lemma 4.2, if H is a triple system with no $K_{s,t}^+$, H has an $(s+1)(t+1)$-full subgraph H' where

$$|H'| \geq |H| - (s+1)(t+1)|\partial H| = |H| - O(n^2).$$

If $|\partial H'| > f(n)$, then we find a $K_{s,t} \subset \partial H'$, which can be greedily expanded to a $K_{s,t}^+ \subset H' \subset H$, a contradiction. Therefore $|\partial H'| \leq f(n)$. If $d = |H|/2f(n)$, then there is a d-full subgraph H'' of H' such that

$$|H''| \geq |H'| - d|\partial H'| \geq \tfrac{1}{2}|H| - O(n^2).$$

Since $|\partial H''| \leq |\partial H'| \leq f(n)$, there is a vertex $v \in V(\partial H'')$ of degree at most $m = f(n)/n$ in $\partial H''$. If $N(v)$ is the neighborhood of v in $\partial H''$, then $N(v)$ induces a subgraph of $\partial H''$ with at most $g(m)$ edges, otherwise we find a $K_{s-1,t}$ in $N(v)$, and adding v we get a $K_{s,t} \subset \partial H'' \subset \partial H'$, a contradiction. However, H'' is d-full, so $N(v)$ induces a subgraph of H'' of minimum degree at least d. This shows $\frac{1}{2}dm \le g(m)$, which implies $|H| = O(g(m)f(n)/m)$. By Theorem 4.3, $f(n) = O(n^{2-1/s})$ which implies $m = O(n^{1-1/s})$. Since $g(m) = O(m^{2-1/(s-1)})$ from Theorem 4.3, we obtain

$$|H| = O(g(m)f(n)/m) = O(n^{3-3/s}).$$

This completes the proof of Theorem 3.12. □

5.2 *Proof of Theorem 3.10 for r ≥ 4*

In this section we show how to use Lemmas 4.4 and 4.11 to prove Theorem 3.10 when $r \ge 4$. Let us assume that $r \ge 4$, G is a forest with v vertices, $\sigma = \sigma(G)$, and $G^+ \not\subset H \subset \binom{[n]}{r}$. We are to prove an upper bound for $|H|$. Put $s = rv$ and let

$$K = \{e \in \partial H : d^*_H(e) \ge s\}.$$

Lemma 4.4 implies that $G \not\subset K$, so the greedy algorithm yields $|K| \le (v-2)n$. Let $V(K) = \{x_1, \dots, x_n\}$ so that $d_K(x_i) \ge d_K(x_{i+1})$. Let $L := \{x_1, \dots, x_p\}$ be the p vertices of highest degree in K, where $p = n^\epsilon$ for some small positive ϵ depending only on G and r. Using $|K| \le (v-2)n$ and an averaging argument yields

$$z := d_K(x_{p+1}) \le \frac{d_K(x_1) + \cdots + d_K(x_{p+1})}{p+1} \le \frac{2|K|}{p+1} < \frac{2(v-2)n}{p}. \tag{3}$$

Consider the partition $H = H_1 \cup H_2$ in Lemma 4.11 and write $H_2 = H_3 \cup H_4 \cup H_5 \cup H_6$ where:

- H_3 is the set of edges in H_2 with center outside L
- $H_4 \subset H$ is the set of edges meeting L in at least two vertices
- $H_5 \subset H_2 \setminus H_3$ is the set of edges f with $|f \cap L| = 1$ and $d_{H_2 \setminus H_3}(f \setminus L) \le \sigma - 1$
- $H_6 := H_2 \setminus (H_3 \cup H_4 \cup H_5)$.

We now show that $|H_3|, |H_4|, |H_6|$ are each $O(n^{r-1-\delta})$ for some $\delta > 0$ while $|H_5| \le (\sigma - 1)\binom{n-1}{r-1}$.

Counting edges of H_3 from their centers, an easy averaging argument, Lemma 4.11 and (3) give

$$|H_3| \le \sum_{i=p+1}^{n} \binom{d_K(x_i)}{r-1} \le \frac{\sum_i d_K(x_i)}{z} \binom{z}{r-1} = O(n^{r-1}/p^{r-2}) = O(n^{r-1-r\epsilon}).$$

Clearly

$$|H_4| \leq \binom{p}{2}\binom{n}{r-2} = O(n^{r-2+2\epsilon}).$$

Counting the edges of H_5 from their portion outside L immediately gives

$$|H_5| \leq (\sigma - 1)\binom{n-p}{r-1}.$$

For $A \in \binom{L}{\sigma}$ let $\mathcal{B}_A := \{B : \{a\} \cup B \in H \text{ for all } a \in A\}$ and

$$H_A := \{f \in H : a \in A, B \in \mathcal{B}_A, \{a\} \cup B = f\}.$$

Each edge $f \in H_6$ can be written in the form $f = \{a\} \cup B$ where $a \in L$, $B \cap L = \emptyset$, such that there are at least σ edges in $H \setminus H_3$ containing B. Moreover, because none of these σ edges are in H_3 their last point is in L. Consequently, $H_6 \subset \bigcup_{A\in\binom{L}{\sigma}} H_A$.

Given a copy of G^+, let Y be a set of σ vertices of this copy that intersect each edge of G^+ in exactly one vertex (here is where we use the definition of σ). Let $\mathcal{C}$ be the $(r-1)$-graph obtained from G^+ by removing the set Y. In other words, every edge of G^+ gets shrunk by removing the unique vertex of Y from it. Since $G^+ \not\subset H$, we conclude that $\mathcal{C} \not\subset \mathcal{B}_A$ for each A. Consequently, $|\mathcal{B}_A| \leq \mathrm{ex}_{r-1}(n, \mathcal{C}) = O(n^{r-2})$ where the bound holds from an easy application of Lemma 4.2. Therefore

$$|H_6| \leq \sum_{A\in\binom{L}{\sigma}} |H_A| \leq \sum_{A\in\binom{L}{\sigma}} |A||\mathcal{B}_A| \leq \sigma\binom{p}{\sigma}O(n^{r-2}) = O(n^{r-2+\epsilon\sigma}).$$

5.3 Proof of Theorems 3.8, 3.10, and 3.11 for r = 3

The proofs of Theorems 3.8, 3.10, and 3.11 for $r = 3$ all use the same template, which we describe as follows. Let G be a tree or a cycle or a graph with $\sigma(G^+) = 2$. Let H be an n-vertex triple system with at least $(\sigma(G^+) - 1 + \varepsilon)\binom{n}{2}$ edges, where $\varepsilon > 0$. If we can show $G^+ \subset H$ when n is large enough, then we have $\mathrm{ex}_3(n, G^+) \sim (\sigma(G^+) - 1)\binom{n}{2}$.

By Lemma 4.2, H has a σ-full subgraph H' with at least $\varepsilon\binom{n}{2}$ edges. Let $|V(G)| = k$. If $|\partial H'| < \frac{\varepsilon}{2k}\binom{n}{2}$, then by Lemma 4.2, H' contains a $2k$-full subgraph H''. In particular, $\partial H''$ has minimum degree at least k, and therefore contains G. Now every edge of this copy of G has degree at least $2k$ in H'', and therefore we may greedily expand G to G^+ in H', as required. Next we consider the case $|\partial H'| \geq \frac{\varepsilon}{2k}\binom{n}{2}$.

By Lemma 4.18, for any t there exists n_0 such that $n > n_0$, then there exists a $K = K_{t,t} \subset \partial H'$ each edge of which has out-codegree at least σ. So if $e \in K$, there exist vertices $v_1(e), v_2(e), \ldots, v_\sigma(e) \in V(H')\backslash V(K)$ such that $e \cup \{v_i(e)\} \in H'$ for $1 \leq i \leq \sigma$. In the language of list colorings, we have σ list colorings $\chi_1, \chi_2, \ldots, \chi_\sigma$ of K, defined by $\chi_i(e) = v_i(e)$ for $e \in K$. By Theorem 4.17, for any s there exists n_1 such that if $n > n_1$, there exists a complete bipartite graph $J \subset \partial H'$ with parts X and Y of size s such that the restriction of χ_i to J is canonical, monochromatic, or rainbow, for $1 \leq i \leq \sigma$.

The remainder of the proof consists in showing that we can embed G^+ in H' using the colorings of J. The proof of the embedding is generally technical for Theorem 3.10 and Theorem 3.11, so we focus on the special case of embedding an expansion of a quadrilateral, i.e., Theorem 3.8 for $k = 4$. In this case $\sigma = 2$, and we let $s = 4$, and we consider the cases one by one:

- *Some χ_i is rainbow.* Then $C_4^+ \subset J^+ \subset H'$.
- *χ_1 and χ_2 are monochromatic.* Suppose $\chi_1(e) = v_1$ for $e \in J$ and $\chi_2(e) = v_2$ for $e \in J$. Let e_1, e_2 and e_3, e_4 form vertex-disjoint paths of length two in J. Then $e_1 \cup \{v_1\}, e_2 \cup \{v_2\}, e_3 \cup \{v_2\}, e_4 \cup \{v_1\}$ is a C_4^+ in H'.
- *χ_1 and χ_2 are both canonical.* Suppose χ_1 and χ_2 are both X-canonical, the other cases are similar. Then pick any quadrilateral $\{e_1, e_2, e_3, e_4\} \subset J$, where $e_1 \cap e_2 \subset X$ and $e_3 \cap e_4 \subset X$. By definition there exist distinct vertices v_1, v_2 and w_1, w_2 such that $\chi_1(e_1) = v_1$, $\chi_2(e_2) = v_2$, $\chi_1(e_3) = w_1$, and $\chi_2(e_4) = w_2$. Then $e_1 \cup \{v_1\}, e_2 \cup \{v_2\}, e_3 \cup \{w_1\}, e_4 \cup \{w_2\}$ is a C_4^+ in H'.
- *χ_1 is canonical and χ_2 is monochromatic.* Suppose χ_1 is X-canonical. Let e_1, e_2, e_3, e_4 form a path of length four in J with $e_2 \cap e_3 \subset Y$. Then there exist distinct vertices v_1, v_2, v_3 such that $\chi_1(e_2) = v_2$, $\chi_1(e_3) = v_3$, and $\chi_2(e_1) = \chi_2(e_4) = v_1$. Now $e_1 \cup \{v_1\}, e_2 \cup \{v_2\}, e_3 \cup \{v_3\}, e_4 \cup \{v_1\}$ is a C_4^+ in H'.

In all cases, $C_4^+ \subset H'$, and we conclude $\mathrm{ex}_3(n, C_4^+) \sim \binom{n}{2}$. The cases of embedding G^+ when G is a cycle or a tree follow the same template, except that the cases are substantially more complicated, and for trees the embedding depends on the structure of the tree relative to crosscuts. □

6 Open problems

When G is a forest or a cycle, or $\sigma(G^+) = 2$, the main theorems in the survey show $\mathrm{ex}_r(n, G^+) \sim (\sigma(G^+) - 1)\binom{n-1}{r}$. It is an interesting general question to determine for which r-graphs F one has $\mathrm{ex}_r(n, F) \sim (\sigma(F) - 1)\binom{n-1}{r}$, and furthermore, for which F the extremal constructions all have a small transversal, or can be transformed by deleting few edges to an r-graph with a small transversal. A general problem is to determine those graphs G with $\mathrm{ex}_r(n, G^+) \sim (\sigma(G^+) - 1)\binom{n}{r-1}$. In the case $r = 3$, the proof template described in Section 5.3 breaks the argument into two broad cases depending on the size of the shadow. If the shadow is very small, then the greedy procedure allows us to embed a rather large class of expansions, including those with treewidth at most two. It is convenient to use the following characterization of graphs with treewidth at most two to see this: start with an edge, and given any edge xy in the current graph, add a new vertex z and the two edges xz and yz. This leads us to pose the following problem:

Problem A. If G has treewidth two, determine whether $\mathrm{ex}_3(n, G^+) \sim (\sigma(G) - 1)\binom{n}{2}$.

By Theorem 3.11, if $\sigma(G^+) = 2$ then $\mathrm{ex}_3(n, G^+) \sim (\sigma(G) - 1)\binom{n}{2}$. The next case $\sigma(G^+) = 3$ is wide open.

Problem B. Determine the order of magnitude of $\mathrm{ex}_3(n, G^+)$ when G is a graph with $\sigma(G^+) = 3$.

The results of [52] show $\mathrm{ex}_3(n, K_{3,t}^+)/n^2 \to \infty$ as $t \to \infty$, so $\mathrm{ex}_3(n, G^+)$ probably depends substantially on the structure of G. However, one may ask if $\mathrm{ex}_3(n, G^+) = O(n^2)$ when $\sigma(G^+) = 3$.

A general problem is to determine for which r-graphs F one has $\mathrm{ex}_r(n, F) = O(n^{r-1})$. Theorem 3.13 shows that if $F = G^+$ and G is a bipartite graph with $\mathrm{ex}_2(n, G) = O(n^{\varphi})$ where $\varphi = \frac{1}{2}(\sqrt{5} - 1)$, then $\mathrm{ex}_3(n, G^+) = O(n^2)$, and this gives $\mathrm{ex}_3(n, Q_3^+) = O(n^2)$ and also $\mathrm{ex}_3(n, G^+) = O(n^2)$ when G is a bipartite graph such that every vertex in some part of G has degree at most two. On the other hand, we saw that $\mathrm{ex}_3(n, G^+)$ is not quadratic in n if in every 3-coloring of G, every pair of color classes induces a cycle, or if $|G| \geq 3|V(G)| - 5$. We pose the following problem. An *acyclic coloring* of a graph G is a proper vertex-coloring of G such that any pair of color classes induce a forest.

Problem C. Determine whether $\mathrm{ex}_3(n, G^+) = O(n^2)$ for every graph G with an acyclic 3-coloring.

In particular, every wheel graph W_k with $k + 1$ vertices, with k even, has an acyclic 3-coloring, and the above problem is open even for W_4. We may phrase the above problem in the language of graphs: if an n-vertex graph has a superquadratic number of triangles, does it contain every graph with a bounded number of vertices and an acyclic 3-coloring, and in particular, every wheel graph W_k with k even?

Recall that a graph is 2-degenerate if each of its subgraphs has minimum degree at most two. Every graph with treewidth two is 2-degenerate, but not vice versa, for example, the graph obtained from K_4 by subdividing an edge is 2-degenerate but has treewidth three. Burr and Erdős [8] conjectured that for every 2-degenerate bipartite graph G, $\mathrm{ex}_2(n, G) = O(n^{3/2})$. Using Theorem 3.13, this would show $\mathrm{ex}_3(n, G^+) = O(n^2)$ for every 2-degenerate bipartite graph G. It is therefore perhaps reasonable to expect that every 2-degenerate graph G has $\mathrm{ex}_3(n, G^+) = O(n^2)$ – note that these graphs all have an acyclic 3-coloring.

Problem D. If G is a 2-degenerate graph, is $\mathrm{ex}_3(n, G^+) = O(n^2)$?

If G is a bipartite graph such that all the vertices in one part have degree at most two, then it is known that $\mathrm{ex}(n, G) = O(n^{3/2})$ by the method of Dependent Random Choice [26], and therefore in this case $\mathrm{ex}_3(n, G^+) = O(n^2)$.

Finally, we mention that one can also consider expansions of hypergraphs: if G is an s-graph then the r-uniform expansion G^+ of G is obtained by adding a set of $r-s$ vertices to each edge of G, with disjoint sets for distinct edges. Recently, Füredi and Jiang [36] extended Theorem 3.10 to a large class of r-uniform expansions of s-graphs forests. Also, the first author and Stading [59] determined the asymptotics for $\text{ex}_r(n, P_t(s)^+)$ for $r \geq 2s$, where $P_t(s)$ is the $2s$-graph consisting of t edges $e_1, \ldots, e_t$ where consecutive edges intersect in s points and nonconsecutive edges are pairwise disjoint.

In general, determining $\text{ex}_r(n, G^+)$ for an s-graph G is a wide open problem with almost no general results known. An interesting case is when G is a tight tree. A *tight tree* is a hypergraph with edge set $\{e_1, e_2, \ldots, e_k\}$ such that for every $i \geq 2$, there exists $j < i$ such that $e_i \cap \bigcup_{h=1}^{i-1} e_h \subset e_j$. In particular, if all the e_i have size two, then we have the usual definition of a graph tree. An important conjecture of Kalai [30], extending the well-known Erdős-Sós conjecture for trees in graphs [17], states that if T is an r-uniform tight tree with k vertices, then $\text{ex}_r(n, T) \leq \frac{k-r}{r}\binom{n}{r-1}$. This conjecture remains open in all but a very few cases, and the proof of the Erdős-Sós conjecture for graphs was only recently claimed. The following general question is raised for expansions:

Problem E. Determine $\lim_{n\to\infty} \frac{\text{ex}_r(n,T^+)}{\binom{n}{r-1}}$ when T is the shadow of an r-uniform tight tree.

Problem A asks whether the limit is $\sigma(T^+) - 1$ in the case $r = 3$, since the shadow of a 3-uniform tight tree is a graph of treewidth two.

Acknowledgements Research of the author Dhruv Mubayi was partially supported by NSF grant DMS-1300138.

Research of the author Jacques Verstraëte was supported by NSF grant DMS-1362650.

References

1. R. Ahlswede, L.H. Khachatrian, The complete intersection theorem for systems of finite sets. Eur. J. Comb. **18**, 125–136 (1997)
2. N. Alon, O. Pikhurko, Personal communication
3. N. Alon, C. Shikhelman, Triangles in H-free graphs. http://arxiv.org/abs/1409.4192
4. N. Alon, L. Rónyai, T. Szabó, Norm-graphs: variations and applications. J. Comb. Theory, Ser. B **76**, 280–290 (1999)
5. B. Bollobás, D.E. Daykin, P. Erdős, Sets of independent edges of a hypergraph. Q. J. Math. Oxford Ser. (2) **27**, 25–32 (1976)
6. J.A. Bondy, U.S.R. Murty, *Graph Theory*. Graduate Texts in Mathematics, vol 244 (Springer, New York, 2008)
7. W.G. Brown, On graphs that do not contain a Thomsen graph. Can. Math. Bull. **9**, 281–285 (1966)
8. S. Burr, P. Erdős, On the magnitude of generalized Ramsey numbers for graphs, in *Infinite and Finite Sets I*, ed. by J. Bolyai, Colloquia Mathematica Societatis, vol. 10 (NorthHolland, Amsterdam, 1975), pp. 214–240
9. N. Bushaw, N. Kettle, Turán Numbers for Forests of Paths in Hypergraphs. arXiv:1303.5022

10. W.Y.C. Chen, J. Liu, L.X. Wang, Families of sets with intersecting clusters. SIAM J. Discrete Math. **23**(3), 1249–1260 (2009)
11. F.R.K. Chung, Unavoidable stars in 3-graphs. J. Comb. Theory (A) **35**, 252–262 (1983)
12. F.R.K. Chung, P. Frankl, The maximum number of edges in a 3-graph not containing a given star. Graphs Comb. **3**(2), 111–126 (1987)
13. R. Csákány, J. Kahn, A homological approach to two problems on finite sets. J. Algebraic Comb. **9**(2), 141–149 (1999)
14. M. Deza, P. Erdős, P. Frankl, Intersection properties of systems of finite sets. Proc. Lond. Math. Soc. (3) **36**, 369–384 (1978)
15. R.A. Duke, P. Erdős, Systems of finite sets having a common intersection, in *Proceedings, 8th S-E Conference on Combinatorics, Graph Theory and Computing* (1977), pp. 247–252
16. P. Erdős, On extremal problems of graphs and generalized graphs. Isr. J. Math. **2**, 183–190 (1964)
17. P. Erdős, Some problems in graph theory, *Theory of Graphs and Its Applications*, ed. by M. Fiedler (Academic Press, New York, 1965), pp. 29–36
18. P. Erdős, A problem on independent r-tuples. Ann. Univ. Sci. Budapest. **8**, 93–95 (1965)
19. P. Erdős, T. Gallai, On maximal paths and circuits of graphs. Acta Math. Acad. Sci. Hungar. **10**, 337–356 (1959)
20. P. Erdős, R. Rado, A combinatorial theorem. J. Lond. Math. Soc. **25**, 249–255 (1950)
21. P. Erdős, M. Simonovits, A limit theorem in graph theory. Studia Sci. Math. Hungar. **1**, 51–57 (1966)
22. P. Erdős, M. Simonovits, Supersaturated graphs and hypergraphs. Combinatorica **3**(2), 181–192 (1983)
23. P. Erdős, M. Simonovits, Cube supersaturated graphs and related problems, in *Progress in Graph Theory, (Waterloo, Ont., 1982)*, ed. by J.A. Bondy, U.S.R. Murty (Academic Press, Toronto, 1984), pp. 203–218
24. P. Erdős, A.H. Stone, On the structure of linear graphs. Bull. Am. Math. Soc. **52**, 1087–1091 (1946)
25. P. Erdős, C. Ko, R. Rado, Intersection theorems for systems of finite sets. Q. J. Math. Oxford Second Series **12**, 313–320 (1961)
26. J. Fox, B. Sudakov, Dependent random choice. Random Struct. Algoritm. **38**, 1–32 (2011)
27. P. Frankl, On families of finite sets no two of which intersect in a singleton. Bull. Aust. Math. Soc. **7**(1), 125–134 (1977)
28. P. Frankl, An extremal problem for 3-graphs. Acta Math. Acad. Sci. Hung. **32**, 157–160 (1978)
29. P. Frankl, Improved bounds for Erdős' matching conjecture. J. Comb. Theory Ser. A **120**(5), 1068–1072 (2013)
30. P. Frankl, Z. Füredi, Exact solution of some Turán-type problems. J. Comb. Theory Ser. A **45**(2), 226–262 (1987)
31. P. Frankl, T. Łuczak, K. Mieczkowska, On matchings in hypergraphs. Electron. J. Comb. **19**, Paper 42, 5 (2012)
32. Z. Füredi, New asymptotics for bipartite Turán numbers. J. Comb. Theory, Ser. A **75**, 141–144 (1996)
33. Z. Füredi, On finite set systems whose every intersection is a kernel of a star. Discrete Math. **47**, 129–132 (1983)
34. Z. Füredi, Linear trees in uniform hypergraphs. Eur. J. Comb. Theory Ser. A **35**, 264–272 (2014)
35. Z. Füredi, T. Jiang, Hypergraph Turán numbers of linear cycles. J. Comb. Theory Ser. A, **123**, 252–270 (2014)
36. Z. Füredi, T. Jiang, Turan number of Hypergraph trees. arxiv1505.03210v1
37. Z. Füredi, M. Simonovits, O. Pikhurko, On triple systems with independent Neighborhoods. Probab. Comput, **14**, 795–813 (2005)
38. Z. Füredi, D. Mubayi, O. Pikhurko, Quadruple systems with independent neighborhoods. J. Comb. Theory, Ser. A **115**(8), 1552–1560 (2008)

39. Z. Füredi, M. Simonovits, Triple systems not containing a Fano configuration. Comb. Probab. Comput **14**, 467–484 (2005)
40. Z. Füredi, T. Jiang, R. Seiver, Exact solution of the hypergraph Turán problem for k-uniform linear paths. Combinatorica **34**(3), 299–322 (2014)
41. R. Gu, X. Li, Y. Shi, Turán numbers of vertex disjoint cycles. arXiv:1305.5372
42. H. Huang, P. Loh, B. Sudakov, The size of a hypergraph and its matching number. Comb. Probab. Comput. **21**, 442–450 (2012)
43. T. Jiang, O. Pikhurho, Z. Yilma, Set systems without a strong simplex. SIAM J. Discrete Math. **24**, 1038–1045 (2010)
44. P. Keevash, D. Mubayi, Stability results for cancellative hypergraphs. J. Comb. Theory Ser. B **92**(1), 163–175 (2004)
45. P. Keevash, D. Mubayi, Set systems without a simplex or a cluster. Combinatorica **30**(2), 175–200 (2010)
46. P. Keevash, D. Mubayi, R. Wilson, Set systems with no singleton intersection. SIAM J. Discrete Math. **20**(4), 1031–1041 (2006)
47. P. Keevash, B. Sudakov, The Turán number of the Fano plane. Combinatorica **25**, 561–574 (2005)
48. P. Keevash, B. Sudakov, On a hypergraph Turán problem of Frankl. Combinatorica **25**, 673–706 (2005)
49. J. Kim, Regular subgraphs of uniform hypergraphs. arXiv:1502.02177 [math.CO]
50. A. Kostochka, D. Mubayi, J. Verstraëte, Turán problems and shadows I: paths and cycles. J. Comb. Theory Ser. A (to appear)
51. A. Kostochka, D. Mubayi, J. Verstraëte, Turán problems and shadows II: trees (submitted)
52. A. Kostochka, D. Mubayi, J. Verstraëte, Turán problems and shadows III: expansions of graphs, SIAM Journal on Discrete Mathematics **29**(2), 868–876 (2015)
53. T. Kövari, V.T. Sós, P. Turán, P, On a problem of K. Zarankiewicz. Colloquium Math. **3**, 50–57 (1954)
54. T. Łuczak, K. Mieczkowska, On Erdős' extremal problem on matchings in hypergraphs. J. Comb. Theory Ser. A **124**, 178–194 (2014)
55. D. Mubayi, A hypergraph extension of Turán's theorem. J. Comb. Theory, Ser. B **96**(1), 122–134 (2006)
56. D. Mubayi, An intersection theorem for four sets. Adv. Math. **215**(2), 601–615 (2007)
57. D. Mubayi, R. Ramadurai, Simplex stability. Comb. Probab. Comput. **18**(3), 441–454 (2009)
58. D. Mubayi, R. Ramadurai, Set systems with union and intersection constraints. J. Comb. Theory Ser. B **99**(3), 639–642 (2009)
59. D. Mubayi, R. Stading, Ph.D thesis of R. Stading, University of Illinois at Chicago (2014)
60. D. Mubayi, J. Verstraëte, Proof of a conjecture of Erdős on triangles in set-systems. Combinatorica **25**(5), 599–614 (2005)
61. D. Mubayi, J. Verstraëte, Minimal paths and cycles in set-systems. Eur. J. Comb. **28**(6), 1681–1693 (2007)
62. D. Mubayi, J. Verstraëte, Two-regular subgraphs of hypergraphs. J. Comb. Theory Ser. B **99**(3), 643–655 (2009)
63. O. Pikhurko, Exact computation of the hypergraph Turán function for expanded complete 2-graphs. J. Comb. Theory Ser. B **103**, 220–225 (2013)
64. M. Simonovits, A method for solving extremal problems in graph theory, stability problems, in *Theory of Graphs (Proc. Colloq., Tihany, 1966)* (Academic Press, New York, 1968), pp. 279–319
65. V.T. Sós, Some remarks on the connection of graph theory, finite geometry and block designs, in Proceedings of the Combinatorial Conference, Rome (1976), pp. 223–233
66. P. Turán, On an extremal problem in graph theory. Mat. Fiz. Lapok **48**, 436–452 (1941)
67. R.M. Wilson, The exact bound in the Erdős-Ko-Rado theorem. Combinatorica **4**, 247–257 (1984)
68. K. Zarankiewicz, Problem of P101. Colloq. Math. **2**, 301 (1951)

Recent advances on Dirac-type problems for hypergraphs

Yi Zhao

Abstract A fundamental question in graph theory is to establish conditions that ensure a graph contains certain spanning subgraphs. Two well-known examples are Tutte's theorem on perfect matchings and Dirac's theorem on Hamilton cycles. Generalizations of Dirac's theorem, and related matching and packing problems for hypergraphs, have received much attention in recent years. New tools such as the absorbing method and regularity method have helped produce many new results, and yet some fundamental problems in the area remain unsolved. We survey recent developments on Dirac-type problems along with the methods involved, and highlight some open problems.

Keywords Perfect matchings • Hypergraphs • Hamilton cycles • Graph packing • Absorbing method

Mathematics Subject Classification (2010): 05C65, 05C70, 05C45, 05C35

Given two (hyper)graphs F and H, which conditions guarantee H contains F as a subgraph? When $|V(F)| = |V(H)|$, the decision problem of whether H contains F is often NP-complete, *e.g.*, deciding if a graph H contains a Hamilton cycle is a well-known NP-complete problem. Therefore it is natural to look for sufficient conditions for such problems. A classical result of Dirac [13] states that every graph on $n \geq 3$ vertices with minimum degree $n/2$ contains a Hamilton cycle. Problems that relate the minimum degree (in general, minimum d-degree in hypergraphs) to the structure of the (hyper)graphs are often referred to as *Dirac-type problems*. The Dirac-type problems for hypergraphs have received much attention in recent years. In this survey we concentrate on three such problems: matching problems (Section 1), packing problems (Section 2), and Hamilton cycles (Section 3). Many problems in this survey were already considered in the survey of Rödl and Ruciński [67].

Y. Zhao (✉)
Department of Mathematics and Statistics, Georgia State University, Atlanta, GA 30303, USA
e-mail: yzhao6@gsu.edu

A. Beveridge et al. (eds.), *Recent Trends in Combinatorics*, The IMA Volumes in Mathematics and its Applications 159, DOI 10.1007/978-3-319-24298-9_6

However, since this is a fast-growing area, there are new developments in the last few years and we will emphasize these new advances. Since we only consider Dirac-type problems, we do not discuss matching, packing, or Hamilton cycles in random or quasi-random hypergraphs. We also omit corresponding results in graphs and digraphs. Many results that we omit can be found in other surveys, *e.g.*, Kühn and Osthus [52, 54, 55], and Gould [23, 24].

1 Matching problems

Given $k \geq 2$, a *k-uniform hypergraph* (*k-graph*) consists of a vertex set V and an edge set E, where each edge is a k-element subset (k-subset) of V. Thus a 2-graph is simply a graph. In this survey a *hypergraph* refers to a k-graph with $k \geq 3$. Given a k-graph H with $k \geq 2$, a *matching of size s* is a collection of s disjoint edges; a *perfect matching* is a matching that covers the vertex set of H (thus it is necessary that k divides $|V(H)|$). Many open problems in combinatorics can be formulated as a problem of finding perfect matchings in hypergraphs, *e.g.*, Ryser's conjecture that every Latin square of odd order has a transversal, and the existence of combinatorial designs (recently solved by Keevash [41]).

A well-known result of Tutte [83] characterized all the graphs with perfect matchings and there are efficient algorithms (e.g., Edmond's algorithm [14]) that determine if a graph has a perfect matching. However, deciding if a 3-partite 3-graph contains a perfect matching is among the first 21 NP-complete problems given by Karp [37]. Therefore it is natural to look for sufficient conditions that guarantee a perfect matching.

1.1 Degree Conditions for Perfect Matchings

There are multiple ways to define degrees in hypergraphs. Given a k-graph H with a set S of d vertices, where $0 \leq d \leq k-1$, the *degree* of S, denoted by $\deg_H(S)$ or simply $\deg(S)$, is the number of edges containing S. The *minimum d-degree* $\delta_d(H)$ of H is the minimum of $\deg(S)$ over all d-subsets S of $V(H)$. Hence $\delta_0(H) = e(H)$ is the number of edges in H. We refer to $\delta_1(H)$ as the *minimum vertex degree* of H and $\delta_{k-1}(H)$ as the *minimum codegree* of H. The simple monotonicity

$$\frac{\delta_0(H)}{\binom{n}{k}} \geq \frac{\delta_1(H)}{\binom{n-1}{k-1}} \geq \cdots \geq \frac{\delta_{k-1}(H)}{n-k+1}$$

suggests that a codegree condition is stronger than other degree conditions. Bollobás, Daykin, and Erdős [5] first related the minimum (vertex) degree to the existence of a large (but far from perfect) matching in k-graphs. Daykin and Häggkvist [12] extended this result by showing that every k-graph with $\delta_1(H) \geq (1-1/k)\binom{n-1}{k-1}$ contains a perfect matching.

Given integers $d < k \le n$ such that k divides n, define the *minimum d-degree threshold* $\mathbf{m_d(k,n)}$ as the smallest integer m such that every k-graph H on n vertices with $\delta_d(H) \ge m$ contains a prefect matching. A simple greedy argument shows that $m_1(2,n) = n/2$ for all $n \in 2\mathbb{N}$. Given $k \ge 3$, a result of Rödl, Ruciński, and Szemerédi [70] on Hamilton cycles implies that $m_{k-1}(k,n) \le n/2 + o(n)$. Kühn and Osthus [51] sharpened this bound to $m_{k-1}(k,n) \le n/2 + 3k^2\sqrt{n\log n}$ by reducing the problem to the one for k-partite k-graphs. Rödl, Ruciński, and Szemerédi [71] improved it further to $m_{k-1}(k,n) \le n/2 + O(\log n)$ by using the absorbing method. Rödl, Ruciński, Schacht, and Szemerédi [72] found a simple proof of $m_{k-1}(k,n) \le n/2 + k/4$. Finally Rödl, Ruciński, and Szemerédi [74] determined $m_{k-1}(k,n)$ exactly for all $k \ge 3$ and sufficiently large n (again by the absorbing method). In order to state this and later results, let us describe a class of extremal configurations that are usually referred to as *divisibility barriers*.

Construction 1.1. [79] *Define $\mathcal{H}_{\text{ext}}(n,k)$ to be the family of all k-graphs $H = (V,E)$, in which there is a partition of V into two parts A,B and $i \in \{0,1\}$ such that $|A| \ne i|V|/k \mod 2$ and $|e \cap A| = i \mod 2$ for all edges $e \in E$.*

It is easy to see that no hypergraph $H \in \mathcal{H}_{\text{ext}}(n,k)$ contains a perfect matching. Indeed, suppose H contains a perfect matching M, then $|A| = \sum_{e\in M} |e\cap A| = i|V|/k \mod 2$, contradicting the definition of H.

Define $\delta(n,k,d)$ to be the maximum of the minimum d-degrees among all the hypergraphs in $\mathcal{H}_{\text{ext}}(n,k)$ and note that $m_d(k,n) > \delta(n,k,d)$. It is easy to see that

$$\delta(n,k,k-1) = \begin{cases} n/2-k+2 & \text{if } k/2 \text{ is even and } n/k \text{ is odd} \\ n/2-k+3/2 & \text{if } k \text{ is odd and } (n-1)/2 \text{ is odd} \\ n/2-k+1/2 & \text{if } k \text{ is odd and } (n-1)/2 \text{ is even} \\ n/2-k+1 & \text{otherwise.} \end{cases} \qquad (1)$$

Theorem 1.2. [74] *For $k \ge 3$, $m_{k-1}(k,n) = \delta(n,k,k-1)+1$ for sufficiently large n.*

With more case analysis, one may determine $\delta(n,k,k-2)$, *e.g.*, it is shown in [79] that

$$\delta(n,4,2) \le \frac{n^2}{4} - \frac{5n}{4} - \frac{\sqrt{n-3}}{2} + \frac{3}{2}$$

and equality holds for infinitely many n. In general, $\delta(n,k,d) = (1/2 + o(1))\binom{n-d}{k-d}$ for any fixed $k > d$ but the general formula of $\delta(n,k,d)$ is unknown – this is related to the open problem of finding the minima of binary Krawtchouk polynomials. Nevertheless, Treglown and the author [79, 80] determined $m_d(k,n)$ in terms of $\delta(n,k,d)$ for all $d \ge k/2$.

Theorem 1.3. [79, 80] *Let $k \ge 3$ and $d \ge k/2$. Then $m_d(k,n) = \delta(n,k,d) + 1$ for sufficiently large n.*

Previously Pikhurko [66] showed that $m_d(k,n) = (1/2+o(1))\binom{n-d}{k-d}$ for all $d \geq k/2$. Independently Czygrinow and Kamat [8] determined $m_2(4,n)$ for sufficiently large n.

Another class of extremal constructions are known as *space barriers*.

Construction 1.4. *Given $s,k,n \in \mathbb{N}$ such that $s \leq \lceil n/k \rceil$ (k may not divide n), let $H_s^0(n,k)$ be the k-graph on n vertices whose vertex set is partitioned into two parts A and B such that $|A| = s-1$, and whose edge set consists of all those edges with at least one vertex in A. When k divides n, let $H^0(n,k) := H_{n/k}^0(n,k)$.*

Since each edge contains at least one vertex from A and $|A| < s$, $H_s^0(n,k)$ contains no matching of size s; in particular, $H^0(n,k)$ contains no prefect matching. Note that

$$\delta_d(H^0(n,k)) = \binom{n-d}{k-d} - \binom{(1-1/k)n-d+1}{k-d} \approx \left(1-\left(\frac{k-1}{k}\right)^{k-d}\right)\binom{n-d}{k-d}.$$

Hàn, Person, and Schacht [33] proved that $m_1(3,n) = (5/9+o(1))\binom{n}{2} \approx \delta_1(H^0(n,3))$. Khan [46] and independently Kühn, Osthus, and Treglown [58] obtained that $m_1(3,n) = \delta_1(H^0(n,3)) + 1$ for sufficiently large n. Khan [45] also proved that $m_1(4,n) = \delta_1(H^0(n,4)) + 1$ for sufficiently large n. Alon, Frankl, Huang, Rödl, Ruciński, and Sudakov [3] determined $m_d(k,n)$ asymptotically for all $d \geq k-4$, including the new cases when $(k,d) = (5,1)$, $(5,2)$, $(6,2)$, and $(7,3)$. Very recently Treglown and the author [81] determined $m_2(5,n)$ and $m_3(7,n)$ exactly for sufficiently large n.

All these results point to the following conjecture, whose asymptotic version (2) has appeared earlier, *e.g.*, [33, 52].

Conjecture 1.5. [81] *Let $k,d \in \mathbb{N}$ such that $d \leq k-1$. Then for sufficiently large $n \in k\mathbb{N}$,*

$$m_d(k,n) = \max\left\{\delta(n,k,d),\ \binom{n-d}{k-d} - \binom{(1-1/k)n-d+1}{k-d}\right\} + 1.$$

In particular,

$$m_d(k,n) = \left(\max\left\{\frac{1}{2},\ 1-\left(\frac{k-1}{k}\right)^{k-d}\right\} + o(1)\right)\binom{n-d}{k-d}. \tag{2}$$

Note that for all $1 \leq d \leq k-1$,

$$\left(\frac{k-1}{k}\right)^{k-d} < \left(\frac{1}{e}\right)^{1-\frac{d}{k}} \quad \text{and} \quad 1-\left(\frac{k-1}{k}\right)^{k\ln 2} \to \frac{1}{2} \quad \text{as} \quad k \to \infty,$$

where *ln* denotes the natural logarithm. Thus, for $1 \ll k \ll n$, if d is significantly bigger than $(1-\ln 2)k \approx 0.307k$, then $\delta(n,k,d) > \binom{n-d}{k-d} - \binom{(1-1/k)n-d+1}{k-d}$. On the other hand, if d is smaller than $(1-\ln 2)k$, then $\delta(n,k,d) < \binom{n-d}{k-d} - \binom{(1-1/k)n-d+1}{k-d}$ for sufficiently large n.

Other than the aforementioned results, no other asymptotic or exact value of $m_d(k,n)$ is known. When $k \geq 3$ and $1 \leq d < k/2$, Hàn, Person, and Schacht [33] gave a general bound: $m_d(k,n) \leq ((k-d)/k + o(1))\binom{n-d}{k-d}$. This was improved by Markström and Ruciński [64] to $m_d(k,n) \leq ((k-d)/k - 1/k^{k-d} + o(1))\binom{n-d}{k-d}$ and by Kühn, Osthus, and Townsend [59] to

$$m_d(k,n) \leq \left(\frac{k-d}{k} - \frac{k-d-1}{k^{k-d}} + o(1)\right)\binom{n-d}{k-d}.$$

Let us discuss proof techniques. Most aforementioned results were obtained by the *absorbing method*, initiated by Rödl, Ruciński, and Szemerédi [70]. Roughly speaking, the absorbing method reduces the task of finding a spanning sub(hyper)graph to that of finding a near spanning sub(hyper)graph by using some *absorbing structure*. Given a k-graph that contains a matching M and a vertex set S such that $V(M) \cap S = \emptyset$, we say that M *absorbs* S if there is another matching M' with $V(M') = V(M) \cup S$. Suppose we want to prove (2) for some $d < k$. Let $a = \max\{\frac{1}{2}, 1 - (\frac{k-1}{k})^{k-d}\}$. Let H be a k-graph with $\delta_d(H) \geq (a + 2\gamma)\binom{n-d}{k-d}$ for some $\gamma > 0$. We first apply the following absorbing lemma of Hàn, Person, and Schacht [33, Lemma 2.4].

Lemma 1.6. [33] *For all $\gamma > 0$ and positive integers $k > d$ there exists n_0 such that the following holds for all $n \geq n_0$. Suppose H is a k-graph on n vertices with $\delta_d(H) \geq (\frac{1}{2}+2\gamma)\binom{n-d}{k-d}$, then H contains a matching M of size $\gamma^k n/k$ that can absorb any vertex set $W \subseteq V(H) \setminus V(M)$ with $|W| \in k\mathbb{N}$ and $|W| \leq \gamma^{2k} n$.*

We next remove $V(M)$ from H and let $H' = H[V(H) \setminus V(M)]$. Then $\delta_d(H) \geq (a+\gamma)\binom{n-d}{k-d}$. If we can show that H' contains a matching that covers all but at most $\gamma^{2k} n$ vertices, then M can absorb these vertices and we obtain the desired perfect matching of H. Thus it suffices to prove that *every k-graph on n vertices with $\delta_d(H) \geq (a + o(1))\binom{n-d}{k-d}$ contains a matching that covers all but $o(n)$ vertices.* We call this assertion the *almost perfect matching lemma*, and note that it is weaker than an asymptotic version of Conjecture 1.10 from Section 1.2. When $d \geq k/2$, this lemma essentially follows from a greedy argument (see, *e.g.*, [33, Theorem 1.3]), but it appears hard to prove the lemma in general when $d < k/2$. As shown in [3, 67], it suffices to find an almost perfect *fractional matching* instead (see Section 1.3 for details). This helps the authors of [3] to obtain $m_d(k,n)$ asymptotically for $d \geq k-4$. However, finding an almost perfect fractional matching for smaller d is still an open problem (see Conjecture 1.12).

Now suppose we want to find $m_d(k,n)$ *exactly*. Naturally we separate the *extremal case* (when H is close to the extremal configuration) from the non-extremal case. The proof for the non-extremal case follows the procedure described above, except that we may assume that H is *not* close to the extremal configuration when proving the absorbing lemma and the almost perfect matching lemma. For example, suppose $1 - (\frac{k-1}{k})^{k-d} < 1/2$, we need the following refinement of Lemma 1.6. Let $\varepsilon > 0$ and H and H' be two k-graphs on the same n vertices. We say that H is *ε-close to H'* if H becomes a copy of H' after adding and deleting at most εn^k edges. Let $\mathcal{B}_{n,k}$ ($\overline{\mathcal{B}}_{n,k}$) denote the k-graph whose vertex set can be partitioned into A, B with

$|A| = \lfloor n/2 \rfloor$ and $|B| = \lceil n/2 \rceil$ such that all its edges intersect A in an odd (even) number of vertices.

Theorem 1.7. [81, Theorem 5] *Given any $\varepsilon > 0$ and integer $k \geq 2$, there exist $0 < \alpha, \xi < \varepsilon$, and $n_0 \in \mathbb{N}$ such that the following holds. Suppose that H is a k-graph on $n \geq n_0$ vertices with $\delta_1(H) \geq \left(\frac{1}{2} - \alpha\right)\binom{n-1}{k-1}$. Then H is ε-close to $\mathcal{B}_{n,k}$ or $\overline{\mathcal{B}}_{n,k}$, or H contains a matching M of size $|M| \leq \xi n/k$ that absorbs any set $W \subseteq V(H) \setminus V(M)$ such that $|W| \in k\mathbb{N}$ with $|W| \leq \xi^2 n$.*

When $1 - (\frac{k-1}{k})^{k-d} < 1/2$, a general extremal case was solved in [79, Theorem 4.1]: *for $1 \leq d < k$, there exists n_0 such that every k-graph H on $n \geq n_0$ vertices contains a perfect matching if H is ε-close to $\mathcal{B}_{n,k}$ or $\overline{\mathcal{B}}_{n,k}$ and $\delta_d(H) \geq \delta(n,k,d) + 1$.*

1.2 Smaller matchings

In this section we discuss the connection between $\nu(H)$, the size of the largest matching in H, and the minimum d-degree $\delta_d(H)$. Given $0 \leq d < k \leq n$ and $s \leq n/k$, let $\mathbf{m}_\mathbf{d}^s(\mathbf{k}, \mathbf{n})$ be the smallest m such that every k-graph H on n vertices with $\delta_d(H) \geq m$ contains a matching of size s.

When $d = k - 1$, Rödl, Ruciński, and Szemerédi [70, 74] noticed a striking contrast between the codegree threshold for perfect matchings (about $n/2$ as seen in Theorem 1.2) and the one for almost perfect matchings (about n/k as shown below). We present the simple proof from [74] because it is a good example of a greedy argument.

Proposition 1.8. [74, Fact 2.1] *Given integers $n \geq k \geq 2$, every k-graph H with $\delta_{k-1}(H) \geq s$ for some $s \leq \lfloor n/k \rfloor - k + 2$ contains a matching of size s.*

Proof. Let M be the largest matching of H and $U = V(H) \setminus V(M)$. Suppose to the contrary, that $|M| < s$ and thus $|V(M)| \leq k(\lfloor n/k \rfloor - k + 1) \leq n - (k-1)k$. Hence $|U| \geq (k-1)k$. Let $S_1, \ldots, S_k$ denote (arbitrary) k disjoint $(k-1)$-subsets of U. By the minimum codegree condition, we have $\sum_{i=1}^k \deg(S_i) \geq ks$. All the neighbors of S_i are in $V(M)$ – otherwise we can enlarge M. Since $|M| < s$, there exists an edge $e \in M$ that contains at least $k+1$ neighbors of $S_1, \ldots, S_k$. Consequently there are two vertices $v_1, v_2 \in e$ and $i \neq j$ such that $v_1 \in N(S_i)$ and $v_2 \in N(S_j)$. By replacing e with $\{v_1\} \cup S_i$ and $\{v_2\} \cup S_j$, we obtain a matching of size $|M| + 1$, contradiction. □

When k does not divide n, the largest matching in a k-graph on n vertices is of size $\lfloor n/k \rfloor$. Such matching is called a *near perfect* matching. Rödl, Ruciński, and Szemerédi [74] proved that every k-graph H on n vertices with $\delta_{k-1}(H) \geq n/k + O(\log n)$ contains a near perfect matching and conjectured that $\delta_{k-1}(H) \geq \lfloor n/k \rfloor$ suffices. Using the absorbing method, Han [26] recently proved this conjecture.

Theorem 1.9. [26] *Let $n \geq k \geq 2$ be integers such that $k \nmid n$ and n is sufficiently large. Then every k-graph H on n vertices with $\delta_{k-1}(H) \geq \lfloor n/k \rfloor$ contains a matching of size $\lfloor n/k \rfloor$.*

A corollary of Theorem 1.9 is $m^s_{k-1}(k,n) \le s$ for *all* $s < n/k$ and sufficiently large n, which generalizes Proposition 1.8. To see this, let $H = (V,E)$ be a k-graph on n vertices with $\delta_{k-1}(H) \ge s$ for some $s < n/k$. By adding a set T of about $\frac{n-ks}{k-1}$ vertices to V and all the k-subsets of $V \cup T$ that intersect T to E, we obtain a k-graph H' on n' vertices with $\delta_1(H') \ge \lfloor n'/k \rfloor$ and $n' \notin k\mathbb{N}$. Applying Theorem 1.9, we obtain a near perfect matching of H'. Removing the vertices of T from this matching, we obtain a matching in H of size at least s.

For any integer $s \le n/k$, the k-graph $H^0_s(n,k)$ defined in Construction 1.4 satisfies $\nu(H^0_s(n,k)) = s-1$ and $\delta_d(H) = \binom{n-d}{k-d} - \binom{n-s-d+1}{k-d}$ for all $0 \le d \le k-1$. This implies that

$$m^s_d(k,n) \ge \binom{n-d}{k-d} - \binom{n-s-d+1}{k-d} + 1 \quad \text{for all } s \le n/k. \tag{3}$$

Combining this with the aforementioned corollary of Theorem 1.9, we obtain that for all $k \ge 2$ and sufficiently large n,

$$m^s_{k-1}(k,n) = s \qquad \text{for all } s < n/k.$$

This prompts us to conjecture that $H^0_s(n,k)$ provides the correct value of $m^s_d(k,n)$ for all $d < k$ and most $s < n/k$. As noted in Section 1.1, an asymptotic version of this conjecture already implies (2), which determines $m_d(k,n)$ asymptotically for all $d < k$.

Conjecture 1.10. *Given $1 \le d \le k-2$, there exist n_0 and C such that*

$$m^s_d(k,n) = \binom{n-d}{k-d} - \binom{n-s-d+1}{k-d} + 1 \tag{4}$$

for all $n \ge n_0$ and all $s \le n/k - C$.

When $d = 1$, a result of Bollobás, Daykin, and Erdős [5] showed that (4) holds for all $n > 2k^3(s+1)$. When $d = 1$ and $k = 3$, Kühn, Osthus, and Treglown [58] proved (4) for all $s \le n/3$ and sufficiently large n (not necessarily divisible by 3). Kühn, Osthus, and Townsend [59, Conjecture 1.3] proposed an asymptotic version of Conjecture 1.10,

$$m^s_d(k,n) = \left(1 - \left(1 - \frac{s}{n}\right)^{k-d} + o(1)\right)\binom{n-d}{k-d},$$

and proved it for all $s \le \min\{\frac{n}{2(k-d)}, \frac{n-o(n)}{k}\}$.

1.3 *Erdős Conjecture and fractional matching*

How many edges of a k-graph guarantee a matching of size s? This question dates back to an old conjecture of Erdős [15] that has received much attention lately.

Conjecture 1.11. [15] *Let s, k, n be integers such that $2 \le k \le n$ and $1 \le s \le n/k$. Then*

$$m_0^s(k,n) = \max\left\{\binom{ks-1}{k}, \binom{n}{k} - \binom{n-s+1}{k}\right\} + 1.$$

The lower bound comes from (3) and the k-graph consisting of a complete k-graph K_{ks-1}^k on $ks-1$ vertices and $n-ks+1$ isolated vertices.

The $s = 2$ case of Conjecture 1.11 is the well-known Erdős-Ko-Rado theorem [17]. A classic theorem of Erdős and Gallai [16] confirms the conjecture for $k = 2$. Erdős [15] proved the conjecture for $n \ge n_0(k,s)$. Bollobás, Daykin, and Erdős [5] proved the conjecture for $n > 2k^3(s-1)$ and Huang, Loh, and Sudakov [36] recently improved it to $n \ge 3k^2 s$. When $k = 3$, Frankl, Rödl, and Ruciński [21] proved the conjecture for $n \ge 4s$ while Łuczak and Mieczkowska [63] proved it for sufficiently large s. Recently Frankl [18] proved the conjecture for $k = 3$. Frankl [19] also proved the conjecture for $s \le n/(2k)$.

Alon et al. [3] considered a fractional version of the Erdős conjecture. Let $H = (V,E)$ be a k-graph on n vertices. A *fractional matching* in H is a function $w : E \to [0,1]$ such that for each $v \in V$ we have $\sum_{e \ni v} w(e) \le 1$. The *size* of w, denoted by $\nu^*(H)$, is $\sum_{e\in E} w(e) = \frac{1}{k}\sum_v \sum_{e\ni v} w(e) \le n/k$. If $\nu^*(H) = n/k$, then we call w a *perfect fractional matching*. Determining $\nu^*(H)$ is a linear programming problem. Its dual problem is finding a minimum *fractional vertex cover* $\tau^*(H) = \sum_{v\in V} w(v)$ over all functions $w : V \to [0,1]$ such that $\sum_{v\in e} w(v) \ge 1$ for all $e \in E$. Correspondingly $\tau(H)$ is the minimum number of vertices in a vertex cover of H.

Given integers $n \ge k > d \ge 0$ (with $k \ge 2$) and a real number $0 < s \le n/k$, define $\mathbf{f_d^s(k,n)}$ to be the smallest integer m such that every k-graph H on n vertices with $\delta_d(H) \ge m$ contains a fractional matching of size s. We let $f_d(k,n) := f_d^{n/k}(k,n)$ (note that n/k may not be an integer). The hypergraph $H^0_{\lceil s\rceil}(n,k)$ defined in Construction 1.4 contains no fractional matching of size s because

$$\nu^*(H^0_{\lceil s\rceil}(n,k)) = \tau^*(H^0_{\lceil s\rceil}(n,k)) \le \tau(H^0_{\lceil s\rceil}(n,k)) = \lceil s\rceil - 1 < s.$$

When s is an integer, the complete k-graph K_{ks-1}^k contains no fractional matching of size s. When $s = n/k$, it was shown in [71] that $f_{k-1}(k,n) = \delta_{k-1}(H^0_{\lceil n/k\rceil}(n,k)) + 1 = \lceil n/k\rceil$. The following two conjectures were given in [3].

Conjecture 1.12. [3]

(i) *For all $1 \le d \le k-1$, $f_d(k,n) = \left(1 - \left(\frac{k-1}{k}\right)^{k-d} + o(1)\right)\binom{n-d}{k-d}$.*

(ii) *For all positive integers n, k, s such that $k \ge 2$ and $s \le n/k$,*

$$f_0^s(k,n) = \max\left\{\binom{ks-1}{k}, \binom{n}{k} - \binom{n-s+1}{k}\right\} + 1.$$

Alon et al. [3] first proved Conjecture 1.12 (ii) asymptotically for $k \in \{3, 4\}$ by reducing it to an old probabilistic conjecture of Samuels [76] and then they applied this result to prove (i) with $k - d \in \{3, 4\}$.

One can convert an almost perfect fractional matching to an almost perfect integer matching by applying either a result of Frankl and Rödl [20] or the weak Regularity Lemma (see [3, Section 4] and [59, Section 5] for details of these two approaches). This implies the following connection between $m_d(k, n)$ and $f_d(k, n)$.

Theorem 1.13. [3] *For* $1 \leq d \leq k - 1$ *if there exists* $c > 0$ *such that* $f_d(k, n) = (c + o(1))\binom{n-d}{k-d}$ *then*

$$m_d(k, n) = \left(\max\left\{\frac{1}{2}, c\right\} + o(1)\right)\binom{n-d}{k-d}.$$

This theorem implies that to determine $m_d(k, n)$, it suffices to prove Conjecture 1.12 (i). Theorem 1.13 was slightly improved in [81] to

$$m_d(k, n) = \max\left\{\delta(n, k, d) + 1, \ (c + o(1))\binom{n-d}{k-d}\right\},$$

where $\delta(n, k, d)$ was defined in Section 1.1.

1.4 Other Matching problems

First we consider matching problems in multipartite hypergraphs. Let $\mathcal{K}_k(n)$ denote the family of all k-partite k-graphs H with parts $V_1, \cdots, V_k$ such that $|V_1| = \cdots = |V_k| = n$. Given $I \subseteq [k]$, a vertex set S is *I-crossing* if $|S \cap V_i| = 1$ for $i \in I$ and $S \cap V_i = \emptyset$ otherwise. Let $\delta_I(H)$ be the minimum of $\deg(S)$ over all I-crossing sets S and let $\delta'_d(H) := \min \delta_I(H)$ over all $I \subseteq [k]$ of size d. Kühn and Osthus [51] proved that every $H \in \mathcal{K}_k(n)$ contains a perfect matching if $\delta'_{k-1}(H) \geq n/2 + \sqrt{2n \log n}$. Aharoni, Georgakopoulos, and Sprüssel [1] found an elegant proof of the following theorem.

Theorem 1.14. [1] *Let* $H \in \mathcal{K}_k(n)$. *If* $\deg(S) > n/2$ *for all* $[k-1]$*-crossing sets* S *and* $\deg(S) \geq n/2$ *for all* $\{2, 3, \ldots, k\}$*-crossing sets* S, *then* H *contains a perfect matching.*

This theorem implies that H contains a perfect matching if $\delta'_{k-1}(H) > n/2$ for all values of n. This is tight when k is even and $n = 2 \pmod 4$ but may not be tight in other cases (off by at most one). Pikhurko [66] proved that every $H \in \mathcal{K}_k(n)$ contains a perfect matching if there exists nonempty $L \subsetneq [k]$ such that

$$\frac{\delta_L(H)}{n^{k-|L|}} + \frac{\delta_{[k]\setminus L}(H)}{n^{|L|}} = 1 + \Omega\left(\sqrt{\frac{\log n}{n}}\right).$$

This implies that H contains a perfect matching if $\delta'_d(H) \geq (1/2 + o(1))n^{k-d}$ for some $d \geq k/2$. Recently Lo and Markström [60] determined the minimum $\delta'_1(H)$ that guarantees a perfect matching when $H \in \mathcal{K}_3(n)$ and n is sufficiently large.

Keevash and Mycroft [44] investigated the structure of k-graphs satisfying $\delta_{k-1}(H) \geq n/k - o(n)$ but containing no perfect matching. They showed that such k-graphs are close to either space or divisibility barriers (see Theorem 2.7 in the next section). Applying the results in [44], Keevash, Knox, and Mycroft recently [42] found a necessary and sufficient condition for the existence of a perfect matching in all k-graphs H with $\delta_{k-1}(H) \geq n/k + o(n)$.

Given $k \geq 3$ and $0 \leq c \leq 1$, let $\mathbf{PM}(k, c)$ denote the decision problem of determining whether a k-graph H contains a perfect matching when $\delta_{k-1}(H) \geq cn$. The result of Karp [37] says that $\mathbf{PM}(k, 0)$ is NP-complete. On the other hand, Theorem 1.2 implies that $\mathbf{PM}(k, 1/2)$ can be decided in constant time. Szymańska [78] proved that for $c < 1/k$ the problem $\mathbf{PM}(k, 0)$ admits a polynomial-time reduction to $\mathbf{PM}(k, c)$ and hence $\mathbf{PM}(k, c)$ is also NP-complete. Karpiński, Ruciński, and Szymańska [38] showed that there exists $\varepsilon > 0$ such that $\mathbf{PM}(k, 1/2 - \varepsilon)$ is in P. Applying the aforementioned structure result, Keevash, Knox, and Mycroft [42] proved that $\mathbf{PM}(k, 1/k + \varepsilon)$ is in P for all $\varepsilon > 0$ (furthermore, their algorithm provides a perfect matching if it exists). Very recently Han [27] solved the remaining case $\mathbf{PM}(k, 1/k)$. In fact, he proved that $\mathbf{PM}(k, c)$ is in P for all $c \geq 1/k$ by using some theory developed in [42] and a lattice-based absorbing method.

2 Packing Problems

2.1 Results

(Hyper)graph packing, alternatively called *(hyper)graph tiling*, is a natural extension of the matching problem. Given two k-graphs H and F, an *F-packing* is a sub(hyper)graph of H that consists of vertex-disjoint copies of F. An F-packing is *perfect* (called an *F-factor*) if it is a spanning sub(hyper)graph of H. In this case it is necessary that $|V(F)|$ divides $|V(H)|$.

Packing problems have been studied extensively for graphs. The celebrated Hajnal-Szemerédi theorem [25] states that for all $n \in t\mathbb{N}$ every n-vertex graph with minimum degree at least $(1 - 1/t)n$ contains a K_t-factor. Given a k-graph F of order f and an integer n divisible by f, we define *the F-packing threshold* $\boldsymbol{\delta_{d(n,F)}}$ as the smallest integer t such that every n-vertex k-graph H with $\delta_d(H) \geq t$ contains an F-factor. We simply write $\delta(n, F)$ for $\delta_{k-1}(n, F)$. When $k = 2$, Kühn and Osthus [53] determined $\delta(n, F)$ up to an additive constant for all graphs F. This improves the earlier results of Alon and Yuster [2], Komlós [47], and Komlós, Sárközy, and Szemerédi [49]. For more results on graph packing, see the survey [52].

It is not surprising that packing problems become harder in hypergraphs. Other than the matching problems mentioned in Section 1, only a few packing thresholds are known. Recall that K_t^k is the complete k-graph on t vertices. The first step towards a hypergraph Hajnal-Szemerédi theorem is determining $\delta(n, K_4^3)$ (K_4^3-packing is also interesting because the corresponding Turán problem is a famous conjecture of Turán [82]). Czygrinow and Nagle [10] showed that $\delta(n, K_4^3) \geq 3n/5 + o(n)$. Keevash and Sudakov observed that $\delta(n, K_4^3) \geq 5n/8 + o(n)$. Pikhurko [66] proved $3n/4 - 2 \leq \delta(n, K_4^3) \leq 0.861n$; independently Keevash and the author (unpublished, see [44]) proved that $2n/3 - 1 \leq \delta(n, K_4^3) \leq 4n/5 + o(n)$. Lo and Markström [61] showed that $\delta(n, K_4^3) = 3n/4 + o(n)$ by the absorbing method. Independently and simultaneously Keevash and Mycroft [44] determined $\delta(n, K_4^3)$ *exactly.*

Theorem 2.1. [44] *For sufficiently large $n \in 4\mathbb{N}$,*

$$\delta(n, K_4^3) = \begin{cases} 3n/4 - 2 & \text{if } n \in 8\mathbb{N} \\ 3n/4 - 1 & \text{otherwise.} \end{cases}$$

We will elaborate on the approaches used in [44, 61] in the next subsection. It is desirable to find the packing threshold $\delta(n, K_t^k)$ for all $3 \leq k < t$. However, at present such a hypergraph Hajnal-Szemerédi theorem seems out of reach. When $t = k + 1$, Lo and Markström [61] showed that $\delta(n, K_{k+1}^k) \leq (1 - 1/2k)n$ for $k \geq 3$. It is plausible that one can prove $\delta(n, K_{k+1}^k) \leq \frac{k}{k+1}n + o(n)$ by applying the approach of [44]. Unfortunately we do not know a matching lower bound (it was shown in [61] that $\delta(n, K_{k+1}^k) \geq 2n/3$ for even k). For arbitrary t, it was shown in [61] that

$$\left(1 - \frac{193\log(t-1)}{(t-1)^2}\right) n \leq \delta(n, K_t^3) \leq \left(1 - \frac{2}{t^2 - 3t + 4} + o(1)\right) n$$

and $\delta(n, K_t^k) \leq (1 - \binom{t-1}{k-1}^{-1} + o(1))n$ for $k \geq 6$ and $t \geq (3 + \sqrt{5})k/2$.

Problem 2.2. *(1) Prove or disprove that $\delta(n, K_{k+1}^k) = \frac{k}{k+1}n + o(n)$ for $k \geq 4$.*
(2) Improve the existing bounds for $\delta(n, F)$ when $F = K_t^3$, $t \geq 5$ and $F = K_t^k$, $4 \leq k < t$.

Let K_4^- be the (unique) 3-graph with four vertices and three edges. Lo and Markström [57] showed that $\delta(n, K_4^-) = n/2 + o(n)$ by using the absorbing method.

All the remaining success was on packing with k-partite k-graphs. Given positive integers $m_1 \leq \cdots \leq m_k$, let $K_{m_1,\ldots,m_k}^k$ denote the complete k-partite k-graph with parts of sizes $m_1, \ldots, m_k$. In particular, let $K_k^k(m) = K_{m,\ldots,m}^k$. It is clear that $\delta_d(n, K_k^k(m)) \geq m_d(k, n)$ but it is possible to have $\delta_d(n, K_{m_1,\ldots,m_k}^k) < m_d(k, n)$ for certain $m_1, \ldots, m_k$. Other than the matching problems, perhaps the earliest result on hypergraph packing was on $K_{1,1,2}^3$*−packing* (note that $K_{1,1,2}^3$ is the unique 3-graph with four vertices and 2 triples). As a corollary of their main result on loose Hamilton cycles, Kühn and Osthus [50] proved that $\delta(n, K_{1,1,2}^3) = n/4 + o(n)$. Recently Czygrinow, DeBiasio, and Nagle [11] determined this threshold exactly for sufficiently large n.

Mycroft [65] recently determined the codegree packing thresholds asymptotically for all complete k-partite k-graphs, as well as a large class of non-complete k-partite k-graphs. Given a complete k-partite k-graph $K = K^k_{m_1,\dots,m_k}$, define $\gcd(K)$ to be $\gcd\{m_j - m_i : 1 \le i < j \le k\}$ and $\sigma(K) = m_1/(m_1 + \dots + m_k)$. We say that K is type 0 if $\gcd(m_1, \dots, m_k) > 1$ or all $m_i = 1$; K is type $d \ge 1$ if $\gcd(m_1, \dots, m_k) = 1$ and $\gcd(K) = d$. It was shown in [65] that

$$\delta(n, K) = \begin{cases} n/2 + o(n) & \text{if } K \text{ is type } 0; \\ \sigma(K)n + o(n) & \text{if } K \text{ is type } 1; \\ \max\{\sigma(K)n, n/p\} + o(n) & \text{if } K \text{ is type } d \ge 2, \end{cases}$$

where p is the smallest prime factor of d. Mycroft [65] also answered a question of Rödl and Ruciński [67] by determining $\delta(n, C^k_s)$ asymptotically for all $k \ge 3$ and $s \ge 2$, where C^k_s is a k-uniform loose cycle with s edges (see Section 3 for its definition). The proof of [65] makes use of hypergraph regularity, including the Regularity Approximation Lemma of Rödl and Schacht [69] and the Blow-up Lemma of Keevash [40].

Let us consider hypergraph packing under vertex degree conditions.[1] Very little is known beyond the matching problem. Lu and Székely [62] used the Local Lemma to derive a general upper bound for $\delta_1(n, F)$ for arbitrary k-graph F.

Theorem 2.3. [62] *Let F be a t-vertex m-edge k-graph in which each edge intersects at most d other edges. Then*

$$\delta_1(n, F) \le \left(1 - \frac{1}{e(d + 1 + \frac{m}{t}k^2)}\right)\binom{n-1}{k-1}, \quad \text{where } e = 2.718\dots.$$

For example, when $F = K^k_t$, we have $m = \binom{t}{k}$, $d < k\binom{t-1}{k-1}$, and consequently $\delta_1(n, K^k_t) \le \left(1 - \frac{1}{2ek\binom{t-1}{k-1}}\right)\binom{n-1}{k-1}$. On the other hand, we have $\delta_1(n, K^k_t) > \binom{n-1}{k-1} - \binom{(k-1)n/t}{k-1}$ by considering $H^0_{(t-k+1)n/t}(n, k)$ defined in Construction 1.4. Therefore, when k is fixed and $t \to \infty$, we have

$$\delta_1(n, K^k_t) = \left(1 - \Theta\left(\frac{1}{t^{k-1}}\right)\right)\binom{n-1}{k-1}. \tag{5}$$

By considering complement r-graphs, we can translate (5) to the following corollary on equitable colorings of hypergraphs. An *equitable ℓ-coloring* of a (hyper)graph is a proper vertex coloring with ℓ colors such that the sizes of any two color classes differ by at most one.

Corollary 2.4. *For every $k \ge 3$ there exists $c_k > 0$ such that every k-graph with maximum vertex degree d has an equitable ℓ-coloring for any $\ell \ge c_k d^{1/(k-1)}$.*

[1] We are not aware of any packing results on d-degree conditions for $1 < d < k - 1$.

Problem 2.5. *Improve the constant c_k in Corollary 2.4,* i.e., *improve the constants hidden in* (5).

Lo and Markström [61] applied the results in [45, 46] to determine $\delta_1(n, K_3^3(m))$ and $\delta_1(n, K_4^4(m))$ asymptotically. Recently Han and the author [30] and independently Czygrinow [7] determined $\delta_1(n, K_{1,1,2}^3)$ exactly for sufficiently large n. More recently Han, Zang, and the author [35] determined $\delta_1(n, K_{a,b,c}^3)$ asymptotically for arbitrary $a \le b \le c$. Given $a \le b \le c$, let $d = \gcd(b-a, c-b)$ and define

$$f(a,b,c) := \begin{cases} 1/4, & \text{if } a = 1, \gcd(a,b,c) = 1 \text{ and } d = 1; \\ 6 - 4\sqrt{2} \approx 0.343, & \text{if } a \ge 2, \gcd(a,b,c) = 1 \text{ and } d = 1; \\ 4/9, & \text{if } \gcd(a,b,c) = 1 \text{ and } d \ge 3 \text{ is odd}; \\ 1/2, & \text{otherwise.} \end{cases} \tag{6}$$

Theorem 2.6. [35]

$$\delta_1(n, K_{a,b,c}) = \left(\max\left\{f(a,b,c), 1 - \left(\frac{b+c}{a+b+c}\right)^2, \left(\frac{a+b}{a+b+c}\right)^2\right\} + o(1)\right)\binom{n}{2}.$$

It is interesting to note that Theorem 2.6 contains a case where the coefficient of the packing threshold is *irrational*. In fact, as far as we know, all the previously known tiling thresholds have rational coefficients. The lower bound in Theorem 2.5 follows from six constructions: three of them are divisibility barriers, two are space barriers, and the last construction was called a *tiling barrier* because there exists a vertex that is not contained in any copy of $K_{a,b,c}$. In general, given a k-graph F, let $\tau_d(n, F)$ denote the minimum integer t such that every k-graph H of order n with $\delta_d(H) \ge t$ has the property that every vertex of H is covered in some copy of F. When F is a graph, it is not hard to see that $\tau_1(n, F) = (1 - 1/(\chi(F) - 1) + o(1))n$ (see the concluding remarks of [35]). Given a k-graph F, trivially

$$\mathrm{ex}_d(n, F) < \tau_d(n, F) \le \delta_d(n, F), \tag{7}$$

where $\mathrm{ex}_d(n, F)$ is the *d-degree Turán number* of F, defined as the smallest integer t such that every r-graph H of order n with $\delta_d(H) \ge t + 1$ contains a copy of F. It was shown [35] that $\tau_1(n, K_{a,b,c}^3) \le (6 - 4\sqrt{2} + o(1))\binom{n}{2}$, and equality holds when $a \ge 2$. It will be interesting to know $\tau_1(n, F)$ for other 3-graphs.

2.2 *Methods*

Most packing thresholds on graphs, *e.g.*, [49, 53], were obtained by the regularity method using the Regularity Lemma of Szemerédi [77] and the Blow-up Lemma of Komlós, Sárközy, and Szemerédi [48]. As the hypergraph versions of these two lemmas are now available, it is possible to attack hypergraph packing problems by the same approach though the proofs become long and technical – [44, 65] are two examples.

Keevash and Mycroft [44] derived a theorem on perfect matchings for simplicial complexes that is very useful for packing problems. To state this result precisely, we need several definitions. A *k-system* J is a set system in which the largest set of J has size k and $\emptyset \in J$. A *k-complex* is a downward closed k-system, namely, every subset of a set in J is also in J. Given a k-system J, let J_r denote the family of r-sets in J for $0 \le r \le k$. The *minimum r-degree* of J, denoted by $d_r(J)$, is the minimum $\deg_{J_{r+1}}(e)$ among all $e \in J_r$. (Note that this is different from $\delta_r(J_{r+1})$, which is the minimum $\deg_{J_{r+1}}(S)$ among all r-sets $S \subseteq V(J)$.) The *degree sequence* of J is $\mathbf{d}(J) = (d_0(J), d_1(J), \dots, d_{k-1}(J))$. Given a vector $\mathbf{a} = (a_0, a_1, \dots, a_{k-1})$, we write $\mathbf{d}(J) \ge \mathbf{a}$ if $d_i(J) \ge a_i$ for $0 \le i \le k-1$.

Let $H = (V, E)$ be a k-graph and let $\mathcal{P}$ be an ordered partition of V into $V_1, \dots, V_t$. The *index vector* $\mathbf{i}_{\mathcal{P}}(S)$ of a set $S \subseteq V$ is defined as $(|S \cap V_1|, \dots, |S \cap V_t|)$. A vector of $\mathbb{Z}^t$ is referred to as a *k-vector* if all its coordinates are non-negative and sum to k (thus all $\mathbf{i}_{\mathcal{P}}(e)$, $e \in E$, are k-vectors). A *lattice* in $\mathbb{Z}^t$ is an additive subgroup of $\mathbb{Z}^t$. We let $L_{\mathcal{P}}(H)$ denote the lattice generated by the index vectors $\mathbf{i}_{\mathcal{P}}(e)$ for all edges e of H. Given $\mu > 0$, let $L^{\mu}_{\mathcal{P}}(H)$ denote the lattice generated by all vectors $\mathbf{x} \in \mathbb{Z}^t$ such that there are at least μn^k edges e of H with $\mathbf{i}_{\mathcal{P}}(e) = \mathbf{x}$. A lattice is *complete* if it contains all k-vectors; otherwise, it is *incomplete*. A lattice is *transferral-free* if it contains no $\mathbf{u}_i - \mathbf{u}_j$ for any $i \ne j$, where $\mathbf{u}_i$ is the 1-vector whose ith coordinate is one.

Theorem 2.7. [44, Theorem 2.9] *Suppose that $1/n \ll \alpha \ll \mu, \beta \ll 1/k$, and k divides n. Let J be a k-complex on n vertices with degree sequence*

$$\mathbf{d}(J) \ge \left(n, (\tfrac{k-1}{k} - \alpha)n, (\tfrac{k-2}{k} - \alpha)n, \dots, (\tfrac{1}{k} - \alpha)n\right). \tag{8}$$

Then at least one of the following properties holds:

1: *J_k contains a perfect matching.*

2 (Space barrier): *There exists a set $S \in V(J)$ with $|S| = jn/k$ for some $1 \le j \le k-1$ such that all but at most βn^k edges of J_k intersect S with at most j vertices.*

3 (Divisibility barrier): *There exists a partition $\mathcal{P}$ of $V(J)$ into $t \le k$ parts of size at least $d_{k-1}(J) - \beta n$ such that $L^{\mu}_{\mathcal{P}}(J_k)$ is incomplete and transferral-free.*

After obtaining this theorem, Keevash and Mycroft [44] attacked the K_4^3-packing problem as follows. Let $H = (V, E)$ be a 3-graph with $\delta(H) \ge 3n/4$. Consider the so-called *clique 4-complex* J of H, in which $J_i = \binom{V}{i}$ for $i \le 2$, $J_3 = H$, and J_4 is the family of all 4-sets that span copies of K_4^3. It is easy to see that

$$d_0(J) = n, \quad d_1(J) = n-1, \quad d_2(J) = \delta_2(H) \ge \frac{3}{4}n, \quad \textit{and} \quad d_3(J) \ge \frac{n}{4} + 3.$$

Indeed, to find $d_3(J)$, we fix a 3-set $abc \in E(H)$; each of the pairs ab, ac, bc has at least $3n/4 - 1$ neighbors in $V \setminus \{a, b, c\}$ and thus at least $n/4 + 3$ vertices $d \in V$

are the neighbors of all ab, ac, bc.[2] Next we apply Theorem 2.7 to conclude that either J_4 contains a perfect matching or there is a space or divisibility barrier. What remains is to show that H contains a K_4^3-factor if it is a space or divisibility barrier satisfying the minimum codegree condition.

Most of the aforementioned packing thresholds were obtained by the absorbing method. As described in Section 1.1 for the matching problems, our goal is first obtaining a small absorbing F-packing that can absorb any smaller set of vertices, and then finding an F-packing that covers most of the remaining vertices.

Given a k-graph F of order f, suppose we want to find an F-factor in an n-vertex k-graph $H = (V, E)$ with certain degree conditions. Given $\varepsilon > 0$, $i \in \mathbb{N}$, and two vertices $x, y \in V(H)$, we say that x and y are (ε, i)-*reachable* if there are at least εn^{if-1} $(if-1)$-subsets of $W \subset V$ such that both $H[\{x\} \cup W]$ and $H[\{y\} \cup W]$ contain F-factors. A vertex set $U \subseteq V$ is (ε, i)-*closed* if any two vertices of U are (ε, i)-reachable. In order to find the desired absorbing F-packing, it suffices to show that V is (ε, i)-closed for some $\varepsilon > 0$ and $i \in \mathbb{N}$ – see [61, Lemma 1.1]. Sometimes this is straightforward but sometimes this is done in two steps, referred to as a *lattice-based absorbing method* in [27, 35]. In Step 1 we find a partition $\mathcal{P} = (V_1, \ldots, V_t)$ of V such that each V_i is not small and (ε, i)-closed for some $\varepsilon > 0$ and $i \in \mathbb{N}$. In Step 2 we show that $L^{\mu}_{\mathcal{P},F}$ is complete for some $\mu > 0$, where $L^{\mu}_{\mathcal{P},F}$ is the lattice generated by all vectors $\mathbf{v} \in \mathbb{Z}^t$ such that there are at least μn^f f-sets S that span copies of F and satisfy $\mathbf{i}_{\mathcal{P}}(S) = \mathbf{v}$. Once these steps are done, we can easily show that V is (ε', i')-closed for some $\varepsilon' > 0$ and $i' \in \mathbb{N}$.

3 Hamilton Cycles

Cycles in hypergraphs have been studied since the 1970s. There are several notions of cycles. A k-graph (V, E) is called a *Berge cycle* if E consists of k distinct edges $e_1, \ldots, e_t$ and V contains distinct vertices $v_1, \ldots, v_t$ (and possibly other vertices) such that each e_i contains v_i and v_{i+1}, where $v_{t+1} = v_1$. Bermond, Germa, Heydemann, and Sotteau [4] proved a Dirac-type theorem for Berge cycles. In recent years a more structured notion of cycles has become more popular. Given $1 \le l < k$, a k-graph C is a called an *l-cycle* if its vertices can be ordered cyclically such that each of its edges consists of k consecutive vertices and every two consecutive edges (in the natural order of the edges) share exactly l vertices. In a k-graph, a $(k-1)$-cycle is often called a *tight* cycle while a 1-cycle is often called a *loose* cycle (sometimes called *linear* cycle). We say that a k-graph contains a Hamilton l-cycle if it contains an l-cycle as a spanning subhypergraph. Note that a k-uniform l-cycle of order n contains exactly $n/(k-l)$ edges, implying that $k - l$ divides n.

[2] Here we see why we need to consider the 4-complex J instead of the 4-graph J_4 alone: $\delta_3(J_4) = 0$ because a 3-set $abc \notin E(H)$ has degree zero in J_4.

We define the threshold $\mathbf{h}_\mathbf{d}^\mathbf{l}(\mathbf{k},\mathbf{n})$ as the smallest integer m such that every k-graph H on n vertices with $\delta_d(H) \geq m$ contains a Hamilton l-cycle, provided that $k-l$ divides n. As before, we may omit the subscript when $d = k-1$. Unless stated otherwise, we assume that n is sufficiently large in this section.

Katona and Kierstead [39] first studied $h^{k-1}(k,n)$ and proved that

$$\left\lfloor \frac{n-k+3}{2} \right\rfloor \leq h^{k-1}(k,n) \leq \left(1 - \frac{1}{2k}\right) n + O(1)$$

The following construction provides the lower bound.

Construction 3.1. *Let $H = (V,E)$, where $V = X \cup Y \cup \{v\}$ such that $|X| = \lfloor \frac{n-1}{2} \rfloor$ and $|Y| = \lceil \frac{n-1}{2} \rceil$, and E consists of all k-subsets of V containing v and all k-sets $S \subset X \cup Y$ such that $|S \cap X| \neq \lfloor k/2 \rfloor$.*

It is easy to check that $\delta_{k-1}(H) \geq \lfloor (n-k+1)/2 \rfloor$. Suppose H contains a tight Hamilton cycle. Then there exists an ordering $v_1, \ldots, v_{n-1}$ of the vertices of $X \cup Y$ such that all $e_i := \{v_i, \ldots, v_{i+k-1}\}$, $1 \leq i \leq n-k$, are edges. Let $a_i = |X \cap e_i|$. Then $\sum_{i=1}^{n-k} a_i$ is about $|X|k$ when n is sufficiently large because all the vertices of X are counted k times except for those v_i with $i < k$ and $i > n-k$. Thus the average a_i is about $k/2$. On the other hand, we have $a_i \neq \lfloor k/2 \rfloor$ and $|a_i - a_{i+1}| \leq 1$ for all i. Thus either all $a_i \leq \lfloor k/2 \rfloor - 1$ or all $a_i \geq \lfloor k/2 \rfloor + 1$, a contradiction.

Using the absorbing method, Rödl, Ruciński, and Szemerédi [70, 73] showed that

$$h^{k-1}(k,n) = \frac{n}{2} + o(n) \tag{9}$$

for all $k \geq 3$. With long and involved arguments, they [75] were able to obtain an exact result when $k = 3$.

Theorem 3.2. [75] *For sufficiently large n, $h^2(3,n) = \lfloor n/2 \rfloor$.*

Assume that $1 \leq l < k$ and $k-l$ divides n. Since every $(k-1)$-cycle of order n contains an l-cycle on the same vertices, (9) implies that $h^l(k,n) \leq \frac{n}{2} + o(n)$. On the other hand, it is not hard to see that $h^l(k,n) \geq \frac{n}{2} - k$ when $k-l$ divides k. In fact, when k divides n, a Hamilton l-cycle contains a perfect matching thus Construction 1.1 provides this bound; when k does not divide n, this was proven by Markström and Ruciński [64, Proposition 2]. Consequently,

$$h^l(k,n) = \frac{n}{2} + o(n) \quad \text{if } k-l \mid k. \tag{10}$$

Very recently Han and the author [29] determined the d-degree threshold $h_d^{k/2}(k,n)$ exactly for all even $k \geq 6$ and all $d \geq k/2$. The value of $h_d^{k/2}(k,n)$ turns out to be close (but not always equal) to $\delta(n,k,d)$ defined in Section 1.1.

When $k-l$ does not divide k, the threshold $h^l(k,n)$ is much smaller. Kühn and Osthus [50] proved that $h^1(3,n) = n/4 + o(n)$. This was generalized to arbitrary k and $l = 1$ by Keevash, Kühn, Mycroft, and Osthus [43] and to arbitrary k and arbitrary $l < k/2$ by Hàn and Schacht [34]. Later Kühn, Mycroft, and Osthus [56] showed that

$$h^l(k,n) = \frac{n}{\lceil \frac{k}{k-l} \rceil (k-l)} + o(n) \quad \text{if } k-l \nmid k. \tag{11}$$

The following simple construction supports the lower bound in (11).

Construction 3.3. *Suppose* $1 \le l < k$, $k-l \nmid k$, *and* $k-l \mid n$. *Let* $t = n/(k-l)$ *and* $s = \lceil k/(k-l) \rceil$. *Let* $H_0 = (V,E)$ *be an* n*-vertex* k*-graph in which* V *is partitioned into sets* A *and* B *such that* $|A| = \lceil t/s \rceil - 1$. *The edge set* E *consists of all the* k*-sets that intersect* A.

We have $\delta_{k-1}(H_0) = |A| = \lceil n/\left(\lceil \frac{k}{k-l} \rceil (k-l)\right) \rceil - 1$. If H_0 contains a Hamilton l-cycle C, then each vertex is contained in at most s edges of C. Since A is a vertex cover of C, $|C| \le |A|s < t$, a contradiction.

Recently Czygrinow and Molla [9] showed that $h^1(3,n) = \lceil n/4 \rceil$. Independently Han and the author [28] proved that $h^l(k,n) = \lceil n/(2k-2l) \rceil$ for all $l < k/2$. It was conjectured [28] that Construction 3.3 is an extremal configuration for Hamilton l-cycles whenever $k-l$ does not divide k. On the other hand, since the extremal cases in [29, 75] require involved work, it seems harder to determine the exact value of $h^l(k,n)$ when $k-l$ divides k.

Problem 3.4. *Determine* $h^l(k,n)$ *exactly for all* $l \ge k/2$, *in particular, prove that* $h^l(k,n) = \lceil n/\left(\lceil \frac{k}{k-l} \rceil (k-l)\right) \rceil$ *when* $k-l \nmid k$.

Much less is known on the value of $h_d^l(k,n)$ when $d \le k-2$. Buß, Hàn, and Schacht [6] showed that $h_1^1(3,n) = (\frac{7}{16} + o(1))\binom{n}{2}$. Recently Han and the author [31] improved this to an exact result.

Theorem 3.5. [31] *There exists* n_0 *such that the following holds. If* H *is a 3-graph* H *on* $n \ge n_0$ *vertices with* $n \in 2\mathbb{N}$ *and* $\delta_1(H) \ge \binom{n-1}{2} - \binom{\lfloor \frac{3}{4}n \rfloor}{2} + c$, *where* $c = 2$ *if* $n \in 4\mathbb{N}$ *and* $c = 1$ *otherwise. Then* H *contains a loose Hamilton cycle.*

It is conjectured in [6] that $h_1^1(k,n) = \delta_1(H_0) + o(n^{k-1})$ for H_0 defined in Construction 3.3.

Problem 3.6. *Determine, asymptotically or exactly, the other values of* $h_d^l(k,n)$, *in particular, prove or disprove that* $h_1^1(k,n) = \delta_1(H_0) + o(n^{k-1})$ *for all* $k \ge 4$.

It was conjectured in [67] that $h_d^{k-1}(k,n) = m_d(k,n) + o(n^{k-d})$ for all $d < k$ and the aforementioned result [74] confirmed this for $d = k-1$. However, Han and the author [32] recently disproved this conjecture. Nevertheless, it might be true that $h_1^2(3,n) \approx m_1(3,n) = (\frac{5}{9} + o(1))\binom{n}{2}$. This problem seems (much) harder than the corresponding matching problem. The best known bound $h_1^2(3,n) \le \frac{1}{3}(5 - \sqrt{5} + o(1))\binom{n}{2}$ was given by Rödl and Ruciński [68] recently. For arbitrary $d < k$, Glebov, Person, and Weps [22] proved that $h_d^{k-1}(k,n) \le (1 - c_{k,d})\binom{n-d}{k-d}$ for some small $c_{k,d} > 0$.

As with the matching and packing problems, the absorbing method is the main tool of finding a Hamilton l-cycle in k-graphs. Since [67, Section 2.2] gave a detailed sketch on this approach (in the case of $l = d = k - 1$), we only highlight the main ideas. In a k-graph, an *l-path* consists of vertices $v_0, v_1, \ldots, v_{t-1}$ for some $t = (k-l)q + l$ and edges $E_0, E_1, \ldots, E_{q-1}$, where $E_i = v_{(k-l)i} \cdots v_{(k-l)i+k-1}$. We call $v_0 v_1 \ldots v_{l-1}$ and $v_{t-l} \ldots v_{t-1}$ two *ends* of the path. Given an l-path P and a vertex set S such that $V(P) \cap S = \emptyset$, we say that P *absorbs* S if there is an l-path on $V(P) \cup S$ with the same ends of P. To find a Hamilton l-cycle in a k-graph H, we proceed in the following steps.

Step 1: Find an *absorbing l-path* P with $|P| \ll n$; namely, P can absorb any vertex set $W \subseteq V(H) \setminus V(P)$ such that $|W| \ll |P|$ and $|W| \in (k-l)\mathbb{N}$. Denote the two ends of P by L_1 and L_2 and let $V' = (V \setminus V(P)) \cup L_1 \cup L_2$.
Step 2: Find a *reservoir set* $R \subset V'$ of size $|R| \ll |P|$ such that any two l-subsets of V' can be connected to an l-path via many s-subsets of R for some constant s (*i.e.*, independent of n). Let $V'' = V \setminus (V(P) \cup R)$.
Step 3: Cover all but $o(n)$ vertices of V'' with constant many vertex-disjoint l-paths.

Once these three steps are done, we connect all the l-paths that we have found by using the vertices of R and finally absorb the leftover vertices in Step 3 and the remaining vertices in R with P. If $\delta_d(H) \geq m_d(k,n) + o(n^{k-d})$, then Step 3 can be easily done by first applying the weak Regularity Lemma and then finding an almost perfect matching in the reduced k-graph. (Thus the difficulty of proving $h_1^2(3,n) \leq \frac{5}{9}\binom{n}{2} + o(n^2)$ resides on the first two steps.) On the other hand, when $l < k-1$, a smaller $\delta_d(H)$ may suffice for Step 3. For example, when $2l < k$ and $\delta_{k-1}(H) \geq n/(2k-2l)$, the authors of [28] accomplished Step 3 by finding an almost $Y_{k,2l}$-factor in the reduced k-graph, where $Y_{k,2l}$ consists of two k-sets that share exactly $2l$ vertices. This is possible because one can convert a copy of $Y_{k,2l}$ in the reduced k-graph to an almost spanning l-path in the original k-graph, due to the fact that, when $2l < k$, every (long) l-path is a k-partite k-graph whose first $2l$ parts are of size about m and the remaining parts are of size about $2m$ for some integer m.

Acknowledgements The author would like to thank Jie Han for valuable discussion when preparing this manuscript. He also thanks Albert Bush, Jie Han, Allan Lo, Richard Mycroft, and Andrew Treglown for their comments that improved the presentation of the manuscript.

The author is partially supported by NSF grant DMS-1400073.

References

1. R. Aharoni, A. Georgakopoulos, P. Sprüssel, Perfect matchings in r-partite r-graphs. Eur. J. Comb. **30**(1), 39–42 (2009)
2. N. Alon, R. Yuster, H-factors in dense graphs. J. Comb. Theory Ser. B **66**(2), 269–282 (1996)
3. N. Alon, P. Frankl, H. Huang, V. Rödl, A. Ruciński, B. Sudakov, Large matchings in uniform hypergraphs and the conjecture of Erdős and Samuels. J. Comb. Theory Ser. A **119**(6), 1200–1215 (2012)

4. J.-C. Bermond, A. Germa, M.-C. Heydemann, D. Sotteau, Hypergraphes Hamiltoniens, in *Problèmes combinatoires et théorie des graphes (Colloq. Internat. CNRS, Univ. Orsay, Orsay, 1976)*. Colloq. Internat. CNRS, vol. 260 (CNRS, Paris, 1978), pp. 39–43
5. B. Bollobás, D.E. Daykin, P. Erdős, Sets of independent edges of a hypergraph. Q. J. Math. Oxford Ser. (2) **27**(105), 25–32 (1976)
6. E. Buß, H. Hàn, M. Schacht, Minimum vertex degree conditions for loose Hamilton cycles in 3-uniform hypergraphs. J. Comb. Theory Ser. B **103**(6), 658–678 (2013)
7. A. Czygrinow, Minimum degree condition for C_4-tiling in 3-uniform hypergraphs (submitted)
8. A. Czygrinow, V. Kamat, Tight co-degree condition for perfect matchings in 4-graphs. Electron. J. Comb. **19**(2), Paper 20, 16 (2012)
9. A. Czygrinow, T. Molla, Tight codegree condition for the existence of loose Hamilton cycles in 3-graphs. SIAM J. Discrete Math **28**(1), 67–76 (2014)
10. A. Czygrinow, B. Nagle, A note on codegree problems for hypergraphs. Bull. Inst. Comb. Appl. **32**, 63–69 (2001)
11. A. Czygrinow, L. DeBiasio, B. Nagle, Tiling 3-uniform hypergraphs with $K_4^3 - 2e$. J. Graph Theory **75**(2), 124–136 (2014)
12. D.E. Daykin, R. Häggkvist, Degrees giving independent edges in a hypergraph. Bull. Aust. Math. Soc. **23**(1), 103–109 (1981)
13. G.A. Dirac, Some theorems on abstract graphs. Proc. Lond. Math. Soc. (3) **2**, 69–81 (1952)
14. J. Edmonds, Paths, trees, and flowers. Canad. J. Math. **17**, 449–467 (1965)
15. P. Erdős, A problem on independent r-tuples. Ann. Univ. Sci. Budapest. Eötvös Sect. Math. **8**, 93–95 (1965)
16. P. Erdős, T. Gallai, On maximal paths and circuits of graphs. Acta Math. Acad. Sci. Hungar. **10**, 337–356 (unbound insert) (1959)
17. P. Erdős, C. Ko, R. Rado, Intersection theorems for systems of finite sets. Q. J. Math. Oxford Ser. (2) **12**, 313–320 (1961)
18. P. Frankl, On the maximum number of edges in a hypergraph with given matching number (2012). arXiv:1205.6847
19. P. Frankl, Improved bounds for Erdős' matching conjecture. J. Comb. Theory Ser. A **120**(5), 1068–1072 (2013)
20. P. Frankl, V. Rödl, Near perfect coverings in graphs and hypergraphs. Eur. J. Comb. **6**(4), 317–326 (1985)
21. P. Frankl, V. Rödl, A. Ruciński, On the maximum number of edges in a triple system not containing a disjoint family of a given size. Comb. Probab. Comput. **21**(1–2), 141–148 (2012)
22. R. Glebov, Y. Person, W. Weps, On extremal hypergraphs for Hamiltonian cycles. Eur. J. Comb **33**(4), 544–555 (2012)
23. R.J. Gould, Advances on the Hamiltonian problem—a survey. Graphs Comb. **19**(1), 7–52 (2003)
24. R.J. Gould, Recent advances on the Hamiltonian problem: survey III. Graphs Comb. **30**(1), 1–46 (2014)
25. A. Hajnal, E. Szemerédi, Proof of a conjecture of P. Erdős, in *Combinatorial Theory and Its Applications, II (Proc. Colloq., Balatonfüred, 1969)* (North-Holland, Amsterdam, 1970), pp. 601–623
26. J. Han, Near perfect matchings in k-uniform hypergraphs. Comb. Probab. Comput. **24**(5), 723–732 (2015)
27. J. Han, Decision problem for perfect matchings in dense k-uniform hypergraphs (2014). arXiv:1409.5931
28. J. Han, Y. Zhao, Minimum codegree threshold for Hamilton ℓ-cycles in k-uniform hypergraphs. J. Comb. Theory Ser. A **132**, 194–223 (2015)
29. J. Han, Y. Zhao, Minimum degree conditions for Hamilton $(k/2)$-cycles in k-uniform hypergraphs (2015, manuscript)
30. J. Han, Y. Zhao, Minimum degree thresholds for C_4^3-tiling. J. Graph Theory **79**(4), 300–317 (2015)

31. J. Han, Y. Zhao, Minimum vertex degree threshold for loose Hamilton cycles in 3-uniform hypergraphs. J. Comb. Theory Ser. B **114**, 70–96 (2015)
32. J. Han, Y. Zhao, Forbidding Hamilton cycles in uniform hypergraphs (2015). arXiv: 1508.05623
33. H. Hàn, Y. Person, M. Schacht, On perfect matchings in uniform hypergraphs with large minimum vertex degree. SIAM J. Discrete Math. **23**, 732–748 (2009)
34. H. Hàn, M. Schacht, Dirac-type results for loose Hamilton cycles in uniform hypergraphs. J. Comb. Theory. Ser. B **100**, 332–346 (2010)
35. J. Han, C. Zang, Y. Zhao, Minimum vertex degree thresholds for tiling complete 3-partite 3-graphs (2015). arXiv:1503.08730
36. H. Huang, P.-S. Loh, B. Sudakov, The size of a hypergraph and its matching number. Comb. Probab. Comput. **21**(3), 442–450 (2012)
37. R.M. Karp, Reducibility among combinatorial problems, in *Complexity of computer computations (Proc. Sympos., IBM Thomas J. Watson Res. Center, Yorktown Heights, N.Y., 1972)* (Plenum, New York, 1972), pp. 85–103
38. M. Karpiński, A. Ruciński, E. Szymańska, Computational complexity of the perfect matching problem in hypergraphs with subcritical density. Int. J. Found. Comput. Sci. **21**(6), 905–924 (2010)
39. G. Katona, H. Kierstead, Hamiltonian chains in hypergraphs. J. Graph Theory **30**(2), 205–212 (1999)
40. P. Keevash, A hypergraph blow-up lemma. Random Struct. Algorithm **39**(3), 275–376 (2011)
41. P. Keevash, The existence of designs (2014). arXiv:1401.3665
42. P. Keevash, F. Knox, R. Mycroft, Polynomial-time perfect matchings in dense hypergraphs. Adv. Math. **269**(0), 265–334 (2015)
43. P. Keevash, D. Kühn, R. Mycroft, D. Osthus, Loose Hamilton cycles in hypergraphs. Discrete Math. **311**(7), 544–559 (2011)
44. P. Keevash, R. Mycroft, A geometric theory for hypergraph matching. Memoirs of the American Mathematical Society 233(Monograph 1908) (2014)
45. I. Khan, Perfect matchings in 4-uniform hypergraphs (2011). arXiv:1101.5675
46. I. Khan, Perfect matchings in 3-uniform hypergraphs with large vertex degree. SIAM J. Discrete Math. **27**(2), 1021–1039 (2013)
47. J. Komlós, Tiling Turán theorems. Combinatorica **20**(2), 203–218 (2000)
48. J. Komlós, G. Sárközy, E. Szemerédi. Blow-up lemma. Combinatorica **17**(1), 109–123 (1997)
49. J. Komlós, G. Sárközy, E. Szemerédi. Proof of the Alon-Yuster conjecture. Discrete Math. **235**(1–3), 255–269 (2001). Combinatorics (Prague, 1998)
50. D. Kühn, D. Osthus, Loose Hamilton cycles in 3-uniform hypergraphs of high minimum degree. J. Comb. Theory Ser. B **96**(6), 767–821 (2006)
51. D. Kühn, D. Osthus, Matchings in hypergraphs of large minimum degree. J. Graph Theory **51**(4), 269–280 (2006)
52. D. Kühn, D. Osthus, Embedding large subgraphs into dense graphs, in *Surveys in Combinatorics 2009*. London Mathematical Society Lecture Note Series, vol. 365 (Cambridge University Press, Cambridge, 2009), pp. 137–167
53. D. Kühn, D. Osthus, The minimum degree threshold for perfect graph packings. Combinatorica **29**(1), 65–107 (2009)
54. D. Kühn, D. Osthus, A survey on Hamilton cycles in directed graphs. Eur. J. Comb. **33**(5), 750–766 (2012)
55. D. Kühn, D. Osthus, Hamilton cycles in graphs and hypergraphs: an extremal perspective, in *Proceedings of the International Congress of Mathematicians 2014, Seoul*, vol. 4 (2014), pp. 381–406
56. D. Kühn, R. Mycroft, D. Osthus, Hamilton ℓ-cycles in uniform hypergraphs. J. Comb. Theory Ser. A **117**(7), 910–927 (2010)
57. A. Lo, K. Markström, Minimum codegree threshold for $(K_4^3 - e)$-factors. J. Comb. Theory Ser. A **120**(3), 708–721 (2013)

58. D. Kühn, D. Osthus, A. Treglown, Matchings in 3-uniform hypergraphs. J. Comb. Theory Ser. B **103**(2), 291–305 (2013)
59. D. Kühn, D. Osthus, T. Townsend, Fractional and integer matchings in uniform hypergraphs. Eur. J. Comb. **38**, 83–96 (2014)
60. A. Lo, K. Markström, Perfect matchings in 3-partite 3-uniform hypergraphs. J. Comb. Theory Ser. A **127**, 22–57 (2014)
61. A. Lo, K. Markström, F-factors in hypergraphs via absorption. Graphs Comb. **31**(3), 679–712 (2015)
62. L. Lu, L. Székely, Using Lovász local lemma in the space of random injections. Electron. J. Comb. **14**(1), Research Paper 63, 13 (2007)
63. T. Łuczak, K. Mieczkowska, On Erdős' extremal problem on matchings in hypergraphs. J. Comb. Theory Ser. A **124**, 178–194 (2014)
64. K. Markström, A. Ruciński, Perfect Matchings (and Hamilton cycles) in Hypergraphs with large degrees. Eur. J. Comb. **32**(5), 677–687 (2011)
65. R. Mycroft, Packing k-partite k-uniform hypergraphs (2014). arXiv:1402.5643
66. O. Pikhurko, Perfect matchings and K_4^3-tilings in hypergraphs of large codegree. Graphs Comb. **24**(4), 391–404 (2008)
67. V. Rödl, A. Ruciński, Dirac-type questions for hypergraphs — a survey (or more problems for endre to solve). An Irregular Mind. Bolyai Soc. Math. Stud. **21**, 561–590 (2010)
68. V. Rödl, A. Ruciński, Families of triples with high minimum degree are Hamiltonian. Discuss. Math. Graph Theory **34**(2), 361–381 (2014)
69. V. Rödl, M. Schacht, Regular partitions of hypergraphs: regularity lemmas. Comb. Probab. Comput. **16**(6), 833–885 (2007)
70. V. Rödl, A. Ruciński, E. Szemerédi, A Dirac-type theorem for 3-uniform hypergraphs. Comb. Probab. Comput. **15**(1–2), 229–251 (2006)
71. V. Rödl, A. Ruciński, E. Szemerédi, Perfect matchings in uniform hypergraphs with large minimum degree. Eur. J. Comb. **27**(8), 1333–1349 (2006)
72. V. Rödl, A. Ruciński, M. Schacht, E. Szemerédi, A note on perfect matchings in uniform hypergraphs with large minimum collective degree. Comment. Math. Univ. Carolin. **49**(4), 633–636 (2008)
73. V. Rödl, A. Ruciński, E. Szemerédi, An approximate Dirac-type theorem for k-uniform hypergraphs. Combinatorica **28**(2), 229–260 (2008)
74. V. Rödl, A. Ruciński, E. Szemerédi, Perfect matchings in large uniform hypergraphs with large minimum collective degree. J. Comb. Theory Ser. A **116**(3):613–636 (2009)
75. V. Rödl, A. Ruciński, E. Szemerédi, Dirac-type conditions for Hamiltonian paths and cycles in 3-uniform hypergraphs. Adv. Math. **227**(3), 1225–1299 (2011)
76. S.M. Samuels, On a Chebyshev-type inequality for sums of independent random variables. Ann. Math. Stat. **37**, 248–259 (1966)
77. E. Szemerédi, Regular partitions of graphs, in *Problèmes combinatoires et théorie des graphes (Colloq. Internat. CNRS, Univ. Orsay, Orsay, 1976)*. Colloq. Internat. CNRS, vol. 260 (CNRS, Paris, 1978), pp. 399–401
78. E. Szymańska, The complexity of almost perfect matchings and other packing problems in uniform hypergraphs with high codegree. Eur. J. Comb. **34**(3), 632–646 (2013)
79. A. Treglown, Y. Zhao, Exact minimum degree thresholds for perfect matchings in uniform hypergraphs. J. Comb. Theory Ser. A **119**(7), 1500–1522 (2012)
80. A. Treglown, Y. Zhao, Exact minimum degree thresholds for perfect matchings in uniform hypergraphs II. J. Comb. Theory Ser. A **120**(7), 1463–1482 (2013)
81. A. Treglown, Y. Zhao, A note on perfect matchings in uniform hypergraphs (2015). arXiv:1503.03357
82. P. Turán, On an extremal problem in graph theory, (Hungarian). Mat. Fiz. Lapok **48**, 436–452 (1941)
83. W.T. Tutte, The factorization of linear graphs. J. Lond. Math. Soc. **22**, 107–111 (1947)

Rainbow Hamilton cycles in random graphs and hypergraphs

Asaf Ferber and Michael Krivelevich

Abstract Let H be an edge-colored hypergraph. We say that H contains a *rainbow* copy of a hypergraph S if it contains an isomorphic copy of S with all edges of distinct colors. We consider the following setting. A randomly edge-colored random hypergraph $H \sim \mathscr{H}_c^k(n,p)$ is obtained by adding each k-subset of $[n]$ with probability p, and assigning it a color from $[c]$ uniformly, independently at random. As a first result we show that a typical $H \sim \mathscr{H}_c^2(n,p)$ (that is, a random edge-colored graph) contains a rainbow Hamilton cycle, provided that $c = (1 + o(1))n$ and $p = \frac{\log n + \log\log n + \omega(1)}{n}$. This is asymptotically best possible with respect to both parameters, and improves a result of Frieze and Loh. Secondly, based on an ingenious coupling idea of McDiarmid, we provide a general tool for tackling problems related to finding "nicely edge-colored" structures in random graphs/hypergraphs. We illustrate the generality of this statement by presenting two interesting applications. In one application we show that a typical $H \sim \mathscr{H}_c^k(n,p)$ contains a rainbow copy of a hypergraph S, provided that $c = (1 + o(1))|E(S)|$ and p is (up to a multiplicative constant) a threshold function for the property of containment of a copy of S. In the second application we show that a typical $G \sim \mathscr{H}_c^2(n,p)$ contains $(1 - o(1))np/2$ edge-disjoint Hamilton cycles, each of which is rainbow, provided that $c = \omega(n)$ and $p = \omega(\log n/n)$.

Keywords Hamilton cycles • rainbow subgraphs • random graphs and hypergraphs

Mathematics Subject Classification (2010): 05C80, 05C15, 05C45

A. Ferber
Department of Mathematics, Yale University, New Haven, CT 06520, USA

Department of Mathematics, MIT, Cambridge, MA 02139, USA
e-mail: asaf.ferber@yale.edu; ferbera@mit.edu

M. Krivelevich (✉)
School of Mathematical Sciences, Raymond and Beverly Sackler Faculty of Exact Sciences, Tel Aviv University, Tel Aviv, 6997801, Israel
e-mail: krivelev@post.tau.ac.il

A. Beveridge et al. (eds.), *Recent Trends in Combinatorics*, The IMA Volumes in Mathematics and its Applications 159, DOI 10.1007/978-3-319-24298-9_7

1 Introduction

We consider the following model of edge-colored random k-uniform hypergraphs. Let $p \in [0,1]$ and let c be a positive integer. Then we define $\mathscr{H}_c^k(n,p)$ to be the probability space of edge-colored k-uniform hypergraphs with vertex set $[n] := \{1,\ldots,n\}$, obtained by first choosing each k-tuple $e \in \binom{[n]}{k}$ to be an edge independently with probability p and then by coloring each chosen edge independently and uniformly at random with a color from the set $[c]$. For example, the case $k = 2$, denoted $\mathscr{G}_c(n,p)$, is the standard binomial graph $\mathscr{G}(n,p)$ whose edges are colored at random in c colors. In the special case where $c = 1$, we write $\mathscr{H}^k(n,p) := \mathscr{H}_c^k(n,p)$, and observe that this is just the standard binomial random hypergraph model. For $H \sim \mathscr{H}_c^k(n,p)$ and a hypergraph S, we say that H contains a *rainbow* copy of S if H contains an isomorphic copy of S with all edges in distinct colors. A frequent theme in recent research is to determine the conditions on p and c under which a random hypergraph $H \sim \mathscr{H}_c^k(n,p)$ contains, with high probability (w.h.p.), a rainbow copy of a given hypergraph S.

Let us first discuss the case $k = 2$ of binomial random graphs. Perhaps two of the most natural properties to address are when S is a perfect matching or S is a Hamilton cycle. Note that in these cases we need $c \geq n/2$ and $c \geq n$, respectively. Moreover, it is well known (see, e.g., [6]) that a perfect matching (respectively a Hamilton cycle) starts to appear in a typical $G \sim \mathscr{G}(n,p)$ whenever $p = \frac{\log n+\omega(1)}{n}$ (respectively, $p = \frac{\log n+\log\log n+\omega(1)}{n}$), and therefore we can restrict ourselves to these regimes. In [7], Cooper and Frieze showed that for $p \approx 42\log n/n$ and $c = 21n$, a graph $G \sim \mathscr{G}_c(n,p)$ typically contains a rainbow Hamilton cycle. Later on, Frieze and Loh [11] improved this to $p = \frac{(1+o(1))\log n}{n}$ and $c = (1+o(1))n$, which is asymptotically optimal with respect to both of the parameters p and c. Recently, Bal and Frieze [3] obtained the optimal c by showing that for $p = \omega(\log n/n)$ and $c = n/2$ (respectively $c = n$), a graph $G \sim \mathscr{G}_c(n,p)$ w.h.p. contains a rainbow perfect matching (respectively a rainbow Hamilton cycle). For general graphs, Ferber, Nenadov, and Peter [10] showed that for every graph S on n vertices with maximum degree $\Delta(S)$ and for $c = (1+o(1))e(S)$, a typical $G \sim \mathscr{G}_c(n,p)$ contains a rainbow copy of S, provided that $p = n^{-1/\Delta(S)}\text{polylog}(n)$ (here, as elsewhere, $e(S)$ denotes the number of edges in S). In this case, the number of colors c is asymptotically optimal, whereas the edge probability p is almost certainly not.

Our first main result improves the main theorem of Frieze and Loh from [11] to $p = \frac{\log n+\log\log n+\omega(1)}{n}$, which is clearly optimal. Our proof technique is completely different, resulting in a shorter proof than the one given in [11].

Theorem 1. *Let $\varepsilon > 0$, let $c = (1+\varepsilon)n$ and let $p = \frac{\log n+\log\log n+\omega(1)}{n}$. Then a graph $G \sim \mathscr{G}_c(n,p)$ w.h.p. contains a rainbow Hamilton cycle.*

Next, building upon an ingenious coupling idea of McDiarmid [18], we give a general statement regarding the problem of finding "nice" structures in randomly edge-colored random hypergraphs. Then, we exhibit its applications to derive interesting corollaries. Before doing so, let us introduce some useful notation. For an integer c, suppose that $\mathscr{C} := \mathscr{C}(c,n,k)$ is a collection of edge-colored k-uniform hypergraphs on the same vertex set $[n]$, whose edge set is colored with colors from $[c]$. We say that $\mathscr{C}$ is *ℓ-rich* if for any $C \in \mathscr{C}$ and for any $e \in E(C)$ there are at least ℓ distinct ways to color e in order to obtain an element of $\mathscr{C}$. For example, consider the case where $k = 2$ and $c = n_1 \geq n$, and let $\mathscr{C}(c, n, k)$ be the set of all possible rainbow Hamilton cycles in K_n. Note that for each $C \in \mathscr{C}$ and for every $e \in E(C)$, there are $n_1 - n + 1$ ways to color e in order to obtain a rainbow Hamilton cycle. Therefore, $\mathscr{C}$ is $(n_1 - n + 1)$-rich. Now, given a collection of edge-colored hypergraphs $\mathscr{C}$, we define $\tilde{\mathscr{C}}$ to be the set of all hypergraphs obtained by taking $C \in \mathscr{C}$ and deleting the colors from its edges. With this notation in hand we can state the following theorem:

Theorem 2. *Let $p := p(n) \in [0, 1]$ and let ℓ, c be positive integers for which $q := \frac{cp}{\ell} \leq 1$. Let k be any positive integer and let $\mathscr{C} := \mathscr{C}(c, n, k)$ be any ℓ-rich set. Then, we have*

$$\Pr\left[H \sim \mathscr{H}^k(n, p) \text{ contains some } C \in \tilde{\mathscr{C}}\right] \leq \Pr\left[H \sim \mathscr{H}_c^k(n, q) \text{ contains some } C \in \mathscr{C}\right].$$

Next we present two interesting applications for Theorem 2, combining it with known results. First, we show that one can find a rainbow Hamilton cycles in a random hypergraph with an optimal (up to a multiplicative constant) edge density when working with the approximately optimal number of colors. Second, we present an application which is somewhat different in nature. We show that one can find "many" edge-disjoint Hamilton cycles, each of which is rainbow in a random graph. Before stating it formally, we need the following definition. Let H be a k-uniform hypergraph on n vertices. For $0 \leq \ell < k$ we define a Hamilton ℓ-cycle as a cyclic ordering of $V(H)$ for which the edges consist of k consecutive vertices, and for each two consecutive edges e_i and e_{i+1} we have $|e_i \cap e_{i+1}| = \ell$ (where we consider $n + 1 = 1$). It is easy to show that a Hamilton ℓ-cycle contains precisely $m_\ell := \frac{n}{k-\ell}$ edges and therefore we cannot expect to have one unless n is divisible by $k - \ell$. (Note that we can consider a perfect matching as a Hamilton 0-cycle.)

The problem of finding the threshold for the existence of Hamilton ℓ-cycles in random hypergraphs has drawn a lot of attention in the last decade. Among the many known results, it is worth mentioning the one of Johansson, Kahn, and Vu [13], who showed that $p = \Theta(\log n/n^{k-1})$ is a threshold for the appearance of a Hamilton 0-cycle (that is, a perfect matching) in a typical $H \sim \mathscr{H}^k(n, p)$, provided that n is divisible by k. In general, for every $\ell < k$, the threshold for the appearance of a Hamilton ℓ-cycle in a typical $H \sim \mathscr{H}^k(n, p)$ (assuming that n is divisible by $k - \ell$) is around $p \approx \frac{1}{n^{k-\ell}}$ (in some cases an extra log factor appears). For more details we refer the reader to [8] and its references.

Now we are ready to state our next result:

Theorem 3. *Let $0 \leq \ell < k$ be two integers, and let $p \in [0, 1]$ be such that*

$$\Pr\left[H \sim \mathscr{H}^k(n, p) \text{ contains a Hamilton } \ell\text{-cycle}\right] = 1 - o(1).$$

Then, for every $\varepsilon > 0$, letting $c = (1 + \varepsilon)m_\ell$ and $q = \frac{cp}{\varepsilon m_\ell + 1}$ we have

$$\Pr\left[H \sim \mathscr{H}_c^k(n, q) \text{ contains a rainbow Hamilton } \ell\text{-cycle}\right] = 1 - o(1).$$

Remark: Note that we allow ε to be a function of n (or even 0) in the statement above. Moreover, if we take ε to be a constant, then in particular we see that by losing a multiplicative constant in the threshold, a rainbow Hamilton ℓ-cycle w.h.p. exists. This for example reproves and extends the first result obtained by Cooper and Frieze [7] in a very concise way.

The second application is regarding the problem of finding many rainbow edge-disjoint Hamilton cycles in a typical $G \sim \mathcal{G}(n,p)$. The analogous problem without the rainbow requirements is well studied and quite recently, completing a long sequence of papers, Knox, Kühn, and Osthus [14], Krivelevich and Samotij [15] and Kühn and Osthus [17] solved this question for the entire range of p. Combining these results with Theorem 2 we can in particular obtain the following:

Theorem 4. *For every $0 < \varepsilon < 1$ there exists $C := C(\varepsilon) > 0$ such that for every $p \geq \omega(\log n/n)$ and $c = Cn$ the following holds:*

$$\Pr\left[G \sim \mathcal{G}_c(n,p) \textit{ contains } (1-\varepsilon)np/2 \textit{ edge-disjoint rainbow Hamilton cycles}\right] = 1 - o(1).$$

Notation. Our graph theoretic notation is quite standard and mainly follows that of [20]. For $p \in [0,1]$ we let $\mathcal{H}^k(n,p)$ denote the probability space of k-uniform hypergraphs on vertex set $[n]$, obtained by adding each possible k-subset of $[n]$ as an edge with probability p, independently at random. In the special case $k = 2$, we denote $\mathcal{G}(n,p) := \mathcal{H}^2(n,p)$, the well-studied binomial random graph model. For an integer c, we let $\mathcal{H}_c^k(n,p)$ be the probability space of edge-colored k-uniform hypergraphs on vertex set $[n]$ obtained as follows. First, take $H \sim \mathcal{H}^k(n,p)$, and then, to each edge, assign a color from $\mathcal{C} := [c]$ uniformly, independently at random. As before, in the case $k = 2$ we write $\mathcal{G}_c(n,p) := \mathcal{H}_c^2(n,p)$. Given a subhypergraph H' of an edge-colored hypergraph H, we say that H is *rainbow* if all its edges receive distinct colors. For a vertex $v \in V(H)$ we denote by $d_H^c(v)$ its *color degree*, that is, the number of distinct colors appearing on edges incident to v. For an edge $e \in E(H)$, we let $c(e)$ denote its color. Given a subset of vertices $W \subseteq V(H)$ and a subset of colors $\mathcal{C}_0 \subseteq \mathcal{C}$, we let $H[W;\mathcal{C}_0]$ denote the subhypergraph of H on a vertex set W for which $e \in \binom{W}{k} \cap E(H)$ is an edge of $H[W;\mathcal{C}_0]$ if and only if $c(e) \in \mathcal{C}_0$. In case that G is a graph, given two disjoint subsets of vertices $S, W \subseteq V(G)$ and a subset of colors $\mathcal{C}_0 \subseteq \mathcal{C}$, we let $G[S,W;\mathcal{C}_0]$ denote the bipartite subgraph of G with parts S and W, and edges $sw \in E(G)$, where $s \in S$, $w \in W$, and $c(sw) \in \mathcal{C}_0$. Moreover, for two disjoint subsets $S, W \subseteq V(G)$ and an integer D, we say that G contains a D-matching from S to W if there exists a rainbow subgraph M of G such that $d_M(s,W) = D$ for every $s \in S$ and $d_M(w) \leq 1$ for every $w \in W$.

We will frequently omit rounding signs for the sake of clarity of presentation.

2 Tools

In this section we introduce some tools and auxiliary results to be used in our proofs.

2.1 *Probabilistic tools*

We will routinely employ bounds on large deviations of random variables. We will mostly use the following well-known bound on the lower and the upper tails of the binomial distribution due to Chernoff (see [2, 12]).

Lemma 1 (Chernoff's inequality). *Let* $X \sim \mathrm{Bin}(n, p)$ *and let* $\mu = \mathbb{E}(X)$. *Then*

- $\Pr[X < (1-a)\mu] < e^{-a^2\mu/2}$ *for every* $a > 0$;
- $\Pr[X > (1+a)\mu] < e^{-a^2\mu/3}$ *for every* $0 < a < 3/2$.

We also make use of the following approximation for the lower tail of a binomially distributed random variable.

Lemma 2. *Let* $\frac{\log n}{n} \le p \le \frac{2\log n}{n}$, *and let* $0 < \delta < 1$ *be such that* $\left(\frac{e^2}{\delta}\right)^{\delta} e^{-1+\delta} \le e^{-0.7}$. *Then*

$$\Pr[\mathrm{Bin}(n, p) \le \delta np] \le n^{-2/3}.$$

Proof. Note that

$$\begin{aligned}\Pr[\mathrm{Bin}(n,p) \le \delta np] &= \sum_{i=0}^{\delta np} \binom{n}{i} p^i (1-p)^{n-i} \le \delta np \left(\frac{e}{\delta}\right)^{\delta np} e^{-(1-\delta)np} \\ &\le \delta np \left[\left(\frac{e}{\delta}\right)^{\delta} e^{-1+\delta}\right]^{np} \le \delta np e^{-0.7np} \\ &\le e^{-(2/3)\log n} = n^{-2/3}.\end{aligned}$$

□

Before introducing the next tool to be used, we need the following definition.

Definition 1. Let $(A_i)_{i=1}^n$ be a collection of events in some probability space. A graph D on a vertex set $[n]$ is called a *dependency graph* for $(A_i)_i$ if A_i is mutually independent of all the events $\{A_j : ij \notin E(D)\}$.

We make use of the following asymmetric version of the Lovász Local Lemma (see, e.g., [2]).

Lemma 3. *(Asymmetric Local Lemma) Let* $(A_i)_{i=1}^n$ *be a sequence of events in some probability space. Suppose that* D *is a dependency graph for* $(A_i)_i$, *and suppose that* *there are real numbers* $(x_i)_{i=1}^n$ *such that for every i the following holds:*

$$\Pr[A_i] \leq x_i \prod_{ij \in E(D)} (1 - x_j).$$

Then, $\Pr[\bigcap_{i=1}^n \bar{A}_i] > 0$.

2.2 Properties of $\mathscr{G}_c(n, p)$

Here we gather fairly standard typical properties of sparse binomial random graphs. Given a graph $G = (V, E)$ with vertex set $V = [n]$ vertices, define the set of vertices $SMALL \subseteq V$ by

$$SMALL := \{v \in [n] : d_G(v) \leq \delta \log n\},$$

where $\delta > 0$ is a small enough absolute constant.

Lemma 4. *Let* $0 < \beta, \varepsilon < 1$ *be absolute constants, let* $c = (1+\varepsilon)n$, *and let* $\frac{\log n}{n} \leq p \leq \frac{2 \log n}{n}$. *Then, w.h.p. a graph* $G \sim \mathscr{G}_c(n, p)$ *satisfies the following properties.*

(*P*1) $\Delta(G) \leq 10 \log n$.
(*P*2) $|SMALL| \leq n^{0.4}$.
(*P*3) *For every* $v \in [n]$, $d_G^c(v) \geq d_G(v) - 2$.
(*P*4) *Let* $E_0 = \{e \in E(G) : e \cap SMALL \neq \emptyset\}$. *Then all the elements of* E_0 *are of distinct colors.*
(*P*5) *No two vertices* $x, y \in SMALL$ *(x and y might be the same) have a path of length at most* 4 *with* x, y *as its endpoints in G.*
(*P*6) *For every two disjoint subsets X and Y of size* $|X| = |Y| = \omega\left(\frac{n}{(\log n)^{1/2}}\right)$, *the number of colors appearing on the edges between X and Y is at least* $(1 + \varepsilon - o(1))n$.
(*P*7) *For every subset* $\mathscr{C} \subseteq [c]$ *of size* βn *and for every subset* $X \subseteq [n]$ *for which* $|X|^2 p = \omega(n)$, *we have that* $\frac{\beta}{3}|X|^2 p \leq e(G[X; \mathscr{C}]) \leq \beta |X|^2 p$.
(*P*8) *For every subset* $\mathscr{C} \subseteq [c]$ *of size* βn *and for every two disjoint subsets* $X, Y \subseteq [n]$ *such that* $|X||Y|p = \omega(n)$, *we have that* $\frac{\beta}{2}|X||Y|p \leq e(G[X, Y; \mathscr{C}]) \leq \beta |X||Y|p$.
(*P*9) *For every* $s \in [c]$, *the number of edges in G which are colored in s is at most* $10 \log n$.
(*P*10) *For every subset* $X \subseteq [n]$, *if* $|X| \leq \frac{n}{\log^{4/3} n}$, *then* $e_G(X) \leq 8|X|$.
(*P*11) *For every* $X \subseteq [n]$ *of size* $|X| \geq \frac{n}{\log^{4/3} n}$, *we have* $e_G(X) \leq |X|^2 p \left(\frac{n}{|X|}\right)^{1/2}$.

Proof. For (*P*1), just note that given a vertex $v \in [n]$, since $d_G(v) \sim \text{Bin}(n-1, p)$, it follows that

$$\Pr[d_G(v) \geq 10\log n] \leq \binom{n}{10\log n} p^{10\log n}$$
$$\leq \left(\frac{enp}{10\log n}\right)^{10\log n} = o(1/n).$$

Therefore, by applying the union bound we obtain that w.h.p. $\Delta(G) \leq 10\log n$.

For $(P2)$ note that by Lemma 2, the expected number of such vertices is at most $n^{1/3}$. Therefore, by applying Markov's inequality, $(P2)$ immediately follows.

For $(P3)$, assume that $d_G^c(v) \leq d_G(v) - 3$. In particular, this means that there are 2 disjoint pairs $\{x_1, y_1\}$ and $\{x_2, y_2\}$ of neighbors of v such that for $i = 1, 2$ both vx_i and vy_i have the same color. The probability of this event is upper bounded by $(d_G(v))^4 c^{-2} = o(n^{-1})$. Therefore, by applying the union bound we obtain $(P3)$.

For $(P4)$, note that by $(P2)$ we have that $|SMALL| = o(n^{0.49})$. Now, since $\Delta(D) \leq 10\log n$, it follows that $|E_0| = o(n^{1/2})$. Therefore,

$$\Pr\left[\exists e, e' \in E_0 \text{ with the same color}\right] \leq \binom{|E_0|}{2} c^{-1} = o(1).$$

For $(P5)$, note that, given two vertices x, y, the probability that there exists a path of length at most 4 between them is upper bounded by $p + np^2 + n^2p^3 + n^3p^4 \leq \frac{17\log^4 n}{n}$. Now, since by $(P2)$ we have $|SMALL| = o(n^{-0.49})$, and since the events "$v \in SMALL$" are "almost independent," by applying the union bound we can easily obtain that the probability for having two such vertices x, y in $SMALL$ is $o(1)$.

For $(P6)$, let X and Y be two disjoint subsets. Note that for a set $\mathscr{C}' \subseteq \mathscr{C}$ of size t, the probability that none of the colors of $\mathscr{C}'$ appears on $E(X, Y)$ is upper bounded by:

$$(1 - p + p(1 - t/c))^{|X||Y|} \leq e^{-pt|X||Y|/c}.$$

Therefore, if $t = \gamma n$ for some fixed constant $\gamma > 0$, then by applying the union bound we obtain that the probability for having a subset $\mathscr{C}'$ of colors of size γn, and two disjoint subsets X and Y of sizes $|X| = |Y| = x = \frac{10n}{(\gamma\log n)^{1/2}}$ for which none of the edges in $E(X, Y)$ uses colors of $\mathscr{C}'$ is at most

$$\binom{n}{x}\binom{n}{x}\binom{(1+\varepsilon)n}{\gamma n} e^{-\gamma npx^2/c} \leq 8^n e^{-\frac{100\gamma pn^2}{c\gamma\log n}} = o(1)\,.$$

For $(P7)$ just note that the expectation of the number of such edges is $\binom{|X|}{2}\cdot p \cdot \frac{\beta n}{c} = \omega(n)$, and therefore, by Chernoff's inequality and the union bound over all choices of X and of $\mathscr{C}$, we easily obtain the desired claim.

For $(P8)$, let $X, Y \subseteq [n]$ be such subsets. Note that since $\mathscr{C}$ is of size βn, the expected number of edges between X and Y which are assigned colors from

$\mathscr{C}$ is $\frac{\beta}{1+\varepsilon}|X||Y|p = \omega(n)$. Therefore, the property follows easily from Chernoff's inequality and the union bound.

For $(P9)$, $s \in [c]$ be some color and let Y denote the random variable which counts the number of edges colored s in G. Clearly, $Y \sim \text{Bin}\left(e(G), \frac{1}{(1+\varepsilon)n}\right)$. Now, it is easy to show that w.h.p. $e(G) = (1+o(1))\binom{n}{2}p \leq (1+o(1))n\log n \leq 2n\log n$, and therefore,

$$\Pr[Y \geq 10\log n] \leq \binom{2n\log n}{10\log n}\left(\frac{1}{(1+\varepsilon)n}\right)^{10\log n}$$
$$\leq \left(\frac{2en\log n}{10(1+\varepsilon)n\log n}\right)^{10\log n} = o(1/n).$$

Next, by applying the union bound we obtain the desired claim.

We leave $(P10)$ and $(P11)$ as an exercise for the reader. □

2.3 Finding rainbow star matchings between appropriate sets

In this subsection we describe the main technical lemma which will be used in the proof of Theorem 1. Informally speaking, this lemma ensures the existence of rainbow star matchings between sets of appropriate sizes.

Lemma 5. *Let $\alpha, \varepsilon > 0$ be constants, let D be a fixed integer, let $c = (1+\varepsilon)n$ and let $\frac{\log n}{n} \leq p \leq \frac{2\log n}{n}$. Then, a graph $G \sim \mathscr{G}_c(n,p)$ is w.h.p. such that the following holds. Suppose that*

(i) $W \subseteq [n]$ of size $(1+o(1))\frac{n}{\log\log n}$, and
(ii) $S \subseteq [n]$ of size $\frac{n}{\log^{0.4} n} \leq |S| \leq \frac{2n}{\log^{0.4} n}$, and
(iii) $\mathscr{C}_1 \subseteq \mathscr{C} := [c]$ of size $|\mathscr{C}_1| = \alpha n$, and
(iv) for every $s \in S$ there are at least $\frac{\log n}{(\log\log n)^2}$ edges $e = sw$ with $w \in W$ and $c(e) \in \mathscr{C}_1$.

Then, there exists a rainbow D-matching from S to W, with all colors from $\mathscr{C}_1$.

The main ingredient in the proof of Lemma 5 is the following powerful tool due to Aharoni and Haxell [1], generalizing Hall's theorem to hypergraphs.

Theorem 5. *Let g and D be positive integers and let $\mathscr{H} = \{\mathscr{H}_1, \ldots, \mathscr{H}_t\}$ be a family of g-uniform hypergraphs on the same vertex set. If, for every $I \subseteq [t]$, the hypergraph $\bigcup_{i\in I}\mathscr{H}$ contains a matching of size greater than $Dg(|I|-1)$, then there exists a function $f : [t] \times [D] \to \bigcup_{i=1}^{t} E(\mathscr{H}_i)$ such that $f(i,j) \in E(\mathscr{H}_i)$ for every i and j, and $f(i,j) \cap f(i',j') = \emptyset$ for $(i,j) \neq (i',j')$.*

When applying Theorem 5, we will distinguish between few cases according to the size of $I \subseteq [t]$. The following lemmas will make our life a bit easier with it.

Lemma 6. *Let $\varepsilon > 0$, let $c = (1+\varepsilon)n$, let $D \in \mathbb{N}$ and let $\frac{\log n}{n} \leq p \leq \frac{2\log n}{n}$. Then a graph $G \sim \mathcal{G}_c(n,p)$ is w.h.p. such that the following holds. For every collection of $j \leq \frac{n}{\log^{0.9} n}$ vertex disjoint stars, each of size $\log^{0.2} n$, the number of colors appearing on their edges is at least $2Dj$.*

Proof. Let $s := \log^{0.2} n$. We show that the probability of having a collection of $j \leq \frac{n}{\log^{0.9} n}$ stars, each of which of size s whose union contains at most $2Dj$ colors is $o(1)$. The following expression is an upper bound for this probability:

$$\sum_{j=1}^{\frac{n}{\log^{0.9} n}} \binom{n}{j}\binom{n}{s}^j p^{js}\binom{(1+\varepsilon)n}{2Dj}\left(\frac{2Dj}{(1+\varepsilon)n}\right)^{js}$$

$$\leq \sum_{j=1}^{\frac{n}{\log^{0.9} n}} \left(\frac{en}{j}\right)^j \left(\frac{e(1+\varepsilon)n}{2Dj}\right)^{2Dj} \left(\frac{2eDjnp}{s(1+\varepsilon)n}\right)^{js} = o(1).$$

Indeed, fix $j \leq \frac{n}{\log^{0.9} n}$ and first choose j vertices to be the "centers" of the stars. For each of these vertices choose s neighbors and multiply by the probability of all these edges to appear. Next, choose a set of $2Dj$ colors from $c = (1+\varepsilon)n$ and multiply by the probability that all the edges of the stars are colored with these colors. This completes the proof of the lemma. □

The following lemma may look at the first glance a bit complicated to understand, but its role will become clear during the proof of Lemma 5.

Lemma 7. *Let $0 < \alpha < \varepsilon < 1$, let $c = (1+\varepsilon)n$, let $D \in \mathbb{N}$ and let $\frac{\log n}{n} \leq p \leq \frac{2\log n}{n}$. Then a graph $G \sim \mathcal{G}_c(n,p)$ is w.h.p. such that for every*

(i) $\frac{n}{\log^{0.9} n} \leq j \leq \frac{2n}{\log^{0.4} n}$,
(ii) $W \subseteq [n]$ *of size* $|W| = (1+o(1))\frac{n}{\log\log n}$, *and*
(iii) $\mathcal{C}_1 \subseteq \mathcal{C} := [c]$ *of size* $|\mathcal{C}_1| = \alpha n$,

the following holds. The probability of having subsets $X \subseteq [n]$ of size j, $W' \subseteq W$ and $\mathcal{C}_2 \subseteq \mathcal{C}_1$ of sizes at most $2Dj$ such that for every edge $xw \in E_G(X,W)$ we have $c(xw) \in \mathcal{C}_2$ or $w \in W'$ or $c(xw) \notin \mathcal{C}_1$ is $o(1)$.

Proof. In order to prove the lemma, note that we can upper bound the probability by

$$\binom{n}{|W|}\binom{|W|}{|W'|}\binom{c}{|\mathcal{C}_1|}\binom{|\mathcal{C}_1|}{|\mathcal{C}_2|}\cdot\sum_{j=\frac{n}{\log^{0.9} n}}^{\frac{n}{\log^{0.4} n}}\left(1-p+p\left(\frac{|\mathcal{C}_2|+|\mathcal{C}\setminus\mathcal{C}_1|}{|\mathcal{C}|}\right)\right)^{j|W\setminus W'|}$$

$$\leq 16^n \sum_{j=\frac{n}{\log^{0.9} n}}^{\frac{n}{\log^{0.4} n}} \left(1-p+p\frac{2Dj}{(1+\varepsilon)n}+p(1-\alpha/(1+\varepsilon)+o(1))\right)^{j|W\setminus W'|}$$

$$\leq 16^n \sum_{j=\frac{n}{\log^{0.9} n}}^{\frac{n}{\log^{0.4} n}} \exp\left(-Cjp\left(\frac{n}{\log\log n} - Dj\right)\right) = o(1).$$

Here, C is some constant that depends on α. This completes the proof. □

The following lemma shows that in a typical random graph $G \sim \mathscr{G}(n,p)$, any bipartite subgraph $B = (S \cup W, E') \subseteq G$ with all the vertices in S of "large" degree contains an s-matching from S to W, for an appropriate choice of parameters.

Lemma 8. *Let* $\frac{\log n}{n} \leq p \leq \frac{2\log n}{n}$. *Then, a graph* $G \sim \mathscr{G}(n,p)$ *is w.h.p. such that the following holds. Suppose that* $B = (S \cup W, E')$ *is a bipartite (not necessarily induced) subgraph of G, with*

(*i*) $|S| \leq \frac{n}{\log^{0.9} n}$, *and*
(*ii*) $|W| = (1+o(1))\frac{n}{\log\log n}$, *and*
(*iii*) $d_B(v) \geq \frac{\log n}{(\log\log n)^2}$ *for every* $v \in S$.

Then there exists a $\log^{0.2} n$*-matching from S to W.*

Proof. Let $B = (S \cup W, E')$ be the subgraph of G as described above. In order to prove the lemma, we use the following version of Hall's Theorem (see, e.g., [20]). A bipartite graph $B = (S \cup W, E')$ contains an s-matching from S to W, if and only if the following holds:

$$\text{For every } X \subseteq S \text{ we have } |N_B(X)| \geq s|X|. \tag{1}$$

Suppose that (1) fails for B with $s = \log^{0.2} n$. Then, there exists a subset $X \subseteq S$ for which $|N_B(X) \cup X| < (s+1)|X|$. In particular, letting $Y = N_B(X) \cup X$, by (*iii*), we conclude that $e_B(Y) \geq |Y|\frac{\log n}{(s+1)(\log\log n)^2} \geq |Y|\frac{\log^{0.8} n}{2(\log\log n)^2}$. Since $|Y| < (s+1)|X| \leq \frac{2n}{\log^{0.7} n}$, and since $|Y|\log^{0.7} n > |Y|^2 p\,(n/|Y|)^{1/2}$ for every $|Y| \leq n/\log^{0.6} n$, we obtain a contradiction to ($P10$) and ($P11$). □

Now we are ready to prove Lemma 5.

Proof (of Lemma 5). Let $\alpha < \varepsilon$, let $D \in \mathbb{N}$ and let $W, S \subseteq [n]$ and $\mathscr{C}_1 \subseteq \mathscr{C}$ as described in the assumptions of the lemma. For every $s \in S$, we define a graph $\mathscr{H}_s$ with vertex set $W \cup \mathscr{C}_1$ in the following way. For every $w \in W$ and $x \in \mathscr{C}_1$, $wx \in E(\mathscr{H}_s)$ if and only if $sw \in E(G)$ and $c(sw) = x$. Consider the family $\mathscr{H} := \{\mathscr{H}_s : s \in S\}$, and note that in order to prove the lemma, we need to show that there exists a function $f : S \times [D] \to \bigcup_{s\in S} E(\mathscr{H}_s)$ such that $f(s,i) \in E(\mathscr{H}_s)$ for every s and i, and $f(s,i) \cap f(s',i') = \emptyset$ for $(s,i) \neq (s',i')$. To this end, we make use of Theorem 5. All we need to show is that for every $T \subseteq S$, the graph $\bigcup_{t\in T} E(\mathscr{H}_t)$ contains a matching of size greater than $2D(|T|-1)$. We distinguish between two cases:

Case 1: $|T| \leq \frac{n}{\log^{0.9} n}$. Consider the bipartite graph $B = (T \cup W, E')$, where $E' := \{tw : t \in T, w \in W \text{ and } c(tw) \in \mathcal{C}_1\}$. By applying Lemma 8 to B, we conclude that there exists an s-matching from T to W in B, where $s = \log^{0.2} n$. Let M be such a matching, and note that by applying Lemma 6 to M, it follows that the number of colors appearing in M is at least $2D|T|$. Now, one can easily deduce that $\bigcup_{t \in T} E(\mathcal{H}_t)$ contains a matching of size at least $2D|T| > 2D(|T| - 1)$.

Case 2: $\frac{n}{\log^{0.9} n} \leq |T| \leq |S|$. Let M be a matching in $\bigcup_{t \in T} E(\mathcal{H}_t)$ of maximum size, let $\mathcal{C}_2 := \{x \in \mathcal{C}_1 : \exists w \in W \text{ s.t } wx \in M\}$ and let $W' := \{w \in W : \exists x \in \mathcal{C}_2 \text{ s.t } wx \in M\}$. Suppose that $|M| \leq 2D(|T| - 1) < 2D|T|$. In particular, it means that for every $v \in T$ and $w \in W$ we have $vw \notin E(G)$, or $c(vw) \in \mathcal{C}_2$, or $w \in W'$, or $c(vw) \notin \mathcal{C}_1$, which contradicts Lemma 7. This completes the proof. □

2.4 Expansion properties of subgraphs of random graphs

In the following lemma we show. that given an edge-colored graph G, one can find two subsets of colors $\mathcal{C}_1, \mathcal{C}_2$ and a vertex subset W which inherits some desired properties from G. The statement of the lemma is adjusted so as to facilitate its application in the proof of Theorem 1.

Lemma 9. *Let $0 < \alpha, \delta, \varepsilon < 1/2$ be constants and let n be an integer. Let G be an edge-colored graph on $m \geq (1 - o(1))n$ vertices, and let $\mathcal{C}^*$ be its set of colors, of size $|\mathcal{C}^*| \geq (1 + \varepsilon/2)n$. Suppose that $\delta \log n \leq \delta(G) \leq \Delta(G) \leq 10 \log n$, that each color appears at most $10 \log n$ times in G, and that for each $v \in V(G)$, $d^c_G(v) \geq d_G(v) - 2$. Then one can find subsets $\mathcal{C}_0, \mathcal{C}_1 \subseteq \mathcal{C}^*$, and $W \subseteq V(G)$ satisfying the following properties:*

(i) $|W| = (1 + o(1))\frac{n}{\log \log n}$, *and*
(ii) $\mathcal{C}_0$ *and* $\mathcal{C}_1$ *are two disjoint subsets of sizes* $(1 + o(1))\alpha n$, *and*
(iii) *for every* $w \in W$, $d_{\mathcal{C}_0}(w, W) \in \left(\frac{\alpha d_G(w)}{2 \log \log n}, \frac{2\alpha d_G(w)}{\log \log n}\right)$, *and*
(iv) *for every* $v \in V(G)$, $d_{\mathcal{C}_1}(v, W) \in \left(\frac{\alpha d_G(v)}{2 \log \log n}, \frac{2\alpha d_G(v)}{\log \log n}\right)$, *and*
(v) *for every* $x \in \mathcal{C}_0$, *x appears on at most* $\frac{100 \log n}{(\log \log n)^2}$ *edges in* $G[W]$.

Proof. Let $\mathcal{C}_0, \mathcal{C}_1 \subseteq \mathcal{C}^*$ be two disjoint random subsets, obtained in the following way: for each element of $\mathcal{C}^*$ toss a coin with probability 2α to decide whether it belongs to $\mathcal{C}_0 \cup \mathcal{C}_1$, then, with probability $1/2$ decide to which of these sets it belongs. All these choices are independent. Let $W \subseteq V(G)$ be a random subset of vertices, obtained by picking each $v \in V(G)$ with probability $\frac{1}{\log \log n}$, independently at random. We wish to show that the obtained sets satisfy (i)–(v) with positive probability. In order to do so, we consider several types of events. First, let A_W denote the event "$|W| \notin (1 \pm o(1))\frac{n}{\log \log n}$". Second, for each $i \in \{0, 1\}$, let C_i denote the event "$|\mathcal{C}_i| \notin (1 \pm o(1))\alpha n$". Third, for each vertex $v \in V(G)$, let $\Gamma_i(v)$

$(i \in \{0,1\})$ denote the event "$d_{\mathscr{C}_i}(v,W) \notin \left(\frac{\alpha d_G(v)}{2\log\log n}, \frac{2\alpha d_G(v)}{\log\log n}\right)$". Lastly, for each $x \in \mathscr{C}^*$, let B_x be the event "more than $\frac{100\log n}{(\log\log n)^2}$ edges in $G[W]$ are colored x". With this notation at hand, we wish to show that

$$\Pr\left[\overline{A_W} \cap \overline{C_1} \cap \overline{C_2} \cap \left(\bigcap_{i\in\{0,1\}, v\in V(G)} \overline{\Gamma_i(v)}\right) \cap \left(\bigcap_{x\in\mathscr{C}^*} \overline{B_x}\right)\right] > 0.$$

First, define $\mathscr{E} := \{A_W, C_1, C_2, \Gamma_i(v), B_x : v \in V(G), i \in \{0,1\}, x \in \mathscr{C}^*\}$, and let us estimate the probabilities of each of the events $X \in \mathscr{E}$. By using Chernoff's bounds we trivially get

(*a*) $\Pr[A_W] = \exp(-\Theta(n/\log\log n))$,
(*b*) $\Pr[C_i] = \exp(-\Theta(\alpha n))$, and
(*c*) $\Pr[\Gamma_i(v)] = \exp(-C\alpha d_G(v)/\log\log n)$, where C is an absolute constant which does not depend on α.

For estimating $\Pr[B_x]$, note that since $d_G^c(v) \geq d_G(v) - 2$ for every $v \in V(G)$, it follows that each color class can be partitioned into at most four matchings, each of size at most $10\log n$ (the maximum number of edges with the same color in G). Fix such a partition (into matchings) for each color class x. It follows that if B_x fails, then in at least one of the four matchings, at least $\frac{25\log n}{(\log\log n)^2}$ edges have been chosen. Since in each matching these choices are independent, and since for a fixed edge e, the probability that $e \in W$ is $\frac{1}{(\log\log n)^2}$, it follows by Chernoff's bounds that

(*d*) $\Pr[B_x] \leq \exp\left(-\frac{\log n}{2(\log\log n)^2}\right)$.

Next, let us define a dependency graph D for $\mathscr{E}$, where the edges of D are as follows:

- All pairs $A_W X$, where $X \in \mathscr{E}$, and
- all pairs $C_i X$, where $i \in \{0,1\}$ and $X \in \mathscr{E}$, and
- all pairs $\Gamma_i(v)\Gamma_j(u)$, where $i \neq j$ and $v = u$, or $v \neq u$ and $N_G(v) \cap N_G(u) \neq \emptyset$, or if the same color $c \in \mathscr{C}^*$ appears on edges incident to both u and v, and
- all pairs $\Gamma_i(v)B_x$ for which there exists an edge uw such that $\{u,w\} \cap (\{v\} \cup N_G(v)) \neq \emptyset$ and $c(uw) = x$, and
- all pairs B_xB_y, for which there exist two edges e and f, of colors x and y, respectively, such that $e \cap f \neq \emptyset$.

Now, for some fixed constant $c_0 > 1$, define $x_W = \sqrt{\Pr[A_W]}$, $y_i = \sqrt{\Pr[C_i]}$ (where $i \in \{0,1\}$), $x_{i,v} = c_0 \Pr[\Gamma_i(v)]$ (where $i \in \{0,1\}$ and $v \in V(G)$), and $z_x = \sqrt{\Pr[B_x]}$ for $x \in \mathscr{C}^*$. Note that

$$\Pr[A_W] = \exp(-\Theta(n/\log\log n)) \leq x_W \prod_{i,v}(1 - x_{i,v}),$$

and

$$\Pr[C_i] = \exp(-\Theta(\alpha n)) \le y_i \prod_{i,v}(1 - x_{i,v}),$$

and

$$\Pr[\Gamma_i(v)] = \exp(-C\alpha d_G(v)/\log\log n) \le c_0 x_{i,v} \prod_{j,u:\Gamma_j(u)\Gamma_i(v)\in E(D)} (1 - x_{j,u}) \prod_{x\in N_G^c(v)} (1 - z_x).$$

and

$$\Pr[B_x] \le z_x \prod_{x\in N_G^c(v),i} (1 - x_{i,v}) \prod_{y:B_xB_y\in E(D)} (1 - z_y).$$

(The last two inequalities hold because the corresponding "degrees of dependencies" are some polylog(n)). All in all, one can apply the Asymmetric Local Lemma (Lemma 3) and obtain the desired claim. □

Now, let $\frac{\log n}{n} \le p \le \frac{2\log n}{n}$, let $c = (1+\varepsilon)n$, and let $G \sim \mathscr{G}_c(n,p)$. We show that w.h.p. G is such that every (not necessarily induced) subgraph $G_1 \subseteq G$ on $(1+o(1))n/\log\log n$ vertices with some degree constraints is also a very good expander.

Lemma 10. *Let $\alpha, \delta, \varepsilon > 0$, let $\frac{\log n}{n} \le p \le \frac{2\log n}{n}$, and let $c = (1+\varepsilon)n$. Then a graph $G \sim \mathscr{G}_c(n,p)$ is w.h.p. such that the following properties hold. Suppose that*

(*i*) *$W \subseteq [n]$ is of size $(1+o(1))n/\log\log n$, and*
(*ii*) *$\mathscr{C}_0 \subseteq \mathscr{C}$ is of size $(1+o(1))\alpha n$.*

Then $H := G[W;\mathscr{C}_0]$ satisfies:

(*a*) *For every subset $X \subseteq W$, if $|X| \le \frac{n}{\log^{4/3} n}$, then $e_H(X) \le 8|X|$, and*
(*b*) *for every $X \subseteq W$ of size $\frac{n}{\log^{4/3} n} \le |X| \le |W|$, we have $e_H(X) \le |X|^2 p \left(\frac{n}{|X|}\right)^{1/2}$, and*
(*c*) *for every $X \subseteq W$, if $|X| \ge n/\log^{0.4} n$, then $e_H(X) \le \alpha|X|^2 p$, and*
(*d*) *for every two subsets $X, Y \subseteq W$ satisfying $|X||Y|p = \omega(n)$, we have $e_H(X,Y) \ge \frac{\alpha}{2}|X||Y|p$.*

Proof. All these properties follow from the properties in Lemma 4 and similar arguments, hence are left as an exercise for the reader. □

Let us define the following useful notion of a (k,d)-expander.

Definition 2. A graph G is called a *(k,d)-expander* if for every subset $X \subseteq V(G)$ of size at most k we have

$$|N_G(X) \setminus X| \ge d|X|.$$

The following lemma is almost identical to Lemma 2.4 in [16] (although with few minor modifications). For the convenience of the reader, we briefly sketch the proof.

Lemma 11. *Let* $0 < \varepsilon, \delta < 1$ *and let* $\alpha < \delta e^{-100}$ *be constants. Let* $\frac{\log n}{n} \leq p \leq \frac{2\log n}{n}$, *and let* $c = (1+\varepsilon)n$. *Then there exists* $d_0 \in \mathbb{N}$ *such that for every* $d \geq d_0$, *a graph* $G \sim \mathcal{G}_c(n,p)$ *is w.h.p. such that the following holds. Suppose that* $W \subseteq [n]$ *is of size* $(1+o(1))n/\log\log n$, $\mathcal{C}_0 \subseteq \mathcal{C}$ *of size* $(1+o(1))\alpha n$ *and* $H := G[W;\mathcal{C}_0]$ *is a subgraph of* G *satisfying* $\frac{\alpha\delta\log n}{2\log\log n} \leq \delta(H) \leq \Delta(H) \leq \frac{20\alpha\log n}{\log\log n}$ *and properties (a)–(d) of Lemma 10. Moreover, assume that no color from* $\mathcal{C}_0$ *appears in* H *more than* $\frac{100\log n}{(\log\log n)^2}$ *times. Then, there exists a subgraph* $R \subseteq H$ *satisfying the following:*

(*a*) *R is rainbow, and*
(*b*) *R is a* $(k,100)$*-expander (where* $k = \alpha\delta|W|/100$*), and*
(*c*) $|E(R)| \leq d|W|$.

Proof (Lemma 2.4 in [16]). Let d be a large enough integer. Condition on G satisfying all the properties of Lemma 4. Now, for every $w \in W$, let w choose d random edges of H incident with w (with repetitions), and let $\Gamma(w)$ be the set of the edges chosen by w. Let R be the graph whose edge set is $\bigcup_{w\in W}\Gamma(w)$. We wish to show that R satisfies $(a) - (c)$ with positive probability.

We consider few types of events. First, the events regarding the rainbow part. For every two edges e_1, e_2 of the same color, let us denote by $A(e_1,e_2)$ the event "both e_1 and e_2 are in R" (in case $e_1 \neq e_2$), and "e_1 is chosen in more than one trial" (in case $e_1 = e_2$). Define

$$\mathcal{A} := \{A(e_1,e_2) : e_1 \text{ and } e_2 \text{ have the same color}\}.$$

Second, we consider the events ensuring the expansion of sets. For a set $X \subseteq W$, let $B(X)$ denote the event that $e_R(X) \geq \frac{d}{101}|X|$, and for every $\frac{n}{\log^{4/3} n} \leq x \leq k$, let

$$\mathcal{B}_x = \{B(X) : |X| = 101x\}.$$

Clearly, if none of the events in $\mathcal{A}$ happens, then (a) and (c) hold. Now, assume in addition that none of the events $\mathcal{B}_x$ happens, and we wish to show that (b) holds. Let $X \subseteq V(H)$ be a subset of size at most k, and we wish to show that $|N_H(X)\setminus X| \geq 100|X|$. Assuming otherwise, we obtain a subset X for which $|X \cup N_H(X)| < 101|X|$. Since none of the events in $\mathcal{A}$ holds, it follows that $|E_H(X \cup N_H(X))| \geq d|X| > d|X \cup N_H(X)|/101$. Now, if $|X \cup N_H(X)| \leq \frac{n}{\log^{4/3} n}$, then for a large enough d it contradicts (a) of Lemma 10. If $\frac{n}{\log^{4/3} n} \leq |X \cup N_H(X)| \leq 101k$, then it contradicts the event $B(X \cup N_H(X))$.

It thus suffices to show that with positive probability none of these events occurs. To this end we make use of the Local Lemma. We estimate the probabilities of each event above, define a dependency graph D and estimate its degrees.

The family $\mathcal{A}$**:** For a fixed pair $e_1, e_2 \in E(G)$ of the same color we have

$$\Pr[A(e_1, e_2)] \le \left(\frac{4d \log\log n}{\alpha\delta \log n}\right)^2.$$

Define $x = c_0 \left(\frac{4d \log\log n}{\alpha\delta \log n}\right)^2$ for some constant $c_0 > 1$. For the "degree of dependency" within $\mathscr{A}$, note that $A(e_1, e_2)$ depends on $A(f_1, f_2)$ if and only if $e_1 \cup e_2$ intersects $f_1 \cup f_2$. Now, recall that $\Delta(H) \le \frac{20\alpha \log n}{\log\log n}$, and that each color class contains at most $\frac{100 \log n}{(\log\log n)^2}$ edges. Therefore, the number of events in $\mathscr{A}$ which are neighbors of $A(e_1, e_2)$ in the dependency graph is at most $4\Delta(H)\frac{100 \log n}{(\log\log n)^2} \le \frac{8000\alpha \log^2 n}{(\log\log n)^3}$.

The family $\mathscr{B}_t$: For a fixed set X of size $101t$, similarly to [16], one can show that

$$\Pr[B(X)] \le \binom{e_H(X)}{dt} \cdot (2d)^{dt} \cdot \left(\frac{2 \log\log n}{\alpha\delta \log n}\right)^{dt},$$

where for $t \le n/(101 \log^{0.4} n)$, by assumption (b) of Lemma 10 we have $e_H(X) \le 101^2 t^2 p \left(\frac{n}{101t}\right)^{1/2}$, and therefore,

$$\Pr[B(X)] \le \left(10^9 \left(\frac{t}{n}\right)^{1/2} \log\log n/\alpha\delta\right)^{dt}.$$

For $t \le n/(101 \log^{0.4} n)$, define $y_t = \left(10^9 e \left(\frac{t}{n}\right)^{1/2} \log\log n/\alpha\delta\right)^{dt/2}$ and note that for an appropriately large d we have

$$\sum_{t=n/(\log^{4/3} n)}^{n/\log^{0.4} n} |\mathscr{B}_t| \cdot y_t = o(1).$$

Now, for $n/(101 \log^{0.4} n) \le t \le k$, we use (c) of Lemma 10 and obtain

$$\Pr[B(X)] \le \left(\frac{10^9 t \log\log n}{\delta n}\right)^{dt}.$$

Define $y_t = e^{C_1 t}\left(\frac{4et \log\log n}{\delta n}\right)^{dt}$ for some constant $C_1 > 0$ and note that for appropriate choices of δ, d, and C_1 we obtain

$$\sum_{t=n/(101 \log^{0.4} n)}^{k} |\mathscr{B}_t| \cdot y_t = \sum_{t=n/(101 \log^{0.4} n)}^{\alpha\delta n/\log\log n} \binom{n/\log\log n}{101t} e^{C_1 t}\left(\frac{4et \log\log n}{\delta n}\right)^{dt} = o(1).$$

To compute the "degree of dependency" with members of $\mathscr{A}$ in D, note that $B(X)$ is correlated with an event $A(e_1, e_2)$ if $(e_1 \cup e_2) \cap X \ne \emptyset$. Therefore, since the maximum degree of H is at most $\frac{20\alpha \log n}{\log\log n}$ and since each color appears at most $\frac{100 \log n}{(\log\log n)^2}$ times, the number of events $A(e_1, e_2)$ correlated with $B(X)$ is upper bounded by

$$|X|\frac{20\alpha \log n}{\log\log n}\frac{100\log n}{(\log\log n)^2} = |X|\frac{2000\alpha \log^2 n}{(\log\log n)^3}.$$

In order to apply the Asymmetric Local Lemma (Lemma 3) we need to show that the following inequalities hold.

$$\Pr[A(e_1, e_2)] \leq x \cdot (1-x)^{\frac{8000\alpha \log^2 n}{(\log\log n)^3}} \left(\prod_t (1-y_t)^{|\mathscr{B}_t|}\right),$$

$$\Pr[B(X)] \leq y_t \cdot (1-x)^{|X|\frac{2000\alpha \log^2 n}{(\log\log n)^3}} \cdot \left(\prod_t (1-y_t)^{|\mathscr{B}_t|}\right).$$

For the first inequality, note that since $x = c_0\left(\frac{4d\log\log n}{\alpha\delta \log n}\right)^2$, it follows that

$$\begin{aligned}
& x \cdot (1-x)^{\frac{8000\alpha \log^2 n}{(\log\log n)^3}} \left(\prod_t (1-y_t)^{|\mathscr{B}_t|}\right) \\
&\geq c_0\left(\frac{4d\log\log n}{\alpha\delta\log n}\right)^2 e^{-2c_0\left(\frac{4d\log\log n}{\alpha\delta\log n}\right)^2 \frac{8000\alpha\log^2 n}{(\log\log n)^3}} e^{-2\sum_t |\mathscr{B}_t| y_t} \\
&= (1+o(1))c_0\left(\frac{4d\log\log n}{\alpha\delta\log n}\right)^2 \\
&\geq \Pr[A(e_1, e_2)].
\end{aligned}$$

We used the facts that $\sum |\mathscr{B}_t| \cdot y_t = o(1)$ and that for small values of x we have $1 - x \geq e^{-2x}$.

The second inequality is even easier to verify and is left as an exercise for the reader. □

2.5 *Finding a long rainbow path*

In this section, we show that a graph $G \sim \mathscr{G}_c(n, p)$ is k-rainbow-pseudorandom in a "robust" way, and that an edge-colored k-pseudorandom graph contains a long rainbow path. Before stating our first lemma, we introduce the following definition:

Definition 3. A graph G on n vertices whose set of edges is colored is called *k-rainbow-pseudorandom*, if for every two disjoint subsets of vertices $A, B \subseteq V(G)$

of size $|A| = |B| = k$, the number of colors appearing on the edges of G between A and B is at least n.

Lemma 12. *Let $0 < \alpha < \varepsilon < 1$ be two constants, let $c = (1+\varepsilon)n$ and let $\frac{\log n}{n} \leq p \leq \frac{2\log n}{n}$. Then a graph $G \sim \mathcal{G}_c(n,p)$ is w.h.p. such that the following holds. For every subset $\mathcal{C}^* \subseteq [c]$ of size $|\mathcal{C}^*| \geq (1+\alpha)n$, the graph $G[[n];\mathcal{C}^*]$ is k-rainbow-pseudorandom for $k = \frac{n}{\log^{0.49} n}$.*

Proof. Follows immediately from Property ($P6$) of Lemma 4. □

Now we state a modification of Lemma 4.4 from [4].

Lemma 13. *Let G be an edge-colored graph on n vertices which is k-pseudorandom. Then G contains a rainbow path of length at least $n - 2k + 1$.*

Proof. The proof is more or less identical to the proof of Lemma 4.4 in [4] with some minor changes which are left to the reader. □

2.6 Expander graphs

Here we show that the union of few expander graphs yields an expander graph as well.

First, we show that adding vertex disjoint stars to an expander graph does not harm the expansion properties too much.

Lemma 14. *Let G be a graph, let $m > 1$, and let k be a positive integer. Let $S \subseteq V(G)$ be a subset of vertices for which there exists an m-matching from S to $V(G)\setminus S$. If $G[V(G)\setminus S]$ is a (k,m)-expander, then G is a $(k,(m-1)/2)$-expander.*

Proof. Let $X \subseteq V(G)$ of size $|X| \leq k$, and we wish to show that $|N_G(X)\setminus X| \geq \frac{m-1}{2}|X|$. Let us distinguish between the following two cases:

Case I: $|X \cap S| \leq |X|/2$. In this case, since $G[V(G)\setminus S]$ is a (k,m), it follows that $X \setminus S$ expands by a factor of c and therefore $|N_G(X)\setminus X| \geq \frac{(m-1)|X|}{2}$.

Case II: $|X \cap S| > |X|/2$. In this case, there exists an m-matching from S to $V(G)\setminus S$, hence $|N_G(X)\setminus X| \geq |N_G(X\cap S)\cap(V(G)\setminus S)| - |X\setminus S| \geq m|X\cap S| - \frac{|X|}{2} \geq \frac{(m-1)|X|}{2}$. □

The following simple lemma is from [5] (Claim 2.8).

Lemma 15. *Let G be a graph, let $m > 0$, and let k be a positive integer. Let $U \subseteq V(G)$ be a subset for which $d_G(u) \geq m-1$ for every $u \in U$, and, moreover, there is no path of length at most 4 in G whose (possibly identical) endpoints lie in U. If $G[V(G)\setminus U]$ is a (k,m)-expander, then G is a $(k,m-1)$-expander.*

2.7 Boosters

In the proof of Theorem 1 we need to find a Hamilton path between two designated vertices x' and y' in a sparse expander subgraph G_1 of a typical $G \sim \mathcal{G}_c(n,p)$. Moreover, we need such a Hamilton path to be rainbow within a prescribed subset of colors. In this section we show how to achieve such a goal.

A routine way to turn a non-Hamiltonian expander graph G_1 into a Hamiltonian graph is by using *boosters*. A booster is a non-edge e of G_1 such that the addition of e to G_1 decreases the number of connected components of G_1, or creates a path which is longer than a longest path of G_1, or turns G_1 into a Hamiltonian graph. In order to turn G_1 into a Hamiltonian graph, we start by adding a booster e of G_1. If the new graph $G_1 \cup \{e\}$ is not Hamiltonian then one can continue by adding a booster of the new graph. Note that after at most $2|V(G_1)|$ successive steps the process must terminate and we end up with a Hamiltonian graph. The main point using this method is that it is well-known (for example, see [19]) that a non-Hamiltonian graph G_1 with "good" expansion properties has many boosters. However, our goal is a bit different. We wish to turn G_1 into a graph that contains a rainbow Hamilton path with x' and y' as its endpoints. In order to do so, we add one (possibly) fake edge $x'y'$ to G_1, color it with a new color (which does not belong to $\mathcal{C}$) and try to find a rainbow Hamilton cycle that contains the edge $x'y'$. Then, the path obtained by deleting this edge from the Hamilton cycle will be the desired path. For that we need to define the notion of *e-boosters*.

Given a graph G_1 and a pair $e \in \binom{V(G_1)}{2}$, consider a path P of $G_1 \cup \{e\}$ of maximal length which contains e as an edge. A non-edge e' of G_1 is called an e-booster if $G_1 \cup \{e, e'\}$ has fewer connected components than $G_1 \cup \{e\}$ has, or contains a path P' which passes through e and which is longer than P, or that $G_1 \cup \{e, e'\}$ contains a Hamilton cycle that uses e. The following lemma is from [9] and shows that every connected and non-Hamiltonian graph G_1 with "good" expansion properties has many e-boosters for every possible e.

Lemma 16. *Let G_1 be a connected graph for which $|N_{G_1}(X) \setminus X| \geq 2|X| + 2$ holds for every subset $X \subseteq V(G_1)$ of size $|X| \leq k$. Then, for every pair $e \in \binom{V(G_1)}{2}$ such that $G_1 \cup \{e\}$ does not contain a Hamilton cycle which uses the edge e, the number of e-boosters for G_1 is at least $(k+1)^2/2$.*

Remark 1. The proof of Lemma 16 can be easily modified (in fact, the same proof holds) for the following case. G_1 is a graph obtained by adding not too many (say, at most polylog(n)) vertices of degree 2, any two of them are far apart (say, of distance at least 3), to a $(k, 3)$ expander, and e is not incident with any of these vertices.

As another remark, note that for a $(k, 2)$-expander, we trivially have that each connected component is of size larger than k, and therefore, if the graph is not connected, then there are at least $(k+1)^2$ boosters which decrease the number of connected components.

Note that in order to turn a rainbow graph G_1 into a graph that contains a rainbow Hamiltonian cycle passing through e, one should repeatedly add e-boosters, one by one, every time adding a booster with an unused color, at most $2|V(G_1)|$ times. Therefore, we wish to show that a graph $G \sim G_c(n,p)$ typically contains "many" e-boosters of "many" colors for every sparse expander subgraph G_1 and every pair $e \in \binom{V(G_1)}{2}$.

Lemma 17. *Let $0 < \varepsilon < 1, \beta > 0$, let $c = (1+\varepsilon)n$ and let $\frac{\log n}{n} \leq p \leq \frac{2\log n}{n}$. Then a graph $G \sim \mathscr{G}_c(n,p)$ is w.h.p. such that the following holds. Suppose that*

(i) $G_1 \subseteq G$ is any subgraph with $\frac{n}{\log\log n} \leq |V(G_1)| \leq \frac{2n}{\log\log n}$ and $|E(G_1)| = \Theta(n/\log\log n)$ which is an $(\beta|V(G_1)|, 2)$-expander, and
(ii) $e \in \binom{V(G_1)}{2}$ is any pair which is not incident with vertices of degree 2 in G_1, and
(iii) $\mathscr{C}_2 \subseteq [c]$ is a subset of size at least $\varepsilon n/100$,

then, G contains e-boosters for G_1 assigned with colors from $\mathscr{C}_2$.

Proof. Note first that by Remark 1 after Lemma 16, there are at least $\beta^2|V(G_1)|^2/2 \geq \frac{\beta^2 n^2}{2(\log\log n)^2}$ e-boosters for every such G_1. Fix a subset $\mathscr{C}_2 \subseteq [c]$ of size at least $\varepsilon n/100$, and observe that the probability of $E(G)$ not to contain any e-booster which is assigned with a color from $\mathscr{C}_2$ is at most

$$\left(1-p+p\frac{1+0.99\varepsilon n}{(1+\varepsilon)n}\right)^{\frac{\beta^2n^2}{2(\log\log n)^2}} \leq (1-p)^{\frac{\varepsilon\beta^2n^2}{300(\log\log n)^2}} \leq \exp\left(-\frac{\varepsilon\beta^2 n\log n}{300(\log\log n)^2}\right).$$

Now, taking the union bound over all subsets $V(G_1) \subseteq [n]$ of size $\frac{n}{\log\log n} \leq |V(G_1)| \leq \frac{2n}{\log\log n}$ and over all subgraphs G_1 of G on vertex set $V(G_1)$ with at most $\frac{Cn}{\log\log n}$ many edges (where C is some fixed constant), and over all subsets of colors $\mathscr{C}_2 \subseteq [c]$ of size at least $\varepsilon n/5$ we obtain that the probability of having a counterexample is upper bounded by

$$\sum_{t=n/\log\log n}^{2n/\log\log n} 2^c\binom{n}{t}\binom{\binom{t}{2}}{Ct}p^{Ct}\exp\left(-\frac{\varepsilon\beta^2n\log n}{300(\log\log n)^2}\right)$$

$$\leq \frac{2n}{\log\log n}2^{(1+\varepsilon)n}2^n\left(\frac{enp}{C\log\log n}\right)^{2Cn/\log\log n}\exp\left(-\frac{\varepsilon\beta^2n\log n}{300(\log\log n)^2}\right)$$

$$\leq 8^n\exp\left(C\frac{2n}{\log\log n}\log\frac{enp}{C\log\log n}\right)\exp\left(-\frac{\varepsilon\beta^2n\log n}{300(\log\log n)^2}\right) = o(1).$$

This completes the proof. □

3 Proof of Theorem 1

In this section we prove Theorem 1.

Proof. Let $G \sim \mathcal{G}_c(n,p)$, and let $\delta > 0$ be a sufficiently small constant (to be specified later). Throughout the proof we assume that G satisfies all the properties of the lemmas from the previous section.

Our proof strategy goes as follows. For each vertex $v \in SMALL$ let us arbitrarily choose a set $A(v) = \{x,y\}$ of exactly two distinct neighbors of v and set $V_0 = SMALL \cup \left(\bigcup_{v\in SMALL} A(v)\right)$, and $E_0 = \{vz : v \in SMALL \text{ and } z \in A(v)\}$. By $(P4)$ and $(P5)$ of Lemma 4 we have that all the $A(v)$'s are disjoint and that E_0 is rainbow. Let $\mathcal{C}_{small} := \{c(e) : e \in E_0\}$ denote the set of colors used in E_0, and let $\mathcal{C}^* := \mathcal{C} \setminus \mathcal{C}_{small}$ be its complement. Observe that by $(P2)$ of Lemma 4 for a small enough δ we have $|\mathcal{C}^*| \geq (1+\varepsilon/2)n$.

Now, note that by $(P5)$ of Lemma 4 it easily follows that $\delta(G[V \setminus V_0]) \geq \delta \log n$. Therefore, letting $\alpha = \min\{\varepsilon/5, \delta e^{-10}\}$, by applying Lemma 9 to $G[V\setminus V_0]$ we find subsets $W \subseteq [n] \setminus V_0$ and $\mathcal{C}_0, \mathcal{C}_1 \subset \mathcal{C}^*$ for which

(i) $|W| = (1+o(1))\frac{n}{\log\log n}$, and
(ii) $\mathcal{C}_0 \cap \mathcal{C}_1 = \emptyset$, and
(iii) $|\mathcal{C}_0|, |\mathcal{C}_1| = (1+o(1))\alpha n$, and
(iv) for every $v \in [n] \setminus SMALL$ we have $d_{\mathcal{C}_1}(v, W) \in \left(\frac{\alpha\delta \log n}{2\log\log n}, \frac{20\alpha \log n}{\log\log n}\right)$, and
(v) the subgraph $H := G[W; \mathcal{C}_0]$ satisfies all the properties of Lemma 10.

In order to find the desired rainbow Hamilton cycle we proceed as follows. First, find a rainbow path P of length $n - |W| - n/\log^{0.4} n$ in $[n] \setminus (V_0 \cup W)$ whose edges receive colors from $\mathcal{C}^* \setminus (\mathcal{C}_0 \cup \mathcal{C}_1)$. The existence of such a path is ensured by Lemmas 12 and 13. Second, let x, y denote P's endpoints, define $S = ([n] \setminus (SMALL \cup V(P) \cup W)) \cup \{x,y\}$ be the set of "unused" vertices, and consider the bipartite graph $B := G[S, W; \mathcal{C}_1]$. Lemma 5 ensures that B contains (say) a rainbow 9-matching M from S to W. Let x' and y' be two neighbors (in M) of x and y, respectively, and define $M' := M \setminus \{e \in M : e \cap \{x,y\} \neq \emptyset\}$.

Next, by applying Lemma 11 to $G[W; \mathcal{C}_0]$, we find a subgraph $R \subseteq G[W; \mathcal{C}_0]$ which satisfies the following:

(a) R is rainbow, and
(b) R is an $(\alpha\delta|W|/100, 100)$-expander, and
(c) $|E(R)| = \Theta(n/\log\log n)$.

Now, let us define G_1 to be the subgraph of G on vertex set $V_1 := [n] \setminus V(P)$, with edge set $M' \cup E_0 \cup E(R)$. Note that since R is an $(\alpha\delta|W|/100, 100)$-expander, and since for $S' := S \setminus \{x,y\}$, there exists a 9-matching from S' to W, it follows by Lemma 14 that adding S' and M' to R yields an $(\alpha\delta|W|/100, 4)$-expander. Now, since by $(P5)$ we have that vertices in $SMALL$ are far apart, by Lemma 15 it follows that G_1 is an $(\alpha\delta|V_1|/200, 2)$-expander with $\Theta(n/\log\log n)$ edges, and is clearly rainbow. Finally, we wish to turn G_1 (in G) into a graph which contains a rainbow

Hamilton path with x' and y' as its endpoints. Note that both x' and y' are not neighbors of vertices of degree 2 in G_1. Now, one can repeatedly apply Lemma 17 to G_1 with respect to the set of available colors to obtain a rainbow Hamilton path of G_1 connecting x' to y' which uses only colors not appearing on P. (Each time we add a booster e whose color $c(e) \in \mathscr{C}^* \setminus (\mathscr{C}_0 \cup \mathscr{C}_1)$ has not been used before we update the set of available colors by excluding $c(e)$. Since $|W| = o(n)$, along the process we still have a linear number of colors available, and thus Lemma 17 applies.) A moment's thought now reveals that such a path, together with P and the edges xx' and yy', yields a rainbow Hamilton cycle in G. This completes the proof. □

4 Proof of Theorem 2

In this section we prove Theorem 2.

Proof. Let us define the following sequence $\Gamma_0, \Gamma_1, \ldots, \Gamma_N$ of random edge-colored k-uniform hypergraphs, where $N = \binom{n}{k}$, in the following way: Let $e_1, \ldots, e_N$ be an arbitrary enumeration of all the elements of $\binom{[n]}{k}$. Now, in Γ_i, for every $j > i$ we add the corresponding edge with probability p, independently at random and assign it all the colors in $[c]$ (these edges can be seen as multiple edges with multiplicity c). For every $j \leq i$, we add e_j to Γ_i with probability q, independently at random and then assign it a unique color from $[c]$ uniformly, independently at random. Note that $\Gamma_0 \sim \mathscr{H}^k(n,p)$ while $\Gamma_N \sim \mathscr{H}_c^k(n,q)$. Therefore, in order to complete the proof it is enough to show that

$$\Pr\left[\Gamma_i \text{ contains some } C \in \mathscr{C}\right] \geq \Pr\left[\Gamma_{i-1} \text{ contains some } C \in \mathscr{C}\right].$$

To this end, expose all edges but e_i and its color(s) in both spaces. There are three possible scenarios:

(*a*) Γ_{i-1} contains some $C \in \mathscr{C}$ not using e_i, or
(*b*) Γ_{i-1} does not contain any member $C \in \mathscr{C}$ even if we add e_i with all the possible colors, or
(*c*) Γ_{i-1} contains a member of $\mathscr{C}$ if we add e_i with all the possible colors.

Note that in (*a*) and (*b*) there is nothing to prove. Therefore, it is enough to consider case (*c*). The crucial observation here is that if e_i is needed for finding a copy of some $C \in \mathscr{C}$, then since $\mathscr{C}$ is ℓ-rich, it follows that at least ℓ colors are valid for e_i in order to obtain such a copy. Now, the probability for Γ_{i-1} to contain a member of $\mathscr{C}$ is precisely p (recall that e_i is crucial for this aim and that we add e_i with all possible colors), where the probability for Γ_i to have such a copy is at least $q\ell/c = p$. This completes the proof. □

5 Applications of Theorem 2

In this section we show how to use Theorem 2 in order to derive Theorems 3 and 4. For Theorem 3 we prove a stronger statement from which the proof immediately follows.

Theorem 6. *Let S be any k-uniform hypergraph on n vertices with m edges, and let p be such that*

$$\Pr\left[H \sim \mathcal{H}^k(n,p) \text{ contains a copy of } S\right] = 1 - o(1).$$

Then, for every $\varepsilon \geq 0$, *letting* $c = (1+\varepsilon)m$ *and* $q = \frac{cp}{\varepsilon m+1}$, *if* $q \leq 1$ *then we have*

$$\Pr\left[H \sim \mathcal{H}_c^k(n,q) \text{ contains a rainbow } S\right] = 1 - o(1).$$

Proof. Let $\mathcal{C}$ be the set of all possible rainbow copies of S on n vertices with colors from $[c]$, where $c = (1+\varepsilon)m$. Note that for any $e \in E(C)$, $C - e$ has exactly $m-1$ edges and since there are $(1+\varepsilon)m$ colors, it follows that there are $\varepsilon m + 1$ ways to color e to obtain a rainbow copy of S. All in all, $\mathcal{C}$ is $(\varepsilon m + 1)$-rich, and therefore by applying Theorem 2 to $\mathcal{C}$ we obtain the desired claim. □

Now we prove Theorem 4 which informally speaking states that for $c = \omega(n)$ and $p = \omega(\log n/n)$, in a typical $G \sim \mathcal{G}_c(n,p)$ one can find $(1-o(1))np/2$ edge-disjoint Hamilton cycles, each of which is rainbow.

Proof. First, observe that for example by the main results of [14, 15], it follows in particular that for $p = \omega(\log n/n)$ we have

$$\Pr\left[G \sim \mathcal{G}(n,p) \text{ contains } (1-o(1))np/2 \text{ edge-disjoint Hamilton cycles}\right] = 1 - o(1).$$

Now, let C be such that $\frac{Cn}{(C-1)n+1} \leq 1 + \varepsilon/2$ and let $c = Cn$. Let us define $\mathcal{C}$ to be the family of all collections C of $(1-\varepsilon/2)np/2$ edge-disjoint Hamilton cycles, each of which is rainbow. Note that for every $C \in \mathcal{C}$ and every $e \in E(C)$, since e belongs to a given rainbow Hamilton cycle, there are at most $n-1$ colors which are forbidden for it. Therefore, there are $Cn - (n-1) = (C-1)n + 1$ ways to color e in order to obtain an element of $\mathcal{C}$ and we conclude that $\mathcal{C}$ is $((C-1)n+1)$-rich. Now, by applying Theorem 2 for $q = \frac{Cnp}{(C-1)n+1}$ we obtain

$$\Pr\left[G \sim \mathcal{G}_c(n,q) \text{ contains } (1-o(1))np/2 \text{ edge-disjoint rainbow Hamilton cycles}\right] = 1 - o(1).$$

All in all, since $q \leq (1+\varepsilon/2)p$ we obtain that $(1-o(1))np \geq \frac{(1-o(1))nq}{1+\varepsilon/2} \geq (1-\varepsilon)nq$ as desired. □

Acknowledgements The research of Michael Krivelevich was supported in part by USA-Israel BSF grant 2010115 and by grant 912/12 from the Israel Science Foundation.

References

1. R. Aharoni, P. Haxell, Hall's theorem for hypergraphs. J. Graph Theory **35**, 83–88 (2000)
2. N. Alon, J.H. Spencer, *The Probabilistic Method*, 3rd edn. (Wiley, New York, 2008)
3. D. Bal, A. Frieze, Rainbow matchings and Hamilton cycles in random graphs (2013). arXiv preprint [arXiv:1311.6423]
4. I. Ben-Eliezer, M. Krivelevich, B. Sudakov, The size Ramsey number of a directed path. J. Comb. Theory Ser. B **102**, 743–755 (2012)
5. S. Ben-Shimon, A. Ferber, D. Hefetz, M. Krivelevich, Hitting time results for Maker-Breaker games, in *Proceedings of the 22nd Symposium on Discrete Algorithms (SODA'11)*, pp. 900–912. Journal version: Random Struct. Alg. **41**, 23–46 (2012)
6. B. Bollobás, *Random Graphs*, 2nd edn. (Cambridge University Press, Cambridge, 2001)
7. C. Cooper, A. Frieze, Multi-coloured Hamilton cycles in random edge-coloured graphs. Comb Probab. Comput. **11**, 129–133 (2002)
8. A. Dudek, A. Frieze, Tight Hamilton cycles in random uniform hypergraphs. Random Struct Alg. **42**, 374–385 (2013)
9. A. Ferber, M. Krivelevich, H. Naves, Generating random graphs in biased Maker-Breaker games. Random Struct. Alg. **47**, 615–634 (2015)
10. A. Ferber, R. Nenadov, U. Peter, Universality of random graphs and rainbow embedding. Random Struct. Alg. (to appear)
11. A. Frieze, P.-S. Loh, Rainbow Hamilton cycles in random graphs. Random Struct. Alg. **44**, 328–354 (2014)
12. S. Janson, T. Łuczak, A. Rucinski, *Random Graphs* (Wiley, New York, 2000)
13. A. Johansson, J. Kahn, V. Vu, Factors in random graphs. Random Struct. Alg. **33**, 1–28 (2008)
14. F. Knox, D. Kühn, D. Osthus, Edge-disjoint Hamilton cycles in random graphs. Random Struct. Alg. **46**, 397–445 (2015)
15. M. Krivelevich, W. Samotij, Optimal packings of Hamilton cycles in sparse random graphs. SIAM J. Disc. Math. **26**, 964–982 (2012)
16. M. Krivelevich, C. Lee, B. Sudakov, Compatible Hamilton cycles in random graphs (2014). arXiv preprint [arXiv:1410.1438]
17. D. Kühn, D. Osthus, Hamilton decompositions of regular expanders: applications. J. Comb. Theory Ser. B **104**, 1–27 (2014)
18. C. McDiarmid, Clutter percolation and random graphs. Math. Program. Stud. **13**, 17–25 (1980)
19. L. Pósa, Hamiltonian circuits in random graphs. Disc. Math. **14**, 359–364 (1976)
20. D.B. West, *Introduction to Graph Theory* (Prentice Hall, Upper Saddle River, 2001)

Further applications of the Container Method

József Balogh and Adam Zsolt Wagner

Abstract Recently, Balogh–Morris–Samotij and Saxton–Thomason proved that hypergraphs satisfying some natural conditions have only few independent sets. Their main results already have several applications. However, the methods of proving these theorems are even more far reaching. The general idea is to describe some family of events, whose cardinality a priori could be large, only with a few certificates. Here, we show some applications of the methods, including counting C_4-free graphs, considering the size of a maximum C_4-free subgraph of a random graph and counting metric spaces with a given number of points. Additionally, we discuss some connections with the Szemerédi Regularity Lemma.

Keywords Container method • enumeration in graph theory

Mathematics Subject Classification (2010): Primary 05C30, Secondary 05C35

1 Introduction

Recently, an important trend in probabilistic combinatorics is to transfer extremal results in a dense environment into a sparse, random environment. The first main breakthrough is due to Conlon–Gowers [8] and Schacht [23]. Not much later, Balogh–Morris–Samotij [5] and Saxton–Thomason [22] had a different approach, which not only proved most of the results of Conlon–Gowers [8] and Schacht [23], but also provided counting versions of these results. Additionally, there are some further applications of the main theorems of [5] and [22]. However, the proof methods have the potential to be more influential than the theorems themselves. The general idea, which can be traced back to the classical paper of Kleitman–

J. Balogh (✉) • A.Z. Wagner
Department of Mathematical Sciences, University of Illinois at Urbana-Champaign, Urbana, IL 61801, USA
e-mail: jozsebal@gmail.com; jobal@math.uiuc.edu; zawagner22@gmail.com; zawagne2@illinois.edu

A. Beveridge et al. (eds.), *Recent Trends in Combinatorics*, The IMA Volumes in Mathematics and its Applications 159, DOI 10.1007/978-3-319-24298-9_8

Winston [13] (and more explicitly in several papers of Sapozhenko) is to describe some family of events, whose cardinality a priori could be large, only with a few certificates.

Here, we try to outline what lies behind this method. Because each of the above-mentioned four papers [5, 8, 22, 23] already gave a nice survey of the field, additionally Conlon [7], Rödl–Schacht [20], and Samotij [21] have survey papers of the topics, we choose a different route. Each of the following four sections discusses the method with some applications. Three of the sections contain new results.

In Section 2 we state an important corollary of the main results of [5] and [22] and will discuss its connection with the Szemerédi Regularity Lemma [24].

In Section 3 we estimate the volume of the convex subset of $[0,1]^{\binom{n}{2}}$ representing metric spaces with n points, and consider a discrete variant of this problem as well. Kozma–Meyerovitch–Peled–Samotij [15] considered the following question: choose randomly and independently $\binom{n}{2}$ numbers from the interval $[0,1]$, labelling the edges of a complete graph K_n, what is the probability that any three of them forming a triangle will satisfy the triangle inequality? They used entropy estimates to derive an upper bound on this probability and noticed that Szemerédi Regularity Lemma can be used to count metric spaces whose all distances belong to a discrete set of a fixed size.

Mubayi and Terry [19] more recently pushed the discrete case into $\{0,1\}$-law type results. Using the general results of [5] and [22], we improve the results implied by the regularity lemma, and obtain good results for the continuous case. Parallel to our work, Kozma, Meyerovitch, Morris, Peled, and Samotij [16] practically solved the continuous version of the problem using more advanced versions of the methods of [5] and [22]. In order to avoid duplicate work, our goal in this section is clarity over pushing the method to its limit.

It was probably Babai–Simonovits–Spencer [1] who first considered extremal problems in random graphs. Most of their proposed problems are resolved, but there was no progress on the following: What is the maximum number of the edges of a C_4-free subgraph of the random graph $G(n,p)$ when $p = 1/2$? In the case $p = o(1)$, strong bounds were given by Kohayakawa–Kreuter–Steger [14]. Here, in Section 4, we improve on the trivial upper bound, noting that an n-vertex C_4-free graph can have at most $(1/2 + o(1))n^{3/2}$ edges.

Theorem 1.1. *For every $p \in (0,1)$, there is a $c > 0$ that the largest C_4-free subgraph of $G(n,p)$ has at most $(1/2 - c)n^{3/2}$ edges w.h.p. In particular, if $p = 0.5$ we can take $c = 0.028$.*

Kleitman–Winston [13] authored one of the first papers in the field whose main idea was to find small certificates of families of sets in order to prove that there are not many of them. They proved that the number of C_4-free graphs on n vertices is at most $2^{cn^{3/2}}$ for $c \approx 1.081919$. In Section 5, we improve the constant in the exponent by using ideas found in Section 3. The improvement is tiny; the main aim is to demonstrate that the method has potential applications for other problems as well.

2 Connections to Szemerédi's Regularity Lemma

In this section we describe some connections between the Szemerédi Regularity Lemma and the new counting method. Originally, without using the regularity lemma, Erdős–Kleitman–Rothschild [9] estimated the number of K_k-free graphs.

Theorem 2.1. *There are* $2^{(1+o(1))\cdot \mathsf{ex}(n,K_k)}$ K_k*-free graphs on n vertices, where* $\mathsf{ex}(n, K_k)$ *is the maximum number of the edges of a* K_k*-free graph on n vertices.*

Here, we sketch a well-known proof of Erdős–Frankl–Rödl [10], using the Szemerédi Regularity Lemma. We apply the Regularity Lemma for a K_k-free graph G_n, which outputs a cluster graph R_t, where $t = O(1)$. Then we clean G_n, i.e. we remove edges inside the clusters, between the sparse and the irregular pairs. Define C_n to be the blow-up of R_t to n vertices (getting back the same preimage vertices of G_n). Observe that C_n contains all but $o(n^2)$ edges of G_n, and C_n are K_k-free, hence $e(C_n) \leq \mathsf{ex}(n, K_k)$. Now we can count the number of choices for G_n: The number of choices for C_n is $O(1) \cdot n^n$, the number of choices for $E(G_n) \cap E(C_n)$ given C_n is at most $2^{\mathsf{ex}(n,K_k)}$, and the number of choices for $E(G_n) - E(C_n)$ given C_n is $2^{o(n^2)}$, completing the proof of the result.

An important corollary of the main results of Balogh–Morris–Samotij [5] and Saxton–Thomason [22] is the following characterisation of K_k-free graphs.

Proposition 2.2. *There is a* $t \leq 2^{O(\log n \cdot n^{2-1/(k-1)})}$ *and a set* $\{G_1, \ldots, G_t\}$ *of graphs, each containing* $o(n^k)$ *copies of* K_k*, such that for every* K_k*-free graph H there is an* $i \in [t]$ *such that* $H \subseteq G_i$.

Note that it follows from standard results in extremal graph theory that each such graph G_i necessarily contains at most $(1 + o(1)) \cdot \mathsf{ex}(n, K_k)$ edges.

This provides an even shorter proof of Theorem 2.1: For each K_k-free graph H there is an $i \in [t]$ such that $H \subseteq G_i$. The number of choices for i is $2^{O(\log n \cdot n^{2-1/(k-1)})}$, and the number of subgraphs of G_i is at most $2^{(1+o(1))\mathrm{ex}(n,K_k)}$.

The proof using the Regularity Lemma yields two variants of Proposition 2.2, one which is weaker, as it gives $t = 2^{o(n^2)}$, though it is strong enough for this particular application, and one that we call the Szemerédi Approximate Container Lemma.

Proposition 2.3. *There is a $t = 2^{o(n^2)}$ and a set $\{G_1, \ldots, G_t\}$ of graphs, each containing at most $(1 + o(1))\mathrm{ex}(n, K_k)$ edges, and $o(n^k)$ copies of K_k, such that for every K_k-free graph H there is an $i \in [t]$ such that $H \subseteq G_i$.*

Proposition 2.4. *There is a $t = O(1)$ and a set $\{G_1, \ldots, G_t\}$ of graphs, each K_k-free, such that for every K_k-free graph H there is an $i \in [t]$ and a permutation of $V(H)$, giving an isomorphic graph H', such that $|E(H') - E(G_i)| = o(n^2)$.*

Most of the tools from this section are likely to be useful in counting *maximal* K_r-free graphs, where there are still no satisfactory bounds when $r \geq 4$.

Problem 2.5 *What is the number of **maximal** K_r-free graphs with vertex set $[n]$? For $r = 3$ this was a question of Erdős, which was settled in [2] and [6].*

3 The number of metric spaces

Our goal is to estimate the number of metric spaces on n points, where the distance between any two points lies in $\{1, \ldots, r\}$ for some $r = r(n)$. This problem was considered first by Kozma–Meyerovitch–Peled–Samotij [15], who, using the regularity lemma gave an asymptotic bound on the number of such metric spaces for a fixed constant r. Recently, Mubayi–Terry [19] provided a characterisation of the typical structure of such metric spaces for a fixed constant r, while $n \to \infty$. We will be more interested in what happens if r is allowed to grow as a function of n. Our main result is the following:

Theorem 3.1. *Fix an arbitrary small constant $\epsilon > 0$. If*

$$r = O\left(\frac{n^{1/3}}{\log^{\frac{4}{3}+\epsilon} n}\right),$$

then the number of such metric spaces is

$$\left\lceil \frac{r+1}{2} \right\rceil^{\binom{n}{2}+o(n^2)}.$$

Kozma–Meyerovitch–Peled–Samotij [15] pointed out that the discrete and the continuous problems are related. They considered the same question in the continuous case with distances in $[0, 1]$. Their entropy-based approach yields Theorem 3.1 for $r < n^{1/8}$.

At the end of this section we show how our results translate to the continuous setting.

For a positive integer r define $m(r) = \lceil \frac{r+1}{2} \rceil$. We will use an easy corollary of Mubayi–Terry ([19] Lemma 4.9):

Lemma 3.2. *Let $A, B, C \subset [r]$, all non-empty. Suppose the triple $\{A, B, C\}$ does not contain a non-metric triangle – that is, every triple $\{a, b, c : a \in A, b \in B, c \in C\}$ satisfies the triangle inequality. Then if r is even we have $|A| + |B| + |C| \leq 3m(r)$, and if r is odd we have $|A| + |B| + |C| \leq 3m(r) + 1$.*

Let $\mathscr{H}$ be the 3-uniform hypergraph with vertex set r rows, one for each colour, and $\binom{n}{2}$ columns, one for each edge of K_n. A vertex (i, f) of $\mathscr{H}$ corresponds to the event that the graph edge f has colour i. Three vertices of $\mathscr{H}$ form a hyperedge when the graph edge coordinates of the vertices form a triangle in K_n while the 'colours' do not satisfy the triangle inequality. With other words, the hyperedges correspond to non-metric triangles, and independent sets having exactly one vertex from each column correspond to points of the metric polytope. Our plan is to prove a supersaturation statement, but first we need two lemmas.

The first lemma we use is due to Füredi [11]. For a graph G, write G^2 for the "proper square" of G, i.e. where xy is an edge if and only if there is a z such that xz and zy are edges in G. Write $e(G)$ for the number of edges in G.

Lemma 3.3. *For any graph G with n vertices, we have*

$$e(G^2) \geq e(G) - \lfloor n/2 \rfloor.$$

The second lemma bounds the size of the largest independent set in $\mathscr{H}$.

Lemma 3.4. *Let $S \subset V(\mathcal{H})$ have no empty columns and contain no edges in $\mathcal{H}$. Then if r is even we have $|S| \le m(r)\binom{n}{2}$, and if r is odd we have $|S| \le m(r)\binom{n}{2} + rn$.*

Proof. The even case follows directly from Lemma 3.2, and we note that this bound is tight – let S contain the interval $[r/2, r]$ from each column. The bound in the odd case is slightly more difficult, and we make no effort to establish a tight bound, which should probably be $|S| \le m(r)\binom{n}{2} + n/2$.

Let r be odd, and let A, B, C be three columns that form a triangle of S. Note that if for some $k \ge 1$ we have $|A| \ge m(r) + k$ and $|B| \ge m(r) + k$ then $|C| \le m(r) - 2k + 1 \le m(r) - k$ by Lemma 3.2. Write B_k for the set of columns in S of order at least $m(r) + k$ and write S_k for the set of columns in S of order at most $m(r) - k$. Let G_k be the graph on $[n]$ with edges B_k. Then by Lemma 3.3 we get

$$|S_k| \ge e(G_k^2) \ge e(G_k) - \lfloor n/2 \rfloor \ge |B_k| - n.$$

Hence

$$|S| - m(r)\binom{n}{2} = \sum_{k=1}^{r} k(|B_k| - |B_{k+1}|) - \sum_{k=1}^{r} k(|S_k| - |S_{k+1}|) = \sum_{k=1}^{r} (|B_k| - |S_k|) \le nr,$$

and the result follows. □

Now we are ready to prove a supersaturation-like result.

Lemma 3.5. *Let $\epsilon > 0$ and let $S \subset V(\mathcal{H})$ with no empty columns.*

1. *If r is even and $|S| \ge (1+\epsilon)\binom{n}{2}m(r)$, then S contains at least $\frac{\epsilon}{10}\binom{n}{3}$ hyperedges.*
2. *If r is odd, $n > n_\epsilon$ sufficiently large and $|S| \ge (1+\epsilon)\binom{n}{2}m(r)$, then S contains at least $\frac{\epsilon^4}{40000}\binom{n}{3}$ hyperedges.*

Proof. Suppose first that r is even. Then there are at least $\frac{\epsilon}{10}\binom{n}{3}$ triangles in G such that the corresponding columns contain at least $(1+\epsilon/10)3m(r)$ vertices from S. Indeed, if this was not the case, then

$$\frac{\frac{\epsilon}{10}\binom{n}{3}3r + \binom{n}{3}3m(r)(1+\epsilon/10)}{n-2} = \binom{n}{2}\left(m(r) + \frac{\epsilon(r+m(r))}{10}\right) > |S|,$$

which is a contradiction. Hence part 1 of the lemma follows from Lemma 3.2.

Now suppose r is odd. Given $T \subset [n]$, write $f_S(T)$ for the set of vertices of S contained in the $\binom{|T|}{2}$ columns of $\mathcal{H}$ corresponding to the edges spanned by T. Set $n_0 = 20/\epsilon$, so that by Lemma 3.4, whenever $T \subset [n]$ with $|T| = n_0$ and $|f_S(T)| \ge m(r)\binom{n_0}{2}(1+\frac{\epsilon}{3})$ then $\mathcal{H}[f_S(T)]$ contains a hyperedge.

First, we claim that there are at least $\frac{\epsilon}{4}\binom{n}{n_0}$ choices of $T \subset [n]$ with $|T| = n_0$ and $|f_S(T)| \ge m(r)\binom{n_0}{2}(1+\frac{\epsilon}{3})$.

Indeed, if this was not the case, then we would have

$$|S| \leq \frac{\frac{\epsilon}{4}\binom{n}{n_0}\binom{n_0}{2}r + \binom{n}{n_0}m(r)\binom{n_0}{2}(1+\frac{\epsilon}{3})}{\binom{n-2}{n_0-2}} = \binom{n}{2}\left(m(r) + \frac{\epsilon r}{4} + \frac{\epsilon m(r)}{3}\right) \\ < \binom{n}{2}m(r)(1+\epsilon), \tag{1}$$

which is not possible. So the number of hyperedges contained in S is at least

$$e(\mathscr{H}[S]) \geq \frac{\epsilon}{4}\binom{n}{n_0}\bigg/\binom{n-3}{n_0-3} \geq \frac{\epsilon}{4}\frac{\epsilon^3}{20^3}\binom{n}{3},$$

and the result follows. □

Write $\bar{d}$ for the average degree of $\mathscr{H}$, and for $j \in [3]$ define the j-th maximum co-degree

$$\Delta_j = \max\{|\{e \in E(\mathscr{H}) : \sigma \subset e\}| : \sigma \subset V(\mathscr{H}) \text{ and } |\sigma| = j\}.$$

Below, we will make use of a version of the container theorem of [5, 22], the way it was formulated by Mousset–Nenadov–Steger [18].

Theorem 3.6. *There exists a positive integer c such that the following holds for every positive integer N. Let $\mathscr{H}$ be a 3-uniform hypergraph of order N. Let $0 \leq p \leq 1/(3^6 c)$ and $0 < \alpha < 1$ be such that $\Delta(\mathscr{H}, p) \leq \alpha/(27c)$, where*

$$\Delta(\mathscr{H}, p) = \frac{4\Delta_2}{\bar{d}p} + \frac{2\Delta_3}{\bar{d}p^2}.$$

Then there exists a collection of containers $\mathscr{C} \subset \mathscr{P}(V(\mathscr{H}))$ such that

(i) every independent set in $\mathscr{H}$ is contained in some $C \in \mathscr{C}$,
(ii) for all $C \in \mathscr{C}$ we have $e(\mathscr{H}[C]) \leq \alpha e(\mathscr{H})$, and
(iii) the number of containers satisfies

$$\log|\mathscr{C}| \leq 3^9 c(1 + \log(1/\alpha))Np\log(1/p).$$

Proof of Theorem 3.1. Let $\mathscr{H}$ be the hypergraph defined earlier, i.e. the 3-uniform hypergraph with vertex set formed by pairs of the r colours and the $\binom{n}{2}$ edges of K_n, with 3-edges corresponding to non-metric triangles. Let $\epsilon, \delta > 0$ be arbitrarily small constants and set $p = \frac{1}{r\log^{2+\delta} n}$ and $\alpha = \frac{10^{10} c \log^{4+2\delta} n}{n}$. In $\mathcal{H}$ we have $\Delta_1 \le nr^2$, $\Delta_2 \le r$, $\Delta_3 = 1$, $\bar{d} \ge r^2 n/64$ and

$$\Delta(\mathcal{H}, p) \le 4\left(\frac{64r^2 \log^{2+\delta} n}{r^2 n} + \frac{64r^2 \log^{4+2\delta} n}{2r^2 n}\right) \le \frac{\alpha}{27c}.$$

Then Theorem 3.6 provides containers with

$$e(\mathcal{H}[C]) \le \alpha e(\mathcal{H}) \le 10^4 c r^3 n^2 \log^{4+2\delta} n,$$

and the number of containers is

$$\log |\mathcal{C}| \le \frac{c3^{10} r n^2 \cdot \log n \cdot \log r \cdot \log\log n}{r \log^{2+\delta} n} = o(n^2).$$

Now assume

$$r = o\left(\frac{n^{1/3}}{\log^{(4+2\delta)/3} n}\right).$$

Then the maximum number of edges in a container is $o(n^3)$, hence by Lemma 3.5, and the fact that a useful container does not have an empty column, we have for n large enough,

$$|V(C)| < (1+\epsilon) m(r) \binom{n}{2}.$$

Hence, the number of colourings in a container is at most $(1+\epsilon)^{\binom{n}{2}} m(r)^{\binom{n}{2}} = m(r)^{\binom{n}{2} + o(n^2)}$. The logarithm of the number of containers is $o(n^2)$. The total number of good colourings is at most the number of containers times the maximum number of colourings in a container. Hence the total number of good colourings is

$$m(r)^{\binom{n}{2} + o(n^2)},$$

as required. □

Now we turn our attention to the continuous setting. The set-up in [15] is as follows. Given a metric space with n points and all distances being in $[0, 1]$, we regard the set of distances as a vector in $[0, 1]^{\binom{n}{2}}$. We will call the union of all such n points in $[0, 1]^{\binom{n}{2}}$ for all finite metric spaces the *metric polytope* M_n. More precisely, the metric polytope M_n is the convex polytope in $\mathbb{R}^{\binom{n}{2}}$ defined by the inequalities $0 < d_{ij} \le 1$ and $d_{ij} \le d_{ik} + d_{jk}$.

Note that if $a + b \ge c$ then

$$\lceil a \rceil + \lceil b \rceil \ge \lceil c \rceil. \tag{2}$$

Theorem 3.7. *Fix $\delta > 0$ constant. Then for $n > n_\delta$ sufficiently large, we have*

$$(vol(M_n))^{1/\binom{n}{2}} \leq \frac{1}{2} + \frac{1}{n^{\frac{1}{6}-\delta}}.$$

Proof. First consider the discrete setting, colouring with r colours, where r is the even integer closest to $n^{\frac{1}{6}-\frac{\delta}{2}}$. W.l.o.g. $\delta < 1/4$ and set

$$1/p = n^{\frac{1}{3}-\frac{\delta}{4}}, \qquad \alpha = 300cn^{\delta-\frac{2}{3}},$$

where c is the constant from Theorem 3.6. Then

$$\Delta(\mathcal{H}, p) < 300\left(\frac{1}{rnp} + \frac{1}{p^2r^2n}\right) \leq 300\left(\frac{n^{1/3-\delta/4}}{n^{7/6-\delta/2}} + \frac{n^{2/3-\delta/2}}{n^{4/3-\delta}}\right) \leq \alpha,$$

and we get containers with

$$e(\mathcal{H}[C]) \leq \alpha n^3 r^3 \leq n^{3-1/6-\delta/4},$$

where the number of containers satisfies

$$\log |\mathcal{C}| \leq n^2 rp \log^3 n \leq n^{2-1/6-\delta/5}.$$

Hence, by Lemma 3.5, the number of vertices in a container is at most

$$|V(C)| \leq (1 + n^{-1/6-\delta/6})\binom{n}{2} m(r).$$

This implies that the number of colourings contained in a container is at most

$$\text{col}(C) \leq \left(\frac{V(C)}{\binom{n}{2}}\right)^{\binom{n}{2}} \leq \left((1 + n^{-1/6-\delta/6})m(r)\right)^{\binom{n}{2}} \leq m(r)^{\binom{n}{2}} e^{n^{2-1/6-\delta/7}}.$$

That is, the total number X of colourings is at most

$$X \leq m(r)^{\binom{n}{2}} e^{n^{2-1/6-\delta/7}} e^{n^{2-1/6-\delta/5}} \leq m(r)^{\binom{n}{2}} e^{n^{2-1/6-\delta/8}}.$$

Now consider colourings in the continuous setting. Cut up each edge of the cube into r pieces. We get by (2) that

$$
\begin{aligned}
(\mathrm{vol}(M_n))^{1/\binom{n}{2}} &\leq \left(\frac{(m(r)+1)^{\binom{n}{2}} e^{n^{2-1/6-\delta/8}}}{r^{\binom{n}{2}}} \right)^{1/\binom{n}{2}} \\
&\leq \left(2^{-\binom{n}{2}} \left(1 + \frac{4}{r}\right)^{\binom{n}{2}} \right)^{1/\binom{n}{2}} \left(1 + \frac{1}{n^{1/6+\delta/9}}\right) \leq \frac{1}{2} + \frac{1}{n^{\frac{1}{6}-\delta}},
\end{aligned}
$$

as required. □

Remark: Kozma, Meyerovitch, Morris, Peled, and Samotij [16] using a stronger supersaturation result and a somewhat different container type of theorem, independently, parallel to our work, improved the error term in Theorem 3.7 to $(\log^2 n)n^{-1/2}$, where the $-1/2$ is best possible as it was pointed out in [15].

It would be interesting to extend the above ideas to generalised metric spaces (note that there are several different definitions of these), hence we propose the following purposely vague question:

Problem 3.8 *It is a natural question to ask: is there an interesting extension of the problem discussed in this section, when instead of requiring metric triangles, one wants metric d-dimensional simplices for a fixed d?*

4 The largest C_4-free subgraph of a random graph

Now we turn our attention to proving Theorem 1.1. As before, for a graph G write G^2 for the "proper square" of G, i.e. where xy is an edge if and only if there is a z such that xz and zy are edges in G.

To simplify the technicality of the proof, we present here only the case when $p = 1/2$; for larger p we can use a properly chosen smaller c and the argument is the same.

Assume that H is the large C_4-free graph that we aim to find in $G(n, 1/2)$. A natural approach would be to do what Füredi [12] did, working on a Ramsey-type problem. Using the Kleitman–Winston method, he gave an upper bound on the number of C_4-free graphs with m edges. If p is sufficiently small, then the expected number of copies of such graphs in $G(n, p)$ is $o(1)$, proving the desired result. His upper bound was, assuming that $m > 2n^{4/3}(\log^2 n)$, that the number of such graphs is at most $(4n^3/m^2)^m$, hence the expected number of them is $(4pn^3/m^2)^m = o(1)$, as long as $m > (1/2 - a)n^{3/2}$ and $p < 1/16 - a$, where $a > 0$ is some small constant.

The idea of our proof is that the union bound in the above argument is too wasteful; hence, instead of considering each H separately, we want to show that for every C_4-free subgraph H with many edges, there exists some certificate in the graph. Each of these certificates will say that in a certain part of the graph one needs to select $(1 - o(1))\delta n^{3/2}$ edges of $G(n,p)$ out of possible $(1 + o(1))\delta n^{3/2}$ pairs. Since the number of certificates is much smaller than 1/(probability that the above unlikely event holds), which is of order $2^{cn^{3/2}}$, we can now show using the union bound that w.h.p. there is no such subgraph.

So our first task is to build such a certificate. Fix a C_4-free graph H with at least $m > (1/2 - c)n^{3/2}$ edges, where for the $p = 1/2$ case we can set $c = 10^{-5}$. First fix a linear order π of the vertex set of H, which we will use as a tiebreaker among vertices. We will also need a second linear ordering, as follows:

Definition 4.1. Given a graph G, a *min-degree ordering* is an ordering $\{v_1, \dots, v_n\}$ of $V(G)$ such that for each $i \in [n]$, the vertex v_i is of minimum degree in $G[v_i, \dots, v_n]$. When there are multiple such vertices, then we let v_i be the first among them in the ordering π.

Fix a min-degree ordering of H and let $Y := \{v_1, \dots, v_s\}$ and $X = V(H) - Y$, where we will choose s such that the minimum degree of $H[X]$ is large. Now we fix $F \subset H[X]$ with the following properties. The first is that F is sparse, i.e. $e(F) \leq n^{3/2}/\log^2 n$. The second is that the independent sets in F^2 approximate the independent sets in $H^2[X]$ rather well. In particular, we choose F such that every large independent set of $H^2[X]$ is in a 'container' determined by F, where the number of containers is at most $r := 2^{n^{2/5}\log^{20} n}$. The key observation is that for every $v_j \in Y$ its neighbourhood in X spans an independent set in $H^2[X]$, as H is C_4-free.

The certificate for H will be the vector $[Y, F, \{d_j\}_{j=1}^{s}, \{r_j\}_{j=1}^{s}]$, where $d_j := |N(v_j) \cap X|$, and $r_j \leq r$ is the index of the container containing $N(v_j) \cap X$, and $s = |Y|$. The number of certificates is at most $2^n \cdot \binom{n^2/2}{n^{3/2}/\log^2 n} \cdot n^n \cdot r^n \leq 2^{2n^{3/2}/\log n}$.

After this preparation, the proof is simple: for every H we fix a certificate. Standard arguments show that if H is dense, then there must be many edges between X and Y, however, the number of places to put the edges, given the certificate of H is w.h.p. not sufficiently big, and here we can apply the union bound using the *number of certificates* to bound the probability that $H \subseteq G(n, 1/2)$. Note that we give a bound on the probability that $H \subseteq G(n, 1/2)$ *simultaneously* for all H with a given certificate.

We shall use some standard properties of C_4-free graphs:

Theorem 4.2. *(i) Let R be a C_4-free graph on n vertices. Then $e(R) \leq 0.5n^{3/2}+n$.*

(ii) Let $d_1, \dots, d_n$ be the degree sequence of R, where R is a C_4-free graph on n vertices. Then $\sum d_i^2 \leq n^2 + 2n^{3/2}$.

(iii) Let R be a C_4-free bipartite graph with class sizes $a \leq b$. Then $e(R) \leq a\sqrt{b} + 2b$.

Next we prove a lower bound on $e_H[X, Y]$. In everything that follows, we fix a C_4-free graph H on n vertices with a min-degree ordering $\{v_1, \dots, v_n\}$, and $e(H) > \frac{1}{2}(1 - c)n^{3/2}$, where $c > 0$ is a sufficiently small constant.

Lemma 4.3. *For any two constants γ, δ with $0 < \gamma < 1/2$ and $0 < \delta < 1/2$ there exist constants $c_0 = c_0(\delta, \gamma)$ and $n_0 = n_0(\delta, \gamma, c)$ such that if $0 < c < c_0$, $n > n_0$ and $Y = \{v_1, \ldots, v_{\delta n}\}$ and $X = V(H)\backslash Y$, then*

$$e(X, Y) > (1-\gamma)\delta n^{3/2}. \tag{3}$$

Proof. Suppose we have the above set-up, yet (3) is false. Let $\gamma' \geq \gamma$ be such that $e(X, Y) = (1-\gamma')\delta n^{3/2}$ and let $\beta \geq 0$ be such that $e(Y) = \frac{1}{2}\beta(\delta n)^{3/2}$. By Theorem 4.2(i) we have $\beta \leq 1 + \frac{1}{\delta\sqrt{n}}$. Now note that

$$\sum_{i=1}^{|Y|} d_i = 2e(Y) + e(X, Y) = \beta(\delta n)^{3/2} + (1-\gamma')\delta n^{3/2}.$$

By the convexity of the function x^2 we get

$$\sum_{i=1}^{|Y|} d_i^2 \geq \delta n \left(\frac{\beta(\delta n)^{3/2} + (1-\gamma')\delta n^{3/2}}{\delta n}\right)^2 \geq \delta n \left(\beta(\delta n)^{1/2} + (1-\gamma')n^{1/2}\right)^2. \tag{4}$$

Moreover, we have

$$\sum_{i=|Y|+1}^{n} d_i = 2e(H) - 2e(Y) - e(X, Y) \geq (1-c)n^{3/2} - \beta(\delta n)^{3/2} - (1-\gamma')\delta n^{3/2}.$$

Hence, again by convexity, we get

$$\sum_{i=|Y|+1}^{n} d_i^2 \geq \frac{\left((1-c)n^{3/2} - \beta(\delta n)^{3/2} - (1-\gamma')\delta n^{3/2}\right)^2}{n - \delta n}. \tag{5}$$

We will derive a contradiction with Theorem 4.2(ii) by showing that

$$\sum_{i=1}^{n} d_i^2 > n^2 + 2n^{3/2}.$$

To achieve this, we combine (4) and (5):

$$\frac{1}{n^2}\sum_{i=1}^{n} d_i^2 \geq \delta\left(\beta\delta^{1/2} + (1-\gamma')\right)^2 + \frac{\left((1-c) - \beta\delta^{3/2} - (1-\gamma')\delta\right)^2}{1-\delta}. \tag{6}$$

We want to show that the right-hand side (denoted by $A = A(c, \delta, \gamma)$) is larger than 1 for c sufficiently small. Observe that A is a continuous function of c, hence it is enough to show that $A > 1$ for $c = 0$. Now since at $c = 0$ we have

$$A - 1 = \frac{\delta}{1-\delta}(\beta\sqrt{\delta} - \gamma')^2,$$

the lemma follows. □

An instant corollary of Lemma 4.3 is that by adding a few vertices to Y we may assume that the minimum degree of $H[X]$ is large:

Corollary 4.4. *For any two constants γ, δ with $0 < \gamma < 1/2$ and $0 < \delta < 1/2$ there exist constants $c_0(\delta, \gamma)$ and $n_0(\delta, \gamma, c)$ such that if $0 < c < c_0$ and $n > n_0$ then there exists an α with $\delta/2 < \alpha < \delta$ such that if we set $Y = \{v_1, \ldots, v_{\alpha n}\}$ and $X = V(H)\setminus Y$ then the minimum degree of $H[X]$ satisfies*

$$\delta(H[X]) \geq (1 - 2\gamma)\sqrt{n}.$$

Proof. For each $i \leq \delta n$ let $d^*(i)$ be the number of neighbours of v_i in $\{v_{\delta n+1}, \ldots, v_n\}$. Then by the ordering and by Theorem 4.2(i) we have $d^*(i) \leq \sqrt{n}$ for all i. Note that by the properties of our vertex ordering, all we need to do is find an index i with $\delta n/2 < i < \delta n$ such that $d^*(i) \geq (1 - 2\gamma)\sqrt{n}$. Indeed, then we could set $X = \{v_i, \ldots, v_n\}$. So assume for contradiction that $d^*(i) < (1 - 2\gamma)\sqrt{n}$ for all i with $\delta n/2 < i < \delta n$. Let $Y' = \{v_1, \ldots, v_{\delta n}\}$ and $X' = V(G)\setminus Y'$. Now

$$e(Y', X') < \frac{\delta n}{2}\sqrt{n} + \frac{\delta n}{2}(1 - 2\gamma)\sqrt{n} \leq (1 - \gamma)\delta n^{3/2},$$

contradicting Lemma 4.3. □

We will need the following container lemma for graphs:

Lemma 4.5. *Let $t = \log^3 n$, let $\epsilon > 0$ and $0 < b < 1/4$ be small constants, and let n be sufficiently large depending on ϵ and b. Let H be an n-vertex C_4-free graph with $X \subset V(H)$ and $|X| \geq n/2$, where every $v \in X$ has $d(v) > (1 - b)n^{1/2}$. Then there is an $F \subset H$ with $e(F) \leq 2n^{3/2}/t$ such that there are $C_1, \ldots, C_r \subset X$ that*

(i) for every independent set in $H^2[X]$ there is a C_i containing it,
(ii) $r < n^{n^{2/5}t^5}$,
(iii) $C_1, \ldots, C_r$ depend only on F,
(iv) $|C_i| \leq (1 + 4b)n^{1/2}$ for every i.

Proof. We first prove, as Kleitman–Winston [13], that $H^2[X]$ does not have a large sparse subset:

Lemma 4.6. *Let $Z \subset X$ with $|Z| = Cn^{1/2}$, where $C = C(n) \gg 1$. Then $e(H^2[Z]) > C^2 n/8$.*

Proof. Observe that

$$\sum_{v \in Z} d(v) \geq (1 - b)n^{1/2}|Z| > Cn/2.$$

Counting cherries (paths of length two), and using that H is C_4-free, this means that

$$e(H^2[Z]) > C^2 n/8.$$

□

Lemma 4.7. *Let $Z \subset X$ with $|Z| = (1+3b)n^{1/2}$, where $b < 1/4$. Then $e(H^2[Z]) > bn$.*

Proof. Using that $H[Z]$ is C_4-free, we know that $H[Z]$ does not span many edges, only $O(n^{3/4})$. Observe that

$$\sum_{v \in Z} d(v) \geq (1-b)n^{1/2}|Z| > (1+b)n.$$

Counting cherries (paths of length two), and using that H is C_4-free, this means that

$$e(H^2[Z]) = \sum_{v \in V(H)} \binom{d_Z(v)}{2} \geq bn\binom{2}{2} = bn.$$

□

Now, we shall choose F as a random subgraph of H, keeping each edge with probability $1/t$. Then $e(F) \leq 2n^{3/2}/t$ w.h.p.

Lemma 4.8. *(i) We have w.h.p. for any $Z \subset X$ of size $(1+3b)n^{1/2}$ that $e(F^2[Z]) > bn/(16t^2)$.*
(ii) We have w.h.p. for any $Z \subset X$ of size $n^{3/5}$, that $e(F^2[Z]) > |Z|^2/(32t^2)$.

Proof. (i) The number of choices for such Z is at most $2^{O(n^{1/2}\log n)}$. The proof of Lemma 4.7 yields that for each given Z there are at least $bn/4$ edge-disjoint cherries contributing to $E(H^2[Z])$. The reason is that writing the degree sequence of the vertices in $X - Z$ toward Z, a degree d contributes $\lfloor d/2 \rfloor$ edge-disjoint cherries and the degree sum is bn more than the number of the vertices. Hence, the expected number of them in $E(F^2[Z])$ is at least $bn/(8t^2)$. By Chernoff's bound, w.h.p. for each Z we have $e(F^2[Z]) > bn/(16t^2)$, as the concentration $\exp(-\frac{bn}{100t^2})$ beats the number of choices for Z.

(ii) Let $C = C(n) = n^{1/10}$. Choose any $Z \subset X$ of size $Cn^{1/2}$. The number of choices for Z is at most $2^{O(n^{3/5}\log n)}$. In the graph F, the degree sum of the vertices in Z is w.h.p. (beating the number of choices for Z) at least $Cn/(4t)$, hence $e(F^2[Z]) > C^2 n/(32t^2) = |Z|^2/(32t^2)$. □

From now on we fix such an F satisfying the conclusions of Lemma 4.8. Now we construct the family of the container sets $\{C_i\}$ in F^2. Fix an independent set I in F^2. Our aim is to construct a pair $(T(I), C(T))$ with

$$T(I) \subset I \subset T(I) \cup C(T),$$

where we call $C := T(I) \cup C(T)$ the *container* containing I. The set $T(I)$ is the (small) certificate of the container, as crucially, $C(T)$ depends only on T, not on I.

We construct the pair $(T(I), C(T))$ algorithmically. First we set $T = \emptyset$ and $A = [n]$, where A is the set of *available* vertices. In each step we choose the largest degree vertex, say v, in $F^2[A]$. In case there is more than one such vertex we choose the one which comes first in the ordering π. If $v \notin I$, then we just set $A := A - \{v\}$ and we iterate this step.

If $v \in I$, then we add v to T, and as I is an independent set, we can remove its neighbourhood from A, i.e. set $A := A - \{v\} - N(v)$.

We stop this process when A shrinks to a 'small' set, and then let $T(I) := T$, $C(T) := A$.

As we build a container for each independent set I in $H^2[X]$, condition (i) is clearly satisfied. For checking whether the other conditions hold, we first give an upper bound on $|T|$. We always add to T from A the largest degree vertex of $F^2[A]$. Until $|A| > n^{3/5}$, by Lemma 4.8 (ii), in each step we remove at least $n^{3/5}/(32t^2)$ vertices from A. Until $|A|$ is at least $(1+3b)n^{1/2}$, by Lemma 4.8 (i), in each step we remove at least always $bn^{2/5}/(16t^2)$ vertices from A. Putting together, we have that $|T| < 32n^{2/5}t^2 + (16/b)n^{1/5}t^2 < n^{2/5}t^3 < b\sqrt{n}$. When we reach this point, set the container to be the union of T and A, where it depends only on F, and the number of choices is bounded by the number of choices on T. Now, conditions (ii)–(iv) clearly hold. □

Now we just have to put together the details in order to prove Theorem 1.1.

Proof (of Theorem 1.1). The certificate for H will be the vector $[Y, F, \{d_j\}_{j=1}^s, \{r_j\}_{j=1}^s]$, where $d_j := |N(v_j) \cap X|$, and $r_j \le r$ is the index of the container containing $N(v_j) \cap X$, and $s = |Y|$.

Assume we are given the degree sequence, X, Y, and F. For each $v \in Y$ we fix the container of $N(v) \cap X$, noting that $N(v) \cap X$ should be an independent set in F^2. The number of choices for v and the container of the neighbourhood of v in X is at most $2^{n^{7/5}\log^{10} n}$ (this bound is for all v simultaneously). The number of $[X, Y]$ edges to be placed is, by Lemma 4.3, at least $(1-\gamma)\delta n^{3/2}$, but the number of pairs of vertices where they could be placed is at most $(1+4\gamma)\delta n^{3/2}$. By Chernoff's bound, for γ sufficiently small it is unlikely that this could be done in the random graph. The concentration $\exp(-c_{\gamma,\delta,p}n^{3/2})$ clearly beats the bound $\exp(o(n^{3/2}))$ for the number of choices, hence this completes the proof of Theorem 1.1. □

In what follows, we will sketch a second proof of Theorem 1.1. This proof is much easier and gives much better constants than the above proof: unfortunately it only works when $p < \frac{9}{16}$.

Proof (of Theorem 1.1 when $p < \frac{9}{16}$). Let $p < \frac{9}{16}$ be a constant, fix an ordering π and a degree sequence, and let H be a C_4-free graph with $\left(\frac{1}{2} - c\right)n^{3/2}$ edges just like in the previous proof. Moreover, for each i we will fix the right-degree d_i^* of v_i - that is, $d_i^* = |N(v_i) \cap \{v_i, \dots, v_n\}|$. Now instead of fixing containers for the neighbourhood of v_i in H, we will fix containers for the *right-neighbourhood*.

That is, for each i the container C_i satisfies $N(v_i) \cap \{v_i, \dots, v_n\} \subset C_i$. Now note that we can make the container C_i have order just barely larger than $\frac{n-i}{d_i^*}$ for all i - the proof of this is similar but easier than Lemma 4.5 (also see Lemma 5.3). Now the crucial idea (which was present in the previous proof as well) is that if for many i we have $d_i^* > |C_i|p$ then the number of edges of H is larger than the expected number of places in $G(n,p)$, hence the Chernoff's bound implies that embedding of H is unlikely to happen in a random graph.

More precisely, fix $I = \{i \in [n] : d_i^* > \frac{n-i}{d_i^*}p\}$, the set of vertices with too small containers. A simple calculation shows that if for a constant ϵ we have $c < \frac{3-4\sqrt{p}}{6}(1-\epsilon)$ then the excess degree sum $D = \sum_I \left(d_i^* - \frac{n-i}{d_i^*}p\right)$ will be at least a constant proportion of the total number of edges. Then H has at least $\sum_I d_i^*$ edges between pairs $\cup_I (i, C_i)$, so

$$\sum_{i \in I} d_i^* \geq \sum_{i \in I} p \cdot |C_i| + D.$$

It means that $G(n,p)$ on $\cup_I (i, C_i)$ needs to have D more edges than the expected number of edges, which has a low chance by the Chernoff's bound. Indeed, the probability of this happening is at most $e^{-c_\epsilon n^{3/2}}$, hence the concentration beats the number of choices and the proof is complete. For $p = \frac{1}{2}$ we can take any c less than $\frac{3-2\sqrt{2}}{6} \approx 0.028$. □

One might think that a maximum C_4-free subgraph of $G(n,p)$ is (close to) a regular graph - but somewhat surprisingly, if we are looking for the largest regular C_4-free subgraph H of $G(n,p)$, then everything is much simpler. Denote by d the degree of H, then the container of each vertex will have size at most $(1+o(1))n/d$, therefore we have to place $dn/2$ edges (each twice) into $(1+o(1))n^2/(d)$ places, which, after some technical argument which we omit, gives the restriction that $d \leq (1+o(1))\sqrt{pn}$.

Note that the natural conjecture, that the largest C_4-free subgraph of $G(n,p)$ has $(1/2 + o(1))pn^{3/2}$ edges, is false in general. The following construction is a slight modification of the one given by Morris–Saxton [17], and provides a counterexample for small p. Take an extremal C_4-free graph G_0 on $n/2$ vertices and let G' be obtained by blowing up each vertex of G_0 to size two and replace every edge by a (not necessarily perfect) matching. Note that every G' obtained in this way is C_4-free. Now consider $G(n,p)$ and try to count how many edges it has in common with a graph obtained as above. Since

$$E(X) \geq \frac{1}{2}\left(\frac{n}{2}\right)^{3/2} \left(4p(1-p)^3 + 2\left(2p^2(1-p)^2 + 4p^3(1-p) + p^4\right)\right) > \frac{n^{3/2}}{2}p$$

holds for $p < p_0 \approx 0.2$, the result follows.

Above we have seen that looking for the largest *regular* C_4-free subgraph of $G(n,p)$ seems much simpler than the general problem. A natural question to ask is, whether there are some other properties (like regularity) whose assumption simplifies the problem.

Problem 4.9 *Is there a natural extra condition, e.g. some property $\mathscr{P}$, such that the maximum number of edges of a C_4-free subgraph satisfying $\mathscr{P}$ of $G(n,p)$ could be determined asymptotically?*

5 The number of C_4-free graphs

Let F_n be the number of n-vertex labelled C_4-free graphs. The magnitude of F_n was upper bounded by Kleitman and Winston [13]. Let the constant γ be defined as follows:

$$\gamma = \frac{2}{3} \max_{x\in(0,1)} \frac{H(x^2)}{x} \approx 1.081919$$

where $H(y) = -y\log_2 y - (1-y)\log_2(1-y)$ is the binary entropy function, and let $c^* \approx 0.49$ be the constant satisfying the following equality:

$$\gamma = \frac{2}{3}\frac{H\left((c^*)^2\right)}{c^*}.$$

Theorem 5.1. *[13] The number of n-vertex labelled C_4-free graphs satisfies*

$$\log_2 F_n \le (1+o(1))\gamma n^{3/2}.$$

Our main result in this section is the improvement of their constant by a tiny amount.

Theorem 5.2. *There exists a $\delta > 0$ such that*

$$\log_2 F_n \le (1+o(1))(\gamma - \delta)n^{3/2}.$$

We make no effort to optimise the value of δ - we suspect $\delta = 2^{-100}$ is small enough to make the proof work. The improvement in our theorem comes from insisting that our containers not only have few vertices in them, but also have not too large degree measure, as described below. The remainder of this section is devoted to proving Theorem 5.2.

Fix an ordering $v_1, \ldots, v_n$ such that v_i has minimum degree in $G_i = G[v_i, \ldots, v_n]$, write d_i for the degree of v_i in G and write d_i^* for the degree of v_i in G_i (the "right-degree" of v_i).

Our approach is similar to that of Kleitman–Winston [13], and to the approach in the previous section. Given this ordering, and degree sequences, we find a small container C_i for each vertex v_i, with the following properties: (i) C_i should contain $N(v_i) \cap \{v_i, \ldots, v_n\}$. (ii) The number of choices should be small, when we consider different graphs having the same vertex ordering and degree sequences.

If for every i, both C_i and the number of choices for C_i are small, then we could obtain an upper bound for the number of choices for $N(v_i) \cap \{v_i, \ldots, v_n\}$, yielding an upper bound for the number of C_4-free graphs. The number of choices for $N(v_i) \cap \{v_i, \ldots, v_n\}$ is $\binom{|C_i|}{d_i^*}$, therefore we improve on the upper bound if d_i^* is smaller than it should be. If for most i it is not smaller, then the degree measure, and hence the size, of an average container is smaller, yielding again an improvement.

Our additional idea is that for every G_n, it cannot be that all the containers have the largest possible sizes. In the first half of the proof, we describe containers of vertices in a fixed graph G_n, so every vertex has only one container, containing its neighbourhood.

Let ϵ be a very small fixed constant. Fix an ordering as above, a degree sequence and a right-degree sequence (at most n^n choices each). Also, for each vertex v_i fix a container C_i for its neighbourhood, and in what follows we will show that one can create the containers such that for $i < (1-\epsilon)n$, the degree measure of C_i in G_i is at most

$$\mu_i(C_i) = \sum_{v \in C_i} d_{G_i}(v) \leq (1+\epsilon^2)(n-i+1). \tag{7}$$

Lemma 5.3. *Let G be a C_4-free graph with degree sequence and the ordering of its vertices given, as above. Then for each $i = 1, \ldots, (1-\epsilon)n$ we can fix a container C_i containing $N(v_i) \cap \{v_i, \ldots, v_n\}$ such that we have $\sum_{v \in C_i} d_{G_i}(v) \leq (1+\epsilon^2)(n-i+1)$.*

Proof. The proof is very similar to the one given by Kleitman–Winston [13] and Lemma 4.5, so we only give a sketch here. Set $m = n - i + 1$. If $d_i^* < \sqrt{m}/\log^2 m$ then the number of choices for $N(v_i) \cap \{v_i, \ldots, v_n\}$ is at most $2^{O(\sqrt{m}/\log m)}$. So assume $d_i^* \geq \sqrt{m}/\log^2 m$. Now we construct a fingerprint in the exact same way as in Lemma 4.5. As long as the set of available vertices has order at least $m^{3/5}$ we always remove at least $m^{3/5}/(100 \log^4 m)$ vertices. Then until we reach order $3\sqrt{m}$ we always remove at least $m^{2/5}/(100 \log^4 m)$ vertices. For the final touch, notice that if $\sum_{v \in C_i} d_{G_i}(v) > (1+\epsilon^2)m$ then $|E(G^2[C_i])| > \epsilon^2 m$ as in Lemma 4.7, and in $G^2[C_i]$ every vertex in the fingerprint T has degree zero. Hence we can always find an available vertex of degree at least $\epsilon^2\sqrt{m}/2$ and add it to our fingerprint T. But we can only add $O(1)$ vertices to the fingerprint this way, as $|C_i| = O(\sqrt{n})$. As before, the order of the certificate of the container will be at most $m^{2/5}t^5$ and the degree measure of the container will be as required. □

Definition 5.4. The vertex v_i is **win**, if at least one of the following two conditions hold:

- $|d_i^* - c^*\sqrt{n-i+1}| > \epsilon\sqrt{n-i+1}$.
- The container of v_i has order at most $(1-\epsilon^2)(n-i+1)/(c^*\sqrt{n-i+1}) = \frac{1-\epsilon^2}{c^*}\sqrt{n-i+1}$.

Denote the set of win vertices by W.

Recall that in the original proof of Kleitman–Winston [13] we add vertices one by one, according to the ordering given above. The final bound came from noting that in the worst case scenario we have $2^{\frac{H((c^*)^2)}{c^*}\sqrt{n-i}}$ choices for the neighbourhood of v_i.

Note that if we have at least ϵn wins then we are done. Indeed, if v_i is a win vertex then following the original proof of Kleitman–Winston, the number of choices for its neighbourhood contributes at most $\frac{H((c^*\pm\epsilon)^2)}{c^*\pm\epsilon}$ to the final sum in the exponent, which is strictly smaller than $\frac{H((c^*)^2)}{c^*}$. A linear number of win vertices then gives us a constant factor improvement in the final bound. Hence from now on we will assume that we have less than ϵn wins, and derive a contradiction.

In what follows we will use the two notions *right-degree* and *degree* quite frequently. Recall that the former always means $d^*(v_i) = |N(v_i) \cap \{v_{i+1}, \ldots, v_n\}|$, and the latter is the usual degree in the whole graph, denoted by $d(v)$ or d_v; unless otherwise specified, the word 'degree' will always mean the latter.

We sketch the proof of Theorem 5.2, then fill in the details. The first observation is that we need to have at least *some* win vertices. Indeed, if none of the vertices were win, then every vertex v_i would have right-degree (roughly) $c^*\sqrt{n-i}$. Hence the total number of edges in the graph would be $\frac{2}{3}c^*n^{3/2}$. But since the vertices early on in the ordering have small degrees ($d(v_i) \approx c^*\sqrt{n}$ for i small), this means that some later vertices would have much large degrees, i.e. $d(v_i) \geq \frac{4}{3}c^*\sqrt{n}$ for some large values of i. Now suppose that the first vertex has a container only containing such large degree vertices! Then the degree measure of the container is a factor of $4/3$ bigger than allowed. Indeed, the key idea to our proof will be that a container can only contain very few of these large degree vertices.

To exploit this observation we will partition our vertices in a few classes, according to the magnitude of their left- and right-degrees. If $d_v \approx c^*\sqrt{n}$ then we are dealing with a nice, everyday, average vertex. We cannot use these to derive contradictions of any sort. But as noted above, the total number of edges in our graph cannot come from these normal vertices only! So we need to have at least *some* larger degree vertices. But why cannot we have, say, $\sqrt{n}$ vertices of degree close to n? This brings us to our next crucial observation.

Suppose v_i is a vertex at position i in our ordering, and it has a *huge* degree (the exact threshold for being 'huge' will be specified later). Since our graph is C_4-free, the right-degree of this vertex is at most $\sqrt{n-i}+1$. Hence it will have quite large left-degree, meaning that it has to be contained in many containers! (This is because if u_iu_j is an edge of the graph with $i < j$ then u_j has to be in the container of u_i.)

However, as mentioned before, a container can only contain very few such large degree vertices. Combining these ideas will tell us that even though there is a huge amount of "excess degree" we have to distribute among our vertices, we cannot get too many huge ones. This in turn will imply that there is a linear proportion (at least $1/7$) of vertices which have large degree.

The finishing blow will come by repeating the same procedure as above. A large degree vertex, if not a win vertex, needs to be contained in many containers. But again, a container can contain only a few large degree vertices to not violate the degree measure condition, which will give us the long sought contradiction that finishes the proof. This bound on the degree measure is the essential ingredient, the mysterious heroine that will make an appearance multiple times throughout our proof and makes all our estimates work smoothly.

With this overview in hand, we now make all above ideas and definitions precise, and prove Theorem 5.2.

Definition 5.5. A vertex v is **large** if $d_v > (1 + 30\sqrt{\epsilon})c^* \sqrt{n}$. Denote the set of large vertices by L. A vertex v is **huge** if $d_v > \sqrt{n}$. Denote the set of huge vertices by H. For $i \in [n]$, say a vertex v_k with $k > i$ is ***i*-alive** if $d(v_k, G_i) > (1 + 10\sqrt{\epsilon})c^* \sqrt{n - i + 1}$, where $d(v_k, G_i)$ is the degree of v_k in G_i. Denote the set of i-alive vertices by L_i. Note that $H \subset L \subset L_1$.

Since the graph is C_4-free, we have by Theorem 4.2 (i) that for all i,

$$d_i^* \le \sqrt{n - i + 1} + 1. \tag{8}$$

By the definitions, we have that

$$2|E(G)| = \sum d_v = 2\sum d_v^* \ge 2 \sum_{i=1}^{(1-\epsilon)n} \left(c^* \sqrt{i} - \epsilon\sqrt{n}\right) \ge \frac{4}{3}c^*(1-5\epsilon)n^{3/2}, \tag{9}$$

where the first inequality comes from excluding the at most ϵn win vertices from our sum. Next, we make the idea "a container can only contain very few large degree vertices" precise.

Proposition 5.6. *Let $i < (1-\epsilon)n$. The container C_i of a non-win vertex v_i contains at most $\sqrt{\epsilon(n - i + 1)}$ vertices that are i-alive.*

Proof. Let $m = n-i+1$. Suppose C_i contains at least $\sqrt{\epsilon m}$ vertices that are i–alive. Since the order of the container is at least $(1 - \epsilon^2)\sqrt{m}/c^*$ (by definition of a win vertex), the degree measure of the container C_i in G_i is at least

$$\mu_i(C_i) \ge \sqrt{\epsilon m}(1+10\sqrt{\epsilon})c^* \sqrt{m} + \left(\frac{1-\epsilon^2}{c^*} - \sqrt{\epsilon}\right) \sqrt{m}(c^* - \epsilon)\sqrt{m} \ge m\,(2\epsilon + 1)\,. \tag{10}$$

Here we used that v_i is the minimum degree vertex in G_i, and that $c^* \approx 0.49$. This contradicts the constraint (7) on the degree measure of the containers. □

We also have a corresponding upper bound for all other vertices:

Lemma 5.7. *For every i, the container C_i contains at most $10\sqrt{n}$ vertices that are i-alive.*

Proof. If this was not the case, then the degree measure of C_i would violate the constraint (7). □

Now we show that the "excess edges" cannot be coming from only very few vertices - instead they must be rather evenly distributed among a linear proportion of $V(G)$.

Lemma 5.8. *We have*

$$|L\backslash W| > \frac{n}{7}.$$

Proof. Let v_k be a huge vertex. Then regardless of whether v_k is win or not, it has to be contained in at least $d_{v_k} - \sqrt{n-k} - 1$ containers C_i with $i < k$, as $d_k^* \leq \sqrt{n-k} + 1$. Consider a bipartite graph G' with vertex sets $A = [n]$ and $B = H = \{h_1, \ldots, h_{|H|}\}$. Add the edge (i, h_j) if the following two conditions hold:

1. C_i contains h_j, and
2. there are at most $d_{h_j} - \sqrt{n} - 1$ vertices v_t with $t < i$ such that C_t contains h_j.

Then in G', every vertex h_i in B has degree at least $d_{h_i} - \sqrt{n}$ (in fact their degrees are $\lceil d_{h_i} - \sqrt{n}\rceil$). So

$$|E(G')| \geq \sum_H (d_v - \sqrt{n}). \tag{11}$$

Now it's time to use Proposition 5.6. If (i, h_j) is an edge in G' and v_i is non-win, then $d(h_j, G_i) \geq \sqrt{n}$. (Why? Because v_j appeared in at most $d_{h_j} - \sqrt{n}$ containers prior to i.) Hence h_j is i-alive. So in G', the degree of a non-win vertex v_i in A is at most $\sqrt{\epsilon n}$ if $i < (1-\epsilon)n$, and at most $10\sqrt{n}$ if v_i is non-win with $i \geq (1-\epsilon)n$ by Lemma 5.7. There are at most ϵn win vertices in total, and similarly each win vertex v_i has a container containing at most $10\sqrt{n}$ vertices that are i-alive. Hence

$$|E(G') \leq \sqrt{\epsilon n}n + 20\epsilon n^{3/2} < 2\sqrt{\epsilon}n^{3/2}. \tag{12}$$

Note that by equation (9) we have that

$$\sum_L d_v = 2|E(G)| - \sum_{[n]\backslash L} d_v \geq \frac{4}{3}c^*(1-5\epsilon)n^{3/2} - (1+30\sqrt{\epsilon})c^* n^{3/2}$$

$$\geq c^* n^{3/2}\left(\frac{1}{3} - 50\sqrt{\epsilon}\right) \tag{13}$$

and by using the definition of a huge vertex, a corollary of equations (11) and (13) is that

$$|E(G')| \geq \sum_L (d_v - \sqrt{n}) \geq c^* n^{3/2} \left(\frac{1}{3} - 50\sqrt{\epsilon} \right) - |L|\sqrt{n}. \tag{14}$$

Putting equations (12) and (14) together, we get

$$2\sqrt{\epsilon} n^{3/2} + |L|\sqrt{n} > c^* n^{3/2} \left(\frac{1}{3} - 50\sqrt{\epsilon} \right).$$

Because

$$|L| > n \left(\frac{c^*}{3} - 100\sqrt{\epsilon} \right) \tag{15}$$

and $|W| < \epsilon n$, we get

$$|L \backslash W| > n \left(\frac{c^*}{3} - 101\sqrt{\epsilon} \right) > \frac{n}{7}. \tag{16}$$

□

Now we can finish the proof of Theorem 5.2. Note that if a large non-win vertex v_p appeared in at most $20\sqrt{\epsilon}c^*\sqrt{n}$ different C_k with $k < i < p$ then v_p is i-alive. As in the proof of Lemma 5.8, let us create a bipartite graph and count the edges in two ways.

Consider a bipartite graph G'' with vertex sets $A = [n]$ and $B = L\backslash W = \{p_1, \ldots, p_{|L\backslash W|}\}$. Add the edge (i, p_j) if the following two conditions hold:

1. C_i contains p_j, and
2. There are at most $20\sqrt{\epsilon}c^*\sqrt{n}$ vertices v_t with $t < i$ such that C_t contains p_j.

As before, we have that

$$(\sqrt{\epsilon} + 20\epsilon)n^{3/2} \geq E(G'') \geq 20\sqrt{\epsilon}c^*\sqrt{n}|L\backslash W| > \frac{20c^*}{7}\sqrt{\epsilon}n^{3/2}, \tag{17}$$

which is a contradiction, since $\frac{20c^*}{7} \approx 1.4 > 1$. This completes the proof. □

The original question still remains open. In general, whether the number of H-free graphs with vertex set $[n]$ is $2^{O(\text{ex}(n,H))}$ is still not known for many bipartite graphs H. When the order of magnitude of $\text{ex}(n, H)$ is known then the situation is better: see [3] and [4] for when H is a complete bipartite graph, and see [17] for when H is an even cycle.

Acknowledgements The authors are grateful to Hong Liu, Maryam Sharifzadeh, and Wojciech Samotij for careful reading of the manuscript.

Research by József Balogh is partially supported by Simons Fellowship, NSA Grant H98230-15-1-0002, NSF CAREER Grant DMS-0745185, NSF Grant DMS-1500121, Arnold O. Beckman Research Award (UIUC Campus Research Board 15006) and Marie Curie FP7-PEOPLE-2012-IIF 327763.

References

1. L. Babai, M. Simonovits, J. Spencer, Extremal subgraphs of random graphs. J. Graph Theory **14**(5), 599–622 (1990)
2. J. Balogh, S. Petrickova, The number of the maximal triangle-free graphs. Bull. Lond. Math. Soc. **46**(5), 1003–1006 (2014)
3. J. Balogh, W. Samotij, The number of $K_{m,m}$-free graphs. Combinatorica **31**, 131–150 (2011)
4. J. Balogh, W. Samotij, The number of $K_{s,t}$-free graphs. J. Lond. Math. Soc. **83**, 368–388 (2011)
5. J. Balogh, R. Morris, W. Samotij, Independent sets in hypergraphs. J. Am. Math. Soc. **28**(3), 669–709 (2015)
6. J. Balogh, H. Liu, S. Petrickova, M. Sharifzadeh, The typical structure of maximal triangle-free graphs. arXiv preprint [arXiv:1501.02849]
7. D. Conlon, Combinatorial theorems relative to a random set, in *Proceedings of the International Congress of Mathematicians* 4, 303–328 (2014)
8. D. Conlon, W.T. Gowers, Combinatorial theorems in sparse random sets (2010). arXiv preprint [arXiv:1011.4310]
9. P. Erdős, D. Kleitman, B. Rothschild, Asymptotic enumeration of K_n-free graphs. Colloquio Internazionale sulle Teorie Combinatorie (Rome, 1973), Tomo II (Accad. Naz. Lincei, Rome, 1976), pp. 19–27. Atti dei Convegni Lincei, No. 17
10. P. Erdős, P. Frankl, V. Rödl, The asymptotic number of graphs not containing a fixed subgraph and a problem for hypergraphs having no exponent. Graphs Comb. **2**, 113–121 (1986)
11. Z. Füredi, The maximum number of edges in a minimal graph of diameter 2. J. Graph Theory **16**(1), 81–98 (1992)
12. Z. Füredi, Random Ramsey graphs for the four-cycle. Discret. Math. **126**, 407–410 (1994)
13. D.J. Kleitman, K.J. Winston, On the number of graphs without 4-cycles. Discret. Math. **41**(2), 167–172 (1982)
14. Y. Kohayakawa, B. Kreuter, A. Steger, An extremal problem for random graphs and the number of graphs with large even-girth. Combinatorica **18**(1), 101–120 (1998)
15. G. Kozma, T. Meyerovitch, R. Peled, W. Samotij, Random points in the metric polytope (in preparation)
16. G. Kozma, T. Meyerovitch, R. Morris, R. Peled, W. Samotij, An almost optimal bound on the volume of the metric polytope (in preparation)
17. R. Morris, D. Saxton, The number of C_{2l}-free graphs. Adv. Math. arXiv preprint [arXiv:1309.2927v3], (2013)
18. F. Mousset, R. Nenadov, A. Steger, On the number of graphs without large cliques. SIAM J. Discret. Math. **28**(4), 1980–1986 (2014)
19. D. Mubayi, C. Terry, Discrete metric spaces: structure, enumeration, and $0-1$ laws. arXiv preprint [arXiv:1502.01212]
20. V. Rödl, M. Schacht, Extremal results in random graphs, in *Erdős Centennial*. Society Mathematical Studies, vol. 25 (János Bolyai Society Mathematical Studies, Budapest, 2013), pp. 535–583
21. W. Samotij, Counting independent sets in graphs. Eur. J. Comb. **48**, 5–18 (2015)
22. D. Saxton, A. Thomason, Hypergraph containers. Invent. Math. **201**(3), 925–992 (2015)
23. M. Schacht, Extremal results for random discrete structures (2009, submitted)
24. E. Szemerédi, *Regular Partitions of Graphs*. No. STAN-CS-75-489 (Department of Computer Science, Stanford University California, Stanford, 1975)

Independent transversals and hypergraph matchings - an elementary approach

Penny Haxell

Abstract We give a self-contained elementary approach to a body of combinatorial results that were previously proved using topology. One example is the following hypergraph version of Hall's Theorem, due to Aharoni and Haxell from 2000. Let $\mathcal{H}$ be an r-uniform hypergraph with a vertex partition (A, X), in which every edge has exactly one vertex in A. Suppose that for every subset $S \subseteq A$, the neighbourhood hypergraph $\Gamma_{\mathcal{H}}(S)$ of S has a matching of size $(r-1)(|S|-1)+1$. Then $\mathcal{H}$ has a matching of size $|A|$.

Keywords Independent transversals • Hypergraph matchings

Mathematics Subject Classification (2010): 05C65, 05D15

1 Introduction

In this article we develop and prove a number of combinatorial theorems that were first proved using a topological approach [2, 5, 6], some of which were later given proofs using homology [29, 30]. It should be emphasised that neither the results nor the ideas covered here are new, but instead of using topology explicitly we re-cast the material in a combinatorial setting. One motivation for this is to make this body of work conveniently accessible to those who are not so familiar with topology or homology. A second motivation is the fact that in the end we derive results that are purely combinatorial in nature, and so it is perhaps natural to make the whole theory self-contained and use only elementary combinatorial arguments.

An example of a combinatorial theorem that can be proved using these methods is concerned with independent transversals in graphs. Let G be a graph and let $\mathcal{P} = \{V_1, \ldots, V_m\}$ be a partition of the vertex set $V(G)$ into m parts. An *independent*

P. Haxell (✉)
Department of Combinatorics and Optimization, University of Waterloo,
Waterloo, ON, Canada N2L 3G1
e-mail: pehaxell@uwaterloo.ca

A. Beveridge et al. (eds.), *Recent Trends in Combinatorics*, The IMA Volumes in Mathematics and its Applications 159, DOI 10.1007/978-3-319-24298-9_9

transversal of G with respect to $\mathcal{P}$ is an independent set $\{v_1, \ldots, v_m\}$ in G such that $v_i \in V_i$ for each i. This is a very general notion, and many combinatorial problems can be formulated by asking whether a certain graph with a certain vertex partition has an independent transversal. (For example, the SAT problem can be expressed in these terms, see, e.g., [21].) We cannot expect a good characterisation of those $(G, \mathcal{P})$ for which G has an independent transversal with respect to $\mathcal{P}$, but we can give some sufficient conditions. A simple example is the following [16, 18].

Theorem 1. *Let G be a graph with maximum degree Δ, and let $\mathcal{P} = \{V_1, \ldots, V_m\}$ be a partition of $V(G)$. If $|V_i| \geq 2\Delta$ for each i, then G has an independent transversal with respect to $\mathcal{P}$.*

Some examples of applications of Theorem 1 can be found in [4, 9, 10, 12, 26–28].

The topological method that is the basis for this article was initiated in [5] and used Sperner's Lemma and special triangulations of simplices to prove statements about hypergraph matching (including the theorem mentioned in the abstract). Both Sperner's Lemma and the relevant properties of triangulated simplices have purely combinatorial proofs, and several later developments using the method can also be described in combinatorial terms. Here we exploit this by removing the explicit use of topology and replacing it with the underlying combinatorics, but still using topological intuition as a guide, and indeed still using some topological terminology (such as "sphere" and "connectedness") for the abstract notions we discuss. Section 2 gives the background on simplicial complexes that we will need. Since we argue there in terms of sets and parity, in essence we develop a small amount of homology from first principles, although we do not use the language of homology. Section 3 is devoted to an abstract version of Sperner's Lemma, and in Sections 4 and 5 we show how to use these ideas (in particular the notion of connectedness) to apply Sperner's Lemma to combinatorial problems about graphs and hypergraphs. In Section 6 we describe a number of applications, including what has been perhaps the most striking success of this method, namely Aharoni's proof [2] in 2000 of the famous conjecture of Ryser [31] from 1967 on matchings and covers in 3-partite 3-uniform hypergraphs.

2 Pure Simplicial Complexes

A *simplicial complex* consists of a set $\mathcal{A}$ of subsets A of a finite set V with the property that if $A \in \mathcal{A}$ and $B \subset A$ then $B \in \mathcal{A}$. We call the sets A the *simplices* of $\mathcal{A}$, and the *dimension* of A is $|A| - 1$. We will identify $V = V(\mathcal{A})$ with its set of 0-dimensional simplices (called 0-simplices) of $\mathcal{A}$, in other words we assume that every element of V is in some $A \in \mathcal{A}$. The *dimension* of $\mathcal{A}$ is the maximum dimension of any $A \in \mathcal{A}$. We say that $\mathcal{A}$ is *pure* if every maximal $A \in \mathcal{A}$ has the same dimension. Note then that a pure d-dimensional simplicial complex (abbreviated d-PSC) is the *closure* of the $(d+1)$-uniform hypergraph $\mathcal{A}^d$ consisting of the d-dimensional simplices of $\mathcal{A}$, that is, we form $\mathcal{A}$ from $\mathcal{A}^d$ by adding all subsets of the edges of $\mathcal{A}^d$. In general $\mathcal{A}^t$ denotes the set of all t-dimensional simplices in $\mathcal{A}$, in particular $V(\mathcal{A}) = \mathcal{A}^0$.

For a d-PSC $\mathcal{A}$, the *boundary* $\partial\mathcal{A}$ of $\mathcal{A}$ is the $(d-1)$-PSC that is the closure of the d-uniform hypergraph

$$\{B : |B| = d, |\{A \in \mathcal{A}^d : B \subset A\}| \equiv 1 \mod 2\}.$$

If $\partial(\mathcal{A})$ is empty, we say that $\mathcal{A}$ is a *d-sphere*. Note in particular that a 0-sphere is any nonempty set of singletons of even size, since the empty set should be contained in an even number of them.

For example, a simple graph G with no isolated vertices is a 1-PSC, its edges are the 1-simplices and its vertices are the 0-simplices. Its boundary is the set of vertices of odd degree. Another helpful example to keep in mind (that we will refer to several times) is the following. A 3-dimensional ball that is triangulated into small tetrahedra is a 3-PSC. Its boundary is a sphere triangulated into triangles. Each pair of vertices of the triangulation T of the sphere is either in exactly two of the triangles of T (if it forms a 1-simplex of T) or in zero (if it does not form a simplex of T).

We will sometimes use the following convenient way to determine the boundary of $\mathcal{A}$. For $0 \leq i \leq d+1$ we take an arbitrary but fixed ordering $I_i(V)$ of all i-subsets of $V = V(\mathcal{A})$. The incidence matrix $\mathcal{A}[d, d+1]$ has rows indexed by $I_d(V)$ and columns indexed by $\mathcal{A}^d$, and its (B, A)-entry is 1 if $B \subset A$ and 0 otherwise. Then $\partial(\mathcal{A})$ is the closure of the d-uniform hypergraph whose characteristic vector is $\mathcal{A}[d, d+1]\mathbf{1}$, where $\mathbf{1}$ is the all-1's vector with $|\mathcal{A}^d|$ coordinates. (All matrix operations in this article will be over $\mathbb{F}_2$.)

First we make explicit some basic properties of spheres and boundaries.

Lemma 2. *Let $\mathcal{A}$ be a d-PSC. Then $\partial(\mathcal{A})$ is a $(d-1)$-sphere.*

***Proof*.** Let $\mathbf{v} = \mathcal{A}[d, d+1]\mathbf{1}$. Let $\mathcal{K}$ denote the $(d-1)$-PSC that is the closure of the complete d-uniform hypergraph on V. Note that the C-entry of $\mathcal{K}[d-1, d]\mathbf{v}$ counts the number of $(d-1)$-simplices in $\partial(\mathcal{A})$ that contain C (mod 2). But

$$\mathcal{K}[d-1, d]\mathbf{v} = \mathcal{K}[d-1, d]\mathcal{A}[d, d+1]\mathbf{1}$$

and the (C, A)-entry of $\mathcal{K}[d-1, d]\mathcal{A}[d, d+1]$ is 0 if $C \not\subset A$, and is $|\{B : |B| = d, C \subset B \subset A\}|$ if $C \subset A$. But this quantity is exactly 2 since $|C| = d-1$ and $|A| = d+1$. Therefore $\mathcal{K}[d-1, d]\mathbf{v}$ is the zero vector, and so $\partial(\mathcal{A})$ is a $(d-1)$-sphere. □

Lemma 3. *Let $\mathcal{A}$ be a $(d-1)$-sphere. Suppose $w \notin V(\mathcal{A})$. Then the closure $\mathcal{C}$ of $\{A \cup \{w\} : A \in \mathcal{A}^{d-1}\}$ is a d-PSC with boundary $\mathcal{A}$.*

***Proof*.** Let $B \subset V \cup \{w\}$ be a set of size d. Since every $(d+1)$-set in $\mathcal{C}$ contains w, the set of $(d+1)$-sets in $\mathcal{C}$ containing B is exactly $\{D \cup \{w\} : D \in \mathcal{D}\}$, where

$$\mathcal{D} = \{A \in \mathcal{A} : B \setminus \{w\} \subseteq A\}.$$

Since $\mathcal{A}$ is a $(d-1)$-sphere, if $|B \setminus \{w\}| = d-1$, then $|\mathcal{D}|$ is even, which implies $B \notin \partial\mathcal{C}$.

Thus if $B \in \partial\mathcal{C}$, then we must have $|B \setminus \{w\}| = d$, in other words $w \notin B$. Thus in this case the only $(d+1)$-set in $\mathcal{C}$ that can contain B is $B \cup \{w\}$. Thus $B \in \mathcal{A}$, and we find $\partial\mathcal{C} \subseteq \mathcal{A}$.

Conversely, note that each $A \in \mathcal{A}$ is in exactly one $(d+1)$-set $A \cup \{w\} \in \mathcal{C}$. Therefore $\mathcal{A} \subseteq \partial\mathcal{C}$ and hence $\mathcal{A} = \partial\mathcal{C}$. □

As an example for Lemma 3, consider $\mathcal{A}$ to be a sphere triangulated into triangles, place a new point w inside the sphere and form $\mathcal{C}$ by "filling up" the space inside the sphere with the set of all tetrahedra that have a triangle in $\mathcal{A}$ as base and w as apex.

Let $\mathcal{A}$ be a d-PSC, and let $U \subset V(\mathcal{A})$ be a subset of size at most d. The *link* of U in $\mathcal{A}$ is $\mathcal{L}_{\mathcal{A}}(U) = \{A \setminus U : A \in \mathcal{A}, U \subset A\}$.

In our graph example, the link of a single vertex is just its neighbourhood in the graph. In the triangulated 3-dimensional ball example, consider U to be an edge xy in the interior of the triangulated ball. The link of U together with U forms a "bicycle wheel", where U is the "axle", the link of U is a cycle (the "rim" of the wheel) and each vertex of the link has an edge ("spoke") to each of x and y.

Lemma 4. *Let $\mathcal{A}$ be a d-PSC and let $U \in \mathcal{A}$ be a simplex of size $u \leq d$ that is not a simplex of $\partial\mathcal{A}$. Then $\mathcal{L}_{\mathcal{A}}(U)$ is a $(d-u)$-sphere.*

Proof. Let $B \subset V(\mathcal{A})$ be a set of size $d-u$. If $B \not\subset V(\mathcal{A}) \setminus U$, then no element of $\mathcal{L}_{\mathcal{A}}(U)$ contains B. Suppose $B \subset V(\mathcal{A}) \setminus U$. Then $B \cup U$ has size d. Since for any set A containing U we have $B \subset A \setminus U$ if and only if $B \cup U \subset A$, and $B \cup U \notin \partial\mathcal{A}$, we find that $|\{A \setminus U \in \mathcal{L}_{\mathcal{A}}(U) : B \subset A \setminus U\}|$ is even. Hence $\mathcal{L}_{\mathcal{A}}(U)$ is a $(d-u)$-sphere. □

In later arguments we will need a method for modifying a given PSC $\mathcal{B}$ to remove a specific simplex $B \in \mathcal{B}$, without altering the boundary of $\mathcal{B}$. When B is a maximal simplex this is easy to do using the following notion. For a $(d+1)$-PSC $\mathcal{B}$, a $(d+1)$-simplex $B \in \mathcal{B}$, and a new vertex $w \notin V(\mathcal{B})$, the *stellar subdivision* of $\mathcal{B}$ with respect to (B, w) is the $(d+1)$-PSC that is the closure of

$$\mathcal{B}^{d+1} \setminus \{B\} \cup \{B' \cup \{w\} : B' \subset B, |B'| = |B| - 1\}.$$

Lemma 5. *Let $\mathcal{B}$ be a $(d+1)$-PSC, let B be a $(d+1)$-simplex in $\mathcal{B}$ and let $w \notin V(\mathcal{B})$ be a new vertex. Let $\mathcal{B}'$ be the stellar subdivision of $\mathcal{B}$ with respect to (B, w). Then $\mathcal{B}'$ satisfies $\partial\mathcal{B}' = \partial\mathcal{B}$ and does not contain B as a simplex.*

Proof. The simplex B is not present in $\mathcal{B}'$ by definition. To check that $\partial\mathcal{B}' = \partial\mathcal{B}$, observe that for each $(d+1)$-set A, if $w \notin A$ then the number of $(d+1)$-simplices in $\mathcal{B}'$ containing A is the same as the number in $\mathcal{B}$. If $w \in A$, then $A \not\subset B$ and A is contained in exactly zero or two of $\{B' \cup \{w\} : B' \subset B, |B'| = |B| - 1\}$. Therefore $\partial\mathcal{B}' = \partial\mathcal{B}$. □

When B is not a maximal simplex then we will use the more technical construction in the following "hole-filling" lemma. Topological versions of this type of argument were used in, e.g., [6, 33].

Lemma 6. *Let $\mathcal{B}$ be a $(d+1)$-PSC and let B be a simplex in $\mathcal{B}$ with $|B| \leq d+1$ that is not a simplex of $\partial\mathcal{B}$. Let $\mathcal{L} = \mathcal{L}_{\mathcal{B}}(B)$ and set $b = |B|$. Then by Lemma 4 we* *know $\mathcal{L}$ is a $(d-b+1)$-sphere. Let $\mathcal{C}$ be a $(d-b+2)$-PSC with $\partial\mathcal{C} = \mathcal{L}$ and $V(\mathcal{C}) \cap B = \emptyset$. Suppose that every $(d-b+2)$-simplex of $\mathcal{C}$ contains a vertex that is not a vertex of $\mathcal{B}$. Let $\mathcal{B}'$ be the $(d+1)$-PSC that is the closure of*

$$\mathcal{B}^{d+1} \setminus \{D \cup B : D \in \mathcal{L}, |D| = d-b+2\} \cup \{C \cup B' : C \in \mathcal{C},$$
$$|C| = d-b+3, B' \subset B, |B'| = |B|-1\}.$$

Then $\mathcal{B}'$ satisfies $\partial\mathcal{B}' = \partial\mathcal{B}$ and does not contain B as a simplex.

Proof. Again it is clear that $\mathcal{B}'$ does not have B as a simplex.

Let A be a $(d+1)$-set. Let $\mathcal{S}_1 = \{D \cup B : D \in \mathcal{L}, |D| = d-b+2, A \subset D \cup B\}$ and $\mathcal{S}_2 = \{C \cup B' : C \in \mathcal{C}, |C| = d-b+3, B' \subset B, |B'| = |B|-1, A \subset C \cup B'\}$. Note that for each $C \cup B' \in \mathcal{S}_2$, the simplex C is a $(d-b+2)$-simplex of $\mathcal{C}$ and hence contains a vertex that is not in $\mathcal{B}$. Therefore $\mathcal{B} \cap \mathcal{S}_2 = \emptyset$. Note also that $\mathcal{S}_1 \subset \mathcal{B}$. Therefore we just need to show that $|\mathcal{S}_2| - |\mathcal{S}_1|$ is even.

Suppose $B \subseteq A$. Since $V(\mathcal{C}) \cap B = \emptyset$ we see that $|\mathcal{S}_2| = 0$. Since $|A \setminus B| = d-b+1$, and $\mathcal{L}$ is a $(d-b+1)$-sphere, the number of simplices $D \in \mathcal{L}^{d-b+1}$ with $A \setminus B \subset D$ is even. This tells us that $|\mathcal{S}_1|$ is even, as required.

Suppose $|A \cap B| = b-1$, say $B_1 = A \cap B$. Since $V(\mathcal{C}) \cap B = \emptyset$, we find $\mathcal{S}_2 = \{C \cup B_1 : C \in \mathcal{C}, |C| = d-b+3, A \setminus B_1 \subset C\}$.
Note that $|A \setminus B_1| = d-b+2$. If $A \setminus B_1 \in \mathcal{L} = \partial\mathcal{C}$, then the unique element of $\mathcal{S}_1$ is $(A \setminus B_1) \cup B$. Also, the number of $C \in \mathcal{C}^{d-b+2}$ with $A \setminus B_1 \subset C$ is odd. Thus $|\mathcal{S}_2|$ is odd as well, and so $|\mathcal{S}_2| - |\mathcal{S}_1|$ is even. If $A \setminus B_1 \notin \mathcal{L} = \partial\mathcal{C}$, then $\mathcal{S}_1 = \emptyset$ and the number of $C \in \mathcal{C}$ with $A \setminus B_1 \subset C$ is even. Thus again $|\mathcal{S}_2| - |\mathcal{S}_1|$ is even.

Suppose $|A \cap B| = b-2$, say $B_2 = A \cap B$. Then $\mathcal{S}_1 = \emptyset$. Suppose $\mathcal{S}_2 \neq \emptyset$, so that some $C \cup B' \in \mathcal{S}_2$. Since $A \subset C \cup B'$ and B' contains a vertex y that is not in B_2, this y must be the unique vertex in $C \cup B'$ that is not in A. Therefore $C = A \setminus B_2$ and $B' = B_2 \cup \{y\}$.
From this we see that there is exactly one other element in $\mathcal{S}_2$, namely $C \cup B''$ where $B'' = B_2 \cup \{z\}$ and $B = B_2 \cup \{y, z\}$. Therefore if $\mathcal{S}_2$ is nonempty, then $|\mathcal{S}_2| = 2$.

Suppose $|A \cap B| \leq b-3$. Then $\mathcal{S}_1 = \mathcal{S}_2 = \emptyset$.
Thus in all cases $|\mathcal{S}_2| - |\mathcal{S}_1|$ is even, completing the proof.

□

As an example of the use of the hole-filling lemma, consider again the triangulated 3-dimensional ball, and the simplex $B = xy$ in its interior. The link of B is the rim $\mathcal{L}$ of the bicycle wheel. Imagine erasing all the tetrahedra of the triangulation that contain xy, forming a "hole" in the ball. Now take a triangulated disk $\mathcal{C}$ whose boundary is identified with the wheel rim $\mathcal{L}$, and "fill up" the hole by adding the tetrahedra $C \cup \{x\}$ and $C \cup \{y\}$ for each triangle C of $\mathcal{C}$. This gives us a new triangulated ball $\mathcal{B}'$ whose boundary sphere is the same as that of $\mathcal{B}$, and we have removed the simplex B.

3 Sperner's Lemma

The familiar form of Sperner's Lemma states that if the points of a triangulated geometric d-simplex are coloured with $d+1$ colours satisfying certain conditions, then there exists a simplex of the triangulation that gets all $d+1$ colours on its vertices. In this section we simply translate this well-known result into a more abstract form, using the language of pure simplicial complexes. Our abstract version of Sperner's Lemma is as follows.

Lemma 7. *Let $S = \{v_1, \ldots, v_m\}$ be a nonempty set and suppose for each $T \subseteq S$ there exists a $(|T|-1)$-PSC $\mathcal{A}_T$ such that*

(i) $\mathcal{A}_\emptyset = \{\emptyset\}$ and $\mathcal{A}^0_{\{v_i\}} = \{v_i\}$ for each i,
(ii) for each $|T| \geq 1$ we have $\partial(\mathcal{A}_T) = \cup_{U\subset T, |U|=|T|-1} \mathcal{A}_U$.

Suppose further that each vertex of $\mathcal{A}_S$ is coloured with an element of S such that for every T, if $\{x\} \in \mathcal{A}_T$, then the colour of x is in T. Then $\mathcal{A}_S$ contains a multicoloured simplex, i.e. a simplex whose $|S|$ vertices are coloured with the whole set S.

Proof. Here we copy a standard proof of Sperner's Lemma, proving the stronger statement that the number of multicoloured simplices is odd. We proceed by induction on m. When $m = 1$ the number of multicoloured simplices is one. Suppose that $m \geq 2$ and the statement is true for smaller m. Let $\mathcal{D}$ denote the set of all $(m-2)$-simplices in $\mathcal{A}_S$ that are multicoloured with $U_0 = \{v_1, \ldots, v_{m-1}\}$. By the colouring condition, any simplex in $\mathcal{D}$ that is in $\partial(\mathcal{A}_S) = \cup_{U\subset S, |U|=|S|-1} \mathcal{A}_U$ must be in $\mathcal{A}_{U_0}$. So by the induction hypothesis applied to $\mathcal{A}_{U_0}$, the number of simplices of $\mathcal{D}$ in $\partial(\mathcal{A}_S)$ is odd. We define a graph H with vertex set $\mathcal{D}$ as follows: we join simplices D and D' in H if and only if there is an $(m-1)$-simplex A of $\mathcal{A}_S$ that contains both D and D'. Observe that if A contains $D \in \mathcal{D}$ then it contains a second $D' \in \mathcal{D}$ (and hence corresponds to an edge of H) if and only if A is not itself multicoloured (with S).

We now consider the degrees of vertices in H. For each $D \in \mathcal{D}$ let $mc(D)$ denote the number of multicoloured $(m-1)$-simplices in $\mathcal{A}_S$ that contain D, and let $t(D)$ denote the total number of $(m-1)$-simplices in $\mathcal{A}_S$ that contain D. Then the degree of D in H is $t(D) - mc(D)$ for each $D \in \mathcal{D}$. Thus the set $\mathcal{M}$ of multicoloured $(m-1)$-simplices $A \in \mathcal{A}_S$ has size

$$\sum_{A\in\mathcal{M}} 1 = \sum_{D\in\mathcal{D}} mc(D) = \sum_{D\in\mathcal{D}} t(D) - d_H(D).$$

But $\sum_{D\in\mathcal{D}} d_H(D)$ is even, and $\sum_{D\in\mathcal{D}} t(D) \equiv \sum_{D\in\mathcal{D}\cap\partial(\mathcal{A}_S)} 1 \mod 2$ by definition of $\partial(\mathcal{A}_S)$. Thus by the induction hypothesis

$$|\mathcal{M}| \equiv \sum_{D\in\mathcal{D}\cap\partial(\mathcal{A}_S)} 1 \equiv 1 \mod 2.$$

□

We will apply Sperner's Lemma to specially constructed PSCs that are built up in order of dimension. The next result is a technical lemma that facilitates this type of construction.

Lemma 8. *Let $S = \{v_1, \dots, v_m\}$ be a set of at least two vertices and suppose for each $T \subset S$ with $0 \leq |T| \leq |S| - 1$ there exists a $(|T| - 1)$-PSC $\mathcal{A}_T$ such that*

(i) $\mathcal{A}_\emptyset = \{\emptyset\}$ and $\mathcal{A}^0_{\{v_i\}} = \{v_i\}$ for each i,
(ii) for each $|T| \geq 1$ we have $\partial(\mathcal{A}_T) = \cup_{U \subset T, |U|=|T|-1} \mathcal{A}_U$,
(iii) for each $T \neq T'$ with $|T| = |T'|$ we have $\mathcal{A}_T^{|T|-1} \cap \mathcal{A}_{T'}^{|T'|-1} = \emptyset$.

Then $\mathcal{A} = \cup_{T \subset S, |T|=|S|-1} \mathcal{A}_T$ is a $(|S| - 2)$-sphere.

Proof. To find the boundary of $\mathcal{A}$ we consider the product $\mathcal{A}[|S| - 2, |S| - 1]\mathbf{1}$. Condition (iii) tells us that no $(|S| - 1)$-set A is a simplex of more than one $\mathcal{A}_T$ with $|T| = |S| - 1$, which implies that the B-entry of $\mathcal{A}[|S| - 2, |S| - 1]\mathbf{1}$ is the sum over all $T \subset S$, $|T| = |S| - 1$ of the B-entries in

$$\mathcal{A}_T[|S| - 2, |S| - 1]\mathbf{1}. \tag{1}$$

If $B \notin \partial(\mathcal{A}_T)$, then the B-entry of (1) is zero. Suppose $B \in \partial(\mathcal{A}_{T_0})$, so the B-entry in (1) for T_0 is 1. Since B is a full-dimensional simplex in $\partial(\mathcal{A}_{T_0})$ and by (ii) $\partial(\mathcal{A}_{T_0}) = \cup_{U \subset T_0, |U|=|T_0|-1} \mathcal{A}_U$, and again by (iii) no two of the $\mathcal{A}_U$ can share a $(|S| - 2)$-set, there is exactly one $U_0 \subset T_0$ such that $B \in \mathcal{A}_{U_0}$. Now since $|U_0| = |S| - 2$, there is exactly one other $T_1 \subset S$ with $U_0 \subset T_1$ and $|T_1| = |S| - 1$. By (ii) we know $\partial(\mathcal{A}_{T_1}) = \cup_{U \subset T_1, |U|=|T_1|-1} \mathcal{A}_U$. Therefore the B-entry in (1) for T_1 is also 1, and for all other T it is zero. Thus the B-entry in $\mathcal{A}[|S| - 2, |S| - 1]\mathbf{1}$ is zero for each B. This completes the proof that $\mathcal{A}$ is a sphere. □

4 Labellings, Simplicial Maps and Connectedness

In order to study problems such as the independent transversal problem mentioned in the introduction, we would like to be able to apply Sperner's Lemma to "difficult" (i.e. complicated) simplicial complexes Σ, such as the independence complex of a graph (defined in the next section). Our approach will be to try to construct a "nice" pure simplicial complex $\mathcal{A}$, associated with Σ, and a map from $\mathcal{A}$ to Σ such that applying Sperner's Lemma in $\mathcal{A}$ "simulates" applying it in Σ. This is the idea of the main result of this section, Theorem 10.

Let Σ be a simplicial complex, and let $\mathcal{A}$ be a d-PSC. A *Σ-labelling* of $\mathcal{A}$ is just a map $f : V(\mathcal{A}) \to V(\Sigma)$. A simplex $A \in \mathcal{A}$ is *properly labelled* by f if $\{f(a) : a \in A\} \in \Sigma$. (Note that $|\{f(a) : a \in A\}| < |A|$ is possible.) If every $A \in \mathcal{A}$ is properly labelled by f, then we say $f : \mathcal{A} \to \Sigma$ is a *simplicial map* from $\mathcal{A}$ to Σ.

The simplicial complex Σ is said to be *k-connected* if for each $-1 \leq d \leq k$, for every d-sphere $\mathcal{A}$ and every simplicial map $f : \mathcal{A} \to \Sigma$, there exist a $(d + 1)$-PSC

$\mathcal{B}$ and a simplicial map $f' : \mathcal{B} \to \Sigma$ such that $\partial\mathcal{B} = \mathcal{A}$ and the restriction $f'|_{\mathcal{A}}$ of f' to $\mathcal{A}$ satisfies $f'|_{\mathcal{A}} = f$. In particular, saying that Σ is (-1)-connected is the same as saying that Σ is nonempty.

For technical reasons (required by Lemma 6) we will need a PSC with an additional property, but this will be implied by the existence of $\mathcal{B}$, as made explicit in the following lemma.

Lemma 9. *Let $\mathcal{A}$ be a d-sphere and let $f : \mathcal{A} \to \Sigma$ be a simplicial map. Suppose there exist a $(d+1)$-PSC $\mathcal{B}$ and a simplicial map $f' : \mathcal{B} \to \Sigma$ such that $\partial\mathcal{B} = \mathcal{A}$ and $f'|_{\mathcal{A}} = f$. Then there exist a $(d+1)$-PSC $\mathcal{B}_1$ and a simplicial map $f_1 : \mathcal{B}_1 \to \Sigma$ such that $\partial\mathcal{B}_1 = \mathcal{A}$ and $f_1|_{\mathcal{A}} = f$, and every $(d+1)$-simplex of $\mathcal{B}_1$ has a vertex that is not in $\partial\mathcal{B}_1$.*

Proof. Suppose $\mathcal{B}$ has a $(d+1)$-simplex B such that each vertex of B is in $\partial\mathcal{B}$. Let us call such a simplex *bad*. Let w be a new vertex not in $V(\mathcal{B})$, and let $\mathcal{B}'$ be the stellar subdivision of $\mathcal{B}$ with respect to (B, w). Then by Lemma 5 we know that $B \notin \mathcal{B}'$ and $\partial\mathcal{B}' = \partial\mathcal{B}$.

To extend the simplicial map f' to $\mathcal{B}'$, we may set $f'(w) = f'(v)$ for an arbitrary vertex v of B. For any $B' \cup \{w\} \in \mathcal{B}'$ we have $\{f'(b) : b \in B' \cup \{w\}\} \subseteq \{f'(b) : b \in B\}$, which is a simplex of Σ since f' was a simplicial map from $\mathcal{B}$ to Σ.

The number of bad simplices in $\mathcal{B}'$ is smaller than the number in $\mathcal{B}$, since we removed B and did not introduce any new bad simplices. By repeating this construction for each bad simplex, we obtain $\mathcal{B}_1$ with the required properties. □

We are ready for the main theorem of this section, stated here in its defect formulation (set $d = 0$ to get the basic formulation). The basic form of this result was first used implicitly in [5], and was stated in an explicit form in [3]. Our proof here follows the ideas of [5]. A proof using homology and the Nerve Theorem was given in [29] (see also [30]).

Theorem 10. *Suppose $d \geq 0$, and let Σ be a simplicial complex whose vertices are coloured from a set S. Suppose that for every $T \subseteq S$, the subcomplex Σ_T of simplices whose vertices are coloured from T is $(|T| - d - 2)$-connected. Then Σ has a multicoloured $(|S| - d - 1)$-simplex.*

Proof.

The case $d = 0$. Let $m = |S|$. The statement is trivially true if $m = 0$ so assume $m \geq 1$. Our aim is to construct a $(|T| - 1)$-PSC $\mathcal{A}_T$ for each $T \subseteq S$ with the following properties:

- there is a simplicial map $f : \mathcal{A}_S \to \Sigma$. This defines a colouring c_f of $V(\mathcal{A}_S)$ where $c_f(v)$ is the element of S that is the colour of $f(v)$, for each $v \in V(\mathcal{A}_S)$.
- $\{\mathcal{A}_T : T \subseteq S\}$ together with c_f satisfy the conditions of Lemma 7.

Then by Lemma 7, there is a simplex in $\mathcal{A}_S$ that is multicoloured by c_f with the set S. Thus by definition of c_f, the labels of this simplex form a simplex of Σ that is multicoloured by S.

We define the $\mathcal{A}_T$ and f for each T in increasing order of the size of T. We set $\mathcal{A}_\emptyset = \{\emptyset\}$. Let $S = \{v_1, \dots, v_m\}$, and let $\mathcal{A}^0_{\{v_i\}} = \{v_i\}$ for each i. The conditions guarantee that Σ_T is nonempty for each T with $|T| = 1$, so we may define f on each v_i by letting $f(v_i)$ be a vertex of $\Sigma_{\{v_i\}}$.

Fix a set $T_0 \subseteq S$ with $|T_0| \geq 2$. Suppose we have defined $\mathcal{A}_T$ for each $T \subset T_0$, $|T| < |T_0|$, and a labelling f of $\bigcup_{T \subset T_0} V(\mathcal{A}_T)$, such that the following hold:

(i) the restriction of f to each $V(\mathcal{A}_T)$ is a simplicial map from $\mathcal{A}_T$ to Σ_T,
(ii) $\{\mathcal{A}_T : T \subset T_0\}$ satisfies the conditions of Lemma 8 (applied with $S := T_0$).

Note that if $|T_0| = 2$ then (i) and (ii) hold.

By Lemma 8 we know by (ii) that $\mathcal{C} = \cup_{T \subset T_0, |T| = |T_0|-1} \mathcal{A}_T$ is a $(|T_0| - 2)$-sphere, and by (i) that f gives a simplicial map from $\mathcal{C}$ to Σ_{T_0}. Since Σ_{T_0} is $(|T_0| - 2)$-connected, there exist a $(|T_0| - 1)$-PSC $\mathcal{A}_{T_0}$ with $\partial \mathcal{A}_{T_0} = \mathcal{C}$, and a simplicial map from $\mathcal{A}_{T_0}$ to Σ_{T_0} that extends f (we will call the new simplicial map f as well). By Lemma 9, we may assume that every $(|T_0| - 1)$-simplex of $\mathcal{A}_{T_0}$ has a vertex that is not in $\mathcal{C}$. Hence we may choose these to be completely new vertices, so that each $(|T_0| - 1)$-simplex in $\mathcal{A}_{T_0}$ has a vertex that is not in any $\mathcal{A}_{T'}$ with $|T'| = |T_0|$.

This defines $\mathcal{A}_{T_0}$ and f on $\mathcal{A}_{T_0}$. Then with these definitions, for any set T_1 of size $|T_0| + 1$, (i) holds, and (ii) holds by our choice of completely new vertices for each $\mathcal{A}_{T_0}$. Therefore we may complete this construction to get a $(|S| - 1)$-PSC $\mathcal{A}_S$ such that the colouring c_f together with $\{\mathcal{A}_T : T \subseteq S\}$ satisfy the conditions of Lemma 7. This completes the proof as described above, for the case $d = 0$.

The case $d > 0$. We derive the result for $d > 0$ from the $d = 0$ case. Let Σ be a simplicial complex whose vertices are coloured from the set S. For convenience assume $S = [m]$. We define a new simplicial complex $\Sigma^{(1)}$ as follows. The vertex set is $V(\Sigma) \cup W^{(1)}$, where $W^{(1)} = \{w_1, \dots, w_m\}$ is a set of m new distinct vertices, and w_i is coloured i. Then $\Sigma^{(1)}$ is defined by

$$\Sigma^{(1)} = \{\sigma \cup \tau : \sigma \in \Sigma, \tau \subset W^{(1)}, |\tau| \leq 1\}.$$

We denote this construction by $\Sigma^{(1)} = \Sigma + W^{(1)}$.

We claim that for any set $T \subseteq S$, if Σ_T is $(|T| - d - 2)$-connected then $\Sigma_T^{(1)}$ is $(|T| - d - 1)$-connected. For $|T| \leq d - 1$ there is nothing to prove. If $|T| = d$ the claim states that $\Sigma_T^{(1)}$ is nonempty, which is true since it contains $\{w_t\}$ for each $t \in T$. Assume $|T_0| > d$. Fix r with $-1 \leq r \leq |T_0| - d - 1$, let $\mathcal{A}$ be an r-sphere and let $f : \mathcal{A} \to \Sigma_{T_0}^{(1)}$ be a simplicial map. Assume without loss of generality that $1 \in T_0$. Let x be a new vertex not in $V(\mathcal{A})$ and let $\mathcal{B}$ be the closure of $\{A \cup \{x\} : A \in \mathcal{A}\}$. Then by Lemma 3 we know that $\mathcal{B}$ is an $(r + 1)$-PSC with boundary $\mathcal{A}$.

We extend f by setting $f(x) = w_1$. If no vertex v of $\mathcal{A}$ has $f(v) \in W^{(1)} \setminus \{w_1\}$, then f is a simplicial map from $\mathcal{B}$ to $\Sigma_{T_0}^{(1)}$ as required. Otherwise let B be a maximal simplex of $\mathcal{B}$ whose label set consists of a subset of $W^{(1)}$ of size at least two. Let us call such a simplex *bad*. Then x is a vertex of B, and the label set of B is $\{w_1, w\}$ for some $w \in W^{(1)} \setminus \{w_1\}$.

Case 1. $|B| = r + 2$.

Let $\mathcal{B}'$ be the stellar subdivision of $\mathcal{B}$ with respect to (B, y), where y is a new vertex not in $V(\mathcal{B})$. Then $\partial\mathcal{B}' = \partial\mathcal{B} = \mathcal{A}$ by Lemma 5. Extend f to $\mathcal{B}'$ by setting $f(y)$ to be a vertex of Σ_{T_0}. This is possible since Σ_{T_0} is $(|T_0|-d-2)$-connected, and $|T_0| - d - 2 \geq -1$ so Σ_{T_0} is nonempty.

Case 2. $|B| \leq r + 1$.

By Lemma 4, the link $\mathcal{L} = \mathcal{L}_{\mathcal{B}}(B)$ is an $(r-b+1)$-sphere, where $b = |B|$. Since B is maximal, the labelling f gives a simplicial map from $\mathcal{L}$ to Σ_{T_0}. Thus since $b \geq 2$ and Σ_{T_0} is $(|T_0| - d - 2)$-connected, and $|T_0| - d - 2 \geq r - b + 1$, there exist a $(r - b + 2)$-PSC $\mathcal{C}$ with $\partial\mathcal{C} = \mathcal{L}$ and an extension of f that is a simplicial map from $\mathcal{C}$ to Σ_{T_0}. By Lemma 9 we may assume that every $(r-b+2)$-simplex of $\mathcal{C}$ has a vertex not in $\mathcal{L}$, and $V(\mathcal{B}) \cap V(\mathcal{C}) = V(\mathcal{L})$. Let us denote the extended labelling by f as well.

Let $\mathcal{B}'$ be the $(r + 1)$-PSC from Lemma 6, so $\mathcal{B}'$ is the closure of

$$\mathcal{B}^{r+1} \setminus \{D \cup B : D \in \mathcal{L}, |D| = r - b + 2\} \cup \{C \cup B' : C \in \mathcal{C},$$
$$|C| = r - b + 3, B' \subset B, |B'| = |B| - 1\}.$$

Then $\partial\mathcal{B}' = \partial\mathcal{B} = \mathcal{A}$ by Lemma 6, and by considering f to be a labelling of all of $\mathcal{B}'$ we find that $\mathcal{B}'$ is labelled by f from $\Sigma^{(1)}_{T_0}$.

In both Case 1 and Case 2, the simplex B is not a simplex of $\mathcal{B}'$. Moreover, by construction, every improperly labelled simplex in $\mathcal{B}' \setminus \mathcal{B}$ is improper only because it contains a simplex labelled with $\{w_1, w\}$ that is a proper subset of B. Thus the set of bad simplices of $\mathcal{B}'$ is properly contained in the set of bad simplices of $\mathcal{B}$. Thus we may repeat this construction for each simplex that is not properly labelled from $\Sigma^{(1)}_{T_0}$ until we obtain a $(|T_0| - d)$-PSC $\mathcal{B}^*$ with $\partial\mathcal{B}^* = \mathcal{A}$ and a labelling of $\mathcal{B}^*$ from $\Sigma^{(1)}_{T_0}$ that extends f, in which every simplex is properly labelled. This gives the required simplicial map from $\mathcal{B}^*$ to $\Sigma^{(1)}_{T_0}$, showing that $\Sigma^{(1)}_{T_0}$ is $(|T_0| - d - 1)$-connected. This completes the proof of the claim.

We may now finish the proof of the case $d > 0$. Given Σ as in the conditions of the theorem, we construct complexes $\Sigma^{(1)}, \ldots, \Sigma^{(d)}$ by setting $\Sigma^{(i)} = \Sigma^{(i-1)} + W^{(i)}$, where $W^{(1)}, \ldots, W^{(d)}$ are disjoint sets of new vertices. Then by the claim, for each $T \subseteq S$ the subcomplex $\Sigma^{(d)}_T$ is $(|T| - 2)$-connected. Thus by the case $d = 0$, there is a multicoloured $(|S| - 1)$-simplex C in $\Sigma^{(d)}$. By construction C contains at most d vertices of $W^{(1)} \cup \ldots \cup W^{(d)}$. Thus C contains a multicoloured $(|S| - d - 1)$-simplex as required. □

5 Connectedness via graph parameters

The simplicial complex that will be of most interest to us is the *independence complex* $\mathcal{I}(G)$ of a graph G, that is, the set of all independent subsets of vertices in G. Note that in a vertex-partitioned graph G, an independent transversal is just a multicoloured simplex of $\mathcal{I}(G)$. Theorem 10 provides the following condition that guarantees the existence of an independent transversal in a vertex-partitioned graph. Again we state the defect version, about *partial* independent transversals that intersect all but d of the vertex classes of the partition.

Theorem 11. *Let $d \geq 0$ be given. Let G be a graph and let $\mathcal{P} = \{V_1, \ldots, V_m\}$ be a partition of $V(G)$. Suppose that for each subset T of $[m]$, the subgraph $G_T = G[\bigcup_{i\in T} V_i]$ of G induced by $\bigcup_{i\in T} V_i$ is such that $\mathcal{I}(G_T)$ is $(|T| - d - 2)$-connected. Then G has a (partial) independent transversal with respect to $\mathcal{P}$ of size $m - d$.*

In order for this theorem to be useful however, we need to relate the connectedness properties of the independence complex of a graph to natural parameters of the graph itself. For example, one simple consequence of Lemma 3 is that if a graph G has an isolated vertex then its independence complex is t-connected for every t. The main tool we use in this section to explore relationships with other more useful parameters will be the following theorem from [29] (see also [4]). Here we give a reformulated statement of the theorem that is more convenient for our purposes. We denote the neighbourhood of a vertex x by $\Gamma(x)$.

Theorem 12. *Let $t \geq 0$ and a graph G be given, and let xy be an edge of G. Suppose $\mathcal{I}(G \setminus xy)$ is t-connected and $\mathcal{I}(G - (\Gamma(x) \cup \Gamma(y)))$ is $(t-1)$-connected. Then $\mathcal{I}(G)$ is t-connected.*

Proof. Let r be fixed, where $-1 \leq r \leq t$. Let $\mathcal{A}$ be an r-sphere and let $f : \mathcal{A} \to \mathcal{I}(G)$ be a simplicial map. Since $\mathcal{I}(G \setminus xy)$ is t-connected, there exist an $(r+1)$-PSC $\mathcal{B}$ and a simplicial map $f' : \mathcal{B} \to \mathcal{I}(G \setminus xy)$ with $\partial\mathcal{B} = \mathcal{A}$ and $f'|\mathcal{A} = f$.

If every simplex of $\mathcal{B}$ is properly labelled from $\mathcal{I}(G)$ by f', then we are finished, so suppose the set $\mathcal{B}_0$ of simplices of $\mathcal{B}$ that have label set $\{x,y\}$ is nonempty. Let B be a maximal simplex in $\mathcal{B}_0$. Note that B is not a simplex of $\mathcal{A}$, since every simplex of $\mathcal{A}$ is properly labelled from $\mathcal{I}(G)$ by f'.

Case 1. $|B| = r + 2$.
Let $\mathcal{B}'$ be the stellar subdivision of $\mathcal{B}$ with respect to (B, w), where w is a new vertex not in $V(\mathcal{B})$. Then $\partial\mathcal{B}' = \partial\mathcal{B} = \mathcal{A}$ by Lemma 6. Extend f' to $\mathcal{B}'$ by setting $f(w)$ to be an element of $V(G - (\Gamma(x) \cup \Gamma(y)))$. This is possible since $\mathcal{I}(G - (\Gamma(x) \cup \Gamma(y)))$ is $(t-1)$-connected, and $t - 1 \geq -1$ so $G - (\Gamma(x) \cup \Gamma(y))$ is nonempty.

Case 2. $|B| \leq r + 1$.
By Lemma 4 the link $\mathcal{L} = \mathcal{L}_{\mathcal{B}}(B)$ is an $(r - b + 1)$-sphere, where $b = |B|$. Since B is maximal, no vertex of $\mathcal{L}$ is labelled from $\{x,y\}$. Moreover since f' is a simplicial map from $\mathcal{B}$ to $\mathcal{I}(G \setminus xy)$, no other element of $\Gamma(x) \cup \Gamma(y)$ appears as a label on $\mathcal{L}$. Thus $\mathcal{L}$ is an $(r - b + 1)$-sphere and f' gives a simplicial map from $\mathcal{L}$ to $\mathcal{I}(G - (\Gamma(x) \cup \Gamma(y)))$. Since $b \geq 2$ we know $r - b + 1 \leq t - 1$, so since $\mathcal{I}(G - (\Gamma(x) \cup \Gamma(y)))$ is $(t-1)$-connected, there exist an $(r - b + 2)$-PSC $\mathcal{C}$ with $\partial\mathcal{C} = \mathcal{L}$ (and $V(\mathcal{C}) \cap B = \emptyset$) and a simplicial map $f'' : \mathcal{C} \to \mathcal{I}(G - (\Gamma(x) \cup \Gamma(y)))$ such that $f''|_{\mathcal{L}} = f'$. By Lemma 9 we may assume that every $(r - b + 2)$-simplex of $\mathcal{C}$ has a vertex not in $\mathcal{L}$, and $V(\mathcal{B}) \cap V(\mathcal{C}) = V(\mathcal{L})$.

Let $\mathcal{B}'$ be the $(r+1)$-PSC from Lemma 6, so $\mathcal{B}'$ is the closure of

$$\mathcal{B}^{r+1} \setminus \{D \cup B : D \in \mathcal{L}, |D| = r - b + 2\} \cup \{C \cup B' : C \in \mathcal{C},$$
$$|C| = r - b + 3, B' \subset B, |B'| = |B| - 1\}.$$

Then $\partial\mathcal{B}' = \partial\mathcal{B} = \mathcal{A}$ by Lemma 6, and by considering f'' to be a labelling of all of $\mathcal{B}'$ by setting $f''(v) = f'(v)$ for all $v \in V(\mathcal{B}') \setminus V(\mathcal{C})$ we find that $\mathcal{B}'$ is labelled by f'' from $\mathcal{I}(G \setminus xy)$.

In both Case 1 and Case 2, the simplex B is not a simplex of $\mathcal{B}'$. Moreover, by construction, every improperly labelled simplex in $\mathcal{B}' \setminus \mathcal{B}$ is improper only because it contains a simplex labelled with $\{x, y\}$ that is a proper subset of B. Thus the set of simplices of $\mathcal{B}'$ that have label set $\{x, y\}$ is properly contained in $\mathcal{B}_0$, and we may repeat this construction for each simplex that is not properly labelled from $\mathcal{I}(G)$ until we obtain an $(r+1)$-PSC $\mathcal{B}^*$ with $\partial\mathcal{B}^* = \mathcal{A}$ and a simplicial map $f^* : \mathcal{B}^* \to \mathcal{I}(G)$, where $f^*|_{\mathcal{A}} = f$. Therefore $\mathcal{I}(G)$ is t-connected. □

We may now identify some graph parameters that relate to connectedness of the independence complex. A set U of vertices of a graph G is said to *(strongly) dominate* G if every vertex of G is adjacent to a vertex in U. (The more usual definition of domination just requires this property for each vertex of $V(G) \setminus U$, but here we mean the stronger notion.) Then as shown in [29], Theorem 12 implies the following.

Theorem 13. *Suppose that the graph G cannot be dominated by the vertices of any set F of edges of G with $|F| \leq t + 1$. Then $\mathcal{I}(G)$ is t-connected.*

Proof. Note that if $V(G)$ is empty then the assumption of the theorem is not satisfied for any $t \geq -1$, so we may assume $V(G) \neq \emptyset$. We prove the contrapositive statement by induction on $|E(G)|$. If $|E(G)| = 0$, then $\mathcal{I}(G)$ is t-connected for every t. Assume $|E(G)| \geq 1$ and that the theorem is true for graphs with fewer edges. Let xy be an edge of G. Suppose that for some $t \geq 0$, the complex $\mathcal{I}(G)$ is not t-connected. Then by Theorem 12 either (i) $\mathcal{I}(G \setminus xy)$ is not t-connected or (ii) $\mathcal{I}(G - (\Gamma(x) \cup \Gamma(y)))$ is not $(t-1)$-connected.

If (i) holds, then by induction $G \setminus xy$ can be dominated by the vertices of a set F of edges of $G \setminus xy$ with $|F| \leq t + 1$. Therefore G is dominated by the same set of vertices. If (ii) holds, then by induction $G - (\Gamma(x) \cup \Gamma(y))$ can be dominated by the vertices of a set F of edges of $G - (\Gamma(x) \cup \Gamma(y))$ with $|F| \leq t$. But then G is dominated by the vertices of $F \cup \{xy\}$. □

This gives the following immediate corollary.

Corollary 14. *Suppose G cannot be dominated by any set of $2t + 2$ vertices. Then $\mathcal{I}(G)$ is t-connected.*

From here it is a short step to prove Theorem 1, which was stated in the introduction. In fact it is just as easy to prove the following slightly stronger version.

Corollary 15. *Let G be a graph with maximum degree Δ, let $\mathcal{P} = \{V_1, \dots, V_m\}$ be a partition of $V(G)$, and let x be a vertex of G. If $|V_i| \geq 2\Delta$ for each i, then G has an independent transversal with respect to $\mathcal{P}$ that contains x.*

***Proof*.** Assume without loss of generality that $x \in V_m$. Let $G' = G-(V_m \cup \Gamma(x))$. For each $T \subseteq [m-1]$, the subgraph G'_T has at least $2|T|\Delta - |\Gamma(x)| \geq (2|T|-1)\Delta$ vertices and therefore it cannot be dominated by any set of $2(|T|-2)+2$ vertices. Therefore by Corollary 14 we find that $\mathcal{I}(G'_T)$ is $(|T|-2)$-connected. Then Theorem 11 implies that G' has an independent transversal with respect to $\{V_1, \dots, V_{m-1}\}$, which together with x forms an independent transversal of G. □

More generally, the same proof shows that if $|\bigcup_{i\in T} V_i| > (2|T|-2)\Delta$ for each $T \subseteq [m]$, then G has an independent transversal.

The bound given in Theorem 1 and Corollary 15 is best possible, since there exist graphs for which $|V_i| = 2\Delta - 1$ for each i that do not have independent transversals [33]. A more precise version of Theorem 1 (taking into account the number of vertex classes m) was proved in [22].

A second parameter that relates to connectedness of the independence complex involves domination of independent sets. For a graph G we define $i(G)$ to be the largest ℓ for which there exists an independent set W in G that is not dominated by any set U of vertices in G with $|U| \leq \ell$. The following forms part of the main theorem in [5]. Again the proof given here is as in [29].

Theorem 16. *Let G be a graph and suppose $i(G) \geq t+1$. Then $\mathcal{I}(G)$ is t-connected.*

***Proof*.** Note that if $V(G)$ is empty then the assumption of the theorem is not satisfied for any $t \geq -1$, so we may assume $V(G) \neq \emptyset$. We prove the statement by induction on $|E(G)|$. If $|E(G)| = 0$, then $\mathcal{I}(G)$ is t-connected for every t. Assume $|E(G)| \geq 1$ and that the theorem is true for graphs with fewer edges. We may also assume that G has no isolated vertices. Suppose that for some $t \geq 0$ we have $i(G) \geq t+1$. Let W be an independent set in G that is not dominated by any set of size at most $t+1$. Then $W \neq \emptyset$. Since no vertex of W is isolated, we may choose an edge vw with $w \in W$.

Consider the graph $G \setminus vw$. The set W is an independent set in this graph, and it cannot be dominated in $G \setminus vw$ by any set U with $|U| \leq t+1$. Therefore $i(G \setminus vw) \geq t+1$, and so by induction $\mathcal{I}(G \setminus vw)$ is t-connected.

Consider the graph $G - (\Gamma(v) \cup \Gamma(w))$. The set $W \setminus \Gamma(v)$ is an independent set in this graph. Suppose $W \setminus \Gamma(v)$ is dominated in $G - (\Gamma(v) \cup \Gamma(w))$ by U' where $|U'| \leq t$. Then W is dominated in G by $U' \cup \{v\}$, contradicting our assumption on W. Therefore $W \setminus \Gamma(v)$ witnesses the fact that $i(G - (\Gamma(v) \cup \Gamma(w))) \geq t$. Hence by induction $\mathcal{I}(G - (\Gamma(v) \cup \Gamma(w)))$ is $(t-1)$-connected. Therefore by Theorem 12 we conclude that $\mathcal{I}(G)$ is t-connected, as required. □

In fact in [5] the result is somewhat stronger, in that the sets U can also be assumed to be independent. A minor modification of the proof above yields this stronger version.

6 Applications

Theorem 11 gives a sufficient condition for the existence of an independent transversal in G, in terms of the connectedness of the independence complexes of various subgraphs of G. In fact this notion of connectedness should not be that effective in general at capturing the property of having an independent transversal, since it is not a *monotone* parameter. By this we mean that it is possible to add to a graph H some new vertices, and some new edges incident to the new vertices, and make the connectedness of $\mathcal{I}(H)$ go down. Thus it can happen that $(G, \mathcal{P})$ satisfies the assumptions of Theorem 11 and therefore has an independent transversal S, but we can add some new vertices and edges to G, so that S is still an independent transversal, but the resulting graph no longer satisfies the assumptions of Theorem 11. In spite of this shortcoming however, there are several problems that can be solved using Theorem 11, and for some of them (for example, the first two below) we do not know a solution by any other means. (We remark that Theorem 1 is not an example of such a problem, as it also has a purely combinatorial proof ([16], see also, e.g., [11, 21, 24]) that does not use the notion of connectedness.) In this section we give several more examples of applications of the results we have seen. Further generalisations and applications to other settings such as matroids have been investigated, see, for example, [3].

6.1 Hall's Theorem for hypergraphs

We begin with a simple application of Theorem 16 to line graphs, which forms the core of the main result in [5]. For a (multi)hypergraph $\mathcal{G}$ (i.e. multiple edges may be present), we define the *line graph* $L(\mathcal{G})$ to be the graph whose vertex set is the edge set of $\mathcal{G}$, in which two edges e and f of $\mathcal{G}$ are joined by an edge of $L(\mathcal{G})$ if and only if $e \cap f \neq \emptyset$. A *matching* in $\mathcal{G}$ is a set of pairwise disjoint edges. The largest size of a matching in $\mathcal{G}$ is denoted by $\nu(\mathcal{G})$.

Lemma 17. *Let $\mathcal{G}$ be an r-uniform (multi)hypergraph. Then the simplicial complex $\mathcal{I}(L(\mathcal{G}))$ is $(\lfloor(\nu(\mathcal{G}) - 1)/r\rfloor - 1)$-connected.*

Proof. Let $\mathcal{M}$ be a matching in $\mathcal{G}$ of size $\nu(\mathcal{G})$. Then $\mathcal{M}$ is an independent set in $L(\mathcal{G})$. Let $\ell = \lfloor(|\mathcal{M}| - 1)/r\rfloor$ and let U be an arbitrary set of vertices in $L(\mathcal{G})$ (i.e. edges in $\mathcal{G}$) of size at most ℓ. Then the total number of vertices of $\mathcal{G}$ contained in the edges of U is at most $r|U| \leq |\mathcal{M}| - 1$. Therefore there exists an edge of $\mathcal{M}$ that is disjoint from every edge of U, and hence U does not dominate $\mathcal{M}$. This shows that $i(L(\mathcal{G})) \geq \ell$, which implies that $\mathcal{I}(L(\mathcal{G}))$ is $(\ell - 1)$-connected by Theorem 16. □

Hall's classical theorem for graphs states that in a bipartite graph H with vertex classes A and X, if every $T \subseteq A$ has the property that the neighbourhood of T in X is big enough (has size at least $|T|$), then H has a matching of size $|A|$. (Of course Hall's Theorem also tells us this condition is necessary.) Here we generalise these notions to hypergraphs as follows. Let $\mathcal{H}$ be a *bipartite* r-uniform hypergraph, which means that the vertex set of $\mathcal{H}$ can be partitioned into parts A and X, and every edge of $\mathcal{H}$ has one vertex in A and $r-1$ vertices in X. We are interested in a condition that guarantees the existence of a matching in $\mathcal{H}$ of size $|A|$. For $T \subseteq A$, the *neighbourhood* $\Gamma(T)$ of T is the $(r-1)$-uniform (multi)hypergraph with vertex set X, that contains one edge $e \setminus \{a\}$ for each edge e of $\mathcal{H}$ that contains a vertex a of T. (Note that multiple edges are indeed possible.) Then Hall's Theorem for hypergraphs [5] is as follows.

Theorem 18. *Let $\mathcal{H}$ be a bipartite r-uniform hypergraph with vertex classes A and X. Suppose that $\nu(\Gamma(T)) \geq (r-1)(|T|-1)+1$ for every $T \subseteq A$. Then $\mathcal{H}$ has a matching of size $|A|$.*

Proof. For the (multi)hypergraph $\mathcal{G} = \Gamma(A)$, we consider the line graph $L(\mathcal{G})$. Then A defines a vertex partition of $L(\mathcal{G})$, in which V_a consists of those edges f of $\mathcal{G}$ for which $f \cup \{a\} \in \mathcal{H}$. With these definitions, the conclusion of Theorem 18 states that $L(\mathcal{G})$ has an independent transversal with respect to this vertex partition. The assumption together with Lemma 17 implies that for each $T \subseteq A$, the subgraph $L(\mathcal{G})_T$ is such that $\mathcal{I}(L(\mathcal{G})_T)$ is $(|T|-2)$-connected. Therefore by Theorem 11 we find that $L(\mathcal{G})$ has an independent transversal, as required. □

In fact Theorem 18 is best possible, since there exist bipartite r-uniform hypergraphs $\mathcal{H}$ as above for which each $\Gamma(T)$ has a matching of size at least $(r-1)(|T|-1)$, but $\mathcal{H}$ has no matching of size $|A|$ (see [1]).

Unlike Hall's Theorem, the condition in Theorem 18 is not necessary. The above methods imply the following if-and-only-if version of the theorem (see [5]).

Theorem 19. *Let $\mathcal{H}$ be a bipartite r-uniform hypergraph with vertex classes A and X. Then $\mathcal{H}$ has a matching of size $|A|$ if and only if for every $T \subseteq A$ there exists a matching $\mathcal{M}_T$ in $\Gamma(T)$, such that $\mathcal{M}_T$ is not dominated by any set of $|T|-1$ edges of $\bigcup_{S \subset T} \mathcal{M}_S$.*

6.2 Ryser's Conjecture for 3-partite hypergraphs

A *cover* of a hypergraph $\mathcal{H}$ is a set of vertices that intersects every edge of $\mathcal{H}$. The size of a smallest cover of $\mathcal{H}$ is denoted by $\tau(\mathcal{H})$. Since the vertex set of any maximal matching in $\mathcal{H}$ is a cover, we see that if $\mathcal{H}$ is r-uniform then $\tau(\mathcal{H}) \leq r\nu(\mathcal{H})$. A famous old conjecture of Ryser [31] from 1967 states that if $\mathcal{H}$ is r-partite then this bound can be improved. The r-uniform hypergraph $\mathcal{H}$ is said to be *r-partite* if its vertices can be partitioned into r parts such that every edge has exactly one vertex in each part.

Conjecture 20 (Ryser's Conjecture). *If $\mathcal{H}$ is an r-partite r-uniform hypergraph, then*

$$\tau(\mathcal{H}) \leq (r-1)\nu(\mathcal{H}).$$

In spite of a substantial amount of work on this problem (see, e.g., [23]) it is wide open in general. For $r = 2$ it is König's Theorem. The only other known case is $r = 3$, which after a series of partial results [17, 34, 35] was proved in [2] as a consequence of Theorem 18. Here we give the proof from [2], in a slightly different formulation.

We begin with the following simple observation. Note that a 3-partite 3-uniform hypergraph with parts A, B and C can be viewed as a bipartite hypergraph (in the sense of the previous subsection) in which $X = B \cup C$.

Lemma 21. *Let $\mathcal{H}$ be a 3-partite 3-uniform hypergraph with parts A, B and C. Then for each $S \subseteq A$, the bipartite graph $\Gamma(S)$ satisfies $\nu(\Gamma(S)) \geq |S| - |A| + \tau(\mathcal{H})$.*

Proof. For each S, a cover of $\mathcal{H}$ is given by $(A \setminus S) \cup T$ where T is a minimum cover of the bipartite graph $\Gamma(S)$. Since $|T| = \nu(\Gamma(S))$ by König's Theorem, we find

$$\tau(\mathcal{H}) \leq |A \setminus S| + |T| = |A| - |S| + \nu(\Gamma(S)).$$

□

Now Ryser's Conjecture for $r = 3$ follows easily from Lemma 17 and Theorem 11.

Theorem 22. *Let $\mathcal{H}$ be a 3-partite 3-uniform hypergraph. Then $\tau(\mathcal{H}) \leq 2\nu(\mathcal{H})$.*

Proof. For each $S \subseteq A$ we know by Lemma 17 that $\mathcal{I}(L(\Gamma(S)))$ is $(\lfloor(\nu(\Gamma(S)) - 1)/2\rfloor - 1)$-connected. Lemma 21 tells us that $\nu(\Gamma(S)) \geq |S| - |A| + \tau(\mathcal{H})$, from which we conclude that $\mathcal{I}(L(\Gamma(S)))$ is $(|S| - (|S| + |A| - \tau(\mathcal{H}))/2 - 2)$-connected. Thus for every S we have that $\mathcal{I}(L(\Gamma(S)))$ is $(|S| - d - 2)$-connected where $d = A - \tau(\mathcal{H})/2$. Therefore Theorem 11 applied to $L(\Gamma(A))$ with vertex partition given by A (as in the proof of Theorem 18) implies that $\mathcal{H}$ has a hypergraph matching of size $|A| - d \geq \tau(\mathcal{H})/2$, as required. □

6.3 The Cycle-plus-triangles problem

Our next application is related to the famous "Cycle-plus-triangles" problem, popularised by Erdős in the 1980s, and finally solved by Fleischner and Stiebitz [15] in 1992 and with a different proof by Sachs [32] in 1993. It asks whether every union of a cycle C_{3k} of length divisible by three, together with a set of k disjoint triangles on the same vertex set, is 3-colourable. This question of Erdős was motivated by an earlier question of Du, Hsu and Hwang [13], who asked whether every C_{3k} with an arbitrary vertex partition into classes of size three has an independent transversal. This problem was also unsolved until the proofs of Fleischner and Stiebitz, and Sachs, of the stronger statement. Both of these proofs are ingenious but quite difficult. Here we can give a solution to the question of Du, Hsu and Hwang that is almost immediate from Theorems 11 and 16, and in fact is more general (see, e.g., [4, 21]).

Theorem 23. *Let G be a graph of maximum degree two, in which each cycle is of length divisible by three. Then G has an independent transversal with respect to any vertex partition into classes of size at least three.*

Proof. Fix a vertex partition, and let T be any subset of classes. Then G_T has at least $3|T|$ vertices, and its components are paths, and cycles of length divisible by three. Let W_T be a maximum independent set in G_T that is spaced three apart, i.e. for each path component take an endvertex and then every third vertex starting from that end. For each cycle component C_{3k} take an independent set of size k by taking every third vertex. We claim that W_T witnesses the fact that $i(G_T) \geq |T| - 1$ for each T, and hence by Theorem 16 it follows that $\mathcal{I}(G_T)$ is $(|T| - 2)$-connected. To see this, let U be an arbitrary set in G_T of size at most $|T| - 1$. Since W_T is spaced three apart, each vertex of U is adjacent to at most one vertex of W_T. Thus since $|W_T| \geq |T|$, there is a vertex $w \in W_T$ that is not a neighbour of any vertex of U. Therefore G_T is not dominated by U, which shows $i(G_T) \geq |T| - 1$ as required. Therefore by Theorem 11 we find that G has an independent transversal. □

6.4 *Strong chromatic number*

In fact the cycle-plus-triangles problem is a question about a general colouring parameter called the *strong chromatic number*. Let r be a positive integer and let G be a graph with n vertices where r divides n. We call G *strongly r-colourable* if for every partition of the vertex set into parts V_k of size r, there exists a proper vertex colouring of G with r colours, with the additional property that each V_k receives all r colours. In other words, any union of G together with a collection of disjoint r-cliques on the same vertex set has chromatic number r. (Thus the cycle-plus-triangles problem asks whether the strong chromatic number of C_{3k} is three). If r does not divide n, then we say G is strongly r-colourable if the graph obtained by adding $r\lceil n/r\rceil - n$ isolated vertices to G is strongly r-colourable. The *strong chromatic number $s\chi(G)$* of a graph G is defined to be the minimum r such that G is strongly r-colourable. The notion of strong chromatic number was introduced in [7] and [14], and has been studied by various authors in a number of different contexts. See, e.g., Problem 4.14 in [25] and [20] for more on the history of this parameter.

A significant open problem concerning the strong chromatic number has been to give a best possible upper bound in terms of the maximum degree Δ of G. The first linear bound for $s\chi(G)$ in terms of $\Delta(G)$ was given in [8], and an explicit estimate was given later in [19], where it was proved that $s\chi(G) \leq 3\Delta(G) - 1$ for every graph G. An asymptotic upper bound of $(11/4 + o(1))\Delta(G)$ was established in [20]. On the other hand, the constructions given in [33] show that the constant factor cannot be less than two, suggesting that $s\chi(G) \leq 2\Delta(G)$ may hold for every graph G. The fractional version of this statement was proved in [4].

Here we give a short proof from [4] of an upper bound that is only slightly weaker than the best known bound for general graphs.

Theorem 24. *For every graph G we have* $s\chi(G) \leq 3\Delta(G)$.

***Proof*.** By adding isolated vertices as necessary we may assume that $|V(G)|$ is divisible by 3Δ. Let $\mathcal{P} = \{V_1, \ldots, V_m\}$ be a partition of $V(G)$ into parts of size 3Δ, and let ϕ be a partial proper colouring of G with 3Δ colours, under which no two vertices in the same part V_i get the same colour, and such that the number of vertices coloured is as large as possible. Suppose that some vertex x is not coloured by ϕ. Then some colour, say red, does not appear on any vertex of the part containing x, say V_m. By possibly renumbering the parts we may assume that there exists $j \geq 1$ such that for $j \leq i \leq m-1$, each V_i contains a vertex y_i that is coloured red by ϕ, and $V_1, \ldots, V_{j-1}$ do not. For $j \leq i \leq m-1$ let

$$V_i' = V_i \setminus \{z : \phi(z) = \phi(w) \text{ for some } w \in \Gamma(y_i)\}.$$

Similarly let $V_m' = V_m \setminus \{z : \phi(z) = \phi(w) \text{ for some } w \in \Gamma(x)\}$. Note that $y_i \in V_i$ for each such i, and $x \in V_m$. Then $|V_i'| \geq |V_i| - \Delta = 2\Delta$ for each $j \leq i \leq m$, and hence by Corollary 15 there exists an independent transversal $U = \{u_j, \ldots, u_m\}$ of $G[\bigcup_{i=j}^{m}]V_i'$ such that $x = u_m$.

We modify ϕ as follows: for each $j \leq i \leq m$ our new colouring ϕ' colours u_i red. For $j \leq i \leq m-1$ we set $\phi'(y_i) = \phi(u_i)$ (if u_i is not coloured by ϕ, then y_i is not coloured by ϕ'). Note that for some i it may happen that $u_i = y_i$. For all remaining vertices $\phi'(z) = \phi(z)$. Then no two vertices in the same part get the same colour under ϕ', and ϕ' colours one more vertex than ϕ does, since $x = u_m$ is coloured by ϕ'. To see that ϕ' is a proper colouring, note that no two adjacent vertices are coloured red because U is an independent set, and no two adjacent vertices get the same (non-red) colour by definition of V_i' and the fact that $\{y_j, \ldots, y_{m-1}\}$ is independent. Thus ϕ' is a better partial colouring than ϕ, contradicting our choice of ϕ. Hence no vertex x can be left uncoloured by ϕ, showing that $s\chi(G) \leq 3\Delta(G)$ as required. □

Acknowledgements This work was partially supported by NSERC. The author is also grateful for the support of the Mittag-Leffler Institute in Djursholm, Sweden, where part of this work was done.

References

1. R. Aharoni, Matchings in n-partite n-graphs. Graphs Comb. **1**, 303–304 (1985)
2. R. Aharoni, Ryser's Conjecture for tripartite 3-graphs. Combinatorica **21**, 1–4 (2001)
3. R. Aharoni, E. Berger, The intersection of a matroid and a simplicial complex. Trans. Am. Math. Soc. **358**, 4895–4917 (2006)
4. R. Aharoni, E. Berger, R. Ziv, Independent systems of representatives in weighted graphs. Combinatorica **27**, 253–267 (2007)
5. R. Aharoni, P. Haxell, Hall's Theorem for hypergraphs. J. Graph Theory **35**, 83–88 (2000)
6. R. Aharoni, M. Chudnovsky, A. Kotlov, Triangulated spheres and colored cliques. Discret. Comput. Geom. **28**, 223–229 (2002)

7. N. Alon, The linear arboricity of graphs. Isr. J. Math. **62**, 311–325 (1988)
8. N. Alon, The strong chromatic number of a graph. Random Struct. Algorithm. **3**, 1–7 (1992)
9. N. Alon, V. Asodi, Edge colouring with delays. Comb. Probab. Comput. **16**, 173–191 (2007)
10. N. Alon, G. Ding, B. Oporowski, D. Vertigan, Partitioning into graphs with only small components. J. Comb. Theory Ser. B **87**, 231–243 (2003)
11. R. Berke, P. Haxell, T. Szabó, Bounded transversals in multipartite graphs. J. Graph Theory **70**, 318–331 (2012)
12. J. Britnell, A. Evseev, R. Guralnick, P. Holmes, A. Maróti, Sets of elements that pairwise generate a linear group. J. Comb. Theory Ser. A **115**, 442–465 (2008)
13. D.Z. Du, D.F. Hsu, F.K. Hwang, The Hamiltonian property of consecutive-d digraphs, in *Graph-Theoretic Models in Computer Science*, vol. II (Las Cruces, NM, 1988–1990). Math. Comput. Model. **17**, 61–63 (1993)
14. M. Fellows, Transversals of vertex partitions in graphs. SIAM J. Discret. Math. **3**, 206–215 (1990)
15. H. Fleischner, M. Stiebitz, A solution to a colouring problem of P. Erdős. Discret. Math. **101**, 39–48 (1992)
16. P. Haxell, A condition for matchability in hypergraphs. Graphs Comb. **11**, 245–248 (1995)
17. P. Haxell, A note on a conjecture of Ryser. Period. Math. Hung. **30**, 73–79 (1995)
18. P. Haxell, A note on vertex list colouring. Comb. Probab. Comput. **10**, 345–348 (2001)
19. P. Haxell, On the strong chromatic number. Comb. Probab. Comput. **13**, 857–865 (2004)
20. P. Haxell, A new bound on the strong chromatic number. J. Graph Theory **58**, 148–158 (2008)
21. P. Haxell, On forming committees. Am. Math. Mon. **118**, 777–788 (2011)
22. P. Haxell, T. Szabó, Odd independent transversals are odd. Comb. Probab. Comput. **15**, 193–211 (2006)
23. P. Haxell, L. Narins, T. Szabó, Extremal hypergraphs for Ryser's Conjecture I. Connectedness of line graphs of bipartite graphs (submitted)
24. P. Haxell, T. Szabó, G. Tardos, Bounded size components—partitions and transversals. J. Comb. Theory Ser. B **88**, 281–297 (2003)
25. T.R. Jensen, B. Toft, *Graph Coloring Problems* (Wiley, New York, 1995)
26. T. Kaiser, D. Král, R. Skrekovski, X. Zhu, The circular chromatic number of graphs of high girth. J. Comb. Theory Ser. B **97**, 1–13 (2007)
27. A. King, Hitting all maximum cliques with a stable set using lopsided independent transversals. J. Graph Theory **67**, 300–305 (2011)
28. M. Krivelevich, Almost perfect matchings in random uniform hypergraphs. Discret. Math. **170**, 259–263 (1997)
29. R. Meshulam, The clique complex and hypergraph matchings. Combinatorica **21**, 89–94 (2001)
30. R. Meshulam, Domination numbers and homology. J. Comb. Theory Ser. A **102**, 321–330 (2003)
31. H.J. Ryser, Neuere Probleme der Kombinatorik, in *Vorträge über Kombinatorik*, Oberwolfach, MFO, Germany, 24–29 July 1967, pp. 69–91
32. H. Sachs, Elementary proof of the cycle-plus-triangles theorem, in *Combinatorics, Paul Erdős is Eighty*, vol. 1, ed. by D. Miklós, V. T. Sós, T. Szönyi (Bolyai Society Mathematical Studies, Budapest, 1993), pp. 347–359
33. T. Szabó, G. Tardos, Extremal problems for transversals in graphs with bounded degree. Combinatorica **26**, 333–351 (2006)
34. E. Szemerédi, Z. Tuza, Upper bound for transversals of tripartite hypergraphs. Period. Math. Hung. **13**, 321–323 (1982)
35. Z. Tuza, On the order of vertex sets meeting all edges of a 3-partite hypergraph. Ars Comb. **24**(A), 59–63 (1987)

Giant components in random graphs

Mihyun Kang

Abstract The phase transition in random graphs was first discussed by Erdős and Rényi who showed that a random graph undergoes a drastic change in the order and structure of the largest component. In view of recent results, one can recognise two main trends that have increased our understanding of random graphs. The first trend is the study of various random graph models (e.g. random hypergraphs, random graph processes, random graphs with degree constraints, random graphs on surfaces). The second trend is finding new simple proofs (and thereby improvement) of classical results on random graphs. This article discusses developments in random graphs in the light of these trends, with focus on giant components, limit theorems, and proof techniques.

Keywords Giant component • Phase transition • Limit theorems • Random graphs • Random hypergraphs

Mathematics Subject Classification (2010): 05C80, 05C65

1 Emergence of giant component

One of the most important results of Erdős and Rényi [18] deals with a *sudden change* in the order and structure of the largest component in a random graph, which is caused by altering the average degree near the critical value of one. Roughly speaking, the "giant component" emerges when the average vertex degree passes through one. Before stating this result more precisely, we introduce some notation.

Given a real number $p \in [0, 1]$ we let $G(n, p)$ be a graph with vertex set $[n] := \{1, \ldots, n\}$ in which each pair of vertices is joined by an edge independently with probability p. Given an integer $m \in \{0, \ldots, \binom{n}{2}\}$ we let $G(n, m)$ denote a graph chosen uniformly at random from the set of all graphs with vertex set $[n]$ and m

M. Kang (✉)
Institute of Discrete Mathematics, Graz University of Technology,
Steyrergasse 30, 8010 Graz, Austria
e-mail: kang@math.tugraz.at

A. Beveridge et al. (eds.), *Recent Trends in Combinatorics*, The IMA Volumes in Mathematics and its Applications 159, DOI 10.1007/978-3-319-24298-9_10

edges. These two models are essentially equivalent when m is about the expected number of edges in $G(n,p)$, which is $p\binom{n}{2}$.

Given a graph G with vertex set $[n]$ we say that a vertex $v \in [n]$ is *reachable* from $w \in [n]$ if there exists a sequence of distinct edges $e_1, \dots, e_k$ such that $v \in e_1$, $w \in e_k$, and $e_i \cap e_{i+1} \neq \emptyset$ for all $1 \leq i < k$. Reachability is an equivalence relation, and the equivalence classes are called the *components* of G. Given a component C of G the number of vertices (resp. edges) of C is called the *order* (resp. *size*) of C. We let $L(G)$ denote the maximum order of a component of G. Since the subsets of $V(G)$ can be ordered lexicographically, we call the lexicographically first component of G that has order $L(G)$ the *largest component* of G. When a component of order $L(G)$ is *unique* in the sense that the second largest component is of order $o(L(G))$, then the largest component is also called the *giant component*. We let $M(G)$ denote the size of the largest component of G.

We are often concerned with properties that hold *with high probability* (in short *whp*), meaning with probability tending to one as the number n of vertices tends to ∞.

The result of Erdős and Rényi and their improvements can now be stated [10, 19].

Theorem 1. *Let $\varepsilon \in \mathbb{R}$ be a fixed constant and let $p = p(n) \in [0, 1]$ be a sequence such that $\varepsilon = (n-1)p - 1$ for all integers $n \in \mathbb{N}$. Then the following hold whp.*

(i) If $\varepsilon < 0$, then $L(G(n,p)) = O(\log n)$, and the largest component is a tree and all other components are either trees or unicyclic components.
(ii) If $\varepsilon > 0$, then $L(G(n,p)) = (1 + o(1))\,\rho n$, where $\rho = \rho(\varepsilon) \in (0, 1)$ is the unique positive solution of the equation $1-\rho = \exp(-(1+\varepsilon)\,\rho)$. Furthermore, the largest component contains more than two cycles, while all but the largest component are trees or unicyclic components of order $O(\log n)$.

The reason why the average vertex degree $(n-1)p$ being equal to one is so crucial in view of the emergence of the giant component can be nicely explained by branching processes. We shall sketch this approach in Section 4.

Bollobás [9] and Łuczak [23] made a breakthrough in the study of the critical phenomena associated with the phase transition. They determined how the order of the largest component changes, depending on how fast the average degree $(n-1)p$ converges to the critical value of one. Their results can be restated as follows.

Theorem 2. *Let $\varepsilon = \varepsilon(n)$ be a sequence satisfying $\varepsilon \to 0$ and $\varepsilon^3 n \to \infty$ as $n \to \infty$ and let $p = p(n) \in [0, 1]$ be a sequence such that*

$$\varepsilon = (n-1)p - 1$$

for all integers $n \in \mathbb{N}$. Then the following hold whp.

(i) If $\varepsilon < 0$, then $L(G(n,p)) = O(\varepsilon^{-2} \log(|\varepsilon|^3 n))$.
(ii) If $\varepsilon > 0$, then $L(G(n,p)) = (1 \pm o(1))\, 2\varepsilon n$.

Results analogous to Theorems 1 and 2 hold for $G(n,m)$ with $m = (1+\varepsilon)\,n/2$.

In the supercritical regime, much stronger results are known for the giant component in $G(n,p)$ and $G(n,m)$, which will be the subject of Sections 1.1–1.2.

1.1 Limit theorems for giant component in $G(n,p)$

Throughout this section we consider the binomial random graph $G(n,p)$ in the supercritical regime, when the average degree $(n-1)p$ is strictly larger than one and bounded. We let $I \subset (1,\infty)$ be a compact (i.e. closed and bounded) interval, let $p = p(n) \in [0,1]$ be a sequence such that

$$d := (n-1)\,p \in I, \quad \forall n \in \mathbb{N}.$$

We let $\rho = \rho(d) \in (0,1)$ be the unique positive solution of the equation

$$1-\rho = \exp(-d\,\rho) \tag{1}$$

and define

$$\mu := \rho n, \qquad \tilde{\mu} := \frac{d(1-(1-\rho)^2)}{2}\,n, \qquad \bar{\sigma}^2 := \frac{\rho(1-\rho)}{(1-d(1-\rho))^2}\,n,$$

$$\tilde{\sigma}^2 := \left(\frac{d^2(1-\rho)^2\,(2+(1-\rho)\,(-2d\rho-2-\rho))}{(1-d(1-\rho))^2} + \frac{d(1-(1-\rho)^2)}{2}\right) n,$$

$$\hat{\sigma} := \frac{d(1-\rho)\left(1-(1-\rho)^2-d\rho(1-\rho)\right)}{(1-d(1-\rho))^2}\,n.$$

In terms of limit theorems, Theorem 1 (ii) provides a strong law of large numbers for the order $L(G(n,p))$ of the giant component in $G(n,p)$.

Theorem 3 (Strong law of large numbers). *For any $\delta_1, \delta_2 > 0$, there exists $n_0 \in \mathbb{N}$ such that for any $n \geq n_0$*

$$1-\delta_2 \leq \mathbb{P}\left[\,\left|\frac{L(G(n,p))-\mu}{n}\right| \leq \delta_1\,\right] \leq 1+\delta_2.$$

Therefore, the typical value of $L(G(n,p))$ can be determined up to fluctuations of order $o(n)$. It can be more precisely characterised as follows.

Theorem 4 (Central limit theorem). *For any $a, b \in \mathbb{R}$ with $a < b$ and any $\delta > 0$, there exists $n_0 \in \mathbb{N}$ such that for any $n \geq n_0$*

$$(1-\delta)\,F(a,b) \leq \mathbb{P}\left[\,a \leq \frac{L(G(n,p))-\mu}{\bar{\sigma}} \leq b\,\right] \leq (1+\delta)\,F(a,b),$$

where

$$F(a,b) := \frac{1}{\sqrt{2\pi}} \int_a^b \exp\left(-\frac{x^2}{2}\right) dx.$$

In other words, $\bar{\sigma}^{-1}(L(G(n,p))-\mu)$ converges in distribution to the standard normal distribution $N(0,1)$.

The central limit theorem for $L(G(n,p))$ was proved first via counting techniques by Stepanov [27] and then by Pittel and Wormald [24]. As a special case of the corresponding result on random k-uniform hypergraph $H_k(n,p)$ for any $k \geq 2$, an alternative proof of Theorem 4 was provided by Behrisch, Coja-Oghlan, and Kang [4]. They introduced a variant of the component-exposure process based on the breadth-first search and analysed random variables related to $L(G(n,p))$ via Martingale techniques. There is yet another proof of Theorem 4 which makes use of Stein's method [3]. Roughly speaking, Stein's method states that a sum of a family of dependent random variables converges to the normal distribution if one can bound the correlations within any constant-sized subfamily sufficiently well. This method was used by Barbour, Karoński, and Ruciński [2] in order to prove that in $G(n,p)$ the number of tree components of a given bounded size is asymptotically normal. This technique was extended by Behrisch, Coja-Oghlan, and Kang [3] to study the total number of vertices *outside* of the giant component in $H_k(n,p)$ for any $k \geq 2$.

Note that Theorem 4 provides the estimation of $L(G(n,p))$ up to an error of $o(\bar{\sigma}) = o(\sqrt{n})$. However, it does not yield the asymptotic probability that $L(G(n,p))$ hits the value s *exactly*. This is provided by the next theorem.

Theorem 5 (Local limit theorem). *For any compact interval $J \subset \mathbb{R}$ and any $\delta > 0$, there exists $n_0 \in \mathbb{N}$ such that for any $n \geq n_0$ and any integer $s \in \mathbb{N}$ satisfying $n^{-1/2}(s-\mu) \in J$, we have*

$$(1-\delta)\,\phi_{\mu,\bar{\sigma}}(s) \leq \mathbb{P}[\,L(G(n,p)) = s\,] \leq (1+\delta)\,\phi_{\mu,\bar{\sigma}}(s),$$

where

$$\phi_{\mu,\bar{\sigma}}(s) := \frac{1}{\bar{\sigma}\sqrt{2\pi}} \exp\left(-\frac{(s-\mu)^2}{2\bar{\sigma}^2}\right).$$

This local limit theorem follows from a general result on random k-uniform hypergraph $H_k(n,p)$ for any $k \geq 2$ by Behrisch, Coja-Oghlan, and Kang [4], who used the sprinkling arguments (also known as the two-round exposure) and smoothing techniques. We first include each possible edge with probability $p_1 = (1-\varepsilon_1)p$ independently, where $\varepsilon_1 > 0$ is small but independent of n and denote the resulting random hypergraph by G_1. Then, we add each possible edge that is not present in G_1 with a small probability $p_2 \sim \varepsilon_1 p$ and investigate closely how these additional random edges attach further vertices to the largest component in G_1. Denote by S the number of these "attached" vertices. The conditional distribution of S *given the value of* $L(G_1)$ is "locally uniform". Because p_1 and p_2 are chosen such that each edge is present with probability p after the second portion of edges has been added, this yields the desired local limit theorem for $L(G(n,p))$.

Stepanov [27] and Pittel and Wormald [24] derived a *local limit theorem for the joint distribution* of the order $L(G(n,p))$ and size $M(G(n,p))$ of the giant component in $G(n,p)$.

Theorem 6 (Local limit theorem for joint distribution). *For any compact set $\mathcal{J} \subset \mathbb{R}^2$ and any $\delta > 0$, there exists $n_0 \in \mathbb{N}$ such that for any $n \geq n_0$ and any integers $s, t \in \mathbb{N}$ satisfying $\left(n^{-1/2}(s-\mu), n^{-1/2}(t-\tilde{\mu})\right) \in \mathcal{J}$, we have*

$$(1-\delta)\,P(s,t) \;\leq\; \mathbb{P}[\,L(G(n,p)) \;=\; s,\;\; M(G(n,p)) \;=\; t\,] \;\leq\; (1+\delta)\,P(s,t),$$

where

$$P(s,t) := \frac{1}{2\pi\sqrt{\bar{\sigma}^2\tilde{\sigma}^2-\hat{\sigma}^2}} \cdot \exp\left(-\frac{\bar{\sigma}^2\tilde{\sigma}^2}{2(\bar{\sigma}^2\tilde{\sigma}^2-\hat{\sigma}^2)}\left(\frac{(s-\mu)^2}{\bar{\sigma}^2}-\frac{2\hat{\sigma}(s-\mu)(t-\tilde{\mu})}{\bar{\sigma}^2\tilde{\sigma}^2}+\frac{(t-\tilde{\mu})^2}{\tilde{\sigma}^2}\right)\right).$$

Behrisch, Coja-Oghlan, and Kang [5] provided a new simple proof of Theorem 6, indeed a proof of the corresponding result on random k-uniform hypergraph $H_k(n,p)$ for any $k \geq 2$. Their proof begins with the observation that *conditioned* on the event that the total number $E(G(n,p))$ of edges in $G(n,p)$ is equal to $m' \in \{0,\ldots,\binom{n}{2}\}$, $G(n,p)$ is distributed exactly as $G(n,m')$, so we have

$$\begin{aligned}&\mathbb{P}[\,L(G(n,p)) \;=\; s,\;\; M(G(n,p)) \;=\; t\,]\\ &= \sum_{m'=0}^{\binom{n}{2}} \mathbb{P}\left[\,\mathrm{Bi}\left(\binom{n}{2},\,p\right) = m'\,\right]\cdot \mathbb{P}[\,L(G(n,m')) \;=\; s,\;\; M(G(n,m')) \;=\; t\,].\end{aligned}$$

Furthermore, the asymptotic estimation for $\mathbb{P}[\,\mathrm{Bi}\left(\binom{n}{2},\,p\right) \;=\; m'\,]$ follows from the Chernoff bound and the local limit theorem for the binomial distribution. The asymptotic estimation for $\mathbb{P}[\,L(G(n,m')) \;=\; s,\;\; M(G(n,m')) \;=\; t\,]$ follows from a local limit theorem for the joint distribution of the order $L(G(n,m'))$ and size $M(G(n,m'))$ of the giant component in $G(n,m')$, which we will see in the next section (Theorem 7).

1.2 *Local limit theorem for giant component in* $G(n,m)$

In this section we consider the uniform random graph $G(n,m)$ in the supercritical regime when the average degree $2m/n$ is strictly larger than one and bounded. We let $I \subset (1,\infty)$ be a compact interval, let $m = m(n)$ be a sequence of integers such that

$$d := \frac{2m}{n} \in I, \qquad \forall n \in \mathbb{N}.$$

We let $\rho = \rho(d) \in (0, 1)$ be the unique positive solution of the equation (1) and define

$$\mu := \rho n, \qquad \tilde{\mu} := \frac{d(1-(1-\rho)^2)}{2} n,$$

$$\bar{\tau}^2 := \frac{1-\rho-(2d+1)(1-\rho)^2+4d(1-\rho)^3-2d(1-\rho)^4}{(1-d(1-\rho))^2} n,$$

$$\tilde{\tau}^2 := \frac{d(1-\rho)^2\left(1-(1-\rho)^2\left(3d^2-2d+1+(4d^2-2d)(1-\rho)-d^2(1-\rho)^2\right)\right)}{2(1-d(1-\rho))^2} n,$$

$$\hat{\tau} := \frac{d(1-\rho)^2\left(1-2d(1-\rho)+(3d-1)(1-\rho)^3-d(1-\rho)^3\right)}{(1-d(1-\rho))^2} n.$$

The following *bivariate* local limit theorem for the order and size of the giant component in $G(n, m)$ can be obtained from the *univariate* local limit theorem for the order of the giant component in $G(n, p)$ (Theorem 5) with help of Fourier analysis [5]. We shall sketch this Fourier analytic approach in Section 5.

Theorem 7 (Local limit theorem for joint distribution). *For any compact set $\mathcal{J} \subset \mathbb{R}^2$ and any $\delta > 0$, there exists $n_0 \in \mathbb{N}$ such that for any $n \geq n_0$ and any integers $s, t \in \mathbb{N}$ satisfying $\left(n^{-1/2}(s-\mu), n^{-1/2}(t-\tilde{\mu})\right) \in \mathcal{J}$, we have*

$$(1-\delta)\, Q(s,t) \leq \mathbb{P}[\, L(G(n,m)) = s,\ M(G(n,m)) = t\,] \leq (1+\delta)\, Q(s,t),$$

where

$$Q(s,t) := \frac{1}{2\pi\sqrt{\bar{\tau}^2\tilde{\tau}^2-\hat{\tau}^2}} \cdot \exp\left(-\frac{\bar{\tau}^2\tilde{\tau}^2}{2(\bar{\tau}^2\tilde{\tau}^2-\hat{\tau}^2)}\left(\frac{(s-\mu)^2}{\bar{\tau}^2}-\frac{2\hat{\tau}(s-\mu)(t-\tilde{\mu})}{\bar{\tau}^2\tilde{\tau}^2}+\frac{(t-\tilde{\mu})^2}{\tilde{\tau}^2}\right)\right).$$

2 Random k-uniform hypergraphs

One of the most natural generalisations of $G(n, p)$ is the random k-uniform hypergraph $H_k(n, p)$ for any integer $k \geq 2$, which is a hypergraph with vertex set $[n]$ in which each of the $\binom{n}{k}$ possible edges (i.e. k-element subsets of $[n]$) is present independently with probability p. Similarly, a natural generalisation of $G(n, m)$ is $H_k(n, m)$, a hypergraph chosen uniformly at random from the set of all k-uniform hypergraphs with vertex set $[n]$ and m edges for a given $m \in \{0, \cdots, \binom{n}{k}\}$.

The notion of connectivity in hypergraphs is however ambiguous, and there are several possible definitions. We shall consider the one suggested by Bollobás and Riordan [12], i.e. the j-tuple connectivity of a k-uniform hypergraph H with vertex set $[n]$, where k, j are fixed integers satisfying $2 \leq k \leq n-1$ and $1 \leq j \leq k-1$. We say that a j-element subset J_1 in a hypergraph H is *reachable* from another j-element subset J_2 if there exists a sequence of edges $(E_1, \ldots, E_\ell)$ such that $J_1 \subseteq E_1$, $J_2 \subseteq E_\ell$, and $|E_i \cap E_{i+1}| \geq j$ for each $i = 1, \ldots, \ell - 1$. The reachability is an equivalence relation on j-element subsets, and the equivalence classes are called j-tuple connected components, or *j-components* in short. We let $L_j(H)$ denote the maximum number of j-element subsets contained in a j-component of H and call the lexicographically first j-component of H that has order $L_j(H)$ the largest component. The case $j = 1$ corresponds to the notion of vertex-connectedness and $L_1(H)$ is simply the number of vertices in the largest component of H, so we let $L(H) = L_1(H)$. In addition we let $M(H)$ denote the number of edges in the largest component of H.

2.1 Vertex-connectivity in $H_k(n,p)$

The critical probability threshold for the emergence of the giant component in $H_k(n,p)$ was first determined by Schmidt-Pruzan and Shamir in [26]. This coincides with the critical probability threshold for the emergence of the giant component in $G(n,p)$ when $k = 2$.

Theorem 8. *Let $k \geq 2$ be a fixed integer, let $\varepsilon \in \mathbb{R}$ be a fixed constant, and let $p = p(n) \in [0,1]$ be a sequence such that*

$$\varepsilon = (k-1)\binom{n-1}{k-1}p - 1$$

for all $n \in \mathbb{N}$. Then the following hold whp.

(i) If $\varepsilon < 0$, then $L(H_k(n,p)) = O(\log n)$.
(ii) If $\varepsilon > 0$, then $L(H_k(n,p)) = (1+o(1))\rho n$, where $\rho = \rho(k,\varepsilon) \in (0,1)$ is the unique positive solution of the equation $1 - \rho = \exp\left(-\frac{1+\varepsilon}{k-1}\left(1-(1-\rho)^{k-1}\right)\right)$.

In the rest of this section we consider $H_k(n,p)$ in the supercritical regime when the "average degree" $(k-1)\binom{n-1}{k-1}p$ is strictly larger than one and bounded. We let $I \subset (1,\infty)$ be a compact interval and let $p = p(n) \in [0,1]$ be a sequence such that

$$d := (k-1)\binom{n-1}{k-1}p \in I$$

for all $n \in \mathbb{N}$. Let $\rho = \rho(k,d) \in (0,1)$ be the unique positive solution of the equation

$$1 - \rho = \exp\left(-\frac{d}{k-1}\left(1-(1-\rho)^{k-1}\right)\right). \tag{2}$$

We define

$$\mu_k := \rho n, \qquad \tilde{\mu}_k := \frac{d(1-(1-\rho)^k)}{k(k-1)}n,$$

$$\bar{\sigma}_k^2 := \frac{(1-\rho)(\rho + d(1-\rho-(1-\rho)^{k-1}))}{(1-d(1-\rho)^{k-1})^2}n,$$

$$\tilde{\sigma}_k^2 := \left(\frac{d^2(1-\rho)^k(2+f(k,d,\rho))}{(k-1)^2(1-d(1-\rho)^{k-1})^2} + \frac{d(1-(1-\rho)^k)}{k(k-1)}\right)n,$$

$$\hat{\sigma}_k := \frac{d(1-\rho)(1-(1-\rho)^k - d(1-\rho)^{k-1}\rho)}{(k-1)(1-d(1-\rho)^{k-1})^2}n,$$

where $f(k,d,\rho) := (1-\rho)^{k-1}\left(d(1-\rho)^{k-1} - d(1+\rho) - 2 + \rho\right)$.

The first limit theorem for the giant component in $H_k(n,p)$ is *a strong law of large numbers* provided by Theorem 8 (ii).

Theorem 9 (Strong law of large numbers). *For any $\delta_1, \delta_2 > 0$, there exists $n_0 \in \mathbb{N}$ such that for any $n \geq n_0$*

$$1-\delta_2 \leq \mathbb{P}\left[\left|\frac{L(H_k(n,p)) - \mu_k}{n}\right| \leq \delta_1\right] \leq 1+\delta_2.$$

Therefore, the typical value of $L(H_k(n,p))$ can be determined up to fluctuations of order $o(n)$. The better estimations of $L(H_k(n,p))$ are provided by the central and local limit theorems for $L(H_k(n,p))$.

Theorem 10 (Central limit theorem). *For any $a, b \in \mathbb{R}$ with $a < b$ and any $\delta > 0$, there exists $n_0 \in \mathbb{N}$ such that for any $n \geq n_0$*

$$(1-\delta)F(a,b) \leq \mathbb{P}\left[a \leq \frac{L(H_k(n,p)) - \mu_k}{\bar{\sigma}_k} \leq b\right] \leq (1+\delta)F(a,b),$$

where

$$F(a,b) := \frac{1}{\sqrt{2\pi}}\int_a^b \exp\left(-\frac{x^2}{2}\right)dx.$$

In other words, $\bar{\sigma}_k^{-1}(L(H_k(n,p)) - \mu_k)$ converges in distribution to the standard normal distribution $N(0,1)$.

This yields the estimation of $L(H_k(n,p))$ up to an error of $o(\bar{\sigma}) = o(\sqrt{n})$, while the following local limit theorem provides the asymptotic probability that $L(H_k(n,p))$ hits exactly the value s.

Theorem 11 (Local limit theorem). *For any compact interval $J \subset \mathbb{R}$ and any $\delta > 0$, there exists $n_0 \in \mathbb{N}$ such that for any $n \geq n_0$ and any integer $s \in \mathbb{N}$ satisfying $n^{-1/2}(s - \mu_k) \in J$, we have*

$$(1-\delta)\,\phi_{\mu_k,\bar{\sigma}_k}(s) \;\leq\; \mathbb{P}[\,L(H_k(n,p)) \;=\; s\,] \;\leq\; (1+\delta)\,\phi_{\mu_k,\bar{\sigma}_k}(s),$$

where

$$\phi_{\mu_k,\bar{\sigma}_k}(s) := \frac{1}{\bar{\sigma}\sqrt{2\pi}}\exp\left(-\frac{(s-\mu_k)^2}{2\bar{\sigma}_k^2}\right).$$

The above central and local limit theorems were proved among others by Behrisch, Coja-Oghlan, and Kang [4], who proved in [5] the following stronger result.

Theorem 12 (Local limit theorem for joint distribution). *For any compact set $\mathcal{J} \subset \mathbb{R}^2$ and any $\delta > 0$, there exists $n_0 \in \mathbb{N}$ such that for any $n \geq n_0$ and any integers $s, t \in \mathbb{N}$ satisfying $\left(n^{-1/2}(s-\mu_k),\, n^{-1/2}(t-\tilde{\mu}_k)\right) \in \mathcal{J}$, we have*

$$(1-\delta)\,P_k(s,t) \;\leq\; \mathbb{P}[\,L(H_k(n,p)) \;=\; s,\;\; M(H_k(n,p)) \;=\; t\,] \;\leq\; (1+\delta)\,P_k(s,t),$$

where

$$P_k(s,t) := \frac{1}{2\pi\sqrt{\bar{\sigma}_k^2\tilde{\sigma}_k^2 - \hat{\sigma}_k^2}}$$

$$\cdot\,\exp\left(-\frac{\bar{\sigma}_k^2\tilde{\sigma}_k^2}{2(\bar{\sigma}_k^2\tilde{\sigma}_k^2 - \hat{\sigma}_k^2)}\left(\frac{(s-\mu_k)^2}{\bar{\sigma}_k^2} - \frac{2\hat{\sigma}_k(s-\mu_k)(t-\tilde{\mu}_k)}{\bar{\sigma}_k^2\tilde{\sigma}_k^2} + \frac{(t-\tilde{\mu}_k)^2}{\tilde{\sigma}_k^2}\right)\right).$$

2.2 Vertex-connectivity in $H_k(n,m)$

In this section we consider $H_k(n,m)$ in the supercritical regime when the "average degree" $k(k-1)\,m/n$ is strictly larger than one and bounded. We let $I \subset (1,\infty)$ be a compact interval, let $m = m(n)$ be a sequence of integers such that

$$d := \frac{k(k-1)\,m}{n} \in I$$

for all $n \in \mathbb{N}$, and let $\rho = \rho(k,d) \in (0,1)$ be the unique positive solution of the equation (2). Further, we define

$$\mu_k := \rho n, \qquad \tilde{\mu}_k := \frac{d(1-(1-\rho)^k)}{k(k-1)}n,$$

$$\bar{\tau}_k^2 := \frac{1-\rho-\left(1+\frac{d}{k-1}\right)(1-\rho)^2 - d(1-\rho)^k + \frac{2dk}{k-1}(1-\rho)^{k+1} - \frac{dk}{k-1}(1-\rho)^{2k}}{(1-d(1-\rho)^{k-1})^2}n,$$

$$\tilde{\tau}_k^2 := \frac{d(1-\rho)^k\,(1-g(k,d,\rho))}{k(k-1)(1-d(1-\rho)^{k-1})^2}\,n,$$

$$\hat{\tau}_k := \frac{d\,(1-\rho)^k\,\left(1-\frac{d}{k-1}(1-\rho)-h(k,d,\rho)\right)}{(k-1)(1-d(1-\rho)^{k-1})^2}\,n,$$

where

$$g(k,d,\rho) := \frac{d(k-2)}{k-1}(1-\rho)^{k-1} + \left(\frac{d^2k}{(k-1)^2} - \frac{dk}{k-1} + 1\right)(1-\rho)^k$$
$$+\frac{d^2}{k-1}(1-\rho)^{2k-2} - \frac{2d}{k-1}\left(\frac{dk-k+1}{k-1}\right)(1-\rho)^{2k-1} + \frac{d^2}{(k-1)^2}(1-\rho)^{3k-2},$$

$$h(k,d,\rho) := d(1-\rho)^{k-1} - \frac{d(k+1)-k+1}{k-1}(1-\rho)^k + \frac{d}{k-1}(1-\rho)^{2k-1}.$$

Analogous to the graph case, Behrisch, Coja-Oghlan, and Kang [5] showed that the *univariate* local limit theorem for the order of the giant component in $H_k(n,p)$ (Theorem 11) can be converted into the following *bivariate* local limit theorem for the order and size of the giant component in $H_k(n,m)$, using Fourier analysis.

Theorem 13 (Local limit theorem for joint distribution). *For any compact set $\mathcal{J} \subset \mathbb{R}^2$ and any $\delta > 0$, there exists $n_0 \in \mathbb{N}$ such that for any $n \geq n_0$ and any integers $s,t \in \mathbb{N}$ satisfying $\left(n^{-1/2}(s-\mu_k),\, n^{-1/2}(t-\tilde{\mu}_k)\right) \in \mathcal{J}$, we have*

$$(1-\delta)\,Q_k(s,t) \leq \mathbb{P}[\,L(H_k(n,m)) = s,\; M(H_k(n,m)) = t\,] \leq (1+\delta)\,Q_k(s,t),$$

where

$$Q_k(s,t) := \frac{1}{2\pi\sqrt{\bar{\tau}_k^2\tilde{\tau}_k^2-\hat{\tau}_k^2}}$$
$$\cdot\exp\left(-\frac{\bar{\tau}_k^2\tilde{\tau}_k^2}{2(\bar{\tau}_k^2\tilde{\tau}_k^2-\hat{\tau}_k^2)}\left(\frac{(s-\mu_k)^2}{\bar{\tau}_k^2} - \frac{2\hat{\tau}_k(s-\mu_k)(t-\tilde{\mu}_k)}{\bar{\tau}_k^2\tilde{\tau}_k^2} + \frac{(t-\tilde{\mu}_k)^2}{\tilde{\tau}_k^2}\right)\right).$$

2.3 High-order connectivity

In view of the emergence of the giant j-component, Cooley, Kang, and Person [15] showed that $H_k(n,p)$ undergoes a phase transition at the critical probability threshold $p = \left(\left(\binom{k}{j}-1\right)\binom{n-j}{k-j}\right)^{-1}$, which coincides with the critical probability threshold for the emergence of the giant component in $H_k(n,p)$ when $j=1$.

Theorem 14. *Let k, j be fixed integers satisfying $k \geq 2$ and $1 \leq j \leq k-1$, let $\varepsilon \in \mathbb{R}$ be a fixed constant, and let $p = p(n) \in [0, 1]$ be a sequence such that*

$$\varepsilon = \left(\binom{k}{j} - 1\right) \binom{n-j}{k-j} p - 1$$

for all $n \in \mathbb{N}$. Then the following hold whp.

(i) If $\varepsilon < 0$, then $L_j(H_k(n, p)) = O(\log n)$.
(ii) If $\varepsilon > 0$, then $L_j(H_k(n, p)) = \Omega(n^j)$.

Cooley, Kang, and Koch [16] took a closer look at the giant j-component in the weakly subcritical and weakly supercritical regimes. Among others they proved the following theorem.

Theorem 15. *Let k, j be fixed integers satisfying $k \geq 2$ and $1 \leq j \leq k-1$, and let $\varepsilon = \varepsilon(n)$ be a sequence satisfying*

$$\varepsilon \to 0, \quad \varepsilon^3 n^j \to \infty, \quad \varepsilon^2 n^{1-2\delta} \to \infty, \quad as \quad n \to \infty, \tag{3}$$

for some constant $\delta > 0$. Let $p = p(n) \in [0, 1]$ be a sequence such that $\varepsilon = p\left(\binom{k}{j} - 1\right)\binom{n-j}{k-j} - 1$ for all $n \in \mathbb{N}$. Then the following hold whp.

(i) If $\varepsilon < 0$, then $L_j(H_k(n, p)) = O(\varepsilon^{-2} \log n)$.
(ii) If $\varepsilon > 0$, then

$$L_j(H_k(n, p)) = (1 \pm o(1)) \frac{2\varepsilon}{\binom{k}{j} - 1} \binom{n}{j}.$$

Note that when $j = 1$ the conditions on ε (3) are simply $\varepsilon \to 0$ and $\varepsilon^3 n \to \infty$. This provides the best possible range for ε. However, for $j \geq 2$, the third condition takes over from the second one and the conditions $\varepsilon \to 0$ and $\varepsilon^2 n^{1-2\delta} \to \infty$ are required. This condition arises from proof methods. When $k = 2, j = 1$, Theorem 15 (ii) matches Theorem 2 (ii).

3 Asymptotic number of connected (hyper)graphs

In this section we derive the asymptotic number of connected graphs and connected k-uniform hypergraphs with given numbers of vertices and edges.

3.1 Asymptotic number of connected graphs

Let $C(n,m)$ denote the number of connected graphs with n vertices and m edges, and let $c(n,m)$ denote the probability that $G(n,m)$ is connected. Obviously we have

$$C(n,m) = c(n,m)\binom{\binom{n}{2}}{m}.$$

Therefore, the asymptotic number of connected graphs with n vertices and m edges can easily be deduced from the asymptotic estimation of the probability $c(n,m)$ that $G(n,m)$ is connected.

In order to estimate $c(n,m)$ we choose $N > n$ and $p \in [0,1]$ in such a way that n and m will be probable outcomes of the order $L(G(N,p))$ and size $M(G(N,p))$ of the giant component in $G(N,p)$. We observe that *conditioned* on the event that $L(G(N,p)) = n$ and $M(G(N,p)) = m$, the giant component in $G(N,p)$ is a *uniformly distributed* connected graph with n vertices and m edges. Therefore, we have

$$\begin{aligned}\mathbb{P}[L(G(N,p)) = n,\ M(G(N,p)) = m]\\ \sim \binom{N}{n}\binom{\binom{N}{2}}{m} c(n,m)p^m(1-p)^{\binom{N}{2}-\binom{N-n}{2}-\binom{n}{2}+m}.\end{aligned} \tag{4}$$

This observation was used by Łuczak in [22], and asymptotic estimations of the terms on the right-hand side of (4) (apart from the desired term $c(n,m)$) were carried out by Coja-Oghlan, Moore, and Sanwalani in [14]. More crucially, the local limit theorem for the order and size of the giant component in $G(N,p)$ (Theorem 6) provides the asymptotic expression of the left-hand side. Putting these together, we obtain the following exact asymptotic expression for the probability $c(n,m)$ that $G(n,m)$ is connected [6, 7].

Theorem 16. *We let $I_2 \subset (2,\infty)$ be a compact interval, let $m = m(n)$ be a sequence of integers such that $d := \frac{2m}{n} \in I_2$ for all $n \in \mathbb{N}$, and let $r = r(d) \in (0,1)$ be a unique solution of the equation*

$$r = \exp\left(-\frac{d(1-r)}{1+r}\right).$$

Then for any $\delta > 0$ there exists $n_0 \in \mathbb{N}$ such that for any $n \geq n_0$ we have

$$(1-\delta)\,R(d,r,n) \leq c(n,m) \leq (1+\delta)\,R(d,r,n),$$

where

$$R(d,r,n) := \frac{1+r-dr}{\sqrt{(1+r)^2-2dr}} \cdot \exp\left(\frac{dr}{1+r} - \frac{dr^2}{2(1+r)}\right)\left(r^{\frac{r}{1-r}}(1-r)^{1-d}(1-r^2)^{\frac{d}{2}}\right)^n. \tag{5}$$

This result is consistent with the result of Bender, Canfield, and McKay [8], who were the first to compute the asymptotic probability that $G(n,m)$ is connected for *any* ratio m/n. There are alternative proofs of this result by van der Hofstad and Spencer [28] using branching process arguments and by Bollobás and Riordan [13] using smoothing techniques.

3.2 Asymptotic number of connected k-uniform hypergraphs

Similarly to (4), we have the relation

$$\mathbb{P}[L(H_k(N,p)) = n,\ M(H_k(N,p)) = m] \sim \binom{N}{n}\binom{\binom{N}{k}}{m} c_k(n,m) p^m (1-p)^{\binom{N}{k}-\binom{N-n}{k}-\binom{n}{k}+m},$$

so we can estimate the probability $c_k(n,m)$ that $H_k(n,m)$ is connected, using Theorem 12.

Theorem 17. *Let $k \geq 3$ be a fixed integer. We let $I_k \subset (k,\infty)$ be a compact interval, let $m = m(n)$ be a sequence of integers such that $d := \frac{k(k-1)m}{n} \in I_k$ for all $n \in \mathbb{N}$, and let $r = r(k,d) \in (0,1)$ be a unique solution of the equation*

$$r = \exp\left(-\frac{d(1-r)(1-r^{k-1})}{(k-1)(1-r^k)}\right).$$

Then for any $\delta > 0$ there exists $n_0 \in \mathbb{N}$ such that for any $n \geq n_0$ we have

$$(1-\delta)R_k(d,r,n) \leq c_k(n,m) \leq (1+\delta)R_k(d,r,n),$$

where

$$R_k(d,r,n) := \frac{1-r^k-d(1-r)r^{k-1}}{\sqrt{(1-r^k+d(r-r^{k-1}))(1-r^k)-\frac{dk}{k-1}r(1-r^{k-1})^2}} \cdot \exp\left(\frac{d(r-2r^k+r^{k-1})}{2(1-r^k)}\right)\left(r^{\frac{r}{1-r}}(1-r)^{1-\frac{d}{k-1}}(1-r^k)^{\frac{d}{k(k-1)}}\right)^n. \tag{6}$$

Note that there is an extra factor $\exp\left(\frac{dr^2}{2(r+1)}\right)$ in the asymptotic expression for $c(n,m)$, in comparison to that for $c_k(n,m)$ when taking $k = 2$. More precisely, we have $R(d,r,n) = \exp\left(\frac{dr^2}{2(r+1)}\right) \cdot R_2(d,r,n)$ from (5) and (6).

Let $C_k(n,m)$ denote the number of connected k-uniform hypergraphs with n vertices and m edges. Because we have

$$C_k(n,m) = c_k(n,m)\binom{\binom{n}{k}}{m},$$

Theorem 17 yields an asymptotic formula for $C_k(n,m)$ in the strictly supercritical regime when $m - \frac{n}{k-1} = \Theta(n)$. Karoński and Łuczak [20] derived an asymptotic formula for $C_k(n,m)$ in the barely supercritical regime when $m - \frac{n}{k-1} \ll \frac{\log n}{\log\log n}$, using combinatorial arguments. Bollobás and Riordan [13] derived an asymptotic formula for $C_k(n,m)$ all the way through the supercritical regime $m - \frac{n}{k-1} = o(n)$. Sato and Wormald [25] also derived an asymptotic formula for $C_3(n,m)$ when $n^{1/3}\log^2 n \ll m - \frac{n}{2} \ll n$, using cores and kernels.

4 Giant component and branching processes

In this section and the next, we describe two techniques for analysing the giant components in random graphs, one classical (branching processes) and one new (Fourier analysis).

The branching process with offspring distribution Z is a random rooted tree constructed as follows. We start with a single root vertex in generation 0. Each vertex in generation $i \in \mathbb{N} \cup \{0\}$ has a random number of children in generation $i+1$, independently drawn from distribution Z. A classical result from the theory of branching processes states that if $\mathbb{E}(Z) < 1$, then the mean total number of vertices in the branching process is $1/(1-\mathbb{E}(Z))$ and in particular, the process dies out with probability one. On the other hand, if $\mathbb{E}(Z) > 1$, then the process continues forever ("survives") with probability $\rho \in (0,1)$, which is the unique positive solution of the equation

$$1 - \rho = \Phi_Z(1-\rho), \tag{7}$$

where Φ_Z is the probability generating function of Z.

Two branching processes that are useful for the analysis of the order of the largest component of $G(n,p)$ are the Poisson branching process with offspring distribution $\mathrm{Po}(d)$ and the binomial branching process with offspring distribution $\mathrm{Bi}(n,p)$ [1]. Let $d := (n-1)p$. The probability generating function of the Poisson branching process with offspring distribution $\mathrm{Po}(d)$ satisfies

$$\Phi_{\mathrm{Po}(d)}(x) = \exp(d(x-1)), \tag{8}$$

and thus, (7) and (8) imply that if $d > 1$, then the survival probability satisfies

$$1 - \rho = \exp(-d\,\rho). \tag{9}$$

The probability generating function of the binomial branching process with offspring distribution $\mathrm{Bi}(n,p)$ satisfies

$$\Phi_{\mathrm{Bi}(n,p)}(x) = (1 - p(1-x))^n, \tag{10}$$

so, (7) and (10) imply that if $d > 1$, then the survival probability $\rho \in (0,1)$ satisfies

$$1 - \rho = (1 - p\rho)^n. \tag{11}$$

Below we demonstrate how the theory of branching processes can be applied to obtain Theorem 1 and Theorem 2 (ii). In order to determine the order of the largest component of $G(n,p)$ we consider the component-exploration process based on the breadth-first search [21]. Given a vertex v, let C_v denote the component containing v. We construct a spanning tree T_v (rooted at v) obtained by exploring C_v as follows. We begin at v and discover all the neighbours of v, say $v_1, \dots, v_k$. Then we discover all the neighbours of v_1 different from the vertices found so far, then the new neighbours of v_2, and so on, ending the first generation with the new neighbours of v_k. We continue this breadth-first exploration until there are no more new vertices left in C_v. At the conclusion of this process, we have constructed a spanning tree T_v of the component C_v. We then couple T_v (from below and above) with properly chosen branching processes.

We now sketch the coupling argument for Theorem 1. In the subcritical regime (i.e. before the giant component emerges), we couple the component-exploration process with the Poisson branching process with offspring distribution $\mathrm{Po}(1+\varepsilon)$ with $\varepsilon < 0$ and obtain a strong upper bound on the size of all components by a standard application of large deviation inequalities. This yields (the first part of) Theorem 1 (i). In the supercritical regime (i.e. after the giant component is formed), we couple the component-exploration process with the Poisson branching process with offspring distribution $\mathrm{Po}(1+\varepsilon)$ with $\varepsilon > 0$ and bound (from below and above) the number of vertices in "large" components (of order, say at least $n^{2/3}$). Then we show that there are no components of order substantially smaller than $n^{2/3}$, but larger than $\Theta(\log n)$, using the component-size gap idea of Bollobás [9]. Finally we show that almost all vertices in large components are in a single large(st) component, which can be done by applying the classical sprinkling argument of Erdős and Rényi [18]. The survival probability ρ satisfying (9) corresponds to the fraction of vertices that lie in the giant component of $G(n,p)$, which leads to (the first part of) Theorem 1 (ii).

Recently, Bollobás and Riordan [11] showed that a variant of the branching process approach works also in the *weakly supercritical* regime, i.e. when

$$\varepsilon = (n-1)p - 1 > 0, \quad \varepsilon \to 0, \quad \varepsilon^3 n \to \infty, \quad \text{as} \quad n \to \infty$$

and reproved Theorem 2 (ii). Their proof uses the binomial branching process with offspring distribution $\mathrm{Bi}(n,p)$, which is considered as a random rooted tree and denoted by $T(n,p)$. Note that (11) implies that for $\varepsilon = (n-1)p - 1 \to 0$, we have

$$\rho \sim 2\varepsilon. \tag{12}$$

A crucial notion in [11] is the dual process of $T(n,p)$ defined as follows. We think of $T(n,p)$ as a random subgraph of the infinite n-ary tree $T(n,1)$ by including each edge independently with probability p and keeping only the component containing the root. Consider the event D that $T(n,p)$ dies out. This event happens if and only if every vertex in the first generation is a neighbour of the root and has only finitely many descendants in $T(n,p)$ (which happens with probability $p(1-\rho)$), or it is not a neighbour of the root in $T(n,p)$ (which happens with probability $1-p$). Thus, the distribution of $T(n,p)$ conditioned on D is the same as the (unconditional) distribution of $T(n,p')$, where

$$p' = p(1-\rho)/(1-p\rho).$$

We call $T(n,p')$ the dual process of $T(n,p)$. Let $|T(n,p)|$ (resp. $|T(n,p')|$) denote the total number of vertices in $T(n,p)$ (resp. $T(n,p')$). Then we have

$$\mathbb{E}(|T(n,p')|) = 1/(1-np') \sim \varepsilon^{-1}.$$

This, (12), and Markov's inequality imply that for any integer $K = K(n)$

$$\mathbb{P}(|T(n,p)| \geq K) \sim 2\varepsilon + \mathbb{P}(D)\,\mathbb{P}(|T(n,p')| \geq K) \lesssim 2\varepsilon + 1/(\varepsilon K). \tag{13}$$

Another important property of $T(n,p)$ is its width $W(T(n,p))$, which is defined as the maximum of the sizes of the generations. Then for any integer $K = K(n)$ satisfying $\varepsilon K \to \infty$,

$$\mathbb{P}(\{W(T(n,p)) \geq K\} \wedge D) = o(\varepsilon). \tag{14}$$

We then couple the rooted tree T_v (constructed by the breadth-first exploration process) with appropriately chosen binomial branching process $T(n,p)$. Using the properties of $T(n,p)$, e.g. (13) and (14), as well as the second moment method, we show that the number of vertices in "large" components (of order K satisfying $\varepsilon^2 K \to \infty$ and $K = o(\varepsilon n)$) is concentrated around $(2+o(1))\varepsilon n$. Finally, using the sprinkling argument we show that there are no two such large components and that whp the number of vertices in the largest component $G(n,p)$ satisfies

$$L(G(n,p)) \sim 2\varepsilon n.$$

This concludes our sketch of the branching process proof of Theorem 2 (ii).

Cooley, Kang, and Koch [16] extended the branching process approach of Bollobás and Riordan [11] to the giant j-component in the random k-uniform hypergraph $H_k(n,p)$ and obtained Theorem 15. Below we briefly sketch the main ideas, difficulties, and the key lemma (smoothness lemma) for the proof.

We explore a j-component via the breadth-first search process as follows. We begin with one j-set, say J. We then find all edges containing J. There are $\binom{n-j}{k-j}$ k-sets containing J, each of which is an edge with probability p. For each edge E, discover $k-j$ new j-sets contained in the edge E. (It could be fewer if some of these j-sets were discovered earlier.) So, the expected number of j-sets discovered from J in one generation is equal to

$$d := \left(\binom{k}{j} - 1\right)\binom{n-j}{k-j} p.$$

Then from each of these new j-sets in turn we look for any more edges containing them and so on. We may therefore approximate the breadth-first search process by a branching process, in which we represent j-sets by vertices and the expected number of children of each vertex is given by d. If the expected number of children d is smaller than 1, then the process always has a tendency to shrink and will therefore die out with probability one. This roughly corresponds to the j-set being in a small j-component. On the other hand, if the expected number of children d is larger than one, the process has a tendency to grow and therefore has a certain positive probability of surviving indefinitely, corresponding to the original j-set being in a large j-component. Note that the critical threshold probability for the emergence of the giant j-component in $H_k(n,p)$ that we have seen in Theorems 14 and 15 is exactly when $d = \left(\binom{k}{j} - 1\right)\binom{n-j}{k-j} p = 1$.

The difficulty with larger j is often that the j-sets of a structure may be "widely distributed" in that they are highly correlated in an uncontrollable way, rather than spread evenly over the hypergraph. Cooley, Kang, and Koch [16] presented a tool, the *smoothness lemma*, which guarantees that in $H_k(n,p)$ whp various collections of j-sets (e.g. large components, unions of small components, boundaries of component-exploration processes) are "evenly distributed" in the sense that for any $1 \leq \ell \leq j-1$ and for any set ℓ-set L, the number of j-sets containing L is about the "right" number. This smoothness lemma has been applied to determine the asymptotic order of the giant j-component close to the phase transition threshold [16] and to determine a threshold and hitting time for the hypergraph to be completely j-connected [17].

5 Giant component and Fourier analysis

In this section we shall briefly sketch how Fourier analysis and the univariate local limit theorem for the order of the giant component in the binomial random graph $G(n,p)$ (Theorem 5) can be used to derive the joint distribution of the order and size of the giant component in the uniform random graph $G(n,m)$ (Theorem 7).

For $p' \in [0, 1]$ and $m' \in \{0, \cdots, \binom{n}{2}\}$ we let $\bar{M}(G(n, p'))$ (resp. $\bar{M}(G(n, m'))$) denote the total number of edges *outside* the largest component in $G(n, p')$ (resp. $G(n, m')$). In addition we let $E(G(n, p'))$ denote the total number of edges in $G(n, p')$, which is a random variable with binomial distribution $\mathrm{Bi}\left(\binom{n}{2}, p'\right)$. A crucial observation for Fourier analytic approach is that for any edge probability p', *conditioned* on the event that $E(G(n, p'))$ is equal to m', $G(n, p')$ is *distributed exactly* as $G(n, m')$. Therefore, for any $s', t' \in \mathbb{N}$

$$\begin{aligned}
&\mathbb{P}[\, L(G(n,p')) = s',\ \bar{M}(G(n,p')) = t' \,] \\
&= \sum_{m'=0}^{\binom{n}{2}} \mathbb{P}[\, E(G(n,p')) = m' \,] \cdot \mathbb{P}[\, L(G(n,m')) = s',\ \bar{M}(G(n,m')) = t' \,] \\
&= \sum_{m'=0}^{\binom{n}{2}} \mathbb{P}\left[\, \mathrm{Bi}\left(\binom{n}{2}, p'\right) = m' \,\right] \cdot \mathbb{P}[\, L(G(n,m')) = s',\ \bar{M}(G(n,m')) = t' \,].
\end{aligned} \tag{15}$$

We recollect the conditions for Theorem 7 and Theorem 5. We let $I \subset (1, \infty)$ and $\mathscr{J} \subset \mathbb{R}^2$ be compact sets, let $m = m(n)$ be a sequence of integers such that

$$d := \frac{2m}{n} \in I, \quad \forall n \in \mathbb{N}. \tag{16}$$

We let $\rho = \rho(d) \in (0, 1)$ be the unique positive solution of the equation $1 - \rho = \exp(-d\rho)$ and set $\mu := \rho n$ and $\tilde{\mu} := \frac{d(1-(1-\rho)^2)}{2} n$. We let $s = s(n)$, $t = t(n)$, $\bar{t} = \bar{t}(n)$ be sequences of integers such that $\bar{t} = m - t$ and

$$\left(\frac{s-\mu}{\sqrt{n}}, \frac{t-\tilde{\mu}}{\sqrt{n}}\right) \in \mathscr{J}, \quad \text{equiv.} \quad \left(\frac{s-\mu}{\sqrt{n}}, \frac{1}{\sqrt{n}}\left(\frac{d(1-\rho)^2 n}{2} - \bar{t}\right)\right) \in \mathscr{J}. \tag{17}$$

Furthermore we let $p = p(n) \in [0, 1]$ be a sequence such that

$$d = (n-1)\, p, \quad \forall n \in \mathbb{N} \tag{18}$$

and set $\sigma := \sqrt{\binom{n}{2} p(1-p)}$. Note that $\frac{2m}{n} = d = (n-1)\, p$, because of (16) and (18).

Our aim is to derive an asymptotic estimation of $\mathbb{P}[\, L(G(n, m)) = s,\ \bar{M}(G(n, m)) = \bar{t} \,]$ for $m, s, \bar{t}$ satisfying (16) and (17). To this end, we shall apply Fourier inversion to (15). We let

$$p_z := p + z\sigma \binom{n}{2}^{-1} \quad \textit{and} \quad m_z := \binom{n}{2} p_z = \binom{n}{2} p + z\sigma = m + z\sigma$$

for $z \in \mathbb{R}$ and set $z^* = \ln^2 n$. Then (15) implies that for any $z \in [-z^*, z^*]$ and for n sufficiently large to ensure that $p_z \in [0, 1]$ for all such z, we have

$$\begin{aligned}&\mathbb{P}[\,L(G(n,p_z)) = s,\ \bar{M}(G(n,p_z)) = \bar{t}\,]\\&= \sum_{m'=0}^{\binom{n}{2}} \mathbb{P}\left[\,\mathrm{Bi}\left(\binom{n}{2}, p_z\right) = m'\,\right] \cdot \mathbb{P}[\,L(G(n,m')) = s,\ \bar{M}(G(n,m')) = \bar{t}\,].\end{aligned} \tag{19}$$

We denote the density of the standard normal distribution by

$$\phi(z) := \frac{1}{\sqrt{2\pi}} \exp\left(-\frac{z^2}{2}\right) \tag{20}$$

and define functions

$$\begin{aligned}f(z) &:= n \cdot \mathbb{P}[\,L(G(n,p_z)) = s,\ \bar{M}(G(n,p_z)) = \bar{t}\,]\\g(z) &:= n \cdot \mathbb{P}[\,L(G(n,m_z)) = s,\ \bar{M}(G(n,m_z)) = \bar{t}\,]\end{aligned}$$

if $z \in [-z^*, z^*]$ and 0 otherwise.

Our goal is to compute $g(0)$ explicitly. If it *were* true that $f = g * \phi$ and an asymptotic expression for $f(z)$ *were* known, then taking the Fourier transform *would* yield $\hat{f} = \hat{g}\hat{\phi}$. So dividing by $\hat{\phi}$ and taking the Fourier transform once more we *would* get the expression for g arising from $\hat{f}/\hat{\phi}$. However, we only know that

$$\|f - g * \phi\|_2 = o(1),$$

which follows essentially from (19): by the Chernoff bound for the binomial distribution and the properties of f, we have for any $z \in [-z^*, z^*]$,

$$f(z) \sim \sum_{m'=m-z^*}^{m+z^*} \mathbb{P}\left[\,\mathrm{Bi}\left(\binom{n}{2}, p_z\right) = m'\,\right] \cdot \mathbb{P}[\,L(G(n,m')) = s,\ \bar{M}(G(n,m')) = \bar{t}\,].$$

For each integer $m - z^* \le m' \le m + z^*$ let $J_{m'}$ be the set of all $z \in \mathbb{R}$ such that $m_z = m + z\sigma = m'$. Then $\int_{J_{m'}} 1du = 1/\sigma$. By definition of g, we get $g(z) = \mathbb{P}[\,L(G(n,m')) = s,\ \bar{M}(G(n,m')) = \bar{t}\,]$ for all $z \in J_{m'}$, and therefore

$$\mathbb{P}[\,L(G(n,m')) = s,\ \bar{M}(G(n,m')) = \bar{t}\,] = \frac{\int_{J_{m'}} g(u)du}{\int_{J_{m'}} 1du} = \sigma \int_{J_{m'}} g(u)du.$$

Applying the local limit theorem for the binomial distribution, we have for any $u \in J_{m'}$

$$\mathbb{P}[\,\mathrm{Bi}\left(\binom{n}{2}, p_z\right) = m'\,] \sim \frac{1}{\sigma\sqrt{2\pi}} \exp\left(-\frac{(u-z)^2}{2}\right) = \sigma^{-1}\phi(z-u).$$

Summing up, we obtain

$$f(z) \sim \int_{\mathbb{R}} g(u)\phi(z-u)du.$$

On the other hand, we can derive an asymptotic expression for $f(z)$, using the duality principle for the largest component: the random graph obtained from $G(n,p_z)$ by removing the largest component is close in distribution to a random graph $G(n',p_z)$ on the remaining vertices, where $n' = n - L(G(n,p_z))$. In particular, conditioning on the event that $L(G(n,p_z)) = s$, the number $\bar{M}(G(n,p_z))$ of edges outside the largest component has approximately a binomial distribution $\mathrm{Bi}\left(\binom{n-s}{2}, p_z\right)$, so we have for all $z \in [-z^*, z^*]$

$$\begin{aligned}\frac{f(z)}{n} &:= \mathbb{P}[\,L(G(n,p_z)) = s,\ \bar{M}(G(n,p_z)) = \bar{t}\,] \\ &\sim \mathbb{P}[\,L(G(n,p_z)) = s\,] \cdot \mathbb{P}[\,\mathrm{Bi}\left(\binom{n-s}{2}, p_z\right) = \bar{t}\,].\end{aligned}$$

This yields an asymptotic formula for $f(z)$, because the local limit theorem for $L(G(n,p_z))$ (Theorem 5) gives an asymptotic expression for $\mathbb{P}[\,L(G(n,p_z)) = s\,]$ and the local limit theorem for the binomial distribution yields an explicit expression for $\mathbb{P}[\,\mathrm{Bi}\left(\binom{n-s}{2}, p_z\right) = \bar{t}]$. Putting these together we see that the function

$$\begin{aligned}F(z) &:= \frac{n}{2\pi(1-\rho)\sigma\,\bar{\sigma}} \cdot \exp\left(-\frac{1}{2}\left(\frac{2\sigma((1-\rho)-(1-\rho)^2)}{\bar{\sigma}(1-d(1-\rho))}\,z - \frac{s-\mu}{\bar{\sigma}}\right)^2\right) \\ &\quad \cdot \exp\left(-\frac{1}{2}\left((1-\rho)\,z + \frac{\tilde{\mu}-\bar{t}}{(1-\rho)\sigma} - \frac{d(s-\mu)}{\sigma}\right)^2\right) \qquad (21)\end{aligned}$$

is such that for all $z \in [-a,a]$ for arbitrarily large but fixed constant $a > 0$,

$$|f(z) - F(z)| = o(1).$$

Furthermore, by the definition of $m_z = m + z\sigma$, we have $m_0 = m$, and because in $G(n,m)$ the total number of edges is fixed to be m, we have

$$\bar{M}(G(n,m)) = m - M(G(n,m)).$$

The choice of parameters (e.g. $t = m - \bar{t}$) ensures that

$$\mathbb{P}[\, L(G(n,m)) = s,\ M(G(n,m)) = t\,] \\ = \mathbb{P}[\, L(G(n,m_0)) = s,\ \bar{M}(G(n,m_0)) = \bar{t}\,] = \frac{g(0)}{n}.$$

Finally, we show that there exists an explicitly defined function $h(z)$ satisfying

$$F = h * \phi, \quad \|f - F\|_2 = o(1), \quad |h(0) - g(0)| = o(1). \tag{22}$$

Indeed, we obtain the explicit expression of $h(z)$ by taking the Fourier transform of $\hat{F}/\hat{\phi}$, in which F and ϕ (and therefore $\hat{F}$ and $\hat{\phi}$) are explicitly given in (20) and (21). In particular, we have

$$h(0) = n\, Q(s,t),$$

where $Q(s,t)$ is defined as in Theorem 7. This and the third relation in (22) yield $|g(0) - n\, Q(s,t)| = o(1)$ and also the desired bivariate local limit theorem (Theorem 7), because $Q(s,t) = \Theta(1/n)$. This concludes the sketch of the proof using Fourier analysis.

This Fourier analytic approach was introduced by Behrisch, Coja-Oghlan, and Kang [5] to obtain the bivariate local limit theorem for the order and size of the giant component in $H_k(n,m)$ for any $k \geq 2$ (Theorem 13).

6 Concluding remarks

Since the seminal work of Erdős and Rényi [18] five decades ago, the giant component has been one of the central themes in the theory of random graphs. We have now extensive results on the birth, order, size, central and local limit theorems of the giant component. These results have been proved and strengthened using various techniques such as combinatorial counting, sprinkling, branching processes, Martingale, Fourier analysis, smoothing techniques, just to mention a few.

In this article we have discussed recent results on giant components in random graphs and random hypergraphs, central and local limit theorems for the giant components, and two proof techniques (branching processes and Fourier analysis). This article is not an exhaustive survey of the topic; our aim and hope is to provide readers with accessible and concise material on some developments and innovations in this ever growing area.

Acknowledgements This article is based on the author's lecture at the IMA Workshop "Probabilistic and Extremal Combinatorics", which took place at the Institute of Mathematics and its Applications (IMA), University of Minnesota, 8–12 September 2014. The author is supported by the Austrian Science Fund (FWF): P26826, W1230.

References

1. N. Alon, J. Spencer, *The Probabilistic Method*, 3rd edn. (Wiley, New York, 1992)
2. A. Barbour, M. Karoński, A. Ruciński, A central limit theorem for decomposable random variables with applications to random graphs. J. Comb. Theory Ser. B **47**, 125–145 (1989)
3. M. Behrisch, A. Coja-Oghlan, M. Kang, The order of the giant component of random hypergraphs (2007). http://arxiv.org/abs/0706.0496
4. M. Behrisch, A. Coja-Oghlan, M. Kang, The order of the giant component of random hypergraphs. Random Struct. Algorithm. **36**, 149–184 (2010)
5. M. Behrisch, A. Coja-Oghlan, M. Kang, Local limit theorems for the giant component of random hypergraphs. Comb. Probab. Comput. **23**, 331–366 (2014)
6. M. Behrisch, A. Coja-Oghlan, M. Kang, The asymptotic number of connected d-uniform hypergraphs. Comb. Probab. Comput. **23**, 367–385 (2014)
7. M. Behrisch, A. Coja-Oghlan, M. Kang, Corrigendum 'The asymptotic number of connected d-uniform hypergraphs'. Comb. Probab. Comput. **24**, 373–375 (2014)
8. E. Bender, E. Canfield, B. McKay, The asymptotic number of labeled connected graphs with a given number of vertices and edges. Random Struct. Algorithm. **1**, 127–169 (1990)
9. B. Bollobás, The evolution of random graphs. Trans. Am. Math. Soc. **286**, 257–274 (1984)
10. B. Bollobás, *Random Graphs*, 2nd edn. (Cambridge University Press, Cambridge, 2001)
11. B. Bollobás, O. Riordan, A simple branching process approach to the phase transition in $G_{n,p}$. Electron. J. Comb. **19**, Paper 21 (2012)
12. B. Bollobás, O. Riordan, Asymptotic normality of the size of the giant component in a random hypergraph. Random Struct. Algorithm. **41**, 441–450 (2012)
13. B. Bollobás, O. Riordan, Counting connected hypergraphs via the probabilistic method (2014). http://arxiv.org/abs/1404.5887
14. A. Coja-Oghlan, C. Moore, V. Sanwalani, Counting connected graphs and hypergraphs via the probabilistic method. Random Struct. Algorithm. **31**, 288–329 (2007)
15. O. Cooley, M. Kang, Y. Person, Giant components in random hypergraphs (2014). http://arxiv.org/abs/1412.6366
16. O. Cooley, M. Kang, C. Koch, The size of the giant components in random hypergraphs (2015). http://arxiv.org/abs/1501.07835
17. O. Cooley, M. Kang, C. Koch, Threshold and hitting time for high-order connectivity in random hypergraphs (2015). http://arxiv.org/abs/1502.07289
18. P. Erdős, A. Rényi, On the evolution of random graphs. Publ. Math. Inst. Hung. Acad. Sci. **5**, 17–61 (1960)
19. S. Janson, T. Łuczak, A. Ruciński, *Random Graphs* (Wiley, New York, 2000)
20. M. Karoński, T. Łuczak, The phase transition in a random hypergraph. J. Comput. Appl. Math. **142**, 125–135 (2002)
21. R. Karp, The transitive closure of a random digraph. Random Struct. Algorithm. **1**, 73–93 (1990)
22. T. Łuczak, On the number of sparse connected graphs. Random Struct. Algorithm. **1**, 171–173 (1990)
23. T. Łuczak, Component behavior near the critical point of the random graph process. Random Struct. Algorithm. **1**, 287–310 (1990)
24. B. Pittel, N. Wormald, Counting connected graphs inside out. J. Comb. Theory Ser. B **93**, 127–172 (2005)
25. C.M. Sato, N. Wormald, Asymptotic enumeration of sparse connected 3-uniform hypergraphs (2014). http://arxiv.org/abs/1401.7381
26. J. Schmidt-Pruzan, E. Shamir, Component structure in the evolution of random hypergraphs. Combinatorica **5**, 81–94 (1985)
27. V.E. Stepanov, On the probability of connectedness of a random graph $g_m(t)$. Theory Prob. Appl. **15**, 55–67 (1970)
28. R. van der Hofstad, J. Spencer, Counting connected graphs asymptotically. Eur. J. Comb. **27**, 1294–1320 (2006)

Infinite random graphs and properties of metrics

Anthony Bonato and Jeannette Janssen

Abstract We give a survey of recent developments in the theory of countably infinite random geometric graphs. Classical results of Erdős and Rényi establish that countably infinite random graphs are isomorphic with probability 1. Infinite random graphs have vertices identified with points in a metric space, and edges are added with a given probability dependent on the relative location of their endpoints. The probability that infinite random geometric graphs are isomorphic is considered. The metric spaces where such a unique isotype emerges are indeed fairly rare, and specifically arise in the context of finite-dimensional normed spaces equipped with the ℓ_∞-metric. We survey negative results for random geometric graphs in the cases of the Euclidean and hexagonal metric. Recent work which considers infinite random geometric graphs in the general setting of normed linear spaces is described. Open problems in the area are provided in the final section.

Keywords Random geometric graph • Infinite graph • Rado set • Step isometry • Graphs in normed spaces

Mathematics Subject Classification (2010): 05C63, 05C80

1 Introduction

Geometric random graph models play an emerging role in the modelling of real-world networks such as on-line social networks [11–13], wireless networks [21], and the web graph [1, 20]. In such stochastic models, vertices of the network are represented by points in a suitably chosen metric space, and edges are chosen by a mixture of relative proximity of the vertices and probabilistic rules. In real-world networks, the underlying metric space is a representation of the hidden reality that

A. Bonato
Department of Mathematics, Ryerson University, Toronto, ON, Canada, M5B 2K3
e-mail: abonato@ryerson.ca

J. Janssen (✉)
Department of Mathematics and Statistics, Dalhousie University, Halifax, NS, Canada, B3H 3J5
e-mail: jeannette.janssen@dal.ca

A. Beveridge et al. (eds.), *Recent Trends in Combinatorics*, The IMA Volumes in Mathematics and its Applications 159, DOI 10.1007/978-3-319-24298-9_11

leads to the formation of edges. Such networks can be viewed as embedded in a *feature space*, where vertices with similar features are more closely positioned. An example arises in the social sciences called Blau space [27]. In Blau space, agents in the social network correspond to points in a metric space, and the relative position of nodes follows the principle of *homophily*: nodes with similar socio-demographics are closer together in the space. The web graph may be viewed in *topic space*, where web pages with similar topics are closer to each other. We note that the theory of random geometric graphs has been extensively developed; see, for example, [3, 18, 23, 25, 33], and the books [28, 29].

While random finite geometric graphs have been closely investigated, the same is not true for the infinite case. One of the most studied examples of an infinite limit graph arising from a stochastic model is the infinite random graph. The probability space $G(\mathbb{N}, p)$ consists of graphs with vertices $\mathbb{N}$, so that each distinct pair of integers is adjacent independently with a fixed probability $p \in (0, 1)$. Erdős and Rényi [19] discovered that with probability 1, all $G \in G(\mathbb{N}, p)$ are isomorphic. A graph G is *existentially closed* (or *e.c.*) if for all finite disjoint sets of vertices A and B (one of which may be empty), there is a vertex $z \notin A \cup B$ adjacent to all vertices of A and to no vertex of B. We say that z is *correctly joined* to A and B. The unique isomorphism type of countably infinite e.c. graph is named the *infinite random graph*, or the *Rado* graph, and is written R. The graph R possesses many properties such as homogeneity, universality, and inexhaustibility. See Chapter 6 of [6] and the surveys [15, 16] for additional background on R.

We considered infinite random geometric graphs first in [8], and studied in that paper the isomorphism types of their associated countable limit structures. The following stochastic model introduced in [8] is sufficiently general for our purposes. Consider a metric space S with distance function

$$d : S \times S \to \mathbb{R},$$

a positive real number δ, a countable subset V of S, and $p \in (0, 1)$. The *Local Area Random Graph* LARG(V, δ, p) has vertices V, and for each pair of vertices u and v with $d(u, v) < \delta$, an edge is added independently with probability p. In other words, we consider a random, constant-radius disk model on a subset of a metric space. Note that V may be either finite or infinite, although we will consider only V infinite in this chapter.

The LARG model generalizes well-known classes of random graphs. For example, special cases of the LARG model include the random geometric graphs (where $p = 1$), and the binomial random graph $G(n, p)$ (where S has finite diameter D, and $\delta \geq D$).

Our focus will be on the case of separable metric spaces (that is, those containing a countable dense set), where S is chosen as a countable dense subset. Such metric spaces are precisely the second-countable ones (that is, containing a countable base). Following notation introduced in [4], we say that a countable dense set S is *Rado* if the resulting graph is with probability 1 unique up to isomorphism, for any $p \in (0, 1)$, and we say it is *strongly non-Rado* if any two such graphs are with probability 1 not isomorphic. Note that these are properties of subsets of metric spaces. A fundamental question when studying graphs generated by the LARG model is to determine which sets are Rado or strongly non-Rado. Perhaps surprisingly, as described in Section 4, there are sets S which are neither Rado nor strongly non-Rado.

In this chapter, we survey results (with proofs omitted or sketched) on the Rado property of dense subsets of normed spaces such as finite-dimensional ℓ_p spaces. A fundamental tool in the analysis of infinite geometric graphs is the geometric e.c. property, discussed in Section 2. We then describe an infinite family of Rado sets in Section 3, working in the normed spaces ℓ_∞^d of dimension d. Strongly non-Rado sets are described in ℓ_2^2 and under the hexagonal metric in the plane in Section 4. The recent results of [10] and [4] are then surveyed in Section 5, which shows that almost all sets are non-Rado in a normed space, unless the space is isomorphic to ℓ_∞^d for some d. Interestingly, the tools used in [4] come from classical results in functional analysis. We finish with a set of open problems.

Throughout, all graphs considered are simple, undirected, and countable unless otherwise stated. The cardinality of the natural numbers is denoted by $\aleph_0$. We will encounter two distinct notions of distance: metric distance and graph distance. In a metric space with metric d, we write $d(u, v)$ for the metric distance of the points. For a graph G, we write $d_G(u, v)$ for the graph distance. For a real number $1 \leq p \leq \infty$ and $d \geq 1$ an integer, the vector space $\mathbb{R}^d$ of dimension d equipped with the metric derived from the p-norm is denoted by ℓ_p^d. Given a metric space S with distance function d, denote the (open) *ball of radius δ around x* by

$$B_\delta(x) = \{u \in S : d(u, x) < \delta\}.$$

We will sometimes just refer to $B_\delta(x)$ as a *δ-ball* or *ball of radius δ*. A subset V is *dense* in S if for every point $x \in S$, every ball around x contains at least one point from V. We refer to $u \in S$ as points or vertices, depending on the context. For a reference on graph theory the reader is directed to [17, 34], while [14] is a reference on metric spaces.

2 Geometrically e.c. graphs

Adjacency properties have an important role in characterizing infinite random geometric graphs. For a survey of adjacency properties, see [7]. A graph G is *existentially closed* or *e.c.* if for all finite sets A and B of disjoint vertices, there is a vertex $z \notin A \cup B$ that is *correctly joined* to A and B; that is, z is joined to each vertex in A and to no vertex in B. Any two countable e.c. graphs are isomorphic; the isomorphism type is named the *infinite random* or *Rado graph*, and is written R. A central result in infinite graph theory was proven by Erdős and Rényi [19] which states that with probability 1, a countably infinite random graph is isomorphic to R.

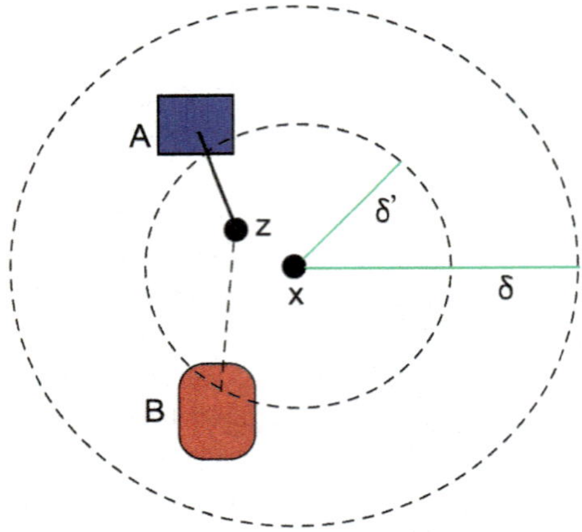

Fig. 1 The δ-g.e.c. property.

We next consider a geometric analogue of the e.c. property. Let $G = (V, E)$ be a graph whose vertices are points in the metric space S with metric d. The graph G is *geometrically e.c. at level* δ (or δ*-g.e.c.*) if for all δ' so that $0 < \delta' < \delta$, for all $x \in V$, and for all disjoint finite sets A and B so that $A \cup B \in B_\delta(x)$, there exists a vertex $z \notin A \cup B \cup \{x\}$ so that

(i) z is correctly joined to A and B,
(ii) for all $u \in A \cup B$, $d(u, z) < \delta$, and
(iii) $d(x, z) < \delta'$.

This definition implies that V is dense in itself, since we may choose A and B to be empty. Further, if G is δ-g.e.c., then G is δ'-g.e.c. for any $\delta' < \delta$.

The g.e.c. property bears clear similarities with the e.c. property defined in the introduction. The important differences are that a correctly joined vertex must exist only for sets A and B which are contained in an open ball with radius δ and centre x, and it must be possible to choose the vertex z correctly joined to A and B arbitrarily close to x; see Fig. 1.

The following result demonstrates that graphs generated by the LARG model are typically g.e.c.

Theorem 1 ([8]). *Let* (V, d) *be a metric space and* S *a countable dense subset of* V. *If* $\delta > 0$ *and* $p \in (0, 1)$, *then with probability* 1, LARG(V, δ, p) *is* δ*-g.e.c.*

A graph $G = (V, E)$ whose vertices are points in the metric space (S, d) has *threshold* δ if for all edges $uv \in E$, $d(u, v) < \delta$. A graph that is g.e.c. at level δ and has threshold δ is called a *geometric* δ*-graph*. By definition, a graph G generated by LARG(V, δ, p) has threshold δ, and, if V is countable and dense in itself, then G is a geometric δ-graph. Thus, this random graph model generates geometric δ-graphs.

Balls with radius δ in δ-g.e.c. graphs contain copies of R, and hence, contain isomorphic copies of all countable graphs.

Theorem 2 ([8]). *Let $U \subseteq S$ be so that $U \subseteq B_\delta(x)$ for some $x \in U$. Then a δ-g.e.c. graph with vertex set U is e.c., and so is isomorphic to R.*

The following important theorem demonstrates that there exists a close relationship between graph distance and metric distance in any graph that is a geometric δ-graph. We denote the closure of set V in S by $\overline{V}$. The set W is *convex* if for every pair of points x and y in W, there exists a point $z \in W$ such that

$$d(x, z) + d(z, y) = d(x, z).$$

Theorem 3 ([8]). *Let $G = (V, E)$ be geometric δ-graph, and let $\overline{V}$ be convex. Let $u, v \in V$ so that $d(u, v) > \delta$. Then we have that*

$$d_G(u, v) = \lfloor d(u, v)/\delta \rfloor + 1.$$

Theorem 3 directly leads to the following corollary, which is an important tool for proofs in the next two sections.

Corollary 1 ([8]). *If $\overline{V}$ and $\overline{W}$ are convex, and there is a geometric δ-graph with vertex set V and a geometric γ-graph with vertex set W which are isomorphic via f, then for every pair of vertices $u, v \in V$,*

$$\lfloor d(u, v)/\delta \rfloor = \lfloor d(f(u), f(v))/\gamma \rfloor.$$

Corollary 1 will prove useful as we survey results in the chapter, and it suggests the following generalization of isometry. Given metric spaces (S, d_S) and (T, d_T), sets $V \subseteq S$ and $W \subseteq T$, and positive real numbers δ and γ, a *step-isometry at level* (δ, γ) from V to W is a bijective map $f : V \to W$ with the property that for every pair of vertices $u, v \in V$,

$$\lfloor d_S(u, v)/\delta \rfloor = \lfloor d_T(f(u), f(v))/\gamma \rfloor.$$

Every isometry is a step-isometry, but the converse is false, in general. For example, consider $\mathbb{R}$ with the Euclidean metric, and let $\delta = \gamma = 1$. Then $f : \mathbb{R} \to \mathbb{R}$ given by $f(x) = \lfloor x \rfloor + (x - \lfloor x \rfloor)^2$ is a step-isometry, but is not an isometry.

Throughout the rest of the chapter, for simplicity and without loss of generality, we assume that $\delta = 1$.

3 Isomorphism results for ℓ_∞

The first success in the theory of infinite random geometric graphs came from the realization that for certain dense sets, the LARG model generates graphs that are, with probability 1, isomorphic, regardless of the choice of threshold δ or link probability p. The metric spaces containing these dense sets are the spaces ℓ_∞^d. In fact, we will see that for each d, there is a unique isomorphism type that arises for "almost all" dense vertex sets in $\mathbb{R}^d$.

3.1 ℓ_∞^d gives unique random infinite graphs

First, we consider ℓ_∞^1. The vertex set V must be dense in $\mathbb{R}$, and it must satisfy the property given by the following definition. A set $V \subseteq \mathbb{R}$ is *integer distance free* (or *idf*) if for any two distinct elements u, v in V, we have that $d(u, v) \notin \mathbb{Z}$.

Theorem 4 ([8]). *Let V and W be two countable dense idf subsets of $\mathbb{R}$. If G is a geometric 1-graph with vertex set V and H is a geometric 1-graph with vertex set W, then $G \cong H$.*

A proof sketch shows how an isomorphism can be constructed. For each $x \in \mathbb{R}$, let $q(x) = \lfloor x \rfloor$ and $r(x) = x - q(x)$. We refer to $r(x)$ as the represenative of x and to $q(x)$ as the *quotient*. The following lemma gives an alternative definition of step-isometries in terms of quotients and representatives.

Lemma 1 ([8]). *Let V and W be subsets of $\mathbb{R}$. Assume V and W are idf. Then a bijective function $f : V \to W$ is a step-isometry if and only if the following two conditions hold.*

1. *For every $u, v \in V$, if $r(u) < r(v)$, then $r(f(u)) < r(f(v))$.*
2. *For every $u \in V$, $q(u) = q(f(u))$.*

The proof of Theorem 4 follows using a variant of the back-and-forth method (used to show that R is the unique isotype of e.c. graph). Let $V = \{v_i : i \geq 0\}$ and $W = \{w_i : i \geq 0\}$ and assume that $v_0 = w_0 = 0$. We inductively construct an isomorphism F between pairs of finite sets (V_i, W_i) where V_i contains v_i and W_i contains w_i.

As an additional induction hypothesis we maintain conditions (1) and (2) from Lemma 1.

We first set $f(0) = 0$ and let $V_0 = W_0 = \{0\}$. For the induction step, fix $i \geq 0$. To extend f, we find an image of v_{i+1} which maintains the conditions. Let $v = v_{i+1}$.

Define

$$a = \max\{r(f(u)) : u \in V_i \text{ and } r(u) < r(v)\} \cup \{0\},$$

$$b = \min\{r(f(u)) : u \in V_i \text{ and } r(u) > r(v)\} \cup \{1\}.$$

Since the sets defining a and b are disjoint, and the idf property guarantees that all remainders in W are distinct, we have that $a < b$.

To maintain the induction hypothesis, $r(f(v))$ must lie in $[a, b)$, and $q(f(v))$ must equal to $q(v)$. Set

$$I = (q(v) + a, q(v) + b).$$

Any vertex in $W \cap I$ will qualify as a candidate for $f(v)$, so that conditions (1) and (2) are maintained, and thus, f remains a step-isometry. We must then find a vertex in I that will also guarantee that f is an isomorphism, by making sure it has the correct neighbors. This can be easily done by invoking the g.e.c condition.

Theorem 4 has the following corollaries.

Corollary 2 ([8]). *A countable dense idf set in $\mathbb{R}$ is Rado.*

The condition that a dense set should be idf is a weak one. In fact, if we select a dense set V from $\mathbb{R}$ randomly under a reasonable model, then V will be idf with probability 1. Examples of reasonable models to select a dense set V are to form V by taking countably many samples of a distribution on $\mathbb{R}$ whose density function is strictly positive almost everywhere, or to take the union of countably many Poisson point processes [32].

We obtain similar unique isomorphism types of graphs in all dimensions, as proven in the following result. For higher dimensions, we need to extend the definition of idf. Given a set $V \subseteq \mathbb{R}^n$, denote the i-th component set of V as:

$$V_i = \{x_i : x \in V\}$$

where $x = (x_1, x_2, \ldots, x_d)$. A set $V \subseteq \mathbb{R}^d$ is *idf* if the coordinate sets $V_1, \ldots, V_d$ are all idf.

Theorem 5 ([8]). *Consider the metric space ℓ_∞^d. Let V and W be two countable dense idf sets in $\mathbb{R}^d$. If G is a geometric 1-graph with vertex set V and H is a geometric 1-graph with vertex set W, then $G \cong H$. In particular, for all choices of dense idf vertex set V, there is a unique isomorphism type of g.e.c. graphs in ℓ_∞^d, written GR_d.*

We inductively construct an isomorphism in a similar way as was sketched for Theorem 4.

For higher dimensions, we do not have a complete characterization of step-isometries as in Lemma 1, but we can use this lemma to obtain necessary conditions. Precisely, a bijective function $f : V \to W$ is a step-isometry if the following two conditions hold for all $u, v \in V$ and for all i, $1 \leq i \leq d$:

$$r(u_i) < r(v_i) \text{ if and only if } r(f(u)_i) < r(f(v)_i),$$
$$q(u_i) = q(f(u)_i).$$

To prove the theorem, we construct an isomorphism between sets V_i and W_i much as in the one-dimensional case. We now sketch how to extend an isomorphism $f : V_i \to W_i$ to a new vertex $v = v_{i+1}$.

For all j, $1 \leq j \leq d$, define

$$a_j = \max\{r(f(u)_j) : u \in V_i \text{ and } r(u_j) \leq r(v_j)\},$$
$$b_j = \min\{r(f(u)_j) : u \in V_i \text{ and } r(u_j) > r(v_j)\}.$$

Note that $a_j < b_j$ for all j. Namely, if not there must exist $1 \leq j \leq d$ and two points $u, w \in W_i$ so that $r(u_j) = r(w_j)$. This contradicts the fact that W is idf.

In order to maintain the induction hypothesis, for all j, $r(f(v))_j$ should lie in interval $[a_j, b_j)$, and $q(f(v)_j)$ should be equal to $q(v_j)$. Let $k_j = q(v_j)$, and consider the product set

$$I = \prod_{1 \leq j \leq d} (q(v_j) + a_j, q(v_j) + b_j).$$

Any vertex in I will qualify as a candidate for $f(v)$ so that f satisfies condition 1. To complete the proof, we use the fact that W is 1-g.e.c. to show that I contains a vertex that is correctly joined to the vertices in W_i so that f remains an isomorphism.

Corollary 3 ([8]). *For each dimension d, there exists a unique isotype of graph, written GR_d, such that for all countable dense subsets V of $\mathbb{R}$, so that V is idf, for all $p \in (0, 1)$, and for all $\delta > 0$, the graph* LARG(V, δ, p) *is isomorphic to GR_d.*

We name GR_d the *infinite random geometric graph of dimension d*. Note that GR_d has infinite diameter for all $d \geq 1$ (unlike R, which has diameter 2). We will explore the structure of the graphs GR_d in the next subsection.

We can apply Theorem 5 to obtain a result about isomorphisms between graphs with vertex sets in $\mathbb{R}^n$ if there exists a special type of map between the sets.

Theorem 6 ([8]). *Consider the metric space ℓ_∞^d. Let V and W be two countable idf sets in $\mathbb{R}^d$. Assume that for all $1 \leq i \leq d$, there exists a step-isometry from V_i to W_i. If G is a geometric 1-graph with vertex set V and H is a geometric 1-graph with vertex set W, then $G \cong H$.*

Note that the statement of this theorem does not require V and W to be dense in $\mathbb{R}^n$, but only in a compact subset of $\mathbb{R}^n$. It is not hard to see, using Lemma 1, that there exists a step-isometry between two intervals (a, b) and (a', b') if their lengths $t_1 = b - a$ and $t_2 = b' - a$ are not integers, and if $\lfloor t_1 \rfloor = \lfloor t_2 \rfloor$. Applying this fact to each dimension, and using the above theorem, we obtain the following corollary.

Corollary 4. *Let V and W be two countable sets in $\mathbb{R}^d$. Suppose that for each $1 \leq i \leq d$, the closures $\overline{V_i}$ and $\overline{W_i}$ are intervals of length $t_{V,i}$, $t_{W,i}$, respectively, where $t_{V,i}$ and $t_{W,i}$ are not integers, and $\lfloor t_{V,i} \rfloor = \lfloor t_{W,i} \rfloor$. Then there exists a step-isometry between V and W. Moreover, if G is a geometric 1-graph with vertex set V and H is a geometric 1-graph with vertex set W, then $G \cong H$.*

Thus, for any d-tuple $(t_1, t_2, \ldots, t_d)$ there is a unique isomorphism type of g.e.c. graphs in ℓ_∞^d with countable vertex set, corresponding to 1-g.e.c. graphs whose vertex sets are dense in a set $I_1 \times \cdots \times I_d$, where each I_i is an interval of length $t_i + \epsilon_i$, and $0 < \epsilon_i < 1$.

3.2 Properties of the unique limit

One question is to determine whether for distinct dimensions d, the infinite random geometric graphs GR_d are non-isomorphic. We settle this question in the affirmative here using a geometric property of metric spaces.

The *equilateral dimension* of a metric space S is the maximum number of equidistant points in S. In [30], it was proven that in a Banach space of dimension d, the equilateral dimension is bounded above by 2^d, and from below by $\min\{4, d+1\}$. The equilateral dimension of ℓ_∞^d equals 2^d (with the lower bound witnessed by the set of binary vectors), while it equals $d+1$ in Euclidean space ℓ_2^d. Interestingly, the equilateral dimension of ℓ_1^d is unknown, and is claimed to be $2d$ in what is referred to as *Kusner's conjecture*; see [24]. Note that the standard basis vectors and their negatives witness the lower bound of $2d$ in the ℓ_1^d case. See [2] for further background on equilateral dimension of the ℓ_p spaces.

We utilize equilateral dimension in our graph theoretic setting via the following definition. For a graph G and positive integer t, define the *equilateral clique number*, written $\omega(G, t)$, to be the supremum of the cardinalities of a set A of vertices so that every pair of distinct vertices in A have graph distance exactly t. We first provide a lower bound on the equilateral clique number for any geometric 1-graph in ℓ_p^d.

Theorem 7. *Let k be the equilateral dimension of ℓ_p^d, where $d \geq 1$ and $1 \leq p \leq \infty$. Then for any geometric 1-graph G with vertex set V dense in ℓ_p^d, and for all $t > 2$, we have that $\omega(G, t) \geq k$.*

Proof. Fix $t \geq 2$. Choose a real number $\epsilon > 0$ so that $5\epsilon < 1$. Let S be a subset of ℓ_p^d so that $|S| = k$ and every pair of distinct points in S have distance exactly $t + 3\epsilon$. For each $x \in S$, choose a vertex $v_x \in V$ such that $d(x, v_x) < \epsilon$. Set $V_S = \{v_x : x \in S\}$. For all $v_x, v_y \in V_S$, we have that

$$t + \epsilon = d(x, y) - 2\epsilon < d(v_x, v_y) < d(x, y) + 2\epsilon = t + 5\epsilon < t + 1.$$

Hence, we have that $\lfloor d(v_x, v_y) \rfloor = t$, and so by Theorem 3, we have that the graph distance of v_x to v_y equals t. It follows that $\omega(G, t) \geq k$. □

The following theorem proves that the graphs GR_d are non-isomorphic for distinct dimensions d.

Theorem 8. *For all $d \geq 1$ and positive integers $t \geq 3$, $\omega(GR_d, t) = 2^d$.*

Proof. Recall that ℓ_∞^d has equilateral dimension 2^d. It follows from Theorem 7 and the facts that GR_d can be realized as a geometric 1-graph that $\omega(GR_d, t) \geq 2^d$.

Now, to prove that $\omega(GR_d, t) \leq 2^d$, we use induction on $d \geq 1$. In fact, we prove a stronger statement that ℓ_∞^d does not contain $2^d + 1$ points whose pairwise distance lies strictly between t and $t + 1$. The result then follows by Theorem 3.

The case $d = 1$, is elementary, since one cannot find real numbers x, y, and z with the property that $|x-y|$, $|x-z|$, and $|y-z|$ are all greater than 3, but differ by at most 1. The proof for $d = 1$ now follows by Theorem 3.

Now, fix $d \geq 1$ and assume the statement is true for $d = 1$. Let A' be a set of maximum cardinality in ℓ_∞^d such that for all pairs $a, b \in A'$ we have that $\lfloor d(a,b) \rfloor = t$. For all $a = (a_1, a_2, \ldots, a_d) \in A$, let a' be the projection onto the first $d-1$ dimensions; that is, $a' = (a_1, a_2, \ldots, a_{d-1})$. Assume, without loss of generality, that the origin is an element of A, and $a_1 \geq 0$ for all $a \in A$.

We set

$$A_1 = \{a \in A : 0 \leq a_d < \frac{t+1}{2}\},$$

and

$$A_2 = \{a \in A : \frac{t+1}{2} \leq a_d < t+1\}.$$

It is evident that A_1 and A_2 partition A. Without loss of generality, assume that $|A_1| \geq |A_2|$, so $|A_1| \geq \frac{|A|}{2}$. For all $a, b \in A_1$, by definition we have that $|a_d - b_d| < \frac{t+1}{2} < t$, and also that $d(a,b) \geq t$.

Therefore, the distance between a and b is not achieved in the last coordinate, so $d(a', b') \geq t$, where $a', b' \in \ell_\infty^{d-1}$. Hence, A_1', the projection of A_1 onto the first $d-1$ coordinates, is a set of vertices in ℓ_∞^{d-1} such that for all $a', b' \in A_1'$, $\lfloor d(a', b') \rfloor = t$. It follows by induction that $|A_1| \leq 2^{d-1}$ and so $|A| \leq 2|A_1| \leq 2^d$. □

We note that it is straightforward to prove that $\omega(GR_d, t) = \aleph_0$, where $t = 1, 2$ (since such graphs contain the infinite random graph as an induced subgraph).

Deleting a point from a dense set S with the idf property in $\mathbb{R}^n$ gives another dense set with the idf property. As a consequence of Corollary 6, we have the following *inexhaustibility* property.

Corollary 5 ([8]). *For all $d > 0$ and vertices x in GR_d, $GR_d - x \cong GR_d$.*

The g.e.c. property may be used to prove a relationship between the graphs GR_d for different d. Suppose V is a dense idf subset of $\mathbb{R}^d$, and G a 1-geometric graph with vertex set V (so $G \cong GR_d$). Let $V^* = \{v \in V : 0 < v_d < 1\}$. Then it is not hard to see that the floor of the distance, and thus, the graph distance, between two vertices is determined by the projection onto the first $d-1$ dimensions. It follows that the subgraph induced by V^* is isomorphic to a 1-geometric graph whose vertex set is the projection of V^* onto the first $d-1$ dimension. By Theorem 5, this graph is isomorphic to GR_{d-1}, and since rounded distances are preserved, so are graph distances. Hence, the embedding of this subgraph into G is isometric. Applying this argument repeatedly, we obtain the following theorem.

Theorem 9. *For $j < k$, GR_k contains GR_j as an isometric induced subgraph.*

A *step-isometric isomorphism* is an isomorphism of graphs that is a step-isometry. The following corollary shows that the graphs GR_n act transitively on step-isometric isomorphic induced subgraphs.

Corollary 6 ([8]). *Let G and H be finite induced subgraphs of GR_d for some positive integer d. A step-isometric isomorphism $f : G \to H$ extends to an automorphism of GR_d.*

4 Non-isomorphism results

The normed space ℓ_∞^d is rather special from the viewpoint of infinite random geometric graphs. This notion will be rigorously explored in the next section. Here, we are content to provide some results demonstrating that for some metric spaces, all dense subsets are strongly non-Rado.

4.1 The Euclidean metric in the plane

We first consider a result proved in [8]. The familiar Euclidean metric space in the plane gives examples of strongly non-Rado sets.

Theorem 10 ([8]). *If S is a countable set dense in ℓ_2^2, then S is strongly non-Rado*

The main tool in proving Theorem 10 is the following geometric lemma.

Lemma 2 ([8]). *Let V and W be dense subsets of ℓ_2^2. Then every step-isometry from V to W is an isometry.*

The lemma is proved by showing, using geometric configurations, that any small "error" in the distances between the vertices leads inevitably to larger errors in the distances between other pairs of vertices, and by repetition of the construction leads to errors that exceed 1. This then contradicts the properties of a step-isometry.

An intuitive interpretation of how this lemma leads to the non-isomorphism result is the following. Any isometry in ℓ_2^2 is determined by the images of three points in general position. Any isomorphism between two geometric 1-graphs G and H must be a step-isometry and thus, an isometry. Suppose that G and H were generated by the LARG process. We have countably many choices for the images, under an isomorphism for G to H, of the first three vertices. Once these images are fixed, the images of the remaining vertices of G and H are completely determined. Thus, if G and H are to be isomorphic, for each of the $2^{\aleph_0}$ pairs of vertices, G and H must agree on the existence of an edge between that pair.

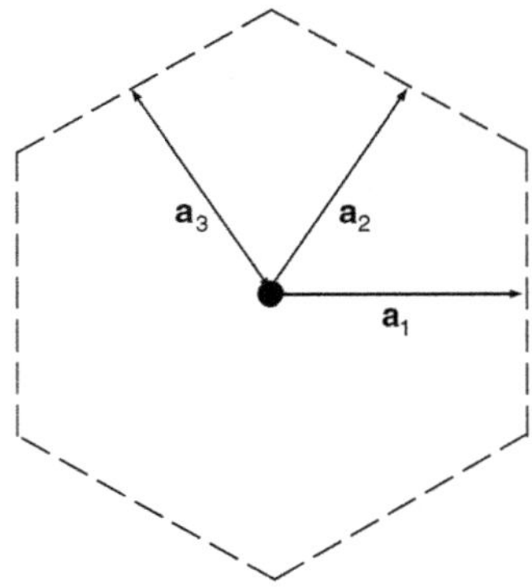

Fig. 2 The vectors a_i are the normal vectors for the regular hexagon.

4.2 Hexagonal metric

The hexagonal metric arises in the study of Voronoi diagrams and period graphs (see [22]) in computational geometry, which in turn has applications to nanotechnology. The hexagonal metric arises as a special case of *convex polygonal (or polygon-offset) distance functions*, where distance is in terms of a scaling of a convex polygon containing the origin; for example, see [5]. A precise definition of this metric is given below. We note that *honeycomb networks* formed by tilings by hexagonal meshes have been studied as, among other things, a model of interconnection networks.

We now formally define the hexagonal metric. Consider the vectors

$$a_1 = \begin{pmatrix} 1 \\ 0 \end{pmatrix}, a_2 = \begin{pmatrix} \frac{1}{2} \\ \frac{1}{2}\sqrt{3} \end{pmatrix}, \text{ and } a_3 = \begin{pmatrix} -\frac{1}{2} \\ \frac{1}{2}\sqrt{3} \end{pmatrix}.$$

These are the normal vectors to the sides of a regular hexagon, as shown in Fig. 2. We will call the three line directions perpendicular to a_1, a_2, a_3 the *principal directions*.

For $x \in \mathbb{R}^2$ define the *hexagonal norm* of x as follows:

$$\|x\|_{\text{hex}} = \max_{i=1,2,3} |a_i \cdot x|,$$

where $u \cdot v$ is the dot product of vectors u, v. The *hexagonal metric* in $\mathbb{R}^2$ is derived from the hexagonal norm, and defined by

$$d_{\text{hex}}(x, y) = \|x - y\|_{\text{hex}}.$$

Note that the unit balls with the hexagonal metric are regular hexagons as in Figure 2. We denote the hexagonal metric by d_{hex}.

Theorem 11 ([9]). *Almost all dense sets in $\mathbb{R}^2$ with the d_{hex}-metric are strongly non-Rado.*

The proof of Theorem 11 uses the fact that any isomorphism between geometric 1-graphs with vertex set V must be a step-isometry. It is evident that for any point $v \in V$, the balls $B(v, z)$ where $z \in \mathbb{N}$, must be *respected* by any isometry f; that is, any points inside the ball $B(v, z)$ must have an image inside the corresponding ball $B(f(v), z)$. This fact can be used to show that any line with a principal direction and going through a point $v \in V$ must also be respected. Moreover, any line at integer distance from such a line must also be respected. Thus, a finite set of points in V generates a grid of lines, all in principal directions, that must be respected. The regions bounded by this grid are of positive area, so this in itself does not lead to a contradiction. However, it can be shown that any line in a principal direction and going through an intersection point of lines of this grid must also be respected. Applying this fact iteratively leads to the formation of a dense grid of lines which must be respected, derived from only a finite number of points. Thus, any step-isometry is determined by the images of a finite number of points. The non-isomorphism result then follows along the same lines as in the previous case.

5 Normed spaces

In [4], the notions of Rado and strongly Rado sets were introduced, along with general results on the existence and non-existence of Rado sets in the setting of normed spaces. One of the key results of [4] is the following theorem, which establishes the special role of ℓ_∞ in the theory of random geometric graphs. A *random dense set* is a countable union of sets each chosen according to a Poisson point process [32].

Theorem 12 ([4]). *Let S be a finite-dimensional normed space not isometric to ℓ_∞^d. Then almost every random dense set V is strongly non-Rado.*

An analogue of Theorem 12 was proven in [10], using a different approach (in particular, segmenting the cases into the polygonal and non-polygonal metric cases).

The results of Theorem 12 correspond to a special case of the following key result. As proved in [4], for a finite-dimensional normed space S, there exists a unique maximal subspace T isometric to ℓ_∞^d for some d, such that there is a subspace U with $S = U \bigoplus T$ and

$$\|u + w\| = \max\{\|u\|, \|w\|\}$$

for all $u \in U$ and $w \in T$. This is referred to as the ℓ_∞-decomposition of V, and is written $(U \bigoplus \ell_\infty^d)_\infty$.

Theorem 13 ([4]). *Let S be a finite-dimensional normed space with ℓ_∞-decomposition $(U \bigoplus \ell_\infty^d)_\infty$, and fix $0 < p < 1$. Then the following hold.*

1. *If $U = 0$ in the ℓ_∞-decomposition, then almost all countable dense sets V are Rado, but there exist countable dense sets which are strongly non-Rado. Further, there exist countable dense sets S for which the probability that two graphs $G, G' \in \mathrm{LARG}(V, 1, p)$ are isomorphic lies strictly between 0 and 1.*
2. *If $S = U$, then all countable dense sets S are strongly non-Rado.*
3. *If $d > 0$ and $U \neq \{0\}$, then almost all countable dense sets V are strongly non-Rado, but there exist countable dense sets V which are Rado. Additionally, there exist countable dense sets for which the probability that two graphs $G, G' \in \mathrm{LARG}(V, 1, p)$ are isomorphic lies strictly between 0 and 1.*

Item (1) in Theorem 13 covers the case where the metric space is ℓ_∞^d, which we discussed in Section 3. Namely, almost all countable dense sets are idf, and thus, the first part of the case follows from Theorem 5 (which proves the stronger result that there is a unique isomorphism type for all countable dense sets that are idf). The remainder of this case can be understood when we consider what happens if V is not idf. In particular, suppose that V contains a subset W of points which have only integer coordinates. Then any isomorphism must be an isometry on W and must fix W. Hence, this restricts the possibilities for an isomorphism. We then have a relationship between the probability that the graphs are isomorphic to the probability that the subgraphs on W are isomorphic. The corresponding probability can be zero, or it can be between 0 and 1, depending on W.

Item (2) is a generalization of the results for ℓ_2^2 and the hexagonal metric discussed above. The proofs rely on the analysis of step-isometries, and their characterization. The proofs employ results from functional analysis such as the Mazur-Ulam theorem, and properties of extreme points in normed spaces. It is shown that any step-isometry f is highly constrained. For example, as part of the proof, it is shown that f generates a lattice derived from the extreme points of the unit ball, and that f must be an isometry on this lattice. The restrictions on step-isometries limit the possibilities for an isomorphism, and the non-isomorphism result then follows by an argument similar to the one sketched in the previous subsections. Item (3) follows similarly. The fact that there exist Rado sets and sets that are neither Rado nor non-Rado follows by construction of special sets where the ℓ_∞ component of the space is somehow dominant for pairs of vertices in the set.

6 Open problems

The most fundamental question in the study of infinite random geometric graphs is to classify Rado sets in all normed spaces. This problem was mentioned in [4], and may be summarized as follows.

Problem 1. [4]: Let V be a normed space with ℓ_∞-decomposition $V = (U \bigoplus \ell_\infty^d)_\infty$ for some $d > 1$. Classify the countable dense Rado sets.

The infinite-dimensional case is wide open. The second problem mentioned in [4] is the following.

Problem 2. [4]: Let V be an infinite-dimensional normed space. Classify the countable dense Rado sets.

The most concrete example to consider here is the space of bounded sequences, written ℓ_∞^∞, which has dimension $\aleph_0$. A challenge with the space ℓ_∞^∞ from our perspective is that it is not separable. Does such a space even contain Rado sets?

Notice that the LARG model easily adapts to any metric space. Theorem 1, which states that graphs generated with the LARG model are g.e.c., holds for any metric space. However, the other results discussed here are all about normed spaces, and the tools developed to prove the results apply only to such spaces. Perhaps the broadest generalization of this line of research is to consider general separable metric spaces.

Problem 3. Let V be a separable metric space. Classify the countable dense Rado sets.

The LARG model extends to the case of uncountable graphs; for example, one may consider the unit disk graph of uncountable dense subsets of the plane equipped with the ℓ_∞-metric. Infinite random geometric graphs where the underlying vertex set is uncountable appear to be more challenging to study, and the classification of such likely relies on various axioms of infinity in set theory. An interesting problem is therefore, the following.

Problem 4. Are there are uncountable Rado sets in normed spaces?

As a final remark, we point out possible connections with model theory. The infinite random graph R is the unique countably infinite graph satisfying almost sure (first-order) theory of graphs (see, for example, [31]). In related work, McColm [26] considered 0–1 laws for *Gilbert* graphs, which in our terminology are finite graphs generated by the LARG model with $p = 1$, where the underlying space is the one-dimensional unit circle S^1 (with metric the distance between reals modulo 2π). As mentioned in [26], it would be interesting to establish whether there exist 0–1 laws for the LARG model in certain metric spaces, and analyze the isomorphism types of countable models of the almost sure theory. In particular, an intriguing question is whether the infinite random geometric graphs GR_d of dimension d play an analogous role to the one played by R in the almost sure theory of geometric graphs.

Acknowledgements The second author wishes to thank IMA, which she visited during the annual thematic program on Discrete Structures. IMA gave her the opportunity to present this work as part of the seminar series, and to discuss it with other visitors. Both authors acknowledge support from grants from NSERC.

References

1. W. Aiello, A. Bonato, C. Cooper, J. Janssen, P. Prałat, A spatial web graph model with local influence regions. Internet Math. **5**, 175–196 (2009)
2. N. Alon, P. Pudlák, Equilateral sets in ℓ_p^n. Geom. Funct. Anal. **13**, 467–482 (2003)
3. P. Balister, B. Bollobás, A. Sarkar, M. Walters, Highly connected random geometric graphs. Discret. Appl. Math. **157**, 309–320 (2009)
4. P. Balister, B. Bollobás, K. Gunderson, I. Leader, M. Walters, Random geometric graphs and isometries of normed spaces (2015, preprint) [arXiv:1504.05324]
5. G. Barequet, M.T. Dickerson, M.T. Goodrich, Voronoi diagrams for convex polygon-offset distance functions. Discret. Comput. Geom. **25**, 271–291 (2001)
6. A. Bonato, *A Course on the Web Graph*. Graduate Studies Series in Mathematics (American Mathematical Society, Providence, 2008)
7. A. Bonato, The search for n-e.c. graphs. Contrib. Discret. Math. **4**, 40–53 (2009)
8. A. Bonato, J. Janssen, Infinite random geometric graphs. Ann. Comb. **15**, 597–617 (2011)
9. A. Bonato, J. Janssen, Infinite random geometric graphs from the hexagonal metric, in *Proceedings of IWOCA'12* (2012)
10. A. Bonato, J. Janssen, Infinite geometric graphs and properties of metrics (2015, preprint)
11. A. Bonato, D.F. Gleich, M. Kim, D. Mitsche, P. Prałat, A. Tian, S.J. Young, Dimensionality matching of social networks using motifs and eigenvalues. PLOS ONE **9**(9), e106052 (2014)
12. A. Bonato, J. Janssen, P. Prałat, Geometric protean graphs. Internet Math. **8**, 2–28 (2012)
13. A. Bonato, M. Lozier, D. Mitsche, X. Perez Gimenez, P. Prałat, The domination number of online social networks and random geometric graphs, in *Proceedings of TAMC'15* (2015)
14. V. Bryant, *Metric Spaces: Iteration and Application* (Cambridge University Press, Cambridge, 1985)
15. P.J. Cameron, The random graph, in *Algorithms and Combinatorics*, vol. 14, ed. by R.L. Graham, J. Nešetřil (Springer, New York, 1997), pp. 333–351
16. P.J. Cameron, The random graph revisited, in *European Congress of Mathematics*, vol. I, ed. by C. Casacuberta, R.M. Miró-Roig, J. Verdera, S. Xambó-Descamps (Birkhauser, Basel, 2001), pp. 267–274
17. R. Diestel, *Graph Theory*, 4th edn. (Springer, New York, 2010)
18. R. Ellis, X. Jia, C.H. Yan, On random points in the unit disk. Random Struct. Algorithm. **29**, 14–25 (2006)
19. P. Erdős, A. Rényi, Asymmetric graphs. Acta Math. Acad. Sci. Hung. **14**, 295–315 (1963)
20. A. Flaxman, A.M. Frieze, J. Vera, A geometric preferential attachment model of networks. Internet Math. **3**, 187–205 (2006)
21. A.M. Frieze, J. Kleinberg, R. Ravi, W. Debany, Line of sight networks. Comb. Probab. Comput. **18**, 145–163 (2009)
22. N. Fu, H. Imai, S. Moriyama, Voronoi diagrams on periodic graphs, in *Proceedings of the International Symposium on Voronoi Diagrams in Science and Engineering* (2010)
23. A. Goel, S. Rai, B. Krishnamachari, Monotone properties of random geometric graphs have sharp thresholds. Ann. Appl. Prob. **15**, 2535–2552 (2005)
24. R. Guy (ed.), Unsolved problems: an Olla-Podrida of open problems, often Oddly posed. Am. Math. Mon. **90**, 196–200 (1983)
25. J. Janssen, Spatial models for virtual networks, in *Proceedings of the 6th Computability in Europe* (2010)
26. G.L. McColm, First order zero-one laws for random graphs on the circle. Random Struct. Algorithm. **14**, 239–266 (1999)
27. J.M. McPherson, J.R. Ranger-Moore, Evolution on a dancing landscape: organizations and networks in dynamic Blau space. Soc. Forces **70**, 19–42 (1991)
28. R. Meester, R. Roy, *Continuum Percolation* (Cambridge University Press, Cambridge, 1996)
29. M. Penrose, *Random Geometric Graphs* (Oxford University Press, Oxford, 2003)
30. C.M. Petty, Equilateral sets in Minkowski spaces. Proc. Am. Math. Soc. **29**, 369–374 (1971)
31. J. Spencer, *The Strange Logic of Random Graphs* (Springer, Berlin, 2001)

32. B. Tsirelson, Brownian local minima, random dense countable sets and random equivalence classes. Electron. J. Probab. **11**, 162–198 (2006)
33. M. Walters, Random geometric graphs, in *Surveys in Combinatorics 2011*, ed. by R. Chapman. London Mathematical Society Lecture Note Series, vol. 392 (Cambridge University Press, Cambridge, 2011)
34. D.B. West, *Introduction to Graph Theory*, 2nd edn. (Prentice Hall, Upper Saddle River, 2001)

Nordhaus-Gaddum problems for Colin de Verdière type parameters, variants of tree-width, and related parameters

Leslie Hogben

Abstract A Nordhaus-Gaddum problem for a graph parameter is to determine a tight lower or upper bound for the sum or product of the parameter evaluated on a graph and on its complement. This article surveys Nordhaus-Gaddum results for the Colin de Verdière type parameters μ, ν, and ξ; tree-width and its variants largeur d'arborescence, path-width, and proper path-width; and minor monotone ceilings of vertex connectivity and minimum degree.

Keywords Nordhaus-Gaddum • Colin de Verdière type parameters • Tree-width • Halin S-functions • Maximum nullity • Minimum rank

Mathematics Subject Classification (2010): 05C50, 05C10, 05C15, 05C40, 05C83

1 Introduction

For a graph parameter ζ, the following are *Nordhaus-Gaddum* problems:

- Determine a (sharp) lower or upper bound on $\zeta(G) + \zeta(\overline{G})$.
- Determine a (sharp) lower or upper bound on $\zeta(G) \cdot \zeta(\overline{G})$.

The name comes from the next theorem (see Section 2 for the definition of the chromatic number $\chi(G)$ of G).

Theorem 1. [30] *For any graph G,*

$$2\sqrt{|G|} \leq \chi(G) + \chi(\overline{G}) \leq |G| + 1$$

L. Hogben (✉)
Department of Mathematics, Iowa State University, Ames, IA 50011, USA

American Institute of Mathematics, 600 E. Brokaw Road, San Jose, CA 95112, USA
e-mail: hogben@aimath.org

A. Beveridge et al. (eds.), *Recent Trends in Combinatorics*, The IMA Volumes in Mathematics and its Applications 159, DOI 10.1007/978-3-319-24298-9_12

and

$$|G| \leq \chi(G) \cdot \chi(\overline{G}) \leq \left(\frac{|G|+1}{2}\right)^2.$$

Each bound is assumed for infinitely many values of n.

An extensive survey of Nordhaus-Gaddum problems in general can be found in [1]. The focus of this survey is on Nordhaus-Gaddum problems for the following parameters: the Colin de Verdière type parameters μ, ν, and ξ; tree-width and its variants largeur d'arborescence la, path-width pw, and proper path-width ppw; and minor monotone ceilings of vertex connectivity $\lceil\kappa\rceil$ and minimum degree $\lceil\delta\rceil$.

These parameters are defined in Section 2, where their background and connections are discussed. Section 3 discusses the Nordhaus-Gaddum sum upper bounds for these parameters, which take the form of a constant times the order of the graph. The Nordhaus-Gaddum sum lower bounds are discussed in Section 4, which for μ, ν, and ξ is known in linear algebra as a Graph Complement Conjecture (GCC), and Section 5 discusses Nordhaus-Gaddum product upper and lower bounds. All of the parameters μ, ν, ξ, tw, la, pw, ppw, $\lceil\kappa\rceil$, $\lceil\delta\rceil$ are minor monotone, which makes the Nordhaus-Gaddum bounds more interesting (see Sections 3 and 4 for comments on the trivial values of the related non-minor monotone parameters).

2 Parameters

In this section we define some graph terminology and then define and discuss parameters. Colin de Verdière type parameters μ, ν, ξ, are discussed in Section 2.1, tree-width tw and variants largeur d'arborescence la, path-width pw, and proper-path-width ppw are discussed in Section 2.2, and the minor monotone ceilings of vertex connectivity $\lceil\kappa\rceil$ and minimum degree $\lceil\delta\rceil$ are discussed in Section 2.4. Section 2.3 reviews the less well-known Halin S-functions, which include tw $+1$ and $\lceil\kappa\rceil + 1$. Relationships between the parameters discussed here are shown in a diagram in Figure 5 in Section 4. A more extensive discussion of the connections between the parameters μ, ν, ξ, tw, la, pw, ppw, $\lceil\kappa\rceil$, $\lceil\delta\rceil$ and the problem of determining maximum nullity among real symmetric matrices described by a graph can be found in [4].

All graphs are simple, undirected, and finite. The *complement* of a graph $G = (V, E)$ is the graph $\overline{G} = (V, \overline{E})$, where $\overline{E}$ consists of all two element sets from V that are not in E. For a graph $G = (V, E)$ and $W \subseteq V$, the *induced subgraph* $G[W]$ is the graph with vertex set W and edge set $\{wu \in E : w, u \in W\}$ (wu and uw denote the same edge $\{w, u\}$). The subgraph induced by $\overline{W} = V \setminus W$ is usually denoted by $G - W$, or in the case W is a single vertex $\{v\}$, by $G - v$. If $G_1 = (V_1, E_1)$ and $G_2 = (V_2, E_2)$ are disjoint graphs, the *join* of G_1 and G_2, denoted by $G_1 \vee G_2$, is the graph having vertex set $V(G_1 \vee G_2) = V(G_1) \cup V(G_2)$ and edge set $E(G_1 \vee G_2) = E(G_1) \cup E(G_2) \cup E$, where E consists of all the edges v_1v_2 with $v_1 \in V(G_1)$ and $v_2 \in V(G_2)$. A graph is a *cograph* or *decomposable* if it can be expressed as a sequence of disjoint unions and joins of isolated vertices. If graphs G and H each contain a clique K_t, then a graph obtained by identifying these two specific copies of K_t is a *clique-sum of G and H on K_t* and is denoted by $G \langle K_t \rangle H$. The graph $\emptyset$ has no vertices, whereas $\overline{K_n}$ has $n \geq 1$ vertices but no edges.

The *vertex connectivity* $\kappa(G)$ of G is the smallest number k such that there is a set S of k vertices such that $G - S$ is disconnected, provided G is not complete. By convention, $\kappa(K_n) = n - 1$. The *minimum degree* of G is $\delta(G) = \min_{v \in V} \deg_G(v)$.

The *contraction* of an edge $e = uv$ of G, denoted by G/e, is obtained by identifying the vertices u and v, deleting any loops that arise in this process, and replacing any multiple edges by a single edge. A *minor* of G arises by performing a (possibly empty) sequence of deletions of edges, deletions of isolated vertices, and/or contractions of edges. A minor H of G obtained only by contractions can also be constructed by partitioning the vertices of G as $W_1 \dot{\cup} \cdots \dot{\cup} W_h = V(G)$ such that the induced subgraph $G[W_i]$ is connected for all $i = 1, \ldots, h$. Then $V(H) = \{W_1, \ldots, W_h\}$ and $W_iW_j \in E(H)$ if and only if $i \neq j$ and there exist $w_i \in W_i, w_j \in W_j$ such that $w_iw_j \in E(G)$. The notation $H \preceq G$ means that H is a minor of G, and $H \prec G$ means that H is a *proper minor* of G, i.e., $H \preceq G$ and $H \neq G$. A graph parameter ζ is *minor monotone* if $H \preceq G$ implies $\zeta(H) \leq \zeta(G)$.

For a graph parameter ζ that takes nonnegative integer values, the *minor monotone ceiling* $\lceil \zeta \rceil$ is defined by

$$\lceil \zeta \rceil(G) = \max\{\zeta(H) : H \preceq G\}$$

(see, for example, [4]). Observe that $\lceil \zeta \rceil$ is minor monotone and for all G, $\zeta(G) \leq \lceil \zeta \rceil(G)$. Minor monotone ceilings have been studied for several parameters by various authors, e.g., [4, 14, 15].

The *clique number* $\omega(G)$ is the maximum order of a clique in G, and the *Hadwiger number* $\eta(G)$ is the maximum order of a clique minor of G, i.e., $\eta = \lceil \omega \rceil$. The *chromatic number* $\chi(G)$ is the minimum number of colors necessary to color the vertices of G so that adjacent vertices have different colors. Since $\chi(K_n) = n$, $\omega(G) \leq \chi(G)$ and thus $\eta(G) \leq \lceil \chi \rceil(G)$ for all graphs G. The *Hadwiger Conjecture* is that for all graphs G, $\chi(G) \leq \eta(G)$, or equivalently, $\lceil \chi \rceil = \eta$.

2.1 Colin de Verdière Type Parameters

In 1990 Colin de Verdière ([8] in English) introduced the graph parameter μ to characterize planarity; $\mu(G)$ is equal to the maximum nullity among all matrices satisfying several conditions including the Strong Arnold Hypothesis (defined below). The Colin de Verdière number μ is the first of several parameters (called *Colin de Verdière type parameters*) that take the maximum nullity among real symmetric matrices that are described by the graph and satisfy the Strong Arnold Hypothesis, and possibly other conditions. All the Colin de Verdière type parameters we discuss have been shown to be minor monotone [2, 8, 9].

All matrices discussed are real and symmetric; the set of $n \times n$ real symmetric matrices is denoted by $S_n(\mathbb{R})$. Define $[n] = \{1, \ldots, n\}$. For $A = [a_{ij}] \in S_n(\mathbb{R})$, the *graph* of A is $\mathscr{G}(A) = (V, E)$ where $V = [n]$ and $E = \{ij : i, j \in [n], i \neq j, \text{ and } a_{ij} \neq 0\}$; the diagonal of A is ignored in determining $\mathscr{G}(A)$. The *set of symmetric matrices described by* G is $\mathscr{S}(G) = \{A \in S_n(\mathbb{R}) : \mathscr{G}(A) = G\}$. The *maximum nullity* of a graph is $\mathrm{M}(G) = \max\{\mathrm{null} A : A \in \mathscr{S}(G)\}$. We note that this parameter is not a Colin de Verdière type parameter and is not minor monotone.

A real symmetric matrix A satisfies the *Strong Arnold Hypothesis* provided there does not exist a nonzero real symmetric matrix X satisfying

- $AX = 0$,
- $A \circ X = 0$, and
- $I \circ X = 0$,

where $\circ$ denotes the entry-wise product, i.e., $(A \circ B)_{ij} = a_{ij} b_{ij}$, and I is the identity matrix. The Strong Arnold Hypothesis is equivalent to the requirement that certain manifolds intersect transversally (see [21]).

The Colin de Verdière number $\mu(G)$ is defined to be the maximum nullity among symmetric matrices $A = [a_{ij}]$ such that:

- $A \in \mathscr{S}(G)$;
- A satisfies the Strong Arnold Hypothesis;
- A has exactly one negative eigenvalue (counting multiplicity); and
- A is a generalized Laplacian (i.e., for all $i \neq j$, $a_{ij} \leq 0$).

Colin de Verdière introduced the parameter ν in [9]; the definition utilizes positive semidefinite matrices. A matrix $A \in S_n(\mathbb{R})$ is *positive semidefinite* if $\mathbf{x}^T A \mathbf{x} \geq 0$ for all $\mathbf{x} \in \mathbb{R}^n$, or equivalently all eigenvalues of A are nonnegative. Then $\nu(G)$ is defined to be the maximum nullity among matrices A such that:

- $A \in \mathscr{S}(G)$;
- A satisfies the Strong Arnold Hypothesis; and
- A is positive semidefinite.

The parameter $\xi(G)$ was introduced in [2]; $\xi(G)$ is the maximum nullity among real symmetric matrices such that:

- $A \in \mathscr{S}(G)$ and
- A satisfies the Strong Arnold Hypothesis.

Remark 1. For every graph G, $\nu(G) \leq \xi(G)$ and $\mu(G) \leq \xi(G)$, but μ and ν are noncomparable (see [4] for examples).

As noted in [19], the following result can be derived from results in [27, 28].

Theorem 2. [19, 27, 28] *For every graph* G, $\kappa(G) \leq \nu(G)$.

Since all Colin de Verdière parameters are minor monotone, the next corollary is immediate.

Corollary 1. [4] *For every graph* G, $\lceil \kappa \rceil(G) \leq \nu(G)$.

2.2 Tree-width and Variants

Tree-width is a well-known parameter with many equivalent characterizations, any one of which may be chosen as the definition; for more information see a standard reference such as [10]. Here we use the k-tree characterization as the definition.

A *k-tree* is constructed inductively by starting with a complete graph on $k + 1$ vertices and connecting each new vertex to the vertices of an existing clique on k vertices. A tree is a 1-tree. A k-tree is a k-connected chordal graph with the maximum order of a clique equal to $k + 1$ (G is *chordal* if G has no induced cycle of length greater than three). A *partial k-tree* is a subgraph of a k-tree. The maximal cliques of a k-tree are of order $k + 1$, and the *facets* of a maximal clique are its k-clique subgraphs. Two maximal cliques are *adjacent* if they share a facet. For a graph G, $\text{tw}(G)$ is the minimum k such that G is a partial k-tree. Note that $\omega(G) - 1 \leq \text{tw}(G)$ and an equivalent way to define tree-width is $\text{tw}(G) = \min\{\omega(H) - 1 : H \text{ is chordal supergraph of } G\}$.

Several related parameters are defined as the minimum k for which the graph is a subgraph of a certain type of k-tree. Examples with $k = 2$ are shown in Figure 1.

A *linear k-tree* is a k-tree constructed with the restriction that at each stage the new vertex is connected to the vertices of an existing K_k that includes a vertex of degree k (without loss of generality it can be assumed that the new vertex is adjacent to the vertex just added). Equivalently, a linear k-tree is a k-tree in which exactly two vertices have degree k (or is the k-tree K_{k+1}). A *partial linear k-tree* is a subgraph of a linear k-tree. The *proper path-width* of a graph G, denoted by $\text{ppw}(G)$, is the minimum k for which G is a partial linear k-tree. Proper path-width was introduced in [34].

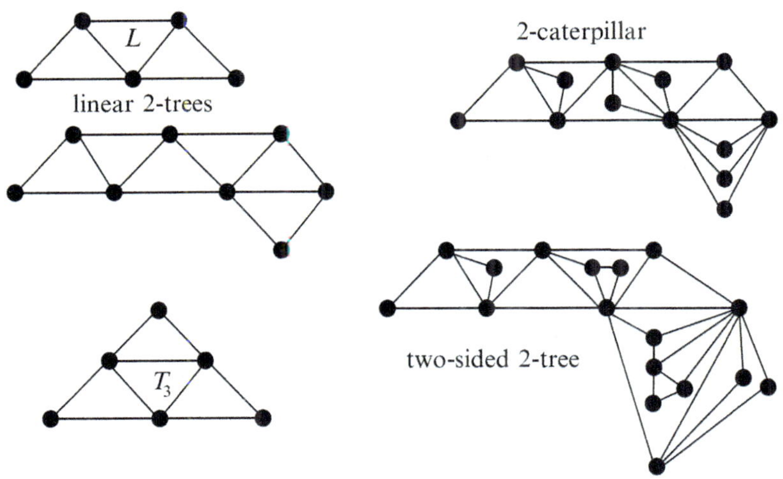

Fig. 1 Examples of 2-trees. Clockwise from top left: two linear 2-trees including the specific 2-tree L; a 2-caterpillar that is not a linear 2-tree; a two-sided 2-tree that is not a 2-caterpillar; the supertriangle T_3, which is not a two-sided 2-tree.

A *k-caterpillar* is a k-tree constructed with the restriction that at each stage the new vertex is adjacent to the k vertices of some facet of the maximal clique that was added at the previous stage; an example with $k = 2$ is shown in Figure 1. A k-caterpillar can also be constructed by first constructing a linear k-tree and then possibly adding extra vertices, with each new vertex adjacent to all the vertices in a facet that is shared by two maximal cliques of the linear k-tree used to construct the k-caterpillar. The *path-width* of a graph G, denoted by $\mathrm{pw}(G)$, is the minimum k for which G is a subgraph of a k-caterpillar.

A *two-sided k-tree* (or *straight k-tree*) is a k-tree constructed with the restriction that at each stage the new vertex is adjacent to the k vertices of an existing K_k that either includes a vertex of degree k or is the same as the K_k to which some previous vertex was connected. Every tree is a two-sided 1-tree. The *largeur d'arborescence* of a graph G is the minimum k for which G is a subgraph of a two-sided k-tree, and is denoted by $\mathrm{la}(G)$ [9, 20].

Any linear k-tree is a k-caterpillar, any k-caterpillar is a two-sided k-tree, and any two-sided k-tree is a k-tree, but not vice versa, as illustrated in Figure 1. Thus $\mathrm{tw}(G) \leq \mathrm{la}(G) \leq \mathrm{pw}(G) \leq \mathrm{ppw}(G)$. There are known close relationships between tree-width and largeur d'arborescence, and between path-width and proper path-width, as described in the next theorem.

Theorem 3. *For every graph G,*

1. [9] $\mathrm{tw}(G) \leq \mathrm{la}(G) \leq \mathrm{tw}(G) + 1$, *and*
2. [35] $\mathrm{pw}(G) \leq \mathrm{ppw}(G) \leq \mathrm{pw}(G) + 1$.

Tree-width, largeur d'arborescence, path-width, and proper path-width are all minor monotone. Alternative characterizations of largeur d'arborescence, path-width, and proper path-width in terms of a two-coloring with color change rule (called zero forcing) were given in [4] and we summarize these here.[1] Let G be a graph.

- A subset $B \subseteq V(G)$ defines a *coloring* by coloring all vertices in B blue and all the vertices not in B white.
- A *color change rule* (CCR) is a rule describing conditions on a vertex u and its neighbors under which u can cause a white vertex w to change color to blue; in this case we say *u forces w* and write $u \rightarrow w$.
- Given a coloring of G and a color change rule CCR-X, a CCR-X *final coloring* is a set of blue vertices obtained by applying CCR-X until no more color changes are possible.
- A CCR-X *zero forcing set* for G is a subset B of vertices such that if initially the vertices in B are colored blue and the remaining vertices are colored white, $V(G)$ is a CCR-X final coloring.

[1] There is also a characterization for tree-width by color change rule but it is more complicated and there are many other useful characterizations for tree-width, so we omit it here.

- The CCR-X *zero forcing number* is the minimum cardinality of B over all CCR-X zero forcing sets $B \subseteq V(G)$. The CCR-X zero forcing number is often denoted by X.

The study of zero forcing arose independently in mathematical physics in the study of control of quantum systems, and in combinatorial matrix theory in the study of maximum nullity/minimum rank of a graph, where it is associated with forcing zeros in a null vector of a matrix. Since our interest is in minor monotone parameters, we skip over the original parameters (see [4] for more information) and proceed directly to the minor monotone floors. The color change rules for $\lfloor Z \rfloor = \text{ppw}$, $\lfloor Z_\ell \rfloor = \text{pw}$, and $\lfloor Z_+ \rfloor = \text{la}$ are:[2]

CCR-$\lfloor Z \rfloor$ If u is blue and w is the only white neighbor of u, then change the color of w to blue. If u is blue, all neighbors of u are blue, and u has not yet performed a force, then change the color of any white vertex w to blue; in this case we say that u *hops* to w. It is not required that u have any neighbors for u to hop.

CCR-$\lfloor Z_\ell \rfloor$ If u is blue and exactly one neighbor w of u is white, then change the color of w to blue. If w is white, w has a neighbor, and every neighbor of w is blue, then change the color of w to blue. If u is blue, all neighbors of u are blue, and u has not yet performed a force, then change the color of any white vertex w to blue (this does not require that u have any neighbors); in this case we say that u hops to w.

CCR-$\lfloor Z_+ \rfloor$ Let B be the set consisting of all the blue vertices. Let $W_1, \ldots, W_k$ be the sets of vertices of the k components of $G - B$ (note that it is possible that $k = 1$). For each component $1 \le i \le k$, let $C_i \subseteq B$ be the subset of blue vertices that are considered to be "active" with regard to that component. (Initially, each $C_i = B$.) If $u \in C_i$, $w \in W_i$, and u has no white neighbors in $G[W_i \cup B] - w$, then change the color of w to blue (this allows for either a normal forcing move, or a hop from u to w). To each connected component of $G[W_i] - w$, associate a new active set equal to $(C_i \setminus \{u\}) \cup \{w\}$.

2.3 *Halin's S-functions and Tree-width*

Halin [15] defined an S-function to be an integer valued graph parameter ζ such that

1. ζ is minor monotone,
2. $\zeta(\emptyset) = 0$ [or equivalently, $\zeta(K_1) = 1$],
3. $\zeta(G \vee K_1) = \zeta(G) + 1$, and
4. $\zeta(G \langle K_t \rangle H) = \max\{\zeta(G), \zeta(H)\}$.

[2] All of these equalities of parameters require the graph to have at least one edge.

He studied the complete lattice of S-functions and observed that η is the minimum element of the lattice. The first definition of tree-width was given by Halin for the parameter tw $+1$, which he called $s(G)$ [15]; tree-width was redefined later by Robertson and Seymour using tree-decompositions. Halin defined $s(G)$ as the maximum element of the lattice of S-functions, and characterized it as follows: A graph H is an *S-graph* if it has a simplicial decomposition in which each member is a simplex, which is another way of saying that H has a decomposition as clique-sums of cliques. As Halin notes, this is a well-known characterization of chordal graphs. An *S-covering of order k* of a graph G is an S-graph H that contains G and such that the maximum order of a member of the simplicial decomposition of H is k, i.e., H is a chordal supergraph of G with $\omega(H) = k$. Then $s(G)$ is the minimum order of an S-covering of G, i.e., $s(G) = \min\{\omega(H) : H \text{ is chordal supergraph of } G\} = \text{tw}(G) + 1$.

Halin [15] also studied other parameters such as $\lceil\chi\rceil(G)$, which he called the modified chromatic number and denoted by $c(G)$; he showed that $\lceil\chi\rceil$ is an S-function (recall that Hadwiger's Conjecture is equivalent to $\lceil\chi\rceil = \eta$). Halin [15] studied $\lceil\kappa\rceil + 1$, which he called the *modified vertex connectivity* and denoted by $d(G)$; he showed that $\lceil\kappa\rceil + 1$ is an S-function.

The next two examples show that $\mu + 1$ and $\nu + 1$ are not S-functions (even ignoring issues of the correct starting value).

Example 1. Observe that $K_{1,3} = P_3 \langle K_1 \rangle P_2$ but $\mu(P_3) = \mu(P_2) = 1$ and $\mu(K_{1,3}) = 2$.

Example 2. Observe that $T_3 = L \langle K_2 \rangle K_3$ but $\nu(L) = \nu(K_3) = 2$ and $\nu(T_3) = 3$ (see Figure 1 for L and T_3 and [4] for values of ν).

2.4 Minor Monotone Ceilings of Vertex Connectivity and Minimum Degree

In this section we discuss the minor monotone ceilings $\lceil\kappa\rceil$ and $\lceil\delta\rceil$ in more detail. For all G, $\kappa(G) \le \delta(G)$, so $\lceil\kappa\rceil(G) \le \lceil\delta\rceil(G)$. It is clear from the definition of a k-tree that if G a subgraph of a k-tree, then $\delta(G) \le k$. Thus for every graph G, $\delta(G) \le \text{tw}(G)$ and since tree-width is minor monotone, $\lceil\delta\rceil(G) \le \text{tw}(G)$. If $K_r \preceq G$, then $r - 1 = \kappa(K_r) \le \lceil\kappa\rceil(G)$, so $\eta(G) - 1 \le \lceil\kappa\rceil(G)$ for all G. Fijavž [13] and Fijavž and Wood [14] studied $\lceil\kappa\rceil$ and $\lceil\delta\rceil$ via minimal forbidden minors.

Example 3. The graph D_3 shown in Figure 2 is defined in [13, 14], where it is shown that $\lceil\kappa\rceil(D_3) = 4$, $\lceil\delta\rceil(D_3) = 5$, and $\lceil\delta\rceil(G_{5,4}) = 4$ for $G_{5,4} \simeq D_3[\{1, 2, 3, 4, 5, 6, 13\}] \simeq D_3[\{7, 8, 9, 10, 11, 12, 13\}]$. Observe that $D_3 = G_{5,4} \langle K_1 \rangle G_{5,4}$, but $\lceil\delta\rceil(G_{5,4}) = 4$ and $\lceil\delta\rceil(D_3) = 5$. This shows that $\lceil\delta\rceil$ is not a Halin S-function.

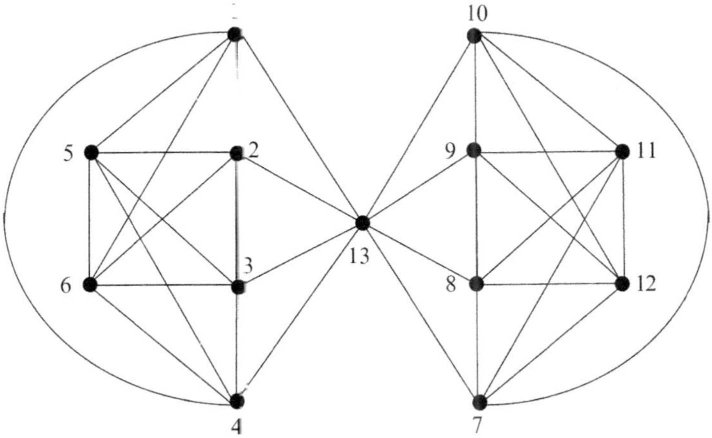

Fig. 2 The graph D_3.

In our study of Nordhaus-Gaddum sum upper bounds (see Section 3) we use a lower bound on the number of edges in G in terms of ξ (see Theorem 7 below). This bound carries over to $\lceil\kappa\rceil \leq \nu \leq \xi$ (and also to μ, although a similar bound for μ was known previously [31]). Here we prove an analogous bound for $\lceil\delta\rceil$.

Proposition 1. *For any graph $G = (V, E)$, $|E| \geq \frac{\lceil\delta\rceil(G)(\lceil\delta\rceil(G)+1)}{2}$.*

Proof. Let $G' = (V', E')$ be a minor of G having $\delta(G') = \lceil\delta\rceil(G)$. Observe that the operations used to produce minors (edge contraction, edge deletion, isolated vertex deletion) cannot increase the number of edges, so $|E| \geq |E'|$. Since each vertex of G' must have degree at least $\delta(G')$ and there must be at least $\delta(G') + 1$ vertices in G', $|E'| \geq \frac{\delta(G')(\delta(G')+1)}{2} = \frac{\lceil\delta\rceil(G)(\lceil\delta\rceil(G)+1)}{2}$. □

3 Nordhaus-Gaddum Sum Upper Bounds

Let $\beta \in \{\eta, \lceil\kappa\rceil, \lceil\delta\rceil, \mu, \nu, \xi, \mathrm{tw}, \mathrm{la}, \mathrm{pw}, \mathrm{ppw}\}$. What can be said about an upper bound for $\beta(G) + \beta(\overline{G})$? In [6], the *NG sum upper multiplier for β*, denoted by b_β, is defined to be the least value of b making $\beta(G) + \beta(\overline{G}) \leq b\,|G|$ true for all G with $|G| \geq 2$. In the spirit of Nordhaus and Gaddum's requirement that the tight bound be realized for arbitrarily large order graphs, here we modify the definition slightly. For $n \geq 2$,

$$b(n, \beta) := \min\{b : \beta(G) + \beta(\overline{G}) \leq b\,|G| \text{ for all } G \text{ with } |G| \geq n\}$$

and

$$b_\beta := \lim_{n\to\infty} b(n,\beta);$$

since $b(n,\beta)$ is nonincreasing and $b(n,\beta) \geq 0$, the limit exists. In fact, K_n immediately shows that $b_\beta \geq 1$.

Since $\beta(G) \leq |G| - 1$ (except $\eta(G) \leq |G|$), $b_\beta \leq 2$. It is known that the value of b_β is less than 2 for some of the parameters and equal to 2 for others; some are known precisely whereas for others we have bounds and open questions. We begin with the parameters β for which b_β is known exactly.

Theorem 4. [23] *For all G*

$$\eta(G) + \eta(\overline{G}) \leq \frac{6}{5}|G|$$

and there exist graphs of arbitrarily large order achieving $\eta(G) + \eta(\overline{G}) = \frac{6}{5}|G|$, *so*

$$b_{\eta-1} = b_\eta = \frac{6}{5} = 1.2.$$

The lower bound on b_η is established by the next example.

Example 4. [23] Construct a graph G on $n = 5s$ vertices partitioned into 5 sets $V_1, \ldots, V_5$ of s vertices each. Include all edges between pairs of vertices u and v with $u, v \in V_i$ for $i \in \{1, 3\}$, and all edges between pairs of vertices u and v with $u \in V_i$ and $v \in V_{i+1}$ (where $V_6 := V_1$). This is illustrated in Figure 3, where the gray areas indicate all edges present. Contracting a perfect matching between edges in V_1 and V_2 and contracting perfect matching between V_4 and V_5 gives a K_{3s} minor of G, so $\eta(G) \geq \frac{3}{5}n$. Since $G \prec \overline{G}$, $\eta(\overline{G}) \geq \frac{3}{5}n$. Thus $\eta(G) + \eta(\overline{G}) \geq \frac{6}{5}|G|$.

The next result, due to Joret and Wood, follows from a recent result of Perarnau and Serra [32] that $\operatorname{tw}(G(n, \frac{1}{2})) = n - o(n)$ as $n \to \infty$.

Theorem 5. [22] *Asymptotically almost surely as* $n \to \infty$,

$$\operatorname{tw}\left(G\left(n, \frac{1}{2}\right)\right) + \operatorname{tw}\left(\overline{G\left(n, \frac{1}{2}\right)}\right) = 2n - o(n),$$

and so $b_{\operatorname{tw}} = 2$.

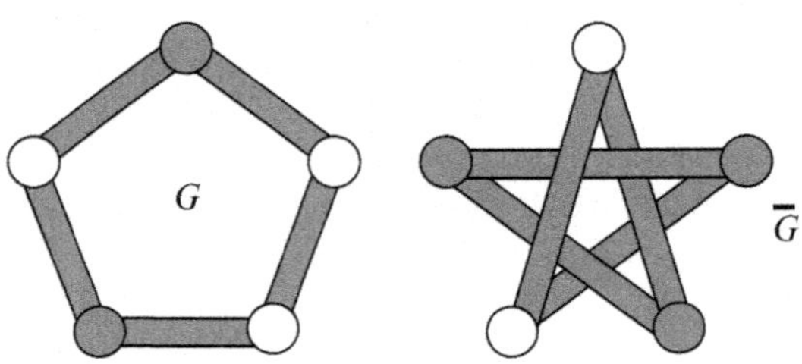

Fig. 3 Schematic diagram of the edges in the graph G and its complement $\overline{G}$.

Corollary 2. $b_{\text{la}} = b_{\text{pw}} = b_{\text{pFw}} = 2$.

Although it is known that for almost all large graphs $\text{tw}(G) + \text{tw}(\overline{G}) \approx 2|G|$, we do not know explicit examples showing $b_{\text{tw}} = 2$.

Question 1. Can we construct explicit examples showing $b_{\text{tw}} = 2$?

Theorem 6. [6]

$$1.333 < \frac{4}{3} \leq b_{\lceil\kappa\rceil} \leq b_{\nu} \leq b_{\xi} \leq \sqrt{2} < 1.415.$$

As for Hadwiger number, the lower bound on $b_{\lceil\kappa\rceil}$ comes from the construction of an example, but the example is considerably more complicated, and we refer the reader to [6]. The upper bound comes from the next theorem.

Theorem 7. [17] *For a connected graph $G = (V, E)$,*

$$|E| \geq \frac{\xi(G)(\xi(G)+1)}{2} - 1$$

(where the -1 is unnecessary unless G is bipartite and every optimal matrix for $\xi(G)$ has zero diagonal).

As noted in [6], $b_{\lceil\kappa\rceil} \leq b_{\lceil\delta\rceil}$, so $\frac{4}{3} \leq b_{\lceil\delta\rceil}$. Then $b_{\lceil\delta\rceil} \leq \sqrt{2}$ follows from Proposition 1 by a simplification of the argument that deduces $b_{\xi} \leq \sqrt{2}$ from Theorem 7. Here we reproduce the brief proof for the convenience of the reader.

Theorem 8.

$$1.333 < \frac{4}{3} \leq b_{\lceil\delta\rceil} \leq \sqrt{2} < 1.415.$$

Proof. Suppose $|G| = n$. By Proposition 1, $2E(G) \geq \lceil\delta\rceil(G)(\lceil\delta\rceil(G)+1) \geq (\lceil\delta\rceil(G))^2$ and similarly for $\overline{G}$. Then $n^2 \geq n(n-1) = 2E(G) + 2E(\overline{G}) \geq \lceil\delta\rceil(G)^2 + \lceil\delta\rceil(\overline{G})^2$. Thus

$$b_{\beta} \leq \max_{\theta}(\cos\theta + \sin\theta) = \sqrt{2}.$$

□

Question 2. Is there an example that increases the lower bound $\frac{4}{3}$ for $b_{\lceil\kappa\rceil}, b_{\lceil\delta\rceil}, b_{\nu}$ or b_{ξ}? Or a way to reduce the upper bound $\sqrt{2}$ for $b_{\xi}, b_{\nu}, b_{\lceil\delta\rceil}$, or $b_{\lceil\kappa\rceil}$?

From prior results for η and ξ,

$$1.2 = \frac{6}{5} = b_{\eta} \leq b_{\mu} \leq b_{\xi} \leq \sqrt{2} < 1.415.$$

Question 3. Is there an example that increases the (Hadwiger) lower bound $\frac{6}{5}$ for b_μ? Or a way to reduce the upper bound $\sqrt{2}$?

Since $\mathrm{M}(K_n) = n-1$ and $\mathrm{M}(\overline{K_n}) = n$, trivially the Nordhaus-Gaddum sum upper bound for M is $2|G| - 1$. (Note that $\mathrm{M}(G) + \mathrm{M}(\overline{G}) \le 2n - 1$ requires $n \ge 2$, as we have assumed throughout this section.)

4 Nordhaus-Gaddum Sum Lower Bound

Within the combinatorial matrix community, the assertion that $|G| - 2$ is the Nordhaus-Gaddum sum lower bound for β, i.e., for all graphs G

$$|G| - 2 \le \beta(G) + \beta(\overline{G}) \tag{1}$$

is referred to as the Graph Complement Conjecture for β or GCC_β, because it has been conjectured by various authors for the Colin de Verdière type parameters (and for other parameters related to the maximum nullity of matrices associated with a graph). Here we discuss Nordhaus-Gaddum sum lower bounds for the parameters $\beta \in \{\lceil\kappa\rceil, \lceil\delta\rceil, \mu, \nu, \xi, \mathrm{tw}, \mathrm{la}, \mathrm{pw}, \mathrm{ppw}\}$ in terms of parameters for which GCC_β is true, conjectured, and false. In Section 4.3 we show that when it is true this bound is tight.

4.1 Parameters β for which GCC_β is True

For the four parameters tw, la, pw, and ppw, (1) is known to be true.

Theorem 9. [11, 22] *For all graphs G,*

$$|G| - 2 \le \mathrm{tw}(G) + \mathrm{tw}(\overline{G}).$$

Corollary 3. [11] *For* $\beta \in \{\mathrm{la}, \mathrm{pw}, \mathrm{ppw}\}$, $|G| - 2 \le \beta(G) + \beta(\overline{G})$ *for all graphs G.*

It is worth noting that for large random graphs (1) is immediate from Perarnau and Serra's result on the expected value of tree-width [32]. However, although suggestive, large random graph results do not imply (1) is true for all graphs. Compare Kostochka's result that $\eta(G(n, \frac{1}{2})) \le \frac{n}{\sqrt{\log n}}$ almost always as $n \to \infty$ [24] and his Nordhaus-Gaddum sum upper bound for η quoted in Theorem 4.

4.2 Parameters β for which GCC_β is Conjectured

For the parameters $\beta \in \{\mu, \nu, \xi\}$, various authors have conjectured that GCC_β is true:

Conjecture 1. [26] For all graphs G,

$$|G| - 2 \leq \mu(G) + \mu(\overline{G}).$$

Conjecture 2. [3] For all graphs G,

$$|G| - 2 \leq \nu(G) + \nu(\overline{G}).$$

Of course, Conjectures 1 and 2 each imply the following:

Conjecture 3. [3] For all graphs G,

$$|G| - 2 \leq \xi(G) + \xi(\overline{G}).$$

Note that for $\beta \in \{\nu, \xi\}$, where GCC_β (1) is conjectured, the information on large random graphs is inconclusive: Hall et al. [17] showed that for $\beta \in \{\nu, \xi\}$,

$$\frac{1}{2} \leq \liminf_{n \to \infty} \frac{\mathbf{E}\left[\beta(G(n, \frac{1}{2}))\right]}{n} \leq \frac{1}{\sqrt{2}}.$$

However, there is some evidence for GCC_μ and GCC_ν (and of course either of those would imply GCC_ξ).

Theorem 10. [26] *If G is planar, then $|G| - 2 \leq \mu(G) + \mu(\overline{G})$.*

Corollary 4. $|G| - 2 \leq \mu(G) + \mu(\overline{G})$ *for a graph G whenever*

1. $\mu(G) \geq |G| - 6$ *or* $\mu(\overline{G}) \geq |G| - 6$*;*
2. $|G| \leq 10$.

Proof. That $\mu(G) \geq |G| - 6$ implies GCC_μ was noted in [12]; here we reproduce the brief explanation. Assume $\mu(G) \geq |G| - 6$. If $\mu(\overline{G}) \geq 4$ then $\mu(G) + \mu(\overline{G}) \geq |G| - 2$. If $\mu(\overline{G}) \leq 3$, then $\overline{G}$ is planar [8], so by Theorem 10, $\mu(G) + \mu(\overline{G}) \geq |G| - 2$.

If $\mu(G) \leq 3$ or $\mu(\overline{G}) \leq 3$, then G or $\overline{G}$ is planar [8], so $\mu(G) + \mu(\overline{G}) \geq |G| - 2$ by Theorem 10. Thus if $\mu(G) + \mu(\overline{G}) < |G| - 2$, necessarily $\mu(G) \geq 4$ and $\mu(\overline{G}) \geq 4$, and thus $10 < |G|$. □

Theorem 11. $|G| - 2 \leq \nu(G) + \nu(\overline{G})$ *for a graph G whenever*

1. [33] $\mathrm{tw}(G) \leq 3$ *or* $\mathrm{tw}(\overline{G}) \leq 3$*;*
2. [33] *G is a k-connected partial k-tree, in particular, if G is a k-tree;*
3. $\nu(G) \geq \mathrm{tw}(G)$ *or* $\nu(\overline{G}) \geq \mathrm{tw}(\overline{G})$*;*
4. $\nu(G) \leq 2$ *or* $\nu(\overline{G}) \leq 2$*;*
5. [12] $\nu(G) \geq |G| - 4$ *or* $\nu(\overline{G}) \geq |G| - 4$*;*

6. [3] $|G| \leq 8$;
7. [3] *G is a cograph;*
8. [29] *G is chordal.*

Furthermore, if G and H are graphs that satisfy GCC$_\nu$*, then the join* $G \vee H$ *and disjoint union* $G \dot{\cup} H$ *satisfy GCC*$_\nu$ [3].

Proof. Most of the statements appear in the literature. For statement (3): By [33, Theorem 5], $\nu(\overline{G}) \geq |G| - \text{tw}(G) - 2$ for every graph G. So if $\nu(G) \geq \text{tw}(G)$,

$$\nu(G) + \nu(\overline{G}) \geq \text{tw}(G) + |G| - \text{tw}(\overline{G}) - 2 = |G| - 2.$$

Statement (4) follows from Statement (3) and the fact that $\nu(G) \leq 2$ implies $\text{tw}(G) \leq \nu(G)$, which follows from the minimal forbidden minor characterizations of tw and ν: The minimal forbidden minor for $\text{tw}(G) \leq 2$ is K_4 [7]. The minimal forbidden minors for $\nu(G) \leq 2$ are K_4 and T_3 [19]. □

4.3 GCC$_\beta$ is Tight if True for β

We show in Proposition 3 below that for $\beta \in \{\mu, \nu, \xi, \text{tw}, \text{la}, \text{pw}, \text{ppw}\}$, $\beta(\overline{P_n}) = n-3$, and thus for any of these parameters for which the inequality (1) is correct (in particular, tw, la, pw, and ppw), it is tight because for $n \geq 4$,

$$\beta(P_n) + \beta(\overline{P_n}) = 1 + (n-3) = n-2.$$

Proposition 2. *Let T be a forest that contains* P_4 *as a subgraph. Then* $\text{ppw}(\overline{T}) \leq n-3$.

Proof. Since $\text{ppw} = \lfloor Z \rfloor$, it suffices to show $\lfloor Z \rfloor(\overline{T}) \leq n-3$. Assume the vertices of T are numbered starting with the P_4 as 1, 2, 3, 4 in path order. Then $B = \{2, 5, 6, \ldots, n\}$ is a zero forcing set of cardinality $n-3$ for $\overline{T}$ (with $2 \to 4, 4 \to 1, 1 \to 3$). □

Proposition 3. *For* $\beta \in \{\kappa, \delta, \lceil\kappa\rceil, \lceil\delta\rceil, \mu, \nu, \xi, \text{tw}, \text{la}, \text{pw}, \text{ppw}\}$ *and* $n \geq 4$, $\beta(\overline{P_n}) = n-3$.

Proof. For all β, $\beta \leq \text{ppw} = \lfloor Z \rfloor$, and $\kappa \leq \beta$ for all β except μ. By Proposition 2, $\lfloor Z \rfloor(\overline{P_n}) \leq n-3$. Since P_n is planar, it is known that $\mu(P_n) + \mu(\overline{P_n}) \geq n-2$ and thus that $\mu(\overline{P_n}) \geq n-3$.

To see that $\kappa(\overline{P_n}) \geq n-3$, assume $P_n = (1, 2, \ldots, n)$ and suppose $S \subset [n]$ is a cut-set of $\overline{P_n}$. Note first that this implies there must exist $x < y \in [n] \setminus S$ that are not joined by a path in $\overline{P_n} - S$. For $x \sim y$ in $\overline{P_n}$, this is impossible. So $x = k, y = k+1$ for some k with $1 \leq k \leq n-1$. Assume first that $2 \leq k \leq n-2$. Then for $z = 1, \ldots, k-2, k+3, \ldots, n$, $(k, z, k+1)$ is a path between k and $k+1$. So the $n-4$ vertices $1, \ldots, k-2, k+3, \ldots, n$ must be in S. But k and $k+1$ are still connected by the path $(k, k+2, k-1, k+1)$, so at least one of $k-1$ or $k+2$ must also in S, and thus $|S| \geq n-3$. The case $k=1$ or $k=n-1$ is even easier, as there are $n-3$ paths on 3 vertices between k and $k+1$ in $\overline{P_n}$. Thus $\kappa(\overline{P_n}) \geq n-3$. □

4.4 Parameters β for which GCC_β is False

In this section we show that GCC_β is false for $\beta \in \{\lceil\delta\rceil, \lceil\kappa\rceil, \lceil\chi\rceil - 1\}$ and reproduce the known Nordhaus-Gaddum lower bound for η.

Theorem 12. [24, Corollary 5][3]

$$\min_{|G|=n} \left(\eta(G) + \eta(\overline{G})\right) = \Theta\left(\frac{n}{\sqrt{\log n}}\right).$$

The tree T shown Figure 4 provides a counterexample to (1) for $\lceil\delta\rceil$ and thus for $\lceil\kappa\rceil$ and $\eta - 1$, and also for $\lceil\chi\rceil - 1$. (A smaller but less structured example was found by Berrera-Cruz and Lin [5].)

Proposition 4. *For the tree T shown in Figure 4, $\lceil\delta\rceil(\overline{T}) = 8$ and $\lceil\chi\rceil(\overline{T}) = 9$ and thus*

$$|T| - 3 \geq \lceil\delta\rceil(T) + \lceil\delta\rceil(\overline{T}) \geq \lceil\kappa\rceil(T) + \lceil\kappa\rceil(\overline{T}),$$

$$|T| - 1 \geq \lceil\chi\rceil(T) + \lceil\chi\rceil(\overline{T}) \geq \eta(T) + \eta(\overline{T}).$$

Proof. We first show $\lceil\delta\rceil(\overline{T}) \leq 8$. Deletion of a vertex in a connected graph cannot result in a greater minimum degree than contraction of an edge incident with that vertex, and deletion of an edge cannot raise minimum degree, so there is a minor realizing $\lceil\delta\rceil(\overline{T})$ is obtained from $\overline{T}$ by contractions. That is, there exists $H \prec \overline{T}$ with $\delta(H) = \lceil\delta\rceil(\overline{T})$ and H is defined by a partition $W_1, \ldots, W_h$. For $v = 1, 2, 3, 4, 5$, $\deg_T v = 3$, so $\deg_{\overline{T}} v = 8$. Thus to achieve $\delta(H) \geq 9$, each of these 5 vertices must be in a W_i of cardinality greater than one. But that would imply $|H| \leq 9$ and thus $\delta(H) \leq 8$.

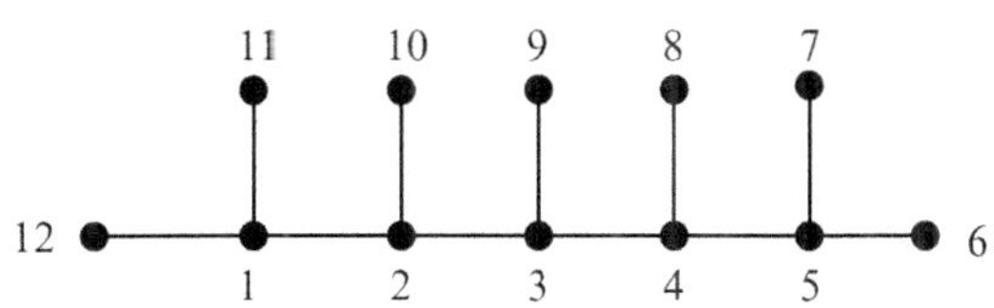

Fig. 4 A tree T for which $GCC_{\lceil\delta\rceil}$ fails.

[3] $f(n) = \Theta(g(n))$ means there exist constants K, C, k, c such that for all $n \geq K$, $f(n) \leq Cg(n)$, and for all $n \geq k$, $f(n) \geq cg(n)$.

For edges $e = \{1,5\}, f = \{2,4\}, g = \{3,6\}$, $\overline{T}/\{e,f,g\} = K_9$, so $\eta(\overline{T}/\{e,f,g\}) = 9$ and $\lceil\delta\rceil(\overline{T}/\{e,f,g\}) = 8$. Note that $9 = \eta(\overline{T}) \leq \lceil\chi\rceil(\overline{T})$. Since any order n graph $G \neq K_n$ has $\chi(G) \leq n-1$, any order 10 minor H of $\overline{T}$ has $\chi(H) \leq 9$. Since $\chi(\overline{T}) = 7$ and for every contraction of a single edge e, $\chi(\overline{T}/e) \leq 8$,[4] $\lceil\chi\rceil(\overline{T}) = 9$. □

The fact that (1) fails for η for large random graphs is clear from Kostochka's sum lower bound for η quoted in Theorem 12. But it is nice to have concrete examples, such as T shown in Figure 4 or the icosahedral graph [3].

We note that the construction in Proposition 4 can be generalized to further lower the bound; see [18] where it is shown that a Nordhaus-Gaddum sum lower bound for $\lceil\delta\rceil$ cannot exceed $n - c\sqrt{n}$ for any $c > 0$.

Question 4. What is the Nordhaus-Gaddum sum lower bound for $\lceil\delta\rceil$? Or what is its form?

Question 5. What is the Nordhaus-Gaddum sum lower bound for $\lceil\kappa\rceil$? Or what is its form?

It is easy to construct a graph G of arbitrarily large order with $\delta(G) = \delta(\overline{G}) = 1$, so trivially the Nordhaus-Gaddum sum lower bound for δ and κ is 2.

4.5 Relationships between Parameters and GCC

Relationships between the parameters discussed here are shown in Figure 5 together with a summary concerning the parameters for which (1) is known to be true, is conjectured, and is false. If two parameters are connected by a line segment, then the upper parameter is greater than or equal to the lower parameter for all graphs that have an edge. A dashed line of small triangles indicates a conjecture. Note that if Hadwiger's Conjecture is true then $\lceil\chi\rceil - 1 = \eta - 1$, but that is not indicated on the diagram (except by the conjecture that $\lceil\chi\rceil - 1 \leq \mu$, which it implies). The horizontal lines delineate the status of (1) as the Nordhaus-Gaddum sum lower bound. This diagram is adapted from [4, Figure 1.1][5], which provides additional information such as proofs of the claimed inequalities, examples showing all the parameters are distinct (except $\lceil\chi\rceil - 1$ and $\eta - 1$), and examples showing noncomparability.

It has been conjectured by various authors that δ is less than several parameters related to maximum nullity (see [12, Section 46.7] for more information). H. Tracy Hall has presented a proof that $\delta(G) \leq \nu(G)$ [16] but it is not yet fully written up, so is still considered a conjecture.

[4]It is easy to verify this in a computer algebra system such as *Sage*.

[5][4, Figure 1.1] does not includes $\lceil\chi\rceil - 1$ nor does it have any information about the Nordhaus-Gaddum lower bound.

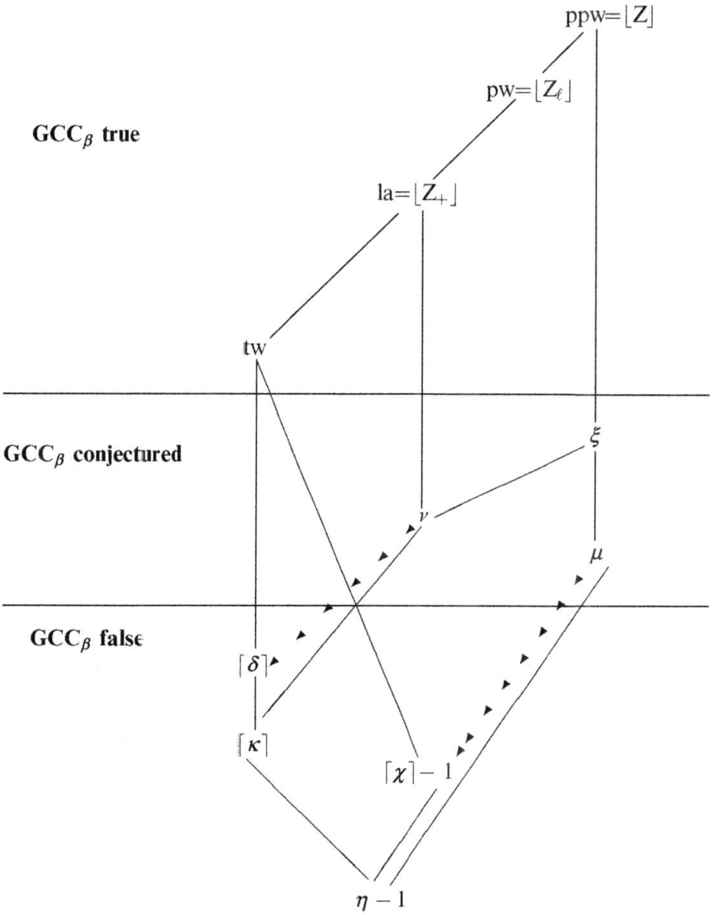

Fig. 5 Relationships between minor monotone parameters and status of GCC_β.

5 Nordhaus-Gaddum Product Bounds

For $n \geq 2$, the *NG product upper multiplier for* β, denoted by c_β, is defined by

$$c(n,\beta) := \min\{c : \beta(G)\cdot\beta(\overline{G}) \leq c\,|G|^2 \text{ for all } G \text{ with } |G| \geq n\}$$

and

$$c_\beta := \lim_{n\to\infty} c(n,\beta).$$

Since $c(n,\beta)$ is nonincreasing and $c(n,\beta) \geq 0$, the limit exists.

As in the case of Nordhaus and Gaddum's original results (cf. Theorem 1), we can often obtain good information about the NG product upper multiplier for β from the NG sum upper multiplier for β.

Remark 2. If $\beta(G)+\beta(\overline{G}) \leq bn$ for all graphs of order n, then $\beta(G)\beta(\overline{G}) \leq x(bn-x) \leq \frac{1}{4}b^2n^2$, so $c_\beta \leq \frac{1}{4}b_\beta^2$. If there exists a construction of graphs G_n of order n such that $\beta(G_n) = \beta(\overline{G_n})$ and $\lim_{n\to\infty} \beta(G_n)/n = d$, then $d^2 \leq c_\beta$.

Since Kostochka's construction in [23] for $\eta(G)+\eta(\overline{G}) = \frac{6}{5}|G|$ realizes $\eta(G) = \frac{3}{5}n = \eta(\overline{G})$, the next corollary is immediate.

Corollary 5. *The NG product upper multiplier for η (or for $\eta - 1$) is*

$$c_{\eta-1} = c_\eta = \frac{9}{25}.$$

The next corollary follows from the fact that the construction in [6] of $H(a)$ that realizes $\lceil\kappa\rceil(H(a)) \approx \frac{2}{3}n$ is self-complementary.

Corollary 6. *We have the following bounds on the NG product upper multiplier for $\lceil\kappa\rceil, \lceil\delta\rceil, \nu, \xi$*

$$0.444 < \frac{4}{9} \leq c_{\lceil\kappa\rceil} \leq c_\nu \leq c_\xi \leq \frac{1}{2} = 0.5 \textit{ and } 0.444 < \frac{4}{9} \leq c_{\lceil\delta\rceil} \leq \frac{1}{2} = 0.5.$$

For a lower bound to $\beta(G)\cdot\beta(\overline{G})$, we assume neither G nor $\overline{G}$ is the empty graph (no edges), because for many of our parameters this would result in a lower bound of zero.

Remark 3. For $\beta \in \{\lceil\kappa\rceil, \lceil\delta\rceil, \nu, \mu, \xi, \mathrm{tw}, \mathrm{la}, \mathrm{pw}, \mathrm{ppw}\}$, $\beta(P_n) = 1$ and $\beta(\overline{P_n}) = n-3$ (cf. Proposition 3), so any lower bound on $\beta(G)\beta(\overline{G})$ cannot exceed $|G|-3$. If $|G|-2 \leq \beta(G)+\beta(\overline{G})$ (as is true for $\beta \in \{\mathrm{tw}, \mathrm{la}, \mathrm{pw}, \mathrm{ppw}\}$), then

$$|G|-3 \leq \beta(G)\cdot\beta(\overline{G})$$

and this bound is tight.

It should be noted that whereas subtracting 1 from a parameter (e.g., $\eta - 1$ vs. η) has no effect on the upper multiplier, it makes a dramatic difference in the NG product lower bound, which Kostochka has determined for η.

Theorem 13. [25] cited in [1] *For all graphs such that G and $\overline{G}$ each have an edge,*

$$\left\lceil \frac{3n-5}{2} \right\rceil \leq \eta(G)\cdot\eta(\overline{G}).$$

Acknowledgements This research was supported in part by the Institute for Mathematics and its Applications with funds provided by the National Science Foundation. I thank IMA for providing both financial support and a wonderful research environment. I thank Shaun Fallat, Gašper Fijavž, and David Wood for their conversations on this topic and/or comments on drafts, which have been a great help.

References

1. M. Aouchiche, P. Hansen, A survey of Nordhaus–Gaddum type relations. Discret. Appl. Math. **161**, 466–546 (2013)
2. F. Barioli, S.M. Fallat, L. Hogben, A variant on the graph parameters of Colin de Verdière: implications to the minimum rank of graphs. Electron. J. Linear Algebra **13**, 387–404 (2005)
3. F. Barioli, W. Barrett, S. Fallat, H.T. Hall, L. Hogben, H. van der Holst, On the graph complement conjecture associated with minimum rank. Linear Algebra Appl. **436**, 4373–4391 (2012)
4. F. Barioli, W. Barrett, S. Fallat, H.T. Hall, L. Hogben, B. Shader, P. van den Driessche, H. van der Holst, Parameters related to tree-width, zero forcing, and maximum nullity of a graph. J. Graph Theory **72**, 146–177 (2013)
5. F. Berrera-Cruz, J.C.-H. Lin, Personal communication
6. W. Barrett, S. Fallat, H.T. Hall, L. Hogben, Note on Nordhaus-Gaddum problems for Colin de Verdière type parameters. Electron. J. Comb. **20**, P56 (2013)
7. R.B. Borie, R.G. Parker, C.A. Tovey, Recursively constructed graphs, in *Handbook of Graph Theory*, ed. by J.L. Gross, J. Yellen (CRC Press, Boca Raton, 2004), pp. 99–118
8. Y. Colin de Verdière, On a new graph invariant and a criterion for planarity, in *Graph Structure Theory* (American Mathematical Society, Providence, RI, 1993), pp. 137–147
9. Y. Colin de Verdière. Multiplicities of eigenvalues and tree-width graphs. J. Comb. Theory B **74**, 121–146 (1998)
10. R. Diestel, *Graph Theory*, Electronic Edition (Springer, New York, 2005)
11. J. Ekstrand, C. Erickson, D. Hay, L. Hogben, J. Roat, Note on positive semidefinite maximum nullity and positive semidefinite zero forcing number of partial 2-trees. Electron. J. Linear Algebra **23**, 79–97 (2012)
12. S. Fallat, L. Hogben, Minimum rank, maximum nullity, and zero forcing number of graphs, in *Handbook of Linear Algebra*, 2nd edn. ed. by L. Hogben (CRC Press, Boca Raton, 2014), pp. 46-1–46-36
13. G. Fijavž, Graphs minors and connectivity. Ph.D. thesis [in Slovene], University of Ljubljana, Slovenia (2001)
14. G. Fijavž, D.R. Wood, Graph minors and minimum degree. Electron. J. Comb. **17**, R151 (2010)
15. R. Halin, S-functions of graphs. J. Geom. **8**, 171–178 (1976)
16. H.T. Hall, The greedegree theorem [proof that $\delta(G) \leq \nu(G)$ for all graphs G]. Colloquium, Iowa State University, November 2010
17. H.T. Hall, L. Hogben, R. Martin, B. Shader, Expected values of parameters associated with the minimum rank of a graph. Linear Algebra Appl. **433**, 101–117 (2010)
18. L. Hogben, J.C.-H. Lin, M. Young, Nordhaus-Gaddum sum lower bounds for $\lceil\delta\rceil$ and $\lceil\kappa\rceil$ (preprint)
19. H. van der Holst, Graphs with magnetic Schrödinger operators of low corank. J. Comb. Theory B **84**, 311–339 (2002)
20. H. van der Holst, On the "Largeur d'Arborescence." J. Graph Theory **41**, 24–52 (2002)
21. H. van der Holst, L. Lovász, A. Schrijver, The Colin de Verdière graph parameter, in *Graph Theory and Computational Biology* (Balatonlelle, 1996). Bolyai Society Mathematical Studies, vol. 7 (János Bolyai Math. Soc., Budapest, 1999), pp. 29–85
22. G. Joret, D.R. Wood, Nordhaus-Gaddum for treewidth. Eur. J. Comb. **33**, 488–490 (2012)

23. A.V. Kostochka, On Hadwiger numbers of a graph and its complement, in *Finite and Infinite Sets*, vol. I, II (Eger, 1981), (North-Holland, Amsterdam, 1984), pp. 537–545
24. A.V. Kostochka, Lower bound of the Hadwiger number of graphs by their average degree. Combinatorica **4**, 307–316 (1984)
25. A.V. Kostochka, A lower bound for the product of the Hadwiger numbers of a graph and its complement. Komb. Anal. **8**, 50–62 (1989) [in Russian]
26. A. Kotlov, L. Lovász, S. Vempala, The Colin de Verdière number and sphere representations of a graph. Combinatorica **17**, 483–521 (1997)
27. L. Lovász, M. Saks, A. Schrijver, Orthogonal representations and connectivity of graphs. Linear Algebra Appl. **114/115**, 439–454 (1989)
28. L. Lovász, M. Saks, A. Schrijver, A correction: orthogonal representations and connectivity of graphs. Linear Algebra Appl. **313**, 101–105 (2000)
29. L.H. Mitchell, On the graph complement conjecture for minimum semidefinite rank. Linear Algebra Appl. **435**, 1311–1314 (2011)
30. E.A. Nordhaus, J. Gaddum, On complementary graphs. Am. Math. Mon. **63**, 175–177 (1956)
31. R. Pendavingh, On the relation between two minor-monotone graph parameters. Combinatorica **18**(2), 281–292 (1998)
32. G. Perarnau, O. Serra, On the tree-depth of random graphs. Discret. Appl. Math. **168**, 119–126 (2014)
33. J. Sinkovic, H. van der Holst, The minimum semidefinite rank of the complement of partial k-trees. Linear Algebra Appl. **434**, 1468–1474 (2011)
34. A. Takahashi, S. Ueno, Y. Kajitani, Minimal acyclic forbidden minors for the family of graphs with bounded path-width. Discret. Math. **127**, 293–304 (1994)
35. A. Takahashi, S. Ueno, Y. Kajitani, Mixed searching and proper-path-width. Theor. Comput. Sci. **137**, 253–268 (1995)

Algebraic aspects of the normalized Laplacian

Steve Butler

Abstract Spectral graph theory looks at the interplay between the structure of a graph and the eigenvalues of a matrix associated with the graph. Many interesting graphs have rich structure which can help in determining the eigenvalues associated with a particular graph matrix. This survey looks at some common techniques in working with and determining the eigenvalues associated with the normalized Laplacian matrix, in addition to some algebraic applications of these eigenvalues.

Keywords Spectral graph theory • Normalized Laplacian • Weighted graphs • Random walks • Cospectral graphs • Graph products • Twins • Characteristic polynomials

Mathematics Subject Classification (2010): 05C50

1 Introduction

Spectral graph theory investigates the relationship between the structure of a graph and the eigenvalues, or spectrum, of some matrix (or collection of matrices) associated with the graph. Researchers consider many different matrices that are considered, including the adjacency matrix A, whose entries indicate when two vertices are adjacent, the Laplacian $L = D - A$, where D is the diagonal degree matrix, the signless Laplacian $L = D + A$, and the normalized Laplacian commonly defined as $\mathscr{L} = D^{-1/2}(D - A)D^{-1/2}$ (this works for graphs without vertices of degree 0, a general definition will be given in Section 2.1). The goal of this paper is to investigate the spectrum of the normalized Laplacian.

One can roughly divide spectral graph theory into two camps. The first camp looks at the structure of graphs and subgraphs, how their eigenvalues relate; how to build smaller graphs into larger graphs and the effect on the spectrum; and equalities between eigenvalues and combinatorial parameters. This is *algebraic spectral graph*

S. Butler (✉)
Iowa State University, Ames, IA 50011, USA
e-mail: butler@iastate.edu

A. Beveridge et al. (eds.), *Recent Trends in Combinatorics*, The IMA Volumes in Mathematics and its Applications 159, DOI 10.1007/978-3-319-24298-9_13

theory. The second camp looks at approximating the structures of a graph, giving bounds for the spectrum using various combinatorial parameters and conversely using various combinatorial parameters to give bounds for the spectrum. This is *analytic spectral graph theory*.

While both are important in understanding and using eigenvalues, this survey focuses on the algebraic aspects of the normalized Laplacian matrix. In particular, we aim to understand how eigenvalues can be computed. In particular, we consider various graph operations and the corresponding impact on the eigenvalues or characteristic polynomials.

We proceed by first going into more detail on the formal definition of the normalized Laplacian (Section 2). We will then look at common ways to combine two graphs together, i.e., graph products, and explore how this affects the spectrum (Section 3). We also consider equitable partitions and duplicated structures in the graph (Section 4). Finally we will consider some properties of the characteristic polynomial for the normalized Laplacian, including its relation to Kemeny's constant (Section 5). Throughout the paper we will give several examples showing how these various techniques can be applied.

More information about the normalized Laplacian with a greater focus on the analytic results can be found in the work of Chung [11] and recent surveys [5, 7]. Finally, we note that there is a large amount of literature on algebraic graph theory which involves spectral graph theory for *regular* graphs (see [3, 13]). We will for the most part avoid results that deal with regular graphs (where possible). This is for the simple reason that if a graph is regular, then knowing the spectrum of any one of the adjacency, Laplacian, signless Laplacian, or normalized Laplacian will give us the spectrum of all the others.

2 Weighted graphs and the normalized Laplacian

Most of the focus in spectral graph theory (and in this paper) is on *simple* graphs, i.e., graphs without loops or multiple edges. This allows us to reduce the edge relationships between vertices to a simple binary 'yes' or 'no.' While this is useful combinatorially, we will see that in some cases it is better to work in the larger context of weighted graphs. These are graphs where there are weight functions $w(u, v)$ on the *edges* of the graph and $w(u)$ on the *vertices* of the graph.

The weight function on the edges satisfies $w(u, v) \geq 0$ and $w(u, v) = w(v, u)$. (There has been some investigation into the case when negative edge weights are allowed which came up in the context of a variation of a "2-lift" (see [4]).) We will follow the convention that $w(u, v) > 0$ if and only if $u \sim v$, i.e., u and v are adjacent. The requirement that $w(u, v) = w(v, u)$ reflects that we want to limit our discussion to undirected graphs. With the weight function on the edges we can define the *weighted* adjacency matrix by letting $A_{uv} = w(u, v)$.

The weight function on the vertices satisfies $w(u) \geq 0$. The weight function affects the degree of a vertex u, denoted $d(u)$ and defined by

$$d(u) = w(u) + \sum_v w(u, v).$$

We now can define the weighted degree matrix D. We note here that $w(u, u)$ and $w(u)$ are *not* the same thing. Namely $w(u, u)$ represents the weight of a loop at the vertex u (which is an edge) while $w(u)$ is some intrinsic weight to the vertex. Put in another way, $w(u, u)$ contributes both to A and D, while $w(u)$ only contributes to D.

Most applications will assume that the weight function on the vertices are all zero. However, these weights can play an important and useful role, particularly when we have several duplicate structures in a graph. They also show up when looking at random walks on a portion of the graph, i.e., a subgraph. In such a situation there are generally three options available as to what to do with the edges between the subgraph we are interested in and the remainder graph.

1. Delete the edges (i.e., work on the "induced" subgraph). This preserves all the weights on the edges of the subgraph, i.e., A, but will affect the degree of vertices incident with deleted edges, i.e., changes D.
2. Change edges to loops (i.e., we take random walks in the whole graph, but if we exit the subgraph then we go right back to the vertex we came from). This will now preserve the degree of vertices, i.e., D, but with added loops we affect A. (This is known as the Neumann boundary condition [11].)
3. Let the edges become escape routes (i.e., we take random walks in the whole graph, but if we exit the subgraph then we stop the walk). This is done by having the weights of any deleted incident edges of a vertex be added to the weight of the vertex, which helps to preserve both D and A. (This is known as the Dirichlet boundary condition [11].)

In particular, a useful way to think about degree weights are as half-edges incident to a vertex. In later sections we will give examples of how we can compute the spectrum of some of these graphs.

2.1 Formal definition of the normalized Laplacian

We can now give the formal definition for the normalized Laplacian. Namely,

$$\mathscr{L}_{uv} = \begin{cases} \frac{-w(u,v)}{\sqrt{d(u)d(v)}} & \text{if } u \neq v \text{ and } u \sim v; \\ \frac{d(u)-w(u,u)}{d(u)} & \text{if } u = v \text{ and } d(u) \neq 0; \\ 0 & \text{else.} \end{cases} \tag{1}$$

In the special case when none of the vertices has degree 0, then this is equivalent to $\mathscr{L} = D^{-1/2}(D - A)D^{-1/2}$ (and in many papers this is how it is defined). When the graph is regular of degree d then $D^{-1/2} = (1/\sqrt{d})I$ so that $\mathscr{L} = I - (1/d)A$. In particular, if we know the spectrum of the normalized Laplacian matrix, we also know the spectrum of the adjacency matrix (as well as the Laplacian and signless Laplacian). Therefore any known results for regular graphs will translate into the normalized Laplacian matrix. Further if a graph is bipartite, i.e., $V = A \cup B$ with edges going between A and B and the $d(u) = s$ for $u \in A$ and $d(u) = t$ for $u \in B$ (i.e., the graph is bi-regular), then $\mathcal{L} = I - (1/\sqrt{st})A$. So again results for such graphs translate between the normalized Laplacian and adjacency matrices.

When the minimum degree is positive, the normalized Laplacian is similar to the matrix $I - D^{-1}A$. The matrix $D^{-1}A$ is the probability transition matrix for random walks. Therefore the study of the normalized Laplacian is intrinsically linked to understanding random walks.

As we mentioned earlier, most of spectral graph theory is concerned with looking at simple graphs, which are the case $w(u, v) \in \{0, 1\}$ and $w(u, u) = w(u) = 0$. When there are no loops and vertex weights are 0 (such as in simple graphs), then the diagonal entries of the normalized Laplacian consists of 0's and 1's, where the 0's correspond to isolated vertices. In particular, the trace is the number of non-isolated vertices, or put another way, we can determine the number of isolated vertices by looking at the sum of the eigenvalues compared to the number of eigenvalues. Once we identify how many isolated vertices there are we can then remove the corresponding number of 0's and have the spectrum of the part of the graph which has no isolated vertices. Therefore when working with simple graphs, we can simply assume that there are no isolated vertices.

We conclude this section by noting the following basic properties of $\mathcal{L}$.

Theorem 1 (Chung [11]). *The matrix $\mathcal{L}$ is positive semi-definite and satisfies the following:*

1. *All eigenvalues lie in the interval* $[0, 2]$.
2. *The multiplicity of* 0 *is the number of connected components of the graph.*
3. *The multiplicity of* 2 *is the number of bipartite components with at least two vertices.*
4. *A graph without isolated vertices is bipartite if and only if the spectrum is symmetric around* 1.

2.2 Scaling and the normalized Laplacian

Given a weighted graph G and a parameter $\alpha > 0$ we will let αG denote the weighted graph where $w_{\alpha G}(u, v) = \alpha w_G(u, v)$ and $w_{\alpha G}(u) = \alpha w_G(u)$, i.e., we have scaled all the weights.[1] (Here and throughout we will use subscripts to indicate which graph is being indicated if there are several graphs being considered.) Scaling all the weights by α also scales the degrees by α. Now considering equation (1) we immediately get the following.

[1]This is different from the usual convention of letting αG denote α disjoint copies of G.

Proposition 1. *For any weighted graph G we have* $\mathscr{L}_{\alpha G} = \mathscr{L}_G$.

In particular, scaling the weights does not affect the matrix, and hence also the spectrum. This is the "normalization" of the normalized Laplacian.

2.3 Harmonic eigenvectors

One of the best ways to understand what is happening with the eigenvalues associated with a matrix is to look at the eigenvectors and how they relate to the structure. For the normalized Laplacian we will find it easier to work with the following variant.

Definition 1. Given a graph with no isolated vertices and $\mathscr{L}\mathbf{x} = \lambda\mathbf{x}$, i.e., $\mathbf{x}$ is an eigenvector of $\mathscr{L}$, then the corresponding *harmonic eigenvector* associated with the eigenvalue λ is $\mathbf{y} = D^{-1/2}\mathbf{x}$.

We can now rewrite $\mathscr{L}\mathbf{x} = \lambda\mathbf{x}$ as follows:

$$\begin{aligned} D^{-1/2}(D-A)D^{-1/2}\mathbf{x} &= \lambda\mathbf{x} \\ (D-A)\mathbf{y} &= \lambda D^{1/2}(D^{1/2}\mathbf{y}) = \lambda D\mathbf{y} \\ (1-\lambda)D\mathbf{y} &= A\mathbf{y} \end{aligned}$$

This relationship becomes a *local* requirement at each vertex. Namely, $\mathbf{y}$ is a harmonic eigenvector for λ if and only if it is nonzero and at each vertex u

$$\sum_{v\sim u} w(u,v)\mathbf{y}(v) = (1-\lambda)d(u)\mathbf{y}(u). \tag{2}$$

It is instructive to compare this to what happens if we use the adjacency matrix, i.e., $A\mathbf{z} = \lambda\mathbf{z}$ if and only if

$$\sum_{v\sim u} w(u,v)\mathbf{z}(v) = \lambda\mathbf{z}(u).$$

From this we see that any eigenvector of the adjacency matrix associated with the eigenvalue 0 is also a harmonic eigenvector of the normalized Laplacian associated with the eigenvalue 1 (i.e., in both cases the right-hand side is 0).

Given we have a graph with no vertex weights, then if we let $\mathbf{z} = \mathbf{1}$ (the all 1's vector), equation (2) becomes

$$\sum_{v\sim u} w(u,v)\mathbf{z}(v) = \sum_{v\sim u} w(u,v) = d(u) = (1-0)d(u)\mathbf{z}(u),$$

showing $\mathbf{1}$ is the harmonic eigenvector for the eigenvalue 0. This places limitations on the remaining (harmonic) eigenvectors. In particular, since the normalized Laplacian is symmetric, the remaining eigenvectors must be perpendicular to

$D^{1/2}\mathbf{1}$.We will say that the harmonic eigenvectors $\mathbf{y}$ and $\mathbf{z}$ are perpendicular if and only if

$$0 = (D^{1/2}\mathbf{y})^T(D^{1/2}\mathbf{z}) = \mathbf{y}^T D\mathbf{z} = \sum_u d(u)\mathbf{y}(u)\mathbf{z}(u),$$

i.e., the corresponding eigenvectors are perpendicular.

3 Graph products

Our goal in this section is to show how to use graphs as building blocks to construct larger graphs. Before we begin, we will find it useful to have a basic collection of simple graphs for which the eigenvalues are known. We will use the complete graph K_n, the complete bipartite graph $K_{m,n}$, the cycle graph C_n, and the path graph P_n. For the normalized Laplacian these graphs have the following spectrum (see Chung [11] for more information):

$$\begin{aligned} K_n: &\quad \left\{0, \tfrac{n}{n-1}^{(n-1)}\right\} \\ K_{s,t}: &\quad \left\{0, 1^{(s+t-2)}, 2\right\} \\ C_n: &\quad \left\{1 - \cos\tfrac{2k\pi}{n} : 0 \le k \le n-1\right\} \\ P_n: &\quad \left\{1 - \cos\tfrac{k\pi}{n-1} : 0 \le k \le n-1\right\}. \end{aligned}$$

Here we use the notation $*^{(i)}$ to indicate that an eigenvalue is repeated i times.

From this simple list of graphs we already see that the normalized Laplacian cannot always determine the number of edges or if a graph is regular. For example, $K_{1,3}$ and $K_{2,2}$ both have spectrum $\{0, 1^{(2)}, 2\}$, but they have differing number of edges and one is regular while the other is not. Note that for the other common matrices studied in spectral graph theory (adjacency, Laplacian, and signless Laplacian), the number of edges and whether a graph is regular can always be determined by the spectrum.

We now want to look at ways to combine our building blocks. This will be done by looking at various products and the join operation which are defined for simple graphs.

Definition 2. The *Cartesian product* of G and H, denoted $G \square H$, the *tensor product* of G and H, denoted $G \times H$, and the *strong product* of G and H, denoted $G \boxtimes H$, all have vertex set $\{(a,b) : a \in V(G), b \in V(H)\}$ and edge sets respectively as follows:

$$\begin{aligned} E(G \square H) &= \big\{\{(a,b),(c,d)\} : a = c \text{ and } b{\sim}d;\text{ or } a{\sim}c \text{ and } b = d\big\}, \\ E(G \times H) &= \big\{\{(a,b),(c,d)\} : a{\sim}c \text{ and } b{\sim}d\big\}, \\ E(G \boxtimes H) &= E(G \square H) \cup E(G \times H). \end{aligned}$$

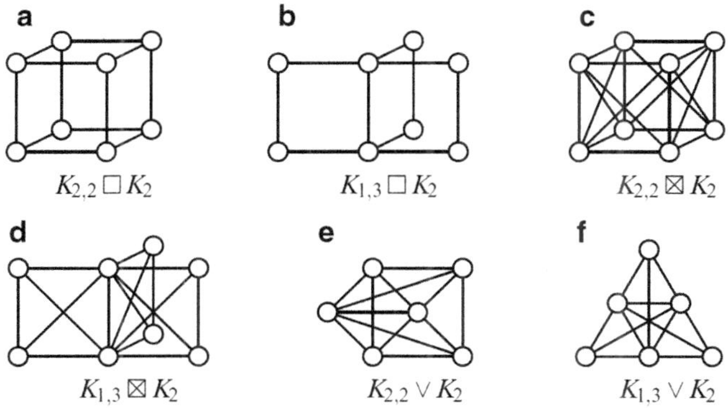

Fig. 1 Examples of graph operations.

Note that the notation used for the different products reflects the result when you take the product of K_2 with itself.

Definition 3. The *join* of G and H, denoted $G \vee H$, is the graph formed by taking the disjoint union of G and H and then adding an edge between every vertex in G and every vertex in H.

Proposition 2. *The spectra of the normalized Laplacians of G and H do not uniquely determine the spectra of the normalized Laplacians for the graphs $G \square H$, $G \boxtimes H$ or $G \vee H$.*

This can be seen by starting with $K_{1,3}$ and $K_{2,2}$, which have the same spectrum, and then looking at the different graphs that result when combined with K_2. These graphs are shown in Fig. 1.

The eigenvalues for the normalized Laplacian are given below, the important thing to notice being that the various graphs differ in their spectrum.

$$K_{2,2} \square K_2\colon \quad \left\{0, \tfrac{2}{3}^{(3)}, \tfrac{4}{3}^{(3)}, 2\right\} \qquad K_{1,3} \square K_2\colon \quad \left\{0, \tfrac{1}{2}^{(2)}, \tfrac{3}{4}, \tfrac{5}{4}, \tfrac{3}{2}^{(2)}, 2\right\}$$

$$K_{2,2} \boxtimes K_2\colon \quad \left\{0, \tfrac{4}{5}^{(2)}, \tfrac{6}{5}^{(4)}, \tfrac{8}{5}\right\} \qquad K_{1,3} \boxtimes K_2\colon \quad \left\{0, \tfrac{2}{3}^{(2)}, \tfrac{8}{7}, \tfrac{4}{3}^{(3)}, \tfrac{32}{21}\right\}$$

$$K_{2,2} \vee K_2\colon \quad \left\{0, 1^{(2)}, \tfrac{6}{5}, \tfrac{13}{10}, \tfrac{3}{2}\right\} \qquad K_{1,3} \vee K_2\colon \quad \left\{0, 1^{(2)}, \tfrac{6}{5}^{(2)}, \tfrac{8}{5}\right\}$$

When the graphs are regular, relationships of the spectrums can be established.

Proposition 3. *Let G be an r-regular simple graph with normalized Laplacian eigenvalues $0 = \lambda_0 \le \lambda_1 \le \cdots \le \lambda_{n-1}$, and let H be an s-regular simple graph with normalized Laplacian eigenvalues $0 = \theta_0 \le \theta_1 \le \cdots \le \theta_{m-1}$. Then the following hold.*

- *If at least one of r and s is nonzero, the eigenvalues of $G \square H$ are*

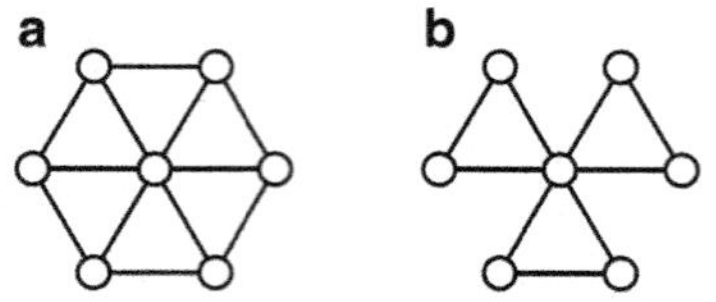

Fig. 2 Examples of graphs formed by joins

$$\left\{\frac{r\lambda_i + s\theta_j}{r+s} : 0 \le i \le n-1, 0 \le j \le m-1\right\}.$$

- *If at least one of r and s is nonzero, the eigenvalues of $G \boxtimes H$ are*

$$\left\{\frac{rs(\lambda_i + \theta_j - \lambda_i\theta_j) + r\lambda_i + s\theta_j}{rs + r + s} : 0 \le i \le n-1, 0 \le j \le m-1\right\}.$$

- *The eigenvalues of $G \vee H$ are*

$$\left\{0, 2 - \frac{r}{m+r} - \frac{s}{n+s}\right\} \cup \left\{\frac{m + r\lambda_i}{m+r} : 1 \le i \le n-1\right\} \cup \left\{\frac{n + s\theta_j}{n+s} : 1 \le j \le m-1\right\}.$$

The results for the Cartesian product and strong product readily follow by noting that the product of regular graphs is regular. Therefore we can translate the eigenvalues of the normalized Laplacian into the eigenvalues of the adjacency matrix, apply known results for these products on the adjacency matrix (see [3]), and then translate the resulting eigenvalues back to the normalized Laplacian.

The result for the join is more interesting because the join of two graphs does not have to be regular so we cannot simply translate results for the adjacency matrix.

Example 1. Two graph families constructed by joins are the wheel graph $W_{n+1} = K_1 \vee C_n$ and the friendship graph F_k on $2k+1$ vertices, formed by taking the join of a single vertex and k disjoint copies of K_2, see Fig. 2. The eigenvalues for W_{n+1} and F_k are respectively

$$\left\{0, \tfrac{4}{3}\right\} \cup \left\{1 - \tfrac{2}{3}\cos\tfrac{2i\pi}{n} : 1 \le i \le n-1\right\} \quad \text{and} \quad \left\{0, \tfrac{1}{2}^{(k-1)}, \tfrac{3}{2}^{(k+1)}\right\}.$$

The proof for the join result can be found in [4]. We sketch the key idea here as it is indicative of a common theme in this area. Consider $G \vee H$ and let us look at what happens with the part that is restricted to G. In addition to the edges of G each vertex in G now has m additional edges (those that connect to the vertices of H) and so now the degree of vertices is $m + r$. Now take a harmonic eigenvector of G which is perpendicular to the harmonic eigenvector $\mathbf{1}$ (there are $n-1$ such eigenvectors). By definition we know that such a harmonic eigenvector has the following two properties in G for some λ:

$$\sum_{v\sim u}\mathbf{y}(v)=(1-\lambda)r\,\mathbf{y}(u)\qquad\text{and}\qquad\sum_{v}\mathbf{y}(v)=0.$$

(The first is from equation (2), the second from the perpendicular requirement.) Now we take this vector and create a new vector $\hat{\mathbf{y}}$ for $G\vee H$ by making the value $\hat{\mathbf{y}}(w)=0$ for all $w\in V(H)$. Then we note the following happens

$$\sum_{v\sim u}\hat{\mathbf{y}}(v)=\begin{cases}\sum_{v\sim u}\mathbf{y}(v)=(1-\lambda)r\mathbf{y}(u)=(1-\lambda')(m+r)\hat{\mathbf{y}}(u) & \text{if } u\in V(G);\\ \sum_{v}\mathbf{y}(v)=0=(1-\lambda')(n+s)\hat{\mathbf{y}}(u) & \text{if } u\in V(H).\end{cases}$$

In particular this will be a harmonic eigenvector for the eigenvalue λ' where

$$(1-\lambda')(m+r)=(1-\lambda)r\quad\text{or}\quad\lambda'=\frac{m+r\lambda}{m+r}.\tag{3}$$

So therefore we can readily find all but two of the eigenvalues. On the other hand, we know that 0 must be an eigenvalue and it comes from none of these, leaving us with one eigenvalue, and then knowing that the sum of the eigenvalues must be $m+n$ allows us to find the last. (Equitable partitions, discussed in Section 4, give another method to find the remaining eigenvalues by "collapsing" G and H.)

The argument boiled down to two facts. First, we knew that the sum of the entries of the harmonic eigenvector was 0 and since we "coned" over the graph we could assign all other values to 0 and still preserve the local requirements. Second, when we start with a regular graph and then perturb the degree, this will perturb the eigenvalue in a predictable way.

In the case that we have a bipartite graph where one part is regular we can cone over the regular side in a similar way and still be able to locally control the eigenvalues. This basic idea was used by Butler and Grout [8] to form cospectral graphs by gluing in such bipartite graphs into generic graphs.

It remains to now look at what happens for the tensor product of two graphs.

Theorem 2. *Let G and H be simple graphs without isolated vertices, further let $0=\lambda_0\le\lambda_1\le\cdots\le\lambda_{n-1}$ and $0=\theta_0\le\theta_1\le\cdots\le\theta_{m-1}$ be the normalized Laplacian eigenvalues for G and H, respectively. Then the normalized Laplacian eigenvalues for $G\times H$ are*

$$\{\lambda_i+\theta_j-\lambda_i\theta_j:0\le i\le n-1,0\le j\le m-1\}.$$

Proof. This follows from combining two simple observations. First, λ is an eigenvalue of $\mathscr{L}$ if and only if $1-\lambda$ is an eigenvalue of $D^{-1/2}AD^{-1/2}$. Second, since $d_{G\times H}\big((u,v)\big)=d_G(u)d_G(v)$ then when $(u,v)\sim(s,t)$ we have

$$\frac{1}{\sqrt{d_{G\times H}\big((u,v)\big)d_{G\times H}\big((s,t)\big)}}=\frac{1}{\sqrt{d_G(u)d_G(s)}\sqrt{d_H(v)d_H(t)}}.$$

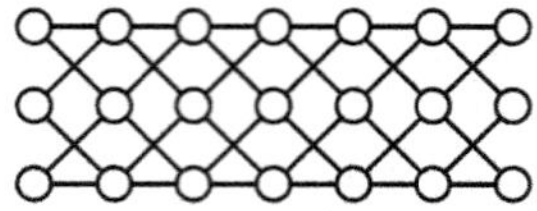

Fig. 3 The graph product $K_3 \times P_7$

And in particular for the tensor product we have

$$D_{G\times H}^{-1/2}A_{G\times H}D_{G\times H}^{-1/2} = \left(D_G^{-1/2}A_GD_G^{-1/2}\right) \otimes \left(D_H^{-1/2}A_HD_H^{-1/2}\right)$$

where "$\otimes$" indicates the tensor product of matrices. But the eigenvalues of a tensor product of matrices is found by taking all possible products of the eigenvalues of the initial matrices. □

Example 2. The spectrum of $K_t \times P_n$ (see Fig. 3) is

$$\left\{1 - \cos\tfrac{k\pi}{n-1}, \left(1 + \tfrac{1}{t-1}\cos\tfrac{k\pi}{n-1}\right)^{(t-1)} : 0 \le k \le n-1\right\}.$$

4 Equitable partitions and twins

Some graphs contain groups of vertices with similar structure. More precisely, we have the following definition.

Definition 4. An *equitable partition* of G is a partition $V(G) = V_1 \cup V_2 \cup \cdots \cup V_k$ so that the sum of edge weights from a particular vertex in V_i to all the vertices in V_j does not depend on which vertex we choose, i.e., for $s, t \in V_i$ we have

$$\sum_{v\in V_j} w(s, v) = \sum_{v\in V_j} w(t, v).$$

Theorem 3 (Butler and Chung [7]). *Given an equitable partition $V = V_1 \cup \cdots \cup V_k$ of a weighted graph G with no vertex weights we can form a new graph H with vertices $v_1, \ldots, v_k$ and edge weights*

$$w_H(v_i, v_j) = \sum_{\substack{s\in V_i \\ t\in V_j}} w_G(s, t).$$

Then the eigenvalues of $\mathscr{L}_H$ are also eigenvalues of $\mathscr{L}_G$ (including multiplicity).

The basic idea is that if we have an eigenvector for H then we can lift it up to an eigenvector for G that is constant on each parts of the equitable partition. (Note by our assumptions that the degrees of the vertices in each part agree.)

Example 3. For the complete tripartite graph $K_{r,s,t}$ we can find $r + s + t - 3$ independent eigenvectors associated with 1, i.e., choose two vertices in the same part and assign one of them the value 1, the other the value -1 and all other vertices 0. For the remaining three eigenvalues we can form an equitable partition of three parts, i.e., one for each part, this equitable partition gives a weighted triangle on three vertices with edge weights rs, rt, and st, for which the eigenvalues can be computed. Therefore the spectrum of $K_{r,s,t}$ is

$$\left\{0, 1^{(r+s+t-3)}, \tfrac{3}{2} \pm \tfrac{1}{2}\sqrt{9 - 8\tfrac{(r+s+t)(rs+rt+st)}{(r+s)(r+t)(s+t)}}\right\}.$$

For example, $K_{5,7,9}$ has spectrum $\{0, 1^{(18)}, \frac{11}{8}, \frac{13}{8}\}$.

One common way to form an equitable partition is to use the orbits of an automorphism. An *automorphism* of a weighted graph G is a mapping $\pi : V(G) \to V(G)$ such that for each $u, v \in V(G)$ we have $w(u, v) = w(\pi(u), \pi(v))$ and $w(u) = w(\pi(u))$.

Another common method which relates to equitable partitions is to use covers. We say that a weighted graph G *covers* a weighted graph H if there is a surjective map $\pi : V(G) \to V(H)$ which preserves edge weights so that for each $v \in V(G)$ the restriction of π to the neighborhood of v is a bijection. The preimages of vertices in H give an equitable partition. A classic example of this are cycles C_n are covered by cycles C_{kn} for $k \geq 1$, and therefore the eigenvalues of C_n are a subset of the eigenvalues of C_{kn}. More information about covers as it relates to the normalized Laplacian is given by Chung and Yau [12].

4.1 Twin vertices

We can use equitable partitions to find *some* of the eigenvalues of a graph, the question then remains on how do we find the rest. This can sometimes be done locally within the parts of the partition.

Theorem 4. *Let V_i be a part in an equitable partition of a graph without vertex weights and with vertices which all have degree k satisfying that for each $u \notin V_i$, $w(u, s) = w(u, t)$ for all $s, t \in V_i$ (i.e., the part* cones *to the remainder of the graph). If $G[V_i]$ (the induced subgraph of G on V_i) is regular of degree d and has as eigenvalues of the normalized Laplacian $0 = \lambda_0 \leq \lambda_1 \leq \cdots \leq \lambda_{m-1}$, then*

$$\left\{\frac{k - d + d\lambda_i}{k} : 1 \leq i \leq m - 1\right\}$$

are eigenvalues of G with eigenvectors whose support lies in V_i.

This follows from the same ideas used in determining the eigenvalues of the join, which is itself a special case of an equitable partition. More precisely, we find the eigenvalues using equation (3).

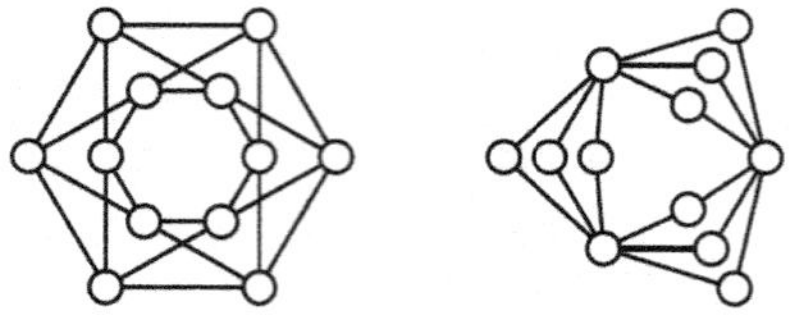

Fig. 4 Two cospectral graphs.

One special case of this result in simple graphs is when we have *twins*, which are vertices which have the same set of (closed) neighbors. (There is also a notion of twins for weighted graphs, see [6].)

Corollary 1. *Let V_i consist of a collection of twins in a simple graph which are mutually non-adjacent. Then there are $|V_i| - 1$ eigenvalues of 1 which come from eigenvectors restricted to V_i.*

Let V_i consist of a collection of twins in a simple graph which are mutually adjacent and have common degree d. Then there are $|V_i| - 1$ eigenvalues of $\frac{d+1}{d}$ which come from eigenvectors restricted to V_i.

By identifying twin vertices we can collapse a graph and deal with smaller graphs. For example, we can readily identify sets of twins for the two graphs in Fig. 4 so that the resulting equitable partitions correspond to weighted copies of C_6. Both graphs will have six eigenvalues of 1 from the twins, and the remaining eigenvalues from the graphs $4C_6$ and $3C_6$, respectively. By Proposition 1 the graphs $4C_6$ and $3C_6$ have the same normalized Laplacian, allowing us to conclude the original graphs are cospectral. This can also be shown by noting that these particular graphs are $C_3 \times K_{2,2}$ and $C_3 \times K_{1,3}$.

More examples of this phenomenon are given in [6], including the following construction which can be used to form cospectral graphs with differing numbers of edges for the normalized Laplacian.

Theorem 5 (Butler [6]). *Given a simple bipartite graph G with $V = A \cup B$ and natural numbers s, t, s', t' satisfying $s|A| + t|B| = s'|A| + t'|B|$, then form a new graph H by replacing each vertex in A by s independent vertices, each vertex in B by t independent vertices and each edge by the complete bipartite graph $K_{s,t}$, similarly form H' but using s' and t' for the respective sizes. Then H and H' are cospectral and have $st|E(G)|$ and $s't'|E(G)|$ edges respectively.*

An example using twins that are mutually adjacent will be given later.

4.2 Twin subgraphs

We can generalize the idea of twin vertices to twin subgraphs. The key element is that we have duplicate structures in the graph which connect in a consistent manner. More precisely we have the following.

Definition 5. Let G be a simple graph and let $V_1, V_2, \ldots, V_k$ be disjoint subsets of the vertices such that $G[V_1], \ldots, G[V_k]$ (i.e., the induced subgraphs on the V_i) are isomorphic with the isomorphism $\pi_i : V_1 \to V_i$. Further, these subgraphs connect in a consistent manner, i.e., for $u \notin V_1 \cup \cdots \cup V_k$ then $u \sim v$ for $v \in V_1$ if and only if $u \sim \pi_i(v)$ for $2 \leq i \leq k$. Then $G[V_1], \ldots, G[V_k]$ are *twin subgraphs*.

The notion of twin subgraphs can be applied to weighted graphs. Information on how this is done can be found in [6].

Proposition 4. *Let G be a simple graph with twin subgraphs coming from $V_1, \ldots, V_k$. Then the eigenvalues of G can be found using the following weighted graphs:*

- *Multiplicity $k-1$ for each of the eigenvalues arising from the graph $G[V_1]$ with vertex weights added so that the degrees agree with the degrees in G.*
- *Multiplicity 1 for each of the eigenvalues arising from the graph obtained by deleting $V_2 \cup \cdots \cup V_k$ and changing the edge weight of each edge incident to a vertex in V_1 to k.*

The idea of twin subgraphs has appeared in the work of Banarjee and Jost [1] and Banarjee and Mehatari [2]. Their approach was to start with a fixed graph and then create twin subgraphs (what they called "motifs") by duplicating an induced subgraph and its connections. They then considered what happens with the eigenvalues, where particular focus was given on the eigenvalue 1.

Example 4. The graph $K_{1,t} \square K_2$ can be viewed as a collection of 4-cycles which have been glued along an edge. We can view this as a collection of t disjoint copies of K_2 that consistently glue into a single copy of K_2. Therefore the eigenvalues of $K_{1,t} \square K_2$ are found by taking $(t-1)$ copies of the eigenvalues arising from the edge K_2 where the vertices have weight 1 (which have eigenvalues $\{\frac{1}{2}, \frac{3}{2}\}$), and the graph C_4 where one edge has weight 1 and the remaining three edges have weight t (which has eigenvalues $\{0, \frac{t+3}{2t+2}, \frac{3t+1}{2t+2}, 2\}$). Therefore the spectrum is

$$\{0, \tfrac{1}{2}^{(t-1)}, \tfrac{t+3}{2t+2}, \tfrac{3t+1}{2t+2}, \tfrac{3}{2}^{(t-1)}, 2\}.$$

Example 5. We construct a cospectral graph for the cycle C_{4k}, as shown in Fig. 5. We start with eigenvalues of some basic graphs which can be found using twins.

1. The path P_{2n+1} can be viewed as two copies of the path P_n (the twins) which connect to a central vertex. In particular, the eigenvalues of P_{2n+1} are the eigenvalues of $2P_{n+1}$ together with a path on n vertices with *one* end having vertex weight 1. Since we know the eigenvalues of P_{2n+1} and $2P_{n+1}$ (which has the same eigenvalues of P_{n+1}), we can conclude that the eigenvalues of P_n with one end having vertex weight 1 are
$$\{1 - \cos \tfrac{(2k+1)\pi}{2n} : 0 \leq k \leq n-1\}.$$
2. The cycle C_{2n} can be viewed as two copies of the path P_{n-1} (the twins) with the ends connecting to two special vertices. In particular, the eigenvalues of C_{2n} are

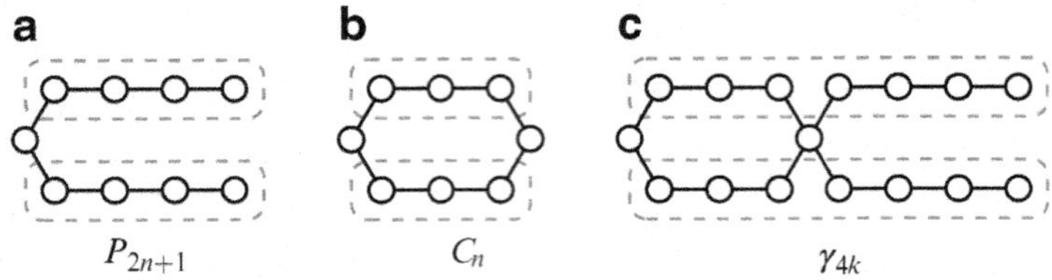

Fig. 5 Twin subgraphs of Example 5. Graphs C_{4k} and γ_{4k} are cospectral.

the eigenvalues of $2P_{n+1}$ together with a path on $n-1$ vertices with *both* ends having vertex weight 1. Since we know the eigenvalues of C_{2n} and $2P_{n+1}$ we can conclude the eigenvalues of P_{n-1} with both ends having vertex weight 1 are

$$\left\{1-\cos\tfrac{2k\pi}{2n} : n+1 \le k \le 2n-1\right\}.$$

3. Let γ_{4k} be the graph formed starting with C_{2n} and P_{2n+1} and then identifying the center vertex of the path to a point on the cycle. This can again be viewed as twin subgraphs gluing onto two special vertices so that the eigenvalues of the graph are the union of the eigenvalues of $2P_{2n+1}$, P_n with one end having vertex weight one, and P_{n-1} with both ends having vertex weight one, i.e.,

$$\begin{aligned}&\left\{1-\cos\tfrac{k\pi}{2n} : 0 \le k \le 2n\right\} \cup \left\{1-\cos\tfrac{(2k+1)\pi}{2n} : 0 \le k \le n-1\right\}\\ &\cup \left\{1-\cos\tfrac{2k\pi}{2n} : n+1 \le k \le 2n-1\right\} = \left\{1-\cos\tfrac{k\pi}{2n} : 0 \le k \le 4n-1\right\}.\end{aligned}$$

As promised above, γ_{4k} is cospectral with C_{4k}.

For the adjacency, Laplacian, and signless Laplacian it is an easy exercise to show that there are no graphs which are cospectral with the cycles. The graph γ_{4k} shows that this is *not* the case for the normalized Laplacian, though no other examples of graphs cospectral with a cycle are known.

Conjecture 1. The only graphs cospectral with a cycle are $K_{1,3}$ and γ_{4k}.

A variant of twin subgraphs comes when we can "fold" the graph in half (including folding an edge to become a loop). A discussion on how the eigenvalues can be found in this setting is given in [4].

5 The characteristic polynomial

The characteristic polynomial of the normalized Laplacian, which we will denote $\phi(x) = \det(xI - \mathscr{L})$, gives the eigenvalues of a graph. Additionally, the coefficients of the polynomial contain important information about the graph. For graphs without isolated vertices, $\mathscr{L}$ is similar to $I - D^{-1}A$ which has rational entries, so that for simple graphs $\phi(x)$ is a polynomial with rational coefficients.

Definition 6. A *decomposition*, $\mathfrak{D}$, of a graph G is a subgraph consisting of disjoint edges and cycles with no isolated vertices.

The determinant can be computed using decompositions in the following way.

Proposition 5 (Butler and Heysse [9]). *For a simple graph G we have*

$$\phi(x) = \sum_{\mathfrak{D}} \frac{(-1)^{\text{even}(\mathfrak{D})} 2^{\text{long}(\mathfrak{D})} (x-1)^{n-|V(\mathfrak{D})|}}{\prod_{v \in V(\mathfrak{D})} d(v)},$$

where even($\mathfrak{D}$) *is the number of cycles of even length (including isolated edges),* long($\mathfrak{D}$) *is the number of cycles of length at least 3, and the sum ranges over all decompositions $\mathfrak{D}$ (including the empty decomposition).*

This proposition is not particularly useful in practice, but Butler and Heysse [9] were able to find a way to show that certain pairs of graphs were cospectral by using decompositions to show that the characteristic polynomials were equal (they also give a generalization of the decomposition formulation for weighted graphs).

Guo, Li, and Shiu [14] gave a reduction formula for the characteristic polynomial for the normalized Laplacian which involves the sum of characteristic polynomials of subgraphs found by deleting vertices but preserving degrees (i.e., by moving edge weights to vertex weights where needed). The formula itself is cumbersome to work with directly, but can be used to prove a coalescing result.

Theorem 6 (Guo, Li and Shiu [14]). *Let G and H be simple graphs with $u \in V(G)$ and $v \in V(H)$, and let $G\,{}_u\!\circ_v\, H$ be the coalescence of G and H, i.e., formed by taking the union of G and H and identifying the vertices u and v into a new vertex uv. Then*

$$\phi(G\,{}_u\!\circ_v\, H) = \frac{d_G(u)\phi(G)\phi(H') + d_H(v)\phi(G')\phi(H)}{d_G(u) + d_H(v)},$$

where G' (resp. H') is the weighted graph obtained by deleting the vertex u (resp. v) and then increasing the degree weights of any adjacent vertices by 1 (so degrees of the remaining vertices are preserved).

Guo, Li, and Shiu also gave two non-isomorphic graphs G_1 and G_2 and vertices u_1 and u_2 for which $\phi(G_1) = \phi(G_2)$ and $\phi(G_1') = \phi(G_2')$. In particular for any graph H and vertex $v \in K$ we have $G_1\,{}_{u_1}\!\circ_v\, H$ is cospectral with $G_2\,{}_{u_2}\!\circ_v\, H$.

5.1 Spanning trees and Kemeny's coefficient

For a (simple) graph G on n vertices with eigenvalues $0 = \lambda_0 \le \lambda_1 \le \cdots \le \lambda_{n-1}$ we have

$$\begin{aligned}\phi(x) &= x(x-\lambda_1)\cdots(x-\lambda_{n-1})\\ &= x^n + c_{n-1}x^{n-1} + \cdots + c_2x^2 + c_1x.\end{aligned}$$

The coefficients of the characteristic polynomial are symmetric functions of the eigenvalues. For example, we have $c_1 = (-1)^{n-1}\lambda_1 \cdots \lambda_{n-1}$.

Kirchoff's Matrix Three Theorem gives a connection between the coefficient of x in the characteristic polynomial for the Laplacian and the number of spanning trees in a graph. A similar weighted version of this result also holds for the normalized Laplacian.

Theorem 7 (Weighted Matrix Tree Theorem [11]). *Let $\tau(G)$ denote the number of spanning trees of a graph G and let $0 = \lambda_0 \le \lambda_1 \le \cdots \le \lambda_{n-1}$ denote the eigenvalues of the normalized Laplacian of G. Then*

$$(-1)^{n-1}c_1 = \lambda_1 \cdots \lambda_{n-1} = \frac{\tau(G)\sum_v d(v)}{\prod_v d(v)}.$$

The coefficient c_2 also has an interesting combinatorial interpretation related to Kemeny's constant.

Definition 7. Given a graph G, *Kemeny's constant*, denoted $K(G)$, is the expected first passage time from a randomly chosen starting point in the graph to a randomly chosen destination point in the graph (see [16] for more information).

There is a way to compute Kemeny's constant by using the eigenvalues of the probability transition matrix. In particular, we have the following.

Proposition 6 (Levene and Loizou [17]). *Let G be a connected simple graph with $\rho_{n-1} \le \cdots \le \rho_1 \le \rho_0 = 1$ the eigenvalues for the probability transition $D^{-1}A$. Then*

$$K(G) = \sum_{i=1}^{n-1} \frac{1}{1-\rho_i}.$$

We note that $D^{-1}A$ is similar to the normalized adjacency matrix $D^{-1/2}AD^{-1/2}$, and so the eigenvalues of the probability transition matrix are connected to the eigenvalues of the normalized Laplacian by the relationship $\lambda_i = 1 - \rho_i$. So we can rewrite the above expressions as

$$K(G) = \sum_{i=1}^{n-1} \frac{1}{\lambda_i}.$$

We also note

$$c_2 = (-1)^{n-2}\lambda_1 \cdots \lambda_{n-1} \sum_{i=1}^{n-1} \frac{1}{\lambda_i} = -c_1 K(G).$$

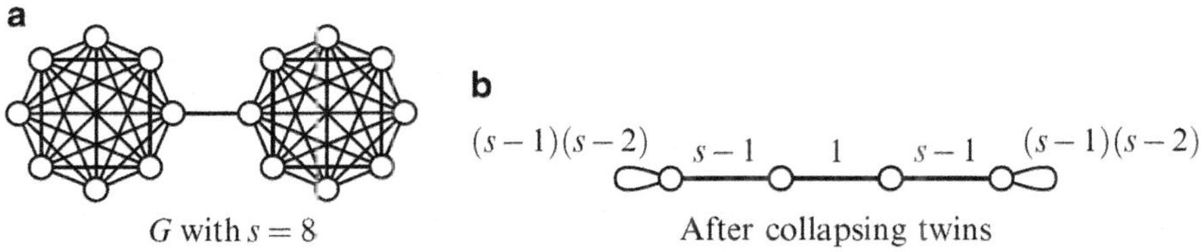

Fig. 6 The graph for Example 6

Allowing us to conclude that

$$K(G) = -\frac{c_2}{c_1}.$$

Example 6. Let G be the simple graph formed by taking two disjoint copies of K_s and joining the two cliques by one edge, see Fig. 6. To calculate $K(G) = -c_2/c_1$, we find the characteristic polynomial using some shortcuts. For example, G has an equitable partition in four parts, two with $s-1$ adjacent twins, and then two additional vertices from the connecting edge. From the twins in the cliques we get eigenvalues $\left\{\frac{s}{s-1}^{(2s-4)}\right\}$. The remaining eigenvalues come from the graph formed from the equitable partition on four vertices with edge weights 1, $s-1$ (twice), and loops $(s-1)(s-2)$ on the ends. (Note that the loops are *not* $\binom{s-1}{2}$, because each edge contributes twice, once as $w(u,v)$ and once as $w(v,u)$.) The characteristic polynomial for the graph on four vertices is

$$x^4 - \tfrac{s}{s-1}x^3 + \tfrac{s^4+s^2-1}{s^2(s-1)^2}x^2 - \tfrac{2(s^2-s+1)}{s^2(s-1)^2}.$$

Therefore the characteristic polynomial for G is

$$\begin{aligned}
\phi(x) &= \left(x-\tfrac{s}{s-1}\right)^{2s-4}\left(x^4-\tfrac{s}{s-1}x^3+\tfrac{s^4+s^2-1}{s^2(s-1)^2}x^2-\tfrac{2(s^2-s+1)}{s^2(s-1)^2}x\right)\\
&= \left(\cdots-(2s-4)\left(\tfrac{s}{s-1}\right)^{2s-5}x+\left(\tfrac{s}{s-1}\right)^{2s-4}\right)\left(\cdots+\tfrac{s^4+s^2-1}{s^2(s-1)^2}x^2-\tfrac{2(s^2-s+1)}{s^2(s-1)^2}x\right)\\
&= \tfrac{s^{2s-7}}{(s-1)^{2s-2}}\left(\cdots-(2s-4)(s-1)x+s\right)\left(\cdots+(s^4+s^2-1)x^2-2(s^2-s+1)x\right)\\
&= \tfrac{s^{2s-7}}{(s-1)^{2s-2}}\left(\cdots+(s^5-4s^4-15s^3+24s^2-21s+8)x^2-2(s^3-s^2+s)x\right),
\end{aligned}$$

so we find that

$$K(G) = \frac{(s^5+4s^4-15s^3+24s^2-21s+8)}{2(s^3-s^2+s)}.$$

There is a related concept called *the degree-Kirchoff index*, denoted $K'(G)$ which is a generalization of the Kirchoff index. It is known that

$$K'(G) = \Big(\sum_v d(v)\Big)\Big(\sum_{i=1}^{n-1}\frac{1}{\lambda_i}\Big) = \Big(\sum_v d(v)\Big)K(G).$$

For more information on this parameter see Chen and Zhang [10].

5.2 Characteristic polynomials and graph operations

Definition 8. For a simple graph G we can construct the following graphs:

- The *line graph*, denoted $\ell(G)$, whose vertex set corresponds to the edges of G and two vertices in $\ell(G)$ are adjacent if and only if the corresponding edges in G were incident to a common vertex.
- The *subdivision graph*, denoted $s(G)$, formed by adding one new vertex v_e for each edge $e = u{\sim}w$ and then deleting edge e and adding in edges $u{\sim}v_e$ and $v_e{\sim}w$.
- The graph $r(G)$ formed by adding back in the deleted edges to $s(G)$.
- The graph $q(G)$ formed by taking $s(G)$ and then adding in the corresponding edges of $\ell(G)$.

The graphs $\ell(G)$ and $s(G)$ are well studied. When the initial graph G is regular, then these graphs are regular ($\ell(G)$) or bi-regular ($s(G)$) graphs, so the known results for the adjacency spectrums will translate to these graphs. The other two graphs are not regular. Huang and Li [15] were able to establish connections between the characteristic polynomial for all of these graphs and the characteristic polynomial of G (through the use of Schur complements and properties of incidence matrices). Using this they then were able to find information on the number of spanning trees in these graphs compared to those of G as well as look at the relationship for the degree-Kirchoff index in regards to G.

Theorem 8 (Huang and Li [15]). *Let G be a k-regular simple graph with n vertices and m edges. Then the following holds:*

- $\phi_{\ell(G)}(x) = \frac{k^n((k-1)x-k)^{m-n}}{2^n(k-1)^m}\phi_G\big(\frac{2(k-1)x}{k}\big)$.
- $\phi_{s(G)}(x) = \big(\frac{-1}{2}\big)^n(x-1)^{m-n}\phi_G\big(2x(2-x)\big)$.
- $\phi_{r(G)}(x) = \frac{(2x-3)^n(x-1)^{m-n}}{4^n}\phi_G(2x)$.
- $\phi_{q(G)}(x) = \frac{(kx-k-1)^m}{2^n k^m}\phi_G(2x)$.

From the above results on the characteristic polynomials, relationships between the eigenvalues of the graph and the graphs obtained by these operations can be obtained.

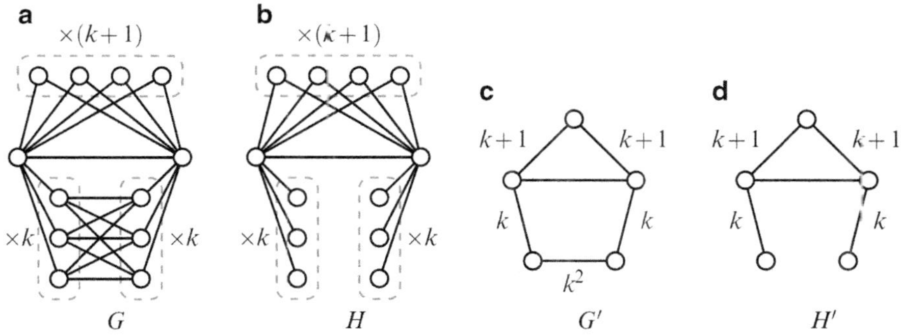

Fig. 7 The graphs from Example 7

5.3 *Verifying cospectral graphs*

One way to show that two graphs are cospectral (have the same eigenvalues) is to use structure to find many eigenvalues which agree on both graphs and then reduce the problem of finding the remaining eigenvalues for a smaller graph which can then be checked using characteristic polynomials. We give here two such examples that illustrate some of the unusual aspects of the spectrum of the normalized Laplacian.

Example 7. Let G and H be the two graphs shown in Fig. 7, where we have put vertices into groups. The number of vertices in each group marked as a function of k. Note that H is a subgraph of G, and the two graphs differ by the edges of the complete bipartite graph $K_{k,k}$. The indicated groups of vertices give an equitable partition of both graphs. In each case, the vertices in the parts are twins and so both graphs have the eigenvalue 1 with multiplicity $3k-2$. We can now use the equitable partitions to form the graphs G' and H' which give the remaining eigenvalues of G and H. These are graphs on five vertices with edge weights as a function of k and in *both* cases the characteristic polynomials are:

$$x^5 - \frac{6k^2+8k+3}{4(k+1)^2}x^3 - \frac{1}{4(k+1)}x^2 + \frac{k(2k+1)}{4(k+1)^2}x.$$

In particular the graphs G and H are cospectral.

This example illustrates that it is possible for a simple graph to be cospectral with a subgraph. More strongly, it shows that it is possible for a *dense* graph to be cospectral with a *sparse* subgraph.

Example 8. Let $B_{s,t}$ be the graph on $2s+2t+2$ vertices formed by taking two copies of K_{s+t} and two special vertices v_s and v_t where v_s connects to s vertices in each of the cliques and v_t connects to the remaining t vertices. Fig. 8 shows $B_{2,2}$ and $B_{1,3}$. The graph $B_{s,t}$ has a natural equitable partition into six parts: v_s, v_t, and two groups for each clique: those vertices that connect to v_s, and those vertices that connect to v_t, see Fig. 8. For the twins in the equitable partitions in the two cliques they will

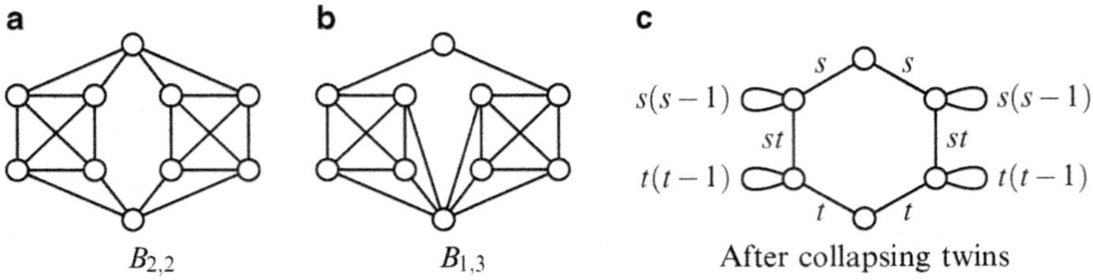

Fig. 8 The graphs from Example 8

contribute $2s+2t-4$ eigenvalues of $\frac{s+t+1}{s+t}$. The remaining eigenvalues comes from a weighted graph on six vertices arising from the equitable partition. This graph has characteristic polynomial

$$x^6 - \frac{4(s+t+1)}{(s+t)}x^5 + \frac{6(s+t+1)^2}{(s+t)^2}x^4 - \frac{(4(s+t)^2+5(s+t)+2)(s+t+2)}{(s+t)^3}x^3 + \frac{((s+t)^3+5(s+t)^2+3(s+t)+1)(s+t+1)}{(s+t)^4}x^2 - \frac{(s+t+1)^2}{(s+t)^3}x.$$

In particular we can conclude that all of the eigenvalues can be expressed in terms of $s+t$, and so the graphs $B_{k-i,k+i}$ for $0 \le i \le k-1$ are all cospectral. Note that $B_{k,k}$ is regular of degree $2k$ while for $1 \le i \le k-1$ the graphs $B_{k-i,k+i}$ are not regular.

This example illustrates that it is possible for a simple graph which is regular to be cospectral with a simple graph which is not regular (in addition to the previous bipartite examples of $K_{s,s}$ and C_{4k}). This stands in stark contrast to the other common matrices in spectral graph theory (adjacency, Laplacian, and signless Laplacian) where this is impossible. Of course this is a common theme in the subject: the normalized Laplacian behaves quite differently, and in interesting and unusual ways. There still remains much to understand for the normalized Laplacian.

Acknowledgements The author thanks Shaun Fallat and Kristin Heysse for many useful discussions. This paper was written while the author was a visitor at the Institute for Mathematics and its Applications and also while the author was supported by an NSA Young Investigator Grant.

References

1. A. Banarjee, J. Jost, The spectrum of the normalized graph Laplacian. Linear Algebra Appl. **428**, 3015–3022 (2008)
2. R. Mehatari, A. Banerjee, Effect on normalized graph Laplacian spectrum by motif attachment and duplication. Appl. Math. Comput. **261**, 382–387 (2015)
3. A. Brouwer, W. Haemers, *Spectra of Graphs* (Springer, New York, 2012)
4. S. Butler, Eigenvalues of 2-edge-coverings. Linear Multilinear Algebra **58**, 413–423 (2010)
5. S. Butler, A gentle introduction to the normalized Laplacian. IMAGE, **53**, 19–27 (Fall 2014)

6. S. Butler, Using twins and scaling to construct cospectral graphs for the normalized Laplacian. Electron. J. Linear Algebra **28**, 54–68 (2015)
7. S. Butler, F. Chung, Spectral graph theory, in *Handbook of Linear Algebra*, ed. by L. Hogben, 2nd edn. (CRC, Boca Raton, 2014)
8. S. Butler, J. Grout, A construction of cospectral graphs for the normalized Laplacian. Electron. J. Comb. **18**, #231, 20 pp (2011)
9. S. Butler, K. Heysse, A cospectral family of graphs for the normalized Laplacian found by toggling. arXiv:1507.01928
10. H. Chen, F. Zhang, Resistance distance and the normalized Laplacian spectrum. Discrete Appl. Math. **155**, 654–661 (2007)
11. F. Chung, *Spectral Graph Theory*, 2nd edn. (AMS, Providence, 1997)
12. F. Chung, S.T. Yau, Coverings, heat Kernels and spanning trees. Electron. J. Comb. **6**, #12, 21 pp (1999)
13. C. Godsil, G. Royle, *Algebraic Graph Theory* (Springer, New York, 2001)
14. J.-M. Guo, J. Li, W.C. Shiu, On the Laplacian, signaless Laplacian and normalized Laplacian characteristic polynomials of a graph. Czech. Math. J. **63**, 701–720 (2013)
15. J. Huang, S. Li, On the normalized Laplacian spectrum, degree-Kirchoff index and spanning trees of graphs. Bull. Aust. Math. Soc. **91**, 353–367 (2015)
16. J. Hunter, The role of Kemeny's constant in properties of Markov chains. Commun. Stat. Theory Methods **43**, 1309–1321 (2014)
17. M. Levene, G. Loizou, Kemeny's constant and the random surfer. Am. Math. Mon. **109**, 741–745 (2002)

Progress on poset-free families of subsets

Jerrold R. Griggs and Wei-Tian Li

Abstract Increasing attention is being paid to the study of families of subsets of an n-set that contain no subposet P. Especially, we are interested in such families of maximum size given P and n. For certain P this problem is solved for general n, while for other P it is extremely challenging to find even an approximate solution for large n. It is conjectured that for any P, the maximum size is asymptotic to a constant times $\binom{n}{\lfloor \frac{n}{2} \rfloor}$, where the constant is a certain integer depending on P. This survey has two purposes. First, we want to bring this exciting line of research to the attention of a wider audience. Second, we want to make experts aware of the broad range of recent progress in the area.

Keywords Poset-free families of subsets • Sperner Theory • Diamond-free families • Lubell function

Mathematics Subject Classification (2010): 05D05, 06A07

1 First Results: Families Without Chains

Problems of forbidding given structures in a larger structure are very popular in extremal combinatorics. The study of forbidden subgraph problems, starting from Mantel's result on triangle-free graphs, is nowadays a well-developed discipline. The Turán theory on hypergraphs contains beautiful theorems and challenging unsolved problems.

A research area that has become fertile in recent years considers forbidding poset structures in families of subsets. Let $P = (P, \leq)$ be a finite poset. We say that another poset Q contains P as a *subposet* (in the weak sense), if there exists an order-

J.R. Griggs (✉)
Department of Mathematics, University of South Carolina, Columbia, SC 29208, USA
e-mail: griggs@math.sc.edu

W.-T. Li
Department of Applied Mathematics, National Chung Hsing University, Taichung, Taiwan 40227
e-mail: weitianli@dragon.nchu.edu.tw

A. Beveridge et al. (eds.), *Recent Trends in Combinatorics*, The IMA Volumes in Mathematics and its Applications 159, DOI 10.1007/978-3-319-24298-9_14

preserving injection i from P to Q. We view a family $\mathcal{F}$ of subsets of a finite set as a poset itself, ordered according to the inclusion relation of sets. More precisely, working in the Boolean lattice $\mathcal{B}_n$ of all 2^n subsets of $[n] := \{1, 2, \cdots, n\}$, ordered by inclusion, we want to understand how large a family of subsets in $\mathcal{B}_n$ can be without containing a given poset P as a subposet. We let $\binom{[n]}{k}$ denote the collection of k-subsets of $[n]$.

The foundational result in the area dates back to 1928. Sperner studied *antichains*, which are families of sets without any set containing another, and obtained the following result.

Theorem 1.1. [47] *Let $\mathcal{F}$ be an antichain of subsets of $[n]$. Then $|\mathcal{F}|$ is at most the largest binomial coefficient, namely $\binom{n}{\lfloor \frac{n}{2} \rfloor}$. Moreover, the only antichains achieving the bound are $\binom{[n]}{\lfloor \frac{n}{2} \rfloor}$ and $\binom{[n]}{\lceil \frac{n}{2} \rceil}$.*

Note that we get two different extremal families, or just one (the two listed are identical), depending on the parity of n.

A generalization of Sperner's theorem was given by Erdős, who developed it in the 1940s to make progress on a question of Littlewood and Offord concerning the distribution of roots of random polynomials. The *size* of a poset is its cardinality. A poset is a *chain* if any two of its elements are comparable. Erdős gave the following result on forbidding chains of a given size:

Theorem 1.2. [18] *Let $k \geq 1$. Let $\mathcal{F}$ be a family of subsets of $[n]$ such that no $k+1$ subsets form a chain. Then $|\mathcal{F}|$ is at most the sum of the k middle binomial coefficients in n. Moreover, the only families achieving the bound are obtained by taking all subsets of the k middle sizes in $[0, n]$.*

Again, the maximum size of such a family $\mathcal{F}$ is given for all n, and there is just one, or two, extremal families. Since these bounds and the families of Erdős's Theorem occur repeatedly in the theory, it is useful to adopt notation from [23]: Let $\mathcal{B}(n, k)$ be a family of subsets of $[n]$ of the k middle sizes, $\binom{[n]}{\lfloor (n-k+1)/2 \rfloor} \cup \cdots \cup \binom{[n]}{\lfloor (n+k-1)/2 \rfloor}$ or $\binom{[n]}{\lceil (n-k+1)/2 \rceil} \cup \cdots \cup \binom{[n]}{\lceil (n+k-1)/2 \rceil}$. So $\mathcal{B}(n, k)$ is one or two possible families, depending on the parity of $n+k$. Also, let $\Sigma(n, k)$ denote $|\mathcal{B}(n, k)|$.

By saying no $k+1$ elements in a family $\mathcal{F}$ form a chain, it means that $\mathcal{F}$ does not contain any $\mathcal{P}_{k+1}$, where $\mathcal{P}_r$ denotes the *chain poset* (also called *path poset*) of size r.

In the early 1980s, Katona started to investigate forbidding posets other than chains in the Boolean lattice. Since then a rapidly growing number of results on such problems have been discovered by Katona, his collaborators, and other researchers who are inspired by the pioneering work. Here we attempt to collect this work, particularly the research papers published in past ten years. We seek to identify the core ideas and prospective directions for continuing study. Interested readers should also see other surveys in recent years, especially [23, 27, 29, 35].

2 Forbidding Other Posets

Following Katona we denote by $\mathrm{La}(n, P)$ the maximum size of a family $\mathcal{F}$ of subsets of $[n]$ containing no subposet P. If a family of subsets does not contain P as a subposet, then we say it is *P-free*. Note that we are forbidding P as a subposet in the weak sense, meaning that we are excluding not only P itself, but also any poset containing P. For instance, let $\mathcal{V}_r$ denote the *fork poset* with one element below the other r elements. When we forbid $\mathcal{V}_r$, we are forbidding any subset being in r others in the family, whether or not those others contain each other. We are interested in the exact value of $\mathrm{La}(n, P)$, when we can determine it, or else we are happy if we can determine its leading asymptotic behavior, for fixed poset P, as $n \to \infty$.

One poset parameter we sometimes need is the *height* of poset P, denoted $h(P)$, which is the largest size of any chain in P.

The first small poset P that was investigated was $\mathcal{V} = \mathcal{V}_2$ (which has a Hasse diagram that looks like the letter V). Katona and Tarján (1981) gave the asymptotic behavior of $\mathrm{La}(n, \mathcal{V})$:

Theorem 2.1. [31] *As $n \to \infty$,*

$$1 + \frac{1}{n} + \Omega\left(\frac{1}{n^2}\right) \le \frac{\mathrm{La}(n, \mathcal{V})}{\binom{n}{\lfloor \frac{n}{2} \rfloor}} \le 1 + \frac{2}{n}.$$

Note that $\mathcal{B}(n, 1)$, the subsets of middle size in n, is a $\mathcal{V}$-free family of asymptotically optimal size. However, optimal families are slightly larger (and not known)–we can add some sets in the level above the middle one while remaining $\mathcal{V}$-free. The lower bound in the theorem adds in a lower order term, based on a construction of Graham and Sloane [19] in coding theory. The error term in the upper bound is twice that of the lower bound. It would be a significant accomplishment to improve the error term and determine constant c such that $\mathrm{La}(n, \mathcal{V}) = (1 + \frac{c}{n} + o(\frac{1}{n}))\binom{n}{\lfloor \frac{n}{2} \rfloor}$.

Thanh (1998) generalized the result of Katona and Tarján on poset $\mathcal{V}$ to the r-fork, and Katona and DeBonis later improved the error term:

Theorem 2.2. [13, 48] *For general r, as $n \to \infty$,*

$$1 + \frac{r-1}{n} + \Omega\left(\frac{1}{n^2}\right) \le \frac{\mathrm{La}(n, \mathcal{V}_r)}{\binom{n}{\lfloor \frac{n}{2} \rfloor}} \le 1 + 2\frac{r-1}{n} + O\left(\frac{1}{n^2}\right).$$

Another way to build up the $\mathcal{V}$ poset is to add a second element below the top two. Called the *butterfly poset* $\bowtie$, it has two minimal elements, each below each of two maximal elements. We have a jump in the possible size of P-free families for this poset compared to $\mathcal{V}$. Particularly surprising is that for the butterfly poset, the value of $\mathrm{La}(n, P)$ can be determined precisely. Observe that for any k, two distinct k-sets have at most one $(k-1)$-subset below them both. It means that the union of two consecutive levels in $\mathcal{B}_n$ is $\bowtie$-free. DeBonis, Katona, and Swanepoel (2005) used the cyclic permutation method to prove that the family $\mathcal{B}(n,2)$ is as large as possible, provided $n \geq 3$:

Theorem 2.3. [14] *For butterfly-free families, when $n \geq 3$,*

$$\mathrm{La}(n, \bowtie) = \Sigma(n, 2).$$

For $n \geq 5$, the only extremal families are $\mathcal{B}(n,2)$.

Omitting one ordered pair in the butterfly poset gives a poset that is intermediate between $\mathcal{V}$ and $\bowtie$. Called $\mathcal{N}$ because of its Hasse diagram, it has elements corresponding to subsets a, b, c, d with $a \subset b, c \subset b, c \subset d$. It immediately satisfies $\mathrm{La}(n, \mathcal{V}) \leq \mathrm{La}(n, \mathcal{N}) \leq \mathrm{La}(n, \bowtie)$. Griggs and Katona (2008) discovered for this poset that $\mathrm{La}(n, P)$ grows like $\mathrm{La}(n, \mathcal{V})$:

Theorem 2.4. [20] *For $\mathcal{N}$-free families, as $n \to \infty$,*

$$1 + \frac{1}{n} + \Omega\left(\frac{1}{n^2}\right) \leq \frac{\mathrm{La}(n, \mathcal{N})}{\binom{n}{\lfloor \frac{n}{2} \rfloor}} \leq 1 + \frac{2}{n} + O\left(\frac{1}{n^2}\right).$$

What if we take the butterfly poset and add more elements to the two levels? Say we have r elements each above each of s elements. Call this the *complete 2-level poset* $\mathcal{K}_{r,s}$, named after the graph that looks like its Hasse diagram. For $r = s = 2$ it is the butterfly. DeBonis and Katona found bounds on $\mathrm{La}(n, K_{r,s})$ that are asymptotically similar to what was found for the butterfly.

Theorem 2.5. [13] *Let $r \geq s \geq 2, r \geq 3$. For $K_{r,s}$-free families, as $n \to \infty$, we have*

$$\Sigma(n,2) + \left(2\,\frac{r+s-4}{n} + \Omega\left(\frac{1}{n^2}\right)\right)\binom{n}{\lfloor \frac{n}{2} \rfloor}$$

$$\leq \mathrm{La}(n, K_{r,s}) \leq \left(2 + 2\,\frac{r+s-3}{n} + O\left(\frac{1}{n^2}\right)\right)\binom{n}{\lfloor \frac{n}{2} \rfloor}.$$

Their upper bound for $\mathrm{La}(\mathcal{K}_{r,s})$ follows from a method of partitioning a $\mathcal{K}_{r,s}$-free families into the union of a $\mathcal{V}_r$-free family and a Λ_s–free family, where the *s-lambda* poset Λ_s is the dual of $\mathcal{V}_s$.

Let us mention that recently Patkós [44] has obtained results for the complete 3-level poset $\mathcal{K}_{r,s,t}$, where the levels have sizes r, s, t, respectively.

Griggs and Lu [24] considered several other families of posets, including batons and height two trees. They also considered a different generalization of the butterfly poset. The *crown poset* $\mathcal{O}_{2k}$, $k \geq 2$, has Hasse diagram that is an up-and-down cycle on $2k$ vertices, namely, $a_1 \leq b_1 \geq a_2 \leq b_2 \cdots \geq a_k \leq b_k \geq a_1$. The smallest crown is $\mathcal{O}_4$, which is the butterfly, while $\mathcal{O}_6$ is the middle two levels of the Boolean lattice $\mathcal{B}_3$. Their 2008 work also introduced more probabilistic methods into the subject. They managed to determine $\mathrm{La}(n, \mathcal{O}_{2k})$ asymptotically for all even $k \geq 2$. Later on (2014), Lu [36] managed to do this as well for odd $k \geq 7$.

Theorem 2.6. [24, 36] *Let $k = 4$ or ≥ 6. For $\mathcal{O}_{2k}$-free families,*

$$1 + \frac{1}{n} + \Omega\left(\frac{1}{n^2}\right) \leq \frac{\mathrm{La}(n, \mathcal{O}_{2k})}{\binom{n}{\lfloor \frac{n}{2} \rfloor}} \leq 1 + \frac{2}{n} + O\left(\frac{1}{n^2}\right).$$

It remains open to determine $\mathrm{La}(n, P)$ asymptotically for $\mathcal{O}_6$ and $\mathcal{O}_{10}$. The best upper bound is larger than above by a factor of $(1 + \frac{1}{\sqrt{2}})$. We see that for $k = 4$ and for all $k \geq 6$, in contrast to the butterfly, $\mathrm{La}(n, \mathcal{O}_{2k})$ is asymptotic to $\binom{n}{\lfloor \frac{n}{2} \rfloor}$ (and likely it also is for $k = 3, 5$).

The next section describes the difficulty determining $\mathrm{La}(n, P)$ when P is a four-element diamond poset. In 2011 Li [35] then weakened the diamond slightly, considering the poset on elements a, b, c, d with $a < b, a < c < d$, denoted by $\mathcal{J}$ (J) due to its Hasse diagram. Forbidding $\mathcal{J}$ is weaker than forbidding the chain $\mathcal{P}_3$, but stronger than forbidding $\mathcal{P}_4$ or the four-element diamond. It was surprising that there is a nice answer here.

Theorem 2.7. [35] *For $\mathcal{J}$-free families,*

$$\mathrm{La}(n, \mathcal{J}) = \Sigma(n, 2).$$

All posets P mentioned so far are *ranked*, which means that there is a function r from P to the nonnegative integers such that whenever element y covers x, $r(y) = r(x) + 1$. For instance, the Boolean lattice $\mathcal{B}_n$ is ranked by taking $r(X) = |X|$ for a subset X of $[n]$. The smallest unranked posets contain five elements. Methuku and Tompkins considered one such poset, with partial order relations: $a_1 < a_2$, $b_1 < b_2 < b_3$, $a_1 < b_3$, $b_1 < a_2$. They named it the *skew-butterfly* and determined $\mathrm{La}(n, P)$ for this poset:

Theorem 2.8. [42] *Let P be the skew-butterfly. For $n \geq 3$,* $\mathrm{La}(n, P) = \Sigma(n, 2)$.

3 Searching for Diamonds

After speaking about the early work on forbidden subposets back in March, 2007, including the butterfly and $\mathcal{N}$ posets, Griggs was asked about another poset with a 4-cycle Hasse diagram, the Boolean lattice $\mathcal{B}_2$. Because of its shape, it makes sense to call it the *diamond poset*. More generally, let $\mathcal{D}_k$ denote the poset with one maximum element above the rest, one minimum element below the rest, and k elements in between. We call this the *k-diamond poset*, so that $\mathcal{D}_2$ denotes the diamond poset. People asked Katona about $\mathcal{D}_2$ at talks before that. He wrote [29] that when he was asked in 2004 about $\mathrm{La}(n, \mathcal{D}_2)$, he answered, "*This is a good question, I think the present method will work on this problem, too.*" To the contrary, rather unexpectedly, it turns out that solving $\mathrm{La}(n, \mathcal{D}_2)$ has become one of the most challenging problems in extremal combinatorics.

Immediately, because $\mathcal{D}_2$ contains a 3-chain, $\mathcal{B}(n, 2)$ is diamond-free. On the other hand, since $\mathcal{D}_2$ is a subposet of a 4-chain, by Erdős's Theorem 1.2, no diamond-free family is larger than $\mathcal{B}(n, 3)$. It means that $\Sigma(n, 2) \leq \mathrm{La}(n, \mathcal{D}_2) \leq \Sigma(n, 3)$, so asymptotically

$$2(1 - o(1)) \leq \frac{\mathrm{La}(n, \mathcal{D}_2)}{\binom{n}{\lfloor \frac{n}{2} \rfloor}} \leq 3.$$

An easy averaging argument reduces the upper bound to 2.5. Griggs and Lu believed that upper bound to be soft– the lower bound of 2 should be the answer asymptotically. While still a student, Li joined the effort to improve the upper bound.

Using averaging arguments and the Lubell function (which we introduce later), Griggs, Li, and Lu brought the upper bound strictly below 2.3. They announced their progress at conferences, and the group of Axenovich, Manske, and Martin responded by reducing the bound. Griggs *et al.* then did more computations to push their argument for an even better bound. More recently, Kramer, Martin, and Young succeeded in reducing it to 2.25, which is important progress, even if it remains well above the value 2 that most researchers expect is the actual value. We summarize these results here:

Theorem 3.1. *For $\mathcal{D}_2$-free families, we have*

(1) *(see* [23]*)* $\mathrm{La}(n, \mathcal{D}_2) \leq 2.296\binom{n}{\lfloor \frac{n}{2} \rfloor}$,
(2) [2] $\mathrm{La}(n, \mathcal{D}_2) \leq 2.283\binom{n}{\lfloor \frac{n}{2} \rfloor}$,
(3) [23] $\mathrm{La}(n, \mathcal{D}_2) \leq 2.273\binom{n}{\lfloor \frac{n}{2} \rfloor}$,
(4) [34] $\mathrm{La}(n, \mathcal{D}_2) \leq 2.25(1 + o(1))\binom{n}{\lfloor \frac{n}{2} \rfloor}$.

Since this problem of determining the largest size of a diamond-free family $\mathcal{F}$ of subsets of $[n]$ is so daunting, how about restricting $\mathcal{F}$ and seeing if the bound of 2 applies? Various researchers independently considered $\mathcal{F}$ restricted to just the three middle sizes, that is, $\mathcal{F} \subseteq \mathcal{B}(n, 3)$. A diamond in such a family consists of a subset A in the lowest level, a subset C in the highest level, and both subsets B obtained by adding an element of $C - A$ to A. Moreover, this problem appeals to graph theorists, who are interested in sets of vertices of the n-cube that induce no 4-cycles. When restricting to just three middle levels, diamonds correspond to the only 4-cycles.

Note that these three middle levels each have size $\sim \binom{n}{\lfloor \frac{n}{2} \rfloor}$, so for a family $\mathcal{F}$ in these levels, $|\mathcal{F}|/\binom{n}{\lfloor \frac{n}{2} \rfloor}$ is essentially just the sum, over the three levels, of the proportions of elements in $\mathcal{F}$. Surely, we can manage this narrower problem on just three levels? In 2012 Axenovich, Manske, and Martin [2] reduced the upper bound for three-level diamond-free families to below the current best bound of

$2.25\binom{n}{\lfloor \frac{n}{2} \rfloor}$ for general diamond-free families. They also observed that a bound on this problem also bounds the size of families in which the sets have just three sizes, not necessarily the middle three. Elegant arguments in 2013 by Manske and Shen [39] reduce it further. More recently Balogh, Hu, Lidický, and Liu [4] employed flag algebras, a sophisticated method for associating inequalities to certain problems. The resulting semidefinite program then requires considerable computer time to produce good bounds. This method has proven successful in Turán theory [3, 45].

Theorem 3.2. *Let $\mathcal{F}$ be a $\mathcal{D}_2$-free family of subsets in $\mathcal{B}(n,3)$, that is, every set in $\mathcal{F}$ has one of the three middle sizes in n. Then we have*

(1) [2] $|\mathcal{F}| \le 2.20\binom{n}{\lfloor \frac{n}{2} \rfloor}$.
(2) [39] $|\mathcal{F}| \le 2.16\binom{n}{\lfloor \frac{n}{2} \rfloor}$.
(3) [4] $|\mathcal{F}| \le 2.15121\binom{n}{\lfloor \frac{n}{2} \rfloor}$.

Here is an even more restricted problem we propose that remains open, as far as we know:

Problem 3.3. *Show that if family $\mathcal{F}$ contains $\sim$ 70% of the subsets from each of the three middle sizes, then $\mathcal{F}$ must contain a diamond $\mathcal{D}_2$.*

Notice that $|\mathcal{F}| \sim 2.1\binom{n}{\lfloor \frac{n}{2} \rfloor}$, which is less than, what we need to use, the bound of Balogh *et al.* That we cannot yet solve this toy problem suggests we do not yet fully understand how multiple levels of the Boolean lattice–even just three levels–interrelate.

Let us mention here additional work on diamond-free families. Czabarka, Dutle, Johnston, and Székely [11] used random families arising from posets based on abelian groups to construct more examples of large diamond-free families. These families can be viewed as additional support for the conjectured asymptotic behavior.

Dove [15] devised examples of diamond-free families $\mathcal{F}$ for all even $n \ge 6$ with $|\mathcal{F}| > \Sigma(n,2)$. It means that $\mathrm{La}(n,\mathcal{D}_2)$ is not likely to have a simple formula for all large n (unlike, for instance, $\bowtie$). These examples are all contained in $\mathcal{B}(n,3)$, the middle three layers. The asymptotic behavior of $\mathrm{La}(n,\mathcal{D}_2)$ must be more complicated than for examples like $\bowtie$, so it is unlikely anyone can determine these values exactly. Sarkis, Shahriari, and students [46] investigated diamond-free families in the subspace lattice, and discovered the analogous asymptotic problem is less formidable.

In a later section we shall say more about $\mathrm{La}(n,P)$ for general diamonds $\mathcal{D}_k$. For now, let us mention one simple way k-diamonds arise. Griggs noticed that early results suggest a dependence of $\mathrm{La}(n,P)$ on the height $h(P)$. Indeed, from the result of DeBonis-Katona on forbidding $\mathcal{K}_{r,s}$ it follows that for every poset P of height at most two, $\mathrm{La}(n,P)/\binom{n}{\lfloor \frac{n}{2} \rfloor} \le 2 + o(1)$ as $n \to \infty$. Griggs wondered whether for general h there is a bound c_h such that $\mathrm{La}(n,P)/\binom{n}{\lfloor \frac{n}{2} \rfloor} \le c_h$ for all posets P of height h. This turned out to be completely false: Jiang and Lu [23] independently noticed that while the k-diamonds $\mathcal{D}_k$ all have height 3, there is no bound on how large $\mathrm{La}(n,\mathcal{D}_k)/\binom{n}{\lfloor\frac{n}{2}\rfloor}$ can get. This is easily seen from the fact that the family of the middle r levels, $\mathcal{B}(n,r)$, contains no diamond $\mathcal{D}_k$ for $k \geq 2^{r-1}-1$. It means that for any r, for large enough k, we have $\mathrm{La}(n,\mathcal{D}_k)/\binom{n}{\lfloor\frac{n}{2}\rfloor} > r$ for all large n.

4 Conjectured Asymptotics for General Posets

For some posets P, such as chains or the butterfly (when $n \geq 3$), $\mathrm{La}(n,P)$ is simply a sum of largest binomial coefficients in n. For many more posets P, this fails, but the asymptotic behavior of $\mathrm{La}(n,P)$ is nice. Examples of this include $\mathcal{V}, \mathcal{N}$, and $\mathcal{K}_{r,s}$. Of course, we have seen that for the diamond $\mathcal{D}_2$, even the asymptotic behavior remains unsettled.

In their study of a variety of posets, Griggs and Lu [24] came to believe that for every finite poset P, as $n \to \infty$, $\mathrm{La}(n,P)$ should be asymptotic to $\binom{n}{\lfloor\frac{n}{2}\rfloor}$ times some integer depending on P. This same belief likely guided the early work of Katona and his collaborators, even if it was not explicitly stated. Even now, some years later, the conjecture continues to hold for all posets for which the asymptotic behavior is known.

For posets P Griggs and Lu introduced the notation

$$\pi(P) := \lim_{n\to\infty} \frac{\mathrm{La}(n,P)}{\binom{n}{\lfloor\frac{n}{2}\rfloor}}.$$

The conjectured limit imitates the Turán density $\pi(H)$ of k-graphs H, which concerns the maximum number $\mathrm{ex}(n,H)$ of edges in a k-graph on n vertices not containing H, $\pi(H) := \lim_{n\to\infty} \frac{\mathrm{ex}(n,H)}{\binom{n}{k}}$. A difference is that the hypergraph limit is known to exist in general, while proving existence remains open for $\pi(P)$. Also, the hypergraph limit is not integer in general, while the poset limit is conjectured to be integer.

When Griggs announced this conjecture at a 2008 conference, two of the experts present, M. Saks and P. Winkler, both pointed out a pattern that explains the known limiting values $\pi(P)$, having to do with finding poset P in Boolean lattices $\mathcal{B}_n$. Griggs and Lu [23] subsequently formulated it by introducing notation: For a poset P, let $e(P)$ denote the maximum k such that the union $\mathcal{B}(n,k)$ of k middle levels in $\mathcal{B}_n$ does not contain P as a subposet, no matter how large n is.

For instance, the butterfly poset is not contained in $\mathcal{B}(n,3)$ when $n=2$, since $\mathcal{B}(2,3)$ is simply the diamond $\mathcal{B}_2$. Starting at $n=3$, $\mathcal{B}(n,3)$ does contain (many) butterflies. On the other hand, $\bowtie$ is not contained in the union of any two consecutive levels in the Boolean lattice. It means that $e(\bowtie)=2$. Since $\Sigma(n,2) \sim 2\binom{n}{\lfloor\frac{n}{2}\rfloor}$, it gives $\pi(\bowtie) = 2 = e(\bowtie)$.

In general, the family $\mathcal{B}(n, e)$ contains no P for $e = e(P)$. So if $\pi(P)$ exists, it must be at least $e(P)$. The Griggs-Lu Conjecture, announced in 2008 [cf. 23], is that $\pi(P)$ exists, and its value is $e(P)$:

Conjecture 4.1. [23, 24] *For any poset P, the limit $\pi(P) := \lim_{n\to\infty} \frac{\mathrm{La}(n,P)}{\binom{n}{\lfloor \frac{n}{2} \rfloor}}$ exists. Moreover, its value is the integer $e(P)$.*

Not long after the conjecture was announced (2009), Bukh [7] proved the existence of $\pi(P)$ for every poset which has a tree (acyclic) Hasse diagram.

Theorem 4.1. [7] *For all tree posets T, $\pi(T) = h(T) - 1 = e(T)$.*

This may be the strongest result yet in support of the Griggs-Lu Conjecture, considering that it includes so many of the posets described already, such as chains, forks, batons, height 2 trees, and $\mathcal{J}$. The proof is an impressive display of probabilistic reasoning.

Moreover, Bukh's result can be used to get a good estimate for other posets. For example, it gives another way to derive $\pi(\bowtie) = 2$: We already have the easy lower bound from $\mathrm{La}(n, \bowtie) \geq \mathcal{B}(n, 2) \sim 2\binom{n}{\lfloor \frac{n}{2} \rfloor}$. For the upper bound, we view the butterfly as a subposet of $\mathcal{X}$, where $\mathcal{X}$ is the poset on five elements with partial order relations $x_1, x_2 \leq x_3 \leq x_4, x_5$. We have that $\mathrm{La}(n, \bowtie) \leq \mathrm{La}(n, \mathcal{X})$. Since the poset $\mathcal{X}$ is a tree of height 3, by Theorem 4.1, $\mathrm{La}(n, \mathcal{X}) = (2 + o(1))\binom{n}{\lfloor \frac{n}{2} \rfloor}$.

5 Lubell Function

We have the general lower bound $e(P)$ on $\pi(P)$, when it exists, directly from the definitions. To verify the conjecture, the key is to get the upper bound. The Lubell function is a tool that was introduced to do just that, at least for some posets. This method is an extension of the heart of Lubell's elegant proof [38] of Sperner's Theorem (the case that $P = \mathcal{P}_2$).

Let $\mathscr{C} := \mathscr{C}_n$ denote the collection of all $n!$ *full (maximal) chains*

$$\emptyset \subset \{i_1\} \subset \{i_1, i_2\} \subset \cdots \subset [n]$$

in the Boolean lattice $\mathcal{B}_n$. Fix a family $\mathcal{F} \subseteq 2^{[n]}$. We say the *height* of $\mathcal{F}$ is

$$h(\mathcal{F}) := \max_{\mathcal{C}\in\mathscr{C}} |\mathcal{F} \cap \mathcal{C}|.$$

It is the same as the height if $\mathcal{F}$ is viewed as a poset under inclusion. Following [23] we define the *Lubell function* of $\mathcal{F}$ by

$$\bar{h}(\mathcal{F}) = \bar{h}_n(\mathcal{F}) := \operatorname*{ave}_{\mathcal{C}\in\mathscr{C}_n} |\mathcal{F} \cap \mathcal{C}|.$$

The bar notation suggests what it is, the average number of times a random full chain meets $\mathcal{F}$, as compared to the height, which is the maximum number of times over all full chains. The Lubell function value bounds the size of a family (extending Lubell's argument for an antichain $\mathcal{F}$):

Lemma 5.1. [23] *Let $\mathcal{F}$ be a collection of subsets of $[n]$. Then $\bar{h}(\mathcal{F}) = \sum_{F\in\mathcal{F}} 1/\binom{n}{|F|} \geq |\mathcal{F}|/\binom{n}{\lfloor\frac{n}{2}\rfloor}$.*

Proof. Let us count the total number of times all full chains meet $\mathcal{F}$ in two different ways. Then the average will be

$$\frac{1}{n!}\sum_{C\in\mathscr{C}} |\mathcal{F}\cap C| = \frac{1}{n!}\sum_{F\in\mathcal{F}} |F|!(n-|F|)! = \sum_{F\in\mathcal{F}} \frac{1}{\binom{n}{|F|}}.$$

The last inequality in the lemma follows from the fact that $\binom{n}{\lfloor\frac{n}{2}\rfloor} \geq \binom{n}{k}$ for all k. □

The Lemma tells us that the ratio we are interested in, $|\mathcal{F}|/\binom{n}{\lfloor\frac{n}{2}\rfloor}$, is bounded above by the Lubell function of $\mathcal{F}$. So upper bounds on the Lubell function for all P-free families $\mathcal{F}$ yield upper bounds on $\mathrm{La}(n,P)/\binom{n}{\lfloor\frac{n}{2}\rfloor}$, the ratio that is conjectured to converge to $e(P)$.

We can view $\bar{h}(\mathcal{F})$ as a weighted sum, where each set F has weight $1/\binom{n}{|F|}$. Given P and n, let us define $\lambda_n(P)$ to be $\max \bar{h}_n(\mathcal{F})$ over all P-free families $\mathcal{F}$ in $\mathcal{B}_n$. For a largest P-free family $\mathcal{F}$, we have $\mathrm{La}(n,P)/\binom{n}{\lfloor\frac{n}{2}\rfloor} = |\mathcal{F}|/\binom{n}{\lfloor\frac{n}{2}\rfloor} \leq \lambda_n(P)$, since $1/\binom{n}{\lfloor\frac{n}{2}\rfloor}$ is the minimum possible weight among all subsets of $[n]$. Therefore, if $\pi(P)$ exists, then it cannot exceed $\lim_{n\to\infty}\lambda_n(P)$. Let us denote this last limit by $\lambda(P)$, if it exists. We have that when the two limits exist,

$$e(P) \leq \pi(P) \leq \lambda(P).$$

How can one estimate $\lambda_n(P)$? Because every poset P is a subposet of a chain on $|P|$ elements, a full chain cannot intersect a P-free family $\mathcal{F}$ more than $|P|-1$ times. Hence, the average number of times a random full chain meets $\mathcal{F}$ is at most $|P|-1$. This gives us a very rough bound, $\lambda_n(P) \leq |P|-1$. To refine it, we may suitably partition the set $\mathscr{C}_n$ of full chains into blocks and find for each block the average number of times a random full chain meets $\mathcal{F}$, and then take the maximum over all the blocks. This will typically overestimate $\bar{h}_n(\mathcal{F})$. Surprisingly, for many posets there is an appropriate choice of partition [21] that provides a practical way to evaluate $\lambda_n(P)$.

For general P, the existence of $\lambda(P)$ remains an open question. Moreover, unlike the lower bound, there are examples P for which $\pi(P) < \lambda(P)$, and this gap can be extremely large. Nevertheless, there are also some "good posets" P for which $\bar{h}_n(\mathcal{F}) \leq e(P)$ for any P-free family $\mathcal{F}$ and any n. For such a poset, not only is the value of $\mathrm{La}(n,P)$ determined, but so are the families achieving it. Griggs and Li [22] say such posets are *uniformly L-bounded*, where L stands for Lubell:

Theorem 5.2. [23] *Let P be a poset that is uniformly L-bounded. Let $e = e(P)$. Then for all n,* $\mathrm{La}(n, P) = \Sigma(n, e)$, *and so* $\pi(P) = e(P)$. *Moreover, if $\mathcal{F}$ is a P-free family of subsets of $[n]$ of maximum size, $\mathcal{F}$ must be the family $\mathcal{B}(n, e)$.*

We see that uniformly L-bounded posets satisfy Conjecture 4.1. In [22], one can see a constructive method to obtain many uniformly L-bounded posets.

One poset of interest that definitely is not uniformly L-bounded is the notorious diamond $\mathcal{D}_2$. In fact, there are examples of diamond-free families $\mathcal{F}_n$ of subsets of $[n]$ for which the Lubell function approaches 2.25 as $n \to \infty$ [23]. Consequently, the sequence $\lambda_n(\mathcal{D}_2)$, which is known to be nonincreasing in n by a simple averaging argument, has a limit $\lambda(\mathcal{D}_2) \geq 2.25$. It means that working purely with the Lubell function, as in [23], one cannot reduce the upper bound on $\pi(\mathcal{D}_2)$ any lower than 2.25.

Kramer *et al.* [34] used the Lubell function in their proof that brought the upper bound on $\pi(\mathcal{D}_2)$ down to the very same value, 2.25. However, they were able to restrict their consideration to diamond-free families of a particular type. As a consequence, the value of $\lambda(\mathcal{D}_2)$ remains open, with lower bound 2.25 and upper bound about 2.273 by the proof of [23].

So why should we still expect that $\pi(\mathcal{D}_2) = 2$? The families $\mathcal{F}_n$ we mentioned, with relatively large Lubell function values, are actually not very large. They consist of very small subsets (size at most 3), or else dually very large subsets (size at least $n-3$): Each set in $\mathcal{F}_n$ makes a rather large contribution to $\bar{h}(\mathcal{F}_n)$.

What happens with the k-diamonds $\mathcal{D}_k$ for general k? Do they get even harder to deal with? Surprisingly, Li discovered that for "almost all" values of k, including k as small as 3 or 4, $\mathcal{D}_k$ is uniformly L-bounded, so satisfies the Griggs-Lu Conjecture, and we know the exact values of $\mathrm{La}(n, P)$ and the extremal families. For other values of k it remains a great challenge to determine $\mathrm{La}(n, P)$:

Theorem 5.3. [23, 35] *Let $k \geq 2$, and define $m := \lceil \log_2(k+2) \rceil$.*

(1) If $k \in [2^{m-1} - 1, 2^m - \binom{m}{\lfloor \frac{m}{2} \rfloor} - 1]$, then $\mathcal{D}_k$ is uniformly L-bounded, and $\pi(\mathcal{D}_k) = e(\mathcal{D}_k) = m$.

(2) If $k \in [2^m - \binom{m}{\lfloor \frac{m}{2} \rfloor}, 2^m - 2]$ then, if $\pi(\mathcal{D}_k)$ exists,

$$m = e(\mathcal{D}_k) \leq \pi(\mathcal{D}_k) \leq m + 1 - \frac{2^m - k - 1}{\binom{m}{\lfloor \frac{m}{2} \rfloor}} < m + 1.$$

Another class of posets that is uniformly L-bounded is the set of suspensions of paths of distinct lengths. Precisely, for $k \geq 1$ let $l_1 \geq \cdots \geq l_k \geq 3$, and define the *harp poset* $\mathcal{H}(l_1, \ldots, l_k)$ to consist of paths $\mathcal{P}_{l_1}, \ldots, \mathcal{P}_{l_k}$ with their top elements identified and their bottom elements identified. For instance, in this notation we have $\mathcal{D}_k$ is the harp $\mathcal{H}(3, \ldots, 3)$ where there are k 3's. Then Griggs, Li, and Lu proved:

Theorem 5.4. [23] *If $l_1 > \cdots > l_k \geq 3$, then* $\mathrm{La}(n, \mathcal{H}(l_1, \ldots, l_k))$ *is uniformly L-bounded, and $\pi = e = l_1 - 1$.*

For example, the unranked poset on five elements called N_5 by lattice theorists is the harp $\mathcal{H}(4,3)$. By the theorem above, it is uniformly L-bounded, so we know that $\mathrm{La}(n, N_5) = \Sigma(n, 3)$ for all n.

The vast majority of subsets of $[n]$ are close to the middle value in size. There are comparatively few subsets that are small or large. More precisely, Shannon's Theorem [1, p.256] gives

$$\sum_{i=0}^{\lfloor \alpha n \rfloor} \binom{n}{i} = o\left(\binom{n}{\lfloor \frac{n}{2} \rfloor}\right),$$

for any positive constant $\alpha < 1/2$, showing there are few small sets. By taking complements there are similarly few large sets.

On the other hand, sets with either few or many elements make major contributions to the weighted sum that is the Lubell function. In using the Lubell function to bound $|\mathcal{F}|/\binom{n}{\lfloor \frac{n}{2} \rfloor}$ asymptotically for P-free families, it is enough to consider such families $\mathcal{F}$ that contain no small or large sets, thereby avoiding the overly heavy contributions of small and large sets to $\bar{h}(\mathcal{F})$. A full hierarchy of poset properties has been devised by Griggs and Li [22], involving the Lubell function, to take advantage of this observation. Each of the properties implies that $\pi(P) = e(P)$, though the values of $\mathrm{La}(n, P)$ may not be determined precisely as with uniformly L-bounded posets.

We say for integer m that poset P is *m-L-bounded*, if for every n every P-free family $\mathcal{F} \subseteq \mathcal{B}_n$, containing only sets of sizes between m and $n-m$, satisfies $\bar{h}_n(\mathcal{F}) \leq e(P)$. For $m = 0$, these are uniformly L-bounded posets. The butterfly poset is not uniformly L-bounded, but it is 1-L-bounded. The m-L-bounded posets satisfy Conjecture 4.1.

As an application Griggs and Li [22] introduced the class of fans, where a *fan poset* is the join of paths, that is, paths (chains) that are identified at the bottom. Fans all satisfy some Lubell boundedness condition, so that $\pi(P) = e(P)$, which we already knew by Bukh's Tree Theorem. However, now we know more. For instance, if there is just one path, or if the path lengths are distinct with the longest path at least 2 longer than the others, the fan is uniformly L-bounded. In this case we know $\mathrm{La}(n, P)$ exactly, as well as the extremal families.

Griggs and Li [22] presented a series of results describing how to construct many more posets satisfying Lubell boundedness properties (and thus the $\pi(P) = e(P)$ conjecture). It is surprising one can build such complicated posets that satisfy the conjecture, when we still seem so far from verifying it for the diamond $\mathcal{D}_2$.

6 Forbidden Induced Subposets

Up to this point, we only considered families that forbid a subposet P in the weak sense, meaning that we are also excluding all posets that contain P. It is perhaps equally natural to investigate what happens if we exclude only families of subsets ordered exactly as P is. A poset P_1 contains poset P_2 as an *induced subposet*, if there is an injection f (a strong embedding) from P_2 to P_1, such that for $x, y \in P_2$, $x \leq_{P_2} y$ if and only if $f(x) \leq_{P_1} f(y)$. For instance, the diamond $\mathcal{D}_2$ is a subposet of the chain $\mathcal{P}_4$, but it is not an induced subposet of any chain, no matter how long.

The notation, $\mathrm{La}^{\#}(n,P)$, introduced by Carroll and Katona [9], is the maximum size of a family $\mathcal{F}$ of subsets of $[n]$ not containing the poset P as an induced subposet. Clearly, $\mathrm{La}(n,P) \leq \mathrm{La}^{\#}(n,P)$, since it is more restrictive to forbid P in the weak sense.

Let us begin again with chains. Since a family $\mathcal{F}$ contains the chain $\mathcal{P}_k$ in the usual weak sense if and only if it contains it in the strong sense, we have $\mathrm{La}(n,\mathcal{P}_k) = \mathrm{La}^{\#}(n,\mathcal{P}_k)$, which is $\Sigma(n,k-1)$, by Erdős's Theorem on chains.

Next, consider $\mathcal{V} = \mathcal{V}_2$, which is what Carroll and Katona did (2008):

Theorem 6.1. [9]

$$1+\frac{1}{n}+\Omega\left(\frac{1}{n^2}\right) \leq \frac{\mathrm{La}^{\#}(n,\mathcal{V}_2)}{\binom{n}{\lfloor \frac{n}{2}\rfloor}} \leq 1+\frac{2}{n}+O\left(\frac{1}{n^2}\right).$$

So far, the results are similar to usual weak containment. However, in general it should be much easier to build families that only avoid containing P as an *induced* subposet, since in order for $\mathcal{F}$ to contain an induced copy of P, not only must we have the containment relations intact, we must also have the non-containments.

Indeed, observe that since a poset P is always a subposet of the chain $\mathcal{P}_{|P|}$, it means that

$$\mathrm{La}(n,P) \leq \mathrm{La}(n,\mathcal{P}_{|P|}) = \Sigma(n,|P|-1) \sim (|P|-1)\binom{n}{\lfloor \frac{n}{2}\rfloor}.$$

By contrast, for general P it is not at all clear how to bound $\mathrm{La}^{\#}(n,P)/\binom{n}{\lfloor \frac{n}{2}\rfloor}$ for all n. Is it even bounded in general? Lu and Milans [37] have conjectured that it is bounded for every poset P (and it is reported Katona also made this conjecture).

In more detail, for the induced subposet problem, Lu and Milans defined the *Turán threshold*

$$\pi^*(P) := \limsup_{n\to\infty} \frac{\mathrm{La}^{\#}(n,P)}{\binom{n}{\lfloor \frac{n}{2}\rfloor}}$$

and the *Lubell threshold*

$$\lambda^*(P) := \limsup_{n\to\infty} \{\bar{h}_n(\mathcal{F}) \mid P \text{ is not as an induced subposet of } \mathcal{F}\}.$$

They conjectured that the two thresholds are finite for all posets. Clearly, if $\lambda^*(P)$ is finite, so is $\pi^*(P)$, since $\pi^*(P) \leq \lambda^*(P)$. Lu and Milans verified that $\lambda^*(P)$ is finite for series-parallel posets and for posets of height at most 2. These innocent-looking claims stymied the experts for some time. (We return to this shortly.)

Given the challenge of working with $\mathrm{La}^{\#}(n, P)$, it is then remarkable that Boehnlein and Jiang (2012) managed to extend Bukh's methods to determine the asymptotics of all tree posets:

Theorem 6.2. [5] *Let T be a tree poset of height $h \geq 2$. Then*

$$\mathrm{La}^{\#}(n, T) = (h-1)\binom{n}{\lfloor \frac{n}{2} \rfloor}(1 + o(1)).$$

We see that $\pi^*(T) = h - 1$ for tree posets of height h.

Next compare $\mathrm{La}^{\#}(n, P)$ with $\mathrm{La}(n, P)$, which in general is smaller. For tree posets P, we have that $\lim_{n\to\infty} \mathrm{La}^{\#}(n, P)/\mathrm{La}(n, P) = 1$. On the other hand, there are examples of posets P such that $\lim_{n\to\infty} \mathrm{La}^{\#}(n, P)/\mathrm{La}(n, P)$ is arbitrarily large [5]. For example, let H_m be a poset consisting of elements $x_1, \ldots, x_m$ and $y_1, \ldots, y_m$ with orderings $x_i \leq y_j$ for $i \leq j \leq m$. This is a poset of height 2, which is contained in $\mathcal{K}_{m,m}$ as a non-induced poset. On the other hand, when $m \geq 3$, H_m contains the butterfly as a subposet. Thus, by Theorems 2.3 and 2.5, $\mathrm{La}(n, H_m) \sim 2\binom{n}{\lfloor \frac{n}{2} \rfloor}$. However, one can show that $\mathrm{La}^{\#}(n, H_m) \geq (m - 1 - o(1))\binom{n}{\lfloor \frac{n}{2} \rfloor}$, which is larger by a factor of $\frac{m-1}{2}$ than $\mathrm{La}(n, H_m)$.

For complete two-level posets, Patkós (2015) gave the following estimates of $\mathrm{La}^{\#}(n, P)$:

Theorem 6.3. [44] *For integers $r, s \geq 2$, we have*

$$\Sigma(n, 2) + \left(\frac{r+s-2}{n} - O_{r,s}\left(\frac{1}{n^2}\right)\right)\binom{n}{\lfloor \frac{n}{2} \rfloor}$$

$$\leq \mathrm{La}^{\#}(n, K_{r,s}) \leq \left(2 + \frac{2(r+s-2)}{n} + o\left(\frac{1}{n}\right)\right)\binom{n}{\lfloor \frac{n}{2} \rfloor}.$$

It is interesting that while H_m is a subposet of $\mathcal{K}_{m,m}$, we have $\mathrm{La}^{\#}(n, H_m) \geq \mathrm{La}^{\#}(n, \mathcal{K}_{m,m})$.

What about the notorious diamond poset $\mathcal{D}_2$? Lu and Milans (2014) gave the best result to date:

Theorem 6.4. [37] *As $n \to \infty$, $\mathrm{La}^{\#}(n, \mathcal{D}_2) \leq (2.583 + o(1))\binom{n}{\lfloor \frac{n}{2} \rfloor}$.*

Later in 2014 there was a breakthrough on the induced subposet problem, when the boundedness conjecture of Katona and of Lu and Milans was proven by Methuku and Pálvőlgyi. They brought the ideas and results from the area of forbidden hypermatrices [32, 40] to the problem and established the next theorem. It is interesting that the parameter, the order dimension of P, has never appeared in the theory before.

Theorem 6.5. [41] *For every poset P, there is a* $C_P = 2^d K$ *such that* $\mathrm{La}^{\#}(n, P) \le C_P\binom{n}{\lfloor \frac{n}{2} \rfloor}$*, where d is the order dimension of P and K is a constant obtained from forbidden hypermatrix problems.*

It means that $\pi^*(P) \le C_P$. Unfortunately, the Lu-Milans conjecture that $\lambda^*(P)$ is finite for all P is not yet proven.[1]

7 General Upper Bounds

Erdős's result, Theorem 1.2, provides an upper bound of $\Sigma(n, k-1)$ on $\mathrm{La}(n, P)$ for any poset P on k elements. This bound is far from accurate when the *width*, the largest size of an antichain in P, is large. In this section, we present ideas for finding upper bounds on $\mathrm{La}(n, P)$ using the invariants of P.

Recall that the Lubell function for a family $\mathcal{F}$ is the average number of times that a full chain $\mathcal{C}$ meets $\mathcal{F}$. We can also count the average number of times other families meet $\mathcal{F}$. In 2013 Burcsi and Nagy [8] called a family of sets in $\mathcal{B}_n$ a *double chain* $\mathcal{D}$, as follows. It consists of a full chain

$$\mathcal{C} = \{I_0 \subset I_1 \subset \cdots \subset I_n\},$$

where $|I_i| = i$ for all i, called the primary chain, together with sets $S_i = I_{i-1} \cup I_{i+1} \setminus I_i$, $1 \le i \le n-1$. The number of double chains containing a specific set is twice the number of full chains containing it, for every subset besides $\emptyset$, $[n]$. Using a double counting argument as in Lubell's method, there is a weighted sum of sets in a P-free family:

$$\frac{1}{n!} \sum_{\substack{F \in \mathcal{F} \\ F \ne \emptyset, [n]}} 2|F|!(n - |F|)! \le \frac{1}{n!} \sum_{\mathcal{D}} |\mathcal{D} \cap \mathcal{F}| \le (|P| + h(P) - 2). \tag{1}$$

This leads to an upper bound that involves not only the size, but also the height of P:

Theorem 7.1. [8] *For any poset P,*

$$\mathrm{La}(n, P) \le \left(\frac{|P| + h(P) - 2}{2}\right) \binom{n}{\lfloor \frac{n}{2} \rfloor}.$$

Note that the height of a poset is at most the size. So, this result indeed improves Theorem 1.2 for general P. While Erdős's upper bound is only tight for chains, there is a series of posets P with $\mathrm{La}(n, P)$ reaching the upper bound, one of them being the butterfly.

[1] After this survey went to press, we learned that Arès Méroueh announced a proof of the conjecture (ArXiv:1506.07056).

Chen and Li (2014) introduced a new type of set family, called *k-linkage*, with a parameter k. They derived a more general inequality that reduces to the previous one when $k = 1$:

Theorem 7.2. [10] *Given a poset P and $k \geq 1$, for sufficiently large n,*

$$\mathrm{La}(n, P) \leq \frac{1}{k+1}\left(|P| + \frac{1}{2}(k^2 + 3k - 2)(h(P) - 1) - 1\right)\binom{n}{\lfloor \frac{n}{2} \rfloor}.$$

By setting $k = \lfloor \sqrt{|P|/h(P)} \rfloor$, it implies $\mathrm{La}(n, P) = O(\sqrt{|P|h(P)})\binom{n}{\lfloor \frac{n}{2} \rfloor}$.

The next inequality is proved by Grosz, Methuku, and Tompkins [26] using the concept of a *k-interval chain* $\mathcal{C}_k$, which is a union of a full chain $\emptyset \subset \{i_1\} \subset \{i_1, i_2\} \subset \cdots \subset [n]$ and the collection of sets S satisfying $\{i_1, \ldots, i_m\} \subset S \subset \{i_1, \ldots, i_{m+k}\}$ for some m. When $k = 2$, a k-interval chain is a double chain $\mathcal{D}$. Counting $|\mathcal{C}_k \cap \mathcal{F}|$ for a P-free family $\mathcal{F}$ is technical and complicated, but it gives a better estimate for $\mathrm{La}(n, P)$.

Theorem 7.3. [26] *Given a poset P, for any integer $k \geq 2$, it holds that*

$$\mathrm{La}(n, P) \leq \frac{1}{2^{k-1}}\left(|P| + (3k - 5)2^{k-2}(h(P) - 1) - 1\right)\binom{n}{\lfloor \frac{n}{2} \rfloor}.$$

As in Theorem 7.2, we get

$$\mathrm{La}(n, P) = O\left(h(P)\log_2\left(\frac{|P|}{h(P)} + 2\right)\right)\binom{n}{\lfloor \frac{n}{2} \rfloor}$$

by an appropriate choice of k. On the other hand, Grosz *et al.* also showed that for some posets P,

$$\mathrm{La}(n, P) \geq \left((h(P) - 2)\log_2\left(\frac{|P|}{h(P)}\right)\right)\binom{n}{\lfloor \frac{n}{2} \rfloor}.$$

8 Packing posets

In their pioneering paper on forbidden subposets (1983), Katona and Tarján [31] proposed the more general problem of forbidding a family. In our notation, given a collection of posets $\{Q_j\}_{j\geq 1}$ (possibly infinite), we seek the maximum size $\mathrm{La}(n, \{Q_j\}_{j\geq 1})$ of a family $\mathcal{F}$ of subsets of $[n]$ that contains no copy of any poset Q_j in the collection. Katona and Tarján, and independently but much later (2015), Dove and Griggs [16], solved an interesting instance of this problem. Recall that Λ is the three-element poset that is the dual of $\mathcal{V} = \mathcal{V}_2$.

Theorem 8.1. [16, 31] *For all n,*

$$\mathrm{La}(n, \{\mathcal{V}, \Lambda\}) = 2\binom{n-1}{\lfloor \frac{n-1}{2} \rfloor}.$$

Dove and Griggs obtained this result by converting it to a packing problem. Instead of trying to pick many subsets of $[n]$ that avoid containing a poset P, we take the entire Boolean lattice $\mathcal{B}_n$ and seek to pack in as many unrelated copies of P as possible. Here, a *copy of P* is meant in the weak sense as usual. So if $\mathcal{F}$ avoids any $\mathcal{V}$ or Λ, it means that $\mathcal{F}$ can be decomposed into unrelated chains, each of size 2 or 1, and this is a problem that can be completely solved.

It depends mainly on how many unrelated copies of the chain $\mathcal{P}_2$ can be packed into $\mathcal{B}_n$. More precisely, following Dove and Griggs [16], let the packing number $\mathrm{Pa}(n, P)$ denote the maximum size of a family of subsets $\mathcal{F}$ constructed from pairwise unrelated copies of P in $\mathcal{B}_n$. Here $\mathcal{F}$ is the union of (weak) copies of P such that no two elements of different copies are equal or even related (by inclusion). We measure the size of such a family, rather than the number of copies of P. The measures are equivalent, of course, but later we will describe a more general model, where we build $\mathcal{F}$ out of unrelated copies of posets P_i coming from a collection. In this case, the family size is more important than the number of copies.

Here is the determination of $\mathrm{Pa}(n, P)$ when P is a chain $\mathcal{P}_k$. It was obtained in 1984 by Griggs, Stahl, and Trotter [25]. It can also be deduced from an earlier paper (1965) of Bollobás [6].

Theorem 8.2. [25] *The packing number for k-paths is* $\mathrm{Pa}(n, \mathcal{P}_k) = k\binom{n-(k-1)}{\lfloor \frac{n-(k-1)}{2} \rfloor}$. *For fixed k, as n goes to infinity, this is asymptotic to* $\frac{k}{2^{k-1}}\binom{n}{\lfloor \frac{n}{2} \rfloor}$.

It is interesting that the packing number for k-paths (chains) is asymptotic to a constant times $\binom{n}{\lfloor \frac{n}{2} \rfloor}$, the same behavior as for $\mathrm{La}(n, \mathcal{P}_k)$, though with a different constant. For general posets P, the asymptotics of $\mathrm{La}(n, P)$ remain rather wide open. It is then surprising that it is possible to determine the leading asymptotic behavior of the packing number $\mathrm{Pa}(n, P)$ for general posets P. Dove and Griggs asked and solved this question in 2013. Only later did they learn that the same problem had been posed first by Katona in 2010 [28]. Moreover, Katona and Nagy also succeeded in solving it, at about the same time in 2013. The papers by the two teams were published together (2015) [16, 30].

Both teams formulated their results using convexity. A family $\mathcal{F} \subseteq \mathcal{B}_n$ generates an *ideal* (or *down-set*) and a *filter* (or *up-set*) denoted as follows:

$$D(\mathcal{F}) = \{S \in \mathcal{B}_n | S \subseteq A \text{ for some } A \in \mathcal{F}\}, \text{ and}$$

$$U(\mathcal{F}) = \{S \in \mathcal{B}_n | A \subseteq S \text{ for some } A \in \mathcal{F}\}.$$

Define the *convex closure operator* on $\mathcal{F}$ to be $\overline{\mathcal{F}} := D(\mathcal{F}) \cap U(\mathcal{F})$. Note that $\mathcal{F} \subseteq \overline{\mathcal{F}}$. It is important that if two subset families are unrelated, then their closures must be unrelated as well. If a family $\mathcal{F}$ satisfies $\mathcal{F} = \overline{\mathcal{F}}$, we say $\mathcal{F}$ is *convex*.

The asymptotic determination of $\mathrm{Pa}(n, P)$ depends on how P can be weakly embedded into a Boolean lattice. There exists a minimum size of the closure of $f(P)$ over all weak embeddings f of P into $\mathcal{B}_k$ over all k. Denote this minimum by $c(P)$.

Theorem 8.3. [16, 30] *For any poset P, as $n \to \infty$,* $\mathrm{Pa}(n, P) \sim \frac{|P|}{c(P)}\binom{n}{\lfloor \frac{n}{2} \rfloor}$.

This result can be proven using Lubell-style chain averaging methods, extending the arguments used to prove Theorem 8.2.

Dove and Griggs expanded their study to allow packing posets from a collection. Specifically, let $\mathrm{Pa}(n, \{P_i\}_{i\geq 1})$ denote the maximum size of a family in $\mathcal{B}_n$ constructed from pairwise unrelated (weak) copies of posets, each chosen from a given collection of posets $\{P_i\}_{i\geq 1}$ (possibly infinite). Note that the sizes of the posets P_i can vary. Dove and Griggs showed that as n goes to infinity,

$$\mathrm{Pa}(n, \{P_1, P_2, \ldots, P_k\}) \sim \max_{1\leq i\leq k}\left(\frac{|P_i|}{c(P_i)}\right)\binom{n}{\lfloor \frac{n}{2} \rfloor}.$$

Similarly, one can investigate families built from *induced* copies of given posets. Denote the maximum size of a family in $\mathcal{B}_n$ constructed from induced copies of P as $\mathrm{Pa}^*(n, P)$. We can also define the more general quantity $\mathrm{Pa}^*(n, \{P_i\}_{i\geq 1})$. Now, for any collection $\{Q_j\}_{j\geq 1}$, the forbidden poset maximum, $\mathrm{La}(n, \{Q_j\}_{j\geq 1})$, is equivalent to $\mathrm{Pa}^*(n, \{P_i\}_{i\geq 1})$, where $\{P_i\}_{i\geq 1}$ is the collection of all possible connected posets that do not contain any of the posets in $\{Q_j\}_{j\geq 1}$ as a subposet. For instance, $\mathrm{La}(n, \mathcal{V}) = \mathrm{Pa}^*(n, \{\Lambda_i\}_{i\geq 0})$. So the problem of determining $\mathrm{Pa}^*(n, \{P_i\}_{i\geq 1})$ can be viewed as more general than the $\mathrm{La}(n, \{Q_j\}_{j\geq 1})$ problem this survey is addressing.

Both teams independently formulated and solved the analogue of the result above for induced subposets. We need to define $c^*(P)$ as the minimum size of the closure of a strong embedding of P in $\mathcal{B}_n$ over all possible n. In general, $c^*(P) \neq c(P)$. Here is the main theorem for packing induced copies of P.

Theorem 8.4. [16, 30] *For any poset P, as $n \to \infty$,* $\mathrm{Pa}^*(n, P) \sim \frac{|P|}{c^*(P)}\binom{n}{\lfloor \frac{n}{2} \rfloor}$.

9 Supersaturation

There has been some effort to obtain results in this area of "supersaturation" or "Erdős-Rademacher type" which concern how many copies of poset P there must be in every family $\mathcal{F} \subseteq \mathcal{B}_n$ of given size greater than $\mathrm{La}(n, P)$.

Kleitman [33] solved this problem in 1966 in the case $P = \mathcal{P}_2$, verifying a conjecture of Erdős-Katona that the minimum number of copies is attained by picking $\mathcal{F}$ to consist of subsets as close to the middle value as possible. He conjectured that for any k, the same families of subsets closest to the middle would achieve the minimum number of copies of chains $\mathcal{P}_k$ for any k.

This conjecture remains open for general k. However, for families of size not much larger than $\mathrm{La}(n, \mathcal{P}_k) = \Sigma(n, k-1)$, it has been verified independently by Dove, Griggs, Kang, and Séreni, and by Das, Gan, and Sudakov. Using different approaches, both groups proved:

Theorem 9.1. [12, 17] *If a family $\mathcal{F}$ of subsets of $[n]$ satisfies $|\mathcal{F}| = \Sigma(n, k-1)+x$, then there must be at least*

$$x \cdot \prod_{i=1}^{k-1}\left(\left\lfloor \frac{n+k}{2} \right\rfloor - i + 1\right)$$

copies of $\mathcal{P}_k$ in $\mathcal{F}$.

Note that this product

$$\prod_{i=1}^{k-1}\left(\left\lfloor \frac{n+k}{2} \right\rfloor - i + 1\right)$$

is the number of copies of $\mathcal{P}_k$ contained in $\mathcal{B}(n, k)$, with one endpoint of the chain being a particular set in the kth middle rank (of ranks in $[0, n]$). Thus, the family that consists of $k-1$ middle ranks $\mathcal{B}(n, k-1)$ and x sets from the kth middle rank witnesses that the bound above is tight for

$$x \le \binom{n}{\lfloor \frac{n}{2} \rfloor + (-1)^k \lfloor \frac{k}{2} \rfloor}.$$

Dove's dissertation [15] further addresses this problem, extending the theorem above to the lattice of subsets of a multiset (the divisor lattice).

The first result we know that goes beyond chains considers P to be the butterfly poset. Recall $\mathrm{La}(n, P) = \Sigma(n, 2)$. In 2014 Patkós [43] considered families of $\Sigma(n, 2) + x$ subsets of $[n]$, $x > 0$. For small values of x, similar to the result above for chains P, he determined the minimum possible number of butterflies. Beyond that, he found a general bound that is asymptotically best-possible.

10 Future Study

There are posets P for which we can readily determine $\mathrm{La}(n, P)$ for all n. In most cases, $\mathrm{La}(n, P)$ is simply a sum of binomial coefficients, often of the form $\Sigma(n, k)$ for some k. It may even be possible to describe all of the extremal P-free families. Examples include chains $\mathcal{P}_k$, the butterfly $\bowtie$, and diamonds $\mathcal{D}_3, \mathcal{D}_4$.

Then there are posets P such as $\mathcal{V}$, for which it seems to be very hard to determine $\mathrm{La}(n, P)$, but for which the asymptotic behavior $\pi(P)$ is known.

However, there seem to be many posets P for which even the asymptotic behavior is challenging or beyond current methods. The most notable example is the diamond $\mathcal{D}_2$. Other examples include the 5-diamond $\mathcal{D}_5$, and the crowns $\mathcal{O}_6, \mathcal{O}_{10}$.

Beyond trying to find $\mathrm{La}(n, P)$, or at least $\pi(P)$, how can we tell which posets we can solve and which are going to be difficult (or hopeless)?

Many researchers continue to be interested in the diamond-free family problem, particularly for families restricted to the middle three levels. We have learned that the upper bound on $\pi(\mathcal{D}_2)$ provided by the Lubell function can only go down to 2.25, which is still well above the conjectured upper bound of 2. If the sets in the family $\mathcal{F}$ are excluded from the smallest and largest sizes–which won't really affect the asymptotics of $|\mathcal{F}|/\binom{n}{\lfloor \frac{n}{2} \rfloor}$ since asymptotically few subsets are very small or very large–it is suspected the Lubell bound can be brought down to the conjectured 2. However, no one has been able to reduce the Lubell bound with such size restrictions.

A smaller problem that remains open is to prove that the Lubell limit $\lambda(\mathcal{D}_2) = 2.25$. We have the lower bound, and know the limit exists, but the upper bound remains around 2.27.

Flag algebra arguments (and related computations) have led to improved bounds for the diamond poset $\mathcal{D}_2$. Can they give the conjectured value of 2, or solve other open subposet problems, as they have in Turán theory?

After the diamond $\mathcal{D}_2$, the most intriguing posets for which the existence of $\pi(P)$ remains open may be the crowns $\mathcal{O}_6$ and $\mathcal{O}_{10}$.

Probability theory has figured into Lubell function arguments and in the proof of Bukh's Tree Theorem. It seems likely that we can do more with probability to answer the asymptotic questions raised in this field.

It feels like we have only gotten started in the study of induced subposet problems. The recent proof of the conjecture of Katona and Lu-Milans, that the Turán threshold $\pi^*(P)$ is finite for all posets P, is encouraging. It remains open to prove the conjecture of Lu-Milans that the Lubell threshold $\lambda^*(P)$ is finite for all posets P.

This survey has not done justice to all of the wonderful ideas developed in the area, but at least we have collected all of the references to date we are aware of, and the interested reader is encouraged to study them. This continues to be a lively research area, and there is a long way to go!

Acknowledgements Research of Jerrold R. Griggs was supported in part by a grant from the Simons Foundation (#282896 to Jerrold Griggs) and by a long-term visiting position at the IMA, University of Minnesota

Wei-Tian Li was supported by Ministry of Science and Technology (No. 103-2115-M-005 -003 -MY2).

References

1. N. Alon, J.H. Spencer, *The Probabilistic Method*, 3rd edn. (Wiley, New York, 2008)
2. M. Axenovich, J. Manske, R. Martin, Q_2-free families in the Boolean lattice. Order **29**, 177–191 (2012)
3. R. Baber, J. Talbot, Hypergraphs do jump, Comb. Probab. Comput. **20**, 161–171 (2011)
4. J. Balogh, P. Hu, B. Lidický, H. Liu, Upper bounds on the size of 4- and 6-cycle-free subgraphs of the hypercube. Eur. J. Comb **35**, 75–85 (2014)
5. E. Boehnlein, T. Jiang, Set families with a forbidden induced subposet. Comb. Probab. Comput. **21**, 496–511 (2012)
6. B. Bollobás, On generalized graphs. Acta Math. Acad. Sci. Hung. **16**, 447–452 (1965)
7. B. Bukh, Set families with a forbidden poset. Electron. J. Comb. **16**, R142, 11 p (2009)
8. P. Burcsi, D.T. Nagy, The method of double chains for largest families with excluded subposets. Electron. J. Graph Theory Appl. **1**, 40–49 (2013)
9. T. Carroll, G.O.H. Katona, Bounds on maximal families of sets not containing three sets with $A \cup B \subset C, A \not\subset B$. Order **25**, 229–236 (2008)
10. H.-B. Chen, W.-T. Li, A note on the largest size of families of sets with a forbidden poset. Order **31**, 137–142 (2014)
11. É. Czabarka, A. Dutle, T. Johnston, L.A. Székely, Abelian groups yield many large families for the diamond problem. Eur. J. Math. **1**, 320–328 (2015)
12. S. Das, W. Gan, B. Sudakov, Sperner's theorem and a problem of Erdős, Katona and Kleitman. Comb. Probab. Comput. **24**, 585–608 (2015)
13. A. DeBonis, G.O.H. Katona, Largest families without an r-fork. Order **24**, 181–191 (2007)
14. A. DeBonis, G.O.H. Katona, K.J. Swanepoel, Largest family without $A \cup B \subset C \cap D$. J. Comb. Theory (Ser. A) **111**, 331–336 (2005)
15. A.P. Dove, *Generalizations of Sperner's Theorem: Packing Posets, Families Forbidding Posets, and Supersaturation*. Ph.D. dissertation, University of South Carolina, 2013
16. A.P. Dove, J.R. Griggs, Packing posets in the Boolean lattice. Order **32**, 429–438 (2015)
17. A.P. Dove, J.R. Griggs, R.J. Kang, J.-S. Sereni, Supersaturation in the Boolean Lattice. Integers **14A**, 1–7 (2014)
18. P. Erdős, On a lemma of Littlewood and Offord. Bull. Am. Math. Soc. **51**, 898–902 (1945)
19. R.L. Graham, N.J.A. Sloane, Lower bounds for constant weight codes. IEEE IT **26**, 37–43 (1980)
20. J.R. Griggs, G.O.H. Katona, No four subsets forming an N. J. Comb. Theory (Ser. A) **115**, 677–685 (2008)
21. J.R. Griggs, W.-T. Li, The partition method for poset-free families. J. Comb. Optim. **25**(4), 587–596 (2013)
22. J.R. Griggs, W.-T. Li, Poset-free families and Lubell-boundedness. J. Comb. Theory (Ser. A) **134**, 166–187 (2015)
23. J.R. Griggs, W.-T. Li, L. Lu, Diamond-free families. J. Comb. Theory (Ser. A) **119**, 310–322 (2012)
24. J.R. Griggs, L. Lu, On families of subsets with a forbidden subposet. Comb. Probab. Comput. **18**, 731–748 (2009)
25. J.R. Griggs, J. Stahl, W.T. Trotter, Jr., A Sperner theorem on unrelated chains of subsets. J. Comb. Theory (Ser. A) **36**, 124–127 (1984)

26. D. Grósz, A. Methuku, C. Tompkins, An improvement of the general bound on the largest family of subsets avoiding a subposet P. ArXiv:1408.5783
27. G.O.H. Katona, Forbidden inclusion patterns in the families of subsets (introducing a method), in *Horizons of Combinatorics*. Bolyai Society Mathematical Studies, vol. 17 (Bolyai Mathematical Society, Budapest; Springer, New York, 2008), pp. 119–140
28. G.O.H. Katona, *On the Maximum Number of Incomparable Copies of a Given Poset in a Family of Subsets*. Lecture, International Conference on Recent Trends in Graph Theory and Combinatorics, Cochin University of Science and Technology, India (2010)
29. G.O.H. Katona, Sperner type theorems with excluded subposets. Discrete Appl. Math. **161**, 1251–1258 (2013)
30. G.O.H. Katona, D. Nagy, Incomparable copies of a poset in the Boolean lattice. Order **32**, 419–427 (2015)
31. G.O.H. Katona, T.G. Tarján, Extremal problems with excluded subgraphs in the n-cube, in *Graph Theory, Łagów*, ed. by M. Borowiecki, J. W. Kennedy, M.M. Sysło. Lecture Notes in Mathematics, vol. 1018 (Springer, Berlin/Heidelberg/New York/Tokyo, 1981/1983), pp. 84–93
32. M. Klazer, A. Marcus, Extensions of the linear bound in the F uredi-Hajnal conjecture. Adv. Appl. Math. **38**, 258–266 (2006)
33. D. Kleitman, A conjecture of Erdős-Katona on commensurable pairs among subsets of an n-set, in *Theory of Graphs, Proceedings of the Colloquium Held at Tihany*, September 1966, ed. by P. Erdős, G.O.H. Katona (Academic Press, New York, 1968), pp. 215–218. *Theory of Graphs, Proceedings of the Colloquium*, Tihany (1966), pp. 215–218
34. L. Kramer, R.R. Martin, M. Young, On diamond-free subposets of the Boolean lattice. J. Comb. Theory (Ser. A) **120**, 545–560 (2013)
35. W.-T. Li, *Extremal Problems on Families of Subsets with Forbidden Subposets*. Ph.D. dissertation, University of South Carolina, 2011
36. L. Lu, On crown-free families of subsets. J. Comb. Theory (Ser. A) **126**, 216–231 (2014)
37. L. Lu, K. Milans, Set families with forbidden subposets. J. Comb. Theory (Ser. A) **136**, 126–142 (2015). arXiv:1408.0646
38. D. Lubell, A short proof of Sperner's lemma. J. Comb. Theory **1**, 299 (1966)
39. J. Manske, J. Shen, Three layer Q_2-free families in the Boolean lattice. Order **30**, 585–592 (2013)
40. A. Marcus, G. Tardos, Excluded permutation matrices and the Stanley-Wilf conjecture, J. Comb. Theory (Ser. A) **107**, 153–160 (2004)
41. A. Methuku, D. Pálvőlgyi, Forbidden hypermatrices imply general bounds on induced forbidden subposet problems. ArXiv:1408.4093
42. A. Methuku, C. Tompkins, Exact forbidden subposet results using chain decompositions of the cycle. ArXiv:1411.0415
43. B. Patkós, Supersaturation and stability for forbidden subposet problems. J. Comb. Theory (Ser. A) **136**, 220–237 (2015) ArXiv:1408.1887
44. B. Patkós, Induced and non-induced forbidden subposet problems. Electron. J. Comb. **22**, #P1.30 (2015)
45. A.A. Razborov, Flag algebras. J. Symb. Logic **72**, 1239–1282 (2007)
46. G. Sarkis, S. Shahriari, PCURC, Diamond-free subsets in the linear lattices. Order **31**, 421–43 (2014)
47. E. Sperner, Ein Satz über Untermengen einer endlichen Menge. Math. Z. **27**, 544–548 (1928)
48. H.T. Thanh, An extremal problem with excluded subposets in the Boolean lattice. Order **15**, 51–57 (1998)

The mathematics of causal sets

Graham Brightwell and Malwina Luczak

Abstract The causal set approach to quantum gravity is based on the hypothesis that the underlying structure of spacetime is that of a random partial order. We survey some of the interesting mathematics that has arisen in connection with the causal set hypothesis, and describe how the mathematical theory can be translated to the application area. We highlight a number of open problems of interest to those working in causal set theory.

Keywords Causal sets • Random partial orders

Mathematics Subject Classification (2010): 06A06, 60C05, 83C45

1 The causal set hypothesis

This article constitutes a survey of the mathematical results related to the causal set approach to quantum gravity. Simply put, the causal set hypothesis is that spacetime is fundamentally a discrete structure, consisting of a set of points equipped with a partial order. On a large scale, such a discrete structure may approximate a Lorentzian spacetime manifold, with the partial order given by "is in the past light cone of", as in the theory of general relativity. On a small scale, the behaviour will be completely different, with an inherent discreteness and "randomness", as in the theory of quantum mechanics. Causal set theory is one of the number of proposed alternatives to string theory; research took off in the late 1980s following work of Bombelli, Lee, Meyer, and Sorkin [15], but the concept occurred to others independently, notably in the preprint of Myrheim [43] from 1978. Although causal

G. Brightwell (✉)
Department of Mathematics, London School of Economics, Houghton Street, London, WC2A 2AE, UK
e-mail: g.r.brightwell@lse.ac.uk

M. Luczak
School of Mathematics, Queen Mary University of London, London, UK
e-mail: m.luczak@qmul.ac.uk

A. Beveridge et al. (eds.), *Recent Trends in Combinatorics*, The IMA Volumes in Mathematics and its Applications 159, DOI 10.1007/978-3-319-24298-9_15

set theory is now well established as a field, it does not enjoy the same status or support as string theory, and progress has accordingly been less fast.

Underlying the theory is the notion that the "correct" model of spacetime is a model of a random partially ordered set. For each of the most common models of random partial orders in the mathematical literature – see Brightwell [17] for a survey – some variant or extension has its attractions as a candidate for a model of spacetime. Our aim in this article is to review the mathematics, explain its relevance to physics, and discuss some potentially important open problems.

There are two main strands to the physics. One is to argue that we know what the discrete set should look like: it should be a set of points embedded "faithfully" (i.e., with about the same density in almost all regions of equal volume) in a 4-dimensional Lorentzian manifold, which on a moderate scale is essentially flat. This can be achieved, at least approximately, by taking the points of our discrete set to be a Poisson process in 4-dimensional Minkowski space $\mathbb{M}^4$, with the order induced on this set from the underlying manifold. We then ask what we can do with such a discrete partially ordered set: can we in any sense recover the structure of the manifold it came from?

The other strand aims to address the question of "why" the spacetime resembles $\mathbb{M}^4$ on large scales. The ultimate prize would be a simple, natural random process, making no use of any underlying manifold, whose samples do resemble discrete sets that are "faithfully" embedded in $\mathbb{M}^4$. There is currently no candidate for such a process, but much interesting mathematics has resulted from various approaches to the issue.

Given that spacetime is to be discrete, one might still ask why it makes sense to use a random structure? The most obvious alternative is to model space as a lattice-like structure, with layers representing time. However, a model of this form will exhibit "preferred directions" along the lattice that we do not observe in practice. In particular, observations suggest that the spacetime structure of the universe is invariant under Lorentz transformations, and so our discrete model ought also to enjoy this property, or a very close approximation of it. More generally, there is no explicit construction known of a discrete subset of $\mathbb{M}^4$ that has approximately the "right" number of points in each interval. As we shall discuss later, taking a Poisson process in $\mathbb{M}^4$ does not completely solve this problem, but at least it provides the basis for a theory that does not discriminate between intervals of equal volume. There is also the intuition coming from quantum mechanics that aspects of reality are best modelled using some "randomness", albeit not necessarily one involving a classical probability measure.

There are a number of surveys of the theory of causal sets from a physics perspective [26, 34, 51, 52], and the reader is referred to those for more background and motivation. The present article is not intended as a full survey of the causal set programme; two particular omissions are (i) the prediction from the theory of a small but non-zero value of the cosmological constant, subsequently supported by observations, and (ii) recent progress on the important project of developing notions of fields and particles on top of the underlying causal set.

A partially ordered set, or *poset*, is a pair $(X, <)$ where $<$ is a transitive, irreflexive relation on the *ground-set* X. A causal set is intended to model a universe with a big bang, meaning that the past light cone of every point has finite measure. The equivalent statement in the discrete setting is that every element x has only finitely many elements below it. It is customary in the physics literature to take this as the definition of a *causal set* $(X, <)$: a poset with countable ground-set X such that, for every $x \in X$, the set $\{y \in P : y \le x\}$ is finite. For many purposes, one might instead work with a *locally finite* poset, where the *intervals* $[x, y] = \{z : x \le z \le y\}$ are finite. We introduce other definitions as we proceed.

Whether or not causal set theory is ultimately a success, even the possibility of an application area enriches the study of random partial orders, and provides an interesting alternative viewpoint regarding what questions are of natural interest.

2 Poisson processes in Lorentzian manifolds

In this section, we examine models arising by taking a continuum structure equipped with a partial order and a compatible measure, generating a discrete set of points by taking a Poisson process in the continuum, and then looking purely at the discrete set with its induced partial order.

The most basic continuum model of spacetime is 4-dimensional Minkowski space $\mathbb{M}^4$. Minkowski space $\mathbb{M}^d$ is a Lorentzian manifold, with $d-1$ dimensions of space and one of time: one can represent $\mathbb{M}^d$ as $\mathbb{R}^{d-1} \times \mathbb{R}$, with the product Lebesgue measure, and an order relation given by $(\mathbf{x}, s) \le (\mathbf{y}, t)$ if $|\mathbf{x}-\mathbf{y}| \le t - s$. Here we have expressed our coordinates in units where the speed of light is 1, so to say that one point is below another means that light may travel from the lower point of spacetime to the higher. The set of points above $\mathbf{a} = (\mathbf{x}, s)$ is the *future light cone* of $\mathbf{a}$, and the set of points below $\mathbf{a}$ is its *past light cone*. A pair of points is said to be *timelike* if they are comparable in the partial order: in this case the distance between $(\mathbf{x}, s)$ and $(\mathbf{y}, t)$ is given by $\sqrt{(t-s)^2 - |\mathbf{x}-\mathbf{y}|^2}$ and is called the *proper time* between the two points. Note that this distance might be zero: this corresponds to one point lying on the boundary of the future light cone of the other. If a pair of points is incomparable, they are said to be *spacelike*, and the distance between them is defined as $\sqrt{|\mathbf{x}-\mathbf{y}|^2 - (t-s)^2}$. The measure, the partial order, and the distance function on this space are then invariant under the group of *Lorentz transformations*. These include isometries of the space $\mathbb{R}^{d-1}$, with the time axis left fixed, and also the *Lorentz boosts*: with respect to any orthogonal coordinate system for the space $\mathbb{R}^{d-1}$, a boost in the direction of the first coordinate is given by mapping $((x_1, \ldots, x_{d-1}), t)$ to $((\gamma(x_1 - \beta t), x_2, \ldots, x_{d-1}), \gamma(t - \beta x_1))$, where $\beta \in (-1, 1)$ is the *velocity coefficient* (measuring the magnitude of the boost) and $\gamma = (1-\beta^2)^{-1/2}$ is the *Lorentz factor*. A boost represents transforming the time axis of one observer to that of another travelling at speed β in the direction of the first coordinate with respect to the first observer.

Although it is natural to consider 4-dimensional models of spacetime, it is also reasonable to think about simpler "toy models" where, for instance, $\mathbb{M}^2$ replaces $\mathbb{M}^4$. It is also desirable to be able to replace $\mathbb{M}^4$, representing a flat spacetime, with a more general Lorentzian manifold with curvature.

There is an alternative visual interpretation of a finite partial order P embedded in $\mathbb{M}^d$. Imagine a "horizontal" plane H to the past of all points of P and, for each point x of P, consider the intersection S_x of the past light cone of x with H. This intersection will be a $(d-1)$-sphere, and we have that $S_x \subseteq S_y$ if and only if x is below y in the order on $\mathbb{M}^d$. Hence the collection of spheres S_x, ordered by containment, is a representation of the partial order P. This construction can be reversed, so that a finite partial order can be embedded in $\mathbb{M}^d$ if and only if it has a representation as a collection of $(d-1)$-spheres ordered by containment. A partial order with such a representation is called a *circle order* if $d = 3$, and a $(d-1)$*-sphere order* for higher d. See, for instance, Meyer [42], or Brightwell and Winkler [23].

For two points $\mathbf{a}, \mathbf{b}$ in Minkowski space with $\mathbf{a} \leq \mathbf{b}$, the interval $[\mathbf{a}, \mathbf{b}]$ is given by $\{\mathbf{z} : \mathbf{a} \leq \mathbf{z} \leq \mathbf{b}\}$. For two intervals of the same positive measure, there is a Lorentz transformation taking one to the other. Thus any two intervals of positive measure are "equivalent" up to a scale factor; for most purposes, we can restrict our attention to one such interval.

Any (discrete) subset of $\mathbb{M}^d$ inherits the order structure from the Minkowski space, and we can then ask questions about the partial orders arising. There are various results concerning how much of the structure of a Lorentzian manifold can be recovered from just the causal (partial order) structure, and the basic conclusion is that the topology and geometry can be recovered completely, up to a conformal factor specifying the volume measure. The most general results in this line are those of Hawking, King, and McCarthy [33] and Malament [41].

We now take a Poisson process P with intensity n in $\mathbb{M}^d$. The partial order on $\mathbb{M}^d$ then induces a partial order on the discrete set of points, and we can look at this purely as a discrete structure, "rubbing out" the manifold used to construct it. The basic principle is that the order relation from the manifold survives in the discrete approximation, and the volume measure can be approximated by counting the points in any given region. Hence one might hope that the discrete structure contains enough of the structure of the original manifold to form a sufficient replacement for it.

From a mathematical perspective, one may fix some interval $[\mathbf{a}, \mathbf{b}]$ in the manifold, let the intensity n of the Poisson process tend to infinity, and look at the limiting distribution of suitably scaled parameters of the partial order $P[\mathbf{a}, \mathbf{b}]$ restricted to the interval. The physical perspective does not allow us the luxury of taking the limit: physical considerations lead to a proposition for a natural value of the intensity, where the expected number of points in a "standard" volume of spacetime, with spatial volume on the order of a cubic metre and time duration on the order of the time for light to travel one metre, is about 10^{136}: see [51] for a detailed argument. Indeed, the point of the theory is that small intervals of spacetime should be unlike the continuum, so that we can see "quantum behaviour" and "relativistic behaviour" in the same model, but at different scales.

A fundamental question is whether, if we go to a large enough scale, parameters of the partial order are good approximations to parameters of the underlying manifold. Versions of this question were framed in the seminal paper of Bombelli, Lee, Meyer, and Sorkin [15], as well as in the earlier preprint of Myrheim [43].

Consider two points **a** and **b** in the Poisson process on the underlying manifold M. They will be at some distance in M, which can be either timelike, if $\mathbf{a} \leq \mathbf{b}$, or spacelike, if the two points are incomparable in the partial order on M. A natural candidate for an approximation of the (scaled) distance between two timelike points is the *height* of the partial order $P[\mathbf{a}, \mathbf{b}]$, i.e., the length h of the longest chain $\mathbf{a} = \mathbf{x}_0 < \mathbf{x}_1 < \cdots < \mathbf{x}_h = \mathbf{b}$. The idea that the height could serve as a discrete version of timelike distance was first proposed by Myrheim [43]. More generally, one can ask about the distribution of the height of the random partial order $P[\mathbf{a}, \mathbf{b}]$.

There is a closely related model of random partial orders that has received much more attention in the mathematical literature. Instead of Minkowski space, consider $\mathbb{R}^d$, with Lebesgue measure and coordinate order, so $(a_1, a_2, \ldots, a_d) \leq (b_1, b_2, \ldots, b_d)$ if and only if $a_i \leq b_i$ for each i. Again we consider a Poisson process of intensity n in $\mathbb{R}^d$, and take the induced partial order on the discrete set of points of the Poisson process. For many purposes, it is convenient to work with the finite unit cube $[0, 1]^d$, i.e., the interval $[(0, \ldots, 0), (1, \ldots, 1)]$ instead.

For $d = 2$, the map $f : \mathbb{M}^2 \to \mathbb{R}^2$ given by $f(x, t) = \frac{1}{\sqrt{2}}(t + x, t - x)$ is an isomorphism, but, for higher d, $\mathbb{M}^d$ and $\mathbb{R}^d$ are different spaces, and the random partial orders have different distributions. In $\mathbb{R}^d$ with the coordinate order, the "future light cone" of x, the set of points above x in $\mathbb{R}^d$, has cross-sections that are simplices. For instance, in $\mathbb{R}^3$, the model has what have been dubbed "triangular light cones", with three distinguished directions.

There is an alternative interpretation of the above model. First, there is little essential difference between taking a Poisson process of intensity n in the unit cube, and taking exactly n (labelled) points uniformly and independently in the cube. The latter amounts to taking n independent d-tuples, with each of the nd numbers uniform in $[0, 1]$. All we are interested in is the partial order on the points, which is the intersection of the d linear orders given by the coordinates, and these coordinate orders are independent, each being uniform over all $n!$ total orders. Therefore an equivalent description is given by taking d linear orders uniformly and independently on an n-element ground-set, and intersecting the orders. The *dimension* of a poset P is the minimum number of linear orders on its ground-set whose intersection is P, or equivalently the minimum d such that P can be embedded in $\mathbb{R}^d$. A random partial order $P_d(n)$ obtained by intersecting d iid uniform linear orders on an n-element ground-set is a *random d-dimensional partial order*. This model was introduced by Winkler [54] in 1985; as we have just seen, essentially the same model is obtained by taking a Poisson process of intensity n in $[0, 1]^d$.

There is a long history to the study of the height $H_2(n)$ of a random 2-dimensional partial order $P_2(n)$, partly because this has a further guise as the length of a longest increasing subsequence in a uniformly random permutation of n points. A number of authors contributed to showing that the height is approximately $2\sqrt{n}$, and then Baik, Deift, and Johansson [6] gave the following impressively precise result.

Theorem 1.

$$\Pr\left(\frac{H_2(n) - 2n^{1/2}}{n^{1/6}} \leq x\right) \to F(x), \qquad \textit{as } n \to \infty,$$

where $F(x)$ is a specified function. In particular, the mean and standard deviation of $H_2(n)$ satisfy

$$\mathbb{E} H_2(n) = 2n^{1/2} - \alpha n^{1/6} + o(n^{1/6}); \quad \sigma(H_2(n)) = \beta n^{1/6} + o(n^{1/6}),$$

where $\alpha \simeq 1.711$, $\beta \simeq 0.902$.

See Aldous and Diaconis [2] for more on the history and context of this result.

The proper time between two timelike points $\mathbf{a}$ and $\mathbf{b}$ in $\mathbb{M}^2$ is, up to a constant factor, $(\mathrm{vol}[\mathbf{a},\mathbf{b}])^{1/2}$, and so the result above can be interpreted as saying that, for a Poisson process of unit intensity in $\mathbb{M}^2$, the height of the partial order in $[\mathbf{a},\mathbf{b}]$ gives, up to a constant factor, a very good approximation to the proper time between the two points.

The degree of precision in Theorem 1 remains far out of reach for the height of a random partial order in more than 2 dimensions, either in the cube or in Minkowski space. Bollobás and Winkler [12] showed that the height $H_d(n)$ of $P_d(n)$ behaves as $n^{1/d}$: they showed that there is a sequence of constants (c_d), with $c_d < e$ for all d and $c_d \to e$ as $d \to \infty$, such that, for each fixed d, $H_d(n)n^{-1/d} \to c_d$ in probability. Even the values of c_d for $d \geq 3$ remain unknown, let alone any finer estimates.

This result was generalised by Bollobás and Brightwell [8] to a setting where, as in both $[0,1]^d$ and $\mathbb{M}^d$, all intervals "look the same". A *partially ordered measure space* is a measure space $(X, \mathcal{F}, \mu)$ equipped with a partial order $<$ so that all intervals $[x, y]$ are measurable. Two such spaces $(X, \mathcal{F}, \mu, <)$ and $(X', \mathcal{F}', \mu', <')$ are said to be *order-isomorphic* if there is a bijection $\lambda : X \to X'$ respecting the partial order and the σ-field; they are said to be *scale-isomorphic* if additionally there is a constant α such that $\mu'(\lambda(A)) = \alpha\mu(A)$ for all $A \in \mathcal{F}$. A partially ordered measure space is *homogeneous* if any two intervals of positive measure are scale-isomorphic. A *box space* is a homogeneous partially ordered measure space that is itself an interval of measure 1.

Both $[0,1]^d$ and an interval in $\mathbb{M}^d$ of measure 1 are box spaces, and Bollobás and Brightwell [8] give a few other examples as well. They proved that every box space has a "dimension" $d \geq 1$ (which may be infinite, or non-integer) so that, for any interval $[\mathbf{a},\mathbf{b}]$ of positive measure, the maximum, over $\mathbf{c} \in [\mathbf{a},\mathbf{b}]$, of the minimum of $\mu[\mathbf{a},\mathbf{c}]$ and $\mu[\mathbf{c},\mathbf{b}]$ is $2^{-d}\mu[\mathbf{a},\mathbf{b}]$.

For a Poisson process of density n in a box space X, let $H_X(n)$ denote the height of the induced partial order. The following result of Bollobás and Brightwell [8] states that the height is asymptotic to $m_X n^{1/d}$ for some constant m_X, depending on the space.

Theorem 2. *Let X be a box space of dimension d. Then there is a constant m_X such that $n^{-1/d}H_X(n) \to m_X$ in probability.*

The only case of interest to us where m_X is known is $X = [0,1]^2$, where $m_X = 2$. For X an interval in $\mathbb{M}^d$, $d \geq 3$, the bounds given in [8] are

$$\frac{2^{1-1/d}}{\Gamma(1+1/d)} \leq m_X \leq \frac{2^{1-1/d}e(\Gamma(d+1))^{1/d}}{d}.$$

For $d = 4$, this gives $1.8554 < m_X < 2.5297$.

In $\mathbb{M}^d$, the proper time between two timelike points **a** and **b** is proportional to $(\mathrm{vol}[\mathbf{a},\mathbf{b}])^{1/d}$, so Theorem 2 shows that again, up to a constant factor, the height of the interval provides a good approximation to the proper time between the two points. As explained in Bachmat [5], this result extends to Poisson processes in Lorentzian manifolds with curvature: if we take a Poisson process in an interval in such a manifold, then the longest chain will follow the maximal curve between the endpoints, and its length, suitably scaled, will be a good approximation to the proper time.

It would be interesting to have better bounds on the constant m_X when X is an interval in $\mathbb{M}^d$: as far as we know, the bounds above of Bollobás and Brightwell are the best known. It would also be interesting to have some tight bounds, upper and lower, on the variance of $H_X(n)$. A straightforward application of Talagrand's inequality gives that the standard deviation is at most of order $n^{1/2d}$, but the example of the Baik-Deift-Johansson Theorem suggests that the actual deviation may be much smaller. This issue is discussed by Bachmat in [5], where it is suggested on the basis of heuristics and numerical evidence that the standard deviation in $\mathbb{M}^4$ might be of order $n^{1/24}$.

The implications of Theorem 2 for causal set theory were spelled out by Brightwell and Gregory [19], where they explained how the height of $P[\mathbf{a},\mathbf{b}]$ could operate as a definition of the distance between two timelike points. What about the distance between two spacelike points **a** and **b**? In the underlying manifold, this distance is well-defined, and we might hope that some simple parameter of the partial order approximates spacelike distance up to a constant factor. For a causal set obtained from a Poisson process in $\mathbb{M}^2$, one can define a concept of two points being "nearest neighbours", and then derive a distance function. Brightwell and Gregory [19] pointed out a phenomenon that makes any such definition problematic in higher dimensions. However exactly the definition is framed, we would want that a *sufficient* condition for a spacelike pair **a**, **b** to be nearest neighbours is that there are points $\mathbf{c} < \mathbf{d}$ of the causal set such that **a** and **b** are the only two points of the causal set in $[\mathbf{c},\mathbf{d}]$. However, given any spacelike pair **a**, **b** in $\mathbb{M}^d$ $(d \geq 3)$, by applying larger and larger boosts in a direction perpendicular to the line between **a** and **b**, one can find an infinite collection of intervals in $\mathbb{M}^d$ containing **a** and **b** in the interior, all of essentially the same volume, each pair of intervals disjoint apart from an arbitrarily small tube around the line between **a** and **b**. Given such a collection, where the tube is small enough to contain no points of the causal set, almost surely one of these intervals will contain exactly two points **c** and **d** of the causal set, with **a** and **b** in $[\mathbf{c},\mathbf{d}]$. Thus every spacelike pair of points would be nearest neighbours.

The existence of these rare "voids", where there are very few points of the Poisson process in an interval of large volume, along with the Lorentz invariance of the manifold, dooms any attempt to define a distance in terms of minimising some quantity over some family of intervals. It is, however, worth noting that this precise objection is of a theoretical nature only: Dowker, Henson and Sorkin [28] calculate that the probability of a void of the size of an atom contained within the observed universe is negligibly small.

There remains no strong candidate for a natural definition of spacelike distance within a causal set. See however Rideout and Wallden [49] for some practical approaches to measuring spacelike distances that yield useful answers.

Another direction of research concerns the characterisation of which finite partial orders can be embedded in $\mathbb{M}^d$ (i.e., of d-sphere orders), for each d. For $d = 2$, the answer is the family of 2-dimensional orders, and this is a well-studied and well-understood family. For higher dimensions, rather little is known. Brightwell and Winkler [23], and independently Meyer [42], constructed, for each d, an example of a poset that is a d-sphere order but is not a $(d-1)$-sphere order. Alon and Scheinerman [3] showed that the number of d-sphere orders on m labelled elements grows as $m^{md(1+o(1))}$ for each fixed d (which provides a non-constructive proof of the result of Brightwell/Winkler and Meyer). This number is comparable to the number of m-element partial orders embeddable in $\mathbb{R}^d$, i.e., the number of m-element partial orders of dimension at most d, and grows slowly compared to the total number of m-element partial orders, which is $2^{m^2/4+O(m)}$ – we shall discuss this further later. For a while, it was an open question whether (a) all 3-dimensional partial orders can be embedded in $\mathbb{M}^3$, and (b) every partial order can be embedded in some $\mathbb{M}^d$. In 1999, Felsner, Fishburn, and Trotter [30] answered both questions by showing that there is some 3-dimensional partial order that cannot be embedded in any $\mathbb{M}^d$. Their example is produced through Ramsey-theoretic arguments, and its size is enormously larger than the supposed number of points of spacetime in the observable universe, but it is very likely that there are examples with fewer than a hundred elements: in [23], Brightwell and Winkler suggested that the partial order 2^6 of subsets of a 6-element set is not a sphere order in any dimension. It is remarkable that this concrete question is apparently not easy to answer computationally; to the best of our knowledge, the complexity of determining whether a given finite partial order is (for instance) a circle order is unknown.

If we are presented with a causal set, and told that it arises as a sample from a Poisson process on some Lorentzian manifold (perhaps with a few local "defects"), can we determine the manifold, up to some rough isometry? This is an ill-specified question, but it is nevertheless important for causal set theory to have some kind of answer.

Bombelli and Meyer [14] showed that, if we are given a countable partial order, consisting of an iid sequence of points selected according to the volume measure on some portion of a Lorentzian manifold of finite volume, then the manifold can be recovered from the causal structure on the points.

If we are given instead a sequence of finite partial orders, sampled separately from a region of finite volume in a fixed manifold according to Poisson processes with densities tending to infinity, then we would hope to be able to recover the manifold via a *poset limit*. The general theory of poset limits has been developed recently by Janson [37] and Hladký, Máthé, Patel, and Pikhurko [36]; we give a brief account here.

Given finite posets P and Q, the *density* $t(Q;P)$ of Q in P is the proportion of $|Q|$-tuples of elements of P that are isomorphic to Q. A sequence (P_n) of finite posets with $|P_n| \to \infty$ is said to be *convergent* if, for every fixed finite poset Q, $t(Q; P_n)$ tends to a limit. An appropriate limit object for a sequence of posets is what has been informally dubbed a *poson*, by analogy with the idea of a graphon in the theory of graph limits. Let $([0, 1], \mathscr{B}, \mu)$ denote the standard Borel probability space on the interval $[0, 1]$, and $\prec$ the standard linear order on $[0, 1]$. A *poson* W is a $\mathscr{B} \times \mathscr{B}$ measurable function from $[0, 1] \times [0, 1]$ to $[0, 1]$ such that (i) if $W(x, y) > 0$, then $x \prec y$, and (ii) if $W(x, y) > 0$ and $W(y, z) > 0$, then $W(x, z) = 1$. To generate a random n-element sample from a poson W, we take n elements $X_1, \ldots, X_n$ uniformly and independently from $[0, 1]$, and we put $X_i < X_j$ with probability $W(X_i, X_j)$, each pair independently. Condition (ii) then guarantees that $<$ is a transitive relation, and condition (i) ensures that the linear order induced by $\prec$ on the sample $(X_1, \ldots, X_n)$ gives a linear extension of the derived poset. The density $t(Q; W)$ of a fixed poset Q in a poson W is the probability that a $|Q|$-element sample from W is isomorphic to Q; the density $t(Q; W)$ can be expressed as a multi-dimensional integral. The sequence (P_n) of posets is then said to converge to the poson W if $t(Q; P_n) \to t(Q; W)$ for every finite poset Q. Extending a result of Janson [37], Hladký, Máthé, Patel, and Pikhurko [36] proved that every convergent sequence of posets converges to some poson.

The implications for causal set theory are indirect. The theory of poset limits highlights the following conjecture of Bombelli [13] as fundamental: if U and V are regions of Lorentzian manifolds of finite volume, such that there is no measure-preserving diffeomorphism between the two, then there is some finite poset Q such that $t(Q; U) \neq t(Q; V)$. Here the density $t(Q; U)$ is the probability that an iid uniform sample of $|Q|$ elements from U yields a poset isomorphic to Q.

In causal set theory, we do not have a sequence of partial orders with sizes tending to infinity, but a single partial order that in most ordinary terms would be thought of as very large (although not large enough to meaningfully apply the poset regularity lemma of [36]), and the sense in which it approximates a manifold will necessarily be quantitative, rather than a statement about limits. One would like a result saying that, if two finite portions of manifolds are both approximated well by the same partial order, then the two spaces are roughly isometric. No specific conjecture has been formulated, but this has nevertheless become known as the *Hauptvermutung* of causal set theory.

3 Growth processes

In the previous section, we discussed the properties of a Poisson process in a Lorentzian manifold. However well such a model might fit observations, it cannot make for a theory with any "explanatory power": we start with a manifold, embed the discrete set of points, and then "rub out" the unwanted manifold at the end. What would be far more satisfying is a random process, following some reasonably natural rules, whose samples typically resemble a Poisson process in $\mathbb{M}^4$.

We know of no contender for such a process, and indeed it is surely too much to expect; any eventual theory will presumably not involve classical probability measures on causal sets at all. In the absence of a fully realised theory, what one can do is start from the other extreme, and ask what form a suitable process might take, and then investigate processes of such a form. If there are processes that share some of the observed properties of spacetime, then all the better: if not, then one gains experience in asking the right questions and identifying the key issues.

One outline structure that has arisen [47] is that of a *growth process*. In this construction, points (of spacetime) are generated one at a time, each new point being maximal among those seen so far, and placed above some subset of the existing points chosen according to some (random) rule. In this way, an infinite causal set is generated. The order in which the elements of the causal set are generated is part of the model, and it is a *linear extension* of the causal set: a linear order on the ground-set extending the partial order.

One example of this process is known as the model of *random graph orders* in the mathematics literature, and *transitive percolation* in the physics literature. Here, we fix some value $p \in (0, 1)$, and let the ground-set be $\{x_1, x_2, \dots\}$. For each n in turn, a random subset S_n of $\{x_1, \dots, x_n\}$ is chosen, with each element of $\{x_1, \dots, x_n\}$ appearing in S_n with probability p, independent of all other choices. Then the new element x_{n+1} is placed above all elements of S_n. Typically, this rule will not obey transitivity, so additionally x_{n+1} is placed above all elements that are less than or equal to some element of S_n. An alternative description of this random construction is as follows: (i) take a random graph on the ground-set, with each edge present independently with probability p, (ii) for $i < j$, interpret an edge from x_i to x_j as putting $x_i < x_j$, and (iii) take the transitive closure. We may stop the process after n elements have been added, in which case we denote the resulting random graph order $P_{n,p}$. This model of random graph orders was first considered as a model of random acyclic directed graphs by Barak and Erdős [7], introduced in its own right by Albert and Frieze [1], and then studied by a number of other authors since, including a series of papers by Bollobás and Brightwell [9–11].

Obviously there is an enormous variety of other growth models, and it is difficult to say anything of interest about growth models in full generality. Rideout and Sorkin [47] identified two extra conditions that they felt were desirable in a growth model representing the universe. The first property they called "general covariance": for every n, and every poset Q on n elements, conditional on the first n points generated forming Q, each linear extension of Q is equally likely to be the order in which the first n points were generated. One needs to be careful with this definition: the partial orders in Rideout and Sorkin's definitions are unlabelled, and they discuss their process in terms of transition probabilities for a Markov chain whose states are unlabelled finite partial orders. Given any unlabelled m-element partial order Q, with $e(Q)$ linear extensions, each linear extension induces a path from the empty partial order to Q, adding one maximal element at a time. Some paths may be induced by more than one linear extension, as there may be several isomorphic copies of some partial order that are stems of Q. The condition of general covariance states that, for every Q, each path from the empty partial order to Q occurs with probability proportional to the number of linear extensions of Q inducing that path.

Rideout and Sorkin [47] dubbed their second property *Bell causality*. It reflects the principle that a change in one part of the "universe" cannot affect some other part of the universe that is not in the future light cone of the change. The translation of this to a growth model is as follows. A transition of the Markov chain involves adding a new maximal element: suppose we are considering two possible transitions from a state Q, adding the new element above either of the two down-sets S_1 or S_2 of the existing elements, with transition probabilities $p(Q; S_1)$ and $p(Q; S_2)$, respectively. Now consider transitions from a state Q' which is the poset Q restricted to $S_1 \cup S_2$: the condition of Bell causality is that

$$p(Q; S_1)p(Q'; S_2) = p(Q; S_2)p(Q'; S_1).$$

Thus the relative probabilities of the two transitions are unchanged on the removal of "spectators" which would be incomparable with the new element in both transitions.

We observe that transitive percolation, for any value of the parameter p, satisfies both of these conditions. To see general covariance, it is enough to show that, for any partial order Q on $\{x_1, x_2, \dots, x_m\}$ with $x_1 < x_2 < \cdots < x_m$ as a linear extension, the probability that the first m steps of the transitive percolation process produces Q does not depend on the labelling of the elements of Q. We see that the probability of Q is given by $p^{c(Q)}(1-p)^{i(Q)}$, where $c(Q)$ is the number of covering pairs (i.e., pairs (x, y) such that $x < y$ in Q but there is no z with $x < z < y$ in Q), and $i(Q)$ is the number of incomparable pairs. Checking that transitive percolation satisfies Bell causality is straightforward. Rideout and Sorkin [47] also imposed an extra condition that, from every state, the probability of the new element being incomparable to every existing element should be positive; we say that the growth model is *gregarious*. They then proved that the only growth models satisfying general covariance, Bell causality, and this extra condition are what they called *classical sequential growth models* or *csg models*.

A csg model is specified by a sequence of non-negative constants $t_0, t_1, \dots$, where $t_0 > 0$. At any state Q, a partial order with m elements, a subset S_m of the existing elements is selected, and the new element x_{m+1} is placed above all elements below or equal to some element of S_m. For each S, the probability that $S_m = S$ is proportional to $t_{|S|}$, so the probability that $S_m = S$ is $t_{|S|}\left(\sum_{k=0}^{m}\binom{m}{k}t_k\right)^{-1}$. For instance, transitive percolation is the example where $t_k = (p/(1-p))^k$ for each k.

To see that a csg model, as above, satisfies general covariance, we calculate the probability of generating a particular labelled partial order on the set $\{x_1, \dots, x_n\}$, where x_{i+1} is above a set B_i of b_i elements, and covers a set $C_i \subseteq B_i$ of c_i elements. Thus the given partial order is generated if each S_i contains C_i and is contained in B_i. Now the probability that the given partial order is generated is the product of ratios $a_i / \sum_{k=0}^{i}\binom{i}{k}t_k$, where $a_i = \sum_{\ell=c_i}^{b_i}\binom{b_i-c_i}{\ell-c_i}t_\ell$, the total "weight" attached to all values of S_i with $C_i \subseteq S_i \subseteq B_i$. In the overall product, the denominators are purely functions of the t_k, while the numerator a_i depends only on the structure of the partial order below x_{i+1}. Hence, on taking a different order of generation, the same denominators appear, and the same numerators appear in a different order, so the ratio is unchanged.

The main theorem of Rideout and Sorkin [47] is as follows.

Theorem 3. *The only growth models that satisfy general covariance and Bell causality, and are gregarious, are csg models.*

What if one drops the extra "technical" condition that the growth model be gregarious? This allows also models of the following form: there are a number of "phases" of the growth process, with each phase of the process developing as a csg model and, with the possible exception of a final infinite phase, stopping after a pre-determined finite number of steps. All elements of each new phase are above all elements of previous phases. The number of steps of the current phase and the parameters of the csg model in effect during the phase are arbitrary functions of the partial order at the end of the previous phase. Varadarajan and Rideout [53] and Dowker and Surya [27] showed that the growth models satisfying general covariance and Bell causality are exactly the processes of this form. Thus dropping the requirement that our growth models be gregarious does not introduce any different possible "local" structures.

Let us then consider whether csg models include some that could form the basis of a satisfactory theory of quantum gravity. First of all, we consider whether random graph orders might be a contender. The large-scale structure of a random graph order $P_{n,p}$ has been extensively investigated, and is now quite well understood. Alon, Bollobás, Brightwell, and Janson [4] showed that, for a fixed p and an infinite ground-set, there are a.s. infinitely many *posts*, which are elements of the partial order comparable to all others. Sharper results about the distribution of the sequence of posts were given by Bollobás and Brightwell [11] and Kim and Pittel [39]. For small p, posts become very rare: the probability that a given element is a post is asymptotically $2\pi e^{\pi^2/6} p^{-1} e^{-\pi^2/3p}$. Bollobás and Brightwell [11] explain how the presence of many posts implies that parameters such as the height of a random graph order are normally distributed, with mean proportional to n.

If we instead take a function $p = p(n)$ such that $np/\log p^{-1} \to \infty$ and $p \log n \to 0$, then the random graph order $P_{n,p}$ exhibits a very different structure. Suppose first that $n = \alpha p^{-1} \log(p^{-1})$, for $\alpha < 1$ a fixed constant. Then the number of elements above x_1 in the random graph order is fairly well concentrated around n^α. This implies that, in a random graph order, the element x_m is typically incomparable to most elements in the set $\{x_k : |k - m| < p^{-1} \log(p^{-1})\}$. On the other hand, if $n \geq \alpha p^{-1} \log(p^{-1})$ for some $\alpha > 1$, and $I(1)$ denotes the number of elements incomparable to x_1, then it can be read out of a result of Pittel and Tungol [44] that, whenever $M = O(p^{-1} \log \log(p^{-1}))$,

$$\Pr(I(1) < p^{-1} \log(p^{-1}) + M) = (1 + o(1)) \exp(-e^{-pM}).$$

This implies that each element x_m is almost surely comparable to all elements outside the range $\{x_k : |k - m| < p^{-1}\log(p^{-1}) + \omega(n)p^{-1}\}$, where $\omega(n)$ is any function tending to infinity. There are similar results in the earlier papers of Simon, Crippa, and Collenberg [50] and Bollobás and Brightwell [10]. For this range of parameters, Bollobás and Brightwell [9] showed that the width of the partial order (i.e., the size of a largest antichain) is of order p^{-1} and, moreover, a.s. every element is in an antichain of size of order p^{-1}. By contrast, if we fix p and let n tend to infinity, then there is the occasional antichain of size much larger than p^{-1}.

This is certainly not a picture that suggests a 4-dimensional Lorentzian manifold, and indeed it is possible to point out various concrete ways in which it differs: for a start, almost surely every finite partial order will appear as an interval in a random graph order, whereas this is not the case in a Poisson process in a manifold.

However, a random graph order does share some properties of interest with a Poisson process in a Lorentzian manifold. Rideout and Sorkin [48] gave experimental evidence that a suitably scaled sequence of finite random graph orders approaches a poset limit. Namely, they looked at the densities $t(Q, P_{n,p(n)})$ of various small posets Q in a random graph order $P_{n,p(n)}$, and observed that, if $p(n)$ is tuned so that the density of comparable pairs remains constant, then the densities of all the other posets Q they looked at also approached limits. Moreover, the densities of some suborders appeared to tend to 0. They pointed out that these properties were what would be expected on taking a sequence of Poisson processes, with increasing intensity, in a manifold.

Brightwell and Georgiou [18] explained this phenomenon in terms of the known large-scale structure of a random graph order, as described above. In the language of poset limits, as set out the previous section, they showed that a sequence $(P_{n,p(n)})$ of random graph orders is convergent if and only if either $p^{-1}\log(p^{-1})/n$ tends to a limit in $[0, 1)$, or $\liminf p^{-1}\log(p^{-1})/n \geq 1$. In the case where $p^{-1}\log(p^{-1})/n$ tends to zero, the proportion of comparable pairs of elements in $P_{n,p(n)}$ tends to 1, as described above, and the poset limit is a chain. In the case where $\liminf p^{-1}\log(p^{-1})/n \geq 1$, the random graph order is very sparse, and the limit is an antichain. In the intermediate case, where $p^{-1}\log(p^{-1})/n \to c$, for $c \in (0, 1)$, one can take the limit poson $W : [0, 1] \times [0, 1] \to [0, 1]$ to be given by:

$$W(x, y) = \begin{cases} 1 & \text{if } y > x + c \\ 0 & \text{if } y \leq x + c \end{cases}.$$

The finite partial orders with positive density in W are all *semiorders*: they can be represented by a family of unit-length intervals in $\mathbb{R}$ with $x < y$ if the interval representing x lies entirely to the left of the interval representing y. An alternative characterisation is that semiorders are partial orders containing neither of the posets H or L in Figure 1 as suborders. It is natural to extend the definition of semiorders to include posons; a poson W is a semiorder if $W(x, y)$ is 0–1 valued, non-increasing in x and non-decreasing in y. Chains and antichains are semiorders, so the limit of any convergent sequence $(P_{n,p(n)})$ of random graph orders is a semiorder.

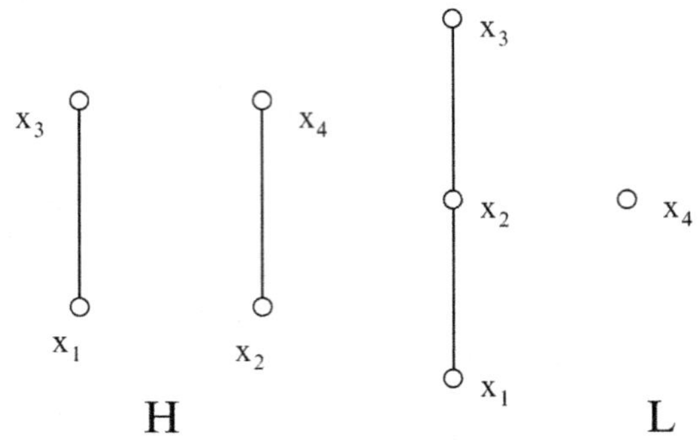

Fig. 1 The partial orders H and L

The causal structure of the observed universe is manifestly not that of a semiorder; a semiorder has a very well-defined temporal order, but no spatial structure at all. The conclusion is that a random graph order, even with a carefully chosen parameter, is not an adequate model.

Might it be that general csg models exhibit very different behaviour? There are certainly some interesting cases where the random partial order produced is sparse, for instance, if all the parameters t_k are zero beyond a certain point. The case where $t_0, t_1 > 0$ and $t_2 = t_3 = \cdots = 0$ produces a random forest: each new element x_m is placed above at most one other, with x_m being minimal with probability $t_0/(t_0+(m-1)t_1)$, and placed above each existing element with probability $t_1/(t_0+(m-1)t_1)$. In the limit, there are a.s. infinitely many connected components, and every element is covered by infinitely many later elements. The next case is that of "random binary orders", where each arriving element is placed above (up to) two existing elements, chosen uniformly at random. The structure of random binary orders was investigated by Georgiou [32]. One of the main results of that paper is that, for any constants $\varepsilon, \eta \in (0, 1)$, there is a constant M such that, for each fixed m, with probability at least $1 - \eta$, the number of elements incomparable with x_m is at most $Mm^{2+\varepsilon}$, and x_m is below all elements x_n with $n > m^{4+\varepsilon}$. There is some delicacy in the statement, as there will be a few exceptional elements that are comparable with very few elements for an unusually long time in the growth process, and indeed there are a.s. infinitely many m such that x_m is incomparable to at least about m^3 elements. Georgiou [32] proves that, a.s., for every m, x_m is incomparable with at most $m^{27/5}$ elements, and is below every x_n with $n \geq m^{12}$. The basic phenomenon here is that, even in this csg model, which is as sparse as possible without being the model of random forests, every element is incomparable with finitely many others. One might conceivably be interested in this behaviour in the context of a model of a very rapidly expanding universe.

However, for our purposes it seems more relevant to consider "dense" csg models, where the proportion of comparable pairs is positive. For comparison purposes, we should consider a sequence (C_n) of finite partial orders, with C_n equal to the induced csg model on $\{x_1, \ldots, x_n\}$ with parameter sequence $\mathbf{t}(n)$. Brightwell and Georgiou [18] showed that, even with this extra generality, with an infinite collection of parameters instead of a single one, the only possible poset limits for a sequence of csg models are semiorders.

Theorem 4. *Let (C_n) be a sequence of finite csg models, where C_n has n elements for each n. If the sequence (C_n) is convergent, then the limit is a semiorder.*

To prove Theorem 4, Brightwell and Georgiou showed directly that, for any sequence of finite csg models with sizes tending to infinity, the expected numbers of copies of the posets H and L in Figure 1 tend to 0.

Moreover, Brightwell and Georgiou [18] determined conditions under which a sequence of csg models does converge to a poset limit. The basic idea is that, in any csg model, the structure is locally quite similar to that of a random graph order. The key sequence of parameters is (p_m), where $p_m = \mathbb{E}\,|S_m|/m$, the probability that any given element x_n, with $n < m$, is selected for the set S_m; at the introduction of the new element x_{m+1}, p_m plays a role analogous to the parameter p of random graph orders. The sequence (p_m) of probabilities can take arbitrary jumps upwards within $(0,1)$: the parameter t_m of the csg model plays no role in generating the sets $S_1, \ldots, S_{m-1}$, but, if it is large enough, then there will be a high probability that $S_m = \{x_1, \ldots, x_m\}$, and so x_{m+1} will be placed above all existing elements. Brightwell and Georgiou showed that $\mathbb{E}\,|S_m|$ is non-decreasing in m, so the sequence (p_m) can only decrease slowly. Therefore, in all interesting cases, (p_m) is slowly varying over long stretches and, if p_m is close to a value p over a long stretch, then the structure of the csg model is, on a large scale, very close to that of a random graph order with parameter p.

Negative results such as Theorem 4 were expected, and there remains interest among physicists in the behaviour of random graph orders, and more generally in csg models. If there is to be a "successful" theory of causal sets, it will not be based on a csg model, indeed most likely not on any classical stochastic process, but nevertheless a csg model may serve as a useful toy model, having some features of the real theory without all of the complexity that will surely prove necessary. For instance, one question that arose in the study of csg models was the nature of the appropriate σ-field of "observable" events. The description of the growth process entails a natural σ-field $\mathscr{G}$, and an event G in $\mathscr{G}$ is deemed to be "observable" if it is label-invariant, i.e., if renumbering the elements of the partial order with a different linear extension does not affect membership in G. Brightwell, Dowker, García, Henson, and Sorkin [24] showed that, up to sets of measure zero, the σ-field of observable events is generated by the *stem* events, which are the events of the form "the causal set contains a down-set isomorphic to L", for each fixed finite partial order L.

One feature that does have resonance for theories of quantum gravity is the existence of posts: recall that, for any fixed p, there are a.s. infinitely many posts in the random graph order with parameter p. This might represent the possibility of a succession of universes, with no information flowing from one to the next. The question then arises of which csg models have posts: this is currently an open problem, but it is possible to suggest an answer. If the sequence (t_k) of parameters of the model behaves as c^k, for any constant c, whether smaller or greater than 1, then the csg model behaves as a random graph order on large scales. If the sequence tends to infinity more rapidly than an exponential function, then on large enough scales the csg model is denser than any random graph order, and the proportion of the elements of $\{x_1, \ldots, x_n\}$ that are posts will tend to 1 as n tends to infinity. It is more interesting to look at sequences of parameters (t_k) that tend to zero more rapidly than any exponential function. For well-behaved sequences, the results of Brightwell and Georgiou [18] imply that the csg model then resembles a random graph order on any scale, but that as the scale increases the effective parameter p_n of the random graph order decreases, implying that posts become rarer as the number of elements increases. Calculations strongly suggest that the threshold for the existence of infinitely many posts is $p_m \simeq \pi^2/3 \log m$. Thus we expect that, if $t_k \simeq (t/\log k)^k$ for some $t < \pi^2/3$, then there are a.s. finitely many posts, while, if $t_k \simeq (t/\log k)^k$ for some $t > \pi^2/3$, then there are a.s. infinitely many posts. Establishing a result along these lines is work in progress.

Sorkin's idea has always been that the eventual theory of causal sets will not involve a model using "classical" probability measures at all. Quantum mechanics can be described using quantum measures, obeying some but not all of the rules of probability measures, and it is thus at least worth considering the possibility that the evolution of spacetime also involves quantum measures.

Another possibility before giving up on classical probability altogether is to challenge one of the hypotheses of Theorem 3, either general covariance or Bell causality. Of the two, general covariance seems mathematically more natural, and also to have more unassailable motivation: it embodies the principle that if we observe the spacetime universe, we see no effects that can be attributed to a specific favoured linear extension of the partial order. For Bell causality, while the principle is appropriate for events taking place in the universe, it does not seem automatic that it should apply to the generation of the universe itself. In the next section, we explore what happens if we drop the condition of Bell causality.

4 Order-invariance

In this section, we give a short account of the work in two papers [20, 21] of the authors. The idea behind these papers is to examine the property of general covariance, introduced in the previous section, in its own right. As we shall explain, measures satisfying general covariance are analogues of the Gibbs measures of statistical physics, and we believe that the theory we develop here will prove to be of interest beyond any relevance it has to the causal set programme.

As mentioned earlier, Rideout and Sorkin [47] expressed the property of general covariance in terms of transitions of a Markov chain whose states are *unlabelled* finite partial orders. It is very often cleaner to work with labelled combinatorial structures, and accordingly we shall discuss Markov chains whose states are labelled partial orders, and in this context we use the term "order-invariance" instead of

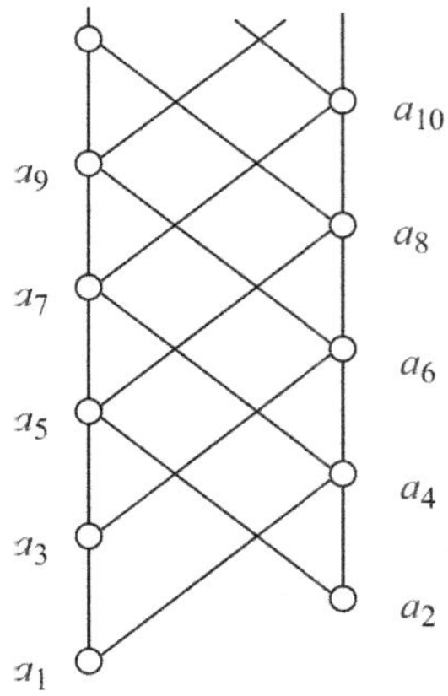

Fig. 2 The causal set $P = (Z, <)$

"general covariance": we give a definition later. As with growth models, we start with the empty poset, and at each stage we add one new maximal element, keeping track of the order in which the elements are generated. The infinite poset $P = (Z, <)$ generated is always a causal set, and the order in which the elements are generated is a *natural extension* of P, i.e., a bijection from $\mathbb{N}$ to Z whose inverse is order-preserving. Alternatively, we can view a natural extension of $(Z, <)$ as an enumeration of Z that respects the partial order $<$.

We begin with an example of a very different nature from that of csg models. Here, we start with a fixed causal set $P = (Z, <)$, and give a prescription for generating a random natural extension of P. Our process will have the property of order-invariance: if we stop after some fixed number of steps, then, conditioned on the structure of the causal set (i.e., the elements included and the comparabilities among them), every possible order of generation is equally likely.

Consider the poset $P = (Z, <)$ in Figure 2.

Here $Z = \{a_1, a_2 \ldots\}$, and $a_j > a_i$ if $j > i + 1$. A bijection $\lambda : \mathbb{N} \to Z$ is then a natural extension if $X_k = \{\lambda(1), \ldots, \lambda(k)\}$ is a down-set in P for each k. We seek a probability measure μ on the set $L(P)$ of natural extensions of P which is "uniform" (in a suitable sense).

For this example, the σ-field of events is generated by events of the form

$$E(a_{i_1} a_{i_2} \cdots a_{i_k}) = \{\lambda : \lambda(j) = a_{i_j} \text{ for } j = 1, \ldots, k\},$$

the set of natural extensions with "initial segment" $a_{i_1} a_{i_2} \cdots a_{i_k}$, for $k \in \mathbb{N}$ and the i_j distinct elements of $\mathbb{N}$. We call $a_{i_1} a_{i_2} \cdots a_{i_k}$ an *ordered stem* if $\{a_{i_1}, a_{i_2}, \ldots, a_{i_j}\}$ is a down-set in P, for $j = 1, \ldots, k$: i.e., if there is a natural extension of P with this initial segment. Our order-invariance condition is that, whenever $a_{i_1} a_{i_2} \cdots a_{i_k}$ and $a_{\ell_1} a_{\ell_2} \cdots a_{\ell_k}$ are ordered stems such that $\{a_{i_1}, a_{i_2}, \ldots, a_{i_k}\} = \{a_{\ell_1}, a_{\ell_2}, \ldots, a_{\ell_k}\}$, we have

$$\mu(E(a_{i_1}a_{i_2}\cdots a_{i_k})) = \mu(E(a_{\ell_1}a_{\ell_2}\cdots a_{\ell_k})). \tag{1}$$

We describe the measure μ via a random process for generating $\lambda(1), \lambda(2), \ldots$ sequentially. Given the set $X_k = \{\lambda(1), \ldots, \lambda(k)\}$, the element $\lambda(k+1)$ has to be one of the minimal elements of $P \setminus X_k$, and there are at most two of these. Our process is defined by the following rules: if there is only one minimal element a_k of $P \setminus X_k$, then we take $\lambda(k+1) = a_k$ with probability 1; if there are two minimal elements a_{k+1} and a_{k+2} of $P \setminus X_k$, then we set $\lambda(k+1) = a_{k+1}$ with probability ϕ and $\lambda(k+1) = a_{k+2}$ with probability $1-\phi$, where $\phi^2 = 1-\phi$.

It is easily seen that $\mu(E(a_2a_1)) = 1-\phi$: we choose $\lambda(1) = a_2$ with probability $1-\phi$, and then a_1 is the only minimal element of $P \setminus \{a_2\}$, so we choose $\lambda(2) = a_1$ with probability 1. Also, $\mu(E(a_1a_2)) = \phi^2$, since we choose $\lambda(1) = a_1$ with probability ϕ, and then $\lambda(2) = a_2$ with probability ϕ. Hence the condition of order-invariance forces the choice of ϕ to satisfy $\phi^2 = 1-\phi$. One can extend the argument above to show that the random process defined above satisfies all the identities of the form (1), so that the process is order-invariant.

We have described a process generating a random natural extension of a fixed P, but we can also think of it as a growth process, growing a causal set by adding one new maximal element at each step: we always get the same infinite causal set P, but the order in which the elements are generated is random.

There is an alternative way to obtain the measure μ in our example. Consider the sets $Z_n = \{a_1, \ldots, a_n\}$, the finite posets $P_n = P_{Z_n}$ induced by P on each Z_n, and the uniform measures ν^{P_n} on their sets of linear extensions, for each n. It can be shown that

$$\nu^{P_n}(E(a_{i_1}a_{i_2}\cdots a_{i_k})) \to \mu(E(a_{i_1}a_{i_2}\cdots a_{i_k})),$$

as $n \to \infty$, for each ordered initial segment $a_{i_1}a_{i_2}\cdots a_{i_k}$.

For another example, we consider growth processes where the causal set generated is a.s. an antichain. If we require our causal sets to be labelled, then a growth process which a.s. generates an antichain is nothing more than a sequence of random variables: the labels of the elements, in the order they are introduced. Order-invariance in this case requires that, if we condition on the set of the first k labels, for any k, then each of the $k!$ orderings of these labels is equally likely. That is, the sequence of labels must be *exchangeable*.

One way to generate a sequence of exchangeable random labels is to take any probability distribution τ on any set X of potential labels, and let the labels be an iid sequence of random elements of X with probability measure τ. Our labels are to be a.s. distinct, so the probability measure τ must be atomless. By the Hewitt-Savage Theorem [35], every sequence of exchangeable random variables is a mixture of sequences of the type above (i.e., there is a probability measure ρ on some space of probability measures on a set X: one measure τ is chosen according to ρ, and then an iid sequence of random elements of X is generated according to τ).

A similar mechanism allows us to incorporate csg models into our labelled context. At each step, as a new element arrives, we generate the down-set below it as in the previous section, and we equip it with a label chosen uniformly, and independently of everything else, from $X = [0, 1]$, with its usual Borel σ-field and Lebesgue measure. We make this, fairly arbitrary, choice of $X = [0, 1]$ as a set of potential labels for all our processes.

We now give the general definition of order-invariance. We specify the outcome of a growth process by giving an infinite string of (labels of) elements, $x_1 x_2 \cdots$ ($x_i \in [0, 1]$) in the order of their generation, together with a suborder $<^{\mathbb{N}}$ of the index set $\mathbb{N}$ with its standard order: $i <^{\mathbb{N}} j$ if and only if $x_i < x_j$ in the causal set $P = (X, <)$ generated. Let Ω be the set of pairs $\omega = (x_1 x_2 \cdots, <^{\mathbb{N}})$, the outcome space. Let π_k be the projection map on Ω given by $\pi_k(x_1 x_2 \cdots, <^{\mathbb{N}}) = (x_1 \cdots x_k, <^{\mathbb{N}}_{[k]})$, where $<^{\mathbb{N}}_{[k]}$ is the restriction of the order $<^{\mathbb{N}}$ to $[k]$. Then we can think of $\pi_k(\omega)$ as the state of the process at step k, corresponding to outcome ω.

Our basic events are subsets of Ω of the form

$$E(B_1, \cdots, B_k, <^{[k]}) = \pi_k^{-1}(B_1, \cdots, B_k, <^{[k]}),$$

where each B_i is a Borel set and $<^{[k]}$ is a suborder of the natural order on $[k]$. We let $\mathscr{F}_k$ be the σ-field generated by the sets $E(B_1, \cdots, B_k, <^{[k]})$, and $\mathscr{F}$ be the σ-field generated by $\bigcup_{k=1}^{\infty} \mathscr{F}_k$. A growth process then gives rise to a probability measure μ on $(\Omega, \mathscr{F})$, which we call a *causal set measure*. A causal set measure μ is *order-invariant* if, for any Borel sets $B_1, \ldots, B_k$, any suborder $<^{[k]}$ of $[k]$, and any linear extension λ of $<^{[k]}$, we have

$$\mu(E(B_1, \cdots, B_k, <^{[k]})) = \mu(E(B_{\lambda(1)}, \cdots, B_{\lambda(k)}, \lambda[<^{[k]}])),$$

where the linear order $\lambda[<^{[k]}]$ on $\{1, \ldots, k\}$ is given by $i(\lambda[<^{[k]}])j$ if and only if $\lambda(i) <^{[k]} \lambda(j)$. This definition incorporates the notion of general covariance from the previous section, as well as that of an order-invariant measure on the space of natural extensions of a fixed order, as in our earlier example.

Ideally we would like to classify order-invariant measures, in a manner similar to Theorem 3 of Rideout and Sorkin. In reality, this is too ambitious a goal, but some partial steps in this direction are taken in [20]. It is noted in that paper that the space of order-invariant measures is a convex subset of the set of all probability measures on $(\Omega, \mathscr{F})$. This means that an arbitrary mixture of order-invariant measures is also order-invariant, and suggests a focus on the *extremal* order-invariant measures: those that cannot be written as a convex combination of two others.

An order-invariant measure that almost surely produces one fixed causal set $P = (Z, <)$, as in our first example, will be called an order-invariant measure on P. The process in that example is the only order-invariant measure on that poset P, and it is extremal. This is however not true in general: there are causal sets with no order-invariant measures on them, and others with infinitely many extremal order-invariant measures.

An example of a causal set admitting no order-invariant measures is an infinite labelled antichain; such an order-invariant measure would give every element of the antichain the same probability of appearing as the bottom element in a random natural extension, and this is not possible. On the other hand, from the Hewitt-Savage Theorem, extremal order-invariant measures a.s. giving rise to an antichain are effectively the same as iid sequences of random elements of $[0, 1]$.

The main result of [20] shows that all extremal order-invariant measures on $(\Omega, \mathcal{F})$ are, in a sense, a combination of extremal order-invariant measures of these two types.

Theorem 5. *Let μ be an extremal order-invariant measure. Then there is a poset $Q = (Z, <)$, either a causal set or a finite poset, with a marked set M of maximal elements such that, if Q' is obtained from Q by replacing each element z of M with a countably infinite antichain A_z, then the poset Π generated by μ is a.s. equal to Q', except for the labels on the antichains A_z.*

Probably the most interesting special case is when the set M of marked maximal elements is empty, so that the extremal measure μ is an order-invariant measure on the fixed (labelled) causal set Q. However, not every order-invariant measure on a fixed causal set is extremal, so the theorem falls short of characterising extremal order-invariant measures. The other extreme case is when Q consists of a single marked element z, and so Q' consists of the single antichain A_z (corresponding to an exchangeable sequence of random labels $[0, 1]$ from an atomless probability distribution on $[0, 1]$).

The concept of order-invariance on fixed causal sets has occurred elsewhere in the literature, often in different guises. Brightwell [16] studied random linear extensions of locally finite posets. His main theorem gives that, if, for some fixed t, a causal set P is such that every element is incomparable with at most t others, then there is a unique order-invariant measure on P. The case when P is the two-dimensional grid $(\mathbb{N} \times \mathbb{N}, <)$ is linked to representation theory and harmonic functions on the Young lattice (the lattice of down-sets of the grid) – see Kerov [38], for instance. The topic of order-invariant measures on fixed causal sets is considered in detail in [21], where there are many more examples and results.

The family of natural extensions of a fixed causal set P can also be viewed as the set of configurations of a (1-dimensional) spin system, and order-invariant measures can then be interpreted as Gibbs measures, so that some of the general results discussed in, e.g., Georgii [31] apply. Specifically, we now give an alternative characterisation of order-invariance.

For $<^{\mathbb{N}}$ a suborder of $\mathbb{N}$, $k \in \mathbb{N}$, and λ a linear extension of $<^{\mathbb{N}}_{[k]}$, let λ^+ be given by

$$\lambda^+(i) = \begin{cases} \lambda(i) & i \le k \\ i & i > k. \end{cases}$$

So λ^+ is the natural extension of $<^{\mathbb{N}}$ obtained from the natural order on $\mathbb{N}$ by permuting the first k natural numbers according to λ. Now, for $\omega = (x_1 x_2 \cdots, <^{\mathbb{N}}) \in \Omega$, $k \in \mathbb{N}$, and λ a linear extension of $<^{\mathbb{N}}_{[k]}$, set $\lambda^+[\omega] = (x_{\lambda^+(1)} x_{\lambda^+(2)} \cdots, \lambda^+[<^{\mathbb{N}}])$; this is the element of Ω, representing the same labelled poset, obtained by permuting the first k elements of the ordered stem according to λ. Now, for $E \in \mathscr{F}$, let $\nu^k(E)(\omega)$ be the proportion of linear extensions λ of $<^{\mathbb{N}}_{[k]}$ such that $\lambda^+[\omega] \in E$.

For any $\omega \in \Omega$ and any k, the function $\nu^k(\cdot)(\omega)$ gives a probability measure on $\mathscr{F}$, namely the uniform measure on $\lambda^+[\omega]$, where λ runs over linear extensions of $<^{\mathbb{N}}_{[k]}$. Now we have our alternative characterisation of order-invariance.

Theorem 6. *Let μ be a causal set measure. Then μ is order-invariant if and only if*

$$\mu(E) = \mathbb{E}_\mu \, \nu^k(E)$$

for every $E \in \mathscr{F}$ and every $k \in \mathbb{N}$.

This is an analogue of the DLR equations from statistical physics about conditional probabilities. It corresponds to specifying a boundary condition outside a finite volume – here this means that we condition on all the information about ω except the order in which the first k elements are generated. We then realise the conditional Gibbs measure, which in our setting is $\nu^k(\cdot)(\omega)$.

We now describe a sequence $(\mathscr{G}_k)_{k\in\mathbb{N}}$ of σ-fields, which corresponds to the *external σ-fields* in interacting particle systems. For $k \in \mathbb{N}$, $\mathscr{G}_k$ is, roughly speaking, the family of sets in $\mathscr{F}$ that are invariant under all permutations of the first k elements producing another natural extension. We also define the *tail σ-field* $\mathscr{G} = \bigcap_{k=1}^\infty \mathscr{G}_k$.

The following result, again from [20], establishes that the functions $\nu^k(\cdot)(\cdot)$ are a family of measure kernels, analogous to Gibbsian specifications in statistical mechanics.

Theorem 7. *Let μ be an order-invariant measure. For any event $E \in \mathscr{F}$, and any $k \in \mathbb{N}$, we have*

$$\mu(E \mid \mathscr{G}_k) = \nu^k(E),$$

almost surely.

We are now in a position to describe some conditions on an order-invariant measure μ that are equivalent to extremality. In general, the limit of the measures ν^k is not necessarily equal to μ, but we do have the following.

Theorem 8. *Let μ be any order-invariant measure, and let E be any event in $\mathscr{F}$. Then the sequence $\nu^k(E)(\omega)$ converges to a $\mathscr{G}$-measurable limit $\nu(E)(\omega)$, μ-a.s. Moreover, $\nu(E) = \mu(E \mid \mathscr{G})$, μ-a.s., and $\mathbb{E}_\mu \, \nu(E) = \mu(E)$.*

We say that an order-invariant measure μ is *essential* if, for every $E \in \mathscr{F}$, $\nu^k(E) \to \mu(E)$ a.s. In other words, μ is essential if, for every E, the limit $\nu(E)$

in Theorem 8 is a.s. equal to $\mu(E)$, or equivalently $\mu(E \mid \mathscr{G}) = \mu(E)$ a.s. We say that μ has *trivial tails* if $\mu(H)$ is equal to 0 or 1 for every $H \in \mathscr{G}$. It is shown in [20] that, for any order-invariant measure μ, the following are equivalent: (i) μ is extremal, (ii) μ is essential, (iii) μ has trivial tails.

We finish this section by returning to the application to causal set theory. We saw in the previous section that there is no csg model whose samples resemble those of a Poisson process in $\mathbb{M}^4$, or in any other manifold. Now that we have relaxed the conditions on our growth processes, can we find an order-invariant causal set process that produces a random causal set whose distribution is that of a Poisson process in (to be specific) the future light cone of a point in $\mathbb{M}^4$? This question was raised in [21], and it remains an open problem. We discuss below the presumably simpler question with $\mathbb{M}^2$ (or, equivalently, $\mathbb{R}^2_+$ with the coordinate order) replacing $\mathbb{M}^4$, but the same principles apply to $\mathbb{M}^4$.

Suppose we take a Poisson process X in the positive quadrant $\mathbb{R}^2_+$, and consider the induced poset $P = (X, <)$: if we can show that there is a.s. an invariant measure on $(X, <)$, then we can represent the overall measure as a mixture of these measures. We now take a sequence of larger and larger squares $S_N = [0, N] \times [0, N]$, and consider the finite partial orders $P_N = P_{X \cap S_N}$, the restrictions of P to the points of the Poisson process in the squares S_N. A natural candidate for an order-invariant measure μ on $(X, <)$ is the limit of the uniform measures ν^{P_N}, as earlier. Now we focus on one minimal element x, say the one closest to the origin in Euclidean distance, and consider $q_N(x) = \nu^{P_N}(E(x))$: the probability, in a uniform random linear extension of P_N, that x is the bottom element. If the limit of measures exists, then the sequence $(q_N(x))$ of probabilities will have to tend to a non-zero limit $q(x)$; indeed, the sum of the $q(x)$ over all minimal elements x will have to be equal to 1. On the other hand, if $q(x) = \lim_{N\to\infty} q_N(x)$ exists and is non-zero, then presumably the same is true of other similar limits, and we should be able to use the values of these limits to define an order-invariant measure on $(X, <)$ (to begin, at the first step of the growth process, we would select each minimal element x with probability $q(x)$). So the question of whether an order-invariant measure on $(X, <)$ exists would seem to boil down to the question of whether the sequence $(q_N(x))$ a.s. tends to a non-zero limit. Unfortunately, our ability to estimate numbers of linear extensions for the partial orders P_N is far too limited to enable a direct attack on this question.

5 Uniform measures

A "path integral" or "sum over histories" approach to quantum gravity would proceed by specifying a space of "potential" histories, and assigning an amplitude (or quantum measure) to sets of those histories. The idea is that most of the "weight" of the measure is on histories similar to the observed universe. In other words, similarity to a Poisson process in a 4-dimensional Lorentzian manifold would be an "emergent property" from some simple and natural random process. A csg model constitutes one example of such an approach, where the probability measure is classical. Another approach is to examine what happens when we take very simple probabilistic models, such as the uniform model on (some class of) partial orders of fixed size.

If we fix the number of elements at some (large) value n, then a uniformly random partial order on n elements is almost surely a 3-layer poset. A *k-layer poset* is one where the elements are partitioned into layers $A_1, A_2, \ldots, A_k$, such that (i) whenever $x \in A_i$ is below $y \in A_j$ in the poset, we have $i < j$, and (ii) all elements of A_i are below all elements of A_{i+2}, for each $i = 1, \ldots, k-2$. The Kleitman-Rothschild Theorem [40] states that almost all n-element partial orders are 3-layer posets in which $|A_1|$ and $|A_3|$ are about $n/4$, and $|A_2|$ is about $n/2$. Notice that, given a partition into sets of the appropriate sizes, and any subset S of $(A_1 \times A_2) \cup (A_2 \times A_3)$, we may form a 3-layer poset by putting $x < y$ only for those pairs in $S \cup (A_1 \times A_3)$. Thus the number of 3-layer partial orders is approximately $2^{n^2/4}$.

Brightwell, Prömel, and Steger [22] gave the following sharper estimate for the number C_n of partial orders on n labelled elements: for some absolute constant $C > 1$,

$$C_n = (1 + O(C^{-n})) \sum_{s=0}^{n} \binom{n}{s} 2^{(s+1)(n-s)} = (1 + O(1/n))\beta 2^{(n+1)^2/4} \binom{n}{\lfloor n/2 \rfloor}.$$

In the sum, the s-term is a very tight estimate of the number of 3-layer posets with s elements in the middle layer, and the result implies that the proportion of n-element posets that are not 3-layer posets is exponentially small. In the second expression, the term β takes one constant value if n is even and a very slightly different one if n is odd.

A typical 3-layer poset has no "spatial structure", and does not resemble a Poisson process in any manifold, so it is unlikely that the uniform model has much direct relevance to the causal set programme. What might however be of interest is to restrict attention to certain classes of partial orders, and/or to assign different "activities" to different n-element partial orders, as is common in statistical physics models.

One simple step is to constrain the number of comparable pairs in our random partial orders. Prömel, Steger, and Taraz produced a series of two papers [45, 46]: in [45], they demonstrated that, whatever constraint is imposed, the typical partial order arising is a k-layer partial order for some k. In [46], they studied the optimisation problem of finding the optimal structure given the target proportion of relations. It follows that, if we give each partial order P a weight proportional to $e^{-cr(P)}$, where c is a positive constant and $r(P)$ is the number of comparable pairs, then the resulting partial order will still almost surely be layered.

One way to assign weights in such a way that the majority of the weight is not supported on the class of layered posets is to give most posets weight zero, i.e., to restrict attention to some much smaller class of partial orders. Specifically, suppose we concentrate on the class of partial orders that can be embedded in $\mathbb{M}^d$, i.e., the $(d-1)$-sphere orders, and then consider the uniform measure on this class. What does a uniformly random n-element $(d-1)$-sphere order look like? By definition, such a partial order would embed into $\mathbb{M}^d$, but there is no a priori reason to suppose that it can be embedded faithfully into any finite region. However, one reason for optimism is that such a result is true in the case $d = 2$.

As we saw in Section 2, the partial orders that can be embedded in $\mathbb{M}^2$ are exactly the 2-dimensional orders, and the model of random partial orders given by a Poisson process of intensity n in $\mathbb{M}^2$ is equivalent to that given by taking two uniformly random linear orders on n labelled points and intersecting them to form a random 2-dimensional partial order. The difference between choosing two linear orders uniformly and choosing a uniform random 2-dimensional partial order is that some 2-dimensional partial orders have more than one representation as the intersection of two linear orders. The extreme case is the antichain, which has $n!$ representations as the intersection of one linear order and its reverse. However, it turns out that most 2-dimensional partial orders, in either model, have rather few representations. Given a pair of linear orders on a fixed n-element ground-set, we say that a pair of elements is *swappable* if they appear consecutively in both linear orders, and they are incomparable in the intersection. A swappable pair can clearly be swapped in each order, and the intersection will be unaffected. Similarly, if a pair of linear orders has k disjoint swappable pairs, then there are (at least) $2^k - 1$ other pairs of linear orders whose intersection is the same as the original pair. If we choose the two linear orders uniformly at random, the expected number of swappable pairs is asymptotic to 1 as $n \to \infty$, and it is easy to see that the limiting distribution is Poisson with mean 1. A result of Winkler [55] states that, up to swappable pairs, most 2-dimensional partial orders have a unique representation. This implies that the two models of random partial orders are *contiguous*, i.e., any event with positive probability in one model also has positive probability in the other.

The situation is even nicer for *unlabelled* 2-dimensional partial orders. The following result was proved by El-Zahar and Sauer [29] and stated in this form by Winkler [55]. We consider two different probability measures on the space D_n of n-element unlabelled 2-dimensional partial orders; one is obtained by intersecting two random linear orders on $\{1, \ldots, n\}$, and the other is the uniform probability measure on D_n. In the unlabelled case, in either model, almost every element of D_n has a unique representation as an intersection of two linear orders, and, if Φ is any isomorphism-invariant statement about partial orders with a limiting probability in either model, a limiting probability exists in the other case as well and the two limits are equal.

The implications of this result for causal sets were explored by Brightwell, Henson, and Surya [25]. The result of El-Zahar and Sauer implies that a 2-dimensional partial order chosen uniformly at random is (effectively) distributed as the partial order induced on a Poisson process in $[0, 1]^2$ (or, equivalently, an interval in $\mathbb{M}^2$), and so such a random partial order can be embedded faithfully in an interval in $\mathbb{M}^2$. This indicates how an approximation of a continuum structure can arise from a probability measure – the uniform measure on D_n – whose definition does not refer to any continuous structure.

For each $d > 2$, it is unknown whether the model of a random d-dimensional partial order obtained by intersecting d uniformly random linear orders is contiguous with the model obtained by choosing a d-dimensional partial order uniformly at random. It is also unknown whether the model of taking a uniformly random order embeddable in $\mathbb{M}^d$ is contiguous with a Poisson process in any subset of $\mathbb{M}^d$.

Just to indicate some of the issues arising, let R be any region of $\mathbb{M}^2$, and consider a partial order chosen uniformly at random from those embeddable in R. Whatever R we choose, these are exactly the 2-dimensional partial orders; we have seen that a uniformly chosen 2-dimensional partial order can be "faithfully" embedded in $[0, 1]^d$, and therefore in any rectangular region of the space, but not in a region of any other shape.

Acknowledgements Malwina Luczak's research was supported by an EPSRC Leadership Fellowship, grant reference EP/J004022/2.

References

1. M. Albert, A. Frieze, Random graph orders. Order **6**, 19–30 (1989)
2. D. Aldous, P. Diaconis, Longest increasing subsequences: from patience sorting to the Baik-Deift-Johansson Theorem. Bull. Am. Math. Soc. **36**, 413–432 (1999)
3. N. Alon, E.R. Scheinerman, Degrees of freedom versus dimension for containment orders. Order **5**, 11–16 (1988)
4. N. Alon, B. Bollobás, G. Brightwell, S. Janson, Linear extensions of a random partial order. Ann. Appl. Probab. **4**, 108–123 (1994)
5. E. Bachmat, Discrete space-time and its applications, in *Random Matrices, Integrable Systems and Applications: A Conference in Honor of Percy Deift's 60th Birthday*, ed. by J. Baik, L-C. Li, T. Kriecherbauer, K. McLaughlin, C. Tomei. Contemporary Mathematics Book Series (American Mathematical Society, Providence, 2008)
6. J. Baik, P. Deift, K. Johansson, On the distribution of the length of the longest increasing subsequence of random permutations. J. Am. Math. Soc. **12**, 1119–1178 (1999)
7. A. Barak, P. Erdős, On the maximal number of strongly independent vertices in a random acyclic directed graph. SIAM. J. Algebraic Discret. Methods. **5**, 508–514 (1984)
8. B. Bollobás, G. Brightwell, Box-spaces and random partial orders. Trans. Am. Math. Soc. **324**, 59–72 (1991)
9. B. Bollobás, G. Brightwell, The width of random graph orders. Math. Sci. **20**, 69–90 (1995)
10. B. Bollobás, G. Brightwell, The dimension of random graph orders, in *The Mathematics of Paul Erdős II*, ed. by R.L. Graham, J. Nešetřil (Springer, New York, 1996), pp. 51–69
11. B. Bollobás, G. Brightwell, The structure of random graph orders. SIAM J. Discret. Math. **10**, 318–335 (1997)
12. B. Bollobás, P. Winkler, The longest chain among random points in Euclidean space. Proc. Am. Math. Soc. **103**, 347–353 (1988)
13. L. Bombelli, Statistical Lorentzian geometry and the closeness of Lorentzian manifolds. J. Math. Phys. **41**, 6944–6958 (2000)
14. L. Bombelli, D.A. Meyer, The origin of Lorentzian geometry. Phys. Lett. A **141**, 226–228 (1989)
15. L. Bombelli, J. Lee, D. Meyer, R.D. Sorkin, Spacetime as a causal set. Phys. Rev. Lett. **59**, 521–524 (1987)
16. G. Brightwell, Linear extensions of infinite posets. Discret. Math. **70**, 113–136 (1988)
17. G. Brightwell, Models of random partial orders, in *Surveys in Combinatorics 1993*, ed. by K. Walker. London Mathematical Society Lecture Notes Series, vol. 187 (Cambridge University Press, Cambridge, 1993), pp. 53–83

18. G. Brightwell, N. Georgiou, Continuum limits for classical sequential growth models. Random Struct. Algorithm **36**, 218–250 (2010)
19. G. Brightwell, R. Gregory, The structure of random discrete spacetime. Phys. Rev. Lett. **66**, 260–263 (1991)
20. G. Brightwell, M. Luczak, Order-invariant measures on causal sets. Ann. Appl. Probab. **21**, 1493–1536 (2011)
21. G. Brightwell, M. Luczak, Order-invariant measures on fixed causal sets. Comb. Prob. Comput. **21**, 330–357 (2012)
22. G. Brightwell, H.-J. Prömel, A. Steger, The average number of linear extensions of a partial order. J. Comb. Theory (A) **73**, 193–206 (1996)
23. G. Brightwell, P. Winkler, Sphere orders. Order **6**, 235–240 (1989)
24. G. Brightwell, H.F. Dowker, R.S. García, J. Henson, R.D. Sorkin, "Observables" in causal set cosmology. Phys. Rev. D **67**, 084031 (2003)
25. G. Brightwell, J. Henson, S. Surya, A 2D model of causal set quantum gravity: the emergence of the continuum. Classical Quantum Gravity **25**, 105025 (2008)
26. F. Dowker, Introduction to causal sets and their phenomenology. Gen. Relativ. Gravit. **45**, 1651–1667 (2013)
27. F. Dowker, S. Surya, Observables in extended percolation models of causal set cosmology. Classical Quantum Gravity **23**, 1381–1390 (2006)
28. F. Dowker, J. Henson, R.D. Sorkin, Quantum gravity phenomenology, Lorentz invariance and discreteness. Mod. Phys. Lett. **A19**, 1829–1840 (2004)
29. M. El-Zahar, N. Sauer, Asymptotic enumeration of two-dimensional posets. Order **5**, 239–244 (1988)
30. S. Felsner, P.C. Fishburn, W.T. Trotter, Finite three dimensional partial orders which are not sphere orders. Discret. Math. **201**, 101–132 (1999)
31. H.-O. Georgii, *Gibbs Measures and Phase Transitions*. de Gruyter Studies in Mathematics, vol. 9 (de Gruyter, Berlin, 1988)
32. N. Georgiou, The random binary growth model. Random Struct. Algorithm. **27**, 520–552 (2005)
33. S.W. Hawking, A.R. King, P.J. McCarthy, A new topology for curved spacetime which incorporates the causal, differential and conformal structures. J. Math. Phys. **17**, 174–181 (1976)
34. J. Henson, The causal set approach to quantum gravity, in *Approaches to Quantum Gravity: Towards a New Understanding of Space and Time*, ed. by D. Oriti (Cambridge University Press, Cambridge, 2009), pp. 26–43
35. E. Hewitt, L.J. Savage, Symmetric measures on Cartesian products. Trans. Am. Math. Soc. **80**, 470–501 (1955)
36. J. Hladký, A. Máthé, V. Patel, O. Pikhurko, Poset limits can be totally ordered. Trans. Am. Math. Soc. **367**, 4319–4337 (2015)
37. S. Janson, Poset limits and exchangeable random posets. Combinatorica **31**, 529–563 (2011)
38. S. Kerov, The boundary of Young lattice and random Young tableaux, in *Formal Power Series and Algebraic Combinatorics*, ed. by L.J. Billera, C. Greene, R. Simion, R.P. Stanley. DIMACS Series, Discrete Mathematics and Theoretical Computer Science, vol. 24 (American Mathematical Society, Providence, 1996), pp. 133–158
39. J.H. Kim, B. Pittel, On tail distribution of interpost distance. J. Comb. Theor. B **80**, 49–56 (2000)
40. D. Kleitman, B. Rothschild, Asymptotic enumeration of partial orders on a finite set. Trans. Am. Math. Soc. **205**, 205–220 (1975)
41. D. Malament, The class of continuous timelike curves determines the topology of spacetime. J. Math. Phys. **18**, 1399–1404 (1977)
42. D.A. Meyer, Spherical containment and the Minkowski dimension of partial orders. Order **10**, 227–237 (1993)
43. J. Myrheim, Statistical geometry. CERN preprint TH-2538 (1978)

44. B. Pittel, R. Tungol, A phase transition phenomenon in a random directed acyclic graph. Random Struct. Algorithm **18**, 164–184 (2001)
45. H.J. Prömel, A. Steger, A. Taraz, Phase transitions in the evolution of partial orders. J. Comb. Theor. Ser. A **94**, 230–275 (2001)
46. H.J. Prömel, A. Steger, A. Taraz, Counting partial orders with a fixed number of comparable pairs. Comb. Probab. Comput. **10**, 159–177 (2001)
47. D. Rideout, R.D. Sorkin, A classical sequential growth dynamics for causal sets. Phys. Rev. D **61**, 024002 (2000)
48. D. Rideout, R.D. Sorkin, Evidence for a continuum limit in causal set dynamics. Phys. Rev. D **63**, 104011 (2001)
49. D. Rideout, P. Wallden, Spacelike distance from discrete causal order. Classical Quantum Gravity **26**, 155013 (2009)
50. K. Simon, D. Crippa, F. Collenberg, On the distribution of the transitive closure in a random acyclic digraph, in *Algorithms ESA'93 (Bad Honnef 1993)*. Lecture Notes in Computer Science, vol. 726 (1993), pp. 345–356
51. R.D. Sorkin, Causal sets: discrete gravity (notes for the Valdivia summer school), in *Proceedings of the Valdivia Summer School*, Valdivia, Chile, January 2002 ed. by A. Gomberoff, D. Marolf. Lectures on Quantum Gravity (Plenum, New York, 2005)
52. S. Surya, Directions in causal set quantum gravity, in *Recent Research in Quantum Gravity*, ed. by A. Dasgupta (Nova Science Publishers, New York, 2012)
53. M. Varadarajan, D. Rideout, A general solution for classical sequential growth dynamics of causal sets. Phys. Rev. D **73**, 104021 (2006)
54. P. Winkler, Random orders. Order **1**, 317–335 (1985)
55. P. Winkler, Random orders of dimension 2. Order **7**, 329–339 (1991)

Part II
Additive and analytic combinatorics

Lectures on approximate groups and Hilbert's 5th problem

Emmanuel Breuillard

Abstract This paper gathers four lectures, based on a mini-course at IMA in 2014, whose aim was to discuss the structure of approximate subgroups of an arbitrary group, following the works of Hrushovski and of Green, Tao and the author. Along the way we discuss the proof of the Gleason–Yamabe theorem on Hilbert's 5th problem about the structure of locally compact groups and explain its relevance to approximate groups. We also present several applications in particular to uniform diameter bounds for finite groups and to the determination of scaling limits of vertex transitive graphs with large diameter.

Keywords Approximate groups • Additive combinatorics • Polynomial growth • Volume doubling • Diameter bounds • Scaling limits

Mathematics Subject Classification (2010): 11B30, 22D05

1 Additive combinatorics on groups

1.1 Introduction

A key idea of Geometric Group Theory, as initiated by Gromov and others in the seventies and eighties, is to study infinite groups as geometric objects via their Cayley graphs. Endowed with the graph distance, the Cayley graphs are naturally metric spaces. Typically we look at a group Γ generated by a finite symmetric set S and define the associated Cayley graph and word metric:

$$d_S(g, h) := \inf\{n \geq 0, g^{-1}h \in S^n\},$$

where $S^n := S \cdot \ldots \cdot S$ is the product set of n copies of S. In particular the ball of radius n around the identity $B_S(e, n)$ is nothing else but S^n. Usually people are interested in

E. Breuillard (✉)
Laboratoire de Mathématiques, Bâtiment 425, Université Paris Sud 11, 91405 Orsay, France
e-mail: emmanuel.breuillard@math.u-psud.fr; emmanuel.breuillard@gmail.com

A. Beveridge et al. (eds.), *Recent Trends in Combinatorics*, The IMA Volumes in Mathematics and its Applications 159, DOI 10.1007/978-3-319-24298-9_16

properties that are independent of the choice of S, and look at very large values of n. So S is of bounded size and n is large.

Additive combinatorics on groups, on the other hand, is concerned with product sets $A^n := A \cdot \ldots \cdot A$, where n is of bounded size, but A is a very large finite subset of an ambient group G. Traditionally, combinatorialists and number theorists have been interested in studying large finite subsets of integers that do not grow much under addition. This is an old and well-developed field of mathematics (much older than Geometric Group Theory) with applications to many number theoretical questions (such as unique factorization in rings of algebraic integers, arithmetic progressions in primes, the Goldbach's problem, Waring's problem, etc.).

Only fairly recently did people start looking at similar questions on non-abelian groups, only to find out that many of the tools developed in the commutative context could also be applied in a non-commutative setting.

1.2 Examples of results from additive combinatorics

To fix ideas, let us recall a sample of landmark results from additive combinatorics:

Theorem 1.1. (Sum–product theorem, Bourgain–Katz–Tao [8], Bourgain–Glibichuk–Konyagin [9]). *Given $\varepsilon > 0$, there is $\delta > 0$ such if p is a prime and $A \subset \mathbb{F}_p$ is a subset of the field $\mathbb{F}_p$ with p elements, then either $|A + A| > |A|^{1+\delta}$ or $|AA| > |A|^{1+\delta}$ or $|A| > p^{1-\varepsilon}$.*

In other words a subset of a prime field must grow either by addition or by multiplication unless it is almost all of the field. This can be viewed as a quantitative version of the fact that there are no non-trivial subrings in $\mathbb{F}_p$. Of course it is easy to construct large subsets A of $\mathbb{F}_p$ that do not grow much under addition or under multiplication: take a finite arithmetic progression or geometric progression; then $|AA| \leq 2|A|$, but $|A|$ can be of arbitrary cardinality inside $\mathbb{F}_p$.

By now pretty good quantitative estimates have been obtained on the exponent $\delta(\varepsilon)$, see, e.g., [36, 37, 66] and the references therein.

Theorem 1.2. (Freiman–Ruzsa–Chang theorem [23, 32, 67]). *Let $A \subset \mathbb{Z}$ be a finite subset such that $|A+A| \leq K|A|$. Then A is $C(K)$-commensurable to a d-dimensional arithmetic progression P with $d = O(\log K)$.*

Here we used the following terminology:

Definition 1.3. (K-commensurability). Two finite subsets A and B of an ambient group G are called K-commensurable if

$$|A \cap B| \geq \frac{1}{K} \max\{|A|, |B|\}.$$

and

Definition 1.4. (Multidimensional arithmetic progression). A d-dimensional arithmetic progression in an ambient group G is a finite subset of the form $P := \pi(B)$, where $\pi : \mathbb{Z}^d \to G$ is a group homomorphism and $B := \prod_{i=1}^d [-L_i, L_i]$ is a "box" with side lengths $L_i \in \mathbb{N}$.

Remarks:

a) It is easy to see that P as above satisfies $|PP| \leq 2^d|P|$ because $B+B$ is contained in at most 2^d translates of B.
b) The proof of the Freiman–Ruzsa–Chang theorem is based on harmonic analysis on $\mathbb{Z}/p\mathbb{Z}$ for some suitably chosen prime p. We refer the reader to [5, 74] and [41] for nice accounts with proofs. The constant $C(K)$ given by this theorem is typically of order $e^{O(K^{O(1)})}$. Tom Sanders recently gave much better, almost polynomial, bounds. See [70] for a recent survey. Conjecturally ("Polynomial Freiman–Ruzsa conjecture," see [42]) one ought to be able to take $C(K)$ to be of order $O(K^{O(1)})$.

In this theorem and its applications, the quality of the bounds (i.e., their dependence on K) is important.

Non-commutative analogues of the above results have started to draw the attention of a number of people in the past ten years. For example, the sum–product theorem can be reformulated as a Freiman type theorem in the affine group of the line (see [51]). The main goal of these lectures is to present a generalization of Freiman's theorem to arbitrary groups, following [54] and [16]. See [17] for another survey on this topic. Proofs were completely avoided in [17]. Our intent in these lectures is not to provide a complete treatment, but rather to give some ideas regarding the proofs, a rather complete outline and even the details of some steps, while remaining as informal as possible. All details can be found in the original paper [16] and full proofs are also given in Tao's recent book [73]. For yet another account, we recommend Van den Dries' Bourbaki report [29]. Also part of our focus here is also on the geometric applications, especially in the last lecture, where the main ones are given full proofs.

We refer the reader to [7, 33, 48, 49, 58, 81] for some older results about the general non-commutative case, mostly dealing with the small K regime. In the arbitrary K regime in the non-commutative context, early papers include [30, 50, 59, 71]. One of the first results in this direction is due to Razborov [64] and deals with finite subsets of the free group. Razborov proved that if F is a non-abelian free group and $A \subset F$ a finite subset, then

$$AAA| \geq |A|^2/(\log |A|)^c$$

for some positive c independent of A and F, unless A lies in a cyclic subgroup. The exponent 2 is best possible here, as is clear by considering, for example, $A := \{x\} \cup \{y^i; -N < i < N\}$ in $F_2 := \langle x, y \rangle$.

S. R. Safin recently improved this and showed:

Theorem 1.5. (Safin [68]). *There is $c \in (0,1)$ such that if F is a free group, then for every $n \geq 3$ and every finite subset A in F*

$$|A^n| \geq c^n |A|^{[\frac{n+1}{2}]}$$

unless A generates a cyclic group.

Safin's proof gives an explicit lower bound on c. One can take $c = \frac{1}{62}$. Again the same example shows that the exponent is sharp for each n. See [22] for further generalizations of the Razborov–Safin results to free products and HNN extensions.

Remark: When G is an abelian group, then $|AA| \leq K|A|$ implies that $|A^n| \leq O_{K,n}(1)|A|$ for every $n \geq 2$, according to the Plünnecke-Ruzsa inequalities (see [74]). This is still the case in the non-abelian context, but when starting with the assumption $|AAA| \leq K|A|$ (see §2.2 below). A set A consisting of a subgroup H and another element x with $H \cap xHx^{-1} = \{1\}$ yields an example with small $|AA|$ and large $|AAA|$. This justifies the consideration of three-fold product sets in non-commutative groups. See §2.2 below for more on this issue.

1.3 The Freiman inverse problem

The general problem ("Freiman inverse problem" as advertised in particular by T. Tao in his blog several years ago) is to describe the structure of large subsets A of an arbitrary ambient group G that satisfy $|AA| \leq K|A|$ for some fixed constant K.

The value $|AA|/|A|$ is usually called the *doubling constant* of A, equivalently we say that A has *doubling at most K*.

For example, we easily have:

Proposition 1.6. *Let G be an arbitrary group. Suppose $|AA| = |A|$. Then there is a finite subgroup H of G and $A = aH = Ha$ for every $a \in A$.*

Proof. Pick $a \in A$ and set $H := Aa^{-1}$. The set H contains the identity, so $A \subset AH$. But $|AH| = |AHa| = |AA| = |A|$, hence $A = AH$. Iterating $A = AH^n$ for every $n \geq 1$ and it follows that the semigroup $\cup_n H^n$ is finite and contained in $a^{-1}A$. But a finite semigroup inside a group is a finite group. So the subgroup generated by H satisfies $\langle H\rangle \subset a^{-1}A$. In particular $|\langle H\rangle| \leq |A| = |H|$. Hence $\langle H\rangle = H$. It follows that $A = Ha$ and that $H = AA^{-1}$. Finally $|H| = |A| = |AA| = |HaHa| = |HaH| = |a^{-1}HaH|$. But $a^{-1}HaH/H \simeq a^{-1}Ha/(a^{-1}Ha \cap H)$. Hence $a^{-1}Ha = H$ and we are done. □

Slightly trickier is the following:

Proposition 1.7. *Let G be an arbitrary group. Suppose $|AA| \leq 1.1|A|$. Then there is a finite subgroup H of G and $a \in A$ such that $aHa^{-1} = H$, $|H| \leq 1.2|A|$ and $A \subset aH$.*

Proof. Let 1_A be the indicator function of A and $\|f\|_1 = \sum_{g \in G} |f(g)|$ the ℓ^1-norm. We will prove that $H := AA^{-1}$ is stable under multiplication using the following observation: if $||1_A - 1_{gA}||_1 < 2|A|$, then $g \in H$. First we show that $H = A^{-1}A$ and that $|H| \leq 1.2|A|$.

Note that $\forall x, y \in G$, $||1_{xA} - 1_{yA}||_1 = |xA \Delta yA| = 2(|A| - |xA \cap yA|)$. But if $x, y \in A$, then $|xA \cup yA| = 2|A| - |xA \cap yA| \leq |A^2| \leq 1.1|A|$. Hence $\forall x, y \in A$, $|xA \cap yA| \geq 0.9|A|$ and

$$||1_A - 1_{x^{-1}yA}||_1 \leq 0.2|A|. \tag{1}$$

Since $|xA \cap yA| > 0$ we conclude that $xa = yb$ for some $a, b \in A$, i.e., $x^{-1}y \in AA^{-1}$. Hence $A^{-1}A = H = AA^{-1}$. In fact $xa = yb$ for at least $0.9|A|$ pairs (a, b) in $A \times A$. Therefore $|H| = |A^{-1}A| \leq |A|^2/0.9|A| \leq 1.2|A|$.

To see that $HH \subset H$ we write:

$$||1_A - 1_{z^{-1}wx^{-1}yA}||_1 \leq ||1_A - 1_{z^{-1}wA}||_1 + ||1_{z^{-1}wA} - 1_{z^{-1}wx^{-1}yA}||_1,$$

we get from (1)

$$||1_A - 1_{z^{-1}wx^{-1}yAA}||_1 \leq 0.4|A|$$

for every $x, y, z, w \in A$. Hence $HH \subset H$. Hence H is a subgroup and $A \subset Ha$ for all $a \in A$, so $H = A^{-1}A \subset a^{-1}Ha$ and a normalizes H. □

Remark 1.8. The above proof can be pushed with little effort to yield that if $|AA| < \frac{3}{2}|A|$, then $H = AA^{-1} = A^{-1}A$ is a finite subgroup of size $< \frac{3}{2}|A|$ which is normalized by A. This is a theorem of G. Freiman, see [34]. Clearly $3/2$ is sharp, take $A = \{0, 1\} \in \mathbb{Z}$.

Remark 1.9. An immediate consequence of the above lemma is the following general fact: if S is a symmetric generating subset of a finite group G with $|S| \geq \alpha|G|$ for some $\alpha \in (0, 1)$, then $S^n = G$ for some $n \leq O(\log(\frac{1}{\alpha}))$. Indeed $|S^{2^n}| > 1.1^n|S|$ unless $|S^{2^{n-1}}| \geq |G|/1.2 > |G|/2$, in which case $S^{2^n} = G$.

Remark 1.10. One may ask: for which values of $K \geq 1$ is it true that the condition $|AA| \leq K|A|$ implies that A is contained in boundedly many (in terms of K) cosets of a finite subgroup ? The answer is : this holds for all $K < 2$. This is due to Yayha Hamidoune. He showed that A is then contained in at most $\frac{1}{2-K}$ cosets of a subgroup of size at most $|A|$. See [47, Prop. 4.1]. Clearly it is sharp because the arithmetic progressions $A := \{0, 1, \ldots, N\} \subset \mathbb{Z}$ have $|A + A| \leq 2|A|$.

What if we increase the parameter K beyond the $K = 2$ threshold ? The next natural question to ask is: for which values of K is it true that the condition $|AA| \leq K|A|$ implies that A is contained in an extension HL, where H is a finite subgroup normalized by L a d-dimensional arithmetic progression ? Clearly such a set has doubling at most 2^d.

However this question is ill-posed, because there are sets of doubling say 2^4 which are not of this form: take a ball S^n of large radius n in the discrete Heisenberg group

$$H_3(\mathbb{Z}) := \left\{ \begin{pmatrix} 1 & x & z \\ 0 & 1 & y \\ 0 & 0 & 1 \end{pmatrix} ; x, y, z \in \mathbb{Z} \right\}$$

Here 4 is the homogeneous asymptotic dimension of the Heisenberg group, namely $|S^n| \simeq n^4$ up to multiplicative constants, for any given finite generating set S of $H_3(\mathbb{Z})$.

In 2008 Elon Lindenstrauss proposed [60] a general conjectural statement, which was subsequently proved by Ben Green and Terry Tao and the author in [16], and is the main subject of these notes:

Theorem 1.11. (Structure theorem, weak version). *Let G be a group, $K \geq 1$ a parameter and A a finite subset with $|AA| \leq K|A|$. Then there is a coset of a virtually nilpotent subgroup of G which intersects A in a subset of size $\geq |A|/C(K)$.*

Remarks.

a) Here $C(K)$ is a number depending on K only, and not on A nor G.
b) The proof also gives a uniform bound on the nilpotency class and the number of generators of the nilpotent subgroup (both in $O(\log K)$ with an absolute implied constant), although no bound can be expected on the index (e.g., take A a large finite simple group). More precisely we can take the virtually nilpotent group to be a finite-by-nilpotent group, where the nilpotent quotient N verifies that $\sum_i d(C^i(N)/C^{(i+1)}(N)) = O(\log K)$, where $(C^i(N))_i$ is the lower central series and $d(\cdot)$ the number of generators.
c) However the proof gives no explicit bound on $C(K)$. The proof is not effective.
d) Stated as above the theorem says nothing about the case when G is finite. And this is enough for a number of geometric applications. However the proof does. We actually show that the virtually nilpotent subgroup can be taken to be finite-by-nilpotent where the finite normal subgroup lies in XA for some X of size $|X| \leq c(K)$ for some constant $c(K)$ depending only on K. We will give a stronger form of the theorem below, which includes this refinement.
e) The theorem had been established earlier by Tao [72] in the special case when G is assumed solvable of bounded derived length.
f) In certain situations, most notably when G is a free group (by the result of Razborov mentioned above) or when $G = GL_n(k)$ for some field k, then pretty good explicit bounds can be found on $C(K)$. For example, in [15] it is shown that $C(K) = O(K^{O_n(1)})$ for $G = GL_n(\mathbb{C})$. See also [39, 63] for the positive characteristic situation.

Theorem 1.11 can be seen as a finitary version of Gromov's theorem on groups with polynomial growth. And indeed, as we now show, it easily implies Gromov's theorem.

Theorem 1.12. (Gromov's polynomial growth theorem). *Let $\Gamma := \langle S \rangle$ be a finitely generated group with polynomial growth (i.e., $|S^n| = O(n^D)$ for some $D > 0$). Then Γ is virtually nilpotent.*

Proof that Theorem 1.11 implies Theorem 1.12. Since $|S^n| = O(n^D)$, there must be arbitrarily large n's for which $|S^{4n}| \leq 5^D|S^n|$. Indeed, suppose not, then $|S^{4n}| > 5^D|S^n|$ for all n large enough, $n \geq n_0$ say, and hence $5^{Dk}|S^{n_0}| < |S^{4^k n_0}| = O(4^{Dk}n_0^D)$ a contradiction for large n.

Pick such an n. By Theorem 1.11 applied to $A := S^n$ and $K = 5^D$, there is a virtually nilpotent subgroup H_n such that S^n has a large intersection with a coset of H_n. In particular $|S^{2n} \cap H_n| \geq |S^n|/C(D) \geq |S^{4n}|/(5^D C(D))$. Theorem 1.12 follows by applying the following lemma with $k = 2n$.

Lemma 1.13. *In a group $\Gamma = \langle S \rangle$ if $|S^k \cap H| > |S^{2k}|/C$ for some subgroup H, some $C > 0$ and some $k \geq C$, then $[\Gamma : H] \leq C$.*

Proof. Look at the Schreier graph of Γ/H. It is connected, so either $S^kH = \Gamma$ or $|S^kH/H| > k$. Write $S^k = \cup_{i=1}^N S^k \cap g_iH$. Clearly $\cup_{i=1}^N (S^k \cap g_iH)(S^k \cap H) \subset S^{2k}$. Hence $N|(S^k \cap H)| \leq |S^{2k}| < C|S^k \cap H|$. And hence $N = |S^kH/H| < C \leq k$. Hence $S^kH = \Gamma$, and $[\Gamma : H] = N < C$. □

□

Remark 1.14. In fact the above proves slightly more than Gromov's theorem. It shows that given $C, D > 0$ there is $n_0 = n_0(C, D)$ such that if $|S^n| \leq Cn^D$ for some $n \geq n_0$, then Γ is virtually nilpotent. Note that this is stronger than Gromov's own finitary version of his result (discussed at the end of his original paper [44]). He showed (basically just by contradiction) that his theorem implies that there is $n_0 = n_0(C, D)$ such that if $|S^n| \leq Cn^D$ holds for all $n \leq n_0$, then Γ is virtually nilpotent. So the added value here is that we need only one scale, where the doubling property occurs.

We end this lecture by giving the following equivalent form of the main theorem:

Theorem 1.15. (Structure theorem, equivalent form). *Let G be a group, $K \geq 1$ a parameter and A a finite subset with $|AA| \leq K|A|$. Then there is a subset $X \subset G$ such that $|X| \leq C(K)$ and a virtually nilpotent subgroup $\Gamma_0 \leq G$ such that $A \subset X\Gamma_0$.*

The equivalence follows immediately from the following basic combinatorial lemma by taking $B = A \cap \Gamma_0 a$, where $\Gamma_0 a$ is the coset obtained in Theorem 1.11.

Lemma 1.16. (Ruzsa covering lemma). *Suppose A, B are subsets in a group G such that $|AB| \leq K|B|$, then there is a subset $X \subset A$ of size $\leq K$ such that $A \subset XBB^{-1}$.*

Proof. Pick a maximal family of disjoint translates of B of the form a_iB for $a_1, \ldots, a_N \in A$. Clearly $N \leq K$. Then for every $a \in A$, aB intersects some a_iB non-trivially and hence $a \in a_iBB^{-1}$. Take $X = \{a_1, \ldots, a_N\}$. □

2 The structure of approximate groups and geometric applications

In the first lecture, we saw that our structure theorem 1.11 easily implies Gromov's polynomial growth theorem. To provide further motivation for our study of the non-commutative inverse Freiman problem and of approximate groups, we begin this lecture with another geometric application.

2.1 A generalized Margulis lemma

In his book on metric structures on Riemannian and non-Riemannian spaces [46], Gromov discusses among many other things the properties of spaces with a lower curvature bound, in particular the Margulis lemma. The Margulis lemma is an important result in Riemannian geometry and hyperbolic geometry, where it is used, for example, to give lower bounds on the volume of closed manifolds and to describe the geometry of cusps. Usually this lemma requires some curvature bounds on the manifold (see, e.g., the books by Buser–Karcher [21] and Burago–Zalgaller [18]). The following result, which was conjectured by Gromov in this book, shows that the lemma holds in much greater generality.

Theorem 2.1. (Generalized Margulis Lemma). *Let X be a metric space and assume that every ball of radius 4 can be covered by at most K balls of radius 1. Now let $\Gamma \leq Isom(X)$ be a discrete subgroup of isometries of X. Then there is $\varepsilon_0 > 0$ depending on K only and not on X such that for every $x \in X$ the subgroup generated by*

$$S_\varepsilon(x) := \{\gamma \in \Gamma; d(\gamma \cdot x, x) < \varepsilon\}$$

is virtually nilpotent if $\varepsilon < \varepsilon_0$.

There is no properness assumption on X. However when we say that Γ is a "discrete group of isometries" we mean that the orbit maps $\Gamma \to X$, $\gamma \mapsto \gamma \cdot x$ are proper for every x, i.e., inverse images of bounded balls are finite. In fact all we need is that $S_2(x)$ is finite for each x.

Proof. Clearly, from the triangle inequality we have: $S_2(x)^2 \subset S_4(x)$. Let us prove that $|S_2(x)^2| \leq K|S_2(x)|$. Write $B(x,4) \subset \cup_{i=1}^K B(x_i,1)$ for a collection of K points $x_i \in X$. If $B(x_i,1)$ intersects $S_4(x)\cdot x$, then pick $\gamma_i \in S_4(x)$ such that $\gamma_i \cdot x \in B(x_i,1)$. Now given any $\gamma \in S_4(x)$, $\gamma \cdot x \in B(x_i,1)$ for some i, hence $d(\gamma\cdot x, \gamma_i \cdot x) \leq 2$. This means that $\gamma_i^{-1}\gamma \in S_2(x)$. Hence we proved $S_4(x) \subset XS_2(x)$ for some set $X \subset \Gamma$ with $|X| \leq K$. In particular $|S_2(x)^2| \leq |S_4(x)| \leq K|S_2(x)|$.

Now we can apply the structure theorem (here Theorem 1.15) to $S_2(x)$ and conclude that there is a virtually nilpotent group Γ_0 such that $S_2(x) \subset X\Gamma_0$ for some finite set X with $|X| \leq C(K)$. Pick an integer n with $C(K) < n < 2C(K)$ and $\varepsilon_0 = \frac{1}{n}$. If $\varepsilon < \varepsilon_0$, then by the triangle inequality $S_\varepsilon(x)^n \subset S_2(x) \subset X\Gamma_0$. But $S_\varepsilon(x)^k\Gamma_0/\Gamma_0$ is a non-decreasing sequence of sets and if for some k $S_\varepsilon(x)^k\Gamma_0/\Gamma_0 = S_\varepsilon(x)^{k+1}\Gamma_0/\Gamma_0$, then the sequence stabilizes and $S_\varepsilon(x)^k\Gamma_0/\Gamma_0 = \langle S_\varepsilon(x)\rangle\Gamma_0/\Gamma_0$. Since $n > |X|$ it follows that $S_\varepsilon(x)^n\Gamma_0/\Gamma_0 = \langle S_\varepsilon(x)\rangle\Gamma_0/\Gamma_0$ and $[\langle S_\varepsilon(x)\rangle : \langle S_\varepsilon(x)\rangle \cap \Gamma_0] \leq n \leq 2C(K)$. Hence $\langle S_\varepsilon(x)\rangle$ is virtually nilpotent. □

Here again the nilpotent subgroup can be taken to have at most $O(\log K)$ generators and be of nilpotency class $O(\log K)$.

The best constant $\varepsilon_0(X)$ is usually called the *Margulis constant* of X.

Yet another application of the structure theorem is the fact, first proved by Gromov, that the fundamental groups of almost flat manifolds are virtually nilpotent. This was later extended by Fukaya–Yamaguchi to almost non-negatively curved manifolds and by Cheeger–Colding and Kapovich–Wilking to Ricci almost non-negatively curved manifolds. We can recover this last result from the structure theorem by a simple application of the Bishop–Gromov volume comparison estimates (see [16])

Theorem 2.2. (Almost non-negatively curved manifolds). *Given d there is $\varepsilon(d) > 0$ such that if M is a Riemannian d-manifold with diameter 1 and Ricci curvature bounded below by $-\varepsilon(d)$, then $\pi_1(M)$ is virtually nilpotent.*

2.2 Approximate groups and the additive combinatorics toolkit

We now return to additive combinatorics and describe a few combinatorial tricks and useful tools that one uses all the time when one deals with sets of small doubling. The first one is the so-called Ruzsa inequality. Given two finite subsets A, B in a group G let

$$d(A, B) := \log \frac{|AB^{-1}|}{\sqrt{|A||B|}}$$

Clearly $d(A, B) = d(B, A)$ and $d(A, B) \geq 0$ because $|AB^{-1}| \geq |A|, |B|$. What is remarkable is the following:

Lemma 2.3. (Ruzsa inequality). *Given three finite subsets A, B, C in G, the triangle inequality holds*

$$d(A, C) \leq d(A, B) + d(B, C)$$

Hence $d(A, B)$ is called the *Ruzsa distance* between A and B. The proof is very easy:

Proof. Consider the map $AC^{-1} \times B \to AB^{-1} \times BC^{-1}$, which sends (x, b) to $(a_x b^{-1}, bc_x^{-1})$, where we made a choice of $a_x \in A$ and $c_x \in C$ for each $x \in AC^{-1}$. Then quite obviously this map is injective. Hence $|B||AC^{-1}| \leq |AB^{-1}||BC^{-1}|$, which is another way to phrase the triangle inequality $d(A, C) \leq d(A, B) + d(B, C)$. □

As a consequence:

$$\frac{|AA^{-1}|}{|A|} = e^{d(A,A)} \leq e^{d(A,A^{-1})+d(A^{-1},A)} = (\frac{|AA|}{|A|})^2.$$

Using the triangle inequality one can also prove [50, 62, 71]:

Lemma 2.4. (Small tripling lemma). *If A is any finite subset in a group G, and $n \geq 3$, then*

$$\frac{|A^n|}{|A|} \leq (\frac{|A^3|}{|A|})^{n-2}$$

Dealing with conditions like $|AA| \leq K|A|$, or $|A^{-1}A| \leq K|A|$ can be cumbersome at times and it is easy to run into unessential technical difficulties that hide the real ones. This is why Terry Tao came up with another set of axioms for a subset A, closely related to the small doubling condition, which is much easier to handle. He coined the following definition [71]:

Definition 2.5. (Approximate subgroup). Let G be a group and $K \geq 1$ a parameter. A subset A of G is called a K-approximate subgroup if

- $A = A^{-1}$ and $1 \in A$,
- $AA \subset XA$ for some subset $X \subset G$ with $|X| \leq K$.

Note: This makes sense for A infinite as well. But we will be mainly concerned with finite ones and by abuse of language A will always be assumed finite in the sequel.

Clearly a K-approximate subgroup has doubling at most K. The converse is not true of course, but we have the following partial converse:

Proposition 2.6. (Tao 2007 [71]). *Suppose A is a finite subset of an ambient group G such that $|AA| \leq K|A|$. Then there is an $O(K^{O(1)})$-approximate subgroup H of G such that $A \subset XH \cap HY$ for some subsets X, Y of size at most $O(K^{O(1)})$ and $|H| \leq O(K^{O(1)})|A|$.*

The implied constants in this result are absolute. This proposition essentially reduces the study of large sets of bounded doubling to that of approximate groups. In a similar flavor one shows that is $|A^3| \leq K|A|$, then $(A \cup A^{-1} \cup \{1\})^2$ is a $O(K^{O(1)})$-approximate subgroup. And that if $|AA| \leq K|A|$, then A has a subset of size $\geq |A|/O(K^{O(1)})$ which has tripling at most $O(K^{O(1)})$.

While the above facts require a bit of thought and are not immediate (see [71]), we can prove here a very easy variant:

Lemma 2.7. *If G is a group and A a subset such that $A = A^{-1}$ and $|A^5| \leq K|A|$, then A^2 is a K-approximate subgroup.*

Proof. Simply apply Ruzsa's covering Lemma 1.16. □

It often happens that group theoretic arguments about finite subgroups of an ambient group can be transferred or adapted to approximate groups. This has been a guiding philosophy in much of our arguments. For example, one has:

Lemma 2.8. *Suppose A is a K-approximate subgroup and B an L-approximate subgroup of an ambient group G (A,B not necessarily finite). Then $A^2 \cap B^2$ is a $(KL)^3$-approximate subgroup of G.*

Proof. Write $A^2 \subset XA$, $B^2 \subset YB$, so that $(A^2 \cap B^2)^2 \subset A^4 \cap B^4 \subset X^3A \cap Y^3B$. But every set of the form $xA \cap yB$ for $x \in X^3$, $y \in Y^3$ is contained in $z_{x,y}(A^2 \cap B^2)$ for some $z_{x,y}$ (pick any $z_{x,y}$ in $xA \cap yB$). So $(A^2 \cap B^2)^2 \subset Z(A^2 \cap B^2)$ for $Z = \{z_{x,y}\}_{x \in X, y \in Y}$ of size $\leq (KL)^3$. □

In particular if H is a subgroup of G and A a K-approximate subgroup, then $A^2 \cap H$ is again a K^3-approximate subgroup.

2.3 Nilprogressions and Tointon's theorem

The structure theorem (Theorem 1.11) gave the key information about sets with bounded doubling, namely that they have a large intersection with a coset of a virtually nilpotent group. In order to get more precise information on the structure of these sets, or equivalently on the structure of approximate groups, we need to introduce the notion of *nilprogression*. This generalizes to nilpotent groups the d-dimensional arithmetic progressions described above.

First we make the following general definition:

Definition 2.9. (Progression in a general group). Let G be a group and $x_1, \ldots, x_r$ be elements in G. Let $L_1, \ldots, L_r$ be non-negative integers. The set $P = P(x_1, \ldots, x_r; L_1, \ldots, L_r)$ of all group elements in G which can be written as a word in the x_i's and x_i^{-1}'s involving at most L_i letters $x_i^{\pm 1}$ for each $i = 1, \ldots, r$ is called a progression of rank r and lengths $\{L_i\}_i$.

For example, if $L_1 = \ldots = L_r = N$, then $P = P(x_1, \ldots, x_r; N, \ldots, N)$ is the ball of radius N for the "ℓ^∞-word metric" with generating set $S := \{x_1^{\pm 1}, \ldots, x_r^{\pm 1}\}$, which we denote by d_S^∞. Note that this is closely related to the usual word metric d_S (or ℓ^1-word metric), because obviously $d_S^\infty \leq d_S \leq r d_S^\infty$.

Similarly we define:

Definition 2.10. (Nilprogression). A progression $P(x_1, \ldots, x_r; N, \ldots, N)$ is said to be a nilprogression with step $\leq s$ and rank $\leq r$ if the x_i's generate nilpotent group with nilpotency class $\leq s$.

and:

Definition 2.11. (Coset nilprogression). A coset nilprogression is the inverse image of a nilprogression by a group homomorphism with finite kernel.

Side remark: The ball–box principle (see [45]) regarding word balls in nilpotent groups (i.e., that they are sandwiched in between two "nilpotent boxes" of comparable size) remains valid for nilprogressions. This and some routine nilpotent algebra yield:

Proposition 2.12. *There is $K = K(r,s) > 0$ such that every coset nilprogression of rank $\leq r$, step $\leq s$ and large enough side length (i.e., $L_i \geq C(r,s)$ for some constant $C(r,s) > 0$) is a K-approximate subgroup.*

For the proof, see either [73, Prop. 2.2.3] or [14, Cor. 3.6]. Following previous work of Green–Ruzsa [43] on the abelian case, Green and the author [13] on torsion-free nilpotent groups, and Tao [71] on 2-step groups, M. Tointon recently generalized the Freiman–Ruzsa–Chang theorem (Theorem 1.2) to an arbitrary nilpotent group. Using a clever lifting of approximate subgroups and induction on nilpotency class, he proved:

Theorem 2.13. (Tointon's theorem [75]). *Let A be a K-approximate subgroup of a nilpotent group G with nilpotency class s. Then there is a coset nilprogression P with rank $\leq r = O_s(K^{C(s)})$ and step $\leq s$ such that $A \subset P \subset XA$ for some $X \subset G$ of size $|X| \leq O_s(K^{C(s)})$.*

2.4 Stronger form of the structure theorem

We now are in a position to state the stronger form of our structure theorem.

Theorem 2.14. (BGT strong form). *Suppose A is a K-approximate subgroup in an ambient group G. Then $A \subset XP$, where $X \subset G$ has size $|X| \leq C(K)$ and P is a coset nilprogression whose rank and step are bounded in terms of K only and such that $|P|/|A|$ is bounded in terms of K only.*

Remarks:

a) Here again $C(K)$ is some constant depending only on K, but the proof gives no explicit estimate in terms of K. It can in fact be shown that $C(K)$ is not upper bounded by K^C for some absolute constant $C > 0$. See [14] for an example.
b) Below we derive the above strong form from the weak form given earlier in Theorem 1.15 using Tointon's theorem. Originally in [16] we proved the strong form directly (thus obtaining as a by-product a new proof of the original abelian Freiman–Ruzsa theorem) and obtained explicit bounds for the rank and step of the nilprogression and its size as well, but not on the constant $C(K)$. In fact we show in [16] that one can find a coset nilprogression P as in Theorem 2.14 with $P \subset A^{12}$ (and hence $|P| \leq K^{11}|A|$) whose rank and step are $O(K^3)$.

Proof. From the weak form of the structure theorem, more precisely Theorem 1.15, we can find a virtually nilpotent subgroup $\Gamma_0 \leq G$ such that $A \subset X\Gamma_0$ for some subset X of size $|X| \leq C(K)$. So $|A^2 \cap \Gamma_0| \geq |A|/C(K)$. In fact (see Remark d) after Theorem 1.11) the proof shows that Γ_0 fits into an exact sequence $1 \to N \to \Gamma_0 \xrightarrow{\pi} L \to 1$, where L is nilpotent with bounded rank and step, and N is finite with $N \subset XA$ for some bounded set X, $|X| \leq c(K)$. By Lemma 2.8 $A^2 \cap \Gamma_0$ is a K^3-approximate subgroup. Then $\pi(A^2 \cap \Gamma_0)$ is a K^3-approximate subgroup of L and Tointon's theorem applies: there is a coset nilprogression Q with bounded rank and step which contains $\pi(A^2 \cap \Gamma_0)$ and is contained in boundedly many translates of $\pi(A^2 \cap \Gamma_0)$. Set $P = \pi^{-1}(Q)$. This is again a coset nilprogression with bounded rank and step. It contains $A^2 \cap \Gamma_0$, so by Ruzsa covering (Lemma 1.16) A lies in XP for some bounded X. Moreover, Q is contained in boundedly many cosets of A, because N is. □

We now give an interesting consequence of the strong form of the theorem and in the last lecture, we will give two geometric applications, one to Cayley graphs of finite simple groups, the other to scaling limits of vertex transitive graphs, that both require this strong form.

Suppose

$$P = P(x_1, \ldots, x_r; L_1, \ldots, L_r)$$

is a nilprogression of rank r and step s, then for every $n \in \mathbb{N}$ the power set P^n is obviously contained in $P(x_1, \ldots, x_r; nL_1, \ldots, nL_r)$. Using some nilpotent algebra it can be shown (by arguments similar to those in the proof of Proposition 2.12) that if the L_i's are large enough, it also contains $P(x_1, \ldots, x_r; \frac{n}{C}L_1, \ldots, \frac{n}{C}L_r)$ for some $C = C(r, s) > 0$ and that the three sets are of comparable size up to multiplicative constants depending only on r and s. In particular $|P^{2n}| \leq K|P^n|$ for some constant $K = K(r, s)$ independent of the L_i's and independent of n. So the bound on the doubling P is inherited to all powers P^n. From this observation (proved in detail in [14]) and the strong form of the theorem it is then an exercise (similar to the derivation of Theorem 1.12 from Theorem 1.11) to derive:

Theorem 2.15. (doubling at one scale ⇒ doubling at all scales, [14]). *Given $K > 0$ there is $n_0, C(K) > 0$ such that if S is a finite subset of a group G such that S^{n_1} is a K-approximate subgroup for some $n_1 > n_0$, then S^n is a $C(K)$-approximate subgroup for every $n \geq n_1$.*

3 Hilbert's fifth problem and approximate groups

In this third lecture, we outline the proof of the structure theorem (Theorem 1.11). A good deal of this lecture is devoted to explaining the key step of the Gleason–Yamabe theorem on the structure theory of locally compact groups.

3.1 Introduction

Approximate subgroups of specific groups (such as linear groups, free groups, finite simple groups, etc...) had been studied by various authors relying on techniques specific to the given class of groups they were considering, often yielding explicit bounds. But no approach to tackle the general problem of understanding the structure of approximate subgroups of arbitrary groups existed prior to the work of Hrushovski [54].

In his paper [54], Hrushovski proposed an approach, based on an analogy between this question and various problems and results from Stability Theory, a branch of Model Theory. His theorem, which we state below in a form that avoids the language of model theory, is the first step of our proof. For readers interested in the connection with model theory, we recommend van den Dries' Bourbaki report [29] as well as Hrushovski's own lecture notes.

Combined with the Gleason–Yamabe structure theorem for locally compact groups (Hilbert's 5th problem) Hrushovski's theorem already gives non-trivial information on the structure of an arbitrary approximate group. Roughly speaking Hrushovski derives from it a weaker structure theorem than Theorem 1.11 in which approximate groups are showed to admit a nested sequence $S_{n+1}^2 \subset S_n$ of large approximate subgroups S_n of A, which exhibits the following weak form of nilpotence: $[S_n, S_n] \subset S_{n+1}$, see [54, Theorem 1.1].

In order to obtain the finer structure given by our Theorem 1.11, one has to go inside the proof of the Gleason–Yamabe theorem and adapt arguments discovered there and then (more than sixty years ago) in the context of topological groups to the setting of approximate groups.

Going through the proof of the Gleason–Yamabe theorem in the original papers of Gleason and of Yamabe or in the subsequent exposition in the book by Montgomery and Zippin can be daunting. Nevertheless the task has been greatly eased by a number of people over the years, including I. Kaplansky [56] and especially J. Hirschfeld [53] who identified and formulated the key claims (the Gleason–Yamabe lemmas) and gave a much streamlined treatment of the proof using ultra-products (recently simplified further by model theorists L. van den Dries and I. Goldbring in a forthcoming article [28]). The whole thing is now pretty demystified, and a clear and complete modern treatment is now available in Tao's book [73].

3.2 Proof strategy

We turn to the proof of the structure theorem. As in Gromov's proof of his theorem [44] on groups with polynomial growth, one needs to pass from a discrete object (in our case a finite approximate group) to an infinite object on which we can perform analysis. While Gromov used his notion of Gromov–Hausdorff convergence to define his limit space (i.e., the asymptotic cone = the Cayley graph viewed from infinity) van den Dries and Wilkie [27] later suggested to construct this space using ultrafilters. We will follow the same route and start with a non-principal ultrafilter $\mathcal{U}$ on the set of positive integers.

Fix $K \geq 1$. Given any family of groups G_n and a family of finite K-approximate subgroups $A_n \leq G_n$, we form the ultra-product:

$$\mathbb{A} := \prod_{\mathcal{U}} A_n, \quad \mathbb{G} := \prod_{\mathcal{U}} G_n.$$

The first observation is that $\mathbb{A}$ is a K-approximate subgroup of $\mathbb{G}$. Indeed the ultra-product of a sequence of sets of cardinal at most K has cardinal at most K. We call such a set $\mathbb{A}$ an *ultra approximate group*.

The proof has essentially three main steps:

Step 1 : Build a topology.
Step 2 : Adapt the Gleason–Yamabe lemmas.
Step 3 : Conclude.

Step 1: This step is due to Hrushovski [54]. While Gromov started with a sequence of metric spaces (Cayley graphs), we do not have any distance given a priori with our sequence of sets A_n. So the first goal here is to define a notion of closeness or nearness for elements of A. One would like to say that an element g is close to the identity if the symmetric difference $gA \Delta A$ is very small in size. While this does not quite work on the nose, a variation of this idea can be implemented successfully. This "distance" being defined for each A_n, it passes to the ultra-product to give rise to a genuine left-invariant pseudo-distance $d(\cdot, \cdot)$ on the group $\langle \mathbb{A} \rangle$. One then shows that $G := \langle \mathbb{A} \rangle / \{\mathfrak{a}; d(\mathfrak{a}, 1) = 0\}$ is a locally compact second countable group.

Step 2: This part consists in adapting the Gleason–Yamabe lemmas from locally compact groups to approximate groups. The Gleason–Yamabe lemmas lie at the heart of the proof of the structure theorem for locally compact groups proved by Gleason and Yamabe in the early 1950s. According to this theorem, every locally compact group has an open subgroup which is a projective limit of Lie groups, i.e., has arbitrarily small normal compact subgroups modulo which it is a Lie group with a finite number of connected components. The Gleason–Yamabe lemmas serve two purposes in the proof: they allow to reduce to the NSS case (NSS= no small subgroups) and to show that the set of one-parameter subgroups of an NSS group is a vector space, thus giving rise to a Lie algebra and a Lie structure. It turns out that under some assumptions on the finite approximate group A (which are anyway satisfied by a large approximate subgroup of A as follows from Step 1), these lemmas can be adapted almost word for word to A and yield a finite subgroup H of A and an element $a \in A \setminus H$ which commutes with A modulo H. This element will serve as a stem from which to build the nilprogression.

Step 3: Performing Step 2 for each A_n, we obtain an ultra finite subgroup $\mathbb{H} := \prod_{\mathcal{U}} H_n \lhd \langle \mathbb{A} \rangle$ and an ultra cyclic subgroup $\langle \mathfrak{a} \rangle \leq \langle \mathbb{A} \rangle$, where $\mathfrak{a}$ is the class of $(c_n)_n$. By Step 1 and the structure theorem for locally compact groups, we may assume

after passing to an open subgroup that G is a projective limit of Lie groups. Moding out by the maximal normal compact subgroup of G, we obtain a connected Lie group L. If $\dim L = 0$, then $\mathbb{A}$ is covered by finitely many translates of a compact open subgroup and this means A_n is covered by boundedly many translates of a finite subgroup for $\mathcal{U}$-almost every n. If $\dim L > 0$, the projection of $\langle \mathfrak{a} \rangle$ to L is a central one-parameter subgroup. The rest of the proof proceeds by induction on $\dim L$, moding out $\langle \mathbb{A} \rangle$ by $\langle \mathfrak{a} \rangle$ and starting all over again. A careful treatment of this step requires the introduction of *local groups*[1]. At the end we obtain that the A_n's are covered by boundedly many translates of a finite-by-nilpotent subgroup, the finite part being itself contained in boundedly many translates of A_n. The theorem then follows by contradiction, picking a sequence of potential counter-examples A_n to the statement of the theorem and carefully negating the quantifiers.
N.B. In Gromov's theorem the induction in the final step is done on the exponent of polynomial growth. In his paper [54], Hrushovski also gave another proof of Gromov's theorem and even a strengthening of it (which is now a simple consequence of Theorem 1.11) using induction on $\dim L$ as in Step 3 above. The novelty of [16], say compare to [54], consists mainly in Step 2.

Below we describe Step 1 and Step 2 and refer the reader to the original paper for Step 3, whose details are a bit more technical.

3.3 Building a locally compact topology (Step 1)

The goal here is to define a non-trivial topology on $\langle \mathbb{A} \rangle$ which, when a certain normal subgroup is quotiented out, turns $\langle \mathbb{A} \rangle$ into a locally compact group. We need to define a base of neighborhoods of the identity. The main tool for this is the following lemma:

Lemma 3.1. (Sanders [69], Croot–Sisask [24], Hrushovksi [54]). *If A is a K-approximate group and $k \geq 1$, then there is $S \subset A^4$ such that*

- $1 \in S$, $S = S^{-1}$,
- $S^k \subset A^4$,
- $|S| \geq |A|/C(K,k)$

In other words, this lemma allows to take a "square root," or k-th root, of any given approximate group. The lemma was proved independently by Sanders using a combinatorial argument (showing the existence of a subset $B \subset A$ such that there is a non-trivial proportion of elements $g \in A^4$ such that $|gAB \Delta AB|$ is at most $\frac{1}{k}|AB|$ say), by Croot–Sisask (who use the probabilistic method to find the subset S) and by Hrushovski (using tools from Model Theory).

[1] They can be avoided if one only shoots for Theorem 1.11, see the proof in [29], but seem necessary for the finer structure described in Remark d) after Theorem 1.11, and hence for the strong form Theorem 2.14.

Using this lemma, we can define a base of neighborhoods of the identity in $\langle \mathbb{A} \rangle$ as follows. For each A_n and each $k \geq 1$ pick a subset $S_{n,k}$ as given by the above lemma. In fact a rather straightforward strengthening of the above lemma ensures that we can pick $S_{n,k}$ in such a way that $S_{n,k+1}^2 \subset S_{n,k}$ and $gS_{n,k+1}g^{-1} \in S_{n,k}$ for all $g \in A$.

Set $\mathbb{S}_k := \prod_{\mathcal{U}} S_{n,k}$. This defines a topology on $\langle \mathbb{A} \rangle$ with the $\mathbb{S}_k$ as a base of neighborhoods of the identity. Moreover the intersection $\Gamma := \cap_{k \geq 1} \mathbb{S}_k$, where $\mathbb{S}_k := \prod_{\mathcal{U}} S_{n,k}$ is a normal subgroup of $\langle \mathbb{A} \rangle$. We define G as the quotient of $\langle \mathbb{A}^4 \rangle / \Gamma$. In this topology $\mathbb{A}$ mod Γ is a compact neighborhood of the identity.

Given a Hausdorff topological group with a nested sequence of symmetric neighborhoods of the identity $S_{i+1}^2 \subset S_i$ a classical construction of Birkhoff and Kakutani [6, 61] yields a left invariant distance compatible with the topology (in which the S_i are essentially the balls of radius $\frac{1}{2^i}$ around the identity). We can do this here with the $\mathbb{S}_k$'s and obtain a left invariant distance d on G.

Then the local compactness of G is proven in the usual way by establishing completeness with respect to d and total boundedness (i.e., $\forall\, 0 < \varepsilon < r$ every ball of radius r can be covered by $\leq C(\varepsilon, r)$ balls of radius ε). This follows from the Ruzsa covering lemma. Completeness follows by the countable saturation property of ultrafilters.

In the Birkhoff–Kakutani construction of the left invariant distance d, balls can be approximated by finite products of the S_i's. Since the $\mathbb{S}_k$'s are ultra-product sets, i.e., given by the ultra-product of subsets of A_n, we obtain that balls in G, and more generally any compact subset of G, can be approximated by ultra-product sets.

We summarize the above discussion in the following statement:

Theorem 3.2. (Hrushovski local model theorem, [54, Theorem 4.2.]) *There exists a locally compact group G and a group homomorphism*

$$\pi : \langle \mathbb{A} \rangle \to G$$

such that $\pi(\mathbb{A})$ is a compact neighborhood of the identity in G, and for every $F \subset O \subset \pi(\mathbb{A}^4)$ with F closed and O open subsets, there are $B_n \subset A_n^4$ such that

$$\pi^{-1}(F) \subset \mathbb{B} \subset \pi^{-1}(O),$$

where $\mathbb{B} := \prod_{\mathcal{U}} B_n$.

Remarks.

a) If G is the trivial group, then the theorem implies that the A_n's are finite groups for almost all indices n w.r.t the ultrafilter $\mathcal{U}$. Indeed take $O = F = \{1\}$. Then according to the theorem there is $\mathbb{B} = \prod_{\mathcal{U}} B_n \subset \mathbb{A}$ which is the kernel of π. Hence all of $\langle \mathbb{A} \rangle$, and we get that $\mathbb{A}$ is stable under multiplication, hence so are the A_n for $\mathcal{U}$-almost every n.

b) Similarly if the "model group" G is totally disconnected, then we can conclude that there is a finite group $H_n \subset A_n^C$ for some bounded C such that A_n is contained in boundedly translates of H_n. Indeed a totally disconnected locally compact group contains an open compact subgroup (van Dantzig). According to the theorem it must be an ultra-product set $\mathbb{H} := \prod_{\mathcal{U}} H_n$ and the H_n must be stable under multiplication for $\mathcal{U}$-almost every n, i.e., be finite subgroups. Since $\mathbb{A}$ is compact in projection to G, it must be finitely covered by $\mathbb{H}$.

c) Recall the Gleason–Yamabe structure theorem for locally compact groups:

Theorem 3.3. (Gleason–Yamabe [40, 80]). *If G is a locally compact group, then G has an open subgroup G' which is a "generalized Lie group" in the sense that every identity neighborhood contains a compact subgroup $H \lhd G'$ such that G'/H is a Lie group with finitely many connected components.*

N.B. If G is compact, the above boils down to the Peter–Weyl theorem (which is itself an application of the spectral theorem for self-adjoint convolution operators acting on $L^2(G)$). It is worth noting that the proof of Theorem 3.3 for non-compact groups is not independent of the Peter–Weyl theorem as it makes use of it in the crucial reduction step to the NSS (No Small Subgroup) case.

A consequence of this is that attached to every non-compact locally compact group there is a canonically defined non-compact connected Lie group L with no non-trivial compact normal subgroups. Indeed L is the connected component of the identity of G'/H modulo its maximal normal compact subgroup. Its isomorphism class is independent of G' and H. The connected component G^0 is compact if and only if L is trivial.

In the setting of Theorem 3.2, we call L the *Hrushovski Lie group* associated to the A_n's. If it is trivial, then the A_n are commensurable to finite groups (this is Remark b) above). In order to understand what happens if L is non-trivial, we need to go into the proof of the Gleason–Yamabe theorem. This is the aim of Step 2. At the end (Step 3) we establish that L is a simply connected nilpotent Lie group.

3.4 The Gleason–Yamabe theorem

Let us digress for a little while and discuss the proof of the Gleason–Yamabe structure theorem for locally compact groups (Theorem 3.3). At the heart of it lies the so-called Gleason–Yamabe lemmas (Lemmas 3.4 and 3.6 below). They serve a dual purpose:

(1) Reduce the structure theorem to the NSS case (i.e., the case when G has an identity neighborhood with no non-trivial subgroup).
(2) Show that if G is NSS, the set of one-parameter subgroups (i.e., homomorphisms from $(\mathbb{R}, +)$ to G) forms a vector space, thus giving rise to the sought after Lie algebra and Lie group structure.

Historically, Gleason [40] used a construction closely related to Lemma 3.6 below to prove Theorem 3.3 in the case of NSS connected locally compact groups of finite topological dimension. It was Yamabe [79, 80] who first understood how to deal with the general (non NSS) case by reducing it to the Peter–Weyl theorem using a variant of Gleason's argument (encapsulated in Lemma 3.4). We stress that there are two important cases in which the theorem is much easier to prove: the compact case (Peter–Weyl), and the totally disconnected case (van Dantzig).

Suppose G is a locally compact group and pick a compact identity neighborhood V in G. Given a subset $Q \subset V$, let $n_V(Q)$ be the largest integer such that $1, Q, \ldots, Q^{n-1}$ are all contained in V. Clearly $n_V(Q) > 1$ iff $Q \subset V$, and $n_V(Q) = +\infty$ iff the subgroup $\langle Q \rangle$ lies in V entirely. Set $n_V(g) = n_V(\{g\})$.

Lemma 3.4. (Gleason–Yamabe 1). *There is $C = C_V > 0$ such that for every $K \geq 1$, there is an identity neighborhood $U \subset V$, such that for every subset $Q \subset U$,*

- *either $n_{V^4}(Q) \geq K n_V(Q)$.*
- *or there is $g \in Q$ such that $n_U(g) \leq K C_V n_V(Q)$.*

In other words, unless it takes a very long time for powers of Q to venture outside of V^4 after they first pop out of V, the time it takes for powers of Q to escape V is comparable to the maximal time needed for powers of individual members of Q to escape U.

A consequence of this is:

Corollary 3.5. (subgroup trapping). *For every identity neighborhood V in G, there is an identity neighborhood $U \subset V$ such that the subgroup H generated by all subgroups contained in U is contained in V.*

Proof. Take for Q the union of all subgroups contained in U and apply Lemma 3.4.

Proof of Theorem 3.3 : reduction to the NSS case. Granting the above corollary and using Peter–Weyl's theorem, it is easy to find an open subgroup G' in G and a normal subgroup $N \lhd G'$ modulo which it is NSS (i.e., G'/N has no small subgroups) thus reducing the proof of the Gleason–Yamabe theorem to the NSS case.

Indeed since V is compact, this subgroup H is compact and we can apply the Peter–Weyl theorem to it. In particular H has arbitrarily small compact normal subgroups N modulo which it is NSS. Picking an identity neighborhood W such that $W^3 \subset U$, we may take $N \subset W$ and assume that every subgroup of H contained in W^3 is contained in N. But since, by definition of H, every subgroup contained in U is contained in H, we conclude that N is normalized by W and that $\langle W \rangle / N$ is NSS.

We could set $G' = \langle W \rangle$, but this would make it dependent on U. To remove this dependence is a simple task however. Indeed van Dantzig's theorem ensures that every totally disconnected locally compact group has an open compact subgroup. So simply define G' as the pre-image of a compact open subgroup of G/G^0, where G^0 is the connected component of G and perform the above inside G' instead of G. Then $\langle W \rangle$ is an open subgroup of G', hence it has finite index in G', and there are only finitely many conjugates $g_i N g_i^{-1}$ of N. So $N' := \cap g_i N g_i^{-1}$ is normal in G' and contained in the neighborhood of the identity $W' := \cap g_i W g_i^{-1}$. Moreover any subgroup contained in W' is contained in N', so G'/N' is NSS.

Proof of Corollary 3.5. We now derive Corollary 3.5 from Lemma 3.4. This uses Peter–Weyl's theorem one more time. Let $B(t)$ be the closed ball of radius t around the identity (for a fixed left invariant distance), let Q_t be the union of all subgroups contained in $B(t)$. Fix $s > 0$. For $t < s$, $Q_t \subset B(s)$, and $n_{B(s)}(g) = +\infty$ for $g \in Q_t$. So Lemma 3.4 implies that either $n_{B(s)}(Q_t) = +\infty$ in which case $\langle Q_t \rangle$ is relatively compact and we are done, or $\frac{n_{B(4s)}(Q_t)}{n_{B(s)}(Q_t)}$ is unbounded as $t \to 0$. This implies that $Q_t^{n_{B(s)}(Q_t)}$ converges along a sequence of $t \to 0$ towards a compact subgroup $H_s \subset B(4s)$ not contained in $B(s/2)$. By diagonal process we obtain this way a nested family of groups $H_s \leq H_{s'}$ if $s \leq s'$. But by Peter–Weyl, since H_1 is compact, it has a closed normal subgroup N_1 such that H_1/N_1 is NSS. Now doing the above construction with $B(s)N_1$ in place of $B(s)$ we obtain a nested family of subgroups $H_s' \subset (B(s)N_1)^4 \cap H_1 \subset B(4s)N_1 \cap H_1$ such that H_s' is not contained in $B(s/2)N_1$. This contradicts the fact that H_1/N_1 is NSS.

We now pass to the second part of the Gleason–Yamabe lemmas and thus assume that G is NSS. It is then easy to see (arguing by contradiction and taking a limit) that n_U and n_V are comparable if U and V are any two small enough identity neighborhoods, i.e.:

$$\frac{1}{C} n_U(Q) \leq n_V(Q) \leq C n_U(Q) \tag{2}$$

for some $C = C(U, V) > 0$ and all subsets Q. We then define for $g \in G$

$$\|g\|_U := \frac{1}{n_U(g)},$$

and call it the *escape norm* associated to U.

Lemma 3.6. (Gleason–Yamabe 2). *Suppose G is a locally compact group which is NSS and V is a neighborhood of the identity. Then V contains a compact identity neighborhood U such that, for some $C = C_U > 0$, and all $g, h \in U$,*

(1) $\|h^{-1}gh\|_U \leq C\|g\|_U$
(2) $\|gh\|_U \leq C(\|g\|_U + \|h\|_U)$
(3) $\|[g,h]\|_U \leq C\|g\|_U \cdot \|h\|_U$

Item 1) is clear from the definition and (2). Item 2) is an immediate consequence of the first Gleason–Yamabe lemma, i.e., Lemma 3.4, and of (2) (take $Q = \{g, h\}$). Item 3) can be proved by an argument similar that used in the proof of Lemma 3.4 (see below).

Lemma 3.6 is used to prove that the set $L(G)$ of one-parameter subgroups of G forms a vector space on which (an open subgroup of) G acts by conjugation with central kernel. From this point the end of the proof of Theorem 3.3 is fairly standard: one shows that $L(G)$ is locally compact, hence finite dimensional, thus (an open subgroup of) G embeds in $\mathrm{GL}(\mathbb{R}^n)$ modulo the center and one concludes by the Cartan-von-Neumann theorems that G is a Lie group.

The NSS assumption easily implies (taking the cyclic group generated by group elements closer and closer to the identity and passing to a limit) that $L(G)$ is non-trivial. In order to prove that $L(G)$ is a vector space, one needs to make sense of the formula:

$$\lim_{n\to+\infty}(X(\frac{t}{n})Y(\frac{t}{n}))^n = (X+Y)(t)$$

given two one-parameter subgroups $X, Y \in L(G)$. This is where Item 2) of Lemma 3.6 comes into play, because it allows to conclude that $(X(\frac{t}{n})Y(\frac{t}{n}))^n$ remains in a fixed compact part of G when n tends to $+\infty$ and t remains bounded. In fact, using Item 3), i.e., the commutator shrinking property, one shows that this is a Cauchy sequence (NSS groups are metrizable) and hence converges.

Finally, using Item 3) again, one proves that every small enough element of G lies on a one-parameter subgroup. Then after passing to an open subgroup of G, we see that the kernel of the adjoint representation is central. This concludes the proof of Theorem 3.3.

The proof of the Gleason–Yamabe lemmas. Having explained how the Gleason–Yamabe theorem is derived from the Gleason–Yamabe lemmas, let us now say a few words about the proof of the lemmas themselves. As said earlier Items 1) and 2) follow easily from the NSS condition and Lemma 3.4, so we are left with proving Lemma 3.4 and Item 3) of Lemma 3.6.

Before we start, it is instructive to notice that if G were a Lie group, then both statements would be easy to prove using the exponential map $\exp : L(G) \to G$, as $\|\exp(X)\|_U$ is then comparable to $||X||$ for any fixed norm on $L(G)$. But we are precisely trying to establish the existence of an exponential map...

We now focus on Lemma 3.4. There are two bright ideas involved here: the first, say in the setting of Lemma 3.4, is to introduce a continuous non-negative bump function ϕ, which vanishes outside V^4 and is equal to 1 at the identity, and compare $\|g\|_U$ to the sup norm $||\partial_g\phi||_\infty$, where $\partial_g\phi(x) := \phi(g^{-1}x) - \phi(x)$. We make the following simple observations:

(1) We have the triangle inequality $||\partial_{gh}\phi||_\infty \leq ||\partial_g\phi||_\infty + ||\partial_h\phi||_\infty$,
(2) If $||\partial_g\phi||_\infty < 1$, then $g \in Supp(\phi) \subset V^4$.

from which it follows immediately that $||g||_V \leq ||\partial_g\phi||_\infty$. We cannot expect that an inequality in the other direction holds for all g, because that would imply that ϕ is invariant under every subgroup contained in V. But, for every $K \geq 1$ we will find U small enough and build a function ϕ (depending on K,V) supported in V^4, such that $\phi(1) = 1$, and for which the following inequality holds for all $Q \subset U$, all $g \in Q$, and some $C_V > 0$

$$||\partial_g\phi||_\infty \leq C_V\frac{1}{n_U(g)} + \frac{1}{2K}\cdot\frac{1}{n_V(Q)}. \tag{3}$$

Lemma 3.4 follows easily. Indeed, by definition of $n_{V^4}(Q)$ there is a word w of length $n_V(Q)$ with letters in Q such that $w \notin V^4$. Hence $||\partial_w\phi||_\infty \geq 1$ by (2) above, and by the triangle inequality (1) there must be $g \in Q$ such that $||\partial_g \phi||_\infty \geq \frac{1}{n_{V^4}(Q)}$. Suppose $n_{V^4}(Q) \leq K n_V(Q)$, then we conclude from (3) that $n_U(g) \leq KC_V n_V(Q)$ as desired.

So things boil down to finding ϕ such that (3) holds. If we were on a Lie group and ϕ were say of class C^2, because the second derivatives are controlled, we would have $\partial_{g^n}\phi \simeq n\partial_g\phi$ as long as $g^n \in U$ if U is small enough. This follows immediately from the Taylor expansion of ϕ near the identity, or from the following abstract form of the Taylor formula, which always holds (direct check):

$$\partial_{g^n}\phi = n\partial_g\phi + \sum_{i=0}^{n-1} \partial_{g^i}\partial_g\phi \tag{4}$$

The second term in the right-hand side would then be negligible and it would follow that $||\partial_g\phi||_\infty \leq C||g||_U$, which is a strong form of (3).

This is where the second bright idea comes in: making use of a (left-) Haar measure, we can define ϕ as the convolution of two bump functions $\phi(x) := \frac{\psi_1 * \psi_2(x)}{\psi_1 * \psi_2(1)}$, where $\psi_1 * \psi_2(x) = \int \psi_1(xy)\psi_2(y^{-1})dy$. The following simple formula allows to control the second derivatives of ϕ in terms of the first derivatives of ψ_1 and ψ_2:

$$\partial_h\partial_g(\psi_1 * \psi_2) = \int (\partial_g\psi_1)(y) \cdot (\partial_{h^y}\psi_2)(y^{-1}x)dy.$$

where $h^y = y^{-1}hy$. We take

$$\psi_1(x) := (1 - \frac{1}{n_V(Q)} \inf\{n \geq 0, x \in Q^n V\})^+$$

and

$$\psi_2(x) := (1 - \frac{1}{L} \inf\{n \geq 0, x \in W^n V\})^+,$$

where $y^+ := \max\{0, y\}$ and W is an identity neighborhood such that $W^L \subset V$ for some integer $L > 2K vol(V^3)/vol(V \cap V^{-1})$.

It is clear that ψ_1 and ψ_2 are equal to 1 on V and their supports are contained in V^2. Hence the support of ϕ lies in V^4, $||\psi_1 * \psi_2|| \leq vol(V^2)$ and $\psi_1 * \psi_2(1) \geq vol(V \cap V^{-1})$. Hence

$$||\phi|| \leq \frac{vol(V^2)}{vol(V \cap V^{-1})}$$

We also have Lipschitz control on both ψ_1 and ψ_2, because if $g \in Q$, then $||\partial_g\psi_1|| \leq \frac{1}{n_V(Q)}$, while if $h \in W$, then $||\partial_h\psi_2|| \leq \frac{1}{L}$. Moreover if $g \in V$, then the support of $\partial_g\psi_1$ is contained in V^3.

Setting $U = \bigcap_{y \in V^3} y^{-1} W y$, we conclude that for all $g \in Q$ and $h \in U$:

$$||\partial_h \partial_g (\psi_1 * \psi_2)|| \leq vol(V^3) \frac{1}{n_V(Q)} \frac{1}{L}. \tag{5}$$

Finally plugging (5) into the Taylor formula (4), and writing $||\partial_{g^n} \phi|| \leq 2||\phi||$, we get that as long as $g^i \in U$ for all $i = 0, \ldots, n$,

$$||\partial_g \phi|| \leq \frac{||\partial_{g^n} \phi||}{n} + \sum_0^{n-1} ||\partial_{g^i} \partial_g \phi|| \leq C_V \frac{1}{n} + \frac{vol(V^3)}{vol(V \cap V^{-1})} \frac{1}{n_V(Q)} \frac{1}{L} \leq C_V \frac{1}{n} + \frac{1}{2Kn_V(Q)}$$

since $L > 2K vol(V^3)/vol(V \cap V^{-1})$, and where $C_V = 2 \frac{vol(V^2)}{vol(V \cap V^{-1})}$. We thus obtain (3) and we are done.

This concludes the proof of Lemma 3.4. The proof of the commutator shrinking property, Item 3) of Lemma 3.6, is similar and we do not include it here. We also leave out the details of the remainder of the proof of Theorem 3.3, namely the proof that the set of one-parameter subgroups is finite dimensional space on which G acts with central kernel. We refer the reader to Tao's book [73] for these missing items.

3.5 *The Gleason–Yamabe lemmas for approximate groups (Step 2)*

We now return to our main theorem (Theorems 1.11 and 2.14) and to the second step of its proof. Here the intent is to adapt the Gleason–Yamabe lemmas to the setting of approximate groups. In much the same way as these lemmas were used in the locally compact setting to exhibit a small normal subgroup modulo which the group is Lie, the adaptation of the Gleason–Yamabe lemmas to the approximate group setting will yield a finite normal subgroup modulo which the approximate group centralizes a non-trivial subgroup.

Given g, let $n_A(g)$ be the largest integer such that $1, g, \ldots, g^n$ belong to A.

In order to implement the Gleason–Yamabe lemmas in the approximate group setting, we need an assumption on the approximate group.

Definition 3.7. (Strong approximate group). A K-approximate group A is said to be a strong approximate group if

(1) if $1, g, \ldots, g^{1000} \in A^{100}$, then $g \in A$.
(2) there is a subset $S = S^{-1} \subset A$ such that $(S^{A^4})^{10^6 K^3} \subset A$ and if $1, g, \ldots, g^{10^6 K^3} \in A$, then $g \in S$.

Observe that, as follows from this definition, if A is a strong K-approximate subgroup of G, then

- $\forall g, n_A(g) \le n_{A^{100}}(g) \le 10^3 n_A(g)$,
- $10^6 K^3 n_S(g) \ge n_A(g)$.

Set

$$||g||_A = \frac{1}{n_A(g)}$$

the *escape norm* of g w.r.t A. Note that $\|g\|_A = 0$ iff the cyclic group $\langle g \rangle$ lies in A entirely.

Lemma 3.8. (Gleason–Yamabe lemmas, approximate group version). *Given $K \ge 1$, there is $C = C(K) > 0$ such that if A is a strong K-approximate subgroup of a group G, then $\forall g, h \in A$,*

(1) $\|h^{-1}gh\|_A \le C\|g\|_A$
(2) $\|gh\|_A \le C(\|g\|_A + \|h\|_A)$
(3) $\|[g,h]\|_A \le C\|g\|_A \cdot \|h\|_A$

We do not repeat the proof, which is essentially identical to the original Gleason–Yamabe lemma that we proved above. The key point here is the second item. For example, it implies that if the $\langle g \rangle$ and $\langle h \rangle$ lie both in A, then the subgroup generated by gh is also contained in A.

Of course in order to be able to apply this lemma, we need to be able to produce a strong approximate subgroup out of the original approximate subgroup A. Indeed we prove:

Lemma 3.9. (existence of strong approximate subgroups). *Given $K > 0$, there is $C(K) > 0$ such that for every K-approximate subgroup A of a group G there is a strong $C(K)$-approximate subgroup $A' \subset A^4$, such that A is contained in a most $C(K)$ left translates of A'.*

This allows to work with A' instead of A, and thus apply the Gleason–Yamabe lemmas to it. Unfortunately, we were not able to give a purely combinatorial proof of this lemma. It is a consequence of Hrushovski's local model theorem (Theorem 3.2) and of the original Gleason–Yamabe structure theorem for locally compact groups (applied to the model group G from Theorem 3.2). Consequently $C(K)$ is ineffective, and there lies the source of ineffectiveness of our main theorem.

Proof. Apply the Gleason–Yamabe structure theorem for locally compact groups to the model group G from Theorem 3.2. Up to passing to an open subgroup, we may assume (up to changing A into a commensurate approximate subgroup) that G is a projective limit of Lie groups. But it is easy to see that sufficiently small balls around the identity in exponential coordinates in a connected Lie group (say for a choice of a Euclidean metric on the Lie algebra) satisfy the strong approximate group axioms. Hence we can pull back those sets to $\mathbb{A}^4$ via Theorem 3.2, and we are done. $\square$

Note that this argument actually shows that $\{g \in A; \|g\|_A < \varepsilon\}$ is a positive proportion of A, for every $\varepsilon > 0$. So up to passing to that subset for $\varepsilon = \frac{1}{C}$, we may assume that $\|g\|_A < \frac{1}{C}$ for all $g \in A$. It is then easy to exploit the approximate group version of the Gleason–Yamabe lemmas and conclude Step 2. Indeed set $H := \{g \in A; \|g\| = 0\}$. By (1) and (2), this forms a subgroup of G, which is normalized by A. Moreover set pick $a \in A$ with $\|a\|_A$ minimal. Then (3) implies that a commutes with every element of A modulo H. This concludes Step 2.

For Step 3 and the induction on $\dim L$, we refer the reader to our original paper [16], or to Tao's book [73].

4 Applications to diameter bounds and scaling limits of finite groups

4.1 Diameter of finite simple groups

A celebrated conjecture of Babai [1] proposes the following universal polylogarithmic upper bound on the diameter of all Cayley graphs of finite simple groups:

$$\mathrm{diam}_S(G) \leq C \log(|G|)^D$$

for every finite simple group G with symmetric generating set S, where $\mathrm{diam}_S(G) := \inf\{n \in \mathbb{N}; S^n = G\}$, and absolute constants $C, D > 0$.

All attempts to this conjecture so far are based on the classification of finite simple groups, by examining case by case various infinite families, such as the alternating groups (for which the conjecture is still open, see [52] for the best available bounds), or finite simple group of Lie type in the q large limit (such as $\mathrm{PSL}_n(q)$ for n fixed). Actually in the latter case, when the rank is bounded, the conjecture has been solved by Helfgott [50] for $\mathrm{PSL}_2(p)$ and by Pyber-Szabo [63], and independently Green, Tao and myself [15] in general. In fact it is conjectured that $D = 1$ for bounded rank groups [11]. It has also been extended to some non simple perfect groups such as $\mathrm{PSL}_n(\mathbb{Z}/q\mathbb{Z})$ with q arbitrary by Varjú [78] (see also [10]).

On the other hand, one can very easily derive from our structure theorem for approximate groups, the following weaker, but very general bound.

Theorem 4.1. (Diameter bound for finite simple groups [14]). *Given $\varepsilon > 0$, there are only finitely many non-abelian finite simple groups admitting a symmetric generating set S with diameter*

$$\mathrm{diam}_S(G) \geq |G|^{\varepsilon}.$$

Proof. We argue as in the proof of Theorem 1.12, but instead of Theorem 1.11, we use the strong form of our structure theorem, i.e., Theorem 2.14. By contradiction, suppose there were infinitely many such groups. Then there would be arbitrarily large integers n and simple groups G with generating sets S, such that $|S^{5n}| \leq K|S^n|$, where $K = 5^{4/\varepsilon}$, and $n \leq \sqrt{\mathrm{diam}_S(G)}$. For otherwise there would be $n_0 \geq 1$ independent of G, S such that $|G| \geq |S^{5^m n_0}| \geq K^m |S^{n_0}| \geq K^m$ for all m with $5^m n_0 \leq \sqrt{\mathrm{diam}_S(G)}$, but this is not compatible with $\mathrm{diam}_S(G) \geq |G|^\varepsilon$, when $|G|$ is large. By Lemma 2.7, S^{2n} is a K^2-approximate subgroup. We may then apply Theorem 2.14 and conclude that $S^n \subset XP$, where $|X| \leq C(K)$ and P is a coset nilprogression of size $|P| \leq C(K)|S^{2n}|$. In particular $|S^{2n} \cap \langle P \rangle| \geq |S^{5n}|/KC(K)^2$. By Lemma 1.13, if n is large enough, $\langle P \rangle$ has index at most $KC(K)^2$ in G. But G is a non-abelian simple group, so it has no non-trivial subgroup of bounded index (for every subgroup of index d contains a normal subgroup of index at most $d!$). Hence $G = \langle P \rangle$. However, by definition, P contains a subgroup H normalized by P, such that $\langle P \rangle / H$ is nilpotent. Since G is simple, we must have $G = H$, and hence $|G| \leq |XP| \leq C(K)^2 |S^{2n}|$, and hence applying Remark 1.9 $\mathrm{diam}_S(G) \leq O_K(n)$, which is eventually a contradiction, for $n \leq \sqrt{\mathrm{diam}_S(G)}$. □

The above argument can be strengthened to arbitrary finite groups. For example, in [14] we prove:

Theorem 4.2. ([14]). *Assume that G is a finite group with symmetric generating set S such that* $\mathrm{diam}_S(G) \geq |G|^\varepsilon$. *Then G has a quotient Q with diameter at least c_ε* $\mathrm{diam}_S(G)$, *which contains a cyclic subgroup of index at most C_ε. Here $c_\varepsilon, C_\varepsilon$ are positive constants depending on ε only.*

On a Cayley graph, the diameter and the first eigenvalue of the Laplacian enjoy the following general inequality (see [26] or [14, Lemma 5.1]:

$$\lambda_1 \geq \frac{1}{8\,\mathrm{diam}_S(G)^2}$$

As a consequence of this and the above theorem, we see that every Cayley graph of a finite group G satisfies a spectral gap lower bound of the form

$$\lambda_1 \geq \frac{1}{|G|^\varepsilon}$$

unless it admits a large virtually cyclic quotient with diameter comparable to that of G. Such bounds are weaker than what is required for expander graphs, but they can be useful (they can be used for example, in the work of Ellenberg-Hall-Kowalski [31]). For additional results in this vein, we refer the reader to [14].

4.2 *Scaling limits of vertex transitive graphs*

In [3] Benjamini, Finucane, and Tessera considered the problem of determining the scaling limits of finite vertex transitive graphs satisfying the volume bound (6) below. It turns that this problem can be analyzed thanks to our structure theorem for approximate subgroups. Given constants $C, d > 0$, consider the family $\mathcal{F}_{C,d}$ of all finite, vertex transitive connected graphs X such that

$$\frac{|X|}{\deg(X)} \leq C \cdot \mathrm{diam}(X)^d \tag{6}$$

where $|X|$ is the number of vertices of X and $\deg(X)$ is the number of neighbors of a given vertex in X (all vertices have the same number of neighbors, since X is assumed vertex transitive). We denote by $\widehat{X}$ the finite metric space defined on the set X by the graph distance renormalized by a factor $\frac{1}{\mathrm{diam}(X)}$ so that $\widehat{X}$ has diameter 1. They prove:

Theorem 4.3. ([3]) *For every sequence $\{X_n\}_{n\geq 1}$ of finite graphs belonging to $\mathcal{F}_{C,d}$ for some fixed C, d, and such that* $\mathrm{diam}(X_n) \to +\infty$, *there is a subsequence $\{X_{n_k}\}$ such that $\widehat{X_{n_k}}$ converges in the Gromov–Hausdorff topology towards a torus $\mathbb{R}^q/\mathbb{Z}^q$, endowed with a left invariant Finsler metric.*

In other words: scaling limits of vertex transitive graphs with (polynomially) large diameter are flat tori equipped with a norm. We give a complete proof of Theorem 4.3 in this section. In [3] the stronger condition $|X| \leq C \cdot \mathrm{diam}(X)^d$ is assumed, but the proof works just as well assuming the weaker bound (6). When $\deg(X)$ is uniformly bounded it is possible to further show that $q \leq d$ and that the limit norm is polyhedral, see [3], but we will not pursue this here.

A left invariant Finsler metric on $\mathbb{R}^k/\mathbb{Z}^k$ is nothing but the choice of a norm $\|\cdot\|$ on $\mathbb{R}^k$, which induces a translation invariant distance on the torus $\mathbb{R}^k/\mathbb{Z}^k$ by setting $d([x],[y]) = \inf_{u\in\mathbb{Z}^k} \|x - y + u\|$.

Here is a graphic way to illustrate the above theorem: the result implies that it is impossible to approximate the Euclidean sphere S^2 by finer and finer finite vertex transitive graphs. The soccer ball (or truncated icosahedron, which is vertex transitive) is perhaps (one of) the best attempts at this game, see [2, 38].

4.2.1 Gromov compactness criterion

The Gromov–Hausdorff metric is the natural distance on the space of compact metric spaces up to isometry extending the Hausdorff topology on compact subsets of a given metric space. We refer the reader to the books [19, 46] for background on this notion. For the moment we only recall the well-known:

Gromov compactness criterion *A family $\mathcal{F}$ of compact metric spaces is relatively compact in the GH-topology if and only if it is uniformly bounded in diameter, and for all $\varepsilon > 0$ there is $N(\varepsilon) \in \mathbb{N}$ such that every $X \in \mathcal{F}$ can be covered by at most $N(\varepsilon)$ balls of radius ε.*

It easily follows from this criterion that the space of all compact metric spaces up to isometry is complete for the GH-metric. In fact this is the non-trivial part of the statement (it is easy to see that the above criterion is equivalent to the total boundedness of the family $\mathcal{F}$, and clearly a totally bounded subset of a complete metric space is relatively compact). Completeness is only proved for path metric spaces in Gromov's book [46, Prop 5.2], but it is folklore that it holds in general, see, e.g., [65], or [73, Prop. 1.8.7] for a proof with ultra-products.

We prove Theorem 4.3 in two stages, first showing the existence of a limit and then identifying it.

First note that pulling back the graph metric on $X \in \mathcal{F}_{C,d}$ to the group of automorphisms of the graph, we get a natural word metric on this group. To this effect pick a base point $x_0 \in X$ and consider the subset $S \subset G := Aut(X)$ of those $s \in S$ which either fix x_0 or send it to a neighbor. Since X is connected and vertex transitive G is generated by S and $S = S^{-1}$. Moreover the preimage under the orbit map $G \to X$, $g \mapsto g \cdot x_0$ of the ball $B(x_0, n)$ of radius n around x_0 is precisely S^n. In particular $\mathrm{diam}(X) = \mathrm{diam}_S(G)$ and

$$\frac{|S^n|}{|S|} = \frac{|B(x_0, n)|}{\deg(X)}$$

for all $n \geq 1$. This allows to easily translate the statements from X to G.

Proof of the existence of a limit in Theorem 4.3. To apply the Gromov compactness criterion above to the family $\mathcal{F} = \{\widehat{X_n}\}_{n\geq 1}$, it is enough to check that given $\varepsilon > 0$, there is $K_\varepsilon > 0$ such that every ball of radius $r \geq \varepsilon$ can be covered by at most K_ε balls of radius $\frac{r}{2}$. In fact we will show the stronger property (needed only for the identification of the limit) that K_ε can be taken to be independent of ε, if n is large enough. This follows easily from Theorem 2.15. Indeed setting $K = 9^d$ in Theorem 2.15 we get an integer $n_0 = n_0(K) \geq 1$. And we see that if $\mathrm{diam}(X_n)$ is large enough, there must be some $n_1 \in [n_0, \varepsilon\,\mathrm{diam}(X_n)]$ such that $|S^{3n_1}| \leq K|S^{n_1}|$, for otherwise $|S^{3^m n_0}| \geq 9^{dm}|S^{n_0}|$ for all m with $3^m n_0 \leq \varepsilon\,\mathrm{diam}(X_n)$, and thus $|G| \geq 9^{dm}|S|$ for such m, which implies $C\,\mathrm{diam}(X_n)^d \geq 9^{dm} \geq (\varepsilon\,\mathrm{diam}(X_n)/n_0)^{2d}$, a contradiction for n large. So we apply Theorem 2.15 and get that $|B(x_0, 4k)| \leq K'|B(x_0, k)|$ for all $k \geq \varepsilon\,\mathrm{diam}(X_n)$ and all n large enough, for some $K' > 0$ independent of n. From this we conclude that every ball of radius $4k$ can be covered by at most K' balls of radius $2k$, and this ends the proof. □

We remember from this argument, that there exists some $K = K(C, d) > 0$ such that if X is a limit of a sequence of $\widehat{X_n}$ with X_n satisfying (6) and with $\mathrm{diam}(X_n) \to +\infty$, then X is K-doubling in the sense that every ball of radius $r > 0$ can be covered by at most K balls of radius $\frac{r}{2}$. It follows immediately from this that X has finite Hausdorff dimension, hence *finite topological dimension.*

4.2.2 Identification of the limit: outline

We now turn to the identification of the limit. This part is almost purely formal in that it follows from a simple combination of known facts (we differ here from the original treatment of [3]). Here is the argument. If $\widehat{X_n}$ converges to some compact metric space X, then $G := Isom(X)$ must act transitively on X, because $G_n := Isom(X_n)$ acts transitively on X_n. Moreover, X is pathwise connected and locally connected (because the balls in the graphs X_n are connected), and has finite topological dimension. From this, a lemma of Montgomery–Zippin, which is a simple consequence of the Peter–Weyl theorem, tells us that G is a (compact) Lie group. Moreover up to passing to a subsequence (equivariant GH-convergence), we may assume that there is an ε_r -representation of G_n in G, which is ε_n-dense, for some $\varepsilon_n \to 0$. A well-known result (the so-called *Ulam stability* of almost representations of finite groups in compact groups, due to Alan Turing in this setting and further generalized by Kazhdan and others) ensures that the ε_n-representation must be close to a genuine representation. Then Jordan's lemma on finite subgroups of Lie groups and the almost density of the representations imply that G^0 must be abelian, hence $X = G^0$ is a torus. The metric on $X = G^0$ is geodesic and left invariant: all such metrics on a torus are Finsler (i.e., given by a norm).

4.2.3 Montgomery–Zippin lemma

We now pass to the details. Set $G := Isom(X)$. We already know that X is pathwise connected, locally connected, G-transitive, and finite dimensional. We have the following well-known lemma:

Lemma 4.4. *If G is a compact group acting faithfully and transitively on a finite-dimensional, connected and locally connected topological space, then G is a Lie group with finitely many connected components.*

Proof. The lemma holds for locally compact groups in general and is due to Montgomery–Zippin (see [61, §6.3, p. 243]). It easily follows from the Gleason–Yamabe theorem. In the case of compact groups, only Peter–Weyl is needed: we include the proof for the reader's convenience.

According to the Peter–Weyl theorem, there is a nested sequence of closed normal subgroups N_i in G such that G/N_i is a Lie group and $\bigcap_i N_i = \{1\}$. We have to show that N_i is trivial for i large. Let $X_i := X/N_i$ be the quotient space under the action of N_i. The space X is naturally a projective limit of the X_i's, i.e., $X = \varprojlim X_i$. Note that the X_i's are manifolds: they are homogeneous spaces of the compact Lie group G/N_i. Since X is finite dimensional as a topological space, we must have $\dim(X_i) = \dim(X_{i_0})$ when i is larger than some i_0. This implies that N_{i_0}/N_i is zero-dimensional if $i \geq i_0$. Being a compact Lie group, it must be finite. Therefore X_i is a finite normal cover of X_{i_0} with finite group of deck transformations N_{i_0}/N_i. The group N_{i_0} is a projective limit of finite groups, i.e., it is profinite.

We claim that N_{i_0} is finite. This follows from the local connectedness of X. Indeed, by covering theory, the universal cover of X_{i_0} is a cover of each X_i, $i \geq i_0$. Therefore, given $x \in X_{i_0}$ there exists an open neighborhood U of x such that $\pi_i^{-1}(U)$ is a disjoint union of open subsets of X_i each homeomorphic to U and permuted by the action of N_{i_0}/N_i for $i \geq i_0$. Hence $\Omega := \varprojlim \pi_i^{-1}(U)$ is homeomorphic to $U \times N_{i_0}$. Note that since $X = \varprojlim X_i$, Ω is also an open subset of X, which is not locally connected at any point, unless N_{i_0} is finite. This ends the proof. □

Note in passing that the hypothesis that X be locally connected is important to avoid examples such as the solenoid $(\mathbb{R} \times \mathbb{Q}_2)/\mathbb{Z}[\frac{1}{2}]$, which is a connected and finite dimensional compact group, but not a Lie group.

Note also that, since X is connected, the connected component of the identity G^0 in G acts transitively on X.

4.2.4 Ulam stability and almost representations

We now recall the notion of almost representation. Given a target group G, endowed with a left invariant distance d making it a metric space, a mapping ϕ from a group Γ to G, is called *an ε-representation* if for every $x, y \in \Gamma$,

$$d(\phi(xy), \phi(x)\phi(y)) \leq \varepsilon.$$

We say moreover that ϕ is *ε-dense* if for each $g \in G$ there is $x \in \Gamma$ such that $d(g, \phi(x)) \leq \varepsilon$.

Ulam stability (see [77]) is the general problem of determining to what extent an ε-representation of Γ is $\delta(\varepsilon)$-close to a genuine group homomorphism, i.e., does there exists a group homomorphism $\rho : \Gamma \to G$ such that $d(\phi(x), \rho(x)) \leq \delta(\varepsilon)$, for some $\delta(\varepsilon)$ tending to 0 as $\varepsilon \to 0$. A little known and recently rediscovered [3, 38] 1938 paper of Alan Turing [76] deals precisely with this problem in the case when Γ is a finite group and G a connected Lie group endowed with a left invariant Riemannian metric:

Theorem 4.5. (Turing's theorem). *Suppose G is a connected compact Lie group, then there is $C > 0$ such that every ε-dense ε-representation of some finite group into G is $C\varepsilon$-close to a group homomorphism. In particular, if there is a sequence of finite groups Γ_n and ε_n-representations $\phi_n : \Gamma_n \to G$ that are ε_n-dense in G for some $\varepsilon_n \to 0$, then G is a compact abelian Lie group.*

The second part of the statement follows immediately from the combination the first and of Jordan's theorem on finite subgroups of Lie groups: that they contain an abelian subgroup of bounded index [12, 25, 55]. The first part actually holds without the assumption of ε-density (we do not need this refinement for the proof of Theorem 4.3), as follows from the following more general result of Kazhdan [57, Theorem 1]. Let $G = \mathcal{U}(\mathcal{H})$ be the unitary group of a separable Hilbert space, endowed with the bi-invariant distance $d(g, h) := \|g - h\|$.

Theorem 4.6. (Ulam stability of amenable groups). *Every ε-representation ϕ of an amenable group Γ into G is ε-close to a genuine representation ρ, i.e., $d(\phi(\gamma), \rho(\gamma)) \leq \varepsilon$ for all $\gamma \in \Gamma$.*

Note that every finite group is amenable, so this applies uniformly to all finite groups Γ, and also note that every connected compact Lie group is a subgroup[2] of $\mathcal{U}(\mathcal{H})$ (via the regular representation) so the result applies to this situation as well and implies Turing's theorem.

For a very short proof of Kazhdan's theorem, as well as a modern view on Ulam stability, we refer the reader to the paper [20], in particular Theorem 3.2 there.

4.2.5 Equivariant GH convergence

When each term of a sequence of compact metric spaces comes equipped with a group action by isometries, one can strengthen the Gromov compactness theorem and argue that under the same hypotheses, one can find a subsequence which converges in an almost equivariant way. Namely we have:

Equivariant Gromov–Hausdorff compactness *Let $(X_n, d_n) \to (X, d)$ be a converging sequence of compact metric spaces. Let $\Gamma_n := Isom(X_n)$ Then there is a closed subgroup of isometries G of X and subsequence $\{X_{n_k}\}$, such that $\{(X_{n_k}, d_{n_k})\}$ equivariantly GH converges to (X, G), in other words, there are maps $f_k : X_{n_k} \to X$ and $\phi_k : \Gamma_{n_k} \to G$ with ε_k-dense image in X and G respectively (where $\varepsilon_k \to 0$), such that*

$$\sup_{x \in X_{n_k}, \gamma \in \Gamma_{n_k}} d(f_k(\gamma x), \phi_k(\gamma) f_k(x)) \xrightarrow[k \to +\infty]{} 0 \tag{7}$$

This is a classical fact, which holds more generally for locally compact metric spaces and pointed GH convergence, see, for example, [35, Prop. 3.6]. We give a proof for convenience.

Proof. By definition there are almost isometries $f_n : X_n \to X$ and $g_n : X \to X_n$ such that $f_n \circ g_n$ and $g_n \circ f_n$ are both uniformly close to the identity. Pick a nested sequence of ε_k-dense finite subsets F_k in X, where $\varepsilon_k \to 0$. The restrictions of elements in $f_n \Gamma_n g_n$ to F_k form a compact set of maps from F_k to X. By the usual diagonal process we can extract a subsequence of $(X_n)_n$ such that these compact sets converge as n tends to infinity for every given k. Let G_k be the limit. The restriction map $G_{k+1} \to G_k$ is surjective. The projective limit $G := \varprojlim G_k$ is a compact set of maps on $\cup_k F_k$. Maps in G_k are distance preserving, hence so are maps in G. Thus maps in G extend by continuity to isometries of X.

Given k there is n_k such that G_k is ε_k-close to the restriction of $f_n \Gamma_n g_n$ to F_k for $n = n_k$. Thus we may find for each $\gamma \in \Gamma_{n_k}$ an element $\phi_k(\gamma)$ in G_k (hence also one

[2] Ulam stablility may fail for certain target groups other that U(H), for example it fails for G the p-adic integers, see [57][Prop. 1.]

in G) such that $f_{n_k}\gamma g_{n_k}$ is $3\varepsilon_k$-close to $\phi_k(\gamma)$ on X. Similarly the image of ϕ_k is $3\varepsilon_k$-dense in G. It is then straightforward to check (7) and that G is indeed a subgroup. □

Note that it follows directly from (7) that ϕ_{n_k} is a δ_k-representation of Γ_{n_k} inside G for some $\delta_k \to 0$. Therefore Kazhdan's theorem 4.6 guarantees that we may in fact choose the ϕ_{n_k} to be a genuine group homomorphism from Γ_{n_k} into G.

By Lemma 4.4, in the setting of Theorem 4.3, the limit group G is a compact Lie group and the groups Γ_n are finite. Therefore Turing's theorem (or the combination of Kazhdan and Jordan) implies that G^0 is an abelian compact Lie group, i.e., a torus $\mathbb{R}^k/\mathbb{Z}^k$. We have $X = G^0/H$ for some closed subgroup H. But since G^0 acts faithfully on X (it is a subgroup of isometries) and G^0 is abelian, we must have $H = 1$, i.e., $X = G^0$.

4.2.6 Invariant length metrics on Lie groups

To conclude the proof of Theorem 4.3 it remains to identify the metric on the limit space X. We know that it is a *length metric* a.k.a intrinsic metric, i.e., the distance between two points coincides with the infimum of the length of paths between two points. This is so, because X is a GH limit of graphs and the graph metric is a length metric. Since $X = \mathbb{R}^k/\mathbb{Z}^k$, the distance on X lifts naturally to a length metric on $\mathbb{R}^k$, which is invariant under translations. So the last part of Theorem 4.3 is now a consequence of the following simple and well-known lemma, of which we give a proof for convenience.

Lemma 4.7. *If d is a translation invariant length metric on $\mathbb{R}^k$, then there is a norm $\|\cdot\|$ on $\mathbb{R}^k$ such that $d(x,y) = \|x-y\|$ for all x,y.*

Proof. Set $\|x\| := \limsup_{t\to 0} \frac{1}{t}d(0,tx)$. Clearly, by translation invariance and the triangle inequality $d(0,x) \le nd(0,\frac{x}{n})$ for every $n \in \mathbb{N}$. It follows that $d(0,x) \le \|x\|$. It is also straightforward to verify that $\|x+y\| \le \|x\| + \|y\|$ and that $\|\lambda x\| = \lambda\|x\|$ if $\lambda > 0$. Hence $\|x\|$ is a norm on $\mathbb{R}^k$. Moreover we also have $\|x\| := \liminf_{t\to 0} \frac{1}{t}d(0,tx)$. Indeed if not there would be some $\beta < 1$ and arbitrarily small s,t such that $\frac{1}{s}d(0,sx) \le \beta\frac{1}{t}d(0,tx)$. If s is chosen much smaller than t, say $t = ns + s'$, $0 \le s' \le s$, $n \ge 1$, then by the triangle inequality and translation invariance $d(0,tx) \le nd(0,sx) + d(0,s'x)$, which eventually contradicts the previous inequality.

So $\|x\| := \lim_{t\to 0} \frac{1}{t}d(0,tx)$. By local compactness of $\mathbb{R}^k$, this limit is uniform over x with $\|x\| = 1$. From this it is easy to conclude that $\|x\| = d(0,x)$. Indeed if not, then $d(0,x) \le (1-\varepsilon)\|x\|$ for some x and $\varepsilon > 0$. Moreover given any $\delta > 0$, there are points $x_0 = 0, \ldots, x_n = x$ such that $d(x_i, x_{i+1}) < \delta$ and $d(x_0,x_1) + \ldots + d(x_{n-1},x_n) \le (1-\frac{\varepsilon}{2})\|x\|$. However $d(x_i,x_{i+1})$ is uniformly very close to $\|x_i - x_{i+1}\|$, but $\|x\| \le \|x_0 - x_1\| + \ldots + \|x_{n-1} - x_n\|$ by the triangle inequality. A contradiction. □

This lemma is a special case of the following much more general result of Berestovski, which gives a complete characterization of left invariant length metrics on Lie groups.

Before we state Berestovski's theorem, let us give a general construction of left invariant length metrics on a Lie group. Pick a norm on the Lie algebra $\mathfrak{g}$ of the Lie group G and use it to define the length of piecewise differential paths in exactly the same way as left invariant Riemannian metrics on G are defined, except that the norm is now arbitrary instead of Euclidean. This obviously gives rise to a left invariant (Finsler) length metric on G. More generally, we may consider a norm on a vector subspace V of $\mathfrak{g}$, which generates $\mathfrak{g}$ as a Lie algebra. Similarly this allows to define the length of piecewise differentiable *horizontal* paths, that is paths whose derivative, once translated back to the origin, everywhere belongs to V. The property of V that it generates $\mathfrak{g}$ can be shown (this is the Chow–Rachevski theorem from sub-Riemannian geometry) to imply that every two points in G can be joined by a horizontal piecewise differentiable continuous path. Hence this yields a well-defined length metric on G. Metrics arising from the above construction are called left invariant Carnot–Caratheodory–Finsler metrics.

When G is abelian, then every subspace of $\mathfrak{g}$ is a subalgebra, so this refinement does not produce new examples of left invariant metrics. But if G is nilpotent (e.g., the Heisenberg group), then many new length metrics arise this way. In fact it can be shown ([4]) that G abelian and $G = \mathbb{R} \ltimes \mathbb{R}^n$, where $\mathbb{R}$ acts by homotheties, are the only Lie groups in which all Carnot–Caratheodory–Finsler metrics are Finsler. We can now state:

Theorem 4.8. (Berestovski [4]). *Every left invariant length metric on a connected Lie group is Carnot–Caratheodory–Finsler.*

This ends our discussion of Theorem 4.3.

Acknowledgements These lectures were prepared for the workshop on *Additive and Analytic Combinatorics* held at the IMA in Minneapolis from September 29 to October 3, 2014. Due to a health problem, I was unfortunately not able to give the lectures. I am very grateful to Terry Tao, who replaced me on the spot. I also thank T. Tao, R. Tessera and M. Tointon for their valuable comments on the text. The author is partially supported by ERC grant no. 617129.

References

1. L. Babai, A. Seress, On the diameter of Cayley graphs of the symmetric group. J. Comb. Theory Ser. A **49**, 175–179 (1988)
2. I. Benjamini, *Coarse Geometry and Randomness*. Lecture Notes from the 41st Probability Summer School Held in Saint-Flour, 2011. Lecture Notes in Mathematics (Springer, Cham, 2013), viii+129 pp.
3. I. Benjamnin, H. Finucane, R. Tessera, On the scaling limit of finite vertex transitive graphs with large diameter. arXiv:1203.5624 (2011)
4. V.N. Berestovski, Homogeneous manifolds with intrinsic metric. Sib. Mat. Z. **29**(6), 17–29 (1988)

5. Y. Bilu, Structure of sets with small sumset, in *Structure Theory of Set Addition*. Astérisque, vol. 258 (1999), pp. 77–108
6. G. Birkhoff, A note on topological groups. Compos. Math. **3**, 427–430 (1936)
7. L.V. Brailovsky, G.A. Freiman, *On a product of finite subsets in a torsion-free group*, J. Algebra **130**(2), 462–476 (1990)
8. J. Bourgain, N. Katz, T. Tao. A sum-product estimate in finite fields and their applications. Geom. Funct. Anal. **14**(1), 27–57 (2004)
9. J. Bourgain, A.A. Glibichuk, S.V. Konyagin, Estimates for the number of sums and products and for exponential sums in fields of prime order. J. Lond. Math. Soc. **73**(2), 380–398 (2006)
10. H. Bradford, New Uniform Diameter Bounds in Pro-p Groups. arXiv preprint 1410.3007
11. E. Breuillard, Diophantine geometry and uniform growth of finite and infinite groups, in *Proceedings of the International Congress of Mathematicians*, ICM, Seoul, 2014
12. E. Breuillard, An exposition of Camille Jordan's original proof of his theorem on finite subgroups of invertible matrices. Notes available at http://www.math.u-psud.fr/~breuilla/cour.html (2011)
13. E. Breuillard, B.J. Green, Approximate groups. I. The torsion-free nilpotent case. J. Inst. Math. Jussieu **10**(1), 37–57 (2011)
14. E. Breuillard, M. Tointon, Nilprogressions and groups with moderate growth. arXiv preprint 1506.00886
15. E. Breuillard, B. Green, T. Tao, Approximate subgroups of linear groups. Geom. Funct. Anal. **21**(4), 774–819 (2011)
16. E. Breuillard, B. Green, T. Tao, The structure of approximate groups. Publ. Inst. Hautes Études Sci. **116**(1), 115–221 (2012)
17. E. Breuillard, B. Green, T. Tao, *Small Doubling in Groups*, Erds centennial. Bolyai Society Mathematical Studies, vol. 25 (János Bolyai Mathematical Society, Budapest, 2013), pp. 129–151
18. Y. Burago, V.A. Zalgaller, *Geometric Inequalities*. Translated from the Russian by A.B. Sosinski. Grundlehren der Mathematischen Wissenschaften. Springer Series in Soviet Mathematics, vol. 285 (Springer, Berlin, 1988), xiv+331 pp.
19. D. Burago, Y. Burago, S. Ivanov, A course in metric geometry. Graduate Studies in Mathematics, vol. 33 (American Mathematical Society, Providence, 2001). MR1835418 (2002e:53053)
20. M. Burger, N. Ozawa, A. Thom, On Ulam stability. Isr. J. Math. **193**(1), 109–129 (2013)
21. P. Buser, H. Karcher, *Gromov's Almost Flat Manifolds*. Astérisque, vol. 81 (Société Mathmatique de France, Paris, 1981), 148 pp.
22. J.O. Button, Explicit Helfgott type growth in free products and in limit groups. J. Algebra **389**, 61–77 (2013)
23. M.C. Chang, A polynomial bound in Freiman's theorem. Duke Math. J. **113**(3), 399–419 (2002)
24. E. Croot, O. Sisask, A probabilistic technique for finding almost-periods of convolutions. Geom. Funct. Anal. **20**(6), 1367–1396 (2010)
25. C.W. Curtis, I. Reiner, *Representation Theory of Finite Groups and Associative Algebras* (Interscience, New York, 1962)
26. P. Diaconis, L. Saloff-Coste, Comparison techniques for random walk on finite groups. Ann. Probab. **21**(4), 2131–2156 (1993)
27. L. van den Dries, A.J. Wilkie, Gromov's theorem on groups of polynomial growth and elementary logic. J. Algebra **89**, 349–374 (1984)
28. L. van den Dries, I. Goldbring, Hilbert 5th problem. L'enseignement Mathematique. **61**, 3–43 (2015)
29. L. van den Dries, *Approximate Groups*. Séminaire Bourbaki, Exposé 1077. Astérisque 367–368 (2015), pp. 79–113
30. G. Elekes, Z. Király, On the combinatorics of projective mappings. J. Algebraic Comb. **14**(3), 183–197 (2001)
31. J.S. Ellenberg, C. Hall, E. Kowalski, Expander graphs, gonality and variation of Galois representations. Duke Math. J. **161**(7), 1233–1275 (2012)

32. G.A. Freiman, *Foundations of a Structural Theory of Set Addition*. Translations of Mathematical Monographs, vol. 37 (American Mathematical Society, Providence, 1973)
33. G.A. Freiman, *Groups and the inverse problems of additive number theory*, Number-theoretic studies in the Markov spectrum and in the structural theory of set addition, Kalinin. Gos. Univ., Moscow, 175–183 (1973)
34. G.A. Freiman, On finite subsets of nonabelian groups with small doubling. Proc. Am. Math. Soc. **140**(9), 2997–3002 (2012)
35. K. Fukaya, T. Yamaguchi, The fundamental groups of almost non-negatively curved manifolds. Ann. Math. (2) **136**(2), 253–333 (1992)
36. M.Z. Garaev, An explicit sum-product estimate in $\mathbb{F}_p$. Int. Math. Res. Not. **11**, 1–11 (2007)
37. M.Z. Garaev, The sum-product estimates for large subsets of prime fields. Proc. Am. Math. Soc. **137**(8), 2735–2739 (2008)
38. T. Gelander, Limits of finite homogeneous metric spaces. Enseign. Math. (2) **59**(1–2), 195–206 (2013)
39. N. Gill, H. Helfgott, Growth in solvable subgroups of $GL_r(\mathbb{Z}/p\mathbb{Z})$. Math. Ann. **360**(1–2), 157–208 (2014)
40. A.M. Gleason, Groups without small subgroups. Ann. Math. (2) **56**, 193–212 (1952)
41. A. Granville, An introduction to additive combinatorics, *Additive Combinatorics*. CRM Proceedings and Lecture Notes, vol. 43,(American Mathematical Society, Providence, 2007), pp. 1–27
42. B.J. Green, The polynomial Freiman-Ruzsa conjecture, in *Open Problems in Mathematics* [Online], vol. 2 (2014). http://opmath.org/index.php/opm/index
43. B. Green, I.Z. Ruzsa, Freiman's theorem in an arbitrary abelian group. J. Lond. Math. Soc. (2) **75**(1), 163–175 (2007)
44. M. Gromov, Groups of polynomial growth and expanding maps. Publ. Inst. Hautes Études Sci. **53**, 53–73 (1981)
45. M. Gromov, Carnot-Carathodory spaces seen from within, in *Sub-Riemannian Geometry*. Progress in Mathematics, vol. 144 (Birkhuser, Basel, 1996), pp. 79–323
46. M. Gromov, Metric structures for Riemannian and non-Riemannian spaces, in *Modern Birkhäuser Classics* (2007). Based on the 1981 French original Birkauser (Springer)
47. Y. Hamidoune, Two inverse results. Combinatorica **33**(2), 217–230 (2013)
48. Y. Hamidoune, A. S. Lladó, O. Serra, *On subsets with small product in torsion-free groups*, Combinatorica **18**(4), 529–540 (1998)
49. Y. Hamidoune, *On small subset product in a group*, Structure theory of set addition. Astèrisque No. **258**, xiv–xv, 281–308 (1999)
50. H.A. Helfgott, Growth and generation in $SL_2(\mathbb{Z}/p\mathbb{Z})$. Ann. Math. (2) **167**(2), 601–623 (2008)
51. H.A. Helfgott, Growth and generation in $SL_3(\mathbb{Z}/p\mathbb{Z})$. J. Eur. Math. Soc. **13**(3), 761–851 (2011)
52. H.A. Helfgott, A. Seress, On the diameter of permutation groups. Ann. Math. (2) **179**(2), 611–658 (2014)
53. J. Hirschfeld, The nonstandard treatment of Hilbert's fifth problem. Trans. Am. Math. Soc. **321**(1), 379–400 (1990)
54. E. Hrushovski, Stable group theory and approximate subgroups. J. Am. Math. Soc. **25**(1), 189–243 (2012)
55. C. Jordan, Mémoire sur les équations différentielles linéaires à intégrale algébrique. J. Reine Angew. Math. **84**, 89–215 (1878)
56. I. Kaplansky, *Lie Algebras and Locally Compact Groups* (The University of Chicago Press, Chicago/London, 1971), xi+148 pp.
57. D. Kazhdan, On ε-representations. Isr. J. Math. **43**, 315–323 (1982)
58. J.H.B. Kemperman, *On complexes in a semigroup*, Nederl. Akad. Wetensch. Proc. Ser. A 59 = Indag. Math. **18**, 247–254 (1956)
59. E. Lindenstrauss, Pointwise theorems for amenable groups. Invent. Math. **146**(2), 259–295 (2001)
60. E. Lindenstrauss, Personal communication (2007)

61. D. Montgomery, L. Zippin, *Topological Transformation Groups* (Interscience Publishers, New York/London, 1955)
62. G. Petridis, *New proofs of Plünnecke-type estimates for product sets in groups*, Combinatorica **32**(6) (2012), 721–733.
63. L. Pyber, E. Szabó, Growth in finite simple groups of Lie type. J. Am. Math. Soc. **29**(1), 95–146 (2016)
64. A.A. Razborov, A product theorem in free groups. Ann. Math. **179**(2), 405–429 (2014)
65. J. Rouyer, Generic properties of compact metric spaces. Topol. Appl. **158**(16), 2140–2147 (2011)
66. M. Rudnev, An improved sum-product inequality in fields of prime order. Int. Math. Res. Not. **2012**(16), 3693–3705 (2012)
67. I.Z. Ruzsa, Generalized arithmetical progressions and sumsets. Acta Math. Hung. **65**(4), 379–388 (1994)
68. S.R. Safin, Powers of subsets of free groups. Mat. Sb. **202**(11), 97–102 (2011); translation in Sb. Math. **202**(11–12), 1661–1666 (2011)
69. T. Sanders, On a nonabelian Balog-Szemerdi-type lemma. J. Aust. Math. Soc. **89**(1), 127–132 (2010)
70. T. Sanders, The structure theory of set addition revisited. Bull. Am. Math. Soc. **50**, 93–127 (2013)
71. T. Tao, Product set estimates for non-commutative groups. Combinatorica **28**(5), 547–594 (2008)
72. T. Tao, Freiman's theorem for solvable groups. Contrib. Discret. Math. **5**(2), 137–184 (2010)
73. T. Tao, *Hilbert'S Fifth Problem and Related Topics*. Graduate Studies in Mathematics, vol. 153 (American Mathematical Society, Providence, 2014), xiv+338 pp.
74. T. Tao, V. Vu, *Additive Combinatorics*. Cambridge Studies in Advanced Mathematics, vol. 105 (Cambridge University Press, Cambridge, 2006)
75. M.C.H. Tointon, Freiman's theorem in an arbitrary nilpotent group. Proc. Lond. Math. Soc. **109**, 318–352 (2014)
76. A.M. Turing, Finite approximations to Lie groups. Ann. Math. (2) **39**(1), 105–111 (1938)
77. S.M. Ulam, *A Collection of Mathematical Problems* (Interscience Publishers, New York, 1960)
78. P.P. Varjú, Random walks in compact groups. Doc. Math. **18**, 1137–1175 (2013)
79. H. Yamabe, A generalization of a theorem of Gleason. Ann. Math. (2) **58**, 351–365 (1953)
80. H. Yamabe, On the conjecture of Iwasawa and Gleason. Ann. Math. (2) **58**, 48–54 (1953)
81. G. Zémor, *A generalisation to noncommutative groups of a theorem of Mann*, Discrete Math. **126**(1-3), 365–372 (1994)

Character sums and arithmetic combinatorics

Mei-Chu Chang

Abstract In this survey we review some old and new results on character sums and their applications to various problems in analytic number theory, e.g., smallest quadratic nonresidues, Dirichlet L-function, Linnik's problem. We also discuss open problems in this area. Some of the techniques involved belong to arithmetic combinatorics. One may hope for these methods to lead to further progress.

Keywords Character sums • Zero-free regions.

Mathematics Subject Classification (2010): 11L40, 11M06.

A function $\chi : \mathbb{Z}/q\mathbb{Z} \to \{z \in \mathbb{C} : |z| = 1\} \cup \{0\}$ is a multiplicative character mod q if it satisfies the properties that $\chi(mn) = \chi(m)\chi(n)$, and $\chi(n) = 0$ if $\gcd(n, q) \neq 1$. We are interested in bounding non-trivially the character sum of χ over an interval of length $H \leq q$. More precisely, we want to study the following problem.

Problem 1. *Assuming χ is non-principal and $q \gg 0$, how small $H = H(q)$ can be such that*

$$\Big| \sum_{n=a+1}^{a+H} \chi(n) \Big| < H^{1-\epsilon} \ ? \tag{1}$$

In 1918, Polya and Vinogradov (Theorem 12.5 in [27]) had the estimate for $H \gg \sqrt{q} \log q$.

Theorem 1 (Polya-Vinogradov). *Let χ be a non-principal Dirichlet character* mod q. *Then*

$$\Big| \sum_{m=a+1}^{a+H} \chi(m) \Big| < Cq^{\frac{1}{2}}(\log q).$$

M.-C. Chang (✉)
Department of Mathematics, University of California, Riverside, CA 92521, USA
e-mail: mcc@math.ucr.edu

A. Beveridge et al. (eds.), *Recent Trends in Combinatorics*, The IMA Volumes in Mathematics and its Applications 159, DOI 10.1007/978-3-319-24298-9_17

Forty four years later Burgess [5] improved Polya and Vinogradov's result to $H > q^{\frac{1}{4}+\varepsilon}$, first for prime moduli, then for cube-free moduli.

Theorem 2 (Burgess). *For $\varepsilon > 0$ there exists $\delta > 0$ such that if $H > p^{\frac{1}{4}+\varepsilon}$, then*

$$\Big|\sum_{m=a+1}^{a+H} \chi(m)\Big| \ll p^{-\delta}H.$$

Using sieving, there is the following

Corollary 1. *The smallest quadratic nonresidue* mod *p is at most $p^{\frac{1}{4\sqrt{e}}+\varepsilon}$.*

At present time, Burgess' estimate is still the best. Davenport and Lewis generalized it to higher dimensions by replacing the interval by a box $B \subset \mathbb{F}_{p^n}$.

Let $\{\omega_1, \dots, \omega_n\}$ be an arbitrary basis for $\mathbb{F}_{p^n}$ over $\mathbb{F}_p$. Then for any $x \in \mathbb{F}_{p^n}$, there is a unique representation of x in terms of the basis.

$$x = x_1\omega_1 + \cdots + x_n\omega_n$$

A *box* $B \subset \mathbb{F}_{p^n}$ is a set such that for each j, the coefficients x_j form an interval.

$$\begin{aligned} B &= \Big\{\sum_{j=1}^{n} x_j\omega_j : x_j \in [a_j+1, a_j+H_j], \quad \forall i\Big\} \\ &= \prod_{j=1}^{n}[a_j+1, a_j+H_j]. \end{aligned} \tag{2}$$

Davenport and Lewis had the following non-trivial estimate for character sums over boxes. (See [6, 28] for work on special boxes.)

Theorem 3. [12, Theorem 2] *Let $H_j = H$ for $j = 1, \dots, n$, with*

$$H > p^{\frac{n}{2(n+1)}+\delta} \text{ for some } \delta > 0 \tag{3}$$

and let $p > p(\delta)$. Then, with B defined as above

$$\Big|\sum_{x\in B} \chi(x)\Big| < (p^{-\delta_1}H)^n,$$

where $\delta_1 = \delta_1(\delta) > 0$.

For $n = 1$ (i.e., $\mathbb{F}_q = \mathbb{F}_p$) this is Burgess' result. But as n increases, the exponent in (3) tends to $\frac{1}{2}$.

Motivated by the work of Burgess and Davenport-Lewis, we obtained the following estimates on incomplete character sums. Our result improves Theorem DL for $n > 4$ [7, 8] and is also uniform in n.

Theorem 4. *Let χ be a non-trivial multiplicative character of $\mathbb{F}_{p^n}$, and let $\varepsilon > 0$ be given. If*

$$B = \prod_{j=1}^{n} [a_j + 1, a_j + H_j]$$

is a box satisfying

$$\prod_{j=1}^{n} H_j > p^{(\frac{2}{5}+\varepsilon)n},$$

then for $p > p(\varepsilon)$

$$\Big|\sum_{x \in B} \chi(x)\Big| \ll_n p^{-\frac{\varepsilon^2}{4}} |B|,$$

unless n is even and $\chi|_{F_2}$ is principal, where F_2 is the subfield of size $p^{n/2}$, in which case

$$\Big|\sum_{x \in B} \chi(x)\Big| \leq \max_{\xi} \big|B \cap \xi F_2\big| + O_n(p^{-\frac{\varepsilon^2}{4}} |B|).$$

The proof of Theorem 4 used ingredients and techniques from sum–product theory, specially multiplicative energy. Theorem 4 was improved by Konyagin [30] for regular boxes.

Theorem 5. [30] *Let χ be a non-trivial multiplicative character of $\mathbb{F}_{p^n}$, and let B be given as in (2) with*

$$H_j = H > p^{\frac{1}{4}+\epsilon}, \quad \forall j.$$

Then

$$\Big|\sum_{x \in B} \chi(x)\Big| \ll_n p^{-\delta} |B|,$$

where $\delta = \delta(\epsilon) > 0$.

Problem 2. *Obtain a non-trivial estimate on $\sum_{x \in B} \chi(x)$, assuming $|B| > q^{1/4+\varepsilon}$ and B not 'essentially' contained in a multiplicative translate of a subfield.*

As in [12] (see p. 131), Theorem 4 has the following application to the distribution of primitive roots of $\mathbb{F}_{p^n}$.

Corollary 2. *Let $B \subset \mathbb{F}_{p^n}$ be as in Theorem 4 and satisfying* $\max_\xi \left|B \cap \xi F_2\right| < p^{-\varepsilon}|B|$ *if n even. The number of primitive roots of $\mathbb{F}_{p^n}$ belonging to B is*

$$\frac{\varphi(p^n-1)}{p^n-1}|B|(1+o(p^{-\tau'})),$$

where $\tau' = \tau'(\varepsilon) > 0$ and assuming $n \ll \log\log p$.

The proof of Corollary 2 is by combining character sums estimate with the following.

$$\frac{\phi(p^n-1)}{p^n-1}\left\{1+\sum_{\substack{d|p^n-1\\ d>1}}\frac{\mu(d)}{\phi(d)}\sum_{\operatorname{ord}(\chi)=d}\chi(x)\right\} = \begin{cases} 1 & \text{if } x \text{ is primitive,}\\ 0 & \text{otherwise.}\end{cases}$$

We also generalize Burgess's inequality in a slightly different direction.

Theorem 6. *Let $\mathcal{P}$ be a proper d-dimensional generalized arithmetic progression in $\mathbb{F}_p$ with*

$$|\mathcal{P}| > p^{\frac{2}{5}+\varepsilon}, \text{ some } \varepsilon > 0$$

If χ is a non-principal multiplicative character of $\mathbb{F}_p$, we have

$$\left|\sum_{x\in\mathcal{P}}\chi(x)\right| < p^{-\tau}|\mathcal{P}|,$$

where $\tau = \tau(\varepsilon, d) > 0$ and assuming $p > p(\varepsilon, d)$.

Remark 6.1. The exponent $\frac{2}{5} < \frac{1}{2}$ does not depend on d.

Remark 6.2. Similar results hold for $\mathbb{F}_{p^n}$ with worse exponent.

Using Freiman's theorem, sum–product, and character sums, we obtained the following.

Corollary 3. *Given $C > 0$ and $\varepsilon > 0$, there is a constant $\kappa = \kappa(C, \varepsilon)$ and a positive integer $k < k(C, \varepsilon)$ such that if $A \subset \mathbb{F}_p$ satisfies*

(i) $|A| > p^{2/5+\varepsilon}$
(ii) $|A+A| < C|A|$.

Then we have

$$|A^k| > \kappa p,$$

where $A^k = A\cdots A$ is the k-fold product set of A.

There is also the study of multi-linear character sums.

Let $(L_j)_{1\le j\le n}$ be n independent linear forms in n variables over $\mathbb{F}_p$, and let χ be a non-principal multiplicative character mod p. Denote a box by $B = \prod_{i=1}^{n} [a_i,\ a_i + H]$. We are interested in the following non-trivial estimates

$$\left| \sum_{x\in B} \chi\Big(\prod_{j=1}^{n} L_j(x)\Big) \right| < p^{-\delta} H^n. \tag{4}$$

Theorem 7 (Burgess). *Assume*

$$H > p^{\frac{1}{2} - \frac{1}{2(n+1)} + \varepsilon}. \tag{5}$$

Then (4) *holds.*

For $n = 1$, condition (5) is the well-known Burgess assumption $H > p^{\frac{1}{4}+\varepsilon}$ for character sum bound. For $n = 2$, it is $H > p^{\frac{1}{3}+\varepsilon}$. We generalized the Burgess assumption to any dimension. (See [4].)

Theorem 8 (Bourgain-Chang). *Assume*

$$H > p^{\frac{1}{4}+\varepsilon}, \quad \textit{for any } n.$$

Then (4) *holds.*

The theorem above has application to character sums of polynomials.

Theorem 9. *Let $f(x_1, \dots, x_d)$ be a homogeneous polynomial of degree d and split over $\overline{\mathbb{F}}_p$, and let $B = \prod_{i=1}^{d}[a_i, a_i + H] \subset \mathbb{F}_p^d$ be a box of length H with*

$$H = p^{\frac{1}{4}+\varepsilon}.$$

Then

$$\left| \sum_{x\in B} \chi(f(x)) \right| < p^{-\delta} H^d.$$

This improves Gillett's condition $H > p^{\frac{d}{2(d+1)}+\varepsilon}$.

Coming back to Problem 1, for special moduli, the condition on H in Burgess' theorem can be improved. One can proved (1) for much smaller intervals. There are two classical results, based on quite different arguments by Graham–Ringrose ([27], Corollary 12.15) and Iwaniec ([27], Theorem 12.16).

First, for the modulus q, we set up the following notations for the largest prime divisor $\mathcal{P}$ of q, and the *core* k of q.

$$\mathcal{P} = \max_{p|q} p \quad \text{and} \quad k = \prod_{p|q} p.$$

Theorem 10. [18] *Let χ be a primitive character of modulus $q \geq 3$ with q square free. Then, for*

$$N = |I| \geq q^{\frac{4}{\sqrt{\log\log q}}} + \mathcal{P}^9$$

we have

$$\Big|\sum_{n \in I} \chi(n)\Big| \ll N e^{-\sqrt{\log q}}.$$

The purpose of the work of Graham and Ringrose is to study the following least quadratic nonresidue problem.

Problem 3. *Let p be a prime, and let $n(p)$ be the least quadratic nonresidue mod p. Find a lower bound on $n(p)$.*

Independent of Burgess' result $n(p) \leq p^{\frac{1}{4\sqrt{e}}+\varepsilon}$, Friedlander [14] and Salié [35] showed that $n(p) = \Omega(\log p)$, i.e., there are infinitely many p such that $n(p) > c \log p$ for some absolute constant c. By assuming the Generalized Riemann Hypothesis, in 1971 Montgomery [33] proved that $n(p) = \Omega(\log p \ \log\log p)$. Using Theorem 10, Graham and Ringrose showed that $n(p) = \Omega(\log p \ \log\log\log p)$ unconditionally.

In 1974, Iwaniec [26] proved the following theorem by generalizing Postnikov's theorem [34].

Theorem 11. [34] *Let χ be a primitive character of modulus q, $2 \nmid q$. Then, for*

$$k^{100} < N < N' \leq 2N \tag{6}$$

we have

$$\Big|\sum_{N < n \leq N'} \chi(n)\Big| \leq C^{s(\log s)^2} N^{1-\frac{c}{s^2 \log s}},$$

where $s = \frac{\log q}{\log N}$.

Theorem 11 is good for q powerful. A related problem is about the digital aspects of the primes. Let $x < N = 2^n$. Write

$$x = x_0 + 2x_1 + 2^2 x_2 + \cdots + 2^{n-1} x_{n-1} \quad \text{with} \quad x_0, x_1, \cdots, x_{n-1} \in \{0, 1\}.$$

For $A \subset \{1, \cdots, n\}$, given $\{\alpha_j \in \{0,1\}\}_{j \in A}$, one expects

$$\big|\{p = x < N : x_j = \alpha_j,\ \forall j \in A\}\big| \sim \frac{N}{\log N} 2^{-|A|}.$$

Problem 4. *How large can A be?*

For related work, see Sierpinski [39], Harman–Katai [22], Bourgain [3].

Theorem 10 is for q with small prime factors (smooth moduli), for which one assumes

$$\log N \gg \log \mathcal{P} + \frac{\log q}{\sqrt{\log \log q}}. \tag{7}$$

On the other hand, Theorem 11 is for q with small core. Condition (6) implies that

$$\log N \gg \log k. \tag{8}$$

If fix k, to get non-trivial result, in Theorem 11, one needs to assume

$$\log N \gg \log k + (\log q)^{\frac{3}{4}+\epsilon}. \tag{9}$$

Both are special cases of the following theorem in [11].

Denote

$$K = \frac{\log q}{\log k}.$$

Theorem 12. *Assume N satisfies*

$$q > N > \mathcal{P}^{10^3}$$

and

$$\log N > (\log q)^{\frac{9}{10}} \ + \ 10^3 \ \frac{\log 2K}{\log \log q} \ \log k\ . \tag{10}$$

Let χ be a primitive multiplicative character modulo q and I an interval of size N. Then

$$\Big| \sum_{x \in I} \chi(x) \Big| \ll N e^{-(\log N)^{3/5}}. \tag{11}$$

Remark 12.1. In the same spirit as assumptions (7) and (9), assumption (10) can be replaced by the stronger and friendlier assumption.

$$\log N \gg \log \mathcal{P} \ + \ \frac{\log q}{\log \log q}. \tag{12}$$

(The second term of condition (12) is clearly bigger than either term of (10).)

Remark 12.2. This result is completely general and gives bounds for very short character sums as soon as $\log \mathcal{P}$ is small compared with $\log q$.

Remark 12.3. To compare Theorem 12 with Theorem 10, we note also that condition (12) gives a better result than (7). Moreover, condition (7) is restricted to square free q. As for comparing Theorem 12 with Theorem 11, we let k be square free and $q = k^m$, where m is large but not too large. More precisely, we assume $\log m = o(\log\log k)$. Theorem 11 certainly requires that $\log N > C \log k$ while condition (10) in Theorem 12 becomes $\log N > (m \log k)^{\frac{9}{10}} + C\frac{\log m}{\log\log q} \log k$, which is clearly better.

Remark 12.4. We did not try to optimize the power of $\log q$ in the first term in (10) nor the saving in (11).

The following is a mixed character sum version of Theorem 12 (see also [10]).

Theorem 13. [15] *Under the assumptions of Theorem 12,*

$$\Big|\sum_{x\in I} \chi(x) e^{if(x)}\Big| < Ne^{-\sqrt{\log N}}$$

assuming $f(x) \in \mathbb{R}[x]$ *of degree at most* $(\log N)^c$ *for some* $c > 0$.

Corollary 4. *Let* $T > 0$. *Assume* N *satisfies*

$$q > N > \mathcal{P}^{10^3}$$

and q *satisfies*

$$\log N > (\log qT)^{\frac{9}{10}} + 10^3 \frac{\log 2K}{\log\log q} \log k.$$

Then for χ *primitive, we have*

$$\Big|\sum_{n\in I} \chi(n) n^{it}\Big| < Ne^{-\sqrt{\log N}} \quad \textit{for } |t| < T.$$

Following the classical arguments going back to Hadamand, de-la-Vallee-Poissin, and Landau, the above estimates lead to zero-free regions for the corresponding Dirichlet L-functions. Denote

$$L(s, \chi) = \sum_n \chi(n) n^{-s}, \qquad s = \rho + it.$$

Iwaniec [26] obtained the following results.

Theorem 14. [26] *Assume* $|L(s, \chi)| < M$ *for* $\rho > 1 - \eta$, $|t| < T^2$. *Then* $L(s, \chi)$ *has no zeros in the region* $\rho > 1 - \frac{\eta}{400 \log M}$, $|t| < T$, *except for possible Siegel zeros.*

Corollary 5. [26] *Assume Y and $\gamma > 0$ satisfy*

$$\Big|\sum_{n\sim N} \chi(n)n^{it}\Big| < N^{1-\gamma} \quad \textit{for } N > Y, |t| < T.$$

Then

$$\rho > 1 - \frac{c}{\log Y + \frac{1}{\gamma}\log\frac{1}{\gamma} + \frac{1}{\gamma}\log\log\frac{qT}{Y}}.$$

From Corollary 4 and Corollary 5, one derives bounds on the Dirichlet L-function $L(s,\chi)$ and zero-free regions the usual way. For a detailed argument, see, for instance, Lemmas 8–11 in [26]. This leads to the following theorem.

Theorem 15. *Let χ be a primitive multiplicative character with modulus q. For $T > 0$, let*

$$\theta = c\min\Big(\frac{1}{\log\mathcal{P}}, \frac{\log\log k}{(\log k)\log 2K}, \frac{1}{(\log qT)^{9/10}}\Big).$$

Then the Dirichlet L-function $L(s,\chi) = \sum_n \chi(n)n^{-s}, s = \rho + it$ has no zeros in the region $\rho > 1-\theta$, $|t| < T$, except for possible Siegel zeros.

It follows in particular that $\theta\log qT \to \infty$ if $\frac{\log\mathcal{P}}{\log q} \to 0$.

In certain range of k, this improves Iwaniec's bound (See [27], Theorem 8.29.)

$$\theta = \min\left\{c\frac{1}{(\log T)^{\frac{2}{3}}(\log\log T)^{\frac{1}{3}}}, \frac{1}{\log k}\right\}.$$

A well-known application of zero-free region is to primes in arithmetic progressions. In 1944, Linnik proved the following theorem about the least prime in an arithmetic progression.

Theorem 16. [31, 32] *There exists c such that if $(a,q) = 1$, then there is a prime $p < q^c$ such that $p \equiv a \bmod q$.*

For explicit value of c in Theorem 16, Xylouris [40] has the best estimate $c = 5.2$ by using Heath-Brown [23, 24] (who obtained $c = 5.5$) proposed improvements.

Theorem 15 gives a lower c for q smooth.

Theorem 17. *In Theorem 16, one may take $c = \frac{12}{5} + \epsilon$, assuming $\log\mathcal{P} < \frac{\log q}{\log\log q}$.*

Estimates on short character sums are also related to Polya–Vinogradov's theorem. Based on the work of Granville and Soundararajan [19] (which characterizes when character sums are large), Goldmakher [17] used Theorem 10 to improve Polya–Vinogradov's bound on $\sum_{n<x}\chi(n)$ and obtained the following.

Theorem 18. [18] *Let* χ mod q *be a primitive character. Then*

$$\Big|\sum_{n<x}\chi(n)\Big| \ll \sqrt{q}\,\log q\,\sqrt{\log^3 q}\,\Big(\frac{1}{\log^2 q}+\frac{\log \mathcal{P}}{\log q}\Big)^{\frac{1}{4}}.$$

When q *is square free, then*

$$\Big|\sum\chi(n)\Big| \ll \sqrt{q}\,\frac{\log q}{(\log\log q)^{\frac{1}{4}}}.$$

Instead of Theorem 10, we use Theorem 17 (and Granville–Soundararajan) and obtain the following.

Corollary 6. *Let* χ mod q *be a primitive character, and let*

$$M = (\log q)^{\frac{9}{10}} + (\log 2K)\frac{\log k}{\log\log k} + \log \mathcal{P}.$$

Then

$$\Big|\sum_{n<x}\chi(n)\Big| \ll \sqrt{q}\,\sqrt{\log q}\,\sqrt{M}\,\sqrt{\log\log\log q}.$$

When q *is square free,*

$$\Big|\sum_{n<x}\chi(n)\Big| \ll \sqrt{q}\,\frac{\log q}{(\log\log q)^{\frac{1}{2}}}.$$

Let $f(x_1,\dots,x_n) \in \mathbb{F}_q[x_1,\dots,x_n]$, $q = p^\ell$ be a polynomial of degree d. We are interested in bounding the exponential sum

$$S = \sum_{x_1,\dots,x_n} e_p\Big(\mathrm{Tr}\big(f(x_1,\dots,x_n)\big)\Big) \tag{13}$$

as well as a certain incomplete sums where the variables are restricted to a 'box' $B \subset \mathbb{F}_q^n$. More specifically, we consider various instances of this question where Deligne type estimates are not applicable, either because f is too singular or the box B is too small. In their work on Gowers' norms, Green and Tao [20] obtain non-trivial bounds in the situation where $\mathbb{F}_q = \mathbb{F}_p$ and d are fixed and n is large, assuming that the value of f is not determined by a few polynomials of lower degree.

Problem 5. *Obtain quantitative version of the Green–Tao result.*

For Problem 5, see the recent result by Forni and Flaminio [13].

Estimates of this type are also particularly relevant to circuit complexity [21].

Using methods from geometry of numbers, W. Schmidt [36] obtained bounds on incomplete sums (13) over boxes, but without exploring the effect of large n or ℓ.

Problem 6. *Investigate the bounds obtained in* [36] *when n or ℓ is large.*

It is possible that techniques from arithmetic combinatorics may be relevant.
For $n = 1$, estimates of this type are obtained in [2].
There are some other problems related to Problem 1.

Problem 7. *Let V be a vector subspace of $\mathbb{F}_{p^n}$ over $\mathbb{F}_p$, not essentially contained in a multiplicative translate of a subfield. Under what assumptions on $\dim_{\mathbb{F}_p} V$ can one obtain non-trivial bounds on $\sum_{x\in V}\chi(x)$?*

The arithmetic combinatorics approach permits to go below the $n/2$ barrier of classical methods. We are particularly interested in the situation where p is fixed and n is large. Assuming $\xi \in \mathbb{F}_{p^n}$ a generator, one may specify further $V = \langle \xi^j : j \in S\rangle$ with $S \subset \{0, 1, \dots, n-1\}$. In the special case $S = \{0, 1, \dots, m\}$, V. Shoup [38] used the Hasse–Weil method to get results for $m = O(\log n)$.

In the paper [8], we also succeeded in improving Karacuba's result on character sums of the type

$$\sum_{x\in I}\Big|\sum_{y\in A}\chi(x+y)\Big|, \tag{14}$$

where χ is a multiplicative character (mod p), I an interval and $A \subset \mathbb{F}_p$ arbitrary. This result was important to the recent work [16] on the distribution of quadratic and higher order residues (mod p). See also [25].

Further improvement on Karacuba's result was obtained by X. Shao [37].

Theorem 19. [39] *Let $q \in \mathbb{Z}_+$ be cube-free and χ mod q be non-principal. If $A \subset [1, q]$ is a union of disjoint intervals $I_1, \dots, I_s$ with $|A| > q^{1/4+\varepsilon}s^{1/2}$ and $|I_j| > q^{\varepsilon}$, $(1 \le j \le s)$ for any $\varepsilon > 0$, then $\sum_{n\in A}\chi(n)$ has a non-trivial bound.*

One could expect that for 'most p' better results are obtainable. However, the following problem seems still open.

Problem 8. *Show that for most p, the largest gap between quadratic residues is $o(p^{1/4})$.*

The following interesting character sum questions were highlighted in Karacuba's survey [29].

Problem 9. *Obtain a non-trivial bound on general sums*

$$\sum_{x\in A,\, y\in B}\chi(x+y),$$

when $|A| \sim \sqrt{p} \sim |B|$.

It is well-known that this question is related to the Paley graph conjecture and also relevant to the theory of 'extractor' in computer science [1].

Problem 10. *Prove that for p large and* $a \not\equiv 0 \pmod p$

$$\sum_{1 \le x < H} \left(\frac{x+a}{p}\right) \Lambda(x) = o(H),$$

when $H \sim \sqrt{p}$.

Results of this type were obtained by Vinogadov, for large values of H.

Problem 11. *Prove that for large p*

$$\min \left\{ 1 \le x \le p : \left(\frac{a+x^2}{p}\right) = -1 \right\} = o(\sqrt{p})$$

uniformly in $1 \le a \le p$.

In [9], the bound $p^{\frac{1}{2\sqrt{e}}+\varepsilon}$ was obtained, but for $a \neq 0$ given.

Acknowledgements Research by Mei-Chu Chang was supported in part by NSF grant DMS 1301608.

References

1. B. Barak, G. Kindler, R. Shaltiel, B. Sudakov, A. Wigderson, Simulating independence: new constructions of condensers, Ramsey graphs, dispersers, and extractors. J. ACM **57**, 52 (2010). Preliminary version in STOC 2005
2. J. Bourgain, On exponential sums in finite Fields, in *An Irregular Mind*, ed. by I. Bárány, J. Solymosi. Bolyai Society Mathematical Studies, Budapest, vol. 21 (Springer, Berlin, 2010)
3. J. Bourgain, Prescribing the binary digits of primes. Isr. J. Math. **194**, 935–955 (2013)
4. J. Bourgain, M.-C. Chang, On a multilinear character sum of Burgess. C. R. Math. Acad. Sci. **348**(3–4), 115–120 (2010)
5. D.A. Burgess, On character sums and primitive roots. Proc. Lond. Math. Soc. (3) **12**, 179–192 (1962)
6. D.A. Burgess, Character sums and primitive roots in finite fields. Proc. Lond. Math. Soc. (3) **37**, 11–35 (1967)
7. M.-C. Chang, On a question of Davenport and Lewis and new character sum bounds in finite fields. Duke Math. J. **145**(3), 409–442 (2008)
8. M.-C. Chang, Character sums in finite fields. AMS Contemp. Math. **518**, 83–98 (2009)
9. M.-C. Chang, On character sums of binary quadratic forms. J. Number Theory **129**(9), 2064–2071 (2009)
10. M.-C. Chang, An estimate of incomplete mixed character sums, in *An Irregular Mind*, ed. by I. Bárány, J. Solymosi. Bolyai Society Mathematical Studies, Budapest, vol. 21 (Springer, Berlin, 2010), pp. 243–250
11. M.-C. Chang, Short character sums for composite moduli. J. d'Anal. Math. **123**, 1–33 (2014)

12. H. Davenport, D. Lewis, Character sums and primitive roots in finite fields. Rend. Circ. Matem. Palermo-Serie II-Tomo XII-Anno **12**, 129–136 (1963)
13. G. Forni, L. Flaminio, On effective equidistribution for higher step nilflows arXiv:1407.3640
14. V.R. Friedlander, On the least n-th power non-residue. Dokl. Akad. Nauk. SSSR **66**, 351–352 (1949)
15. P.X. Gallagher, Primes in progressions to prime-power modulus. Invent. Math. **16**, 191–201 (1972)
16. M.Z. Garaev, S.V. Konyagin, Y. Malykhin, Asymptotics for the sum of powers of distances between power residues modulo a prime. Proc. Steklov Inst. Math. **276**(1), 77–89 (2012)
17. L. Goldmakher, Character sums to smooth moduli are small. Can. J. Math. **62**, 1099 (2010)
18. S.W. Graham, C.J. Ringrose, Lower bounds for least quadratic nonresidues, in *Analytic Number Theory* (Allerton Park, IL, 1989). Progress in Mathematics, vol. 85 (Birkhauser, Boston, 1990), pp. 269–309
19. A. Granville, K. Soundararajan, Large character sums: pretentious characters and the *Pólya*-Vinogradov theorem. J. Am. Math. Soc. **20**, 357–384 (2007)
20. B. Green, T. Tao, The distribution of polynomials over finite fields, with applications to the Gowers norms. Contrib. Discret. Math.**4**(2), 1–36 (2009)
21. F. Green, A. Roy, H. Straubing, Bounds on an exponential sum arising in Boolean circuit complexity. C. R. Math. Acad Sci. **341**(5), 279–282 (2005)
22. G. Harman, I. Katai, Primes with preassigned digits II. Acta. Arith. **133**(2), 171–184 (2008)
23. D.R. Heath-Brown, Siegel zeros and the least prime in an arithmetic progression. Q. J. Math. Oxf. Ser. (2) **41**, 405–418 (1990)
24. D.R. Heath-Brown, Zero-free regions for Dirichlet L-functions, and the least prime in an arithmetic progression. Proc. Lond. Math. Soc. (3) **64**(2), 265–338 (1992)
25. D.R. Heath-Brown, Burgess's bounds for character sums. arXiv:1203.5219
26. H. Iwaniec, On zeros of Dirichlet's L series. Invent. Math. **23**, 97–104 (1974)
27. H. Iwaniec, E. Kowalski, *Analytic Number Theory* (American Mathematical Society, Providence, 2004)
28. A.A. Karacuba, A certain arithmetic sum. Sov. Math. Dokl. **12**(4), 1172–1174 (1971)
29. A.A. Karatsuba, Arithmetic problems in the theory of Dirichlet characters. (Russian, English) Russ. Math. Surv. **63**(4), 641–690 (2008); translation from Usp. Mat. Nauk **63**(4), 43–92 (2008)
30. S.V. Konyagin, Estimates of character sums in finite fields. Mat. Z. **88**(4), 529–542 (2010)
31. Y.V. Linnik, On the least prime in an arithmetic progression I. The basic theorem. Rec. Math. (Mat. Sbornik) N.S. **15**(57), 139–178 (1944)
32. Y.V. Linnik, On the least prime in an arithmetic progression II. The Deuring-Heilbronn phenomenon. Rec. Math. (Mat. Sbornik) N.S. **15**(57), 347–368 (1944)
33. H.L. Montgomery, *Topics in Multiplicative Number Theory*. Lecture Notes in Mathematics, vol. 227 (Springer, New York, 1971)
34. A.G. Postnikov, On Dirichlet L-series with the character modulus equal to the power of a prime number. J. Indian Math. Soc. (N.S.) **20**, 217–226 (1956)
35. H. Salié, Uber den kleinsten positiven quadratischen Nichtrest nach einer Primzahl. Math. Nachr. **3**, 7–8 (1949)
36. W. Schmidt, Bounds for exponential sums. *Acta Arith.* **44**(3), 281–297 (1984)
37. X. Shao, Character sums over unions of intervals, *Forum Math.* **27**(5), 3017–3026 (2015)
38. V. Shoup, Searching for primitive roots in finite fields. Math. Comput. **58**, 369–380 (1992)
39. W. Sierpinski, Sur un probléme concernment les nombres $k \cdot 2n + 1$. Elem. Math. **15**, 63–74 (1960)
40. T. Xylouris, On Linnik's constant (2009). arXiv:0906.2749

Sum-product formulae

Andrew Granville and József Solymosi

Abstract This is a survey on sum-product formulae and methods. We state old and new results. Our main objective is to introduce the basic techniques used to bound the size of the product and sumsets of finite subsets of a field.

Keywords Additive Combinatorics • Sum-Product problems

Mathematics Subject Classification (2010): 11B75, 05B10

1 Introduction

1.1 A Few Definitions

For an additive or multiplicative group G and subsets $A, B \subset G$ we define

$$A + B = \{g \in G : \text{ There exist } a \in A, b \in B \text{ such that } g = a + b\}; \text{ and}$$
$$A \cdot B = \{g \in G : \text{ There exist } a \in A, b \in B \text{ such that } g = a \cdot b\}.$$

We let $r_{A+B}(n) := \#\{a \in A, b \in B : \; n = a + b\}$, $r_{AB}(n) := \#\{a \in A, b \in B : n = ab\}$, and note that $0 \le r_{A+B}(n) \le \min\{|A|, |B|\}$ since $r_{A+B}(n) = |A \cap (n-B)| \le |A|$ and $r_{A+B}(n) = |B \cap (n - A)| \le |B|$. We write $\hat{A}(t) = \sum_{a\in A} e(at)$ where $e(u) = e^{2i\pi u}$.

A. Granville
Département de mathématiques et de statistiques, Université de Montréal, CP 6128 Succursale Centre-Ville, Montréal, QC, Canada H3C 3J7
e-mail: andrew@dms.umontreal.ca

J. Solymosi (✉)
Department of Mathematics, University of British Columbia, 1984 Mathematics Road, Vancouver, BC, Canada V6T 1Z2
e-mail: solymosi@math.ubc.ca

A. Beveridge et al. (eds.), *Recent Trends in Combinatorics*, The IMA Volumes in Mathematics and its Applications 159, DOI 10.1007/978-3-319-24298-9_18

1.2 Multiplication Tables

We learnt to multiply by memorizing the multiplication tables; that is, we wrote down a table with the rows and columns indexed by the integers between 1 and N and the entries in the table were the row entry times the column entry.[1] Paul Erdős presumably learnt his multiplication tables rather more rapidly than the other students, and was left wondering: How many distinct integers are there in the N-by-N multiplication table? Note that if we take $A = \{1, 2, \ldots, N\}$, then we are asking how big is $A \cdot A$? Or, more specifically, since the numbers in the N-by-N multiplication table are all $\leq N^2$, what proportion of the integers up to N^2 actually appear in the table? That is,

Does $|A \cdot A|/N^2$ tend to a limit as $N \to \infty$?

Erdős showed that the answer is, yes, and that the limit is 0. His proof comes straight from "*The Book*".[2] Erdős's proof is based on the celebrated result of Hardy and Ramanujan that "almost all" positive integers $n \leq N$ have $\sim \log\log N$ (not necessarily distinct) prime factors (here "almost all" means for all but $o(N)$ values of $n \leq N$): Hardy and Ramanujan's result implies that "almost all" products ab with $a, b \leq N$ have $\sim 2\log\log N$ prime factors, whereas "almost all" integers $\leq N^2$ have $\sim \log\log(N^2) \sim \log\log N$ prime factors! The result follows from comparing these two statements.

1.3 The Motivating Conjectures

In fact one can show that $|A \cdot A|$ is large whenever A is an arithmetic progression or, more generally, when A is a generalized arithmetic progression of not-too-large dimension. [3]

This led Erdős and Szemerédi to the conjecture that for any $\epsilon > 0$, there exists $c_\epsilon > 0$ such that for any finite set of integers A,

$$|A + A| + |A \cdot A| \geq c_\epsilon |A|^{2-\epsilon}. \tag{1}$$

---

[1] A.G.: In my primary school we took $N = 12$ which was the basic multiple needed for understanding U.K. currency at that time.

[2] Erdős claimed that the Supreme Being kept a book of all the best proofs, and only occasionally would allow any mortal to glimpse at "The Book."

[3] A *generalized arithmetic progression* is the image of a lattice, that is:

$$C := \{a_0 + a_1 n_1 + a_2 n_2 + \cdots + a_k n_k : \ 0 \leq n_j \leq N_j - 1 \text{ for } 1 \leq j \leq k\},$$

where $N_1, N_2, \ldots, N_k$ are integers ≥ 2. This generalized arithmetic progression is said to have *dimension* k and *volume* $N_1 N_2 \ldots N_k$; and is *proper* if its elements are distinct.

Even more, one can conjecture that if $|A| = |B| = |C|$ then

$$|A + B| + |A \cdot C| \geq c_\epsilon |A|^{2-\epsilon}. \tag{2}$$

The above conjectures might hold for complex numbers even. Perhaps the most general version is

$$\textit{Either } |A + B| \gg (|A||B|)^{1-\epsilon} \textit{ or } |A \cdot C| \gg (|A||C|)^{1-\epsilon}$$

with no restrictions on the sizes of A, B, and C. The thinking in these conjectures is that if $A + B$ is small then A must be "structured," more precisely that it must look like a largish subset of a generalized arithmetic progression, and similarly if AC is small then $\log A$ must look like a largish subset of a generalized arithmetic progression, and that these two structures are incompatible.

2 Sum-Product for Real Numbers

2.1 Results Via Discrete Geometry

The second author proved (2) for $\epsilon = 8/11$ [29] (see Theorem 1 below). We now prove (2) for $\epsilon = 3/4$. We begin by stating the

Szemerédi–Trotter Theorem. *We are given a set $\mathscr{C}$ of m curves in $\mathbb{R}^2$ such that*

- *Each pair of curves meet in $\leq \kappa_1$ points;*
- *Any pair of points lie on $\leq \kappa_2$ curves.*

For any given set $\mathscr{P}$ of n points, there are $\leq m + 4\kappa_2 n + 4\kappa_1\kappa_2^{1/3}(mn)^{2/3}$ pairs (π, γ) with point $\pi \in \mathscr{P}$ lying on curve $\gamma \in \mathscr{C}$.

Székely provided a gorgeous proof of this result, straight from *The Book*, via geometric and random graph theory [27]. From this Elekes elegantly deduced the following in [7]:

Theorem 1. *If $A, B, C \subset \mathbb{R}$, then*

$$|A + B| + |A \cdot C| \geq \frac{1}{2}(|A| - 1)^{3/4}(|B||C|)^{1/4}. \tag{3}$$

Proof. If $|A + B||A \cdot C| \geq (\frac{1}{2^4}|B||C|)^2$, then at least one of $|A + B|$ and $|A \cdot C|$ is $\geq \frac{1}{2^4}|B||C|$, and they are both $\geq |A|$, so that their product is $\geq \frac{1}{2^4}|A|^3|B||C|$. Hence

$$|A + B| + |A \cdot C| \geq \frac{1}{2}|A|^{3/4}(|B||C|)^{1/4},$$

which implies the result. Hence we may assume that $|A + B||A \cdot C| < \frac{1}{2^8}(|B||C|)^2$.

Let $\mathscr{P}$ be the set of points $(A+B)\times(A\cdot C)$; and $\mathscr{C}$ the set of lines $y=c(x-b)$ where $b\in B$ and $c\in C$. In the Szemerédi–Trotter Theorem we have $\kappa_1=\kappa_2=1$ with

$$m=|B||C| \text{ and } n\le N:=|A+B|\,|A\cdot C|,$$

since the set of points is $\cup_{a\in A}(a+B,aC)$. For fixed $b\in B$ and $c\in C$, all of the points $\{(a+b,ac):a\in A\}$ in $\mathscr{P}$ lie on the line $y=c(x-b)$, so that

$$\#\{(\pi,\gamma):\pi\in\mathscr{P} \text{ on } \gamma\in\mathscr{C}\}\ge|A|m.$$

Substituting this into the Szemerédi–Trotter Theorem we obtain

$$(|A|-1)m\le 4n+4(mn)^{2/3}\le 4N+4(mN)^{2/3}.$$

We assumed that $N<m^2/2^8$, so that $N<(mN)^{2/3}/2^{8/3}<(2^{2/3}-1)(mN)^{2/3}$, and hence $(|A|-1)m^{1/3}\le 4(2N)^{2/3}$. This implies that $N>(|A|-1)^{3/2}(|B||C|)^{1/2}/16$. □

Corollary 2.1. *If $A\subset\mathbb{R}$, then*

$$|A+A|+|A\cdot A|\ge\frac{1}{2}|A|^{5/4}.$$

Next we give an argument that improves this. It is still not the best result currently known in the direction of (1) but it only uses the Szemerédi–Trotter Theorem which has several advantages. The most important advantage is that incidence bounds between points and lines on a plane over any field K provide sum-product bounds in K. Even better, the point set where the incidence bounds are needed have a special Cartesian product structure. For example, on the complex plane, $\mathbb{C}^2$, it is quite easy to give a Szemerédi-Trotter type bound (with the same exponents) for lines and points of a Cartesian product like $(A+B)\times(A\cdot C)$ above. The incidence bound for this special case appeared in [28]. Another example is Vinh's work [32] who used a Szemerédi-Trotter type bound to obtain a different proof of Garaev's sum-product estimate in finite fields (see Theorem 4 below).

Theorem 2. *[29] If $A,B,C,D\subset\mathbb{R}$ with $0\notin C$, then*

$$|A/C||A+B||C+D|\gg(|A||C|)^{3/2}(|B||D|)^{1/2}\min\left\{1,\frac{|A/C|}{|B||D|}\right\}^{1/4}.$$

Part of this follows from

Theorem 3. *If $A, B, C, D \subset \mathbb{R}$ with $0 \notin C$ and $|A/C| \le |B||D|$, then*

$$|A/C|^3 (|A+B||C+D|)^4 > \frac{1}{2 \cdot 10^{10}} (|A||C|)^6 (|B||D|), \tag{4}$$

and

$$|AC|^3 (|A+B||C+D|)^4 \ge \frac{1}{10^9} \frac{(|A||C|)^6 (|B||D|)}{\log^3(4 \min\{|A|, |C|\})}.$$

We note some consequences:

Corollary 2.2. *If $A, C \subset \mathbb{R}$ with $0 \notin C$, then*

$$|A/C|^3\ |A+C|^8 > \frac{1}{2 \cdot 10^{10}} (|A||C|)^7 \tag{5}$$

and

$$|AC|^3\ |A+C|^8 \ge \frac{1}{10^9} \frac{(|A||C|)^7}{\log^3(4 \min\{|A|, |C|\})}.$$

Hence

$$|AA|^3\ |A+A|^8 \gg \frac{|A|^{14}}{\log^3(4|A|)};$$

in particular if $|A+A| \le \kappa |A|$, then $|AA| \gg \kappa^{-8/3} |A|^2 / \log(4|A|)$.

Remark. If $A = \{1, \ldots, N\}$, then $|AA| \asymp N^2/(\log N)^\delta (\log\log N)^{3/2}$ for some $\delta = 1 - \frac{1+\log\log 2}{\log 2} = 0.08607\ldots$. Hence some power of log in the denominator in this last result is unavoidable.

Proof of Theorem 3. Let $V(k)$ denote the set of $m \in C/A$ for which $2^k \le r_{C/A}(m) < 2^{k+1}$, for $k = 0, 1, 2, \ldots$ Consider $\mathscr{C} = \mathscr{C}_k$ the set of lines $y = mx + e$ with $m \in V(k)$ which contain at least one point $(x, y) \in B \times D$. Each $(x, y) \in B \times D$ lies on exactly $|V(k)|$ lines of $\mathscr{C}_k$ (as may be seen by taking $e = b - dm$ for each $m \in V(k)$), so that

$$|V(k)||B||D| \le |\mathscr{C}_k| + 4|B||D| + 4(|\mathscr{C}_k||B||D|)^{2/3}$$

by the Szemerédi–Trotter Theorem. Hence either $|V(k)| \le 14$ or $\frac{11}{15}|V(k)||B||D| \le |\mathscr{C}_k| + 4(|\mathscr{C}_k||B||D|)^{2/3}$ which implies that

$$|\mathscr{C}_k| \ge \min\{\frac{1}{4}|V(k)||B||D|, \frac{1}{27}|V(k)|^{3/2}(|B||D|)^{1/2}\}.$$

Now $|V(k)| \le |C/A| \le |B||D|$ so that $|\mathscr{C}_k| \ge \frac{1}{27}|V(k)|^{3/2}(|B||D|)^{1/2}$.

Now consider the set of points $(A + B) \times (C + D)$. If $y = mx + e$ is a line in $\mathscr{C}_k$ containing the point (b, d), then it also contains the points $(a + b, c + d)$ whenever $c/a = m$ with $a \in A, c \in C$. Hence each such line contains at least 2^k points from $(A + B) \times (C + D)$ and the Szemerédi–Trotter Theorem then yields $2^k|\mathscr{C}_k| \leq |\mathscr{C}_k| + 4|A + B||C + D| + 4(|\mathscr{C}_k||A + B||C + D|)^{2/3}$. Hence

$$(2^k - 1)|\mathscr{C}_k| \leq \max\{80|A + B||C + D|, 4.2\,(|\mathscr{C}_k||A + B||C + D|)^{2/3}\}$$

which implies that

$$|\mathscr{C}_k| \leq 80 \max\left\{\frac{|A+B||C+D|}{(2^k-1)}, \frac{(|A+B||C+D|)^2}{(2^k-1)^3}\right\} = 80\frac{(|A+B||C+D|)^2}{(2^k-1)^3},$$

where this last inequality follows since $r_{C/A}(m) \leq \min\{|A|, |C|\}$ which implies that

$$(2^k - 1)^2 < r_{C/A}(m)^2 \leq |A||C| \leq |A + B||C + D|.$$

Combining the deductions at the end of the last two paragraphs gives

$$|V(k)| \leq 6^2 10^{2/3} \frac{(|A+B||C+D|)^{4/3}}{(2^k-1)^2(|B||D|)^{1/3}}. \tag{6}$$

Therefore

$$\sum_{r_{C/A}(m) \geq 2^K} r_{C/A}(m) \leq \sum_{k \geq K} \sum_{m \in V(k)} 2^{k+1} \leq 335 \sum_{k \geq K} \frac{2^k}{(2^k-1)^2} \cdot \frac{(|A+B||C+D|)^{4/3}}{(|B||D|)^{1/3}}$$

$$\leq 670 \frac{2^K}{(2^K-1)^2} \cdot \frac{(|A+B||C+D|)^{4/3}}{(|B||D|)^{1/3}},$$

and this is $\leq \frac{1}{2}|A||C|$ provided $2^K \geq 671(|A + B||C + D|)^{4/3}/(|A||C|)(|B||D|)^{1/3}$. So select the smallest K for which this holds, so that

$$\frac{1}{2}|A||C| \leq \sum_{r_{C/A}(m) < 2^K} r_{C/A}(m) < 2^K|C/A| < 1342|C/A|\,\frac{(|A+B||C+D|)^{4/3}}{(|A||C|)(|B||D|)^{1/3}},$$

and the first result follows.

Let us also note that, by the Cauchy–Schwarz inequality

$$(|A||C|)^2 = \left|\sum_n r_{AC}(n)\right|^2 \leq |AC| \sum_n r_{AC}(n)^2 = |AC| \sum_m r_{C/A}(m)^2 \leq 4|AC| \sum_k V(k) 2^{2k}.$$

By (6), and the fact that $2^k \leq r_{C/A}(m) \leq \min\{|A|, |C|\}$ we deduce the second result. □

Proof of Theorem 2 when $|C/A| > |B||D|$. We may assume that

$$|A+B||C-D| \leq (|A||C|)^{3/2}/(118(|B||D|)^{1/2}),$$

else we obtain the result by multiplying through by $|A/C|^{3/4} > (|B||D|)^{3/4}$.

In the proof of Theorem 3 we note that if $V(k) > |B||D|$ then

$$V(k) \leq \frac{4|\mathscr{C}_k|}{|B||D|} \leq 320\frac{(|A+B||C+D|)^2}{(2^k-1)^3|B||D|},$$

so we obtain $\sum_{r_{C/A}(m)\leq 2^K} r_{C/A}(m) \geq \frac{1}{2}|A||C|$ provided

$$2^K \gg \max\left\{\frac{(|A+B||C-D|)^{4/3}}{(|A||C|)(|B||D|)^{1/3}}, \frac{|A+B||C+D|}{(|A||B||C||D|)^{1/2}}\right\} \asymp \frac{|A+B||C+D|}{(|A||B||C||D|)^{1/2}},$$

the last equality following from our assumption, and the result follows as in the proof of Theorem 3. □

2.2 Some Easier Ideas

The *multiplicative energy* of two finite sets A, B is defined as

$$E_\times(A,B) = \#\{a_1,a_2 \in A,\ b_1,b_2 \in B:\ a_1b_1 = a_2b_2\} = \sum_{a,b\in B} |aA \cap bA|.$$

By the Cauchy–Schwarz inequality we have $E_\times(A,B)^2 \leq E_\times(A,A)E_\times(B,B)$. We also can write

$$E_\times(A,B) = \sum_m r_{AB}(m)^2 = \sum_n r_{A/B}(n)^2 = \sum_n r_{A/A}(n)r_{B/B}(n),$$

and hence, by the Cauchy–Schwarz inequality

$$(|A||B|)^2 = \left(\sum_m r_{AB}(m)\right)^2 \leq |AB|\sum_m r_{AB}(m)^2 = |AB|E_\times(A,B). \tag{7}$$

Similarly $(|A||B|)^2 \leq |A/B|E_\times(A,B)$. Finally, if $A_1 \cap A_2 = \emptyset$, then $r_{(A_1\cup A_2)B}(n) = r_{A_1B}(n) + r_{A_2B}(n)$ and so, by the Cauchy–Schwarz inequality,

$$\begin{aligned} E_\times(A_1\cup A_2, B) &= \sum_m r_{(A_1\cup A_2)B}(m)^2 \\ &\leq 2\sum_m (r_{A_1B}(n)^2 + r_{A_2B}(n)^2) = 2(E_\times(A_1,B) + E_\times(A_2,B)). \end{aligned} \tag{8}$$

Proposition 2.3. (Solymosi, [30]) *If A and B are finite sets of real numbers, not containing $\{0\}$, then*

$$E_\times(A,B) \le 12|A+A||B+B|\log(3\min\{|A|,|B|\}), \tag{9}$$

and hence, by (7),

$$|A+A||B+B|\min\{|A/B|,|AB|\} \ge (|A||B|)^2/(12\log(3\min\{|A|,|B|\}).$$

Remarks. Note that there are examples with $0 \in A \cup B$ where this bound cannot hold. For example, if $0 \in B$, then $E_\times(A,B) \ge \#\{a,a' \in A : a0 = 0 = a'0\} = |A|^2$ whereas the bound in Proposition 2.3 is smaller than $|A|^2$ if A and B are both arithmetic progressions with $|B| \ll |A|/\log|A|$.

Note also that this bound is, more-or-less, best possible in any example with $|A+A| \ll |A|$ and $|B+B| \ll |B|$ since, trivially, $E_\times(A,B) \ge |A||B|$.

Proof. We begin by proving this result when A and B are both finite sets of positive real numbers. Let $m := \min\{|A|,|B|\}$. If $m = 1$, then $E_\times(A,B) = \max\{|A|,|B|\}$ and the result is easy, so we may assume $m \ge 2$.

Let $R_{B/A}(\ell) = \{(a,b) \in A \times B : b = \ell a\}$ which has size $r_{B/A}(\ell)$, and note that $r_{B/A}(\ell) \le m$. Let $L_k := \{\ell : 2^k \le r_{B/A}(\ell) < 2^{k+1}\}$ and $K = [\log m/\log 2]+1$. Then

$$\sum_{k=0}^{K-1}\sum_{\ell \in L_k} r_{B/A}(\ell)^2 = \sum_{\ell} r_{B/A}(\ell)^2 = E_\times(A,B),$$

and

$$\sum_{\substack{k=0\\|L_k|=1}}^{K-1}\sum_{\ell \in L_k} r_{B/A}(\ell)^2 \le \sum_{k=0}^{K-1} 2^{2k+2} < \frac{2^{2K+2}}{3} \le \frac{16m^2}{3}.$$

Hence there exists k with $|L_k| \ge 2$ for which

$$\sum_{\ell \in L_k} r_{B/A}(\ell)^2 \ge \frac{1}{K}\left(E_\times(A,B) - \frac{16m^2}{3}\right).$$

Let $L_k = \{\ell_1 < \ell_2 < \ldots < \ell_r\}$ with $r \ge 2$; we claim that the elements of

$$\bigcup_{i=1}^{r-1}(R_{B/A}(\ell_i) + R_{B/A}(\ell_{i+1})) \subset A \times B + A \times B$$

are distinct. For if $i < j$ with $a_i + a_{i+1} = a_j + a_{j+1}$, then

$$\ell_i a_i + \ell_{i+1}a_{i+1} < \ell_{i+1}(a_i + a_{i+1}) \le \ell_j(a_j + a_{j+1}) < \ell_j a_j + \ell_{j+1}a_{j+1};$$

and if $(a+a', \ell_i a + \ell_{i+1}a') = (x,y)$, then a and a' are determined, and so unique.

Now, as $A \times B + A \times B = (A + A) \times (B + B)$, we deduce that

$$|A + A||B + B| = |A \times B + A \times B| \geq \sum_{i=1}^{r-1} r_{B/A}(\ell_i) r_{B/A}(\ell_{i+1})$$

$$\geq (r-1)2^{2k} \geq \frac{r}{2} \cdot 2^{2k} \geq \frac{1}{8} \sum_{i=1}^{r} r_{B/A}(\ell_i)^2$$

$$\geq \frac{1}{8K} \left(E_\times(A, B) - \frac{16m^2}{3} \right).$$

From this we deduce that

$$E_\times(A, B) \leq \frac{8 \log 2m}{\log 2} |A + A||B + B| + \frac{16}{3} \min\{|A|, |B|\}^2,$$

and then (9) follows for $m \geq 2$ after a little calculation, using the fact that $|C + C| \geq 2|C| - 1$.

Now if A only has positive real numbers and $0 \notin B$, then write $B = B_+ \cup B_-$ where $B_\pm = \{b \in B : \pm b > 0\}$. Now $E_\times(A, B_-) = E_\times(A, -B_-)$ by definition so, by (8) and then the case in which we have already proved (9), we have

$$E_\times(A, B) \leq 2E_\times(A, B_-) + 2E_\times(A, -B_-)$$
$$\leq 12|A + A|(|B_+ + B_+| + |B_- + B_-|) \log(3 \min\{|A|, |B|\})$$

which implies (9), since $B_+ + B_+$ and $B_- + B_-$ are evidently disjoint subsets of $B + B$ (as their elements are of different signs) (Fig. 1).

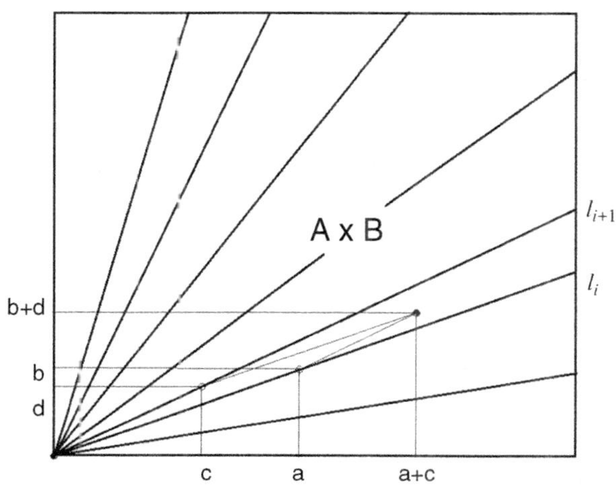

Fig. 1 The sum of $(a, b) \in \ell_i$ and $(c, d) \in \ell_{i+1}$ is a unique point of the Cartesian product $(A + A) \times (B + B)$.

Finally, if $0 \notin A \cup B$, then (9) follows similarly from this last result by partitioning A as $A_+ \cup A_-$. □

Remark. We can deduce bounds on $E_\times(A,B)$, when $0 \in A \cup B$, from Proposition 2.3, using the following

If $0 \notin A$ but $0 \in B = B_0 \cup \{0\}$ then, by definition, $E_\times(A,B) = E_\times(A,B_0) + |A|^2$ and $B + B = (B_0 + B_0) \cup B$.

If $0 \in A = A_0 \cup \{0\}$ and $0 \in B = B_0 \cup \{0\}$, then $E_\times(A,B) = E_\times(A_0,B_0) + (|A| + |B| - 1)^2$ with $A + A = (A_0 + A_0) \cup A$ and $B + B = (B_0 + B_0) \cup B$.

Corollary 2.4. *If A is any finite set of real numbers, then*

$$E_\times(A,B) \leq 12|A + A|^2 \log(3|A|), \tag{10}$$

and hence, by (7),

$$|A + A|^2 \min\{|A/A|, |AA|\} \geq |A|^4/(12\log(3|A|).$$

Proof. If $0 \notin A$, then our bound follows from setting $B = A$ in (9). If $0 \in A$, then we use the information in the previous remark, together with (9), to obtain

$$\begin{aligned} E_\times(A,A) &= E_\times(A_0,A_0) + (2|A| - 1)^2 \leq 12|A_0 + A_0|^2 \log(3|A|) + (2|A| - 1)^2 \\ &= 12(|A + A| - |A|)^2 \log(3|A|) + (2|A| - 1)^2 \\ &= 12|A{+}A|^2 \log(3|A|) {+} 12\log(3|A|)(|A|^2 {-} 2|A||A{+}A|) {+} (2|A| {-} 1)^2 \\ &\leq 12|A + A|^2 \log(3|A|) + 12\log(3|A|)(2|A| - 3|A|^2) + (2|A| - 1)^2 \end{aligned}$$

as $|A + A| \geq 2|A| - 1$, which yields (10). □

A similar bound for complex numbers was obtained by Konyagin and Rudnev in [22]. Very recently Konyagin and Shkredov announced an improvement on the sum-product bound. They proved in [23] that

$$|A + A| + |A \cdot A| \gg |A|^{4/3+c}$$

where $\frac{1}{20598} > c > 0$ is an absolute constant.

2.3 Small Product Sets

From Corollary 2.2 it follows that if the sumset is very small then the product set is almost quadratic. The opposite statement is surprisingly hard to prove. It was Chang's observation [6] that one can use a powerful tool, the Subspace Theorem, to obtain such bound. For the history and more details about the Subspace Theorem we refer to the excellent survey paper of Yuri Bilu [3].

An important variant of the Subspace Theorem was proved by Evertse, Schlickewei, and Schmidt [9]. We present the version with the best known bound due to Amoroso and Viada [1].

Theorem 2.5. *Let K be a field of characteristic 0, Γ a subgroup of K^* of rank r, and $a_1, a_2, \ldots, a_n \in K^*$. Then the number of solutions of the equation*

$$a_1 z_1 + a_2 z_2 + \cdots + a_n z_n = 1 \tag{11}$$

with $z_i \in \Gamma$ and no subsum on the left-hand side vanishing is at most

$$A(n, r) \leq (8n)^{4n^4(n+nr+1)}.$$

We are going to use the following result of Freiman (Lemma 1.14 in [11]).

Proposition 1. *Let $A \subset \mathbb{C}$. If $|AA| \leq C|A|$, then A is a subset of a multiplicative subgroup of $\mathbb{C}^*$ of rank at most r, where r is a constant depending on C.*

Theorem 2.6. *Let $A \subset \mathbb{C}$ with $|A| = n$. Suppose $|AA| \leq Cn$. Then there is a constant C' depending only on C such that*

$$|A + A| \geq \frac{n^2}{2} + C'n.$$

Proof. We consider solutions of $x_1 + x_2 = x_3 + x_4$ with $x_i \in A$. A solution of this equation corresponds to two pairs of elements from A that give the same element in $A + A$. Let us suppose that $x_1 + x_2 \neq 0$ (there are at most $|A| = n$ solutions of the equation $x_1 + x_2 = 0$ with $x_1, x_2 \in A$).

First we consider the solutions with $x_4 = 0$. Then by rearranging we get

$$\frac{x_1}{x_3} + \frac{x_2}{x_3} = 1. \tag{12}$$

By Proposition 1 and Theorem 2.5 there are at most $s_1(C)$ solutions of $y_1 + y_2 = 1$ with no subsum vanishing. Each of these gives at most n solutions of (12) since there are n choices for x_3. There are only two solutions of $y_1 + y_2 = 1$ with a vanishing subsum, namely $y_1 = 0$ or $y_2 = 0$, and each of these gives n solutions of (12). So we have a total of $(s_1(C) + 2)n$ solutions of (12).

For $x_4 \neq 0$ we get

$$\frac{x_1}{x_4} + \frac{x_2}{x_4} - \frac{x_3}{x_4} = 1. \tag{13}$$

Again by Proposition 1 and Theorem 2.5, the number of solutions of this with no vanishing subsum is at most $s_2(C)n$. If we have a vanishing subsum, then $x_1 = -x_2$ which is a case we excluded earlier or $x_1 = x_3$ and then $x_2 = x_4$, or $x_2 = x_3$ and then $x_1 = x_4$. So we get at most $2n^2$ solutions of (13) with a vanishing subsum (these are the $x_1 + x_2 = x_2 + x_1$ identities).

So, in total, we have at most $2n^2 + s(C)n$ solutions of $x_1 + x_2 = x_3 + x_4$ with $x_i \in A$. Suppose $|A + A| = k$ and $A + A = \{\alpha_1, \ldots, \alpha_k\}$. We may assume that $\alpha_1 = 0$. Recall that we ignore sums $a_i + a_j = 0$. Let

$$P_i = \{(a, b) \in A \times A : a + b = \alpha_i\}, \quad 2 \leq i \leq k.$$

Then

$$\sum_{i=2}^{k} |P_i| \geq n^2 - n = n(n-1).$$

Also, a solution of $x_1 + x_2 = x_3 + x_4$ corresponds to picking two values from P_i where $x_1 + x_2 = \alpha_i$. Thus

$$2n^2 + s(C)n \geq \sum_{i=2}^{k} |P_i|^2 \geq \frac{1}{k-1}\left(\sum_{i=2}^{k} |P_i|\right)^2 \geq \frac{n^2(n-1)^2}{k-1}$$

by the Cauchy-Schwarz inequality. The bound for $k = |A + A|$ follows.

2.4 Upper Bounds in the Sum-Product Inequality

One obvious way to obtain upper bounds is to select A to be a largish subset of $\{1, \ldots, x\}$ with lots of multiplicative structure. For example, we could let A be the set of integers $\leq x$ all of whose prime factors are $\leq y$, so that $|A| = \Psi(x, y)$, $|AA| \leq \Psi(x^2, y)$, and $|A+A| \leq 2x$. Roughly $\Psi(x, y) = x((e+o(1))/u \log u)^u$ when $x = y^u$; so that $|AA|/|A|^2 = (1/2 + o(1))^{2u}$ and $|A + A|/|A|^2 = (u \log u/(e + o(1)))^{2u}/x$. We select u so that these are roughly equal, that is, $u = \frac{\log x}{2 \log\log x}\left(1 + \frac{1+o(1)}{\log\log x}\right)$, and thus $y \asymp (\log x)^2$. Therefore $|A| = x^{1/2} 2^{(1+o(1))u}$. Hence we have an infinite family of examples in which, if $|A| = N$ then

$$\max\{|A + A|, |AA|\} \leq N^{2 - \frac{\log 4 + o(1)}{\log\log N}}.$$

We can obtain this result without using any "machinery": Let A be the set of $N := \binom{\pi(y)+u}{u}$ integers composed of no more than u not necessarily distinct prime factors $\leq y$. Then $A \subset [1, y^u]$ so that $|A + A| \leq 2y^u$, whereas $|AA| = \binom{\pi(y)+2u}{2u}$. We select $u = [ey^{1/2}/2 \log y]$ so that, by Stirling's formula and the prime number theorem,

$$N = \left(\frac{e(y/\log y)}{u}\right)^u e^{O(u/\log y)} = \left(2y^{1/2}\right)^u e^{O(u/\log y)} = (u(\log u)^{O(1)})^u,$$

and therefore $u \sim \log N / \log\log N$. Now, by similar calculations we find that

$$|AA| \text{ and } |A+A| = |A|^2/2^{(2+o(1))u} e^{O(u/\log y)} = N^{2-\frac{\log 4+o(1)}{\log\log N}}.$$

3 Sum-Product Inequalities over Finite Fields

The Szemerédi–Trotter Theorem does not hold over $\mathbb{F}_q$ which renders all of the above results moot in this setting. However such results in finite fields are the most applicable, so we will now pursue this. The first thing to note is that we must modify (1) when the set A is large, hence Garaev conjectured that if $A \subset \mathbb{F}_p$ then

$$|A+A| + |A \cdot A| \gg \min\{|A|^2/\sqrt{p}, \sqrt{p|A|}\}/|A|^{o(1)}. \tag{14}$$

3.1 *Upper Bounds in* $\mathbb{F}_p$

We begin by showing that the lower bound in Garaev's conjecture cannot, in general, be increased:

Proposition 3.1. *For any given integers I, J, N with $1 \le N \le I, J \le p$, and $N \le \lceil IJ/p \rceil$, there exist $A \subset B, C \subset \mathbb{F}_p$, with $|A| = N, |B| = J, |C| = I$ such that*

$$|A+B| < 2|B| \text{ and } |A \cdot C| < 2|C|.$$

In particular, for any given N, $1 \le N \le p$, there exists $A \subset \mathbb{F}_p$ with $|A| = N$ such that

$$\max\{|A+A|, |A \cdot A|\} \le \min\{|A|^2, 2\sqrt{p|A|} + 1\}.$$

Remark. If we have $\max\{|A+A|, |A \cdot A|\} \gg |A|^{2-o(1)}$ for all sets $A \subset \mathbb{F}_p$ of size N, then the second part of Proposition 3.1 implies that $N \ll p^{1/3-o(1)}$.

Proof. Let $C := \{g^1, \ldots, g^I\}$ where g is a primitive root mod p, and $A_x := C \cap B_x$ for each $x \in \mathbb{F}_p$ where $B_x := x + \{1, \ldots, J\}$. Now

$$\sum_x |A_x| = \sum_{i=1}^{I} \sum_{j=1}^{J} \#\{x \in \mathbb{F}_p : x = g^i - j\} = IJ,$$

so that there exists x with $|A_x| \ge IJ/p$. Let A be any subset of A_x of size N, and $B = B_x$. Therefore $A + A \subset A + B \subset B + B = \{2x+2, \ldots, 2x+2J\}$, $A \cdot A \subset C \cdot C \subset \{g^2, \ldots, g^{2I}\}$ so that $|A+A| \le |A+B| \le |B+B| < 2J$ and $|A \cdot A| \le$

$|A \cdot C| \leq |C \cdot C| < 2I$, which completes the proof of the first part. Now, taking $I = J = \lceil \sqrt{pN} \rceil$ we find that $|A + A|, |A \cdot A| \leq 2\lceil \sqrt{pN} \rceil - 1$, which implies the second part. □

3.2 A Little Cauchying

Let us make note of a couple of inequalities, for characteristic functions of sets: By Cauchy we obtain

$$\left(\sum_j \left| A\left(\frac{j}{p}\right) \right| \left| \hat{B}\left(\frac{-j}{p}\right) \right| \right)^2 \leq \sum_j \left| A\left(\frac{j}{p}\right) \right|^2 \sum_j \left| B\left(\frac{-j}{p}\right) \right|^2 = p|A| \cdot p|B|. \tag{15}$$

By Cauchy we obtain

$$\begin{aligned} \left| \sum_{a \in A} \sum_{b \in B} e\left(\frac{kab}{p}\right) \right|^2 &\leq \sum_{a \in A} 1 \cdot \sum_{a \in A} \left| \sum_{b \in B} e\left(\frac{kab}{p}\right) \right|^2 \\ &\leq |A| \cdot \sum_{a \in \mathbb{F}_p} \left| \sum_{b \in B} e\left(\frac{kab}{p}\right) \right|^2 = |A| \cdot p|B|, \end{aligned} \tag{16}$$

by Parseval.

3.3 Lower Bounds in $\mathbb{F}_p$

Theorem 4. (Garaev) *If $A, B, C \subset \mathbb{F}_p$ with $0 \notin C$, then*

$$|A + B| \cdot |A \cdot C| \geq \frac{|A|}{4} \cdot \min\left\{ \frac{|A||B||C|}{p}, 2p \right\}.$$

Remark. Taking $I = J = [\sqrt{pN}]$ in Proposition 3.1, we obtain examples with $|A + B|\ |A \cdot C| \leq 4p|A|$. Therefore Theorem 4 is best possible, up to a factor of 8, when $|A||B||C| \geq 2p^2$. In particular for $|A| = |B| = |C| \geq 2p^{2/3}$.

Theorem 4 and its proof remain valid, with suitable modifications, in $\mathbb{F}_p \times \mathbb{F}_p$ (changing both occurrences of p to $q = p^2$ in the lower bound). If we select a set $D \subset \mathbb{F}_p$ such that $|D + D|, |DD| \asymp \min\{|D|^2, p\}$, then taking $A = B = C = D \times \mathbb{F}_p$ we have $|A + B|, |AC| \asymp p \min\{|D|^2, p\} = \min\{|A|^2/p, p^2\}$ so that $|A + A||AA| \asymp \min\{|A|^4/q, q^2\}$. Therefore Theorem 4 is best possible up to a constant factor when $q^{1/2} \leq |A| \leq q^{3/4}$, in this setting.

First by letting $C \to 1/C$ above, and then by taking $A = B = C$ we deduce:

Corollary 3.2. *If $A, B, C \subset \mathbb{F}_p$ with $0 \notin C$, then*

$$|A+B| \cdot |A/C| \geq \frac{|A|}{4} \cdot \min\left\{ \frac{|A||B||C|}{p}, 2p \right\} .$$

If $A \subset \mathbb{F}_p$ with $0 \notin A$, then

$$|A+A| \cdot |AA|,\ |A+A| \cdot |A/A| \geq \frac{|A|}{4} \cdot \min\left\{ \frac{|A|^3}{p}, 2p \right\} .$$

If A is a multiplicative subgroup of $\mathbb{F}_p^*$, then

$$|A+A| \geq \min\left\{ \frac{|A|^3}{4p}, p/2 \right\} .$$

Proof of Theorem 4. For any $a \in A, b \in B, c \in C$ we have a distinct solution to

$$u/c + b = v \tag{17}$$

with $u \in AC, c \in C, b \in B,\ v \in V = A + B$, where $u = ac$ and $v = a + b$. Hence $|A||B||C|$ is no more than the total number of solutions of (3.4), which equals

$$\sum_{u\in AC}\sum_{c\in C}\sum_{b\in B}\sum_{v\in A+B} \frac{1}{p}\sum_{j=0}^{p-1} e\left(\frac{j(u/c+b-v)}{p}\right)$$

$$= \frac{1}{p}\sum_{j=0}^{p-1} \hat{B}\left(\frac{j}{p}\right) \hat{V}\left(\frac{-j}{p}\right) \sum_{u\in AC}\sum_{c\in C} e\left(\frac{ju/c}{p}\right)$$

$$\leq \frac{|B||C||A+B||AC|}{p} + \frac{1}{p}\max_{k\neq 0}\left|\sum_{u\in AC}\sum_{c\in C} e\left(\frac{ku/c}{p}\right)\right| \sum_{j\neq 0}\left|\hat{B}\left(\frac{j}{p}\right)\right| \left|\hat{V}\left(\frac{-j}{p}\right)\right|$$

$$\leq \frac{|B||C||A+B||AC|}{p} + \sqrt{p|B||C||A+B||AC|}$$

by (15) with A replaced by V, and by (16) with A replaced by AC and B replaced by $1/C$, and the result follows. ❑

Theorem 5. *Suppose that $A, B, C, D \subset \mathbb{F}_p$, and C does not contain 0.*
If $|A+B||C+D||A/C|^2|B||D| \leq p^4$, then

$$|A+B||C+D| \geq \frac{(|A||C|)^2|B||D|}{4p^2}.$$

If $|A+B||C+D||A/C|^2|B||D| > p^4$, *then*

$$|A+B||C+D||A/C| \geq \frac{p}{2}|A||C|.$$

Corollary 3.3. *If* $0 \notin A \subset \mathbb{F}_p$ *and* $|A+A||A/A||A| \leq p^2$, *then* $|A+A| \geq |A|^3/2p$.
If $|A+A| \leq \kappa|A|$ *and* $|A| \geq p^{2/3}$, *then* $|AA|, |A/A| \geq p/2\kappa$ *(by Corollary 3.2).*
If $|A+A| \leq \kappa|A|$ *and* $(2\kappa p)^{1/2} < |A| \leq p^{2/3}$, *then* $|A/A| > p/2\kappa^2$.

Remark. The first part is stronger than Garaev's $|AA||A+A| \geq |A|^4/2p$ in this range.

Proof of Theorem 5. Let us look at solutions to $u-b = m(v-d)$ with $b \in B, d \in D, u \in U = A+B, v \in V = C+D, m \in M = A/C$. For each $(a,b,c,d) \in A \times B \times C \times D$ we have the (distinct) solution $(b,d,u,v,m) = (b,d,a+b,c+d,a/c)$, so there are at least $|A||B||C||D|$ solutions. On the other hand we can give an exact count via the exponential sum

$$\sum_{b,d,u,v,m} \frac{1}{p} \sum_{j=0}^{p-1} e\left(\frac{j(u-b-m(v-d))}{p}\right) = \frac{|B||D||U||V||M| + \text{Error}}{p},$$

where

$$|\text{Error}| \leq \max_{i \neq 0} \left| \sum_{m \in M} \hat{D}\left(\frac{im}{p}\right) \hat{V}\left(\frac{-im}{p}\right) \right| \cdot \left| \sum_{j=1}^{p-1} \hat{U}\left(\frac{j}{p}\right) \hat{B}\left(\frac{-j}{p}\right) \right|.$$

By Cauchy–Schwarz this gives

$$|\text{Error}|^2 \leq \sum_{m \in M} \left|\hat{D}\left(\frac{im}{p}\right)\right|^2 \cdot \sum_{m \in M} \left|\hat{V}\left(\frac{-im}{p}\right)\right|^2 \cdot \sum_{j=1}^{p-1} \left|\hat{U}\left(\frac{j}{p}\right)\right|^2 \cdot \sum_{j=1}^{p-1} \left|\hat{B}\left(\frac{-j}{p}\right)\right|^2$$

which is $\leq p|D|p|V|p|U|p|B|$ by (15). Hence we have proved

$$|A||B||C||D| \leq \frac{|B||D||U||V||M|}{p} + p\sqrt{|B||D||U||V|},$$

and the result follows. □

Theorem 6 ([17]). *Suppose that* $A, B, C, D \subset \mathbb{F}_p$, *and* A, C *do not contain* 0.
If $|A+B||C+D||B||D| \ll p^3$, *then*

$$|AC|^2|A+B||C+D| \gg (|A||C|)^2|B||D|/p.$$

If $|A+B||C+D||B||D| \gg p^3$, *then*

$$|AC||A+B||C+D| \gg p|A||C|.$$

Remark. We claim that these bounds can be obtained trivially if $|A||C|(|B||D|)^2 \ll p^3$: The second case cannot hold since

$$|A||C|(|B||D|)^2 = |A||B||C||D||B||D| \geq |A+B||C+D||B||D| \gg p^3,$$

but then $|AC|^2|A+B||C+D| \geq |A||C|\ |A|^{2/3}|B|^{1/3}\ |C|^{2/3}|D|^{1/3} = (|A||C|)^2|B||D|/p \cdot (p^3/(|A||C|(|B||D|)^2)^{1/3} \gg (|A||C|)^2|B||D|/p$.

Proof. There exists $m \in AC$ such that $r_{AC}(m) \geq |A||C|/|AC|$. Now in the set

$$\{(u,v) \in (A-B)\times(C+D),\ (b,d) \in B \times D:\ (u-b)(v-d) = m\}$$

we evidently have the distinct points $((a+b, c+d), (b,d))$ for every $b \in B, d \in D$, and $a \in A, c \in C$ with $ac = m$; a total of $|B||D|r_{AC}(m) \geq |A||B||C||D|/|AC|$ points. To get an exact count, write $U = A+B,\ V = C+D$, to obtain

$$\sum_{b,d,u,v}\sum_{\substack{r,s\\ rs=m}} \frac{1}{p}\sum_i e\left(\frac{i(u-b-r)}{p}\right)\cdot\frac{1}{p}\sum_j e\left(\frac{j(v-d-s)}{p}\right)$$

$$= \frac{1}{p^2}\sum_{i,j}\hat{U}\left(\frac{i}{p}\right)\hat{B}\left(\frac{-i}{p}\right)\hat{V}\left(\frac{j}{p}\right)\hat{D}\left(\frac{-j}{p}\right)\sum_{\substack{r,s\\ rs=m}} e\left(\frac{-(ir+js)}{p}\right).$$

The $i = j = 0$ term yields $\frac{p-1}{p^2}|U||B||V||D|$, since there are exactly $p-1$ solutions to $rs = m$. If $j = 0$ but $i \neq 0$, then our final sum equals -1, so that the sum over $i \neq 0$ is $\frac{1}{p^2}|V||D|(|U||B| - p\,|U \cap B|)$. Similarly with $i = 0$. Finally if $i \neq 0$ and $j \neq 0$, then the final term is $\leq 2\sqrt{p}$ in absolute value by a well-known result on Kloosterman sums, and the total contribution is therefore

$$\leq \frac{2}{p^{3/2}}\sum_{i\neq 0}\left|\hat{U}\left(\frac{i}{p}\right)\hat{B}\left(\frac{-i}{p}\right)\right|\sum_{j\neq 0}\left|\hat{V}\left(\frac{j}{p}\right)\hat{D}\left(\frac{-j}{p}\right)\right|,$$

and by Cauchying the square of this is

$$\leq \frac{4}{p^3}\sum_i\left|\hat{U}\left(\frac{i}{p}\right)\right|^2\sum_i\left|\hat{B}\left(\frac{-i}{p}\right)\right|^2\sum_j\left|\hat{V}\left(\frac{j}{p}\right)\right|^2\sum_j\left|\hat{D}\left(\frac{-j}{p}\right)\right|^2 = 4p|U||B||D||V|.$$

Putting this altogether we obtain

$$\frac{|A||B||C||D|}{|AC|} \leq \frac{p+1}{p^2}|U||B||V||D| + 2\sqrt{p|U||B||D||V|},$$

which implies the result. □

Corollary 3.4. *Suppose that* $0 \notin A \subset \mathbb{F}_p$. *If* $|A+A||A| \ll p^{3/2}$, *then*

$$|AA||A+A| \gg |A|^3/\sqrt{p}.$$

If $|A+A||A| \gg p^{3/2}$, *then*

$$|AA||A+A|^2 \gg p|A|^2.$$

This is only non-trivial if $|A| \gg p^{1/2}$.

4 Ruzsa–Plunnecke Type Inequalities

We begin with a key result of Ruzsa:

Proposition 4.1. *If* $X, A_1, \ldots, A_k \subset \mathbb{F}_p$, *then there exists a non-empty* $Y \subset X$ *such that*

$$\frac{|Y + A_1 + \ldots + A_k|}{|Y|} \leq \prod_{i=1}^{k} \frac{|X + A_i|}{|X|}.$$

Corollary 4.2. *If* $A, B, C \subset \mathbb{F}_p$, *then*

$$|A \pm B| \leq \frac{|A+C||B+C|}{|C|}.$$

Proof. We can define an injective map $\phi : (A-B) \times C \to (A+C) \times (B+C)$ as follows, so that the inequality $|A-B| \leq |A+C||B+C|/|C|$ holds: If $\lambda \in A-B$, fix $a_\lambda \in A$, $b_\lambda \in B$ such that $\lambda = a_\lambda - b_\lambda$ and then define $\phi(\lambda, c) = (a_\lambda + c, b_\lambda + c)$. The map is injective since if $u = a_\lambda + c$ and $v = b_\lambda + c$ then $\lambda = u - v$ and then $c = u - a_\lambda$.

For the other case take $k = 2, A_1 = A, A_2 = B, X = C$ in Proposition 4.1 to obtain that there exists non-empty $Y \subset C$ such that

$$|A+B| \leq |Y+A+B| \leq \frac{|A+C|}{|C|} \cdot \frac{|B+C|}{|C|} \cdot |Y| \leq \frac{|A+C|}{|C|} \cdot \frac{|B+C|}{|C|} \cdot |C|.$$

□

Corollary 4.3. *If* $X, A_1, \ldots, A_k \subset \mathbb{F}_p$, *then there exists* $Z \subset X$ *such that* $|Z| \geq \frac{1}{2}|X|$ *and*

$$\frac{|Z + A_1 + \ldots + A_k|}{|Z|} \leq 2^k \prod_{i=1}^{k} \frac{|X + A_i|}{|X|}.$$

Proof. By Proposition 4.1 we know that there exists a set $Z \subset X$ for which the inequality holds, so let Z' be the largest subset of X for which this inequality is satisfied and suppose that $|Z'| \leq \frac{1}{2}|X|$. Apply Corollary 4.2 with $X' = X \setminus Z'$ in place of X. Noting that $|X'| > |X|/2$, and each $|X' + A_i| \leq |X + A_i|$ we deduce that there exists a non-empty $Y \subset X'$ such that $|Y + A_1 + \ldots + A_k| < 2^k |Y| \prod_{i=1}^k (|X + A_i|/|X|)$. Now let $Z = Z' \cup Y$ so that $|Z + A_1 + \ldots + A_k| \leq |Z' + A_1 + \ldots + A_k| + |Y + A_1 + \ldots + A_k|$, which is $\leq 2^k \prod_{i=1}^k (|X + A_i|/|X|)$ times $|Y| + |Z'| = |Z|$, and thus our inequality is satisfied by Z which is larger than Z', contradicting the hypothesis. □

Corollary 4.4. *For any $a, b \in \mathbb{F}_p^*$ and $A, B \subset \mathbb{F}_p$ we have*

$$|aA \pm bB| \leq \frac{|A + A||B + B|}{|aA \cap bB|},$$

and

$$|aA \pm bB| \leq \frac{|A + A||B + B|}{\max_{n \in \mathbb{F}_p} r_{aA+bB}(n)}.$$

Proof. In Corollary 4.2 replace A by aA, B by bB, and take $C = (x + aA) \cap bB$ for some $x \in \mathbb{F}_p$. We note that $aA + C \subset x + aA + aA$ which has the same size as $A + A$ and, similarly, $bB + C \subset bB + bB$ which has the same size as $B + B$. The first result follows taking $x = 0$. Now $|C| = r_{bB-aA}(x)$, so writing $x = -n$ and changing b to $-b$, we get our second result. □

Corollary 4.5. *For any $a, b \in \mathbb{F}_p^*$ and $A, B \subset \mathbb{F}_p$ we have*

$$aA \pm bB| \leq \frac{|A + B|^2}{|bA \cap aB|},$$

and

$$|aA \pm bB| \leq \frac{|A + B|^2}{\max_{n \in \mathbb{F}_p} r_{aB+bA}(n)}.$$

Proof. In Corollary 4.2 now replace A by aA, B by bB, and take $C = (x + bA) \cap aB$ for some $x \in \mathbb{F}_p$. We note that $aA + C \subset aA + aB$ which has the same size as $A + B$ and, similarly, $bB + C \subset x + bB + bA$ which also has the same size as $A + B$. The first result follows taking $x = 0$. Now $|C| = r_{aB-bA}(x)$, so changing b to $-b$, we get our second result □

5 Lower Bounds on the Size of $A + tB$

Lemma 5.1. *If $A, B \subset \mathbb{F}_p$ with $|A||B| > p$, then $\frac{A-A}{B-B} = \mathbb{F}_p \cup \{\infty\}$.*

Proof. As $|A||B| > p$ and each of $|A|, |B| \le p$ hence $|A|, |B| > 1$, that is, $|A|$, $|B| \ge 2$. Hence $0 = (a-a)/(b_1-b_2)$ and $\infty = (a_1-a_2)/(b-b)$. If $t \ne 0$, then there are $|A||B| > p$ numbers $a + tb$ so that two must be congruent mod p. Taking their difference implies the result. □

Remark. One might expect that if $A, B \subset \mathbb{F}_p$ with $|A||B| > p$ then $AB + AB = \mathbb{F}_p$. However if $A = B = \{m \pmod p) : (m/p) = 1\}$ where p is a prime $\equiv 3 \pmod 4$, then evidently $0 \notin AB + AB$ (and here $|A|, |B| = (p-1)/2$). Hence the best we can hope for is that if $|A||B| > p$ then $AB + AB + AB = \mathbb{F}_p$, and perhaps $AB + AB = \mathbb{F}_p^*$.

Glibichuk [14] proved that if $|A||B| \ge 2p$ then $8AB = \mathbb{F}_p$ (so that if $|A||B| \ge p$ then $8AB = \mathbb{F}_p$, since then $|A+A||B| \ge 2p$ so that $16AB \supseteq 8(A+A)B = \mathbb{F}_p$, unless A is an arithmetic progression, which can be handled).

Let $T = \frac{A-A}{B-B} \setminus \{0, \infty\}$. We are interested in the size of $A + tB$ when $|A|, |B| > 1$. Evidently $|A + tB| \le |A|\,|B|$ with equality if and only if $t \notin T \cup \{0\}$.

Let $R(t) = R_{A,B}(t)$ denote the number of solutions $a, c \in A,\ b, d \in B$ to $a + tb = c + td$. We always have the the "diagonal solutions" where $a = c$ and $b = d$ to $a + tb = c + td$, so that $R(t) \ge |A||B|$. Equality holds, that is, $R(t) = |A||B|$, if and only if $t \notin T \cup \{0\}$. Hence

$$|A + tB| = |A|\,|B| \iff t \notin T \cup \{0\} \iff R(t) = |A||B|. \tag{18}$$

There is a link between $|A + tB|$ that holds no matter what, which is given by the Cauchy–Schwarz inequality: Let $r_t(n) = \#\{a \in A,\ b \in B :\ n = a + tb\}$ so that

$$(|A|\,|B|)^2 = \left(\sum_{n \in A+tB} r_t(n)\right)^2 \le |A + tB|\, R_{A,B}(t), \tag{19}$$

since $R_{A,B}(t) = \sum_n r_t(n)^2$.

Proposition 5.2. *For any finite sets A, B, S with $0 \notin S$, there exists $t \in S$ for which*

$$|A + tB| > \frac{1}{2} \min\{|S|, |A|\,|B|\}.$$

If $A, B \in \mathbb{F}_p$, then we also have

$$|A + tB| > \frac{1}{2} \min\left\{p, \frac{|A|\,|B|\,|S|}{p}\right\}.$$

Proof. Note that $|A + 0B| = |A|$ and $R(0) = |A|\,|B|^2$. Since $R(t) \ge |A||B|$ for all t hence

$$\sum_{t \in S} (R(t) - |A||B|) \le \sum_{t \ne 0} (R(t) - |A||B|) = \#\{a, c \in A,\ b, d \in B :\ b \ne d \text{ and } a \ne c\}$$
$$= (|A|^2 - |A|)(|B|^2 - |B|). \tag{20}$$

Therefore there exists $t \in S$ with

$$R(t) \leq \frac{(|A|^2 - |A|)(|B|^2 - |B|)}{|S|} + |A|\,|B| < 2|A||B| \max\left\{\frac{|A||B|}{|S|}, 1\right\},$$

whence, by (19),

$$|A|\,|B| \leq 2|A + tB| \ \max\left\{\frac{|A||B|}{|S|}, 1\right\}$$

and so the first result follows.

When we are working mod p, we have

$$R(t) = \sum_n r_t(n)^2 = \frac{1}{p}\sum_{j=0}^{p-1} |\hat{A}(j/p)|^2 |\hat{B}(jt/p)|^2 \geq \frac{(|A|\,|B|)^2}{p},$$

taking the $j = 0$ term, since every term is non-negative. If $|A||B| > p$, then $R(t) > |A|\,|B|$ and so $T = \mathbb{F}_p \setminus \{0\}$ by (18) giving another proof of the lemma above.

Now, rearranging (20) we obtain

$$\sum_{t\in S}\left(R(t) - \frac{(|A|\,|B|)^2}{p}\right) \leq \sum_{t\neq 0}\left(R(t) - \frac{(|A|\,|B|)^2}{p}\right) = p|A||B|\left(1 - \frac{|A|}{p}\right)\left(1 - \frac{|B|}{p}\right),$$

so there exists $t \in S$ with

$$|A|\,|B| \leq 2|A + tB| \ \max\left\{\frac{p}{|S|}, \frac{|A|\,|B|}{p}\right\}$$

by (19), and the result follows. □

Corollary 5.3. *Suppose that $|A|, |B| \geq 2$. If $\frac{A-A}{B-B} = \mathbb{F}_p$, then there exist $a_1, a_2 \in A, b_1, b_2 \in B$ such that*

$$|(a_1 - a_2)B + (b_1 - b_2)A| > \frac{1}{2} \min\{p, |A|\,|B|\}.$$

If $\frac{A-A}{B-B} \neq \mathbb{F}_p$, then there exist $a_1, a_2 \in A, b_1, b_2 \in B$ such that

$$|(a_1 - a_2 + b_1 - b_2)B + (b_1 - b_2)A| = |A|\,|B|.$$

In other words, the elements $(a_1 - a_2 + b_1 - b_2)b + (b_1 - b_2)a$ with $a \in A, b \in B$, are distinct.

Proof. For any $t \in T$, there exist $a_1, a_2 \in A$, $b_1, b_2 \in B$, for which $(a_2 - a_1) + (b_1 - b_2)t = 0$. The first case follows immediately from Proposition 5.2 by multiplying through by $b_1 - b_2$.

Suppose that $T \neq \mathbb{F}_p^*$. Now T contains a non-zero element, say t, since A has at least two elements. Moreover we may assume $1 \leq t < p/2$ since $T = -T$ (as may be seen by swapping a_1 and a_2). Hence there exists $t \in T$ such that $t + 1 \notin T$. Therefore $|A + (t + 1)B| = |A||B|$ by (18) and the result follows by multiplying through by $b_1 - b_2$. □

Lemma 5.4. *Let* $I(A, B) := (B - B)A + (A - A)B$.

(i) *If* $t \in \frac{A-A}{B-B}$, *then* $|I(A, B)| \geq |A + tB|$.
(ii) $AB - AB \subset I(A, B)$, *so that* $|I(A, B)| \geq \min\{p, 2|AB| - 1\}$.

Proof. There exist $a_1, a_2 \in A$, $b_1, b_2 \in B$ for which $(a_2 - a_1) + (b_1 - b_2)t = 0$. Each element of $S = (b_1 - b_2)(A + tB)$ can be written as $(b_1 - b_2)a + (b_1 - b_2)tb = (b_1 - b_2)a + (a_1 - a_2)b \in I(A, B)$ and (i) follows. Also if $a_1, a_2 \in A$, $b_1, b_2 \in B$, then $a_1b_1 - a_2b_2 = (b_1 - b_2)a_1 + (a_1 - a_2)b_2 \in I(A, B)$, that is, $AB - AB \subset I(A, B)$, and (ii) follows from the Cauchy–Davenport theorem. □

Corollary 5.5. *If* $|A||B| > p$, *then there exists* $t \in T$ *for which* $|A + tB| > p/2$. *Hence* $|I(A, B)| > p/2$. *We can rephrase this as: There exist* $b_1, b_2 \in B, a_1, a_2 \in A$ *such that* $|(b_1 - b_2)A + (a_1 - a_2)B| > p/2$.

Proof. By Lemma 5.1 we know that $\frac{A-A}{B-B} = \mathbb{F}_p$. Taking $S = T$ in Proposition 5.2 we deduce that there exists $t \in T$ with $|A + tB| > p/2$. The result then follows from Lemma 5.4. □

Proposition 5.6. *Let* $R_k(B)$ *be the set of* $n \in \mathbb{F}_p$ *for which* $r_{B/B}(n) \geq k$ *for* $1 \leq k \leq |B|$; *note that* $1 \in R_k(B)$. *Let* $G_k(B)$ *be the multiplicative group generated by* $R_k(B)$, *and then* $H_k(B) = G_k(B)\frac{A-A}{B-B}$. *There exists* $t \in T$ *for which* $|A + tB| \gg \min\{k|A|, |H_k|\}$.

Proof. If $H_k(B) = \frac{A-A}{B-B}$, then the result follows from Proposition 5.2 (since $|B| \geq k$). Otherwise $\frac{A-A}{B-B} \subsetneq H_k(B)$ so there exist $g \in G_k(B)$ and $t_0 \in T$ such that $gt_0 \notin T$. Now any $g \in G_k(B)$ can be written as $g = n_1 n_2 \ldots n_\ell$ where each $n_j \in R_k(B)$. Define $t_j = n_j t_{j-1}$ for each j, so that $t_0 \in T$ and $t_\ell = gt_0 \notin T$: hence there exist $t = t_{j-1} \in T$ and $n = n_j \in R_k(B)$ such that $nt = t_j \notin T$. But then $|A + ntB| = |A||B|$ by (18); that is, the elements of $A + ntB$ are all distinct. Now $r_{B/B}(n) \geq k$ by the definition of $R_k(B)$, and so there are at least k values of $b \in B$ for which nb is also in B, and hence $A + tB$ contains at least $|A|k$ distinct elements. □

Lemma 5.7. *Let* $B = B_1 \cup B_2$ *be a partition of* B *where* $b_1/b_2 \notin G_k$ *for any* $b_1 \in B_1, b_2 \in B_2$. *Then* $|B_1||B_2| \leq (k - 1)|B_1B_2|$.

Proof. If $s \notin B_1/B_2$, then $r_{B_1/B_2}(s) = 0$. If $s \in B_1/B_2$, then $s \notin G_k$ by hypothesis, so that $s \notin R_k$ and hence $r_{B_1/B_2}(s) \leq r_{B/B}(s) < k$. Therefore

$$(|B_1||B_2|)^2 = \left(\sum_s r_{B_1B_2}(s)\right)^2 \le |B_1B_2| \sum_s r_{B_1B_2}(s)^2 = |B_1B_2| \sum_s r_{B_1/B_2}(s)^2$$
$$\le |B_1B_2|(k-1) \sum_s r_{B_1/B_2}(s) = |B_1B_2|(k-1)|B_1||B_2|.$$

□

Lemma 5.8. *Let $k = |B|^2/100|BB|$. There exists $h \neq 0$ for which $|B \cap hG_k(B)| \ge \frac{49}{50}|B|$.*

Proof. Let H be the set of cosets of G_k in $\mathbb{F}_p^*$. For any partition $H = H_1 \cup H_2$ let $B_j := \cup_{h\in H_j}(B \cap hG_k)|$ for $j = 1, 2$ so that $B_1 \cup B_2$ is a partition of B; note that $b_1/b_2 \in (h_1/h_2)G_k$ for some $h_1 \in H_1, h_2 \in H_2$ so that $h_1 \neq h_2$ and thus $b_1/b_2 \notin G_k$. Now $|B_1|(|B| - |B_1|) = |B_1||B_2| < k|B_1B_2| < k|BB| = \frac{|B|^2}{100}$ by Lemma 5.7, and so either $|B_1|$ or $|B_2|$ is $> \frac{49}{50}|B|$.

Now let H_1 be a maximal subset of H such that $|B_1| < |B|/50$. Therefore for any $h \in H_2$ we must have $|B_1 \cup (B \cap hG_k)| \ge |B|/50$ and hence $> \frac{49}{50}|B|$ by the previous paragraph, so that $|B \cap hG_k| \ge \frac{24}{25}|B|$. We deduce H_2 has no more than one element, and thus exactly one element (since $|B_2| > 0$). The result follows. □

Lemma 5.9. *Let $C \subset G$, a subgroup of $\mathbb{F}_p^*$, with $1 < |C| < \sqrt{p}$. Then we have $|G(C-C)| \gg |C|^{3/2}$ and $|G| \cdot |C-C| \gg |C|^{5/2}$.*

Proof. First note that $|G(C-C)| \ge |G|$ and $|C-C| \ge |C|$ so the results follow unless $|C| \ge 2$ and $|G| \le |C|^{3/2}$, which we now assume.

Now, GS is a union of cosets of G for any set $S \in \mathbb{F}_p^*$, so we can write

$$G(C-C) = \{0\} \cup \cup_{i=1}^m t_iG,$$

where we order these cosets of G so that $r_{G-G}(t_1) \ge r_{G-G}(t_2) \ge \ldots \ge r_{G-G}(t_m)$ (note that $r_{G-G}(t_i)$ takes the same value for any choice of t_i inside a fixed coset of G). Since $|G(C-C)| = |G|m+1$, the first result follows unless $m \le M := [|C|^{3/2}/|G|]$.

If $J \le M$, then $J\,G|^4 \le M|G|^4 \le |G|^3|C|^{3/2} \le |C|^{9/2}|C|^{3/2} = |C|^6 \le p^3$. Therefore

$$Jr_{G-G}(t_J) \le \sum_{i=1}^J r_{G-G}(t_i) \le 4(J|G|)^{2/3} \tag{21}$$

by Lemma 5 of [18] (with the constant '4' made explicit) and so

$$|G(C-C)| > m|G| \ge \frac{1}{8}\left(\sum_{i=1}^m r_{G-G}(t_i)\right)^{3/2}. \tag{22}$$

For any fixed $c_0 \in C$ the solutions to $h_1 - h_2 = t_i$ with $h_1, h_2 \in G$ are in 1–1 correspondence with the solutions $h_3 - c_0 = t_i h_4$ with $h_3, h_4 \in G$, as may be seen by taking $h_3 = h_1 c_0/h_2$ and $h_4 = c_0/h_2$. Hence

$$r_{G-G}(t_i) = \sum_{t \in t_i G} r_{G-c_0}(t) \geq \sum_{t \in t_i G} r_{C-c_0}(t). \tag{23}$$

We then deduce

$$\sum_{i=1}^{m} r_{G-G}(t_i) \geq \sum_{t \in (C-C)G \setminus \{0\}} r_{C-c_0}(t) = |C| - 1,$$

and the first result follows from (22).

We now prove the second result, no longer assuming that $m \leq M$: Since $r_{C-C}(t) \leq r_{G-G}(t) = r_{G-G}(t_i)$ for all $t \in t_i G$, we have

$$\sum_{t \in t_i G} r_{C-C}(t)^2 \leq r_{G-G}(t_i) \sum_{t \in t_i G} r_{C-C}(t) \leq r_{G-G}(t_i) \sum_{c_0 \in C} \sum_{t \in t_i G} r_{C-c_0}(t) \leq |C| r_{G-G}(t_i)^2$$

by (23). Therefore, using (21) for the bound $r_{G-G}(t_J) \leq 4|G|^{2/3}/\min\{J, M\}^{1/3}$, we obtain

$$\begin{aligned}
\sum_{t \neq 0} r_{C-C}(t)^2 &= \sum_{i=1}^{M} \sum_{t \in t_i G} r_{C-C}(t)^2 + \sum_{i=M+1}^{m} \sum_{t \in t_i G} r_{C-C}(t)^2 \\
&\leq \sum_{i=1}^{M} |C| r_{G-G}(t_i)^2 + \sum_{i=M+1}^{m} r_{G-G}(t_i) \sum_{t \in t_i G} r_{C-C}(t) \\
&\leq |C| \sum_{i=1}^{M} 16 |G|^{4/3} i^{-2/3} + 4|G|^{2/3} M^{-1/3} \sum_{i=M+1}^{m} \sum_{t \in t_i G} r_{C-C}(t) \\
&\leq 48 |C| |G|^{4/3} M^{1/3} + 4|C|^2 |G|^{2/3} M^{-1/3} \asymp 52 |C|^{3/2} |G|.
\end{aligned}$$

Hence

$$(|C|^2 - |C|)^2 = \left(\sum_{t \neq 0} r_{C-C}(t) \right)^2 \leq |C - C| \sum_{t \neq 0} r_{C-C}(t)^2 \ll |C - C| |C|^{3/2} |G|,$$

and the result follows. □

Theorem 7. *If $|A| < \sqrt{p}$, then $|I(A, A)| \gg |A|^{3/2}$.*

Proof. We have $|I(A,A)| \geq 2|AA| - 1$ by Lemma 5.4(ii), and the result follows unless $|AA| \ll |A|^{3/2}$, which we now assume.

Let $k = |A|^2/100|AA|$ ($\gg |A|^{1/2}$), and define $R_k(A), G_k(A), H_k(A)$ as above. By Lemma 5.8 there exists $h \neq 0$ such that if $C = \{g \in G_k(A) : gh \in A\} = G_k \cap h^{-1}A$, then

$$|C| = |A \cap hG_k(A)| \geq \frac{49}{50}|A|.$$

Therefore, using the fact that $H = G(A-A)/(A-A)$ we have

$$|H| + 1 \geq |G(A-A)| \geq |G(hC - hC)| = |G(C-C)| \gg |C|^{3/2} \gg |A|^{3/2}$$

by Lemma 5.9. The result follows by Lemma 5.4 and Proposition 5.6 since now $|I(A,A)| \gg \min\{k|A|, |H_k|\} \gg |A|^{3/2}$. □

Theorem 8. *We have*

$$E_\times(A,A) \leq 4|A+A|^2 \max\left\{\frac{|A|^2}{p}, \frac{p}{|A|}\right\} \log|A|,$$

and

$$E_\times(A,A)^4 < 32|A+A|^8|A|^2 \max\left\{\frac{|A|^3}{p}, 2|A+A|\right\} (\log|A|)^4.$$

If $|A| \leq \sqrt{p}$, then $|A|^3/p \leq |A| \leq |A+A|$, so the above becomes $E_\times(A,A)^4 < 64|A+A|^9|A|^2(\log|A|)^4$. This yields a sum-product bound which is non-trivial for all $|A| \leq \sqrt{p}$:

Corollary 5.10. *We have*

$$E_\times(A,B) \leq 4|A+A||B+B| \log(|A||B|) \left(\max\left\{\frac{|A|^2}{p}, \frac{p}{|A|}\right\} \max\left\{\frac{|B|^2}{p}, \frac{p}{|B|}\right\}\right)^{1/2},$$

and

$$E_\times(A,B)^8 < 2^{10}(|A+A||B+B| \log|A||B|)^8(|A||B|)^2 \max\left\{\frac{|A|^3}{p}, 2|A+A|\right\} \times \max\left\{\frac{|B|^3}{p}, 2|B+B|\right\},$$

which implies that if $|A|, |B| \leq p^{1/2}$ then

$$|AB|^8(|A+A||B+B|)^9 > \frac{(|A||B|)^{14}}{2^{12}(\log|A||B|)^8},$$

Proof. By the Cauchy–Schwarz inequality we have

$$E_\times(A,B)^2 = \left(\sum_n r_{A/A}(n) r_{B/B}(n)\right)^2 \le \sum_m r_{A/A}(m)^2 \sum_n r_{B/B}(n)^2 = E_\times(A,A)E_\times(B,B)$$

and then the first two results follow from Theorem 8. Now, if $|A|, |B| \le p^{1/2}$, then $|A|^3/p \le |A| < |A+A|$. Therefore, by the second inequality and (7), we obtain our third and final inequality □

If $|A|, |B| \ge p^{2/3}$, then by the first inequality in Corollary 5.10, and (7), we obtain

$$|A+A||B+B||AB| \ge \frac{p|A||B|}{4\log(|A||B|}.$$

which is weaker than Theorem 4.

Corollary 5.11. *If* $4p^4/|A|^6 \ge |A+A| > |A|^3/2p$ *and* $|A| \le 2p^{5/9}$, *then*

$$|AA|^4|A+A|^9 \ge \frac{|A|^{14}}{2^6(\log|A|)^4}$$

If $|A+A| \le |A|^3/2p$ *and* $p^{1/2} \le |A| \le p^{5/9}$, *then*

$$|AA||A+A|^2 \ge \frac{|A|^{11/4}p^{1/4}}{2^{5/4}\log|A|}$$

Proof. This follows by Theorem 8 and (7) with $B = A$, which gives

$$|A|^4 \le |AA|E_\times(A,A).$$

□

Proof of Theorem 8. We begin by noting that

$$E_\times(A,A) = \sum_{b\in A}\sum_{k=0}^{[\log|A|]} \sum_{\substack{a\in A\\ 2^k \le |aA\cap bA| < 2^{k+1}}} |aA\cap bA|$$

where the logarithm here is in base 2. Hence there exists $2^k \le |A|$ for which there is $A_1 \subset A$ and $b_0 \in A$ such that

$$2^k \le |aA \cap b_0A| < 2^{k+1}$$

for every $a \in A_1$, where $|A_1|2^{k+1} \ge E_\times(A,A)/(|A|\log|A|)$.

By Proposition 5.2 with $S = A_1$ there exists $a \in A_1$ such that

$$|aA - b_0A| \geq \frac{1}{2}\min\{p, |A|^2|A_1|/p\};$$

and $|aA - b_0A| \leq |A + A|^2/|aA \cap b_0A|$ by Corollary 4.4. Hence

$$A + A|^2 \geq \frac{E_\times(A,A)}{4\log|A|} \min\left\{\frac{p}{|A_1||A|}, \frac{|A|}{p}\right\},$$

and the first result follows.

If $\frac{A_1-A_1}{A_1-A_1} = \mathbb{F}_p$, then by Corollary 5.3 there exist $a_1, a_2, a_3, a_4 \in A_1$ such that

$$|(a_1 - a_2)A_1 + (a_3 - a_4)A_1| > \frac{1}{2}\min\left\{p, |A_1|^2\right\};$$

By Proposition 4.1 we have $Y \subseteq b_0A$ such that

$$|(a_1 - a_2)A + (a_3 - a_4)A| \leq |Y + a_1A - a_2A + a_3A - a_4A| \leq |Y|\prod_{i=1}^{4}\frac{|a_iA \pm b_0A|}{|b_0A|}$$

$$\leq |b_0A|\prod_{i=1}^{4}\frac{|A + A|^2}{|b_0A|\ |a_iA \cap b_0A|} \leq \frac{|A + A|^8}{|A|^3\ 2^{4k}}$$

using Corollary 4.4. Hence

$$|A + A|^8 > \frac{(|A_1|2^{k+1})^4}{32}\min\left\{p/|A|, |A|\right\},$$

and so $E_\times(A,A)^4 \leq 32|A + A|^8|A|^3(\log|A|)^4\max\left\{1, |A|^2/p\right\}$.

If $\frac{A_1-A_1}{A_1-A_1} \neq \mathbb{F}_p$, then by Corollary 5.3 there exist $a_1, a_2, a_3, a_4 \in A_1$ such that if $A_2 \subset A_1$ then

$$|(a_1 - a_2)A_2 + (a_1 - a_2 + a_3 - a_4)A_1| = |A_1|\ |A_2|.$$

By Corollary 4.3 there exists $(a_1 - a_2)A_2 \subset (a_1 - a_2)A_1$ with $|A_2| \geq \frac{1}{2}|A_1|$ such that

$$|(a_1 - a_2)A_2 + (a_1 - a_2)A_1 + (a_3 - a_4)A_1|$$
$$\leq 4|A_2|\frac{|A_1 + A_1|}{|(a_1 - a_2)A_1|}\cdot\frac{|(a_1 - a_2)A_1 + (a_3 - a_4)A_1|}{|(a_1 - a_2)A_1|}.$$

Bounding the last term as above we obtain $|A + A|^9 > (|A_1|2^{k+1})^4|A|^2/64$, so that $E_\times(A,A)^4 \leq 64|A + A|^9|A|^2(\log|A|)^4$. □

Remark (A few ideas). In the case that $\frac{A-A}{A-A} = \mathbb{F}_p$, we get in the above proof that there exist $a_1, a_2, a_3, a_4 \in A$ such that

$$\frac{1}{2}\min\{p, |A|^2\} \le |(a_1 - a_2)A + (a_3 - a_4)A| \le |bA| \prod_{i=1}^{4} \frac{|A+A|^2}{|bA|\,|a_iA \cap bA|}.$$

Now note that $\sum_{a,b\in A} |aA \cap bA| = E_\times(A, A) \ge |A|^4/|AA|$ so $|aA \cap bA|$ is $|A|^2/|AA|$ on average. If we somehow get that, even with the loss of a constant (or even $|A|^\epsilon$) for our $|a_iA \cap bA|$ then our bound here would be $|A+A|^8|AA|^4 \gg |A|^{11}\min\{p, |A|^2\}$ which is what we get in Corollary 5.11, but in a less complicated way. If we could take $b = a_1$ so we can replace one term in our product by $|A+A|/|A|$. Then we would get the bound $|A+A|^7|AA|^3 \gg |A|^9\min\{p, |A|^2\}$; this improves the exponent from $\frac{13}{12}$ when $|A| \le \sqrt{p}$ to $\frac{11}{10}$.

If $\frac{A-A}{A-A} \ne \mathbb{F}_p$, then we can change $\min\{p, |A|^2\}$ to $\min\{\left|\frac{A-A}{A-A}\right|, |A|^2\}$ using the same argument.

Combining all the results to this point, here are the results we obtained on sum-product in finite fields:

Corollary 5.12. *If $A \subseteq \mathbb{F}_p$, then*

$$\max\{|AA|, |A+A|\} \gg (\log|A|)^{O(1)} \cdot \begin{cases} \sqrt{p|A|} & \text{if } |A| \ge p^{2/3} \\ |A|^2/\sqrt{p} & \text{if } p^{2/3} > |A| \ge p^{7/13} \\ |A|^{11/12}p^{1/12} & \text{if } p^{7/13} > |A| \ge p^{13/25} \\ |A|^{14/13} & \text{if } p^{13/25} > |A|. \end{cases}$$

As one can conjecture there is room for improvements. Indeed Rudnev recently published [26] a new bound

$$\max\{|AA|, |A+A|\} \gg (\log|A|)^{O(1)} \cdot |A|^{12/11} \text{ if } p^{1/2} > |A|.$$

More strikingly after completing this survey we have learned that Roche-Newton, Rudnev, and Shkredov announced [24] a fantastic bound

$$\max\{|AA|, |A+A|\} \gg |A|^{6/5} \text{ if } p^{5/8} > |A|.$$

In their proof they use incidence bounds in "Elekes style." Misha Rudnev [25] used a result of Guth and Katz on point-line incidences in space (see in [15] and in [21]) to obtain an unexpectedly strong point-plain incidence bound in K^3 for arbitrary field K. This beautiful result led to new sum-product bounds in $\mathbb{F}_p$ even for composite p.

5.1 Getting the Full Field

We now give a result of Glibichuk discussed at the start of this section:

Theorem 9. *If* $|A||B| \geq 3p/2 + \sqrt{p}$, *then* $8AB = \mathbb{F}_p$.

Proof. Suppose $|B| \geq |A|$, let $B_+ = \{b \in B : -b \in B\}$ and $B_- = \{b \in B : -b \notin B\} \cup \{b \in B_+ : 1 \leq b \leq \frac{p-1}{2}\}$. By definition B_+ is symmetric ($b \in B_+ \Leftrightarrow -b \in B_+$) and B_- is anti-symmetric ($b \in B_- \implies -b \notin B_-$). Let B_* be the larger of the two. By a simple counting argument we know that $|B_*| = \max\{|B_+|, |B_-|\} \geq \max\{\frac{2|B|-1}{3}, 2|B| - p\}$. Note that $|A||B_*| \geq |A| \cdot \frac{2|B|-1}{3} \geq p + 2$.

Noting that $a + tb = c + td$ iff $a - td = c - tb$ we have $R(t) = R(-t)$ in Proposition 5.2 so there exists $t \neq 0$ such that $R(t) = R(-t) \leq |A||B_*| + |A|^2|B_*|^2/(p-1)$ so that

$$|A + tB_*|, |A - tB_*| \geq \frac{|A|^2|B_*|^2}{R(t)} \geq \frac{|A|^2|B_*|^2}{|A||B_*| + |A|^2|B_*|^2/(p-1)} > p/2$$

as $|A||B_*| \geq p + 2$. If $B_* = B_+$, then $I(A, B) = A(B + B) + B(A + A) \subset 4AB$ and so, by Lemma 5.4(i), we have $|4AB_+| \geq |I(A, B_+)| \geq |A + tB_+| > p/2$, and thus $8AB_+ = \mathbb{F}_p$ by the pigeonhole principle. If $B_* = B_-$, then, by the pigeonhole principle, there exist $a_1, a_2 \in A, b_1, b_2 \in B_-$ such that $a_1 - tb_1 = -(a_2 - tb_2)$ so that $t = (a_1 + a_2)/(b_1 + b_2)$ (and $b_1 + b_2 \neq 0$ as B_- is anti-symmetric). But then $|4AB_-| \geq |(b_1 + b_2)A + (a_1 + a_2)B_-| = |A + tB| > p/2$, and thus $8AB_- = \mathbb{F}_p$ by the pigeonhole principle.

The result follows. □

Corollary 5.13. *If* $|A||B| \geq 3p/4 + \sqrt{p}$, *then* $16AB = \mathbb{F}_p$.

Proof. Suppose $|B| \geq |A|$ so that $|A||B+B| \geq |A|(2|B|-1) \geq 2|A||B| - \sqrt{|A||B|} \geq 3p/2 + \sqrt{p}$, and the result follows by applying Theorem 9 with B replaced by $B+B$, so that $16AB \subset 8A(B + B) = \mathbb{F}_p$. □

We now give a result of Hart and Iosevitch [16]:

Theorem 10. *If* $\prod_{j=1}^{m} |A_j||B_j|/p > (p-1)$, *then* $\sum_{i=1}^{m} A_iB_i \supseteq \mathbb{F}_p^*$. *In particular, if* $|A||B| > p(p-1)^{1/m}$, *then* $mAB \supseteq \mathbb{F}_p^*$. *If* $\prod_{j=1}^{m} |A_j||B_j|/p > (p-1)^2$, *then* $\sum_{i=1}^{m} A_iB_i \supseteq \mathbb{F}_p$.

Proof. Let $r(t)$ be the number of representations of t as $\sum_{i=1}^{m} a_ib_i$, so that

$$r(t) = \sum_{\substack{a_i \in A_i \\ b_j \in B_j}} \frac{1}{p} \sum_k e\left(\frac{k(\sum_i a_ib_i - t)}{p}\right) = \frac{\prod_i |A_i||B_i|}{p} + \frac{\text{Error}}{p},$$

where, by the Cauchy–Schwarz inequality, and writing $u \equiv k/l \pmod{p}$

$$|\text{Error}|^2 = \left|\sum_{a_i\in A_i}\sum_{k\neq 0} e\left(\frac{-kt}{p}\right)\prod_j \sum_{b_j\in B_j} e\left(\frac{ka_jb_j}{p}\right)\right|^2$$

$$\leq \sum_{a_i\in A_i}\left|\sum_{k\neq 0} e\left(\frac{-kt}{p}\right)\prod_j \sum_{b_j\in B_j} e\left(\frac{ka_jb_j}{p}\right)\right|^2$$

$$\leq \prod_i |A_i|\cdot \sum_{a_i}\sum_{k,l\neq 0} e\left(\frac{(l-k)t}{p}\right)\prod_j \sum_{b_j,b_j'\in B_j} e\left(\frac{a_j(kb_j-lb_j')}{p}\right)$$

$$\leq \prod_i |A_i|\cdot \sum_{u,l\neq 0} e\left(\frac{(1-u)lt}{p}\right)\prod_j p \sum_{b_j,ub_j\in B_j} 1.$$

Assume, for now, that $t\neq 0$. If $u\neq 1$, then the sum over l equals -1, otherwise it equals $p-1$. Hence the above is $\leq (p-1)\prod_i |A_i|\cdot\prod_j p\sum_{b_j\in B_j}1 = (p-1)p^m$ $\prod_i |A_i||B_i|$. We deduce that $pr(t)\geq \prod_i |A_i||B_i| - ((p-1)p^m\prod_i |A_i||B_i|)^{1/2}$ and the result follows.

If $t=0$, the sum over l is $p-1$ and it is feasible that $ub_j\in B_j$ for all u,j, so the above is $\leq (p-1)^2p^m\prod_i |A_i||B_i|$ and the result follows (one can also prove this bound more directly using (16)). □

6 Helfgott's More General Bounds

Theorem 11 (Helfgott's Theorem). *Let G be a group and Γ be an abelian group of automorphisms. Let $S\subset\Gamma$ with the property*

$$\textit{If } g^\sigma = g \textit{ for some } g\in G, \sigma\in S^{-1}S, \textit{ then } g=1 \textit{ or } \sigma=1. \tag{24}$$

Then for any $A\subset G$ we have one of the following:

(i) *There exists $g\in A$ such that $|Ag^S| = |A|\,|S|$*
Or there exist $c\in A^{-1}A$ and $\lambda\neq\tau\in S$ such that
(ii) *There exists $b\in A\cup A^{-1}$ such that $\{(ab^\sigma)^\tau c^\sigma (ab^\sigma)^{-\lambda} : a\in A,\ \sigma\in S\}$ contains $|A|\,|S|$ distinct elements.*

Or (iii) *There exists $\eta\in S\cup S^{-1}$ such that $\{a^\tau (c^\eta)^\sigma a^{-\lambda} : a\in A,\ \sigma\in S\}$ contains $|A|\,|S|$ distinct elements;*
or (iv) *$\{a^\tau c^\sigma a^{-\lambda} : a\in A,\ \sigma\in S\}$ contains $\geq \frac{1}{2}\min\{|A||S|, |\mathcal{O}|\}$ distinct elements, where $\mathcal{O}$ is the union of the orbits of elements of A under the two maps $a\to ba$ for any $b\in A\cup A^{-1}$, and $a\to a^\eta$ for any $\eta\in S\cup S^{-1}$.*

Proof. Define $U := \{g \in G :$ There exists $\lambda \neq \tau \in S$ such that $g^{\tau-\lambda} \in A^{-1}A\}$. For any $g \in G$ define $\phi_g : A \times S \to G$ by $\phi_g(a, \sigma) = ag^{\sigma}$. Note that ϕ_g is injective if and only if $g \notin U$, and in this case $|Ag^S| = |A|\,|S|$.

Next define $\delta_{\lambda,\tau}(g) = g^{\tau-\lambda}$ for $g \in G$, for each $\lambda \neq \tau \in S$. This is always injective, for if $\delta_{\lambda,\tau}(g_1) = \delta_{\lambda,\tau}(g_2)$, then $(g_1^{-1}g_2)^{\lambda\tau^{-1}} = g_1^{-1}g_2$, and so $g_1 = g_2$ by (24). Hence if $g \notin U$, then $|\delta_{\lambda,\tau}(Ag^S)| = |A|\,|S|$. We observe that

$$\delta_{\lambda,\tau}(ag^{\sigma}) = (ag^{\sigma})^{\tau-\lambda} = a^{\tau}g^{\sigma(\tau-\lambda)}a^{-\lambda}, \tag{25}$$

using the fact that Γ is abelian.

Suppose that $\mathcal{O} \not\subset U$. By following how the orbits are created from A by applying the two maps, we consider the first element of $\mathcal{O}$ that is not in U. Then one of the following must be true:

(i) There exists $g \in A$ such that $g \notin U$;
(ii) There exists $u \in \mathcal{O} \cap U$ such that $g = bu \notin U$ with $b \in A \cup A^{-1}$;
(iii) There exists $u \in \mathcal{O} \cap U$ such that $g = u^{\eta} \notin U$ with $\eta \in S \cup S^{-1}$, where

In case (i) we have that ϕ_g is injective so that $|Ag^S| = |A|\,|S|$.

In cases (ii) and (iii) we have $u \in \mathcal{O} \cap U$ and so there exist $\lambda \neq \tau \in S$ and $c \in A^{-1}A$ such that $u^{\tau-\lambda} = c$.

In case (ii) we then have $g^{\tau-\lambda} = (bu)^{\tau-\lambda} = b^{\tau}u^{\tau-\lambda}b^{-\lambda} = b^{\tau}cb^{-\lambda}$, and so $\delta_{\lambda,\tau}(ag^{\sigma}) = (ab^{\sigma})^{\tau}c^{\sigma}(ab^{\sigma})^{-\lambda}$ by (25) and the commutativity of Γ. Hence

$$\{(ab^{\sigma})^{\tau}c^{\sigma}(ab^{\sigma})^{-\lambda} : a \in A,\ \sigma \in S\} = \delta_{\lambda,\tau}(Ag^S)$$

which has size $|\delta_{\lambda,\tau}(Ag^S)| = |A|\,|S|$.

In case (iii) we then have $g^{\tau-\lambda} = u^{(\tau-\lambda)\eta} = c^{\eta}$ and so, proceeding as above,

$$\{a^{\tau}(c^{\eta})^{\sigma}a^{-\lambda} : a \in A,\ \sigma \in S\} = \delta_{\lambda,\tau}(Ag^S)$$

which has size $|\delta_{\lambda,\tau}(Ag^S)| = |A|\,|S|$.

Now suppose that $\mathcal{O} \subset U$ and define $R_g := \{a, b \in A, \lambda \neq \tau \in S : ag^{\lambda} = bg^{\tau}\}$. Note these are disjoint for if $(a, b, \lambda\tau) \in R_g \cap R_h$, then $g^{\tau-\lambda} = b^{-1}a = h^{\tau-\lambda}$ which is impossible as $\delta_{\lambda,\tau}$ is injective. Therefore,

$$\min_{g\in U}|R_g| \leq \frac{1}{|U|}\#\{a, b \in A, \lambda \neq \tau \in S\} < \frac{(|A||S|)^2}{|\mathcal{O}|}$$

and, since

$$\sum_n r_{Ag^S}(n)^2 = \#\{a, b \in A, \lambda, \tau \in S : ag^{\lambda} = bg^{\tau}\} = |R_g| + |A||S|,$$

we deduce that for some $g \in U$,

$$|A|^2|S|^2 = \left(\sum_n r_{Ag^S}(n)\right)^2 \leq |Ag^S| \sum_n r_{Ag^S}(n)^2 \leq |Ag^S| \left(\frac{(|A||S|)^2}{|\mathcal{O}|} + |A||S|\right)$$

by the Cauchy–Schwarz inequality. As $g \in U$, there exist $\lambda \neq \tau \in S$ and $c \in A^{-1}A$ such that $g^{\tau-\lambda} = c$, and so, by (25),

$$\{a^\tau c^\sigma a^{-\lambda} : a \in A, \sigma \in S\} = \delta_{\lambda,\tau}(Ag^S)$$

which has size

$$|\delta_{\lambda,\tau}(Ag^S)| = |Ag^S| \geq \frac{|A||S||\mathcal{O}|}{|A||S| + |\mathcal{O}|} \geq \frac{1}{2} \min\{|A||S|, |\mathcal{O}|\}.$$

□

For further readings on the subject see [2, 4, 5, 8, 10, 12, 13, 19, 20, 31].

Acknowledgements Thanks are due to Todd Cochrane, Ernie Croot, Harald Helfgott, Chris Pinner. Andrew Granville is partially supported by an NSERC Discovery Grant, as well as a Canadian Research Chair. Jozsef Solymosi is partially supported by ERC Advanced Research Grant no 267165 (DISCONV), by Hungarian National Research Grant NK 104183, and by an NSERC Discovery Grant.

References

1. F. Amoroso, E. Viada, Small points on subvarieties of a torus. Duke Math. J. **150**(3), 407–442 (2009)
2. A. Ayyad, T. Cochrane, Z. Zheng, The congruence $x_1x_2 \equiv x_3x_4 \pmod p$, the equation $x_1x_2 = x_3x_4$, and mean values of character sums. J. Number Theory **59**, 398–413 (1996)
3. Y. Bilu, The Many Faces of the Subspace Theorem (after Adamczewski, Bugeaud, Corvaja, Zannier...). Séminaire Bourbaki, Exposé 967, 59ème année (2006–2007); Astérisque **317**, 1–38 (2007/2008)
4. J. Bourgain, A.A. Glibichuk, S.V. Konyagin, Estimates for the number of sums and products and for exponential sums in fields of prime order. J. Lond. Math. Soc. **73**, 380–398 (2006)
5. J. Bourgain, N. Katz, T. Tao, A sum-product estimate in finite fields, and applications. GAFA **14**, 27–57 (2004)
6. M.-C. Chang, Sum and product of different sets. Contrib. Discrete Math. **1**(1), (2006)
7. G. Elekes, On the number of sums and products. Acta Arith. **81**(4), 365–367 (1997)
8. P. Erdős, E. Szemerédi, *On Sums and Products of Integers*. Studies in Pure Mathematics (Birkhäuser, Basel, 1983), pp. 213–218
9. J.-H. Evertse, H.P. Schlickewei, W.M. Schmidt, Linear equations in variables which lie in a multiplicative group. Ann. Math. **155**(3), 807–836 (2002)
10. K. Ford, The distribution of integers with a divisor in a given interval. Ann. Math. **168**, 367–433 (2008)

11. G.A. Freiman, *Foundations of a Structural Theory of Set Addition*. Translations of Mathematical Monographs (American Mathematical Society, Providence, 1973)
12. M.Z. Garaev, An explicit sum-product estimate in $\mathbb{F}_p$. IMRN **35**, 11 (2007)
13. M.Z. Garaev, The sum-product estimate for large subsets of prime fields. Proc. Am. Math. Soc. **136**, 2735–2739 (2008)
14. A. Glibichuk, M. Rudnev, Additive properties of product sets in an arbitrary finite field. J. d'Analyse Mathématique **108**(1), 159–170 (2009)
15. L. Guth, N.H. Katz. On the Erdős distinct distances problem in the plane. Ann. Math. **181**(1), 155–190 (2015)
16. D. Hart, A. Iosevitch, Sums and products in finite fields: an integral geometric viewpoint, in *Radon Transforms, Geometry, and Wavelets: AMS Special Session*, January 7–8, 2007, New Orleans. Contemporary Mathematics, vol. 464 (2008), pp. 129–135
17. D. Hart, A. Iosevitch, J. Solymosi, Sums and products in finite fields via Kloosterman sums. IMRN Art. ID rnm007, 1–14 (2007)
18. D.R. Heath-Brown, S.V. Konyagin, New bounds for Gauss sums derived from k-th powers, and for Heilbronn's exponential sums. Q. J. Math. **51**, 221–235 (2000)
19. H. Helfgott, Growth and generation in $SL\_2(Z/pZ)$. Ann. Math. **167**, 601–623 (2008)
20. H.N. Katz, C.-Y. Shen, A slight improvement to Garaev's sum-product estimate. Proc. Am. Math. Soc. **136**, 2499–2504 (2008)
21. J. Kollár, Szemerédi-Trotter-type theorems in dimension 3. Adv. Math. **271**, 30–61 (2015)
22. S.V. Konyagin, M. Rudnev, On new sum-product type estimates. SIAM J. Discrete Math. **27**(2), 973–990
23. S.V. Konyagin, I.D. Shkredov. On sum sets, having small product set. In *Proceedings of the Steklov Institute of Mathematics* **290**(1), 288–299 (2015)
24. O. Roche-Newton, M. Rudnev, I.D. Shkredov, New sum-product type estimates over finite fields. arXiv:1408.0542 [math.CO]
25. M. Rudnev, On the number of incidences between planes and points in three dimensions. arXiv:1407.0426 [math.CO]
26. M. Rudnev, An improved sum-product inequality in fields of prime order. Int. Math. Res. Not. **16**, 3693–3705 (2012)
27. L. Székely, Crossing numbers and hard Erdős problems in discrete geometry. Comb. Probab. Comput. **6**(3), 353–358 (1997)
28. J. Solymosi, G. Tardos, On the number of k-rich transformations, in *Proceedings of the twenty-third annual symposium on Computational geometry (SCG '07)* (ACM, New York, 2007), pp. 227–231
29. J. Solymosi, On the number of sums and products. Bull. Lond. Math. Soc **37**, 491–494 (2005)
30. J. Solymosi, Bounding multiplicative energy by the sumset. Adv. Math. **222**(2), 402–408 (2009)
31. T. Tao, V.H. Vu, *Additive Combinatorics*. Cambridge Studies in Advanced Mathematics, vol. 105 (Cambridge University Press, Cambridge, 2006)
32. L.A. Vinh, The Szemerédi-Trotter type theorem and the sum-product estimate in finite fields. Eur. J. Comb. **32**(8), 1177–1181 (2011)

Ajtai–Szemerédi Theorems over quasirandom groups

Tim Austin

Abstract Two versions of the Ajtai–Szemerédi Theorem are considered in the Cartesian square of a finite non-abelian group G. In case G is sufficiently quasirandom, we obtain strong forms of both versions: if $E \subseteq G \times G$ is fairly dense, then E contains a large number of the desired patterns for most individual choices of 'common difference'. For one of the versions, we also show that this set of good common differences is syndetic.

Keywords Ajtai-Szemerédi theorem • Quasirandom groups • Corners • Syndeticity

Mathematics Subject Classification (2010): 05D99, 20C15, 37A25

1 Introduction

A classical result of Ajtai and Szemerédi [1] asserts the following. For every $\delta > 0$ there is an $N_0 < \infty$ such that, if G is a finite cyclic group with $|G| \geq N_0$, and $E \subseteq G \times G$ satisfies

$$|E| \geq \delta |G|^2, \tag{1}$$

then there are $x, y \in G$ and $r \in G \setminus \{0\}$ such that

$$E \supseteq \{(x, y), (x + r, y), (x, y + r)\}. \tag{2}$$

A set satisfying (1) is called **δ-dense** in $G \times G$. Subsets of $G \times G$ of the kind on the right-hand side of (2) are called **abelian corners**.

T. Austin (✉)
Courant Institute, New York University, New York, NY 10012, USA
e-mail: tim@cims.nyu.edu

A. Beveridge et al. (eds.), *Recent Trends in Combinatorics*, The IMA Volumes in Mathematics and its Applications 159, DOI 10.1007/978-3-319-24298-9_19

Several proofs of this are now known, and it has been generalized to subsets of $G \times G$ for arbitrary finite abelian groups G: see, for instance, the discussions around [21, Propositions 10.47 and 11.28].

All known proofs actually give a stronger conclusion: that there is some $c > 0$ depending only on δ with the property that any δ-dense set $E \subseteq G \times G$ must satisfy

$$\left|\{(x, y, r) \in G^3 : \{(x, y), (x + r, y), (x, y + r)\} \subseteq E\}\right| \geq c|G|^3. \tag{3}$$

This paper considers two candidate generalizations of these results to non-abelian groups G. The most obvious non-abelian analogs of abelian corners are triples of the form

$$\{(x, y), (gx, y), (x, gy)\}, \quad g \neq e, \tag{4}$$

where e is the identity of G. We refer to these as **naïve corners**. A second possibility is triples of the form

$$\{(x, y), (gx, y), (gx, gy)\}, \quad g \neq e. \tag{5}$$

Triples of this second kind first appeared in [6] (in the related setting of infinite amenable groups and their probability-preserving actions), so we refer to them as **Bergelson–McCutcheon–Zhang ('BMZ') corners**. Solymosi [20] has studied BMZ corners in finite groups in; he writes them as

$$\{(x, y), (xg, y), (x, gy)\}, \quad g \neq e,$$

but these may be identified with BMZ corners by first writing in terms of y, g, and $x' := xg$ and then applying the transformation $(a, b) \mapsto (a^{-1}, b)$.

When G is abelian, naïve corners are abelian corners, and BMZ corners are equivalent to them by a simple change of variables, but this is not true for non-abelian G. As discussed in [6], when searching for BMZ corners inside a δ-dense subset of $G \times G$, some methods from the abelian setting have useful generalizations which do not seem to apply to naïve corners.

Either of these patterns, and indeed many others, must appear inside a δ-dense subset of $G \times G$ for any sufficiently large finite group G. This is simply because such a G must contain a fairly large abelian subgroup [11], to whose cosets one can apply the known results for abelian groups.

Our first theorem states this result formally. Here and for the rest of the paper, we also allow G to be a non-finite compact metrizable group, since our methods handle these with little change. We generally refer to these simply as 'compact' groups, suppressing the assumption that they are metrizable. If G is a compact group then its Haar probability measure is denoted by m_G; if $H \leq G$ is a closed subgroup and $g \in G$, then m_{gH} and m_{Hg} denote respectively the left- and right-translates of m_H, regarded as a measure on G, by the element g.

Theorem A. *For any $\delta > 0$ and any $k \in \mathbb{N}$ there is an $N_0 < \infty$ for which the following holds. If G is a compact group with $|G| \geq N_0$ (in particular, if G is not finite), and $E \subseteq G^k$ is Borel and satisfies $m_{G^k}(E) \geq \delta$, then there are*

$$(x_1, \ldots, x_k) \in G^k \quad \textit{and} \quad g \in G \setminus \{e\}$$

such that

$$E \supseteq \{(x_1^{\eta_1}, \ldots, x_k^{\eta_k}) : (\eta_1, \ldots, \eta_k) \in \{0, 1\}^k\},$$

where for each $i \leq k$ we set $x_i^0 := x_i$ and $x_i^1 := gx_i$.

The analogous result in which we set $x_i^1 := x_i g$ follows immediately, simply by applying Theorem A to the set

$$\{(y_1^{-1}, \ldots, y_k^{-1}) : (y_1, \ldots, y_k) \in E\}.$$

The simple proof of Theorem A will be given in Section 2. However, the proof does not give a generalization of the 'counting' result of inequality (3) to non-abelian G, because it counts only those patterns for which the parameter g lies in some abelian subgroup of G, which may be very small compared to G itself.

It seems important to begin this paper with Theorem A, but our main results are of a different kind. These assert that, if G is sufficiently quasirandom (as defined in [13]) in terms of δ, and $E \subseteq G \times G$ is δ-dense, then there is a large set of possible 'common differences' g such that E must contain both many naïve corners and many BMZ corners with that specific choice of g. In contrast to the abelian setting, the assumption of quasirandomness allows one to obtain reasonable and explicit bounds. The current best bounds in the abelian case are those of Shkredov [18, 19].

Theorem B. *Let $\varepsilon > 0$ and suppose that G is a D-quasirandom compact group. Let $E \subseteq G \times G$ be Borel and set $\delta := m_{G\times G}(E)$. Then the set*

$$A := \big\{g \in G : m_{G\times G}\big\{(x,y) : \{(x,y), (gx,y), (x,gy)\} \subseteq E\big\} \geq \delta^3 - \varepsilon\big\}$$

has $m_G(A) \geq 1 - 2\sqrt{3}D^{-1/4}/\varepsilon$. In particular, if $D \gg (1/\varepsilon)^4$ then A is most of G.

This is proved in Section 4. As far as I know, this is the first result for naïve corners in non-abelian groups, although the proof uses only some simple estimates and classical representation theory.

Similar methods give an analogous result for BMZ corners, although the bounds we obtain are more complicated, and it seems unlikely that they are optimal. In formulating the results for BMZ corners, we no longer take care over the exact values of constants: universal constants are simply denoted by O(1).

Theorem C. *For any $\varepsilon \in (0, 1/2)$ there is a*

$$D = \exp\big((1/\varepsilon)^{O(1)}\big) \tag{6}$$

for which the following holds. Let G be a D-quasirandom compact group, let $E \subseteq G \times G$ *be Borel, and set* $\delta := m_{G\times G}(E)$. *Then the set*

$$B := \big\{g \in G :\ m_{G\times G}\big\{(x,y) :\ \{(x,y),(gx,y),(gx,gy)\} \subseteq E\big\} \geq \delta^4 - \varepsilon\big\}$$

has $m_G(B) \geq 1 - \varepsilon$.

In the setting of Theorem C, a closely related argument proves another fact about the set B of 'good common differences'. Recall that a subset S of a group G is K-**left-syndetic** for an integer K if there are K left-translates of S that cover the whole of G. We henceforth abbreviate 'K-left-syndetic' to 'K-syndetic'.

Theorem D. *For any* $\varepsilon \in (0, 1/2)$ *there is a*

$$D = \exp\big(\exp\big((1/\varepsilon)^{O(1)}\big)\big) \tag{7}$$

for which the following holds. If G is a D-quasirandom compact group and $E \subseteq G \times G$ *is Borel, then the set B defined in Theorem C is K-left-syndetic for some*

$$K = \exp\big((1/\varepsilon)^{O(1)}\big).$$

This gives a different sense in which the set B is 'large'.

Qualitative versions of Theorems C and D have recently been proved by Bergelson, Robertson and Zorin-Kranich [7]. Their work uses ergodic theory for actions of certain ultraproducts of sequences of finite groups, which are of course highly infinite; it gives no explicit control on D or K in terms of ε. Our proof below uses only elementary inequalities and representation theory, so should be simpler for the reader not versed in non-standard analysis. In this respect, the present paper relates to [7] as did the earlier work [3] to [5]. We prove Theorems C and D in Section 5.

Theorems B, C and D together raise the following.

Question 1. Is it true that, if G is sufficiently quasirandom in terms of $\varepsilon \in (0, 1/2)$, then the set A from Theorem B is K-syndetic for some K depending only on ε? $\lhd$

Intuitively, quasirandom groups are very far from abelian groups. Taken together, the Ajtai–Szemerédi Theorem for abelian groups and our Theorems B and C make it natural to ask whether a lower bound of the kind in (3) holds for either naïve or BMZ corners and for all finite groups G that are sufficiently large.

In fact, this is true for BMZ corners, since the original graph-theoretic proof of Ajtai and Szemerédi generalizes fairly easily to those: see [20]. This observation has a counterpart in ergodic theory, where related questions about multiple recurrence have been studied [4, 6, 10]. This is why the pattern (5) was first introduced into ergodic theory by Bergelson, McCutcheon and Zhang in [6].

However, I do not see a way to apply that graph-theoretic argument in looking for naïve corners, so for those the following question may be of interest.

Question 2. Is it true that for every $\delta > 0$ there is a $c > 0$ such that, whenever G is a compact group and $E \subseteq G \times G$ satisfies $m_{G^2}(E) \geq \delta$, one also has

$$m_{G^3}\{(x, y, g) \in G^3 : \{(x, y), (gx, y), (x, gy)\} \subseteq E\} \geq c?$$

◁

2 Proof of Theorem A

The special case of Theorem A for naïve corners already appears as [6, Corollary 6.4]. The general case needs only a slight expansion of their argument.

Proof (Theorem A).

Step 1: finite groups. The theorem is well-known among finite abelian groups: see, for instance, [21, Proposition 11.28]. Fix $\delta > 0$ and $k \in \mathbb{N}$, and let n be so large that, if A is any finite abelian group of order at least n and $F \subseteq A^k$ satisfies $m_{A^k}(F) \geq \delta$, then F must contain a pattern of the kind in Theorem A.

By a classical result of Erdős and Straus [11], we may now choose N so large that any finite group G of order at least N contains an abelian subgroup H of order at least n (see also [15] for an essentially optimal estimate of N in terms of n). Suppose that $E \subseteq G^k$ has $m_{G^k}(E) \geq \delta$. Then

$$m_{G^k}(E) = \int \cdots \int m_{Hg_1 \times \cdots \times Hg_k}(E)\, m_{H\backslash G}(\mathrm{d}(Hg_k)) \cdots m_{H\backslash G}(\mathrm{d}(Hg_1)),$$

so there are some cosets $Hg_1, \ldots, Hg_k$ for which

$$m_{Hg_1 \times Hg_2 \times \cdots \times Hg_k}(E) \geq \delta.$$

Define $F \subseteq H^k$ by

$$E \cap (Hg_1 \times Hg_2 \times \cdots \times Hg_k) = F \cdot (g_1, \ldots, g_k).$$

Applying the abelian case of the theorem to F gives some

$$(x_1, \ldots, x_k) \in H^k \quad \text{and} \quad h \in H \setminus \{e\}$$

such that

$$F \supseteq \{(x_1^{\eta_1}, \ldots, x_k^{\eta_k}) : (\eta_1, \ldots, \eta_k) \in \{0, 1\}^k\},$$

with the obvious analog of the notation in the statement of Theorem A. Translating back to E, this gives

$$E \supseteq \{((x_1 g_1)^{\eta_1}, \dots, (x_k g_k)^{\eta_k}) : (\eta_1, \dots, \eta_k) \in \{0,1\}^k\}.$$

Step 2: other compact Lie groups. If G is a non-finite compact Lie group, then it contains a nontrivial toral subgroup. This, in turn, contains finite cyclic subgroups of arbitrarily large cardinality. Letting H such a subgroup of cardinality at least n, we may complete the proof as in Step 1.

Step 3: general compact groups. Finally, let G be an arbitrary compact group which is not finite or a Lie group. As a standard consequence of the Peter–Weyl Theorem [8, Section III.3], there are a continuous surjective homomorphism $\pi : G \longrightarrow \overline{G}$ to a compact Lie group such that the set

$$\overline{E} := \{(\overline{x}_1, \dots, \overline{x}_k) \in \overline{G}^k : m_{\pi^{-1}\{\overline{x}_1\}\times\cdots\times\pi^{-1}\{\overline{x}_k\}}(E) > 1 - 2^{-k-1}\} \tag{8}$$

satisfies

$$m_{G^k}(E \triangle (\pi \times \cdots \times \pi)^{-1}(\overline{E})) < \delta/2,$$

and hence $m_{\overline{G}^k}(\overline{E}) > \delta/2$.

Since G is not finite or a Lie group, we may choose $\overline{G}$ to have cardinality as large as we please (allowing infinity if necessary). Having done so, either Step 1 or Step 2 gives some

$$(\overline{x}_1, \dots, \overline{x}_k) \in \overline{G}^k \quad \text{and} \quad \overline{g} \in \overline{G} \setminus \{\overline{e}\}$$

such that

$$\overline{E} \supseteq \{(\overline{x}_1^{\eta_1}, \dots, \overline{x}_k^{\eta_k}) : (\eta_1, \dots, \eta_k) \in \{0,1\}^k\},$$

with the obvious analog of the notation in the statement of Theorem A.

Finally, consider lifts $x_1, \dots, x_k$ and g chosen independently at random from the Haar measures $m_{\pi^{-1}\{\overline{x}_1\}}, \dots, m_{\pi^{-1}\{\overline{x}_k\}}$ and $m_{\pi^{-1}\{\overline{g}\}}$. Observe that each gx_i is then a random lift of $\overline{gx}_i$ with distribution $m_{\pi^{-1}\{\overline{gx}_i\}}$. Define each $x_i^{\eta_i}$ using these lifts in the usual way. Then it follows from the definition (8) that each of the events

$$(x_1^{\eta_1}, \dots, x_k^{\eta_k}) \in E \quad \text{for } (\eta_1, \dots, \eta_k) \in \{0,1\}^k$$

has probability at least $1 - 2^{-k-1}$. Therefore, by the first-moment bound, there is some choice of $x_1, \dots, x_k$ and g for which all of these events occur simultaneously.

□

3 Preliminary discussion of the results for quasirandom groups

The proofs of Theorems B, C and D have several common elements. This section introduces some of those.

It will be convenient to use some simple notation and terminology from ergodic theory. Given a compact group G, a **probability** $(G \times G)$**-space** is a tuple (Y, Σ, ν, S, T) in which (Y, Σ, ν) is a probability space and S and T are two commuting, ν-preserving G-actions on that space. Since they commute, S and T together define an action of $G \times G$, hence the name. Given such a probability $(G \times G)$-space, we will write Σ^S, Σ^T and $\Sigma^{S,T}$ for the σ-subalgebras of sets in Σ that are invariant under S, T, or the whole $(G \times G)$-action, respectively.

For example, let $X := G \times G$ with the measure $\mu = m_X$, let Σ_X be the Borel σ-algebra of X, and define

$$S^g(x, y) := (gx, y) \quad \text{and} \quad T^g(x, y) := (x, gy).$$

This turns (X, Σ_X, μ, S, T) into a probability $(G \times G)$-space. It will appear repeatedly below.

Now consider measurable functions $f_1, f_2, f_3 : X \longrightarrow [-1, 1]$, and suppose G is highly quasirandom. For Theorems B, C and D we need to estimate the values taken by

$$\text{either} \quad \int f_1 \cdot f_2 S^g \cdot f_3 T^g \, d\mu \quad \text{or} \quad \int f_1 \cdot f_2 S^g \cdot f_3 S^g T^g \, d\mu$$

for 'typical' group elements g. We ultimately need to do this in case $f_1 = f_2 = f_3 = 1_E$, but allowing more flexibility will be important for the proofs. The strategy for these estimates takes a form that has become well-known in additive combinatorics: each of the functions f_i will be decomposed as

$$f_i = f_i^\circ + f_i^\perp$$

in such a way that the functions $f_i^\perp$ contribute very little to the integrals of interest for 'most' group elements g, while the functions f_i° have some extra 'structure' which makes the estimate of that integral easier. In such a decomposition, the negligible terms $f_i^\perp$ are often called 'quasirandom'. Here, 'most' group elements will mean those lying in a very large subset of G in the case of Theorem B or C, or those lying in a suitable syndetic subset in the case of Theorem D. A nice discussion of this methodology can be found in [14].

Different decompositions are required for studying naïve and BMZ corners. For the former, the required decompositions are fairly simple, and will be introduced within the proof of Theorem B. For BMZ corners we need a rather more complicated construction, based on the regularity lemma of Frieze and Kannan [12]; this will be explained separately in Subsection 5.1.

3.1 Estimates from probability

The following simple lemma plays the rôle of the classical van der Corput estimate in the present paper. It will be the basis for several other estimates later.

Lemma 1. *Let (Y, Σ, ν) be a probability space, let V be a real or complex Hilbert space with inner product $\langle\cdot,\cdot\rangle$ and corresponding norm $\|\cdot\|$, and let $y \mapsto u_y$ be a strongly measurable function $Y \longrightarrow V$. Let v be a unit vector in V. Then*

$$\int |\langle v, u_y\rangle|\, \nu(\mathrm{d}y) \leq \sqrt{\iint |\langle u_y, u_{y'}\rangle|\, \nu(\mathrm{d}y)\, \nu(\mathrm{d}y')}.$$

Proof. Define

$$\varphi(y) := \overline{\langle v, u_y\rangle}/|\langle v, u_y\rangle|,$$

using the convention that $\varphi(y) = 1$ if $\langle v, u_y\rangle = 0$. So φ takes values in the unit circle, and more specifically in $\{-1, 1\}$ if V is real. This is a measurable function of y, and

$$\begin{aligned}\int |\langle v, u_y\rangle|\, \nu(\mathrm{d}y) &= \int \varphi(y)\langle v, u_y\rangle\, \nu(\mathrm{d}y) = \Big\langle v, \int \varphi(y)u_y\, \nu(\mathrm{d}y)\Big\rangle \\ &\leq \Big\| \int \varphi(y)u_y\, \nu(\mathrm{d}y)\Big\| = \sqrt{\iint \langle \varphi(y)u_y, \varphi(y')u_{y'}\rangle\, \nu(\mathrm{d}y)\, \nu(\mathrm{d}y')} \\ &\leq \sqrt{\iint |\langle u_y, u_{y'}\rangle|\, \nu(\mathrm{d}y)\, \nu(\mathrm{d}y')}.\end{aligned}$$

□

We will also need the following very general probabilistic estimate. It can be found as [9, Lemma 1.6], and then [7] cites it for a similar purpose to ours.

Lemma 2. *Let (Y, Σ, ν) be a probability space, let $\Sigma_1, \ldots, \Sigma_k$ be σ-subalgebras of Σ, and let f be a bounded non-negative measurable function on Y. Then*

$$\int f \cdot \prod_{i=1}^{k} \mathsf{E}(f \mid \Sigma_i)\, \mathrm{d}\nu \geq \Big(\int f\, \mathrm{d}\nu\Big)^{k+1}.$$

□

3.2 Estimates from representation theory

All of the representation theory in this paper concerns either unitary representations on complex Hilbert spaces or orthogonal representations on real Hilbert spaces.

The results we need from representation theory are all standard, and can be found in many textbooks. A good reference for our purposes is [8, Chapters II and III]: that book actually focuses on compact Lie groups, but all the facts we cite from it clearly hold for arbitrary compact groups.

If $\pi : G \curvearrowright V$ is a unitary or orthogonal representation, then P^{π} denotes the orthogonal projection from V onto the subspace of $\pi(G)$-fixed vectors.

A unitary (resp. orthogonal) representation is **D-quasirandom** if it has no irreducible subrepresentations of complex (resp. real) dimension less than D. Following [13], the group G itself is D-quasirandom if and only if all its non-identity irreducible unitary representations are D-quasirandom. Quasirandomness will be exploited in this paper by way of the following two lemmas.

Lemma 3. *If G is a D-quasirandom compact group, $\pi : G \curvearrowright V$ is a unitary representation, and $u, v \in V$, then*

$$\left\| P^{\pi\otimes\pi}(u \otimes v) - P^{\pi} u \otimes P^{\pi} v \right\|_{V\otimes V} \leq D^{-1/2}\|u\|_V\|v\|_V,$$

where $\|\cdot\|_V$ denotes the norm on V and $\|\cdot\|_{V\otimes V}$ denotes the Hilbertian tensor product of that norm on $V \otimes V$.

Proof. This is a routine consequence of Schur's Lemma: it can be found as [3, Lemma 2]. □

Unlike the above, the next lemma will be needed for both unitary and orthogonal representations.

Lemma 4. *If G is a D-quasirandom compact group, $\pi : G \curvearrowright V$ is a unitary or orthogonal representation, and $u, v \in V$, then*

$$\int |\langle u, \pi^g v\rangle_V - \langle P^{\pi} u, P^{\pi} v\rangle_V|^2 \,\mathrm{d}g \leq D^{-1}\|u\|_V^2\|v\|_V^2.$$

Proof. In the unitary case, this is [3, Corollary 3] (although the proof given there contained an error, corrected in [2]).

Now suppose that π is an orthogonal representation. Then its complexification $\pi^{\mathbb{C}} : G \curvearrowright V \otimes_{\mathbb{R}} \mathbb{C}$ is a unitary representation, which may also be regarded as an orthogonal representation isomorphic to $\pi \oplus \pi$ (see, for instance, [8, Section II.6]). A simple calculation shows that $P^{\pi^{\mathbb{C}}}$ is simply the complexification of P^{π}. Hence the desired inequality for π follows from its counterpart for $\pi^{\mathbb{C}}$. □

4 Naïve corners

This section proves Theorem B. The key to this proof is the following proposition, which roughly asserts that 'correlations' across naïve corners almost vanish unless one starts with functions that have some nontrivial structure.

Let (X, Σ_X, μ, S, T) be as in Section 3.

Proposition 1. *If* $f_1, f_2, f_3 : X \longrightarrow [-1, 1]$, *and either* $\mathsf{E}(f_2 \mid \Sigma_X^S) = 0$ *or* $\mathsf{E}(f_3 \mid \Sigma_X^T) = 0$, *then*

$$\int \Big| \int f_1 \cdot f_2 S^g \cdot f_3 T^g \, \mathrm{d}\mu \Big| \, \mathrm{d}g \leq \sqrt{3} D^{-1/4}.$$

Proposition 1 will in turn be deduced from a kind of 'mixing' estimate, formulated in the next lemma. Before stating the lemma, it will be helpful to have some notation for representations of product groups. In this section the representations will all be unitary, even though the functions in Proposition 1 are real-valued, since we will need to make another direct appeal to the Schur Orthogonality Relations. Given any two unitary representation $\pi : G \curvearrowright V$ and $\theta : G \curvearrowright W$, one obtains a representation $G \times G \curvearrowright V \otimes W$ by defining

$$(\pi \boxtimes \theta)^{(g,h)} := \pi^g \otimes \theta^h.$$

It is a standard result that $\pi \boxtimes \theta$ is an irreducible $(G \times G)$-representation if each of π and θ is irreducible, and that all irreducibles of $G \times G$ arise this way [8, Proposition II.4.14]. (On the other hand, as far as I know there is no standard notation for $\pi \boxtimes \theta$.) It follows at once that if G is D-quasirandom, then so is $G \times G$.

Lemma 5. *Let G be D-quasirandom, let $\rho : G \times G \curvearrowright U$ be a unitary representation, and let $u \in U$. Then*

$$\Big\| \int \rho^{(g,g^{-1})} u \, \mathrm{d}g - P^\rho u \Big\| \leq D^{-1} \|u\|.$$

Proof. By decomposing ρ into irreducibles, it suffices to prove this when ρ is itself a nontrivial irreducible. According to the discussion above, this means that $\rho = \pi \boxtimes \theta$ for some irreducible unitary representations $\pi : G \curvearrowright V$ and $\theta : G \curvearrowright W$, not both the identity.

In this case we will prove that

$$\int \rho^{(g,g^{-1})}(v \otimes w) \, \mathrm{d}g = \int (\pi^g v \otimes \theta^{g^{-1}} w) \, \mathrm{d}g = \begin{cases} \frac{1}{\dim(V)}(w \otimes v) & \text{if } \pi = \theta \\ 0 & \text{else} \end{cases} \tag{9}$$

for any $v \in V$ and $w \in W$. Since π and θ cannot both be the identity, if they are equal then $\dim(V) \geq D$, by the assumption of quasirandomness. Therefore (9) implies that

$$\left\| \int \rho^{(g,g^{-1})} u \, \mathrm{d}g \right\| \leq D^{-1} \|u\| \quad \forall u \in U.$$

To prove (9), it suffices to check the inner products of the two sides of this equation against another element of U, which can also be of tensor product form. Thus the desired equality becomes

$$\int \left\langle \pi^g v \otimes \theta^{g^{-1}} w, v' \otimes w' \right\rangle \mathrm{d}g = \int \langle \pi^g v, v' \rangle \langle \theta^{g^{-1}} w, w' \rangle \, \mathrm{d}g$$
$$= \begin{cases} \frac{1}{\dim(V)} \langle w, v' \rangle \langle v, w' \rangle & \text{if } \pi = \theta, \\ 0 & \text{else.} \end{cases}$$

This is now one of the standard Schur Orthogonality Relations: see [8, Theorem II.4.5(ii)]. □

Corollary 1. *If G is D-quasirandom and $F_1, F_2 \in L^2_{\mathbb{C}}(G \times G)$, then*

$$\int \left| \int F_1 \cdot \overline{F_2} T^g S^{g^{-1}} \, \mathrm{d}\mu - \int F_1 \, \mathrm{d}\mu \int \overline{F_2} \, \mathrm{d}\mu \right| \mathrm{d}g \leq 2D^{-1/2} \|F_1\|_2 \|F_2\|_2.$$

Remark 1. The use of $\overline{F_2}$ rather than F_2 on the left-hand side here is only for the sake of convenience. With this choice, the integral $\int F_1 \cdot \overline{F_2} T^g S^{g^{-1}} \, \mathrm{d}\mu$ is the Hermitian inner product in $L^2_{\mathbb{C}}(G \times G)$, which leads more easily to an application of Lemma 5. ◁

Proof. By replacing each F_i with $F_i - \int F_i \, \mathrm{d}\mu$, we may assume that they are both orthogonal to the constant functions. In this case we will show that

$$\int \left| \int F_1 \cdot \overline{F_2} T^g S^{g^{-1}} \, \mathrm{d}\mu \right|^2 \mathrm{d}g \leq 2D^{-1} \|F_1\|^2 \|F_2\|^2,$$

from which the result follows by the Cauchy–Bunyakowksi–Schwartz inequality.

This follows using a standard tensor-product trick. Let $\tilde{X} := X \times X$, $\tilde{\mu} := \mu \otimes \mu$, $\tilde{S} := S \times S$ and $\tilde{T} := T \times T$. Then the integral above is equal to

$$\int_G \int_{\tilde{X}} \tilde{F}_1 \cdot \overline{\tilde{F}_2} \tilde{T}^g \tilde{S}^{g^{-1}} \, \mathrm{d}\tilde{\mu} \, \mathrm{d}g = \int_{\tilde{X}} \tilde{F}_1 \Big(\int \overline{\tilde{F}_2} \tilde{T}^g \tilde{S}^{g^{-1}} \, \mathrm{d}g \Big) \, \mathrm{d}\tilde{\mu},$$

where $\tilde{F}_i(x, y, x', y') := F_i(x, y) \overline{F_i(x', y')}$. Applying Lemma 5 to the inner integral on the right here, this is at most

$$\int \mathsf{E}(\tilde{F}_1 \mid \Sigma_{\tilde{X}}^{\tilde{S},\tilde{T}})\overline{\mathsf{E}(\tilde{F}_2 \mid \Sigma_{\tilde{X}}^{\tilde{S},\tilde{T}})}\,\mathrm{d}\tilde{\mu} + D^{-1}\|\tilde{F}_1\|_{L^2_{\mathbb{C}}(\tilde{\mu})}\|\tilde{F}_2\|_{L^2_{\mathbb{C}}(\tilde{\mu})}$$
$$\leq \big\|\mathsf{E}(\tilde{F}_1 \mid \Sigma_{\tilde{X}}^{\tilde{S},\tilde{T}})\big\|_{L^2_{\mathbb{C}}(\tilde{\mu})}\big\|\mathsf{E}(\tilde{F}_2 \mid \Sigma_{\tilde{X}}^{\tilde{S},\tilde{T}})\big\|_{L^2_{\mathbb{C}}(\tilde{\mu})} + D^{-1}\|F_1\|^2\|F_2\|^2.$$

Finally, in view of the product form of $\tilde{F}_i$ and the fact that $G \times G$ is still D-quasirandom, an appeal to Lemma 3 gives

$$\big\|\mathsf{E}(\tilde{F}_i \mid \Sigma_{\tilde{X}}^{\tilde{S},\tilde{T}})\big\|_{L^2_{\mathbb{C}}(\tilde{\mu})} \leq D^{-1/2}\|F_i\|^2 \quad \text{for } i = 1, 2.$$

Substituting this into the bound above completes the proof. □

Proof (Proposition 1). We give the proof in case $\mathsf{E}(f_2 \mid \Sigma_X^S) = 0$, the other case being analogous.

For each $g \in G$, let

$$u_g := f_2 S^g \cdot f_3 T^g.$$

By Lemma 1, it suffices to prove that

$$\iint |\langle u_g, u_{g'}\rangle|\,\mathrm{d}g\,\mathrm{d}g' = \iint |\langle u_g, u_{hg}\rangle|\,\mathrm{d}g\,\mathrm{d}h \leq 3D^{-1/2}.$$

The second integrand here may be re-written as

$$\begin{aligned}
|\langle u_g, u_{hg}\rangle| &= \Big|\int f_2 S^g \cdot f_2 S^{hg} \cdot f_3 T^g \cdot f_3 T^{hg}\,\mathrm{d}\mu\Big| \\
&= \Big|\int f_2 \cdot f_2 S^h \cdot f_3 T^g S^{g^{-1}} \cdot f_3 T^{hg} S^{g^{-1}}\,\mathrm{d}\mu\Big| \\
&= \Big|\int F_{2,h} \cdot F_{3,h} T^g S^{g^{-1}}\,\mathrm{d}\mu\Big|,
\end{aligned}$$

where $F_{2,h} := f_2 \cdot f_2 S^h$ and $F_{3,h} := f_3 \cdot f_3 T^h$.

Both $F_{2,h}$ and $F_{3,h}$ are real-valued and bounded by 1 in absolute value, so Corollary 1 applies to give

$$\int |\langle u_g, u_{hg}\rangle|\,\mathrm{d}g \leq \Big|\int f_2 \cdot f_2 S^h\,\mathrm{d}\mu\Big|\Big|\int f_3 \cdot f_3 T^h\,\mathrm{d}\mu\Big| + 2D^{-1/2}$$

for any h. Now integrating over h, and using again that $|f_3 \cdot f_3 T^h| \leq 1$, the right-hand side above turns into the bound

$$\int \Big|\int f_2 \cdot f_2 S^h\,\mathrm{d}\mu\Big|\,\mathrm{d}h + 2D^{-1/2} \leq \Big(\int \Big(\int f_2 \cdot f_2 S^h\,\mathrm{d}\mu\Big)^2\,\mathrm{d}h\Big)^{-1/2} + 2D^{-1/2}.$$

By Lemma 4 and the fact that $\mathsf{E}(f_2 \mid \Sigma_X^S) = 0$, this is at most $3D^{-1/2}$. This completes the proof. □

Proof (Theorem B). We will prove that, for any function $f : G \times G \longrightarrow [0,1]$, the set

$$A := \left\{ g \in G : \iint f(x,y)f(gx,y)f(x,gy)\,\mathrm{d}x\,\mathrm{d}y \geq \left(\int f\,\mathrm{d}\mu \right)^3 - \varepsilon \right\}$$

has measure at least $1 - 2\sqrt{3}D^{-1/4}/\varepsilon$. Applying this to $f = 1_E$ gives Theorem B.

Define new $[-1,1]$-valued functions $f_2^\perp$ and $f_3^\perp$ by the decompositions

$$f = \mathsf{E}(f \mid \Sigma_X^S) + f_2^\perp = \mathsf{E}(f \mid \Sigma_X^T) + f_3^\perp.$$

In the present setting, these are the decompositions of $f_2 = f$ and $f_3 = f$ into 'structured' and 'quasirandom' parts, as promised at the beginning of Section 3. It turns out that no related decomposition $f_1^\circ + f_1^\perp$ is needed here.

Substituting the first decomposition into the middle position of the relevant integral, and then the second decomposition into the last position, we obtain

$$\begin{aligned}
&\iint f(x,y)f(gx,y)f(x,gy)\,\mathrm{d}x\,\mathrm{d}y \\
&= \iint f(x,y)\mathsf{E}(f \mid \Sigma_X^S)(gx,y)\mathsf{E}(f \mid \Sigma_X^T)(x,gy)\,\mathrm{d}x\,\mathrm{d}y \\
&\quad + \iint f(x,y)f_2^\perp(gx,y)f(x,gy)\,\mathrm{d}x\,\mathrm{d}y \\
&\quad + \iint f(x,y)\mathsf{E}(f \mid \Sigma_X^S)(gx,y)f_3^\perp(x,gy)\,\mathrm{d}x\,\mathrm{d}y. \qquad (10)
\end{aligned}$$

Since $\mathsf{E}(f \mid \Sigma_X^S)$ is S-invariant and $\mathsf{E}(f \mid \Sigma_X^T)$ is T-invariant, the first integral on the right-hand side of (10) is equal to

$$\iint f(x,y)\mathsf{E}(f \mid \Sigma_X^S)(x,y)\mathsf{E}(f \mid \Sigma_X^T)(x,y)\,\mathrm{d}x\,\mathrm{d}y$$

for any g, and this is bounded below by $\left(\int f\,\mathrm{d}\mu \right)^3$ by Lemma 2. Re-arranging (10), it follows that

$$\begin{aligned}
A \supseteq A' := \Big\{ g \in G : &\left| \iint f(x,y)f_2^\perp(gx,y)f(x,gy)\,\mathrm{d}x\,\mathrm{d}y \right| \\
&+ \left| \iint f(x,y)\mathsf{E}(f \mid \Sigma_X^S)(gx,y)f_3^\perp(x,gy)\,\mathrm{d}x\,\mathrm{d}y \right| \leq \varepsilon \Big\}.
\end{aligned}$$

On the other hand, since $\mathsf{E}(f_2^\perp \mid \Sigma_X^S) = \mathsf{E}(f_3^\perp \mid \Sigma_X^T) = 0$, two appeals to Proposition 1 give

$$\int \left| \iint f(x,y) f_2^\perp(gx,y) f(x,gy)\,\mathrm{d}x\,\mathrm{d}y \right| \mathrm{d}g + \int \left| \iint f(x,y)\mathsf{E}(f \mid \Sigma_X^S)(gx,y) f_3^\perp(x,gy)\,\mathrm{d}x\,\mathrm{d}y \right| \mathrm{d}g \le 2\sqrt{3}D^{-1/4}.$$

Hence $m_G(A') \ge 1 - 2\sqrt{3}D^{-1/4}/\varepsilon$, by Chebyshev's Inequality, so the proof is complete. □

5 BMZ corners

This section proves Theorems C and D. The two proofs have much in common. We will explain the overarching structure of both proofs first, and then separate their finer details.

In this section it now makes more sense to work with orthogonal (real) representations than with unitary ones, since we will not need any representation theoretic results beyond Lemma 4.

Again let (X, Σ_X, μ, S, T) be as in Section 3.

5.1 Decomposition into structured and quasirandom functions

We now describe the decompositions into 'structured' and 'quasirandom' functions appropriate to the analysis of the family of triple forms

$$\int f_1 \cdot f_2 S^g \cdot f_3 S^g T^g \,\mathrm{d}\mu, \quad g \in G. \tag{11}$$

The functions f_1, f_2 and f_3 play different rôles here, so each will need its own notion of 'quasirandom' and 'structured' summands, even though our ultimate interest is in the case $f_1 = f_2 = f_3 = 1_E$.

As is common in this area, the appropriate notions of 'quasirandomness' are measured as smallness in certain norms. We introduce these next.

For a bounded function $f : X \longrightarrow \mathbb{R}$, define

$$\|f\|_{\check{\otimes}_{1,2}} := \sup\left\{ \int f(x,y)g(x)h(y)\,\mathrm{d}\mu : \|g\|_\infty, \|h\|_\infty \le 1 \right\}.$$

This is a classical construction in Banach space theory: the injective tensor norm $\|\cdot\|_{L^1(m_G)\check{\otimes}L^1(m_G)}$. It originates in Schatten's work on tensor products of Banach spaces, where it is referred to as the 'bound cross-norm'. See [17, Section II.3], or [16, Chapter 3] for a more modern treatment. It has now become popular in extremal combinatorics, where it is called the 'box norm'.

Similarly, define

$$\|f\|_{\check{\otimes}_{1,12}} := \sup\left\{\int f(x,y)g(y)h(x^{-1}y)\,\mathrm{d}\mu\ :\ \|g\|_\infty, \|h\|_\infty \leq 1\right\}$$

and

$$\|f\|_{\check{\otimes}_{12,2}} := \sup\left\{\int f(x,y)g(x^{-1}y)h(x)\,\mathrm{d}\mu\ :\ \|g\|_\infty, \|h\|_\infty \leq 1\right\}.$$

These may also be viewed as injective tensor norms, once one chooses appropriate 'coordinate axes' on $G \times G$.

For general functions f_1, f_2 and f_3 on $G \times G$, we will make use of Frieze and Kannan's weak version of the Szemerédi Regularity Lemma [12] (see also [14, Proposition 2.11] for a formulation closer to the present paper). The next proposition gives the specific instance of this result that we need. Recall that if $\mathcal{P}$ is a partition of a set S and $s \in S$, then $\mathcal{P}(s)$ denotes the cell of $\mathcal{P}$ which contains s.

Proposition 2 (Weak Regularity Lemma). *Given a measurable function $f : X \longrightarrow [-1,1]$, and also $\eta > 0$, there are partitions $\mathcal{P}_{1,1}$ and $\mathcal{P}_{1,12}$ of G, each into at most $\exp((1/\eta)^{O(1)})$ cells, for which the following holds.*

Define a new partition of $G \times G$ cell-wise by setting

$$\mathcal{Q}_1(x,y) := \mathcal{P}_{1,1}(y) \cap \mathcal{P}_{1,12}(x^{-1}y), \tag{12}$$

and let $\mathsf{E}_1 : L^\infty(\mu) \longrightarrow L^\infty(\mu)$ be the operator of conditional expectation onto $\mathcal{Q}_1$. Then

$$\|f - \mathsf{E}_1 f\|_{\check{\otimes}_{1,12}} \leq \eta.$$

□

This is not the formulation of the Frieze–Kannan Regularity Lemma given in [12] or [14], but it is an easy consequence. The methods of [12] or [14] give instead a function

$$f^\circ(x,y) = \sum_{m=1}^{m_0} \lambda_m h'_m(y) h''_m(x^{-1}y)$$

which approximates f in the sense that

$$\|f - f^\circ\|_{\check{\otimes}_{1,12}} \leq \eta,$$

and where

- $m_0 = (1/\eta)^{O(1)}$,
- each $\lambda_m \in [-1, 1]$,
- and each h'_m and h''_m is an indicator function on G.

For our purposes it is important to approximate f by a function that still takes values in $[0, 1]$, hence our preference for approximating by a conditional expectation of f itself. To obtain suitable partitions $\mathcal{P}_{1,1}$ and $\mathcal{P}_{1,12}$ from the function f° above, one simply lets $\mathcal{P}_{1,1}$ be generated by the level sets of all the functions h'_m, $m = 1, 2, \ldots, m_0$, and similarly for $\mathcal{P}_{1,12}$ using h''_m.

In recent years, techniques for decomposing a function into structured and quasirandom parts have become quite sophisticated: see [14], for example. A more careful argument than ours might enable one to be more efficient in the above proposition, and perhaps ultimately improve the bound in Theorem C to $D = (1/\varepsilon)^{O(1)}$. However, it is not the purpose of the present paper to explore this kind of enhancement, and we content ourselves with the partitions obtained above.

We will also need two variants of Proposition 2 that are obtained from it by simple changes of variables in $G \times G$. Given f and η as above, there are also partitions $\mathcal{P}_{2,1}$, $\mathcal{P}_{2,2}$, $\mathcal{P}_{3,2}$ and $\mathcal{P}_{3,12}$, all into at most $\exp((1/\eta)^{O(1)})$ cells, for which the following hold: defining two new partitions of $G \times G$ by

$$\mathcal{Q}_2(x, y) := \mathcal{P}_{2,2}(x) \cap \mathcal{P}_{2,1}(y) \tag{13}$$

$$\text{and} \quad \mathcal{Q}_3(x, y) := \mathcal{P}_{3,12}(x^{-1}y) \cap \mathcal{P}_{3,2}(x), \tag{14}$$

and letting $\mathsf{E}_i : L^\infty(\mu) \longrightarrow L^\infty(\mu)$ be the operator of conditional expectation onto $\mathcal{Q}_i$ for $i = 2, 3$, one has

$$\|f - \mathsf{E}_2 f\|_{\check{\otimes}_{1,2}} \le \eta \quad \text{and} \quad \|f - \mathsf{E}_3 f\|_{\check{\otimes}_{12,2}} \le \eta.$$

At some points below, we will need to write out the functions $\mathsf{E}_i f$, $i = 1, 2, 3$ in terms of more elementary summands. Using the individual cells of $\mathcal{Q}_i$, one may always express

$$(\mathsf{E}_1 f)(x, y) = \sum_{m_1=1}^{M_1} h_{1,1,m_1}(y) h_{1,12,m_1}(x^{-1}y), \tag{15}$$

$$(\mathsf{E}_2 f)(x, y) = \sum_{m_2=1}^{M_2} h_{2,1,m_2}(y) h_{2,2,m_2}(x) \tag{16}$$

$$\text{and} \quad (\mathsf{E}_3 f)(x, y) = \sum_{m_3=1}^{M_3} h_{3,2,m_3}(x) h_{3,12,m_3}(x^{-1}y), \tag{17}$$

where $M_i = |\mathcal{Q}_i|$ for $i = 1,2,3$, and each $h_{\bullet,\bullet,\bullet}$ is a measurable function $G \longrightarrow [-1,1]$.

5.2 *Estimates for structured functions*

This subsection analyses the triple form (11) when each f_i is replaced by a structured approximant $\mathsf{E}_i f_i$ as given by Proposition 2. For these approximants, we can exert very precise control over that triple form: it turns out that it hardly depends on g at all. This fact will result from the following.

Lemma 6. *Let $h_1, h_2, h_{12} : G \longrightarrow [-1,1]$ be measurable, and suppose that G is D-quasirandom. Then*

$$\Big|\iint h_1(x)h_2(y)h_{12}(x^{-1}y)\,\mathrm{d}x\,\mathrm{d}y - \Big(\int h_1\Big)\Big(\int h_2\Big)\Big(\int h_{12}\Big)\Big| \le D^{-1/2}.$$

Proof. This is essentially the implication (v) $\Longrightarrow$ (iv) in [13, Theorem 4.5]. Let R be the left-action of G on itself. The left-hand side above may be re-written as

$$\begin{aligned}\Big|\int h_1(x)\Big(\int h_2(y)h_{12}(x^{-1}y)\,\mathrm{d}y - \Big(\int h_2\Big)\Big(\int h_{12}\Big)\Big)\,\mathrm{d}x\Big| \\ = \Big|\int h_1(x)\Big(\langle h_2, h_{12}R^{x^{-1}}\rangle_{L^2(G)} - \Big(\int h_2\Big)\Big(\int h_{12}\Big)\Big)\,\mathrm{d}x\Big|.\end{aligned}$$

By the Cauchy–Bunyakowski–Schwartz inequality, this is at most

$$\Big(\int \Big(\langle h_2, h_{12}R^{x^{-1}}\rangle_{L^2(G)} - \Big(\int h_2\Big)\Big(\int h_{12}\Big)\Big)^2\,\mathrm{d}x\Big)^{1/2},$$

and Lemma 4 bounds this by $D^{-1/2}$. □

Corollary 2. *Suppose that G is D-quasirandom, and that*

$$f_1(x,y) = h_{1,1}(y)h_{1,12}(x^{-1}y), \quad f_2(x,y) = h_{2,1}(y)h_{2,2}(x)$$
$$\text{and} \quad f_3(x,y) = h_{3,2}(x)h_{3,12}(x^{-1}y)$$

for some measurable functions $h_{\bullet,\bullet} : G \longrightarrow [-1,1]$. Then the quantity

$$\varphi(g) := \int f_1 \cdot f_2S^g \cdot f_3S^gT^g\,\mathrm{d}\mu$$

satisfies

$$|\varphi(g) - \varphi(g')| \leq 2D^{-1/2} \quad \forall g, g' \in G.$$

Proof. For this choice of functions f_i, one has

$$\varphi(g) = \iint (h_{1,1}h_{2,1})(y) \cdot (h_{1,12}h_{3,12})(x^{-1}y) \cdot (h_{2,2}h_{3,2})(gx)\,\mathrm{d}x\,\mathrm{d}y.$$

Applying Lemma 6 to this integral, we find that it lies within $D^{-1/2}$ of

$$\Big(\int h_{1,1}h_{2,1}\Big)\Big(\int h_{1,12}h_{3,12}\Big)\Big(\int h_{2,2}h_{3,2}\Big)$$

for all g. □

Proposition 3. *Let $\varepsilon > 0$, and let $\mathcal{Q}_i$, $i = 1, 2, 3$, be partitions as in (12), (13) and (14). Assume that G is D-quasirandom for some*

$$D \geq \frac{16|\mathcal{Q}_1|^2|\mathcal{Q}_2|^2|\mathcal{Q}_3|^2}{\varepsilon^2}. \tag{18}$$

Let the sets E and B be as in Theorem C. Then B contains the set

$$C := \Big\{g \in G :\ \mu\big(E \cap (g^{-1}, e) \cdot E \cap (g^{-1}, g^{-1}) \cdot E\big)$$
$$\geq \int (\mathsf{E}_1 1_E) \cdot (\mathsf{E}_2 1_E)S^g \cdot (\mathsf{E}_3 1_E)S^g T^g\,\mathrm{d}\mu - \varepsilon/2\Big\}.$$

Proof. According to the decompositions (15)–(17), the quantity

$$\psi(g) := \int (\mathsf{E}_1 1_E) \cdot (\mathsf{E}_2 1_E)S^g \cdot (\mathsf{E}_3 1_E)S^g T^g\,\mathrm{d}\mu$$

is a sum of at most $|\mathcal{Q}_1||\mathcal{Q}_2||\mathcal{Q}_3|$ quantities having the form φ treated by Corollary 2. Therefore, by that corollary, $\psi(g)$ varies by at most

$$2|\mathcal{Q}_1||\mathcal{Q}_2||\mathcal{Q}_3|D^{-1/2} \leq \varepsilon/2$$

as g varies in G. Therefore

$$\mu\big(E \cap (g^{-1}, e) \cdot E \cap (g^{-1}, g^{-1}) \cdot E\big) \geq \psi(e) - \varepsilon = \int \mathsf{E}_1 1_E \cdot \mathsf{E}_2 1_E \cdot \mathsf{E}_3 1_E\,\mathrm{d}\mu - \varepsilon$$

for all $g \in C$.

Finally, an appeal to Lemma 2 gives

$$\int \mathsf{E}_1 1_E \cdot \mathsf{E}_2 1_E \cdot \mathsf{E}_3 1_E \,\mathrm{d}\mu \geq \int 1_E \cdot \mathsf{E}_1 1_E \cdot \mathsf{E}_2 1_E \cdot \mathsf{E}_3 1_E \,\mathrm{d}\mu \geq \mu(E)^4.$$

Hence $g \in B$ for all $g \in C$. □

In the remainder of the proofs of Theorems C and D, we show that the set C from the above corollary is large in the required senses, rather than handle the set B from Theorem C itself.

5.3 *Completed proof of Theorem C*

In order to use the approximants $\mathsf{E}_i f_i$ to the functions f_i in estimating the triple forms (11), we need to convert a bound on the norms such as $\|\cdot\|_{\check{\otimes}_{1,2}}$ into some more direct control on those triple forms. That control is given by the next proposition.

Proposition 4. *Let $f_1, f_2, f_3 : X \longrightarrow [-1,1]$ be measurable. Then*

$$\int \Big| \int f_1 \cdot f_2 S^g \cdot f_3 S^g T^g \,\mathrm{d}\mu \Big| \,\mathrm{d}g$$
$$\leq \sqrt{2D^{-1/4} + \sqrt{\min\{\|f_1\|_{\check{\otimes}_{1,12}}, \|f_2\|_{\check{\otimes}_{1,2}}, \|f_3\|_{\check{\otimes}_{12,2}}\}}}.$$

This may be viewed as the analog of Proposition 1 for BMZ corners. It will be deduced from the following intermediate estimates.

Lemma 7. *For any $f : X \longrightarrow [-1,1]$, the following inequalities hold:*

$$\int \|\mathsf{E}(f \cdot f S^h \mid \Sigma_X^T)\|_2^2 \,\mathrm{d}h \leq D^{-1/2} + \|f\|_{\check{\otimes}_{1,2}}, \tag{19}$$

$$\int \|\mathsf{E}(f \cdot f (ST)^h \mid \Sigma_X^T)\|_2^2 \,\mathrm{d}h \leq D^{-1/2} + \|f\|_{\check{\otimes}_{12,2}}, \tag{20}$$

and

$$\int \|\mathsf{E}(f \cdot f (ST)^h \mid \Sigma_X^S)\|_2^2 \,\mathrm{d}h \leq D^{-1/2} + \|f\|_{\check{\otimes}_{1,12}}. \tag{21}$$

Proof. We first prove (19). Define a new probability $(G \times G)$-space as follows: let $Y := G \times G \times G$, let $\nu := m_Y$, and let the two generating actions be

$$\tilde{S}^g(x,y,z) = (gx, y, z) \quad \text{and} \quad \tilde{T}^g(x,y,z) = (x, gy, gz).$$

(In ergodic-theory terms, this is the relative product of two copies of (X,μ,S,T) over Σ_X^T.) Then a simple calculation shows that

$$\|\mathsf{E}(f\cdot fS^h\mid\Sigma_X^T)\|_2^2=\int F\cdot F\tilde{S}^h\,\mathrm{d}\nu=\langle F,F\tilde{S}^h\rangle_{L^2(\nu)},$$

where

$$F(x,y,z):=f(x,y)f(x,z).$$

Let $H:=\mathsf{E}(F\mid\Sigma_Y^{\tilde{S}})$. Applying Lemma 4 to F and $\tilde{S}$, it follows that

$$\begin{aligned}\int\|\mathsf{E}(f\cdot fS^h\mid\Sigma_X^T)\|_2^2\,\mathrm{d}h&=\int\langle F,F\tilde{S}^h\rangle_{L^2(\nu)}\,\mathrm{d}h\\&\leq\int\left|\langle F,F\tilde{S}^h\rangle_{L^2(\nu)}-\langle F,H\rangle_{L^2(\nu)}\right|\mathrm{d}h+\left|\langle F,H\rangle_{L^2(\nu)}\right|\\&\leq D^{-1/2}+\left|\langle F,H\rangle_{L^2(\nu)}\right|.\end{aligned}$$

On the other hand, H is a function of only the coordinates y and z for a point $(x,y,z)\in Y$, and therefore

$$\begin{aligned}\left|\langle F,H\rangle_{L^2(\nu)}\right|&=\left|\iiint f(x,y)f(x,z)H(y,z)\,\mathrm{d}x\,\mathrm{d}y\,\mathrm{d}z\right|\\&\leq\int\left|\iint f(x,y)\cdot(f(x,z)H(y,z))\,\mathrm{d}x\,\mathrm{d}y\right|\mathrm{d}z\\&\leq\int\|f\|_{\check{\otimes}_{1,2}}\,\mathrm{d}z=\|f\|_{\check{\otimes}_{1,2}}.\end{aligned}$$

The proof of (20) is very similar. One uses the same auxiliary $(G\times G)$-space $(Y,\nu,\tilde{S},\tilde{T})$ and function F as before, but now one proceeds from the estimate

$$\begin{aligned}\int\|\mathsf{E}(f\cdot f(ST)^h\mid\Sigma_X^T)\|_2^2\,\mathrm{d}h&=\iint F\cdot F(\tilde{S}\tilde{T})^h\,\mathrm{d}\nu\,\mathrm{d}h\\&=\int\langle F,F(\tilde{S}\tilde{T})^h\rangle_{L^2(\nu)}\,\mathrm{d}h\leq D^{-1/2}+\left|\int F\cdot\mathsf{E}(F\mid\Sigma_Y^{\tilde{S}\tilde{T}})\,\mathrm{d}\nu\right|.\end{aligned}$$

The function $H':=\mathsf{E}(F\mid\Sigma_Y^{\tilde{S}\tilde{T}})$ can be written as a function of only $x^{-1}y$ and $x^{-1}z$ for a point $(x,y,z)\in Y$. Using this, the change of variables $w:=x^{-1}z$ gives

$$\begin{aligned}\left|\int F\cdot H'\,\mathrm{d}\nu\right|&=\left|\iiint f(x,y)f(x,z)H'(x^{-1}y,x^{-1}z)\,\mathrm{d}x\,\mathrm{d}y\,\mathrm{d}z\right|\\&=\left|\iiint f(x,y)f(x,xw)H'(x^{-1}y,w)\,\mathrm{d}x\,\mathrm{d}y\,\mathrm{d}w\right|\end{aligned}$$

$$\leq \int \left| \iint f(x,y) \cdot (f(x,xw)H'(x^{-1}y,w))\, \mathrm{d}x\, \mathrm{d}y \right| \mathrm{d}w$$
$$\leq \int \|f\|_{\check{\otimes}_{12,2}}\, \mathrm{d}w = \|f\|_{\check{\otimes}_{12,2}}.$$

Finally, inequality (21) is simply (20) with the rôles of S and T reversed. □

Proof (Proposition 4). We first prove the bound that uses either $\|f_2\|_{\check{\otimes}_{1,2}}$ or $\|f_3\|_{\check{\otimes}_{12,2}}$. Let $u_g := f_2 S^g \cdot f_3 S^g T^g$. By Lemma 1, it suffices to prove that

$$\iint |\langle u_g, u_{g'}\rangle|\, \mathrm{d}g\, \mathrm{d}g' = \iint |\langle u_g, u_{hg}\rangle|\, \mathrm{d}g\, \mathrm{d}h$$
$$\leq 2D^{-1/4} + \sqrt{\min\{\|f_2\|_{\check{\otimes}_{1,2}}, \|f_3\|_{\check{\otimes}_{12,2}}\}}.$$

For any $g, h \in G$, one has

$$|\langle u_g, u_{hg}\rangle| = \left| \int f_2 S^g \cdot f_2 S^{hg} \cdot f_3 S^g T^g \cdot f_3 S^{hg} T^{hg}\, \mathrm{d}\mu \right|$$
$$= \left| \int f_2 \cdot f_2 S^h \cdot f_3 T^g \cdot f_3 S^h T^{hg}\, \mathrm{d}\mu \right|$$
$$= |\langle F_h, F'_h T^g\rangle|,$$

where $F_h := f_2 \cdot f_2 S^h$ and $F'_h := f_3 \cdot f_3 (ST)^h$, both of which take values in $[-1, 1]$. Therefore Lemma 4 and the Cauchy–Bunyakowski–Schwartz Inequality give

$$\iint |\langle u_g, u_{hg}\rangle|\, \mathrm{d}g\, \mathrm{d}h \leq \int \left|\left\langle \mathsf{E}(F_h \mid \Sigma_X^T), \mathsf{E}(F'_h \mid \Sigma_X^T)\right\rangle\right| \mathrm{d}h + D^{-1/2}$$
$$\leq \int \|\mathsf{E}(F_h \mid \Sigma_X^T)\|_2 \|\mathsf{E}(F'_h \mid \Sigma_X^T)\|_2\, \mathrm{d}h + D^{-1/2}$$
$$\leq \sqrt{\int \|\mathsf{E}(F_h \mid \Sigma_X^T)\|_2^2\, \mathrm{d}h \cdot \int \|\mathsf{E}(F'_h \mid \Sigma_X^T)\|_2^2\, \mathrm{d}h} + D^{-1/2}.$$

Each of the integrals inside this last square root is certainly at most 1. Therefore, by Lemma 7, the last line above may be bounded by

$$\text{either} \quad \sqrt{\|f_2\|_{\check{\otimes}_{1,2}} + D^{-1/2}} + D^{-1/2} \leq \sqrt{\|f_2\|_{\check{\otimes}_{1,2}}} + 2D^{-1/4}$$
$$\text{or} \quad \sqrt{\|f_3\|_{\check{\otimes}_{12,2}} + D^{-1/2}} + D^{-1/2} \leq \sqrt{\|f_3\|_{\check{\otimes}_{12,2}}} + 2D^{-1/4}.$$

Finally, the proof of the bound using $\|f_1\|_{\check{\otimes}_{1,12}}$ is exactly analogous to the case of $\|f_3\|_{\check{\otimes}_{12,2}}$ once one makes the substitution $g' := g^{-1}$ to write

$$\int\Big|\int f_1\cdot f_2S^g\cdot f_3S^gT^g\,\mathrm{d}\mu\Big|\,\mathrm{d}g = \int\Big|\int f_3\cdot f_2T^{g'}\cdot f_1S^{g'}T^{g'}\,\mathrm{d}\mu\Big|\,\mathrm{d}g'.$$

□

Proof (Theorem C). Let $f := 1_E$ and $\eta := \varepsilon^8/(4\cdot 6^4)$. Then there are certainly values of D satisfying (6) for which

$$\sqrt{2D^{-1/4}+\sqrt{\eta}} \leq \varepsilon^2/6. \tag{22}$$

For this function f and error tolerance η, let the partitions $\mathcal{Q}_i$, $i = 1,2,3$, and corresponding operators E_i be given by Proposition 2 and its two variants. Then $|\mathcal{Q}_i| \leq \exp(2(1/\eta)^{O(1)})$ for each i, and so there are values of D satisfying (6) for which (18) also holds. Therefore Proposition 3 applies, and so it suffices to show that the set C from that proposition has measure at least $1-\varepsilon$. We will in fact prove that

$$\int\Big|\int f\cdot fS^g\cdot fS^gT^g\,\mathrm{d}\mu - \int(\mathsf{E}_1f)\cdot(\mathsf{E}_2f)S^g\cdot(\mathsf{E}_3f)S^gT^g\,\mathrm{d}\mu\Big|\,\mathrm{d}g \leq \varepsilon^2/2.$$

The desired lower bound on $m_G(C)$ follows from this by Chebyshev's Inequality.

Let $f_i^{\perp} = f - \mathsf{E}_if$ for each i. By the triangle inequality, the above estimate is a consequence of the following:

$$\int\Big|\int f\cdot fS^g\cdot f_3^{\perp}S^gT^g\,\mathrm{d}\mu\Big|\,\mathrm{d}g \leq \varepsilon^2/6,$$

$$\int\Big|\int f\cdot f_2^{\perp}S^g\cdot(\mathsf{E}_3f)S^gT^g\,\mathrm{d}\mu\Big|\,\mathrm{d}g \leq \varepsilon^2/6$$

and

$$\int\Big|\int f_1^{\perp}\cdot(\mathsf{E}_2f)S^g\cdot(\mathsf{E}_3f)S^gT^g\,\mathrm{d}\mu\Big|\,\mathrm{d}g \leq \varepsilon^2/6.$$

These are all now implied by Proposition 4, together with (22). □

5.4 Anti-neighbourhoods and syndeticity

The proof of Theorem D will be based on having a large supply of fairly syndetic subsets of a quasirandom group ready to hand. These subsets will be obtained from a simple construction in terms of representations.

Let $\pi : G \curvearrowright V$ be an orthogonal representation, let $u, v \in V$, and let $\varepsilon > 0$. If $P^\pi = 0$ (that is, π contains no copy of the identity representation), then let

$$A(\pi, u, v, \varepsilon) := \{g \in G : |\langle u, \pi^g v\rangle| < \varepsilon\},$$

and call this the (π, u, v, ε)-**anti-neighbourhood**. For general π, let

$$A(\pi, u, v, \varepsilon) := \{g \in G : |\langle u, \pi^g v\rangle - \langle P^\pi u, P^\pi v\rangle| < \varepsilon\}.$$

If G is D-quasirandom for some large D, and u, v are unit vectors, then the corresponding anti-neighbourhoods are quite large: Lemma 4 and Chebyshev's Inequality imply that

$$m_G(A(\pi, u, v, \varepsilon)) \geq 1 - \frac{D^{-1}}{\varepsilon^2}. \tag{23}$$

This is very intuitive. If π is irreducible with large dimension d, then the orbit points $\pi^g v$ should be fairly evenly spread around the high-dimensional unit sphere in V, and so most of them will be nearly orthogonal to any fixed direction u.

The present section shows that, if G is highly quasirandom, then anti-neighbourhoods are also fairly syndetic. Moreover, one can intersect a controlled number of anti-neighbourhoods, and that intersection is still fairly syndetic. This is not implied solely by the largeness of those intersections, but it will follow from some simple inner-product estimates. In the next section, the syndeticity of the set in Theorem D will be proved by showing that it contains such an intersection of anti-neighbourhoods.

Let us begin with a rough sketch of why a single anti-neighbourhood should be fairly syndetic, before carefully proving the result we need about intersections. The key point is that, if d is very large and we choose a moderately large number of group elements $h_1, \ldots, h_k$ independently at random, then with high probability the image-points $\pi^{h_i} u$ will all be nearly orthogonal to one another. The intuition here is the same as that above: each of the random points $\pi^{h_i} u$ should be fairly evenly distributed around the high-dimensional unit sphere of V, independently of the others. However, having obtained some h_i's for which the vectors $\pi^{h_i} u$ are all nearly orthogonal, this then forces any other unit vector to be nearly orthogonal to at least some of them. In particular, for any $g \in G$, there is an i such that $\pi^{h_i} u$ and $\pi^g v$ are nearly orthogonal. This implies that u is nearly orthogonal to $\pi^{h_i^{-1} g} v$, and hence that g is in $h_i \cdot A(\pi, u, v, \varepsilon)$ for some small ε. Crucially, after fixing the right choice of $h_1, \ldots, h_k$, this argument works for every $g \in G$.

Such 'near orthogonality' will be deduced using the following.

Lemma 8. *Let V be a real Hilbert space and let v and $u^1, \ldots, u^m$ be unit vectors in V. Suppose that*

$$|\langle u^i, u^j\rangle| \leq 1/m^2 \quad \textit{whenever } i \neq j.$$

Then

$$\sum_{i\leq m}\langle v,u^i\rangle^2 \leq 2.$$

Proof. Let $a_i := \langle v, u^i\rangle$ for each i, so one always has $|a_i| \leq 1$, and let

$$w := \sum_{i\leq m} a_i u^i.$$

The assumed inequalities give

$$\|w\|^2 = \sum_{i,j\leq m} a_i a_j \langle u^i, u^j\rangle \leq \sum_{i\leq m} a_i^2 + (m^2-m)/m^2 \leq \sum_{i\leq m} a_i^2 + 1,$$

and hence

$$\sum_{i\leq m} a_i^2 = \langle v, w\rangle \leq \|v\|\,\|w\| \leq \sqrt{\sum_{i\leq m} a_i^2 + 1}.$$

This implies that $\sum_{i\leq m} a_i^2 \leq 2$. □

In order to study intersections of anti-neighbourhoods, we will actually need the following crude corollary which concerns several Hilbert spaces simultaneously.

Corollary 3. *Let $V_1, \ldots, V_k$ be real Hilbert spaces, and let v_ℓ and $u_\ell^1, \ldots, u_\ell^m$ be unit vectors in V_ℓ for each ℓ. Suppose that*

$$|\langle u_\ell^i, u_\ell^j\rangle| \leq 1/m^2 \quad \textit{whenever } \ell \leq k,\ i \neq j.$$

Then there is some $i \leq m$ such that $|\langle v_\ell, u_\ell^i\rangle| \leq \sqrt{2k/m}$ for all $\ell \leq k$.

Proof. Summing the inequalities proved in the preceding lemma gives

$$\sum_{i\leq m}\Big(\sum_{\ell\leq k}\langle v_\ell, u_\ell^i\rangle^2\Big) \leq 2k.$$

□

Corollary 4. *Let $\eta > 0$ and $k \geq 1$, and set $K := \lceil 2k/\eta^2 + 1\rceil$ and $D := K^6 k + 1$. Let $\pi_\ell : G \curvearrowright V_\ell$ for $\ell = 1, 2, \ldots, k$ be orthogonal representations that are all D-quasirandom, and let $u_\ell, v_\ell \in V_\ell$ be unit vectors for each ℓ. Then*

$$A := \bigcap_{\ell=1}^{k} A(\pi_\ell, u_\ell, v_\ell, \eta)$$

is K-syndetic.

Proof. Let

$$A' := \bigcap_{\ell=1}^{k} A(\pi_\ell, u_\ell, u_\ell, 1/K^2).$$

Let $h_1, \ldots, h_K$ be a K-tuple of elements of G drawn at random from the measure $m_G^{\otimes K}$. Then the estimate (23) and a first-moment bound give

$$m_G^{\otimes K}\{h_i^{-1}h_j \in A' \ \forall i \neq j \text{ in } \{1, 2, \ldots, K\}\}$$
$$\geq 1 - \sum_{i \neq j} \sum_{\ell=1}^{k} m_G\big(G \backslash A(\pi_\ell, u_\ell, u_\ell, 1/K^2)\big) \geq 1 - K^6 k D^{-1} > 0.$$

This implies that there exists a K-tuple $h_1, \ldots, h_K$ in G witnessing the above event, hence such that

$$|\langle \pi_\ell^{h_i} u_\ell, \pi_\ell^{h_j} u_\ell \rangle| \leq 1/K^2 \quad \text{whenever } \ell \leq k,\ i \neq j.$$

Therefore, for any $g \in G$, Corollary 3 promises some $i \leq K$ for which

$$|\langle \pi_\ell^{h_i} u_\ell, \pi_\ell^{g} v_\ell \rangle| = |\langle u_\ell, \pi_\ell^{h_i^{-1}g} v_\ell \rangle| \leq \sqrt{2k/K} < \eta \quad \forall \ell \leq k,$$

so $g \in \{h_1, \ldots, h_K\} \cdot A$. □

5.5 *Completed proof of Theorem D*

The next step is the following rather technical proposition.

Proposition 5. *Let $\varepsilon > 0$ and $n \in \mathbb{N}$. Set $k := \lceil 4/\varepsilon^2 \rceil$, and now set $\eta := 1/(3k)^8$. Suppose that G is D-quasirandom for some $D > 4k^4/\eta^4$.*

For $\ell = 2, 3$, let f_1^ℓ, f_2^ℓ and f_3^ℓ be three $[-1, 1]$-valued functions, and suppose that

$$\|f_2^2\|_{\breve{\otimes}_{1,2}} \leq \eta \quad \textit{and} \quad \|f_3^3\|_{\breve{\otimes}_{12,2}} \leq \eta.$$

Finally, let

$$C_\ell := \left\{ g \in G : \left| \int f_1^\ell \cdot f_2^\ell S^g \cdot f_3^\ell S^g T^g \, \mathrm{d}\mu \right| < \varepsilon \right\} \qquad \textit{for } \ell = 2, 3.$$

Then there are elements $h_1, \dots, h_k \in G$ *and some auxiliary* $[-1,1]$*-valued functions*

$$F^2_{2,i,j},\quad F^2_{3,i,j},\quad F^3_{2,i,j}\quad \textit{and}\quad F^3_{3,i,j}\quad \textit{for } 1 \le i < j \le k$$

such that the set

$$E := \bigcap_{1\le i<j\le k} \big(A(T, F^2_{2,i,j}, F^2_{3,i,j}, \eta) \cap A(T, F^3_{2,i,j}, F^3_{3,i,j}, \eta)\big)$$

satisfies

$$E \subseteq \{h_1^{-1}, \dots, h_k^{-1}\} \cdot (C_2 \cap C_3).$$

Remark 2. Letting $m := \lceil 2k^2/\eta^2 + 1 \rceil$, the set E above is m-syndetic by Corollary 4 provided D large enough, and so the above conclusion implies that the intersection $C_2 \cap C_3$ is (mk)-syndetic. However, this fact alone is not quite what we need for the proof of Theorem D: ultimately, that will require the syndeticity of the smaller intersection $C_1 \cap C_2 \cap C_3$ for some other set C_1. In order to prove that, it will be important to have an explicit 'witness' to the syndeticity of $C_2 \cap C_3$ in the form of an intersection of anti-neighbourhoods, such as E above. This is why the above proposition is formulated as it is. $\lhd$

Proof. We will prove that if $h_1, \dots, h_k$ are chosen independently at random from the Haar measure m_G, then with positive probability one obtains a tuple for which the remaining objects required by the proposition also exist.

Step 1. For such a random choice of $h_1, \dots, h_k$, each difference $h_jh_i^{-1}$ for $i \neq j$ is also a uniform random element of G, and so Lemma 7 gives the estimates

$$\iint \cdots \int \|\mathsf{E}(f_2^2 \cdot f_2^2 S^{h_jh_i^{-1}} \mid \Sigma_X^T)\|_2^2 \, \mathrm{d}h_1 \cdots \mathrm{d}h_{k-1}\, \mathrm{d}h_k \le D^{-1/2} + \eta$$

and

$$\iint \cdots \int \|\mathsf{E}(f_3^3 \cdot f_3^3 (ST)^{h_jh_i^{-1}} \mid \Sigma_X^T)\|_2^2 \, \mathrm{d}h_1 \cdots \mathrm{d}h_{k-1}\, \mathrm{d}h_k \le D^{-1/2} + \eta$$

for each $i \neq j$.

Our assumptions imply that $D^{-1/2} < \eta$, and so the right-hand sides above are all less than 2η. Therefore, if $h_1, \dots, h_k$ are chosen randomly as described above, then, by Chebyshev's Inequality and a first-moment bound, the event that

$$\|\mathsf{E}(f_2^2 \cdot f_2^2 S^{h_jh_i^{-1}} \mid \Sigma_X^T)\|_2^2 < \sqrt{2\eta} \quad \text{for all } i \neq j$$
$$\text{and} \quad \|\mathsf{E}(f_3^3 \cdot f_3^3 (ST)^{h_jh_i^{-1}} \mid \Sigma_X^T)\|_2^2 < \sqrt{2\eta} \quad \text{for all } i \neq j \qquad (24)$$

has probability at least

$$1 - 2k^2\sqrt{2\eta} = 1 - 2\sqrt{2}k^2/(3k)^4 > 0.$$

Therefore there exists a tuple $h_1, \ldots, h_k \in G$ for which the inequalities in (24) all hold simultaneously.

Step 2. We now define the required auxiliary functions as follows:

$$F^2_{2,i,j} := f_2^2 S^{h_i} \cdot f_2^2 S^{h_j} \quad \text{and} \quad F^2_{3,i,j} := f_3^2 S^{h_i} T^{h_i} \cdot f_3^2 S^{h_j} T^{h_j}$$
$$F^3_{2,i,j} := f_2^3 S^{h_i} \cdot f_2^3 S^{h_j} \quad \text{and} \quad F^3_{3,i,j} := f_3^3 S^{h_i} T^{h_i} \cdot f_3^3 S^{h_j} T^{h_j}$$

whenever $1 \le i < j \le k$.

Step 3. Having chosen those auxiliary functions, let E be as in the statement of the proposition. Suppose that $g \in E$. We must show that $h_i g \in C_2 \cap C_3$ for some $i \le k$.

For each $g \in G$ and $\ell = 2, 3$, define

$$u^\ell_g := f_2^\ell S^g \cdot f_3^\ell S^g T^g.$$

For any $i < j$ in $\{1, 2, \ldots, k\}$ and $\ell = 2, 3$, we have

$$\begin{aligned} |\langle u^\ell_{h_i g}, u^\ell_{h_j g}\rangle| &= \left| \int f_2^\ell S^{h_i g} \cdot f_2^\ell S^{h_j g} \cdot f_3^\ell S^{h_i g} T^{h_i g} \cdot f_3^\ell S^{h_j g} T^{h_j g} \, \mathrm{d}\mu \right| \\ &= \left| \int f_2^\ell S^{h_i} \cdot f_2^\ell S^{h_j} \cdot f_3^\ell S^{h_i} T^{h_i g} \cdot f_3^\ell S^{h_j} T^{h_j g} \, \mathrm{d}\mu \right| \\ &= \left| \int F^\ell_{2,i,j} \cdot F^\ell_{3,i,j} T^g \, \mathrm{d}\mu \right|. \end{aligned}$$

Since $g \in E$, in particular $g \in A(T, F^\ell_{2,i,j}, F^\ell_{3,i,j}, \eta)$, and so the above is at most

$$\left| \int \mathsf{E}(F^\ell_{2,i,j} \mid \Sigma^T_X) \mathsf{E}(F^\ell_{3,i,j} \mid \Sigma^T_X) \, \mathrm{d}\mu \right| + \eta.$$

If $\ell = 2$ then this expression is bounded by

$$\begin{aligned} \|\mathsf{E}(F^2_{2,i,j} \mid \Sigma^T_X)\|_2 + \eta &= \|\mathsf{E}(f_2^2 S^{h_i} \cdot f_2^2 S^{h_j} \mid \Sigma^T_X)\|_2 + \eta \\ &= \|\mathsf{E}(f_2^2 \cdot f_2^2 S^{h_j h_i^{-1}} \mid \Sigma^T_X)\|_2 + \eta < \sqrt[4]{2\eta} + \eta < 3\sqrt[4]{\eta} = 1/k^2, \end{aligned}$$

by (24). Similarly, if $\ell = 3$ then it is bounded by

$$\|\mathsf{E}(F^3_{3,i,j} \mid \Sigma^T_X)\|_2 + \eta = \|\mathsf{E}(f_3^3 \cdot f_3^3 (ST)^{h_j h_i^{-1}} \mid \Sigma^T_X)\|_2 + \eta < 1/k^2.$$

Thus, we have shown that

$$|\langle u^{\ell}_{h_i g}, u^{\ell}_{h_j g}\rangle| < 1/k^2 \quad \text{whenever } i \neq j,\ \ell = 2, 3.$$

We may therefore apply Corollary 3 to the inner products

$$\langle f^{\ell}_1, u^{\ell}_{h_i g}\rangle, \quad i = 1, 2, \ldots, k,\ \ \ell = 2, 3$$

to conclude that, for any $g \in E$, there is at least one $i \leq k$ for which

$$|\langle f^2_1, u^2_{h_i g}\rangle| \leq \sqrt{4/k} < \varepsilon \quad \text{and} \quad |\langle f^3_1, u^3_{h_i g}\rangle| < \varepsilon.$$

For this choice of i one has $h_i g \in C_2 \cap C_3$, as required. □

Proof (Theorem D). Let $f := 1_E$. As for Theorem C, this proof will be based on three different decompositions of f as given by Proposition 2. However, a new complication here is that the partition $\mathcal{Q}_1$ will need to be chosen after $\mathcal{Q}_2$ and $\mathcal{Q}_3$, and considerably finer than those others.

In the course of the proof, we will meet three points at which we require a lower bound on D. All of these lower bounds will be satisfied for some D as in (7), so there is a choice of D of the form in (7) for which the whole proof can be carried out.

Step 1. Set $k := \lceil 36/\varepsilon^2 \rceil$ and $\eta := 1/(3k)^8$. Let $\mathcal{Q}_2$ and $\mathcal{Q}_3$ be partitions as given by the two variants of Proposition 2 for this error tolerance η, and consider the resulting decompositions

$$f = f^{\perp}_2 + \mathsf{E}_2 f = f^{\perp}_3 + \mathsf{E}_3 f.$$

Let

$$C_2 := \left\{ g \in G : \left| \int f \cdot f^{\perp}_2 S^g \cdot f S^g T^g \, \mathrm{d}\mu \right| < \varepsilon/3 \right\}$$

and

$$C_3 := \left\{ g \in G : \left| \int f \cdot (\mathsf{E}_2 f) S^g \cdot f^{\perp}_3 S^g T^g \, \mathrm{d}\mu \right| < \varepsilon/3 \right\}.$$

Given the above choice of k and η, there is a D as in (7) which satisfies $D > 4k^4/\eta^4$ (indeed, at this point (7) leaves vastly more room than we need). Therefore Proposition 5 applies to give $h_1, \ldots, h_k \in G$ and some auxiliary $[-1, 1]$-valued functions

$$F^2_{2,i,j}, \quad F^2_{3,i,j}, \quad F^3_{2,i,j} \quad \text{and} \quad F^3_{3,i,j} \quad \text{for } 1 \leq i < j \leq k$$

such that $E \subseteq \{h_1^{-1}, \ldots, h_k^{-1}\} \cdot (C_2 \cap C_3)$, where

$$E = \bigcap_{1 \le i < j \le k} \big(A(T, F^2_{2,i,j}, F^2_{3,i,j}, \eta) \cap A(T, F^3_{2,i,j}, F^3_{3,i,j}, \eta)\big).$$

Let $M_i := |\mathcal{Q}_i|$ for $i = 2, 3$.

Step 2. Now set $\eta' := \varepsilon/6M_2M_3$, let $\mathcal{Q}_1$ be as given by Proposition 2 for this error tolerance η', and consider the decomposition

$$f = f_1^{\perp} + \mathsf{E}_1 f.$$

Let

$$C_1 := \Big\{g \in G : \Big| \int f_1^{\perp} \cdot (\mathsf{E}_2 f) S^g \cdot (\mathsf{E}_3 f) S^g T^g \, \mathrm{d}\mu \Big| < \varepsilon/3\Big\}.$$

Observe that

$$\eta' \ge (\varepsilon/6) \exp\big(-4(1/\eta)^{\mathrm{O}(1)}\big) \ge \exp\big(-(1/\varepsilon)^{\mathrm{O}(1)}\big)$$

for $\varepsilon \in (0, 1/2)$. Therefore

$$\frac{16|\mathcal{Q}_1|^2|\mathcal{Q}_2|^2|\mathcal{Q}_3|^2}{\varepsilon^2} \le \frac{16 \exp\big(4(1/\eta')^{\mathrm{O}(1)} + 8(1/\eta)^{\mathrm{O}(1)}\big)}{\varepsilon^2} \le \exp\big(\exp\big((1/\varepsilon)^{\mathrm{O}(1)}\big)\big)$$

for all $\varepsilon \in (0, 1/2)$, and so there are values of D satisfying (7) for which (18) holds for these partitions $\mathcal{Q}_1$, $\mathcal{Q}_2$ and $\mathcal{Q}_3$. (This is the only point at which we need the double exponential in (7).) Therefore Proposition 3 applies, and so it suffices to show that the set C from that proposition is K-syndetic. Moreover, that set C clearly contains $C_1 \cap C_2 \cap C_3$, so it suffices to show that this triple intersection is K-syndetic.

Step 3. We now make use of the decompositions (16) and (17). Substituting these into the integral that appears inside the definition of C_1, we obtain

$$\sum_{m_2, m_3} \iint \big(f_1^{\perp}(x, y) h_{2,1,m_2}(y) h_{3,12,m_3}(x^{-1}y)\big)(h_{2,2,m_2} h_{3,2,m_3})(gx) \, \mathrm{d}x \, \mathrm{d}y. \tag{25}$$

Let

$$\psi_{m_2,m_3}(x, y) := f_1^{\perp}(x, y) h_{2,1,m_2}(y) h_{3,12,m_3}(x^{-1}y),$$

let

$$\varphi_{m_2,m_3}(x, y) := (h_{2,2,m_2} h_{3,2,m_3})(x)$$

(so φ depends only nominally on y), and let

$$E' := \bigcap_{m_2,m_3} A\big(S, \psi_{m_2,m_3}, \varphi_{m_2,m_3}, \eta'\big).$$

We will now show that $E' \subseteq C_1$, so suppose that $g \in E'$. Then the definition of E' and the expression (25) give

$$\begin{aligned}\Big|\int f_1^{\perp} \cdot (\mathsf{E}_2 f)S^g \cdot (\mathsf{E}_3 f)S^g T^g \,\mathrm{d}\mu\Big| &\leq \sum_{m_2,m_3} \Big|\int \psi_{m_2,m_3} \cdot \varphi_{m_2,m_3} S^g \,\mathrm{d}\mu\Big| \\ &\leq \sum_{m_2,m_3} \Big(\Big|\int \psi_{m_2,m_3}\,\mathrm{d}\mu\Big|\Big|\int \varphi_{m_2,m_3}\,\mathrm{d}\mu\Big| + \eta'\Big) \\ &\leq \sum_{m_2,m_3} \Big|\int \psi_{m_2,m_3}\,\mathrm{d}\mu\Big| + \varepsilon/6.\end{aligned}$$

Substituting from the definition of ψ_{m_2,m_3}, this is

$$\begin{aligned}&\sum_{m_2,m_3} \Big|\iint f_1^{\perp}(x,y) h_{2,1,m_2}(y) h_{3,12,m_3}(x^{-1}y)\,\mathrm{d}x\,\mathrm{d}y\Big| + \varepsilon/6 \\ &\qquad \leq M_2 M_3 \|f_1^{\perp}\|_{\check{\otimes}_{1,12}} + \varepsilon/6 \leq M_2 M_3 \eta' + \varepsilon/6 = \varepsilon/3,\end{aligned}$$

so $g \in C_1$.

Step 4. Finally, letting

$$E'' = \bigcap_{i=1}^{k} h_i^{-1} E' = \bigcap_{i=1}^{k} \bigcap_{m_2,m_3} A\big(S, \psi_{m_2,m_3}, \varphi_{m_2,m_3} S^{h_i}, \eta'\big),$$

it follows that

$$E \cap E'' \subseteq \{h_1^{-1}, \dots, h_k^{-1}\} \cdot (C_1 \cap C_2 \cap C_3) \subseteq \{h_1^{-1}, \dots, h_k^{-1}\} \cdot C.$$

On the other hand,

$$\begin{aligned}E \cap E'' = &\bigcap_{1 \leq i < j \leq k} \big(A(T, F^2_{2,i,j}, F^2_{3,i,j}, \eta) \cap A(T, F^3_{2,i,j}, F^3_{3,i,j}, \eta)\big) \\ &\cap \bigcap_{i=1}^{k} \bigcap_{m_2,m_3} A\big(S, \psi_{m_2,m_3}, \varphi_{m_2,m_3} S^{h_i}, \eta'\big).\end{aligned}$$

Since $\varepsilon < 1/2$, this is an intersection of at most

$$2k^2 + kM_2M_3 \le 2k^2 + k\exp(4(1/\eta)^{O(1)}) \le \exp\left((1/\varepsilon)^{O(1)}\right)$$

anti-neighbourhoods. On the other hand,

$$\eta, \eta' \ge \exp(-(1/\varepsilon)^{O(1)}).$$

Let

$$K_0 := \lceil 2(2k^2 + kM_2M_3)/\min\{\eta, \eta'\}^2 + 1 \rceil \le \exp\left((1/\varepsilon)^{O(1)}\right).$$

There is a D satisfying (7) for which

$$D \ge K_0^6(2k^2 + kM_2M_3) + 1.$$

Therefore Corollary 4 can be applied to deduce that this intersection is still K_0-syndetic. Therefore C is K-syndetic for

$$K = K_0 k = \exp\left((1/\varepsilon)^{O(1)}\right),$$

as required. □

Remark 3. The above proof uses three different decompositions of f, one for each the three positions in the triple form (11). However, these decompositions do not have equal status: the decomposition $f = f_1^{\perp} + \mathsf{E}_1 f$ (corresponding to the first position) requires a much finer partition $\mathcal{Q}_1$, and depends on having already obtained the partitions $\mathcal{Q}_2$ and $\mathcal{Q}_3$ corresponding to the second and third positions. This contrasts with the proof of Theorem C, where the three positions in the triple form (11) have roughly equal status.

The reason for this difference can be seen in the proof of Proposition 4. First, we proved the bound that uses the norm $\|f_2\|_{\check{\otimes}_{1,2}}$ or $\|f_3\|_{\check{\otimes}_{12,2}}$; then, the bound that uses $\|f_1\|_{\check{\otimes}_{1,12}}$ was obtained by the same argument upon changing variables to $g' := g^{-1}$. However, this change of variables converts left-syndeticity to *right*-syndeticity, so we cannot use it in the same way to analyse the set C_1 in the proof of Theorem D. Instead, we must first replace f with its structured approximations $\mathsf{E}_2 f$ (in the second position) and $\mathsf{E}_3 f$ (in the third position), and then use the special structure of those approximations to analyse the contribution of f in the first position of the triple form, without changing variables.

This discussion also suggests why our methods fail to answer Question 1 from the Introduction (about syndeticity in the setting of Theorem B). The current version of Theorem B involves an estimate of the triple form for naïve corners which uses an integral of the kind appearing in Corollary 1. That integral features both T^g and $S^{g^{-1}}$. Once again, the appearance of g^{-1} converts left-syndetic sets into right-syndetic sets, and so it is not clear how to obtain control on this integral on any particular anti-neighbourhood. ◁

Acknowledgements Research supported by a fellowship from the Clay Mathematics Institute. I am grateful to Vitaly Bergelson for sharing [7] with me, to Ben Green for pointing me to the references [11, 15], to Sean Eberhard for pointing me to the reference [20] and to Julia Wolf for suggesting some useful clarifications.

References

1. M. Ajtai, E. Szemerédi, Sets of lattice points that form no squares. Stud. Sci. Math. Hung. **9**, 9–11 (1974/1975)
2. T. Austin, Quantitative equidistribution for certain quadruples in quasi-random groups: Erratum. Comb. Probab. Comput. http://dx.doi.org/10.1017/S0963548315000243 (the print version hasn't appeared yet).
3. T. Austin, Quantitative equidistribution for certain quadruples in quasi-random groups. Comb. Probab. Comput. **24**(2), 376–381 (2015). doi:10.1017/S0963548314000492. http://dx.doi.org/10.1017/S0963548314000492
4. V. Bergelson, R. McCutcheon, Central sets and a non-commutative Roth theorem. Am. J. Math. **129**(5), 1251–1275 (2007). doi:10.1353/ajm.2007.0031. http://dx.doi.org/10.1353/ajm.2007.0031
5. V. Bergelson, T. Tao, Multiple recurrence in quasirandom groups. Geom. Funct. Anal. **24**(1), 1–48 (2014). doi:10.1007/s00039-014-0252-0. http://dx.doi.org/10.1007/s00039-014-0252-0
6. V. Bergelson, R. McCutcheon, Q. Zhang, A Roth theorem for amenable groups. Am. J. Math. **119**(6), 1173–1211 (1997)
7. V. Bergelson, D. Robertson, P. Zorin-Kranich, Triangles in Cartesian squares of quasirandom groups. Available online at `arXiv.org`: 1410.5385 (preprint)
8. T. Bröcker, T. tom Dieck, *Representations of Compact Lie Groups* (Springer, New York, 1985)
9. Q. Chu, Convergence of weighted polynomial multiple ergodic averages. Proc. Am. Math. Soc. **137**, 1363–1369 (2009)
10. Q. Chu, P. Zorin-Kranich, Lower bound in the Roth Theorem for amenable groups. Ergod. Theory Dyn. Syst. 35(6), 1746–1766 (2015)
11. P. Erdős, E.G. Straus, How abelian is a finite group? Linear Multilinear Algebra **3**(4), 307–312 (1975/1976)
12. A. Frieze, R. Kannan, Quick approximation to matrices and applications. Combinatorica **19**(2), 175–220 (1999). doi:10.1007/s004930050052. http://dx.doi.org/10.1007/s004930050052
13. W.T. Gowers, Quasirandom groups. Comb. Probab. Comput. **17**(3), 363–387 (2008). doi:10.1017/S0963548307008826. http://dx.doi.org/10.1017/S0963548307008826
14. W.T. Gowers, Decompositions, approximate structure, transference, and the Hahn-Banach theorem. Bull. Lond. Math. Soc. **42**(4), 573–606 (2010). doi:10.1112/blms/bdq018. http://dx.doi.org/10.1112/blms/bdq018
15. L. Pyber, How abelian is a finite group? in *The Mathematics of Paul Erdős, I*. Algorithms and Combinatorics, vol. 13 (Springer, Berlin, 1997), pp. 372–384. doi:10.1007/978-3-642-60408-9_27. http://dx.doi.org/10.1007/978-3-642-60408-9_27
16. R.A. Ryan, *Introduction to Tensor Products of Banach Spaces*. Springer Monographs in Mathematics (Springer, London, 2002). doi:10.1007/978-1-4471-3903-4. http://dx.doi.org/10.1007/978-1-4471-3903-4
17. R. Schatten, *A Theory of Cross-Spaces*. Annals of Mathematics Studies, vol. 26 (Princeton University Press, Princeton, 1950)
18. I.D. Shkredov, On a problem of Gowers. Dokl. Akad. Nauk **400**(2), 169–172 (2005) (Russian)
19. I.D. Shkredov, On a problem of Gowers. Izv. Ross. Akad. Nauk Ser. Mat. **70**(2), 179–221 (2006) (Russian)
20. J. Solymosi, Roth-type theorems in finite groups. Eur. J. Comb. **34**(8), 1454–1458 (2013). doi:10.1016/j.ejc.2013.05.027. http://dx.doi.org/10.1016/j.ejc.2013.05.027
21. T. Tao, V. Vu, *Additive Combinatorics* (Cambridge University Press, Cambridge, 2006)

Part III
Enumerative and geometric combinatorics

Some algorithmic applications of partition functions in combinatorics

Alexander Barvinok

Abstract Given a combinatorially defined family $\mathscr{F} \subset 2^X$ of subsets of a finite set X and non-negative weights $w : X \longrightarrow \mathbf{R}_+$, we define the *partition function* as $P(\mathscr{F}, w) = \sum_{S \in \mathscr{F}} \prod_{x \in S} w(x)$. We describe various situations when $P(\mathscr{F}, w)$ can be computed efficiently, concentrating on two methods: one relating computability to complex zeros of $P(\mathscr{F}, w)$ as a polynomial in $w(x)$ and the other based on scaling. Examples include enumeration of Hamiltonian cycles in graphs, perfect matchings in hypergraphs, and dense subgraphs in graphs.

Keywords Partition function • Permanent • Algorithm • Graph homomorphism • Scaling

Mathematics Subject Classification (2010): 68Q25, 68C25, 15A15, 82B20, 05C30, 05C85, 65F35

1 An introductory example: Hamiltonian cycles in a graph

Let $G = (V, E)$ be an undirected graph with set $V = \{1, \ldots, n\}$ of vertices and set $E \subset \binom{V}{2}$ of edges, without loops or multiple edges. A *Hamiltonian cycle* in G is a cyclic ordering $1 = i_1 \mapsto i_2 \mapsto \ldots \mapsto i_n \mapsto i_1$ of the vertices of G such that $\{i_k, i_{k+1}\} \in E$ for $k = 1, \ldots, n$, where we agree that $i_{n+1} = i_1$. As is well known, the problem of testing whether a given graph contains a Hamiltonian cycle is NP-complete.

Since we cannot solve the problem, we generalize it. Given an $n \times n$ matrix $A = (a_{ij})$, we define its *Hamiltonian permanent* as

$$\operatorname{ham} A = \sum_{1=i_1 \mapsto i_2 \mapsto \ldots \mapsto i_n \mapsto 1} \prod_{k=1}^{n} a_{i_k i_{k+1}}, \tag{1}$$

A. Barvinok (✉)
Department of Mathematics, University of Michigan, Ann Arbor, MI 48109-1043, USA
e-mail: barvinok@umich.edu

A. Beveridge et al. (eds.), *Recent Trends in Combinatorics*, The IMA Volumes in Mathematics and its Applications 159, DOI 10.1007/978-3-319-24298-9_20

where the sum is taken over all $(n-1)!$ cycles. In particular, if A is the adjacency matrix of G, that is, if

$$a_{ij} = \begin{cases} 1 & \text{if } \{i,j\} \in E \\ 0 & \text{if } \{i,j\} \notin E, \end{cases} \tag{2}$$

then ham A is twice the number of Hamiltonian cycles in G (since the same cycle can be traversed in two directions). Now we want to compute ham A for a non-negative matrix A.

At first glance, generalizing the problem does not help us at all: it is still an NP-complete problem to test whether $\operatorname{ham} A \neq 0$ for a given 0–1 matrix A and it is an even harder #P-complete problem to compute ham A for a given non-negative matrix A. However, introducing the quantity (1) gives us a lot of flexibility: now we can try to compute ham A for matrices A that mimic the matrices we are really interested in.

Let us choose a $0 \leq \delta \leq 1$ and, given a graph G, let us modify the matrix A of (2) by

$$a_{ij} = \begin{cases} 1 & \text{if } \{i,j\} \in E \\ 1-\delta & \text{if } \{i,j\} \notin E. \end{cases} \tag{3}$$

Thus if $\delta = 1$, we get the matrix of (2) and ham A is the number of Hamiltonian cycles in G. If $\delta = 0$, then A is the adjacency matrix of the complete graph and $\operatorname{ham} A = (n-1)!$. For $0 < \delta < 1$, the value of ham A still enumerates the Hamiltonian cycles in the complete graph, only that the non-edges of G are weighed down by a factor of $1-\delta$ each, so that the Hamiltonian cycles in G have each weight 1 and the weight of a general Hamiltonian cycle in the complete graph is exponentially small in the number of the non-edges of G it uses.

1.1 Computing the Hamiltonian permanent of a nearly constant matrix

In [9], we prove that for some absolute constant $\delta > 0$ (one can choose $\delta = 0.05$), the value of ham A for any (real or complex) $n \times n$ symmetric matrix $A = (a_{ij})$ satisfying

$$|1 - a_{ij}| \leq \delta \quad \text{for all} \quad i,j \tag{4}$$

can be computed within any given relative error of $0 < \epsilon < 1$ in $n^{O(\ln n - \ln \epsilon)}$ time (this kind of complexity is called *quasi-polynomial*).

1.2 Approximating the Hamiltonian permanent of a positive matrix

In [4], we prove that for any $0 < \delta < 1$, fixed in advance, the value of ham A for any real (not necessarily symmetric) matrix $A = (a_{ij})$ satisfying (4) can be computed in polynomial time within a factor of $n^{O(\ln n)}$, where the implicit constant in "O" depends on δ.

What do the algorithms of Section 1.1 and Section 1.2 buy us? One immediate corollary is that they allow us to distinguish efficiently graphs that are sufficiently far from having a Hamiltonian cycle from graphs that have sufficiently many Hamiltonian cycles, even when "sufficiently many" still means that the probability to hit a Hamiltonian cycle at random is exponentially small. Indeed, let us fix some $0 < \alpha, \beta \leq 1$. Given a graph G on n vertices, let us define A by (3) for some $\delta = \delta(\alpha, \beta) > 0$ to be specified later. If G does not acquire a Hamiltonian cycle unless at least some new βn edges are introduced, then

$$\operatorname{ham} A \leq (1-\delta)^{\beta n}(n-1)!.$$

On the other hand, if G contains at least $\frac{1}{2}\alpha^n(n-1)!$ Hamiltonian cycles, then

$$\operatorname{ham} A \geq \alpha^n(n-1)!.$$

If we choose δ so that

$$(1-\delta)^{\beta} < \alpha$$

then, applying the algorithm of Section 1.2 we can distinguish between these two cases in polynomial time. Interestingly, for some fixed $\beta > 0$ (one can choose $\beta = 1/321$) it is an NP-complete problem to test whether one can introduce at most βn new edges to a given graph so that the graph acquires a Hamiltonian cycle, cf. [31] and [16] and for any fixed $0 < \alpha < 1$ it is an NP-hard problem to test whether a given graph contains at least $\alpha^n n!$ Hamiltonian cycles, cf. [4]. We note some vague similarity with the property testing paradigm [18].

The Hamiltonian permanent (1) is an example of a *partition function*. For A defined by (3), we can write

$$\operatorname{ham} A = \sum_{\sigma} e^{w(\sigma)}, \tag{5}$$

where the sum is taken over all one-cycle permutations ("configurations") σ and $w(\sigma)$ is the "energy" of the configuration, which is just $\ln(1-\delta)$ times the number of non-edges of G in σ. In this form, (5) fits the profile of a partition function in statistical physics, see, for example, [2]. In Section 5, we talk more about parallels with statistical physics, and, in particular, link the ability to compute the partition function efficiently with the absence of a phase transition. In this paper, we consider two general computational methods, one, underlying the algorithm of Section 1.1, is linked to the complex zeros of the partition function and the other, underlying the algorithm of Section 1.2, is linked to scaling. We do not discuss other approaches, such as the one based on the idea of "correlation decay," see [3, 10, 36].

In Section 2 we discuss computing the permanent of a matrix, which demonstrates what forces are at work in Section 1.1 and Section 1.2 in the most transparent way. Meanwhile, we finish this section with an open question.

1.3 Open question

Let us fix a $0 < \delta < 1$. Is there an algorithm, which, given an $n \times n$ real matrix $A = (a_{ij})$ satisfying (4) and $0 < \epsilon < 1$, computes $\operatorname{ham} A$ within relative error ϵ in polynomial or quasi-polynomial time?

2 A case study: computing the permanent of a matrix

Let $A = (a_{ij})$ be an $n \times n$ real or complex matrix. The *permanent* of A is defined as

$$\operatorname{per} A = \sum_{\sigma \in S_n} \prod_{i=1}^{n} a_{i\sigma(i)}, \tag{6}$$

where S_n is the symmetric group of all $n!$ permutations σ of the set $\{1, \ldots, n\}$. If A is the adjacency matrix of a directed graph G on n vertices, then $\operatorname{per} A$ is the number of directed cycle covers. If $G = (V, E)$ is a bipartite graph with set $V = \{1_L, \ldots, n_L; 1_R, \ldots, n_R\}$ of $2n$ vertices (where "L" stands for "left" and "R" stands for "right") and

$$a_{ij} = \begin{cases} 1 & \text{if } \{i_L, j_R\} \in E \\ 0 & \text{otherwise,} \end{cases}$$

then $\operatorname{per} A$ is the number of perfect matchings (also known as dimer covers) in G. This explains the popularity of permanents of 0–1 matrices in combinatorics, cf. [30] and [27].

Permanents of complex matrices are of interest to physics, in connection with boson statistics, see [1]. The greatest success in efficient approximation of the permanent of a non-negative matrix was achieved via the Markov chain Monte Carlo approach, where a fully polynomial randomized approximation scheme was found [21]. For a general $n \times n$ complex matrix the algorithm of Ryser computes the permanent exactly in $O(n^2 2^n)$ time, see Chapter 7 of [30].

Next, we present two theorems.

Theorem 1. *There is an absolute constant $\delta_0 > 0$ (one can choose $\delta_0 = 0.275$) such that if $A = (a_{ij})$ is an $n \times n$ complex matrix satisfying*

$$|1 - a_{ij}| < \delta_0 \quad \textit{for all} \quad i, j, \tag{7}$$

then

$$\operatorname{per} A \neq 0.$$

Let $\mathcal{U}_n(\delta_0) \subset \mathbf{C}^{n\times n}$ be the polydisc of $n \times n$ matrices $A = (a_{ij})$ satisfying (7). Since $\operatorname{per} A \neq 0$ for all $A \in \mathcal{U}_n(\delta_0)$, we can choose a branch of the function $A \longmapsto \ln \operatorname{per} A$ for $A \in \mathcal{U}_n(\delta_0)$ in such a way that $\ln \operatorname{per} A$ is a real number if $A = J_n$ is the $n \times n$ matrix filled with 1s. It turns out that $\ln \operatorname{per} A$ is well-approximated by a polynomial in the entries of A of a small (logarithmic) degree.

Theorem 2. *Let us fix $0 < \delta < \delta_0$ where δ_0 is the constant of Theorem 1. Then there exists a constant $\gamma = \gamma(\delta) > 0$ and for any $\epsilon > 0$ and integer $n \geq 1$ there exists a polynomial $p = p_{n,\delta,\epsilon}$ in the entries of an $n \times n$ matrix $A = (a_{ij})$ such that*

$$\deg p \leq \gamma(\ln n - \ln \epsilon)$$

and

$$|\ln \operatorname{per} A - p(A)| \leq \epsilon$$

provided

$$|1 - a_{ij}| \leq \delta \quad \textit{for all} \quad i, j.$$

We will see that the polynomial p in Theorem 2 is efficiently computable and hence we get a quasi-polynomial algorithm to approximate $\operatorname{per} A$. Theorem 1 was proved (with a weaker bound of $\delta_0 = 0.195$) in [5]. The enhancement to $\delta_0 = 0.275$ follows from a suggestion of Boris Bukh [12], who noticed that one geometric lemma in [5] can be sharpened. Theorem 2 is also proved in [5], though in a different language. First, we establish that Theorem 1 implies Theorem 2.

2.1 *Theorem 1 implies Theorem 2*

The key argument is provided by the following simple lemma.

Lemma 1. *Let $g(z)$ be a univariate complex polynomial with $\deg g = n$ and let $\beta > 1$ be a real number such that*

$$g(z) \neq 0 \quad \textit{provided} \quad |z| \leq \beta.$$

Let us choose a branch of

$$f(z) = \ln g(z) \quad \textit{for} \quad |z| \leq 1.$$

For an integer $m > 0$, let

$$T_m(z) = f(0) + \sum_{k=1}^{m} \frac{z^k}{k!} \frac{d^k}{dz^k} f(z)\Big|_{z=0}$$

be the degree m Taylor polynomial of $f(z)$ computed at $z = 0$. Then

$$|f(1) - T_m(1)| \leq \frac{n}{(m+1)\beta^m(\beta - 1)}.$$

Proof. We can write

$$g(z) = g(0) \prod_{i=1}^{n} \left(1 - \frac{z}{\alpha_i}\right),$$

where $g(0) \neq 0$ and $\alpha_1, \ldots, \alpha_n$ are the roots of $g(z)$. In particular,

$$|\alpha_i| > \beta \quad \text{for} \quad i = 1, \ldots, n.$$

Then

$$f(1) = f(0) + \sum_{i=1}^{n} \ln\left(1 - \frac{1}{\alpha_i}\right).$$

Expanding the logarithm, we get

$$\ln\left(1 - \frac{1}{\alpha_i}\right) = -\sum_{k=1}^{m} \frac{1}{k\alpha_i^k} + \eta_i \quad \text{where} \quad |\eta_i| \leq \frac{1}{(m+1)\beta^m(\beta - 1)}.$$

Since

$$T_m(1) = -\sum_{i=1}^{n} \sum_{k=1}^{m} \frac{1}{k\alpha_i^k},$$

the proof follows. □

Let J_n denote the $n \times n$ matrix filled with 1s. To deduce Theorem 2 from Theorem 1, we apply Lemma 1 to the polynomial

$$g(z) = \operatorname{per}\big(J_n + z(A - J_n)\big).$$

Choosing $\beta = \delta_0/\delta > 1$ in Lemma 1, we conclude that to approximate $f(1) = \ln \operatorname{per} A$ within an additive error of $\epsilon > 1$, it suffices to compute the degree m Taylor polynomial $T_m(z)$ of $f(z) = \ln g(z)$ for some $m = O(\ln n - \ln \epsilon)$. It remains to show that $T_m(1)$ is in fact a polynomial of degree m in the matrix entries a_{ij}.

We have $f'(z) = g'(z)/g(z)$ and hence $g'(z) = f'(z)g(z)$. Thus

$$g''(z) = f''(z)g(z) + f'(z)g'(z), \quad g^{(3)}(z) = f^{(3)}(z)g(z) + 2f''(z)g'(z) + f'(z)g''(z)$$

and generally

$$g^{(k)}(z) = \sum_{j=1}^{k} \binom{k-1}{j-1} f^{(j)}(z) g^{(k-j)}(z). \tag{8}$$

Since $g(0) = n!$, equations (8) represent a non-singular triangular system of linear equations in the derivatives $f^{(k)}(0)$ and all we need is to write $g^{(k)}(0)$ as a polynomial of degree k in the matrix entries a_{ij}. However,

$$\begin{aligned} g^{(k)}(0) &= \frac{d^k}{dz^k} \sum_{\sigma \in S_n} \prod_{i=1}^{n} \left(1 + z\left(a_{i\sigma(i)} - 1\right)\right) \Big|_{z=0} \\ &= \sum_{\sigma \in S_n} \sum_{i_1,\ldots,i_k} \left(a_{i_1\sigma(i_1)} - 1\right) \ldots \left(a_{i_k\sigma(i_k)} - 1\right) \\ &= (n-k)! \sum_{i_1,\ldots,i_k; j_1,\ldots,j_k} \left(a_{i_1j_1} - 1\right) \ldots \left(a_{i_kj_k} - 1\right), \end{aligned}$$

where the last sum is taken over all pairs of ordered k-tuples of distinct indices $1 \leq i_1, \ldots, i_k \leq n$ and $1 \leq j_1, \ldots, j_k \leq n$. Hence we conclude that $g^{(k)}(0)$ and, therefore, $f^{(k)}(0)$ is a polynomial of degree k in the entries a_{ij} of the matrix A and so $T_m(1)$ is a polynomial of degree m in a_{ij}.

2.2 The idea of the proof of Theorem 1

Our goal is to repeatedly split the terms of the monomial expansion (6) into several groups in such a way so that the sum over each group is a non-zero complex number, and, moreover, the angle between different sums (interpreted as non-zero vectors in $\mathbf{R}^2 = \mathbf{C}$) is sufficiently small, so that the total sum over all groups is also non-zero. We obtain such a splitting via the row expansion

$$\operatorname{per} A = \sum_{j=1}^{n} a_{1j} \operatorname{per} A_j, \tag{9}$$

where A_j is the matrix obtained from A by crossing out the first row and the j-th column, and use the induction on n.

Let us fix some $\delta > 0$ and let $\mathcal{U}_n(\delta) \subset \mathbf{C}^{n\times n}$ be the set of $n \times n$ matrices $A = (a_{ij})$ satisfying (7) with δ_0 replaced by δ. We note that

$$|a_{ij}| \geq 1 - \delta \quad \text{and} \quad |\arg a_{ij}| \leq \arcsin \delta \quad \text{for all} \quad i, j \tag{10}$$

and $A \in \mathcal{U}_n(\delta)$.

Let us fix some real $0 < \tau \leq 1$ and $\theta \geq 0$. We attempt to prove by induction on n the following three statements 2.2.1 – 2.2.3:

Statement 2.2.1 *For any $A \in \mathcal{U}_n(\delta)$ we have* per $A \neq 0$.

Statement 2.2.2 *For any two matrices $A, B \in \mathcal{U}_n(\delta)$ that differ in one row or in one column only, the angle between two non-zero complex numbers* per A *and* per B, *interpreted as vectors in $\mathbf{R}^2$, does not exceed θ.*

Statement 2.2.3 *For any $A \in \mathcal{U}_\delta(A)$, we have*

$$|\operatorname{per} A| \geq \tau \sum_{j=1}^{n} |a_{1j}| \left|\operatorname{per} A_j\right|,$$

where A_j is the matrix obtained from A by crossing out the first row and the j-th column.

Note that if $\tau = 1$ then from (9) it follows that all vectors $a_{1j}\operatorname{per} A_j$ are collinear.

Obviously, Statement 2.2.1 for $\mathcal{U}_n(\delta)$ together with Statement 2.2.3 for $\mathcal{U}_n(\delta)$ implies Statement 2.2.1 for $\mathcal{U}_{n+1}(\delta)$. Next, suppose that Statement 2.2.2 holds for $\mathcal{U}_n(\delta)$ and let us consider a matrix $A \in \mathcal{U}_{n+1}(\delta)$. Since any two matrices A_i and A_j in the row expansion (9) differ in not more than one column, the angle between any two complex numbers per A_i and per A_j does not exceed θ and hence by the second inequality of (10) the angle between any two complex numbers $a_i\operatorname{per} A_i$ and $a_j\operatorname{per} A_j$ in (9) does not exceed $\theta + 2\arcsin\delta$. If $\theta + 2\arcsin\delta < 2\pi/3$, then Statement 2.2.3 holds for $\mathcal{U}_{n+1}(\delta)$ provided

$$\tau \leq \cos\left(\frac{\theta}{2} + \arcsin\delta\right). \tag{11}$$

This observation is due to Boris Bukh [12], in [5] a weaker bound is used. Indeed, one can argue that 0 is not in the convex hull of the vectors $a_j\operatorname{per} A_j$ and projecting orthogonally each of those vectors onto the bisector of the angle enclosed by them, we get a lower bound on $|\operatorname{per} A|$ in terms of $\sum_{j=1}^{n} |a_{1j}||\operatorname{per} A_j|$ and θ. Finally, if Statement 2.2.3 holds for $\mathcal{U}_n(\delta)$, then Statement 2.2.2 holds for $\mathcal{U}_n(\delta)$ provided

$$\theta \geq \frac{2\delta}{(1-\delta)\tau}. \tag{12}$$

Indeed, the inequality of Statement 2.2.3 together with the first inequality of (10) implies that

$$\sum_{j=1}^{n} \left| \frac{\partial}{\partial z_{1j}} \ln \operatorname{per} Z \right| \leq \frac{1}{(1-\delta)\tau}$$

for all $Z \in \mathcal{U}_n(\delta)$. We note that for any two $A, B \in \mathcal{U}_n(\delta)$, where $A = (a_{ij})$ and $B = (b_{ij})$ we have $|a_{1j} - b_{1j}| \leq 2\delta$ for $j = 1, \ldots, n$.

To complete the induction, apart from checking that Statements 2.2.1–2.2.3 hold for $\mathcal{U}_1(\delta)$, we need to find a fixed point (θ, τ) where the inequalities (11)–(12) become equations. In other words, we would like to find a $\delta > 0$ (the larger the better) such that the equation

$$\theta = \frac{2\delta}{(1-\delta)\cos\left(\frac{\theta}{2} + \arcsin\delta\right)}$$

has a solution $\theta < 2\pi/3 - 2\arcsin\delta$. Clearly, it has such a solution for a sufficiently small $\delta > 0$ and numerical computations show that one can choose $\delta = 0.275$.

2.3 Open question

What is the optimal value of δ_0 in Theorem 1? We note that for

$$A = \begin{pmatrix} \frac{1+i}{2} & \frac{1-i}{2} \\ \frac{1-i}{2} & \frac{1+i}{2} \end{pmatrix} \quad \text{we have} \quad \operatorname{per} A = 0$$

and hence we must have

$$\delta_0 < \frac{\sqrt{2}}{2} \approx 0.707.$$

Moreover, as Boris Bukh noticed [12], for A as above we have

$$\operatorname{per}\ (A \otimes J_m) = 0,$$

where J_m is an $m \times m$ matrix filled with 1s and m is odd, so we cannot beat the bound of $\sqrt{2}/2$ by allowing the size of the matrix to grow.

A very useful approach to computing permanents is provided by matrix scaling.

2.4 Matrix scaling

Let $A = (a_{ij})$ be an $n \times n$ matrix with non-negative entries. We say that A is obtained by *scaling* from an $n \times n$ matrix $B = (b_{ij})$ provided for some positive $\lambda_1, \ldots, \lambda_n$ and positive $\mu_1, \ldots, \mu_n$ we have

$$a_{ij} = \lambda_i \mu_j b_{ij} \quad \text{for all} \quad i,j.$$

We note that in this case

$$\operatorname{per} A = \left(\prod_{i=1}^{n} \lambda_i\right)\left(\prod_{j=1}^{n} \mu_j\right) \operatorname{per} B. \tag{13}$$

Any $n \times n$ matrix A with strictly positive entries can be obtained by scaling from a *doubly stochastic* matrix B, that is, a non-negative $n \times n$ matrix with all row and column sums equal to 1. Such a matrix B can be found, for example, as a solution to a convex optimization problem

$$\text{Minimize} \quad \sum_{i,j=1}^{n} x_{ij} \ln \frac{x_{ij}}{a_{ij}} \tag{14}$$

$$\text{Subject to:} \quad X = (x_{ij}) \quad \text{is doubly stochastic,} \tag{15}$$

see [28, 33]. Indeed, it is easy to see that the minimum is attained at a positive doubly stochastic matrix and then the Lagrange multipliers give rise to the scaling factors λ_i and μ_j. Given A, the matrix B is unique, while the scaling factors are unique up to a rescaling $\lambda_i \longmapsto \lambda_i \tau$, $\mu_j \longmapsto \mu_j \tau^{-1}$ for some $\tau > 0$.

The problem (14)–(15) in fact leads to a polynomial time algorithm to find a doubly stochastic B and the scaling factors λ_i and μ_j [22]. An even faster (combinatorial) algorithm is obtained in [25].

As is observed in [25], the identity (13) allows one to estimate the permanent of an $n \times n$ positive (and hence, by continuity, non-negative) matrix within a factor of e^n. Indeed,

$$\frac{n!}{n^n} \leq \operatorname{per} B \leq 1 \quad \text{provided} \quad B \quad \text{is doubly stochastic,} \tag{16}$$

where the first inequality is the van der Waerden bound, see [19] for a short proof and a powerful generalization. A more sophisticated upper bound allows one to estimate the permanent within a factor of 2^n (and conjectured factor $2^{n/2}$) [20].

Scaling provides a much more accurate estimate of the permanent if the matrix entries are within a constant factor of each other.

Definition 1. Let $0 < \epsilon < 1$. We say that a matrix $A = (a_{ij})$ with positive entries is ϵ-balanced if

$$\epsilon a_{i_1 j_1} \leq a_{i_2 j_2} \quad \text{for all} \quad i_1, j_1, i_2, j_2.$$

Theorem 3. *Let $A = (a_{ij})$ be an ϵ-balanced $n \times n$ matrix and let $B = (b_{ij})$ be the doubly stochastic matrix obtained from A by scaling. Then B is ϵ^3-balanced and*

$$b_{ij} \leq \frac{1}{n\epsilon^2} \quad \textit{for all} \quad i, j.$$

Proof. We have

$$b_{ij} = \lambda_i \mu_j a_{ij} \quad \text{for all} \quad i, j \tag{17}$$

and some positive $\lambda_1, \ldots, \lambda_n$ and $\mu_1, \ldots, \mu_n$. Then

$$\frac{b_{ij_1}}{b_{ij_2}} = \frac{\mu_{j_1}}{\mu_{j_2}} \frac{a_{ij_1}}{a_{ij_2}} \leq \frac{\mu_{j_1}}{\mu_{j_2}} \frac{1}{\epsilon} \quad \text{for all} \quad i, j_1, j_2. \tag{18}$$

Since

$$\sum_{i=1}^{n} b_{ij_1} = \sum_{i=1}^{n} b_{ij_2} = 1,$$

we conclude that

$$\frac{\mu_{j_1}}{\mu_{j_2}} \geq \epsilon \quad \text{for all} \quad j_1, j_2. \tag{19}$$

Similarly we prove

$$\frac{\lambda_{i_1}}{\lambda_{i_2}} \geq \epsilon \quad \text{for all} \quad i_1, i_2. \tag{20}$$

Since A is ϵ-balanced, combining (17), (19), and (20), we conclude that B is ϵ^3-balanced. Since the row sums of B are 1s, for every i there is a j such that $b_{ij} \leq 1/n$. It follows by (18) and (19) that $b_{ij} \leq 1/n\epsilon^2$ for all i and j. □

The permanent of an $n \times n$ doubly stochastic matrix with uniformly small entries strongly concentrates about e^{-n}.

Theorem 4. *Let B be an $n \times n$ stochastic matrix (non-negative matrix with row sums 1) such that*

$$b_{ij} \leq \frac{1}{m} \tag{21}$$

for some integer $m > 1$ and all i, j. Then

$$\operatorname{per} B \leq \frac{(m!)^{n/m}}{m^n} \leq (9m)^{\frac{n}{2m}} e^{-n}.$$

It follows from Theorem 3 and Theorem 4 that scaling allows one to approximate the permanent of an ϵ-balanced matrix A within a factor of $n^{O(1/\epsilon^2)}$. Indeed, if B is the doubly stochastic matrix obtained from A by scaling, then in Theorem 4 we can choose $m = \lfloor n\epsilon^2 \rfloor$ and then the lower bound of (16) together with the upper bound of Theorem 4 provides an approximation of per B and hence of per A to within the desired factor.

The proof of Theorem 4 is based on the Bregman - Minc upper bound for permanents of 0–1 matrices,

$$\operatorname{per} A \leq \prod_{i=1}^{n} (r_i!)^{1/r_i}, \tag{22}$$

where $r_1, \ldots, r_n$ are the row sums of a 0–1 matrix A, see, for example, Chapter 11 of [35]. Since the function $B \longmapsto \operatorname{per} B$ is linear in each row of B, the maximum of the function on the polyhedron of stochastic matrices satisfying (21) is attained at a vertex of the polyhedron, that is, where $b_{ij} \in \{0, 1/m\}$ for all i, j. The inequality of Theorem 4 can be extended to the case of non-integer m [34].

We conclude this section by noting that a certain combination of scaling and the Taylor polynomial approximation to compute permanents is explored in [29].

3 Back to Hamiltonian cycles

Let us go back to Section 1 and see how the methods developed in Section 2 can be applied to the Hamiltonian permanent. One can obtain results mirroring Theorem 1 and Theorem 2 for the Hamiltonian permanent. Namely, one can prove that ham $A \neq 0$ if $A = (a_{ij})$ is an $n \times n$ symmetric complex matrix such that

$$\left|1 - a_{ij}\right| \leq \delta_1 \quad \text{for all} \quad i, j,$$

where $\delta_1 > 0$ is an absolute constant (one can choose $\delta_1 = 0.051$). Then, using Lemma 1, we obtain a quasi-polynomial algorithm to approximating ham A as long as A satisfies (4) for any $0 < \delta < \delta_1$, fixed in advance. This is how we obtain the algorithm of Section 1.1. More generally, in [9] we consider the following functional.

Let $G = (V, E)$ be an undirected graph with set V of vertices, set E of edges, without loops or multiple edges. Let $m = (\mu_1, \ldots, \mu_k)$ be a vector of positive integers such that

$$\mu_1 + \ldots + \mu_k = |V|,$$

and let $A = (a_{ij})$ be a $k \times k$ complex symmetric matrix. We define the *partition function of graph homomorphisms of G with multiplicities m*:

$$P_{G,m}(A) = \sum_{\substack{\phi : V \longmapsto \{1, \ldots, k\} \\ |\phi^{-1}(i)| = \mu_i \quad \text{for} \quad i = 1, \ldots, k}} \prod_{\{u,v\} \in E} a_{\phi(u)\phi(v)}. \tag{23}$$

Here the sum is taken over all maps $\phi : V \longrightarrow \{1, \ldots, k\}$ that map precisely μ_i vertices of G into i for all $i = 1, \ldots, k$ and the product is taken over all edges of G. Hence $P_{G,m}(A)$ is a polynomial in the matrix entries and $\deg P_{G,m}(A) = |E|$. As is shown in [9], the polynomial $P_{G,m}(A)$ specializes to various interesting quantities, such as the number of independent sets of a given cardinality in G or the number of proper colorings of G with prescribed number of vertices colored in each color. It is a refinement of a much studied graph homomorphism partition function, see [13, 14], and [26].

In particular, if $|V| = k$, $\mu_1 = \ldots = \mu_k = 1$, and G consists of a single cycle, we have $P_{G,M}(A) = k(\text{ham } A)$.

Theorem 5. *There exists an absolute constant $\alpha > 0$ (one can choose $\alpha = 0.107$) such that for any undirected graph $G = (V, E)$ with maximum vertex degree $\Delta(G)$, for any vector $m = (\mu_1, \ldots, \mu_k)$ of positive integers such that $\mu_1 + \ldots + \mu_k = |V|$, and for any $k \times k$ complex symmetric matrix $A = (a_{ij})$ satisfying*

$$|1 - a_{ij}| < \frac{\alpha}{\Delta(G)} \quad \textit{for all} \quad 1 \leq i, j \leq k$$

one has

$$P_{G,m}(A) \neq 0.$$

Theorem 5 is proved in [9]. While the proof is much more cumbersome than that of Theorem 1 outlined in Section 2.2, by and large it follows a similar scheme. As in Section 2.1, we obtain the following corollary, cf. [9].

Theorem 6. *Let $G = (V, E)$, $m = (\mu_1, \ldots, \mu_k)$, $A = (a_{ij})$, and α be as defined in Theorem 5. Let us fix*

$$0 < \delta < \frac{\alpha}{\Delta(G)}.$$

Then there exists $\gamma = \gamma(\delta) > 0$ and for any $\epsilon > 0$, G, m, and k there exists a polynomial $p = p_{\epsilon,\delta,G,m}$ in the entries of a $k \times k$ symmetric matrix A such that

$$\deg p \leq \gamma(\ln |E| - \ln \epsilon)$$

and

$$|\ln P_{G,m}(A) - p(A)| \leq \epsilon$$

provided

$$|1 - a_{ij}| \leq \delta \quad \textit{for all} \quad i, j.$$

Moreover, for any δ, fixed in advance, given G and $\epsilon > 0$, the polynomial p can be computed in $(k|E|)^{O(\ln |E| - \ln \epsilon)}$ time (the constant implicit in the "O" notation depends on δ). The proof more or less follows the method of Section 2.1.

The algorithm of Section 1.2 is based on the following result proved in [4] for ϵ-balanced matrices, see Definition 1.

Theorem 7. *Let $A = (a_{ij})$ be an ϵ-balanced $n \times n$ matrix and let*

$$r = \left\lfloor \frac{4 \ln n}{\epsilon^2} \right\rfloor + 6.$$

Then

$$\frac{1}{2r}\left(\frac{\epsilon}{n}\right)^r \operatorname{per} A \leq \operatorname{ham} A \leq \operatorname{per} A.$$

Thus, for any fixed $\epsilon > 0$, the value of ham A is within a factor of $n^{O(\ln n)}$ from the value of per A and hence to approximate ham A within a factor of $n^{O(\ln n)}$, we can use the matrix scaling algorithm of Section 2.4.

The proof of Theorem 7 is based on the following combinatorial lemma.

Lemma 2. *Let A be an ϵ-balanced $n \times n$ matrix. Let us consider the symmetric group S_n as a finite probability space endowed with probability measure*

$$\mathbf{P}(\sigma) = (\operatorname{per} A)^{-1} \prod_{i=1}^{n} a_{i\sigma(i)} \quad \textit{for} \quad \sigma \in S_n.$$

For $i = 1, \ldots, n$ let $l_i(\sigma)$ be the length of the cycle of σ that contains i. Then

$$\mathbf{P}\big(\sigma \in S_n : \quad l_i(\sigma) = m\big) \leq \frac{1}{\epsilon^2(n-m)} \quad \textit{for} \quad i = 1, \ldots, n$$

and $m = 1, \ldots, n-1$.

Proof. Without loss of generality, we assume that $i = 1$. Let $X \subset S_n$ be the set of permutations such that $l_1(\sigma) = m$. We construct a set $Y \subset S_n$ so that each $\sigma \in X$ contributes $n - m$ permutations to Y. To that effect, given a $\sigma \in X$, we write the cycle of σ containing 1 as $1 = j_1 \mapsto j_2 \mapsto \ldots \mapsto j_m \mapsto 1$, pick any of the $n - m$ numbers, say r, not in the cycle, write the cycle containing r as $r = j_{m-1} \mapsto j_{m+2} \mapsto \ldots \mapsto j_{m+k} \mapsto r$, and produce a permutation $\tau \in Y$ by merging the two cycles together: $1 = j_1 \mapsto \ldots \mapsto j_m \mapsto j_{m+1} \mapsto j_{m+2} \mapsto \ldots \mapsto j_{m+k} \mapsto 1$. We observe that $\mathbf{P}(\tau) \geq \epsilon^2\mathbf{P}(\sigma)$ and that from each permutation $\tau \in Y$ we can reconstruct $\sigma \in X$ in a unique way: we pick the cycle of τ that contains 1, write it as $1 = j_1 \mapsto j_2 \mapsto \ldots \mapsto j_{m+k} \mapsto 1$, and cut it in two, $1 = j_1 \mapsto j_2 \mapsto \ldots \mapsto j_m \mapsto j_1$ and $j_{m+1} \mapsto j_{m+2} \mapsto \ldots j_{m+k} \mapsto j_{m+1}$. This shows that

$$\mathbf{P}(X) \leq \frac{1}{\epsilon^2(n-m)}\mathbf{P}(Y) \leq \frac{1}{\epsilon^2(n-m)}.$$

□

Lemma 2 implies that

$$\mathbf{E}\left(l_i^{-1}\right) = O\left(\frac{\ln n}{\epsilon^2 n}\right) \quad \text{for} \quad i = 1, \ldots, n.$$

Since the number of cycles of $\sigma \in S_n$ can be written as $l_1^{-1}(\sigma) + \ldots + l_n^{-1}(\sigma)$, we conclude that the expected number of cycles of $\sigma \in S_n$ is $O\left(\epsilon^{-2} \ln n\right)$, and hence by the Markov inequality, the bulk of per A is concentrated on permutations σ with $O\left(\epsilon^{-2} \ln n\right)$ cycles. Patching the cycles of such a permutation σ into a Hamiltonian cycle, we complete the proof of Theorem 7.

4 Higher-dimensional permanents

Let $A = (a_{i_1 i_2 \ldots i_d})$ be a d-dimensional $n \times \ldots \times n$ array of complex numbers. Similarly to the case of $d = 2$, we define the *d-dimensional permanent* of A by

$$\text{PER } A = \sum_{\sigma_2, \ldots, \sigma_d \in S_n} \prod_{i=1}^{n} a_{i\sigma_2(i_2)\ldots\sigma_d(i_d)}.$$

If A is a 0–1 array, then PER A enumerates perfect matchings in the uniform d-partite hypergraph encoded by A. Consequently, for $d \geq 3$, it is an NP-complete problem to tell the permanent of a 0–1 array from 0.

Some of the results and methods developed for permanents readily extend to d-dimensional permanents, see [7] and [5].

Theorem 8. *For an integer $d \geq 2$, there is $\delta_d > 0$ such that if $A = (a_{i_1 \ldots i_d})$ is an array of complex numbers satisfying*

$$|1 - a_{i_1 \ldots i_d}| < \delta_d \quad \textit{for all} \quad i_1, \ldots, i_d,$$

then

$$\mathrm{PER}\, A \neq 0.$$

For any $0 < \delta < \delta_d$ there is $\gamma = \gamma(\delta, d) > 0$ and for any $\epsilon > 0$ and any n there is a polynomial $p = p_{d,n,\epsilon,\delta}$ in the entries of an array A such that

$$\deg p \leq \gamma \left(\ln n - \ln \epsilon\right)$$

and

$$\left|\ln \mathrm{PER}\, A - p(A)\right| \leq \epsilon$$

provided

$$|1 - a_{i_1 \ldots i_d}| < \delta \quad \textit{for all} \quad i_1, \ldots, i_d.$$

The proof follows the scheme outlined in Section 2.1 and Section 2.2.

Let us choose a pair of integers $1 \leq k \leq d$ and $1 \leq j \leq n$. The set of entries of A with the k-th index equal to j we call the *(k, j) slice* of A.

Instead of the row expansion (9), we use the "(1, 1) slice expansion"

$$\mathrm{PER}\, A = \sum_{1 \leq i_2, \ldots, i_d \leq n} a_{1 i_2 \ldots i_d} \mathrm{PER}\, A_{i_2 \ldots i_d}, \tag{24}$$

where $A_{i_2 \ldots i_d}$ is the d-dimensional $(n-1) \times \ldots \times (n-1)$ array obtained by crossing out all d slices containing $a_{1 i_2 \ldots i_d}$.

The value of δ_d can be found as the largest $\delta > 0$ for which the equation

$$\theta = \frac{2\delta}{(1-\delta)\cos\left(\frac{(d-1)\theta}{2} + \arcsin \delta\right)}$$

has a solution $\theta \geq 0$ such that $(d-1)\theta + 2\arcsin\delta < 2\pi/3$ (note that any two different subarrays $A_{i_2\ldots i_d}$ and $A_{j_2\ldots j_d}$ in (24) differ in $(d-1)$ slices). Thus we can choose $\delta_2 = 0.275$, $\delta_3 = 0.181$, and $\delta_4 = 0.135$. As in Section 2.1, for any fixed $\delta < \delta_d$, the polynomial p can be constructed in $n^{O(\ln n - \ln \epsilon)}$ time.

We say that A is obtained from B by *scaling* if

$$a_{i_1\ldots i_d} = \lambda_{1i_1}\ldots\lambda_{di_d} b_{i_1\ldots i_d} \tag{25}$$

for some real $\lambda_{kj} > 0$ for $1 \leq k \leq d$ and $1 \leq j \leq n$. In that case,

$$\text{PER}\, A = \left(\prod_{k=1}^{d}\prod_{j=1}^{n}\lambda_{kj}\right)\text{PER}\, B.$$

We say that A is *d-stochastic* if the entries of A are non-negative and the sum of entries in every slice is 1. Again, it is not hard to show that every cubical array A with positive entries can be scaled to a d-stochastic array B, which can be obtained as a solution to the convex optimization problem:

$$\text{Minimize} \quad \sum_{1\leq i_1,\ldots,i_d\leq n} x_{i_1\ldots i_d}\ln\frac{x_{i_1\ldots i_d}}{a_{i_1\ldots i_d}} \tag{26}$$

$$\text{Subject to:} \quad X = (x_{i_1\ldots i_d}) \quad \text{is } d\text{-stochastic}, \tag{27}$$

see also [17].

For $d \geq 3$ one can construct d-stochastic arrays B such that PER B $=$ 0. However, the situation changes dramatically for ϵ-balanced arrays (we adapt Definition 1 for arrays in the obvious way). The following two results are proved in [7].

Theorem 9. *Let A be an ϵ-balanced d-dimensional array and let B be a d-stochastic array obtained from A by scaling. Then B is ϵ^{d+1}-balanced.*

The proof is a straightforward extension of that of Theorem 3.

Theorem 10. *For any $0 < \epsilon < 1$ and any integer $d > 1$ there exists $\gamma = \gamma(d,\epsilon) >$ 0 such that if B is an ϵ-balanced d-stochastic $n \times \ldots \times n$ array then*

$$n^{-\gamma}e^{-n(d-1)} \leq \text{PER}\, B \leq n^{\gamma}e^{-n(d-1)}.$$

Hence for any $0 < \epsilon < 1$, fixed in advance, scaling allows us to approximate permanents of ϵ-balanced d-dimensional $n \times \ldots \times n$ arrays within a polynomial in n factor, just as in the case of $d = 2$ considered in Section 2.4.

The proof of Theorem 10 is based on the observation that if an ϵ-balanced array is close to d-stochastic, then the product of scaling factors is very close to 1.

Lemma 3. *For any* $0 < \epsilon < 1$ *and any integer* $d > 1$ *there exists* $\beta = \beta(d, \epsilon) > 0$ *such that the following holds. Suppose that* A *is an* ϵ*-balanced* d*-dimensional* $n \times \ldots \times n$ *array with the total sum of entries equal to* n. *Suppose further that the sum of entries in the* (k, j) *slice of* A *is* $1 - \delta_{kj}$ *where*

$$|\delta_{kj}| \leq \frac{1}{\epsilon n} \quad \textit{for} \quad k = 1, \ldots, d \quad \textit{and} \quad j = 1, \ldots, n.$$

Let B *be the* d*-stochastic array obtained from* A *by scaling and let* $\lambda_{kj} > 0$ *be the scaling factors for* $k = 1, \ldots, d$ *and* $j = 1, \ldots, n$ *so that (25) holds. Then*

$$0 \leq \sum_{k=1}^{d} \sum_{j=1}^{n} \ln \lambda_{kj} \leq \frac{\beta}{n}. \tag{28}$$

Sketch of proof. We observe that the optimal value (26) in the optimization problem (26)–(27) is $d \sum_{k=1}^{d} \sum_{j=1}^{n} \ln \lambda_{kj}$. Since the sum of the entries of A is n, we have

$$\sum_{j=1}^{n} \delta_{kj} = 0 \quad \text{for} \quad k = 1, \ldots, d.$$

Therefore the sum of entries in every slice of the array $X = (x_{i_1 \ldots i_d})$ defined by

$$x_{i_1 \ldots i_d} = a_{i_1 \ldots i_d} + \frac{1}{n^{d-1}} \sum_{k=1}^{d} \delta_{k i_k}$$

is equal to 1. Since A is ϵ-balanced, for all sufficiently large n, the entries of X are positive and hence X is a feasible point in (27). Substituting X into the objective function (26), we obtain the upper bound in (28).

To get the lower bound of 0, we observe that $\sum_{k=1}^{d} \sum_{j=1}^{n} \ln \lambda_{kj}$ is the value of the convex dual optimization problem

$$\text{Maximize} \quad \sum_{k=1}^{d} \sum_{j=1}^{n} x_{kj}$$

$$\text{Subject to:} \quad \sum_{1 \leq i_1, \ldots, i_d \leq n} a_{i_1 \ldots i_d} \exp\left\{ \sum_{k=1}^{d} x_{k i_k} \right\} \leq n.$$

Since $x_{kj} = 0$ is a feasible point, we get the lower bound in (28). □

Sketch of proof of Theorem 10. All constants implicit in the "O" notation below depend on d and ϵ only. We prove by induction on n that

$$\text{PER } B = \exp\left\{-n(d-1) + O\left(\sum_{j=1}^{n} \frac{1}{j}\right)\right\}. \tag{29}$$

We use the expansion (24). Namely, let $B_{i_2 \ldots i_d}$ be the $(n-1) \times \ldots \times (n-1)$ array obtained from B by crossing out all d slices containing $b_{1 i_2 \ldots i_d}$. Then the sum of entries in every slice of $B_{i_2 \ldots i_d}$ is $1 + O(1/n)$ while the total sum of entries is $s = n - d + O(1/n)$. Let us scale $B_{i_2 \ldots i_d}$ to the total sum of entries $n-1$:

$$A_{i_2 \ldots i_d} = \frac{n-1}{s} B_{i_2 \ldots i_d},$$

and hence

$$\text{PER } B_{i_2 \ldots i_d} = \exp\left\{-(d-1) + O\left(\frac{1}{n}\right)\right\} \text{PER } A_{i_2 \ldots i_d}. \tag{30}$$

We note that the sum of entries in every slice of $A_{i_2 \ldots i_d}$ is still $1 + O(1/n)$. Scaling $A_{i_2 \ldots i_d}$ to a d-stochastic array (which will be ϵ^{d+1}-balanced by Theorem 9) and applying the induction hypothesis and Lemma 3, we obtain

$$\text{PER } A_{i_2 \ldots i_d} = \exp\left\{-n(d-1) + O\left(\sum_{j=1}^{n-1} \frac{1}{j}\right)\right\}. \tag{31}$$

Since B is d-stochastic, by (24) we conclude that PER B is a convex combination of PER $B_{i_2 \ldots i_d}$. Combining (31) and (30) we get (29). Note that at any given moment we work with arrays obtained from subarrays of B by scaling. By Theorem 9 all those arrays are ϵ^{d+1}-balanced and the induction can proceed. □

We conclude this section by noting that while lower bounds for permanents of d-dimensional arrays for $d \geq 3$ are usually very hard to come by, there were successful applications of the "entropy method" to obtain upper bounds, similar to the Bregman - Minc bound (22), see [15, 32], and [24].

5 More examples, connections to physics and the limits of computability

The following partition function was considered in [6]. Let $A = (a_{ij})$ be an $n \times n$ non-negative symmetric matrix. For an integer $2 \leq m \leq n$, we define

$$P_m(A) = \sum_{|S|=m} \prod_{\{i,j\} \subset S} a_{ij},$$

where the sum is taken over all m-subsets $S \subset \{1, \ldots, n\}$ and the product over all unordered pairs $\{i,j\}$ of distinct indices $i, j \in S$. In particular, if A is the adjacency matrix of an undirected graph G on n vertices, without multiple edges or loops, then $P_m(W)$ is the number of *cliques* of size m in G, that is, the number of m-subsets S of the set of vertices, every two of which are connected by an edge of G. It is shown in [6] that if

$$\left|1 - a_{ij}\right| < \frac{\gamma}{m-1} \quad \text{for all} \quad i, j,$$

then $P_m(A)$ can be computed within an absolute error ϵ in $n^{O(\ln n - \ln \epsilon)}$ time. Here $\gamma > 0$ is an absolute constant, one can choose $\gamma = 0.07$ and if $m \geq 10$ and $n \geq 8m$, one can choose $\gamma = 0.27$ (one can also obtain this result, with weaker constants though, from Theorem 6). Given a graph $G = (V, E)$ with set $V = \{1, \ldots, n\}$ of vertices and set $E \subset \binom{V}{2}$ of edges, let us define A by

$$a_{ij} = \begin{cases} 1 + \frac{\gamma}{m-1} & \text{if } \{i,j\} \in E \\ 1 - \frac{\gamma}{m-1} & \text{otherwise.} \end{cases}$$

Then one can write

$$e^{\gamma m} \left(1 + \frac{\gamma}{m-1}\right)^{-\binom{m}{2}} P_m(A) = \sum_{|S|=m} \exp\left\{\gamma m \sigma(S) + O\left(\frac{1}{m}\right)\right\}, \tag{32}$$

where $0 \leq \sigma(S) \leq 1$ is the density of the subgraph induced by S, that is, the ratio of the number of edges of S with both endpoints in S to the maximum number $\binom{m}{2}$ of edges spanned by an m-set of vertices.

The right-hand side of (32) looks like a classical partition function from statistical physics, with "$-\sigma(S)$" playing the role of the energy of a configuration S and "γm" playing the role of the inverse temperature, cf. [2]. From this point of view, an algorithm of the type we discussed in Section 2.1 is interpreted as an interpolation of the logarithm of the partition function from higher (in fact, infinitely high) to lower temperatures. Lemma 1 asserts that as long as we stay away from complex zeros of the partition function, the interpolation works very well. Since [23] and [37], complex zeros of the partition function are linked to *phase transition*. Thus we claim that the logarithm of the partition function at lower temperatures is well-approximated by a low-degree Taylor polynomial of the logarithm of the partition function at higher temperatures as long as we avoid phase transition.

The bound for the constant γ above, as well as bounds in Theorem 1 and Theorem 5 appears to be more artifacts of the method of proof rather than inherent to the problem. For instance, there seems to be no reason why (32) cannot be efficiently computable for any $\gamma > 0$, fixed in advance. Nevertheless, there are examples where computational complexity imposes some bounds.

Let $G = (V, E)$ be a graph as above and let us consider the partition function $P_{G,m}(A)$ of graph homomorphisms with multiplicities, see (23). We restrict ourselves to the class of graphs with $\Delta(G) \leq 3$, in which case Theorem 6 asserts that $P_{G,m}(A)$ is efficiently computable as long as the entries of A are within a certain constant $\delta > 0$ from 1. We claim that there are computational complexity barriers that do not allow us to choose δ arbitrarily close to 1. Let us choose $k = 2$, $m = (\mu_1, \mu_2)$, and for $0 < \tau < 1$ let us define a 2×2 matrix A_τ by

$$a_{ij} = \begin{cases} \tau & \text{if } i = j \\ 1 & \text{if } i \neq j. \end{cases} \tag{33}$$

The *size of the cut* associated with a set $S \subset V$ of vertices is the number of edges in G with precisely one endpoint in S. It is shown in [8] and indeed not hard to see that the value of

$$\frac{\ln P_{G,m}(A_\tau)}{\ln(1/\tau)} + |E|$$

approximates the size of the maximum cut in G associated with a set S of cardinality μ_1 within an additive error of $|V| \ln 2 / \ln(1/\tau)$. Approximating the maximum size of a cut in connected graphs with $\Delta(G) \leq 3$ to within a certain absolute constant factor $\kappa > 1$ is NP-hard [11]. It follows that computing $P_G(A)$ is hard when τ in (33) is sufficiently close to 0.

Acknowledgements This work was partially supported by NSF Grant DMS 1361541.

References

1. S. Aaronson, A. Arkhipov, The computational complexity of linear optics. Theory Comput. **9**, 143–252 (2013)
2. R. Azencott, Simulated annealing. Astérisque **1987/88**(161–162), 223–237 (1988)
3. A. Bandyopadhyay, D. Gamarnik, Counting without sampling: asymptotics of the log-partition function for certain statistical physics models. Random Struct. Algorithm **33**(4), 452–479 (2008)
4. A. Barvinok, On testing Hamiltonicity of graphs. Discret. Math. **338**(1), 53–58 (2015)
5. A. Barvinok, Computing the permanent of (some) complex matrices. Found. Comput. Math. doi:10.1007/s10208-014-9243-7 (2015, to appear)
6. A. Barvinok, Computing the partition functions for cliques in a graph. Theory Comput., **11**, 339–355 (2015) doi:10.4086/toc.2015.v011a013

7. A. Barvinok, A. Samorodnitsky, Computing the partition function for perfect matchings in a hypergraph. Comb. Probab. Comput. **20**(6), 815–835 (2011)
8. A. Barvinok, P. Soberón, Computing the partition function for graph homomorphisms. Combinatorica, to appear. Preprint `arXiv:1410.1842` (2014)
9. A. Barvinok, P. Soberón, Computing the partition functions for graph homomorphisms with multiplicities. Journal of Combinatorial Theory, Series A, **137**, 1–26 (2016)
10. M. Bayati, D. Gamarnik, D. Katz, C. Nair, P. Tetali, Simple deterministic approximation algorithms for counting matchings, in *STOC07 Proceedings of the 39th Annual ACM Symposium on Theory of Computing* (ACM, New York, 2007), pp. 122–127
11. P. Berman M. Karpinski, On some tighter inapproximability results (extended abstract), in *Automata, languages and programming (Prague, 1999)*. Lecture Notes in Computer Science, vol. 1644 (Springer, Berlin, 1999), pp. 200–209
12. B. Bukh, personal communication (2015)
13. A. Bulatov, M. Grohe, The complexity of partition functions. Theor. Comput. Sci. **348**(2–3), 148–186 (2005)
14. J.-Y. Cai, X. Chen, P. Lu, Graph homomorphisms with complex values: a dichotomy theorem. SIAM J. Comput. **42**(3), 924–1029 (2013)
15. B. Cuckler, J. Kahn, Entropy bounds for perfect matchings and Hamiltonian cycles. Combinatorica **29**(3), 327–335 (2009)
16. L. Engebretsen, M. Karpinski, TSP with bounded metrics. J. Comput. Syst. Sci. **72**(4), 509–546 (2006)
17. S. Friedland, Positive diagonal scaling of a nonnegative tensor to one with prescribed slice sums. Linear Algebra Appl. **434**, 1615–1619 (2011)
18. O. Goldreich, S. Goldwasser, D. Ron, Property testing and its connection to learning and approximation. J. ACM **45**(4), 653–750 (1998)
19. L. Gurvits, Van der Waerden/Schrijver-Valiant like conjectures and stable (aka hyperbolic) homogeneous polynomials: one theorem for all. With a corrigendum. Electron. J. Comb. **15**(1), Research Paper 66, 26 pp (2008)
20. L. Gurvits, A. Samorodnitsky, Bounds on the permanent and some applications, in *2014 IEEE 55th Annual Symposium on Foundations of Computer Science (FOCS)* (2014), pp. 90–99. doi:10.1109/FOCS.2014.18
21. M. Jerrum, A. Sinclair, E. Vigoda, A polynomial-time approximation algorithm for the permanent of a matrix with nonnegative entries. J. ACM **51**(4), 671–697 (2004)
22. B. Kalantari, L. Khachiyan, On the complexity of nonnegative-matrix scaling. Linear Algebra Appl. **240**, 87–103 (1996)
23. T.D. Lee, C.N. Yang, Statistical theory of equations of state and phase transitions. II. Lattice gas and Ising model. Phys. Rev. (2), **87**, 410–419 (1952)
24. N. Linial, Z. Luria, An upper bound on the number of high-dimensional permutations. Combinatorica **34**(4), 471–486 (2014)
25. N. Linial, A. Samorodnitsky, A. Wigderson, A deterministic strongly polynomial algorithm for matrix scaling and approximate permanents. Combinatorica **20**(4), 545–568 (2000)
26. L. Lovász, *Large Networks and Graph Limits*. American Mathematical Society Colloquium Publications, vol. 60 (American Mathematical Society, Providence, 2012)
27. L. Lovász, M.D. Plummer, *Matching Theory* (AMS Chelsea Publishing, Providence, 2009). Corrected reprint of the 1986 original
28. A.W. Marshall, I. Olkin, Scaling of matrices to achieve specified row and column sums. Numerische Mathematik **12**, 83–90 (1968)
29. P. McCullagh, An asymptotic approximation for the permanent of a doubly stochastic matrix. J. Stat. Comput. Simul. **84**(2), 404–414 (2014)
30. H. Minc, *Permanents, Encyclopedia of Mathematics and Its Applications*, vol. 6 (Addison-Wesley, Reading, 1978)
31. C.H. Papadimitriou, M. Yannakakis, The traveling salesman problem with distances one and two. Math. Oper. Res. **18**(1), 1–11 (1993)

32. J. Radhakrishnan, An entropy proof of Bregman's theorem. J. Comb. Theory Ser. A **77**(1), 161–164 (1997)
33. U.G. Rothblum, H. Schneider, Scalings of matrices which have prespecified row sums and column sums via optimization. Linear Algebra Appl. **114/115**, 737–764 (1989)
34. G.W. Soules, New permanental upper bounds for nonnegative matrices. Linear Multilinear Algebra **51**(4), 319–337 (2003)
35. J.H. van Lint, R.M. Wilson, *A Course in Combinatorics*, 2nd edn. (Cambridge University Press, Cambridge, 2001)
36. D. Weitz, Counting independent sets up to the tree threshold, in: *STOC06: Proceedings of the 38th Annual ACM Symposium on Theory of Computing* (ACM, New York, 2006), pp. 140–149
37. C.N. Yang, T.D. Lee, Statistical theory of equations of state and phase transitions. I. Theory of condensation. Phys. Rev. (2), **87**, 404–409 (1952)

Partition analysis, modular functions, and computer algebra

Peter Paule and Silviu Radu

Abstract This article describes recent developments connecting problems of enumerative combinatorics, constrained by linear systems of Diophantine inequalities, with number theory topics like partitions, partition congruences, and q-series identities. Special emphasis is put on the role of computer algebra algorithms. The presentation is intended for a broader audience; to this end, elementary introductions to notions like modular functions and to algorithmic aspects of algebra are given.

Keywords MacMahon's partition analysis • Linear systems of Diophantine inequalities and equations • Modular functions • Ramanujan congruences • Discrete geometry • Computer algebra • Symbolic combinatorics

Mathematics Subject Classification (2010): 05A17, 05A20, 05E45, 11D75, 11P83, 11P81, 68W30

1 Introduction

As indicated by the title, this article has a relatively wide topical range which reaches from enumerative combinatorics and linear systems of Diophantine inequalities to number theoretic themes like partitions, partition congruences, and q-series identities. From the methods point of view, despite relying also on analytic concepts like modular functions, special emphasis is put on transforming the analytic framework into algebra, in particular, into computer algebra tools like the Ramanujan-Kolberg package to compute q-identities as witnesses for divisibility properties of partition numbers. The underlying mathematics of the Omega package is more on the algebraic side: semigroups, posets, etc. Omega is an implementation of MacMahon's method of partition analysis, having strong connections also to aspects of discrete

P. Paule (✉) • S. Radu
Research Institute for Symbolic Computation (RISC),
Johannes Kepler University, 4040 Linz, Austria
e-mail: Peter.Paule@risc.uni-linz.ac.at; Silviu.Radu@risc.uni-linz.ac.at

A. Beveridge et al. (eds.), *Recent Trends in Combinatorics*, The IMA Volumes in Mathematics and its Applications 159, DOI 10.1007/978-3-319-24298-9_21

geometry. The objective of this article is to provide an introduction to several recent developments and trends in these areas. The explanatory style of the exposition is chosen to attract also non-expert readers.

To illustrate the possible scope of applications, we quote a problem from Polya [21, Example 5]: "The three sides of a triangle are of lengths l, m, and n, respectively. The numbers l, m, and n are positive integers, $l \leq m \leq n$. Find the number of different triangles of the described kind for a given n. Find a general law governing the dependence of the number of triangles on n." The answer to this problem can be easily extracted from

$$\sum_{\substack{1\leq a\leq b\leq c\\ \text{s.t. } a+b>c}} x^a y^b z^c = \frac{xyz}{(1-yz)(1-xyz)(1-xyz^2)},$$

a relation which, as explained with other examples below, can be easily computed with the Omega package; see also [5]. Apart from elementary problems like this, partition analysis can be used in far more challenging contexts, for instance, as we shall see in Section 3 for the construction of combinatorial objects having modular forms as generating functions.

As a second example consider $p(n)$, the number of partitions of n; for instance, $p(4) = 5$ since $4 = 3+1 = 2+2 = 2+1+1 = 1+1+1+1$. In view of $p(9) = 30, p(14) = 135, p(19) = 490$, etc., Ramanujan conjectured that all these numbers are divisible by 5. Ramanujan also discovered a beautiful identity from which this divisibility is immediate:

$$\sum_{n=0}^{\infty} p(5n+4)q^n = 5\prod_{j=1}^{\infty}\frac{(1-q^{5j})^5}{(1-q^j)^6}. \tag{1}$$

With regard to (1), Hardy [13, xxi–xxxvi] wrote, "It would be difficult to find more beautiful formulae than the 'Rogers-Ramanujan' identities [...]; but here Ramanujan must take second place to Prof. Rogers; and, if I had to select one formula from all Ramanujan's work, I would agree with Major MacMahon in selecting (1)." In this article we outline the structure of Radu's Ramanujan-Kolberg package which computes (1) and many other identities of this kind in automatic fashion.

The rest of this article is structured as follows. In Section 2 we give a brief account of MacMahon's partition analysis and of the "Omega project" devoted to its algorithmic revitalization. Section 3 illustrates the usage of the Omega package as a tool for mathematical discovery. We will see how Omega computations led to a new class of partitions ("broken diamond partitions") having generating functions in the form of quotients of Dedekind eta functions. One observes that arithmetic subsequences of coefficient sequences of such quotients satisfy various divisibility properties. Section 4 presents q-series identities (identities of "Ramanujan-Kolberg type") witnessing divisibility of this kind in a direct fashion. The rest of the article explains how such identities can be found in automatic fashion by the Ramanujan-Kolberg package. In Section 8 the steps of Radu's algorithm are presented.

As a preparation, basic facts from modular functions are given in Section 5 and Section 6 (zero recognition of modular functions). Underlying algorithmics based on elementary facts from monoid theory is discussed in Section 7.

2 Partition Analysis

The beginning of the algorithmic revitalization of partition analysis, the "Omega project," is described by Andrews in [1, Sect. 2.10]: "The number of partitions of N of the form $b_1 + \cdots + b_n$ satisfying

$$\frac{b_n}{n} \geq \frac{b_{n-1}}{n-1} \geq \cdots \geq \frac{b_2}{2} \geq \frac{b_1}{1} \geq 0 \tag{2}$$

equals the number of partitions of N into odd parts each $\leq 2n - 1$. This problem cried out for MacMahon's Partition Analysis [...] Given that Partition Analysis is an algorithm for producing partition generating functions, I was able to convince Peter Paule and Axel Riese to join an effort to automate this algorithm."

Note. A video account of this project of Andrews can be found in [27, 28].

To illustrate MacMahon's method, consider the partition problem constrained by (2) in the special case $n = 3$ but in a slightly more general setting: Find a "good" closed form of

$$L(x_1, x_2, x_3) := \sum_{b_1, b_2, b_3 \in \mathbb{N} \text{ s.t. } 2b_3 - 3b_2 \geq 0, b_2 - 2b_1 \geq 0} x_1^{b_1} x_2^{b_2} x_3^{b_3}.$$

To remove the inequality constraints on the summation variables, MacMahon introduced the "Omega" operator. This operator acts on additionally introduced slack variables which in their exponents carry the inequality information.

$$L(x_1, x_2, x_3) = \underset{\geqq}{\Omega} \sum_{b_1, b_2, b_3 \geq 0} \lambda_1^{2b_3 - 3b_2} \lambda_2^{b_2 - 2b_1} x_1^{b_1} x_2^{b_2} x_3^{b_3} = \underset{\geqq}{\Omega} \frac{1}{1 - \frac{x_1}{\lambda_2^2}} \frac{1}{1 - \frac{\lambda_2 x_2}{\lambda_1^3}} \frac{1}{1 - \lambda_1^2 x_3}.$$

After geometric series summation, the original problem is transformed into the problem of eliminating the slack variables; here, λ_1, λ_2, and λ_3. To this end, in [18] MacMahon compiled tables of elimination rules like

$$\underset{\geqq}{\Omega} \frac{1}{(1 - \lambda^2 A)\left(1 - \frac{B}{\lambda^3}\right)} = \frac{1 + A^2 B}{(1 - A)(1 - A^3 B^2)}$$

or

$$\underset{\geq}{\Omega} \frac{\lambda^{\delta}}{(1-\lambda^2 A)\left(1-\frac{B}{\lambda^2}\right)} = \frac{1}{(1-A)(1-AB)}, \quad \delta = 0, 1.$$

With the Omega package (written in Mathematica) at hand, all these steps are executed automatically as follows.

```
In[1] := « Omega.m
```

Omega Package by Axel Riese (in cooperation with George E. Andrews and Peter Paule) - ©RISC, JKU Linz - V 2.47

```
In[2] := LCrude = OSum[ x1^b1 x2^b2 x3^b3,
 {2 b3 - 3 b2 ≥ 0, b2 - 2 b1 ≥ 0 , b1 ≥ 0}, λ]
```

Out[2] = $\underset{\lambda_1,\lambda_2}{\underset{\geq}{\Omega}} \frac{1}{\left(1-\frac{x1}{\lambda_2^2}\right)\left(1-\frac{\lambda_2 x2}{\lambda_1^3}\right)\left(1-\lambda_1^2 x3\right)}$

```
In[3] := L=OR[LCrude]
```

Out[3] = $\frac{1+x2\,x3^2}{(1-x3)\left(1-x2^2\,x3^3\right)\left(1-x1\,x2^2\,x3^3\right)}$

```
In[4] := L /. {x1->q, x2->q, x3->q}
```

Out[4] = $\frac{1+q^3}{(1-q)\left(1-q^5\right)\left(1-q^6\right)}$

Note. Out[4] presents the generating function for "Lecture Hall" partitions into maximally three parts.

It might be illuminating to put partition analysis into a somewhat more general context. To this end, consider the problem of solving linear Diophantine systems of inequalities, resp. equations, over non-negative integers. More concretely, given integers $a_{i,j}$ and c_k, find $b_1, \ldots, b_n \in \mathbb{N}$ such that

$$\begin{pmatrix} a_{1,1} & \cdots & a_{1,n} \\ a_{2,1} & \cdots & a_{2,n} \\ \vdots & \ddots & \vdots \\ a_{m,1} & \cdots & a_{m,n} \end{pmatrix} \begin{pmatrix} b_1 \\ \vdots \\ b_n \end{pmatrix} \geq \begin{pmatrix} c_1 \\ c_2 \\ \vdots \\ c_m \end{pmatrix}. \tag{3}$$

It is easy to see that exchanging "$\geq$" with "$=$" results in an equivalent problem. In both cases the algebraic structure of the set of solutions of the homogenous version of the problem (i.e., where the c_k are all 0) is that of an (additive) monoid. It is a well-known fact that this submonoid of $\mathbb{N}^n$ is finitely generated; for instance, see the classical book by Grace and Young where this is proved as a consequence of a version of the celebrated Hilbert Basis Theorem [12].

A connection to combinatorics is made by translating things to generating functions. Given an $m \times n$ integer matrix $A = (a_{ij})$, an integer vector $c =$

$(c_1,\ldots,c_m)^t$, consider the non-negative integer solutions $b = (b_1,\ldots,b_n)^t$ to $A\,b \geq c$ presented in the form of a multivariate formal power series

$$L(x_1,\ldots,x_n) = \sum_{\substack{b=(b_1,\ldots,b_n)^t\in\mathbb{N}^n\\ Ab\geq c}} x_1^{b_1}\ldots x_n^{b_n}.$$

It turns out that such $L(x_1,\ldots,x_n)$ arising from linear Diophantine systems of inequalities (resp. equations) always are rational functions of the form

$$L(x_1,\ldots,x_n) = \frac{p(x_1,\ldots,x_n)}{(1-x_1^{\alpha_1}\ldots x_n^{\alpha_n})\ldots(1-x_1^{\gamma_1}\ldots x_n^{\gamma_n})}$$

where $p(x_1,\ldots,x_n)$ is a polynomial in $x_1,\ldots,x_n$.

Note. The exponent vectors $(\alpha_1,\ldots,\alpha_n)^t,\ldots,(\gamma_1,\ldots,\gamma_n)^t$ consisting of non-negative integers are called the "fundamental" generators of the respective monoid of non-negative integer solutions to the associated homogeneous problem. Other generators come from the exponent vectors of the summand monomials in $p(x_1,\ldots,x_n)$.

Example. The example above in matrix form reads as

$$\begin{pmatrix} -2 & 1 & 0 \\ 0 & -3 & 2 \end{pmatrix}\begin{pmatrix} b_1 \\ b_2 \\ b_3 \end{pmatrix} \geq \begin{pmatrix} 0 \\ 0 \end{pmatrix};$$

and according to `Out[3]` the solution monoid is generated as follows:

$$\alpha\begin{pmatrix}0\\0\\1\end{pmatrix}+\beta\begin{pmatrix}0\\2\\3\end{pmatrix}+\gamma\begin{pmatrix}1\\2\\3\end{pmatrix}+\delta\begin{pmatrix}0\\1\\2\end{pmatrix};\quad \alpha,\beta,\gamma\in\mathbb{N},\delta\in\{0,1\}.$$

The first three vectors correspond to the exponent vectors of the monomials in the denominator of `Out[3]`; the exponent vectors of the numerator monomials in `Out[3]` give the choices with $\delta = 0$ and $\delta = 1$, respectively.

In his pioneering book [18] MacMahon describes on more than a hundred pages how partition analysis is put into action to compute $L(x_1,\ldots,x_n)$ for a large variety of problems. In particular, he points to a complete algorithmic method, Elliott's algorithm, for executing this task mechanically. But without having a computer algebra system at hand, MacMahon had to use other methods - often combinations of ingenious manipulatorics and table look-up techniques.

Note. MacMahon's main interest in developing partition analysis was to find a proof for his conjectured form $\prod_{n=1}^{\infty}(1-q^n)^{-n}$ of the generating function for plane partitions. Towards the end of his exposition of partition analysis in [18] MacMahon confessed that his insight into the method was not sufficient to achieve this goal. Finally such a proof was accomplished in the course of the Omega project; see [2, 4]. Another goal of the partition analysis project was to turn algorithms into concrete software; for the resulting Omega package see [5, 6]. Also this algorithmic type of research has attracted renewed interest in the field. A very recent development is the new algorithm "Polyhedral Omega" by Breuer and Zafeirakopoulos [7] which combines strategies a la MacMahon with methods from polyhedral geometry such as Brion decompositions and Barvinok's short rational function representations. In [7] one also finds careful complexity analysis and comparisons to other methods, for instance, to the work of Xin [26]. Despite the development of computer algebra packages like Omega, the primary goal of the Omega project by Andrews, Paule, and Riese was not the improvement of computational complexity but the usage of such packages in the process of mathematical discovery.

3 Omega and Mathematical Discovery

As already mentioned, MacMahon generalized partitions of numbers arranged "on a line" like $3 = 2 + 1 = 1 + 1 + 1$, to plane partitions arranged "in the plane," like

$$3,\ 2+1,\ \begin{array}{c}2\\+\\1\end{array},\ 1+1+1,\ \begin{array}{c}1\\+\\1\\+\\1\end{array},\ \begin{array}{cc}1 & 1+1\\ & +\\ & 1\end{array}.$$

Alternatively, plane partitions which, for example, are arranged in maximally two rows can be described by posets, respectively, directed graphs, as follows:

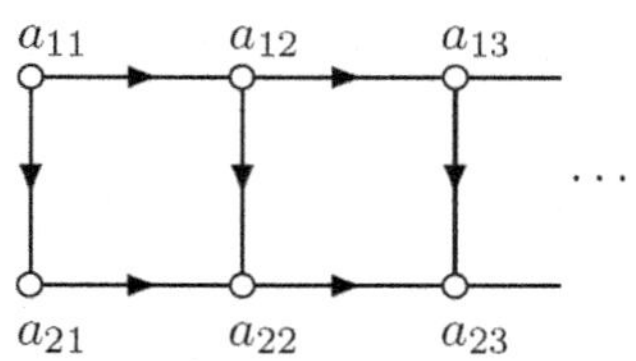

Here the a_{ij} represent non-negative integers following order conditions prescribed by the arrows. For instance, the arrow from a_{11} to a_{21} means $a_{11} \geq a_{21}$. Using the Omega package it is easy to compute corresponding partition generating functions. For example, consider the following poset

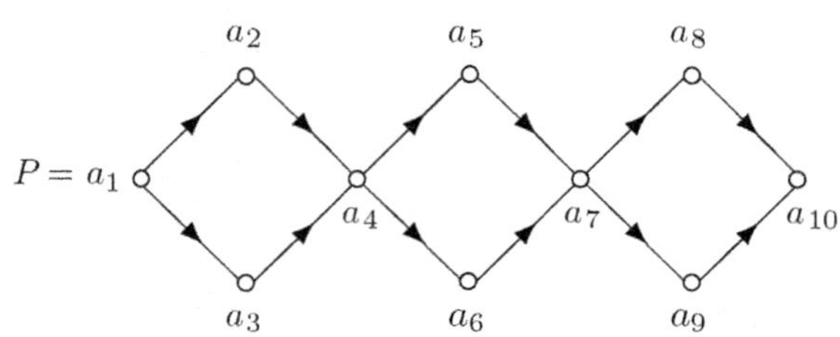

giving rise to the associated generated function

$$L(q) := \sum_{a_1,\ldots,a_{10}\geq 0 \text{ s.t. } P} q^{a_1+\cdots+a_{10}}.$$

One computes the rational function presentation of $L(q)$ just as in the Omega example above, and obtains

$$L(q) = \frac{1+q^8}{(1-q)(1-q^2)^2(1-q^3)(1-q^5)^2(1-q^6)(1-q^7)(1-q^8)(1-q^9)}.$$

Computational experiments with the Omega package led to replacing the poset P by a k-elongated diamond of length 1:

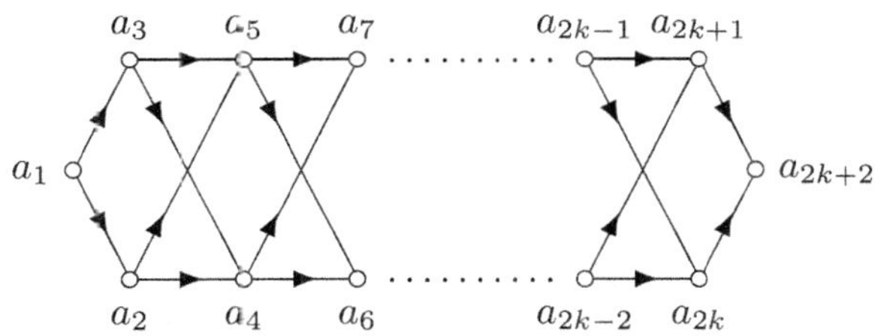

More generally, one can glue n such diamonds together to obtain a k-elongated partition diamond of length n:

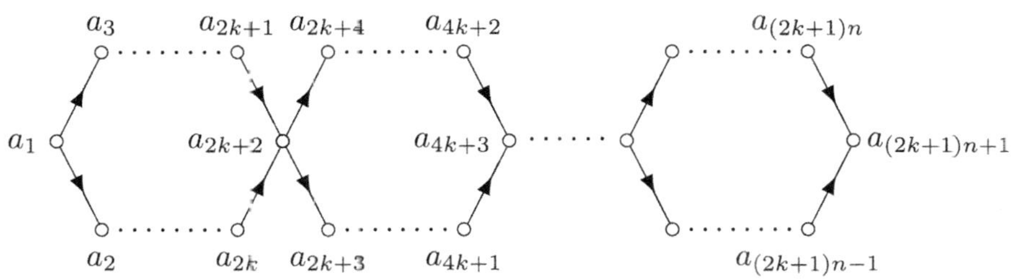

In [3] it is shown that the generating function for k-elongated diamonds of length n is

$$h_{n,k}(q) = \frac{\prod_{j=0}^{n-1}(1+q^{(2k+1)j+2})(1+q^{(2k+1)j+4})\ldots(1+q^{(2k+1)j+2k})}{\prod_{j=1}^{(2k+1)n+1}(1-q^j)}.$$

But one does not need to stop here. Andrews ingeniously suggested to "delete the source"; this means, to remove the a_1-vertex together with its outgoing edges. The result is a surprise; namely, for the generating function $h^*_{n,k}(q)$ over the resulting poset one obtains

$$h^*_{n,k}(q) = \frac{\prod_{j=0}^{n-1}(1+q^{(2k+1)j+1})(1+q^{(2k+1)j+3})\dots(1+q^{(2k+1)j+2k-1})}{\prod_{j=1}^{(2k+1)n}(1-q^j)}.$$

This leads us to consider the poset which results after gluing these diamonds together:

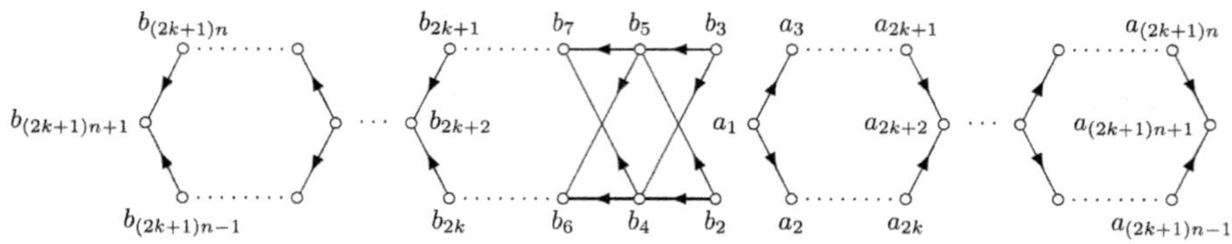

Why considering this? In the limit $n \to \infty$ the corresponding generating function becomes:

$$\begin{aligned}\sum_{m=0}^{\infty}\Delta_k(m)q^m &:= \lim_{n\to\infty} h_{n,k}(q)h^*_{n,k}(q)\\ &= \frac{\prod_{j=1}^{\infty}(1+q^j)}{\prod_{j=1}^{\infty}(1-q^j)^2\prod_{j=1}^{\infty}(1+q^{(2k+1)j})}\\ &= \frac{\prod_{j=1}^{\infty}(1+q^j)(1-q^j)}{\prod_{j=1}^{\infty}(1-q^j)^3\prod_{j=1}^{\infty}(1+q^{(2k+1)j})}\\ &= \prod_{j=1}^{\infty}\frac{(1-q^{2j})(1-q^{(2k+1)j})}{(1-q^j)^3(1-q^{(4k+2)j})}.\end{aligned}$$

Consequently, we have constructed combinatorial objects whose generating function is a non-trivial eta quotient:

$$\sum_{m=0}^{\infty}\Delta_k(m)q^m = q^{\frac{k+1}{12}}\frac{\eta(2\tau)\eta((2k+1)\tau)}{\eta(\tau)^3\eta((4k+2)\tau)}$$

with η the Dedekind eta function defined as usual as

$$\eta(\tau) := q^{\frac{1}{24}}\prod_{n=1}^{\infty}(1-q^n) \tag{4}$$

where $q = e^{2\pi i\tau}$ for $\tau \in \mathbb{H}$, $\mathbb{H} := \{\tau \in \mathbb{C} : \text{Im} > 0\}$ the upper half complex plane. Note that when presenting the generating function in terms of the Dedekind eta function, we moved from viewing q as an indeterminate arising in a formal power series to $q = q(\tau)$ presenting a function in the complex plane. In the following we will move freely between these two worlds; issues like convergence can be settled easily in each individual case.

Most relevant for our context, q-series defined by quotients of η-functions often possess remarkable number theoretic properties. Such properties can be studied most comfortably with the help of computer algebra systems.

For example, let us input a truncated product version of the generating function for k-elongated diamonds:

In[5] := bd[N_, k_] $:=\prod_{j=1}^{N} \frac{(1-q^{2j})(1-q^{(2k+1)j})}{(1-q^{j})^3(1-q^{(4k+2)j})}$

Already the case $k = 1$ will turn out to be interesting. We will inspect the coefficients of the Taylor series expansion up to that of q^{30}.

```
In[6] := bd1 = Normal[Series[bd[30,1], {q,0,30}]]
```

Out[6] = $1 + 3q + 8q^2 + 18q^3 + 38q^4 + 75q^5 + 142q^6 + 258q^7 + 455q^8 + 780q^9 + 1308q^{10}$
$+ 2148q^{11} + 3467q^{12} + 5505q^{13} + 8168q^{14} + 13314q^{15} + 20327q^{16} + 30693q^{17}$
$+ 45882q^{18} + 67944q^{19} + 99745q^{20} + 145239q^{21} + 209882q^{22} + 301128q^{23}$
$+ 429148q^{24} + 607710q^{25} + 855414q^{26} + 1197228q^{27} + 1666585q^{28}$
$+ 2308014q^{29} + 3180668q^{30}$

First we take all the coefficients modulo 2.

```
In[7] := Mod[CoefficientList[bd1, q], 2]
Out[7] = {1, 1, 0, 0, 0, 1, 0, 0, 1, 0, 0, 0, 1, 1, 0,
     0, 1, 1, 0, 0, 1, 1, 0, 0, 0, 0, 0, 0, 1, 0, 0}
```

Next we take the coefficients modulo 3 and 4.

```
In[8] := Mod[CoefficientList[bd1, q], 3]
Out[8] = {1, 0, 2, 0, 2, 0, 1, 0, 2, 0, 0, 0, 2, 0, 2,
     0, 2, 0, 0, 0, 1, 0, 2, 0, 1, 0, 0, 0, 1, 0, 2}
In[9] := Mod[CoefficientList[bd1, q], 4]
Out[9] = {1, 3, 0, 2, 2, 3, 2, 2, 3, 0, 0, 0, 3, 1, 2,
     2, 3, 1, 2, 0, 1, 3, 2, 0, 0, 2, 2, 0, 1, 2, 0}
```

In contrast to the cases modulo 2 and modulo 4, a quick inspection suggests a clear pattern modulo 3:

$$\Delta_1(2n+1) \equiv 0 \pmod{3}, \quad n \in \mathbb{N}. \tag{5}$$

Proof. We proceed as in [3] by recalling the freshman's dream relation

$$(1-q^j)^3 \equiv 1-q^{3j} \pmod 3, \quad j \in \mathbb{N}.$$

Here "$\equiv$" is considered coefficient-wise with respect to powers of q. Then

$$\sum_{m=0}^{\infty} \Delta_1(m)q^m = \prod_{j=1}^{\infty} \frac{(1-q^{2j})(1-q^{3j})}{(1-q^j)^3(1-q^{6j})}$$

$$\equiv \prod_{j=1}^{\infty} \frac{(1-q^{2j})(1-q^{3j})}{(1-q^{3j})(1-q^{6j})} \pmod 3.$$

Hence the coefficients of odd powers of q have to be zero. □

Hirschhorn and Sellers [14] found a proof of (5) which reveals the divisibility by 3 in beautifully direct fashion. Namely, they established the identity

$$\sum_{n=0}^{\infty} \Delta_1(2n+1)q^n = 3\prod_{j=1}^{\infty} \frac{(1-q^{2j})^2(1-q^{6j})^2}{(1-q^j)^6}. \tag{6}$$

With Radu's algorithmic method [23] such identities, including (6), can be established automatically. We will discuss its underlying mathematics in the following sections.

Before doing so, we want to point out that broken diamond partition numbers satisfy a whole variety of identities similar to (5). For example, in [3] it was conjectured that for all $n \in \mathbb{N}$,

$$\Delta_2(10n+2) \equiv 0 \pmod 2 \tag{7}$$

and

$$\Delta_2(25n+14) \equiv 0 \pmod 5. \tag{8}$$

Chan [8] proved this and also

$$\Delta_2(10n+6) \equiv 0 \pmod 2 \tag{9}$$

and

$$\Delta_2(25n+24) \equiv 0 \pmod 5. \tag{10}$$

It should be noted that the cases $10n+2$ and $10n+6$ were first proved by Hirschhorn and Sellers [14]. Pointers to further congruences, variants, and generalizations can be found in [9].

Summarizing, in this section we have seen that partition analysis can be used to construct combinatorial objects ("partition diamonds") with generating functions being eta quotients of number theoretic interest. More precisely, subsequences of the coefficients of Taylor series expansions of such eta quotients satisfy a variety of divisibility properties. In Section 4 we shall see how such divisibilities can be explained by establishing so-called "Ramanujan-Kolberg identities" which by Radu's package can be derived automatically. In Sections 5 and 6 the necessary modular function background is provided. In Section 7 we shall see that a fundamental task in Radu's algorithmic approach is to find generators of a monoid, a theme strongly related to linear Diophantine systems like (3).

4 Ramanujan-Kolberg Identities

In this section we will exemplify how proofs of congruences like (5), (7), (8), (9), and (10) can be derived automatically by using Radu's package. The main task for the package is to establish an underlying Ramanujan-Kolberg identity. Historically, this idea traces back to Ramanujan's "most beautiful" identity. More precisely, in [24] Ramanujan stated without proof that for the partition numbers $p(n)$, satisfying the generating function identity

$$\sum_{n=0}^{\infty} p(n)q^n = \prod_{j=1}^{\infty} \frac{1}{1-q^j},$$

one has (1) and

$$\sum_{n=0}^{\infty} p(7n+5)q^n = 7\prod_{j=1}^{\infty} \frac{(1-q^{7j})^3}{(1-q^j)^4} + 49q\sum_{j=1}^{\infty} \frac{(1-q^{7j})^7}{(1-q^j)^8}. \tag{11}$$

Like (6) above, these identities give a direct explanation for the first two of Ramanujan's classical congruences

$$p(5n+4) \equiv 0 \pmod 5, \quad p(7n+5) \equiv 0 \pmod 7, \tag{12}$$

and

$$p(11n+6) \equiv 0 \pmod{11}. \tag{13}$$

For the third one, Ramanujan gave no such identity, and we will come back to this issue in Section 8.

Radu's package computes identities like (6), (1), and (11) automatically. The input-output specification of his algorithm is as follows.

INPUT: integers $\ell, m, M, N \in \mathbb{N}$ with $0 \leq \ell < m$, and a sequence $(a_r(n))_{n\geq 0}$ defined with respect to a given integer tuple $r = (r_\delta)_{\delta|M}$ by

$$\sum_{n=0}^{\infty} a_r(n)q^n := \prod_{\delta|M} \prod_{n=1}^{\infty} (1-q^{\delta n})^{r_\delta}.$$

Note. In view of eta quotients we have for $\tau \in \mathbb{H}$ and $q = e^{2\pi i\tau}$ that

$$\sum_{n=0}^{\infty} a_r(n)q^n = q^{-\frac{\sum_{\delta|M} \delta r_\delta}{24}} \prod_{\delta|M} \eta(\delta\tau)^{r_\delta}.$$

OUTPUT: a rational number $\sigma \in \mathbb{Q}$, an integer tuple $s = (s_\delta)_{\delta|N}$, a finite set $P_{m,r}(\ell)$ of integers such that $\ell \in P_{r,m}(\ell)$, and eta quotients $e_1, \ldots, e_k$ together with $c_1, \ldots, c_k \in \mathbb{Q}$ such that

$$q^\sigma \prod_{\delta|N} \prod_{n=1}^{\infty} (1-q^{\delta n})^{s_\delta} \prod_{\ell' \in P_{m,r}(\ell)} \sum_{n=0}^{\infty} a_r(mn+\ell')q^n = c_1 e_1 + \cdots + c_k e_k. \tag{14}$$

An identity of the format as in (14) is called a Ramanujan-Kolberg identity.

Note. Apart from computational complexity, Radu's algorithm returns such an identity if such an identity exists under conditions specified more precisely in Section 8.

For example, the identities (1) and (11) rewritten in this output format read as

$$q\prod_{j=1}^{\infty}(1-q^{5j})\sum_{n=0}^{\infty} p(5n+4)q^n = 5\left(\frac{\eta(5\tau)}{\eta(\tau)}\right)^6 \tag{15}$$

and

$$q\prod_{j=1}^{\infty}(1-q^{7j})\sum_{n=0}^{\infty} p(7n+5)q^n = 7\left(\frac{\eta(7\tau)}{\eta(\tau)}\right)^4 + 49\left(\frac{\eta(7\tau)}{\eta(\tau)}\right)^8, \tag{16}$$

respectively. In both cases $P_{m,r}(\ell) = \{\ell\}$ with $\ell = 4$ and $\ell = 5$, respectively.

Note. Identities of the form (14) where $P_{m,r}(\ell)$ is bigger than $\{\ell\}$ go back to Kolberg [16].

To give a concrete example in the context of broken partition diamonds, recall the congruences (8) and (10):

$$\Delta_2(25n+14) \equiv \Delta_2(25n+24) \equiv 0 \pmod 5, \quad n \in \mathbb{N}. \tag{17}$$

Using $(1-q^j)^5 \equiv 1-q^{5j} \pmod 5$ we observe that

$$\sum_{m=0}^{\infty} \Delta_2(n)q^n = \prod_{j=1}^{\infty} \frac{(1-q^{2j})(1-q^{5j})}{(1-q^j)^3(1-q^{10j})} \equiv \prod_{j=1}^{\infty} \frac{(1-q^{2j})(1-q^j)^2}{(1-q^{10j})} \pmod 5$$
$$=: \sum_{n=0}^{\infty} d(n)q^n.$$

Hence $\Delta_2(n) \equiv d(n) \pmod 5$, and we will prove (17) with $d(n)$ instead of $\Delta_2(n)$.

Radu's program "Ramanujan-Kolberg" delivers

$$q^{\frac{3}{2}} \frac{\eta(2\tau)^{12}\eta(5\tau)^{10}}{\eta(\tau)^{6}\eta(10\tau)^{20}} \left(\sum_{m=0}^{\infty} d(25n+14)q^{n}\right)\left(\sum_{m=0}^{\infty} d(25n+24)q^{n}\right)$$
$$= 25(2t^4 + 28t^3 + 155t^2 + 400t + 400) \tag{18}$$

where

$$t = \frac{\eta(\tau)^3\eta(5\tau)}{\eta(2\tau)\eta(10\tau)^3}. \tag{19}$$

Rewriting the η-quotients in terms of q-products of the form

$$\prod_{k=1}^{\infty}(1-q^{\delta k})^{t_k}$$

makes the divisibility by 5 for each of the two classes $d(25n+14)$ and $d(25n+24)$ explicit.

Here, in view of (14), $\ell = 14$, $m = 25$, $M = N = 10$, $r = (r_1, r_2, r_5, r_{10}) = (2, 1, 0, -1)$, $s = (s_1, s_2, s_5, s_{10}) = (-6, 12, 10, -20)$, $\sigma = -4$, and $P_{m,r}(\ell) = \{14, 24\}$.

Note. The program computes a similar identity also for $\Delta_2(n)$ instead of $d(n)$, but the output is much bigger.

For the automatic derivation of Ramanujan-Kolberg identities of the form (14) Radu has chosen a particular setting in order to deal with modular functions in an algebraic fashion. The basic ingredients to this setting are given in the Sections 5, 6, and 7. The description of the steps of Radu's Ramanujan-Kolberg Algorithm can be found in Section 8.

5 Modular Functions: Basic Notions

As defined by (4) at the end of Section 3, eta functions are holomorphic functions defined on the upper half of the complex plane. Obviously, for $f(\tau) := \eta(\tau)^{24}$ we have the periodicity $f(\tau+1) = f(\tau)$ for $\tau \in \mathbb{H}$.

Note. In view of q-series representation we recall a fundamental, but important fact. Namely, for a given holomorphic function $f(\tau)$ on $\mathbb{H}$ with period $N \in \mathbb{N} \setminus \{0\}$ (i.e., $f(\tau+N) = f(\tau)$, $\tau \in \mathbb{H}$), there exists (uniquely) a holomorphic function $h(\tau)$ on the open unit disk, punctured at 0, such that for all $\tau \in \mathbb{H}$:

$$f(\tau) = h(e^{2\pi i\tau/N}). \tag{20}$$

For example, for $f(\tau) := \eta(\tau)^{24}$ (i.e., $N = 1$) one has

$$h(q) = q\left(\sum_{n=1}^{\infty} p(n)q^n\right)^{-24}$$

for all q from the punctured open unit disc.

But more is true. Namely, $f(\tau) := \eta(\tau)^{24}$ satisfies

$$f\left(\frac{a\tau+b}{c\tau+d}\right) = (c\tau+d)^k f(\tau), \quad \tau \in \mathbb{H}, \tag{21}$$

for $k = 12$ and all $\begin{pmatrix} a & b \\ c & d \end{pmatrix} \in \mathrm{SL}_2(\mathbb{Z}) = \left\{\begin{pmatrix} a & b \\ c & d \end{pmatrix} \in \mathbb{Z}^{2\times 2} : ad - bc = 1\right\}$. Holomorphic functions on $\mathbb{H}$ with this property plus suitable asymptotic behavior at $\tau \in \mathbb{Q} \cup \{\infty\}$ are called modular forms of weight k for $\mathrm{SL}_2(\mathbb{Z})$.

Note. A standard reference for modular forms and the related arithmetic of their q-series coefficients is [20].

Taking quotients $g(\tau) := f_1(\tau)/f_2(\tau)$ of such modular forms $f_1(\tau)$ and $f_2(\tau)$ results in a stronger symmetry:

$$g\left(\frac{a\tau+b}{c\tau+d}\right) = g(\tau), \quad \begin{pmatrix} a & b \\ c & d \end{pmatrix} \in \mathrm{SL}_2(\mathbb{Z}), \quad \tau \in \mathbb{H}. \tag{22}$$

Note:

$$\begin{pmatrix} a & b \\ c & d \end{pmatrix}(\tau) := \frac{a\tau+b}{c\tau+d}$$

defines a group action of $\mathrm{SL}_2(\mathbb{Z})$ on $\mathbb{H}$. Often one writes $\gamma\tau$ instead of $\gamma(\tau)$ where $\gamma \in \mathrm{SL}_2(\mathbb{Z})$.

Sometimes one needs to restrict property (22) to $\begin{pmatrix} a & b \\ c & d \end{pmatrix}$ from subgroups of $\mathrm{SL}_2(\mathbb{Z})$, for instance, $\begin{pmatrix} a & b \\ c & d \end{pmatrix} \in \Gamma_0(N)$, $N \in \mathbb{N} \setminus \{0\}$, where

$$\Gamma_0(N) := \left\{\begin{pmatrix} a & b \\ c & d \end{pmatrix} \in \mathrm{SL}_2(\mathbb{Z}) : N|c\right\}.$$

Modular forms g of weight $k = 0$ with symmetry (22) for $\begin{pmatrix} a & b \\ c & d \end{pmatrix} \in \Gamma_0(N)$ (i.e., (23) below) and being holomorphic on $\mathbb{H}$ are called modular functions for $\Gamma_0(N)$. It is obvious that such functions for fixed N form $\mathbb{C}$-algebras; i.e., they are commutative rings with 1 and vector spaces over $\mathbb{C}$.

Notation. The $\mathbb{C}$-algebra of modular functions for $\Gamma_0(N)$ will be denoted by $M(N)$.

For example, one can show that the functions $(\eta(5\tau)/\eta(\tau))^6$ from (15), and $f(\tau)$ and $f(\tau)^2$ with $f(\tau) = (\eta(7\tau)/\eta(\tau))^4$ from (16) are elements from $M(5)$ and $M(7)$, respectively. To this end, one needs to verify for $N = 5$, resp. $N = 7$, the following variant of (22) :

$$g\left(\frac{a\tau + b}{c\tau + d}\right) = g(\tau), \begin{pmatrix} a & b \\ c & d \end{pmatrix} \in \Gamma_0(N), \ \tau \in \mathbb{H}, \tag{23}$$

and also to check the suitable asymptotic behavior at all $\tau \in \mathbb{Q} \cup \{\infty\}$. This "suitable asymptotic behavior" is described by using the Laurent series expansion of h, as in (20), around 0. Consider $\tau \in \mathbb{H}$ close to ∞ or to a point $\frac{a}{c} \in \mathbb{Q}$. Allowing $c = 0$ we include the case $\infty = \frac{a}{0}$. For $\gamma = \begin{pmatrix} a & b \\ c & d \end{pmatrix} \in \mathrm{SL}_2(\mathbb{Z})$ we have $\gamma(\infty) = \frac{a}{c}$. Because of periodicity the following theorem holds [15, Thm 4].

Theorem 5.1. *Let $\gamma(\infty) = \frac{a}{c}$ for $\gamma = \begin{pmatrix} a & b \\ c & d \end{pmatrix} \in \mathrm{SL}_2(\mathbb{Z})$ and g be a holomorphic function on $\mathbb{H}$ satisfying (23). Then there exists a Laurent series expansion such that for all $\tau \in \mathbb{H}$ sufficiently close to $\frac{a}{c} \in \mathbb{Q} \cup \{\infty\}$:*

$$g(\tau) = \sum_{n=-\infty}^{\infty} c_n(\gamma) e^{2\pi i n(\gamma^{-1}\tau)/w_\gamma} \tag{24}$$

where

$$w_\gamma := \min\left\{h \in \mathbb{N} \setminus \{0\} : \begin{pmatrix} 1 & h \\ 0 & 1 \end{pmatrix} \in \gamma^{-1}\Gamma_0(N)\gamma\right\}.$$

Now we can give a precise definition of modularity. Namely, g as in the theorem is called a modular function if in addition to (23):

$$\text{for all } \gamma \in \mathrm{SL}_2(\mathbb{Z}) : c_n(\gamma) = 0 \text{ for almost all negative } n. \tag{25}$$

(Instead of (25) one also says that g is meromorphic at $\tau = \frac{a}{c} = \gamma\infty$.) If (25) holds and if $m = m(\gamma)$ is the smallest index such that $c_m(\gamma) \neq 0$, then we call m the γ-order of g at $\tau = \frac{a}{c}$; notation: $m = \mathrm{ord}^{\gamma}_{a/c}(g)$.

Note. If $c_n(\gamma) = 0$ for all $n \in \mathbb{Z}$, we set $\mathrm{ord}^{\gamma}_{a/c}(g) = \infty$.

Using the fact that $\gamma_2^{-1}\gamma_1\infty = \infty$ iff $\gamma_2^{-1}\gamma_1 = \begin{pmatrix} 1 & h \\ 0 & 1 \end{pmatrix}$ for some $h \in \mathbb{Z}$, it is not too difficult to check that if $\frac{a}{c} = \gamma_1\infty = \gamma_2\infty$ for $\gamma_1, \gamma_2 \in \mathrm{SL}_2(\mathbb{Z})$, then $w_{\gamma_1} = w_{\gamma_2}$ and

$$\mathrm{ord}^{\gamma_1}_{a/c}(g) = \mathrm{ord}^{\gamma_2}_{a/c}(g). \tag{26}$$

Thus we can define the order of a modular function g at $\frac{a}{c} \in \mathbb{Q} \cup \{\infty\}$ by

$$\mathrm{ord}_{a/c}(g) := \mathrm{ord}^{\gamma}_{a/c}(g)$$

for some $\gamma \in \mathrm{SL}_2(\mathbb{Z})$ such that $\gamma\infty = \frac{a}{c}$.

On the same line, let g be a modular function, and let $\gamma\infty = \frac{a}{c} \in \mathbb{Q} \cup \{\infty\}$ for $\gamma = \begin{pmatrix} a & b \\ c & d \end{pmatrix} \in \mathrm{SL}_2(\mathbb{Z})$ with expansion at $\frac{a}{c}$ being

$$g(\tau) = \sum_{n \geq \mathrm{ord}_{a/c}(g)} c_n(\gamma) e^{2\pi i (\gamma^{-1}\tau)/w_\gamma}.$$

Furthermore, let $\gamma' = \begin{pmatrix} a' & b' \\ c' & d' \end{pmatrix} \in \mathrm{SL}_2(\mathbb{Z})$ be such that $\gamma' = \gamma_0\gamma$ for some $\gamma_0 \in \Gamma_0(N)$. Then

$$w_{\gamma'} = w_\gamma$$

owing to $(\gamma')^{-1}\Gamma_0(N)\gamma' = \gamma^{-1}\Gamma_0(N)\gamma$. In addition, g has an expansion at $\frac{a'}{c'} = \gamma'\infty$ as follows

$$\begin{aligned} g(\tau) &= \sum_{n \geq \mathrm{ord}_{a'/c'}(g)} c_n(\gamma') e^{2\pi i ((\gamma')^{-1}\tau)/w_{\gamma'}} \\ &= \sum_{n \geq \mathrm{ord}_{a'/c'}(g)} c_n(\gamma') e^{2\pi i (\gamma^{-1}\tau)/w_\gamma} \end{aligned}$$

where the second equality is by $g(\tau) = g(\gamma_0\tau)$. Hence, by uniqueness of Laurent expansion, for all $n \in \mathbb{Z}$:

$$c_n(\gamma') = c_n(\gamma); \text{ in particular, } \mathrm{ord}^{\gamma'}_{a'/c'}(g) = \mathrm{ord}^{\gamma}_{a/c}(g). \tag{27}$$

In other words, the (24) expansions at points $\frac{a'}{c'} = \gamma_0\frac{a}{c}$ with $\gamma_0 \in \Gamma_0(N)$ are all the same.

Example. For $g(\tau) := (\eta(5\tau)/\eta(\tau))^6$ property (23) for $N = 5$ follows from a refined version of the symmetry (21) for the η-function. To show property (25) we need to show that for all $\gamma \in \mathrm{SL}_2(\mathbb{Z})$ the expansion (24) has only a finite sum as its principal part. In view of property (27), this task can be reduced to a finite number of inspections. To this end, we make use of the coset decomposition [15].

$$\mathrm{SL}_2(\mathbb{Z}) = \Gamma_0(5) \cup \Gamma_0(5)T \cup \Gamma_0(5)TS \cup \cdots \cup \Gamma_0(5)TS^4 \tag{28}$$

where

$$S = \begin{pmatrix} 1 & 1 \\ 0 & 1 \end{pmatrix} \text{ and } T = \begin{pmatrix} 0 & -1 \\ 1 & 0 \end{pmatrix};$$

this means, we just need to check for $\gamma \in \{\mathrm{id}, T, TS, \ldots, TS^4\}$.

A further reduction of the number of inspections comes from the fact that $S(\infty) = \infty$ and thus

$$\mathrm{id}(\infty) = \infty \text{ and } TS^j(\infty) = 0 \ (j = 0, \ldots, 4).$$

This means, we need to inspect the Laurent expansion (24) for all τ close to ∞ (i.e., choosing $\gamma = \mathrm{id}$) and close to 0 (i.e., for $\gamma = T$). For $\gamma = \mathrm{id}$ we can invoke (4) with $q = e^{2\pi i\tau}$

$$g(\tau) = q \prod_{j=1}^{\infty} \left(\frac{1 - q^{5j}}{1 - q^j} \right)^6;$$

i.e., $\mathrm{ord}_\infty(g) = 1$. For $\gamma = T$ we have to use [15, Thm. 9]:

$$\eta\Big(-\frac{1}{\tau}\Big) = (-i)^{1/2} \tau^{1/2} \eta(\tau), \ \tau \in \mathbb{H}, \tag{29}$$

(taking that branch of the square root function $\tau^{1/2}$ which is positive for real $\tau > 0$) and obtain for τ close to 0:

$$g(\tau) = \frac{1}{5^3} \frac{1}{Q} \prod_{j=1}^{\infty} \Big(\frac{1 - Q^j}{1 - Q^{5j}} \Big)^6 \tag{30}$$

where $Q = e^{2\pi i (T^{-1}\tau)/5}$. This means, $\mathrm{ord}_0(g) = -1$.

Note 1. Relation (30) corresponds to the following relation:

$$g(\tau) g\Big(-\frac{1}{5\tau}\Big) = \frac{1}{5^3}, \quad \tau \in \mathbb{H}.$$

Note 2. Matching the right-hand side of (30) to the series expansion as in (24) one sees that $w_T = 5$. This is in accordance with the general fact [17, Lemma 3.2.4]:

$$w_\gamma = \frac{N}{\gcd(c^2, N)} \text{ if } \gamma = \begin{pmatrix} a & b \\ c & d \end{pmatrix} \in \mathrm{SL}_2(\mathbb{Z}). \tag{31}$$

Summarizing, with these considerations we have sharpened our understanding of modular functions for a given group $\Gamma_0(N)$. Recall that these objects form a $\mathbb{C}$-algebra which we denoted by $M(N)$. We also note that if $f(\tau) \in M(N)$ without having roots in the upper half complex plane, then also

$$\frac{1}{f(\tau)} \in M(N).$$

6 Modular Functions: Zero Recognition

Our task is to establish and to prove algebraic relations between functions from $M(N)$. To illustrate various points, consider the following functions on $\mathbb{H}$ using $q = e^{2\pi i\tau}$:

$$f(\tau) := q\prod_{j=1}^{\infty}(1-q^{5j})\sum_{n=0}^{\infty}p(5n+4)q^n \tag{32}$$

and

$$g(\tau) := q\prod_{j=1}^{\infty}\left(\frac{1-q^{5j}}{1-q^j}\right)^6 = \left(\frac{\eta(5\tau)}{\eta(\tau)}\right)^6. \tag{33}$$

Following the example above we know that $g \in M(5)$. Suppose we also know that $f \in M(5)$. Equipped with this knowledge: how does one prove (15); i.e.,

$$f(\tau) - 5g(\tau) = 0, \ \tau \in \mathbb{H}? \tag{34}$$

Since both functions can be expressed as power series in q, one way to proceed would be comparing the (infinitely) many coefficients in these expansions. However, in the given context, it is non-trivial to turn this strategy into a feasible (finitary) argument. Rather than that, one looks at expansions at points $\frac{a}{c} \in \mathbb{Q} \cup \{\infty\}$, where poles arise. In view of (24), the representations in (32) and (33) correspond to expansions at $\gamma\infty = \frac{1}{0} = \infty$ with $\gamma = \begin{pmatrix} 1 & 0 \\ 0 & 1 \end{pmatrix}$ and where $w_{\mathrm{id}} = 1$ by (31). Next consider the expansion at $\gamma\infty = \frac{0}{1} = 0$ with $\gamma = \begin{pmatrix} 0 & -1 \\ 1 & 0 \end{pmatrix} = T$ and $w_T = 5$ by (31). If we expand (30) in powers of $Q = e^{2\pi i(T^{-1}\tau)/5}$ we obtain

$$g(\tau) = \frac{1}{5^3}Q^{-1} - \frac{6}{5^3} + \frac{9}{5^3}Q^1 + \dots.$$

As we shall see below, to prove (15) it is sufficient to check for the Q-expansion

$$f(\tau) = \sum_{n=-\infty}^{\infty} c_n(T)Q^n,$$

whether $\mathrm{ord}_0(f) = -1$ and whether

$$c_{-1}(T) = \frac{1}{5^2} \text{ and } c_0 = -\frac{6}{5^2}. \tag{35}$$

This check is done as follows. First, rewrite $f(\tau)$ as defined in (32) as

$$f(\tau) = q^{\frac{19}{24}}\eta(5\tau)\sum_{n=0}^{\infty} p(5n+4)q^n$$

and define

$$H(\tau) := \frac{1}{5}\eta(5\tau) \text{ and } F(\tau) := 5q^{\frac{19}{24}}\sum_{n=0}^{\infty} p(5n+4)q^n.$$

By (29) we have

$$H(T\tau) = \frac{1}{5}(-i)^{1/2}\left(\frac{\tau}{5}\right)^{1/2}(q^{\frac{1}{5}})^{24}\prod_{j=1}^{\infty}(1-(q^{\frac{1}{5}})^j).$$

To express $F(T\tau)$ in terms of powers of $q^{\frac{1}{5}}$ is more involved. Using properties of the eta function, Rademacher [22] derived that

$$\begin{aligned} F(T\tau) &= \sum_{\lambda=0}^{4} \eta\left(\frac{\tau+24\lambda}{5}\right)^{-1} \\ &= (-i)^{-1/2}(5\tau)^{-1/2}(q^{\frac{1}{5}})^{-\frac{25}{24}} \\ &\quad \times \left(\prod_{n=1}^{\infty}\left(1-(q^{\frac{1}{5}})^{25n}\right)^{-1} - 5\sum_{n=1}^{\infty}\left(\frac{n}{5}\right)p(n-1)(q^{\frac{1}{5}})^n\right) \end{aligned} \tag{36}$$

where $\left(\frac{n}{5}\right)$ is the Legendre symbol. Consequently,

$$\begin{aligned} f(\tau) &= f(T(T^{-1}\tau)) = H(T(T^{-1}\tau))F(T(T^{-1}\tau)) \\ &= \frac{1}{5^2}Q^{-1}\prod_{j=1}^{\infty}(1-Q^j) \\ &= \times\Big(\sum_{n=0}^{\infty}p(n)Q^{25n} - 5\sum_{n=1}^{\infty}\left(\frac{n}{5}\right)p(n-1)Q^n\Big) \\ &= \frac{1}{5^2}Q^{-1} - \frac{6}{5^2} + \frac{9}{5^2}Q + \dots, \end{aligned}$$

which confirms (35). Why is this sufficient to prove (34)?

Recall the following classical fact from complex analysis:

Theorem 6.1 (MMT). *Let f be a holomorphic function on a connected open subset $U \subseteq \mathbb{C}$. Suppose there is a point $p \in U$ such that $|f(z)| \le |f(p)|$ for all $z \in U$. Then f is constant on U.*

This theorem (Maximum Modulus Theorem) generalizes word by word replacing $\mathbb{C}$ by a connected Riemann surface X; see [19, Thm. 1.36]. As a corollary one obtains a fundamental tool for zero testing of modular functions:

Theorem 6.2 (ZT). *Let f be a holomorphic function on a compact Riemann surface X. Then f is a constant function.*

Proof. The function $|f|$ is continuous on the compact space X, hence taking on a maximum at some point in X. Consequently, the Riemann surface version of the MMT implies that f is a constant function. □

Note. Taking $U = \mathbb{C}$ in MMT gives Liouville's theorem. In view of ZT this would correspond to choosing the Riemann sphere for X.

Where in our context is the compact Riemann surface X to apply Theorem ZT for zero testing of modular functions?

To answer this question we extend our group action from $\mathbb{H}$ to $\mathbb{H} \cup \mathbb{Q} \cup \{\infty\}$:

$$\Gamma_0(N) \times \mathbb{H} \cup \mathbb{Q} \cup \{\infty\} \to \mathbb{H} \cup \mathbb{Q} \cup \{\infty\},$$
$$\left(\begin{pmatrix} a & b \\ c & d \end{pmatrix}, \tau\right) \mapsto \begin{pmatrix} a & b \\ c & d \end{pmatrix}\tau := \frac{a\tau + b}{c\tau + d}.$$

Notation. We write $[\tau] := \{\gamma\tau : \gamma \in \Gamma_0(N)\}$ for τ-orbits and $X_0(N)$ for the set of all orbits.

We remark that orbits of $\tau \in \mathbb{Q} \cup \{\infty\}$ contain only elements from $\mathbb{Q} \cup \{\infty\}$. Thus

$$X_0(N) = \{[\tau] : \tau \in \mathbb{H}\} \cup \{[\tau] : \tau \in \mathbb{Q} \cup \{\infty\}\}$$

as a disjoint union of orbit sets.

Note. There are only finitely many orbits in $\{[\tau] : \tau \in \mathbb{Q} \cup \{\infty\}\}$. These orbits are called cusps of $X_0(N)$; the underlying geometrical motivation and the connection to Riemann surfaces can be found in books like [10]. Sometimes, by abuse of language, also a representative τ of a cusp $[\tau]$ is called cusp.

Example. For $N = 5$ let us identify the elements of $\{[\tau] : \tau \in \mathbb{Q} \cup \{\infty\}\}$: Either $\tau = \frac{a}{c} = \frac{1}{0} = \infty$ or $\tau = \frac{a}{c}$ for relatively prime integers a and c. In any case there exists $\gamma = \begin{pmatrix} a & b \\ c & d \end{pmatrix} \in \mathrm{SL}_2(\mathbb{Z})$ such that $\gamma\infty = \frac{a}{c}$. According to (28), $\gamma \in \Gamma_0(5)$ or $\gamma = \gamma_0 TS^j$ with $\gamma_0 \in \Gamma_0(5)$ and $j \in \{0, \dots, 4\}$. In the first case $\left[\frac{a}{c}\right] = [\infty]$, in the second case $\left[\frac{a}{c}\right] = [0]$ because of $TS^j\infty = 0$. Hence

$$X_0(5) = \{[\tau] : \tau \in \mathbb{H}\} \cup \{[0], [\infty]\}.$$

Modular functions on $\mathbb{H}$ can be turned into meromorphic functions on $X_0(N)$ as follows: Suppose $f \in M(N)$. For $\tau \in \mathbb{H}$ define

$$\tilde{f}([\tau]) := f(\tau).$$

Because of (23) this is well defined. For $\frac{a}{c} \in \mathbb{Q} \cup \{\infty\}$ let $\gamma\infty = \frac{a}{c}$ with $\gamma = \begin{pmatrix} a & b \\ c & d \end{pmatrix} \in \mathrm{SL}_2(\mathbb{Z})$. Consider the Laurent expansion as in (24) for $\tau \in \mathbb{H}$ close to $\frac{a}{c}$:

$$f(\tau) = \sum_{n=\mathrm{ord}_{a/c}(f)}^{\infty} c_n(\gamma) e^{2\pi i n(\gamma^{-1}\tau)/w_\gamma}$$

with $w_\gamma = \frac{N}{\gcd(c^2, N)}$ as in (31). If now $\mathrm{ord}_{a/c}(f) < 0$, define

$$\tilde{f}\left(\left[\frac{a}{c}\right]\right) := \infty.$$

If $\mathrm{ord}_{a/c}(f) = 0$, define

$$\tilde{f}\left(\left[\frac{a}{c}\right]\right) := c_0(\gamma);$$

if $\mathrm{ord}_{a/c}(f) > 0$, define

$$\tilde{f}\left(\left[\frac{a}{c}\right]\right) := 0.$$

Because of (27) all these function values at $\frac{a}{c} \in \mathbb{Q} \cup \{\infty\}$ are well defined. Consequently, with the definitions above any $f \in M(N)$ gives rise to a function $\tilde{f}$ on $X_0(N)$.

Without going into detail, the set of orbits $X_0(N)$ can be equipped with a (natural) topology to make it a compact Hausdorff space. In addition, by introducing suitable charts (local homeomorphisms from $X_0(N)$ to $\mathbb{C}$) $X_0(N)$ can be turned into a Riemann surface.

Therefore, modular functions $f \in M(N)$ can be viewed as functions $\tilde{f}$ on the compact Riemann surface $X_0(N)$. From the definitions it is easy to see that the functions $\tilde{f}$ are meromorphic functions on $X_0(N)$ which, owing to f being holomorphic on $\mathbb{H}$, have possible poles only at the cusps; i.e., at the points $\left[\frac{a}{c}\right]$ with $\frac{a}{c} \in \mathbb{Q} \cup \{\infty\}$.

As concrete representations at these cusps, one has that for all $\tau \in \mathbb{H}$ from a suitably chosen open neighborhood of $\frac{a}{c} = \gamma\infty \in \mathbb{Q} \cup \{\infty\}$ with $\gamma = \begin{pmatrix} a & b \\ c & d \end{pmatrix} \in \mathrm{SL}_2(\mathbb{Z})$:

$$\tilde{f}([\tau]) = f(\tau) = \sum_{n=\mathrm{ord}_{a/c}(f)}^{\infty} c_n(\gamma) e^{2\pi i(\gamma^{-1}\tau)/w_\gamma}.$$

Note. As "suitable neighborhoods" of $\frac{a}{c}$ for $\frac{a}{c} \in \mathbb{Q}$ one can take open discs in $\mathbb{H}$ which are tangent to the real axis and to which $\frac{a}{c}$ is adjoined; for $\frac{a}{c} = \infty$ one can choose the "degenerated discs" $\{\tau \in \mathbb{H} : \mathrm{Im}(\tau) > c\}$, $c \geq 0$.

If $\mathrm{ord}_{a/c}(f) \geq 0$ for all cusps $\left[\frac{a}{c}\right]$, then $\tilde{f}$ is holomorphic and, by Theorem ZT, $\tilde{f}$ is even a constant function on $X_0(N)$.

As an obvious consequence, f is constant function on $\mathbb{H}$. And as another obvious consequence we obtain the following zero test for a modular function $f \in M(N)$:

MF-ZeroTest.

(T1) Determine all different cusps $\{\left[\frac{a}{c}\right] : \frac{a}{c} \in \mathbb{Q} \cup \{\infty\}\}$.

(T2) For each cusp representative $\frac{a}{c} = \gamma\infty$ with $\gamma = \begin{pmatrix} a & b \\ c & d \end{pmatrix} \in \mathrm{SL}_2(\mathbb{Z})$ determine whether all the coefficients of the principal part

$$\sum_{n=\mathrm{ord}_{a/c}(f)}^{-1} c_n(\gamma) e^{2\pi i(\gamma^{-1}\tau)/w_\gamma}$$

are zero.

(T3) If the answer to (T2) in each instance is yes: choose a suitable cusp $\left[\frac{a}{c}\right]$ to test whether $c_0(\gamma) = 0$.

7 Interlude: Monoids and Modular Functions

Before describing the Ramanujan-Kolberg Algorithm in Section 8, we have to introduce a special $\mathbb{C}$-algebra of modular functions. This algebra and a related (commutative) monoid structure allow to simplify the MF-ZeroTest and to carry out also other algebraic/algorithmic tasks.

To this end, we consider those functions $f \in M(N)$ for which $\tilde{f}$ as a meromorphic function on $X_0(N)$ has a pole only at the cusp $\left[\frac{a}{c}\right] = [\infty]$; i.e.,

$$M^{\infty}(N) := \{f \in M(N) : \mathrm{ord}_{a/c}(f) \geq 0 \text{ for all } \frac{a}{c} \in \mathbb{Q}\}.$$

As its ambient space $M(N)$, also $M^{\infty}(N)$ forms a $\mathbb{C}$-algebra which, in addition, gives rise to a naturally defined additive monoid induced by the pole orders.

The natural numbers $\mathbb{N} = \{0, 1, \dots\}$ form a commutative monoid with respect to both multiplication (with identity element 1) and addition (with identity element 0). Let $M \subseteq \mathbb{N}$ be an additive submonoid. It is easily seen that M is finitely generated. Namely, fix some $m \in M \setminus \{0\}$ and consider in M the equivalence classes modulo m. If $a, a' \in M$ such that $a' = a + \ell m$ for some $\ell \in \mathbb{N}$, then a' can be discarded as a generator. Hence, choosing the minimal element $u_i \in M$ from each equivalence class gives a set of generators $\{u_0 = 0, u_1, \dots, u_{k-1}, m\}$, $k \leq m$, that generate all of M.

Note. It may happen that $k < m$ because some of the equivalence classes may have empty intersection with M. However, this cannot happen in the case of numerical semigroups M which have the additional defining property that $1 \in G_M$ (i.e., $G_M = \mathbb{Z}$) where

$$G_M := \{\alpha_1 m_1 + \dots + \alpha_\ell m_\ell : \ell \in \mathbb{N}, \alpha_i \in \mathbb{Z}, m_j \in M\}$$

is the (cyclic) subgroup of $\mathbb{Z}$ generated by M. Because then one can find a representation of 1 as

$$1 + \alpha_1 m_1 + \dots + \alpha_i m_i = \alpha_{i+1} m_{i+1} + \dots + \alpha_j m_j \in M$$

with $\alpha_k \in \mathbb{N}$. For $m' := \alpha_1 m_1 + \dots + \alpha_i m_i \in M$ we have $1 + m' \in M$; this implies $u_i := i(1 + m') \in M$ are solutions of the congruences $u_i \equiv i \pmod{m'}$ for each $i \in \{0, \dots, m' - 1\}$. Hence $\mathbb{N} \setminus M$ is a finite set, and also modulo any other non-zero element $m \in M$ no congruence class in M can remain empty. Vice versa, if $\mathbb{N} \setminus M$ is a finite set for a submonoid $M \subseteq \mathbb{N}$, then there exists an $m' \in M$ such that $1 + m' \in M$; hence $(1 + m') - m' = 1 \in G_M$ and M is a numerical semigroup. Finally we note that considering numerical semigroups is no serious restriction of generality because for each submonoid M of $\mathbb{N}$ there is a monoid homomorphism $\phi : M \to M_d, m \mapsto m/d$, where $M_d := \{m/d : m \in M\}$ is a numerical semigroup with d being the generator of G_M. Further information on numerical semigroups with connections to Diophantine systems of inequalities like (3) can be found in [25].

Given non-constant modular functions $f, f_1, \ldots, f_n \in M^\infty(N)$, our goal is to develop an algorithm for checking membership in the $\mathbb{C}$-algebra R generated by $\{f_1, \ldots, f_n\}$; i.e., to decide whether

$$f \in R := \mathbb{C}[f_1, \ldots, f_n]?$$

For simplicity we assume $\gcd(\operatorname{ord}_\infty(f_1), \ldots, \operatorname{ord}_\infty(f_n)) = 1$. In this case

$$M := \{-\operatorname{ord}_\infty(g) : g \in R\}$$

is a numerical semigroup. Define $m := -\operatorname{ord}_\infty(f_1)$ and $t := f_1$, and let $u_0 = 0$, $u_1, \ldots, u_{m-1}$, and m be the generators of M as explained above. Let $g_0 = 1$, $g_1, \ldots, g_{m-1} \in R$ be such that $-\operatorname{ord}_\infty(g_i) = u_i$ for $i \in \{0, \ldots, m-1\}$, then we have

Theorem 7.1. *For every $f \in R$ there exist polynomials $p_0(x), \ldots, p_{m-1}(x) \in \mathbb{C}[x]$ such that*

$$f = p_0(t) + p_1(t)\, g_1 + \cdots + p_{m-1}(t)\, g_{m-1}.$$

Proof. If $d = -\operatorname{ord}_\infty(f)$, then $d = u_i + \ell m$ for some index $i \in \{0, \ldots, m-1\}$ and some $\ell \in \mathbb{N}$. In particular, $\operatorname{ord}_\infty(g_i t^\ell) = \operatorname{ord}_\infty(f)$; therefore there exists $c \in \mathbb{C}$ such that $\operatorname{ord}_\infty(f - c g_i t^\ell) > \operatorname{ord}_\infty(f)$. Continuing this process with $h_1 := f - c g_i t^\ell \in R$ instead of f, one obtains an $h_2 \in R$, and so on. Finally, after k steps one obtains an $h_k \in R$ with $\operatorname{ord}_\infty(h_k) > 0$, and thus $h_k = 0$ by the MF-ZeroTest. By back substitution we obtain the desired form of f. □

Theorem 7.2. *With the assumptions as in the previous theorem we have that*

$$\mathbb{C}[f_1, \ldots, f_n] = \langle g_0 = 1, g_1, \ldots, g_{m-1} \rangle_{\mathbb{C}[t]}, \tag{37}$$

the right side denoting the $\mathbb{C}[t]$-module generated by $g_0 = 1, g_1, \ldots, g_{m-1} \in M^\infty(N)$ where the g_i are computed according to the Algorithm MODULE GENERATORS.

Proof. The following algorithm description shows how to obtain $g_1, \ldots, g_{m-1}$ from $f_1, \ldots, f_n$ in an algorithmic fashion. The rest of the statement is only a reformulation of Theorem 7.1. □

Algorithm MODULE GENERATORS.

INPUT: non-constant $t := f_1, f_2, \ldots, f_n \in M^\infty(N)$ with $m := -\operatorname{ord}_\infty(t)$ and

$$\gcd(\operatorname{ord}_\infty(t), \operatorname{ord}_\infty(f_2), \ldots, \operatorname{ord}_\infty(f_n)) = 1;$$

OUTPUT: $g_0 = 1, g_1, \ldots, g_{m-1} \in M^\infty(N)$ such that (37).

STEP 0: Set $E := \{\}$. Set $m_k := -\operatorname{ord}_\infty(f_k)$, $k = 2, \ldots, n$. For each $1 \leq i \leq m-1$: if there exists $f_j \in \{f_2, \ldots, f_n\}$ such that $-\operatorname{ord}_\infty(f_j) \equiv i \pmod{m}$ then do nothing. Suppose such f_j for i does not exist. Then there exist $\alpha_{i,k} \in \mathbb{N}$ such that

$$\alpha_{i,2}m_2 + \alpha_{i,3}m_3 + \ldots \alpha_{i,n}m_n \equiv i \pmod{m}$$

owing to $\gcd(m, m_2, m_3, \ldots, m_n) = 1$; in this case, add to the set E the element $f_2^{\alpha_{i,2}} f_3^{\alpha_{i,3}} \cdots f_n^{\alpha_{i,n}}$. — Remark. After STEP 0 the set $\{h_1, \ldots, h_\ell\} := \{f_2, \ldots, f_n\} \cup E$ has the property

$$\{-\operatorname{ord}_\infty(h_1), \ldots, -\operatorname{ord}_\infty(h_\ell)\} \equiv \{1, \ldots, m-1\} \pmod{m}.$$

STEP 1: Define as input list $L := \{h_1, \ldots, h_\ell\} := \{f_2, \ldots, f_n\} \cup E$. To do an "elementary step" consider a pair $\{h_i, h_j\} \subseteq L$ with $-\operatorname{ord}_\infty(h_j) \geq -\operatorname{ord}_\infty(h_i)$. If

$$-\operatorname{ord}_\infty(h_j) \not\equiv -\operatorname{ord}_\infty(h_i) \pmod{m},$$

do nothing; otherwise, set

$$F := h_j - c\, t^{\frac{-\operatorname{ord}_\infty(h_j)+\operatorname{ord}_\infty(h_i)}{m}} h_i$$

where $c \in \mathbb{C}$ is chosen such that $-\operatorname{ord}_\infty(F) < -\operatorname{ord}_\infty(h_j)$. If $F = 0$, delete h_j from L; otherwise, replace h_j by F.

Repeat this "elementary step" until the list of functions remains unchanged.

Note 1. STEP 1 turns L into $\{F_1, \ldots, F_{m-1}\} \subseteq M^\infty(N)$ such that

$$\mathbb{C}[f_1, f_2, \ldots, f_n] = \mathbb{C}[t, f_2, \ldots, f_n] = \mathbb{C}[h_1, h_2, \ldots, h_\ell] = \mathbb{C}[t, F_1, \ldots, F_{m-1}]$$

and

$$\{-\operatorname{ord}_\infty(F_1), \ldots, -\operatorname{ord}_\infty(F_{m-1})\} \equiv \{1, \ldots, m-1\} \pmod{m}.$$

Note 2. In view of the finiteness of $\mathbb{N} \setminus M'$ the monoid M' generated by the orders $-\operatorname{ord}_\infty(F_1), \ldots, -\operatorname{ord}_\infty(F_{m-1})$ is a numerical semigroup. But, in general, M' is only a submonoid of $M = \{-\operatorname{ord}_\infty(g) : g \in \mathbb{C}[F_1, \ldots, F_{m-1}]$. STEP 2 now takes care of products of the F_i.

STEP 2: To start with STEP 2, take the output $L := \{F_1, \ldots, F_{m-1}\}$ from STEP 1. To do an "elementary step," we now consider $L^* := L \cup \{F_iF_j\}$ with fixed i, j such that $1 \leq i \leq j \leq m-1$, and apply STEP 1 to L^* to obtain $\{G_1, \ldots, G_{m-1}\} \subseteq M^\infty(N)$. Then these G_i have the property that

$$\mathbb{C}[t, F_1, \ldots, F_{m-1}] = \mathbb{C}[t, F_1, \ldots, F_{m-1}, F_iF_j] = \mathbb{C}[t, G_1, \ldots, G_{m-1}]$$

and

$$\{-\operatorname{ord}_\infty(G_1), \dots, -\operatorname{ord}_\infty(G_{m-1})\} \equiv \{1, \dots, m-1\} \pmod{m}.$$

Moreover, one can easily check that either $\{G_1, \dots, G_{m-1}\} = \{F_1, \dots, F_{m-1}\}$ or

$$-\sum_{i=1}^{m-1} \operatorname{ord}_\infty(G_i) < -\sum_{i=1}^{m-1} \operatorname{ord}_\infty(F_i). \tag{38}$$

The next "elementary step" is made with choosing $G_i, G_j \in \{G_1, \dots, G_{m-1}\}$ such that $F_iF_j \neq G_iG_j$ and applying STEP 1 to $L^* := \{G_1, \dots, G_{m-1}\} \cup \{G_iG_j\}$. Such "elementary steps" are continued until the set $\{G_1, \dots, G_{m-1}\}$ does not change anymore, and the Algorithm MODULE GENERATORS returns the output

$$(g_0, g_1, \dots, g_{m-1}) := (1, G_1, \dots, G_{m-1}).$$

Termination is guaranteed by (38).

Note. There are more efficient, but less transparent strategies for this algorithm.

8 The Ramanujan-Kolberg Algorithm

To describe the steps of Radu's Ramanujan-Kolberg Algorithm [23], in short called "RK Algorithm" below, we need to specify certain sets of eta quotients:

$$E(N) := \left\{\prod_{\delta|N} \eta(\delta\tau)^{r_\delta} : (r_\delta)_{\delta|N} \text{ is proper}\right\},$$

where an integer sequence $(r_\delta)_{\delta|N}$ is called proper if

$$\sum_{\delta|N} r_\delta = 0, \ \sum_{\delta|N} \delta r_\delta \equiv \sum_{\delta|N} \frac{N}{\delta} r_\delta \equiv 0 \pmod{24}, \text{ and } \prod_{\delta|N} \delta^{|r_\delta|/2} \in \mathbb{N}.$$

Using transformation properties of the η-function one sees that $E(N) \subseteq M(N)$. The set $E(N)$ and also

$$E^\infty(N) := E(N) \cap M^\infty(N)$$

clearly form a (multiplicative) monoid. The following, algorithmically important lemma traces back to Newman; see [23].

Lemma 8.1. *For each $N \in \mathbb{N}^*$ there exists $\mu \in E^\infty(N)$ such that*

$$\mathrm{ord}_{a/c}(\mu) > 0 \text{ for all } \frac{a}{c} \in \mathbb{Q}.$$

As a consequence, multiplying functions $f \in M(N)$ with suitable powers μ^k of such eta quotients μ, one can remove all the poles sitting at the cusps $\left[\frac{a}{c}\right]$ whenever $\frac{a}{c} \in \mathbb{Q}$; i.e., resulting in $\mu^k f \in M^\infty(N)$.

RK Algorithm Step 1. For the INPUT as specified in Section 4, compute $\sigma \in \mathbb{Q}$ and an integer sequence $(s_\delta)_{\delta|N}$ together with the finite set $P_{m,r}(\ell)$ of integers such that

$$q^\sigma \prod_{\delta|N} \prod_{n=1}^{\infty} (1 - q^{\delta n})^{s_\delta} \prod_{\ell' \in P_{m,r}(\ell)} \sum_{n=0}^{\infty} a_r(mn + \ell') q^n \in M^\infty(N). \tag{39}$$

In [23] one finds conditions for the existence of such σ and (s_δ) together with algorithms to compute them.

Example. In the case of (16) the algorithm determines

$$q^{\frac{17}{24}} \frac{\eta(\tau)^8}{\eta(7\tau)^7} \sum_{n=0}^{\infty} p(7n + 5) q^n \in M^\infty(7). \tag{40}$$

Define

$$\langle E(N) \rangle_{\mathbb{C}} := \text{the } \mathbb{C}\text{-vector space generated by the elements of } E(N).$$

Note. This vector space is infinite dimensional and it contains the constant functions; i.e., $1 \in \langle E(N) \rangle_{\mathbb{C}}$.

Suppose $f \in M^\infty(N)$ is given in a form as on the left side of (39). The goal of the RK Algorithm is to find $e_1, \ldots, e_k \in E(N)$ such that for $c_1, \ldots, c_k \in \mathbb{C}$:

$$f = c_1 e_1 + \cdots + c_k e_k \in \langle E(N) \rangle_{\mathbb{C}} \cap M^\infty(N). \tag{41}$$

Suppose that

$$\langle E(N) \rangle_{\mathbb{C}} \cap M^\infty(N) = \langle E^\infty(N) \rangle_{\mathbb{C}} \tag{42}$$

where

$$\langle E^\infty(N) \rangle_{\mathbb{C}} := \text{the } \mathbb{C}\text{-vector space generated by the elements of } E^\infty(N).$$

Then the $e_1, \ldots, e_k$ in (41) can be chosen from $E^\infty(N)$. Owing to the next lemma, and to Theorem 8.3 below, there is a finite presentation of the infinite vector space $\langle E^\infty(N)\rangle_{\mathbb{C}}$.

Lemma 8.2. *The (multiplicative) monoid $E^\infty(N)$ is finitely generated.*

RK Algorithm Step 2. Compute monoid generators $E_1, \ldots, E_m$ of $E^\infty(N)$ according to [23, Lemma 25].

Note. The generators can be computed using algorithms that solve Diophantine problems like (3).

Example. The RK Algorithm tells that the monoid $E^\infty(7)$ is generated by $E_1 := \left(\frac{\eta(\tau)}{\eta(7\tau)}\right)^4 \in E^\infty(7)$. Consequently,

$$\begin{aligned}\langle E^\infty(7)\rangle_{\mathbb{C}} &= \{c_1e_1 + \cdots + c_je_j : j \in \mathbb{N}, c_i \in \mathbb{C},\ e_i \in E^\infty(7)\}\\ &= \mathbb{C}[E_1] = \text{polynomials in } E_1 \text{ with coefficients in } \mathbb{C}.\end{aligned}$$

As an alternative to a polynomial ring presentation, $\langle E^\infty(7)\rangle_{\mathbb{C}}$ can be also viewed as $\mathbb{C}[E_1]$-module over the trivial generator $1 \in E^\infty(7)$; in notation:

$$\langle E^\infty(7)\rangle_{\mathbb{C}} = \langle 1\rangle_{\mathbb{C}[E_1]}.$$

Indeed, this is possible in general by the following theorem extracted from [23]:

Theorem 8.3. *Suppose the multiplicative monoid $E^\infty(N)$ is generated by $E_1, \ldots, E_m$. Then there are elements t and $z_1, \ldots, z_n \in \langle E^\infty(N)\rangle_{\mathbb{C}}$ such that the $\mathbb{C}$-vector space $\langle E^\infty(N)\rangle_{\mathbb{C}} = \mathbb{C}[E_1, \ldots, E_m]$ can be represented as a $\mathbb{C}[t]$-module freely generated by $z_1, \ldots, z_n$; i.e.,*

$$\langle E^\infty(N)\rangle_{\mathbb{C}} = \langle 1, z_1, \ldots, z_n\rangle_{\mathbb{C}[t]}.$$

RK Algorithm Step 3. Apply Algorithm MODULAR GENERATORS from Section 7 to $\{E_1, \ldots, E_m\}$.

Note. All these computations rely on the fact that we are dealing with modular functions in $M^\infty(N)$. As already mentioned, for such functions it is sufficient to handle the principal part of the Laurent series expansions in powers of $q = e^{2\pi i\tau}$ at $\frac{a}{c} = \infty$. Also the last step benefits from this fact.

RK Algorithm Step 4. Let $f(\tau)$ denote the left-hand side of (39) and let

$$\langle E^\infty(N)\rangle_{\mathbb{C}} = \langle 1, z_1, \ldots, z_n\rangle_{\mathbb{C}[t]}$$

be the output of Step 3. Following the steps as described in the proof of Theorem 7.1, compute polynomials $p_0(t), \ldots, p_n(t) \in \mathbb{C}[t]$ such that

$$f = p_0(t) + p_1(t)z_1 + \cdots + p_n(t)z_n.$$

Example. Define

$$\sum_{k=0}^{\infty} L(k)q^k := \prod_{j=1}^{\infty} \frac{1}{1-q^{2j-1}} = \prod_{j=1}^{\infty} \frac{1-q^{2j}}{1-q^j}.$$

Note. This is the generating function of partitions into odd parts; we started with the Lecture Hall partition problem (2) where such partitions arise but with a restriction on the size of the parts.

RK Step 1 with $M = 2$, $r = (r_1, r_2) = (-1, 1)$, $m = 7$, and $\ell = 3$ delivers $P_{m,r}(3) = \{3, 4, 6\}$ and

$$\begin{aligned} &q^{-5} \prod_{j=1}^{\infty} \frac{(1-q^j)^{13}(1-q^{7j})^8}{(1-q^{2j})^5(1-q^{14j})^{16}} \\ &\times \sum_{n=0}^{\infty} L(7n+3)q^n \cdot \sum_{n=0}^{\infty} L(7n+4)q^n \cdot \sum_{n=0}^{\infty} L(7n+6)q^n \in M^{\infty}(14). \end{aligned} \tag{43}$$

RK Step 2 gives that the multiplicative monoid $E^{\infty}(14)$ is generated by

$$E_1 = \left(\frac{\eta(2\tau)}{\eta(\tau)}\right)^{1} \left(\frac{\eta(7\tau)}{\eta(\tau)}\right)^{7} \left(\frac{\eta(14\tau)}{\eta(\tau)}\right)^{-1} \in E^{\infty}(14),$$

$$E_2 = \left(\frac{\eta(2\tau)}{\eta(\tau)}\right)^{8} \left(\frac{\eta(7\tau)}{\eta(\tau)}\right)^{4} \left(\frac{\eta(14\tau)}{\eta(\tau)}\right)^{-8} \in E^{\infty}(14),$$

$$E_3 = \left(\frac{\eta(2\tau)}{\eta(\tau)}\right)^{-5} \left(\frac{\eta(7\tau)}{\eta(\tau)}\right)^{5} \left(\frac{\eta(14\tau)}{\eta(\tau)}\right)^{-13} \in E^{\infty}(14),$$

$$E_4 = \left(\frac{\eta(2\tau)}{\eta(\tau)}\right)^{1} \left(\frac{\eta(7\tau)}{\eta(\tau)}\right)^{3} \left(\frac{\eta(14\tau)}{\eta(\tau)}\right)^{-7} \in E^{\infty}(14),$$

$$E_5 = \left(\frac{\eta(2\tau)}{\eta(\tau)}\right)^{5} \left(\frac{\eta(7\tau)}{\eta(\tau)}\right)^{7} \left(\frac{\eta(14\tau)}{\eta(\tau)}\right)^{-11} \in E^{\infty}(14),$$

and

$$E_6 = \left(\frac{\eta(2\tau)}{\eta(\tau)}\right)^{-2} \left(\frac{\eta(7\tau)}{\eta(\tau)}\right)^{6} \left(\frac{\eta(14\tau)}{\eta(\tau)}\right)^{-10} \in E^{\infty}(14).$$

RK Step 3 computes that

$$\langle E^{\infty}(14)\rangle_{\mathbb{C}} = \langle 1, E_4\rangle_{\mathbb{C}[E_1]};$$

i.e., $t = E_1$, $z_1 = E_4$. Denoting the function in (43) by f, RK Step 4 outputs

$$f = 8(p_0(t) + p_1(t)z_1)$$

where $p_0(t) = -16t + 9t^2$ and $p_1(t) = 2t$.

Note. It is easily checked that this relation implies for all $n \in \mathbb{N}$:

$$L(7n+3) \equiv L(7n+4) \equiv L(7n+6) \equiv 0 \pmod 2. \tag{44}$$

In a different setting, Gordon and Ono [11] have shown the strikingly general result that almost all values of $L(n)$ are divisible by 2^k for any $k \in \mathbb{N}$.

Radu's "Ramanujan-Kolberg" package also delivers that

$$\sum_{n=0}^{\infty} L(7n)q^n \cdot \sum_{n=0}^{\infty} L(7n+1)q^n \cdot \sum_{n=0}^{\infty} L(7n+5)q^n$$
$$= q^6 \prod_{j=1}^{\infty} \frac{(1-q^{2j})^5(1-q^{14j})^{16}}{(1-q^j)^{13}(1-q^{7j})^8}(3E_1^3 + 24E_1^2 + 64E_1)$$

and

$$\sum_{n=0}^{\infty} L(7n+2)q^n = q^3 \prod_{n=1}^{\infty} \frac{(1-q^{14j})^8}{(1-q^j)^3(1-q^{2j})(1-q^{7j})^4}(8E_1 + E_4 - 8).$$

The summands on the right-hand sides have no common integer factor, so no divisibility results from these identities.

In Section 4 we remarked that so far in the literature one could not find any Ramanujan-Kolberg identity presenting a witness for the congruence (13). In the next example we show what the Ramanujan-Kolberg package produces for this case.

Example. RK Step 1 computes that

$$q^{-14} \prod_{j=1}^{\infty} \frac{(1-q^j)^{10}(1-q^{2j})^2(1-q^{11j})^{11}}{(1-q^{22j})^{22}} \sum_{n=0}^{\infty} p(11n+6)q^n \in M^{\infty}(22).$$

Note. Step 1 is quite technical; it is explained in detail in [23].

RK Step 2 computes $E_1, \ldots, E_8$ as generators of $E^{\infty}(22)$ where

$$E_1(\tau) = \frac{\eta(2\tau)^8\eta(11\tau)^4}{\eta(\tau)^4\eta(22\tau)^8},\ E_2(\tau) = \frac{\eta(2\tau)\eta(11\tau)^{11}}{\eta(\tau)\eta(22\tau)^{11}},\ E_3(\tau) = \frac{\eta(\tau)^7\eta(11\tau)^3}{\eta(2\tau)^3\eta(22\tau)^7},$$

and the other E_i as in [23].

RK Step 3 computes

$$\langle E^{\infty}(22)\rangle_{\mathbb{C}} = \langle 1, z_1, z_2\rangle_{\mathbb{C}[t]}$$

where

$$t = \frac{1}{11}E_1 - \frac{1}{8}E_2 + \frac{3}{88}E_3,\ z_1 = -3 + \frac{2}{11}E_1 - \frac{1}{8}E_2 - \frac{5}{88}E_3,$$
$$z_2 = -\frac{3}{11}E_1 + \frac{5}{4}E_2 + \frac{1}{44}E_3.$$

Note. When describing the Algorithm MODULE GENERATORS we have chosen t as one of the monoid generators of $E^{\infty}(N)$ and kept it fixed when running the steps of the algorithm. In practice, however, to obtain a most economical description of the resulting module, one is led to update the t. More precisely, whenever a step produces an element with an order smaller than $-\mathrm{ord}_{\infty}(t)$, t is replaced by this element and all other elements are updated according to this new t.

RK Step 4 computes the following Ramanujan-Kolberg identity witnessing the congruence (13):

$$\begin{aligned}\sum_{n=0}^{\infty} p(11n+6)q^n = q^{14}\prod_{j=1}^{\infty}&\frac{(1-q^{22j})^{22}}{(1-q^j)^{10}(1-q^{2j})^2(1-q^{11j})^{11}}\\ &\times\{(1078t^4+13893t^3+31647t^2+11209t-21967)\\ &+z_1(187t^3+5390t^2+594t-9581)\\ &+z_2(11t^3+2761t^2+5368t-6754)\}.\end{aligned}\tag{45}$$

Notice that all integer coefficients on the right-hand side are divisible by 11. But in contrast to Ramanujan's identities (1) and (11), the divisibility of $p(11n+6)$ by 11 follows from (45) only after showing that no denominator of the rational number coefficients in the q-series expansions of t, z_1, and z_2 is divisible by 11. But, using the freshman's dream relation $(1-x)^{11} \equiv 1-x^{11} \pmod{11}$, it is easily seen that $88t$ equals

$$8\prod_{k=1}^{\infty}\frac{(1-q^{2k})^8(1-q^{11k})^4}{(1-q^k)^4(1-q^{22k})^8} - 11\prod_{k=1}^{\infty}\frac{(1-q^{2k})(1-q^{11k})^{11}}{(1-q^k)(1-q^{22k})^{11}} + 3\prod_{k=1}^{\infty}\frac{(1-q^k)^7(1-q^{11k})^3}{(1-q^{2k})^3(1-q^{22k})^7}$$
$$\equiv 8\prod_{k=1}^{\infty}\frac{(1-q^k)^{40}}{(1-q^{2k})^{80}} - 11\prod_{k=1}^{\infty}\frac{(1-q^k)^{120}}{(1-q^{2k})^{120}} + 3\prod_{k=1}^{\infty}\frac{(1-q^k)^{40}}{(1-q^{2k})^{80}} \pmod{11}.$$

The cases z_1 and z_2 work analogously. It should be noticed that even more is true. Namely, the q-series expansions of t, z_1, and z_2 in fact have *integer* coefficients. Using the relations $(1 \pm x)^8 \equiv (1+x^2)^4 \pmod 8$ and $(1-x^2)^8 \equiv (1+x^2)^8 \pmod 8$, this can be verified as follows:

$$\prod_{k=1}^{\infty}\frac{(1-q^k)^{40}}{(1-q^{2k})^{80}}-\prod_{k=1}^{\infty}\frac{(1-q^k)^{120}}{(1-q^{2k})^{120}}\equiv\prod_{k=1}^{\infty}\frac{(1+q^{2k})^{20}}{(1+q^{2k})^{80}}-\prod_{k=1}^{\infty}\frac{1}{(1+q^k)^{120}} \pmod 8$$
$$\equiv\prod_{k=1}^{\infty}\frac{1}{(1+q^{2k})^{60}}-\prod_{k=1}^{\infty}\frac{1}{(1+q^{2k})^{60}} \pmod 8.$$

Note. Despite t, z_1, and z_2 having integer coefficients in their q-expansions, the sufficiently involved structure of this witness identity might explain why Ramanujan (and others) did not come up with such a relation.

Finally we want to come back to the assumption (42) made above. Actually we do not know whether this is true in general. Nevertheless, this "blind spot" is no restriction to the framework in which the RK Algorithm works owing to the following lemma in [23].

Lemma 8.4. *Let* $\mu \in E^{\infty}(N)$ *be as in Lemma* (8.1). *Then there exists an* $n \in \mathbb{N}$, *computable in finitely many steps, such that*

$$\mu^n(\langle E(N)\rangle_{\mathbb{C}} \cap M^{\infty}(N)) \subseteq \langle E^{\infty}(N)\rangle_{\mathbb{C}}.$$

If the left side of (39) has a presentation of the type (41), then the Lemma 8.4 tells us that we have to apply a preprocessing step before executing the RK Algorithm. Namely, multiplying the input (39) by μ^n allows us to find the e_j in (41) as elements in $E^{\infty}(N)$ as desired.

Acknowledgements The research of Radu was supported by the strategic program "Innovatives OÖ 2010 plus" by the Upper Austrian Government in the frame of project W1214-N15-DK6 of the Austrian Science Fund (FWF).

References

1. G.E. Andrews, *The Selected Works of George E. Andrews*. ICP Selected Papers. vol. 3 (Imperial College Press, London, 2013). With commentary, ed. by Andrew V. Sills
2. G.E. Andrews, P. Paule, MacMahon's partition analysis XII: plane partitions. J. Lond. Math. Soc. **76**, 647–666 (2007)
3. G.E. Andrews, P. Paule, MacMahon's partition analysis XI: broken diamonds and modular forms. Acta Arith. **126**, 281–294 (2007)
4. G.E. Andrews, P. Paule, MacMahon's dream, in *Partitions, q-series, and Modular Forms*, Developments in Mathematics, vol. 23 (Springer, New York, 2012), pp. 1–12
5. G.E. Andrews, P. Paule, A. Riese, MacMahon's partition analysis III: the Omega package. Europ. J. Comb. **22**, 887–904 (2001)
6. G.E. Andrews, P. Paule, A. Riese, MacMahon's partition analysis VI: a new reduction algorithm. Ann. Comb. **5**, 251–270 (2001)
7. F. Breuer, Z. Zafeirakopoulos, Polyhedral Omega: a new algorithm for solving linear Diophantine systems. arXiv:1501.07773 (2015)

8. S.H. Chan, Some congruences for Andrews-Paule's broken 2-diamond partitions. Discrete. Math. **308**, 5735–5741 (2008)
9. W.Y.C. Chen, A.R.B. Fan, R.T. Yu, Ramanujan-type congruences for broken 2-diamond partitions modulo 3. Sci. China Math. **57**, 1553–1560 (2014)
10. F. Diamond, J. Shurman, *A First Course in Modular Forms* (Springer, New York, 2005)
11. B. Gordon, K. Ono, Divisibility properties of certain partition functions by powers of primes. Ramanujan J. **1**, 25–35 (1997)
12. J.H. Grace, A. Young, *The Algebra of Invariants* (Cambridge University Press, Cambridge, 1903)
13. G.H. Hardy, P.V.S. Aiyar, B.M. Wilson, *Collected Papers of Srinivasa Ramanujan*. AMS Chelsea Publishing Series (AMS Chelsea Publishing, Providence, 1927)
14. M.D. Hirschhorn, J. Sellers, On recent congruence results of Andrews and Paule for broken k-diamonds. Bull. Aust. Math. Soc. **75**, 121–126 (2007)
15. M.I. Knopp, *Modular Functions in Analytic Number Theory*, vol. 337 (AMS Chelsea Publishing, Providence, 1993)
16. O. Kolberg, Some identities involving the partition function. Mathematica Scandinavica **5**, 77–92 (1957)
17. G. Ligozat, Courbes modulaires de genre 1. Mémoires de la S.M.F **43**, 5–80 (1975)
18. P.A. MacMahon, *Combinatory Analysis*, vols. I, II, 3 edn. (Chelsea, Providence, 1984)
19. R. Miranda, *Algebraic Curves and Riemann Surfaces*. Graduate Studies in Mathematics, vol. 5 (American Mathematical Society, Providence, 1995)
20. K. Ono, The web of modularity: arithmetic of the coefficients of modular forms and q-series, in *CBMS Regional Conference Series in Mathematics*, Washington, DC, vol. 102, 2004
21. G. Pólya, *Mathematics and Plausible Reasoning: Induction and Analogy in Mathematics*, vol. 1 (Princeton University Press, Princeton, 1954)
22. H. Rademacher, The Ramanujan identities under modular substitutions. Trans. Am. Math. Soc. **51**, 609–636 (1942)
23. C.-S. Radu, An algorithmic approach to Ramanujan-Kolberg identities. J. Symb. Comput. **68**, 1–33 (2014)
24. S. Ramanujan, Some properties of $p(n)$, the number of partitions of n. Proc. Camb. Philos. Soc. **19**, 207–210 (1919)
25. J.C. Rosales, P. García-Sánchez, J.I. García-García, M.B. Branko, Systems of inequalities and numerical semigroups. J. Lond. Math. Soc. **65**, 611–623 (2002)
26. G. Xin, An Euclid style algorithm for MacMahon's partition analysis. J. Comb. Theory A **318**, 32–60 (2015)
27. D. Zeilberger, George Eyre Andrews (b. Dec. 4, 1938): A Reluctant REVOLUTIONARY (Part 1). vimeo.com/81396551 in *Rutgers Experimental Mathematics Seminar*, Dec 5, 2013
28. D. Zeilberger, George Eyre Andrews (b. Dec. 4, 1938): A Reluctant REVOLUTIONARY (Part 2). vimeo.com/81410565 in *Rutgers Experimental Mathematics Seminar*, Dec 5, 2013

Moments of orthogonal polynomials and combinatorics

Sylvie Corteel, Jang Soo Kim, and Dennis Stanton

Abstract This paper is a survey on combinatorics of moments of orthogonal polynomials and linearization coefficients. This area was started by the seminal work of Flajolet followed by Viennot at the beginning of the 1980s. Over the last 30 years, several tools were conceived to extract the combinatorics and compute these moments. A survey of these techniques is presented, with applications to polynomials in the Askey scheme.

Keywords Orthogonal polynomials • Moments • Linearization coefficients • Askey Wilson polynomials • Motzkin paths • Staircase tableaux

Mathematics Subject Classification (2010): 33C45, 05A15, 33D45, 05A19

1 Introduction

In this article we will consider polynomials $P_n(x)$ in one variable x with coefficients in a commutative ring $\mathbb{K}$ indexed by a non-negative integer n. In general $\mathbb{K}$ will be $\mathbb{C}$ or $\mathbb{R}$. Sometimes these polynomials will depend on formal variables a, b, c, d, and q.

Given a sequence $\{\mu_n\}_{n\geq 0}$ of elements of $\mathbb{K}$, there exists a unique linear functional $\mathcal{L}$ on the space of polynomials, $\mathcal{L} : \mathbb{K}[x] \to \mathbb{K}$ such that $\mu_n = \mathcal{L}(x^n)$. The sequence $\{\mu_n\}_{n\geq 0}$ is called the *moment sequence* of this linear functional.

S. Corteel (✉)
LIAFA, CNRS et Université Paris Diderot, Case 7014, 75205 Paris, Cedex 13, France
e-mail: Sylvie.Corteel@liafa.univ-paris-diderot.fr

J.S. Kim
Department of Mathematics, Sungkyunkwan University (SKKU), 2066 Seobu-ro, Jangan-gu, Gyeonggi-do, Suwon, 440-746, South Korea
e-mail: jangsookim@skku.edu

D. Stanton
School of Mathematics, University of Minnesota, Minneapolis, MN 55455, USA
e-mail: stanton@math.umn.edu

A. Beveridge et al. (eds.) *Recent Trends in Combinatorics*, The IMA Volumes in Mathematics and its Applications 159, DOI 10.1007/978-3-319-24298-9_22

Definition 1.1. A sequence $\{P_n(x)\}_{n\geq 0}$ is called an *orthogonal polynomial sequence* (OPS) with respect to a linear functional $\mathcal{L}$ if

1. $P_n(x)$ is a polynomial of degree n for each $n \geq 0$,
2. $\mathcal{L}(P_n(x)P_m(x)) = K_n\delta_{n,m}$ for all $n, m \geq 0$, where $K_n \neq 0$.

In this case we also call $P_n(x)$ an *orthogonal polynomial.*

Note that if $\mathcal{L}$ is a linear function for an OPS, then any linear functional $\mathcal{L}'$ which is obtained by multiplying a nonzero constant to $\mathcal{L}$ is also a linear functional for the OPS. Conversely, if $\mathcal{L}$ and $\mathcal{L}'$ are linear functionals for an OPS, then $\mathcal{L} = c \cdot \mathcal{L}'$ for a nonzero constant c. Thus, there is a unique linear functional $\mathcal{L}$ for an OPS such that $\mathcal{L}(1) = 1$. In this case we say that $\mathcal{L}$ and its corresponding moments are *normalized.* Several books were written on the subject [2, 5, 27, 31]. Here we focus on the moments of these polynomials and their relation to combinatorics started by Flajolet [12] and Viennot [33].

The *Hankel determinant* of the moment sequence $\{\mu_n\}_{n\geq 0}$ is

$$\Delta_n = \begin{vmatrix} \mu_0 & \mu_1 & \cdots & \mu_n \\ \mu_1 & \mu_2 & \cdots & \mu_{n+1} \\ \vdots & \vdots & \ddots & \vdots \\ \mu_n & \mu_{n+1} & \cdots & \mu_{2n} \end{vmatrix}.$$

Theorem 1.2. [31] *A linear functional $\mathcal{L}$ has an OPS if and only if $\Delta_n \neq 0$ for all $n \geq 0$. Moreover, an OPS is uniquely determined up to a constant multiple. Indeed, the polynomials $P_n(x)$ in monic form are given explicitly by :*

$$P_n(x) = \frac{1}{\Delta_{n-1}} \begin{vmatrix} \mu_0 & \mu_1 & \cdots & \mu_n \\ \mu_1 & \mu_2 & \cdots & \mu_{n+1} \\ \vdots & \vdots & \ddots & \vdots \\ \mu_{n-1} & \mu_n & \cdots & \mu_{2n-1} \\ 1 & x & \cdots & x^n \end{vmatrix}.$$

One of the most important theorems in the classical theory of orthogonal polynomials is the three-term recurrence relation.

Theorem 1.3. *Any monic OPS $\{P_n(x)\}_{n\geq 0}$ satisfies a three-term recurrence relation*

$$P_{n+1}(x) = (x - b_n)P_n(x) - \lambda_n P_{n-1}(x),$$

for $n \geq 1$, for some sequences $b_0, b_1, \ldots$ and $\lambda_1, \lambda_2, \ldots$, where $\lambda_n \neq 0$.

For non-monic orthogonal polynomials we can always rescale them to get monic ones. The three-term recurrence relation and the moments can be easily obtained as follows.

Proposition 1.4. *Suppose that $P_n(x)$ is an orthogonal polynomial with a three-term recurrence relation*

$$P_{n+1}(x) = (A_n x - B_n)P_n(x) - C_n P_{n-1}(x).$$

Then the rescaled orthogonal polynomial $\tilde{P}_n(x) = a_n P_n(tx)$ satisfies

$$\tilde{P}_{n+1}(x) = (\tilde{A}_n x - \tilde{B}_n)\tilde{P}_n(x) - \tilde{C}_n \tilde{P}_{n-1}(x),$$

where

$$\tilde{A}_n = \frac{a_{n+1}}{a_n} t A_n, \qquad \tilde{B}_n = \frac{a_{n+1}}{a_n} B_n, \qquad \tilde{C}_n = \frac{a_{n+1}}{a_{n-1}} C_n.$$

Moreover, if $\{\mu_n\}_{n\geq 0}$ and $\{\tilde{\mu}_n\}_{n\geq 0}$ are moment sequences of $P_n(x)$ and $\tilde{P}_n(x)$, respectively, then there exists a constant c such that for all $n \geq 0$,

$$\mu_n = c \cdot t^n \tilde{\mu}_n.$$

The orthogonality of orthogonal polynomials with real coefficients is defined by a measure $w(x)$, that is a functional $\mathcal{L} : \mathbb{R}[x] \to \mathbb{R}$ defined by the Stieltjes integral

$$\mathcal{L}(P(x)) = \int_{-\infty}^{\infty} P(x)\mathrm{d}w(x).$$

The most general classical orthogonal polynomials are called the Askey-Wilson polynomials [3]. All of the other classical polynomials can be obtained as a specialization or a limit of these polynomials. The Askey-Wilson polynomials depend upon five parameters: a, b, c, d, and q and are defined by basic hypergeometric series. We use the usual notation

$$(a;q)_k = \prod_{i=0}^{k-1}(1 - aq^i), \quad (a_1, \ldots, a_m; q)_k = \prod_{i=1}^{m}(a_i; q)_k.$$

Definition 1.5. The *Askey-Wilson polynomials* are defined by

$$p_n(x; a, b, c, d|q) = \frac{(ab, ac, ad; q)_n}{a^n} \sum_{k=0}^{n} \frac{(q^{-n}, abcdq^{n-1}, az, a/z; q)_k}{(q, ab, ac, ad; q)_k} q^k \tag{1}$$

with $x = \frac{z+z^{-1}}{2}$.

These functions are polynomials in x of degree n due to the relation

$$(az, a/z; q)_k = \prod_{j=0}^{k-1}(1 - 2axq^j + a^2 q^{2j}).$$

A representing measure for the Askey-Wilson polynomials may be given for $0 < q < 1$ and $\max\{|a|, |b|, |c|, |d|\} < 1$. The orthogonality relation is

$$\int_0^{\pi} p_n(\cos\theta; a, b, c, d|q) p_m(\cos\theta; a, b, c, d|q) w(\cos\theta; a, b, c, d|q) d\theta = 0, \quad n \neq m,$$

where $x = \cos\theta$ and the measure is given by

$$w(x; a, b, c, d|q) = \frac{(z^2, z^{-2}; q)_\infty}{(az, a/z, bz, b/z, cz, c/z, dz, d/z; q)_\infty}.$$

The enumeration formula and the combinatorics of the moments of these polynomials were recently presented in [8, 10, 24, 25]. A recent catalog on known enumeration formulas for moments appeared in [29].

In this paper, we will consider the moments of orthogonal polynomials of the Askey scheme, and emphasize their relationship to combinatorial enumeration. In Section 2, we give a general interpretation of the orthogonal polynomials and their moments using paths [33]. In Section 3, we introduce some basic combinatorics related to the moments of Hermite, Charlier, and Laguerre polynomials and their q-analogues. This was first started in the memoir of Viennot [33] and then in a series of papers by several authors. In Section 4, we present an odd-even trick which is often useful to compute moments and can also give non-trivial combinatorial results. In Section 5, we focus on the case of the Askey-Wilson moments and their enumeration formula. In Section 6, we present some modified moments coming from the Askey-Wilson basis. We end in Section 7 with linearization coefficients.

This paper is a survey and contains a collection of diverse results; so that the interested reader can learn a series of important tools built for the study of moments of orthogonal polynomials.

2 Path interpretation of the polynomials and the moments

For general orthogonal polynomials which satisfy Theorem 1.3, the moments μ_n are always polynomials in the three-term recurrence coefficients with non-negative integral coefficients. For example,

$$\mu_0 = 1, \quad \mu_1 = b_0, \quad \mu_2 = b_0^2 + \lambda_1, \quad \mu_3 = b_0^3 + 2b_0\lambda_1 + b_1\lambda_1.$$

The orthogonal polynomials $p_n(x)$ may also be given as polynomials in x and the three-term recurrence coefficients, but not necessarily positive. For example,

$$p_0(x) = 1, \quad p_1(x) = x - b_0, \quad p_2(x) = x^2 - (b_0 + b_1)x + (b_0 b_1 - \lambda_1).$$

The purpose of this section is to give combinatorial interpretations for these two phenomena using paths.

Paths live in the quarter plane $\Pi = \mathbb{N} \times \mathbb{N}$. A *path* ω is a sequence $\omega = (s_0, \ldots, s_n)$ where $s_i = (x_i, y_i) \in \Pi$. The s_i's are the *vertices* of the path. The starting vertex is s_0, the ending vertex is s_n, and (s_i, s_{i+1}) is an *elementary step* of the path.

The step (s_i, s_{i+1}) is called

- *North* if $x_i = x_{i+1}$ and $y_i + 1 = y_{i+1}$;
- *North-North* if $x_i = x_{i+1}$ and $y_i + 2 = y_{i+1}$;
- *North-East* if $x_i + 1 = x_{i+1}$ and $y_i + 1 = y_{i+1}$;
- *South-East* if $x_i + 1 = x_{i+1}$ and $y_i - 1 = y_{i+1}$; and
- *East* if $x_i + 1 = x_{i+1}$ and $y_i = y_{i+1}$.

The paths are weighted. There is a partial application $\mathrm{wt} : \Pi \times \Pi \to \mathbb{K}$ which associates to each step (s_i, s_{i+1}) a *weight* $\mathrm{wt}(s_i, s_{i+1}) \in \mathbb{K}$. The *weight* of the path ω is the product of the weight of the steps :

$$\mathrm{wt}(\omega) = \prod_{i=0}^{n-1} \mathrm{wt}(s_i, s_{i+1}).$$

Definition 2.1. [33] A *Favard path* is a path $\eta = (s_0, \ldots, s_n)$ starting at $s_0 = (0, 0)$ with three types of elementary steps: North, North-North, and North-East. The weight of the elementary step (s_i, s_{i+1}) is x if the step is North-East and $-b_k$ (resp. $-\lambda_{k+1}$) if the step is North (resp. North-North) and the y-coordinate of s_i is k. The *length* of the path is the y-coordinate of s_n.

Let Fav_n be the set of Favard paths of length n.

Lemma 2.2. [33] *Let $\{P_n(x)\}$ be a sequence of polynomials which satisfies the recurrence of Theorem 1.3 and with boundary conditions $P_0(x) = 1$ and $P_1(x) = x - b_0$. Then*

$$P_n(x) = \sum_{\omega \in \mathrm{Fav}_n} \mathrm{wt}(\omega).$$

A *Motzkin path* of length n is a path consisting of North-East steps, East steps, and South-East steps which lies in the first quadrant. Let $\mathrm{Mot}_{n,k,\ell}$ be the set of Motzkin paths from $(0, k)$ to (n, ℓ), and let $\mathrm{Mot}_n = \mathrm{Mot}_{n,0,0}$.

Suppose that two sequences $b = \{b_n\}_{n\geq 0}$ and $\lambda = \{\lambda_n\}_{n\geq 1}$ are given. For a Motzkin path P, we define $\mathrm{wt}_{b,\lambda}(P)$ to be the product of the weight of every step in P, where the weight of each step is defined as follows.

(1) A North-East step has weight 1.
(2) An East step has weight b_k, where k is the y-coordinate of starting point.
(3) A South-East step has weight λ_k, where k is the y-coordinate of starting point.

For the remainder of this section we assume that $P_n(x)$ are orthogonal polynomials given by $P_{-1}(x) = 0$, $P_0(x) = 1$, and for $n \geq 1$,

$$P_{n+1}(x) = (x - b_n)P_n(x) - \lambda_n P_{n-1}(x),$$

with the normalized linear functional $\mathcal{L}$ and the nth moment $\mu_n = \mathcal{L}(x^n)$.

Theorem 2.3. [33] *We have*

$$\mu_n = \sum_{\omega \in \mathrm{Mot}_n} \mathrm{wt}_{b,\lambda}(\omega).$$

We will sketch a proof of a generalization of Theorem 2.3 that also gives a combinatorial proof of the orthogonality.

Theorem 2.4. [33] *For all n, k, ℓ,*

$$\mathcal{L}(x^n P_k(x) P_\ell(x)) = \lambda_1 \cdots \lambda_\ell \sum_{\omega \in \mathrm{Mot}_{n,k,\ell}} \mathrm{wt}(\omega). \tag{2}$$

Proof. (Sketch [33]) The proof of this theorem uses the combinatorial interpretation of the polynomials as Favard paths. Let $E_{n,k,\ell}$ be the set of triplets made of

- A Favard path f of length k
- A Favard path g of length ℓ
- A Motzkin path ω of length $n + p(f) + p(g)$ where $p(f)$ is the number of North-East steps of f.

Let $F_{n,k,\ell}$ be the subset of $E_{n,k,\ell}$ where $p(f) = k$, $p(g) = \ell$ and ω starts with k North-East steps and ends with ℓ South-East steps. Proving the theorem is equivalent to defining a sign-reversing involution θ on $E_{n,k,\ell}$ whose fixed points are exactly the elements in the set $F_{n,k,\ell}$. Let j be the minimal index such that (s_j, s_{j+1}) is an East step or a South-East step in ω. If $j < k$, then let $h(\omega) = y_j$ otherwise $h(\omega) = \infty$. Let f_j be the first step of f which is a North or a North-North step. Let $h(f) = j$ if such a j exists otherwise $h(f) = \infty$. If $h(\omega)$ or $h(f)$ is finite, then if $h(f) - 1$ is smaller than or equal to $h(\omega)$, if the $h(f)^{th}$ step of f is North-North, we add a North-East step and a South-East step to ω in position $h(f) - 1$ and change the North-North step to two North-East steps; otherwise (it is a North step), we add an East step to ω in position $h(f)$. The step in f is changed to one North-East step. Otherwise ($h(f) - 1 > h(\omega)$), we take off the first East step or the first South-East step and its previous North-East step from ω and change two North-East steps in f by a North-North step or one North-East step in f by a North step. One can easily check that this changes the sign. If both h's are infinite, we apply a similar algorithm on g and the last ℓ steps of ω. The only triplets where the algorithm cannot be applied are exactly the ones in $F_{n,k,\ell}$. □

Note that this implies that

$$\mathcal{L}(P_k(x)P_\ell(x)) = \lambda_1 \ldots \lambda_\ell \delta_{k\ell}.$$

and this gives a proof of the orthogonality relation.

Let

$$C_i(z) = \cfrac{1}{1 - b_{0+i}z - \cfrac{\lambda_{1+i}z^2}{1 - b_{1+i}z - \cfrac{\lambda_{2+i}z^2}{\ddots}}}.$$

Using the theory of Flajolet [12] we have

Corollary 2.5. *We have*

$$\sum_{n\ge 0} \mu_n z^n = \cfrac{1}{1 - b_0 z - \cfrac{\lambda_1 z^2}{1 - b_1 z - \cfrac{\lambda_2 z^2}{\ddots}}} = C_0(z);$$

$$\sum_{n\ge 0} \mathcal{L}(x^n P_\ell(x))z^n = \lambda_1 \cdots \lambda_\ell z^\ell \prod_{i=0}^{\ell} C_i(z);$$

$$\sum_{n\ge 0} \mathcal{L}(x^n P_\ell(x)P_k(x))z^n = \sum_{j=0}^{\min(k,\ell)} \lambda_1 \cdots \lambda_j C_j(z) \prod_{i=j+1}^{\ell} \lambda_i z C_i(z) \prod_{i=j+1}^{k} \lambda_i z C_i(z).$$

The last equation in Corollary 2.5 can be proved by observing the fact that we can uniquely decompose a Motzkin path $\omega \in \mathrm{Mot}_{n,k,\ell}$ as

$$\omega = \omega_{k-j} D \cdots \omega_1 D \omega_0 U \omega'_1 \cdots U \omega'_{\ell-j},$$

where j is a non-negative integer, each ω_i and ω'_i is a Motzkin path in which the starting and ending vertices have the smallest y-coordinates, D is a South-East step, and U is a North-East step.

3 Combinatorics

In this section we show that the moments of Hermite, Charlier, and Laguerre polynomials are equal to the numbers of perfect matchings, set partitions, and permutations, respectively.

Fig. 1 The diagram representing the set partition $\{\{1,4,7\},\{2,5,9\},\{3\},\{6,7\}\}$.

A *set partition* of a set X is a collection of mutually disjoint nonempty subsets of X whose union is X. We denote by Π_n the set of set partitions of $[n] = \{1,2,\ldots,n\}$. Let $\pi = \{B_1, B_2, \ldots, B_k\} \in \Pi_n$. Each element B_i of π is called a *block*. The *size* of a block B_i is the cardinality $|B_i|$. An *edge* of π is a pair (i,j) of integers such that $i < j$, both i and j are contained in the same block, and the block containing i and j does not contain any integer k with $i < k < j$.

We will represent a set partition of $[n]$ as a diagram by placing integers $1,2,\ldots,n$ in a row and connecting i and j for each edge (i,j) of the set partition, see Figure 1. In this pictorial representation, the following definitions are natural.

Let $\pi \in \Pi_n$. An integer $i \in [n]$ is called a *singleton*, an *opener*, a *closer*, or a *continuation* of π, if i is the unique element in a block of size 1, the minimum of a block of size at least 2, the maximum of a block of size at least 2, or none of these, respectively. For example, if π is the set partition in Figure 1, then 3 is a singleton, $1,2,6$ are openers, $7,8,9$ are closers, and $4,5$ are continuations.

A *crossing* of π is a set of two edges (a,b) and (c,d) such that $a < c < b < d$. Let $\mathrm{crossing}(\pi)$ denote the number of crossings of π. For example, if π is the set partition in Figure 1, then $\mathrm{crossing}(\pi) = 4$ since there are four crossings $\{(1,4),(2,5)\}, \{(2,5),(4,7)\}, \{(4,7),(5,9)\}, \{(4,7),(6,8)\}$.

A *matching* is a set partition in which every block has one or two elements. A matching is called *perfect* if every block has 2 elements. Let $\mathcal{M}_n$ denote the set of perfect matchings of $[n]$. One can easily see that $|\mathcal{M}_{2n+1}| = 0$ and

$$|\mathcal{M}_{2n}| = (2n-1)!! = 1 \cdot 3 \cdots (2n-1).$$

Since $\pi \in \mathcal{M}_{2n}$ is also a set partition, we can represent π using a diagram, and crossings of π are defined.

3.1 Hermite polynomials

The *Hermite polynomials* $H_n(x)$ are defined by

$$H_{-1}(x) = 0, \quad H_0(x) = 1, \text{ and}$$

$$H_{n+1}(x) = xH_n(x) - nH_{n-1}(x), \text{ for } n \geq 0.$$

A *Dyck path* of length n is a Motzkin path in Mot_n without East steps. A *Hermite history of length* n is a Dyck path of length n in which every South-East step between the lines $y = i$ and $y = i-1$ is labeled by an integer in $\{0,1,\ldots,i-1\}$.

By Theorem 2.3, the nth moment μ_n of the Hermite polynomials is

$$\mu_n = \sum_{P \in \mathrm{Mot}_n} \mathrm{wt}_{b,\lambda}(P),$$

where $b = \{b_n\}_{n\ge 0}$ and $\lambda = \{\lambda_n\}_{n\ge 1}$ are given by $b_n = 0$ for all $n \ge 0$ and $\lambda_n = n$ for $n \ge 1$. Let $P \in \mathrm{Mot}_n$. Then $\mathrm{wt}_{b,\lambda}(P) = 0$ if P has an East step. Otherwise, P is a Dyck path and $\mathrm{wt}_{b,\lambda}(P)$ is equal to the number of Hermite histories whose underlying Dyck path is P. Thus μ_n is equal to the number of Hermite histories of length n.

We now give a bijection $\psi_H : \mathcal{M}_n \to \mathcal{H}_H(n)$ from the set $\mathcal{M}_n$ of perfect matchings of $[n]$ to the set $\mathcal{H}_H(n)$ of Hermite histories of length n, which appears in [11, Section 2]. It suffices to consider the case when n is even since both sets are empty otherwise. Let $\pi \in \mathcal{M}_{2n}$ be a perfect matching of $[2n]$. Then the corresponding Hermite history $\psi_H(\pi)$ is defined as follows. For $i = 1, 2, \dots, 2n$, if i is an opener of π, then the ith step of $\psi_H(\pi)$ is a North-East step. If i is a closer of π, then the ith step of $\psi_H(\pi)$ is a South-East step labeled by the number of crossings $\{(a,b),(c,d)\}$ of π with $a < c < b < d$ and $b = i$. See Figure 2 for an example. One can check that ψ_H is a bijection. Thus we have the following theorem.

Theorem 3.1. *The nth moment of the Hermite polynomial is equal to the number of perfect matchings of* $[n]$. *In other words,* $\mu_{2n+1} = 0$ *and* $\mu_{2n} = (2n-1)!!$.

Classically, Theorem 3.1 may be derived using the representing measure for the Hermite linear functional

$$\mathcal{L}_{\text{Hermite}}(p(x)) = \frac{1}{\sqrt{2\pi}} \int_{-\infty}^{\infty} p(x) e^{-x^2/2} dx.$$

The *q-Hermite polynomials* $H_n(x|q)$ are defined by $H_{-1}(x|q) = 0$, $H_0(x|q) = 1$, and for $n \ge 0$,

$$H_{n+1}(x|q) = 2xH_n(x|q) - (1-q^n)H_{n-1}(x|q).$$

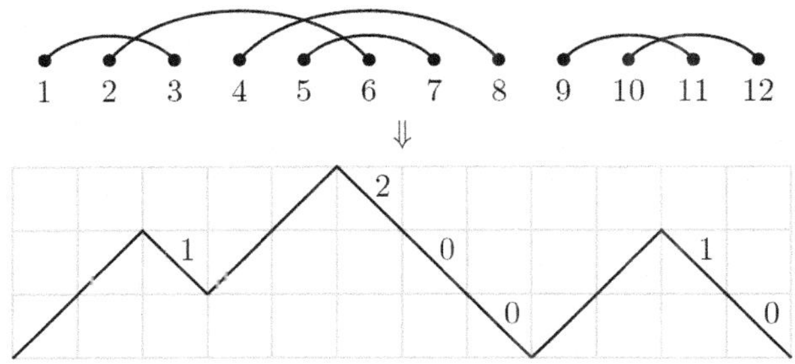

Fig. 2 An example of the bijection ψ_H sending perfect matchings to Hermite histories.

For our purposes, it is more convenient to consider the *rescaled q-Hermite polynomials*

$$\tilde{H}_n(x|q) = \frac{H_n(\frac{\sqrt{1-q}}{2}x|q)}{(1-q)^{n/2}}.$$

Then by Proposition 1.4, we have $\tilde{H}_{-1}(x|q) = 0$, $\tilde{H}_0(x|q) = 1$, and for $n \geq 1$,

$$\tilde{H}_{n+1}(x|q) := x\tilde{H}_n(x|q) - [n]_q\tilde{H}_{n-1}(x|q).$$

where we have used

$$[n]_q = \frac{1-q^n}{1-q} = 1 + q + \cdots + q^{n-1}.$$

We see that the nth moment of $\tilde{H}_n(x|q)$ is

$$\mu_n = \sum_{\eta \in \mathcal{H}_H(n)} q^{|\eta|},$$

where $|\eta|$ is the sum of labels of South-East steps. The bijection $\psi_H : \mathcal{M}_n \to \mathcal{H}_H(n)$ satisfies $\mathrm{crossing}(\pi) = |\psi_H(\eta)|$. Thus we obtain the following theorem.

Theorem 3.2. [18, Eq. (3.8)] *The nth moment of the rescaled q-Hermite polynomials $\tilde{H}_n(x|q)$ is*

$$\mu_n = \sum_{\pi \in \mathcal{M}_n} q^{\mathrm{crossing}(\pi)}.$$

The rescaled *discrete q-Hermite polynomials* $\tilde{h}_n(x;q)$ are defined by $\tilde{h}_{-1}(x;q) = 0$, $\tilde{h}_0(x;q) = 1$, and for $n \geq 0$,

$$\tilde{h}_{n+1}(x;q) = x\tilde{h}_n(x;q) - q^{n-1}[n]_q\tilde{h}_{n-1}(x;q).$$

Again the odd moments are 0. The analogous version to Theorem 3.2 is the next result.

Theorem 3.3. [30, p. 310, (5.4)] *The $2n^{th}$ moment of the rescaled discrete q-Hermite polynomials $\tilde{h}_n(x;q)$ is*

$$\mu_{2n} = [1]_q[3]_q \cdots [2n-1]_q = \sum_{\pi \in \mathcal{M}_{2n}} q^{\mathrm{crossing}(\pi) + 2\,\mathrm{nesting}(\pi)},$$

where $\mathrm{nesting}(\pi)$ *is the number of pairs of two edges (a,b) and (c,d) of π such that $a < c < d < b$.*

3.2 Charlier polynomials

The *Charlier polynomials* $C_n(x)$ are defined by

$$C_{n+1}(x) = (x - n - 1)C_n(x) - nC_{n-1}(x),$$

for $n \geq 1$ with initial conditions $C_{-1}(x) = 0$ and $C_0(x) = 1$.

Theorem 3.4. *The nth moment of the Charlier polynomials is equal to the number of set partitions of* $[n]$.

Theorem 3.4 may be derived using the representing measure for the Charlier linear functional

$$\mathcal{L}_{\text{Charlier}}(p(x)) = \frac{1}{e}\sum_{n=0}^{\infty} p(n)\frac{1}{n!}.$$

We will instead prove a generalization of the above theorem using the *q-Charlier polynomials* $C_n(x, a; q)$ given by

$$C_{n+1}(x, a; q) = (x - a - [n]_q)C_n(x, a; q) - a[n]_q C_{n-1}(x, a; q).$$

A *Charlier history of length n* is a Motzkin path of length n in which every South-East step between the lines $y = i$ and $y = i - 1$ is labeled by an integer in $\{0, 1, \ldots, i-1\}$ and every East on the line $y = i$ is either unlabeled or labeled by an integer in $\{0, 1, \ldots, i-1\}$.

By Theorem 2.3, the nth moment μ_n of the Charlier polynomials is

$$\mu_n = \sum_{P \in \text{Mot}_n} \text{wt}_{b,\lambda}(P),$$

where $b = \{b_n\}_{n\geq 0}$ and $\lambda = \{\lambda_n\}_{n\geq 1}$ are given by $b_n = n + a$ for all $n \geq 0$ and $\lambda_n = an$ for $n \geq 1$. Let $P \in \text{Mot}_n$. Then

$$\text{wt}_{b,\lambda}(P) = \sum_{\eta} a^{\text{unlabeled}(\eta)} q^{|\eta|},$$

where the sum is over all Charlier histories η with underlying Motzkin path P, unlabeled(η) is the number of unlabeled steps of η, and $|\eta|$ is the sum of labels in η.

We now give a bijection $\psi_C : \Pi_n \to \mathcal{H}_C(n)$ from the set Π_n of set partitions of $[n]$ to the set $\mathcal{H}_C(n)$ of Charlier histories of length n. This map can be considered as a special case of Foata and Zeilberger's map [13], which will be described in the next subsection. Let $\pi \in \Pi_n$ be a set partition of $[n]$. Then the corresponding Charlier history $\psi_C(\pi) = \eta$ is defined as follows. For $i = 1, 2, \ldots, n$, if i is an

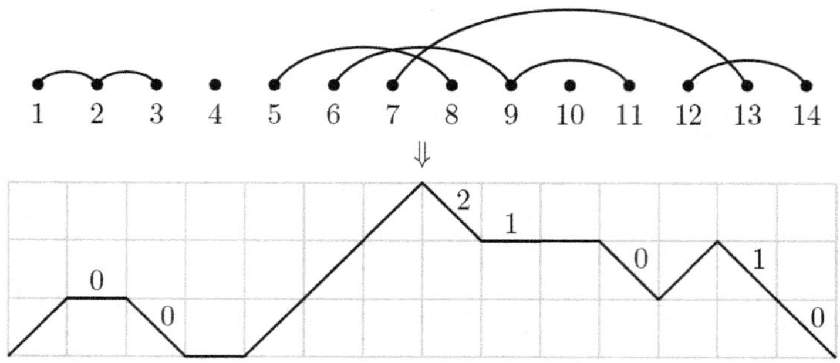

Fig. 3 An example of the bijection ψ_C sending set partitions to Charlier histories.

opener of π, then the ith step of η is a North-East step. If i is a closer of π, then the ith step of η is a South-East step labeled by the number of crossings $\{(a,b),(c,d)\}$ of π with $a < c < b < d$ and $b = i$. If i is a singleton of π, then the ith step of η is an unlabeled East step. If i is a continuation of π, then the ith step of η is an East step labeled by the number of crossings $\{(a,b),(c,d)\}$ of π with $a < c < b < d$ and $b = i$. See Figure 3 for an example.

It is easy to see that ψ_C is a bijection such that if $\psi_C(\pi) = \eta$ then $\text{unlabeled}(\eta) = \text{block}(\pi)$ and $|\eta| = \text{crossing}(\pi)$. This implies the following theorem due to Kim, Stanton, and Zeng [26, Theorem 4].

Theorem 3.5. *The* q-Charlier polynomials $C_n(x,a;q)$ *given by*

$$C_{n+1}(x,a;q) = (x - a - [n]_q)C_n(x,a;q) - a[n]_q C_{n-1}(x,a;q)$$

have the nth moment

$$\mu_n = \sum_{\pi \in \Pi_n} a^{\text{block}(\pi)} q^{\text{crossing}(\pi)}.$$

3.3 Laguerre polynomials

The *Laguerre polynomials* $L_n(x)$ are defined by

$$L_{n+1}(x) = (x - 2n - 1)L_n(x) - n^2 L_{n-1}(x).$$

Theorem 3.6. *The nth moment of the Laguerre polynomials* $L_n(x)$ *is equal to the number of permutations of* $[n]$, *that is,* $\mu_n = n!$.

Theorem 3.6 may be derived using the representing measure for the Laguerre linear functional

$$\mathcal{L}_{\text{Laguerre}}(p(x)) = \int_0^\infty p(x)e^{-x}dx.$$

We will consider the following generalization of the above theorem.

The *q-Laguerre polynomials* $L_n(x;q)$ are defined by

$$L_{n+1}(x;q) = (x - [n]_q - y[n+1]_q)L_n(x;a) - y[n]_q^2 L_{n-1}(x;q).$$

A *Laguerre history of length n* is a Motzkin path of length n in which every North-East or South-East step between the lines $y = i$ and $y = i-1$ is labeled by an integer in $\{0, 1, \dots, i-1\}$ and every East step on the line $y = i$ is either unlabeled or labeled by an integer in $\{-i, -i+1, \dots, -1, 0, 1, \dots, i-1\}$.

By Theorem 2.3, the nth moment μ_n of the Laguerre polynomials is

$$\mu_n = \sum_{P \in \text{Mot}_n} \text{wt}_{b,\lambda}(P),$$

where $b = \{b_n\}_{n\geq 0}$ and $\lambda = \{\lambda_n\}_{n\geq 1}$ are given by $b_n = y[n+1]_q + [n]_q$ for all $n \geq 0$ and $\lambda_n = y[n]_q^2$ for $n \geq 1$. Let $P \in \text{Mot}_n$. Then it is easy to see that

$$\text{wt}_{b,\lambda}(P) = \sum_\eta y^{w(\eta)} q^{|\eta|},$$

where the sum is over all Laguerre histories η with underlying Motzkin path P. $w(\eta)$ is the total number of North-East steps, unlabeled East steps, and East steps labeled by a non-negative integer, and $|\eta|$ is the sum of labels in η.

We now give a bijection $\psi_L : S_n \to \mathcal{H}_L(n)$ from the set S_n of permutations of $[n]$ to the set $\mathcal{H}_L(n)$ of Laguerre histories of length n, which is due to Foata and Zeilberger [13]. We need the following definitions to describe $\psi_L(\pi)$.

Let $\pi \in S_n$ be a permutation of $[n]$. An *upper edge* of π is a pair (i,j) such that $i < \pi(i) = j$. A *lower edge* of π is a pair (i,j) such that $i = \pi(j) < j$. We represent π using the diagram obtained by putting n vertices labeled by $1, 2, \dots, n$ and placing an upper or lower arc connecting i and j for each upper or lower edge (i,j). For example, the upper diagram in Figure 4 represents the following permutation in the two-line notation:

$$\pi = \begin{pmatrix} 1 & 2 & 3 & 4 & 5 & 6 & 7 & 8 & 9 & 10 & 11 & 12 & 13 & 14 \\ 7 & 9 & 4 & 3 & 5 & 11 & 8 & 1 & 2 & 14 & 10 & 6 & 12 & 13 \end{pmatrix},$$

where the entry in the second row and in the ith column means $\pi(i)$. We say that i is an *opener* of π if $i < \pi(i)$ and $i < \pi^{-1}(i)$, a *closer* of π if $\pi^{-1}(i) < i$ and $\pi(i) < i$, a *singleton* of π if $\pi(i) = i$, and a *transient* of π otherwise. An *upper crossing* of π is

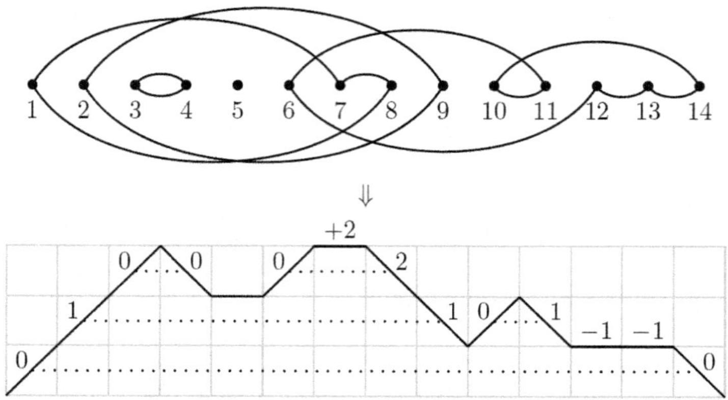

Fig. 4 An example of the bijection ϕ_L sending permutations to Laguerre histories. The matching North-East steps and South-East steps are connected by dotted lines.

a set $\{(a,b),(c,d)\}$ of two upper edges of π with $a < c < b < d$ and $b = i$. A *lower crossing* of π is a set $\{(a,b),(c,d)\}$ of two lower edges of π with $a < c \le b < d$ and $b = i$.

Let $\pi \in S_n$ be a permutation of $[n]$. Then the corresponding Laguerre history $\psi_L(\pi) = \eta$ is defined as follows.

For $i = 1, 2, \ldots, n$, if i is an opener of π, then the ith step of η is a North-East step whose label will be determined later. If i is a closer of π, then the ith step of η is a South-East step. In this case suppose that the jth step and ith step are the matching North-East step and South-East step. Then the label of the jth step (resp. ith step) is the number of upper crossings (resp. lower crossings) $\{(a,b),(c,d)\}$ of π with $a < c < b < d$ and $b = i$. If i is a singleton, then the ith step of η is an unlabeled East step. Finally suppose that i is a transient of π. Then the ith step of η is an East step whose labeled is determined as follows. If the vertex i is incident to two upper edges, then the label is $+k$, where k is the number of upper crossings $\{(a,b),(c,d)\}$ of π with $a < c < b < d$ and $b = i$. If the vertex i is incident to two lower edges, then the label is $-k$, where k is the number of lower crossings $\{(a,b),(c,d)\}$ of π with $a < c \le b < d$ and $b = i$. See Figure 4 for an example. One can check that ϕ_L is a bijection such that if $\phi_L(\pi) = \eta$ then $\mathrm{wex}(\pi) = w(\eta)$ and $\mathrm{crossing}(\pi) = |\eta|$. Here $\mathrm{wex}(\pi)$ is the number of *weak excedances*, which are integers i such that $i \le \pi(i)$. Thus we obtain the following theorem, which is shown in [23, Theorem 2]. An equivalent version of this theorem in terms of continued fractions has been proved earlier, see [6, 30].

Theorem 3.7. *The q-Laguerre polynomials $L_n(x;q)$ have the nth moment*

$$\mu_n = \sum_{\pi \in S_n} y^{\mathrm{wex}(\pi)} q^{\mathrm{crossing}(\pi)}.$$

Corteel et al. [9] showed that the moment has several interpretations in terms of permutations, permutation tableaux, matrix ansatz, etc. Simion and Stanton [30] consider octabasic Laguerre polynomials (whose recurrence relation has eight independent q's) and express their moments using various statistics of permutations. There is an explicit formula for the nth moment of q-Laguerre polynomials, see [9, 21, 22].

4 The odd-even trick

In this section we give a relation between moments for two different sets of orthogonal polynomials. We call this occurrence *the odd-even trick.* It occurs for the three cases considered in Section 3, and offers alternative combinatorial interpretations for the moments. This idea appears in [5, p. 40].

Definition 4.1. For an orthogonal polynomial satisfying the Theorem 2.4, let

$$\mu_n = \mu_n(\{b_k\}_{k\geq 0}, \{\lambda_k\}_{k\geq 1})$$

denote the nth moment.

From Theorem 2.4, μ_n is a positive polynomial function of the three-term recurrence coefficients. The odd-even trick is the observation that if $b_k = 0$ for all k, the polynomials $p_n(x)$ are alternately even and odd. Moreover if

$$p_{2n}(x) = e_n(x), \quad p_{2n+1}(x) = xo_n(x^2),$$

the even and odd polynomials $e_n(x)$ and $o_n(x)$ are themselves orthogonal polynomials. The moment sequences for $e_n(x)$ and $o_n(x)$ are related to the moment sequence of $p_n(x)$.

Proposition 4.2. *Given a sequence λ_k with $\lambda_0 = 0$ we have*

$$\begin{aligned}\mu_{2n}(0, \{\lambda_k\}_{k\geq 1}) &= \mu_n(\{\lambda_{2k} + \lambda_{2k+1}\}_{k\geq 0}, \{\lambda_{2k-1}\lambda_{2k}\}_{k\geq 1}),\\ \mu_{2n+2}(0, \{\lambda_k\}_{k\geq 1}) &= \lambda_1\mu_n(\{\lambda_{2k+1} + \lambda_{2k+2}\}_{k\geq 0}, \{\lambda_{2k}\lambda_{2k+1}\}_{k\geq 1}).\end{aligned}$$

Proof. There are several ways to prove this. We give a sketch of the combinatorial proof due to Viennot [34]. The first statement is equivalent to a bijection between

- Dyck paths of length $2n$ where South-East steps starting at y-coordinate k are weighted by λ_k,
- Motzkin paths of length n where the East steps (resp. South-East steps) starting at y-coordinate k are weighted by $\lambda_{2k} + \lambda_{2k+1}$ (resp. $\lambda_{2k}\lambda_{2k-1}$).

Starting with a Dyck path of length $2n$, $(s_0, \ldots, s_{2n})$. For i from 0 to $n-1$, suppose that $s_{2i} = (2i, 2k)$

- If (s_{2i}, s_{2i+1}) and (s_{2i+1}, s_{2i+2}) are North-East steps, change them to one North-East step;
- If (s_{2i}, s_{2i+1}) and (s_{2i+1}, s_{2i+2}) are South-East steps, change them to one South-East step weighted by $\lambda_{2k}\lambda_{2k-1}$;
- If (s_{2i}, s_{2i+1}) is a South-East step and (s_{2i+1}, s_{2i+2}) is a North-East step, change them to one East step weighted by λ_{2k};
- If (s_{2i}, s_{2i+1}) is a North-East step and (s_{2i+1}, s_{2i+2}) is a South-East step, change them to one East step weighted by λ_{2k+1}.

This gives a Motzkin path of length n with the correct weights. The second statement is equivalent to a bijection between

- Dyck paths of length $2n+2$ where South-East steps starting at y-coordinate k are weighted by λ_k
- Motzkin paths of length n where the East steps (resp. South-East step) starting at y-coordinate k are weighted by $\lambda_{2k+1}+\lambda_{2k+2}$ (resp. $\lambda_{2k+1}\lambda_{2k}$).

One needs to delete the first and last step of the Dyck path and apply the same bijection and adjust the weights accordingly. □

Example 4.3. The Hermite polynomials have $b_k = 0$, $\lambda_k = k$, so

$$(2n-1)!! = \mu_{2n}(0, \{k\}_{k\geq 1}) = \mu_n(\{4k+1\}_{k\geq 0}, \{2k(2k-1)\}_{k\geq 1}).$$

Example 4.4. The Laguerre polynomials have $b_k = 2k+1$, $\lambda_k = k^2$, so

$$n! = \mu_n(\{2k+1\}_{k\geq 0}, \{k^2\}_{k\geq 1}) = \mu_{2n}(0, \{\lfloor (k+1)/2 \rfloor\}_{k\geq 1}).$$

Example 4.5. The q-Charlier polynomials have $b_k = a + [k]_q$, $\lambda_k = a[k]_q$, so in this case

$$\mu_n(\{b\}_{k\geq 0}, \{\lambda_k\}_{k\geq 1}) = \mu_{2n}(0, \{\Lambda_k\}_{k\geq 1}),$$

where

$$\Lambda_k = \begin{cases} [k/2]_q, & \text{if } k \text{ is even,} \\ a, & \text{if } k \text{ is odd.} \end{cases}$$

Example 4.6. The renormalized Askey-Wilson polynomials $Q_n(y)$ (see Proposition 5.1)

$$y = a + a^{-1} - 2x, \quad Q_n(y) = (-1)^n P_n(x)$$

satisfy Theorem 1.3 with

$$b_k = A_{k-1} + C_k, \quad \lambda_k = A_k C_k.$$

In this case

$$\mu_n(\{b_k\}_{k\geq 0}, \{\lambda_k\}_{k\geq 1}) = \mu_{2n}(0, \{\Lambda_k\}_{k\geq 1}),$$

where

$$\Lambda_k = \begin{cases} C_{k/2}, & \text{if } k \text{ is even}, \\ A_{(k-1)/2}, & \text{if } k \text{ is odd}. \end{cases}$$

The next Theorem follows from Example 4.6.

Theorem 4.7. *Let*

$$\theta_{2n} = \mu_{2n}(0, \{\Lambda_k\}_{k\geq 1}).$$

The Askey-Wilson moments satisfy

$$2^n \mu_n(a, b, c, d|q) = \sum_{s=0}^{n} \binom{n}{s} (a + 1/a)^s (-1)^{n-s} \theta_{2n-2s}.$$

Using the Proposition 4.2 we see

Corollary 4.8. *Given a sequence a_k for $k \geq 0$ with $a_0 = 0$ we have*

$$\mu_n(\{a_{2k+1} + a_{2k}\}_{k\geq 0}, \{a_{2k}a_{2k-1}\}_{k\geq 1}) = a_1 \mu_{n-1}(\{a_{2k+2} + a_{2k+1}\}_{k\geq 0}, \{a_{2k+1}a_{2k}\}_{k\geq 1}).$$

Example 4.9. The Laguerre polynomials have $b_k = 2k+1$, $\lambda_k = k^2$, so $a_k = \lceil k/2 \rceil$

$$(n+1)! = \mu_{n+1}(\{2k+1\}_{k\geq 0}, \{k^2\}_{k\geq 1}) = \mu_n(\{2k+2\}_{k\geq 0}, \{k(k+1)\}_{k\geq 1}).$$

5 Askey-Wilson moments

In this section we consider the Askey-Wilson polynomials and explicit forms for their moments. Recall that these polynomials in x have five parameters: a, b, c, d, and q. In Subsection 5.1 we give explicit formulas for the moments. In Subsection 5.2 we use a non-linear change of variables on the parameters to obtain four new parameters α, β, γ, and δ. Using these parameters we give a combinatorial interpretation for the moments as weights of special tableaux.

Recall from (1.5) that the Askey-Wilson polynomial $p_n(x; a, b, c, d|q)$ is not monic and has the leading term $(abcd; q)_n(2x)^n$. It is true, but not obvious, that the Askey-Wilson polynomial $p_n(x; a, b, c, d|q)$ is symmetric in all four parameters a, b, c, and d, not just b, c, and d.

Let

$$P_n(x) = \frac{1}{(abcd; q)_n} p_n(x; a, b, c, d|q)$$

be the normalized Askey-Wilson (AW) polynomial whose leading term is $(2x)^n$. The moment sequences for $P_n(x)$ and $p_n(x; a, b, c, d|q)$ are the same. For historical reasons, and to agree with the literature, we keep the $2x$.

Proposition 5.1. *The normalized Askey-Wilson polynomials $P_n(x)$ satisfy the three-term recurrence relation :*

$$P_{n+1}(x) = (2x - b_n)P_n(x) - \lambda_n P_{n-1}(x)$$

for $n \geq 0$ with $P_{-1}(x) = 0$ and $P_0(x) = 1$,

$$b_n = (a + a^{-1} - A_n - C_n), \quad \lambda_n = A_{n-1}C_n,$$

where

$$A_n = \frac{(1 - abq^n)(1 - acq^n)(1 - adq^n)(1 - abcdq^{n-1})}{a(1 - abcdq^{2n-1})(1 - abcdq^{2n})},$$

$$C_n = \frac{a(1 - q^n)(1 - bcq^{n-1})(1 - bdq^{n-1})(1 - cdq^{n-1})}{(1 - abcdq^{2n-2})(1 - abcdq^{2n-1})}.$$

In this section, we will start by giving enumeration formulas for the moments of these polynomials. We will then present the combinatorics coming from staircase tableaux. We will end the section with some special cases.

5.1 Enumeration formulas

The moments of the AW polynomials, denoted $\mu_n(a, b, c, d|q)$, are not polynomials in a, b, c, d, and q. But it was recently proven [24, Proposition 2.1] that $2^n(abcd; q)_n\mu_n(a, b, c, d|q)$ are polynomials in a, b, c, d, q with integer coefficients. Several explicit formulas for $\mu_n(a, b, c, d|q)$ are known. The following is the simplest known expression as a double sum.

Theorem 5.2. [10, Theorem 1.13] *We have*

$$\mu_n(a, b, c, d|q) = \frac{1}{2^n}\sum_{m=0}^{n}\frac{(ab, ac, ad; q)_m}{(abcd; q)_m}q^m\sum_{j=0}^{m}\frac{q^{-j^2}a^{-2j}(aq^j + q^{-j}/a)^n}{(q, q^{1-2j}/a^2; q)_j(q, q^{2j+1}a^2; q)_{m-j}}.$$

Theorem 5.2 was proved using techniques built in [16, 17]. The formula has two clear defects. It does not demonstrate the polynomiality of $2^n(abcd; q)_n\mu_n(a, b, c, d|q)$ in a, and it is not obviously symmetric in the four parameters a, b, c, and d.

A fivefold sum formula was found where the polynomiality is clear.

Corollary 5.3. [24, Theorem 5.6] *We have*

$$2^n\mu_n(a,b,c,d|q)=\sum_{k=0}^{n}\left(\binom{n}{\frac{n-k}{2}}-\binom{n}{\frac{n-k}{2}-1}\right)\sum_{u+v+w+x+2t=k}a^ub^vc^wd^x$$
$$\times\frac{(ac;q)_v(bd;q)_w}{(abcd;q)_{v+w}}(-1)^tq^{\binom{t+1}{2}}\begin{bmatrix}u+v+w+t\\u\end{bmatrix}_q\begin{bmatrix}v+w+x+t\\v,w,x+t\end{bmatrix}_q\begin{bmatrix}u+x+t\\x\end{bmatrix}_q,$$

where the second sum is over all integers $u,v,w,x\geq 0$, and $-k\leq t\leq k/2$ satisfying $u+v+w+x+2t=k$.

A combinatorial approach to the proof of Corollary 5.3 is given in Section 7. When $d=0$, the moments $2^n\mu_n(a,b,c,0|q)$ are polynomials and there is an explicit polynomial formula which also establishes symmetry.

Corollary 5.4. [24, Theorem 2.3] *The Askey-Wilson moments for $d=0$ are*

$$2^n\mu_n(a,b,c,0|q)=\sum_{k=0}^{n}\left(\binom{n}{\frac{n-k}{2}}-\binom{n}{\frac{n-k}{2}-1}\right)$$
$$\times\sum_{u+v+w+2t=k}a^ub^vc^w(-1)^tq^{\binom{t+1}{2}}\begin{bmatrix}u+v+t\\v\end{bmatrix}_q\begin{bmatrix}v+w+t\\w\end{bmatrix}_q\begin{bmatrix}w+u+t\\u\end{bmatrix}_q,$$

where the second sum is over all integers $u,v,w\geq 0$, and $-k\leq t\leq k/2$ satisfying $u+v+w+2t=k$.

There is a result showing symmetry using very-well poised basic hypergeometric series.

Theorem 5.5. [24, Theorem 2.10] *For an arbitrary A,*

$$2^n\mu_n(a,b,c,d|q)=\sum_{m=0}^{n}\frac{(aA,bA,cA,dA;q)_m}{(A^2,abcd;q)_m}(-q)^m{}_8W_7(m)\sum_{s=0}^{n+1}\left(\binom{n}{s}-\binom{n}{s-1}\right)$$
$$\times\sum_{p=0}^{n-2s-m}A^{-n+2s+2p}\begin{bmatrix}m+p\\m\end{bmatrix}_q\begin{bmatrix}n-2s-p\\m\end{bmatrix}_q q^{m(-n+2s+p)+\binom{m}{2}},$$

where

$${}_8W_7(m)={}_8W_7(A^2/q;A/a,A/b,A/c,A/d,q^{-m};q,abcdq^m),$$

and [14, Chap. 2.1]

$$_8W_7(a_0; a_1, \ldots, a_5; q, z) = \sum_{k=0}^{\infty} \frac{(1 - a_0 q^{2k})}{(1 - a_0)} \frac{(a_0; q)_k}{(q; q)_k} \prod_{i=1}^{5} \frac{(a_i; q)_k}{(qa_0/a_i; q)_k} z^k.$$

The expression for $2^n(abcd; q)_n \mu_n(a, b, c, d|q)$ in Theorem 5.5 is clearly symmetric in a, b, c, and d. From Theorem 5.5 it may be shown that it is a polynomial in these four parameters. We do not give the details here. Instead we offer another representation which shows polynomiality but breaks symmetry

$$\frac{(aA, bA, cA, dA; q)_m}{(A^2; q)_m} {}_8W_7(m) = \sum_{j=0}^{m} \begin{bmatrix} m \\ j \end{bmatrix}_q (cd)^j (A/c, A/d; q)_j (ab; q)_j \\ \times (Aaq^j, Abq^j, cd; q)_{m-j}. \quad (3)$$

The details appear in [24].

The results in Corollaries 5.3 and 5.4, and Theorem 5.5, all involve a product of differences of binomial coefficients and q-binomial coefficients. The reason for this unusual behavior is the form of Theorem 5.2. The power in the numerator leads to binomial coefficients, while the denominator terms lead to differences of q-binomial coefficients. This difference can be switched to the binomial coefficients, giving the differences that are displayed.

The proofs of these results are analytic and use properties of the Askey-Wilson functional $\mathcal{L}_{AW}$. There is a combinatorial proof of the $c = 0$ case of Corollary 5.4 using weighted Motzkin paths in [24]. Josuat-Vergès also did this case (which is the Al-Salam-Chihara polynomials), see [20, Theorem 6.1.1]. The differences of binomial coefficients occur naturally in these combinatorial approaches, see (14). A combinatorial proof of a general result remains open.

5.2 *Combinatorics of the moments*

We define here a combinatorial object that was defined to study the stationary distribution of the asymmetric exclusion process with open boundaries [32].

Definition 5.6. [8] A *staircase tableau* of size n is a Young diagram of "staircase" shape $(n, n-1, \ldots, 2, 1)$ such that boxes are either empty or labeled with α, β, γ, or δ, subject to the following conditions:

- no box along the diagonal is empty;
- all boxes in the same row and to the left of a β or a δ are empty;
- all boxes in the same column and above an α or a γ are empty.

See the left entry of Figure 5 for an example of a staircase tableau.

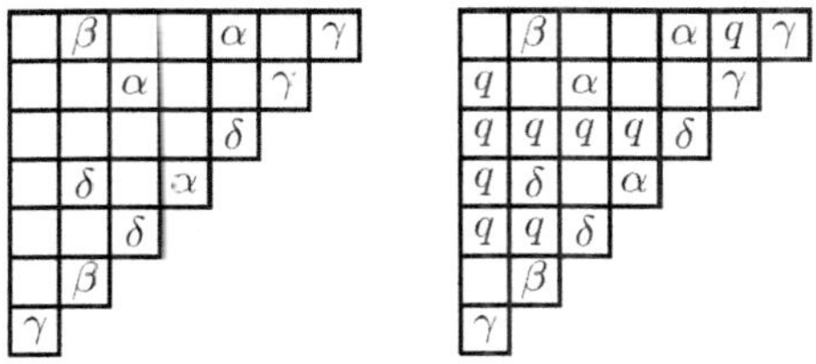

Fig. 5 A staircase tableau of size 7

Definition 5.7. [8] The *weight* $\mathrm{wt}(\mathcal{T})$ of a staircase tableau $\mathcal{T}$ is a monomial in $\alpha, \beta, \gamma, \delta$, and q, which we obtain as follows. Some blank boxes of $\mathcal{T}$ are assigned a q, based on the label of the closest labeled box to its right in the same row and the label of the closest labeled box below it in the same column, such that:

- every blank box which sees a δ to its right gets assigned a q;
- every blank box which sees an α or γ to its right, and a β or γ below it, gets assigned a q.

After this assignment, the *weight* of $\mathcal{T}$, $\mathrm{wt}(\mathcal{T})$ is then defined as the product of all labels in all boxes.

The right entry of Figure 5 shows that this staircase tableau has weight $\alpha^3\beta^2\gamma^3\delta^3q^9$. The *type* of the tableau $\mathcal{T}$ is the number of α's plus the number of δ's on the diagonal. We denote it by $t(\mathcal{T})$.

Let

$$Z_n(y,\alpha,\beta,\gamma,\delta,q) = \sum_{\mathcal{T}} \mathrm{wt}(\mathcal{T})y^{t(\mathcal{T})}, \tag{4}$$

where the sum is taken on the staircase tableaux of size n.

We now link this generating polynomial to the moments of the Askey-Wilson polynomials. Let

$$a = \frac{1-q-\alpha+\gamma+\sqrt{(1-q-\alpha+\gamma)^2+4\alpha\gamma}}{2\alpha},$$

$$c = \frac{1-q-\alpha+\gamma-\sqrt{(1-q-\alpha+\gamma)^2+4\alpha\gamma}}{2\alpha},$$

$$b = \frac{1-q-\beta+\delta+\sqrt{(1-q-\beta+\delta)^2+4\beta\delta}}{2\beta},$$

$$d = \frac{1-q-\beta+\delta-\sqrt{(1-q-\beta+\delta)^2+4\beta\delta}}{2\beta}.$$

Proposition 5.8. (see [10]; Theorem 1.11). *The generating polynomial* $Z_n(y;\alpha,\beta,\gamma,\delta;q)$ *is*

$$(abcd;q)_n\sqrt{y}^n(\alpha\beta)^n\times\mu_n$$

where μ_n *are the moments of the orthogonal polynomials defined by*

$$G_{n+1}(x)=(x-B_n)G_n(x)-\Lambda_nG_{n-1}(x),$$

where

$$B_n=\frac{1/\sqrt{y}+\sqrt{y}+b_n}{1-q},\qquad \Lambda_n=\frac{A_{n-1}C_n}{(1-q)^2}$$

are given by the Askey-Wilson recurrence with the substitutions $a\to a/\sqrt{y}$, $b\to b\sqrt{y}$, $c\to c/\sqrt{y}$, $d\to d\sqrt{y}$.

Remark 5.9. This proposition uses a result of [8] that relates the partition function of the ASEP and the generating polynomial of the staircase tableaux. The proof is quite complicated. It is still an open problem to find a simple combinatorial proof.

Remark 5.10. The Askey-Wilson moments $2^n(abcd;q)_n\mu_n(a,b,c,d|q)$ do not have positive coefficients as polynomials in a, b, c, d, and q. The change of variables in the parameters miraculously does make these moments positive.

We also know an enumeration formula:

Theorem 5.11. [10, Theorem 1.14]

$$\begin{aligned}Z_n(y;\alpha,\beta,\gamma,\delta;q)&=(abcd;q)_n\left(\frac{\alpha\beta}{1-q}\right)^n\sum_{k=0}^{n}\frac{(ab,ac/y,ad;q)_k}{(abcd;q)_k}q^k\\&\times\sum_{j=0}^{k}q^{-(k-j)^2}(a^2/y)^{j-k}\frac{(1+y+q^{k-j}a+q^{j-k}y/a)^n}{(q,q^{2j-2k+1}y/a^2;q)_{k-j}(q,a^2q^{1-2j+2k}/y;q)_j}.\end{aligned}$$

This formula is a polynomial in $y,\alpha,\beta,\gamma,\delta,q$ with positive coefficients whose sum of coefficients is $4^nn!$. This is not easy to see from Theorem 5.11. The special case $y=q=1$ does make this clear.

Proposition 5.12. *The generating polynomial of the staircase tableau of size n satisfies*

$$Z_n(1,\alpha,\beta,\gamma,\delta,1)=\prod_{i=0}^{n-1}(\alpha+\beta+\gamma+\delta+i(\alpha+\gamma)(\beta+\delta)).\tag{5}$$

We give here a simple bijective proof based on [7]. A *colored inversion table* of size n is a table T such that for $i \in \{1,\dots,n\}$, $T[i] = (i-1)_x$ with $x \in \{\alpha,\beta,\gamma,\delta\}$ or $T[i] = j_{x,y}$ with $0 \le j < i-1$, $x \in \{\alpha,\gamma\}$, and $y \in \{\beta,\delta\}$. The weight of a colored inversion table is the product of its colors in $\{\alpha,\beta,\gamma,\delta\}$. Computing the generating polynomial of the colored inversion table is trivial and it is equal to :

$$\prod_{i=0}^{n-1}(\alpha+\beta+\gamma+\delta+i(\alpha+\gamma)(\beta+\delta)).$$

Proposition 5.13. *There exists a weight preserving bijection between staircase tableaux (with $q = 1$) of size n and colored inversion tables of size n.*

Proof. Start with a staircase tableau of size n and number the columns from 1 to n from left to right and the rows from 1 to n from top to bottom. The bijection is obtained doing the following steps. For each column i, look at the topmost Greek letter in column i and count the number of cells j directly to its left that does not contain any Greek letter.

- If this letter, say x, is topmost and leftmost, record $T[i] = i_x$.
- Otherwise let y be the first Greek letter to the left of x and let z be the first Greek letter under y. Then $T[i] = j_{x,z}$.

We present the inverse of the algorithm. Start from a colored inversion table T of size n. Mark all the cells of the tableau as free. For i from n down to 1,

- if $T_i = (i-1)_x$, put an x in the ith column, as high as possible. Mark all the cells to its left as reserved.
- Otherwise, T_i is equal to some $j_{x,y}$. Put an x in the ith column, as high as possible. Mark j boxes to its immediate left as reserved.
- Mark all the free boxes in the ith column as β/δ and mark all the boxes to their left as reserved. □

Then choose between the β/δ by using the fact that y in under the letter to the left of x if and only if x is in column i, y in column $i+j+1$ and $T_i = j_{x,y}$.

For example, we obtain the inversion table $T = (0_\gamma, 1_\beta, 2_\alpha, 1_{\alpha,\beta}, 2_{\alpha,\delta}, 2_{\gamma,\delta}, 1_{\gamma,\delta})$ from the tableau on the left of Figure 5. Note that the proof of Proposition 5.13 gives a proof of (5).

5.3 Special cases

We give alternative proofs of special cases for $Z_n(y;\alpha,\beta,\gamma,\delta;q)$ proven in [10, Table 1]. These were proven using the staircase tableaux and the matrix ansatz. We use here the moments and the explicit three-term recurrence relation. The first result is new and does not have (yet) a combinatorial proof.

Proposition 5.14. *If* $\delta = -\beta/y$*, then*

$$Z_n(y;\alpha,\beta,\gamma,\delta;q) = \prod_{j=0}^{n-1}(\alpha y + \gamma q^j).$$

Proof. First let's check that the choice $\delta = -\beta/y$ forces $\Lambda_1 = 0$, so that in Proposition 5.8 $\mu_n = B_0^n$.

In general we have $bd = -\delta/\beta$. So Proposition 5.1 and Proposition 5.8 imply that Λ_1 has the numerator factor $(1 - bdy) = (1 + y\delta/\beta) = 0$.

Since $abcd = \gamma\delta/\alpha\beta$, we have

$$Z_n(y;\alpha,\beta,\gamma,\delta;q) = (\gamma\delta/\alpha\beta;q)_n\sqrt{y}^n(\alpha\beta)^n \times B_0^n.$$

Take the version of Proposition 5.1 which uses b instead of a. In this case $A(0) = C(0) = 0$, and

$$B_0 = \frac{1/\sqrt{y} + \sqrt{y} + b\sqrt{y} + 1/b\sqrt{y}}{1-q} = \frac{\sqrt{y}}{\beta}.$$

□

Proposition 5.12 may be shown using the recurrence relation instead of the bijection of Proposition 5.13. We need the fact, which follows from Proposition 1.4, that the rescaled Laguerre polynomials which have

$$b_n = A(2n + \theta + 1), \quad \lambda_n = A^2 n(n + \theta),$$

have moments of

$$\mu_n = A^n \prod_{i=1}^{n}(\theta + i).$$

We do not give these details.

The next result can be proven bijectively by adapting the bijection between staircase tableaux and inversion tables given in the preceding subsection, see [7].

Proposition 5.15. *If* $\alpha = \delta = 0$*, then*

$$Z_n(y;\alpha,\beta,\gamma,\delta;q) = \prod_{j=0}^{n-1}(\beta + \beta\gamma[j]_q + \gamma q^j).$$

Proof. This analogously comes from the q-Laguerre polynomials studied by de Médicis and Viennct [11] which have

$$b_n = Aq^n\left([n]_q + B + q + \cdots + q^n\right), \quad \lambda_n = A^2q^{2n-1}[n]_q\left(B + q + \cdots + q^{n-1}\right),$$

$$\mu_n = A^n\prod_{j=1}^{n}(B + q + \cdots + q^{j-1}).$$

□

6 Modified moments

The Askey-Wilson polynomials have the polynomial representation in Definition 1.5 using the Askey-Wilson basis of polynomials,

$$\phi_n(x;a) = \prod_{k=0}^{n-1}(1 - 2axq^k + a^2q^{2k}) = (az, a/z; q)_n, \quad z = e^{i\theta}, \quad x = \cos\theta.$$

However it is not clear how Theorem 1.2 and Theorem 5.2 lead to the explicit formula in Definition 1.5. One may ask if an analogue of Theorem 1.2 exists for the Askey-Wilson basis $\phi_n(x;a)$ instead of x^n. We shall see that for the Askey-Wilson polynomials it does exist. It was given by Wilson in his 1978 thesis and is implicit in [35].

Suppose that polynomial sequences $\{\psi_k(x)\}_{k=0}^{\infty}$ and $\{\phi_k(x)\}_{k=0}^{\infty}$ are given with

$$\deg(\psi_k(x)) = \deg(\phi_k(x)) = k.$$

For the linear functional $\mathcal{L}$ define a matrix M by

$$M_{i,j} = \mathcal{L}(\psi_i\phi_j).$$

Proposition 6.1. *The orthogonal polynomial $p_n(x)$ for $\mathcal{L}$ is*

$$p_n(x) = \frac{1}{B_n}\begin{vmatrix} M_{0,0} & M_{0,1} & \cdots & M_{0,n} \\ M_{1,0} & M_{1,1} & \cdots & M_{1,n} \\ \vdots & \vdots & \ddots & \vdots \\ M_{n-1,0} & M_{n-1,1} & \cdots & M_{n-1,n} \\ \phi_0(x) & \phi_1(x) & \cdots & \phi_n(x) \end{vmatrix},$$

where

$$B_n = \begin{vmatrix} M_{0,0} & M_{0,1} & \cdots & M_{0,n-1} \\ M_{1,0} & M_{1,1} & \cdots & M_{1,n-1} \\ \vdots & \vdots & \ddots & \vdots \\ M_{n-1,0} & M_{n-1,1} & \cdots & M_{n-1,n-1} \end{vmatrix}.$$

The determinant B_n in Proposition 6.1 is nonzero and $p_n(x)$ is a polynomial of exact degree n with leading term $\phi_n(x)$. To prove this, let D_n be the matrix of the determinant Δ_n in Theorem 1.2. Let Y and Z be the $n \times n$ non-singular lower triangular matrices defined by $\psi_i(x) = \sum_{s=0}^{i} Y_{is}x^s$, $\phi_j(x) = \sum_{t=0}^{j} Z_{jt}x^t$. Then $M_{i,j}$ is the ij-entry of the matrix YD_nZ^T, which is invertible.

Theorem 1.2 is the special case $\psi_k(x) = \phi_k(x) = x^k$ of Proposition 6.1.

Whenever we have two bases of polynomials for which $M_{i,j}$ may be explicitly computed, we have a determinantal formula for $p_n(x)$. The explicit representation for the polynomials is reduced to evaluating the determinant

$$\det(M_{i,j})_{\substack{0\le i\le n-1 \\ 0\le j\le n, j\neq k}},$$

$$p_n(x) = \frac{1}{B_n}\sum_{k=0}^{n}(-1)^k\phi_k(x)\det(M_{ij})_{\substack{0\le i\le n-1 \\ 0\le j\le n, j\neq k}} \tag{6}$$

The Askey-Wilson bases provide such an example for the Askey-Wilson polynomials. We use a normalized representing measure for $0 < q < 1$, $0 \le |a|, |b|, |c|, |d| < 1$, which is [3]

$$\mathcal{L}_{\mathrm{AW}}(p(x)) = \frac{1}{2\pi}\frac{(ab, ac, ad, bc, bd, cd; q)_\infty}{(abcd; q)_\infty} \times \int_0^\pi p(\cos(\theta))\frac{(e^{2i\theta}, e^{-2i\theta}; q)_\infty \quad d\theta}{(ae^{i\theta}, ae^{-i\theta}, be^{i\theta}, be^{-i\theta}, ce^{i\theta}, ce^{-i\theta}, de^{i\theta}, de^{-i\theta}; q)_\infty}.$$

In this form we have the Askey-Wilson integral

$$\mathcal{L}_{\mathrm{AW}}(1) = 1.$$

Proposition 6.2. *If* $\psi_i(x) = \phi_i(x; d)$, $\phi_j(x) = \phi_j(x; a)$, *then*

$$\mathcal{L}_{\mathrm{AW}}(\psi_i\phi_j) = M_{i,j} = \frac{(ab, ac; q)_j(bd, cd; q)_i(ad; q)_{i+j}}{(abcd; q)_{i+j}}.$$

Proof. Finding $\mathcal{L}_{\mathrm{AW}}(\psi_i\phi_j)$ amounts to shifting a to aq^j and d to dq^i in the Askey-Wilson integral. □

To find the coefficient of $\phi_k(x;a)$ for the Askey-Wilson polynomials in Definition 1.5, the following determinant evaluation of Wilson [35, p. 1155] is used in (6).

Proposition 6.3. *If $0 \le k \le n$, we have*

$$\det\left(\frac{(Aq^i;q)_j}{(Dq^i;q)_j}\right)_{0\le i\le n-1, 0\le j\le n, j\ne k} = \Delta_{nn}\pi_k/\pi_n,$$

where

$$\Delta_{na} = \prod_{r=0}^{n-1} \frac{(q,D/A;q)_r}{(Dq^r;q)_{n-1}} A^r q^{r(r-1)}$$

$$\pi_k = \frac{(q^{-n},Dq^{n-1};q)_k}{(q,A;q)_k}(-q)^k$$

Proposition 6.3 with $A = ad$ and $D = abcd$ and (6) give the explicit formula in Definition 1.5 for the Askey-Wilson polynomials.

Another example occurs with the Hahn polynomials $Q_n(x;\alpha,\beta,N)$, $0 \le n \le N$, whose measure is purely discrete (the hypergeometric distribution), and is located at the integers $x = 0, 1, \cdots, N$,

$$w(x;\alpha,\beta,N) = \frac{\binom{\alpha+x}{x}\binom{\beta+N-x}{N-x}}{\binom{\alpha+\beta+1+N}{N}}.$$

In this case the choices of

$$\psi_i(x) = (-x)_i = (-x)(-x+1)\cdots(-x+i-1),$$
$$\phi_j(x) = (-N+x)_j = (-N+x)(-N+x+1)\cdots(-N+x+j-1)$$

yield

$$M_{i,j} = \frac{(\alpha+1)_i(\beta+1)_j}{(\alpha+\beta+2)_{i+j}} \frac{N!}{(N-i-j)!}(-1)^{i+j}, \quad \text{for } i+j \le N.$$

Here another determinant evaluation leads to the explicit hypergeometric representation [27] for $Q_n(x;\alpha,\beta,N)$.

7 Linearization coefficients

Suppose that $\{P_n(x)\}_{n\ge 0}$ is an orthogonal polynomial sequence with respect to a linear functional $\mathcal{L} : \mathbb{K}[x] \to \mathbb{K}$. Since $\{P_n(x)\}_{n\ge 0}$ is a basis of the ring $\mathbb{K}[x]$, we can express the product $P_n(x)P_m(x)$ of two polynomials in the basis as follows:

$$P_n(x)P_m(x) = \sum_{k=0}^{n+m} c_{n,m}^k P_k(x). \tag{7}$$

If we multiply both sides of (7) by $P_\ell(x)$ and apply the linear functional $\mathcal{L}$, then by the orthogonality we obtain

$$c_{n,m}^\ell = \frac{\mathcal{L}(P_n(x)P_m(x)P_\ell(x))}{\mathcal{L}(P_\ell(x)^2)}.$$

By Theorem 2.4, $\mathcal{L}(P_\ell(x)^2)$ is obtained immediately once we know the three-term recurrence relation. Thus computing the coefficients $c_{n,m}^\ell$ is equivalent to computing the quantity $\mathcal{L}(P_n(x)P_m(x)P_\ell(x))$. Although it is more common to call $c_{n,m}^\ell$ the linearization coefficient, for brevity, we will instead call the quantity $\mathcal{L}(P_{n_1}(x)P_{n_2}(x)\cdots P_{n_k}(x))$ the *linearization coefficient* of the orthogonal polynomials $P_n(x)$.

Using the three-term recurrence relation, Lemma 2.2 allows us to consider orthogonal polynomials $P_n(x)$ as generating functions of Favard paths. By Theorem 2.3, there is a combinatorial meaning to the moments $\mu_n = \mathcal{L}(x^n)$ of $P_n(x)$. Therefore it is possible to understand the linearization coefficients of $P_n(x)$ combinatorially. When $P_n(x)$ are q-Hermite, q-Charlier, or q-Laguerre polynomials, there is a nice combinatorial expression for the linearization coefficients.

In this section we will consider the linearization coefficients of the Hermite polynomials. Then we will see a connection between the linearization coefficients of the q-Hermite polynomials and the moments of Askey-Wilson polynomials.

Recall that the Hermite polynomials $H_n(x)$ are defined by $H_{-1}(x) = 0$, $H_0(x) = 1$, and for $n \geq 0$,

$$H_{n+1}(x) = xH_n(x) - nH_{n-1}(x), \tag{8}$$

and the moment $\mu_n = \mathcal{L}(x^n)$ is equal to the number of perfect matchings of $[n]$. Using the three-term recurrence relation (8), one can easily see that

$$H_n(x) = \sum_{\pi \in \mathrm{Matching}(n)} (-1)^{\mathrm{edge}(\pi)} x^{\mathrm{fix}(\pi)}, \tag{9}$$

where $\mathrm{fix}(\pi)$ is the number of singletons in π, which are also called *fixed points*.

Recall that $\mathcal{M}_n$ is the set of perfect matchings of $[n]$. Suppose that $n = n_1 + n_2 + \cdots + n_k$ and for $1 \leq i \leq k$, let

$$S_i = \{a_{i-1}+1, a_{i-1}+2, \ldots, a_{i-1}+n_i\}, \tag{10}$$

where $a_{i-1} = n_1 + n_2 + \cdots + n_{i-1}$ for $i \geq 2$ and $a_0 = 0$. Then $[n]$ is a disjoint union of S_i's. An edge (i, j) of $\pi \in \mathcal{M}_n$ is called *homogeneous* if $i, j \in S_r$ for some $r \in [k]$, and *inhomogeneous* otherwise. If every edge is inhomogeneous in $\pi \in \mathcal{M}_n$, then we call π an *inhomogeneous perfect matching*. We denote by $\mathcal{M}(n_1, n_2, \ldots, n_k)$ the set of inhomogeneous perfect matchings of $[n]$.

The following theorem is due to Azor, Gillis, and Victor [4] and Godsil [15].

Theorem 7.1. *The linearization coefficient of the Hermite polynomials is equal to the number of inhomogeneous perfect matchings:*

$$\mathcal{L}(H_{n_1}(x) \cdots H_{n_k}(x)) = |\mathcal{M}(n_1, n_2, \ldots, n_k)|.$$

Proof. By (9), $H_{n_1}(x) \cdots H_{n_k}(x)$ is equal to

$$\sum_{\pi_1 \in \text{Matching}(n_1)} \cdots \sum_{\pi_k \in \text{Matching}(n_k)} (-1)^{\text{edge}(\pi_1) + \cdots + \text{edge}(\pi_k)} x^{\text{fix}(\pi_1) + \cdots + \text{fix}(\pi_k)}.$$

Since $\mathcal{L}(x^m)$ is the number of perfect matchings of $[m]$, we have

$$\mathcal{L}(H_{n_1}(x) \cdots H_{n_k}(x)) = \sum_{(\pi_1, \ldots, \pi_k, \pi_0) \in X} (-1)^{\text{edge}(\pi_1) + \cdots + \text{edge}(\pi_k)},$$

where X is the set of all $(k+1)$-tuples $(\pi_1, \ldots, \pi_k, \pi_0)$ such that

$$\pi_1 \in \text{Matching}(n_1), \ldots, \pi_k \in \text{Matching}(n_k),$$

and π_0 is a perfect matching of the union of the sets of fixed points in $\pi_1, \pi_2, \ldots, \pi_k$. As before we say that an edge (a, b) is *homogeneous* if $(a, b) \in S_i$ for some $1 \leq i \leq k$, and *inhomogeneous* otherwise, where S_i is given in (10). Note that all edges in $\pi_1, \ldots, \pi_k$ are homogeneous and π_0 may have both homogeneous and inhomogeneous edges.

We will construct a sign-reversing involution ρ on X. For $(\pi_1, \ldots, \pi_k, \pi_0) \in X$, we define $\rho((\pi_1, \ldots, \pi_k, \pi_0)) = (\pi_1', \ldots, \pi_k', \pi_0')$ as follows. If $\pi_1, \ldots, \pi_k$ are all empty and π_0 has only inhomogeneous edges, then $\pi_i' = \pi_i$ for all $0 \leq i \leq k$. Otherwise, there is a homogeneous edge (a, b) in one of $\pi_1, \pi_2, \ldots, \pi_k$ or π_0. Take the homogeneous edge (a, b) such that b is minimal. If $(a, b) \in \pi_i$ for some $1 \leq i \leq k$, then let $\pi_i' = \pi_i \setminus \{(a, b)\}$ and $\pi_0' = \pi_0 \cup \{(a, b)\}$, and $\pi_j' = \pi_j$ for $j \neq 0, i$. If $(a, b) \in \pi_0$, then let i be the index for which $a, b \in S_i$ and let $\pi_i' = \pi_i \cup \{(a, b)\}$ and $\pi_0' = \pi_0 \setminus \{(a, b)\}$, and $\pi_j' = \pi_j$ for $j \neq 0, i$.

It is not hard to check that ρ is a sign-reversing involution on X whose fixed points are the $(k+1)$-tuples $(\pi_1, \ldots, \pi_k, \pi_0) \in X$ such that $\pi_1, \ldots, \pi_k$ are all empty and π_0 is an inhomogeneous perfect matchings of $S_1 \cup \cdots \cup S_k$. This completes the proof. □

From now on we put $x = \cos\theta$.

The generating function for the q-Hermite polynomials $H_n(x|q)$ is given by

$$H(\cos\theta, z) := \sum_{n=0}^{\infty} \frac{H_n(\cos\theta|q)}{(q;q)_n} z^n = \frac{1}{(ze^{i\theta};q)_\infty (ze^{-i\theta};q)_\infty}.$$

The orthogonality for the q-Hermite polynomials is

$$\int_0^\pi H_n(\cos\theta|q) H_m(\cos\theta|q) v(\cos\theta|q) d\theta = 0, \qquad n \neq m,$$

where

$$v(\cos\theta|q) = \frac{(q;q)_\infty}{2\pi} (e^{2i\theta};q)_\infty (e^{-2i\theta};q)_\infty.$$

We now look at the Askey-Wilson polynomials defined in (1.5). The total mass

$$I_0(a,b,c,d) = \frac{(q;q)_\infty}{2\pi} \int_0^\pi w(\cos\theta, a, b, c, d; q) d\theta = \frac{(abcd;q)_\infty}{(ab, ac, ad, bc, bd, cd; q)_\infty} \tag{11}$$

of the measure is called the *Askey-Wilson integral.* Observe that the Askey-Wilson integral is the generating function for linearization coefficients of the q-Hermite polynomials

$$\begin{aligned} I_0(a,b,c,d) &= \int_0^\pi H(\cos\theta, a) H(\cos\theta, b) H(\cos\theta, c) H(\cos\theta, d) v(\cos\theta|q) d\theta \\ &= \sum_{n_1,n_2,n_3,n_4=0}^{\infty} \mathcal{L}(H_{n_1}(x|q) H_{n_2}(x|q) H_{n_3}(x|q) H_{n_4}(x|q)) \\ &\quad \times \frac{a^{n_1} b^{n_2} c^{n_3} d^{n_4}}{(q;q)_{n_1} (q;q)_{n_2} (q;q)_{n_3} (q;q)_{n_4}}, \end{aligned}$$

where $\mathcal{L}$ is the linear function for $H_n(x|q)$ defined by

$$\mathcal{L}(p(x)) = \int_0^\pi p(\cos\theta) v(\cos\theta|q) d\theta,$$

Using this observation Ismail, Stanton, and Viennot [18] evaluated the Askey-Wilson integral combinatorially. More precisely, they considered the rescaled q-Hermite polynomials, which are more suitable to work with combinatorially, and showed the following generalization of Theorem 7.1.

Theorem 7.2. [18, Theorem 3.2] *Let $\tilde{\mathcal{L}}$ be the normalized linear functional for $\tilde{H}_n(x|q)$. Then*

$$\tilde{\mathcal{L}}(\tilde{H}_{n_1}(x|q)\cdots\tilde{H}_{n_k}(x|q)) = \sum_{\pi\in\mathcal{M}(n_1,\ldots,n_k)} q^{\mathrm{crossing}(\pi)},$$

where $\mathcal{M}(n_1,\ldots,n_k)$ is the set of inhomogeneous perfect matchings of $[n_1]\uplus\cdots\uplus[n_k]$.

This idea can be extended to compute the moments of the Askey-Wilson polynomials as follows. Let

$$I_n(a,b,c,d) = \frac{(q;q)_\infty}{2\pi}\int_0^\pi (\cos\theta)^n w(\cos\theta,a,b,c,d;q)d\theta. \tag{12}$$

Then the normalized moment $\mu_n(a,b,c,d;q)$ of the Askey-Wilson polynomials is

$$\mu_n(a,b,c,d;q) = I_n(a,b,c,d)/I_0(a,b,c,d). \tag{13}$$

By the same observation as above we have

$$\begin{aligned} I_n(a,b,c,d) &= \int_0^\pi (\cos\theta)^n H(\cos\theta,a)H(\cos\theta,b)H(\cos\theta,c)H(\cos\theta,d)v(\cos\theta|q)d\theta \\ &= \sum_{n_1,n_2,n_3,n_4=0}^{\infty} \mathcal{L}(x^n H_{n_1}(x|q)H_{n_2}(x|q)H_{n_3}(x|q)H_{n_4}(x|q)) \\ &\quad\times \frac{a^{n_1}b^{n_2}c^{n_3}d^{n_4}}{(q;q)_{n_1}(q;q)_{n_2}(q;q)_{n_3}(q;q)_{n_4}}. \end{aligned}$$

Using the above formula Kim and Stanton [24] found the formula for $\mu_n(a,b,c,d;q)$ in Corollary 5.3. In what follows we briefly explain the idea of their proof.

First, in [24] they consider the rescaled q-Hermite polynomials and show the following theorem.

Theorem 7.3. [24, Theorem 5.1] *Let $\tilde{\mathcal{L}}$ be the normalized linear functional for $\tilde{H}_n(x|q)$. Then*

$$\tilde{\mathcal{L}}(x^n\tilde{H}_{n_1}(x|q)\cdots\tilde{H}_{n_k}(x|q)) = \sum_{\pi\in\mathcal{M}(n;n_1,\ldots,n_k)} q^{\mathrm{crossing}(\pi)},$$

where $\mathcal{M}(n;n_1,\ldots,n_k)$ is the set of perfect matchings π of $S_1\cup\cdots\cup S_k\cup S_{k+1}$ such that if π contains a homogeneous edge $(a,b)\in S_i$, then $i=k+1$. Here, S_i is given in (10) *for $1\le i\le k$, and*

$$S_{k+1} = \{n_1+\cdots+n_k+1, n_1+\cdots+n_k+2,\ldots,n_1+\cdots+n_k+n\}.$$

In order to evaluate the sum in the left-hand side of the above theorem, they [24] decompose $\pi \in \mathcal{M}(n; n_1, n_2, n_3, n_4)$ as a pair (π_0, π_1) of a matching π_0 of $[n]$ with m fixed points and an inhomogeneous perfect matching $\pi_1 \in \mathcal{M}(m, n_1, n_2, n_3, n_4)$ for some integer m. In this decomposition we have $\text{crossing}(\pi) = \text{crossing}^*(\pi_0) + \text{crossing}(\pi_1)$, where $\text{crossing}^*(\pi_0)$ is the number of pairs (e_1, e_2) such that

- $e_1 = \{a, b\}$ and $e_2 = \{c, d\}$ are edges of π_0 with $a < c < b < d$, or
- $e_1 = \{a, b\}$ is an edge of π_0 and $e_2 = \{c\}$ is a singleton of π_0 with $a < c < b$.

Then they use the following formula due to Josuat-Vergès [21, Proposition 5.1]:

$$(1-q)^{(n-m)/2} \sum_{\pi \in \mathcal{M}^*(n,m)} q^{\text{crossing}^*(\pi)} = \sum_{k \geq 0} \left(\binom{n}{\frac{n-k}{2}} - \binom{n}{\frac{n-k}{2} - 1} \right) (-1)^{(k-m)/2} q^{\binom{(k-m)/2+1}{2}} \begin{bmatrix} \frac{k+m}{2} \\ \frac{k-m}{2} \end{bmatrix}_q, \quad (14)$$

where $\mathcal{M}^*(n, m)$ is the set of matchings of $[n]$ with m fixed points.

The appearance of the difference of binomial coefficients in (14) can be explained as follows. A *Dyck prefix* is a path obtained from a Dyck path by taking the first m steps for some m. By a similar argument using Hermite histories, the left-hand side of (14) is the generating function for Dyck prefixes whose South-East steps are labeled by 1 or $-q^i$, where i is the y-coordinate of the starting point. Then we can use Penaud's idea [28] which decomposes such a labeled Dyck prefix into a Dyck prefix and a certain labeled Dyck path. The number of Dyck prefixes is given by a difference of binomial coefficients.

There are analogous combinatorial interpretations for the linearization coefficients of q-Charlier polynomials and q-Laguerre polynomials.

Theorem 7.4. [1, p. 127] *Let $\mathcal{L}_C$ be the normalized linear functional for the q-Charlier polynomials $C_n^a(x; q)$. Then*

$$\mathcal{L}_C(C_{n_1}^a(x; q) \cdots C_{n_k}^a(x; q)) = \sum_{\pi \in \Pi(n_1, \ldots, n_k)} a^{\text{block}(\pi)} q^{\text{crossing}(\pi)},$$

where $\Pi(n_1, \ldots, n_k)$ is the set of partitions of $S_1 \cup \cdots \cup S_k$ which do not have homogeneous edges. Here, S_i is given in (10) *for $1 \leq i \leq k$.*

Kim, Stanton, and Zeng [26] found a combinatorial proof of Theorem 7.4. Kasraoui, Stanton, and Zeng [23] showed the following theorem using recurrence relations.

Theorem 7.5. [23, Theorem 5] *Let $\mathcal{L}_L$ be the normalized linear functional for the q-Laguerre polynomials $L_n(x; q)$. Then*

$$\mathcal{L}_L(L_{n_1}(x; q) \cdots L_{n_k}(x; q)) = \sum_{\pi \in D(n_1, \ldots, n_k)} y^{\text{wex}(\pi)} q^{\text{crossing}(\pi)},$$

where $D(n_1, \ldots, n_k)$ is the set of permutations π of $S_1 \cup \cdots \cup S_k$ such that if $\pi(a) = b$ then a and b are in different S_i's. Here, S_i is given in (10) *for* $1 \le i \le k$.

Note that in Theorem 7.5, if $n_1 = \cdots = n_k = 1$, then $D(n_1, \ldots, n_k)$ becomes the set of *derangements* of $[k]$, i.e., permutations with no fixed points. Permutations in $D(n_1, \ldots, n_k)$ are called multi-derangements. There is a nice generating function expression for the number of multi-derangements, see [19, p. 563] and references therein.

Ismail, Kasraoui, and Zeng [19] found a general approach to find the linearization coefficients using recurrence relations.

Acknowledgements Research by Sylvie Corteel is funded by the IDEX Université Sorbonne Paris Cité Project "ALEA Sorbonne" and by the city of Paris: project Emergences "Combinatoire à Paris." Jang Soo Kim was partially supported by Basic Science Research Program through the National Research Foundation of Korea (NRF) funded by the Ministry of Education (NRF-2013R1A1A2061006). Dennis Stanton is partially supported by NSF grant DMS-1148634. The authors want to thank Greg Musiker for his constructive comments on the manuscript.

References

1. M. Anshelevich, Linearization coefficients for orthogonal polynomials using stochastic processes. Ann. Probab. **33**(1), 114–136 (2005)
2. R. Askey, *Orthogonal Polynomials and Special Functions* (Society for Industrial and Applied Mathematics, Philadelphia, 1975)
3. R. Askey, J. Wilson, Some basic hypergeometric orthogonal polynomials that generalize Jacobi polynomials. Mem. Am. Math. Soc. **54**(319), iv+55 (1985)
4. R. Azor, J. Gillis, J.D. Victor, Combinatorial applications of Hermite polynomials. SIAM J. Math. Anal. **13**(5), 879–890 (1982)
5. T. Chihara, *An Introduction to Orthogonal Polynomials* (Gordon and Breach Science Publishers, New York/London/Paris, 1978)
6. S. Corteel, Crossings and alignments of permutations. Adv. Appl. Math. **38**(2), 149–163 (2007)
7. S. Corteel, S, Dasse-Hartaut, Statistics on staircase tableaux, Eulerian and Mahonian statistics, in *23rd International Conference on Formal Power Series and Algebraic Combinatorics (FPSAC 2011)*. Discrete Mathematics and Theoretical Computer Science, AO (Assoc. Discrete Math. Theor. Comput. Sci., Nancy, 2011), pp. 245–255
8. S. Corteel, L.K. Williams, Tableaux combinatorics for the asymmetric exclusion process and Askey-Wilson polynomials. Duke Math. J. **159**(3), 385–415 (2011)
9. S. Corteel, M. Josuat-Vergès, T. Prellberg, M. Rubey, Matrix ansatz, lattice paths and rook placements. DMTCS proc. **AK**, 313–324 (2009)
10. S. Corteel, R. Stanley, D. Stanton, L. Williams, Formulae for Askey-Wilson moments and enumeration of staircase tableaux. Trans. Am. Math. Soc. **364**(11), 6009–6037 (2012)
11. A. de Médicis, X.G. Viennot, Moments des q-polynômes de Laguerre et la bijection de Foata-Zeilberger. Adv. Appl. Math. **15**(3), 262–304 (1994)
12. P. Flajolet, Combinatorial aspects of continued fractions. Discret. Math. **32**(2), 125–161 (1980)
13. D. Foata, D. Zeilberger, Denert's permutation statistic is indeed Euler-Mahonian. Stud. Appl. Math. **83**(1), 31–59 (1990)
14. G. Gasper, M. Rahman, *Basic Hypergeometric Series*, 2nd edn. (Cambridge University Press, Cambridge, 2004)

15. C. Godsil, Hermite polynomials and a duality relation for matching polynomials. Combinatorica **1**(3), 257–262 (1981)
16. M.E.H. Ismail, D. Stanton, Classical orthogonal polynomials as moments. Can. J. Math. **49**(3), 520–542 (1997)
17. M.E.H. Ismail, D. Stanton, Expansions in the Askey-Wilson polynomials. J. Math. Anal. Appl. **424**(1), 664–674 (2015)
18. M.E.H. Ismail, D. Stanton, G. Viennot, The combinatorics of q-Hermite polynomials and the Askey-Wilson integral. Eur. J. Comb. **8**(4), 379–392 (1987)
19. M.E.H. Ismail, A. Kasraoui, J. Zeng, Separation of variables and combinatorics of linearization coefficients of orthogonal polynomials. J. Comb. Theory Ser. A **120**(3), 561–599 (2013)
20. M. Josuat-Vergès, Combinatorics of the three-parameter PASEP partition function. Electron. J. Comb. **18**, #P22 (2011)
21. M. Josuat-Vergès, Rook placements in Young diagrams and permutation enumeration. Adv. Appl. Math. **47**, 1–22 (2011)
22. M. Josuat-Vergès, M. Rubey, Crossings, Motzkin paths and moments. Discret. Math. **311**, 2064–2078 (2011)
23. A. Kasraoui, D. Stanton, J. Zeng, The combinatorics of Al-Salam–Chihara q-Laguerre polynomials. Adv. Appl. Math. **47**(2), 216–239 (2011)
24. J.S. Kim, D. Stanton, Moments of Askey-Wilson polynomials. J. Comb. Theory Ser. A **125**, 113–145 (2014)
25. J.S. Kim, D. Stanton, Bootstrapping and Askey-Wilson polynomials. J. Math. Anal. Appl. **421**(1), 501–520 (2015)
26. D. Kim, D. Stanton, J. Zeng, The combinatorics of the Al-Salam-Chihara q-Charlier polynomials. Sém. Lothar. Comb. **54**, Art. B54i, 15 pp. (electronic) (2005/2007)
27. R. Koekoek, P.A. Lesky, R.F. Swarttouw, *Hypergeometric Orthogonal Polynomials and Their q-Analogues*. Springer Monographs in Mathematics (Springer, Berlin, 2010) With a foreword by Tom H. Koornwinder.
28. J.-G. Penaud, Une preuve bijective d'une formule de Touchard-Riordan. Discret. Math. **139**, 347–360 (1995)
29. P.N. Sadjang, W.A. Koepf, M. Foupouagnigni, On moments of classical orthogonal polynomials. J. Math. Anal. Appl. **424**(1), 122–151 (2015)
30. R. Simion, D. Stanton, Octabasic Laguerre polynomials and permutation statistics. J. Comput. Appl. Math. **68**(1–2), 297–329 (1996)
31. G. Szegõ, *Orthogonal Polynomials,* 4th edn. American Mathematical Society, Colloquium Publications, vol. XXIII. (American Mathematical Society, Providence, 1975), xiii+432 pp.
32. M. Uchiyama, T. Sasamoto, M. Wadati, Asymmetric simple exclusion process with open boundaries and Askey-Wilson polynomials. J. Phys. A **37**(18), 4985–5002 (2004)
33. G. Viennot, Une théorie combinatoire des polynômes orthogonaux. Lecture Notes, UQÀM (1984), http://www.xavierviennot.org/xavier/livres.html
34. X.G. Viennot, A combinatorial interpretation of the quotient-difference algorithm, in *Formal Power Series and Algebraic Combinatorics* (Moscow, 2000) (Springer, Berlin, 2000), pp. 379–390
35. J.A. Wilson, Orthogonal functions from Gram determinants. SIAM J. Math. Anal. **22**(4), 1147–1155 (1991)

The combinatorics of knot invariants arising from the study of Macdonald polynomials

Jim Haglund

Abstract This chapter gives an expository account of some unexpected connections which have arisen over the last few years between Macdonald polynomials, invariants of torus knots, and lattice path combinatorics. The study of polynomial knot invariants is a well-known branch of topology which originated in the 1920s with the one-parameter Alexander polynomial [1]. In the early 1980s Jones [41] introduced a different one-parameter polynomial invariant, with important connections to physics. Shortly thereafter a number of authors more or less simultaneously discovered the HOMFLY polynomial, a two-parameter invariant which includes both the Alexander and Jones polynomials as special cases. The HOMFLY polynomial can be calculated recursively through skein relations. In the late 1980s Witten showed that the Jones polynomial and related invariants have an interpretation in terms of Chern–Simons theory, which is central to string theory.

In 2006 Dunfield, Gukov, and Rasmussen [14] hypothesized the existence of a three-parameter knot invariant, now known as the "superpolynomial knot invariant" of a knot K, denoted $\mathcal{P}_K(a,q,t)$, which includes the HOMFLY polynomial as a special case. Since then various authors proposed different possible definitions of the superpolynomial, which are conjecturally all equivalent. These definitions typically involve homology though, and are difficult to compute.

In the case of torus knots an accepted definition of the superpolynomial has recently emerged from work of Aganagic and Shakirov [6, 7] (using refined Chern–Simons theory) and Cherednik [12] (using the double affine Hecke algebra). Gorsky and Negut [24] showed that these two different constructions yield the same three-parameter knot invariant which is now accepted as the definition of the superpolynomial for torus knots. These constructions involve symmetric functions in a set of variables X known as Macdonald polynomials, which also depend on two extra parameters q,t. These symmetric functions are important in algebraic combinatorics and other areas, and play a central role in various character formulas for S_n modules connected to the Hilbert scheme from algebraic geometry. In particular, Haiman's formula for the bigraded character of the space DH_n of

J. Haglund (✉)
Department of Mathematics, University of Pennsylvania, Philadelphia, PA 19104-6395, USA
e-mail: jhaglund@math.upenn.edu

A. Beveridge et al. (eds.), *Recent Trends in Combinatorics*, The IMA Volumes in Mathematics and its Applications 159, DOI 10.1007/978-3-319-24298-9_23

diagonal harmonics under the diagonal action of the symmetric group is expressed in terms of Macdonald polynomials.

E. Gorsky [22], [47, Appendix] noticed that the coefficient of a^j in the superpolynomial of the $(n+1, n)$ torus knot equals the bigraded multiplicity of a certain hook shape in the character of DH_n. This polynomial is known as the (q, t)-Schröder polynomial since the author showed it can be expressed as a weighted sum over Schröder lattice paths in the $n \times n + 1$ rectangle. Gorsky and Negut have shown that the coefficient of a^j in the superpolynomial of the (m, n) torus knot can be viewed as the coefficient of a certain hook Schur function in a symmetric function expression involving Macdonald polynomials, and they have derived many explicit identities for this object. In addition Oblomkov, Rasmussen, and Shende [47] have introduced a conjectured extension of the q, t-Schröder polynomial to general (m, n) giving a positive strictly combinatorial expression for the superpolynomial of the (m, n) torus knot. This conjecture connects nicely with an important conjecture in algebraic combinatorics called the rational shuffle conjecture. In the following pages we will describe these developments in more detail.

Keywords Macdonald polynomials • Diagonal harmonics • Knot invariants • Superpolynomial • Rational shuffle conjecture

Mathematics Subject Classification (2010): 05A15, 05E05

1 Knot Invariants

A *knot* is an embedding of a circle in $\mathbf{R}^3$. Informally, we will think of a knot as a closed, elastic string in $\mathbf{R}^3$. Two knots are equivalent (isotopic) if one string can be deformed into the other without cutting the string. Our knots will carry an orientation, i.e., a forward direction indicated by an arrow. Knots are typically represented by planar diagrams called knot diagrams, as in Figure 1.

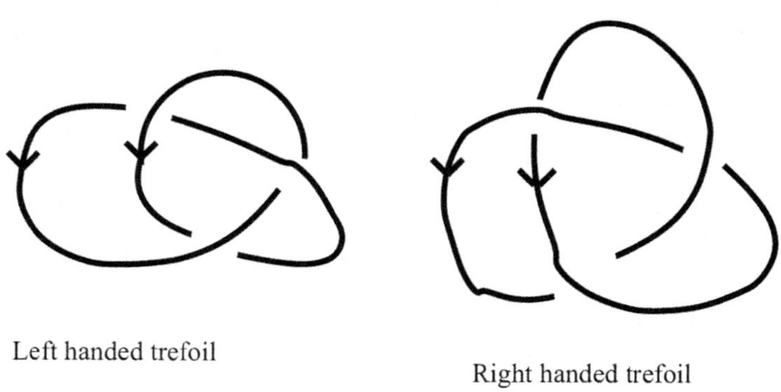

Fig. 1 Knot diagrams of the two trefoil knots

The trefoil knot is the simplest example of a knot which is not isotopic to the unknot O (a closed string with no crossings). The right-handed trefoil is the mirror image of the left-handed trefoil. It is known that the trefoil knot is not isotopic to its mirror image, i.e., is chiral. An interesting family of knots that will be important to us are torus knots. Given two relatively prime positive integers (m,n), we let $T_{(m,n)}$ denote the corresponding torus knot, which can be obtained by wrapping a string around a torus, at such an angle that by the time you return to where you started from you have wrapped around the torus n times in one direction and m times in the other. It is known that $T_{(m,n)}$ is isotopic to $T_{(n,m)}$, and that $T_{(3,2)}$ is the right-handed trefoil. (The left-handed trefoil can be viewed as the $(2,-3)$ torus knot.)

A (polynomial) knot invariant is a polynomial in (possibly) several variables which is the same for isotopic knots. In the 1920s Alexander [1] introduced the first polynomial knot invariant, which he constructed from a knot diagram by taking the determinant of a certain $n \times n$ matrix, where n is the number of crossings in the diagram of the knot. In 1969 Conway [13] showed that a version of this polynomial could be computed recursively using skein relations. Specifically, the Alexander–Conway polynomial $\mathrm{AC}_L(z)$ of the knot L satisfies

$$\mathrm{AC}_{L_+}(z) = \mathrm{AC}_{L_-}(z) + z\mathrm{AC}_{L_0}(z), \tag{1}$$

where L_+, L_-, and L_0 indicate knots which are the same everywhere except at a specific crossing, see Figure 2. We use the initial condition $\mathrm{AC}_O(z) = 1$, where O is the "unknot" consisting of a circle with no crossings. We note that our recursive definition actually applies to oriented links, which can be represented by knot diagrams consisting of possibly more than one string.

Although the same knot can be represented by a knot diagram in many different ways, the polynomial $\mathrm{AC}_L(z)$ turns out to be independent of the diagram.

Example 1. *Let T denote the left-handed trefoil knot of Figure 1. In this example we calculate $AC_T(z)$ using the skein relations. (We follow the procedure carried out in [50] for the right-handed trefoil. The polynomial $AC_T(z)$ turns out to be the same for both trefoil knots, which shows this polynomial cannot always distinguish between a knot and its mirror image). Now*

$$AC_T(z) = AC_O(z) + zAC_H(z), \tag{2}$$

where H is the Hopf link as in Figure 3.

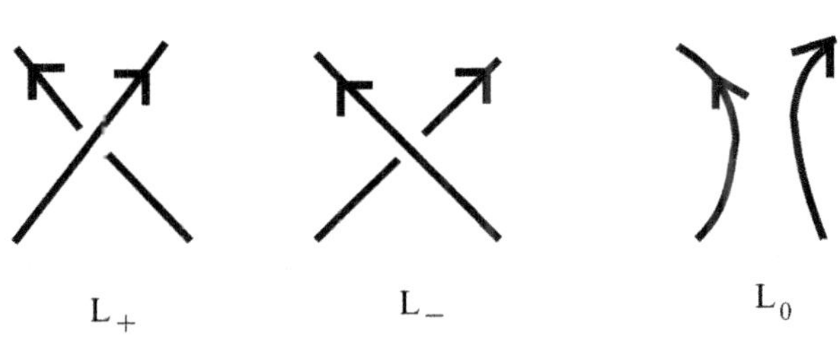

Fig. 2 The three types of crossings in a knot diagram

Fig. 3 One of the two oriented Hopf links, which occurs in the computation of $\mathrm{AC}_T(z)$.

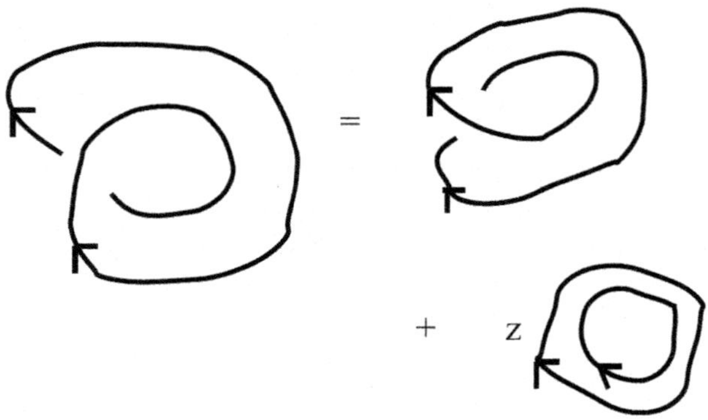

Fig. 4 A recursive relation involving the unlink O_2

Furthermore, we have

$$AC_H = AC_{O_2} + zAC_O, \tag{3}$$

where O_2 is the union of two unknots (called the unlink of two components). To find $AC_{O_2}(z)$ we use the identity in Figure 4, which shows $AC_{O_2}(z) = 0$, since the two other knots are both the unknot O. Putting everything together we have

$$AC_T(z) = 1 + z(0 + z) = 1 + z^2. \tag{4}$$

Jones [41] introduced a different knot invariant $V_L(z)$, which can also be defined recursively via the relation

$$zV_{L_+}(z) = \frac{1}{z}V_{L_-}(z) + (\frac{1}{\sqrt{z}} - \sqrt{z})V_{L_0}(z), \tag{5}$$

together with the initial condition $V_O(z) = 1$. The two trefoil knots have different corresponding Jones polynomials.

Shortly after the introduction of the Jones polynomial, three different pairs of authors simultaneously discovered the HOMFLY polynomial $P_K(\alpha, z)$, a two-variable knot invariant which contains the Jones and Alexander–Conway polynomials as special cases. These three pairs of authors each submitted papers on this to the

Bulletin of the AMS, and the editor convinced them to combine their articles into one [15]. (The word HOMFLY is a combination of the first letters of the last names of the six authors.) Their polynomial can be defined recursively as above by

$$\alpha P_{L+}(\alpha, z) = \frac{1}{\alpha} P_{L-}(\alpha, z) + z P_{L_0}(\alpha, z). \tag{6}$$

If we compute $P_T(\alpha, z)$ for the left-handed trefoil T, as in Example 1 but using (6), we get

$$P_T(\alpha, z) = P_O(\alpha, z)\left(2\alpha^2 + \alpha^2 z^2 - \alpha^4\right). \tag{7}$$

Different initial values for the HOMFLY polynomial $P_O(\alpha, z)$ are used by various authors.

In 2006 Dunfield, Gukov, and Rasmussen [14] hypothesized the existence of a three-parameter knot invariant, the so-called superpolynomial of a knot K, denoted $\mathcal{P}_K(a, q, t)$, which includes the HOMFLY polynomial as a special case. They were not able to define the superpolynomial in general, but listed various desirable properties which it should satisfy, and showed that for many small knots there was a unique polynomial with these properties. Over the next several years various authors proposed different possible definitions of the superpolynomial, which are conjecturally all equivalent. These definitions typically involve homology though, and are difficult to compute.

In 2011 Aganagic and Shakirov [6, 7] used refined Chern–Simons theory to construct knot invariants of torus knots. Cherednik [12] gave an alternate way to construct knot invariants for these knots using the double affine Hecke algebra. Gorsky and Negut [24] showed that these two different constructions gave the same knot invariants. A certain stable limit of these invariants yields a three-parameter knot invariant which is now accepted as the definition of the superpolynomial for torus knots.

E. Gorsky [22] made the exciting discovery that the superpolynomial of the $(n+1, n)$ torus knot has an interpretation in terms of the character of DH_n. To be precise, if we let $\tilde{P}_{T_{(m,r)}}(a, q, t)$ be the modified superpolynomial (obtained from the superpolynomial by replacing t by $1/t$) then he showed [47, Appendix] that the coefficient of a^k in $\tilde{\mathcal{P}}_{T_{(n+1,n)}}(a, q, t)$ equals the scalar product

$$(-1)^k \langle \nabla e_n, e_{n-k} h_k \rangle. \tag{8}$$

Here ∇e_n is Haiman's formula for the character of DH_n, described in more detail in Section 3, e_k is the kth elementary symmetric function, h_k is the kth homogeneous symmetric function, and $\langle , \rangle$ is the Hall scalar product, with respect to which the Schur functions are orthonormal. There is a known combinatorial formula for (8) due to the author [29], which is expressed in terms of weighted lattice paths.

Inspired by the study of the Hilbert scheme of the plane curve $x^n = y^m$, Oblomkov, Rasmussen, and Shende [47] have introduced a conjectured expression for a certain Poincaré polynomial associated to knot homology. Their formula gives a candidate positive combinatorial formula for the superpolynomial of the (m, n) torus knot, which reduces to (8) when $m = n + 1$.

In a parallel development, a natural generalization of the symmetric function ∇e_n (the character of DH_n) has recently emerged from work of Burban, Schiffman, Vasserot, Negut, and others [10, 46, 48, 49] on the Elliptic Hall algebra and other related objects in algebraic geometry and string theory. These symmetric functions are denoted in this article by $Q_{(m,n)}(\mathbf{-1})^{\mathbf{n}}$ (we place $(-1)^n$ in bold characters to indicate we are viewing it as a symmetric function, which is acted upon by the operator $Q_{(m,n)}$, as opposed to the constant $(-1)^n$)). They depend on a pair of relatively prime positive integers (m, n), and can be defined recursively (see [9] for an explicit description of the symmetric function operators $Q_{(m,n)}$). There is a nice conjecture for their expansion into monomials [8, 24] called the rational shuffle conjecture, which reduces to the shuffle conjecture in [34] when $m = n+1$. Gorsky and Negut [24] have shown that

$$\tilde{P}_{T_{(m,n)}}(a, q, t) = \sum_{k=0}^{n} (-a)^k \langle Q_{(m,n)}(\mathbf{-1})^{\mathbf{n}}, e_{n-k} h_k \rangle. \tag{9}$$

The conjectured combinatorial formula for $\tilde{\mathcal{P}}_{T_{(m,n)}}(a, q, t)$ in [47] can be viewed as the "hook case" of the rational shuffle conjecture, since the scalar product of $e_{n-k}h_k$ with a symmetric function F is the sum of the coefficients of the two consecutive hook shapes $s_{k,1^{n-k}}$ and $s_{k+1,1^{n-k-1}}$ in F. (Eq. (9) implies a corresponding formula for the coefficient of a single hook shape in $Q_{(m,n)}(\mathbf{-1})^{\mathbf{n}}$).

All of these constructions involve advanced symmetric functions with two extra parameters q, t known as Macdonald polynomials, which are the subject of the next section.

2 Macdonald polynomials

Let $p_k(X) = \sum_i x_i^k$ be the kth power sum, and for any polynomial $E(t_1, t_2, \ldots, t_n)$ define the plethystic substitution of E into p_k as

$$p_k[E] = E(t_1^k, \ldots, t_n^k). \tag{10}$$

For example,

$$p_k[X(1-t)] = \sum_i x_i^k (1 - t^k) = p_k(X)(1 - t^k) \tag{11}$$

$$p_k[X/(1-t)] = p_k(X)/(1 - t^k). \tag{12}$$

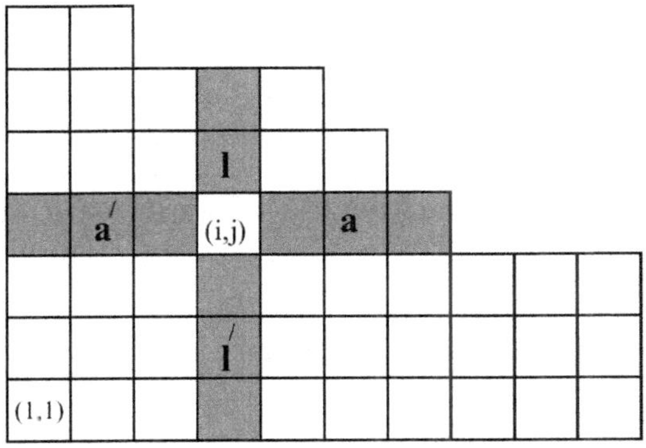

Fig. 5 The arm a, coarm a', leg l, and coleg l' of a cell

So when plethystically substituting E into p_k, we replace any variables by their kth powers, and leave any constants like -1 alone. To plethystically substitute E into an arbitrary symmetric function f, we first express f as a polynomial in the p_k, then substitute E into each p_k in this expression. See [30, pp.19–20] for more background on plethystic substitution.

In 1988 Macdonald [44, 45] introduced symmetric functions $P_\mu(X;q,t)$ which satisfy an orthogonality condition, and include many previously studied bases of symmetric functions as special cases. For example, $P_\mu(X;q,q) = s_\mu(X)$ (for any q), and $P_\mu(X;0,t)$ is the Hall–Littlewood polynomial. Macdonald also defined the integral form J_μ as

$$J_\mu(X;q,t) = \prod_{x\in\mu}(1 - q^a t^{l+1}) P_\mu(X;q,t), \tag{13}$$

with a, l, a', l' denoting the arm, leg, coarm, coleg of the cell, as in Figure 5.

Macdonald conjectured that when J_μ is expanded in terms of the "plethystic Schur basis" $\{s_\lambda[X(1-t)]\}$ the coefficients are all positive, i.e., are in $\mathbb{N}[q,t]$. Garsia and Haiman [16] introduced the modified Macdonald polynomials $\tilde{H}_\mu(X;q,t)$;

$$\tilde{H}_\mu[X;q,t] = t^{n(\mu)} J_\mu\left[\frac{X}{1-1/t}; q, 1/t\right] \tag{14}$$

and gave a conjectured interpretation for $\tilde{H}_\mu(X;q,t)$ as the bigraded character of a certain S_n module V_μ. This refined Macdonald's positivity conjecture since it implies the coefficient of s_λ in $\tilde{H}_\mu$ counts multiplicities of irreducible representations and is hence positive, and this implies a similar interpretation for the coefficient of $s_\lambda[X(1-t)]$ in J_μ. The conjecture of Garsia and Haiman became known as the $n!$ conjecture since it reduced to the problem of showing that the dimension of V_μ is $n!$, for μ any partition of n.

After ten years of intensive research, Haiman [38] proved the $n!$ conjecture using the Hilbert scheme from algebraic geometry. Finding a positive, combinatorial formula for the Schur coefficients in $\tilde{H}_\mu$ is still an important open problem. Haiman's interpretation of $\tilde{H}_\mu$ in terms of the Hilbert scheme has had a growing influence on algebraic geometry and string theory, particularly in the construction of character formulas expressed in terms of Macdonald polynomials.

3 Diagonal Harmonics and Rational Catalan Combinatorics

The space of diagonal harmonics DH_n is defined as

$$DH_n = \{f \in \mathbb{C}[X_n, Y_n] : \sum_{i=1}^{n} \frac{\partial^h}{x_i^h} \frac{\partial^k}{y_i^k} f = 0, \ \forall\, h+k > 0.\}. \tag{15}$$

The symmetric group S_n acts "diagonally" on DH_n by permuting the X and Y variables in the same way;

$$\sigma f(x_1, \ldots x_n, y_1, \ldots, y_n) = f(x_{\sigma_1}, \ldots, x_{\sigma_n}, y_{\sigma_1}, \ldots, y_{\sigma_n}) \quad \forall \sigma \in S_n. \tag{16}$$

Example 2. *An explicit basis for DH_2 is given by $\{1, x_2 - x_1, y_2 - y_1\}$. Taking into account the S_n-action, we see that $x_2 - x_1$ forms a sign-character component, as does $y_2 - y_1$, and 1 forms a trivial character component. If we weight our components by $q^{x\text{-degree}} t^{y\text{-degree}}$, then the Hilbert series of DH_2 is $1 + q + t$, and the Frobenius series (where as usual the Schur function s_λ is used to mark the irreducible component corresponding to λ) is $s_2 + (q+t)s_{1^2}$.*

Garsia and Haiman initiated the study of DH_n [17, 36], and conjectured that the Frobenius series DH_n can be expressed compactly as $\nabla e_n(X)$. Here ∇ is a linear operator on symmetric functions defined on the $\tilde{H}_\mu$ basis via

$$\nabla \tilde{H}_\mu = t^{n(\mu)} q^{n(\mu')} \tilde{H}_\mu, \tag{17}$$

where $n(\mu) = \sum_i (i-1)\mu_i$. Haiman [39] proved this conjecture. His proof uses the Hilbert scheme and other elements of his proof of the $n!$ Theorem.

It follows from Macdonald's original work that

$$\nabla e_n = \sum_{\mu \vdash n} \frac{t^{n(\mu)} q^{n(\mu')} M \tilde{H}_\mu \Pi_\mu B_\mu}{w_\mu}, \tag{18}$$

where $B_\mu = \sum_{s \in \mu} q^{a'} t^{l'}$, $M = (1-q)(1-t)$, $\Pi_\mu = \prod_{s \in \mu, s \neq (1,1)} (1 - q^{a'} t^{l'})$, and $w_\mu = \prod_{s \in \mu} (q^a - t^{l+1})(t^l - q^{a+1})$. Although everything in (18) is explicit, the formula is difficult to interpret combinatorially.

Let $C_n(q,t) = \langle \nabla e_n, s_{1^n} \rangle$, the sign character of DH_n. This polynomial is particularly interesting combinatorially, since $C_n(1,1) = C_n$, the nth Catalan number. We now give a purely combinatorial description of $C_n(q,t)$.

Definition 1. *Let $L^+(m,n)$ denote the set of lattice paths in the first quadrant of the xy-plane, consisting of unit North N and East E steps, which start at $(0,0)$, end at (m,n), and never go below the main diagonal $nx = my$.*

Now assume $m = n$. Given $\pi \in L^+_{n,n}$, define the bounce path of π to be the path described by the following algorithm.

Start at $(0,0)$ and travel North along π until you encounter the beginning of an E step. Then turn East and travel straight until you hit the diagonal $y = x$ Then turn North and travel straight until you again encounter the beginning of an E step of π, then turn East and travel to the diagonal, etc. Continue in this way until you arrive at (n,n).

We can think of our bounce path as describing the trail of a billiard ball shot North from $(0,0)$, which "bounces" right whenever it encounters a horizontal step and "bounces" up when it encounters the line $y = x$. The bouncing ball will strike the diagonal at places

$$(0,0), (j_1, j_1), (j_2, j_2), \ldots, (j_{b-1}, j_{b-1}), (j_b, j_b) = (n,n).$$

We define the bounce statistic bounce(π) to be the sum

$$bounce(\pi) = \sum_{i=1}^{b-1} n - j_i. \tag{19}$$

Also, let $a_i = a_i(\pi)$ denote the number of squares in the ith row (from the bottom) which are to the right of the path π and strictly above the diagonal $x = y$, and set $area(\pi) = \sum_{i=1}^{n} a_i$ For the path in Figure 6, we have bounce $= 19$, area $= 15$, and

$$(a_1, a_2, \ldots a_{11}) = (0,1,2,1,2,0,1,1,2,3,2).$$

Theorem 1. *[18, 19]*

$$C_n(q,t) = \sum_{\pi \in L^+_{n,n}} q^{area(\pi)} t^{bounce(\pi)}. \tag{20}$$

There is another, equivalent way of describing $C_n(q,t)$ which was discovered by M. Haiman [37]. Given a path π as above, we say rows i and j form an inversion if

$$1 \leq i < j \leq n \text{ and either } a_i = a_j \text{ or } a_i = a_j + 1. \tag{21}$$

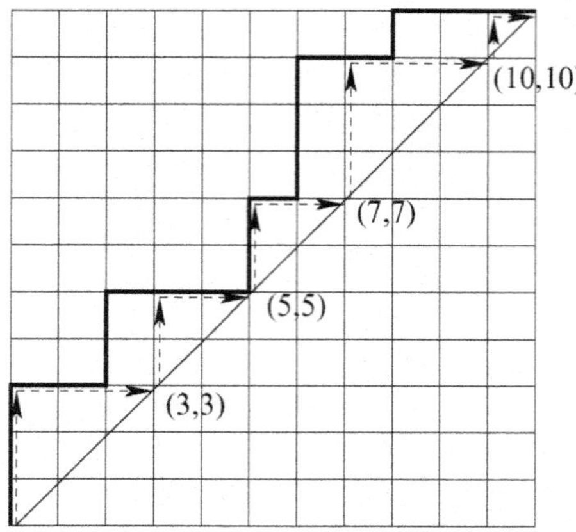

Fig. 6 The bounce path (dotted line) of a Dyck path (solid line). The bounce statistic equals $(11-3)+(11-5)+(11-7)+(11-10)=8+6+4+1=19$, and area $=15$.

Letting $\operatorname{dinv}(\pi)$ denote the number of such inversion pairs, then

$$C_n(q,t) = \sum_{\pi \in L^+_{(n,n)}} q^{\operatorname{dinv}(\pi)} t^{\operatorname{area}(\pi)}. \tag{22}$$

To prove that (22) and (20) are equivalent, we give an abbreviated description of the "zeta map" between Dyck paths, which has the property that

$$\begin{aligned} \operatorname{dinv}(\pi) &= \operatorname{area}(\zeta(\pi)) \\ \operatorname{area}(\pi) &= \operatorname{bounce}(\zeta(\pi)). \end{aligned} \tag{23}$$

This map is described in more detail in [30, pp. 50–51].

Say $b-1$ is the maximal value of the a_i, $1 \le i \le n$. The lengths of the bounce steps of ζ will be $\alpha_1, \ldots, \alpha_b$, where α_i is the number of rows of length $i-1$ in π. To construct the actual path ζ, place a pen at the lattice point with (x, y)-coordinates $(0, 0)$, and draw a line straight up to $(0, \alpha_1)$. Now look at the subsequence of the a_i consisting of only those a_i which are either 0 or 1. Travel through this subsequence, and each time you encounter a 0, draw an E step, and each time you encounter a 1, draw a N step. You should now be at the lattice point $(\alpha_1, \alpha_1+\alpha_2)$. Next look at the subsequence of the a_i consisting of only those a_i which are either 1 or 2. Travel through this subsequence, and each time you encounter a 1, draw an E step, and each time you encounter a 2, draw a N step. You should now be at the lattice point $(\alpha_1+\alpha_2, \alpha_1+\alpha_2+\alpha_3)$. Next, consider the subsequence of 2's and 3's, etc., and continue in this way until you arrive at (n, n). See Figure 7.

It is easy to see this map is a bijection: given $\zeta(\pi)$, from the bounce path we can determine the multiset of row lengths of π. We can then build up the area sequence

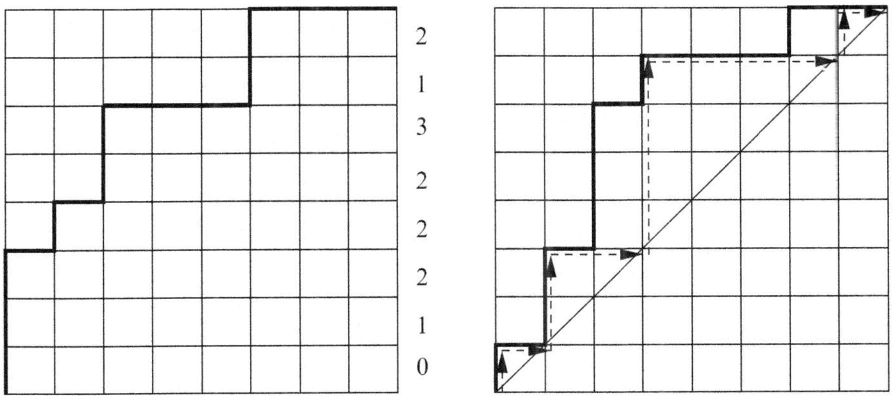

Fig. 7 A path π with row lengths to the right, and the image $\zeta(\pi)$.

of π; from the portion of the path between $(\alpha_1, \alpha_1+\alpha_2)$ we can see how to interleave the rows of lengths 0 and 1, and then we can insert the rows of length 2 into the area sequence, etc. We leave it as an exercise for the reader to verify (23) holds.

There is a more general form of Theorem 1 due to the author. Define the *reading order* of the rows of π to be the order in which the rows are listed by decreasing value of a_i, where if two rows have the same a_i-value, the row above is listed first. For the path on the left in Figure 7, the reading order is

$$\text{row } 6, \text{ row } 8, \text{ row } 5, \text{ row } 4, \text{ row } 3, \text{ row } 7, \text{ row } 2, \text{ row } 1. \tag{24}$$

Finally let $b_k = b_k(\pi)$ be the number of inversion pairs as in (21) which involve the kth row in the reading order and rows that occur before it in the reading order, and set $b_0 = -1$. For the path on the left in Figure 7, we have

$$(b_0, b_1, \ldots, b_8) = (-1, 0, 1, 1, 2, 3, 3, 1, 0). \tag{25}$$

Let

$$C_n(q,t,z) = \sum_{d=0}^{n} z^d \langle \nabla e_n, e_{n-d} h_d \rangle. \tag{26}$$

Theorem 2. *[29]*

$$C_n(q,t,z) = \sum_{\pi \in L_{n,n}^+} q^{dinv(\pi)} t^{area(\pi)} \prod_{\substack{1 \le i \le n \\ b_i > b_{i-1}}} (1 + z/q^{b_i}). \tag{27}$$

Note that in the product on the right-hand side of (27), there is always a factor of $1+z$ corresponding to $i=1$. Furthermore, since by construction dinv $=\sum_{i\geq 1} b_i$, the coefficient of z^d in (27) is clearly in $\mathbb{N}[q,t]$.

We mention that (26) implies

$$\sum_{d=0}^{n-1} z^d \langle \nabla e_n, s_{d+1,1^{n-d-1}} \rangle = C_n(q,t,z)/(1+z). \tag{28}$$

Corollary 1.

$$C_n(q,t,z) = \sum_{\pi \in L^+_{(n,n)}} q^{area(\pi)} t^{bounce(\pi)} \prod_{\substack{a_i > a_{i-1} \\ 1 \leq i \leq n}} (1 + z/q^{a_i}), \tag{29}$$

where we set $a_0 = -1$.

Proof. (Sketch) For $1 \leq i \leq n$, Let $a'_i(\pi)$ be the number of squares in the ith column (from the right) which are below the path π and strictly above the diagonal $x = y$, and set $a'_0 = -1$. It is easy to see that

$$b_i(\pi) = a'_i(\zeta(\pi)).$$

Note that terms in (29) where $a_i > a_{i-1}$ correspond to consecutive N steps of π. By an inductive argument you can identify such pairs of N steps with corresponding pairs of E steps, showing that

$$\prod_{\substack{a_i > a_{i-1} \\ 1 \leq i \leq n}} (1 + z/q^{a_i}) = \prod_{\substack{a'_i > a'_{i-1} \\ 1 \leq i \leq n}} (1 + z/q^{a'_i}).$$

□

Theorem 2 is often called the q,t-Schröder Theorem, since values of i in (27) for which $b_i > b_{i-1}$ correspond to the tops of columns in π. Hence choosing a given term in (27) which contributes to the overall coefficient of z^d corresponds to a selection of d column tops. If we replace each of these column tops by a diagonal step, we get a so-called Schröder path, with q,t-weights.

The polynomial $C_n(q,t,-a)$ has received more attention recently after E. Gorsky noticed that it equals the $m = n+1$ case of the modified superpolynomial $\tilde{P}_{T(m,n)}(a,q,t)$, as mentioned in Section 1. Gorsky and Negut recently obtained a few explicit formulas for $\tilde{P}_{T(m,n)}(a,q,t)$ which we now discuss.

3.1 An extension of ∇e_n

As mentioned in Section 1, given any relatively prime pair (m,n) of positive integers, there is an associated operator $Q_{(m,n)}$ on symmetric functions which has arisen from the study of the Elliptic Hall algebra and associated objects. We refer the reader to [8] and [9] for a concrete, recursive description of these operators. By $Q_{(m,n)}(-\mathbf{1})^{\mathbf{n}}$ we mean the action of $Q_{(m,n)}$ on the constant $(-1)^n$. Then

$$Q_{(n+1,n)}(-\mathbf{1})^{\mathbf{n}} = \nabla e_n,$$

and the symmetric functions $Q_{(m,n)}(-\mathbf{1})^{\mathbf{n}}$ form a natural generalization of ∇e_n. There is a growing family of conjectured interpretations and identities for these functions. For example, in [26] it is conjectured that $Q_{(m,n)}(-\mathbf{1})^{\mathbf{n}}$ equals the bigraded Frobenius series of the finite-dimensional, irreducible module $L_{m/n}$ of the rational Cherednik algebra with parameter $c = m/n$. For $m = kn \pm 1$ this conjecture follows from results of Gordon and Stafford [27, 28]. We now turn to what is known or conjectured about the combinatorics of $Q_{(m,n)}(-\mathbf{1})^{\mathbf{n}}$, and relations to the superpolynomial.

3.2 Tesler Matrices

There is an explicit formula for $Q_{(m,n)}(-\mathbf{1})^{\mathbf{n}}$ due to Gorsky and Negut, expressed in terms of Tesler matrices. These matrices have found previous application to the study of diagonal harmonics [3, 20, 31].

Theorem 3. *Gorsky, Negut [24] For any pair of positive, relatively prime integers (m,n),*

$$Q_{(m,n)}(-\mathbf{1})^{\mathbf{n}} = \sum_{C\in Tes(m,n)} \prod_{i=1}^{m} e_{c_{i,i}} \prod_{\substack{1\le i<m \\ c_{i,i+1}>0}} ([c_{i,i+1}+1]_{q,t} - [c_{i,i+1}]_{q,t}) \prod_{\substack{2\le i+1<j\le m \\ c_{i,j}>0}} (-M)[c_{i,j}]_{q,t}. \tag{30}$$

Here $Tes(m,n)$ is the set of $m\times m$ upper-triangular matrices C of nonnegative integers satisfying

$$c_{i,i} + \sum_{j>i} c_{i,j} - \sum_{j<i} c_{j,i} = \left\lfloor \frac{in}{m} \right\rfloor - \left\lfloor \frac{(i-1)n}{m} \right\rfloor, \quad 1 \le i \le m, \tag{31}$$

$[k]_{q,t} = (t^k - q^k)/(t-q)$, and as before $M = (1-q)(1-t)$.

Example 3. *Note that when $m = n+1$ the conditions (31) reduce to all the "hook sums" $\sum_{j\ge i} c_{ij} - \sum_{j<i} c_{ji}$ equal 1 except for the first which equals 0. Having a first hook sum equal to 0 forces the first row to be all zeros, and so*

$$Tes(3,2) = \{\begin{bmatrix} 0 & 0 & 0 \\ 0 & 1 & 0 \\ 0 & 0 & 1 \end{bmatrix}, \begin{bmatrix} 0 & 0 & 0 \\ 0 & 0 & 1 \\ 0 & 0 & 2 \end{bmatrix}\}.$$

For the leftmost matrix above, the corresponding weight from (30) is e_1e_1 while the weight for the rightmost matrix is $e_2(q+t-1)$. Adding these together we get

$$Q_{(3,2)}(-\mathbf{1})^2 = e_1e_1 + e_2(q+t-1) = s_2 + (q+t)s_{1^2}$$

in agreement with Example 2.

Example 4. *The modified superpolynomial $\tilde{P}_{(m,n)}(q,t,a)$ can be defined as*

$$\tilde{P}_{(m,n)}(q,t,a) = \sum_{d=0}^{n}(-a)^d \langle Q_{(m,n)}(-\mathbf{1})^{\mathbf{n}}, e_{n-d}h_d\rangle, \tag{32}$$

i.e., the generating function for the hook shapes in $Q_{(m,n)}(-\mathbf{1})^{\mathbf{n}}$. Now given a symmetric function f expressed as a polynomial in the e_k, the scalar product $\langle f, e_{n-d}h_d\rangle$ can be found by replacing each e_k by $1+z$, and taking the coefficient of z^d in the resulting expression. Hence a formula for $\tilde{P}_{(m,n)}(a,q,t)$ can be obtained by replacing each $e_{c_{i,i}}$ in (30) by $1-a$.

In [47] and [26] a conjectured combinatorial expression for the superpolynomial of the (m,n) torus knot is given, which we now describe (in terms of the modified superpolynomial). In the appendix of [47] it is shown that this conjecture reduces to the q,t-Schröder theorem when $m = n+1$.

Let Grid(m,n) be the $n \times m$ grid of labelled squares whose upper-left-hand corner square is labelled with $(n-1)(m-1)-1$, and whose labels decrease by m as you go down columns and by n as you go across rows. For example,

Grid(3, 7) = (33)

11	4	−3
8	1	−6
5	−2	−9
2	−5	−12
−1	−8	−15
−4	−11	−18
−7	−14	−21

To the corners of the squares of Grid(m,n) we associate Cartesian coordinates, where the lower-left-hand corner of the grid has coordinates $(0,0)$, and the upper-right-hand-corner of the grid (m,n). One finds the set of paths $L^+_{(m,n)}$ which never go below the diagonal $my = nx$ is the same as the set of π for which none of the squares with negative labels are above π. For a given π, we let area(π) denote the number of squares in Grid(m,n) with positive labels which are below π. Furthermore, let

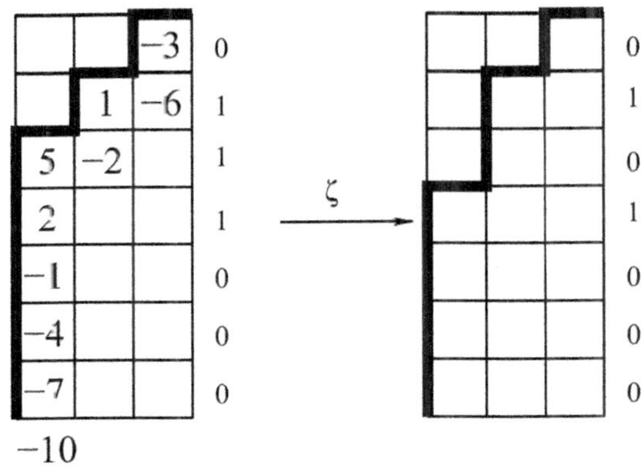

Fig. 8 The sweep map.

dinv(π) denote the number of squares in Grid(m,n) which are above π and whose arm and leg lengths satisfy

$$\frac{a}{l+1} < \frac{m}{n} < \frac{a+1}{l}. \tag{34}$$

Here by the arm and leg lengths of a square s we mean the distances from s to π in the easterly and southerly directions, respectively. For example, if $(m,n) = (3,7)$ and $\pi = NNNNNEENNE$, then area$(\pi) = 2$ (corresponding to the squares with labels 2 and 5). Also, dinv$(\pi) = 2$; the squares with labels 11, 8, 4, 1 have $a = l = 1; a = 1, l = 0; a = 0, l = 1; a = l = 0$, respectively, and so the squares with labels 8 and 11 do not satisfy (34), while the squares with labels 1 and 4 do.

Given $\pi \in L^+_{(m,n)}$, let $R(\pi)$ denote the set of labels of squares which are at the top of some column of π. Say these labels occur in columns $c_1, c_2, \ldots, c_k$ as we move left to right. Then for $1 \le i \le k$, let t_i denote the label of the square which is in the same row as the square at the top of column c_i, and is also in column c_{i+1}, and set $T(\pi) = \{t_1, t_2, \ldots, t_{k-1}\}$. For example, if π is the path on the left of Figure 8, then

$$R(\pi) = \{-3, 1, 5\} \tag{35}$$

$$T(\pi) = \{-6, -2\}. \tag{36}$$

Now form a vector $\alpha(\pi) = (\alpha_1, \ldots, \alpha_k)$ consisting of the elements of $R(\pi)$ in decreasing order, and let $\gamma_i(\pi)$ denote the number of elements of $R(\pi)$ which are larger than α_i, minus the number of elements of $T(\pi)$ which are larger than α_i. For the example of (35), we have $\alpha = (5, 1, -3)$, and so $\gamma_1 = 0 - 0 = 0, \gamma_2 = 1 - 0 = 1, \gamma_3 = 2 - 1 = 1$.

Conjecture 1. *[47] For any pair (m,n) of positive, relatively prime integers,*

$$\tilde{P}_{T(m,n)}(a,q,t) = \sum_{\pi \in L^+_{m,n}} q^{dinv(\pi)} t^{area(\pi)} \prod_{1 \le i \le k} (1 - a/q^{\gamma_i}). \quad (37)$$

3.3 An Extension of the Zeta Map

There is a candidate extension of the zeta map of Figure 7 which can be described as follows. Given a path $\pi \in L^+_{(m,n)}$, call the set of corners of grid squares which are touched by π the "vertices" of π. Next define $S(\pi)$ to be the set consisting of the labels of those squares whose upper-left-hand corners are vertices of π. A given label in $S(\pi)$ is called an N label if the vertex associated to it is the start of an N step, otherwise it is called an E label. For example, if π is the path on the left in Figure 8, then

$$\pi = NNNNNENENENE \quad S(\pi) = \{-10, -7, -4, -1, 2, 5, -2, 1, -6, -3\}. \quad (38)$$

We now define the "sweep map" of [5], denoted ζ, from $L^+_{(m,n)}$ to $L^+_{(m,n)}$ as follows: order the elements of $S(\pi)$ in increasing order to create a vector of labels $D(\pi) = (d_1, d_2, \ldots, d_{m+n})$. Then create a path $\phi(\pi)$ by defining the ith step of $\phi(\pi)$ to be an N step if d_i is an N label, and an E step if d_i is an E label. For the example in (38), we have

$$D(\pi) = (-10, -7, -6, -4, -3, -2, -1, 1, 2, 5) \quad \phi(\pi) = NNNNENNENE. \quad (39)$$

We note that when $m = n + 1$, paths in $L^+_{(m,n)}$ are in bijection with paths in $L^+_{(n,n)}$, and that the sweep map reduces to the ζ map of Figure 7.

4 The Rational Shuffle Conjecture

There is a nice conjecture for the expansion of the symmetric function $Q_{(m,n)}(\mathbf{-1})^{\mathbf{n}}$ into monomials, which is described in [24] and also [8]. If $m = n + 1$ it reduces to the shuffle conjecture from [34]. Let an (m,n)-parking function be a path $\pi \in L^+_{(m,n)}$ together with a placement of the integers 1 through n (called *cars*) just to the right of the N steps of π, with strict decrease down columns. For such a pair P, we let rank(j) be the label of the square that contains j, and we set

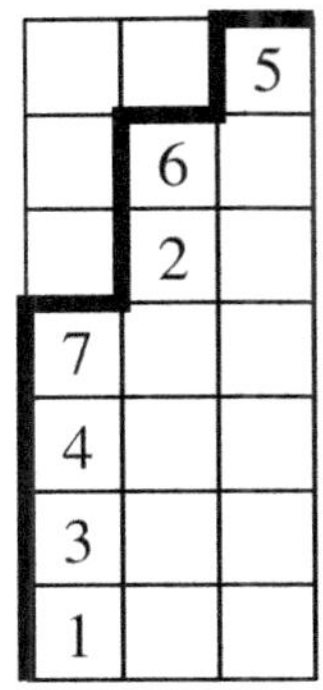

Fig. 9 A (3, 7)-parking function.

$$\mathrm{tdinv}(P) = |\{(i,j) : 1 \le i < j \le n \text{ and } \mathrm{rank}(i) < \mathrm{rank}(j) < \mathrm{rank}(i) + m\}|. \tag{40}$$

Furthermore we let the reading word read(P) be the permutation obtained by listing the cars by decreasing order of their ranks. For example, for the (3, 7)-parking function of Figure 9, tdinv $= 3$, with inversion pairs formed by pairs of cars $(6,7)$, $(4,6)$, and $(2,4)$, and the reading word is 7642531.

Let maxtdinv(π) be tdinv of the parking function for π whose reading word is the reverse of the identity, and for any parking function P for π set

$$\mathrm{dinv}(P) = \mathrm{dinv}(\pi) + \mathrm{tdinv}(P) - \mathrm{maxtdinv}(\pi). \tag{41}$$

Then one formulation of the rational shuffle conjecture is the following

Conjecture 2. *For any pair of relatively prime positive integers (m,n) and any pair of compositions α, β with $\sum_i \alpha + \sum_j beta = n$, we have*

$$\langle Q_{(m,n)}(-\mathbf{1})^{\mathbf{n}}, e_\alpha h_\beta \rangle = \sum_{(m,n) \text{ parking functions } P} q^{dinv(P)} t^{area(\pi)}, \tag{42}$$

where the sum is over all (m,n) parking functions P whose reading word is a shuffle of decreasing sequences of lengths $\alpha_1, \alpha_2, \ldots$ and increasing sequences of lengths $\beta_1, \beta_2, \ldots$. Here $e_\alpha = \prod_i e_{\alpha_i}$, and $h_\beta = \prod_i h_{\beta_i}$.

We note that the values of $\langle Q_{(m,n)}(-\mathbf{1})^{\mathbf{n}}, h_\beta \rangle$ for all partitions β define $Q_{(m,n)}(-\mathbf{1})^{\mathbf{n}}$ as a symmetric function, so at first glance it seems the conditions in (42) over determine $Q_{(m,n)}(-\mathbf{1})^{\mathbf{n}}$. See [30, p. 99] for an explanation of why the case of $\alpha = \emptyset$ in (42) implies the case of general α. We also mention that the motivation for the right-hand-side of (42) comes largely from work of Hikita [40]. He showed that the bigraded Frobenius series of a certain module arising from affine Springer fibers is the same as the bigraded Frobenius series which the rational shuffle conjecture predicts for $Q_{(m,n)}(-\mathbf{1})^{\mathbf{n}}$. Hence one approach to the rational shuffle conjecture is to try and connect the $Q_{(m,n)}(-\mathbf{1})^{\mathbf{n}}$ to Hikita's affine Springer fiber modules directly.

Example 5. *If $\alpha = (n-d)$ and $\beta = (d)$, the left-hand-side of (42) becomes*

$$\langle Q_{(m,n)}(-\mathbf{1})^{\mathbf{n}}, e_{n-d}h_d \rangle. \tag{43}$$

We claim the right-hand-side of (42) reduces to the coefficient of $(-a)^d$ in the right-hand-side of (37). To see why, fix an (m,n) path π, and let our increasing sequence be $(n-d+1,\ldots,n)$ and our decreasing sequence be $(n-d,\ldots,2,1)$. Let P be a parking function for π whose reading word is a shuffle of these sequences. Note the elements of $(n-d+1,\ldots,n)$ of P must occur at the tops of columns. Call these column tops $c_1,\ldots,c_k$, and say their rank numbers are $r_1,\ldots,r_k$. How does dinv(P) differ from dinv(π)? From (41), to answer this question we need to determine $-$tdinv(P) + maxtdinv(π). It is not hard to see that this difference is the sum, over all column tops c_i for $1 \le i \le k$, of the number of cars of P which reside in a column top square whose rank r satisfies $r_i < r < r_i + m$, and which are also in the increasing sequence. The calculation of the γ_i from (37) accomplishes this same purpose. Hence the coefficient of $(-a)^d$ in the right-hand-side of (37) is the same as that which the rational shuffle conjecture predicts for (43), in agreement with (32).

5 Open Problems

Drew Armstrong has coined the phrase "rational Catalan combinatorics" for combinatorial issues connected to paths in $L^+_{(m,n)}$ and the rational shuffle conjecture [2]. (These slides also introduce the right-hand-side of (42), independently of Hikita's work). In this section we list a few open problems involving rational Catalan combinatorics.

Open Problem 1. *Prove that for general coprime (m,n) the sweep map is a bijection from $L^+_{(m,n)} \to L^+_{(m,n)}$.*

This problem has been studied by Gorsky, Mazin, and Vazirani [25] and Armstrong, Loehr, and Warrington [5]. See also [4]. A related question is to find a bounce statistic on paths in $L^+_{(m,n)}$ which, when combined with area, generates $\langle Q_{(m,n)}(-\mathbf{1})^{\mathbf{n}}, s_{1^n}\rangle$. In [25] it is shown that the sweep map is a bijection whenever $m = kn+1$ or $m = kn-1$ for some positive integer k. We note that in the case $m = kn+1$ Loehr [42, 43] (see also [30][pp. 108–109]) has defined an extension of the bounce statistic, which when combined with area generates the q,t-Catalan for $m = kn+1$. For this "paths in a $n \times kn$ rectangle" case there is also an interpretation for the rational shuffle conjecture in terms of a generalization of diagonal harmonics (see [34]).

Open Problem 2. *Prove that*

$$\sum_{\pi \in L^+_{(m,n)}} q^{dinv(\pi)+(n-1)(m-1)/2-area(\pi)} = \frac{1}{[m]}\begin{bmatrix} n+m-1 \\ n \end{bmatrix}. \tag{44}$$

Here $[m] = (1-q^m)/(1-q)$ *and* $\begin{bmatrix} n+m-1 \\ n \end{bmatrix}$ *is the q-binomial coefficient.*

There is currently no known way to generate the right-hand-side of (44) as a sum over weighted lattice paths. It is known [24, p. 28], [21] that

$$q^{(n-1)(m-1)/2}\langle Q_{(m,n)}(-\mathbf{1})^{\mathbf{n}}, s_{1^n}\rangle|_{t=1/q} = \frac{1}{[m]}\begin{bmatrix} n+m-1 \\ n \end{bmatrix}, \tag{45}$$

so the sign-character case of the rational shuffle conjecture would imply (44).

Open Problem 3. *Refine the rational shuffle conjecture by finding a combinatorial prediction for the coefficient of an arbitrary Schur function in* $Q_{(m,n)}(-\mathbf{1})^{\mathbf{n}}$.

We mention that one consequence of the rational shuffle conjecture is that $Q_{(m,n)}(-\mathbf{1})^{\mathbf{n}}$ can be decomposed into a sum of symmetric functions, one for each (m,n) path. These symmetric functions are special cases of LLT polynomials, specifically they are LLT products of vertical strips - see [23], and [30, Chapter 6] for an explanation of why this is true for the $m = n+1$ case. Hence understanding the Schur coefficients of $Q_{(m,n)}(-\mathbf{1})^{\mathbf{n}}$ is embedded in the famous problem of understanding the Schur coefficients of LLT polynomials. (It is also known [32, 33], [30, Appendix A] that the modified Macdonald polynomials $\tilde{H}_\mu(X;q,t)$ are a sum of LLT products of ribbon shapes. It is still an open question to understand the Schur coefficients in this case combinatorially.

Open Problem 4. *The constructions of Aganagic and Shakirov and of Cherednik for torus knot invariants actually depend on a "coloring" by a partition* λ *in addition to the parameters* a, q, t *and* m, n. *The partition* λ *is identified with an irreducible representation. The case we have considered in this chapter corresponds to the case* $\lambda = (1)$, *consisting of a single part of size* 1. *There is no known explicit formula for any other* λ, *although there is an algorithm to compute them [24]. It would be quite interesting to have a conjectured combinatorial formula for the superpolynomial for more general* λ.

Open Problem 5. *In [8], the "compositional rational shuffle conjecture" is introduced, which includes both the rational shuffle conjecture and the compositional shuffle conjecture from [35] (which involves* (n,n)*-paths which hit the diagonal* $x = y$ *in specified places) as special cases. This conjecture is based on the construction of a version of the* $Q_{(m,n)}$ *operators for general non-coprime* (m,n). *Can this be used to define a superpolynomial invariant for* (m,n) *torus links? Is there a Tesler matrix expression for this more general case?*

Note added in Proof: In a recent article on the arXiv [51], N. Williams appears to solve Open Problem 1. In another recent arXiv article, Carlsson and Mellit [11] have made substantial progress on the rational shuffle conjecture, by proving the original shuffle conjecture (i.e. the case $m = n + 1$).

References

1. J.W. Alexander, Topological invariants of knots and links. Trans. Am. Math. Soc. **30**(2), 275–306 (1928)
2. D. Armstrong, Rational Catalan combinatorics 1. Slides from a talk at the JMM 2012 meetings (2012)
3. D. Armstrong, A. Garsia, J. Haglund, B. Sagan, B. Rhoades, Combinatorics of Tesler matrices in the theory of parking functions and diagonal harmonics. J. Comb. **3**, 451–494 (2012)
4. D. Armstrong, C.R.H. Hanusa, B.C. Jones, Results and conjectures on simultaneous core partitions. Eur. J. Comb. **41**, 205–220 (2014)
5. D. Armstrong, N. Loehr, G. Warrington, Rational parking functions and Catalan numbers (2014). Preprint on the math arXiv:1403.1845
6. M. Aganagic, S. Shakirov, Knot homology from refined Chern-Simons theory (2011). Preprint on the math arXiv:1105.5117
7. M. Aganagic, S. Shakirov, Refined Chern-Simons theory and knot homology, in *String-Math 2011*. Proceedings of Symposia in Pure Mathematics, vol. 85 (American Mathematical Society, Providence, 2012), pp. 3–31
8. F. Bergeron, A.M. Garsia, E. Leven, G. Xin, Compositional (km, kn)-shuffle conjectures (2014). Preprint on the math arXiv:1404.4616
9. F. Bergeron, A.M. Garsia, E. Leven, G. Xin, Some remakable new plethystic operators in the thoery of Macdonald polynomials (2014). Preprint on the math arXiv:1405.0316
10. I. Burban, O. Schiffmann, On the Hall algebra of an elliptic curve, I. Duke Math. J. **161**(7), 1171–1231 (2012)
11. E. Carlsson, A. Mellit, A proof of the shuffle conjecture. arXiv:1508.06239 (2015)
12. I. Cherednik, Jones polynomials of torus knots via DAHA. Int. Math. Res. Not. **2013**(23), 5366–5425 (2013)
13. J.H. Conway, An enumeration of knots and links, and some of their algebraic properties, in *Computational Problems in Abstract Algebra* (Proc. Conf., Oxford, 1967) (Pergamon, Oxford, 1970), pp. 329–358
14. N.M. Dunfield, S. Gukov, J. Rasmussen, The superpolynomial for knot homologies. Exp. Math. **15**(2), 129–159 (2006)
15. P. Freyd, D. Yetter, J. Hoste, W.B.R. Lickorish, K. Millett, A. Ocneanu, A new polynomial invariant of knots and links. Bull. Am. Math. Soc. (N.S.) **12**(2), 239–246 (1985)
16. A.M. Garsia, M. Haiman, A graded representation model for Macdonald polynomials. Proc. Nat. Acad. Sci. U.S.A. **90**(8), 3607–3610 (1993)
17. A.M. Garsia, M. Haiman, A remarkable q, t-Catalan sequence and q-Lagrange inversion. J. Algebraic Comb. **5**(3), 191–244 (1996)
18. A.M. Garsia, J. Haglund, A positivity result in the theory of Macdonald polynomials. Proc. Nat. Acad. Sci. U.S.A. **98**, 4313–4316 (2001)
19. A.M. Garsia, J. Haglund, A proof of the q, t-Catalan positivity conjecture. Discret. Math. **256**, 677–717 (2002)
20. A.M. Garsia, J. Haglund, G. Xin, Constant term methods in the theory of Tesler matrices and Macdonald polynomial operators. Ann. Comb. **18**, 83–109 (2014)

21. A. Garsia, E. Leven, N. Wallach, G. Xin, A new plethystic symmetric function operator and the rational compositional shuffle conjecture at $t = 1/q$ (2015). Preprint on the math arXiv: http://arxiv.org/pdf/1501.00631
22. E. Gorsky, Private communication (2012)
23. E. Gorsky, M. Mazin, Rational parking functions and LLT polynomials (2015). Preprint on the math arXiv:1503.04181
24. E. Gorsky, A. Negut, Refined knot invariants and Hilbert schemes (2013). Preprint on the math arXiv: http://arxiv.org/pdf/1304.3328
25. E. Gorsky, M. Mazin, M. Vazirani, Affine permutations and rational slope parking functions (2014). Preprint on the math arXiv:1403.0303
26. E. Gorsky, A. Oblomkov, J. Rasmussen, V. Shende, Torus knots and the rational DAHA. Duke Math. J. **163**(14), 2709–2794 (2014)
27. I. Gordon, J.T. Stafford, Rational Cherednik algebras and Hilbert schemes. Adv. Math. **198**(1), 222–274 (2005)
28. I. Gordon, J.T. Stafford, Rational Cherednik algebras and Hilbert schemes. II. Representations and sheaves. Duke Math. J. **132**(1), 73–135 (2006)
29. J. Haglund, A proof of the q,t-Schröder conjecture. Int. Math. Res. Not. **11**, 525–560 (2004)
30. J. Haglund, *The q,t-Catalan Numbers and the Space of Diagonal Harmonics*. University Lecture Series, vol. 41 (American Mathematical Society, Providence, 2008). With an appendix on the combinatorics of Macdonald polynomials
31. J. Haglund, A polynomial expression for the Hilbert series of the quotient ring of diagonal coinvariants. Adv. Math. **227**, 2092–2106 (2011)
32. J. Haglund, M. Haiman, N. Loehr, A combinatorial formula for Macdonald polynomials. J. Am. Math. Soc. **18**, 735–761 (2005)
33. J. Haglund, M. Haiman, N. Loehr, Combinatorial theory of Macdonald polynomials I: Proof of Haglund's formula. Proc. Nat. Acad. Sci. U.S.A. **102**(8), 2690–2696 (2005)
34. J. Haglund, M. Haiman, N. Loehr, J.B. Remmel, A. Ulyanov, A combinatorial formula for the character of the diagonal coinvariants. Duke Math. J. **126**, 195–232 (2005)
35. J. Haglund, J. Morse, M. Zabrocki, A compositional shuffle conjecture specifying touch points of the Dyck path. Can. J. Math. **64**, 822–844 (2012)
36. M. Haiman, Conjectures on the quotient ring by diagonal invariants. J. Algebraic Comb. **3**, 17–76 (1994)
37. M. Haiman, Private communication (2000)
38. M. Haiman, Hilbert schemes, polygraphs, and the Macdonald positivity conjecture. J. Am. Math. Soc. **14**, 941–1006 (2001)
39. M. Haiman, Vanishing theorems and character formulas for the Hilbert scheme of points in the plane. Invent. Math. **149**, 371–407 (2002)
40. T. Hikita, Affine Springer fibers of type A and combinatorics of diagonal coinvariants. Adv. Math. **263**, 88–122 (2014)
41. V.F.R. Jones, A polynomial invariant for knots via von Neumann algebras. Bull. Am. Math. Soc. (N.S.) **12**(1), 103–111 (1985)
42. N. Loehr, Multivariate analogues of Catalan numbers, parking functions, and their extensions. Ph.D. Thesis, University of California at San Diego, 2003
43. N.A. Loehr, Conjectured statistics for the higher q,t-Catalan sequences. Electron. J. Comb. **12**(Research Paper 9), 54 pp. (electronic) (2005)
44. I.G. Macdonald, A new class of symmetric polynomials. Actes du 20e Séminaire Lotharingien. Publ. Inst. Rech. Math. Av. **372**, 131–171 (1988)
45. I.G. Macdonald, *Symmetric Functions and Hall Polynomials*, 2nd edn. Oxford Mathematical Monographs. Oxford Science Publications (The Clarendon Press/Oxford University Press, New York, 1995)
46. A, Negut, The shuffle algebra revisited. Int. Math. Res. Not. IMRN **2014**(22), 6242–6275 (2014)
47. A. Oblomkov, J. Rasmussen, V. Shende, The Hilbert scheme of a plane curve singularity and the HOMFLY homology of its link (2012). Preprint on the math arXiv:1201.2115

48. O. Schiffmann, E. Vasserot, The elliptic Hall algebra, Cherednik Hecke algebras and Macdonald polynomials. Compos. Math. **147**(1), 188–234 (2011)
49. O. Schiffmann, E, Vasserot, The elliptic Hall algebra and the K-theory of the Hilbert scheme of $\mathbb{A}^2$. Duke Math. J. **162**(2), 279–366 (2013)
50. Wikipedia, Knot theory. https://en.wikipedia.org/wiki/Knot_theory
51. N. Williams, Sweeping up zeta. arXiv:1512.01483 (2015, Preprint)

A survey of consecutive patterns in permutations

Sergi Elizalde

Abstract A consecutive pattern in a permutation π is another permutation σ determined by the relative order of a subsequence of contiguous entries of π. Traditional notions such as descents, runs, and peaks can be viewed as particular examples of consecutive patterns in permutations, but the systematic study of these patterns has flourished in the last 15 years, during which a variety of different techniques have been used. We survey some interesting developments in the subject, focusing on exact and asymptotic enumeration results, the classification of consecutive patterns into equivalence classes, and their applications to the study of one-dimensional dynamical systems.

Keywords Consecutive pattern • Pattern avoidance • Permutation • c-Wilf-equivalence • Cluster method • Dynamical system

Mathematics Subject Classification (2010): 05A05, 05A15, 05A16

1 Introduction

Patterns in permutations are implicit already in the work of MacMahon from one century ago, but interest in them has grown in the last few decades, inspired by work of Knuth [53] on sorting permutations using data structures such as stacks and double-ended queues. He showed that permutations sortable with these devices can be characterized as those not containing subsequences of length 3 whose entries are in a prescribed order (we call such a subsequence a classical pattern), and that they are counted by the Catalan numbers. A few years later, the systematic enumeration of permutations avoiding classical patterns was continued by Simion and Schmidt [65]. Interest in permutation patterns has grow steadily since then, becoming a very active area of research, with connections with other areas of

S. Elizalde (✉)
Department of Mathematics, Dartmouth College, Hanover, NH 03755, USA
e-mail: sergi.elizalde@dartmouth.edu

A. Beveridge et al. (eds.), *Recent Trends in Combinatorics*, The IMA Volumes in Mathematics and its Applications 159, DOI 10.1007/978-3-319-24298-9_24

mathematics, such as algebraic geometry, and other fields such as computer science, biology, and physics.

In this survey we are interested in an important variation to the definition of permutation patterns used in [53, 65], namely the notion of *consecutive* patterns. In an occurrence of a consecutive pattern in a permutation, the positions of the entries are required to be adjacent. Long before this concept was introduced and systematically studied in [36] by Elizalde and Noy, there had been implicit appearances of consecutive patterns in the combinatorics literature. For instance, occurrences of 21 are descents, whose distribution is given by the Eulerian numbers; occurrences of 321 are sometimes called double descents [40]; occurrences of 132 and 231 are called peaks and play an important role in algebraic combinatorics; maximal occurrences of $12\dots m$ are called runs; and permutations avoiding 123 and 321 are the well-known alternating permutations [8], counted by the Euler numbers.

In the last 15 years, a growing number of authors have made significant progress on the study of consecutive patterns in permutations by using an array of different techniques. Despite this ongoing research, many important questions in the field remain unanswered. In this article we survey some of these results and discuss how they are related, without the goal of being exhaustive.

After introducing the basic definitions and notation in Section 2, Section 3 discusses an equivalence relation on patterns which is analogous to Wilf-equivalence for classical patterns. Section 4 surveys exact enumeration formulas that have been obtained using approaches such as the symbolic method, binary trees, the cluster method, symmetric functions, homological algebra, and spectral theory. In Section 5 we summarize some results concerning the asymptotic behavior of the number of permutations avoiding a consecutive pattern. Finally, Section 6 presents a recently discovered application of consecutive patterns to the study of time series.

We will not discuss the extensive literature on classical (i.e., non-consecutive) patterns, for which we refer the reader to the surveys [67] and [49]. We point out that even though some of the questions in both areas are similar, most of the techniques that have been used are significantly different, and so are the results that have been obtained, from the classification into equivalence classes to the description of the asymptotic behavior of permutations avoiding a pattern. Exact enumeration generally seems easier in the consecutive case: the only infinite family of classical patterns for which a generating function is known is the monotone family, whereas for consecutive patterns the distribution of occurrences for several families of patterns is known. On the other hand, some small classical patterns give rise to simple algebraic generating functions, which does not happen in the consecutive case.

2 Definitions

Let $\mathcal{S}_n$ denote the set of permutations of $\{1, 2, \dots, n\}$. For a sequence τ of k distinct positive integers, let $\mathrm{st}(\tau)$ denote the permutation in $\mathcal{S}_k$ obtained by replacing the smallest entry of τ with 1, the second smallest with 2, and so on. For example, st(394176) = 263154. Given $\pi = \pi_1\pi_2\cdots\pi_n \in \mathcal{S}_n$ and $\sigma = \sigma_1\sigma_2\cdots\sigma_k \in \mathcal{S}_k$, an occurrence of σ in π as a *consecutive pattern* is a subsequence of k contiguous entries of π such that $\mathrm{st}(\pi_i\pi_{i+1}\cdots\pi_{i+k-1}) = \sigma_1\sigma_2\cdots\sigma_k$. For example, in $\pi = 132546$, the subsequences 1325 and 2546 are two occurrences of the pattern $\sigma = 1324$.

Denote by $c_\sigma(\pi)$ the number of occurrences of σ in π as a consecutive pattern. If $c_\sigma(\pi) = 0$, we say that π *avoids* σ. For example, 25134 avoids 132. Let $\alpha_n(\sigma)$ be the number of permutations in $\mathcal{S}_n$ that avoid σ as a consecutive pattern. The notions of occurrence, containment, and avoidance in this survey will refer to consecutive patterns unless explicitly stated otherwise, in which case we will use the term *classical pattern* to mean that the entries in an occurrence have no adjacency requirement.

Let

$$P_\sigma(u,z) = \sum_{n\geq 0}\sum_{\pi\in\mathcal{S}_n} u^{c_\sigma(\pi)}\frac{z^n}{n!}$$

be the exponential generating function for occurrences of σ. Setting $u = 0$, we get the generating function for σ-avoiding permutations, $P_\sigma(0,z) = \sum_{n\geq 0}\alpha_n(\sigma)\frac{z^n}{n!}$.

A useful invariant of a pattern σ is its overlap set O_σ, defined as the set of indices i with $1 \leq i < m$ such that $\mathrm{st}(\sigma_{i+1}\sigma_{i+2}\ldots\sigma_m) = \mathrm{st}(\sigma_1\sigma_2\ldots\sigma_{m-i})$. Equivalently, $i \in O_\sigma$ if two occurrences of σ in a permutation can have starting positions at distance i from each other. For example, $O_{1423} = \{2,3\}$. Note that $m-1 \in O_\sigma$ for every $\sigma \in \mathcal{S}_m$. If $m \geq 3$, a pattern $\sigma \in \mathcal{S}_m$ for which $O_\sigma = \{m-1\}$ is said to be *non-overlapping* (sometimes also called *minimally overlapping* or *non-self-overlapping*). Equivalently, σ is non-overlapping if two occurrences of σ in a permutation cannot overlap in more than one position. For example, the patterns 132, 1243, 1342, 21534, and 34671285 are non-overlapping. We denote by $\mathcal{N}_m$ the set of non-overlapping patterns in $\mathcal{S}_m$. Non-overlapping patterns have been studied by Duane and Remmel [25] and by Bóna [17], who showed that $\lim_{m\to\infty}|\mathcal{N}_m|/m! \approx 0.364$.

3 Consecutive Wilf-equivalence classes

In analogy with the definition of Wilf-equivalence for classical patterns, we say that two patterns σ and τ are *c-Wilf-equivalent* if $P_\sigma(0,z) = P_\tau(0,z)$ (that is, $\alpha_n(\sigma) = \alpha_n(\tau)$ for all n), and we say that they are *strongly c-Wilf-equivalent* if $P_\sigma(u,z) = P_\tau(u,z)$. For example, every pattern $\sigma = \sigma_1\cdots\sigma_m$ is strongly c-Wilf-equivalent to its reversal $\sigma_m\cdots\sigma_1$ and its complement $(m+1-\sigma_1)\cdots(m+1-\sigma_m)$. Some non-trivial examples of strong c-Wilf-equivalent pairs are $1342 \sim 1432$ and $154263 \sim 165243$. An open problem in consecutive patterns (whose analogue in the classical case is wide open as well) is to classify patterns into these equivalence classes. It was conjectured by Nakamura that the word *strongly* in the above definition is superfluous, in sharp contrast with the case of classical patterns:

Conjecture 1 ([56, Conjecture 5.6]). If two patterns are c-Wilf-equivalent, then they are also strongly c-Wilf-equivalent.

This conjecture is known to hold in the special case of non-overlapping patterns (see [32, Lemma 3.2] and [55, Theorem 11]). In fact, Dwyer and Elizalde [26] have recently proved that if two non-overlapping patterns σ and τ are c-Wilf-equivalent, then they are *super-strongly* c-Wilf-equivalent, meaning that for every n and for every set of positions S, the number of permutations in $\mathscr{S}_n$ having occurrences of σ exactly at positions S is the same as for τ. For example, 1342 and 2431 are super-strongly c-Wilf-equivalent, but 1423 and 3241 are not. It is worth mentioning that Conjecture 1 has an analogue for words [59, Conjectiure 1.2], which, as Pantone and Vatter showed, would imply the so-called Rearrangement Conjecture from [52].

So far, (strong) c-Wilf-equivalence classes have been characterized for patterns of length up to 6. There are 2 equivalence classes of patterns of length 3, represented by the patterns 123 and 132; 7 classes of patterns of length 4, represented by 1234, 2413, 2143, 1324, 1423, 1342, and 1243 (see [36]); 25 classes for patterns of length 5; and 92 for patterns of length 6 (see [37, 56]).

For non-overlapping patterns of any given length m, it was shown in [24, 25] — proving a conjecture from [28]— that the first and last entry of a pattern determine the strong c-Wilf-equivalence class where it belongs. The converse, once we account for the fact that the reversal and the complement of a pattern preserve its c-Wilf-equivalence class, is conjectured to be true in [32]. The above result has been generalized to sets of (possibly overlapping) consecutive patterns by Khoroshkin and Shapiro [45], by giving a list of sufficient conditions for such sets to be strongly c-Wilf-equivalent. As shown in [26], the same conditions imply super-strong c-Wilf-equivalence.

4 Exact enumeration

4.1 The origins

As mentioned in the introduction, some traditional results on permutations can be interpreted as statements about consecutive patterns, such as the well-known generating function for the Eulerian numbers,

$$P_{21}(u,z) = \frac{1-u}{e^{(u-1)z} - u},$$

and André's generating function $\sec z + \tan z$ for the Euler numbers, which counts up-down permutations [8].

Almost one century later, the first published formula for the number of permutations with no increasing runs of length m or more appears in a 1962 book of David and Barton [22, pg. 156–157]:

$$P_{12\dots m}(0,z)=\left(\sum_{j\geq 0}\frac{z^{jm}}{(jm)!}-\sum_{j\geq 0}\frac{z^{jm+1}}{(jm+1)!}\right)^{-1}. \tag{1}$$

Soon after, we find Entringer's enumeration of permutations according to their number of cyclic peaks (which he called maxima) [38], and Carlitz and Scoville's enumeration of permutations according to rising maxima and falling maxima [21], which are essentially (up to the definition at the beginning and at the end of the permutation) occurrences of 132 and 231.

It was not until 2001 that the systematic study of consecutive patterns in permutations was tackled. In [36], Elizalde and Noy introduced the notion of consecutive patterns and found expressions for the generating functions $P_\sigma(u,z)$ for certain families of patterns of arbitrary length, which include those of length 3 and three of the seven c-Wilf-equivalence classes of patterns of length 4. Letting

$$\omega_\sigma(u,z)=\frac{1}{P_\sigma(u,z)},$$

they gave differential equations satisfied by $\omega_\sigma(u,z)$. The proofs are based on representations of permutations as increasing binary trees, together with the symbolic method of Flajolet and Sedgewick [39]. In the following statements, the derivatives are taken with respect to z.

Theorem 1 ([36]). *Let $\sigma = 12\dots m$, where $m\geq 3$. Then $\omega=\omega_\sigma(u,z)$ is the solution of*

$$\omega^{(m-1)}+(1-u)(\omega^{(m-2)}+\cdots+\omega'+\omega)=0 \tag{2}$$

with $\omega(0)=1, \omega'(0)=-1, \omega^{(i)}(0)=0$ for $2\leq i\leq m-2$.

When $u=0$, the solution of the above differential equation can be expressed as a series, recovering Equation (1).

Another family enumerated in [36] consists of certain non-overlapping patterns that begin with 1. The above observation that the first and last entry determine the (strongly) c-Wilf-equivalence class of a non-overlapping pattern allows us to state the theorem more generally as follows.

Theorem 2 ([36]). *Let $\sigma\in\mathcal{N}_m$ with $\sigma_1=1$, and let $b=\sigma_m$. Then $\omega=\omega_\sigma(u,z)$ is the solution of*

$$\omega^{(b)}+(1-u)\frac{z^{m-b}}{(m-b)!}\omega'=0 \tag{3}$$

with $\omega(0)=1, \omega'(0)=-1, \omega^{(i)}(0)=0$ for $2\leq i\leq b-1$.

When $b = 2$, the above differential equation can be solved to obtain

$$P_\sigma(u,z) = \left(1 - \int_0^z e^{(u-1)t^{m-1}/(m-1)!}dt\right)^{-1}$$

for any $\sigma \in \mathcal{S}_m$ with $\sigma_1 = 1$ and $\sigma_m = 2$ (note that such a pattern is always non-overlapping).

A related problem considered in [36] is the enumeration of permutations that simultaneously avoid several consecutive patterns. Many more results in this direction, which we do not include here, have been obtained by Kitaev [46], Kitaev and Mansour [50, 51], and Aldred, Atkinson, and McCaughan [2].

In [47], Kitaev considered the maximum number of non-overlapping occurrences (i.e., not sharing any entries) of a pattern σ in a permutation π, which we denote by $\mathrm{nlap}_\sigma(\pi)$. He showed that the distribution of this statistic depends only on the number of σ-avoiding permutations:

Theorem 3 ([47]). *For every* σ,

$$\sum_{n\geq 0}\sum_{\pi\in\mathcal{S}_n} u^{\mathrm{nlap}_\sigma(\pi)}\frac{z^n}{n!} = \frac{P_\sigma(0,z)}{(1-u)+u(1-z)P_\sigma(0,z)}.$$

4.2 The cluster method

A powerful method to obtain expressions for $P_\sigma(u,z)$ is the *cluster method* of Goulden and Jackson [43], which is based on inclusion-exclusion (see also [44, Ch. 2.8], its restatement in [66, Ex. 4.40], and its extensions in [58]). Originally used for counting words according to occurrences of factors (i.e., consecutive subwords), it can be formulated in terms of permutations as stating that

$$P_\sigma(u,z) = \frac{1}{1-z-R_\sigma(u-1,z)}, \tag{4}$$

where $R_\sigma(t,z)$ is an exponential generating function where the coefficient of $t^k z^n/n!$ is the number of k-clusters with respect to σ, that is, permutations of length n with k marked occurrences of σ with the property that each marked occurrence overlaps with the next one, and the whole permutation is covered with these marked occurrences. Note that a cluster can have other occurrences of σ that are not marked. For example, 1 6 2 8 3 11 4 9 5 10 7 is a 3-cluster with respect to 14253, where the horizontal lines indicate the three marked occurrences.

To prove Equation (4), note that the generating function $1/(1-z-R_\sigma(t,z))$ counts permutations obtained as sequences of single entries and clusters, where the exponent of t is the number of marked occurrences of σ in these clusters. Thus, a permutation of length n with ℓ occurrences of σ contributes a term $t^k z^n/n!$ for each of the $\binom{\ell}{k}$ ways to mark k of these ℓ occurrences, for every $k \le \ell$. Making the substitution $t = u - 1$, the contribution of such a permutation to the right-hand side of (4) is

$$\sum_{k=0}^{\ell} \binom{\ell}{k} (u-1)^k \frac{z^n}{n!} = u^\ell \frac{z^n}{n!},$$

and thus equal to the contribution to the left-hand side.

Equation (4) is useful because it reduces the enumeration of occurrences of σ in all permutations to the enumeration of clusters, which are typically more tractable. For example, when $\sigma = 12\ldots m$, clusters are simply increasing permutations with marked overlapping occurrences of σ. When $u = 0$, Equation (1) follows from this observation with a little work. When σ is a non-overlapping pattern as in the statement of Theorem 2, computing the number of clusters and applying Equation (4) we get

$$P_\sigma(u,z) = \left(1 - z - \sum_{k\ge 1} \prod_{j=2}^{k} \binom{j(m-1)-b+1}{m-b} \frac{(u-1)^k z^{(m-1)k+1}}{((m-1)k+1)!}\right)^{-1}. \tag{5}$$

In [47], Kitaev used inclusion-exclusion arguments to deduce Equation (5) when $u = 0$.

Rawlings [62] noticed the power of the cluster method and used it to obtain enumeration formulas for the patterns $12\ldots m$, $12\ldots(m-2)m(m-1)$, and $1m(m-1)\ldots 2$ with an additional variable that keeps track of the number of inversions in the permutation, thus generalizing some of the results in [36]. Although his proof relies on a bijection of Fédou, it is in fact possible to directly generalize the above argument proving Equation (4) by adding a variable q that marks the number of inversions (both in the permutations on the left-hand side and on the clusters on the right-hand side), and using generating functions where the $n!$ in the exponential generating functions is replaced by its q-analogue $[n]_q! = \prod_{i=1}^{n}(1+q+\cdots+q^{i-1})$.

For example, refining Equation (5) by applying this q-analogue of the cluster method, we get the following generalization of Rawlings' formulas, where $\mathrm{inv}(\pi)$ denotes the number of inversions of π, and $\binom{n}{k}_q = \frac{[n]_q!}{[k]_q![n-k]_q!}$ is the q-binomial coefficient:

Theorem 4. *Let $\sigma \in \mathcal{N}_m$ with $\sigma_1 = 1$, and let $b = \sigma_m$. Then*

$$\sum_{n\ge 0} \sum_{\pi \in \mathcal{S}_n} q^{\mathrm{inv}(\pi)} u^{c_\sigma(\pi)} \frac{z^n}{[n]_q!}$$
$$= \left(1 - z - \sum_{k\ge 1} \prod_{j=2}^{k} \binom{j(m-1)-b+1}{m-b}_q \frac{q^{k\,\mathrm{inv}(\sigma)}(u-1)^k z^{(m-1)k+1}}{[(m-1)k+1]_q!}\right)^{-1}.$$

Similarly, denoting by $\mathcal{S}_n(12\ldots m)$ the set of permutations in $\mathcal{S}_n$ that avoid $12\ldots m$, the q-analogue of Equation (1) is

$$\sum_{n\geq 0}\sum_{\pi\in\mathcal{S}_n(12\ldots m)} q^{\mathrm{inv}(\pi)}\frac{z^n}{[n]_q!} = \left(\sum_{j\geq 0}\frac{z^{jm}}{[jm]_q!} - \sum_{j\geq 0}\frac{z^{jm+1}}{[jm+1]_q!}\right)^{-1}.$$

Dotsenko and Khoroshkin [24] developed a variation of the cluster method that they call the chain method. What makes their approach particularly interesting is that it is motivated by homological algebra, and more specifically by the computation of free resolutions of Anick type for shuffle algebras. As it turns out, shuffle algebras with monomial relations have bases that can be naturally described via generalized colored permutations avoiding consecutive patterns. From a combinatorial perspective, the chain method is also based on inclusion-exclusion, and the role of clusters is played by chains, which are clusters with additional restrictions. This reduces the number of objects to consider by eliminating some of the clusters whose contributions to $R_\sigma(-1,z)$ cancel out (for example, permutations that are simultaneously a k-cluster and a $k+1$-cluster). Recurrences for the number of clusters and chains are given in [24] for some patterns. Further applications of this homological algebra approach to consecutive patterns are studied by Khoroshkin and Shapiro in [45], and generalizations to patterns in labeled trees are considered by Dotsenko [23].

In [37], Elizalde and Noy observed that counting clusters is equivalent to counting linear extensions of certain posets, which they call *cluster posets*. For example, the cluster posets for the monotone pattern are chains, and the cluster posets for non-overlapping patterns have a simple tree-like structure. Another example is the poset describing the order relations satisfied by the entries of a cluster $\pi \in \mathcal{S}_{11}$ with marked occurrences of $\sigma = 14253$ starting at positions $1, 3, 7$, which is shown in Figure 1. The interpretation in terms of linear extensions is exploited in [37] to obtain differential equations satisfied by $\omega_\sigma(u,z)$ for several patterns. For example, for all patterns for which the cluster posets are chains (this includes 12435 and 124536, for instance), an equation similar to Theorem 1 is derived, where the derivatives that appear depend on the overlap set O_σ. Differential equations are given for other patterns of length 5 and 6 as well, and for the pattern 1324, which has also been studied in [24] and [54].

4.3 Symmetric functions and brick tabloids

Before the cluster method became a common tool in the study of consecutive patterns, Mendes and Remmel [55] developed in 2006 a remarkable technique to express generating functions for permutations and words avoiding consecutive patterns in terms of the reciprocal of another power series. Their method combines symmetric function identities expressed in terms of what they call *brick*

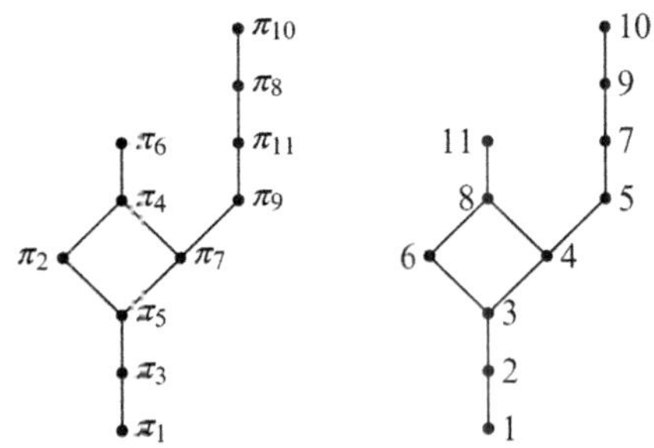

Fig. 1 A cluster poset for $\sigma = 14253$ and a linear extension corresponding to the 3-cluster 1 6 2 8 3 11 4 9 5 10 7.

tabloids, ring homomorphisms from symmetric functions to polynomials (inspired by Brenti's work [20]), and sign-reversing involutions on brick tabloids.

The first application in [55] is an alternative proof of Theorem 3, which is generalized to avoidance of sets of patterns, and refined by the inversion number. Another important result from [55] is an expression of $P_\sigma(0, z)$ as the reciprocal of a power series whose coefficients are signed sums of permutations having occurrences of σ at prescribed positions determined by a certain set associated to σ. This power series is reminiscent of the cluster generating function $R_\sigma(-1, z)$, and in the case of non-overlapping patterns it actually coincides with it [25]. Subsequently, Liese and Remmel [54] gave explicit expressions for this power series in some cases where σ is a shuffle of an increasing sequence with another pattern.

One advantage of the Mendes–Remmel method is that it often yields refinements by statistics such as the number of descents, the number of inversions or the major index, and it extends naturally to permutations avoiding sets of patterns, to tuples of permutations, and to colored permutations (elements of the wreath product $C_k \wr \mathcal{S}_n$). An example of such a refinement is the following, where $\operatorname{des}(\pi)$ denotes the number of descents of π.

Theorem 5 ([55]). *For $m \geq 2$,*

$$\sum_{n\geq 0} \sum_{\pi \in \mathcal{S}_n(m\ldots 21)} x^{\operatorname{des}(\pi)} \frac{z^n}{n!} = \left(\sum_{n\geq 0} \sum_{i\geq 0} (-1)^{n-i} r_{n,i,m}\, x^i \frac{z^n}{n!} \right)^{-1},$$

where $r_{n,i,m}$ is the number of rearrangements of i zeroes and $n-1-i$ ones such that $m-1$ zeroes never appear consecutively.

Setting $x = 1$, Equation (1) follows by applying a sign-reversing involution to the above rearrangements.

On the other hand, most of the formulas resulting from this method do not keep track of the number of occurrences (*matches*, in the terminology from [55]) of the pattern, unlike with the cluster method. Nonetheless, there is a special type of patterns for which Mendes and Remmel [55, Theorem 11] express $P_\sigma(u,z)$ in terms of $P_\sigma(0,z)$. These are precisely the non-overlapping patterns, as shown in [26]. We remark that for $\sigma \in \mathcal{N}_m$, the cluster method can also be used to show that

$$\omega_\sigma(u,z) = 1 + \frac{\omega_\sigma(0,(1-u)^{\frac{1}{m-1}}z) - 1}{(1-u)^{\frac{1}{m-1}}}.$$

4.4 Other questions and approaches

Even though formulas for $P_\sigma(u,z)$ are known for several patterns, the nature of these generating functions for an arbitrary σ is not well understood. For the case of patterns without the adjacency requirement, it was conjectured by Noonan and Zeilberger [57], extending a speculation of Gessel [42], that the generating function for permutations avoiding a classical pattern is always D-finite, that is, it satisfies a linear differential equation with polynomial coefficients. This conjecture has very recently been disproved in [41]. In the case of consecutive patterns, the interesting question is not whether $P_\sigma(0,z)$ is D-finite (since it clearly is not, already for $\sigma = 123$) but whether its reciprocal $\omega_\sigma(0,z)$ is D-finite. Even though it is for the patterns for which the enumeration is known, Elizalde and Noy conjecture that this is not the case in general, and they suggest that 1423 is a counterexample.

Conjecture 2 ([37]). The generating function $\omega_{1423}(0,z)$ is not D-finite.

An automated approach to the enumeration of permutations avoiding consecutive patterns has been carried out by Baxter, Nakamura, and Zeilberger [13]. Their algorithms, implemented in Maple, automatically derive functional equations for the generating functions, and use them to generate the corresponding sequences in polynomial time. In related work, Baxter and Pudwell [12] provide an algorithm that, given any set of consecutive patterns, returns a recurrence (called an *enumeration scheme*) that can be used to efficiently compute the number of permutations avoiding them.

In a different direction, some recent papers study the infinite poset of all permutations $\bigcup_{n\geq 1}\mathcal{S}_n$ where one defines $\sigma \leq \pi$ if π contains σ as a consecutive pattern. In [14], Bernini, Ferrari, and Steingrímsson gave a recursive formula for the Möbius function of this poset. In [63], Sagan and Willenbring used discrete Morse theory to determine its homotopy type. Further properties of this poset are currently being studied in [35].

In recent years, a number of variations and generalizations of consecutive and classical patterns have been studied in the literature. These include *vincular* patterns, originally called generalized patterns when first defined by Babson and Steingrímsson [10]; *bivincular* patterns, introduced by Bousquet-Mélou et al. [18]; *mesh* patterns, introduced by Brändén and Claesson [19]; and *partially ordered* patterns, proposed by Kitaev [47, 48].

5 Asymptotic enumeration

5.1 *Growth rates*

Even in cases where no formula or generating function for the numbers $\alpha_n(\sigma)$ is known, it is natural to ask how fast this sequence grows. The first general result in this direction is the following.

Proposition 1 ([29]). *For every $\sigma \in \mathcal{S}_m$ with $m \geq 3$, the limit*

$$\lim_{n\to\infty} \left(\frac{\alpha_n(\sigma)}{n!}\right)^{1/n}$$

exists, and it is strictly between 0 *and* 1.

This limit, which we denote by ρ_σ, is called the *growth rate* of σ. It is a well-known fact in analytic combinatorics [39, Theorem IV.7] that ρ_σ^{-1} equals the radius of convergence of $P_\sigma(0,z)$ as a function of complex variable, and that $P_\sigma(0,z)$ has a real singularity at $z = \rho_\sigma^{-1}$ [39, Theorem IV.6] and no singularities in $|z| < \rho_\sigma^{-1}$.

The constants ρ_σ were computed in [36] for some small patterns. For example, $\rho_{123} = \frac{3\sqrt{3}}{2\pi} \approx 0.8269933$ and $\rho_{132} \approx 0.7839769$, the reciprocal of the unique positive root of $\int_0^z e^{-t^2/2}dt = 1$. For monotone patterns, it can be shown using Equation (1) and the above discussion that, letting m go to infinity,

$$\rho_{12\ldots m} = 1 - \frac{1}{m!} + \frac{1}{(m+1)!} + O\left(\frac{1}{m!^2}\right).$$

In general, approximate values for ρ_σ can now be found quickly using the Maple packages from [13]. For the patterns of length 3 and 4 solved in [36], it was also shown that as n tends to infinity, $\alpha_n(\sigma) \sim \gamma_\sigma\, \rho_\sigma^n\, n!$ for some constant γ_σ. Warlimont [68] conjectured that this asymptotic behavior holds for all patterns σ, and Ehrenborg, Kitaev, and Perry [27] proved this fact using methods from spectral theory of integral operators:

Theorem 6 ([27]). *For every σ,*

$$\frac{\alpha_n(\sigma)}{n!} = \gamma_\sigma \rho_\sigma^n + O(r_\sigma^n)$$

as n tends to infinity, for some constants γ_σ and $r_\sigma < \rho_\sigma$.

The techniques in [27] give detailed asymptotic expansions. An alternative proof of Theorem 6 seems to be within reach of the standard methods from singularity analysis of generating functions. It was shown in [32] that ρ_σ^{-1} is the smallest positive zero of $\omega_\sigma(0,z)$, and that this zero is simple. Proving that this is the unique zero of $\omega_\sigma(0,z)$ of minimum modulus would imply Theorem 6 by [39, Theorem IV.10].

It is not known whether $\omega_\sigma(0,z)$ is always an entire function in the complex plane. It was shown in [37] that it is entire for a large family of patterns, but that $\omega_{2413}(2,z)$ is not.

5.2 Comparisons between patterns

Another interesting direction of research aims at comparing the growth rates of different patterns. Phrased in a different way, one question is to determine, given two patterns σ and τ, whether $\alpha_n(\sigma) > \alpha_n(\tau)$ for large enough n. Two instances of this question were answered already in [36], where it is shown that

$$\alpha_n(123) > \alpha_n(132) \text{ for } n \geq 4, \quad \text{and} \quad \alpha_n(1342) > \alpha_n(1243) \text{ for } n \geq 7. \tag{6}$$

Even though the question remains wide open in general, the following result describes which patterns of any given length maximize and minimize ρ_σ.

Theorem 7 ([32]). *For every $\sigma \in \mathscr{S}_m$ there exists n_0 such that*

$$\alpha_n(12\dots(m{-}2)m(m{-}1)) \leq \alpha_n(\sigma) \leq \alpha_n(12\dots m)$$

for all $n \geq n_0$.

It is shown in [32] that the above inequalities are strict unless σ is a trivial symmetry of one of the above two patterns. The fact that $\alpha_n(\sigma) \leq \alpha_n(12\dots m)$ for all $\sigma \in \mathscr{S}_m$ had been conjectured in [36], and proved for $\sigma \in \mathscr{N}_m$ in [37]. The fact that $\alpha_n(12\dots(m{-}2)m(m{-}1)) \leq \alpha_n(\sigma)$ for all $\sigma \in \mathscr{S}_m$ had been conjectured in [56].

It is interesting to note that, perhaps counter-intuitively, the analogous statements for classical patterns do not hold, as shown by Bóna [15]. In fact, $12\dots m$ and $12\dots(m{-}2)m(m{-}1)$ are Wilf-equivalent as classical patterns. A different analogue of the upper bound for another definition of pattern avoidance, which requires adjacent positions and adjacent values, was considered in [16].

The proof of Theorem 7 combines singularity analysis of generating functions with combinatorial arguments involving linear extensions of cluster posets. The inequality $\alpha_n(\sigma) < \alpha_n(12\dots m)$ is proved by showing that $\rho_\sigma < \rho_{12\dots m}$ if σ is not monotone. After arguing that ρ_σ^{-1} is the smallest zero of $\omega_\sigma(0,z)$, the proof can be reduced to showing that $\omega_{12\dots m}(0,z) < \omega_\sigma(0,z)$ for $0 < z < C$, for a certain constant C. This is then done by carefully bounding the terms of the cluster generating functions $R_\sigma(-1,z)$.

A version of Theorem 7 for non-overlapping patterns also appears in [32]. It states that for every $\sigma \in \mathscr{N}_m$, there exists n_0 such that

$$\alpha_n(12\dots(m{-}2)m(m{-}1)) \leq \alpha_n(\sigma) \leq \alpha_n(134\dots m2) \tag{7}$$

for all $n \geq n_0$.

A probabilistic proof of the right inequality in Theorem 7 for $m \geq 5$ has been found by Perarnau [61] using Suen's inequality, which gives an upper bound on the probability that none of the events in a certain collection occur simultaneously. The probabilistic method is suitable for consecutive patterns because the events that describe occurrences of σ in a large permutation have few dependencies. It is also shown in [61] that for $\sigma \in \mathcal{S}_m$, as m goes to infinity,

$$\rho_\sigma \geq 1 - \frac{1}{m!} - O\left(\frac{m}{m!^2}\right),$$

and that the growth rate of most patterns is close to this lower bound.

We point out that even though the inequalities in Equation (6) were proved combinatorially by explicitly constructing injective maps [36], no combinatorial proofs of Theorem 7 and Equation (7) are known.

6 Applications to dynamical systems

Aside from their interest from a purely combinatorial perspective, consecutive patterns arise naturally in the study of time series, where they play a role in distinguishing deterministic from random sequences (see [6] for a survey). Given a map f from a linearly ordered set (such as an interval in the real line) to itself, consider the finite sequences (orbits) that are obtained by iterating f, starting from different initial points: $x, f(x), f(f(x)), \dots, f^{n-1}(x)$. If these values are all different, their relative order determines a permutation

$$\mathrm{st}(x, f(x), f(f(x)), \dots, f^{n-1}(x)) \in \mathcal{S}_n.$$

Permutations obtained in this way, for varying x in the domain of f, are called *allowed patterns* of f, or patterns *realized* by f, and denoted by Allow(f). For example, 3241 is an allowed pattern of the *logistic map* $L(x) = 4x(1-x)$ defined in $[0, 1]$, obtained with initial value $x = 0.8$.

Bandt, Keller, and Pompe [11] proved that if f is a piecewise monotone map on a one-dimensional interval, there are some permutations that are not realized by f. These permutations (i.e., those not in Allow(f)) are called *forbidden patterns*. For example, Figure 2 shows that 321 is a forbidden pattern of the logistic map $L(x)$ As shown in [4], $L(x)$ and the so-called *tent map* have the same allowed and forbidden patterns.

It is easy to see that, for every f, the set Allow(f) is closed under consecutive pattern containment: if $\pi \in$ Allow(f) and π contains σ, then $\sigma \in$ Allow(f) [30]. This property allows us to apply combinatorial tools to the study of dynamical systems. For example, it was shown in [11] that for a piecewise monotone map f, the limit

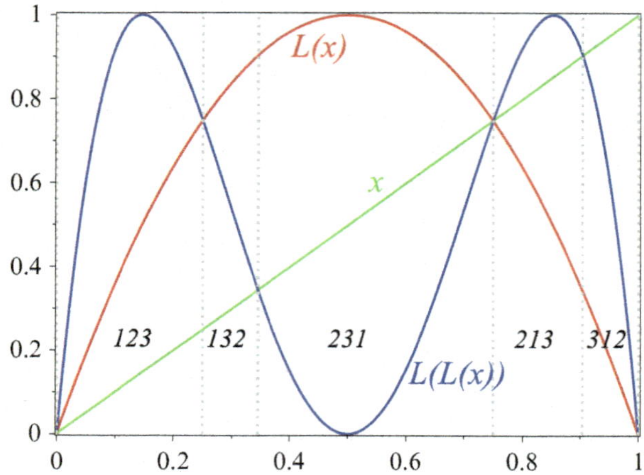

Fig. 2 The allowed patterns of length 3 of the map $L(x) = 4x(1-x)$ on the unit interval.

$$\lim_{n\to\infty} |\operatorname{Allow}(f) \cap \mathcal{S}_n|^{1/n} \tag{8}$$

exists, and its logarithm equals the *topological entropy* of f, which is a measure of the complexity of the dynamical system. Additionally, the fact that Allow(f) is closed under pattern containment allows us to characterize it by the set MF(f) of minimal forbidden patterns, i.e., forbidden patterns where any proper pattern that they contain is allowed. Thus, one can determine the topological entropy of a map by studying the asymptotic behavior of a set of (consecutive) pattern-avoiding permutations. Note that MF(f) is an antichain in the consecutive pattern poset mentioned in Section 4.4. It was shown in [33] that the logistic map defined above and, more generally, the family of maps $L_r(x) = rx(1-x)$, where $1 < r \leq 4$, have infinitely many minimal forbidden patterns.

One of the goals of this combinatorial approach to dynamical systems is to obtain a better understanding of the sets of allowed and forbidden patterns of a map, and in particular how the properties of these sets are related to the properties of the map. Forbidden patterns are useful in certain tests to distinguish random from deterministic time series. These tests are based on the fact that in a sequence of values chosen independently at random from some continuous probability distribution, every permutation appears with positive probability. On the other hand, if the sequence has been generated by iterating a piecewise monotone map, then some patterns will never appear. In fact, since $n!$ grows faster than C^n for any constant C, the fact that the limit in Equation (8) exists implies that most long patterns are forbidden. Such permutation-based tests and their robustness against noisy data have been studied in [3, 7], and they have been applied to the analysis of the stock market in [69, 70].

Even though determining the allowed patterns of a map is a difficult problem in general, such patterns are relatively well understood for the so-called *shift maps*. The shift Σ_N is defined on the set of infinite words over the alphabet $\{0, 1, \ldots, N-1\}$, ordered lexicographically, by $\Sigma_N(w_1w_2w_3\ldots) = w_2w_3w_4\ldots$. For example, $4217536 \in \text{Allow}(\Sigma_3)$ because this permutation describes the relative order of the first few shifts of any word starting as $2102212210\ldots$. By considering words as expansions in base N of real numbers in $[0, 1]$, the shift Σ_N is equivalent to the map that sends x to the fractional part of Nx. It was shown in [4] that the shortest forbidden patterns of Σ_N have length $N + 2$, that there are exactly six of this length, and that there are minimal forbidden patterns of each length $\ell \geq N + 2$. For example, the patterns 1423, 2134, 2314, 3241, 3421, and 4132 are the shortest ones that cannot be realized by shifts of infinite binary words. In [30], the allowed patterns of Σ_N were characterized and enumerated, by means of a formula that determines the minimum number of symbols needed in the alphabet in order for a given permutation to be realized by a shift:

Theorem 8 ([30]). *For* $\pi \in \mathcal{S}_n$, *let* $N(\pi) = \min\{k : \pi \in \text{Allow}(\Sigma_k)\}$. *Then*

$$N(\pi) = 1 + \text{des}(w) + \epsilon,$$

where w *is the word* $w_1 \ldots w_{\pi(n)-1}w_{\pi(n)+1} \ldots w_n$ *defined by* $w_i = \pi(\pi^{-1}(i) + 1)$ *for all* $i \neq \pi(n)$, *and* ϵ *equals* 1 *if* $\pi(n-1)\pi(n) \in \{21, (n-1)n\}$ *and* 0 *otherwise.*

Replacing N with an arbitrary real number $\beta > 1$, one gets the *β-shift* Σ_β. Its combinatorial description is more elaborate [60], but its counterpart on the unit interval is simply the map $x \mapsto \{\beta x\}$. In [31], a method is given to compute, for any permutation π, the smallest real number β such that $\pi \in \text{Allow}(\Sigma_\beta)$, as a root of a polynomial whose coefficients depend on π. This value of β is called the *shift-complexity* of π. For example, the shift-complexity of $\pi = 893146275$ is $\beta \approx 3.343618091$, which is a root of $x^7 - 3x^6 - x^5 - 2x^3 + x^2 + x - 2$.

If instead of considering an arbitrary initial point in the domain of f one restricts to periodic points, the permutations realized by the relative order of the entries in the corresponding orbits (up until the first repetition) are called periodic patterns. For continuous maps, Sharkovskii's theorem [64] describes the possible periods of these orbits, and there is significant literature that studies which periodic patterns are forced by others. From a more combinatorial perspective, Archer and Elizalde [9] characterize the periodic patterns of signed shifts, which are a large family of maps that includes both shifts and the tent map [5], in terms of the structure of the descent set of a certain cyclic permutation associated to the pattern. They also obtain enumeration formulas in some cases.

Another interesting direction of research considers, rather than iterates of a map f, random sequences given by a one-dimensional random walk that at time i is at $X_1 + X_2 + \cdots + X_i$, where the X_j are i.i.d. continuous random variables. It is natural to ask what is the probability distribution on the permutations given by the relative order of the first n values. In general, the probability of obtaining a given permutation depends on the distribution of the X_j, but there are pairs of permutations that are always equally likely to each other, such as 1432 and 2143. Elizalde and Martinez [34] define an equivalence relation on permutations under which equivalent patterns appear with equal frequency in any random walk, and characterize the resulting equivalence classes applying combinatorial methods.

We end by listing a few general questions in this area in the interface of permutation patterns and dynamical systems. They are mostly wide open, although some partial progress has been made in special cases [1, 4, 30, 31, 33]:

1. Characterize the maps f for which MF(f) is finite.
2. Find a method to determine the length of the shortest forbidden pattern of a map.
3. Characterize the sets of permutations that can be MF(f) for some (piecewise monotone) f.
4. Enumerate and/or characterize Allow(f) or MF(f) for particular maps f.
5. Find the limit (8) for particular piecewise monotone maps f, as a combinatorial way to determine their topological entropy.

Acknowledgements Partially supported by grant #280575 from the Simons Foundation and by grant H98230-14-1-0125 from the NSA.

References

1. A. Abrams, E. Babson, H. Landau, Z. Landau, J. Pommersheim, Distributions of order patterns of interval maps. Comb. Probab. Comput. **22**, 319–341 (2013)
2. R.E.L. Aldred, M.D Atkinson, D.J. McCaughan, Avoiding consecutive patterns in permutations. Adv. Appl. Math. **45**, 449–461 (2010)
3. J.M. Amigó, S. Zambrano and M.A.F. Sanjuán, True and false forbidden patterns in deterministic and random dynamics. Europhys. Lett. **79**, 50001 (2007)
4. J.M. Amigó, S. Elizalde, M. Kennel, Forbidden patterns and shift systems. J. Comb. Theory Ser. A **115**, 485–504 (2008)
5. J.M. Amigó, The ordinal structure of the signed shift transformations. Int. J. Bifurcation Chaos **19**, 3311–3327 (2009)
6. J.M. Amigó, *Permutation Complexity in Dynamical Systems*, Springer Series in Synergetics (Springer, Berlin, 2010)
7. J.M. Amigó, S. Zambrano, M.A.F. Sanjuán, Detecting determinism in time series with ordinal patterns: a comparative study. Int. J. Bifurcation Chaos Appl. Sci. Eng. **20**, 2915–2924 (2010)
8. D. André, Mémoire sur les permutations alternées. J. Math. **7**, 167–184 (1881)
9. K. Archer, S. Elizalde, Cyclic permutations realized by signed shifts. J. Comb. **5**, 1–30 (2014)
10. E. Babson, E. Steingrí msson, Generalized permutation patterns and a classification of the Mahonian statistics. Séminaire Lotharingien de Combinatoire **44**, Article B44b (2000)
11. C. Bandt, G. Keller, B. Pompe, Entropy of interval maps via permutations. Nonlinearity **15**, 1595–1602 (2002)
12. A. Baxter, L. Pudwell, Enumeration schemes for vincular patterns. Discret. Math. **312**, 1699–1712 (2002)
13. A. Baxter, B. Nakamura, D. Zeilberger, Automatic Generation of Theorems and Proofs on Enumerating Consecutive-Wilf classes. *Advances in Combinatorics: Waterloo Workshop in Computer Algebra, W80* (Springer, New York, 2011),pp. 121–138
14. A. Bernini, L. Ferrari, E. Steingrímsson, The Möbius function of the consecutive pattern poset. Electron. J. Comb. **18**, Paper 146, 12 pp (2011)
15. M. Bóna, Permutations avoiding certain patterns: the case of length 4 and some generalizations. Discret. Math. **175**, 55–67 (1997)
16. M. Bóna, Where the monotone pattern (mostly) rules. Discret. Math. **308**, 5782–5788 (2008)

17. M. Bóna, Non-overlapping permutation patterns. Pure Math. Appl. (PU.M.A.) **22**, 99–105 (2011)
18. M. Bousquet-Mélou, A. Claesson, M. Dukes, S. Kitaev, (2 + 2)-free posets, ascent sequences and pattern avoiding permutations. J. Comb. Theory Ser. A **117**, 884–909 (2010)
19. P. Brändén, A. Claesson, Mesh patterns and the expansion of permutation statistics as sums of permutation patterns. Electron. J. Comb. **18**, 5 (2011)
20. F. Brenti, Permutation enumeration symmetric functions, and unimodality. Pacific J. Math. **157**, 1–28 (1993)
21. L. Carlitz, R. Scoville, Enumeration of permutations by rises, falls, rising maxima and falling maxima. Acta Mathematica Academiae Scientiarum Hungaricae **25**, 269–277 (1974)
22. F.N. David, D.E. Barton, *Combinatorial Chance* (Lubrecht & Cramer Ltd, New York 1962)
23. V. Dotsenko, Pattern avoidance in labelled trees. Sém. Lothar. Combin. **67**, Art. B67b, 27 (2011/2012)
24. V. Dotsenko, A. Khoroshkin, Shuffle algebras, homology, and consecutive pattern avoidance. Algebra Number Theory **7**, 673–700 (2013)
25. A. Duane, J. Remmel, Minimal overlapping patterns in colored permutations. Electron. J. Comb. **18**, 25 (2011)
26. T. Dwyer, S. Elizalde, A stronger notion of c-Wilf equivalence on permutations. preprint
27. R. Ehrenborg, S. Kitaev, P. Perry, A spectral approach to consecutive pattern avoiding permutations. J. Comb. **2**, 305–353 (2011)
28. S. Elizalde, Consecutive patterns and statistics on restricted permutations, Ph.D. thesis, Universitat Politècnica de Catalunya, 2004
29. S. Elizalde, Asymptotic enumeration of permutations avoiding generalized patterns. Adv. Appl. Math. **36**, 138–155 (2006)
30. S. Elizalde, The number of permutations realized by a shift. SIAM J. Discret. Math. **23**, 765–786 (2006)
31. S. Elizalde, Permutations and β-shifts. J. Comb. Theory Ser. A **118**, 2474–2497 (2011)
32. S. Elizalde, The most and the least avoided consecutive patterns. Proc. Lond. Math. Soc. **106**, 957–979 (2013)
33. S. Elizalde, Y. Liu, On basic forbidden patterns of functions. Discret. Appl. Math. **159**, 1207–1216 (2011)
34. S. Elizalde, M. Martinez, The frequency of pattern occurrence in random walks, in *Proceedings of the Discrete Mathematics and Theoretical Computer Science (FPSAC'15)* (2015), pp. 217–228. http://fpsac2015.sciencesconf.org/70988/document
35. S. Elizalde, P. McNamara, The structure of the consecutive pattern poset. Preprint (2015) [arXiv:1508.05963]
36. S. Elizalde, M. Noy, Consecutive patterns in permutations. Adv. Appl. Math. **30** 110–123 (2003) [Formal power series and algebraic combinatorics (Scottsdale, AZ, 2001)]
37. S. Elizalde, M. Noy, Clusters, generating functions and asymptotics for consecutive patterns in permutations. Adv. Appl. Math. **49**, 351–374 (2012)
38. R. Entringer, Enumeration of permutations of $(1, \ldots, n)$ by number of maxima. Duke Math. J. **36**, 575–579 (1969)
39. P. Flajolet, R. Sedgewick, *Analytic Combinatorics* (Cambridge University Press, Cambridge, 2009)
40. J. Françon, X.G. Viennot, Permutations selon les pics, creux, doubles montées, doubles descentes, nombres d'Euler, nombres de Genocchi. Discret. Math. **28**, 21–35 (1979)
41. S. Garrabrant, I. Pak, Pattern avoidance is not P-recursive, `arXiv:1505.06508` preprint
42. I. Gessel, Symmetric Functions and P-recursiveness. J. Comb. Theory Ser. A **53**, 257–285 (1990)
43. I.P. Goulden, D.M. Jackson, An inversion theorem for cluster decompositions of sequences with distinguished subsequences. J. Lond. Math. Soc. **2**(20), 567–576 (1979)
44. I.P. Goulden, D.M. Jackson, *Combinatorial Enumeration* (Wiley, New York, 1983)
45. A. Khoroshkin, B. Shapiro, Using homological duality in consecutive pattern avoidance. Electron. J. Comb. **18**, 9 (2011)

46. S. Kitaev, Multi-avoidance of generalised patterns. Discret. Math. **260**, 89–100 (2003)
47. S. Kitaev, Partially ordered generalized patterns. Discret. Math. **298**, 212–229 (2005)
48. S. Kitaev, A survey on partially ordered patterns, *Permutation Patterns*, London Mathematical Society Lecture Note Series, vol. 376 (Cambridge University Press, Cambridge, 2010), pp. 115–135
49. S. Kitaev, *Patterns in Permutations and Words* (Springer, Heidelberg, 2011)
50. S. Kitaev, T. Mansour, Simultaneous avoidance of generalized patterns. Ars Combinatoria **75**, 267–288 (2005)
51. S. Kitaev, T. Mansour, On multi-avoidance of generalized patterns. Ars Combinatoria **76**, 321–350 (2005)
52. S. Kitaev, J. Liese, J. Remmel, B.E Sagan, Rationality, irrationality, and Wilf equivalence in generalized factor order. Electron. J. Comb. **16**, 22 (2009)
53. D. Knuth, *The Art of Computer Programming, Vol. I* (Addison-Wesley, Boston, 1968)
54. J. Liese, J. Remmel, Generating functions for permutations avoiding a consecutive pattern. Ann. Comb. **14**, 123–141 (2010)
55. A. Mendes, J. Remmel, Permutations and words counted by consecutive patterns. Adv. Appl. Math. **37**, 443–480 (2006)
56. B. Nakamura, Computational approaches to consecutive pattern avoidance in permutations. Pure Math. Appl. (PU.M.A.) **22**, 253–268 (2011)
57. J. Noonan, D. Zeilberger, The enumeration of permutations with a prescribed number of "forbidden" patterns. Adv. Appl. Math. **17**, 381–407 (1996)
58. J. Noonan, D. Zeilberger, The Goulden–Jackson cluster method: extensions, applications and implementations. J. Differ. Equ. Appl. **5**, 355–377 (1999)
59. J. Pantone, V. Vatter, On the rearrangement conjecture for generalized factor order over $\mathbb{P}$, in *Proceedings of the Discrete Mathematics and Theoretical Computer Science (AT) (2014)*, pp. 217–228. http://www.dmtcs.org/dmtcs-ojs/index.php/proceedings/article/view/dmAT0120/4468
60. W. Parry, On the β-expansions of real numbers. Acta Mathematica Academiae Scientiarum Hungaricae **11**, 401–416 (1960)
61. G. Perarnau, A probabilistic approach to consecutive pattern avoiding in permutations. J. Comb. Theory Ser. A **120**, 998–1011 (2013)
62. D. Rawlings, The q-exponential generating function for permutations by consecutive patterns and inversions. J. Comb. Theory Ser. A **114**, 184–193 (2007)
63. B.E. Sagan, R. Willenbring, Discrete Morse theory and the consecutive pattern poset. J. Algebraic Combinatorics **36**(4), 501–514 (2012)
64. A.N. Sarkovskii, Coexistence of cycles of a continuous map of a line into itself. Ukrainian Math. J. **16**, 61–71 (1964)
65. R. Simion, F.W. Schmidt, Restricted permutations. Eur. J. Comb. **6**, 383–406 (1985)
66. R.P. Stanley, *Enumerative Combinatorics, Vol. 1*, 2nd edn. (Cambridge University Press, Cambridge, 2012)
67. V. Vatter, Permutation classes, in *Handbook of Enumerative Combinatorics*, ed. by M. Bóna (CRC Press, Boca Raton 2015)
68. R. Warlimont, Permutations avoiding consecutive patterns. Ann. Univ. Sci. Budapest. Sect. Comput. **22**, 373–393 (2003)
69. M. Zanin, Forbidden patterns in financial time series. Chaos **18**, 013119 (2008)
70. L. Zunino, M. Zanin, B.M. Tabak, D.G. Perez O.A. Rosso, Forbidden patterns, permutation entropy and stock market inefficiency. Phys. A Stat. Mech. Appl. **388**, 2854–2864 (2009)

Dynamical algebraic combinatorics and the homomesy phenomenon

Tom Roby

Abstract We survey recent work within the area of algebraic combinatorics that has the flavor of discrete dynamical systems, with a particular focus on the *homomesy phenomenon* codified in 2013 by James Propp and the author. In these situations, a group action on a set of combinatorial objects partitions them into orbits, and we search for statistics that are *homomesic*, i.e., have the same average value over each orbit. We give a number of examples, many very explicit, to illustrate the wide range of the phenomenon and its connections to other parts of combinatorics. In particular, we look at several actions that can be defined as a product of *toggles*, involutions on the set that make only local changes. This allows us to lift the well-known poset maps of rowmotion and promotion to the piecewise-linear and birational settings, where periodicity becomes much harder to prove, and homomesy continues to hold. Some of the examples have strong connections with the representation theory of semisimple Lie algebras, and others to cluster algebras via Y-systems.

Keywords antichains • cluster algebras • homomesy • minuscule heaps • noncrossing partitions • orbits • order ideals • Panyushev complementation • permutations • poset • product of chains • promotion • root systems • rowmotion • Suter's symmetry • toggle groups • tropicalization • Y-systems • Young's Lattice • Young tableaux • Zamolodchikov periodicity.

Mathematics Subject Classification (2010): 05E18, 06A11

1 Introduction

The term *Dynamical Algebraic Combinatorics* is meant to convey a range of phenomena involving actions on sets of discrete combinatorial objects, many of which can be built up by small local changes. Schützenberger's operations of promotion and evacuation on Young tableaux are well-known classic examples [13, 27, 42], but

---

T. Roby (✉)
Department of Mathematics, University of Connecticut, Storrs, CT 06269-3009, USA
e-mail: tom.roby@uconn.edu

A. Beveridge et al. (eds.), *Recent Trends in Combinatorics*, The IMA Volumes in Mathematics and its Applications 159, DOI 10.1007/978-3-319-24298-9_25

there are many others. Questions concerning periodicity and orbit structure naturally arise in this setting, and frequently there is a surprising interplay between algebraic and bijective methods of discovery and proof. Connections with representation theory arise even when the original motivations for studying a certain action were purely combinatorial.

The term **homomesy** (Greek for "same middle") was coined by James Propp and the author to describe the following situation. Given a group action on a set of combinatorial objects, a statistic on these objects is called *homomesic* if its average value is the same over all orbits. There are many examples of this phenomenon, spread over large swaths of combinatorics, and of varying degrees of difficulty.

This survey discusses the work of a number of authors, attempting to convey the range of situations in which homomesy occurs. The field is rapidly evolving at present and has already grown too large for this paper to contain all the generalizations and extensions that might appropriately have been included. Many concrete examples and diagrams are provided to aid the reader in quickly grasping the results and key ideas.

1.1 Acknowledgments

This exposition discusses recent work of a number of different authors. The author is grateful to James Propp for the initial ideas that led to the formalization of homomesy and for continuing conversations as it has developed. Collaborations with Propp, Darij Grinberg, Mike LaCroix, Roger Behrend, and Gregg Musiker have been great extended-learning opportunities.

The American Institute of Mathematics hosted an NSF-sponsored workshop on this subject in March 2015 (http://aimath.org/pastworkshops/dynalgcomb.html), which greatly facilitated the author's understanding of the scope of this area. Thanks go to the extremely efficient AIM staff, the organizers (James Propp, Jessica Striker, Nathan Williams, and the author), and the enthusiastic group of participants. The notes taken by Sam Hopkins [32] were particularly helpful. A number of active projects were presented or jump-started at this workshop, and some of the results are mentioned here. Dynamical algebraic combinatorics and homomesy are currently relatively new and active research areas, so much of what is discussed here is (at the time of this writing, June 2015) in arXiv preprints or still "in preparation."

Mike LaCroix wrote fantastic postscript code to generate animations and pictures, which are available on my website http://www.math.uconn.edu/~troby/research.html and in earlier publications. Code contributed to Sage [39, 40] by various authors has been crucial for understanding many examples.

Several people carefully read draft versions and made numerous helpful comments, particularly Karen Edwards, Darij Grinberg, Michael Joseph, Gregg Musiker, James Propp, Victor Reiner, and Elizabeth Sheridan-Rossi. Undoubtedly errors remain, for which the author takes full responsibility.

For productive conversations and valuable insights, thanks go to Drew Armstrong, Anders Björner, Barry Cipra, Keith Conrad, Karen Edwards, Robert Edwards, Miriam Farber, Sam Hopkins, Michael Joseph, Svante Linusson, Gregg Musiker, Alexander Postnikov, Pavlo Pylyavskyy, Vic Reiner, Ralf Schiffler, Richard Stanley, Dennis Stanton, Jessica Striker, Hugh Thomas, Nathan Williams, Pete Winkler, and Ben Young.

2 Homomesy

The notion of homomesy was originally motivated by the work of James Propp and others on chip-firing and roter-routing [21, 22]. Another early example in the context of an action (rowmotion) on antichains in posets was the conjecture (Theorem 2) of Dmitri Panyushev [28], later proved by Drew Armstrong, Christian Stump, and Hugh Thomas [1]. James Propp and the author [30] first codified the notion and coined the term, giving a wide-ranging collection of examples and studying in detail a situation similar to that of Panyushev. The initial discussion here and choice of examples unavoidably has some overlap with [30], although different numerical examples are given where possible.

Definition 1. Given a set $\mathscr{S}$, an invertible map τ from $\mathscr{S}$ to itself such that each τ-orbit is finite, and a function (or "statistic") $f : \mathscr{S} \to \mathbb{K}$ taking values in some field $\mathbb{K}$ of characteristic zero, we say the triple $(\mathscr{S}, \tau, f)$ exhibits **homomesy** iff there exists a constant $c \in \mathbb{K}$ such that for every τ-orbit $\mathscr{O} \subset \mathscr{S}$

$$\frac{1}{\#\mathscr{O}} \sum_{x \in \mathscr{O}} f(x) = c. \tag{1}$$

In this situation we say that the function $f : \mathscr{S} \to \mathbb{K}$ is **homomesic** under the (cyclic) action of τ on $\mathscr{S}$, or more specifically ***c*-mesic**.

When $\mathscr{S}$ is a finite set, homomesy can be restated equivalently as all orbit averages being equal to the global average:

$$\frac{1}{\#\mathscr{O}} \sum_{x \in \mathscr{O}} f(x) = \frac{1}{\#\mathscr{S}} \sum_{x \in \mathscr{S}} f(x). \tag{2}$$

In upcoming sections we will also see how to relax the definition of homomesy to include actions of general (not merely cyclic) groups (Section 2.1), or even non-invertible actions (Section 2.3).

Example 1 (Number of inversions under cyclic rotation of binary strings). Let $\mathscr{S} = \binom{[n]}{k}$ thought of as length n binary strings with exactly k 1's. Define $\tau := C_R : \mathscr{S} \to \mathscr{S}$ by $s = s_1 s_2 \cdots s_n \mapsto s_n s_1 s_2 \cdots s_{n-1}$ (rightward cyclic shift). Let $f(s) := \mathrm{inv}(s) := \#\mathrm{inversions}(s) := \#\{i < j : s_i > s_j\}$ (the usual inversion statistic on multiset permutations).Then it is straightforward to show (though not immediately obvious) that with respect to this action, the statistic inv is $\frac{k(n-k)}{2}$-mesic [30, § 2.3]. For example, if $n = 4$ and $k = 2$, we get two orbits: $(0011, 1001, 1100, 0110)$ with inversion statistic $(0, 2, 4, 2)$ and $(0101, 1010)$, with inversion statistic $(1, 3)$. So over each orbit, the average is $2 = \frac{2(4-2)}{2}$. Similarly, the reader can easily check that among the $\binom{6}{2} = 15$ bitstrings of length 6 with two 1's, this cyclic action gives three orbits with sizes 6, 6, and 3, each with average number of inversions equal to 4.

One important feature of homomesy is that we get constant averages over orbits even though the orbits in general may be of different lengths. It is sometimes useful to think of aggregating a shorter orbit multiple times into a *superorbit* whose length is the order of the action on the entire set (i.e., the size of the cyclic group it generates). This does not affect homomesy, since the average size of a statistic over such a superorbit will be the same as the average over the actual (shorter) orbit. For example, for purposes of homomesy we could replace the orbit $(0101, 1010)$ above with $(0101, 1010, 0101, 1010)$.

Example 2 (Indicator functions of coordinates under cyclic rotation of binary strings). There are other statistics that are more obviously homomesic with respect to the above action. For example, let $\mathbb{1}_i(s) := s_i$ denote the indicator function of the bitstring at position i. Then for any i, $\mathbb{1}_i$ is $\frac{k}{n}$-mesic. This is simply because each value in the string cycles around to each position exactly once over n iterations of the action; thus, in each superorbit, location i is occupied by a 1 k times.

It is instructive to compare the above examples with the "protoexample" given in [34]. See Section 3.1 for more on this example and the relation between homomesy and cyclic sieving.

Example 3 (Sums of centrally symmetric entries under promotion of SSYT). For a fixed Young diagram λ, let $\mathrm{SSYT}_k(\lambda)$ denote the set of **semistandard Young tableaux** of shape λ and ceiling k, i.e., fillings of the cells of λ with elements of $[k] := \{1, 2, \ldots, k\}$ that are *weakly* increasing in each row and *strictly* increasing in each column. These objects and their variants are used to describe and count various things of interest in representation theory and algebraic geometry, such as irreducible representations of GL_n and cohomology classes of flag varieties. (See, e.g., [45, § 7.10] and [12] for more information about these objects and their relationship to symmetric functions.) Schützenberger defined an interesting operation on $\mathrm{SSYT}_k(\lambda)$, which can be generalized to an action on the set of all linear extensions of any finite poset [4, 46, and the references therein]. Readers unfamiliar with promotion will find a self-contained definition in terms of simpler operations at the beginning of Section 3.2.

In the particular case where $\lambda = (n^m) := (n, n, \ldots, n)$ is a rectangular shape with m parts, all equal to n, the Schützenberger promotion operator π satisfies $\pi^k = \mathrm{id}$ [35, Cor. 5.6]. (Simpler proofs are available for *standard Young tableaux*;

see, e.g., [46, Thm. 4.1(a)].) Now fix any subset R of the cells of (n^m), and for $T \in \mathrm{SSYT}_k(n^m)$ set $\sigma_R(T)$ to be the sum of the entries of T whose cells lie in R.

Theorem 1 (Bloom-Pechenik-Saracino). *Let k be a positive integer and suppose that $R \subseteq (n^m)$ is symmetric with respect to 180-degree rotation about the center of (n^m). Then the statistic σ_R is c-mesic with respect to the action of promotion on $\mathrm{SSYT}_k(n^m)$, with $c = |R| \binom{k+1}{2}$.*

For example, consider the following promotion orbit within $\mathrm{SSYT}_5(3^2)$ (where our tableaux are here drawn "English" style, using matrix coordinates, and the symbol ↰ points back to the start of the current orbit):

$$\begin{array}{|c|c|c|}\hline 1 & 1 & 2 \\ \hline 2 & 3 & 4 \\ \hline\end{array} \overset{\pi}{\mapsto} \begin{array}{|c|c|c|}\hline 1 & 1 & 3 \\ \hline 2 & 5 & 5 \\ \hline\end{array} \overset{\pi}{\mapsto} \begin{array}{|c|c|c|}\hline 1 & 2 & 4 \\ \hline 4 & 5 & 5 \\ \hline\end{array} \overset{\pi}{\mapsto} \begin{array}{|c|c|c|}\hline 1 & 3 & 4 \\ \hline 3 & 4 & 5 \\ \hline\end{array} \overset{\pi}{\mapsto} \begin{array}{|c|c|c|}\hline 2 & 2 & 3 \\ \hline 3 & 4 & 5 \\ \hline\end{array} \overset{\pi}{\hookleftarrow}$$

Then the sum of the values in the upper left and lower right cells (shown in red) across the orbit is $(5, 6, 6, 6, 7)$, which averages to $6 = 2\binom{5+1}{2}$. Similarly, the sum of the blue entries in the lower left and upper right corners across the orbit is $(4, 5, 8, 7, 6)$, with average 5, and the sum of the black entries in the middles of the two rows across the orbit is $(4, 6, 7, 7, 6)$, with average 6.

This result was stated as a conjecture in several talks given by the authors, and was proved by Jonathan Bloom, Oliver Pechenik, and Dan Saracino [4]. The latter also proved a version of the result for *cominuscule posets*. For the action of *K-promotion* on *increasing tableaux* of rectangular shapes, they prove an analogous result for two-rowed shapes, and show that it fails in general when λ is a rectangle with more than two rows.

2.1 General group actions

Although the original definition called for the action to be given by an invertible map τ, which is equivalent to the action of a *cyclic* group, the definition of homomesy makes perfect sense if one considers the action of *any* finite group, cyclic or not. There are examples of homomesies under the action of non-cyclic groups G, and homomesy can always be "lifted" from a cyclic subgroup of G to all of G.

Example 4 (Number of inversions under 90-degree rotation of permutation matrices). Let S_n denote the set of permutation matrices, and $\mathcal{Q}$ the ("quarter turn") map which rotates a matrix W by ninety degrees clockwise. Set $\operatorname{inv} W :=$ the number of inversions of the corresponding permutation, which can also be written as

$$\operatorname{inv} W = \sum_{\substack{1 \le i < i' \le n \\ 1 \le j' \le j \le n}} W_{ij} W_{i'j'}.$$

Then inv is homomesic with respect to this action, with average $\frac{n(n-1)}{4}$. For example, when $n = 3$ we get two orbits;

$$\left(\begin{bmatrix}0&1&0\\1&0&0\\0&0&1\end{bmatrix}, \begin{bmatrix}0&1&0\\0&0&1\\1&0&0\end{bmatrix}, \begin{bmatrix}1&0&0\\0&0&1\\0&1&0\end{bmatrix}, \begin{bmatrix}0&0&1\\1&0&0\\0&1&0\end{bmatrix}\right) \text{ and } \left(\begin{bmatrix}1&0&0\\0&1&0\\0&0&1\end{bmatrix}, \begin{bmatrix}0&0&1\\0&1&0\\1&0&0\end{bmatrix}\right)$$

with respective inversion numbers $(1, 2, 1, 2)$ and $(0, 3)$.

The proof of homomesy is easy: $\mathcal{Q}$ takes inversions to non-inversions, and vice-versa. A more interesting homomesy with respect to the action of $\mathcal{Q}$ on the full set of alternating sign matrices (generalizing the above result) is given in [3].

The question then naturally arises: What happens when one considers the full dihedral group acting on the set of matrices? Do we still get homomesy? The answer is yes, and it is straightforward to show this directly. But we also have the following general lemma that shows the relationship between homomesy for the action of a group G and for the action of H a subgroup of G.

Lemma 1. *Let G be a group acting on the set S, and let H be a subgroup of G. If the triple (S, H, f) exhibits homomesy, then so does the triple (S, G, f).*

The proof is simple: aggregating together orbits where a statistic has the same average always gives a larger orbit with that property.

2.2 Suter's action on Young diagrams

Recall the poset *Young's Lattice*, whose elements are all integer partitions (thought of as Young diagrams), ordered by inclusion of diagrams [45, § 7.2]. In [54], Ruedi Suter described an action of the dihedral group D_n ($n \geq 1$) on a particular subgraph Y_n of the Hasse diagram of Young's lattice, defined as follows. Let the **hull** of a Young diagram be the smallest rectangular diagram that contains it, and let Y_n denote the set of all Young diagrams whose hulls are contained in the staircase diagram $(n-1, n-2, \ldots, 1)$. (Equivalently, $Y_n = \{\text{partitions } \lambda : \lambda_1 + \ell(\lambda) \leq n\}$, where $\ell(\lambda)$ denotes the *length*, i.e., number of (nonzero) parts, of λ.) It is not hard to see that there are exactly 2^{n-1} such diagrams.

Given a Young diagram $\lambda \in Y_n$ (which, following Suter, we draw "French" style as rows of boxes in the first quadrant) we discard the boxes in the bottom (first) row (let us say there are k of them), move all the remaining boxes one step downward and to the right, then insert a column of $n-1-k$ boxes at the left. Suter shows that the resulting diagram μ is again in Y_n and that the map $\text{rot}_n : \lambda \mapsto \mu$ is invertible of order n. Thus, rot_n generates a cyclic action on Y_n, and the full Suter action is generated by rot_n along with the *conjugation* map that takes a Young diagram to its transpose in the usual way.

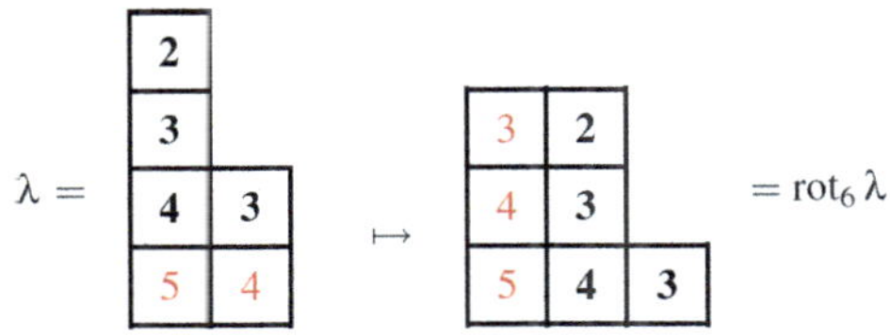

Fig. 1 An example of Suter's map rot_n for $n = 6$ (with $k = 2$); each cell (i, j) is labeled with its weight, $n + 1 - i - j$.

Figure 1 shows an example with $n = 6$ and $k = 2$, where boldface black numbers correspond to boxes that get shifted when one passes from $\lambda = (2, 2, 1, 1)$ to $\mathrm{rot}_6(\lambda) = (3, 2, 2)$.

Example 5 (Suter rotation on Young diagrams). The action of rot_6 on Y_6 produces the following six orbits:

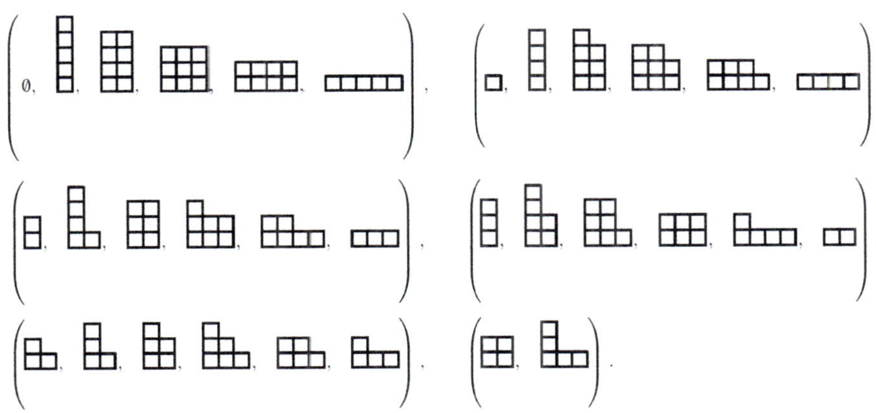

Let f be the statistic on Y_n that sends each Young diagram to the sum of the weights of its constituent boxes, where the box at the lower left has weight $n - 1$, its two neighbors have weight $n-2$, and generally the cell (i, j) is assigned weight $f(i, j) = n+1-i-j$. The boxes in Figure 1 have been marked with their weights, so we can see that $f(\lambda) = 5+4+4+3+3+2 = 21$ while $f(\mathrm{rot}_6(\lambda)) = 5+4+4+3+3+3+2 = 24$. David Einstein and James Propp have shown (unpublished) that this statistic is homomesic with respect to Suter's action.

Proposition 1 (Einstein-Propp). *Let Y_n denote the subset of Young's lattice consisting of all shapes whose hulls fit inside the staircase shape $(n-1, n-2, \ldots, 1)$. Under the action of the cyclic group generated by rot_n on Y_n, the function f is c-mesic with $c = (n^3 - n)/12$.*

Example 6 (A homomesic weight function for Suter rotation). The values of f corresponding to each orbit in Example 5 are

$$(0,15,24,27,24,15), \qquad (5,14,23,26,23,14),$$
$$(9,18,21,23,21,12), \qquad (12,21,24,21,18,9),$$
$$(13,16,19,22,19,16), \qquad (16,19)$$

each of which has average $17.5 = (6^3 - 6)/12$.

As is often the case, there are more refined homomesies lurking underneath the one displayed above. Given positive integers i, j with $i + j = n$, we define statistics $f_{i,j}$ whose sum is f as follows:

$$f_{i,j}(\lambda) := \text{sum of all weights in } \lambda \text{ equal to } i \text{ or } j.$$

Claim. The statistic $f_{i,j}$ is homomesic with average ij with respect to the action of rot_n on Y_n.

Then since $f = \frac{1}{2}(f_{1,n-1} + f_{2,n-2} + \cdots + f_{n-1,1})$, it follows that f is homomesic with average

$$\frac{1}{2}((1)(n-1) + (2)(n-2) + \cdots + (n-1)(1)) = \frac{1}{12}(n-1)(n)(n+1) = \frac{1}{12}(n^3 - n).$$

See [30, § 2.8] for the proof of the claim.

For the example in Figure 1, we get $f_{1,5}(\lambda) = 5$, $f_{2,4}(\lambda) = 10$, and $f_{3,3}(\lambda) = 6$, while $f_{1,5}(\mathrm{rot}_6\,\lambda) = 5$, $f_{2,4}(\mathrm{rot}_6\,\lambda) = 10$, and $f_{3,3}(\mathrm{rot}_6\,\lambda) = 9$. Also, the values of $f_{2,4}$ corresponding to each orbit of Example 5 are

$$(0,6,12,12,12,6), \qquad (0,6,12,12,12,6),$$
$$(4,8,10,10,10,4), \qquad (4,8,10,10,10,4)$$
$$(8,8,8,8,8,8), \qquad (8,8)$$

each of which has average $8 = 2 \cdot 4$. The idea that homomesies of interest can be built up as linear combinations of simpler homomesies is very useful.

Finally, we can apply Lemma 1 to immediately obtain

Proposition 2. *Consider the full Suter action of the dihedral group D_n on Y_n. Define the weight of box (i,j) in λ to be $n + 1 - i - j$, and the statistic $f(\lambda)$ to be the sum of the weights of the boxes in λ. Then the triple (Y_n, D_n, f) is homomesic with average $\frac{n(n^2-1)}{12}$. Furthermore, for positive integers i, j with $i + j = n$, the statistic $f_{i,j}(\lambda) :=$ sum of the weights equal to i or to j in λ is ij-mesic under this action.*

Example 7 (Full dihedral action of Suter on Young diagrams). Note that in Example 5 above, all orbits are closed under the action of conjugation, except for the third and fourth, which combine into a single orbit of size ten under the full D_6 action.

2.3 Homomesy for self-maps & Bulgarian solitaire

In certain situations it is reasonable to generalize the notion of homomesy to non-invertible actions on finite sets, although this comes at a cost of algebraic simplicity. An interesting example is the discrete dynamical system known as *Bulgarian solitaire* [14, 23, 56].

Consider a partition $\lambda \vdash n$ as representing a collection of n identical chips (or uninteresting playing cards) separated into $\ell = \ell(\lambda)$ (unordered) piles containing $\lambda_1, \lambda_2, \ldots, \lambda_\ell$ chips. The game proceeds by taking one chip from each pile and placing them together to form a new pile. We set $\mathfrak{b}(\lambda)$ to be the partition obtained in this way, whose parts are the nonzero elements among $\ell, \lambda_1 - 1, \lambda_2 - 1, \ldots, \lambda_\ell - 1$. Note that the newly created part of size ℓ can range in size from 1 to n, making it hard to write a concise formula for $\mathfrak{b}(\lambda)$ in terms of the parts of λ.

Example 8 (Bulgarian solitaire). For $n = 15$, one trajectory of Bulgarian solitaire is:

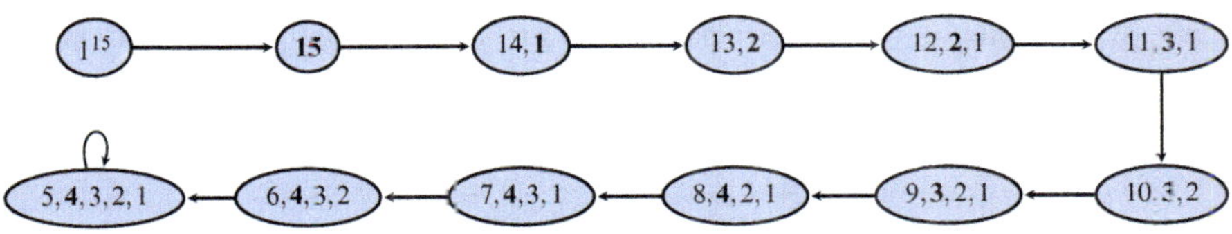

This process first surfaced as a puzzle in Russia around 1980, and a solution by Andrei Toom was published in *Kvant* [56]. A few years later it was popularized in one of Martin Gardner's *Mathematical Games* columns [14]. The puzzle was to show that no matter which of the 176 partitions of 15 one selects for the initial sizes of the piles, one always eventually ends up at the "staircase" partition $(5, 4, 3, 2, 1)$, which is a fixed point of the action (as in the above example). It turns out that if n is a triangular number (so such a staircase partition exists), then any sequence of moves eventually leads to this fixed point of the action; however, in general the action can exhibit more complex dynamical behavior. (See Figure 2.) Some pointers to more recent literature and more information about the history of this problem, including the fanciful, inaccurate (but easily googlable) name, are available in Brian Hopkins's expository survey [23].

Definition 2. Let $\mathscr{S}$ be a *finite* set with a (not necessarily invertible) map $\tau : \mathscr{S} \to \mathscr{S}$ (called a **self-map**). Applying the map iteratively to any $x \in \mathscr{S}$ eventually yields a *recurrent cycle*, and the *recurrent set* is the union of these cycles. (See Figure 2.) We call a statistic $f : \mathscr{S} \to \mathbb{K}$ **homomesic** if the average of f is the same over every recurrent cycle. It is clear that if τ is an invertible action on a finite set S, then this definition specializes to the original one.

Example 9 (Number of parts under Bulgarian solitaire on partitions of $\boldsymbol{n}$*).* Consider the example of Bulgarian solitaire for $n = 8$ as displayed in Figure 2. Let the

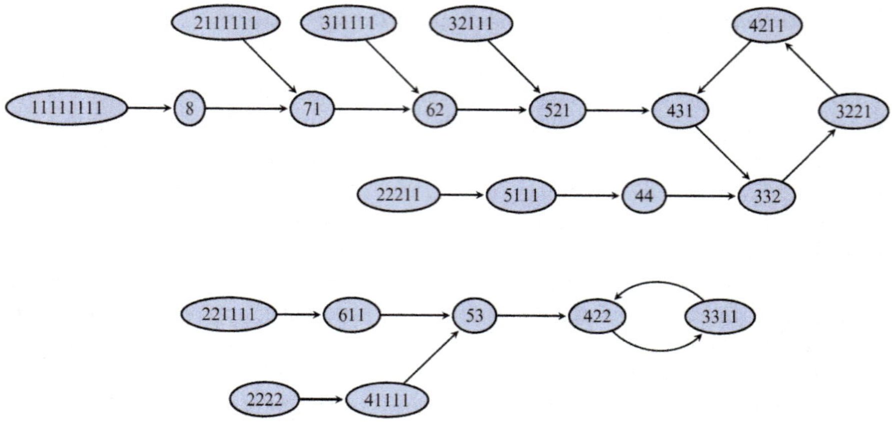

Fig. 2 The action of Bulgarian solitaire on partitions of $n = 8$

statistic $f(\lambda) := \ell(\lambda)$, the number of parts. We claim that this is homomesic in the sense of Definition 2. No matter where one starts, one eventually ends up in one of two recurrent cycles, namely $(431, 332, 3221, 4211)$ or $(422, 3311)$, with average statistics

$$\frac{3+3+4+4}{4} = \frac{7}{2} \text{ and } \frac{3+4}{2} = \frac{7}{2}.$$

So for Bulgarian solitaire acting on partitions of 8, ℓ is $\frac{7}{2}$-mesic.

It is not hard to show that the general situation is as follows:

Proposition 3. *Let $n = k(k-1)/2 + j$ with $0 \leq j < k$, and consider the action of Bulgarian solitaire on the set of partitions of n. Then the length statistic ℓ which computes the number of parts of λ is homomesic with average $(k-1) + j/k$.*

Note that in Example 9, $n = 8$ corresponds to $k = 4, j = 2$, while in the situation that $n = k(k-1)/2$ is a triangular number (so $j = 0$), all paths lead to looping on the shape $\kappa = (k-1, k-2, \ldots, 2, 1)$.

Other statistics homomesic with respect to this action include $f_i(\lambda) := \lambda_i$, the size of the ith largest part of λ, for any $i \geq 1$. For example, when $n = 8$ one sees easily from Figure 2 that f_1 is $\frac{7}{2}$-mesic, f_2 is $\frac{5}{2}$-mesic, f_3 is $\frac{3}{2}$-mesic, and f_4 is $\frac{1}{2}$-mesic.

3 Rowmotion in posets

A number of actions that exhibit homomesy turn out to be defined as compositions of various bijections between sets, particularly involutions. A key example that has been well studied is the action of *rowmotion* on a finite partially ordered set.

Let $\mathscr{A}(P)$ denote the set of antichains, $F(P)$ the set of order filters, and $J(P)$ the set of order ideals of a finite poset P. (See Stanley [44, Ch. 3] for standard definitions and terminology about posets.) There are elementary bijections between each pair of these objects: given an order ideal I, the set of maximal elements of I forms an antichain A, while any antichain generates an order ideal by including all elements smaller or equal to some element of the antichain. A similar bijection between order filters and antichains is found by standing on one's head. Finally, complementation provides a bijection between $J(P)$ and $F(P)$, which happens also to be an involution on the collection of *all* subsets of P. Rowmotion is simply a three-step composition of these bijections.

Definition 3. Let P be a finite poset, and $I \in J(P)$. Set $\rho(I)$ to be the order ideal generated by the minimal elements of the complement of I. If A is an antichain, set $\rho(A)$ to be the set of minimal elements of the complement of the order ideal generated by A.

$$\rho_J : J(P) \to F(P) \to \mathscr{A}(P) \to J(P)$$
$$\rho_A : \mathscr{A}(P) \to J(P) \to F(P) \to \mathscr{A}(P).$$

Here the subscript, which we will generally omit when confusion is unlikely, indicates whether we consider this map to act on $\mathscr{A}(P)$ or $J(P)$. We call either map **rowmotion**. Both actions appear in existing literature, often depending on whether there's a natural representation of antichains or order ideals in terms of other combinatorial objects. In Example 11, the antichains represent *nonnesting partitions*.

Example 10 (The rowmotion operator on order ideals and on antichains). The result of applying rowmotion to the 3-element order ideal I on the left is the 4-element order ideal on the right. In between we have taken the complement of I, and then the minimal elements of that result, which generate the new order ideal.

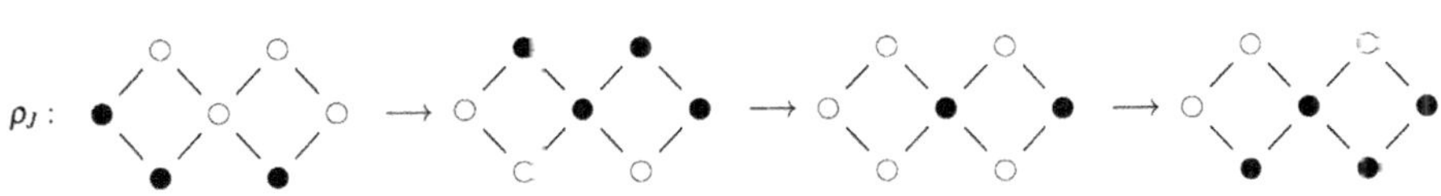

Similarly, the result of applying rowmotion to the 2-element antichain below on the left is the 2-element antichain on the right. Note that if we think about iterating either of these maps, then the resulting actions are the same up to what we consider to be "beat 1" of this 3-beat waltz.

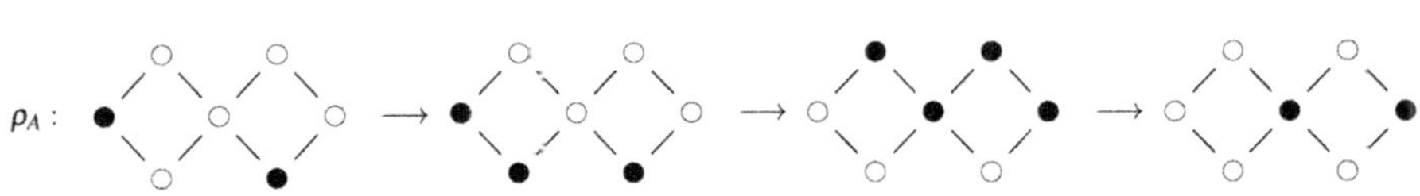

Rowmotion appears first to have been defined by in Andries Brouwer and Lex Schrijver [5], as a map on antichains. It was later studied by Dmitry Fon-der-Flaass [11] and by Peter Cameron and Fon-der-Flaass [6], who also considered it as a permutation of the *monotone Boolean functions* (i.e., indicator functions of order filters of P). In all of this work, the question of the order (aka period) of this operator and the sizes of its orbits was the primary focus, and the main examples were posets that are the product of two or three chains. Jessica Striker and Nathan Williams [53, § 3] give a nice summary of the early history of this map, which has been rediscovered several times, and appears in the literature under a variety of names ("Fon-Der-Flaass action," "Panyushev Complementation," etc.); conventions also differ so that some authors' maps are the inverses of the ones discussed here.

Dmitri Panyushev [28] studied this map for certain graded posets associated with irreducible root systems. In particular, he made the following conjecture, which is one of the earliest statements of an action being homomesic (without using the term). It was proved by Drew Armstrong, Christian Stump, and Hugh Thomas [1].

Theorem 2 (Conjectured [28, Conj. 2.1(iii)] ; proved [1, Thm. 1.2]). *Let W be a finite Weyl group of rank r, with corresponding positive root poset $\Phi^+(W)$. Then for any orbit $\mathcal{O}$ under the action of ρ_A on $\mathcal{A}(\Phi^+(W))$, we have*

$$\frac{1}{|\mathcal{O}|}\sum_{A\in\mathcal{O}}|A| = r/2.$$

In other words, the cardinality statistic is homomesic with respect to rowmotion acting on antichains of the positive root poset, with average half the rank.

Example 11. For the root poset of type A_3 we have the following three rowmotion orbits of sizes 8, 4, and 2:

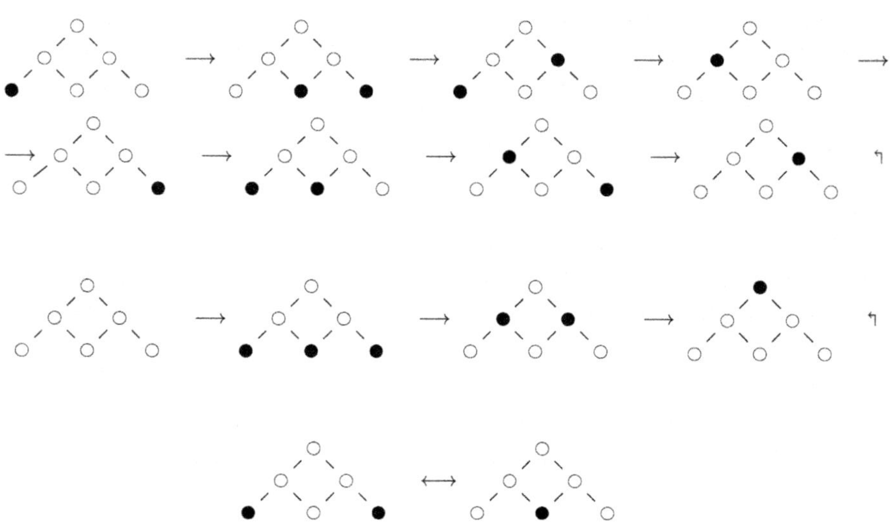

Checking the average cardinality for each orbit we find that

$$\frac{1+2+2-1+1-2+2+1}{8} = \frac{0+3+2+1}{4} = \frac{2+1}{2} = \frac{3}{2},$$

in agreement with the theorem.

For pictures of the other two classes of posets that arise in classical types, see [53, Fig. 2]. Armstrong, Stump, and Thomas get this result by constructing an equivariant bijection (uniformly, though the proof is still type-by-type) between noncrossing partitions under Kreweras complementation and nonnesting partitions under rowmotion. They also show that a certain generalized q-analogue of Catalan numbers defined for any W exhibits cyclic sieving with respect to the action of rowmotion on $J(\Phi^+(W))$.

Example 12. Reconsider Example 11 as an action on $J(P)$, where each order ideal is represented by its maximal elements. Checking for homomesy of the statistics "order-ideal cardinality" we find that the averages are:

$$\frac{1+2+4+3+1+2+4+3}{8} = \frac{5}{2}, \quad \frac{0+3+5+6}{4} = \frac{7}{2}, \quad \text{and} \quad \frac{2+1}{2} = \frac{3}{2},$$

so this statistic is **not** homomesic.

In general, it can be difficult *a priori* to guess statistics that might be homomesic with respect to a certain action, even when there may be many of them. In this situation, Shahrzad Haddadan found an interesting answer.

Theorem 3 ([20, Cor. 36]). *Let P denote the root poset $\Phi^+(A_n)$, and consider the action of rowmotion on $J(P)$ For $I \in J(P)$, set $f(I) := \sum_{x \in I} (-1)^{\mathrm{rk}\,x}$, where* rk x *is the rank of the element, i.e., f is a rank-alternating cardinality statistic. Then $(J(P), \rho_J, f)$ exhibits homomesy.*

Example 13. The averages of the rank-alternating cardinality statistic for Example 11 are

$$\frac{1+2+(3-1)+(2-1)+1+2+(3-1)+(2-1)}{8} = \frac{0+3+(3-2)+(4-2)}{4} = \frac{2+1}{2} = \frac{3}{2}.$$

The example shows that finding interesting homomesic statistics is not always straightforward. In some cases trying to find the right weights for certain indicator functions can be fruitful. (See Section 3.4.)

3.1 Relationship with Cyclic Sieving

Comparisons naturally arise between homomesy and the *cyclic sieving phenomenon (CSP)*. The latter was identified by Victor Reiner, Dennis Stanton, and Dennis White in a seminal paper from 2004 [33]; see also [34] and the excellent exposition of Bruce Sagan [38]. They consider an instance of the CSP to be a triple $(X, X(q), C)$, where $C = \langle c \rangle$ is a cyclic group of order n acting on X, and $X(q)$ is a generating function whose evaluation at roots of unity counts symmetry classes under the action. More specifically, for every $d \in \mathbb{Z}$, the number of elements fixed by c^d equals $X(\zeta^d)$, where $\zeta = e^{2\pi i/n}$ is a root of unity. This generalizes John Stembridge's $q = -1$ phenomenon [50], where $X(1)$ counts the total number of certain combinatorial objects and $X(-1)$ counts a symmetry class of these objects.

An equivalent definition given in [33, Sec. 1] relates more clearly to orbit structures. Given C acting on X, we get a CSP $(X, X(q), C)$ exactly when

$$X(q) \equiv \sum_{i=0}^{n-1} a_i q^i \bmod (q^n - 1),$$

where a_i is the number of orbits of C on X for which the stabilizer cardinality divides i. (The *stabilizer cardinality* aka *stabilizer-order* of an orbit is defined to be the order of the stabilizer subgroup of any (each) element of the orbit.)

Example 14. Recall Example 1, where the cyclic group of order 4 acted on binary strings with exactly two 1's (in obvious bijection with 2-element subsets of $[4]$), giving two orbits: $(0011, 1001, 1100, 0110)$ with stabilizer cardinality 1 and $(0101, 1010)$ with stabilizer cardinality 2. Here the cyclic sieving polynomial is the Gaussian binomial coefficient

$$X(q) = \begin{bmatrix} 4 \\ 2 \end{bmatrix}_q = 1 + q + 2q^2 + q^3 + q^4 \equiv 2 + q + 2q^2 + q^3 \left(\bmod q^4 - 1\right).$$

One reads off from this that the total number of orbits is $a_0 = 2$, and the number of *free orbits* (those whose size is equal to $n = \#C$) is $a_1 = 1$. Both stabilizers-orders divide 2, so $a_2 = 2$, etc.

The two phenomena are similar in that both involve group actions on sets and are intimately bound up with the orbit structure of the action. Interesting examples where both are present are common. And in each case the naturalness of the polynomial (CSP) or statistic (homomesy) is an important consideration.

Cyclic sieving works for cyclic groups, and in some cases for direct products of cyclic groups, i.e., finite abelian groups, but there is no straightforward way to generalize the notion to nonabelian groups. Also, if C' is a subgroup of a cyclic group C, then cyclic sieving of $(X, X(q), C')$ follows from that of $(X, X(q), C)$. Homomesy makes sense for any group action, including non-commutative ones; and by Lemma 1, if H is a subgroup of G, then homomesy for the G-action on X follows from homomesy of the H-action. So homomesies can be lifted to larger groups, and cyclic sieving specializes to smaller groups.

Until recently it was an open problem to find interesting homomesies for a nonabelian group G that are not implied by a homomesy for a commutative subgroup of G. Such homomesies have been proved in recent work of Anne Schilling, Nicolas Thiéry, Graham White, and Nathan Williams [41, Thm. 1.6,

Thm. 5.1]. Another possible avenue is described by Sam Hopkins and Ingrid Zhang [24], who define a natural statistic for oscillating tableaux, whose average value has a surprisingly simple formula. Here the missing piece of the puzzle is finding the correct dihedral action on tableaux that would explain the simplicity of their formula.

Homomesy seems to provide somewhat more flexibility in choices of statistic. There are examples of homomesy in situations where CSP appears to be unlikely, because the order of the action is very large relative to the size of X. See the discussion after Theorem 9 for a concrete example of this. (Conjectural examples were noticed earlier [30, § 4.2]).

3.2 Toggle Operations

Many interesting actions that exhibit homomesy can be expressed as a composition of simpler involutions on the set. For example, the action of promotion π on SSYTs (Example 3) can be expressed as the product of *Bender-Knuth involutions* β_i as follows. Given $T \in \mathrm{SSYT}_k(\lambda)$ and fixed $i \in [k-1]$, consider all entries $\substack{i \\ i+1}$ paired within the same column to be **married**; the involution ignores these pairs. Then in a row with r unmarried i's and s unmarried $i+1$'s, β_i replaces these with s copies of i and r copies of $i+1$:

$$\begin{matrix} & & & & & & & i \\ i & i & i & i & i+1 & i+1 & i+1 & i+1 \\ i+1 & i+1 & \end{matrix} \quad \text{with } \underbrace{i\ i}_{r=2}\ \underbrace{i+1\ \ i+1\ \ i+1}_{s=3} \quad \overset{\beta_i}{\mapsto} \quad \begin{matrix} & & & & & & & i \\ i & i & i & i & i & i+1 & i+1 & i+1 \\ i+1 & i+1 & \end{matrix} \quad \text{with } \underbrace{i\ i\ i}_{r=3}\ \underbrace{i+1\ \ i+1}_{s=2}.$$

By a theorem of Gansner [13, Thm. 4.1], we can write $\pi = \beta_{k-1}\beta_{k-2}\dots\beta_2\beta_1$ on $\mathrm{SSYT}_k(\lambda)$. The first step in the promotion orbit shown in Example 3 can be written as follows. The married pairs at each step are shown as blue (since these values are frozen). Note that $\beta_i(T)$ may be T itself, and that in any case $\beta_i^2(T) = T$.

$$\begin{array}{|c|c|c|}\hline 1 & 1 & 2 \\ \hline 2 & 3 & 4 \\ \hline \end{array} \overset{\beta_1}{\mapsto} \begin{array}{|c|c|c|}\hline 1 & 1 & 2 \\ \hline 2 & 3 & 4 \\ \hline \end{array} \overset{\beta_2}{\mapsto} \begin{array}{|c|c|c|}\hline 1 & 1 & 3 \\ \hline 2 & 3 & 4 \\ \hline \end{array} \overset{\beta_3}{\mapsto} \begin{array}{|c|c|c|}\hline 1 & 1 & 3 \\ \hline 2 & 4 & 4 \\ \hline \end{array} \overset{\beta_4}{\mapsto} \begin{array}{|c|c|c|}\hline 1 & 1 & 3 \\ \hline 2 & 5 & 5 \\ \hline \end{array}$$

Bender-Knuth involutions provide the standard method for proving combinatorially that Schur functions are symmetric [45, Thm. 7.10.2]. But they also provide a useful example of *toggling*. Informally, we consider a toggle to be an *involution* on a collection of objects that changes an object minimally (possibly not at all). More formally, we define the *toggle group* $\mathcal{T}(P)$ of a finite poset as follows:

Definition 4. Let P be a poset. Given $x \in P$, we define the toggle operation $\mathbf{t}_x : J(P) \to J(P)$ ("toggling at x") via

$$\mathbf{t}_x(I) = \begin{cases} I \triangle \{x\} & \text{if } I \triangle \{x\} \in J(P); \\ I & \text{otherwise,} \end{cases}$$

where $A \triangle B$ denotes the symmetric difference $(A \setminus B) \cup (B \setminus A)$. The *toggle group* $\mathcal{T}(P)$ is the group of $\operatorname{Aut} J(P)$ generated by $\{\mathbf{t}_x : x \in P\}$.

Toggling an order ideal I at x means to add x to I if $x \notin I$ or take x away from I if $x \in I$, *provided* that the result is an order ideal; otherwise, do nothing. An easy argument shows the following proposition.

Proposition 4 ([6]). *Let P be a poset. (a) For every $x \in P$, $\mathbf{t}_x$ is an involution, i.e., $\mathbf{t}_x^2 = 1$.*

(b) For every $x, y \in P$ where neither x covers y nor y covers x, the toggles commute, i.e., $\mathbf{t}_x\mathbf{t}_y = \mathbf{t}_y\mathbf{t}_x$.

Thus, $\mathcal{T}(P)$ is the quotient of some Coxeter group. The commutativity of non-adjacent (in the Hasse diagram) toggles means composing them in different orders often leads to the same overall result.

Proposition 5 ([6]). *Let $x_1, x_2, \ldots, x_n$ be any linear extension (i.e., any order-preserving listing of the elements) of a poset P with n elements. Then the composite map $\mathbf{t}_{x_1}\mathbf{t}_{x_2}\cdots\mathbf{t}_{x_n}$ coincides with the rowmotion operation ρ_J.*

The following proposition helps explain Striker and Williams use of the term "rowmotion" for ρ_J.

Corollary 1 ([53], Cor. 4.9). *Let P be a graded poset of rank r, and set $T_k := \prod_{x \text{ has rank } k} \mathbf{t}_x$, the product of all the toggles of elements of fixed rank k. (This is well-defined by Proposition 4.) Then the composition $T_0T_1T_2\cdots T_r$ coincides with ρ_J, i.e., rowmotion is the same as toggling by ranks from top to bottom.*

Example 15. Starting with the order ideal $S = \{(1,1), (2,1)\}$, we toggle successively at the top, right, left, and bottom elements of $P = [2] \times [2]$. Note that in the first step, $\mathbf{t}_{(2,2)}$ leaves the order ideal unchanged, since the set $S' = \{(1,1), (2,1), (2,2)\} \notin J(P)$.

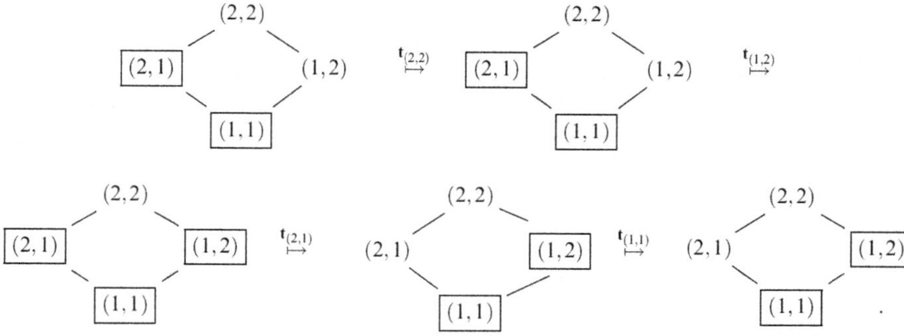

So $\rho(S) = \{(1,1),(1,2)\}$. The reader can easily check that one gets the same result via Definition 3.

Beyond rowmotion, there are several other examples of combinatorial objects with interesting actions that can be realized as some product of toggles acting on linear extensions (e.g., [2]) or on order ideals in a graded poset. It often appears fruitful to look at products where every toggle is used exactly once, which we will call *Coxeter elements* of the toggle group. Striker and Williams [53] provide a number of examples, including ones involving set partitions and alternating sign matrices (ASMs). Many graded posets have a natural notion of *files* (called *columns* in [53]), which can be used to define an operator on $J(P)$ by toggling at each element in files from left to right. (Such posets used to be called *rc-posets*, but now it is understood that any finite graded poset has an **rc-embedding**, which allows for a well-defined notion of files.) The operator, denoted here by ∂, is known as *promotion* (because in some cases it can be related to the Schützenberger promotion operator π). A third operation of this type is called *gyration*, denoted γ, which toggles first at all the even ranks, then at all the odd ranks. Striker and Williams are able to realize Wieland's gyration action on (objects in bijection with) ASMs as toggle-group gyration on the ASM poset. They also prove a general theorem that gives conditions under which various Coxeter elements will be conjugate elements in $\mathscr{T}(P)$, which means that they will have the same orbit structure when acting on $J(P)$. Striker and Williams exploit this to reduce the problem of showing that cyclic sieving holds for *rowmotion* on $J(P)$ to the more easily understood action of *promotion*, obtaining another proof of the cyclic sieving phenomenon for rowmotion acting on $J(\Phi^+(W))$, conjectured by David Bessis and Victor Reiner and proved in [1].

Although results on the order/periodicity of actions and CSP which only depend on the orbit structure will be the same for promotion and rowmotion, this is not true for results on homomesy. There are statistics that are homomesic for rowmotion on a product of two chains, but not for promotion; see the remark after Theorem 4.

3.3 Homomesies in products of two chains

In the papers where they first isolated the notation of homomesy [30, 31], Propp and the author studied cardinality homomesies for rowmotion ρ and promotion ∂ acting on a product of two chains.

Example 16. Figures 3 and 4 show the three orbits for the homomesic triple $(J([4]\times[2]), \rho, \mathrm{card})$, where card denotes the cardinality of an order ideal. Note that here, as opposed to other pictures in this article, we are using *squares* to represent elements of the poset, so the cardinality of the order ideal corresponds to the *area* of the order ideal, which is cut out by a lattice path. (See [30, Fig. 4–5] for more explanation about these different representations.) The homomesic average $c = \frac{4\cdot 2}{2} = 4$ (half the area of the rectangle) as claimed in Theorem 4.

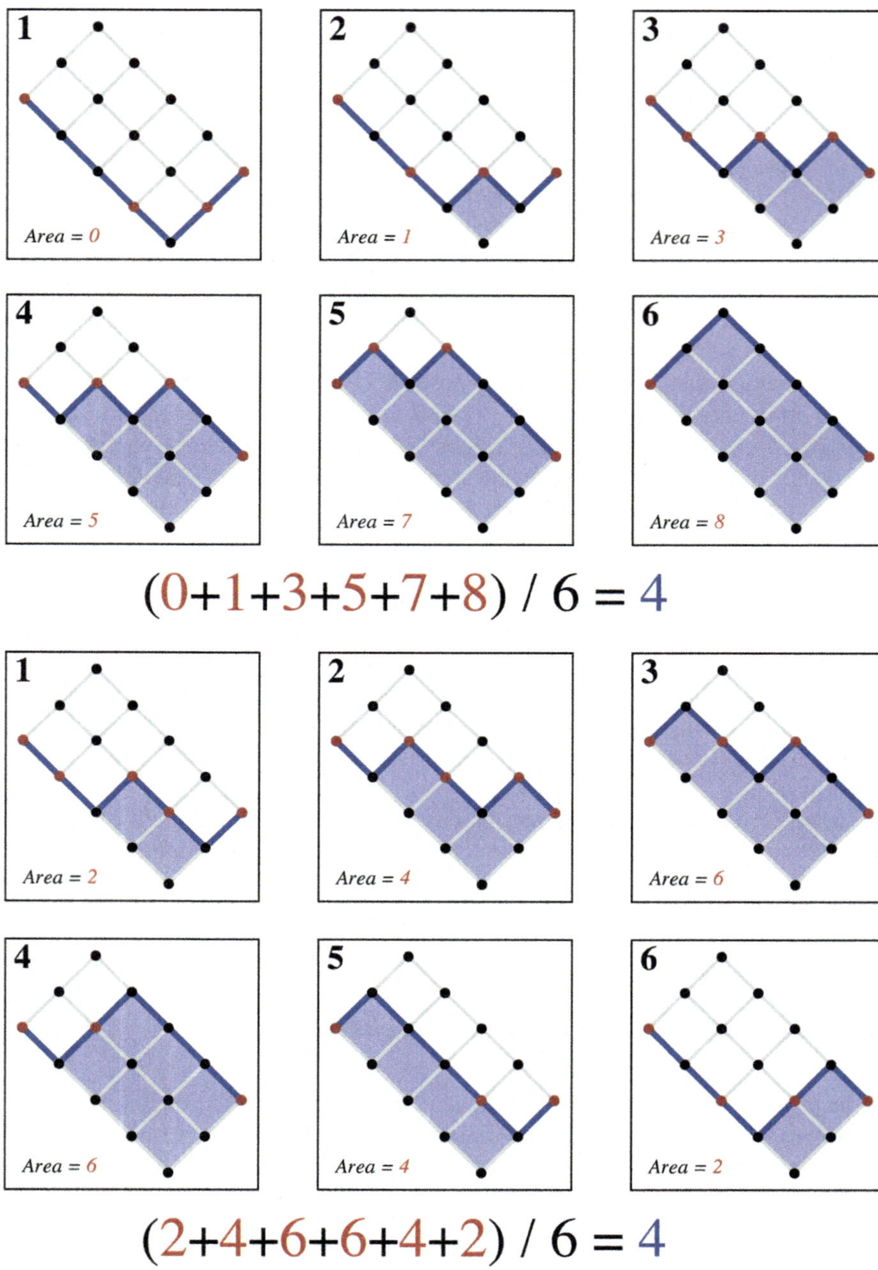

Fig. 3 Two of the three rowmotion orbits in $J([4] \times [2])$. Here the elements of the poset are represented by the filled in squares and order ideals by the shaded regions.

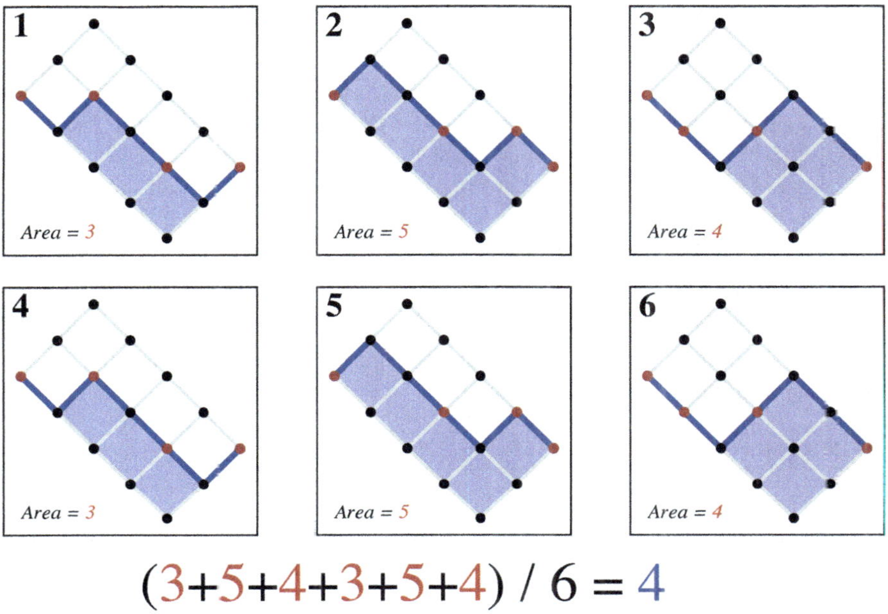

Fig. 4 The third of the three rowmotion orbits in $J([4] \times [2])$, shown as a superorbit, repeating twice the orbit of size three.

Figures 3 and 4 are the final pictures taken from excellent animation by Michael LaCroix. The pictures were created in raw postscript, with the interpreter taking as input a binary string that indicates the lattice path cutting out an order ideal. The code takes care of doing all the computations as well as the drawings. The LaTeX `animate` package takes advantage of a "feature" in Adobe Acrobat Reader to display these. Full animations are available at the author's research webpage, http://www.math.uconn.edu/~troby/research.html.

Theorem 4 ([30]). *Let $P = [p] \times [q]$ be the poset which is the product of two chains. Then the following statistics are c-mesic with respect to the corresponding actions on the set of order ideals, $J(P)$, or the set of antichains $\mathscr{A}(P)$:*

1. *The cardinality of $I \in J(P)$ with respect to rowmotion on $J(P)$, where $c = pq/2$;*
2. *The cardinality of $A \in \mathscr{A}(P)$ with respect to rowmotion on $\mathscr{A}(P)$, where $c = pq/(p+q)$;*
3. *The cardinality of $I \in J(P)$ with respect to promotion on $J(P)$, where $c = pq/2$.*

The cardinality of $A \in \mathscr{A}([p] \times [q])$ with respect to promotion on $\mathscr{A}([p] \times [q])$ is not generally homomesic [30, Example 21], even though the orbit structures for ρ and ∂ are the same.

3.4 Refined homomesies and indicator functions

As we saw earlier in Example 6, one can often think of a homomesy as being built up from simpler or more refined homomesies. For example, define a **file** F of $P = [p] \times [q]$ to be a set of (k, ℓ) with constant $\ell - k$, i.e., a "vertical slice" of a poset, consisting of elements in the standard rc-embedding (as shown in all the pictures). If one restricts attention to a particular file F of $P = [p] \times [q]$, then the cardinality function $f := \#(I \cap F)$ for $I \in J(P)$ is homomesic with respect to ρ.

Example 17 (File cardinalities under rowmotion acting on a product of two chains). Consider the 2-element file F_4 that contains the smallest element within $P = [4]\times[2]$ (adjacent to the rightmost file). Looking at Figures 3 and 4, we can read off the following orbit averages for f_4, all equal to $4/3$:

$$\frac{0+1+1+2+2+2}{6}, \qquad \frac{1+1+2+2+1+1}{6}, \qquad \text{and} \qquad \frac{1+1+2}{3}.$$

The reader can easily check in this example that for each $i \in [5]$, the statistic f_i counting the number of elements in the intersection of an order ideal and the ith file is homomesic. Clearly card $= f_1 + \cdots + f_5$.

It is worth mentioning that in general if f and g are homomesic statistics, then so are $f + g$ and kf, for any $k \in \mathbb{K}$. In other words, given a fixed set and action, the set of homomesic statistics forms a vector space. This naturally leads to questions of dimension, which in general are nontrivial to answer. In the particular case of ρ acting on $\mathscr{S} = J(P)$, we can look for homomesies within the space spanned by the *indicator functions*

$$\mathbb{1}_x(I) := \begin{cases} 1 & \text{if } x \in I \\ 0 & \text{if } x \notin I \end{cases}$$

for every $x \in P$. Then the above example is stating that for any file F (in the obvious rc-embedding of $P = [4] \times [2]$), the statistic $f_F := \sum_{x \in F} \mathbb{1}_x$ is homomesic. It is also true that $\mathbb{1}_x + \mathbb{1}_{x'}$ is homomesic whenever x' is obtained from x by rotating the poset 180 degrees (cf. Example 3). It is possible (though not easy) to show that these homomesies generate the full subspace of homomesic statistics within the vector space $\mathbb{K}^P$ generated by $\{\mathbb{1}_x : x \in P\}$ when $P = [p] \times [q]$. Similarly, for promotion acting on order ideals, the subspace of homomesic statistics is generated by sums of indicator functions along rows and along columns, along with $\mathbb{1}_x + \mathbb{1}_{x'}$ for each $x \in P$. For more information see the discussions at [30, Sec .4.1 and Proof of Thm. 19]. (A later version of [9] should also contain details of these results.) Theorem 3 is an example of successfully finding suitable weights on the indicator functions to give a homomesic statistic.

The situation for ρ_A acting on $\mathscr{A}(P)$ is somewhat different. Define a **positive fiber** of $[p]\times[q]$ to be a set $(k, \ell) \in [p]\times[q]$ with k constant, and a **negative fiber** to be a set with ℓ fixed [30, DEf. 15]. (These are just the natural rows and columns of $[p] \times [q]$, but we use this terminology to avoid confusion with the "rows" and "columns" in the sense of "rc-embedding.") Within the vector space generated by the indicator functions that check whether $x \in P$ belongs to an antichain (defined analogously to those above), the homomesic statistics are generated by the following: $f_F := \sum_{x \in F} \mathbb{1}_x$, where F is any fiber of P, and $\mathbb{1}_x - \mathbb{1}_{x'}$, with x' obtained from x by rotating the poset 180 degrees.

A probabilistic generalization that fiber-cardinalities are homomesic for ρ_A on $\mathscr{A}(P)$ was unearthed by Melody Chan, Shahrzad Haddadan, Sam Hopkins, and Luca Moci [7], whose collaboration began at the aforementioned AIM workshop. They define the *jaggedness* of $I \in J(P)$ to be the number of maximal elements in I plus the number of minimal elements of P not in I. A probability distribution on $J(P)$ is called *toggle-symmetric* if for every $x \in P$, the probability that x is maximal in I equals the probability that x is minimal not in I. They give a formula for the expected jaggedness of an order ideal of P under any toggle-symmetric probability distribution when P is the poset of boxes in a skew Young diagram. Their results depend heavily on results of Striker that a certain "toggleability statistic" is 0-mesic for the actions of rowmotion ρ_J or gyration γ on any finite ranked poset [51, Lemma 6.2, Thm. 6.7].

As corollaries they rederive that fiber-cardinality statistics are homomesic for ρ_A acting on $\mathscr{A}([p]\times[q])$, and also prove that these statistics are homomesic for *gyration* γ acting on antichains. This is noteworthy, since these statistics are not generally homomesic for the promotion operator ∂ on the same space, so the homomesy is not merely a consequence of having the action defined as a Coxeter element of toggles.

3.5 *Refined homomesies for minuscule posets*

The product of two chains is an example of a *minuscule poset* aka *minuscule heaps*, a class that arises naturally in the representation theory of complex semisimple Lie algebras [15, 29, 48, 49]. These are classified into three infinite families (types A, B, and D, with subcases depending on the choice of minuscule element), and a few exceptional types (E_6 and E_7). For a complete classification of these and more pictures, see Stembridge [49, Appendix] or Rush and Shi [36, End of § 1]. Readers unfamiliar with the underlying representation theory can view them simply as a special class of posets with a structure that can be described purely combinatorially, though we do not take the space to do so here. Example 18 should convey the main point of this section

Following up on an idea of Stanley [46], Striker and Williams were able to give elegant proofs via equivariant bijections that the action of rowmotion on minuscule posets in types A and B exhibits cyclic sieving [53, Cor. 6.3]. Rush and Shi then gave a uniform proof that cyclic sieving holds for rowmotion acting on minuscule posets in all types [36].

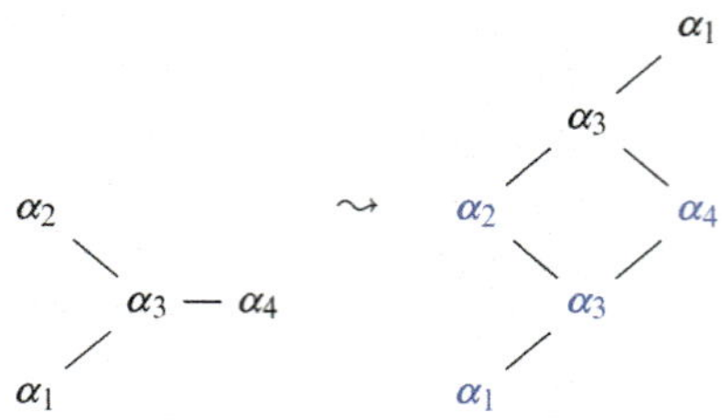

Fig. 5 The Dynkin diagram of type D_4 and the minuscule heap corresponding to the weight α_1.

Since the type A minuscule posets are always products of chains, it is natural to wonder whether the homomesy results extend to other types. This has recently been accomplished by David Rush and Kelvin Wang [37]. Recall that for products of chains, not only is the cardinality of an antichain homomesic, but so are all the statistics defined by counting intersections of the order ideals with any particular file (Example 17). What is exceptionally nice about their work is that they correctly identify what generalizes the notion of files, namely subsets of elements that are labeled by the same simple root within the minuscule poset.

Example 18 (Cardinality of weight-labeled elements under rowmotion acting on a minuscule posets). Consider the Dynkin diagram D_4 and minuscule heap P_{α_1} shown in Figure 5. Note that the Dynkin diagram is embedded in the bottom part of the heap. Although this particular poset happens to be isomorphic to a heap of type B_4, the labeling is different, and this is just a coincidence for n so small. For larger values of n, heaps of type D_n labeled with the weight α_1 (assumed to be a leaf of the trivalent vertex) are not isomorphic to heaps of type B_n.

The two rowmotion orbits on P_{α_1} are shown below.

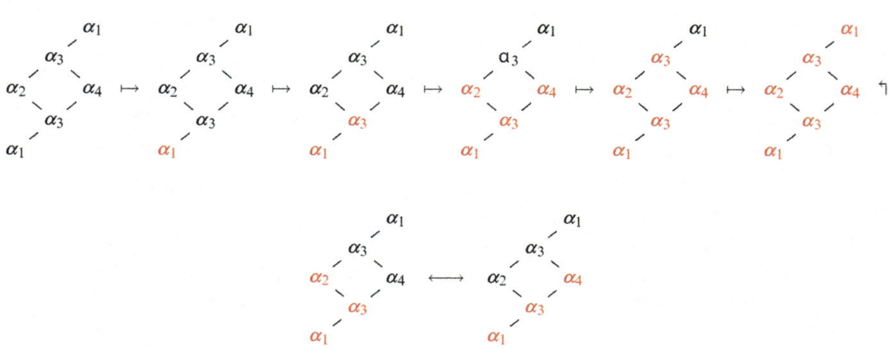

In the first orbit, the average number of times an element labeled α_1 occurs in the order ideal is $\frac{0+1+1+1+1+2}{6} = 1$, while in the second orbit it is $\frac{1+1}{2} = 1$. The same is true for any weight α_i.

The general situation is as follows.

Theorem 5 ([37]). *Let V be a minuscule $\mathfrak{g}$-representation with minuscule weight λ and minuscule heap P_λ. Let α be a simple root of $\mathfrak{g}$ with corresponding fundamental weight ω, and let $P_\lambda^\alpha \subseteq P_\lambda$ be the set of elements of P_λ labeled by α. Let $f^\alpha : J(P_\lambda) \to \mathbb{R}$ be the statistic $|I \cap P_\lambda^\alpha|$ that counts the number of elements in the order ideal labeled by α. Then with respect to the action of rowmotion on P_λ, the statistic f^α is c-mesic with $c = 2\frac{(\lambda,\omega)}{(\alpha,\alpha)}$. Hence, the statistic $f = \text{card } I$ is also homomesic.*

Rush and Wang also give explicit formulae for the homomesic average in terms of standard Lie-theoretic parameters.

4 Piecewise-linear and birational rowmotion

Another advantage of the toggling approach to rowmotion is its flexibility. By defining appropriate generalized toggles, we can lift rowmotion and similar maps to the piecewise-linear and birational categories. We outline here the process for doing this, which is explained in detail in a preprint of Einstein and Propp [9, 10]. Note that in these settings, following standard practice, f will be used to denote a poset labeling, i.e., an object upon which rowmotion acts, rather than a statistic, as earlier in this paper. Context should prevent any serious confusion.

4.1 The order polytope and piecewise-linear toggling

Definition 5. Let P be a finite poset. We define $\widehat{P}$ to be the poset obtained by adjoining to P a new global minimum element $\widehat{0}$ and new global maximum element $\hat{1}$. The **order polytope** $\mathcal{O}(P)$ (introduced by R. Stanley [43]) is the set of functions $f : \widehat{P} \to [0, 1]$ with $f(\hat{0}) = 0$, $f(\hat{1}) = 1$, and $f(x) \leq f(y)$ whenever $x \leq_{\widehat{P}} y$.

Informally, we think of an element of $\mathcal{O}(P)$ as being a labeling of the nodes of the Hasse diagram of $\widehat{P}$ with real numbers that respects the partial order. In this sense they naturally generalize order ideals, which are represented by $\{0, 1\}$-labelings, with 0's labeling elements of the order ideal, and 1's labeling the complementary order filter. In fact, $J(P)$ naturally corresponds to the set of vertices of $\mathcal{O}(P)$.

Definition 6. For each $x \in P$, define the **piecewise-linear toggle** $\sigma_x : \mathcal{O}(P) \to \mathcal{O}(P)$ sending f to the unique f' satisfying

$$f'(y) = \begin{cases} f(y) & \text{if } y \neq x, \\ \min_{z \gtrdot x} f(z) + \max_{w \lessdot x} f(w) - f(x) & \text{if } y = x, \end{cases} \tag{3}$$

where $z \gtrdot x$ means z covers x and $w \lessdot x$ means x covers w.

Note that the interval $[\max_{w \lessdot x} f(w), \min_{z \gtrdot x} f(z)]$ is precisely the set of values that $f'(x)$ could possibly take on while satisfying the order-preserving condition (as long as $f'(y) = f(y)$ for all $y \neq x$), and the map that sends $f(x)$ to $\min_{z \gtrdot x} f(z) + \max_{w \lessdot x} f(w) - f(x)$ is just the affine involution that swaps the endpoints of this interval.

Example 19. Consider the snippet of a poset shown at the left below, where the element x is covered by exactly three elements z_1, z_2, z_3 and covers two elements, w_1, w_2. In order to toggle the labeling f at element x, we first identify the minimum label assigned to an element covering x, and the maximum label below. Then σ_x changes the labeling *only* at the element x:

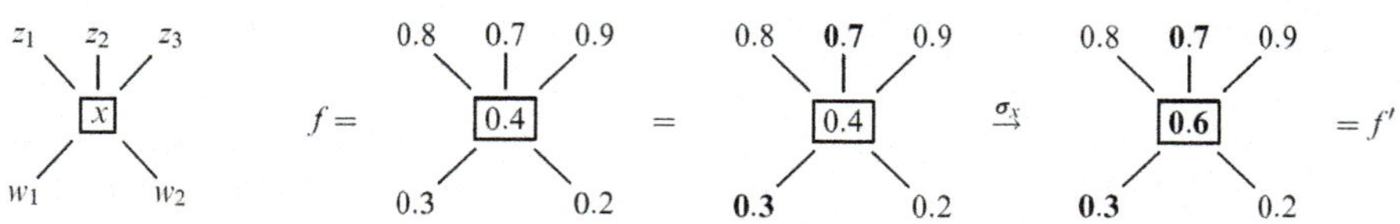

Note that here we have

$$\min_{z \gtrdot x} f(z) + \max_{w \lessdot x} f(w) = 0.7 + 0.3 = 1.0 = 0.4 + 0.6 = f(x) + f'(x)$$

It is straightforward to show that an analogue of Proposition 4 holds for the piecewise-linear toggles, so as before, we can define **piecewise-linear rowmotion** via any linear extension $x_1, \ldots, x_n$ of P to be

$$\rho_{PL} := \sigma_{x_1} \sigma_{x_2} \ldots \sigma_{x_n} : \mathcal{O}(P) \to \mathcal{O}(P).$$

Example 20. Consider the poset $P = [2] \times [2]$, labeled as in Example 15. We show the step-by-step process of toggling at each element of P by rows from top to bottom to obtain $\rho_{PL}(f)$, where f is the element of $\mathcal{O}(P)$ shown at the left.

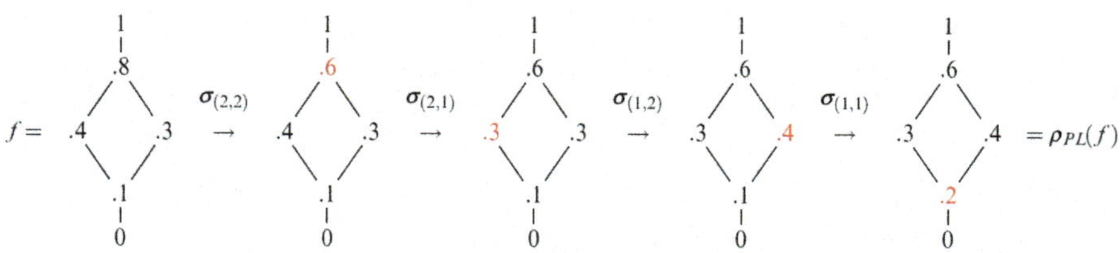

What happens if we apply rowmotion repeatedly? *A priori*, there is no reason to expect the orbits to even be finite since rowmotion is acting on an infinite set. For the f above, we find:

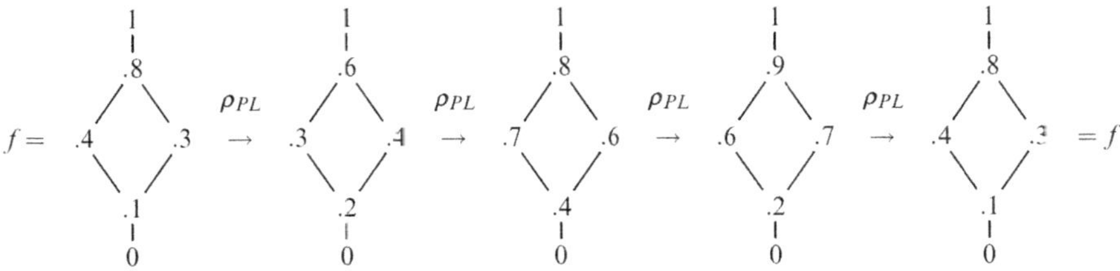

This periodicity turns out to hold in general: the order of the map ρ_{PL} on $\mathcal{O}([p]\times[q])$ is $p+q$, which is the same as the order of ρ on $J([p]\times[q])$. The only proof currently known of this result relies on results from the birational setting. It is an interesting open problem to prove this more directly.

4.2 The birational setting

Much work has been done in the *tropical semiring*, where the usual ring operations $+$ and $\cdot$ are replaced by the tropical operations max and $+$, respectively (see e.g., [26]). The nature of the formula for piecewise-linear toggles allows them to be *detropicalized* to **birational toggles** as follows. First rewrite each occurrence of min in (1) in terms of max using $\min(z_i) = -\max(-z_i)$; then replace each instance of max with $+$, each instance of $+$ with $\cdot$, and each negation with taking reciprocals. This yields the following definition.

Definition 7. Let $\mathbb{K}$ be any field, and $f \in \mathbb{K}^{\widehat{P}}$ any labeling of the elements of $\widehat{P}$ by elements of $\mathbb{K}$. We define the **birational toggle** $T_x : \mathbb{K}^{\widehat{P}} \dashrightarrow \mathbb{K}^{\widehat{P}}$ at $x \in P$ by

$$(T_x f)(y) = \begin{cases} f(y), & \text{if } y \neq x; \\ \dfrac{1}{f(x)} \cdot \dfrac{\sum\limits_{w\in\widehat{P};\; w\lessdot x} f(w)}{\sum\limits_{z\in\widehat{P};\; z\gtrdot x} \dfrac{1}{f(z)}}, & \text{if } y = x \end{cases}$$

for all $y \in \widehat{P}$. Note that this rational map T_x is well-defined, because the right-hand side is well-defined on a Zariski-dense open subset of $\mathbb{K}^{\widehat{P}}$. Finally, define birational rowmotion by $\rho_B := T_{x_1} T_{x_2} \cdots T_{x_n} : \mathbb{K}^{\widehat{P}} \dashrightarrow \mathbb{K}^{\widehat{P}}$, where $x_1, x_2, \ldots, x_n$ is any linear extension of P.

In words, toggling at x changes only the label at x and does this by (a) *inverting* the label at x, (b) multiplying by the *sum* of the labels at vertices *covered by* x, and (c) multiplying by the *parallel sum* of the labels at vertices *covering* x. (By **parallel sum** here, we mean the associative operation $\|$ defined by $a \parallel b := \frac{1}{\frac{1}{a}+\frac{1}{b}}$.)

Example 21. Let $P = [2] \times [2]$, our running example. One iteration of birational rowmotion acting on $\mathbb{K}^{\widehat{P}}$ is shown, toggling step-by-step along the usual linear extension.

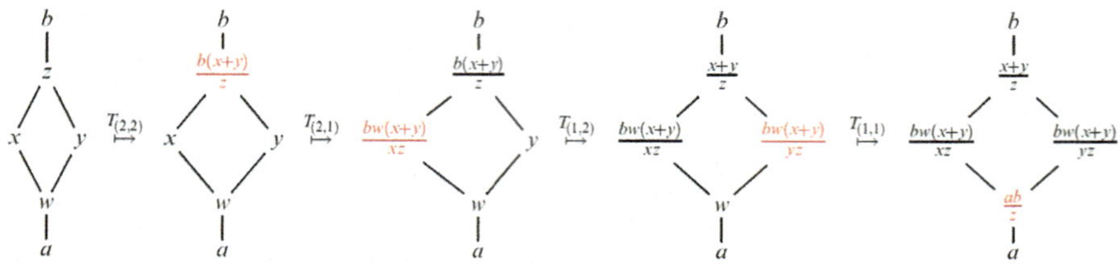

And here is a (generic) orbit of birational rowmotion:

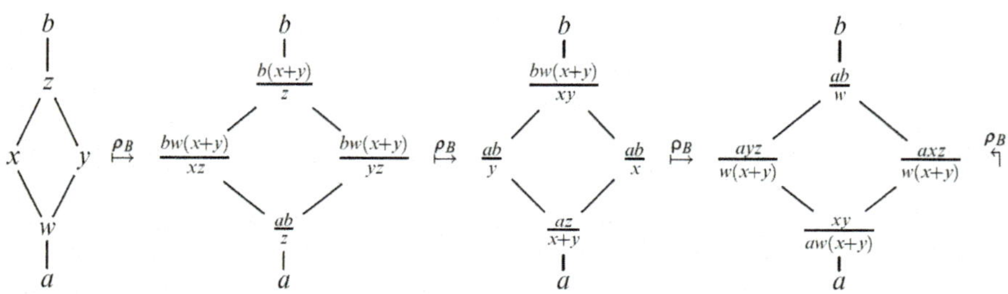

It seems almost miraculous that the orbit of birational rowmotion ρ_B is finite, let alone the same size as for rowmotion ρ_J on order ideals, which we now dub **combinatorial rowmotion**. This appears to be true only for relatively rare types of posets. Indeed for many simple posets such as the three shown in Figure 6, the order of ρ_B can be proven to be infinite [19, § 20].

Einstein and Propp [9, 10] were able to generalize homomesy results for $P = [p] \times [q]$ to the actions of ρ_{PL} and ρ_B by lifting techniques used for combinatorial rowmotion to the birational case, then specializing to the piecewise-linear case. However, their work did not show that the order of these maps on P was finite. The periodicity question loomed large, particularly since they preferred to avoid

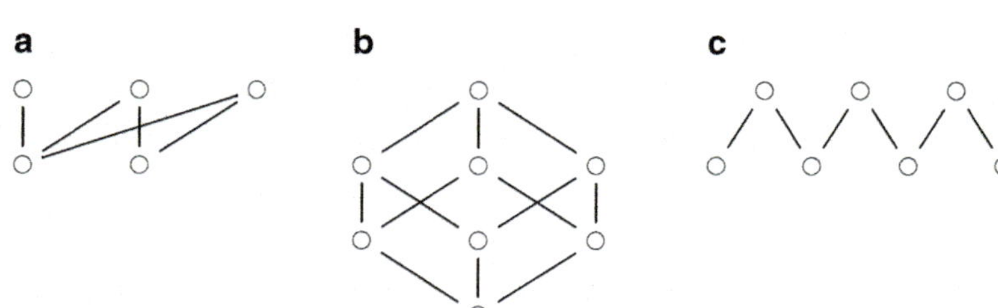

Fig. 6 Three posets for which birational rowmotion ρ_B is **not** periodic

using some sort of asymptotic definition of homomesy in this situation, where the orbits in fact always appeared to be finite. They describe how periodicity results for ρ_B imply the same results for ρ_{PL}, but not vice-versa. This led Darij Grinberg and the author to tackle the periodicity question for birational rowmotion.

Theorem 6 ([16–19]). *The action of ρ_B on generic $\mathbb{K}$-labelings of a finite poset P has finite order when:*

a) $P = [p] \times [q]$, is the product of two chains of lengths p and q;
b) $P = \Phi^+(A_n)$, the positive root poset of type A_n (cf. Example 11);
c) P is a graded rooted forest, with every component oriented towards its root and having the same rank;
d) and more generally, P is a skeletal *poset (a class generalizing graded forests), built up inductively by grafting antichains of various sizes.*

There are a few other posets for which the order of ρ_B is proven to be finite or conjecturally finite, mostly ones coming from Lie theory (root posets and minuscule posets) or built up inductively from simple structures. But getting a complete classification currently seems unlikely.

4.3 Homomesy for piecewise-linear and birational rowmotion

The notion of homomesy is easily lifted to the piecewise-linear and birational settings. In the latter, we need roughly to replace arithmetic means with geometric means. But branching issues with the function $z \mapsto z^{1/n}$ make it more convenient to just compare products across superorbits of the same length, rather than geometric means.

For P a product of two chains, Einstein and Propp are able to lift the homomesy results of [30] as follows.

Theorem 7 (Einstein-Propp [9, Thm. 2, Thm. 4]). *Let $P = [p] \times [q]$ be a product of two chains. Then*

1. *The statistic $G(f) := \sum_{x \in P} f(x)$ is c-mesic with respect to the action of ρ_{PL} on $\mathcal{O}(P)$ with $c = pq/2$.*
2. *For any $\mathbb{K}$-labeling f of $\widehat{P}$ with $f(0) = f(1) = 1$ for which iterations of birational rowmotion $\rho_B\colon \mathbb{K}^{\widehat{P}} \dashrightarrow \mathbb{K}^{\widehat{P}}$ are well-defined, we have*

$$\prod_{k=0}^{p+q-1} \prod_{x \in P} \rho_B^k(f)(x) = 1$$

In fact, even the more refined homomesies seem to lift to these settings. For example, for any *file* F_i in P, the statistic $G(f) := \sum_{x \in F_i} f(x)$ is claimed to be homomesic with respect to the action of ρ_{PL} on $\mathcal{O}(P)$ [9].

Example 22. Consider the orbit of ρ_{PL} shown in (2) and the orbit of ρ_B in 3. For each element $x \in P$, we write the *sum* of the ρ_{PL}-labels at x in the left poset, and the *product* of the ρ_B-labels at x in the right one.

$$f_\Sigma = \begin{array}{ccc} & 3.1 & \\ 2.0 & & 2.0 \\ & .9 & \end{array} \qquad f_\Pi = \begin{array}{ccc} & \frac{(x+y)^2}{xy} & \\ 1 & & 1 \\ & \frac{xy}{(x+y)^2} & \end{array}$$

If we start with a different point in the order polytope and construct the ρ_{PL}-orbit, we might get different sums (other than 3.1 and 0.9) across a ρ_{PL}-orbit at the top and bottom elements, but the sum of these sums will always be 4, just as the sums at the left and right elements will always be 2. And the average value across all elements will be $8/4 = 2 = \frac{2 \cdot 2}{2}$. For the birational orbit, the product of the orbit-products in the middle files is 1, as is the product of all labels at all elements.

4.4 Connection with Y-systems

The proof that ρ_B is periodic on the posets $[p] \times [q]$ and $\Phi^+(A_n)$ (Theorem 1) used a parameterization of $\mathbb{K}$-labelings of P as a ratio of two determinants and a simple three-term Plücker relation. This was modeled on Volkov's approach to proving the type $A_m \times A_n$ Zamolodchikov periodicity conjecture [57], leading a number of people to suspect a possible connection between Y-systems and birational rowmotion. It was only at the aforementioned AIM workshop in March 2015 that an explicit connection was uncovered by Arkady Berenstein, Max Glick, Darij Grinberg, Gregg Musiker, and Hugh Thomas. This gives another path to proving the periodicity of birational rowmotion on these posets and connects it with the theory of cluster algebras. Readers unfamiliar with Y-systems may wish to read this section lightly or seek necessary background and historical information in the excellent survey by Lauren Williams [58].

Informally, a Y-system is a dynamical system of rational functions defined on a graph coming from root systems. The setup is as follows. Let Δ, Δ' be *Dynkin diagrams* (of any type A, B, C, D, E, F, or G) on vertex sets I, I' and let C, C' be the corresponding *Cartan matrices*. Set the graph $\Gamma := I \times I'$.

Define $A = (a_{i,j}) := 2J_{\#I} - C$ and $A' = (a'_{i',j'}) := 2J_{\#I'} - C'$, where J_n denotes the $n \times n$ identity matrix. The matrices A and A' control how this dynamical system updates.

The only example we consider here is Type A, where, for example,

$$\Delta = A_4 \implies C = \begin{bmatrix} 2 & -1 & 0 & 0 \\ -1 & 2 & -1 & 0 \\ 0 & -1 & 2 & -1 \\ 0 & 0 & -1 & 2 \end{bmatrix} \implies A = \begin{bmatrix} 0&1&0&0 \\ 1&0&1&0 \\ 0&1&0&1 \\ 0&0&1&0 \end{bmatrix}.$$

Definition 8. Let h, h' denote the *Coxeter numbers* of Δ, Δ' (i.e., the order of the product of all the simple reflections in any order). The $\Delta \times \Delta'$ **Y-system** is then the collection $\{Y_{i,i',t} : (i, i') \in I \times I', t \in \mathbb{Z}\}$ satisfying the relations

$$Y_{i,i',t+1} Y_{i,i',t-1} = \frac{\prod_{j \in I} (1 + Y_{j,i',t})^{a_{i,j}}}{\prod_{j' \in I'} (1 + Y_{i,j',t}^{-1})^{a_{i',j'}}}.$$

(We can think of the $Y_{i,i',t}$ as positive real numbers or rational functions.)

Simplifying a more complex history, the main conjecture was that this system was periodic, with period given in terms of the Coxeter numbers.

Theorem 8 (Zamolodchikov periodicity conjecture). *With the setup as above, the type (Δ, Δ') Y-system is periodic with period $2(h + h')$, i.e.,*

$$Y_{i,i',t+2(h+h')} = Y_{i,i',t}.$$

Special cases were proved by a number of authors, including Volkov as described above using determinants and cross ratios, Fomin and Zelevinsky using cluster algebras. The theorem in full generality was finally resolved by Bernard Keller [25], using cluster algebra theory and categorification. See [58, § 4] for more details. In type A_n, $h = n + 1$, so the $A_{p-1} \times A_{q-1}$ Y-system has order $2(p + q)$.

The connection between Y-systems and birational rowmotion is then given as follows.

Proposition 6 (Berenstein-Glick-Grinberg-Musiker-Thomas). *Let f be a $\mathbb{K}$-labeling of $\widehat{P}$, where $P = [p] \times [q]$, for which iterations of birational rowmotion $\rho_B : \mathbb{K}^{\widehat{P}} \dashrightarrow \mathbb{K}^{\widehat{P}}$ are well-defined. Assume $f(0) = f(1) = 1$. If we set*

$$Y_{i,i',i+i'-2k} = \frac{\rho_B^k f(i, i' + 1)}{\rho_B^k f(i + 1, i')},$$

then the collection $\{Y_{i,i',j}\}$ form a type $A_{p-1} \times A_{q-1}$ Y-system. Furthermore, almost all type $A_m \times A_n$ Y-systems can be obtained in this way.

Although it takes a certain amount of work, this pathway eventually allows one to obtain periodicity of ρ_B on a product of two chains from periodicity of the corresponding Y-system, and vice-versa. More importantly, it shows a very strong connection between these two phenomena which is yet to be fully explored. In particular, perhaps homomesy results for ρ_B might translate into some interesting property of Y-systems.

5 Homomesy in other toggling situations

A useful abstract generalization of the toggle group of a poset has recently been proposed by Jessica Striker [52]. In this framework, the poset P is replaced with any ground set E, and $J(P)$ with any fixed subset $\mathscr{L}$ of 2^E. Then one defines the **toggle** $t_e : \mathscr{L} \to \mathscr{L}$ by

$$t_e(X) := \begin{cases} X \cup \{e\} & \text{if } e \notin X \text{ and } X \cup \{e\} \in \mathscr{L} \\ X \backslash \{e\} & \text{if } e \in X \text{ and } X \backslash \{e\} \in \mathscr{L} \\ X & \text{otherwise} \end{cases}.$$

It is easy to see that $t_e^2 = 1$ for all $e \in E$, allowing one to define a **generalized toggle group** $T(\mathscr{L})$ to be the subgroup of the symmetric group $\mathfrak{S}_{\mathscr{L}}$ generated by $\{t_e : e \in E\}$. This definition significantly broadens the range of examples to which the ideas of toggling can be applied. Striker gives a few general structure theorems, and describes a number of examples where this approach appears to be fruitful.

One particularly interesting example came out of a working group at the aforementioned AIM workshop in March 2015.

Definition 9. A *noncrossing partition* is a partition $\{B_1, \ldots, B_s\}$ of $[n] := \{1, \ldots, n\}$ such that if $a < b < c < d$ with both $a, c \in B_i$ and $b, d \in B_j$, then $i = j$. Let $NC(n)$ denote the set of all noncrossing partitions of $[n]$.

A noncrossing partition can be represented pictorially by a collection of arcs connecting nodes labeled by $[n]$. For each block B_i, arrange the $|B_i|$ vertices in increasing order and include $|B_i| - 1$ arcs, one for each adjacent pair. Even though the arcs are undirected, we always describe one as an ordered pair (i, j) with $i < j$. Within the set of all possible such graphs on $[n]$ given by connecting any two distinct vertices, the ones representing noncrossing partitions are characterized by the following disallowed types (shown in Figure 7). Two distinct arcs $(i, j), (k, \ell)$ in a noncrossing partition must not satisfy:

1. $i < k < j < \ell$ (crossing),
2. $i = k$ (left half-nesting), or
3. $j = \ell$ (right half-nesting).

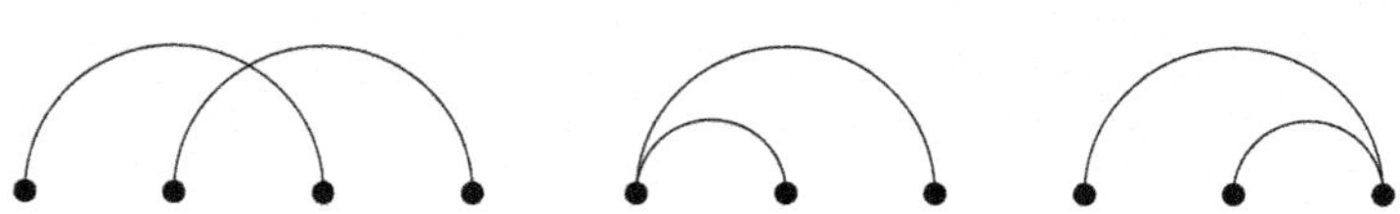

Fig. 7 Disallowed pairs of arcs in a noncrossing partition: crossing, left half-nesting, and right half-nesting, respectively.

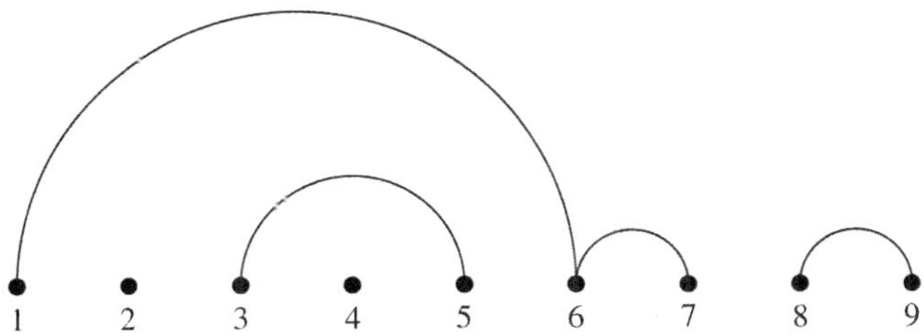

Fig. 8 The noncrossing partition $\{(1,6),(3,5),(6,7),(8,9)\} \in NC(9)$.

Figure 8 shows the noncrossing partition $\{\{1,6,7\},\{2\},\{3,5\},\{4\},\{8,9\}\}$ of size 9, drawn using arcs. Note that nesting arcs are allowed, but by construction, they must describe vertices in different blocks.

Now in Striker's framework, take $E = \{(i,j) \in [n] \times [n] : i < j\}$, and consider $NC(n)$ as a subset of 2^E. The toggle $t_{(i,j)}$ applied to $\nu \in NC(n)$ simply adds or removes the arc (i,j) whenever the result still lies in $NC(n)$, and otherwise does nothing.

The original problem proposed by James Propp was to show that under the action of certain "Coxeter elements" on $NC(n)$, the *arc count statistic* $\alpha :=$ the number of pairs $(i,j) \in \nu$ is $\frac{n-1}{2}$-mesic. He noted that toggling in the "nice" order:

$$\tau_{\text{nice}} = (t_{(n-1,n)})(t_{(n-2,n)}t_{(n-2,n-1)}) \cdots (t_{(2,n)}t_{(2,n-1)} \cdots t_{(2,4)}t_{(2,3)})(t_{(1,n)} \cdots t_{(1,3)}t_{(1,2)})$$

appears to be the inverse of Kreweras complementation on $NC(n)$; hence, this map has order $2n$ and the homomesy follows easily. On the other hand, consider toggling each element of E in the "naughty" order, where all toggles of nodes one apart are applied first, followed by toggles of nodes two apart, etc. In this situation the orbit structure is unpredictable and the period of the map is much greater, but the same statistic appeared to be homomesic. The group working on this problem in fact showed a much more general result.

Theorem 9 (EFGJMPR [8]). *Let τ be any product of toggles $t_{(i,j)}$ (where $i < j$) that contains no toggle more than once (a "partial Coxeter element") and that contains $t_{(i,i+1)}$ for every $i \in [n-1]$. Then the arc count statistics α is $\frac{n-1}{2}$-mesic.*

This theorem provides a good example of homomesy where cyclic sieving seems to be unlikely. For $n = 7$ consider the Coxeter element

$$\tau_{\text{naughty}} = (t_{(1,7)})(t_{(2,7)}t_{(1,6)}) \cdots (t_{(4,7)}t_{(3,6)}t_{(2,5)}t_{(1,4)})(t_{(5,7)}t_{(4,6)}t_{(3,5)}t_{(2,4)}t_{(1,3)})(t_{(6,7)} \cdots t_{(3,4)}t_{(2,3)}t_{(1,2)})$$

Theorem 9 applies, so we get homomesy. But the orbit sizes are 3, 3, 3, 3, 3, 5, 9, 9, 11, 13, 15, 31, 39, 39, 109, 133, with least common multiple 5783868090, which is the order of τ. So an interesting example of cyclic sieving seems to be unlikely for this action, since the size of the cyclic group is so large relative to the number of objects, which in this case is the Catalan number $C_7 = 429$.

References

1. D. Armstrong, C. Stump, H. Thomas, A uniform bijection between nonnesting and noncrossing partitions. Trans. Am. Math. Soc. **365**(8), 4121–4151 (2013). Also available as `arXiv:1101.1277v2`
2. A. Ayyer, S. Klee, A. Schilling, Combinatorial Markov chains on linear extensions. J. Algebraic Comb. **39(4)**, 853–881 (2014). Also available at `arXiv:1205.7074v3`
3. R. Behrend, T. Roby, Dihedral homomesy for Alternating Sign Matrices, in preparation
4. J. Bloom, O. Pechenik, D. Saracino, Proofs and generalizations of a homomesy conjecture of Propp and Roby (2013), Disc. Math. **339**, 194–206 (2016). `arXiv:1308.0546v3`
5. A.E. Brouwer, A. Schrijver, *On the Period of an Operator, Defined on Antichains*, Math. Center report ZW24, Amsterdam (Jun. 1974), http://www.win.tue.nl/~aeb/preprints/zw24.pdf
6. P.J. Cameron, D.G. Fon-der-Flaass, Orbits of Antichains Revisited. Eur. J. Comb. **16**(6), 545–554 (1995), http://www.sciencedirect.com/science/article/pii/0195669895900365
7. M. Chan, S. Haddadan, S. Hopkins, L. Moci, The expected jaggedness of order ideals (2015), `arXiv:1507.00249v1`
8. D. Einstein, M. Farber, E. Gunawan, M. Joseph, M. Macauley, J. Propp, S. Rubinstein-salzedo, Noncrossing partitions, toggles, and homomesies, `arXiv:1510.06362v1`
9. D. Einstein, J. Propp, Combinatorial, piecewise-linear, and birational homomesy for products of two chains (2013), `arXiv:1310.5294`
10. D. Einstein, J. Propp, Piecewise-linear and birational toggling (Extended abstract), DMTCS proceedings of FPSAC 2014 (2014), http://www.dmtcs.org/dmtcs-ojs/index.php/proceedings/article/view/dmAT0145/4518. Also available at `arXiv:1404.3455v1`
11. D.G. Fon-der-Flaass, Orbits of Antichains in Ranked Posets. Eur. J. Comb. **14**(1), 17–22 (1993). http://www.sciencedirect.com/science/article/pii/S0195669883710036
12. W. Fulton, *Young Tableaux: With Applications to Representation Theory and Geometry* (Cambridge University Press, Cambridge, 1997)
13. E. Gansner, On the equality of two plane partition correspondences. Discret. Math. **30**, 121–132 (1980)
14. M. Gardner, Mathematical games: Tasks you cannot help finishing no matter how hard you try to block finishing them. Sci. Am. **249**, 12–21 (1983)
15. R.M. Green, *Combinatorics of Minuscule Representations* (Cambridge University Press, Cambridge 2013)
16. D. Grinberg, T. Roby, The order of birational rowmotion (Extended abstract), DMTCS proceedings of FPSAC 2014, (2014), http://www.dmtcs.org/pdfpapers/dmAT0165.pdf
17. D. Grinberg, T. Roby, Iterative properties of birational rowmotion I: generalities and skeletal posets. Electron. J. Comb. **23**(1), #P1.33 (2016), to appear
18. D. Grinberg, T. Roby, Iterative properties of birational rowmotion II: rectangles and triangles. Electron. J. Comb. **22**(3), #P3.40, (2015). http://www.combinatorics.org/ojs/index.php/eljc/article/view/v22i3p40
19. D. Grinberg, T. Roby, Iterative properties of birational rowmotion (2014), `arXiv:1402.6178v4`. (Full version of above papers with detailed proofs)
20. S. Haddadan, Some instances of homomesy among ideals of posets (2014), `arXiv:1410.4819v2`
21. A.E. Holroyd, J. Propp, Rotor walks and Markov chains, in *Algorithmic Probability and Combinatorics* (American Mathematical Society, Providence, RI, 2010), pp. 105–126
22. A.E. Holroyd, L. Levine, K. Meszaros, Y. Peres, J. Propp, D.B. Wilson, Chip-firing and rotor-routing on directed graphs. Prog. Probability **60**, 331–364 (2008), Updated version at `arXiv:0801.3306v4`
23. B. Hopkins, 30 years of Bulgarian solitaire. Coll. Math. J. **43**, #2, 135–140 (2012)
24. S. Hopkins, I. Zhang, A note on statistical averages for oscillating tableaux. Electron. J. Comb. **22**(2), #P2.48 (2015), http://www.combinatorics.org/ojs/index.php/eljc/article/view/v22i2p48. Also at `arXiv:1408.6183v1`

25. B. Keller, The periodicity conjecture for pairs of Dynkin diagrams. Ann. Math. **177**(1), 111–170 (2013)
26. A.N. Kirillov, *Introduction to tropical combinatorics*, Physics and combinatorics: Proceedings of the Nagoya 2000 International Workshop, held 21 - 26 August 2000 in Nagoya University. ed. by A.N. Kirillov (Nagoya University) & Nadejda Liskova. Published (World Scientific Publishing, Singapore, 2001), pp. 82–150. ISBN #9789812810007
27. A.N. Kirillov, A.D. Berenstein. Groups generated by involutions, Gelfand-Tsetlin patterns, and combinatorics of Young tableaux. Algebra i Analiz **7**(1), 92–152 (1995). Preprint available at http://math.uoregon.edu/~arkadiy/bk1.pdf
28. D.I. Panyushev, On orbits of antichains of positive roots. Eur. J. Comb. **30**(2), 586–594 (2009). Also available at `arXiv:0711.3353v2`
29. R.A. Proctor, Bruhat lattices, plane partition generating functions, and minuscule representations. Eur. J. Comb. **5**, 331–350 (1984)
30. J. Propp, T. Roby, Homomesy in products of two chains. Electron. J. Comb. **22**(3), #P3.4, (2015), http://www.combinatorics.org/ojs/index.php/eljc/article/view/v22i3p4. Also available at `arXiv:1310.5201v5`
31. J. Propp, T. Roby, Homomesy in Products of Two Chains, (extended abstract) DMTCS proceedings of FPSAC 2013 (2013), http://www.dmtcs.org/dmtcs-ojs/index.php/proceedings/article/view/dmAS0180/4290
32. J. Propp, T. Roby, J. Striker, N. Williams, (organizers), S. Hopkins (notetaker), *Notes from the AIM Workshop on Dynamical Algebraic Combinatorics* (American Institute of Mathematics, San Jose, 2015), pp. 23–27, http://aimath.org/pastworkshops/dynalgcomb.html, http://mit.edu/~shopkins/docs/aim_dyn_alg_comb_notes.pdf
33. V. Reiner, D. Stanton, D. White, The cyclic sieving phenomenon. J. Comb. Theory Ser. A **108**, 17–50 (2004)
34. V. Reiner, D. Stanton, D. White, What is... cyclic sieving?. AMS Notices **61**, #2 169–171 (2014)
35. B. Rhoades, Cyclic sieving, promotion, and representation theory. J. Comb. Theory Ser. A **117**(1), 38–76 (2010), http://www.sciencedirect.com/science/article/pii/S0097316509000703 Also available at `arXiv:1005.2568v1`
36. D.B. Rush, X. Shi, On orbits of order ideals of minuscule posets. J. Algebraic Comb. **37**(3), 545–569 (2013), Also available at `arXiv:1108.5245v2`
37. D.B. Rush, K. Wang, On orbits of order ideals of minuscule posets II: Homomesy (2015), `arXiv:1509.08047`
38. B.E. Sagan, *The Cyclic Sieving Phenomenon: A Survey*, London Mathematical Society Lecture Note Series, vol. 392 (Cambridge University Press, Cambridge, 2011). Also available at `arXiv:1008.0790v3`
39. W.A. Stein et. al., *Sage Mathematics Software (Version 6.2.beta2)*, The Sage Development Team (2014), http://www.sagemath.org
40. The Sage-Combinat community, *Sage-Combinat: Enhancing Sage as a Toolbox for Computer Exploration in Algebraic Combinatorics* (2008), http://combinat.sagemath.org
41. A. Schilling, N.M. Thiéry, G. White, N. Williams, Braid moves in commutation classes of the symmetric group (2015), `arXiv:1507.00656v2`
42. M.-P. Schützenberger, Promotion des morphisms d'ensemble ordonnes. Discret. Math. **2**, 73–94 (1972)
43. R.P. Stanley, Two poset polytopes. Discret. Comput. Geom. **1**(1), 9–23 (1986)
44. R.P. Stanley, in *Enumerative Combinatorics, Volume 1, 2nd Edition*, Cambridge Studies in Advanced Mathematics, vol. 49 (Cambridge University Press, Cambridge, 2011). Also available at http://math.mit.edu/~rstan/ec/ec1/
45. R.P. Stanley, in *Enumerative Combinatorics Volume 2*, Cambridge Studies in Advanced Mathematics, vol. 62 (Cambridge University Press, Cambridge, 1999)
46. R.P. Stanley, Promotion and evacuation. Electron. J. Comb. **16**(2), #R9, (2009), http://www.combinatorics.org/ojs/index.php/eljc/article/download/v16i2r9/pdf

47. J.R. Stembridge, Trapezoidal chains and antichains. Eur. J. Comb. **7(4)**, 377–387 (1986), http://www.sciencedirect.com/science/article/pii/S0195669886800099
48. J.R. Stembridge, Minuscule elements of Weyl groups. J. Algebra **235** 722–743 (2001)
49. J.R. Stembridge, On minuscule representations, plane partitions and involutions in complex lie groups. Duke Math. J. **73**(2), 469–490 (1994)
50. J.R. Stembridge, Some hidden relations involving the ten symmetry classes of plane partitions. J. Comb. Theory Ser. A **68**, 372–409 (1994)
51. J. Striker, The toggle group, homomesy, and the Razumov-Stroganov correspondence. Electron. J. Comb. **22**(2), #P2.57 (2015), http://www.combinatorics.org/ojs/index.php/eljc/article/view/v22i2p57/3248. Also available at `arXiv:1503.08898v1`
52. J. Striker, Rowmotion and generalized toggle groups (2016), `arXiv:1601.03710`
53. J. Striker, N. Williams, Promotion and rowmotion. Eur. J. Comb. **33**, 1919–1942 (2012), http://www.sciencedirect.com/science/article/pii/S0195669812000972. Also available at `arXiv:1108.1172v3`
54. R. Suter, Young's lattice and dihedral symmetries. Eur. J. Comb. **23**, 233–238 (2002)
55. H. Thomas, A. Yong, Cominuscule tableau combinatorics (2007), `arXiv:0701215v2`
56. A. Toom, Problem solution M655. Kvant (Quantum) **7**, 28–30 (1981), http://kvant.mccme.ru/1981/07/p28.htm
57. A.Y. Volkov, On the periodicity conjecture for Y-systems. Commun. Math. Phys. **276**, 509–517 (2007). Preprint also available, under the name *On Zamolodchikov's Periodicity Conjecture*, at `arXiv:hep-th/0606094v1`
58. L. Williams, Cluster algebras: an introduction. Bull. AMS **51**(1), 1–26 (2014), http://www.ams.org/journals/bull/2014-51-01/S0273-0979-2013-01417-4/S0273-0979-2013-01417-4.pdf.
59. N. Williams, Cataland, Dissertation, University of Minnesota, 2013, https://conservancy.umn.edu/bitstream/159973/1/Williams_umn_0130E_14358.pdf

Face enumeration on simplicial complexes

Steven Klee and Isabella Novik

Abstract In this chapter we survey many exciting developments on the face numbers of simplicial complexes from the past two decades. We focus on simplicial complexes whose geometric realizations are (homology) manifolds, as well as manifolds with additional combinatorial structure such as balanced manifolds or flag manifolds. The discussed results range from the Upper Bound Theorem for manifolds to the balanced Generalized Lower Bound Theorem for balanced polytopes.

Keywords f-vector • h-vector • g-vector • Triangulated manifold • Upper and lower bound theorems • Stanley-Reisner ring

Mathematics Subject Classification (2010): 05E45, 13F55, 57Q15

1 Introduction

Let M be a closed triangulable manifold, and let Δ be a triangulation of M. What is the smallest number of vertices that Δ can have? How big or small can the number of edges of Δ be as a function of the number of vertices? More generally, what are the possible face numbers (f-numbers, for short) that Δ can have? In other words, what restrictions does the topology of M place on the possible f-numbers of triangulations of M?

To make things even more interesting, we can add some combinatorial conditions on the triangulations we are considering (e.g., flagness, balancedness, etc.) and ask what additional restrictions these combinatorial conditions impose. While only a few theorems in this area of combinatorics were known a couple of decades ago,

S. Klee
Department of Mathematics, Seattle University, 901 12th Avenue, Seattle, WA 98122, USA
e-mail: klees@seattleu.edu

I. Novik (✉)
Department of Mathematics, University of Washington, Box 354350, Seattle, WA 98195, USA
e-mail: novik@math.washington.edu

A. Beveridge et al. (eds.), *Recent Trends in Combinatorics*, The IMA Volumes in Mathematics and its Applications 159, DOI 10.1007/978-3-319-24298-9_26

in the last ten years or so, the field simply exploded with new results and ideas. Thus we feel that a survey paper is long overdue. As new theorems are being proved while we are typing this chapter, and as we have only a limited number of pages, we apologize in advance to our friends and colleagues, some of whose results will not get mentioned here.

The paper is structured as follows. In Section 2 we recall basic definitions pertaining to simplicial complexes. In Section 3 we review classical results on the f-numbers of polytopes and spheres. In Section 4 we survey many recent results on the f-vectors of triangulations of manifolds and pseudomanifolds. Section 5 is devoted to the face numbers of flag and balanced complexes. Finally, in Section 6 we discuss some of the tools and techniques used in the proofs.

2 Definitions

A **simplicial complex** Δ on a (finite) vertex set $V = V(\Delta)$ is a collection of subsets of $V(\Delta)$ that is closed under inclusion. The elements $F \in \Delta$ are called **faces** and the maximal faces of Δ under inclusion are called **facets**. We say that Δ is **pure** if all of its facets have the same cardinality. The **dimension** of a face $F \in \Delta$ is $\dim(F) = |F| - 1$ and the dimension of Δ is $\dim(\Delta) = \max\{\dim(F) : F \in \Delta\}$; we refer to i-dimensional faces as ***i*-faces**. If Δ is a $(d-1)$-dimensional simplicial complex and $k \leq d-1$, the ***k*-skeleton** of Δ is $\mathrm{Skel}_k(\Delta) := \{F \in \Delta : \dim(F) \leq k\}$.

Although simplicial complexes are defined as purely combinatorial objects, each simplicial complex Δ admits a geometric realization $\|\Delta\|$ that contains a geometric i-simplex for each i-face of Δ. We typically do not distinguish between the combinatorial object Δ and the geometric object $\|\Delta\|$ and often say that Δ has certain geometric or topological properties in addition to certain combinatorial properties.

Let Δ be a $(d-1)$-dimensional simplicial complex. The main object of our study is the ***f*-vector** of Δ, $f(\Delta) := (f_{-1}(\Delta), f_0(\Delta), \ldots, f_{d-1}(\Delta))$, where $f_i = f_i(\Delta)$ denotes the number of i-faces of Δ; the numbers f_i are called the ***f*-numbers** of Δ.

For several algebraic reasons, it is often more natural to study a certain (invertible) integer transformation of the f-vector called the ***h*-vector** of Δ, $h(\Delta) = (h_0(\Delta), h_1(\Delta), \ldots, h_d(\Delta))$; it is defined by the following polynomial relation:

$$\sum_{j=0}^{d} h_j(\Delta) \cdot t^{d-j} = \sum_{j=0}^{d} f_{j-1}(\Delta) \cdot (t-1)^{d-j}.$$

Thus, for all $0 \leq j \leq d$,

$$h_j(\Delta) = \sum_{i=0}^{j} (-1)^{j-i} \binom{d-i}{d-j} f_{i-1}(\Delta) \quad \text{and} \quad f_{i-1}(\Delta) = \sum_{j=0}^{i} \binom{d-j}{d-i} h_j(\Delta).$$

In particular, knowing the f-numbers of Δ is equivalent to knowing its h-numbers. Moreover, since the f-numbers are *nonnegative* integer combinations of the h-numbers, any upper/lower bounds on the h-numbers of Δ automatically imply upper/lower bounds on the f-numbers of Δ.

The motivating question in this field of research is the following: given a topological manifold M, characterize the set of f- or h-vectors of simplicial complexes Δ whose geometric realization is homeomorphic to M. This question is incredibly difficult, and there are only a few manifolds for which this question is completely answered. Instead, we relax the question in a number of ways.

Rather than study a manifold of given topological type, we study the family of all manifold triangulations or even more relaxed families of triangulations. Second, rather than seek complete characterizations of the sets of f- or h-vectors of the simplicial complexes in these families, we seek restrictions that can be placed on the f- or h-vectors that connect the combinatorics of the triangulation to the underlying topology of the simplicial complex. We begin by describing some families of simplicial complexes that extend the notion of a manifold triangulation.

If Δ is a simplicial complex and $F \in \Delta$ is a face, the local structure of Δ around F is described by the (closed) **star** of F in Δ: $\mathrm{st}_\Delta(F) := \{G \in \Delta \ : \ F \cup G \in \Delta\}$. Morally, when Δ triangulates a manifold, one should view the star of a face as the closed ball of faces that surround that face (however, even for manifolds, the star of a face need not be a topological ball). Similarly, the **link** of F in Δ is defined as

$$\mathrm{lk}_\Delta(F) := \{G \in \mathrm{st}_\Delta(F) \ : F \cap G = \emptyset\}.$$

When $F = \{v\}$ consists of a single vertex, we write $\mathrm{lk}_\Delta(v)$ in place of $\mathrm{lk}_\Delta(\{v\})$.

If $\mathbf{k}$ is a field (or the ring of integers), the **(reduced) Betti numbers** of Δ are $\beta_i(\Delta;\mathbf{k}) := \dim_{\mathbf{k}} \tilde{H}_i(\Delta;\mathbf{k})$. Here $\tilde{H}_i(-;\mathbf{k})$ stands for the i-th simplicial reduced homology computed with coefficients in $\mathbf{k}$. The **(reduced) Euler characteristic** of Δ is $\tilde{\chi}(\Delta) = \sum_{i=0}^{d}(-1)^{i-1}f_{i-1}(\Delta) = \sum_{i=0}^{d-1}(-1)^i\beta_i(\Delta;\mathbf{k})$. Below we denote by $\mathbb{S}^j$ and $\mathbb{B}^j$ the j-dimensional sphere and ball, respectively.

On the one hand, we can relax the structure of a manifold triangulation (also called a **simplicial manifold**) by studying the families of simplicial homology manifolds and simplicial pseudomanifolds. A $(d-1)$-dimensional simplicial complex Δ is a **k-homology sphere** if $\tilde{H}_*(\mathrm{lk}_\Delta(F);\mathbf{k}) \cong \tilde{H}_*(\mathbb{S}^{d-|F|-1};\mathbf{k})$ for all faces $F \in \Delta$ (including $F = \emptyset$). Similarly, Δ is a **k-homology manifold** (without boundary) if for all nonempty faces $F \in \Delta$, the link of F is a $\mathbf{k}$-homology $(d-|F|-1)$-sphere.

Homology manifolds with boundary are defined in an analogous way: a simplicial complex Δ is a **k-homology manifold with boundary** if (1) the link of each nonempty face F of Δ has the homology of either $\mathbb{S}^{d-|F|-1}$ or $\mathbb{B}^{d-|F|-1}$, and (2) the set of all **boundary faces**, that is,

$$\partial\Delta := \left\{F \in \Delta \ : \ \tilde{H}_*(\mathrm{lk}_\Delta(F);\mathbf{k}) \cong \tilde{H}_*(\mathbb{B}^{d-|F|-1};\mathbf{k})\right\} \cup \{\emptyset\},$$

is a $\mathbf{k}$-homology $(d-2)$-manifold without boundary.

Why do these definitions extend the definitions of manifold triangulations? Applications of the standard homology axioms (see [59, Lemma 3.3]) show that if $F \in \Delta$ is nonempty and p is a point in the relative interior of F, then $H_j(\|\Delta\|, \|\Delta\| - p; \mathbf{k}) \cong \tilde{H}_{j-|F|}(\mathrm{lk}_\Delta(F); \mathbf{k})$. When $\|\Delta\|$ is a $(d-1)$-manifold, $H_j(\|\Delta\|, \|\Delta\| - p; \mathbf{k})$ is nontrivial (and isomorphic to $\mathbf{k}$) if and only if p is an interior point and $j = d - 1$.

A $(d-1)$-dimensional simplicial complex Δ is a **normal pseudomanifold** if (1) it is pure, (2) each $(d-2)$-face (or **ridge**) of Δ is contained in exactly two facets of Δ, and (3) the link of each nonempty face of dimension at most $d-3$ is connected. In particular, every homology manifold without boundary is a normal pseudomanifold.

On the other hand, we can impose extra structure on our simplicial complexes by studying the families of simplicial polytopes and piecewise linear (PL) manifold triangulations. A ***d*-polytope** is the convex hull of a finite set of points in $\mathbb{R}^d$ that affinely span $\mathbb{R}^d$. A d-polytope is simplicial if all of its (proper) faces are simplices. While the boundary of any simplicial d-polytope is a simplicial $(d-1)$-sphere, not every simplicial sphere can be realized as the boundary of a polytope. In fact, for $d \geq 4$, almost all simplicial $(d-1)$-spheres are not polytopal, see [35, 69, 76].

One feature of a simplicial d-polytope is that all of its faces are realized linearly in $\mathbb{R}^d$. A simplicial manifold Δ is a **piecewise linear** (PL) manifold if the link of each nonempty face of Δ is PL-homeomorphic to a simplex or to the boundary of a simplex of the appropriate dimension.

The canonical example of a simplicial complex that distinguishes all of these definitions is the Poincaré 3-sphere. If Δ is a triangulation of the Poincaré 3-sphere, then Δ is a $\mathbb{Z}$-homology 3-sphere, but not a topological 3-sphere. Cannon [15] showed that the double suspension of any $\mathbb{Z}$-homology sphere is a topological sphere, and hence $\Sigma^2(\Delta)$, the double suspension of Δ, is a triangulation of $\mathbb{S}^5$. However, if e is an edge in $\Sigma^2(\Delta)$ connecting suspension vertices, then $\mathrm{lk}_{\Sigma^2(\Delta)}(e) = \Delta$ is not homeomorphic to $\mathbb{S}^3$. Therefore, $\Sigma^2(\Delta)$ is not a PL manifold. This also shows that the link of any face in a sphere triangulation has the homology of a sphere, but need not be homeomorphic to a sphere.

Throughout this paper, Δ denotes a simplicial complex; often, it is a homology manifold. Unless we say that Δ is a homology manifold with boundary, we assume that $\partial\Delta = \emptyset$. Also, by the f- and h-numbers of a simplicial polytope P we mean the f- and h-numbers of the boundary of P.

3 Classical results for polytopes and spheres

One of the first classical results on the structure of f-vectors of triangulations of manifolds is known as Klee's Dehn-Sommerville equations [41].

Theorem 1 ([41]). *Let Δ be a $(d-1)$-dimensional homology manifold (or more generally, any semi-Eulerian simplicial complex). Then*

$$h_{d-j}(\Delta) - h_j(\Delta) = (-1)^j \binom{d}{j} \left[\tilde{\chi}(\Delta) - \tilde{\chi}(\mathbb{S}^{d-1})\right], \quad \text{for all } 0 \leq j \leq d.$$

Thus, the h-vector of a homology sphere is symmetric, and hence determined by $h_0, h_1, \dots, h_{\lfloor d/2 \rfloor}$. For a $(d-1)$-dimensional simplicial complex Δ, $h_0(\Delta) = 1$, while $h_d(\Delta) = \sum_{i=0}^{d}(-1)^{d-i} f_{i-1}(\Delta) = (-1)^{d-1}\tilde{\chi}(\Delta)$. Hence the Dehn-Sommerville equations provide a vast generalization of the Euler-Poincaré formula.

Many early results in the study of face enumeration focused on the properties of f-vectors of (simplicial) polytopes. A first natural question is to ask how large or how small the f-numbers of a simplicial polytope could be as a function of its dimension and number of vertices.

For the question of lower bounds, Barnette [7] showed that among all simplicial d-polytopes with n vertices, **stacked polytopes** minimize the number of faces of each dimension. A stacked d-polytope on n vertices, denoted $ST(n, d)$, is defined inductively as follows. The polytope $ST(d+1, d)$ is a d-simplex. For $n > d+1$, $ST(n, d)$ is constructed by gluing a d-simplex to $ST(n-1, d)$ along a common facet.

Theorem 2 (The Lower Bound Theorem). *Let Δ be a $(d-1)$-dimensional connected homology manifold on n vertices with $d \geq 3$.*

1. [7] *Then $h_2(\Delta) \geq h_1(\Delta)$. It follows that $f_{i-1}(\Delta) \geq ST(n, d)$ for all i.*
2. [34] *Moreover, if $d \geq 4$, then $h_2(\Delta) = h_1(\Delta)$ if and only if Δ is the boundary complex of a stacked d-polytope.*

Part (2) of the Lower Bound Theorem is due to Kalai [34]; Kalai also found an alternative proof of part (1). This theorem was later generalized to all normal pseudomanifolds by Fogelsanger [20] and Tay [89]. The condition that $d \geq 4$ in part (2) of the Lower Bound Theorem is necessary since $h_2 = h_1$ for all 2-spheres, but not all 2-spheres are stacked (e.g., the boundary of the octahedron is not stacked).

For the question of upper bounds, McMullen [54] showed that the **cyclic polytope** is a simplicial polytope with the maximal number of faces of each dimension. A cyclic d-polytope on n vertices, denoted $\mathscr{C}_d(n)$, is defined by taking the convex hull of n distinct points on the moment curve $\varphi(t) := (t, t^2, \dots, t^d) \subseteq \mathbb{R}^d$.

Theorem 3 (The Upper Bound Theorem [54]). *Let P be a simplicial d-polytope on n vertices. Then $h_j(P) \leq h_j(\mathscr{C}_d(n))$ for all j.*

Stanley [80] extended the Upper Bound Theorem to all simplicial homology spheres, and Novik [70] extended it to all odd-dimensional simplicial manifolds and certain even-dimensional simplicial manifolds (see Section 4.2 for more details).

A complete characterization of h-vectors of simplicial polytopes was given in the celebrated g-theorem of Stanley [82] (necessity), and Billera and Lee [10] (sufficiency).

Theorem 4 (The g-theorem [10, 82]). *A vector $\mathbf{h} = (h_0, h_1, \dots, h_d) \in \mathbb{Z}^{d+1}$ is the h-vector of a simplicial d-polytope if and only if*

1. *$h_j = h_{d-j}$ for all j*
2. *$1 = h_0 \leq h_1 \leq \cdots \leq h_{\lfloor d/2 \rfloor}$, and*
3. *the numbers $(g_0, g_1, \dots, g_{\lfloor d/2 \rfloor})$, where $g_j := h_j - h_{j-1}$, form an M-sequence.*

See Section 6.1 for the definition of an M-sequence. The g-theorem shows not only that the h-numbers of a simplicial polytope are unimodal, but also that their rate of growth is bounded. We also mention that Lee's interpretation [47] of Kalai's proof of the Lower Bound Theorem implies that the portion $(g_0(\Delta), g_1(\Delta), g_2(\Delta))$ of the g-vector is an M-sequence for any connected normal pseudomanifold Δ.

The unimodality of the h-numbers of simplicial polytopes, together with the treatment of the cases of equality, is given by the Generalized Lower Bound Theorem (GLBT, for short):

Theorem 5 (The GLBT [57, 65, 82]). *Let P be a simplicial d-polytope. Then*

1. $h_0(P) \leq h_1(P) \leq \cdots \leq h_{\lfloor d/2 \rfloor}(P)$, *and*
2. $h_r(P) = h_{r-1}(P)$ *for some* $r \leq \lfloor d/2 \rfloor$ *if and only if P is* $(r-1)$*-stacked.*

A simplicial d-polytope P is called $(r-1)$**-stacked** if there exists a triangulation, Δ, of P such that $\partial\Delta = \partial P$ and $\mathrm{Skel}_{d-r}(\Delta) = \mathrm{Skel}_{d-r}(P)$. In other words, Δ extends the triangulation of the boundary of P to the interior of P without introducing any interior faces of dimension $\leq d-r$. When $r = 2$, this says that P admits a triangulation with all interior faces having dimension $d-1$ or d. This is precisely what it means for P to be stacked in the sense of the Lower Bound Theorem (Theorem 2).

4 Face enumeration for manifolds

In this section, we discuss extensions of the results from Section 3 from spheres and polytopes to arbitrary manifolds, both with and without boundary. Here we focus on presenting as many results as possible; the tools and techniques that are used will be presented later in the paper.

4.1 *Modified h-numbers and the manifold g-conjecture*

The Dehn-Sommerville equations (Theorem 1) and the Generalized Lower Bound Theorem (Theorem 5) show that the h-numbers of a simplicial polytope are symmetric, nonnegative, and unimodal. In contrast, the h-numbers of a simplicial manifold need not have any of these properties – in particular, they may be negative. For this reason, we modify the definition of the h-numbers in the following way.

Definition 1. Let Δ be a $(d-1)$-dimensional simplicial complex and $\mathbf{k}$ a field. The h''**-numbers** of Δ are defined as

$$h''_j(\Delta) = \begin{cases} h_j(\Delta) - \binom{d}{j}\sum_{i=0}^{j}(-1)^{j-i}\beta_{i-1}(\Delta;\mathbf{k}) & \text{if } 0 \leq j < d, \\ \beta_{d-1}(\Delta;\mathbf{k}) & \text{if } j = d. \end{cases}$$

Note that if Δ is a **k**-homology sphere, then $h''_j(\Delta) = h_j(\Delta)$ for all j. The following result, which generalizes the Dehn-Sommerville equations and provides evidence towards a manifold g-theorem, further indicates that the h''-numbers are the "correct" generalization of the h-numbers for manifolds.

Theorem 6. *Let Δ be a $(d-1)$-dimensional connected, orientable, **k**-homology manifold. Then*

1. ([70, Lemma 5.1]) $h''_j(\Delta) = h''_{d-j}(\Delta)$ *for all* $0 \leq j \leq d$.
2. ([74, Theorem 3.5]) $h''_j(\Delta) \geq 0$ *for all* $0 \leq j \leq d$.
3. ([73, Theorem 3.2], [87, Theorem 2.4]) *If all but (at most)* $d+1$ *of the vertex links in* Δ *have the weak Lefschetz property over* **k**, *then*

   a. $h''_0(\Delta) \leq h''_1(\Delta) \leq \cdots \leq h''_{\lfloor d/2 \rfloor}(\Delta)$ *and*

   b. *the numbers* $(g''_j(\Delta) := h''_j(\Delta) - h''_{j-1}(\Delta))_{j=0}^{\lfloor d/2 \rfloor}$ *form an M-sequence.*

See Section 6.2 for the basics on the weak Lefschetz property (WLP, for short). Specifically, Theorem 27 below asserts that the boundary of any simplicial polytope has the WLP over $\mathbb{Q}$. Thus, part (3) of Theorem 6 applies if $\mathbf{k} = \mathbb{Q}$ and most of vertex links of Δ are polytopal.

If Δ is disconnected with connected components $\Delta_1, \ldots, \Delta_s$, it is easy to show that $h''_0(\Delta) = 1$ and $h''_j(\Delta) = \sum_{i=1}^{s} h''_j(\Delta_i)$ for $1 \leq j \leq d$. Therefore parts (2) and (3) of Theorem 6 continue to hold for disconnected manifolds, and part (1) holds as long as $0 < j < d$. Further, part (2) of Theorem 6 continues to hold for non-orientable manifolds and manifolds with boundary.

The celebrated **g-conjecture for spheres** posits that the conditions of the g-theorem (Theorem 4) hold for *all* simplicial (homology) spheres. Kalai proposed a far-reaching generalization of this conjecture: he conjectured that part (3) of Theorem 6 continues to hold for *all* orientable, **k**-homology manifolds. We refer to this latter conjecture as the **manifold g-conjecture**.

Part (3) of Theorem 6 shows that if **k**-homology spheres have the WLP, then the manifold g-conjecture also holds. In general, very little is known about either version of the g-conjecture. Kubitzke and Nevo [44], building on work of Brenti and Welker [13], proved that the barycentric subdivision of a simplicial homology sphere satisfies the g-conjecture for spheres. Murai [62] extended this result to show that the barycentric subdivision of a simplicial homology manifold satisfies the manifold g-conjecture. Some other classes of spheres for which the g-conjecture is known to hold include spheres with few vertices or few edges [87, Theorem 4.1], as well as **k**-homology spheres that can be obtained from any **k**-homology sphere with the WLP by certain (in fact, most) bistellar flips [87, Theorem 3.1, Theorem 3.2]. We refer our readers to Swartz's recent survey paper [87] for the state-of-the-art on the g-conjectures.

4.2 The Upper Bound Theorem for Manifolds

In 1957, Motzkin proposed the Upper Bound Conjecture (UBC), which states that a cyclic d-polytope on n vertices has the componentwise maximal f-vector among all d-polytopes on n vertices. We use $\mathscr{C}_d(n)$ to denote a cyclic d-polytope on n vertices. Since a small perturbation of the vertices of a polytope will (1) result in a simplicial polytope and (2) only increase the number of faces, it suffices to verify Motzkin's conjecture for simplicial polytopes.

In 1964, V. Klee [42] extended the UBC, proposing that it continues to hold for all **Eulerian** simplicial complexes; i.e., those simplicial complexes Δ with the property that $\tilde{\chi}(\mathrm{lk}_\Delta(F)) = (-1)^{d-|F|-1}$ for all $F \in \Delta$ (including the empty face). Moreover, he proved the UBC for all Eulerian complexes with sufficiently many vertices.

McMullen [54] proved the UBC for polytopes in 1970, and Stanley [80] proved the UBC for all simplicial homology spheres in 1975[1]. Novik [70] extended these results further to homology manifolds that are also Eulerian simplicial complexes.

Theorem 7. *Let Δ be an orientable* **k***-homology manifold of dimension $d-1$ with n vertices. If either*

1. ([70, Theorem 1.4]) $d-1$ *is odd, or*
2. ([70, Theorem 1.5], [71, Theorem 5.3]) $d-1=2k$ *is even and*

$$\beta_k(\Delta;\mathbf{k}) \le 2\beta_{k-1} + 2\sum_{i=0}^{k-3} \beta_i(\Delta;\mathbf{k}), \tag{1}$$

then $h_j(\Delta) \le h_j(\mathscr{C}_d(n))$ for all j. In particular, $f_{i-1}(\Delta) \le f_{i-1}(\mathscr{C}_d(n))$ for all i.

Therefore, the Upper Bound Theorem (Theorem 3) continues to hold for all simplicial odd-dimensional manifolds and for any even-dimensional manifolds for which (1) holds over $\mathbb{Z}/2\mathbb{Z}$. When $\dim(\Delta) = 2k$ is even, (1) is satisfied by all **k**-homology manifolds for which either $\beta_k(\Delta;\mathbf{k}) = 0$ or $(-1)^k(\tilde{\chi}(\Delta)-1) \le 0$.

It is worth adding that the "in particular" part (i.e., the statement on f-vectors) of Theorem 7 was extended to all odd-dimensional Eulerian pseudomanifolds with isolated singularities in [29]; moreover, Novik and Swartz [75], proved the UBT for all Eulerian pseudomanifolds with so-called homologically isolated singularities as long as $n \ge 3d-4$. However, Klee's conjecture [42] that the UBC holds for all Eulerian complexes remains wide open.

[1] We further refer to Stanley's recent article [84] which gives his personal account of how he came to the proof of the Upper Bound Theorem.

4.3 Kühnel's conjectures

By Poincaré duality, the Euler characteristic of any odd-dimensional simplicial manifold is equal to zero. Motivated by the UBT, Kühnel [45] conjectured an upper bound on the Euler characteristic of even-dimensional PL manifolds. He also proved this conjecture for all such manifolds with sufficiently many vertices (see [46, p. 66]). Novik [70] verified Kühnel's conjecture for all $2k$-dimensional simplicial manifolds with $n \geq 4k+3$ or $n \leq 3k+3$ vertices, and Novik and Swartz [74] proved the conjecture in full generality:

Theorem 8 ([74, Theorem 4.4]). *Let Δ be a $2k$-dimensional orientable* **k***-homology manifold with n vertices. Then*

$$(-1)^k \binom{2k+1}{k} (\tilde{\chi}(\Delta) - 1) \leq \binom{n-k-2}{k+1}.$$

Moreover, equality holds if and only if Δ is $(k+1)$-neighborly.

A simplicial complex Δ is s**-neighborly** if every s-vertex subset of $V(\Delta)$ is a face of Δ. For instance, the cyclic polytope $\mathscr{C}_d(n)$ is $\lfloor d/2 \rfloor$-neighborly.

Kühnel also conjectured upper bounds on the individual Betti numbers; these conjectures are recorded in [49, Conjecture 18]. To this end, Murai [63] showed:

Theorem 9. *Let Δ be a $(d-1)$-dimensional* **k***-homology manifold with n vertices.*

1. [63, Theorem 5.1] *If $d = 2k+1$, then $\binom{2k+1}{k}\beta_k(\Delta; \mathbf{k}) \leq \binom{n-k-2}{k+1}$.*
2. [63, Theorem 5.2] *If all vertex links of Δ are polytopal, then*

$$\binom{d+1}{j+1}\beta_j(\Delta; \mathbb{Q}) \leq \binom{n-d+j-1}{j+1} \quad \text{for all } 0 \leq j \leq \lfloor d/2 \rfloor - 1. \tag{2}$$

The same result, but restricted to the class of orientable homology manifolds, was proved in [72].

Note that if $\beta_j(\Delta; \mathbb{Q}) \neq 0$, then equation (2) implies that

$$\binom{d+1}{j+1} \leq \binom{n-d+j-1}{j+1}, \text{ hence } d+1 \leq n-d+j-1, \text{ and so } n \geq 2d+2-j.$$

In fact, for PL-manifolds, Brehm and Kühnel [12] proved the following stronger result: if $n = 2d+1-j$, then the homotopy groups $\pi_0(\Delta), \pi_1(\Delta), \ldots, \pi_j(\Delta)$ are all trivial. Their proof relies on PL Morse theory.

4.4 The Lower Bound Theorem for manifolds

According to the Lower Bound Theorem (Theorem 2), any connected homology manifold (or even a normal pseudomanifold) Δ of dimension $d-1 \geq 2$ satisfies $g_2(\Delta) \geq 0$. In an effort to understand how the topology of $\|\Delta\|$ affects the combinatorics of Δ, Kalai [34] conjectured that for manifolds of dimension $d-1 \geq 3$ this inequality can be strengthened to $g_2(\Delta) \geq \binom{d+1}{2}\beta_1(\Delta;\mathbf{k})$.

Theorem 10 ([63, Theorem 5.3]). *Let Δ be a normal pseudomanifold of dimension $d-1 \geq 3$. Then*

1. $g_2(\Delta) \geq \binom{d+1}{2}(\beta_1(\Delta;\mathbf{k}) - \beta_0(\Delta;\mathbf{k}))$.
2. Moreover, equality holds if and only if Δ is a stacked simplicial manifold.

Note that Theorem 10 strengthens part (2) of Theorem 9 in the case that $j = 1$. Theorem 10 is also due to Murai [63]. In the special case that Δ is an orientable $\mathbf{k}$-homology manifold, part (1) of this result, along with the $d \geq 5$ case of part (2), was proved in [74, Theorem 5.2]. The case of equality for 3-dimensional homology manifolds (both orientable and non-orientable) is due to Bagchi [5].

We will discuss **stacked manifolds**, and more generally, ***r*-stacked manifolds** in Section 4.7. For now, it suffices to say that stacked manifolds form a natural generalization of **stacked spheres** – the boundary complexes of stacked polytopes. In particular, for $d \geq 5$, a $(d-1)$-dimensional simplicial manifold is stacked if and only if all its vertex links are stacked spheres.

4.5 The σ- and μ-vectors

Bagchi and Datta [5, 6], motivated by Morse theory (see [6, Remark 2.11]), introduced the notion of μ-numbers. The following definition is Murai's [63] slight modification of their original definition. Denote by Δ_W the subcomplex of Δ induced by vertices in $W \subseteq V(\Delta)$ and recall that $\beta_0(\{\emptyset\};\mathbf{k}) = 0$, while $\beta_{-1}(\{\emptyset\};\mathbf{k}) = 1$.

Definition 2. Let Δ be a $(d-1)$-dimensional simplicial complex and $\mathbf{k}$ a field. The σ- and μ-vectors of Δ, $(\sigma_{-1}, \sigma_0, \ldots, \sigma_{d-1})$ and $(\mu_0, \ldots, \mu_{d-1})$, are defined by

$$\sigma_j = \sigma_j(\Delta;\mathbf{k}) := \sum_{W \subseteq V} \frac{\beta_j(\Delta_W;\mathbf{k})}{\binom{|V|}{|W|}} \text{ and } \mu_j = \mu_j(\Delta;\mathbf{k}) := \sum_{v \in V} \frac{\sigma_{j-1}(\mathrm{lk}_\Delta(v);\mathbf{k})}{f_0(\mathrm{lk}_\Delta(v)) + 1}.$$

A few remarks are in order. First, note that the μ-numbers are rational, but not necessarily integer numbers. An important property of the μ-numbers is that they satisfy the following inequalities, just as the usual Morse numbers do:

Theorem 11 ([5]). *Let Δ be any simplicial complex. Then for all j,*

1. $\mu_j(\Delta;\mathbf{k}) \geq \beta_j(\Delta;\mathbf{k})$ *if* $j > 0$, *and* $\mu_0(\Delta;\mathbf{k}) \geq \beta_0(\Delta;\mathbf{k}) + 1$.
2. $\sum_{k=0}^{j}(-1)^{j-k}\mu_k(\Delta;\mathbf{k}) \geq \sum_{k=0}^{j}(-1)^{j-k}\beta_k(\Delta;\mathbf{k}) + (-1)^j$.

Murai [63] used Hochster's formula to interpret the μ-numbers of Δ in terms of the graded Betti numbers of the Stanley–Reisner ring of Δ (see Section 6.7 for more details on the graded Betti numbers). He then used this interpretation, along with some recent results from commutative algebra, to prove Theorems 9 and 10. In fact, his proofs show that these two theorems continue to hold even if one replaces the Betti numbers of Δ with the μ-numbers of Δ. Specifically, one obtains the following strengthening of Theorem 10.

Theorem 12. *Let Δ be a normal pseudomanifold of dimension $d-1 \geq 3$. Then* $g_2(\Delta) \geq \binom{d+1}{2}(\mu_1(\Delta;\mathbf{k}) - \mu_0(\Delta;\mathbf{k}) + 1)$.

One of Kalai's long-standing conjectures posits that for any connected homology manifold Δ of dimension $d-1$, $g_2(\Delta)/\binom{d+1}{2}$ is at least as large as the minimum number of generators of the fundamental group of Δ. Currently, the only progress on this conjecture is a result of Klee [38] stating that if Δ is balanced (see Section 5.1 for a definition), then $h_2(\Delta) \geq \binom{d}{2}m(\Delta)$, where $m(\Delta)$ is the minimum number of generators of $\pi_1(\Delta)$. Kalai's conjecture along with Theorem 12 prompted Swartz [88] to ask if $\mu_1(\Delta)$ provides a lower bound for the number of generators of $\pi_1(\Delta)$.

4.6 Stronger manifold g-conjectures

In view of Theorem 10, it is natural to define the following invariants for $r \leq \lfloor d/2 \rfloor$:

$$\tilde{g}_r(\Delta) := g_r(\Delta) - \binom{d+1}{r}\sum_{j=1}^{r}(-1)^{r-j}\beta_{j-1}(\Delta;\mathbf{k}); \quad \text{equivalently,}$$

$$\tilde{g}_r(\Delta) = h''_r(\Delta) - h''_{r-1}(\Delta) - \binom{d}{r-1}\beta_{r-1}(\Delta;\mathbf{k}).$$

In particular, $\tilde{g}_r(\Delta) \leq g''_r(\Delta)$. The following strengthening of part (3) of Theorem 6 holds. See Section 6.2 for the basics on the weak Lefschetz property.

Theorem 13. *Let Δ be an orientable $(d-1)$-dimensional, $\mathbf{k}$-homology manifold and assume that all vertex links of Δ have the WLP over $\mathbf{k}$. Then*

1. *The numbers $\tilde{g}_r(\Delta)$ are nonnegative for all $r \leq d/2$.*
2. *Moreover, $(\tilde{g}_0(\Delta), \tilde{g}_1(\Delta), \ldots, \tilde{g}_{\lfloor d/2 \rfloor}(\Delta))$ is an M-sequence.*

A straightforward computation shows that if Δ is disconnected, then for $r \neq 0$, $\tilde{g}_r(\Delta)$ equals the sum of the $\tilde{g}_r$-numbers of the connected components of Δ, while $\tilde{g}_0(\Delta) = 1$. Hence the above statement for disconnected manifolds follows easily from the connected case. In the connected case, part (1) of Theorem 13 is implicit in [72, Eq. (9)], and part (2) is due to Murai and Nevo [66, Theorem 5.4].

Bagchi [5, Conjecture 2] proposed the following strengthening of part (1) of Theorem 13 and proved it for the case of $d = 4$:

Conjecture 1. Let Δ be a $(d-1)$-dimensional $\mathbf{k}$-homology manifold. Then

$$g_r(\Delta) \geq \binom{d+1}{r} \left[(-1)^r + \sum_{j=1}^{r} (-1)^{r-j} \mu_{j-1}(\Delta; \mathbf{k}) \right] \quad \text{for all } r \leq d/2.$$

Note that Theorem 12 proves the $r = 2$ case of this conjecture even for normal pseudomanifolds. Presently, this conjecture is wide open for $r > 2$, as is a weaker version of the conjecture positing that if Δ is any $\mathbf{k}$-homology manifold, then $\tilde{g}_r(\Delta) \geq 0$ for all $r \leq d/2$.

4.7 *r-stacked manifolds and the cases of equality*

In the previous sections we discussed many inequalities on the f-numbers of manifolds. The next natural question is whether these inequalities are sharp, and if so, when equality is achieved. To provide the answers we need to let manifolds with boundary enter the scene.

Definition 3. Let $0 \leq r \leq d-1$ be an integer.

1. A d-dimensional $\mathbf{k}$-homology manifold with boundary Δ is **r-stacked** if it has no interior faces of dimension $\leq d-r-1$, that is, $\text{Skel}_{d-r-1}(\Delta) = \text{Skel}_{d-r-1}(\partial\Delta)$.
2. A $(d-1)$-dimensional $\mathbf{k}$-homology manifold without boundary is **r-stacked** if it is the boundary of an r-stacked $\mathbf{k}$-homology d-manifold with boundary.
3. A $\mathbf{k}$-homology manifold without boundary is **locally r-stacked** if all of its vertex links are r-stacked.

The 0-stacked manifolds are precisely the (boundaries of) simplices; 1-stacked manifolds are also known as stacked manifolds. The family of r-stacked manifolds was introduced in [66] as a natural generalization of the family of r-stacked polytopes (mentioned at the end of Section 3). It is easy to see that being r-stacked implies being locally r-stacked. In fact, for small r, these two notions are equivalent:

Theorem 14 ([66, Theorem 4.6]). *For $1 \leq r < d/2$, a $\mathbf{k}$-homology $(d-1)$-manifold without boundary is $(r-1)$-stacked if and only if it is locally $(r-1)$-stacked.*

Below we summarize several results on the cases of equality. The first two are due to Murai and Nevo [66], and the last one is due to Bagchi [5].

Theorem 15 ([66, Theorem 3.1]). *Let $1 \leq r \leq d$, and let Δ be a $\mathbf{k}$-homology $(d-1)$-manifold with boundary. Then $h''_r(\Delta) = 0$ if and only if Δ is $(r-1)$-stacked.*

Theorem 16 ([66, Corollary 5.8]). *Let $1 \le r < d/2$, and let Δ be an orientable* $\mathbf{k}$*-homology $(d-1)$-manifold without boundary. If all vertex links of Δ have the WLP, then $\tilde{g}_r(\Delta) = 0$ if and only if Δ is $(r-1)$-stacked.*

If $d = 2k$, does $\tilde{g}_k(\Delta) = 0$ imply that Δ is $(k-1)$-stacked?

Theorem 17 ([5, Theorem 1.11]). *Let $1 \le r \le d/2$ and let Δ be a locally $(r-1)$-stacked* $\mathbf{k}$*-homology $(d-1)$-manifold without boundary. Then*

$$g_r(\Delta) = \binom{d+1}{r}\left[(-1)^r + \sum_{j=1}^{r}(-1)^{r-j}\mu_{j-1}(\Delta;\mathbf{k})\right]. \tag{3}$$

Bagchi [5] also conjectured that the converse holds, namely, if Equation (3) is satisfied as equality, then Δ is locally $(r-1)$-stacked; he verified this conjecture for 3-dimensional homology manifolds.

Modulo the results mentioned in this section, the next natural question is whether r-stacked manifolds exist. The 1-stacked manifolds without boundary (of dimension ≥ 2) are precisely the elements of the Walkup's class introduced in [90] (see also [18]). Each such manifold is obtained by starting with several disjoint boundary complexes of the d-simplex and repeatedly forming connected sums and/or handle additions. Similarly, each 1-stacked manifold with boundary (of dimension ≥ 2) is obtained by starting with a number of d-simplices and repeatedly forming connected unions and/or handle additions; see [18] for more details. Therefore, all 1-stacked manifolds without boundary are spheres or connected sums of sphere bundles over $\mathbb{S}^1$. An analogous statement holds for 1-stacked manifolds with boundary.

A construction of r-stacked manifolds with or without boundary for all r and d is due to Klee and Novik [39]. Specifically, for each pair $0 \le r \le d-1$, they construct a certain d-manifold with boundary, $B(r, d+1)$, that is r-stacked and whose boundary triangulates $\mathbb{S}^r \times \mathbb{S}^{d-1-r}$. The construction of $B(r, d+1)$ is so simple that we sketch it here. The vertex set is $X \sqcup Y$, where $X = \{x_1, \dots, x_{d+1}\}$ and $Y = \{y_1, \dots, y_{d+1}\}$, and the facets of $B(r, d+1)$ are all $(d+1)$-subsets of $X \sqcup Y$ of the form $\{z_1, \dots, z_{d+1}\}$, where $z_i \in \{x_i, y_i\}$ and the number of indices k such that z_k and z_{k+1} were drawn from different sides does not exceed r.

As we saw above, being 1-stacked imposes severe restrictions on the topology of the underlying manifold. More generally, Swartz [87] proved the following:

Theorem 18 ([87, Section 6]). *If Δ is an r-stacked PL manifold with boundary, then Δ has a handle decomposition consisting of handles of index $\le r$.*

Swartz [87] also asked if every PL d-manifold with boundary which has a handle decomposition using only handles of index $\le r$ admits an r-stacked triangulation. While the answer is yes for $r = 0, 1, d-1$, for all other values of r this question is wide open. We refer to [9, Section 1.6] for the definition and discussion of some properties of handle decompositions.

Similarly, Swartz observed (personal communication) that under the assumptions that (1) Theorem 16 continues to hold for the middle $\tilde{g}$-number of odd-dimensional manifolds and (2) all PL spheres have the WLP, the following result would also hold: if $2 \leq r \leq d/2$ and Δ is a PL $(d-1)$-manifold without boundary with $\tilde{g}_r(\Delta) = 0$, then Δ has a surgery presentation with surgeries of index at most $r-2$.

4.8 The Dehn-Sommerville equations for manifolds with boundary

Are there analogs of Dehn-Sommerville equations for manifolds with boundary? To this end, Gräbe [26] (see also [72, Theorem 3.1]) established the following result.

Theorem 19 ([26]). *Let Δ be a* **k**-*homology $(d-1)$-manifold with boundary. Then*

$$h_{d-i}(\Delta) - h_i(\Delta) = \binom{d}{i}(-1)^{d-i-1}\tilde{\chi}(\Delta) - g_i(\partial\Delta) \quad \text{for all } 0 \leq i \leq d.$$

In particular, if Δ is a **k**-*homology ball, then $h_i(\Delta) - h_{d-i}(\Delta) = g_i(\partial\Delta)$.*

In fact, versions of the same result but in the f-vector form go back at least to Macdonald [53]. The above result, along with known inequalities on the h-numbers of manifolds with boundary (such as the fact that $h''_i \geq 0$ for all i), may lead to a "correct" version of the h''-numbers for manifolds with boundary. One such definition was proposed in [72, Definition 3.3].

4.9 The Lower Bound Theorem for manifolds with boundary

What inequalities do the f-vectors of manifolds with boundary satisfy in addition to the nonnegativity of h''-numbers? Not much is known at present. One existing result is an analog of the Lower Bound Theorem. In the following, we use $f_0^\circ(\Delta) := f_0(\Delta) - f_0(\partial\Delta)$ to denote the number of interior vertices of Δ.

Theorem 20. *Let Δ be a connected $(d-1)$-dimensional* **k**-*homology manifold with boundary.*

1. *If $d \geq 5$, then $h_2(\Delta) \geq f_0^\circ(\Delta) + \binom{d}{2}\beta_1(\partial\Delta;\mathbf{k}) + d\beta_0(\partial\Delta;\mathbf{k})$.*
2. *If $d = 4$ and* char **k** $= 2$, *then $h_2(\Delta) \geq f_0^\circ(\Delta) + 3\beta_1(\partial\Delta;\mathbf{k}) + 4\beta_0(\partial\Delta;\mathbf{k})$.*

Theorem 20 provides a strengthening of Kalai's result [34, Theorem 1.3] asserting that connected manifolds with boundary satisfy $h_2(\Delta) \geq f_0^\circ(\Delta)$. Theorem 20 with an additional assumption that Δ has an orientable boundary was proved in [72, Theorem 5.1]. However, the orientability assumption was only needed when $d \geq 5$ to invoke the inequality $g_2(\partial\Delta) \geq \binom{d}{2}\big(\beta_1(\partial\Delta;\mathbf{k}) - \beta_0(\partial\Delta;\mathbf{k})\big)$ of Theorem 10. As Theorem 10 has since been established for non-orientable manifolds of dimension ≥ 3, this additional assumption is no longer required. All 2-dimensional homology manifolds without boundary are disjoint unions of closed surfaces; hence using the field of characteristic two in part (2) of the theorem maximizes the relevant Betti numbers.

Theorem 20 is sharp (see discussion at the end of Section 5 in [72]). Thus the next question to address is the case of equality. A conjecture to this end was proposed in [72, Conjecture 5.7]. However, there are counterexamples to this conjecture that will be discussed in some future paper, and the question of when equality holds remains wide open.

Are there any lower bounds on the higher h-numbers of manifolds with boundary that are analogous to those in Theorem 20? At the moment we do not have even conjectural answers to this question.

4.10 Characterizing f-vectors of manifolds

At present we have only a *conjectural* characterization of the f-vectors of simplicial spheres of dimension five and above. As such, it would be overly optimistic to hope for a complete characterization of the f-vectors of other manifolds of higher dimensions. However, in dimensions up to four, complete characterizations are available for a few manifolds. These manifolds are the topic of this subsection.

For 2-manifolds, it follows from the Dehn-Sommerville equations that the entire f-vector is determined by the number of vertices and the Euler characteristic of a manifold in question. The f-vectors of **all** (closed) 2-manifolds were characterized by Ringel [78] for non-orientable manifolds and by Jungerman and Ringel [32] for orientable manifolds.

In dimension three, Walkup [90] characterized all possible f-vectors of $\mathbb{S}^3$, $\mathbb{R}P^3$, $\mathbb{S}^2 \times \mathbb{S}^1$, and $\mathbb{S}^2 \tilde{\times} \mathbb{S}^1$, the non-orientable $\mathbb{S}^2$-bundle over $\mathbb{S}^1$. Moreover, Lutz, Sulanke, and Swartz [51] determined possible f-vectors of 20 additional 3-manifolds, namely, $(\mathbb{S}^2 \times \mathbb{S}^1)^{\#k}$ and $(\mathbb{S}^2 \tilde{\times} \mathbb{S}^1)^{\#k}$ for $k \in \{2, 3, \dots, 8, 10, 11, 14\}$.

In dimension four, Walkup [90] characterized all possible f-vectors of $\mathbb{S}^4$, while Swartz [86] characterized all possible f-vectors of $\mathbb{C}P^2$, $(\mathbb{S}^2 \times \mathbb{S}^2)^{\#2}$, $\mathbb{S}^3 \times \mathbb{S}^1$, and $K3$-surfaces. The case of the non-orientable $\mathbb{S}^3$-bundle over $\mathbb{S}^1$ was added in [17].

By the Dehn-Sommerville equations, if M is a manifold of dimension 3 or 4 and Δ is a triangulation of M, then $g_1(\Delta)$ and $g_2(\Delta)$ determine the entire f-vector of Δ. Since, in addition, $(1, g_1, g_2)$ must form an M-sequence, it is perhaps not so surprising that all known characterizations of the f-vectors for 3- and 4-dimensional manifolds are of the form $a \leq g_1$ and $b \leq g_2 \leq \binom{g_1+1}{2}$.

What can be said about characterizing possible f-vectors of triangulations of manifolds with boundary? Here the situation seems even more hopeless: at present there is not even a conjecture for characterizing the set of f-vectors of balls of dimension six and above, see [43].

It follows from the g-conjecture for spheres and Theorem 19 that if Δ is a simplicial ball with h-vector $(h_0, h_1, \ldots, h_{d-1}, h_d)$, then $(h_0 - h_d, h_1 - h_{d-1}, \ldots, h_{\lfloor d/2 \rfloor} - h_{\lceil d/2 \rceil})$ is an M-sequence, and moreover, so is

$$(h_0-h_{d+k}, h_1-h_{d+k-1}, \ldots, h_m-h_{d+k-m}) \ \ \forall k = 0, 1, \ldots, d+1, \ m = \lfloor (d+k-1)/2 \rfloor, \tag{4}$$

where we take $h_i = 0$ if $i > d$. Billera and Lee [10] conjectured that these conditions are also sufficient. In other words, they conjectured that every nonnegative integer vector $(h_0, \ldots, h_{d-1}, h_d)$ with the property that all difference vectors as in Equation (4) are M-sequences provides the h-vector of a simplicial ball. While Lee and Schmidt [48] confirmed this conjecture in dimensions three and four, Kolins [43] disproved it in dimensions five and higher.

In fact, at present there is not even a conjecture to characterize the f-vectors of *any* manifold with boundary in dimensions six and above.

5 Manifolds with additional structure

In this section we discuss two combinatorial conditions that can be imposed on a simplicial complex and explore the ways in which these conditions affect the f- or h-vectors of the resulting complexes. While these two families of simplicial complexes seem very different on the surface, the sets of f- and h-vectors arising from them are surprisingly similar and oftentimes directly related.

5.1 Balanced complexes

Definition 4. A $(d-1)$-dimensional simplicial complex Δ is **balanced** if its 1-skeleton (viewed as a graph) admits a proper d-coloring.

Since the graph of a $(d-1)$-simplex is the complete graph on d vertices, balanced complexes are the $(d-1)$-dimensional complexes with the smallest possible chromatic number.

The family of balanced simplicial complexes was introduced by Stanley [81]. In that paper what we will call balanced complexes are called completely balanced simplicial complexes, as he allows more general colorings in which every facet could have, for example, two blue vertices, one green vertex, and three red vertices.

The canonical example of a balanced simplicial complex is the barycentric subdivision of a simplicial complex (or a regular CW complex), Δ, which is denoted as $\mathrm{Sd}(\Delta)$. The vertices of $\mathrm{Sd}(\Delta)$ correspond to the faces of Δ, and the faces of $\mathrm{Sd}(\Delta)$ correspond to chains of nonempty faces in Δ under inclusion. The barycentric subdivision of Δ is balanced by assigning color k to a vertex in $\mathrm{Sd}(\Delta)$ if its corresponding face in Δ has cardinality k.

5.1.1 Upper bounds for balanced complexes

In general the h-vector of a sphere or ball is an M-sequence, meaning there is a family of monomials, closed under divisibility, with h_j monomials of degree j for all j (see Section 6.1 for further details). The following result imposes stronger restrictions on the h''-vectors of balanced manifolds.

Theorem 21. *Let Δ be a balanced manifold, with or without boundary. Then there exists a simplicial complex Γ such that $h''(\Delta) = f(\Gamma)$.*

In particular, it follows that for a balanced manifold, there is a family of *squarefree* monomials, closed under divisibility, with h''_j monomials of degree j for all j.

Recall that if Δ is a simplicial (homology) sphere or ball, then $h''(\Delta) = h(\Delta)$. In that case, Theorem 21 is due to Stanley [81, Corollary 4.5]. For manifolds, Babson and Novik [4, Theorem 6.6] proved a weaker statement than the one in Theorem 21. However, the ideas of their proof, combined with the proof of [74, Theorem 4.6], imply Theorem 21.

In view of the above result, it is natural to extend Kalai's manifold g-conjecture for balanced manifolds in the following way.

Conjecture 2. Let Δ be a balanced, orientable **k**-homology manifold without boundary. Then $(g''_j(\Delta) := h''_j(\Delta) - h''_{j-1}(\Delta))_{j=0}^{\lfloor d/2 \rfloor}$ is the f-vector of a simplicial complex.

Murai [62, Theorem 1.3] proved this conjecture in the specific case that Δ arises as the barycentric subdivision of a homology manifold with or without boundary.

It is also natural to ask if there is a balanced analog of the cyclic polytope; i.e., a balanced simplicial sphere with the componentwise maximal f- or h-numbers among all balanced spheres with a given dimension and number of vertices.

5.1.2 Lower bounds for balanced complexes

The following result extends the Lower Bound Theorem to balanced simplicial complexes.

Theorem 22. *Let Δ be a balanced, connected normal pseudomanifold of dimension $d-1$ with $d \geq 3$. Then*

$$2h_2(\Delta) \geq (d-1)h_1(\Delta). \tag{5}$$

Moreover, when $d \geq 4$, equality holds in Equation (5) if and only if Δ is a stacked cross-polytopal sphere.

Recall that a stacked $(d-1)$-sphere can be defined by taking the connected sum of several disjoint copies of the boundary of a d-simplex. Such spheres are never balanced as the graph of a d-simplex is the complete graph on $d+1$ vertices. Instead, a stacked cross-polytopal sphere is defined by taking the connected sum of several disjoint copies of the boundary of a d-dimensional cross-polytope. The boundary of a d-dimensional cross-polytope is balanced, and so when taking connected sums, vertices of the same color are identified in order to preserve the balanced condition. We refer to this operation as the *balanced connected sum*.

The inequality (5) was proved for balanced spheres by Goff, Klee, and Novik [23, Theorem 5.3] and for balanced manifolds by Browder and Klee [14, Theorem 4.1]. Klee and Novik extended the inequality to the family of balanced normal pseudomanifolds [40, Theorem 3.4] and treated the case of equality [40, Theorem 4.1].

The following conjecture seems to be a natural extension of Theorem 10.

Conjecture 3 ([40, Conjecture 3.8]). Let Δ be a balanced normal pseudomanifold of dimension $d-1 \geq 3$. Then

$$2h_2(\Delta) - (d-1)h_1(\Delta) \geq 4\binom{d}{2}\left(\beta_1(\Delta;\mathbf{k}) - \beta_0(\Delta;\mathbf{k})\right).$$

The case of equality in this conjecture can be achieved by the complexes in the balanced analog of the Walkup class; that is, the family of complexes defined by beginning with disjoint copies of the boundary of the d-dimensional cross-polytope, then applying the operations of balanced connected sums and balanced handle additions. At the moment, we only know that if $\beta_1(\Delta;\mathbb{Q}) \neq 0$, then $2h_2(\Delta) - (d-1)h_1(\Delta) + 4\binom{d}{2}\beta_0(\Delta) \geq 4\binom{d}{2}$ (see [40, Theorem 3.9]).

The inequality (5) is equivalent to the inequality $\binom{d}{1}h_2(\Delta) \geq \binom{d}{2}h_1(\Delta)$. This leads to the following balanced version of the Generalized Lower Bound Conjecture.

Conjecture 4 (Balanced GLBC [40, Conjecture 5.6]). Let Δ be a balanced d-polytope (or more generally, a balanced $\mathbf{k}$-homology $(d-1)$-sphere). Then

$$1 = \frac{h_0(\Delta)}{\binom{d}{0}} \leq \frac{h_1(\Delta)}{\binom{d}{1}} \leq \frac{h_2(\Delta)}{\binom{d}{2}} \leq \cdots \leq \frac{h_{\lfloor d/2\rfloor}(\Delta)}{\binom{d}{\lfloor d/2\rfloor}}. \tag{6}$$

Very recently, Juhnke-Kubitzke and Murai [31] established the inequalities (6) for all balanced polytopes. Klee and Novik also proposed a treatment of the case of equality at any step of (6), but it requires additional terminology that we do not wish to introduce here. See [40, Section 5.2] for further details.

5.2 Flag complexes

Definition 5. Let Δ be a simplicial complex. A subset $S \subseteq V(\Delta)$ is a **missing face** in Δ if $S \notin \Delta$ but $T \in \Delta$ for all proper subsets $T \subset S$. We say that Δ is **flag** if all of its missing faces have cardinality 2.

A flag simplicial complex is completely determined by its graph: $F \subseteq V(\Delta)$ spans a face in Δ if and only if all pairs of vertices in F are connected by edges. For this reason, flag simplicial complexes are also called clique complexes of graphs (or dually, independence complexes of graphs). The barycentric subdivision of a simplicial complex is another canonical example of a flag complex.

The h-vector of a simplicial sphere is palindromic by the Dehn-Sommerville equations. Therefore, it is natural to expand the h-polynomial of a simplicial sphere in a symmetric polynomial basis as

$$\sum_{j=0}^{d} h_j(\Delta)t^j = \sum_{i=0}^{\lfloor d/2 \rfloor} \gamma_i(\Delta) \cdot t^i(1+t)^{d-2i}.$$

The coefficients $\gamma_0, \gamma_1, \dots, \gamma_{\lfloor d/2 \rfloor}$ are called the **γ-numbers** of Δ. The γ-numbers are weighted alternating sums of the h-numbers; in particular, they are integers.

5.2.1 Lower bounds for flag complexes

Gal [22] conjectured the following strengthening of the Generalized Lower Bound Conjecture to the class of flag homology spheres.

Conjecture 5 (Gal's Conjecture [22, Conjecture 2.1.7]). If Δ is a flag homology sphere, then $\gamma_i(\Delta) \geq 0$ for all i.

In the case that $i = \lfloor d/2 \rfloor$, this conjecture is originally due to Charney and Davis [16]. In the language of h-vectors, their conjecture states that if Δ is a flag homology sphere of dimension $2k-1$, then

$$(-1)^k \sum_{j=0}^{2k} (-1)^j h_j(\Delta) \geq 0.$$

In the even-dimensional case, the sum on the left-hand side of the above inequality always equals zero by the Dehn-Sommerville equations.

Presently, Gal's conjecture is known to hold for all flag $\mathbb{Q}$-homology 3-spheres [19] and also for barycentric subdivisions of regular CW spheres [37].

If Gal's conjecture is true, then the inequalities of Equation (6) (the balanced GLBC inequality) continue to hold for flag homology spheres as follows. Proving the balanced GLBC is equivalent to verifying that $j \cdot h_j(\Delta) \geq (d-j+1) \cdot h_{j-1}(\Delta)$ for all $j \leq \lfloor d/2 \rfloor$. And, indeed,

$$\begin{aligned} j \cdot h_j(\Delta) &= \sum_{i=0}^{j} \gamma_i(\Delta) \cdot j \cdot \binom{d-2i}{j-i} \geq \sum_{i=0}^{j-1} \gamma_i(\Delta) \cdot j \cdot \binom{d-2i}{j-i} \qquad (\text{ since } \gamma_j \geq 0) \\ &\geq \sum_{i=0}^{j-1} \gamma_i(\Delta) \cdot (d-j+1) \binom{d-2i}{j-1-i} = (d-j+1) \cdot h_{j-1}(\Delta). \end{aligned}$$

The final inequality uses the fact that $j\binom{d-2i}{j-i} \geq (d-j+1)\binom{d-2i}{j-i-1}$ for $j \leq \lfloor d/2 \rfloor$.

Thus the following conjecture of Nevo (personal communication) may be viewed as a relaxation of Gal's conjecture.

Conjecture 6. The cone generated by the set of h-vectors of balanced d-polytopes (or balanced homology $(d-1)$-spheres) contains the cone generated by the set of h-vectors of flag d-polytopes (or flag homology $(d-1)$-spheres).

At present, this conjecture is completely open, even in the polytopal case.

5.2.2 Attempting to characterize f-vectors of flag complexes

While Gal's conjecture can be viewed as a flag analog of the GLBC, the following increasingly strong conjectures of Nevo and Petersen [67] can be viewed as flag analogs of the necessity part of the g-conjecture.

Conjecture 7. Let Δ be a flag homology sphere. Then there exists a simplicial complex Γ such that

1. [67, Conjecture 1.4] $\gamma(\Delta) = f(\Gamma)$;
2. [67, Conjecture 6.3] Γ is balanced and $\gamma(\Delta) = f(\Gamma)$;
3. [67, Problem 6.4] Γ is flag and $\gamma(\Delta) = f(\Gamma)$.

It follows from [22] that up to dimension four, the second and the third part of Conjecture 7 hold and are equivalent. The reason that the third part of this conjecture is, in general, stronger than the second is due to a result of Frohmader [21], which states that for every flag simplicial complex, there exists a balanced simplicial complex with the same f-vector.

Nevo, Petersen, and Tenner [68] proved that the version of this conjecture asking that Γ should be balanced holds when Δ is the barycentric subdivision of a homology sphere; subsequently, Murai and Nevo [64] extended this result to barycentric subdivisions of *all* (i.e., not only simplicial) polytopes. Furthermore, Nevo and Petersen verified that the version of this conjecture asking that Γ should be flag holds for Coxeter complexes and for flag $(d-1)$-spheres with at most $2d+3$ vertices (i.e., when $\gamma_1 \leq 3$).

5.2.3 Upper bounds on flag complexes

If the second part of Conjecture 7 is true, then standard facts about Turán graphs would imply upper bounds on the γ-numbers of flag spheres. The following conjecture was proposed by Lutz and Nevo [50, Conjecture 6.3].

Conjecture 8. Let $d = 2k$ be even and suppose $d \geq 4$. Let Δ be a flag homology $(d-1)$-sphere on n vertices. Then

$$\gamma_i(\Delta) \leq \gamma_i(\mathcal{J}_k(n)),$$

where $\mathscr{J}_k(n)$ is the flag sphere on n vertices obtained as the join of k (graph) cycles, each of which has $\lfloor \frac{n}{k} \rfloor$ or $\lceil \frac{n}{k} \rceil$ vertices. Moreover, equality holds for some $2 \leq i \leq k$ if and only if $\Delta = \mathscr{J}_k(n)$.

Very recently, Adamaszek and Hladký announced a proof of this conjecture for complexes with a sufficient number of vertices. For $n \gg 0$, they proved the following result:

Theorem 23 ([1]). *For every even $d = 2k$ with $d \geq 4$, there is a constant N_0 for which the following holds: if M is a flag $\mathbb{Z}$-homology manifold of dimension $d - 1$ with $n \geq N_0$ vertices, then*

- $f_i(M) \leq f_i(\mathscr{J}_k(n))$ *for all* $1 \leq i \leq d-1$;
- $h_i(M) \leq h_i(\mathscr{J}_k(n))$ *for all* $2 \leq i \leq d-2$;
- $g_i(M) \leq g_i(\mathscr{J}_k(n))$ *for all* $2 \leq i \leq k$; *and*
- $\gamma_i(M) \leq \gamma_i(\mathscr{J}_k(n))$ *for all* $2 \leq i \leq k$.

Moreover, if equality holds in any of these cases, then M is isomorphic to $\mathscr{J}_k$

6 Tools and techniques

In this section we will introduce some of the key tools used in the proofs of the results discussed in Sections 3–5.

6.1 *The Stanley–Reisner ring*

Throughout this section, we assume that Δ is a $(d-1)$-dimensional simplicial complex with n vertices and $\mathbf{k}$ is an infinite field of an arbitrary characteristic. We identify the vertices of Δ with $[n] = \{1, 2, \ldots, n\}$. The standard reference to this material is Stanley's book [83].

We consider a polynomial ring $S := \mathbf{k}[x_1, x_2, \ldots, x_n]$ with one generator for each vertex in Δ. The **Stanley–Reisner ideal** of Δ is

$$I_\Delta := (x_{i_1} x_{i_2} \cdots x_{i_k} \ : \ \{i_1, i_2, \ldots, i_k\} \notin \Delta) \, .$$

The **Stanley–Reisner ring** (or face ring) of Δ is the quotient $\mathbf{k}[\Delta] := S/I_\Delta$. Since I_Δ is a monomial ideal, the quotient ring $\mathbf{k}[\Delta]$ is graded by degree. The definition of I_Δ ensures that, as a $\mathbf{k}$-vector space, each graded piece of $\mathbf{k}[\Delta]$, denoted $\mathbf{k}[\Delta]_i$, has a natural basis of monomials whose supports correspond to faces of Δ. The following result shows that the h-numbers arise naturally in this algebraic setting.

Theorem 24 ([80, Proposition 3.2]). *The Poincaré-Hilbert series of* $\mathbf{k}[\Delta]$ *can be expressed as a rational function of the form*

$$F(\mathbf{k}[\Delta];t) := \sum_{i\geq 0} \dim_{\mathbf{k}}(\mathbf{k}[\Delta])_i \cdot t^i$$

$$= \sum_{i=0}^{d} \frac{f_{i-1}\cdot t^i}{(1-t)^i} = \frac{h_0(\Delta)+h_1(\Delta)\cdot t+\cdots+h_d(\Delta)\cdot t^d}{(1-t)^d}. \tag{7}$$

As a consequence, the Krull dimension of $\mathbf{k}[\Delta]$ *is* $d = \dim(\Delta)+1$.

Definition 6. A sequence of linear forms, $\theta_1, \theta_2, \ldots, \theta_d \in S$ is called a **linear system of parameters** (or l.s.o.p.) if the ring $\mathbf{k}[\Delta]/(\Theta)$ is a finite-dimensional $\mathbf{k}$-vector space; here $(\Theta) := (\theta_1, \ldots, \theta_d)$.

It is a general fact that if $\mathbf{k}$ is an infinite field, then $\mathbf{k}[\Delta]$ admits an l.s.o.p.; moreover, any choice of generic linear forms $\theta_1, \ldots, \theta_d \in S$ is an l.s.o.p. If $\theta_1, \ldots, \theta_d$ is an l.s.o.p. for Δ, we write $\mathbf{k}(\Delta;\Theta) = \mathbf{k}(\Delta) := \mathbf{k}[\Delta]/(\Theta)$.

Stanley [80] used a result of Reisner [77] to build a first bridge between the algebraic structure of the Stanley–Reisner ring and the combinatorial structure of sphere triangulations.

Theorem 25 ([80, Section 4]). *Let* Δ *be a* $\mathbf{k}$*-homology sphere or* $\mathbf{k}$*-homology ball, and let* $\theta_1, \ldots, \theta_d$ *be any l.s.o.p. for* $\mathbf{k}[\Delta]$*. Then the Poincaré-Hilbert series of* $\mathbf{k}(\Delta)$ *is*

$$F(\mathbf{k}(\Delta);t) = \sum_{j=0}^{d} h_j(\Delta)t^j \tag{8}$$

In other words, $\dim_{\mathbf{k}} \mathbf{k}(\Delta)_j = h_j(\Delta)$ *for all* j.

The key connection between Equations (7) and (8) is that, for homology spheres and homology balls, the face ring $\mathbf{k}[\Delta]$ is free as a $\mathbf{k}[\theta_1, \ldots, \theta_d]$-module. (Simplicial complexes with this property are called **Cohen-Macaulay** complexes.) Equation (8) then follows from the fact that the Poincaré-Hilbert series of $\mathbf{k}[\theta_1, \ldots, \theta_d]$ is $\frac{1}{(1-t)^d}$.

One immediate consequence of Theorem 25 is that the h-numbers of a simplicial homology sphere or ball are nonnegative. In fact, a classical theorem of Macaulay [52] can be used to get much stronger inequalities from Theorem 25. First, we require two definitions.

A **multicomplex**, $\mathcal{M}$, is a family of monomials that is closed under divisibility; i.e., if $\mu \in \mathcal{M}$ and ν divides μ, then $\nu \in \mathcal{M}$. We write $\mathcal{M}_i := \{\mu \in \mathcal{M} : \deg(\mu) = i\}$.

Next, it is easy to verify that for all natural numbers m and i, there exists a unique decomposition of the form

$$m = \binom{a_i}{i} + \binom{a_{i-1}}{i-1} + \cdots + \binom{a_j}{j}, \qquad \text{where } a_i > a_{i-1} > \cdots > a_j \geq j > 0.$$

Define

$$m^{\langle i\rangle} := \binom{a_i+1}{i+1} + \binom{a_{i-1}+1}{i} + \cdots + \binom{a_j+1}{j+1}.$$

A sequence of integers $(1, F_1, F_2, \ldots)$ is an **M-sequence** if $0 \leq F_{i+1} \leq F_i^{\langle i\rangle}$ for all i.

Theorem 26 (Macaulay [52]). *The following conditions are equivalent.*

1. *The sequence $(1, F_1, F_2, \ldots)$ of nonnegative integers is an M-sequence.*
2. *There exists a multicomplex $\mathcal{M}$ such that $|\mathcal{M}_i| = F_i$ for all i.*
3. *There exists a standard graded $\mathbf{k}$-algebra R such that $\dim_{\mathbf{k}} R_i = F_i$ for all i.*

Corollary 1 ([80, Corollary 5.2]). *If Δ is a homology sphere or homology ball, then $h(\Delta)$ is an M-sequence.*

6.2 Lefschetz properties and the g-theorem

When Δ is a homology sphere, Theorem 25 and the Dehn-Sommerville equations imply that

$$\dim_{\mathbf{k}} \mathbf{k}(\Delta)_j = h_j(\Delta) = h_{d-j}(\Delta) = \dim_{\mathbf{k}} \mathbf{k}(\Delta)_{d-j}.$$

The inner equality is a statement that is both combinatorial and topological in nature, while the outer equalities are algebraic. As such, it is natural to ask whether the equality $\dim_{\mathbf{k}} \mathbf{k}(\Delta)_j = \dim_{\mathbf{k}} \mathbf{k}(\Delta)_{d-j}$ can be seen as a sort of algebraic duality.

Definition 7. 1. Let Δ be a homology $(d-1)$-sphere. We say that $\mathbf{k}[\Delta]$ (or simply Δ) has the **strong Lefschetz property** (SLP) if there exists an l.s.o.p. $\theta_1, \ldots, \theta_d$ for $\mathbf{k}[\Delta]$ and a linear form ω such that the multiplication map

$$\times\omega^{d-2j} : \mathbf{k}(\Delta;\Theta)_j \to \mathbf{k}(\Delta;\Theta)_{d-j}$$

is an isomorphism for all $j \leq \lfloor d/2 \rfloor$.

2. Let Δ be a homology $(d-1)$-sphere or ball. We say that $\mathbf{k}[\Delta]$ has the **weak Lefschetz property** (WLP) if there exists an l.s.o.p. $\theta_1, \ldots, \theta_d$ for $\mathbf{k}[\Delta]$ and a linear form ω such that the multiplication map $\times\omega : \mathbf{k}(\Delta;\Theta)_i \to \mathbf{k}(\Delta;\Theta)_{i+1}$ is either injective or surjective for all $i < d$.

A few comments are in order. First, it is well-known that if $\mathbf{k}[\Delta]$ has the SLP, then the desired multiplication map is an isomorphism for any generic choice of $\theta_1, \ldots, \theta_d$ and ω (see [85, Proposition 3.6]). Second, it follows easily from the definition that if $\mathbf{k}[\Delta]$ has the SLP, then it also has the WLP. Third, if Δ is a homology sphere or ball with the WLP, then $h(\Delta)$ is unimodal. Indeed, suppose the multiplication map $\times\omega : \mathbf{k}(\Delta;\Theta)_s \to \mathbf{k}(\Delta;\Theta)_{s+1}$ is surjective for some s.

Then $(\mathbf{k}(\Delta;\Theta)/(\omega))_{s+1} = 0$, so $(\mathbf{k}(\Delta;\Theta)/(\omega))_i = 0$ for all $i \geq s+1$, and hence $\times\omega : \mathbf{k}(\Delta;\Theta)_{s+1} \to \mathbf{k}(\Delta;\Theta)_{s+2}$ is also surjective. In particular, if Δ is a homology sphere with the WLP, then the Dehn-Sommerville equations imply that $h(\Delta)$ is unimodal with its peak at $h_{\lfloor d/2 \rfloor}(\Delta)$.

In fact, the necessity of the conditions of the g-theorem for simplicial polytopes is an easy consequence of the (strong or weak) Lefschetz property: the injectivity of the map $\times\omega : \mathbf{k}(P)_i \to \mathbf{k}(P)_{i+1}$ for all $i \leq \lfloor d/2 \rfloor$ implies that the quotient ring $\mathbf{k}(P)/(\omega)$ is a standard graded $\mathbf{k}$-algebra whose i-th graded piece has dimension $g_i(P) = h_i(P) - h_{i-1}(P)$ for all $i \leq \lfloor d/2 \rfloor$. Thus the g-vector of P is an M-sequence thanks to the following theorem.

Theorem 27 (Stanley [82], McMullen [55, 56]). *A simplicial d-polytope has the strong Lefschetz property over* $\mathbb{Q}$.

More generally, to prove the g-conjecture for homology spheres, it would be sufficient to verify that any homology sphere Δ has the WLP. In fact, it is even sufficient to prove that there exists a one-form ω such that $\times\omega : \mathbf{k}(\Delta)_{\lfloor d/2 \rfloor} \to \mathbf{k}(\Delta)_{\lfloor d/2 \rfloor + 1}$ is surjective (see [73, Theorem 3.2]).

6.3 *From spheres to manifolds: think global, act local*

The main philosophy of many papers on the f-numbers of manifolds and pseudomanifolds is that even though such a space can be very complicated globally, its local structure controls the f-numbers (and other invariants) to a large degree. In this section, we demonstrate several examples of this principle.

One such example is given by the μ-numbers of Section 4.5. These numbers are defined as sums of certain invariants of the vertex links, yet they provide some information on the Betti numbers of the original complex (see Theorem 11).

In a similar spirit, if Δ is a pure $(d-1)$-dimensional simplicial complex, then the **short simplicial h-vector** of Δ, $\hat{h}(\Delta) = (\hat{h}_0(\Delta), \hat{h}_1(\Delta), \ldots, \hat{h}_{d-1}(\Delta))$, is defined by $\hat{h}_i(\Delta) := \sum_{v \in V} h_i(\mathrm{lk}_\Delta(v))$ for $i = 0, 1, \ldots, d-1$. This vector was introduced in [29]. It has a few convenient and easy-to-check properties:

Proposition 1. *The f-numbers of a simplicial complex Δ are nonnegative linear combinations of $\hat{h}$-numbers of Δ. Moreover, if Δ is a homology $(d-1)$-manifold, then $\hat{h}_i(\Delta) = \hat{h}_{d-1-i}(\Delta)$ for all $i = 0, 1, \ldots, d-1$.*

One immediate consequence of these two simple facts, together with Stanley's Upper Bound Theorem for spheres (the h-vector version), is the proof of the f-vector version of the Upper Bound Theorem for all odd-dimensional manifolds (see [29] for a stronger result).

Yet another example of this phenomenon is the following deep observation of Swartz [86, Theorem 4.26] on the existence of Lefschetz-like elements. We denote by $L^i_s(\Delta)$ the set of all pairs (ω, Θ) such that ω is a linear form, Θ is an l.s.o.p., and the multiplication map $\times\omega : \mathbf{k}(\Delta;\Theta)_i \to \mathbf{k}(\Delta;\Theta)_{i+1}$ is a surjection.

Theorem 28. *Let Δ be a* **k***-homology manifold with* **k** *an infinite field. If $L_s^i(\mathrm{lk}_\Delta(v))$ is nonempty for (almost) all of the vertices v of Δ, then $L_s^{i+1}(\Delta)$ is nonempty.*

6.4 Schenzel's formula and socles

Many results on the face numbers of manifolds bound the f-numbers or h-numbers of a simplicial manifold in terms of its Betti numbers. The proofs of most of these results rely on the following theorem of Schenzel [79], which relates the face ring of a simplicial manifold to its topological Betti numbers.

Theorem 29 ([79]). *Let Δ be a* **k***-homology manifold (with or without boundary) of dimension $d-1$, and let $\theta_1, \dots, \theta_d$ be an l.s.o.p. for $\mathbf{k}[\Delta]$. Then*

$$\dim_{\mathbf{k}} \mathbf{k}(\Delta)_j = h_j(\Delta) + \binom{d}{j} \sum_{i=1}^{j-1} (-1)^{j-i-1} \beta_{i-1}(\Delta; \mathbf{k}). \tag{9}$$

In light of this result, we define the h'-numbers of Δ as

$$h'_j(\Delta) := h_j(\Delta) + \binom{d}{j} \sum_{i=1}^{j-1} (-1)^{j-i-1} \beta_{i-1}(\Delta; \mathbf{k}).$$

Note that if Δ is a **k**-homology sphere or ball, then $h'(\Delta) = h(\Delta)$. In particular, Theorem 29 provides a vast generalization of Theorem 25. Also since $\mathbf{k}(\Delta)$ is a standard graded algebra, Schenzel's formula, together with Macaulay's Theorem (Theorem 26) implies that if Δ is a **k**-homology manifold, then $h'(\Delta) := (h'_0(\Delta), \dots, h'_d(\Delta))$ is an M-sequence (cf. Corollary 1). In order to derive Equation (9), Schenzel [79] proved the following result.

Theorem 30 ([79]). *Let Δ be a* **k***-homology manifold (with or without boundary) of dimension $d-1$, and let $\theta_1, \dots, \theta_d$ be an l.s.o.p. for $\mathbf{k}[\Delta]$. For all $1 \le s \le d$ and $0 \le j \le d$, the kernel of the multiplication map*

$$\times\theta_s : (\mathbf{k}[\Delta]/(\theta_1, \dots, \theta_{s-1}))_j \to (\mathbf{k}[\Delta]/(\theta_1, \dots, \theta_{s-1}))_{j+1}$$

has dimension $\binom{s-1}{j}\beta_{j-1}(\Delta; \mathbf{k})$.

Recall that if R is a commutative ring and I, J are ideals of R, then the *ideal quotient* $I : J$ is defined as $I : J = \{x \in R \ : \ xJ \subseteq I\}$.

Definition 8. The **socle** of $\mathbf{k}(\Delta)$ is the ideal

$$\mathrm{Soc}(\mathbf{k}(\Delta)) = 0 : \mathfrak{m} = \{f \in \mathbf{k}(\Delta) \ : \ x_i \cdot f = 0 \text{ for all } i\}.$$

Here we use $\mathfrak{m}$ to denote the maximal ideal $(x_1, \dots, x_n)$.

Assume Δ is a **k**-homology manifold and ω is a generic linear form. In view of Theorem 30, one natural question to ask is what can be said about the dimension of the kernel of $\times\omega : \mathbf{k}(\Delta)_j \to \mathbf{k}(\Delta)_{j+1}$ or at least about the dimension of the socle of $\mathbf{k}(\Delta)$. To this end, we have:

Theorem 31 ([74, Theorem 2.2]). *Let Δ be a **k**-homology manifold, with or without boundary. Then*

$$\dim_{\mathbf{k}} \mathrm{Soc}(\mathbf{k}(\Delta))_j \geq \binom{d}{j}\beta_{j-1}(\Delta;\mathbf{k}).$$

This theorem is due to Novik and Swartz; its proof relies upon the local cohomology module of $\mathbf{k}[\Delta]$ and the following result of Hochster, which describes the structure of $H^i_{\mathfrak{m}}(\mathbf{k}[\Delta])$ as a **k**-vector space in terms of the simplicial (co)homology of Δ and its links.

Theorem 32. *Let Δ be a simplicial complex on n vertices. Let $\mathbf{a} = (a_1, \ldots, a_n) \in \mathbb{Z}^n$ and let $F = \{j \in [n] : a_j \neq 0\}$. Then for any $i \geq 0$, the ith local cohomology module of $\mathbf{k}[\Delta]$ satisfies*

$$\dim_{\mathbf{k}} H^i_{\mathfrak{m}}(\mathbf{k}[\Delta])_{\mathbf{a}} = \begin{cases} 0 & \text{if } F \notin \Delta \text{ or } a_j > 0 \text{ for some } j, \\ \beta_{i-|F|-1}(\mathrm{lk}_\Delta(F);\mathbf{k}) & \text{otherwise.} \end{cases}$$

Let Δ be a **k**-homology manifold (with or without boundary). As we noted earlier, the vector $h'(\Delta)$ is then an M-sequence. One easy but important consequence of Theorem 31 is the following strengthening of this fact:

Proposition 2. *If Δ is a **k**-homology manifold (with or without boundary), then for all $j \leq d-1$, the numbers*

$$\left(h'_0(\Delta), h'_1(\Delta), \ldots, h'_{j-1}(\Delta), h'_j(\Delta) - \binom{d}{j}\beta_{j-1}(\Delta;\mathbf{k}), h'_{j+1}(\Delta)\right)$$

form an M-sequence.

This proposition, in turn, leads to the proof of several theorems discussed in Section 4, among them (i) the Upper Bound Theorem for various classes of homology manifolds [70] (for this result even a certain weaker version of Proposition 2 is enough), and (ii) Kühnel's conjecture on the Euler characteristic of even-dimensional **k**-homology manifolds (see [74, Theorem 4.4] for further details). Theorem 31 along with Theorem 28 also provides a key ingredient of the proof of part (1) of Theorem 10 for orientable **k**-homology manifolds (see [74, Theorem 5.2]).

It is worth mentioning that Theorem 31 can be stated more generally for Buchsbaum graded modules of Krull dimension d over $\mathbf{k}[x_1, \ldots, x_n]$ ([74, Theorem 2.2]). For Buchsbaum modules over local Noetherian rings, a similar result was established by Goto [24, Proposition 3.6].

6.5 *The Gorenstein property*

A graded $\mathbf{k}$-algebra of Krull dimension zero is called *Gorenstein* if its socle is a 1-dimensional $\mathbf{k}$-vector space (see [83, p. 50] for many other equivalent definitions). If Δ is a $\mathbf{k}$-homology sphere, then $\mathbf{k}(\Delta)$ is a Gorenstein ring [83, Theorem II.5.1]. One major consequence of being Gorenstein is the symmetry of the Hilbert function. This naturally leads to the question of how far is $\mathbf{k}(\Delta)$ from being Gorenstein if Δ is a homology manifold other than a sphere. The answer due to Novik and Swartz turns out to be surprisingly simple. We require one definition and a bit of discussion.

Definition 9. Let Δ be a $\mathbf{k}$-homology $(d-1)$-manifold without boundary. Define

$$\overline{\mathbf{k}(\Delta)} := \mathbf{k}(\Delta) / \bigoplus_{i=1}^{d-1} \mathrm{Soc}(\mathbf{k}(\Delta))_i.$$

In the above formula, $\mathrm{Soc}(\mathbf{k}(\Delta))_d$ is not included in the direct sum because that graded piece of the socle is all of $\mathbf{k}(\Delta)_d$. Further, by the definition of the socle, each individual graded piece $\mathrm{Soc}(\mathbf{k}(\Delta))_i$ is an ideal of $\mathbf{k}(\Delta)$ on its own, and so $\overline{\mathbf{k}(\Delta)}$ is a quotient of $\mathbf{k}(\Delta)$ by an ideal.

Recall also the definition of the h''-numbers (see Definition 1). Observe that Schenzel's formula (Theorem 29) and the socle theorem (Theorem 31) imply that for a $\mathbf{k}$-homology $(d-1)$-manifold Δ, $\dim_{\mathbf{k}} \overline{\mathbf{k}(\Delta)}_i \leq h''_i(\Delta)$.

Theorem 33. *Let Δ be a $(d-1)$-dimensional, orientable, $\mathbf{k}$-homology manifold without boundary. Then*

1. [73, Theorem 1.3] $\dim_{\mathbf{k}} \mathrm{Soc}(\mathbf{k}(\Delta))_i = \binom{d}{i}\beta_{i-1}(\Delta)$ *for all $i \leq d$. In particular,* $\dim_{\mathbf{k}} \overline{\mathbf{k}(\Delta)}_i = h''_i(\Delta)$.
2. [73, Theorem 1.4] *Moreover, if Δ is connected, then $\overline{\mathbf{k}(\Delta)}$ is Gorenstein.*

The major ingredient in the proof of Theorem 33 is Gräbe's generalization [25] of Hochster's theorem (Theorem 32) that provides an explicit description of the structure of the local cohomology module $H^i_{\mathfrak{m}}(\mathbf{k}[\Delta])$ as a $\mathbf{k}[\Delta]$-module in terms of the simplicial (co)homology of the links of faces of Δ and maps between them.

In turn, Theorem 33 is one of the main ingredients in the proof of Theorem 6. Indeed, along with some standard properties of Gorenstein rings (see, for instance, [83, Theorems I.12.5 and I.12.10]), Theorem 33 implies that if Δ is a $(d-1)$-dimensional connected, orientable, $\mathbf{k}$-homology manifold, then

$$\mathrm{Hom}_{\mathbf{k}}(\overline{\mathbf{k}(\Delta)}_i, \mathbf{k}) \text{ is naturally isomorphic to } \overline{\mathbf{k}(\Delta)}_{d-i}. \tag{10}$$

In particular, it follows that $h''_i(\Delta) = h''_{d-i}(\Delta)$. A much more substantial consequence of (10) along with Theorem 28 is the injectivity of the map $\times\omega : \overline{\mathbf{k}(\Delta)}_i \to \overline{\mathbf{k}(\Delta)}_{i+1}$ for any $i < \lfloor d/2 \rfloor$ and a generic linear form ω, under an additional assumption that (most of) vertices of Δ have the WLP (see proof of [73, Theorem 3.2]). This injectivity, in turn, yields part (3) of Theorem 6. In fact, Theorems 28 and 33 are also among the key ingredients in the proof of Theorem 13, which gives a strengthening of Theorem 6.

6.6 Generic initial ideals

Generic initial ideals can be traced at least to Hartshorne [28]. They entered the study of the face numbers of simplicial complexes via Kalai's theory of algebraic shifting [11, 33, 36]. Since then they have become an indispensable tool, see, for instance, [3, 60, 61, 70] and many other papers. Perhaps the most spectacular recent application of generic initial ideals is due to Murai and Nevo [65] in their proof of the equality part of the GLBT for simplicial polytopes.

Below we sketch the definition of generic initial ideals and discuss some of their properties. We order the variables of $S = \mathbf{k}[x_1, \ldots, x_n]$ by $x_1 > x_2 > \cdots > x_n$, and let $\succ$ be the rev-lex order on the monomials in $x_1, \ldots, x_n$ of the same degree. For instance, in degree 2 we have: $x_1^2 \succ x_1x_2 \succ x_2^2 \succ x_1x_3 \succ x_2x_3 \succ x_3^2 \succ \cdots$. Furthermore, we refine this partial order by declaring that the lower degree monomials are larger than the higher degree monomials.

For a polynomial A in S, denote by $\mathrm{in}_\succ(A)$ the rev-lex largest monomial that appears in A with a non-zero coefficient. If I is an ideal, then the **rev-lex initial ideal of** I, $\mathrm{in}_\succ(I)$, is defined as $\mathrm{Span}_\mathbf{k}\{\mathrm{in}_\succ(A) : A \in I\}$. It is not hard to see that $\mathrm{in}_\succ(I)$ is in fact an ideal.

Let $\phi = (\phi_{ij}) \in \mathrm{GL}_n(\mathbf{k})$ be a generic matrix. Then ϕ acts on S_1 by $\phi(x_i) = \sum_{k=1}^{n} \phi_{ki}x_k$ and this action can be extended uniquely to a ring automorphism on S. If I is a homogeneous ideal, then the **(rev-lex) generic initial ideal of** I is

$$\mathrm{Gin}_\succ(I) := \mathrm{in}_\succ(\phi(I)).$$

It is a result of Galligo (see, for instance, [27, Theorem 1.27]) that $\mathrm{in}_\succ(\phi(I))$ is independent of ϕ for generic choices of ϕ, and so $\mathrm{Gin}_\succ(I)$ is well-defined.

One beautiful property of generic initial ideals is that they have a very simple structure. For instance, if $\mathbf{k}$ is a field of characteristic zero, I is a homogeneous ideal in S, and $\mu \in \mathrm{Gin}_\succ(I)$ is a monomial, then for any x_t that divides μ and any $1 \le s < t$, the monomial $\mu \cdot x_s/x_t$ is also an element of $\mathrm{Gin}_\succ(I)$. (This result is also due to Galligo [30, Theorem 4.2.1].) In fact, the main reason that generic initial ideals are so useful is that while the degeneration of I to $\mathrm{Gin}_\succ(I)$ considerably simplifies the ideal, it preserves many important properties:

Theorem 34. *Let I be a homogeneous ideal in S.*

1. *The ring S/I is Cohen-Macaulay if and only if $S/\operatorname{Gin}_{\succ}(I)$ is.*
2. *The ring S/I has the WLP if and only if $S/\operatorname{Gin}_{\succ}(I)$ has the WLP.*
3. (Crystallization Principle) *Suppose that $\mathbf{k}$ is a field of characteristic zero and that I is generated by elements of degree $\leq m$. If $\operatorname{Gin}_{\succ}(I)$ has no minimal generators of degree $m+1$, then $\operatorname{Gin}_{\succ}(I)$ is generated by elements of degree $\leq m$.*

The first part is due to Bayer and Stillman, see [30, Corollary 4.3.18]. (In fact, a remarkable extension of this part, due to Bayer, Charalambous, and S. Popescu, asserts that even the values of certain graded Betti numbers are preserved under generic initial ideals; see [8] and [2] for more details.) The second part follows from [30, Lemma 4.3.7] and relies on the properties of the rev-lex order. The third part is due to Green [27, Proposition 2.28]. This part played a crucial role in the Murai–Nevo proof [65] of the equality part of the GLBT.

6.7 Graded Betti numbers of the Stanley–Reisner ring

Let $S = \mathbf{k}[x_1, \ldots, x_n]$ be a polynomial ring. It is a well-known fact that if $I \subseteq S$ is a homogeneous ideal, then there is a unique graded minimal free resolution of S/I (viewed as an S-module) of the following form:

$$0 \to \bigoplus_j S(-j)^{\beta_{\ell,j}} \xrightarrow{\varphi_\ell} \cdots \xrightarrow{\varphi_2} \bigoplus_j S(-j)^{\beta_{1,j}} \xrightarrow{\varphi_1} \bigoplus_j S(-j)^{\beta_{0,j}} \xrightarrow{\varphi_0} S \to S/I \to 0,$$

where $\ell \leq n$. The notation $S(-j)$ denotes a shift in the degree of S: $(S(-j))_k = S_{k-j}$ for all k. The resolution being *graded* means that all maps $\varphi_0, \varphi_1, \ldots, \varphi_\ell$ preserve the degrees of elements of S (after the grading on S is shifted).

The numbers $\beta_{i,j}$ are called the **graded Betti numbers** of S/I. When I is the Stanley–Reisner ideal of a simplicial complex, the following result of Hochster establishes a beautiful relationship between the graded (algebraic) Betti numbers of the minimal free resolution of S/I and the (topological) Betti numbers of the underlying simplicial complex.

Theorem 35 (Hochster's formula). *Let Δ be a simplicial complex. Then*

$$\beta_{i,i+j}(\mathbf{k}[\Delta]) = \sum_{\substack{W \subseteq V(\Delta) \\ |W|=i+j}} \beta_{j-1}(\Delta_W; \mathbf{k}).$$

Kolins [43, Proposition 8] used the graded Betti numbers to place sufficient conditions on whether a homology ball can be split as the boundary connected sum of two other homology balls along a codimension-one face. This allowed him to show that several integer vectors could not be realized as h-vectors of homology balls, despite satisfying very natural algebraic conditions [43, Example 10].

More recently, Murai [63] made a tremendous connection between Hochster's formula and the Bagchi-Datta σ-numbers by noticing that if Δ has n vertices, then

$$\sigma_{j-1}(\Delta;\mathbf{k}) = \sum_{k=j}^{n} \frac{\beta_{k-j,k}(\mathbf{k}[\Delta])}{\binom{n}{k}}.$$

Murai then used several results from commutative algebra to provide upper bounds on the graded Betti numbers of $\mathbf{k}[\Delta]$ and hence also the σ-numbers and μ-numbers. This, along with the results of [58], led to Murai's proof of Theorems 9 and 10.

We are looking forward to many future results on the face numbers of simplicial complexes and new connections between commutative algebra, algebraic topology, and combinatorics!

Acknowledgements We are grateful to Bhaskar Bagchi, Basudeb Datta, Satoshi Murai, Eran Nevo, and Ed Swartz for many helpful conversations and comments on the previous version of this paper.

Novik's research is partially supported by NSF grant DMS-1361423.

References

1. M. Adamaszek, J. Hladký, *Upper bound theorem for odd-dimensional flag manifolds* (2015), http://arxiv.org/abs/1503.05961
2. A. Aramova, J. Herzog, Almost regular sequences and Betti numbers. Am. J. Math. **122**(4), 689–719 (2000). MR 1771569 (2001i:13029)
3. E. Babson, E. Nevo, Lefschetz properties and basic constructions on simplicial spheres. J. Algebraic Comb. **31**(1), 111–129 (2010). MR 2577480 (2011a:55021)
4. E. Babson, I, Novik, Face numbers and nongeneric initial ideals. Electron. J. Comb. **11**(2), Research Paper 25, 23 (2004/06). MR 2195431 (2007c:05202)
5. B. Bagchi, The mu vector, Morse inequalities and a generalized lower bound theorem for locally tame combinatorial manifolds. Eur. J. Comb. **51**, 69–83 (2016). http://dx.doi.org/10.1016/j.ejc.2015.04.007
6. B. Bagchi, B. Datta, On stellated spheres and a tightness criterion for combinatorial manifolds. Eur. J. Comb. **36**, 294–313 (2014). MR 3131896
7. D. Barnette, A proof of the lower bound conjecture for convex polytopes. Pac. J. Math. **46**, 349–354 (1973). MR 0328773 (48 #7115)
8. D. Bayer, H. Charalambous, S. Popescu, Extremal Betti numbers and applications to monomial ideals. J. Algebra **221**(2), 497–512 (1999). MR 1726711 (2001a:13020)
9. B. Benedetti, Smoothing discrete Morse theory. Ann. Sc. Norm. Super. Pisa Cl. Sci. (2014, to appear). http://arxiv.org/pdf/1212.0885v4.pdf
10. L.J. Billera, C.W. Lee, A proof of the sufficiency of McMullen's conditions for f-vectors of simplicial convex polytopes. J. Comb. Theory A **31**(3), 237–255 (1981). MR 635368 (82m:52006)
11. A. Björner, G. Kalai, An extended Euler-Poincaré theorem. Acta Math. **161**(3–4), 279–303 (1988). MR 971798 (89m:52009)
12. U. Brehm, W. Kühnel, Combinatorial manifolds with few vertices. Topology **26**(4), 465–473 (1987). MR 919730 (89c:57022)

13. F. Brenti, V. Welker. f-vectors of barycentric subdivisions. Math. Z. **259**(4), 849–865 (2008). MR 2403744 (2009k:52027)
14. J. Browder, S. Klee, Lower bounds for Buchsbaum complexes. Eur. J. Comb. **32**(1), 146–153 (2011). MR 2727469 (2012a:05350)
15. J.W. Cannon, Shrinking cell-like decompositions of manifolds. Codimension three. Ann. Math. (2) **110**(1), 83–112 (1979) MR 541330 (80j:57013)
16. R. Charney, M. Davis, The Euler characteristic of a nonpositively curved, piecewise Euclidean manifold. Pac. J. Math. **171**(1), 117–137 (1995). MR 1362980 (96k:53066)
17. J. Chestnut, J. Sapir, E. Swartz, Enumerative properties of triangulations of spherical bundles over S^1. Eur. J. Comb. **29**(3), 662–671 (2008). MR 2397347 (2009b:52048)
18. B. Datta, S. Murai, On stacked triangulated manifolds (2014), http://arxiv.org/pdf/1407.6767.pdf
19. M.W. Davis, B. Okun, Vanishing theorems and conjectures for the ℓ^2-homology of right-angled Coxeter groups. Geom. Topol. **5**, 7–74 (2001). MR 1812434 (2002e:58039)
20. A.L. Fogelsanger, The generic rigidity of minimal cycles, Ph.D. Thesis, Cornell University, ProQuest LLC, Ann Arbor, 1988. MR 2636638
21. A. Frohmader, Face vectors of flag complexes. Isr. J. Math. **164**, 153–164 (2008). MR 2391144 (2009a:05150)
22. Ś.R. Gal, Real root conjecture fails for five- and higher-dimensional spheres. Discret. Comput. Geom. **34**(2), 269–284 (2005). MR 2155722 (2006c:52019)
23. M. Goff, S. Klee, I. Novik, Balanced complexes and complexes without large missing faces. Ark. Mat. **49**(2), 335–350 (2011). MR 2826947 (2012k:55026)
24. S. Goto, On the associated graded rings of parameter ideals in Buchsbaum rings. J. Algebra **85**(2), 490–534 (1983). MR 725097 (85d:13032)
25. H.-G. Gräbe, The canonical module of a Stanley-Reisner ring. J. Algebra **86**(1), 272–281 (1984). MR 727379 (85a:13010)
26. H.-G. Gräbe, Generalized Dehn-Sommerville equations and an upper bound theorem. Beitr. Algebra Geom. **25**, 47–60 (1987). MR 899384 (88k:52008)
27. M.L. Green, Generic initial ideals. *Six Lectures on Commutative Algebra (Bellaterra, 1996)*. Progress in Mathematics, vol. 166 (Birkhäuser, Basel, 1998), pp. 119–186. MR 1648665 (99m:13040)
28. R. Hartshorne, Connectedness of the Hilbert scheme. Inst. Hautes Études Sci. Publ. Math. **29**, 5–48 (1966). MR 0213368 (35 #4232)
29. P. Hersh, I. Novik, A short simplicial h-vector and the upper bound theorem. Discret. Comput. Geom. **28**(3), 283–289 (2002). MR 1923232 (2003f:52007)
30. J. Herzog, T. Hibi, *Monomial Ideals*. Graduate Texts in Mathematics, vol. 260 (Springer, London, 2011). MR 2724673 (2011k:13019)
31. M. Juhnke-Kubitzke, S. Murai, Balanced generalized lower bound inequality for simplicial polytopes (2015), http://arxiv.org/abs/1503.06430
32. M. Jungerman, G. Ringel, Minimal triangulations on orientable surfaces. Acta Math. **145**(1–2), 121–154 (1980). MR 586595 (82b:57012)
33. G. Kalai, Characterization of f-vectors of families of convex sets in $\mathbf{R}^d$. Part I: Necessity of Eckhoff's conditions. Isr. J. Math. **48**(2–3), 175–195 (1984). MR 770700 (86c:52003)
34. G. Kalai, Rigidity and the lower bound theorem. I. Invent. Math. **88**(1), 125–151 (1987). MR 877009 (88b:52014)
35. G. Kalai, Many triangulated spheres. Discret. Comput. Geom. **3**(1), 1–14 (1988). MR 918176 (89b:52025)
36. G. Kalai, Algebraic shifting. *Computational Commutative Algebra and Combinatorics (Osaka, 1999)*. Advanced Studies in Pure Mathematics, vol. 33 (Mathematical Society of Japan, Tokyo, 2002), pp. 121–163. MR 1890098 (2003e:52024)
37. K. Karu, *The cd-index of fans and posets*. Compos. Math. **142**(3), 701–718 (2006). MR 2231198 (2007e:14083)

38. S. Klee, The fundamental group of balanced simplicial complexes and posets. Electron. J. Comb. **16**(2), 12 (2009) [Special volume in honor of Anders Björner, Research Paper 7.]. MR 2515770 (2011a:05380)
39. S. Klee, I. Novik, Centrally symmetric manifolds with few vertices. Adv. Math. **229**(1), 487–500 (2012). MR 2854181 (2012j:57047)
40. S. Klee, I. Novik, Lower bound theorems and a generalized lower bound conjecture for balanced simplicial complexes. Mathematika (2014, to appear). http://arxiv.org/abs/1409.5094
41. V. Klee, A combinatorial analogue of Poincaré's duality theorem. Can. J. Math. **16**, 517–531 (1964). MR 0189039 (32 #6466)
42. V. Klee, On the number of vertices of a convex polytope. Can. J. Math. **16**, 701–720 (1964). MR 0166682 (29 #3955)
43. S. Kolins, f-vectors of triangulated balls. Discret. Comput. Geom. **46**(3), 427–446 (2011). MR 2826961 (2012h:52028)
44. M. Kubitzke, E. Nevo, The Lefschetz property for barycentric subdivisions of shellable complexes. Trans. Am. Math. Soc. **361**(11), 6151–6163 (2009). MR 2529927 (2010f:13029)
45. W. Kühnel, Triangulations of manifolds with few vertices. *Advances in Differential Geometry and Topology* (World Science Publisher, Teaneck, 1990), pp. 59–114. MR 1095532 (92k:57043)
46. W. Kühnel, *Tight Polyhedral Submanifolds and Tight Triangulations*. Lecture Notes in Mathematics, vol. 1612 (Springer, Berlin, 1995). MR 1439748 (98b:52018)
47. C.W. Lee, Generalized stress and motions. *Polytopes: Abstract, Convex and Computational (Scarborough, ON, 1993)*. NATO Advanced Science Institutes Series C: Mathematical and Physical Sciences, vol. 440 (Kluwer Academic Publishers, Dordrecht, 1994), pp. 249–271. MR 1322066 (96c:52039)
48. C.W. Lee, L. Schmidt, On the numbers of faces of low-dimensional regular triangulations and shellable balls. Rocky Mt. J. Math. **41**(6), 1939–1961 (2011). MR 2854745 (2012j:52028)
49. F.H. Lutz, Triangulated manifolds with few vertices: combinatorial manifolds (2005), http://arxiv.org/pdf/math/0506372v1.pdf
50. F.H. Lutz, E, Nevo, Stellar theory for flag complexes. Math. Scand. (2014, to appear). http://arxiv.org/abs/1302.5197
51. F.H. Lutz, T. Sulanke, E. Swartz, f-vectors of 3-manifolds. Electron. J. Comb. **16**(2), 33 (2009) [Special volume in honor of Anders Björner, Research Paper 13.]. MR 2515776(2010k:57052)
52. F.S. Macaulay, Some properties of enumeration in the theory of modular systems. Proc. Lond. Math. Soc. **S2-26**(1), 531 (1926). MR 1576950
53. I.G. Macdonald, Polynomials associated with finite cell-complexes. J. Lond. Math. Soc. (2) **4**, 181–192 (1971). MR 0298542 (45 #7594)
54. P. McMullen, The maximum numbers of faces of a convex polytope. Mathematika **17**, 179–184 (1970). MR 0283691 (44 #921)
55. P. McMullen, On simple polytopes. Invent. Math. **113**(2), 419–444 (1993). MR 1228132 (94d:52015)
56. P. McMullen, Weights on polytopes. Discret. Comput. Geom. **15**(4), 363–388 (1996) MR 1384882 (97b:52012)
57. P. McMullen, D.W. Walkup, A generalized lower-bound conjecture for simplicial polytopes. Mathematika **18**, 264–273 (1971). MR 0298557 (45 #7609)
58. J. Migliore, U. Nagel, Reduced arithmetically Gorenstein schemes and simplicial polytopes with maximal Betti numbers. Adv. Math. **180**(1), 1–63 (2003). MR 2019214 (2004k:14082)
59. J.R. Munkres, Topological results in combinatorics. Mich. Math. J. **31**(1), 113–128 (1984). MR 736476 (85k:13022)
60. S. Murai, Generic initial ideals and squeezed spheres. Adv. Math. **214**(2), 701–729 (2007). MR 2349717 (2008h:13037)
61. S. Murai, Algebraic shifting of strongly edge decomposable spheres, J. Comb. Theory A **117**(1), 1–16 (2010). MR 2557878 (2011b:05290)

62. S. Murai, On face vectors of barycentric subdivisions of manifolds. SIAM J. Discret. Math. **24**(3), 1019–1037 (2010). MR 2680231 (2012b:52022)
63. S. Murai, Tight combinatorial manifolds and graded Betti numbers. Collect. Math. **66**(3), 367–386 (2015). http://link.springer.com/article/10.1007 [MR 3384014]
64. S. Murai, E. Nevo, On the cd-index and γ-vector of S*-shellable CW-spheres. Math. Z. **271**(3–4), 1309–1319 (2012). MR 2945609
65. S. Murai, E. Nevo, On the generalized lower bound conjecture for polytopes and spheres. Acta Math. **210**(1), 185–202 (2013). MR 3037614
66. S. Murai, E. Nevo, On r-stacked triangulated manifolds. J. Algebraic Comb. **39**(2), 373–388 (2014). MR 3159255
67. E. Nevo, T.K. Petersen, On γ-vectors satisfying the Kruskal-Katona inequalities. Discret. Comput. Geom. **45**(3), 503–521 (2011). MR 2770549 (2012d:52026)
68. E. Nevo, T.K. Petersen, B.E. Tenner, The γ-vector of a barycentric subdivision. J. Comb. Theory A **118**(4), 1364–1380 (2011). MR 2755087 (2012h:52041)
69. E. Nevo, F. Santos, S. Wilson, Many triangulated odd-spheres. Math. Annalen, to appear (2014), http://arxiv.org/pdf/1408.3501.pdf
70. I. Novik, Upper bound theorems for homology manifolds. Isr. J. Math. **108**, 45–82 (1998). MR 1669325 (2000e:52013)
71. I. Novik, On face numbers of manifolds with symmetry. Adv. Math. **192**(1), 183–208 (2005). MR 2122283 (2006a:52009)
72. I. Novik, E. Swartz, Applications of Klee's Dehn-Sommerville relations. Discret. Comput. Geom. **42**(2), 261–276 (2009). MR 2519879 (2010h:52033)
73. I. Novik, E. Swartz, Gorenstein rings through face rings of manifolds. Compos. Math. **145**(4), 993–1000 (2009). MR 2521251 (2011a:13049)
74. I. Novik, E. Swartz, Socles of Buchsbaum modules, complexes and posets. Adv. Math. **222**(6), 2059–2084 (2009). MR 2562774 (2010j:13041)
75. I. Novik, E. Swartz, Face numbers of pseudomanifolds with isolated singularities. Math. Scand. **110**(2), 198–222 (2012). MR 2943717
76. J. Pfeifle, G.M. Ziegler, Many triangulated 3-spheres. Math. Ann. **330**(4), 829–837 (2004). MR 2102314 (2005j:52017)
77. G.A. Reisner, Cohen-Macaulay quotients of polynomial rings. Adv. Math. **21**(1), 30–49 (1976). MR 0407036 (53 #10819)
78. G. Ringel, Wie man die geschlossenen nichtorientierbaren Flächen in möglichst wenig Dreiecke zerlegen kann. Math. Ann. **130**, 317–326 (1955). MR 0075591 (17,774b)
79. P. Schenzel, On the number of faces of simplicial complexes and the purity of Frobenius. Math. Z. **178**(1), 125–142 (1981) MR 627099 (82k:13022)
80. R.P. Stanley, The upper bound conjecture and Cohen-Macaulay rings. Stud. Appl. Math. **54**(2), 135–142 (1975). MR 0458437 (56 #16640)
81. R.P. Stanley, Balanced Cohen-Macaulay complexes. Trans. Am. Math. Soc. **249**(1), 139–157 (1979). MR 526314 (81c:05012)
82. R.P. Stanley, The number of faces of a simplicial convex polytope. Adv. Math. **35**(3), 236–238 (1980). MR 563925 (81f:52014)
83. R.P. Stanley, *Combinatorics and Commutative Algebra*, 2nd edn. Progress in Mathematics, vol. 41 (Birkhäuser, Boston, 1996). MR 1453579 (98h:05001)
84. R.P. Stanley, How the upper bound conjecture was proved. Ann. Comb. **18**(3), 533–539 (2014). MR 3245897
85. E. Swartz, g-elements, finite buildings and higher Cohen-Macaulay connectivity. J. Comb. Theory A **113**(7), 1305–1320 (2006). MR 2259062 (2007i:05192)
86. E. Swartz, Face enumeration—from spheres to manifolds. J. Eur. Math. Soc. (JEMS) **11**(3), 449–485 (2009). MR 2505437 (2010m:05346)
87. E. Swartz, Thirty-five years and counting (2014), http://arxiv.org/pdf/1411.0987.pdf
88. E. Swartz, What's next?. Oberwolfach Workshop on Geometric and Algebraic Combinatorics (2015), Oberwolfach Report available at http://www.mfo.de/occasion/1506

89. T.-S. Tay, Lower-bound theorems for pseudomanifolds. Discret. Comput. Geom. **13**(2), 203–216 (1995). MR 1314963 (96d:52009)
90. D.W. Walkup, The lower bound conjecture for 3- and 4-manifolds. Acta Math. **125**, 75–107 (1970). MR 0275281 (43 #1038)
91. S. Murai, I. Novik, Face numbers of manifolds with boundary (2015), http://arxiv.org/abs/1509.05115
92. H. Zheng, The flag upper bound theorem for 3- and 5-manifolds (2015), http://arxiv.org/pdf/1512.06958.pdf

Note added in proof

While this Chapter awaited its publication, Murai and Novik [91] established suitable analogs of Theorem 12 and part 1 of Theorem 10 for normal pseudomanifolds with boundary. These results, in particular, led to a strengthening of Theorem 20. Murai and Novik also proved that part 1 of Theorem 13 continues to hold without the orientability assumption and established a suitable extension of this result for homology manifolds with boundary all of whose vertex links have the WLP. Furthermore, Zheng [92] proved that Conjecture 8 holds for all flag 3-manifolds.

Unimodality problems in Ehrhart theory

Benjamin Braun

Abstract Ehrhart theory is the study of sequences recording the number of integer points in non-negative integral dilates of rational polytopes. For a given lattice polytope, this sequence is encoded in a finite vector called the Ehrhart h^*-vector. Ehrhart h^*-vectors have connections to many areas of mathematics, including commutative algebra and enumerative combinatorics. In this survey we discuss what is known about unimodality for Ehrhart h^*-vectors and highlight open questions and problems.

Keywords Ehrhart theory • Lattice polytopes • Unimodality • Log concavity

Mathematics Subject Classification (2010): 52B20, 05A20

1 Introduction

For a sequence $a_0, a_1, a_2, \ldots, a_n$ of real numbers, we define the sequence to be:

- *unimodal* if there is a k such that $a_i \leq a_{i+1}$ when $i \leq k-1$ and $a_i \geq a_{i+1}$ for $i \geq k$.
- *log-concave* if $a_j^2 \geq a_{j-1}a_{j+1}$ for all $1 \leq j < n$.
- *real-rooted* if the generating polynomial $a_0 + a_1x + a_2x^2 + \cdots + a_nx^n$ has all real zeros. (We define constant polynomials to be real-rooted.)

It is well known that real-rootedness implies log-concavity, and that if all the a_i are positive then log-concavity implies unimodality. Further, these implications are strict as shown by the sequences $(1, 4, 6, 4, 1)$, $(1, 1, 1, 1, 1)$, and $(1, 1, 2, 1, 1)$.

Unimodal sequences occur frequently in mathematics, and a wide range of tools from algebra, analysis, probability, and combinatorics are available for establishing unimodality. One reason combinatorialists are interested in unimodality results is that their proofs often point to interesting and unexpected properties of associated

B. Braun (✉)
University of Kentucky, Lexington, KY 40506–0027, USA
e-mail: benjamin.braun@uky.edu

A. Beveridge et al. (eds.), *Recent Trends in Combinatorics*, The IMA Volumes in Mathematics and its Applications 159, DOI 10.1007/978-3-319-24298-9_27

combinatorial, geometric, and algebraic objects. There are three important survey articles regarding these properties in combinatorics, due to Stanley [65], Brenti [19], and Brändén [15]. While the surveys by Stanley and Brenti were written over twenty years ago, they remain relevant and central to the study of unimodal sequences in combinatorics. The recent survey by Brändén is intended to complement the other two works; it covers a range of important newer techniques and developments.

One source of interesting sequences of non-negative integers is Ehrhart theory, the study of enumerating integer points in integer dilates of polytopes whose vertices lie in $\mathbb{Q}^n$. Our focus in this survey is on sequences arising as Ehrhart h^*-vectors for *lattice polytopes*, also known as *integer polytopes*, which are polytopes whose vertices lie in $\mathbb{Z}^n$. While there are many interesting classes of integer polytopes giving rise to unimodal h^*-vectors in Ehrhart theory, fascinating questions and conjectures remain open.

Because of the close connection that exists between lattice polytopes and Cohen-Macaulay semigroup algebras, many of the techniques that have been used in attempts to establish unimodality of h^*-vectors are algebraic in nature. The three surveys mentioned above contain a wealth of additional techniques for establishing unimodality, log-concavity, and real-rootedness. Our hope is that this survey will contribute to increased interest in unimodality problems in Ehrhart theory, and inspire the application of a broader range of tools to these problems.

2 Lattice Polytopes and Ehrhart Theory

In this section we review some well-known background regarding Ehrhart theory for lattice polytopes, including a collection of motivating examples for the study of unimodality. We will frame this discussion in the more general context of the study of integer points in rational cones, of which Ehrhart theory is a special case.

2.1 Integer-Point Transforms and Ehrhart Series

Further discussion of the ideas in this section, including proofs of stated results, can be found in [7, 23, 35, 54, 67].

The study of integer points in rational polyhedra is ubiquitous in mathematics, which is not surprising given that this topic is fundamentally about the tension between arithmetic properties of the lattice $\mathbb{Z}^n$ and geometric convexity in $\mathbb{R}^n$. For any subset $K \subset \mathbb{R}^n$, we define the *integer-point transform* of K to be the formal power series

$$\sigma_K(z) = \sigma_K(z_1, \ldots, z_n) := \sum_{m \in K \cap \mathbb{Z}^n} z^m,$$

where we use the shorthand $z^m := z_1^{m_1} \cdots z_n^{m_n}$. Given a pointed rational cone

$$C = \{x \in \mathbb{R}^n : Ax \geq 0\},$$

where *pointed* requires that C does not contain any non-trivial linear subspace of $\mathbb{R}^n$ and *rational* requires that A have integral (or rational) entries, it is well known that $\sigma_C(z)$ is equal to a rational function in the field of fractions $\mathbb{C}((z_1, z_2, \ldots, z_n))$.

We will assume for the rest of this survey that all cones C are pointed and rational. The rationality of $\sigma_C(z)$ for pointed rational cones is a powerful tool in enumerative combinatorics, and the structure underlying many well-known rational generating function identities. We will soon see that even relatively simple geometric objects like cubes and simplices can have complicated arithmetic properties.

Our key example of a rational cone is the cone over a rational polytope, defined as follows. Let $P \subset \mathbb{R}^n$ be a *rational* polytope, that is, P is the convex hull $P = \text{conv}(v^1, \ldots, v^k)$ of a collection of vectors $v^1, \ldots, v^k \in \mathbb{Q}^n$. (We will use upper indices to index vectors, and lower indices to index their components.) Assume that P has dimension d, though often we will be interested in situations where $d = n$. The *cone over P* is

$$\text{cone}(P) := \text{span}_{\mathbb{R}_{\geq 0}}\{(1, p) : p \in P\} \subset \mathbb{R} \times \mathbb{R}^n,$$

where we consider the new variable to be indexed at 0, i.e. $x \in \mathbb{R} \times \mathbb{R}^n = \mathbb{R}^{1+n}$ is written $x = (x_0, x_1, \ldots, x_n)$. We define the *Ehrhart series* of P to be the generating function

$$E_P(z) := \sigma_{\text{cone}(P)}(z, 1, 1, \ldots, 1) = 1 + \sum_{t \in \mathbb{Z}_{\geq 1}} |tP \cap \mathbb{Z}^n| z^t,$$

where tP is the notation for the t-th dilation of P. That the middle- and right-hand terms displayed above are equal follows from the observation that the points in cone(P) with 0-th coordinate t form a copy of tP.

We will frequently encounter polytopes P that are *lattice polytopes*, also known as *integer polytopes* (we will use both terms freely), which are defined by the condition that $P = \text{conv}(v^1, \ldots, v^k)$ for some $v^1, \ldots, v^k \in \mathbb{Z}^n$. For lattice polytopes, we may combine two well-known theorems due to Ehrhart [29] and Stanley [64] and conclude that there exist values $h_0^*, \ldots, h_d^* \in \mathbb{Z}_{\geq 0}$ with $h_0^* = 1$ such that

$$E_P(z) = \frac{\sum_{j=0}^d h_j^* z^j}{(1-z)^{d+1}}.$$

We say the polynomial $h_P^*(z) := \sum_{j=0}^d h_j^* z^j$ is the h^**-polynomial* of P and the vector of coefficients is called the h^**-vector* of P, denoted h_P^*. That $E_P(z)$ is of this rational form is equivalent to $|tP \cap \mathbb{Z}^n|$ being a polynomial function of t of degree d; the non-negativity of the h^*-vector is an even stronger property. Some authors denote h_i^* by δ_i, and often the h^*-polynomial is referred to as the δ-polynomial of P [35]. We follow the convention of using h^* as introduced by Stanley [66] to emphasize the connections between Ehrhart h^*-polynomials and h-polynomials for simplicial polytopes.

When studying integer-point transforms of cones, one often considers the following properties. An n-dimensional cone is *simplicial* if it is generated by n linearly independent vectors; an n-dimensional polytope is a *simplex* if it is the convex hull of $n+1$ vertices. A pointed rational simplicial cone C is *unimodular* if the minimal length lattice points on the rays generating C form a parallelepiped of volume ± 1 in the lattice obtained by intersecting the affine span of C with $\mathbb{Z}^n$; when C is an n-dimensional cone in $\mathbb{R}^n$, this lattice is $\mathbb{Z}^n$ itself. A lattice simplex P is *unimodular* if cone(P) is unimodular. A *triangulation* of an n-dimensional cone C (resp. a d-dimensional polytope P) is a finite collection T of simplicial n-cones (resp. d-simplices) such that C (resp. P) is the union of the elements of T and for any $\Delta_1, \Delta_2 \in T$, $\Delta_1 \cap \Delta_2$ is a face common to both Δ_1 and Δ_2. The triangulation is *unimodular* if every cone (resp. simplex) in the triangulation is unimodular.

Due to their role in triangulations and the following lemma, simplicial cones are fundamental objects in the study of integer-point transforms.

Lemma 1. *Let $C = \{\sum_{j=0}^{n} \lambda_j v^j : \lambda_j \in \mathbb{R}_{\geq 0}\}$ be a simplicial cone in $\mathbb{R}^{1+n}$ with linearly independent generators $v^0, v^1, \dots, v^n \in \mathbb{Z}^{1+n}$. Then*

$$\sigma_C(z_0, z_1, \dots, z_n) = \frac{\sigma_\Pi(z_0, z_1, \dots, z_n)}{\prod_{j=0}^{n}\left(1 - z^{v^j}\right)}$$

where $\Pi := \{\sum_{j=0}^{n} \lambda_j v^j : \lambda_j \in [0,1)\}$. Furthermore, the number of integer points in Π (and thus the number of monomials in $\sigma_\Pi(z_0, z_1, \dots, z_n)$) is given by the determinant of the matrix with column vectors $v^0, v^1, \dots, v^n$.

We refer to Π as the *fundamental parallelepiped* of C; note that it depends on the choice of generators of C. When $C = \text{cone}(P)$ for a lattice simplex P and Lemma 1 is specialized to the case of the Ehrhart series $E_P(z)$, one obtains the following geometric interpretation of the h^*-vector of the simplex P.

Lemma 2. *For any lattice simplex $P \subseteq \mathbb{R}^n$, with Π the fundamental parallelepiped of* cone(P), *the h^*-vector of P satisfies*

$$h_i(P) = \#\{m \in \Pi \cap \mathbb{Z}^{1+n} : m_0 = i\}.$$

The h^*-vector of a lattice polytope P is a fascinating partial invariant; obtaining a general understanding of h^*-vectors of lattice polytopes and their geometric/combinatorial implications is currently of great interest. A large amount of recent research [4, 6, 13, 15, 31, 36, 55, 58] has focused on the class of reflexive lattice polytopes. Recall that the *dual polytope* of a polytope P containing the origin in its interior is

$$P^{\Delta} := \{x \in \mathbb{R}^n : x \cdot p \leq 1 \text{ for all } p \in P\}.$$

An n-dimensional lattice polytope $P \subset \mathbb{R}^n$ is *reflexive* if P^{Δ} is also a lattice polytope. For example, crosspolytopes and ± 1-cubes (defined below) form dual pairs of reflexives. Reflexive polytopes were named by Batyrev [4] in the context of mirror symmetry and theoretical physics, and independently discovered in the context of Ehrhart theory by Hibi [36], who proved the following theorem.

Theorem 1 (Hibi, [36]). *A lattice polytope $P \subset \mathbb{R}^n$ is reflexive (up to unimodular equivalence) if and only if P is n-dimensional, P contains an interior lattice point, and h_P^* satisfies $h_i^* = h_{n-i}^*$.*

It is known from work of Lagarias and Ziegler [47] that there are only finitely many reflexive polytopes (up to unimodular equivalence) in each dimension, with one reflexive in dimension one, 16 in dimension two, 4,319 in dimension three, and 473,800,776 in dimension four according to computations by Kreuzer and Skarke [46]. The number of five-and-higher-dimensional reflexives is unknown. In addition to Ehrhart theory, reflexive polytopes play important roles in combinatorial mirror symmetry, toric geometry, and the theory of error-correcting codes and random walks.

Reflexive polytopes are a special case of the more general family of Gorenstein polytopes. A lattice polytope $P \subset \mathbb{R}^n$ is *Gorenstein of index k* if kP is a reflexive polytope, up to translation by an element of $\mathbb{Z}^n$. More generally, a pointed rational cone C is *Gorenstein* if there exists an integer point c such that

$$c + (\mathbb{Z}^{1+n} \cap C) = \mathbb{Z}^{1+n} \cap \operatorname{int}(C)$$

where $\operatorname{int}(C)$ denotes the interior of the cone C. This definition of a Gorenstein cone is motivated by results of Stanley and Danilov for Gorenstein semigroup algebras, see [23, Section 6.3] for a textbook exposition. It can be shown that P is Gorenstein if and only if cone(P) is Gorenstein. Gorenstein polytopes may also be identified using the following extension of Hibi's theorem, which is a consequence of the Gorenstein property for semigroup algebras.

Theorem 2 (Stanley, [63]). *A d-dimensional lattice polytope P with*

$$h_P^* = (h_1^*, \ldots, h_s^*, 0, 0, \ldots, 0) \in \mathbb{Z}^d,$$

where $h_s^* \neq 0$, is Gorenstein if and only if $h_i^* = h_{s-i}^*$ for all i.

2.2 Examples

We now consider a range of examples of h^*-vectors along with polytopes that realize them. Throughout this subsection we will denote the i-th standard basis vector in $\mathbb{R}^n$ by e_i. We will begin with examples of polytopes that have well-behaved h^*-vectors.

Example 1. The h^*-polynomial $h_P^*(z) = 1$ is unique to simplices that are unimodularly equivalent to $\Delta_n := \mathrm{conv}(0, e_1, \ldots, e_n)$, the *standard unimodular simplex*. One can easily show that $E_{\Delta_n}(z) = \dfrac{1}{(1-z)^{n+1}}$ using Lemma 1.

Example 2. The polynomial $h_P^*(z) = 1 + z + z^2 + \cdots + z^d$ is the h^*-polynomial of the reflexive simplex of minimal volume in $\mathbb{R}^d$, defined by

$$S_d := \mathrm{conv}\{e_1, e_2, \ldots, e_d, -\sum_{i=1}^{d} e_i\}.$$

This computation is again a straightforward exercise using Lemma 1. Note that the finite geometric series is unimodal but is not real-rooted.

Example 3. The polynomial $(1+z)^d$ is the h^*-polynomial for many different lattice polytopes. Two examples are the *crosspolytopes*

$$C_d := \mathrm{conv}\{e_1, \ldots, e_d, -e_1, \ldots, -e_d\}$$

and the family of reflexive simplices given by

$$T_d := \mathrm{conv}\{-\sum_{i=1}^{d} e_i, e_1 - \sum_{i=2}^{d} e_i, e_2 - \sum_{i=3}^{d} e_i, \cdots, e_{d-1} - e_d, e_d\}.$$

The h^*-polynomial for C_d can be computed in many different ways, see [7, Chapter 2] for one example. The derivation of $h_{T_d}^*(z)$ can be done using either Lemma 1 or by recognizing T_d as an example of an affine free sum of intervals of length 2 and applying techniques from [9].

It is immediate that the polynomial $(1+z)^d$ is real-rooted, hence the binomial coefficients are both log-concave and unimodal. Many additional proofs of unimodality for this polynomial are known. Observe that the symmetry of the binomial coefficients implies that both C_d and T_d are reflexive; the dual polytope to C_d is the cube $[-1,1]^d$.

Example 4. Let S_n denote the symmetric group on n objects. For an element $\pi \in S_n$, the *descent set* of π is $\mathrm{Des}(\pi) := \{i \in [n-1] : \pi(i) > \pi(i+1)\}$ and the *descent statistic* for π is $\mathrm{des}(\pi) := |\mathrm{Des}(\pi)|$. The *n-th Eulerian polynomial* is $A_n(z) := \sum_{\pi \in S_n} z^{\mathrm{des}(\pi)}$. The coefficients of the Eulerian polynomials are called the Eulerian numbers. Eulerian polynomials are known to have symmetric coefficients and to be real-rooted. This family of polynomials has inspired many subsequent developments regarding real-rootedness, log-concavity, and unimodality in enumerative combinatorics.

The *unit cube* $[0, 1]^n$ satisfies $h^*_{[0,1]^n}(z) = A_{n-1}(z)$. While there are many proofs of this, a geometric proof involving triangulations can be found in [5]. Another family of polytopes with Eulerian polynomials as h^*-polynomials are the simplices

$$L_d := \operatorname{conv}\left\{\begin{matrix} e_d, (d-1)e_{d-1} + de_d, \\ (d-2)e_{d-2} + (d-1)e_{d-1} + de_d, \ldots, \\ 2e_2 + \cdots + de_d, e_1 + 2e_2 + \cdots + de_d \end{matrix}\right\}.$$

That L_d is a $(d-1)$-dimensional simplex in $\mathbb{R}^d$ satisfying $h^*_{L_d}(z) = A_{d-1}(z)$ is implicit in work of Corteel, Lee, and Savage [26]. Observe that the symmetry of the Eulerian numbers implies that $[0, 1]^n$ is Gorenstein of index 2 (a straightforward verification) and L_d is reflexive.

Example 5. There is an analogous definition of the descent statistic for the hyperoctahedral group, i.e. the group B_n of signed permutations, which gives rise to the *type B Eulerian polynomial* $\sum_{\pi\in B_n} z^{\operatorname{des}(\pi)}$. The type B Eulerian polynomials are also known to be real-rooted with symmetric coefficients. The cube $[-1, 1]^n$ satisfies

$$h^*_{[-1,1]^n}(z) = \sum_{\pi\in B_n} z^{\operatorname{des}(\pi)},$$

and a geometric proof can be found in [5]. As with the Eulerian numbers, the type B Eulerian polynomial can also be obtained as the h^*-polynomial of a simplex, namely

$$L_n^2 := \{x \in \mathbb{R}^n \mid 0 \le \frac{x_1}{2} \le \frac{x_2}{4} \le \frac{x_3}{6} \le \cdots \le \frac{x_n}{2n} \le 1\}.$$

This was first shown by Savage and Schuster [60].

Again, the symmetry of the type B Eulerian polynomials implies that both the ± 1-cube and L_n^2 are reflexive.

Example 6. Another polynomial with symmetric log-concave coefficients is the *Narayana polynomial* $N_n(z) := \sum_{j=1}^n N(n,j) z^{j-1}$, where $N(n,j) = \frac{1}{n}\binom{n}{j}\binom{n}{j-1}$. It is well known [68, Chapter 6] that $N_n(1)$ is equal to the n-th Catalan number, and that $N(n,j)$ arises in many situations involving natural refinements of Catalan objects. Let $P_n \subset \mathbb{R}^n$ denote the convex hull of the positive simple roots of type A together with the origin, i.e.

$$P_n := \operatorname{conv}\{0, e_i - e_j : n \ge i > j \ge 1\}.$$

It follows from work of Gelfand, Graev, and Postnikov [30] combined with techniques due to Stanley as described in [68, Solution to Exercise 6.31(b)] that $h^*_{P_n}(z) = N_n(z)$, hence P_n is Gorenstein.

In case the preceding examples have painted an overly optimistic portrait of the world of Ehrhart h^*-vectors, we end this subsection with two examples of situations where h^*-polynomials behave in pathological ways.

Example 7. Let $m, d, k \in \mathbb{Z}_{\geq 0}$ satisfy $m \geq 1$, $d \geq 2$, and $1 \leq k \leq \lfloor (d+2)/2 \rfloor$. For each such triple of integers, Higashitani [39] has constructed a d-dimensional lattice simplex with h^*-polynomial $1 + mz^k$. The construction is relatively simple; verifying the h^*-polynomial requires only the use of Lemma 1.

Example 8. Payne [58] gives an explicit construction showing that for any positive integers m and n, there exists a reflexive polytope P and indices

$$i_1 < j_1 < i_2 < j_2 < \cdots < i_m < j_m < i_{m+1}$$

such that the h^*-vector for P satisfies

$$h^*_{i_\ell} - h^*_{j_\ell} \geq n \text{ and } h^*_{i_{\ell+1}} - h^*_{j_\ell} \geq n.$$

Thus, reflexive polytopes exist whose h^*-vectors violate unimodality arbitrarily badly with arbitrary length. For all dimensions greater than or equal to 6, Payne also constructs explicit examples of reflexive simplices with non-unimodal h^*-vectors.

2.3 Integral Closure, Normality, and the Integer Decomposition Property (IDP)

Many of the connections between polytopes and unimodality involve the notions of integral closure and normality. See Bruns and Gubeladze [22] for a complete discussion of the distinction between these two ideas. A lattice polytope P in $\mathbb{R}^n$ is *normal* if

$$\text{span}_{\mathbb{Z}_{\geq 0}}\{(1, P) \cap \mathbb{Z}^{1+n}\} = \text{cone}(P) \cap \text{span}_{\mathbb{Z}}\{(1, P) \cap \mathbb{Z}^{1+n}\}.$$

P is said to be *integrally closed*, or to satisfy the *integer decomposition property (IDP)*, if

$$\text{span}_{\mathbb{Z}_{\geq 0}}\{(1, P) \cap \mathbb{Z}^{1+n}\} = \text{cone}(P) \cap \mathbb{Z}^{1+n}.$$

Thus, normality for lattice polytopes involves integer points in cone(P) that are in the $\mathbb{Z}$-linear span of integer points in $(1, P)$, while integral closure/IDP for lattice polytopes involves all integer points in cone(P). It follows directly from the definitions that IDP implies normal.

In the research literature on this topic, it is common for authors to assume that $\mathbb{Z}^{1+n} = \text{span}_{\mathbb{Z}}\{(1, P) \cap \mathbb{Z}^{1+n}\}$, i.e. to consider the right-hand lattice as the lattice of reference for P, and it is also common for authors to use these terms as synonyms. As a result, the terms "normal" and "integrally closed" are potential causes of confusion in the geometric combinatorics literature, see [27, Remark 0.1], for example, where this arises. For polytopes where the two possible lattices of reference are not equal, one must be careful to specify which is being used for lattice point enumeration; we will assume throughout that $\mathbb{Z}^{1+n}$ is being considered unless specified otherwise.

Several authors [27] have proposed to replace the name "integrally closed" with "IDP" for two reasons: first, to alleviate the existing confusion in the literature between integrally closed and normal polytopes, and second to align with the standard use of IDP as the nomenclature in integer programming and optimization. For these reasons, in this survey we will use the term IDP when discussing integrally closed polytopes.

2.4 Connections to Commutative Algebra

The definitions of integral closure and normality are closely related to connections between Ehrhart theory and Hilbert series for finitely generated graded algebras. Given a finitely generated $\mathbb{C}$-algebra $A = \oplus_{i=0}^{\infty} A_i$, we say A is *standard* if A is generated by A_1 and *semistandard* if A is integral over the subalgebra generated by A_1. Given a lattice polytope P in $\mathbb{R}^n$, the *semigroup algebra associated to P* is

$$\mathbb{C}[P] := \mathbb{C}[x^v : v \in \text{cone}(P) \cap \mathbb{Z}^{1+n}].$$

We grade $C[P]$ by $\deg(x_0^{m_0} \cdots x_n^{m_n}) = m_0$, i.e. by the "height" of the exponent vector in cone(P). With this grading, $\mathbb{C}[P]$ is a semistandard semigroup algebra, where the algebra generated by $\mathbb{C}[P]_1$ is $\mathbb{C}[x^v : v \in (1, P) \cap \mathbb{Z}^{1+n}]$, i.e. the semigroup algebra generated by integer points in $(1, P)$. The role played by the IDP property is captured by the following proposition, whose proof is immediate.

Proposition 1. *A lattice polytope P is IDP if and only if $\mathbb{C}[P]$ is standard.*

Example 9. For the standard unimodular simplex Δ_n, we have

$$\mathbb{C}[\Delta_n] = \mathbb{C}[x_0, x_0x_1, x_0x_2, \ldots, x_0x_n],$$

where every generator of the algebra is of degree 1. Thus, $\mathbb{C}[\Delta_n]$ is isomorphic to a polynomial ring in $n+1$ variables.

Example 10. The interval $P = [0, 2]$ is IDP, thus we have

$$\mathbb{C}[P] = \mathbb{C}[x_0, x_0x_1, x_0x_1^2].$$

Note that this algebra is not a polynomial ring, due to the existence of relations among the generators, e.g., $(x_0)(x_0x_1^2) - (x_0x_1)^2 = 0$. However, $\mathbb{C}[P]$ is integral over the subalgebra $\mathbb{C}[x_0, x_0x_1^2]$, which is isomorphic to a polynomial ring.

Remark 1. While $\mathbb{C}[P]$ is always integral over $\mathbb{C}[x^v : v \in (1, P) \cap \mathbb{Z}^{1+n}]$, the algebra generated by $\mathbb{C}[P]_1$, it is not necessarily the normalization of the latter algebra. In general, the normalization of $\mathbb{C}[x^v : v \in (1, P) \cap \mathbb{Z}^{1+n}]$ is $\mathbb{C}[x^v : v \in \text{cone}(P) \cap \text{span}_{\mathbb{Z}}\{(1, P) \cap \mathbb{Z}^{1+n}]$. If $\mathbb{Z}^{1+n} = \text{span}_{\mathbb{Z}}\{(1, P) \cap \mathbb{Z}^{1+n}\}$, then these two algebras are the same.

Recall that the Hilbert series of a graded algebra A is the generating function

$$H(A; z) := \sum_{i=0}^{\infty} \dim_{\mathbb{C}}(A_i) z^i.$$

It is a straightforward consequence of the definition of $\mathbb{C}[P]$ that

$$H(\mathbb{C}[P]; z) = E_P(z).$$

By a theorem of Hochster [40], $\mathbb{C}[P]$ is a Cohen-Macaulay integral domain. Further, $\mathbb{C}[P]$ always admits a linear system of parameters [35, Lemma 33.7]. Thus, if $\dim(P) = d$, there exist algebraically independent elements $\theta_1, \ldots, \theta_{d+1} \in \mathbb{C}[P]_1$ such that $\mathbb{C}[P]$ is a finitely generated free module over the subalgebra $\mathbb{C}[\theta_1, \ldots, \theta_{d+1}]$. Many important properties of Ehrhart series were first established using this property of $\mathbb{C}[P]$, e.g., the non-negativity of $h_P^*(z)$ [67]. In this context, the symmetry of the h^*-vectors for reflexive polytopes implies that $\mathbb{C}[P]$ are Gorenstein when P is reflexive. Gorenstein algebras form an important subclass of Cohen-Macaulay algebras, and the connection with our previous definition of Gorenstein lattice polytope is that P is Gorenstein if and only if $\mathbb{C}[P]$ is Gorenstein.

One of the open problems of interest in Ehrhart theory is a special case of the following conjecture due to Stanley [65, Conjecture 4(a)], also appearing in Hibi [34, Conjecture 1.5].

Conjecture 1 (Stanley, [65]). For a standard graded Gorenstein integral domain A of Krull dimension d, with

$$H(A; z) = \frac{h_0 + h_1 z + \cdots + h_s z^s}{(1 - z)^d},$$

the sequence $h_0, h_1, \ldots, h_s$ is unimodal.

Stanley's original conjecture was that standard graded Cohen-Macaulay integral domains have unimodal h-vectors, but this is false in general, see [19, Section 5] for further discussion.

3 Unimodality for Polytopes with the Integer Decomposition Property

When Conjecture 1 is restricted to algebras of the form $\mathbb{C}[P]$, we encounter our first major open problem regarding unimodality for Ehrhart h^*-vectors. This problem was proposed by Hibi and Ohsugi in the context of normal polytopes [57].

Conjecture 2 (Hibi and Ohsugi, [57]). If P is Gorenstein and IDP, then h_P^* is unimodal.

Our second major open problem is a relaxation of Conjecture 2 due to Schepers and Van Langenhoven [62], arising as a subconjecture of Stanley's original unimodality conjecture for standard graded Cohen-Macaulay integral domains.

Question 1 (Schepers and Van Langenhoven, [62]). If P is IDP, is it true that h_P^* is unimodal?

These are the outstanding open problems regarding unimodality in Ehrhart theory. A first indication that these problems might be difficult is that they both fail when considering the real-rooted property instead of unimodality, as demonstrated by the simplex S_d in Example 2.

There are two other well-known results regarding h^*-polynomials that further motivate this line of investigation. First, Conjecture 2 is rooted in an earlier conjecture due to Hibi that all reflexive polytopes have unimodal h^*-vectors. This was shown to be false in even dimensions by Mustaţă and Payne [55] and in all dimensions by Payne [58]. While their counterexamples arise as reflexive simplices, these simplices are not normal, hence not IDP. Second, Stanley had conjectured that the h^*-vector of the Birkhoff polytope, i.e. the polytope of doubly stochastic matrices, is unimodal – this was proved to be true by Athanasiadis [3]. The Birkhoff polytope is both Gorenstein and IDP, and remains a source of interesting problems regarding unimodality [28]. A key property of the Birkhoff polytope that Athanasiadis used is that it admits a regular unimodular triangulation. Subsequent work in this area exploited this property further, so we will elaborate on such triangulations next.

3.1 Regular Unimodular Triangulations

We will assume that all triangulations of a lattice polytope $P \subset \mathbb{R}^n$ have a vertex set contained in $\mathbb{Z}^n$. If P admits a unimodular triangulation (or even a covering by unimodular subsimplices), then P is IDP because cone(P) is a union of unimodular cones with lattice-point generators of degree 1. A triangulation of P is regular if it arises as the projection of the lower hull of a lifting of the lattice points of P into $\mathbb{R}^{1+n}$. The following theorem due to Athanasiadis, which had been previously discovered by Hibi and Stanley but never published, places strong restrictions on the "tail" of the h^*-vector for any polytope admitting a regular unimodular triangulation.

Theorem 3 (Athanasiadis, [2]). *Let P be a d-dimensional lattice polytope with $h_P^* = (h_0^*, \ldots, h_d^*)$. If P admits a regular unimodular triangulation, then $h_i^* \geq h_{d-i+1}^*$ for $1 \leq i \leq \lfloor (d+1)/2 \rfloor$,*

$$h_{\lfloor (d+1)/2 \rfloor}^* \geq \cdots \geq h_{d-1}^* \geq h_d^*,$$

and

$$h_i^* \leq \binom{h_1^* + i - 1}{i}$$

for $0 \leq i \leq d$.

As a corollary of Theorem 3, it follows that if P is reflexive and admits a regular unimodular triangulation, then h_P^* is unimodal. Athanasiadis's proof of Theorem 3 relies on lemmas due to Stanley and Kalai on the Hilbert h-vectors of Cohen-Macaulay subcomplexes of the boundary complex of a simplicial polytope.

Athanasiadis's result was further extended by Bruns and Römer [24], who proved that all h^*-vectors of IDP Gorenstein polytopes are h^*-vectors for IDP reflexive polytopes.

Theorem 4 (Bruns and Römer, [24]). *If P is Gorenstein and IDP, then h_P^* is the h^*-vector of an IDP reflexive polytope. Further, if P admits a regular unimodular triangulation, then there exists a simplicial polytope Q such that h_P^* is the h-vector of Q, and hence h_P^* is unimodal as a consequence of the g-theorem.*

Thus, the presence of a regular unimodular triangulation forces strong restrictions on the h^*-vector of a Gorenstein polytope.

In general, the h^*-vectors of lattice polytopes satisfy an array of interesting linear inequalities. Stapledon [69] found combinatorial proofs for many of these, including analogues of Theorem 3. In order to establish these inequalities, for a d-dimensional lattice polytope P with $\deg(h_P^*(z)) = s$ Stapledon introduces the polynomial

$$\overline{h_P^*}(z) := (1 + z + z^2 + \cdots + z^{d-s})h_P^*(z)$$

and proves that $\overline{h_P^*}(z) = a(z) + z^{d+1-s}b(z)$ for symmetric polynomials $a(z)$ and $b(z)$ of degree d and $s-1$, respectively, with integer coefficients. Many of the inequalities for h_P^* that Stapledon produces are consequences of properties of $\overline{h_P^*}(z)$, $a(z)$, and $b(z)$; the following theorem is a representative example of the role played by these polynomials.

Theorem 5 (Stapledon, [69]). *If P is a d-dimensional lattice polytope such that the boundary of P admits a regular unimodular triangulation, then $a(t)$ is unimodal.*

P is reflexive if and only if $h_P^*(z) = a(z)$. Because a reflexive polytope P admits a regular unimodular triangulation if and only if the boundary of P admits a regular unimodular triangulation, we see that this is an extension of the result of Athanasiadis.

While regular unimodular triangulations are a powerful tool [51], not all lattice polytopes admit such decompositions, even when P is normal [56]. Thus, we turn our attention to approaches to establishing unimodality that do not rely on this property.

3.2 Lefschetz Elements

Let $A = \oplus_{i=0}^{s} A_i$ be a finitely generated graded algebra of Krull dimension zero. A useful technique for establishing unimodality for sequences arising as

$$\dim_{\mathbb{C}}(A_0), \dim_{\mathbb{C}}(A_1), \ldots, \dim_{\mathbb{C}}(A_s)$$

is to find a Lefschetz element in A, defined as follows. A linear form $l \in A_1$ is called a *weak Lefschetz element* if the multiplication map

$$\times l : A_i \to A_{i+1}$$

has maximal rank, that is, is either injective or surjective, for each i. By Remark 3.3 of [32], if A has a weak Lefschetz element, then the Hilbert series $H(A; z)$ has unimodal coefficients. For a d-dimensional lattice polytope P, the general theory of Cohen-Macaulay algebras implies that if $\theta_1, \ldots, \theta_{d+1} \in \mathbb{C}[P]_1$ is a linear system of parameters for $\mathbb{C}[P]$, then

$$h_P^*(z) = H(\mathbb{C}[P]/(\theta_1, \ldots, \theta_{d+1}); z),$$

i.e. $h_i^* = \dim_{\mathbb{C}}[\mathbb{C}[P]/(\theta_1, \ldots, \theta_{d+1})]_i$. Thus, the existence of a Lefschetz element in $\mathbb{C}[P]/(\theta_1, \ldots, \theta_{d+1})$ implies h^*-unimodality for P.

Several important recent results and counterexamples in Ehrhart theory [37, 39] have involved only simplices. Also, as shown in the examples given previously, many h^*-polynomials can be realized using simplices. This suggests that lattice simplices are richer objects than they first appear. In the case of a simplex P with vertices $v^0, \ldots, v^n$, we saw in Lemma 2 that h_i^* counts the number of lattice points of degree i in the fundamental parallelepiped for P. Further, the monomials $x_0x^{v^0}, \ldots, x_0x^{v^n}$ form a linear system of parameters in $\mathbb{C}[P]$. This motivates the study of the zero-(Krull)-dimensional algebra

$$R_P := \mathbb{C}[P]/(x_0x^{v^0}, \ldots, x_0x^{v^n})$$

graded by the exponent on x_0, which has a basis given by the lattice points in the fundamental parallelepiped for P. Because of the explicit connection between h^*-vectors, Hilbert series, and R_P, the author and Davis [17] have initiated the investigation of h^*-unimodality for IDP reflexive simplices using Lefschetz techniques.

While experimental data suggests that a weak Lefschetz element exists for many of the algebras R_P when P is an IDP reflexive simplex, such an element need not exist.

Theorem 6 (Braun and Davis, [17]). *Denote by Lef_d the convex hull of the vectors*

$$e_1, \ldots, e_d, -de_1 - \sum_{k=2}^{d}(d+1)e_k.$$

Then

$$h^*_{Lef_d}(z) = 1 + (d+2)z + (d+2)z^2 + \cdots + (d+2)z^{d-1} + z^d$$

and R_{Lef_d} does not have a weak Lefschetz element.

An important observation is that for other choices of linear system of parameters, the quotient of $\mathbb{C}[Lef_d]$ by such a system does have a weak Lefschetz element. However, the failure of the weak Lefschetz approach when using the vertices of P as a system of parameters complicates the situation, as most of the techniques in Ehrhart theory for studying h^*-vectors of simplices correspond to studying the algebra R_P with this choice of system of parameters. Thus, we pose the following questions.

Question 2. For which IDP reflexive simplices does R_P admit a Lefschetz element?

Question 3. For P an IDP reflexive simplex, is there a canonical choice of linear system of parameters such that $\mathbb{C}[P]/(\theta_1, \ldots, \theta_{d+1})$ admits a Lefschetz element?

Some evidence in support of h^*-unimodality for arbitrary IDP reflexive simplices is that this property is preserved under affine free sums. Suppose $P \subset \mathbb{R}^n$ and $Q \subset \mathbb{R}^m$ are full-dimensional simplices with $0 \in P$ and $\{v^0, \ldots, v^m\}$ denoting the vertices of Q. Then for each $i = 0, 1, \ldots, m$ we define

$$P *_i Q := \operatorname{conv}(P \times 0) \cup (0 \times Q - v^i) \subseteq \mathbb{R}^{n+m}.$$

The polytope $P *_i Q$ is an example of an *affine free sum* [9].

Theorem 7 (Braun and Davis, [17]). *If P and Q are IDP reflexive simplices with $0 \in int(P)$, then so is $P *_i Q$ for each i. If, in addition, h^*_P and h^*_Q are unimodal, then so is $h^*_{P*_iQ}$.*

Remark 2. We are not aware of examples of IDP reflexive simplices that do not admit a regular unimodular triangulation. It would be interesting to know if simplices with these properties exist, and if so, to have explicit constructions of infinite families.

3.3 Box Unimodal Triangulations

Another approach to proving h^*-unimodality for IDP reflexive polytopes was proposed by Schepers and Van Langenhoven [62]. For a lattice simplex

$$S = \mathrm{conv}(v^0, v^1, \ldots, v^k) \subset \mathbb{R}^n,$$

define the *open fundamental parallelepiped of S* to be

$$\Pi^\circ := \left\{ \sum_{i=0}^{k} \lambda_i (1, v^i) : \lambda_i \in (0,1) \right\}$$

and the *box polynomial of S* to be

$$B_S(z) := \sum_{m \in \Pi^\circ \cap \mathbb{Z}^{1+n}} z^{m_0},$$

where $B_\varnothing(z) := 1$.

Theorem 8 (Betke and McMullen, [12]). *Let P be a reflexive polytope and let T be a triangulation of the boundary of P. Then*

$$h_P^*(z) = \sum_{F \text{ face of } T} B_F(z) h_F(z)$$

where $h_F(z)$ is the h-polynomial of the link of F in T.

It is well known that if P is d-dimensional and T is regular, then the polynomials $h_F(z)$ are symmetric, unimodal, have non-negative integer coefficients, and have degree $d-1-\dim(F)$. A triangulation T of the boundary of P is called *box unimodal* if T is regular and $B_F(z) = \sum_{i=0}^{\dim(F)} b_i z^i$ is symmetric and unimodal for all non-empty faces of T. It follows from Theorem 8 that if P admits a box unimodal triangulation, then $h_P^*(z)$ is also symmetric and unimodal. An affirmative answer to the following question would therefore imply h^*-unimodality for IDP reflexive polytopes.

Question 4 (Schepers and Van Langenhoven, [62]). Does the boundary of every IDP reflexive polytope have a box unimodal triangulation?

One piece of supporting evidence for the existence of box unimodal triangulations is that for every lattice simplex S, $B_S(z)$ is symmetric. This is a straightforward consequence of the existence of an involution on the lattice points in Π° that sends

$$\sum_{i=0}^{k} \lambda_i (1, v^i) \in \Pi^\circ \cap \mathbb{Z}^{1+n}$$

to

$$\sum_{i=0}^{k}(1-\lambda_i)(1,v^i) \in \Pi^\circ \cap \mathbb{Z}^{1+n}.$$

However, there are at least three complications that arise when considering box unimodal triangulations. First, there exist simplices with non-unimodal box polynomials [62, Example 4.3]. Hence, if the box unimodal triangulation approach is to eventually succeed, one must show that all such simplices are avoided by at least one regular triangulation of the boundary of every IDP reflexive polytope. Second, a result of Haase and Melnikov [31] states that every lattice polytope is a face of some reflexive polytope. Thus, it isn't possible to simply ignore examples of simplices with non-unimodal box polynomials – one must make arguments for why such simplices do not appear in triangulations of IDP reflexives. Third, it isn't clear what geometric or arithmetic conditions on lattice simplices lead to unimodality for their box polynomials, which leads to the following questions.

Question 5. Which lattice simplices have unimodal box polynomials?

Question 6. What are necessary conditions for a sequence to be the coefficients of a box polynomial for a lattice simplex? Sufficient conditions?

3.4 Zonotopes

One of the reasons that Gorenstein lattice polytopes are a focus for questions about unimodality is that there are many techniques for establishing unimodality that only apply to symmetric sequences, e.g., the representation theory of sl_2 [65]. By omitting the requirement that P be Gorenstein, with Question 1 Schepers and Van Langenhoven remove both a strong geometric constraint on the lattice polytopes under consideration and many established techniques for demonstrating unimodality.

Partial answers to Question 1 have been obtained. We call P a *closed lattice parallelepiped* if there exist linearly independent vectors $v^1, \dots, v^r \in \mathbb{Z}^n$ such that

$$P = \left\{ \sum_{i=1}^{r} \lambda_i v^i : \lambda_i \in [0,1] \right\},$$

i.e. if P is a Minkowski sum of linearly independent lattice segments. Lattice parallelepipeds are known to be IDP.

Theorem 9 (Schepers and Van Langenhoven, [62]). *Closed lattice parallelepipeds have unimodal h^*-vectors.*

Lattice parallelepipeds are a subclass of *lattice zonotopes*, which by definition are lattice polytopes arising as Minkowski sums of arbitrary lattice segments [53]. Lattice zonotopes have a covering by lattice parallelepipeds, hence are IDP. A natural question is if the h^*-vectors of lattice zonotopes are unimodal, which the following theorem answers positively.

Theorem 10 (Beck, Jochemko, and McCullough, [11, 41]). *If P is a lattice zonotope, then P has a unimodal h^*-vector.*

In general, IDP lattice polytopes can behave in counterintuitive ways, as shown by the existence of normal lattice polytopes (which are IDP with respect to the lattice $\mathrm{span}_{\mathbb{Z}}\{(1, P) \cap \mathbb{Z}^{1+n}\}$) that do not admit a regular unimodular triangulation [56]. While it seems reasonable to replace the general problem of h^*-unimodality for IDP P by the problem of demonstrating h^*-unimodality for special subfamilies of IDP polytopes, even this can create significant challenges as shown in the next subsection.

3.5 Matroid Polytopes

Recall that a *matroid* M is a system $\mathscr{F}$ of subsets of $\{1, 2, \ldots, n\}$ called *independent sets* such that the following hold:

- $\varnothing \in \mathscr{F}$
- if $X \in \mathscr{F}$ and $Y \subseteq X$, then $Y \in \mathscr{F}$
- if $X, Y \in \mathscr{F}$ and $|X| = |Y| + 1$, there exists $x \in X \setminus Y$ such that $Y \cup x \in \mathscr{F}$

The *bases* of M are the inclusion-maximal independent sets in M, and the set of bases for M is denoted $\mathscr{B}$. The *matroid polytope* of M is

$$P(M) := \mathrm{conv}\{e_B : B \in \mathscr{B}\},$$

where $e_B := \sum_{i \in B} e_i$ with e_i denoting the i-th standard basis vector in $\mathbb{R}^l$. The following conjecture due to De Loera, Haws, and Köppe is based on extensive computational evidence.

Conjecture 3 (De Loera, Haws, and Köppe, [50]). For any matroid M, the h^*-vector of $P(M)$ is unimodal.

Matroid polytopes have many nice properties, for example, Ardila, Benedetti, and Doker proved that they are generalized permutohedra [1], a class of polytopes defined by Postnikov [59]. A geometric variant of an algebraic conjecture due to White [70] regarding the toric ideal of $P(M)$ is the following conjecture of Haws.

Conjecture 4 (Haws, [33]). For any matroid M, the polytope $P(M)$ has a regular unimodular triangulation.

Haws also proposed the following weaker conjecture.

Conjecture 5 (Haws, [33]). For any matroid M, the polytope $P(M)$ is the union of its unimodular subsimplices.

We will focus in this subsection on the special case of Conjecture 3 when M is a uniform matroid. Recall that the *uniform matroid of rank k on n elements*, denoted $U^{k,n}$, is the matroid with bases given by the set of all k-element subsets of $\{1, 2, \ldots, n\}$. The polytope $P(U^{k,n})$ is known in the geometric combinatorics literature as the *(n, k)-hypersimplex*, denoted $\Delta_{n,k}$, and we will use this notation throughout. Several results regarding lattice point enumeration for hypersimplices are known [48, 49]. The unimodality of $h^*_{\Delta_{n,2}}$ follows from work of Katzman [44], but for all $k \geq 3$ the unimodality of the h^*-vectors for hypersimplices remains an open problem. While not all hypersimplices are Gorenstein, $\Delta_{n,k}$ always admits a regular unimodular triangulation, which we will refer to as the *circuit triangulation*. While different descriptions of this triangulation have been found, Lam and Postnikov proved that these descriptions all correspond to the same triangulation [48].

The author and Solus [18] have introduced the following polytopes as part of an attempt to prove that $\Delta_{n,k}$ is h^*-unimodal. Label the vertices of a regular n-gon embedded in $\mathbb{R}^2$ in a clockwise fashion from 1 to n. We define the *circular distance* between two elements i and j of $\{1, 2, \ldots, n\}$, denoted $\mathrm{cd}(i, j)$, to be the number of edges in the shortest path between the vertices i and j of the n-gon. We also denote the path of shortest length from i to j by $\mathrm{arc}(i, j)$. A subset $S \subset \{1, 2, \ldots, n\}$ is called *r-stable* if each pair $i, j \in S$ satisfies $\mathrm{cd}(i, j) \geq r$. The *r-stable n, k-hypersimplex*, denoted $\Delta_{n,k}^{stab(r)}$, is the convex hull of the vectors $\sum_{i \in S} e_i$ where S ranges over all r-stable k-subsets of the n-set. For fixed n and k, these polytopes form the nested chain

$$\Delta_{n,k} \supset \Delta_{n,k}^{stab(2)} \supset \Delta_{n,k}^{stab(3)} \supset \cdots \supset \Delta_{n,k}^{stab(\lfloor \frac{n}{k} \rfloor)}.$$

An approach to establishing unimodality for $h^*_{\Delta_{n,k}}$ is to attempt to inductively prove unimodality for $h^*_{\Delta_{n,k}^{stab(r)}}$ for all r, possibly using the following theorem due to Stanley.

Theorem 11 (Stanley, [66]). *If $P \subseteq Q$ are lattice polytopes, then $h^*_{P,i} \leq h^*_{Q,i}$ for all i, i.e. the h^*-vector for Q dominates the h^*-vector for P in each entry.*

Jochemko and Sanyal [42] have recently proved that this monotonicity property of h^*-vectors is equivalent to the non-negativity of h^*-vectors.

The following geometric properties of $\Delta_{n,k}^{stab(r)}$ are directly related to establishing h^*-unimodality, though they are also independently interesting.

Theorem 12 (Braun and Solus, [18]). *The regular unimodular triangulation of $\Delta_{n,k}$ given by the circuit triangulation restricts to a regular unimodular triangulation of $\Delta_{n,k}^{stab(r)}$ for all r. Further, when n is odd and $k = 2$ there is a shelling of the circuit triangulation of $\Delta_{n,2}$ that first builds the r-stable hypersimplex and then builds the $(r-1)$-stable hypersimplex for every $1 \leq r \leq \lfloor \frac{n}{2} \rfloor$.*

Theorem 13 (Hibi and Solus, [38]). *Let $1 < k < n-1$, and let H denote the hyperplane in $\mathbb{R}^n$ defined by the equation $x_1 + \cdots + x_n = k$. For $1 \leq r < \lfloor \frac{n}{k} \rfloor$, the facets of $\Delta_{n,k}^{stab(r)}$ are supported by the hyperplanes*

$$\{x \in \mathbb{R}^n : x_\ell = 0\} \cap H$$

and

$$\left\{ x \in \mathbb{R}^n : \sum_{i=\ell}^{\ell+r-1} x_i = 1 \right\} \cap H.$$

Further, for $1 \leq r < \lfloor \frac{n}{k} \rfloor$, the polytope $\Delta_{n,k}^{stab(r)}$ is Gorenstein if and only if $n = kr+k$.

The facet descriptions of the stable hypersimplices not provided here have also been obtained, see [38, Remark 2.3].

In [18], the h^*-polynomials of $\Delta_{n,2}^{stab(r)}$ are studied, and connections are obtained with Lucas and Fibonacci polynomials, independence polynomials of graphs, and monomial CR mappings of Lens spaces. Since independence polynomials of graphs have been well-investigated with regard to unimodality, these connections yield additional tools which might be applied in this setting. The vectors $h^*_{\Delta_{n,2}^{stab(r)}}$ are also determined to be unimodal for $r = 2$, $r = 3$, and $r = \lfloor \frac{n}{2} \rfloor - 1$. Many open questions about these polytopes remain, such as the following.

Conjecture 6 (Braun and Solus, [18]). The h^*-vector of $\Delta_{n,2}^{stab(r)}$ is unimodal for all r.

Question 7 (Braun and Solus, [18]). Is the h^*-vector of $\Delta_{n,k}^{stab(r)}$ unimodal for all r?

4 Log-Concavity and Real-Rootedness

We conclude this survey with a brief discussion of log-concavity and real-rootedness for h^*-polynomials. While a large amount of effort has focused on h^*-unimodality, the study of these two properties for lattice polytopes has also been a fruitful area of research. We will discuss two such lines of investigation.

4.1 s-Lecture Hall Polytopes

Motivated by the study of lecture hall partitions [14], given a finite sequence $s = (s_1, \ldots, s_n)$ Savage and Schuster [60] defined the s-lecture hall polytope to be

$$P_n^s := \left\{ x \in \mathbb{R}^n : 0 \leq \frac{x_1}{s_1} \leq \frac{x_2}{s_2} \leq \cdots \leq \frac{x_n}{s_n} \leq 1 \right\}.$$

When $(s_1, s_2, \ldots, s_n) = (1, 2, \ldots, n)$, this polytope contains a subset of the classical lecture hall partitions. It is straightforward to verify that P_n^s is a lattice polytope with vertex set

$$\{(0, 0, \ldots, 0, s_k, s_{k+1}, \ldots, s_n : 1 \leq k \leq n+1\}.$$

Savage and Schuster derived a combinatorial interpretation of $h^*_{P_n^s}(z)$ using ascent statistics for s-inversion sequences. Savage and Visontai [61] used this combinatorial interpretation to prove the following.

Theorem 14 (Savage and Visontai, [61]). *The polynomials $h^*_{P_n^s}(z)$ are real-rooted, hence the h^*-vector of P_n^s is log-concave.*

The proof given by Savage and Visontai uses only the combinatorial interpretation of these polynomials in terms of ascent sequences and *compatible polynomial* techniques due to Chudnovsky and Seymour [25].

Question 8. Is there a proof of real-rootedness, log-concavity, or unimodality for $h^*_{P_n^s}$ that fundamentally relies on the lattice point geometry of $\text{cone}(P_n^s)$?

The following question regarding P_n^s is natural in the context of unimodality.

Question 9. For which sequences s is P_n^s Gorenstein?

The lattice point geometry of $\text{cone}(P_n^s)$ is related to the lattice point geometry of the s-lecture hall cone

$$C_n^s := \left\{ x \in \mathbb{R}^n : 0 \leq \frac{x_1}{s_1} \leq \frac{x_2}{s_2} \leq \cdots \leq \frac{x_n}{s_n} \right\}$$

for which the following is known.

Theorem 15 (Beck, Braun, Köppe, Savage, and Zafeirakopoulos, [10]). *Let $\ell > 0$ and $b \neq 0$ be integers satisfying $\ell^2 + 4b \geq 0$. Let $s = (s_1, s_2, \ldots)$ be defined by*

$$s_n = \ell s_{n-1} + b s_{n-2},$$

with initial conditions $s_1 = 1$, $s_0 = 0$. For each n, let C_n^s be defined by the truncation $(s_1, \ldots, s_n)$. Then C_n^s is Gorenstein for all $n \geq 0$ if and only if $b = -1$. If $b \neq -1$, there exists $n_0 = n_0(b, \ell)$ such that C_n^s fails to be Gorenstein for all $n \geq n_0$.

For example, when $\ell = 2$ and $b = -1$, we obtain the sequence $(1, 2, 3, 4, \ldots)$, corresponding to the classical lecture hall partition cone.

4.2 Dilations of Polytopes and Multivariate Techniques

Our final topic regards the behavior of h^*-vectors for integral dilates nP of a lattice polytope P. This work has generally been motivated by the following theorem.

Theorem 16 (Knudsen, Mumford, and Waterman, [45]). *For every lattice polytope P, there exists an integer n such that nP admits a regular unimodular triangulation.*

Identifying the values of n that correspond to nP with regular unimodular triangulations is a deep and subtle open problem. Bruns and Gubeladze [21] proved that for each d, there exists a natural number c_d such that for all $c > c_d$ and all d-dimensional lattice polytopes P, cP is equal to the union of the unimodular subsimplices of cP. It is unknown whether this is true for the property of cP admitting a regular unimodular triangulation.

Recently there has been interest in the study of the transformations taking a power series

$$\sum_{n\geq 0} a_n z^n = \frac{h(z)}{(1-z)^d}$$

to

$$\sum_{n\geq 0} a_{rn} z^n = \frac{h^{<r>}(z)}{(1-z)^d}$$

for $r \geq 1$. For a lattice polytope P, this corresponds to studying h^*-polynomials for integer dilates of P. The following result strengthens a previous theorem due to Brenti and Welker [20] regarding the Veronese construction for graded algebras and associated Hecke operators.

Theorem 17 (Beck and Stapledon, [8]). *Fix a positive integer d and let $\rho_1 < \rho_2 < \cdots < \rho_d = 0$ denote the roots of $zA_d(z)$, where $A_d(z)$ is the Eulerian polynomial. There exist positive integers m_d and n_d such that, if P is a d-dimensional lattice polytope and $n \geq n_d$, then $h^*_{nP}(z)$ has negative real roots $\beta_1(n) < \beta_2(n) < \cdots < \beta_{d-1}(n) < \beta_d(n) < 0$ with $\beta_i(n) \to \rho_i$ as $n \to \infty$, and the coefficients $\{h^*_i(n)\}_{1\leq i\leq d}$ of $h^*_{nP}(z)$ are positive, strictly log-concave, and satisfy*

$$\begin{aligned} 1 = h^*_0(n) < h^*_d(n) < h^*_1(n) < \cdots \\ < h^*_i(n) < h^*_{d-i}(n) < h^*_{i+1}(n) < \cdots \\ < h^*_{\lfloor \frac{d+1}{2} \rfloor}(n) < m_d h^*_d(n). \end{aligned}$$

A strong conjecture regarding real-rootedness for integral dilates of lattice polytopes is the following.

Conjecture 7 (Beck and Stapledon, [8]). If P is d-dimensional, then for $n \geq d$ the polynomial $h^*_{nP}(z)$ has distinct, negative real roots.

In support of this conjecture, Higashitani (personal communication) has recently proved that for a lattice polytope P, if $n \geq \deg(h_P^*(z))$ then the h^*-vector of nP is strictly log-concave.

McCabe and Smith [52] extended these investigations to the setting of multivariate log-concavity for multigraded Hilbert series, and introduced new techniques for establishing log-concavity. They use these techniques to provide a new proof of the log-concavity of the Eulerian polynomials. The study of log-concavity and unimodality for multivariate h^*-polynomials deserves further investigation, see [52, Section 3] for further discussion and open problems.

Remark 3. Another multivariate approach to Ehrhart h^*-polynomials are *local* h^*-polynomials. Katz and Stapledon [43, Theorem 9.4] have recently established various unimodality results for these objects.

Acknowledgements The author is partially supported by the National Security Agency through award H98230-13-1-0240. Thanks to Christos Athanasiadis, Matthias Beck, Robert Davis, Jesus De Loera, David Haws, Takayuki Hibi, Akihiro Higashitani, Carla Savage, and Liam Solus for their helpful comments and suggestions.

References

1. F. Ardila, C. Benedetti, J. Doker, Matroid polytopes and their volumes. Discret. Comput. Geom. **43**(4), 841–854 (2010). doi:10.1007/s00454-009-9232-9. http://dx.doi.org/10.1007/s00454-009-9232-9
2. C.A. Athanasiadis, h^*-vectors, Eulerian polynomials and stable polytopes of graphs. Electron. J. Comb. **11**(2), Research Paper 6, 13 p. (electronic) (2004/2006)
3. C.A. Athanasiadis, Ehrhart polynomials, simplicial polytopes, magic squares and a conjecture of Stanley. J. Reine Angew. Math. **583**, 163–174 (2005). doi:10.1515/crll.2005.2005.583.163. http://dx.doi.org/10.1515/crll.2005.2005.583.163
4. V.V. Batyrev, Dual polyhedra and mirror symmetry for Calabi-Yau hypersurfaces in toric varieties. J. Algebraic Geom. **3**(3), 493–535 (1994)
5. M. Beck, B. Braun, Euler-Mahonian statistics via polyhedral geometry. Adv. Math. **244**(0), 925–954 (2013). doi:http://dx.doi.org/10.1016/j.aim.2013.06.002. http://www.sciencedirect.com/science/article/pii/S0001870813002065
6. M. Beck, S. Hoşten, Cyclotomic polytopes and growth series of cyclotomic lattices. Math. Res. Lett. **13**(4), 607–622 (2006)
7. M. Beck, S. Robins, *Computing the Continuous Discretely*. Undergraduate Texts in Mathematics (Springer, New York, 2007)
8. M. Beck, A. Stapledon, On the log-concavity of Hilbert series of Veronese subrings and Ehrhart series. Math. Z. **264**(1), 195–207 (2010). doi:10.1007/s00209-008-0458-7. http://dx.doi.org/10.1007/s00209-008-0458-7
9. M. Beck, P. Jayawant, T.B. McAllister, Lattice-point generating functions for free sums of convex sets. J. Comb. Theory A **120**(6), 1246–1262 (2013). doi:10.1016/j.jcta.2013.03.007. http://dx.doi.org/10.1016/j.jcta.2013.03.007
10. M. Beck, B. Braun, M. Köppe, C.D. Savage, Z. Zafeirakopoulos, s-lecture hall partitions, self-reciprocal polynomials, and Gorenstein cones. Ramanujan J. **36**(1–2), 123–147 (2015). doi:10.1007/s11139-013-9538-3. http://dx.doi.org/10.1007/s11139-013-9538-3
11. M. Beck, K. Jochemko, E. McCullough, Manuscript in preparation

12. U. Betke, P. McMullen, Lattice points in lattice polytopes. Monatsh. Math. **99**(4), 253–265 (1985). doi:10.1007/BF01312545. http://dx.doi.org/10.1007/BF01312545
13. C. Bey, M. Henk, J.M. Wills, Notes on the roots of Ehrhart polynomials. Discret. Comput. Geom. **38**(1), 81–98 (2007)
14. M. Bousquet-Mélou, K. Eriksson, Lecture hall partitions. Ramanujan J. **1**(1), 101–111 (1997). doi:10.1023/A:1009771306380. http://dx.doi.org/10.1023/A:1009771306380
15. P. Brändén, Unimodality, log-concavity, real-rootedness, and beyond. in *Handbook of Enumerative Combinatorics* (CRC, Boca Raton, 2014). Preprint available at arXiv:1410.6601v1
16. B. Braun, An Ehrhart series formula for reflexive polytopes. Electron. J. Comb. **13**(1), Note 15, 5 p. (electronic) (2006)
17. B. Braun, R. Davis. Ehrhart series, unimodality, and integrally closed reflexive polytopes to appear in Annals of Combinatorics
18. B. Braun, L. Solus, A shelling of the odd hypersimplex (Preprint), http://arxiv.org/abs/1408.4713
19. F. Brenti, Log-concave and unimodal sequences in algebra, combinatorics, and geometry: an update, in *Jerusalem combinatorics '93*. Contemporary Mathematics, vol. 178 (American Mathematical Society, Providence, 1994), pp. 71–89. doi:10.1090/conm/178/01893. http://dx.doi.org/10.1090/conm/178/01893
20. F. Brenti, V. Welker, The Veronese construction for formal power series and graded algebras. Adv. Appl. Math. **42**(4), 545–556 (2009). doi:10.1016/j.aam.2009.01.001. http://dx.doi.org/10.1016/j.aam.2009.01.001
21. W. Bruns, J. Gubeladze, Unimodular covers of multiples of polytopes. Doc. Math. **7**, 463–480 (electronic) (2002)
22. W. Bruns, J. Gubeladze, *Polytopes, Rings, and K-Theory*. Springer Monographs in Mathematics (Springer, Dordrecht, 2009). doi:10.1007/b105283. http://dx.doi.org/10.1007/b105283
23. W. Bruns, J. Herzog, *Cohen-Macaulay Rings*. Cambridge Studies in Advanced Mathematics, vol. 39 (Cambridge University Press, Cambridge, 1993)
24. W. Bruns, T. Römer, *h*-vectors of Gorenstein polytopes. J. Comb. Theory A **114**(1), 65–76 (2007). doi:10.1016/j.jcta.2006.03.003. http://dx.doi.org/10.1016/j.jcta.2006.03.003
25. M. Chudnovsky, P. Seymour, The roots of the independence polynomial of a clawfree graph. J. Comb. Theory B **97**(3), 350–357 (2007). doi:10.1016/j.jctb.2006.06.001. http://dx.doi.org/10.1016/j.jctb.2006.06.001
26. S. Corteel, S. Lee, C.D. Savage, Enumeration of sequences constrained by the ratio of consecutive parts. Sém. Lothar. Comb. **54A**, Art. B54Aa, 12 (2005/2007)
27. D.A. Cox, C. Haase, T. Hibi, A. Higashitani, Integer decomposition property of dilated polytopes. Electron. J. Comb. **21**(4), Paper 4.28, 17 (2014)
28. R. Davis, Ehrhart series of polytopes related to symmetric doubly-stochastic matrices. Electron. J. Comb. **22**(2), P2.17, (electronic) (2015)
29. E. Ehrhart, Sur les polyèdres rationnels homothétiques à *n* dimensions. C. R. Acad. Sci. Paris **254**, 616–618 (1962)
30. I.M. Gelfand, M.I. Graev, A. Postnikov, Combinatorics of hypergeometric functions associated with positive roots, in *The Arnold-Gelfand mathematical seminars* (Birkhäuser Boston, Boston, 1997), pp. 205–221
31. C. Haase, I.V. Melnikov, The reflexive dimension of a lattice polytope. Ann. Comb. **10**(2), 211–217 (2006)
32. T. Harima, J.C. Migliore, U. Nagel, J. Watanabe, The weak and strong Lefschetz properties for Artinian k-algebras. J. Algebra **262**(1), 99–126 (2003). doi:http://dx.doi.org/10.1016/S0021-8693(03)00038-3. http://www.sciencedirect.com/science/article/pii/S0021869303000383
33. D. Haws, Matroid polytopes: algorithms, theory, and applications, Ph.D Dissertation available at http://arxiv.org/abs/0905.4405
34. T. Hibi, Flawless *O*-sequences and Hilbert functions of Cohen-Macaulay integral domains. J. Pure Appl. Algebra **60**(3), 245–251 (1989). doi:10.1016/0022-4049(89)90085-6. http://dx.doi.org/10.1016/0022-4049(89)90085-6

35. T. Hibi, *Algebraic Combinatorics on Convex Polytopes* (Carslaw Publications, Glebe, 1992)
36. T. Hibi, Dual polytopes of rational convex polytopes. Combinatorica **12**(2), 237–240 (1992)
37. T. Hibi, A. Higashitani, N. Li, Hermite normal forms and δ-vectors. J. Comb. Theory A **119**(6), 1158–1173 (2012). doi:10.1016/j.jcta.2012.02.005. http://dx.doi.org/10.1016/j.jcta.2012.02.005
38. T. Hibi, L. Solus, The facets of the r-stable n,k-hypersimplex. To appear in Annals of Combinatorics
39. A. Higashitani, Counterexamples of the conjecture on roots of Ehrhart polynomials. Discret. Comput. Geom. **47**(3), 618–623 (2012). doi:10.1007/s00454-011-9390-4. http://dx.doi.org/10.1007/s00454-011-9390-4
40. M. Hochster, Rings of invariants of tori, Cohen-Macaulay rings generated by monomials, and polytopes. Ann. Math. (2) **96**, 318–337 (1972)
41. K. Jochemko, On the combinatorics of valuations. PhD Thesis available at http://www.diss.fu-berlin.de/diss/receive/FUDISS_thesis_000000098742
42. K. Jochemko, R. Sanyal, Combinatorial positivity of translation-invariant valuations. To appear in J. Eur. Math. Soc. (JEMS)
43. E. Katz, A. Stapledon, Local h-polynomials, invariants of subdivisions, and mixed Ehrhart theory. Adv. Math. **286**, 181–239 (2016)
44. M. Katzman, The Hilbert series of algebras of the Veronese type. Commun. Algebra **33**(4), 1141–1146 (2005). doi:10.1081/AGB-200053828. http://dx.doi.org/10.1081/AGB-200053828
45. G. Kempf, F.F. Knudsen, D. Mumford, B. Saint-Donat, *Toroidal Embeddings. I*. Lecture Notes in Mathematics, vol. 339 (Springer, Berlin/New York, 1973)
46. M. Kreuzer, H. Skarke, Complete classification of reflexive polyhedra in four dimensions. Adv. Theor. Math. Phys. **4**(6), 1209–1230 (2000)
47. J.C. Lagarias, G.M. Ziegler, Bounds for lattice polytopes containing a fixed number of interior points in a sublattice. Can. J. Math. **43**(5), 1022–1035 (1991). doi:10.4153/CJM-1991-058-4. http://dx.doi.org/10.4153/CJM-1991-058-4
48. T. Lam, A. Postnikov, Alcoved polytopes. I. Discret. Comput. Geom. **38**(3), 453–478 (2007). doi:10.1007/s00454-006-1294-3. http://dx.doi.org/10.1007/s00454-006-1294-3
49. N. Li, Ehrhart h^*-vectors of hypersimplices. Discret. Comput. Geom. **48**(4), 847–878 (2012). doi:10.1007/s00454-012-9452-2. http://dx.doi.org/10.1007/s00454-012-9452-2
50. J.A. De Loera, D.C. Haws, M. Köppe, Ehrhart polynomials of matroid polytopes and polymatroids. Discret. Comput. Geom. **42**(4), 670–702 (2009). doi:10.1007/s00454-008-9080-z. http://dx.doi.org/10.1007/s00454-008-9080-z
51. J.A. De Loera, J. Rambau, F. Santos, *Triangulations*. Algorithms and Computation in Mathematics, vol. 25 (Springer, Berlin, 2010). doi:10.1007/978-3-642-12971-1. http://dx.doi.org/10.1007/978-3-642-12971-1. Structures for algorithms and applications
52. A. McCabe, G.G. Smith, Log-concavity of asymptotic multigraded Hilbert series. Proc. Am. Math. Soc. **141**(6), 1883–1892 (2013). doi:10.1090/S0002-9939-2012-11808-8. http://dx.doi.org/10.1090/S0002-9939-2012-11808-8
53. P. McMullen, On zonotopes. Trans. Am. Math. Soc. **159**, 91–109 (1971)
54. E. Miller, B. Sturmfels, *Combinatorial Commutative Algebra*. Graduate Texts in Mathematics, vol. 227 (Springer, New York, 2005)
55. M. Mustaţă, S. Payne, Ehrhart polynomials and stringy Betti numbers. Math. Ann. **333**(4), 787–795 (2005)
56. H. Ohsugi, T. Hibi, A normal $(0, 1)$-polytope none of whose regular triangulations is unimodular. Discret. Comput. Geom. **21**(2), 201–204 (1999). doi:10.1007/PL00009415. http://dx.doi.org/10.1007/PL00009415
57. H. Ohsugi, T. Hibi, Special simplices and Gorenstein toric rings. J. Comb. Theory A **113**(4), 718–725 (2006). doi:10.1016/j.jcta.2005.06.002. http://dx.doi.org/10.1016/j.jcta.2005.06.002
58. S. Payne, Ehrhart series and lattice triangulations. Discret. Comput. Geom. **40**(3), 365–376 (2008)
59. A. Postnikov, Permutohedra, associahedra, and beyond. Int. Math. Res. Not. IMRN. **6**, 1026–1106 (2009). doi:10.1093/imrn/rnn153. http://dx.doi.org/10.1093/imrn/rnn153

60. C.D. Savage, M.J. Schuster, Ehrhart series of lecture hall polytopes and Eulerian polynomials for inversion sequences. J. Comb. Theory A **119**(4), 850–870 (2012). doi:10.1016/j.jcta.2011.12.005. http://dx.doi.org/10.1016/j.jcta.2011.12.005
61. C.D. Savage, M. Visontai, The s-Eulerian polynomials have only real roots. Trans. Am. Math. Soc. **367**(2), 1441–1466 (2015). doi:10.1090/S0002-9947-2014-06256-9. http://dx.doi.org/10.1090/S0002-9947-2014-06256-9
62. J. Schepers, L. van Langenhoven, Unimodality questions for integrally closed lattice polytopes. Ann. Comb. **17**(3), 571–589 (2013). doi:10.1007/s00026-013-0185-6. http://dx.doi.org/10.1007/s00026-013-0185-6
63. R.P. Stanley, Hilbert functions of graded algebras. Adv. Math. **28**(1), 57–83 (1978)
64. R.P. Stanley, Decompositions of rational convex polytopes. Ann. Discret. Math. **6**, 333–342 (1980). Combinatorial mathematics, optimal designs and their applications (Proc. Sympos. Combin. Math. and Optimal Design, Colorado State Univ., Fort Collins, Colo., 1978)
65. R.P. Stanley, Log-concave and unimodal sequences in algebra, combinatorics, and geometry, in *Graph Theory and Its Applications: East and West (Jinan, 1986)*. Annals of the New York Academy of Sciences, vol. 576 (New York Academy of Sciences, New York, 1989), pp. 500–535. doi:10.1111/j.1749-6632.1989.tb16434.x. http://dx.doi.org/10.1111/j.1749-6632.1989.tb16434.x
66. R.P. Stanley, A monotonicity property of h-vectors and h^*-vectors. Eur. J. Comb. **14**(3), 251–258 (1993). doi:10.1006/eujc.1993.1028. http://dx.doi.org/10.1006/eujc.1993.1028
67. R.P. Stanley, *Combinatorics and Commutative Algebra*. Progress in Mathematics, vol. 41, 2nd edn. (Birkhäuser, Boston, 1996)
68. R.P. Stanley, *Enumerative Combinatorics*, vol. 2. Cambridge Studies in Advanced Mathematics, vol. 62 (Cambridge University Press, Cambridge, 1999). doi:10.1017/CBO9780511609589. http://dx.doi.org/10.1017/CBO9780511609589. With a foreword by Gian-Carlo Rota and appendix 1 by Sergey Fomin
69. A. Stapledon, Inequalities and Ehrhart δ-vectors. Trans. Am. Math. Soc. **361**(10), 5615–5626 (2009). doi:10.1090/S0002-9947-09-04776-X. http://dx.doi.org/10.1090/S0002-9947-09-04776-X
70. N.L. White, A unique exchange property for bases. Linear Algebra Appl. **31**, 81–91 (1980). doi:10.1016/0024-3795(80)90209-8. http://dx.doi.org/10.1016/0024-3795(80)90209-8

Simplicial and cellular trees

Art M. Duval, Caroline J. Klivans, and Jeremy L. Martin

Abstract Much information about a graph can be obtained by studying its spanning trees. On the other hand, a graph can be regarded as a 1-dimensional cell complex, raising the question of developing a theory of trees in higher dimension. As observed first by Bolker, Kalai, and Adin, and more recently by numerous authors, the fundamental topological properties of a tree — namely acyclicity and connectedness — can be generalized to arbitrary dimension as the vanishing of certain cellular homology groups. This point of view is consistent with the matroid-theoretic approach to graphs, and yields higher-dimensional analogues of classical enumerative results including Cayley's formula and the matrix-tree theorem. A subtlety of the higher-dimensional case is that enumeration must account for the possibility of torsion homology in trees, which is always trivial for graphs. Cellular trees are the starting point for further high-dimensional extensions of concepts from algebraic graph theory including the critical group, cut and flow spaces, and discrete dynamical systems such as the abelian sandpile model.

Keywords Tree • Forest • Spanning tree • Matrix-tree theorem • Matroid • Simplicial complex • Cell complex • Combinatorial laplacian • Critical group

Mathematics Subject Classification (2010): 05E45; 05B35, 05C05, 05C30, 05C50, 55U10, 57M15

A.M. Duval (✉)
Department of Mathematical Sciences, University of Texas at El Paso, El Paso, TX, USA
e-mail: aduval@utep.edu

C.J. Klivans (✉)
Division of Applied Mathematics and Department of Computer Science,
Brown University, Providence, RI, USA
e-mail: klivans@brown.edu

J.L. Martin (✉)
Department of Mathematics, University of Kansas, Lawrence, KS, USA
e-mail: jlmartin@ku.edu

A. Beveridge et al. (eds.), *Recent Trends in Combinatorics*, The IMA Volumes
in Mathematics and its Applications 159, DOI 10.1007/978-3-319-24298-9_28

1 Introduction

How can redundancy be eliminated from a network, and what can be said about the resulting substructures? In graph-theoretic terms, these substructures are spanning trees. The collection of all spanning trees of a graph has the structure of a matroid basis system; this observation connects trees to algebraic combinatorics and explains why many graph algorithms can be made computationally efficient. The number of spanning trees of a graph measures its complexity as a network, and there are classical and efficient linear-algebraic tools for calculating this number, as well as for enumerating trees more finely. The algebra of trees is key in studying dynamical systems on graphs, notably the abelian sandpile model (known in other forms as the chip-firing game or dollar game), whose possible states are encoded by a group of size equal to the complexity of the underlying graph.

This chapter is about the more recent theory of spanning trees in *cell complexes*, which are natural higher-dimensional analogues of graphs. The theory for cell complexes parallels the graph-theoretical version in many ways, including the connection to matroid theory. However, higher-dimensional spaces can have much richer topology, which complicates the algebraic and enumerative parts of the story in a very concrete way. It turns out that the simple number of spanning trees of a cell complex is not a well-behaved invariant, and it is better to account for topological complexity in the enumeration. On the other hand, many essential algebraic tools for working with spanning trees of a graph do extend well to arbitrary dimension.

Sections 1 and 2 give an overview of the subject, including the history of the subject of higher-dimensional trees and emphasizing the geometry and topology of trees. As far as possible, we have attempted to make these sections self-contained and accessible to readers not familiar with the machinery of algebraic topology, although a certain amount of technical material is unavoidable. The next two sections are more technical: Section 3 describes the various extensions of the matrix-tree theorem from graphs to cell complexes, and Section 4 examines individual classes of cell complexes whose tree enumerators are understood. Finally, Section 5 covers new directions, including extensions of algebraic graph theory and connections to critical groups, cuts and flows, simplicial decompositions, and matroids.

1.1 Higher-dimensional trees

The higher-dimensional trees considered here were first studied by Bolker [21], in the context of transportation polytopes, which are solution sets to certain combinatorial optimization problems. The vertices of transportation polytopes are spanning trees of *complete colorful complexes*, which are higher-dimensional generalizations of complete bipartite graphs. Bolker pinpointed the basic homological conditions on spanning trees (essentially Definition 2.6 below). Kalai [68] considered spanning trees of *complete simplicial complexes*, which generalize complete graphs, and obtained a beautiful enumerative formula generalizing Cayley's formula. Soon after, Adin [3] gave a corresponding formula for counting spanning trees in complete colorful complexes. More recently, many authors, including Duval, Klivans, and Martin [43, 44, 47], Lyons [79] and Petersson [94], independently observed that Bolker and Kalai's definitions could be extended to general cell complexes to develop a broad theory of higher-dimensional trees.

Algebraically speaking, one can define a spanning tree of a suitably connected cell complex to be any subcomplex generated by faces corresponding to a linearly independent set of columns of a certain matrix (see Definition 2.5). However, this definition does not really reflect the topology of higher-dimensional trees. By analogy, a spanning tree T of a connected graph G can be defined as a collection of edges which correspond to a column basis of the oriented incidence matrix of G. But this definition does not directly capture what a tree looks like, and the most familiar definitions are more topological and combinatorial in nature:

- T is connected and acyclic;
- T is connected and $|T| = n - 1$ (where n is the number of vertices);
- T is acyclic and $|T| = n - 1$;
- T is a maximal acyclic subgraph of G;
- T is a minimal connected subgraph of G.

Similarly, for a d-dimensional cell complex X with codimension-1 Betti number $\beta_{d-1}(X) = 0$ (the analogue of connectedness for graphs), a subcomplex $T \subseteq X$ is a spanning tree if it satisfies any of the following equivalent properties:

- $\beta_d(T) = \beta_{d-1}(T) = 0$;
- $\beta_d(T) = 0$ and $|T_d| = |X_d| - \beta_d(X)$;
- $\beta_{d-1}(T) = 0$ and $|T_d| = |X_d| - \beta_d(X)$;
- T is maximal among the spanning subcomplexes of X with $\beta_d(T) = 0$;
- T is minimal among the spanning subcomplexes of X with $\beta_{d-1}(T) = 0$.

A more complete list of characterizations, including the more general notion of spanning forests when $\beta_{d-1}(X) > 0$, is given in Proposition 2.13. Definitions and notation for the terms above are also given in Section 2.

Many cellular spanning trees "look like" trees. For instance, a 2-dimensional simplicial complex whose dual graph forms a tree automatically satisfies the conditions above; see the first complex in Figure 1. On the other hand, many higher-dimensional trees look less familiar. For instance, if X is any cellular sphere (such as the boundary of a polytope), then its spanning trees are the contractible subcomplexes formed by removing a single maximal face; this is the higher-dimensional generalization of the fact that the spanning trees of a cycle graph are the paths formed from it by deleting one edge. Of particular interest (and the source of much difficulty) is the existence of trees with torsion homology, which can occur in any dimension ≥ 2. Already in dimension 2, trees can have finite non-trivial homology. The smallest simplicial example is the standard triangulation of the real projective plane $\mathbb{R}P^2$, which arises as a spanning tree of the 2-dimensional skeleton of the 6-vertex simplex. See Examples 2.11 and 2.12.

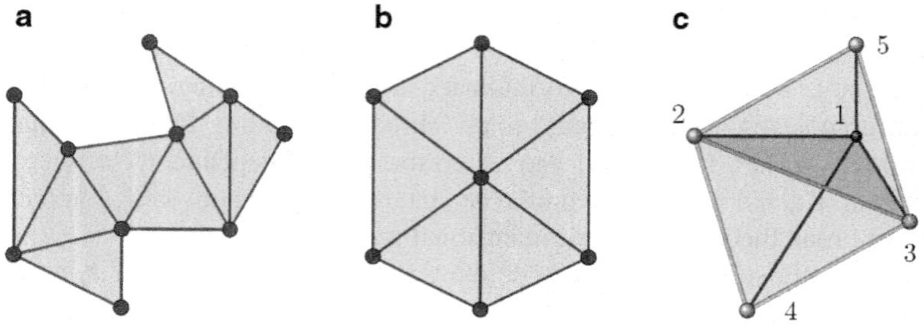

Fig. 1 Some two-dimensional trees.

1.2 Tree numbers

The subject of enumerating spanning trees traditionally dates back to Kirchhoff's 1847 work on electrical circuits [70]. Excellent general references with more historical details include Moon [90] and Stanley [104, pp. 54–69]. In modern terms, let G be a connected graph on n vertices whose Laplacian matrix $L = L(G)$ has non-zero eigenvalues $\lambda_1, \ldots, \lambda_{n-1}$, and let $\mathscr{T}(G)$ denote its set of spanning trees. Then the *matrix-tree theorem* (or *Kirchhoff's theorem*) states that

$$\tau(G) = \det \tilde{L} = \frac{\lambda_1 \cdots \lambda_{n-1}}{n} \tag{1}$$

where $\tau(G) = |\mathscr{T}(G)|$ and $\tilde{L}$ denotes the reduced Laplacian obtained from L by deleting the ith row and column (for any i). Thus tree enumeration is closely tied in with the broad subject of spectral graph theory; see, e.g., [27, 32]. A most important special case is the formula

$$\tau(K_n) = n^{n-2} \tag{2}$$

for the number of spanning trees of the complete graph K_n, known as *Cayley's formula*, although it was in fact first observed by Sylvester [107] and first proved, in an equivalent form, by Borchardt [22]. Cayley's formula can be deduced from the matrix-tree theorem by finding an explicit eigenbasis for $L(K_n)$. Independent combinatorial proofs include the well-known Prüfer code [96] and Joyal's bijection using permutations [67, pp. 15–16]; see also [104, pp. 25–28].

There are other families of graphs with nice spanning tree counts which can be obtained either algebraically (using the matrix-tree theorem) or combinatorially (via a bijection). For example, the number of spanning trees of the complete bipartite graph $K_{m,n}$ is

$$\tau(K_{m,n}) = n^{m-1} m^{n-1} \tag{3}$$

as discovered by several authors [4, 51, 99, 100], and arises in the transportation problem of combinatorial optimization [71].

Moving to higher dimensions, in Bolker's original work, he observed that it is "tempting on numerological grounds" to generalize the well-known formulas (2) and (3), but also was aware that the obvious generalizations were incorrect because of **torsion trees**, i.e.. d-dimensional trees T with $H_{d-1}(T;\mathbb{Z})$ finite but non-trivial. Kalai [68] was the first to pinpoint the role played by torsion. He showed that the correct analogue of the tree count for a d-dimensional simplicial complex X is not simply the cardinality of the set $\mathscr{T}(X)$ of simplicial spanning trees, but rather the quantity

$$\tau(X) = \sum_{T\in\mathscr{T}(X)} |H_{d-1}(T)|^2 \tag{4}$$

where $H_{d-1}(T)$ denotes the $(d-1)$-dimensional reduced homology group of T over $\mathbb{Z}$. The summands in (4) arise naturally in the expansion of the combinatorial Laplacian matrix of X using the Binet–Cauchy formula. Note that torsion has no effect on spanning tree enumeration in graphs; every topological space has torsion-free 0th homology group, so when X is a graph the right-hand side of (4) reduces to the number of spanning trees (On the other hand, torsion does arise in enumerative problems equivalent or related to tree enumeration; see, e.g., [11, 12, 33, 82].) Additional evidence that $\tau(X)$ is the right way to count trees is Kalai's striking generalization of Cayley's formula:

$$\tau(K_{n,d}) = n^{\binom{n-2}{d}} \tag{5}$$

where $K_{n,d}$ denotes the d-dimensional skeleton of an n-vertex simplex. Subsequently, Adin [3] used the same ideas to give a formula for the torsion-weighted count of a complete colorful complex, generalizing (3). Both formulas were precisely those predicted by Bolker. Kalai also gave a weighted version of (5) enumerating spanning trees by their degree sequences [68, Thm. 3′] and showed that torsion accounts for most of the quantity $\tau(X)$: as Kalai put it, spanning trees of $K_{n,d}$ are "on the average, far from being [$\mathbb{Z}$-]acyclic" [68, §5].

The torsion-weighted spanning tree count of a cell complex X^d can be computed from its combinatorial Laplacian operator L. This essential fact, the *cellular matrix-tree theorem*, has appeared in various forms in [25, 43, 44, 79, 94], all building on the ideas originating in [68] and [3]. We will describe the different guises of this theorem, and their interconnections, in Section 3 below. As in the graphical case, $\tau(X)$ may be expressed via a determinant or via eigenvalues. Specifically, for a d-dimensional complex X, the torsion-weighted tree count of d-dimensional trees is given by

$$\tau_d(X) = c\cdot\det\tilde{L} = c'\cdot\frac{\lambda_1\cdots\lambda_k}{\tau_{d-1}(X)}$$

where c and c' are certain correction factors arising from torsion homology; $\tilde{L}$ is the reduced Laplacian formed from L by removing the rows and columns corresponding to a codimension-1 spanning tree; and the λ_i are the non-zero eigenvalues of L.

In practice, for many families of complexes of combinatorial interest (e.g., Cohen-Macaulay complexes), the homological correction factors c and c' disappear. In this case, the weighted tree count is given precisely by a determinant, as in Kalai's enumeration of skeletons of simplices. In general, the eigenvalue formulation of the matrix-tree theorem is recursive; the count for d-dimensional trees is expressed in terms of the count for $(d-1)$-dimensional trees. In the case that $c' = 1$, the recursion yields the alternating-product expression

$$\tau_d(X) = \frac{\boldsymbol{\lambda}_1 \cdot \boldsymbol{\lambda}_3 \cdot \boldsymbol{\lambda}_5 \cdots}{\boldsymbol{\lambda}_0 \cdot \boldsymbol{\lambda}_2 \cdot \boldsymbol{\lambda}_4 \cdots}$$

where $\boldsymbol{\lambda}_i$ is the product of the non-zero eigenvalues of the ith Laplacian.

Laplacians of simplicial complexes first appeared in the work of Eckmann [48] and Dodziuk and Patodi [39]. They were studied as combinatorial objects by Friedman [55] and Friedman and Hanlon [56], who in particular considered the phenomenon of integral Laplacian eigenvalues. Subsequently, many combinatorially significant classes of complexes have been shown to have all integer Laplacian eigenvalues, such as chessboard complexes [56], matroid complexes [72], shifted complexes [41], and hypercubes [44]. The strong connection between Laplacian eigenvalues and tree enumeration suggests that these families might support nice spanning tree counts with combinatorial interpretations. Indeed, there are many successes in this direction including enumerators for shifted complexes in terms of critical pairs [43] and for matroid complexes in terms of flats and the β-invariant [73]. We will look at many of these families individually in Section 4.

1.3 Beyond basic definitions and enumeration

The theory of trees has recently emerged as a natural aspect of the combinatorics of cellular spaces. We describe several new directions in Section 5. Many fundamental objects of algebraic graph theory have been generalized to higher dimension, including critical groups and generalized sandpile/chip-firing models (§5.1) and cuts and flows (§5.2). Simplicial trees provide an inroad to studying theorems and conjectures about decompositions of simplicial complexes (§5.3). Finally, the study of cellular matroids has led to new topological invariants on cellular manifolds and applications to probability (§5.4).

2 Trees and forests: from graphs to cell complexes

The definitions of higher-dimensional trees and forests may be approached from various directions, including linear algebra and matroid theory, but ultimately some topological machinery is necessary to appreciate the ideas fully. We will attempt to give an informal, self-contained description of those concepts that we need such as cell complexes and homology, that will be accessible to a reader willing to endure a little black-box algebra. For the precise details, the reader should consult a standard reference such as Hatcher [63].

2.1 *Trees and forests in graphs*

Let $G = (V, E)$ be a finite graph, with $|V| = n$ and $|E| = m$. For each edge e on vertices v and w, fix an orientation $e = v \to w$. The **signed incidence matrix** ∂_G is then defined as the $n \times m$ matrix with entries

$$\partial_{v\,e} = \begin{cases} 1 & \text{if } e = u \to v \text{ for some } u \neq v, \\ -1 & \text{if } e = v \to w \text{ for some } w \neq v, \\ 0 & \text{otherwise.} \end{cases}$$

In particular, the eth column of ∂ is zero if and only if e is a loop. We now recall several well-known equivalent definitions of trees and forests from graph theory.

Definition 2.1. The graph G is a **tree** if it satisfies any of the following equivalent properties:

(a) G is connected and acyclic.
(b) G is connected and $m = n - 1$.
(c) G is acyclic and $m = n - 1$.
(d) Every pair of vertices in G is connected by exactly one path.
(e) G is a maximal acyclic graph.
(f) G is a minimal connected graph.
(g) For any (hence every) orientation of G, the columns of ∂_G are a basis for the space $\mathbb{R}^n_0 = \{v \in \mathbb{R}^n \mid v_1 + \cdots + v_n = 0\}$.

Meanwhile, G is a **forest** if it satisfies any of the following equivalent properties:

(h) G is acyclic.
(i) Every pair of vertices in G is connected by at most one path.
(j) The columns of the incidence matrix ∂_G of G are linearly independent.
(k) Every connected component of G is a tree.

In order to extend the ideas of trees and forests, we first have to describe the objects that will serve as higher-dimensional analogues of graphs.

2.2 Cell complexes

A **cell complex** X is a topological space that is the disjoint union of *cells*, each homeomorphic to the interior of a k-dimensional disk for some k, attached together in a locally reasonable manner. In particular, the closure of any cell is a union of cells. The set of k-cells is denoted X_k, and the **k-skeleton** $X_{\leq k}$ is the union of all cells of dimension $\leq k$. A cell not contained in the closure of any other cell is called a **facet**, and a complex whose facets all have the same dimension is called **pure**. Every subset S of cells generates a subcomplex $X_S = \overline{\bigcup_{\sigma \in S} \sigma}$. (We often blur the distinction between a set of cells and the subcomplex it generates.) The dimension of X is the largest dimension of one of its cells; the notation X^d indicates a cell complex of dimension d. In particular, a graph is exactly a cell complex of dimension 1.

The combinatorial data about how cells are attached is recorded by a sequence of linear maps $\partial_k = \partial_k(X) : \mathbb{R}^{X_k} \to \mathbb{R}^{X_{k-1}}$ for $0 \leq k \leq d$. The map ∂_1 is just the (signed) incidence matrix of the 1-skeleton graph, and the other maps are higher-dimensional analogues of the signed incidence matrix of a graph. For $k > 0$, each ∂_k may be represented as a matrix with rows indexed by $(k-1)$-cells ρ and columns indexed by k-cells σ; the entry $\partial_{\rho,\sigma}$ is an integer that records, roughly speaking, the number of times that σ is wrapped around ρ, counted with orientation. If ρ is not contained in the closure of σ, then $\partial_{\rho,\sigma} = 0$. In particular, if some $(k-1)$-cell ρ is not in the closure of any k-cell, then the corresponding row of ∂ is zero. We formally define $\partial_0 : \mathbb{R}^{X_0} \to \mathbb{R}$ to be the map that sends every point to $1 \in \mathbb{R}$.

It is a standard fact that $\operatorname{im} \partial_{k+1} \subseteq \ker \partial_k$ for all k, and the quotient $H_k(X; \mathbb{R}) = \ker \partial_k / \operatorname{im} \partial_{k+1}$ is a topological invariant of X called its kth **(reduced) homology group**. The kth **(reduced) Betti number** is $\beta_k(X) = \dim_{\mathbb{R}} H_k(X; \mathbb{R})$. (All homology groups and Betti numbers appearing here are reduced and we will use the notation H_k and β_k in place of the usual $\tilde{H}_k$ and $\tilde{\beta}_k$ for ease of readability.) The 0th reduced Betti number is one less than the number of connected components, so in particular $\beta_0(X) = 0$ if and only if X is connected. If $G = (V, E)$ is a graph with c connected components, then $\beta_1(G) = |E| - (|V| - c)$ is the size of the complement of a maximal spanning forest (sometimes called the *cyclomatic number*).

Simplicial complexes are the cell complexes most familiar to combinatorialists. An (abstract) simplicial complex is a set Δ of subsets of $[n] = \{1, 2, \ldots, n\}$, such that whenever $\sigma \in \Delta$ and $\rho \subseteq \sigma$, then $\rho \in \Delta$. Thus a simplicial complex is completely determined by its set of facets. An abstract simplicial complex can be regarded as a topological space by associating $\sigma = \{i_1, \ldots, i_k\} \in \Delta$ with the interior of the convex hull of $\{e_{i_1}, \ldots, e_{i_k}\}$ in $\mathbb{R}^n$, where e_j is the jth standard basis vector. For readers familiar with hypergraphs, a simplicial complex is just a hypergraph in which every subset of a hyperedge is also a hyperedge.

Simplicial complexes are **regular** cell complexes, which means that every attaching map is locally a homeomorphism; i.e., one-to-one. In particular, for a regular cell complex, every entry of every matrix ∂_k is in $\{0, \pm 1\}$.

Example 2.2. A graph G is a 1-dimensional cell complex whose attaching map ∂_1 is just the signed vertex-edge incidence matrix. As a cell complex, G is regular if G has no loops, and simplicial if in addition G has no parallel edges.

Example 2.3. Gluing two hollow tetrahedra along a common triangle 123 produces a cell complex, the **equatorial bipyramid**, with seven 2-cells (triangles), nine 1-cells (edges), and five 0-cells (vertices). This is an example of a simplicial complex, with facets $\{123, 124, 125, 124, 134, 234, 235\}$ where 1,2,3 are the "equatorial" vertices and 4,5 are the "poles." The collection of facets would often be considered a 3-uniform hypergraph; i.e., a hypergraph where all edges have size 3.

Example 2.4. The real projective plane $\mathbb{R}P^2$ can be realized as a cell complex with one cell σ_k of each dimension k for $k \in \{0, 1, 2\}$. The 1-skeleton $\sigma_1 \cup \sigma_0$ forms the line at infinity; the attaching map ∂_1 is the matrix $[0]$ because σ_1 is a loop. Meanwhile, the attaching map ∂_2 is the matrix $[2]$, representing the fact that σ_2 is wrapped twice around σ_1. The smallest triangulation of $\mathbb{R}P^2$ (i.e., the smallest simplicial complex homeomorphic to $\mathbb{R}P^2$) can be obtained by identifying antipodal faces on an icosahedron. See Figure 3.

2.3 Trees and forests in cell complexes

We can now define the cellular analogues of trees and forests.

Definition 2.5. A d-dimensional cell complex X is a d-**forest** if the columns of $\partial_d(X)$ are linearly independent. A **spanning** d-**forest** of X is a d-forest $F \subseteq X$ such that $F_{\leq d-1} = X_{\leq d-1}$.

Here the term "spanning" refers to the requirement that the $(d-1)$-skeleton of F equals that of X, just as a spanning subgraph of a graph G is a subgraph that includes all vertices. From a homological perspective, an equivalent condition for X to be a d-forest is that $H_d(X; \mathbb{R}) = 0$, because $H_d(X_{\leq d}; \mathbb{R}) = \ker \partial_d(X)$.

Definition 2.6. A d-dimensional cell complex X is a d-**tree** if the columns of $\partial_d(X)$ are linearly independent and $\operatorname{rank} \partial_d(X) = \operatorname{nullity} \partial_{d-1}(X)$. A **spanning** d-**tree** of X is a d-tree $T \subseteq X$ such that $T_{\leq d-1} = X_{\leq d-1}$.

Equivalently, X is a tree if $H_d(X; \mathbb{R}) = H_{d-1}(X; \mathbb{R}) = 0$.

Note that every cell complex has at least one spanning forest (since its boundary matrix has at least one column basis), but a spanning forest of X is a spanning tree if and only if $H_{d-1}(X; \mathbb{R}) = 0$. In general, the conditions $H_d(X; \mathbb{R}) = 0$ and $H_{d-1}(X; \mathbb{R}) = 0$ should be regarded as the analogues of acyclicity and connectedness for a d-dimensional cell complex. Thus Definitions 2.5 and 2.6 say that d-forests and d-trees are respectively "d-acyclic" and "d-acyclic and d-connected" cell complexes.

What do cellular trees and forests look like? A 1-dimensional forest is a graph that contains no cycles (subgraphs homeomorphic to a circle) and a 1-dimensional tree contains no cycles and is connected (i.e., every pair of vertices can be seen as the endpoints of some 1-dimensional path). Correspondingly, a 2-dimensional cell complex X is a forest if X contains no subcomplex homeomorphic to a compact oriented 2-manifold. Imagine a 2-dimensional cell complex as a network of bubbles: in order to make it into a forest, one must remove enough 2-cells to pop all the bubbles, without removing any lower-dimensional cells. Furthermore, X is a 2-tree if every 1-cycle is the boundary of some disk consisting of 2-cells. Definitions 2.5 and 2.6 make sense even when $d = 0$: a 0-dimensional forest is any set of vertices, and a 0-dimensional tree is just a single vertex, which is a 0-spanning tree of X if and only if $X_{\leq 1}$ is a connected graph.

Example 2.7. Figure 1 shows three 2-dimensional simplicial trees. Complex (a) is a triangulation of the disk whose dual graph is a tree. Complex (b) is also a triangulation of the disk, but the dual graph forms a cycle, so it is a 2-tree with no "leaf." Complex (c) is a contractible simplicial complex (hence a tree) with facets 123, 124, 134, 125, 135. It is a spanning tree of the equatorial bipyramid.

Example 2.8. If X^d is homeomorphic to a compact oriented manifold (such as a d-sphere) then $\beta_d(X) = 1$, and the kernel of ∂_d is one-dimensional, generated by a linear relation that involves all its columns non-trivially. Informally, X has exactly one bubble, and it can be popped by removing any facet. Deleting subsequent facets pokes extra holes in the surface. Hence any spanning subcomplex obtained from X by removing one facet (resp., one or more facets) is a maximal spanning forest (resp., a spanning forest).

Example 2.9. Consider the equatorial bipyramid B of Example 2.3. Topologically, it is a "double bubble," recorded homologically by its second Betti number $\beta_2(B) = 2$. In order to pop both bubbles, we can remove either (a) two triangles, one containing vertex 4 and one containing vertex 5, or (b) the middle triangle 123 plus any other triangle. Thus B has a total of $(3 \times 3) + 6 = 15$ spanning trees, all of which are contractible simplicial complexes with 5 facets. See Figure 1(c) and Example 2.7 for one example.

Example 2.10. Figure 2 shows two 2-dimensional simplicial forests, which triangulate the annulus and the Möbius band. In both cases $\beta_2 = 0$ but $\beta_1 = 1 > 0$ (since neither complex is simply connected), so each complex has itself as a spanning 2-forest, but has no spanning tree — just like a non-connected acyclic graph. On the other hand, unlike a 1-dimensional forest, neither complex is the disjoint union of 2-trees.

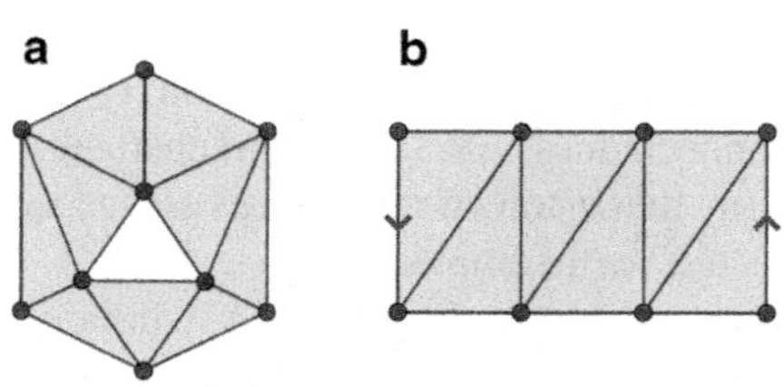

Fig. 2 Two-dimensional forests.

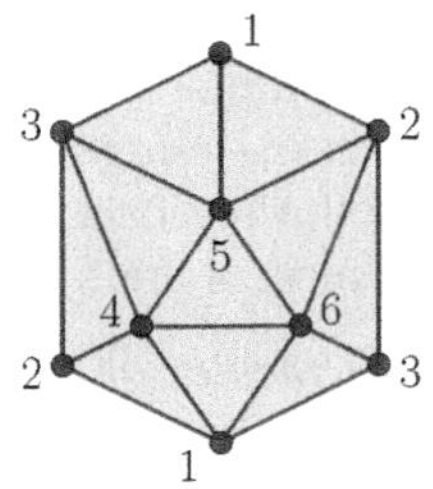

Fig. 3 The triangulation of the real projective plane is a 2-tree with non-trivial torsion.

The connectedness condition $H_{d-1}(X;\mathbb{R}) = 0$ does not preclude the existence of *torsion*. Suppose that we regard the boundary matrix ∂_k as a map on $\mathbb{Z}$-modules $\partial_k^{\mathbb{Z}} : \mathbb{Z}^{X_k} \to \mathbb{Z}^{X_{k-1}}$ (rather than on $\mathbb{R}$-vector spaces), and accordingly work with the integer homology groups $H_k(X;\mathbb{Z}) = \ker \partial_k^{\mathbb{Z}} / \operatorname{im} \partial_{k+1}^{\mathbb{Z}}$. While working over $\mathbb{Z}$ does not change the ranks of the matrices, the condition $\operatorname{rank} \partial_d(X) = \operatorname{nullity} \partial_{d-1}(X)$ no longer implies that $H_{d-1}(X;\mathbb{Z})$ is the trivial group, merely that it is *finite*. This subtlety is concealed in the graph case $d = 1$, since it is a fact that $H_0(X;\mathbb{Z})$ is free abelian for any space X, but is of central importance in higher dimension.

Example 2.11. While every tree graph is a contractible topological space, this is not always true in higher dimension. The simplest example is the real projective plane. In the standard cell structure on $\mathbb{R}P^2$ described in Example 2.4, $\operatorname{im} \partial_2 = 2\mathbb{R} = \ker \partial_1(X)$. So $H_1(\mathbb{R}P^2;\mathbb{R}) = 0$ and $\mathbb{R}P^2$ is a 2-tree (hence the only spanning 2-tree of itself). On the other hand, $H_1(\mathbb{R}P^2;\mathbb{Z}) \cong \mathbb{Z}/2\mathbb{Z} \neq 0$, so $\mathbb{R}P^2$ is not contractible. In the triangulation of $\mathbb{R}P^2$ shown in Figure 3, the cycle consisting of edges 12, 13, and 23 is a representative of the non-trivial torsion homology class.

Example 2.12. Let $K_{6,2}$ be the 2-dimensional skeleton of the 6-vertex simplex, i.e., the 2-dimensional simplicial complex on vertex set $V = [6]$ whose facets are the twenty 3-element subsets of V. Let T_1 be the spanning subcomplex of $K_{6,2}$ whose facets are the ten 3-element sets containing v_1. Then T_1 is contractible (it is a cone), hence is a simplicial spanning tree of $K_{6,2}$ with $H_1(T_1;\mathbb{Z}) = 0$. On the other hand, $\mathbb{R}P^2$ can be triangulated with six vertices (as in Figure 3), and any such triangulation T_2 is also a spanning tree of $K_{6,2}$, with $H_1(T_2;\mathbb{Z}) \cong \mathbb{Z}/2\mathbb{Z}$.

Just as deleting an edge from a graph either decrements the cyclomatic number or increments the number of components, deleting facets of a simplicial complex X^d either decrements β_d or increments β_{d-1}. Thus a d-forest is a subcomplex $T \subset X$ obtained by removing enough facets so that $\beta_d(T) = 0$, and a maximal d-forest is one in which no facets have been removed unnecessarily, so that $\beta_{d-1}(T) = \beta_{d-1}(X)$. If in addition $\beta_{d-1}(X) = 0$, then T is a spanning tree. Thus the vanishing of $\beta_d(T)$ and $\beta_{d-1}(T)$ are the higher-dimensional analogues of the graph-theoretic concepts of acyclicity and connectedness, respectively.

The number of facets of every maximal d-forest $T \subseteq X$ is $|T_d| = |X_d| - \beta_d(X)$, which is the analogue of the statement that a maximal spanning forest of a graph with n vertices and c components has $n - c$ edges.

This discussion leads to the following characterizations of cellular spanning forests and trees, which generalize most of the conditions of Definition 2.1.

Proposition 2.13. *Let X be a cell complex of dimension d and let T contain all faces of dimension $(d-1)$ or less of X. Then T is a maximal spanning forest of X if it satisfies any of the following equivalent properties:*

(a) $\beta_d(T) = 0$ *and* $\beta_{d-1}(T) = \beta_{d-1}(X)$.
(b) $\beta_{d-1}(T) = \beta_{d-1}(X)$ *and* $|T_d| = |X_d| - \beta_d(X)$.
(c) $\beta_d(T) = 0$ *and* $|T_d| = |X_d| - \beta_d(X)$.
(d) Every element of $\ker \partial_{d-1}(X, T)$ *is the boundary of exactly one d-chain in T.*
(e) T is maximal among the spanning subcomplexes of X with $\beta_d(T) = 0$.
(f) T is minimal among the spanning subcomplexes of X with $\beta_{d-1}(T) = \beta_{d-1}(X)$.
(g) The columns of $\partial_d(T)$ are a vector space basis for colspace $\partial_d(X)$.

Meanwhile, T is a spanning forest (not necessarily maximal) if

(h) $\beta_d(T) = 0$.
(i) Every element of $\ker \partial_{d-1}(X, T)$ *is the boundary of at most one d-chain in T.*
(j) The columns of $\partial_d(T)$ are linearly independent.

A maximal spanning forest is a spanning tree when $\beta_{d-1}(X) = 0$. In this case $\ker \partial_{d-1}(X, T)$ becomes $\ker \partial_{d-1}(X)$ in (d) and (i).

When $d = 1$, these properties reduce to the familiar equivalent definitions of trees and forests given in Definition 2.1, letter by letter. One condition that does not carry over to higher dimension is (k) of Definition 2.1, which defines a forest as a graph in which every connected component is a tree. This condition relies on the fact that the incidence matrix $\partial_1(G)$ of a graph G breaks up as a block sum of the incidence matrices of its components, of which there are $\beta_0(G) + 1$. On the other hand, there is no canonical way to decompose a d-dimensional cell complex X, or the matrix $\partial_d(X)$, into $\beta_{d-1}(X) + 1$ pieces. Hence, d-forests are not simply disjoint unions of d-trees, as in the annulus and Möbius band of Example 2.10.

Remark 2.14. If $T \subseteq X$ is a spanning tree, then there is a surjection of finite groups $H_{d-1}(T; \mathbb{Z}) \to H_{d-1}(X; \mathbb{Z})$. In particular, if X has a $\mathbb{Z}$-acyclic spanning tree then $H_{d-1}(X; \mathbb{Z}) = 0$, but the converse does not hold. For example, consider a cell complex with top boundary map $\begin{bmatrix} p & q \end{bmatrix}$ for any relatively prime integers $p, q > 1$.

2.4 Rooted trees and forests

What is the higher-dimensional analogue of a *rooted* tree or forest? In dimension 1, a root of a tree graph is a vertex, while a root of a forest graph is a choice of a vertex from each connected component. As before, in order to generalize these concepts to

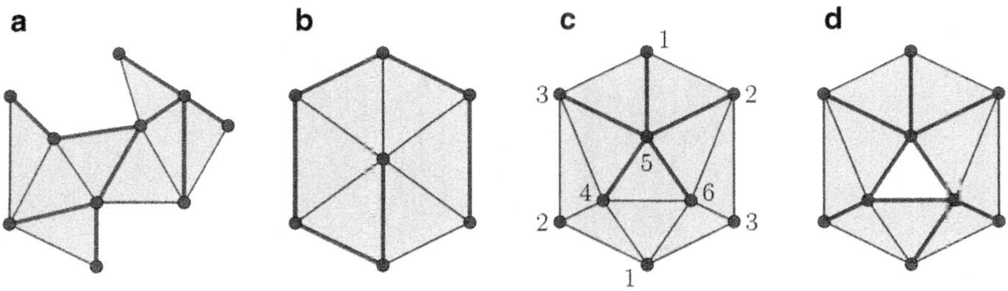

Fig. 4 Rooted trees and forests.

arbitrary dimension, we first state them in terms of linear algebra. It is a standard fact that the nullvectors of a graph G are just the characteristic vectors of its connected components; therefore, a root of a maximal spanning forest of G can be regarded as the complement of a column basis of ∂_G, which motivates the following definition.

Definition 2.15. Let X be a d-dimensional cell complex. A **rooted spanning forest** of X is a pair (F, R), where F is a spanning forest of X; R is a $(d-1)$-dimensional subcomplex of X such that $R_{\leq d-2} = X_{\leq d-2}$; and $\partial_{X \setminus R, F}$ is square and nonsingular, where $\partial_{X \setminus R, F}$ is the submatrix of $\partial_d(X)$ with rows $X \setminus R$ and columns F.

Figure 4 shows rootings of some of the 2-trees encountered previously: the triangulations of the 2-disk (a,b), the projective plane (c), and the annulus (d). The roots are the 1-dimensional subcomplexes indicated with thick lines. In the first three cases $\beta_2(X) = 0$, so a root of X is simply a spanning tree of the codimension-1 skeleton. The annulus has $\beta_1 = 1$, and its root is a unicyclic graph rather than a spanning 1-tree.

In the case that (F, R) is a maximal rooted spanning forest, the root R was called a "relatively acyclic complex" in [47], because the inclusion map $R \hookrightarrow X$ induces isomorphisms $H_i(R; \mathbb{R}) \to H_i(X; \mathbb{R})$ for all $i < d$, a condition equivalent to the vanishing of the relative homology groups $H_i(X, R; \mathbb{R})$ for all $i < d$. Lyons [79] used the term "cobase" for the complement of a root, referring to the fact that a i-cobase is a row basis for the coboundary map ∂^*_{i+1}.

Definition 2.16. Let X be a d-dimensional cell complex. A **directed rooted spanning forest** of X is a triple $(F, R, \mathcal{O})$ where (F, R) is a rooted spanning forest of X, and $\mathcal{O} : X_{d-1} \setminus R_{d-1} \to F_d$ is a bijection such that $\rho \subset \mathcal{O}(\rho)$ for all f. That is, $\mathcal{O}$ pairs each codimension-1 face $\rho \notin R$ with a facet that contains it.

When $d = 1$, this formalism regards an oriented edge $u \to v$ not as an ordered pair of vertices, but as the pairing of vertex u with edge uv. It is a standard fact that every rooting of a tree graph induces a unique orientation of its edges so that each edge is directed towards the root. In higher dimensions, a root can admit more than one orientation; this subject is discussed more fully in §5.3.

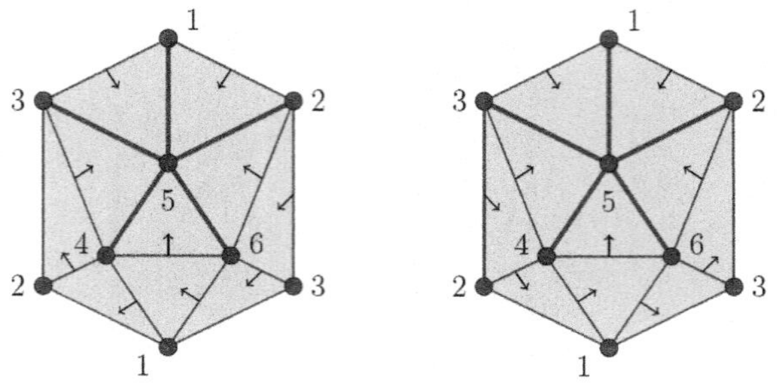

Fig. 5 Two orientations of the rooted spanning tree of $\mathbb{R}P^2$ shown in Fig. 4 (c).

Example 2.17. The rooted trees (a), (b), and (d) in Figure 4 each induces a unique orientation. The rooting of $\mathbb{R}P^2$ shown in (c) induces two distinct orientations: for example, edge 23 can be paired with either triangle 236 or triangle 234. See Figure 5. In general, the number of valid orientations is at least $|H_{d-1}(Y;\mathbb{Z})|$, where Y is the relative complex (F,R), but equality need not hold, as shown by Bernardi and Klivans [16].

2.5 Other formulations of higher-dimensional trees

The definition of a d-forest fits naturally into matroid theory (for a general reference, see, e.g., [92]). The dth **cellular matroid** of a cell complex X is the matroid $M_d(X)$ represented over $\mathbb{R}$ by the columns of $\partial_d(X)$. This is an extension of the definition of a *simplicial matroid*; see [28]. Proposition 2.13 then asserts that the spanning d-forests of X are precisely the independent sets in $M_d(X)$, and the maximal spanning forests are its bases. For a graph, this approach would define spanning trees as the bases of the graphical matroid.

Higher-dimensional trees originate in the work of Bolker [21], Kalai [68], and Adin [3], who studied spanning trees of skeletons of simplices and joins of 0-dimensional complexes, respectively. The foregoing definitions, in varying generality and essentially equivalent but for terminological differences, were formulated by Duval, Klivans, and Martin [43, 44, 47], Lyons [79], and Petersson [94]. The present definition of a maximal spanning forest of a cell complex X matches the definition of "spanning tree" given by Petersson [94, Defn. 3.13]. In addition, Petersson [94, Defn. 3.14] defined a pure connected complex to be a tree if its top homology vanishes (i.e., if it is a forest according to Definition 2.5); by contrast, here Definition 2.6 emphasizes that the higher-dimensional analogue of connectedness is the vanishing of codimension-1 homology. Lyons [79] did not use the terms "tree" and "forest," instead using matroid terminology.

The literature contains many other generalizations of trees to higher dimension: see, e.g., [13, 24, 37, 50, 64, 84]. In general these do not have the same topological properties as the cellular trees considered here, but are defined so as to extend other useful properties of trees. For instance, Faridi's definition [50] extends the recursive description of a tree as a graph formed from a smaller tree by attaching a leaf. Faridi's trees are simplicial complexes whose dual graphs are trees (such as the first, but not the second, tree in Figure 1); the presence of a leaf is important in the study of their facet ideals. In another direction, Masbaum and Vaintrob [84] studied 3-uniform hypergraphs (which can be regarded as pure 2-dimensional simplicial complexes) and found a Pfaffian matrix-tree theorem enumerating their spanning trees.

3 Enumerating spanning trees and forests

This section will be somewhat more technical and algebraic, with the goal of giving precise statements of the various forms of the cellular matrix-tree theorem that have appeared in the literature and explaining how they are interrelated.

3.1 The classical matrix-tree theorem

Let $G = (V, E)$ be a loopless graph with $n = |V|$ and $m = |E|$. Let $e(v, w)$ denote the number of edges between vertices i and j, let $\deg_G(i)$ be the degree of i, and let ∂ be the signed incidence matrix of G (choosing an arbitrary orientation). The **Laplacian matrix** of G is $L(G) = \partial\partial^T$. Explicitly, $L(G) = [\ell_{ij}]_{i,j=1}^n$, where

$$\ell_{ij} = \begin{cases} \deg_G(i) & \text{if } i = j, \\ -e(i,j) & \text{if } i \neq j. \end{cases}$$

The matrix-tree theorem enumerates spanning trees of a graph using its Laplacian matrix.

Theorem 3.1 (Matrix-Tree Theorem). *Let G be a connected graph, let $\mathscr{T}(G)$ be its set of spanning trees, and let $\tau(G) = |\mathscr{T}(G)|$. Then*

$$\tau(G) = \frac{\lambda_1 \cdots \lambda_{n-1}}{n} \tag{6}$$

where the λ_i are the non-zero eigenvalues of $L(G)$. Alternatively, for any i, j,

$$\tau(G) = \det \tilde{L} \tag{7}$$

where $\tilde{L}$ is the **reduced Laplacian** *obtained by deleting the ith row and column for some vertex i (the choice of i does not matter).*

The theorem can be proved by expanding the Laplacian or reduced Laplacian using the Binet–Cauchy formula and observing that the non-vanishing terms correspond exactly to spanning trees. It is also possible to prove (7) using the deletion/contraction recurrence for $\tau(G)$.

Corollary 3.2. *Let G be any graph (not necessarily connected), let $\mathcal{T}(G)$ be its set of maximal spanning forests, and let $\tau(G) = |\mathcal{T}(G)|$. Suppose that G has non-zero Laplacian eigenvalues $\lambda_1, \ldots, \lambda_{n-c}$ and connected components $G_1, \ldots, G_c$ of orders $n_1, \ldots, n_c$. Then*

$$\tau(G) = \frac{\lambda_1 \cdots \lambda_{n-c}}{n_1 \cdots n_c} = \det L_S \tag{8}$$

where L_S is the $(n-c) \times (n-c)$ matrix obtained by choosing a vertex in each G_i and deleting the corresponding c rows and columns from L. Moreover, the number of rooted forests of G is simply the product of the non-zero eigenvalues of $L(G)$.

Indeed, $L(G)$ is the block-diagonal sum $L(G_1) \oplus \cdots \oplus L(G_c)$, so its spectrum is the multiset union of their spectra. Meanwhile, a maximal spanning forest of $L(G)$ is just the union of a spanning tree of each component, and the denominator in (8) is the number of possible choices of roots (since a root consists of a vertex chosen from each component of the graph).

Let $\mathbf{w} = \{w_e\}$ be a family of commuting indeterminates indexed by the edges of K_n. Let G be a graph on $[n]$ and set $w_e = 0$ if $e \notin E(G)$. The **weighted Laplacian matrix** of G is $L(G; \mathbf{w}) = [\hat{\ell}_{i,j}]_{i,j=1}^{n}$, where

$$\hat{\ell}_{ij} = \begin{cases} \sum_{k \neq i} w_{ik} & \text{if } i = j, \\ -w_{ij} & \text{if } i \neq j. \end{cases}$$

The matrix-tree theorem may be extended to the weighted case, producing polynomial tree enumerators that carry finer combinatorial meaning. For example, let $x_1, \ldots, x_n$ be indeterminates corresponding to the vertices of G and set $w_{ij} = x_i x_j$. Then the "Cayley-Prüfer" weighted enumeration formula is

$$\sum_{T \in \mathcal{T}(K_n)} \prod_{i=1}^{n} x_i^{\deg_T(i)} = x_1 \cdots x_n (x_1 + \cdots + x_n)^{n-2}. \tag{9}$$

3.2 Extending the matrix-tree theorem to arbitrary cell complexes

The maximal spanning forests of a cell complex can be enumerated using a higher-dimensional analogue of the matrix-tree theorem. As described above, this idea originated with Kalai [68] and Adin [3] and was extended further by others [25, 43, 44, 47, 79, 94].

Throughout, let X be a d-dimensional pure cell complex. Recall that the kth boundary map $\partial_k : \mathbb{R}^{X_k} \to \mathbb{R}^{X_{k-1}}$ may be regarded as the incidence matrix from k-cells to $(k-1)$-cells. Its adjoint map is the kth **coboundary map** ∂_k^*, which may be regarded as the transpose $[\partial_k^*] = [\partial_k]^T$. The kth **up-down**, **down-up**, and **total Laplacians** are respectively

$$L_k^{\text{ud}}(X) = \partial_{k+1}\partial_{k+1}^*, \qquad L_k^{\text{du}}(X) = \partial_k^*\partial_k, \qquad L_k^{\text{tot}}(X) = L_k^{\text{ud}} + L_k^{\text{du}},$$

all of which may be viewed either as linear operators on $\mathbb{R}^{X_k}$, or equivalently as square matrices of size $|X_k|$. They are all real, diagonalizable, positive semi-definite matrices. The spectrum of $L_k^{\text{tot}}(X)$ is the multiset union of the spectra of $L_k^{\text{ud}}(X)$ and $L_k^{\text{du}}(X)$ (because the up-down and down-up Laplacians annihilate each other). Moreover, the spectra of $L_k^{\text{ud}}(X)$ and $L_{k+1}^{\text{du}}(X)$ coincide up to the multiplicity of the zero eigenvalue. Therefore, the spectra of any of the families $\{L_k^{\text{ud}}(X)\}$, $\{L_k^{\text{du}}(X)\}$, $\{L_k^{\text{tot}}(X)\}$ determines the spectra of the other two. The results here are primarily phrased in terms of the up-down Laplacian, so we abbreviate $L_k^{\text{ud}}(X)$ by $L_k(X)$, or simply L_k when the complex X is clear from the context.

Higher-dimensional tree and forest enumeration requires keeping track of torsion. Accordingly, let $\text{Tor}(A)$ denote the torsion summand of a finitely generated abelian group A, and define

$$t_k(X) = |\,\text{Tor}(H_k(X;\mathbb{Z}))|.$$

We extend this notation: if S is any set of cells of X, we set $t_k(S) = |\,\text{Tor}(H_k(X_S;\mathbb{Z}))|$, where X_S is the subcomplex generated by S.

Let $\mathcal{T}_k(X)$ denote the set of maximal spanning k-forests in X. As in (4) above, the kth **forest count** of X is

$$\tau_k(X) = \sum_{T\in\mathcal{T}_k(X)} t_k(T)^2.$$

Note that when $H_{k-1}(X;\mathbb{R}) = 0$ (i.e., the spanning forests are spanning trees), we have $t_k(T) = |\,\text{Tor}(H_k(T;\mathbb{Z}))| = |H_k(T;\mathbb{Z})|$ for each T. Moreover, when $X = G$ is a graph, every summand $t_0(T)^2$ is 1, so $\tau_1(G)$ is just the number of maximal spanning forests.

3.3 Weightings

As for graphs, one can study weighted analogues of $\tau(X)$ in order to obtain more refined information about its spanning trees and forests.

Let $\mathbf{w} = \{w_F \mid F \in X\}$ be a family of commuting indeterminates indexed by the faces of X. The **weighted forest enumerator** of X is defined as

$$\tau(X;\mathbf{w}) = \tau_k(X;\mathbf{w}) \stackrel{\text{def}}{=} \sum_{T\in\mathcal{T}_k(X)} t_{k-1}(T)^2 \prod_{F\in T} w_F. \tag{10}$$

When X is a simplicial complex, we frequently work with variables v_i indexed by the vertices of X, and set $w_\sigma = \prod_{i\in\sigma} v_i$ for every face σ. We will refer to this weighting as the **vertex weighting** for X, and denote it by $\mathbf{v}$ throughout.

Let D_k denote the diagonal matrix with entries $(w_\sigma : \sigma \in X_k)$. As in [25, 44], one can define[1] a "combinatorial" weighted boundary map $\partial_k D_k^{1/2}$, giving rise to the weighted Laplacian

$$L_{k-1}(X;\mathbf{w}) = \partial_k\, D_k\, \partial_k^*. \tag{11}$$

Note that the combinatorial weighted boundary maps do not form a chain complex. Another approach [2, 74, 83] is to define the weighted boundary map as $D_{k-1}^{-1/2}\partial_k D_k^{1/2}$. These weighted boundaries *do* give rise to a chain complex, and to the "algebraic" weighted Laplacian:

$$L_{k-1}^{\text{alg}}(X;\mathbf{w}) = D_{k-1}^{-1/2}\,\partial_k\, D_k\, \partial_k^*\, D_{k-1}^{-1/2}. \tag{12}$$

3.4 Cellular matrix-forest formulas

The forest counts of a cell complex X can be computed from the spectra of its Laplacians. This relationship appears in several different forms in the literature, all of which share the same underlying principle. We refer to these results collectively as *cellular matrix-forest theorems.*

The relationship between forest counts and the Laplacian becomes simplest when certain rational homology groups of X vanish. In [43, 44] it was assumed that X was **$\mathbb{R}$-acyclic in positive codimension** ($\mathbb{R}$-APC), i.e., that $H_k(X;\mathbb{R}) = 0$ for all $k < d$. This requirement is not necessary and so we weaken the hypotheses slightly. In fact, many complexes of combinatorial interest satisfy a stronger condition: they are **$\mathbb{Z}$-acyclic in positive codimension** ($\mathbb{Z}$-APC), i.e., $H_k(X;\mathbb{Z}) = 0$ for all $k < \dim X$.

[1] This weighted boundary was defined as $\partial_k D_k$ in [44]. Here we adopt the equivalent convention of [25] in order to eliminate squares on the indeterminates in weighted enumerators.

For $S \subseteq X_k$, the notation L_S indicates the square submatrix of $L_k(X)$ restricted to rows and columns indexed by the cells in S. Define the **pseudodeterminant** $\boldsymbol{\lambda}(M)$ of any square matrix M to be the product of its non-zero eigenvalues.

Theorem 3.3 (Cellular Matrix-Forest Theorem). *Let X^d be a cell complex with $H_{d-1}(X;\mathbb{R}) = H_{d-2}(X;\mathbb{R}) = 0$. Let $R \in \mathscr{T}_{d-1}(X)$ and $S = X_{d-1} \setminus R_{d-1}$. Then*

$$\tau_d(X;\mathbf{w}) = \frac{t_{d-2}(X)^2}{t_{d-2}(R)^2} \det L_S(X;\mathbf{w}) \tag{13a}$$

$$= \frac{t_{d-2}(X)^2}{\tau_{d-1}(X)} \boldsymbol{\lambda}(L(X;\mathbf{w})). \tag{13b}$$

Furthermore, for a general cell complex X^d with root R (see Definition 2.15),

$$\tau_d(X;\mathbf{w}) = \frac{t_{d-1}(X)^2}{t_{d-1}(X,R)^2} \det L_S(X;\mathbf{w}). \tag{13c}$$

These three formulas are respectively [43, Thm. 1.4], [44, Thm. 2.8], and [47, Prop. 3.5].[2] In simple terms, (13a) and (13c) state that the forest counts of a complex can be computed from its reduced combinatorial Laplacian, and (13b) states that the forest counts can be computed via a product of eigenvalues of the combinatorial Laplacian.

When X is $\mathbb{Z}$-APC, the correction factors in Theorem 3.3 (e.g., $t_{d-1}(X)$) are equal to 1 and therefore may be omitted.

Example 3.4. Let X be the equatorial bipyramid of Example 2.3; this complex is $\mathbb{Z}$-APC. Let R be the root consisting of all edges that contain vertex 1. Then Theorem 3.3 states that the number of spanning trees equals the determinant of the 5×5 reduced Laplacian formed by removing all rows and columns corresponding to edges in R; this determinant is 15.

The formula (13b) can be iterated to obtain a closed-form expression for the tree count as an alternating product [44, Cor. 2.10]. For simplicity, we state only the $\mathbb{Z}$-APC case.

Corollary 3.5. *Suppose that $H_k(X;\mathbb{Z}) = 0$ for all $k < d$. Then*

$$\tau_d(X) = \prod_{i=0}^{d} \boldsymbol{\lambda}(L_{i-1})^{(-1)^{d-i}}. \tag{14}$$

[2]There is a slight notational error in [47], where the L_Γ in Prop. 3.5 should refer to the reduced Laplacian formed by *removing* the rows and columns corresponding to cells of Γ.

Example 3.6. Let X be the triangulation of the projective plane as given in Figure 3. The pseudodeterminants $\boldsymbol{\lambda}_i = \boldsymbol{\lambda}(L_i)$ of its various Laplacians are:

$$\boldsymbol{\lambda}_2 = 3^4(3+\sqrt{5})^3(3-\sqrt{5})^3 = 3^4 4^3, \qquad \boldsymbol{\lambda}_1 = 6^5, \qquad \boldsymbol{\lambda}_0 = 6.$$

Hence $\tau_2(X) = \boldsymbol{\lambda}_2\boldsymbol{\lambda}_0/\boldsymbol{\lambda}_1 = 4$.

The right-hand side of Equation 14 is essentially the R-torsion (or Reidemeister torsion) of X with respect to the bases of the chain groups given by cells, in the sense of Milnor [89, §8]. Catanzaro, Chernyak, and Klein showed more generally [25, Thm. F] that the square of the R-torsion is the quantity

$$\prod_{k\geq 0}\left(\frac{\operatorname{covol} H_k(X;\mathbb{Z})\operatorname{covol} B_k(X;\mathbb{Z})}{t_k(X)^2}\tau_{k+1}(X)\right)^{(-1)^k}$$

where $B_k(X;\mathbb{Z}) = \operatorname{im}\partial_{k+1}(X)$.

Catanzaro, Chernyak, and Klein [25] obtained a closely related determinant formula for $\tau_d(X)$ in terms of the restriction L_B of $L_{d-1}(X)$ to the space $B = B_{d-1} = \operatorname{im}\partial_{d-1}(X)$ (which is invariant under the action of L_{d-1}). For a lattice (discrete subgroup) $\mathscr{L} \subset \mathbb{R}^n$, define its **covolume**, $\operatorname{covol}\mathscr{L}$, to be the volume of the parallelotope spanned by any (hence every) set of lattice generators. Then:

Theorem 3.7 (Cellular Matrix-Forest Theorem, covolume expansion [25, Thms. C and D]).

$$\tau_d(X;\mathbf{w}) = \frac{t_{d-1}(X)^2}{(\operatorname{covol} B)^2}\det L_B(X;\mathbf{w}).$$

Theorems 3.3 and 3.7 are very close in spirit. Both express the torsion-weighted tree counts of X in terms of determinants of reduced Laplacians. Theorem 3.3 is useful in combinatorial applications, particularly where there is a natural choice of root R (such as when X is a shifted simplicial complex; see §4.3 below). Theorem 3.7 is less combinatorial but has the advantage of expressing $\tau(X)$ purely in terms of invariants of X, with no choice required.

Lyons [79] gave another general formula for $\tau_k(X)$ (which he denoted by $h_{k-1}(X)$) without the requirements that $\beta_{d-1}(X) = \beta_{d-2}(X) = 0$. For a set of k-cells $S \subseteq X_k$, let $t'_k(S)$ denote the size of the group $(\ker_{\mathbb{Z}}\partial_k)/((\ker_{\mathbb{Z}}\partial_k \cap \operatorname{im}_{\mathbb{Q}}\partial_{k+1}) + \ker_{\mathbb{Z}}(\partial_k|_{X_k\setminus S}))$, and define a torsion-weighted enumerator for cobases by

$$h'_k(X) = \sum_{(k+1)\text{-cobases } S} t_k(X_{k+1}\setminus S)\,t'_k(S)^2. \tag{15}$$

Theorem 3.8 (Cellular Matrix-Forest Theorem, cobase expansion [79, Prop. 6.2]). *Let S be a $(d-1)$-cobase of X, i.e., the complement of a $(d-1)$-root R. Then*

$$\tau_d(X) = \frac{t_{d-2}(X)^2}{t_{d-2}(R)^2\, t'_{d-1}(S)^2} \det L_S(X) \tag{16a}$$

$$= \frac{t_{d-2}(X)^2}{h'_{d-2}(X)} \boldsymbol{\lambda}(L(X)). \tag{16b}$$

When $H_{d-1}(X;\mathbb{R}) = H_{d-2}(X;\mathbb{R}) = 0$, these formulas reduce to (13b) and (13b), respectively.

The algebraic weighting (12) can be used to obtain a slightly different matrix-forest formula. Versions of the following result appear in Aalipour, Duval, and Martin [2], Kook and Lee [74, Thm. 4] and Martin, Maxwell, Reiner, and Wilson [83, Thm. 5.3].

Theorem 3.9 (Cellular Matrix-Forest Theorem with algebraic weighting). *Let X be a cell complex of dimension d with weighting $\mathbf{w}$, and suppose that $H_{d-1}(X;\mathbb{R}) = H_{d-2}(X;\mathbb{R}) = 0$. Let $L^{\mathrm{alg}}_{d-1}(X)$ denote the algebraically weighted Laplacian of X as defined in (12). Then*

$$\tau_d(X;\mathbf{w}) = \frac{t_{d-2}(X)^2}{\tau_{d-1}(X;\mathbf{w})} \boldsymbol{\lambda}(L^{\mathrm{alg}}_{d-1}(X)) \prod_{\sigma\in X_{d-1}} w_\sigma . \tag{17}$$

The index d may be replaced with any k for which $H_{k-1}(X;\mathbb{R}) = H_{k-2}(X;\mathbb{R}) = 0$. Observe that setting all the weights w_i to 1 recovers (13b).

Just as iterating (13b) gives (14), iterating Theorem 3.9 yields an alternating-product formula for weighted tree enumerators. For simplicity, we state only the $\mathbb{Z}$-APC case.

Corollary 3.10. *Suppose that X is $\mathbb{Z}$-APC of dimension d, i.e., $H_i(X;\mathbb{Z}) = 0$ for all $i < \dim X$. Then*

$$\tau_d(X;\mathbf{w}) = \prod_{\substack{\sigma\in X\\ \dim\sigma<d}} (w_\sigma)^{(-1)^{d-\dim\sigma-1}} \prod_{k=-1}^{d-1} \boldsymbol{\lambda}(L^{\mathrm{alg}}_k(X))^{(-1)^{d-k-1}} .$$

It is also possible to enumerate rooted forests in terms of products of Laplacian eigenvalues, as observed by Bernardi and Klivans.

Theorem 3.11 (Rooted Cellular Matrix-Forest Theorem [16]). *Let X be a d-dimensional cell complex. Then the characteristic polynomial of the Laplacian $L_d(X)$ gives a generating function for the rooted forests of X:*

$$\sum_{(F,R)} t_{d-1}(X,R)^2 z^{|R|} = \det(L_d(X) + z\cdot \mathrm{Id}). \tag{18}$$

Here Id is the identity matrix with rows and columns indexed by X_{d-1}. The result may also be extended to include indeterminate weights on the faces of the forest and root; see [16, Thm. 18] and [83, Prop. 2.1].

4 Enumeration of trees for specific complexes

We next describe how the theorems of Section 3 have been used to obtain enumerative results for several families of cell complexes. In each case we have provided the full statements of the tree enumeration formulas but refer the reader to the references throughout for detailed treatments of these results.

4.1 *Skeletons of simplices*

Fix integers $n > d \geq 0$, and let $V = [n] = \{1, 2, \ldots, n\}$. The **complete d-dimensional complex on n vertices** is the simplicial complex

$$\Delta = \Delta_{n,d} = \{\sigma \subseteq V \mid \dim \sigma \leq d\}.$$

Equivalently, Δ is the d-dimensional skeleton of the simplex on vertices $[n]$. The complex Δ is $\mathbb{Z}$-APC, with top Betti number $\binom{n-1}{d+1}$.

Kalai [68, Thm. 1] proved that

$$\tau_d(\Delta_{n,d}) = n^{\binom{n-2}{d}} \tag{19}$$

confirming Bolker's [21] guess that this formula was correct and generalizing Cayley's formula $\tau(K_n) = n^{n-2}$. As described in §1.2, Kalai's crucial observation was that the invariant τ_d is the right way to enumerate higher-dimensional trees. The actual *number* of spanning trees probably does not have a closed-form expression; it deviates from Kalai's formula first when $n = 6$ and $d = 2$, since the real projective plane can be triangulated with six vertices; see Example 2.12. To obtain formula (19), Kalai showed that the eigenvalues of the reduced Laplacian are 1 (with multiplicity $\binom{n-2}{d-1}$) and n (with multiplicity $\binom{n-2}{d}$), from which (19) follows immediately.

The weighted Cayley-Prüfer formula (9) also extends nicely to higher dimensions. Kalai [68, Thm. 3′] proved that

$$\tau_d(\Delta_{n,d}; \mathbf{v}) = (v_1 v_2 \cdots v_n)^{\binom{n-2}{d-1}} (v_1 + v_2 + \cdots + v_n)^{\binom{n-2}{d}} \tag{20}$$

where $\mathbf{v}$ is the vertex weighting. It is an open problem to give a bijective proof of this formula. It is not clear how to extend the Prüfer code or Joyal's bijection to arbitrary dimension, or how such a bijection should account for trees with torsion.

4.2 Complete colorful complexes

Let r and $n_1, \ldots, n_r$ be positive integers, and let $V_1, \ldots, V_r$ be pairwise-disjoint vertex sets with $|V_i| = n_i$. The **complete colorful complex** is the $(r-1)$-dimensional complex defined as

$$K_{n_1,\ldots,n_r} = \{\sigma \subset V_1 \cup \cdots \cup V_r \mid \#(\sigma \cap V_i) \leq 1 \ \ \forall i\}.$$

The complexes $K_{n_1,\ldots,n_r}$ are natural generalizations of complete bipartite graphs, which arise as the case $r = 2$. As described in §1.1, the history of higher-dimensional trees owes its origins to the appearance of complete colorful complexes in the transportation problems considered by Bolker [21]. The pure full-dimensional subcomplexes of $K_{n_1,\ldots,n_r}$ are precisely the *balanced simplicial complexes*, which are significant in combinatorial commutative algebra; see, e.g., [103, pp. 95–106].

As observed by Bolker, the complete colorful complex $K_{n_1,\ldots,n_r}$ is $\mathbb{Z}$-APC, with top Betti number $\prod_{i=1}^{r-1}(n_i - 1)$. Adin [3, Thm. 1.5] gave the following closed formula for their tree counts. For $D \subseteq [r]$, set $\sigma_D = \sum_{i \in [r] \setminus D} n_i$ and $\pi_D = \prod_{i \in D}(n_i - 1)$. Then for all $0 \leq k \leq r - 1$,

$$\tau_k(K_{n_1,\ldots,n_r}) = \prod_{\substack{D \subseteq [r] \\ |D| \leq k}} \sigma_D^{\binom{r-2-|D|}{k-|D|}\pi_D}. \tag{21}$$

In the special case $k = r - 1$, Adin's formula becomes

$$\tau_{r-1}(K_{n_1,\ldots,n_r}) = \prod_{i=1}^{r} n_i^{\left(\prod_{j \neq i}(n_j - 1)\right)}. \tag{22}$$

Bolker [21, Thms. 27, 28] observed that (22) gives the exact count of trees if and only if at most two of the n_i's exceed 2 (because the smallest balanced torsion tree is a 9-vertex triangulation of the real projective plane $\mathbb{R}P^2$).

Some additional special cases are worth mentioning. First, setting $r = 2$ in (22) recovers the formula (3) for the number of spanning trees of a complete bipartite graph. Second, if $n_i = 1$ for all i, then $K_{n_1,\ldots,n_r}$ is just the simplex on r vertices, so $\pi_D = 0$ for all D except $D = \emptyset$, when $\pi_\emptyset = 1$, and $\sigma_\emptyset = r$, so (21) specializes to Kalai's formula (19). Third, if $n_i = 2$ for all i, then $K_{n_1,\ldots,n_r}$ is the boundary of the r-dimensional cross-polytope. Now $\pi_D = 1$ for all D and $\sigma_D = 2(r - |D|)$, so (21) becomes

$$\tau_k(K_{2,2,\ldots,2}) = \prod_{d=0}^{k}(2(r-d))^{\binom{r}{d}\binom{r-2-d}{k-d}}.$$

We will have more to say about this case in §4.6 below.

Aalipour, Duval, and Martin [2, Thm. 1.2] used the weighted matrix-tree theorem (Thm. 3.9) to prove a vertex-weighted version of Adin's formula (21), as follows. Let $\mathbf{v} = \{v_{j,t} \mid j \in [r],\ t \in [n_j]\}$ be a family of variables indexed by the vertices of $K_{n_1,\ldots,n_r}$. For $j \in [r]$ and $0 \leq k \leq r-1$, define

$$p_j = \prod_{t=1}^{n_j} v_{j,t}, \qquad s_j = \sum_{t=1}^{n_j} v_{j,t}, \qquad e_{k,j} = \sum_{\substack{D \subseteq [r] \\ j \notin D,\ |D| \leq k-1}} (-1)^{k-1-|D|} \prod_{t \in D} n_t.$$

Then

$$\tau_k(K_{n_1,\ldots,n_r};\mathbf{v}) = \left(\prod_{j=1}^{r} p_j^{e_{k,j}}\right) \prod_{\substack{D \subseteq [r] \\ |D| \leq k}} \left(\sum_{j \in [r]\setminus D} s_j\right)^{\binom{r-2-|D|}{k-|D|}\pi_D}. \tag{23}$$

Note that setting all $v_{j,t}$ to 1 specializes p_j to 1 and $\sum_{j \notin D} s_j$ to σ_D, recovering (21).

Complete colorful complexes play a significant role in the theory of *cyclotomic matroids*, as studied by Martin and Reiner [82]. In the irreducible case, M_n is isomorphic to the dual of the simplicial matroid of the complete colorful complex $K(n) := K_{p_1,\ldots,p_r}$ where n is the product of distinct primes $p_1, \ldots, p_r$. That is, the bases of the matroid are precisely the complements of the facet sets of spanning trees of $K(n)$. Musiker and Reiner [91] extended this work into a beautiful topological interpretation of the coefficients of the nth cyclotomic polynomial Φ_n. In particular, Musiker and Reiner's results explain why $n = 105 = 3 \cdot 5 \cdot 7$ is the first value for which the coefficients of Φ_n are not all ± 1: it is the smallest n for which the complete colorful complex $K(n)$ contains a spanning tree with torsion. In related work, Beck and Hoşten [11] studied *cyclotomic polytopes*, which are dual to the generalized transportation polytopes considered by Bolker, and observed that patterns for their growth series break down for similar reasons involving torsion trees.

4.3 Shifted simplicial complexes

Let $\sigma = \{i_0 < \cdots < i_s\}$ and $\rho = \{j_0 < \cdots < j_r\}$ be sets of positive integers, listed in increasing order. Write $\sigma \preceq \rho$ if $\sigma \subseteq \rho$, or if $s = r$ and $i_k \leq j_k$ for every k. This relation is a partial order, called the **componentwise order**. A simplicial complex Δ^d is called **shifted** if its vertices can be labeled $1, \ldots, n$ so that in each dimension its faces form an order ideal with respect to this order: that is, if $\rho \in \Delta$ and $\sigma \preceq \rho$, then $\sigma \in \Delta$ as well. Shifted complexes are an important class of extremal complexes; every simplicial complex can be transformed into a shifted complex in such a way that many combinatorial properties are preserved while the overall structure is simplified. A signal success of the study of shifted complexes was the classification of all f-vector/Betti number pairs for all simplicial complexes, by Björner and Kalai [19]. Shifted complexes of dimension 1 are known as *threshold graphs* and have generated much study in their own right; see [80].

The homology groups of a shifted complex Δ are free abelian; the Betti number $\beta_k(\Delta)$ is the number of facets of Δ of dimension k not containing vertex 1. To see this, observe that contracting the subcomplex $S = \mathrm{star}_\Delta(1)$ produces a wedge of spheres, one for each facet not containing vertex 1. In particular, if Δ is pure then it is $\mathbb{Z}$-APC, and S is a $\mathbb{Z}$-acyclic spanning tree. For the remainder of this section, we assume that Δ is pure of dimension d on vertex set $[n]$.

Duval and Reiner [41] showed that shifted simplicial complexes are Laplacian integral. Specifically, let $\deg_\Delta(i)$ denote the number of facets containing vertex i, so that $(\deg_\Delta(1), \dots, \deg_\Delta(n))$ is a partition. Then the conjugate (transpose) partition is precisely the spectrum of the top-dimensional Laplacian $L_{d-1}(\Delta)$, up to the multiplicity of the zero eigenvalue. This generalizes the corresponding result for threshold graphs, first observed by Hammer and Kelmans [62, Thm. 5.3] and stated in this form by Merris [88].

Weighted enumerators for spanning trees of threshold graphs were calculated by Martin and Reiner [81, §6], and can be deduced as a special case of a more general bijective result of Remmel and Williamson [98, Thm. 2.4]. Weighted tree enumeration for pure shifted complexes was studied by Duval, Klivans, and Martin [43], as we now summarize. Introduce doubly indexed indeterminates $\mathbf{w} = \{w_{i,j} \mid 0 \le i \le d,\ 1 \le j \le n\}$ and for each face $\sigma = \{a_0 < \cdots < a_k\}$, set $w_\sigma = \prod_{i=0}^{k} w_{i,a_i}$. In other words, w_σ records not just which vertices occur in σ, but their order within σ. Theorem 1.5 of [43] states that the eigenvalues of the weighted Laplacian $L_k(\Delta; \mathbf{w})$ are polynomials in the weights, so that setting all weights to 1 gives an explicit proof of Laplacian integrality for shifted complexes. Specifically, the eigenvalues are in bijection with **critical pairs** of Δ: pairs (σ, ρ) of $(k+1)$-sets of vertices such that $\sigma \in \Delta$, $\rho \notin \Delta$, and σ is covered by ρ in the componentwise order.

Let $\Gamma = \{\rho \in \Delta \mid 1 \notin \rho\}$ and $\Lambda = \{\rho \in \Gamma \mid \rho \cup \{1\} \in \Delta\}$ denote respectively the deletion and link of vertex 1 (which themselves are shifted complexes). Observe that for every critical pair $(\sigma, \rho) = ((a_0, \dots, a_d), (b_0, \dots, b_d))$, there is a unique index j for which $b_j = a_j + 1$ (and $b_i = a_i$ for all $i \neq j$). Define the **signature** of (σ, ρ) to be (S, T), where $S = \{a_0, \dots, a_{j-1}\}$ and $T = \{1, 2, \dots, a_j\}$. Then [43, Thm. 1.6] states

$$\tau_d(\Delta; \mathbf{w}) = \left(\prod_{F \in \Lambda_{d-1}} w_{F \cup \{1\}} \right) \left(\prod_{(S,T) \in \mathrm{Sig}(\Gamma)} \frac{\sum_{j \in T \cup \{1\}} w_{S \cup j}}{w_{S \cup \{1\}}} \right) \tag{24}$$

where $\mathrm{Sig}(\Gamma)$ denotes the set of signatures of critical pairs of Γ. The case $d = 1$ (i.e., Δ is a threshold graph) is equivalent to [81, Thm. 4]. Specializing $w_{i,j} \mapsto v_j$ for every i, j to obtain the vertex weighting $\mathbf{v}$ (called the "coarse weighting" in [43]) produces the formula

$$\tau_d(\Delta; \mathbf{v}) = v_1^{|\Lambda_{d-1}|} \left(\prod_{j=2}^{n} v_j^{\deg_\Lambda(j)} \right) \left(\prod_{(S,T) \in \mathrm{Sig}(\Gamma)} v_1^{-1} \sum_{j \in T \cup \{1\}} v_j \right). \tag{25}$$

4.4 Color-shifted complexes

As in §4.2, let $K = K_{n_1,\dots,n_r}$ be the complete colorful complex with vertex set $V_1 \cup \dots \cup V_r$, where $V_i = \{v_{i,1}, \dots, v_{i,n_i}\}$. A subcomplex $\Delta \subseteq K$ is called **color-shifted** if it satisfies the following property:

$$\text{if } \sigma = \{v_{1,a_1}, v_{2,a_2}, \dots, v_{r,a_r}\} \in \Delta \text{ and } 1 \le b < a_i, \text{ then } \sigma \setminus \{v_{i,a_i}\} \cup \{v_{i,b}\} \in \Delta. \tag{26}$$

Color-shifted complexes were introduced by Babson and Novik [5]. Structurally, color-shifted complexes are to complete colorful complexes as shifted complexes are to simplices. On the other hand, color-shifted complexes (even in dimension 1) are not in general Laplacian integral.

Let G be a color-shifted complex with $r = 2$, i.e., a graph. Assume that G is connected, and let $n = n_1$, $m = n_2$. For $1 \le i \le n$, let $\lambda_i = \max\{j : v_{1,i}v_{2,j} \in E(G)\}$. Then $\lambda = (\lambda_1, \dots, \lambda_n)$ is a partition, and it determines G completely: the edges of G correspond to the boxes in the Ferrers diagram of λ, with V_1 and V_2 corresponding to its rows and columns, respectively. For this reason, 1-dimensional color-shifted complexes were studied under the name of *Ferrers graphs* by Ehrenborg and van Willigenburg, who found an elegant formula for their weighted tree enumerators [49, Thm. 2.1]. Let $G = G_\lambda$ be the Ferrers graph corresponding to a partition λ, and weight the vertices of V_1 and V_2 by $\mathbf{x} = \{x_1, \dots, x_n\}$ and $\mathbf{y} = \{y_1, \dots, y_m\}$, respectively. Then

$$\tau(G_\lambda; \mathbf{x}, \mathbf{y}) = x_1 \cdots x_n y_1 \cdots y_m \prod_{p=1}^{n} (y_1 + \dots + y_{\lambda_p - 1}) \prod_{q=1}^{m} (x_1 + \dots + x_{\tilde{\lambda}_q - 1}) \tag{27}$$

where $\tilde{\lambda}$ is the conjugate partition to λ. The proof of this formula in [49] is recursive, proceeding by calculating $\tau(G_\lambda; \mathbf{x}, \mathbf{y}) / \tau(G_{\lambda'}; \mathbf{x}, \mathbf{y})$, where λ' is a partition obtained from λ by removing a corner of the Ferrers diagram. Another proof, expressing $\tau(G_\lambda; \mathbf{x}, \mathbf{y})$ in terms of the weighted tree enumerator of a related threshold graph, was given in [43, §10.3].

For a color-shifted complex Δ with r colors, let $\mathbf{v} = \{v_{q,i}\}$ be the vertex weighting, where $q \in [r]$ indexes the colors and i indexes the vertices of each color. Let Δ' be the subcomplex of Δ whose faces are $\{\rho \in \Delta \mid v_{q,1} \notin \rho \text{ for any } q\}$. Note that every $(r-2)$-face $\sigma \in \Delta$ is missing a unique color $m = m(\sigma)$; set $k(\sigma) = \max\{j \mid \sigma \cup \{x_{m,j}\} \in \Delta\}$. Aalipour and Duval conjecture [1] that

$$\tau_{r-1}(\Delta; \mathbf{v}) = \prod_{q,i} x_{q,i}^{e(q,i)} \prod_{\sigma \in \Delta'_{r-2}} (x_{m,1} + \cdots + x_{m,k(\sigma)}) \tag{28}$$

where $e(q,i) = \#\{\sigma \in \Delta_{r-1} \mid x_{q,i} \in \sigma \text{ and } x_{q',1} \in \sigma \text{ for some } q' \neq q\}$. Aalipour and Duval have proven this formula in the case $r = 3$ (where a color-shifted complex corresponds to a plane partition). It reduces to (27) when $r = 2$, and to the weighted tree enumerator formula (23) when Δ is a complete colorful complex.

Equation (28) resembles formulas (24) and (25) for weighted tree enumerators of shifted complexes. Specifically, the first factor on the right-hand side is a product of monomials indexed by faces *with* vertices numbered 1, and the second factor is a product of sums indexed by faces *without* vertices numbered 1.

4.5 Matroid complexes

Let M be a matroid on ground set V. The family $\mathscr{I}(M)$ of independent sets of M forms a simplicial complex called a **matroid complex**. These complexes can be characterized in many ways; perhaps the simplest is that a simplicial complex is a matroid complex if and only if every induced subcomplex is pure. Matroid complexes are Cohen-Macaulay and $\mathbb{Z}$-APC. (See Oxley [92] for a general reference on matroids, and Stanley [105, Chap. III] for facts about matroid complexes.)

Every matroid complex is Laplacian integral. This fact was proved first by Kook, Reiner, and Stanton [72], and subsequently generalized by Denham [36], who found an explicit eigenbasis for the Laplacian. An expression for $\tau(\mathscr{I}(M))$ can be extracted from Denham's result (with difficulty) by applying (13b) or (14). One application of the Kook-Reiner-Stanton formula, given in [44, §6.1], is an alternate proof of Adin's formula (21), because the complete colorful complex $K_{n_1,\ldots,n_r}$ is the independence complex of a direct sum of rank-one uniform matroids.

Kook and Lee [73, 74] gave a simpler expression for weighted tree enumerators of matroid complexes. Let $\alpha(M) = T_M(0,1)$, where T_M is the Tutte polynomial of M (see [23]) and $\beta(M) = (-1)^{r(M)} \sum_{A \subseteq E} (-1)^{|A|} r(A)$ (the Crapo beta invariant [30], where r is the rank function; see also [105, pp. 447–448]). Then

$$\tau(\mathscr{I}(M); \mathbf{v}) = \prod_{e \in E} v_e^{|\mathscr{T}(M/e)| - \alpha(M/e)} \prod_{\text{flats } F} \left(\sum_{e \in E \setminus F} v_e \right)^{\alpha(F)\beta(M/F)}$$

[74, Thm. 9], where $\mathbf{v}$ is the vertex weighting. Consequently,

$$\tau(\mathscr{I}(M)) = \prod_{\text{flats } F} (|E \setminus F|)^{\alpha(F)\beta(M/F)}. \tag{29}$$

4.6 Hypercubes

The **hypercube** Q_n is the space $[0,1]^n \subseteq \mathbb{R}^n$, naturally regarded as a cell complex with cells $\{\{0\},\{1\},[0,1]\}^n$. Note that the hypercube is not a simplicial complex, but it has a tight combinatorial structure that enables enumeration of its spanning trees.

The hypercube is contractible, hence $\mathbb{Z}$-APC. The eigenvalues of $L_0(Q_n, r)$ are $0, 2, \ldots, 2j, \ldots, 2n$, with multiplicities $\binom{n}{j}$; this can be shown by directly constructing an eigenbasis as in [104, pp. 61–62]. Applying formula (6) of the classical matrix-tree theorem yields the formula

$$\tau_1(Q_n) = 2^{-n} \prod_{A \subseteq [n]} 2|A| = 2^{2^n-n-1} \prod_{k=2}^{n} k^{\binom{n}{k}}. \tag{30}$$

This formula was discovered (in greater generality) by Cvetković [31, (4.10)], using spectral techniques; see also [32, pp. 75, 218]. The first purely combinatorial proof was recently found by Bernardi [15]. A generalization to higher dimension was obtained by Duval, Klivans, and Martin in [44, Thm. 3.4], who showed that the non-zero eigenvalues of $L_k(Q_n)$ are $2(k+1), \ldots, 2(j+1), \ldots, 2n$, with multiplicities $\binom{j-1}{k}\binom{n}{j}$. Applying the alternating-product formula (14) gives the tree count

$$\tau_k(Q_n) = \prod_{j=k+1}^{n} (2j)^{\binom{n}{j}\binom{j-2}{k-1}}. \tag{31}$$

Weighted tree enumeration of hypercubes proceeds similarly. Introduce indeterminates $\{q_i, x_i, y_i \mid i \in [n]\}$ and weight each face $f = (f_1, \ldots, f_n) \in \{\{0\},\{1\},[0,1]\}^n$ by the monomial

$$w_f = \left(\prod_{i:\, f_i=[0,1]} q_i\right)\left(\prod_{i:\, f_i=0} x_i\right)\left(\prod_{i:\, f_i=1} y_i\right).$$

Now consider the algebraically weighted Laplacian $L_k^{\text{alg}}(Q_n; \mathbf{w})$ (see (12)) for $k \geq 0$. By [44, Thm. 4.2], its non-zero eigenvalues are the rational functions

$u(A) = \sum_{i \in A} q_i/x_i + q_i/y_i$, indexed by subsets $A \subseteq [n]$ with $|A| > k$, with multiplicities $\binom{|A|-1}{k}$. Applying the weighted alternating-product formula[3] (Corollary 3.10) yields

$$\tau_k(Q_n; \mathbf{w}) = (q_1 \cdots q_n)^{\sum_{i=k-1}^{n-1} \binom{n-1}{i}\binom{i-1}{k-2}} \prod_{\substack{A \subseteq [n] \\ |A| > k}} \left(u(A) \prod_{j \in A} x_j y_j \right)^{\binom{|A|-2}{k-1}} \tag{32}$$

as conjectured in [44, Conj. 4.3]. For a complete proof of this formula, see [2]. The case $k = 1$ gives the weighted enumerator for spanning trees of cube graphs [81, Thm. 3].

4.7 *Duality and self-duality*

Let X, Y be cell complexes of dimension d. We say that X and Y are **dual** if there is an inclusion-reversing bijection $\sigma \mapsto \sigma^*$ between their face posets that induces a duality between their cellular chain complexes. On the level of chain complexes, this means that $\partial_{X,k} = \partial^*_{Y,d-k+1}$ for every k, so[4] $L^{\mathrm{ud}}_k(X) = L^{\mathrm{du}}_{d-k}(Y)$. Indeed, $T \subseteq X$ is a spanning k-forest in X if and only if $T^\vee = \{\sigma^* \mid \sigma \in X \setminus T\}$ is a spanning k-forest in Y [44, Prop. 6.1]. Duality carries over to the weighted setting: given a weighting $\mathbf{w}$ on X, define the dual weighting $\mathbf{w}^*$ on Y by $w^*(\sigma^*) = w(\sigma)^{-1}$. Then $L^{\mathrm{ud}}_k(X; \mathbf{w}) = L^{\mathrm{du}}_k(Y; \mathbf{w}^*)$, and likewise for the algebraically weighted Laplacians (see §3.3). By the weighted cellular matrix-tree theorem (Thm. 3.3), $\tau_k(X; \mathbf{w}) = \tau_{d-k}(Y; \mathbf{w}^*)$ for all k. As an application, the $(n-1)$-skeleton of the hypercube Q_n is dual to the complete colorful complex $K_{2,2,\ldots,2}$ (with n 2's), so the formulas (31) and (32) for spanning tree enumerators of Q_n are dual to special cases of (21) and (23), respectively.

A **central reflex** is an embedding of a self-dual planar graph G on the sphere S^2 so that the isomorphism $G \to G^*$ (where G denotes the planar dual) is realized by the antipodal map on $\mathbb{S}^2$. Tutte [108] proved that if G is a central reflex, then $\tau(G)$ is the square of the number of self-dual spanning trees of G. Maxwell [85] generalized Tutte's result to higher dimension. Say that a matroid on ground set $E = \{e_1, \ldots, e_n, \tilde{e}_1, \ldots, \tilde{e}_n\}$ is **involutively self-dual** if it can be represented by an $n \times 2n$ integer matrix M of the form $[I|N]$, where I is the $n \times n$ identity matrix and N is skew-symmetric (so that the fixed-point free involution $E \to E$ given by

---

[3] One needs the convention that $w_\emptyset = 1$, so that $L^{\mathrm{alg}}_{-1}(Q_n)$ is just the 1×1 matrix with entry $\sum_{v \in V(Q_n)} w_v = \prod_{j=1}^n (x_j + y_j)$.

[4] In [44], it was erroneously stated that $L^{\mathrm{ud}}_k(X) = L^{\mathrm{ud}}_{d-k}(Y)$. In fact, duality interchanges up-down and down-up Laplacians.

$e_i \longleftrightarrow \tilde{e}_i$ gives an isomorphism $M \to M^*$). If Δ is the independence complex of such a matroid (so $\dim \Delta = d = n-1$), then [85, Thm. 1.3]

$$\tau_d(\Delta) = \sum_{T \in \mathcal{T}_d(\Delta)} t_{d-1}(T)^2 = \left(\sum_{\text{self-dual } T \in \mathcal{T}_d(\Delta)} t_{d-1}(T) \right)^2 . \tag{33}$$

As an application, let k be a positive integer and let Δ be a simplex on vertex set $[2k+2]$. Then $\tau_k(\Delta_{2k+2,k}) = (2k+2)^{\binom{2k}{k}}$ by Kalai's formula (19). The number $\binom{2k}{k}$ is even for all $k > 0$, so this tree enumerator is a perfect square. The simplicial matroid of the k-skeleton of Δ is involutively self-dual via the map $\sigma \mapsto [2k+2] \setminus \sigma$ so this perfect square phenomenon is explained by (33).

On the level of weighted tree enumerators, the formula for $\tau_k(\Delta_{2k+2,k}; \mathbf{w})$ given by (20) is a perfect square when $n = 2k+2$ and k is odd (since in that case $\binom{n}{k-1}$ must be even). Maxwell [85, Thm. 1.8] proved that this perfect square phenomenon is combinatorial, in the sense that the square root of $\tau_k(\Delta_{2k+2,k}; \mathbf{w}^2)$ has the combinatorial interpretation one would expect from (33):

$$\sum_{T \in \mathcal{T}_k(\Delta)} t_{k-1}(T)^2 w_T^2 = \left(\sum_{\text{self-dual } T \in \mathcal{T}_k(\Delta)} t_{k-1}(T) w_T \right)^2 .$$

The proof identifies the left-hand side with the determinant of a skew-symmetric matrix, and the weighted enumerator for self-dual trees with its Pfaffian.

Self-dual balls of *even* dimension were considered by Martin, Maxwell, Reiner, and Wilson in [83]. If X is a self-dual cellular ball of dimension $d = 2k$ and $\partial = \partial_k(X)$, then $\boldsymbol{\lambda}(L_{k-1}(X)) = \tau_{k-1}(X)\tau_k(X) = \tau_k(X)^2$. More generally, consider a weighting $\mathbf{w}$ such that $w_\sigma = w_{\sigma^*}$, where σ^* is the dual cell to σ. Then $\boldsymbol{\lambda}(L_{k-1}^{\text{alg}}(X; \mathbf{w})) = \tau_k(X; \mathbf{w})^2$, a consequence of [83, Cor. 6.10]. If in addition the self-duality on X is induced by an antipodal map, then it can be shown that X has an orientation giving rise to a symmetric or skew-symmetric boundary matrix ∂ (depending on the parity of k). In either case $\boldsymbol{\lambda}(L_{k-1}) = \boldsymbol{\lambda}(\partial\partial^*) = \boldsymbol{\lambda}(\partial)^2$, so in fact $\tau_k(X) = \tau_{k-1}(X) = \boldsymbol{\lambda}(\partial)$ and more generally $\tau_k(X; \mathbf{w}) = \boldsymbol{\lambda}(D_k\partial)$ by [83, Thm. 7.4].

5 Connections, further directions, and open problems

5.1 Critical groups of cell complexes

The theory of *critical groups* emerges from a discrete dynamical system that has been studied in mathematics, physics, probability, and even economics in several closely related forms, including the *abelian sandpile model*, the *chip-firing game*, and the *dollar game*. Original sources include [8, 17, 20, 38, 77]. See [76] for a short accessible introduction to the subject, and [87] for an overview.

The critical group can be defined purely algebraically from the Laplacian of a graph. In higher dimension, the algebraic definitions have natural analogues using the tools developed earlier in the chapter.

Let $G = (V, E)$ be a graph and let $L = L(G)$ be the graph Laplacian. The **critical group** $K(G)$ is the quotient group

$$K(G) = \text{Tor}(\mathbb{Z}^V / \text{im } L) = \ker \partial_0 / \text{im } L \cong \mathbb{Z}^{|V|-1} / \text{im } \tilde{L}$$

where $\tilde{L}$ is the reduced Laplacian (see Theorem 3.1). The isomorphism is due to Biggs [17, Thm. 4.2]. Regarding a vector $\mathbf{c} \in \mathbb{Z}^V$ as a configuration in which c_i is the number of grains of sand (or chips, or dollars) on vertex v_i, the elements of the critical group can be thought of as equivalence classes of configurations modulo a dispersive Laplacian action. A special system of coset representatives, the *critical configurations*, encodes the possible long-term stable behaviors of the system.

By the matrix-tree theorem, the order of the critical group, and hence the number of distinct long-term configurations, both equal the number of spanning trees of G. There are deep connections between critical configurations and spanning trees, as well as the combinatorics of parking functions, G-parking functions, and hyperplane arrangements; see, e.g., [10, 14, 18, 26, 29, 59, 93].

Accordingly, the ith **critical group** $K_i(X)$ of a cell complex X is defined as

$$K_i(X) = \text{Tor}(\mathbb{Z}^{X_i} / \text{im } L_i) = \ker \partial_i / \text{im } L_i \cong \mathbb{Z}^{|S|} / \text{im } L_S$$

where L_S is the reduced Laplacian obtained by removing a torsion-free i-tree, provided one exists. The isomorphism was proven by Duval, Klivans, and Martin [45, Thm. 3.4] in the simplicial case; the proof carries over to the more general cellular setting. In analogy to the graph case, the size of $K_i(X)$ is the forest count $\tau_{i+1}(X)$.

These higher critical groups model discrete dynamical systems that generalize the sandpile model [45, §5]. A state of the system described by $K_i(X)$ consists of a number for each i-cell; this quantity can be regarded as "charge" ($i = 0$), "current" ($i = 1$), "circulation" ($i = 2$), etc. The Laplacian action redistributes this quantity between i-cells that lie in a common $(i + 1)$-cell. When $i = 0$, this system is the sandpile model described briefly above. When $i = 1$, each state assigns a current to each edge, and redistribution takes the form of circulation around 2-cells. Because the configurations are contained in $\ker \partial_i$, the i-dimensional flow is **conservative**: the sum of oriented flow at every i-dimensional face containing a fixed $(i - 1)$-dimensional face is zero. When $i = 1$, the oriented flow on the edges neither accumulates nor depletes at any vertex. When $i = 2$, the face circulation at each edge adds up to zero.

The model can be regarded as having a sink consisting of an i-dimensional spanning tree. By specifying flow values on all i-faces that are *not* in the sink, the values on the sink needed to make the flow conservative are uniquely defined; this is why the configurations can be described by the cokernel of the reduced Laplacian. Guzmán and Klivans [60] have recently identified meaningful systems of representatives of the higher critical group which reflect the long-term behavior of the Laplacian action.

5.2 Cut and Flow Spaces

The critical group of a cell complex is closely related to its cuts and flows, which generalize the corresponding objects in graph theory. In a connected graph G, an **(edge) cut** is a collection of edges whose removal disconnects the graph, and a **bond** is a minimal cut. In matroid terms, a bond is a cocircuit: a minimal set of edges that meets every basis. The support of every row of the incidence matrix ∂ is a cut, namely the set of all edges incident to a particular vertex, and more generally the **cut space** is defined as im ∂^*, that is, the row space of ∂.

Meanwhile, a **flow** on G is an assignment of a "current" to each edge so that the total flow into each vertex equals the total flow out (called a conservative flow above); that is, a flow is an element of the **flow space** $\ker \partial = (\operatorname{im} \partial^*)^{\perp}$. In matroid terms, the minimal supports of flows are circuits of the graphic matroid (i.e., simple cycles in G). The **cut lattice** $\mathscr{C}(G)$ and **flow lattice** $\mathscr{F}(G)$ are defined in the same way as the cut and flow spaces, except regarding ∂ as a map of $\mathbb{Z}$-modules rather than vector spaces.

Accordingly, for a cell complex X^d, the **cut space**, **cut lattice**, **flow space**, and **flow lattice** of X in dimension k are defined as

$$\begin{aligned} \mathrm{Cut}_k(X) &= \operatorname{im}_{\mathbb{R}} \partial_k^*(X), & \mathscr{C}_k(X) &= \operatorname{im}_{\mathbb{Z}} \partial_k^*(X), \\ \mathrm{Flow}_k(X) &= \ker_{\mathbb{R}} \partial_k(X), & \mathscr{F}_k(X) &= \ker_{\mathbb{Z}} \partial_k(X). \end{aligned}$$

An k-dimensional **bond** of X is a minimal set of k-faces that support a non-zero vector in $\mathrm{Cut}_k(X)$, and a k-dimensional **circuit** of X is a minimal set of k-faces that support a non-zero vector of $\mathrm{Flow}_k(X)$. (These are exactly the cocircuits and circuits, respectively, of the cellular matroid of X; see §5.4 and [106].) Flows and cuts can also be defined homologically: e.g., vectors in $\mathrm{Flow}_d(X)$ are just elements of $H_d(X;\mathbb{R})$, while vectors in $\mathrm{Cut}_d(X)$ are sets of facets whose deletion increases codimension-1 homology.

Given a spanning tree T of a connected graph G, the (signed) characteristic vectors of the fundamental bonds and fundamental cycles of T form bases of the cut and flow spaces, respectively, and in fact they are $\mathbb{Z}$-module bases for the cut and flow lattices. This is a standard fact of algebraic graph theory [57, Chap. 14] that carries over well to general cell complexes [47, §§4–6]. Predictably, it is necessary to account for torsion in higher dimension. The entries of the characteristic vectors of the fundamental bonds and circuits of a cellular spanning tree $T \subseteq X$ can be interpreted as cardinalities of certain torsion groups [47, Thms. 4.11 and 5.3], and in order to form $\mathbb{Z}$-module bases of $\mathscr{C}_d(X)$ or $\mathscr{F}_d(X)$, the tree T must be $\mathbb{Z}$-acyclic.

For an integral lattice $\mathscr{L}$ (that is, a subgroup of $\mathbb{Z}^n$), the **dual lattice** is $\mathscr{L}^\sharp = \{\mathbf{w} \in \mathbb{R}^n \mid \langle \mathbf{w}, \mathbf{v}\rangle \in \mathbb{Z} \ \ \forall \mathbf{v} \in \mathscr{L}\}$, where $\langle\cdot,\cdot\rangle$ is the standard inner product. Thus $\mathscr{L} \subseteq \mathscr{L}^\sharp$, and $\mathscr{L}^\sharp/\mathscr{L}$ is called the **discriminant group** (or determinantal group) of $\mathscr{L}$. Bacher, de la Harpe and Nagnibeda [6] proved that the groups $K(G)$, $\mathscr{C}^\sharp/\mathscr{C}$, $\mathscr{F}^\sharp/\mathscr{F}$, and $\mathbb{Z}^{E(G)}/(\mathscr{C} \oplus \mathscr{F})$ are mutually isomorphic. (See also [57, Chap. 14].) The relationship between the cut and flow lattices, their discriminant groups, and the critical group of X is more complicated in arbitrary dimension. The isomorphisms are replaced with short exact sequences

$$0 \to \mathbb{Z}^n/(\mathscr{C} \oplus \mathscr{F}) \to K(X) \cong \mathscr{C}^\sharp/\mathscr{C} \to E \to 0,$$
$$0 \to E \to \mathbb{Z}^n/(\mathscr{C} \oplus \mathscr{F}) \to K^*(X) \cong \mathscr{F}^\sharp/\mathscr{F} \to 0$$

where E is the "error term" $\mathrm{Tor}(H_{d-1}(X;\mathbb{Z}))$ [47, Thms. 7.6 and 7.7], and $K^*(X)$ is the **cocritical group**, defined roughly by dualizing the definition of the critical group [47, Defn. 7.4]. In particular, the groups $K(X)$, $K^*(X)$, $\mathscr{C}^\sharp/\mathscr{C}$, $\mathscr{F}^\sharp/\mathscr{F}$, and $\mathbb{Z}^{E(G)}/(\mathscr{C} \oplus \mathscr{F})$ are all isomorphic if and only if $H_{d-1}(X;\mathbb{Z})$ is torsion-free.

Bacher, de la Harpe and Nagnibeda's work in [6] arose from viewing a graph as a combinatorial analogue of a Riemann surface, where the critical group appears as the discrete analogue of the Picard group of divisors, or the Jacobian group of holomorphic forms. This point of view also informs the graphical Riemann–Roch theorem of Baker and Norine [9] and subsequent work connected to tropical geometry [61, 66, 78]. An open problem posed in [45, §6] is to understand how the combinatorics of critical and cocritical groups on cell complexes parallels the geometric theory of group invariants such as Chow groups on algebraic varieties.

5.3 *Trees and simplicial complex decomposition*

Recall from Definitions 2.15 and 2.16 that a **directed rooted spanning forest** of a cell complex X^d is a rooted spanning forest (F, R), together with an orientation $\mathscr{O}$ that pairs each $(d-1)$-face ρ outside the root R with a facet $\phi \in F$ that contains ρ. As seen in Example 2.17, rooted forests of dimension ≥ 2 can admit more than one orientation: as shown by Bernardi and Klivans [16], the number of valid orientations is at least $|H_{d-1}(F, R)|$. Equality need not hold: for instance, rooted triangulations of the (torsion-free) dunce cap can admit multiple orientations. In an appropriate linear-algebraic sense, these excess orientations cancel in pairs to the size of the torsion subgroup. It is an open problem to give a combinatorial explanation of this cancellation; see [16].

Iterating the construction of a directed rooted spanning forest gives a discrete (although not gradient) vector field in the sense of discrete Morse theory [53, 54]. To be precise, let $(F, R, \mathscr{O})$ be a directed rooted d-forest of X^d. Then R is a $(d-1)$-dimensional subcomplex of X^d. Choose a directed rooted forest of R which further pairs faces of X^d, this time of dimensions $(d-2)$ and $(d-1)$. Continuing this process, the vector field is then given by the union of the pairings, and the unmatched cells are enumerated by the Betti numbers of X. For example, consider the directed spanning 2-tree of $\mathbb{R}P^2$ shown in (a) or (b) of Figure 5. This is the 1-dimensional root, and the arrows indicate the bijection $\mathscr{O} : F \to X_1 \setminus R$. In the next iteration, choose a single vertex as the 0-dimensional root, and orient all of the edges of R away from the root to match them with nonroot vertices. Finally, match the empty face with the root vertex.

These recursive orientations arise in theorems and conjectures about decompositions of simplicial complexes, with applications to their f- and h-vectors. In particular, certain results about decompositions of simplicial complexes can be usefully restated in the language of simplicial forests and orientations, as we now describe.

Suppose that Δ is a simplicial complex that is acyclic over at least one coefficient ring (it does not matter which), so that the f-polynomial $f(\Delta, x) = \sum_{\sigma \in \Delta} x^{|\sigma|}$ is divisible by $x - 1$. Stanley explained this algebraic fact combinatorially by showing that the face poset of Δ can be decomposed as $\Delta' \dot\cup \Omega$, where Δ' is a subcomplex and there is a bijection $\eta : \Delta' \to \Omega$ such that every $\sigma \in \Delta'$ is a maximal proper subface of $\eta(\sigma)$ [102, Thm. 1.2]. Duval [40, Thm. 1.1] showed more generally that every simplicial complex Δ admits a decomposition $B \dot\cup \Delta' \dot\cup \Omega$, where B contains $\beta_k(\Delta)$ faces for each k and Δ', Ω are as just described. Duval's result (which in fact holds for all regular cell complexes) led to a new proof of Björner and Kalai's characterization of f-vector/Betti number pairs of arbitrary simplicial complexes [19], without using algebraic shifting. In Stanley's and Duval's constructions, the set Ω is a **flag of forests**: an order filter in the face poset of Δ such that for every k, the set $\Omega_k = \{\sigma \in \Omega \mid \dim \sigma = k\}$ is the set of facets of a spanning k-forest of Δ.

More generally, Stanley conjectured [102, Conj. 2.4] that if Δ is **k-uply acyclic**, i.e., if the link of every face of dimension $\leq k-2$ is acyclic, then its face poset can be decomposed into Boolean algebras of rank k. The $k = 1$ case is Stanley's result cited above. The general conjecture remains open, although Duval and Zhang proved a weakening of Stanley's conjecture in [42]. Stanley introduced this conjecture in the same context as the *partitionability conjecture*, which stated that every Cohen-Macaulay simplicial complex Δ can be partitioned into Boolean intervals whose tops are its facets, with ranks given by the h-vector. In light of Remark 2.14, there are implications

every link in X has a $\mathbb{Z}$-acyclic spanning tree	$\Longrightarrow$	X is Cohen-Macaulay	$\Longrightarrow$	every link in X has a spanning tree

because by Reisner's criterion [97], a simplicial complex is Cohen-Macaulay if the link of every face is $\mathbb{Z}$-APC.

The present authors envisioned simplicial spanning trees as an inroad to the partitionability conjecture, culminating in its disproof in [46]. Meanwhile, in a different direction, Katthän [69] had discovered that the *depth conjecture* (proposed by Stanley [101]; see also [65, 95]) could be reduced to the study of certain lattices associated to simplicial spanning trees. While the disproof of the partitionability conjecture implies that the depth conjecture is also false in general, Katthän's methods reveal a deeper connection between trees and combinatorial commutative algebra: e.g., he showed that spanning trees of simplex skeletons (which he called *stoss complexes*) arise as the Scarf complexes of certain extremal lattices [69, Thm. 4.3].

5.4 Trees and Matroids

As mentioned in §2.5, the collection of all spanning forests of a cell complex X forms the collection of bases of a matroid, the *cellular matroid* $M(X)$, which generalizes the simplicial matroids of [28] and coincides with the well-known *graphic matroid* in dimension 1. Matroid structures such as cocircuits are useful in studying cellular trees (as in §5.2), so it is natural to ask what other matroid structures and invariants mean in the context of trees.

Krushkal and Renardy [75] studied a polynomial invariant on cell complexes that generalizes the Tutte polynomial of the cellular matroid. (For a general reference on the Tutte polynomial, see [23].) Let X be a cell complex whose geometric realization is an orientable manifold of even dimension. Its Krushkal–Renardy invariant $P_X(x, y, a, b)$ is a generating function for spanning d-dimensional subcomplexes $Y \subseteq X$, in which the exponents of x, y record combinatorial information about the cellular matroid and those of a, b record topological information about the embedding of Y in X. Setting $a = b = 1$ recovers the Tutte polynomial of $M(X)$, and P satisfies a duality relation $P_X(x, y, a, b) = P_{X^*}(y, x, b, a)$, where X^* is the dual cell complex [75, Thm. 14], generalizing the usual Tutte duality relation $T_M(x, y) = T_{M^*}(y, x)$ on matroids.

Evaluating the Tutte polynomial $T_M(x, y)$ of a matroid M at $(x, y) = (1, 1)$ gives the number of bases of M. If M is the graphic matroid of a graph G then $T_M(1, 1) = \tau(G)$, but this equality does not hold for most cellular matroids because of torsion. On the other hand, torsion can be modeled combinatorially via the theory of *arithmetic matroids*, introduced by d'Adderio and Moci [33–35] and further developed by Fink and Moci [52]. An arithmetic matroid is a matroid equipped with a *multiplicity function* m that models the number of torsion elements in an abelian group. The arithmetic Tutte polynomial studied in [33, 35] records this multiplicity. For a cellular matroid $M(X^d)$, the multiplicity of a set A of facets is $|\operatorname{Tor}(H_{d-1}(A \cup X_{\leq d-1}; \mathbb{Z}))|^2$, and evaluating the arithmetic Tutte polynomial at $(x, y) = (1, 1)$ gives the torsion-weighted tree count of X. Bajo, Burdick and Chmutov [7] similarly modified the Krushkal–Renardy polynomial to include a torsion factor, again obtaining the torsion-weighted tree count as a suitable evaluation of their polynomial invariant.

The matroid structure for trees suggests that graph algorithms could carry over well to the general setting of cell complexes. Of particular interest is the problem of sampling random trees, proposed by Russell Lyons [79] and Igor Pak. In the graph case, there exist efficient algorithms for tree sampling. The best known is Wilson's loop-erased random walk algorithm [109], which generates a uniformly distributed random spanning tree. Extending Wilson's algorithm to general cell complexes is likely to be hard, as there is no known efficient algorithm to sample random bases of an arbitrary matroid. (It is possible for some classes, including equitable matroids; see [86]. Gorodezky and Pak [58] generalized Wilson's algorithm to a restricted class of hypergraphs that do not coincide with cellular forests.) One could look for an algorithm that samples cellular trees either uniformly, or according to a distribution that includes torsion information. Indeed, this problem informs Lyons' motivation in [79], in which he defined determinantal probability measures on cellular trees based on the Laplacian.

Acknowledgements We thank our colleagues Amy Wagler and Jennifer Wagner for valuable suggestions regarding exposition and clarity.

References

1. G. Aalipour, A.M. Duval, Weighted spanning tree enumerators of color-shifted complexes, in preparation
2. G. Aalipour, A.M. Duval, J.L. Martin, A weighted cellular matrix-tree theorem, with applications to complete colorful and cubical complexes, arXiv:1510.00033[math.CO]
3. R.M. Adin, Counting colorful multi-dimensional trees. Combinatorica **12**(3), 247–260 (1992) [MR 1195888 (93j:05036)]
4. T.L. Austin, The enumeration of point labelled chromatic graphs and trees. Can. J. Math. **12**, 535–545 (1960) [MR 0139544 (25#2976)]
5. E. Babson, I. Novik, Face numbers and nongeneric initial ideals. Electron. J. Comb. **11**(2), Research Paper 25, 23 pp. (electronic) (2004/2006) [MR 2195431 (2007c:05202)]
6. R. Bacher, P. de la Harpe, T. Nagnibeda, The lattice of integral flows and the lattice of integral cuts on a finite graph. Bull. Soc. Math. France **125**(2), 167–198 (1997) [MR 1478029(99c:05111)]
7. C. Bajo, B. Burdick, S. Chmutov, On the Tutte-Krushkal-Renardy polynomial for cell complexes. J. Comb. Theory Ser. A **123**, 186–201 (2014) [MR 3157807]
8. P. Bak, C. Tang, K. Wiesenfeld, Self-organized criticality. Phys. Rev. A (3) **38**(1), 364–374 (1988) [MR 949160 (89g:58126)]
9. M. Baker, S. Norine, Riemann-Roch and Abel-Jacobi theory on a finite graph. Adv. Math. **215**(2), 766–788 (2007) [MR 2355607 (2008m:05167)]
10. M. Baker, F. Shokrieh, Chip-firing games, potential theory on graphs, and spanning trees. J. Comb. Theory Ser. A **120**(1), 164–182 (2013) [MR 2971705]
11. M. Beck, S. Hoşten, Cyclotomic polytopes and growth series of cyclotomic lattices. Math. Res. Lett. **13**(4), 607–622 (2006) [MR 2250495 (2007h:52018)]
12. M. Beck, F. Breuer, L. Godkin, J.L. Martin, Enumerating colorings, tensions and flows in cell complexes. J. Comb. Theory Ser. A **122**, 82–106 (2014) [MR 3127679]
13. L.W. Beineke, R.E. Pippert, Properties and characterizations of k-trees. Mathematika **18** 141–151 (1971) [MR 0288046 (44 #5244)]
14. B. Benson, D. Chakrabarty, P. Tetali, G-parking functions, acyclic orientations and spanning trees. Discret. Math. **310**(8), 1340–1353 (2010) [MR 2592488 (2011i:05152)]
15. O. Bernardi, On the spanning trees of the hypercube and other products of graphs. Electron. J. Comb. **19**(4), Paper 51, 16 pp. (2012) [MR 3007186]
16. O. Bernardi, C.J. Klivans, Directed rooted forests in higher dimension, arXiv:1512.07757 [math.CO]
17. N.L. Biggs, Chip-firing and the critical group of a graph. J. Algebraic Comb. **9**(1), 25–45 (1999) [MR 1676732 (2000h:05103)]
18. N. Biggs, P. Winkler, Chip-firing and the chromatic polynomial. Technical Report LSE-CDAM-97-03, London School of Economics, Center for Discrete and Applicable Mathematics, 1997
19. A. Björner, G. Kalai, An extended Euler-Poincaré theorem. Acta Math. **161**(3-4), 279–303 (1988) [MR 971798 (89m:52009)]

20. A. Björner, L. Lovász, P.W. Shor, Chip-firing games on graphs. European J. Comb. **12**(4), 283–291 (1991) [MR 1120415 (92g:90193)]
21. E.D. Bolker, Simplicial geometry and transportation polytopes. Trans. Am. Math. Soc. **217** 121–142 (1976) [MR 0411983 (54 #112)]
22. C.W. Borchardt, Über eine Interpolationsformel für eine Art Symmetrischer Functionen und über Deren Anwendung. Math. Abh. der Akademie der Wissenschaften zu Berlin 1–20 (1860)
23. T. Brylawski, J. Oxley, The Tutte polynomial and its applications, *Matroid applications, Encyclopedia of Mathematics and its Applications*, vol. 40 (Cambridge University Press, Cambridge, 1992), pp. 123–225 [MR 1165543 (93k:05060)]
24. Y. Burman, A. Ploskonosov, A. Trofimova, Matrix-tree theorems and discrete path integration. Linear Algebra Appl. **466**, 64–82 (2015) [MR 3278240]
25. M.J. Catanzaro, V.Y. Chernyak, J.R. Klein, Kirchhoff's theorems in higher dimensions and Reidemeister torsion. Homology Homotopy Appl. **17**(1), 165–189 (2015) [MR 3338546]
26. D. Chebikin, P. Pylyavskyy, A family of bijections between G-parking functions and spanning trees. J. Comb. Theory Ser. A **110**(1), 31–41 (2005) [MR 2128964 (2005m:05010)]
27. F.R.K. Chung, *Spectral Graph Theory*, in CBMS Regional Conference Series in Mathematics, vol. 92. Published for the Conference Board of the Mathematical Sciences, Washington, DC (American Mathematical Society, Providence, RI, 1997) [MR 1421568 (97k:58183)]
28. R. Cordovil, B. Lindström, *Simplicial Matroids*, Combinatorial geometries, Encyclopedia of Mathematical and its Application, vol. 29 (Cambridge Univ. Press, Cambridge, 1987), pp. 98–113 [MR 921070]
29. R. Cori, Y. Le Borgne, The sand-pile model and Tutte polynomials. Adv. Appl. Math. **30** (1-2), 44–52 (2003), Formal power series and algebraic combinatorics (Scottsdale, AZ, 2001) [MR 1979782 (2004d:05095)]
30. H.H. Crapo, A higher invariant for matroids. J. Comb. Theory **2**, 406–417 (1967) [MR 0215744 (35 #5579)]
31. D.M. Cvetković, The spectral method for determining the number of trees. Publ. Inst. Math. (Beograd) (N.S.) **11**(25), 135–141 (1971) [MR 0309772 (46 #8877)]
32. D.M. Cvetković, M. Doob, H. Sachs, *Spectra of Graphs: Theory and Application*. Pure and Applied Mathematics, vol. 87 (Academic, New York-London, 1980) [Harcourt Brace Jovanovich, Publishers] [MR 572262 (81i:05054)]
33. M. D'Adderio, L. Moci, Ehrhart polynomial and arithmetic Tutte polynomial. Eur. J. Comb **33**(7), 1479–1483 (2012) [MR 2923464]
34. M. D'Adderio, L. Moci, Arithmetic matroids, the Tutte polynomial and toric arrangements. Adv. Math. **232**, 335–367 (2013) [MR 2989987]
35. M. D'Adderio, L. Moci, Graph colorings, flows and arithmetic Tutte polynomial. J. Comb. Theory Ser. A **120**(1), 11–27 (2013) [MR 2971693]
36. G. Denham, The combinatorial Laplacian of the Tutte complex. J. Algebra **242**(1), 160–175 (2001) [MR 1844702 (2002h:05039)]
37. A.K. Dewdney, Higher-dimensional tree structures. J. Combinatorial Theory Ser. B **17**, 160–169 (1974) [MR 0369115 (51 #5351)]
38. D. Dhar, Self-organized critical state of sandpile automaton models. Phys. Rev. Lett. **64**(14), 1613–1616 (1990) [MR 1044086 (90m:82053)]
39. J. Dodziuk, V.K. Patodi, Riemannian structures and triangulations of manifolds. J. Indian Math. Soc. (N.S.) **40**(1-4), 1–52 (1977/1976) [MR 0488179 (58 #7742)]
40. A.M. Duval, A combinatorial decomposition of simplicial complexes. Israel J. Math. **87**(1-3), 77–87 (1994) [MR 1286816 (96e:52023)]
41. A.M. Duval, V. Reiner, Shifted simplicial complexes are Laplacian integral. Trans. Am. Math. Soc. **354**(11), 4313–4344 (2002) [MR 1926878 (2003j:15017)]
42. A.M. Duval, P. Zhang, Iterated homology and decompositions of simplicial complexes. Israel J. Math. **121**, 313–331 (2001) [MR 1818393 (2003a:52013)]
43. A.M. Duval, C.J. Klivans, J.L. Martin, Simplicial matrix-tree theorems. Trans. Am. Math. Soc. **361**(11), 6073–6114 (2009) [MR 2529925 (2011a:05385)]

44. A.M. Duval, C.J. Klivans, J.L. Martin, Cellular spanning trees and Laplacians of cubical complexes. Adv. Appl. Math. **46**(1-4), 247–274 (2011) [MR 2794024 (2012e:05182)]
45. A.M. Duval, C.J. Klivans, J.L. Martin, Critical groups of simplicial complexes. Ann. Comb. **17**(1), 53–70 (2013) [MR 3027573]
46. A.M. Duval, B. Goeckner, C.J. Klivans, J.L. Martin, A non-partitionable Cohen-Macaulay simplicial complex, arXiv:1504.04279 (2015, preprint)
47. A.M. Duval, C.J. Klivans, J.L. Martin, Cuts and flows of cell complexes. J. Algebraic Combinatorics **41**, 969–999 (2015) [MR 3342708]
48. B. Eckmann, Harmonische Funktionen und Randwertaufgaben in einem Komplex. Comment. Math. Helv. **17**, 240–255 (1945) [MR 0013318 (7,138f)]
49. R. Ehrenborg, S. van Willigenburg, Enumerative properties of Ferrers graphs. Discret. Comput. Geom. **32**(4), 481–492 (2004) [MR 2096744 (2005j:05076)]
50. S. Faridi, The facet ideal of a simplicial complex. Manuscripta Math. **109**(2), 159–174 (2002) [MR 1935027 (2003k:13027)]
51. M. Fiedler, J. Sedláček, Über Wurzelbasen von gerichteten Graphen. Časopis Pěst. Mat. **83**, 214–225 (1958) [MR 0097071 (20 #3551)]
52. A. Fink, L. Moci, Matroids over a ring. J. Eur. Math. Soc., arXiv:1209.6571, (to appear, 2012)
53. R. Forman, Morse theory for cell complexes. Adv. Math. **134**(1), 90–145 (1998) [MR 1612391 (99b:57050)]
54. R. Forman, A user's guide to discrete Morse theory. Sém. Lothar. Combin. **48**, Art. B48c, 35 (2002) [MR 1939695 (2003j:57040)]
55. J. Friedman, Computing Betti numbers via combinatorial Laplacians. Algorithmica **21**(4), 331–346 (1998) [MR 1622290 (99c:52022)]
56. J. Friedman, P. Hanlon, On the Betti numbers of chessboard complexes. J. Algebraic Combinatorics **8**(2), 193–203 (1998) [MR 1648484 (2000c:05155)]
57. C. Godsil, G. Royle, *Algebraic Graph Theory*, Graduate Texts in Mathematics, vol. 207 (Springer, New York, 2001) [MR 1829620 (2002f:05002)]
58. I. Gorodezky, I. Pak, Generalized loop-erased random walks and approximate reachability. Random Struct. Algoritm. **44**(2), 201–223 (2014) [MR 3158629]
59. J. Guzmán, C. Klivans, Chip-firing and energy minimization on M-matrices. J. Comb. Theory Ser. A **132**, 14–31 (2015) [MR 3311336]
60. J. Guzman, C. Klivans, Chip-firing on general invertible matrices, arXiv:1508.04262 [math.CO]
61. C. Haase, G. Musiker, J. Yu, Linear systems on tropical curves. Math. Z. **270**(3-4), 1111–1140 (2012) [MR 2892941]
62. P.L. Hammer, A.K. Kelmans, Laplacian spectra and spanning trees of threshold graphs. Discret. Appl. Math. **65**(1-3), 255–273 (1996). First International Colloquium on Graphs and Optimization (GOI), 1992 (Grimentz) [MR 1380078 (97d:05205)]
63. A. Hatcher, *Algebraic Topology* (Cambridge University Press, Cambridge, 2002) [MR 1867354 (2002k:55001)]
64. J. Herzog, E.M. Li Marzi, Bounds for the Betti numbers of shellable simplicial complexes and polytopes, in *Commutative Algebra and Algebraic Geometry (Ferrara)*. Lecture Notes in Pure and Application Mathematic, vol. 206 (Dekker, New York, 1999), pp. 157–167 [MR 1702104 (2001b:13017)]
65. J. Herzog, A survey on Stanley depth, in *Monomial Ideals, Computations and Applications*, Lecture Notes in Mathematics, vol. 2083 (Springer, Heidelberg, 2013), pp. 3–45 [MR 3184118]
66. J. Hladký, D. Kráľ, S. Norine, Rank of divisors on tropical curves. J. Comb. Theory Ser. A **120**(7), 1521–1538 (2013) [MR 3092681]
67. A. Joyal, Une théorie combinatoire des séries formelles. Adv. Math. **42**(1), 1–82 (1981) [MR 633783]
68. G. Kalai, Enumeration of **Q**-acyclic simplicial complexes. Israel J. Math. **45**(4), 337–351 (1983) [MR 720308 (85a:55006)]

69. L. Katthän, Stanley depth and simplicial spanning trees. J. Algebraic Combin. **42**(2), 507–536 (2015) [MR 3369566]
70. G. Kirchhoff, ÜIJber die Auflösung der Gleichungen, auf welche man bei der Untersuchung der linearen Verteilung galvanischer Ströme geführt wird. Ann. Phys. Chem. **72** 497–508 (1847)
71. V. Klee, C. Witzgall, Facets and vertices of transportation polytopes, *Mathematics of the Decision Sciences, Part I* (Seminar, Stanford, Calif., 1967), (American Mathematical Society, Providence, RI, 1968), pp. 257–282 [MR 0235832]
72. W. Kook, V. Reiner, D. Stanton, Combinatorial Laplacians of matroid complexes. J. Am. Math. Soc. **13**(1), 129–148 (2000) [MR 1697094 (2001e:05028)]
73. W. Kook, K.-Ju Lee, A formula for simplicial tree numbers of matroid complexes, Europ. J. Combin., to appear (2015)
74. W. Kook, *Weighted Tree-Numbers of Matroid Complexes*, 27th International Conference on Formal Power Series and Algebraic Combinatorics (FPSAC 2015), Discrete Mathematics & Theoretical Computer Science Proceedings, AS, Association of Discrete Mathematics & Theoretical Computer Science, Nancy, pp. 709–720 (to appear, 2015)
75. V. Krushkal, D. Renardy, A polynomial invariant and duality for triangulations. Electron. J. Combin. **21**(3), Paper 51, 16 pp. (2012) [MR 3262279]
76. L. Levine, J. Propp, What is ... a sandpile?. Not. Am. Math. Soc. **57**(8), 976–979 (2010) [MR 2667495]
77. D.J. Lorenzini, A finite group attached to the Laplacian of a graph. Discret. Math. **91**(3), 277–282 (1991) [MR 1129991 (93a:05091)]
78. Y. Luo, Rank-determining sets of metric graphs. J. Comb. Theory Ser. A **118**(6), 1775–1793 (2011) [MR 2793609 (2012d:05122)]
79. R. Lyons, Random complexes and l^2-Betti numbers. J. Topol. Anal. **1**(2), 153–175 (2009) [MR 2541759 (2010k:05130)]
80. N.V.R. Mahadev, U.N. Peled, *Threshold Graphs and Related Topics*, Annals of Discrete Mathematics, vol. 56 (North-Holland Publishing Co., Amsterdam, 1995) [MR 1417258 (97h:05001)]
81. J.L. Martin, V. Reiner, Factorization of some weighted spanning tree enumerators. J. Comb. Theory Ser. A **104**(2), 287–300 (2003) [MR 2019276 (2004i:05070)]
82. J.L. Martin, V. Reiner, Cyclotomic and simplicial matroids. Israel J. Math. **150**, 229–240 (2005) [MR 2255809 (2007g:05040)]
83. J.L. Martin, M. Maxwell, V. Reiner, S.O. Wilson, Pseudodeterminants and perfect square spanning tree counts. J. Comb. **6**(3), 295–325 (2015) [MR 3357126]
84. G. Masbaum, A. Vaintrob, A new matrix-tree theorem. Int. Math. Res. Not. **27**, 1397–1426 (2002) [MR 1908476 (2003a:05107)]
85. M. Maxwell, Enumerating bases of self-dual matroids. J. Comb. Theory Ser. A **116**(2), 351–378 (2009) [MR 2475022 (2010a:05048)]
86. D. Mayhew, Equitable matroids. Electron. J. Comb. **13**(1), Research Paper 41, 8 pp. (electronic) (2006) [MR 2223516 (2007c:05047)]
87. C. Merino, The chip-firing game. Discret. Math. **302**(1-3), 188–210 (2005) [MR 2179643 (2007c:91036)]
88. R. Merris, Degree maximal graphs are Laplacian integral. Linear Algebra Appl. **199**, 381–389 (1994) [MR 1274427 (95e:05083)]
89. J. Milnor, Whitehead torsion. Bull. Am. Math. Soc. **72**, 358–426 (1966) [MR 0196736 (33 #4922)]
90. J.W. Moon, *Counting Labelled Trees*, From lectures delivered to the Twelfth Biennial Seminar of the Canadian Mathematical Congress (Vancouver, vol. 1969, (Canadian Mathematical Congress, Montreal, Quebec, 1970) [MR 0274333 (43 #98)]
91. G. Musiker, V. Reiner, The cyclotomic polynomial topologically. J. Reine Angew Math. **687**, 113–132 (2014) [MR 3176609]
92. J. Oxley, *Matroid Theory*, Oxford Graduate Texts in Mathematics, 2nd ed. vol. 21 (Oxford University Press, Oxford, 2011) [MR 2849819 (2012k:05002)]

93. D. Perkinson, J. Perlman, J. Wilmes, *Primer for the Algebraic Geometry of Sandpiles*, Tropical and non-Archimedean geometry, Contemporary Mathematics, vol. 605 (American Mathematical Society, Providence, RI, 2013), pp. 211–256 [MR 3204273]
94. A. Petersson, Enumeration of spanning trees in simplicial complexes. Master's thesis, Uppsala Universitet (2009)
95. M.R. Pournaki, S.A. Seyed Fakhari, M. Tousi, S. Yassemi, What is ... Stanley depth?. Not. Am. Math. Soc. **56**(9), 1106–1108 (2009) [MR 2568497 (2010k:05346)]
96. H. Prüfer, Neuer beweis eines satzes über permutationen. Arch. Math. Phys. **27**, 142–144 (1918)
97. G.A. Reisner, Cohen-Macaulay quotients of polynomial rings. Adv. Math. **21**(1), 30–49 (1976) [MR 0407036 (53 #10819)]
98. J.B. Remmel, S. Gill Williamson, Spanning trees and function classes. Electron. J. Comb. **9**(1), Research Paper 34, 24 pp. (electronic) (2002) [MR 1928786 (2003g:05067)]
99. H.I. Scoins, The number of trees with nodes of alternate parity. Proc. Cambridge Philos. Soc. **58**, 12–16 (1962) [MR 0136554 (25 #24)]
100. M.A. Simmonard, G.F. Hadley, The maximum number of iterations in the transportation problem. Naval Res. Logistics Q. **6**, 125–129 (1959) [MR 0108401 (21 #7117)]
101. R.P. Stanley, Linear Diophantine equations and local cohomology. Invent. Math. **68**(2), 175–193 (1982) [MR 666158 (83m:10017)]
102. R.P. Stanley, A combinatorial decomposition of acyclic simplicial complexes. Discret. Math. **120**(1-3), 175–182 (1993) [MR 1235904 (94k:55027)]
103. R.P. Stanley, *Combinatorics and Commutative Algebra*, 2nd ed., Progress in Mathematics, vol. 41 (Birkhäuser Boston, Inc., Boston, MA, 1996) [MR 1453579 (98h:05001)]
104. R.P. Stanley, *Enumerative Combinatorics, Vol. 2*. Cambridge Studies in Advanced Mathematics, vol. 62 (Cambridge University Press, Cambridge, 1999). With a foreword by Gian-Carlo Rota and appendix 1 by Sergey Fomin. [MR 1676282 (2000k:05026)]
105. R.P. Stanley, *An Introduction to Hyperplane Arrangements*, Geometric combinatorics, IAS/Park City Math. Ser., vol. 13, (American Mathematical Society, Providence, RI, 2007), pp. 389–496 [MR 2383131]
106. Y. Su, D.G. Wagner, The lattice of integer flows of a regular matroid. J. Comb. Theory Ser. B **100**(6), 691–703 (2010) [MR 2718687 (2012a:05066)]
107. J.J. Sylvester, On the change of systems of independent variables. Quart. J. Math. **1** 42–56 (1857). [Collected Mathematical Papers, vol. 2, Cambridge, 1908, pp. 65–85]
108. W. T. Tutte, *On the Spanning Trees of Self-Dual Maps*, Second International Conference on Combinatorial Mathematics (New York, 1978), Annals of the New York Academy of Sciences, vol. 319, (New York Academy of Sciences, New York, 1979), pp. 540–548 [MR 556066 (81c:05031)]
109. D.B. Wilson, Generating Random Spanning Trees More Quickly than the Cover Time, in *Proceedings of the Twenty-eighth Annual ACM Symposium on the Theory of Computing (Philadelphia, PA, 1996)*, (ACM, New York, 1996), pp. 296–303 [MR 1427525]

Parametric polyhedra with at least k lattice points: Their semigroup structure and the k-Frobenius problem

Iskander Aliev, Jesús A. De Loera, and Quentin Louveaux

Abstract Given an integral $d \times n$ matrix A, the well-studied affine semigroup $\mathrm{Sg}(A) = \{b : Ax = b,\ x \in \mathbb{Z}^n, x \geq 0\}$ can be stratified by the number of lattice points inside the parametric polyhedra $P_A(b) = \{x : Ax = b, x \geq 0\}$. Such families of parametric polyhedra appear in many areas of combinatorics, convex geometry, algebra, and number theory. The key themes of this paper are: (1) A structure theory that characterizes precisely the subset $\mathrm{Sg}_{\geq k}(A)$ of all vectors $b \in \mathrm{Sg}(A)$ such that $P_A(b) \cap \mathbb{Z}^n$ has *at least* k solutions. We demonstrate that this set is finitely generated, it is a union of translated copies of a semigroup which can be computed explicitly via Hilbert bases computations. Related results can be derived for those right-hand-side vectors b for which $P_A(b) \cap \mathbb{Z}^n$ has *exactly* k solutions or *fewer than* k solutions. (2) A computational complexity theory. We show that, when n, k are fixed natural numbers, one can compute in polynomial time an encoding of $\mathrm{Sg}_{\geq k}(A)$ as a multivariate generating function, using a short sum of rational functions. As a consequence, one can identify all right-hand-side vectors of bounded norm that have at least k solutions. (3) Applications and computation for the k-Frobenius numbers. Using generating functions we prove that for fixed n, k the k-Frobenius number can be computed in polynomial time. This generalizes a well-known result for $k = 1$ by R. Kannan. Using some adaptation of dynamic programming we show some practical computations of k-Frobenius numbers and their relatives.

I. Aliev
Cardiff University, Cardiff, UK
e-mail: AlievI@cardiff.ac.uk

J.A. De Loera (✉)
University of California, Davis, USA
e-mail: deloera@math.ucdavis.edu

Q. Louveaux
Université de Liège, Liège, Belgium
e-mail: q.louveaux@ulg.ac.be

A. Beveridge et al. (eds.), *Recent Trends in Combinatorics*, The IMA Volumes in Mathematics and its Applications 159, DOI 10.1007/978-3-319-24298-9_29

Keywords Lattice points in polyhedra • Linear Diophantine equations • Hilbert bases • Combinatorial commutative algebra • Combinatorial number theory • Affine semigroups • Frobenius numbers

Mathematics Subject Classification (2010): 52C07; 52B; 11D07; 05E40; 05A15

1 Introduction

An *affine semigroup* is a semigroup (always containing a zero element) which is finitely generated and can be embedded in $\mathbb{Z}^n$ for some n. This paper studies special polyhedral affine semigroups that appear naturally in many interesting problems in combinatorics, convexity, commutative algebra, and number theory and that can be described in very explicit terms.

Given an integer matrix $A \in \mathbb{Z}^{d \times n}$ and a vector $b \in \mathbb{Z}^d$, we study the semigroup $\mathrm{Sg}(A) = \{b : b = Ax,\ x \in \mathbb{Z}^n, x \geq 0\}$. Geometrically $\mathrm{Sg}(A)$ can be described as *some* of the lattice points inside the convex polyhedral cone cone(A) of nonnegative linear combinations of the columns of A. It is well-known that $\mathrm{Sg}(A) \subset \mathrm{cone}(A) \cap \mathbb{Z}^d$, but the equality is not always true. It is also well-known that $\mathrm{cone}(A) \cap \mathbb{Z}^d$ is a finitely generated semigroup, this time with generators given by the *Hilbert bases* of cone(A) [40, 61]. The study of the difference between $\mathrm{Sg}(A)$ and $\mathrm{cone}(A) \cap \mathbb{Z}^n$ is quite interesting (e.g., it has been part of many papers in commutative algebra about semigroups and their rings. See [12, 51, 63, 64] and the references therein).

Practically speaking, membership of b in the semigroup $\mathrm{Sg}(A)$ reduces to the challenge, given a vector b, to find whether the linear Diophantine system

$$Ax = b, \quad x \geq 0, \quad x \in \mathbb{Z}^n, \tag{1}$$

has a solution or not. Geometrically, problem (1) asks whether there is at least one lattice point inside the *parametric polyhedron* $P_A(b) = \{x : Ax = b, x \geq 0\}$.

Now, for a given integer k, there are three variations of the classical feasibility problem above that in a natural way measure the number of integer points in $P_A(b)$:

- Are there *at least* k distinct solutions for $P_A(b) \cap \mathbb{Z}^n$? If yes, we say that the polyhedron $P_A(b)$ is $\geq$ *k-feasible.*
- Are there *exactly* k distinct solutions for $P_A(b) \cap \mathbb{Z}^n$? If yes, we say that the polyhedron $P_A(b)$ is $=$ *k-feasible.*
- Are there *less than* k distinct solutions for $P_A(b) \cap \mathbb{Z}^n$? If yes, we say that the polyhedron $P_A(b)$ is $<$ *k-feasible.*

The standard integer feasibility problem is just the problem of deciding whether $P_A(b)$ is $\geq$ 1-feasible. We call these three problems, *the fundamental problems of k-feasibility*, thus these problems are already NP-hard in complexity for $k = 1$. For convenience $IP_A(b)$ will denote the integer points inside $P_A(b)$. Similarly, we say that b is $\geq$ k-feasible, or respectively, $=$ k-feasible or $<$ k-feasible if the corresponding polyhedron $P_A(b)$ is.

This paper investigates the question of determining, given an integral matrix A, which right-hand-side vectors b are $\geq k$-feasible, $= k$-feasible, or $< k$-feasible and the structure of the corresponding sets of all such vectors. The first part of this paper is about a decomposition or stratification of the semigroup $\mathrm{Sg}(A)$ by values of k. The decomposition structure we unveil is as translated semigroups that group together all elements $b \in \mathrm{Sg}(A)$ that are $\geq k$-feasible (similarly for the other cases). In the second part we consider the computational complexity of deciding whether a vector b is $\geq k$-feasible (and similarly for $< k$ and $= k$ feasibility). The final answer is a clean nice application of the theory of combinatorial commutative algebra. The algorithms we propose in the second part are rather interesting theoretically, yet not practical beyond a few dozen variables, thus the third part focuses in computer experimentation with the k-Frobenius number, an invariant of interest in combinatorial number theory which is a special case for $1 \times n$ matrices.

1.1 Motivation, prior, and related work

Why study k-feasibility? The three k-feasibility problems appear directly in a wide range of situations and are also special cases of important algebraic and geometric problems we describe now.

To begin, k-feasibility is strongly connected to a classical problem in combinatorial number theory. Let a be a positive integral n-dimensional primitive vector, i.e., $a = (a_1, \ldots, a_n)^T \in \mathbb{Z}^n_{>0}$ with $\gcd(a_1, \ldots, a_n) = 1$. For a positive integer k, the *k-Frobenius number* $\mathrm{F}_k(a)$ is the largest number which cannot be represented in *at least* k different ways as a nonnegative integral combination of the a_i's. Thus, putting $A = a^T$,

$$\mathrm{F}_k(a) = \max\{b \in \mathbb{Z} : IP_A(b) \text{ is } < k\text{-feasible}\}. \tag{2}$$

When $k = 1$ this was studied by a large number of authors and both the structure and algorithmic properties are well-understood. Computing $\mathrm{F}_1(a)$ when n is not fixed is an NP-hard problem (Ramirez Alfonsin [57]). On the other hand, for any fixed n the classical Frobenius number can be found in polynomial time by sophisticated deep algorithms due to Kannan [41] and later Barvinok and Woods [17]. The general problem of finding $\mathrm{F}_1(a)$ has been traditionally referred to as the *Frobenius problem*. There is a rich literature on the various aspects of this question. For a comprehensive and extensive survey we refer the reader to the book of Ramirez Alfonsin [58].

The k-feasibility framework yields a generalization of the Frobenius number which was introduced and studied by Beck and Robins in [18]. Now one is interested on finding the largest number b that cannot be represented in more than k ways. Beck and Robins gave formulas for $n = 2$ of the k-Frobenius number, but for general n and k only bounds on the k-Frobenius number $\mathrm{F}_k(a)$ are available (see [5, 6] and [36] for prior work). Since then, many number theorists have contributed to the topic (e.g., see [23, 62] and the references therein). Note also that Frobenius problem is highly related to what optimizers call the knapsack problem and it has many applications (see [1, 2, 6, 42] and the references therein).

The structure of k-feasibility is also closely related to the properties of affine semigroups and the associated commutative rings. Fix a semigroup S and a field K. The *semigroup algebra* over S with coefficients in K, denoted $K[S]$, is the set of finite formal sums of elements of the form cx^q for $q \in S$ and $c \in K$. Here the multiplication is given by the structure of the semigroup addition $x^a \cdot x^b = x^{a+b}$. The ring theoretic properties of $K[S]$ and the structure of S are interconnected, e.g., the Krull dimension of $K[S]$ equals the rank of S. There are a number of questions about semigroup rings which play an important role in the theory of toric varieties (as the coordinate rings of general toric varieties) and the structure of lattice polyhedra, rational cones, and linear Diophantine systems (see, e.g., [38, 51, 64] and the references therein). For example, the question of when is the semigroup algebra normal, is equivalent to the case of determining 1-feasibility (see, e.g., [24]). There are many interesting affine semigroups in algebraic combinatorics, e.g., through the work of several authors, the values of Littlewood–Richardson coefficients, or more general Clebsch–Gordan coefficients, can be interpreted (using the lattice point interpretation of those numbers) as the values of $IP_A(b)$ (see [44, 47, 54] and the references therein).

Given a finite list of integers $S = [a_1, \ldots, a_n]$ the numerical semigroup $\langle S \rangle$ with minimal generating set S is the set of all linear nonnegative combinations. For a fixed element $b \in \langle S \rangle$ a *factorization* of b is an expression $b = \lambda_1 a_1 + \cdots + \lambda_n a_n$. The integer vector $\bar{\lambda} = (\lambda_1, \ldots, \lambda_n)$ uniquely defines the factorization. The *factorization set* of b is $Z(b)$ is the set of all its factorizations of b (note this is exactly what we call $IP_A(b)$ for $A = [a_1 a_2 \ldots a_n]$). Algebraists study the factorization properties in monoids and integral domains. Numerical monoids provide an excellent venue to explore various measurements of non-unique factorization (see [25, 26, 28, 29, 37] and the references therein). An interesting connection to k-feasibility is that the cardinality of $Z(b)$ is the multivariate Hilbert function of certain graded rings [53]. The gaps of semigroups (see, e.g., [59]) correspond to the $<$ 1-feasible elements. Here we describe the rest of the stratification of the elements in the semigroup.

Naturally k-feasibility problems have also interesting applications in combinatorics and statistics: A *vector partition function* of the $d \times n$ matrix A is

$$\phi_A(b) \quad = \quad \#\{x : Ax = b, x \geq 0, \ x \text{ integral}\}.$$

It is well-known [65] the function ϕ_A is piecewise polynomial of degree $n - rank(A)$. Its domains of polynomiality are convex polyhedral cones, these are the so-called *chambers* of A [49]. Vector partition functions for special matrices are crucial in several topics of mathematics (see [10, 22, 30, 60, 65, 66] and the references therein). In the particular case that b is fixed and one considers dilations of b, one has an Ehrhart function enumerating the lattice points on a dilated polyhedron dilations $P_A(Nb)$ as N grows to infinity. The Ehrhart functions of parametric polytopes of the form $P_A(b)$ has been studied in depth by many researchers (see [15, 35] and the references therein. See also the article [21] in this volume). Our investigations aim to describe those vectors b for which the number of lattice points has at least (or at most or exactly) a value k and uncover their structure.

Consider as concrete example the widely popular recreational puzzle *sudoku*, each instance can be thought of as an integer linear program where the hints provided in some of the entries are the given right-hand sides of the problem. Of course in that case newspapers wish to give readers a puzzle where the solution is unique ($k = 1$), but several authors have studied recently conditions on which there are k possible solutions. It is not difficult to see that this is a special case of a *3-dimensional transportation problem*, that is, the question to decide whether the set of *integer* feasible solutions of the $r \times s \times t$-transportation problem

$$\left\{ x \in \mathbb{Z}^{rst} : \sum_{i=1}^{r} x_{ijk} = u_{jk}, \sum_{j=1}^{s} x_{ijk} = v_{ik}, \sum_{k=1}^{t} x_{ijk} = w_{ij}, x_{ijk} \geq 0 \right\}$$

has a unique solution given right-hand sides u, v, w. Another application of k-feasibility appears in statistics, concretely in application in the data security problem of *multi-way contingency tables*, because when the number of solutions is small, e.g., unique, the margins of the statistical table may disclose personal information which is illegal [32]. The study of k-feasibility appears in explicitly in [39].

Besides the directly related problems above, k-feasibility questions are special cases (for parametric polyhedra of the form $\{x : Ax = b, x \geq 0\}$) of the problem to classify polyhedra with k lattice points. Polyhedra, specially lattice polyhedra, with fixed number of (interior) lattice points play a role in many areas of pure mathematics including representation theory, algebraic geometry and combinatorial geometry. One of the challenges has been to classify them. For this purpose there has been a lot of work, going back to classical results of Minkowski and van der Corput, to show that the volume of a lattice polytope P with $k = \text{card}(\mathbb{Z}^n \cap \text{int} P) \geq 1$ is bounded above by a constant that only depends in n and k (see, e.g., [45, 55] and the references therein). Similarly, the supremum of the possible number of points of $\mathbb{Z}^n$ in a lattice polytope in $\mathbb{R}^n$ containing precisely k points of $\mathbb{Z}^n$ in its interior can be bounded by a constant that only depends in n and k. For each positive integer k, there exist only a finite number of non-equivalent lattice polytopes with k interior points. There have been efforts to create a census of polytopes with k interior points. For the plane this was done in [56] Wei and Din for $k = 2$ [69]. W. Castryk [27] recently provided a census for lattice polygons with given number of interior points for $k \leq 30$. A convex body is *hollow* if $k = 0$, i.e., it does not contain any lattice points in its interior. The classification of hollow lattice polytopes is a much more difficult problem, but there are some results toward a classification too [9, 52]. The study of the asymptotic bounds of the number of non-equivalent polyhedra with given number of interior lattice points was initiated by V. A Arnol'd [8] and continued by many mathematicians [11, 46].

1.2 Our Results

This paper has six new contributions to the study of k-feasibility for the semigroup $\mathrm{Sg}(A)$, and the associated polyhedral geometry. Throughout the paper we assume that the cone $\mathrm{cone}(A)$ is pointed. The six contributions guide the structure of this paper:

1. First, we prove a structural result that implies that the set $\mathrm{Sg}_{\geq k}(A)$ of b's, inside the semigroup $\mathrm{Sg}(A)$, that provide $\geq k$-feasible polytopes $P_A(b)$ is finitely generated.
   Let $\mathrm{Sg}_{\geq k}(A)$ (respectively $\mathrm{Sg}_{=k}(A)$ and $\mathrm{Sg}_{<k}(A)$) be the set of right-hand-side vectors $b \in \mathrm{cone}(A) \cap \mathbb{Z}^d$ that make $P_A(b) \geq k$-feasible (respectively $= k$-feasible, $< k$-feasible). Note that $\mathrm{Sg}_{\geq 1}(A)$ is equal to $\mathrm{Sg}(A)$, the semigroup generated by the column vectors of the matrix A.
   Our first structural theorem gives an algebraic description of the sets $\mathrm{Sg}_{\geq k}(A)$ and $\mathrm{Sg}_{<k}(A)$. Let $e_1, \ldots, e_n$ be the standard basis vectors in $\mathbb{Z}^n_{\geq 0}$. Following the book [31] we define the *coordinate subspace of* $\mathbb{Z}^n_{\geq 0}$ of dimension $r \geq 1$ determined by $e_{i_1}, \ldots, e_{i_r}$ with $i_1 < \cdots < i_r$ as the set $\{e_{i_1}z_1 + \cdots + e_{i_r}z_r : z_j \in \mathbb{Z}_{\geq 0} \text{ for } 1 \leq j \leq r\}$. By the 0-dimensional coordinate subspace of $\mathbb{Z}^n_{\geq 0}$ we understand the origin $0 \in \mathbb{Z}^n_{\geq 0}$.

   Theorem 1. *(i) There exists a monomial ideal $I^k(A) \subset \mathbb{Q}[x_1, \ldots, x_n]$ such that*

   $$\mathrm{Sg}_{\geq k}(A) = \{A\lambda : \lambda \in E^k(A)\}, \tag{3}$$

   where $E^k(A)$ is the set of exponents of monomials in $I^k(A)$.
   (ii) The set $\mathrm{Sg}_{<k}(A)$ can be written as a finite union of translates of the sets $\{A\lambda : \lambda \in S\}$, where S is a coordinate subspace of $\mathbb{Z}^n_{\geq 0}$.

   By the Gordan–Dickson lemma, the ideal $I^k(A)$ is finitely generated, thus one can conclude

   Corollary 1. *$\mathrm{Sg}_{\geq k}(A)$ is a finite union of translated copies of the semigroup $A\mathbb{Z}^n_{\geq 0}$.*

   The proof of Theorem 1 relies on some basic facts on lattice points when we think of them as generators of monomial ideals. The basic tool is a characterization of the complement of a monomial ideal (see [31]). Some of the arguments are of interest for the study of affine semigroups and toric varieties [25, 64].
   Theorem 1 extends the earlier decomposition theorem of Hemmecke, Takemura and Yoshida [39] for $k = 1$. They investigated the semigroup $\mathrm{Sg}(A)$ and the integral vectors that are *not* in the semigroup, but still lie within the cone $\mathrm{cone}(A)$ generated by the columns of A. Those authors studied $Q_{\mathrm{sat}} = \mathrm{cone}(A) \cap \mathrm{lattice}(A)$, where $\mathrm{lattice}(A)$ is the lattice generated by the columns of A. They called $H = Q_{\mathrm{sat}} \setminus \mathrm{Sg}(A)$ the set of *holes* of $\mathrm{Sg}(A)$. The set of holes H may be finite or infinite, but their main result is to give a finite description of the

holes as a finitely generated set. Our Theorem 1 was inspired by theirs. The elements of the set $\mathrm{Sg}_{<k}(A)$ can be viewed as *k-holes* (in the context of numeric semigroups and the Frobenius number, 1-holes have also been called *gaps*, see [59, 67]), namely those right-hand sides b for which $Ax = b$ has *less than* k nonnegative integral solutions. Section 2 gives a proof of Theorem 1 that relies on basic commutative algebra.

2. Second, although traditionally the Frobenius problem has been studied for $1 \times n$ matrices, in Sections 3–5 we discuss $\mathrm{F}_k(A)$, a generalization of k-Frobenius number, but this time applicable to all matrices. We explain the meaning of this generalized k-Frobenius number to the structure of the set $\mathrm{Sg}_{\geq k}(A)$ when seen far away from the origin moving toward asymptotic directions inside the $\mathrm{cone}(A)$. Essentially, under some natural assumptions for the matrix A, the set $\mathrm{Sg}_{\geq k}(A)$ can be decomposed into the set of all integer points in the interior of a certain translated cone and a smaller complex complementary set. We discuss the location of such a cone along a given direction c in the interior of $\mathrm{cone}(A)$. The key goal of Sections 3–5 is to derive the lower and upper bounds for $\mathrm{F}_k(A)$. Based on the results obtained in [3, 6] we show that $\mathrm{F}_k(A)$ is bounded from above in terms of $\det(AA^T)$ and k.

Theorem 2. *Let A be a matrix in $\mathbb{Z}^{d\times n}$, $1 \leq d < n$, satisfying*

$$\begin{aligned} &\text{i) } \gcd\left(\det(A_{I_d}) : A_{I_d} \textit{ is an } d\times d \textit{ minor of } A\right) = 1, \\ &\text{ii) } \{x \in \mathbb{R}^n_{\geq 0} : Ax = 0\} = \{0\}. \end{aligned} \tag{4}$$

Then the k-Frobenius number associated with A satisfies the inequality

$$\mathrm{F}_k(A) \leq \frac{n-d}{2(n-d+1)^{1/2}} \det(AA^T) + \frac{(k-1)^{1/(n-d)}}{2(n-d+1)^{1/2}} (\det(AA^T))^{1/2+1/(2(n-d))}. \tag{5}$$

In addition, the structural Theorem 1 allows us to obtain the lower and upper bounds for $\mathrm{F}_k(A)$ in terms of a finite basis of the monomial ideal $I^k(A)$. Let $||\cdot||_\infty$ denote the maximum norm.

Theorem 3. *Let $x^{g_1}, \ldots, x^{g_t}$ be a finite basis for the ideal $I^k(A)$ and let $m(A) = \min_{1\leq i\leq t} ||g_i||_\infty$. The number $\mathrm{F}_k(A)$ satisfies the inequalities*

$$m(A) - 1 \leq \mathrm{F}_k(A) \leq \frac{n-d}{2(n-d+1)^{1/2}} \det(AA^T) + m(A) \left(\frac{\det(AA^T)}{n-d+1}\right)^{1/2}. \tag{6}$$

3. Third. In Section 6, we propose a way to compute the k-holes, i.e., $\mathrm{Sg}_{<k}(A)$, of the semigroup $\mathrm{Sg}(A)$. We give a natural generalization of the proof techniques used by Hemmecke et al.[39] that relies on Hilbert bases to obtain the following theorem:

Theorem 4. *There exists an algorithm that computes for an integral matrix A a finite explicit representation for the set $\mathrm{Sg}_{<k}(A)$ of k-holes of the semigroup $\mathrm{Sg}(A)$. The algorithm computes (finitely many) vectors $h_i \in \mathbb{Z}^d$ and (affine) semigroups M_i, $i \in I$, each given by a finite set of generators in $\mathbb{Z}^d$, such that*

$$\mathrm{Sg}_{<k}(A) = \bigcup_{i \in I} (h_i + M_i) .$$

Here M_i could be trivial, that is, $M_i = \{0\}$. In the special case $k = 1$, Theorem 4 was proved in [39].

4. Fourth. While it is known that computing k-holes is NP-hard, even for the original Frobenius number case, here we show that for n and k fixed integer numbers, there is an efficient algorithm to detect all the $\geq k$-feasible vectors b's, not explicitly one by one, but rather the entire set of k-feasible vectors is encoded as a single multivariate *generating function*, $\sum_{\geq k-\text{feasible}} t^b$.

Theorem 5. *Let n and k be fixed positive integers. Let $A \in \mathbb{Z}^{d \times n}$ and let M be a positive integer. Then there is a polynomial time algorithm to compute a short sum of rational functions $G(t)$ which efficiently represents a formal sum*

$$\sum_{b \in \mathrm{Sg}_{\geq k}(A),\ b \in [-M,M]^d} t^b .$$

Moreover, from the algebraic formula, one can perform the following tasks in polynomial time:

a. Count how many such b's are there (finite because M provides a box).
b. Extract the lexicographic-smallest such b, $\geq k$-feasible vector.
c. Find the $\geq k$-feasible vector b that maximizes the dot product $c^T b$.

Let us explain a bit the philosophy of such a theorem using generating functions for those not familiar with this point of view: In 1993 A. Barvinok [13, 16] gave an algorithm for counting the lattice points inside a polyhedron P in polynomial time when the dimension of P is a constant. The input of the algorithm is the inequality description of P, the output is a polynomial-size formula for the multivariate generating function of all lattice points in P, namely $f(P) = \sum_{a \in P \cap \mathbb{Z}^n} x^a$, where x^a is an abbreviation of $x_1^{a_1} x_2^{a_2} \ldots x_n^{a_n}$. Hence, a long polynomial with exponentially many monomials is encoded as a much shorter sum of rational functions of the form

$$f(P) \quad = \quad \sum_{i \in I} \pm \frac{x^{u_i}}{(1 - x^{c_{1,i}})(1 - x^{c_{2,i}}) \ldots (1 - x^{c_{s,i}})}. \tag{7}$$

Later on Barvinok and Woods [17] developed a set of powerful manipulation rules for using these short rational functions in Boolean constructions on various sets of lattice points, as well as a way to recover the lattice points inside the image of a linear projection of a convex polytope.

It must be remarked that from the results of Barvinok [13] for fixed n, but not necessarily fixed k, one can decide whether a particular b is k-feasible in polynomial time. Recently Eisenbrand and Hänhle [34] showed that the problem of finding the right-hand-side vector b that maximizes the number of lattice points solutions, when b is restricted to take values in a polyhedron, is NP-hard. One can prove a nice theorem for the computation of the k-Frobenius number as a corollary of Theorem 5. Recall that given a vector $(a_1, \ldots, a_n)$ and a positive integer b a *knapsack problem* is a linear Diophantine problem of the form $a_1x_1 + \cdots + a_nx_n = b$ with $x_i \geq 0$. The question about finding the k-Frobenius number is a query over a parametric family of knapsack problems.

Corollary 2. *Let n, k be two fixed positive integers. Consider the parametric knapsack problem $a^Tx = b$, $x \geq 0$ associated with the vector $a = (a_1, \ldots, a_n)^T \in \mathbb{Z}^n_{>0}$ with $\gcd(a_1, \ldots, a_n) = 1$. Then the k-Frobenius number $F_k(a)$ can be computed in polynomial time.*

Corollary 2 greatly generalizes a similar celebrated theorem of R. Kannan [41] and extends the proof of Barvinok and Woods for $k = 1$. Section 7 gives proofs of Theorem 5 and its Corollary 2.

5. The fifth contribution of our article is concerned with practical computations of $=k$-feasible knapsack problems and k-Frobenius problems. Given $a = (a_1, \ldots, a_n) \in \mathbb{Z}^n_{>0}$ with $\gcd(a_1, \ldots, a_n) = 1$, let $g_k = g_k(a)$ denote the largest positive integer b such that $a_1x_1 + \cdots + a_nx_n = b$ has *exactly* k nonnegative integral solutions if such b exists and zero otherwise. The numbers g_k were studied by many authors, see, e.g., [18, 19, 23] and [62]. In Section 8 we prove the inequality $g_0 < g_1$ for $n = 3$. This answers in affirmative a question proposed in [23].
6. From Corollary 2 one further can ask: What is the computational complexity of computing the k-Frobenius number when the dimension n is fixed but k is part of the input? On the basis of the results of [34] we suspect that this is an NP-hard problem, but we do not know of the answer. It is then of interest to experiment with the values of $F_k(a)$ to see the growth for fixed values of n and possibly predict a formula. The sixth and final contribution of our article is about practical computation and experimental exploration on the behavior of the k-Frobenius numbers. In Section 9 we give an algorithm for fast practical computation of the k-Frobenius numbers using ideas from dynamic programming. Our experiments with knapsacks of three variables ($n = 3$) provided support for a new conjecture on the asymptotic properties of the average value of the k-Frobenius numbers.

In what follows we assume that the reader is familiar with polyhedral convexity, monomial ideals, toric ideals, semigroup rings, linear Diophantine equations, and Gröbner bases as presented in [17, 31, 64].

2 Proof of Theorem 1: Monomial ideals and $\geq k$-feasibility

For $f \in \operatorname{cone}(A) \cap \mathbb{Z}^d$ define

$$L^k_{A,f} = \{\lambda \in \mathbb{Z}^n_{\geq 0} : IP_A(f + A\lambda) \text{ is } \geq k \text{ feasible}\},$$

so that $\mathrm{Sg}_{\geq k}(A) = \{A\lambda : \lambda \in L^k_{A,0}\}$. Define then the monomial ideal $I^k(A)$ as follows (with its set of exponent vectors denoted by $E^k(A)$).

$$I^k(A) := \langle x^\lambda : \lambda \in L^k_{A,0}\rangle .$$

To see that the equation $\mathrm{Sg}_{\geq k}(A) = \{A\lambda : \lambda \in E^k(A)\}$ is satisfied it is enough to check that for any $\lambda_0 \in L^k_{A,0}$ the inclusion $\lambda_0 + \mathbb{Z}^n_{\geq 0} \subset L^k_{A,0}$ holds. We will prove the following more general statement.

Lemma 1. *For any $f \in \operatorname{cone}(A) \cap \mathbb{Z}^d$ and $\lambda_0 \in L^k_{A,f}$ we have the inclusion*

$$\lambda_0 + \mathbb{Z}^n_{\geq 0} \subset L^k_{A,f} . \tag{8}$$

Proof. Let $\lambda_0 \in L^k_{A,f}$, so that there exist k distinct vectors $\lambda_1, \ldots, \lambda_k \in \mathbb{Z}^n_{\geq 0}$ with

$$f + A\lambda_0 = A\lambda_1 = \cdots = A\lambda_k .$$

Take any vector $\mu \in \mathbb{Z}^n_{\geq 0}$ and set $\nu = \lambda_0 + \mu$. Then, clearly, we have

$$f + A\nu = A(\lambda_1 + \mu) = \cdots = A(\lambda_k + \mu),$$

where all vectors $\lambda_1 + \mu, \ldots, \lambda_k + \mu \in \mathbb{Z}^n_{\geq 0}$ are distinct. Consequently, $IP_A(f + A\nu)$ is $\geq k$ feasible and, thus, $\nu \in L^k_{A,f}$. Hence (8) holds and the lemma is proved.

Lemma 1 with $f = 0$ clearly implies the first claim of Theorem 1. Let us now prove the second claim. Recall that the elements of the set $\mathrm{Sg}_{<k}(A)$ are also called *k-holes*. A k-hole f is *fundamental* if there is no other k-hole $h \in \mathrm{Sg}_{<k}(A)$ such that $f - h \in \mathrm{Sg}_{\geq 1}(A)$. In other words, for any $u \in \mathrm{Sg}(A)$ $f - u \notin \mathrm{Sg}_{<k}(A)$.

Lemma 2. *The set of fundamental k-holes is a subset of the zonotope*

$$P = \{A\lambda : \lambda \in [0, 1)^n\} .$$

Proof. Let $f \in \mathrm{Sg}_{<k}(A)$ be a fundamental hole. We can write

$$f = A\lambda , \ \lambda \in \mathbb{Q}^n_{\geq 0} .$$

Suppose $f \notin P$. Then for some j we must have $\lambda_j \geq 1$. Thus, denoting by A_j the jth column vector of A, the element $f' = f - A_j$ is a k-hole as any k distinct solutions for $IP_A(f')$ would correspond to k distinct solutions for $IP_A(f)$. Thus we get a contradiction with our choice of f as a fundamental k-hole. This implies $\lambda_j < 1$ for all j and, consequently, $f \in P$. The lemma is proved.

Lemma 2 shows, in particular, that the number of fundamental k-holes is finite. Let us fix a fundamental k-hole f. If the set $L^k_{A,f}$ is empty then $f + A\lambda$ is a k-hole for all $\lambda \in \mathbb{Z}^n_{\geq 0}$. Assume now that $L^k_{A,f}$ is not empty and consider the monomial ideal $I^k_{A,f} \subset \mathbb{Q}[x_1, \ldots, x_n]$ defined as

$$I^k_{A,f} = \langle x^\lambda : \lambda \in L^k_{A,f} \rangle .$$

Then, in view of (8), $f + A\lambda$ is not a k-hole if and only if $x^\lambda \in I^k_{A,f}$.

Thus we need to write down the set $C(I^k_{A,f})$ of exponents of *standard monomials* (the monomials not in the ideal) for $I^k_{A,f}$. Any such exponent $\lambda \in C(I^k_{A,f})$ corresponds to the k-hole $f + A\lambda$.

By Theorem 3 in Chapter 9 of [31], the set $C(I^k_{A,f})$ can be written as a finite union of translates of coordinate subspaces of $\mathbb{Z}^n_{\geq 0}$. Since the number of fundamental k-holes is finite, the second claim of Theorem 1 is proved.

3 Asymptotic structure of $\mathrm{Sg}_{\geq k}(A)$

In this section we assume that $A \in \mathbb{Z}^{d \times n}$, $1 \leq d < n$, is an integral $d \times n$ matrix satisfying

$$\begin{aligned} &\text{i)}\ \gcd(\det(A_{I_d}) : A_{I_d} \text{ is an } d \times d \text{ minor of } A) = 1, \\ &\text{ii)}\ \{x \in \mathbb{R}^n_{\geq 0} : Ax = 0\} = \{0\} . \end{aligned} \tag{9}$$

In the important special case $d = 1$ the matrix $A = a^T$ is just a row vector with $a = (a_1, \ldots, a_n)^T \in \mathbb{Z}^n$ and (9) i) says that $\gcd(a_1, \ldots, a_n) = 1$. Due to the second assumption (9) ii) we may assume that all entries of the vector a are positive. It follows that the largest integral value b such that the problem $IP_A(b)$ is $< k$-feasible, the k-Frobenius number $\mathrm{F}_k(a)$, is well-defined. Clearly, for $d = 1$ we have the inclusion

$$\mathrm{int}(\mathrm{F}_k(a) + \mathbb{R}_{\geq 0}) \cap \mathbb{Z} \subset \mathrm{Sg}_{\geq k}(a^T) , \tag{10}$$

where $\mathrm{int}(\cdot)$ denotes the interior of the set.

In general, the structure of the set $\mathrm{Sg}_{\geq k}(A)$, apart from a few special cases, is not well understood. It is known that, in analogy with (10), $\mathrm{Sg}_{\geq k}(A)$ can be decomposed into the set of all integer points in the interior of a certain translated cone and a complex complementary set. More recent results (see [6]) attempt to estimate the location of such a cone along the fixed direction $v = A\mathbf{1}$, where $\mathbf{1}$ is the all-1-vector, in the interior of cone(A). The choice of v as the direction vector is dated back to the paper of Khovanskii [43] for $k = 1$. In general, given any rational vector u in the interior of cone(A), the multiple tu will, for large enough $t \in \mathbb{Z}$, lie in $\mathrm{Sg}_{\geq k}(A)$. Hence the name *asymptotic structure*.

In this paper we consider a generalization of the Frobenius number that reflects $\geq k$-feasibility properties of the whole family of the problems $IP_A(b)$, when b runs over all integer vectors in the interior of the cone cone(A). Given a direction vector $b \in \mathrm{int}(\mathrm{cone}(A)) \cap \mathbb{Z}^d$ put

$$\mathrm{g}_k(A, b) = \min\{t \geq 0 : \mathrm{int}(tb + \mathrm{cone}(A)) \cap \mathbb{Z}^d \subset \mathrm{Sg}_{\geq k}(A)\}.$$

We define the *k-Frobenius number associated with A* as

$$\mathrm{F}_k(A) = \max\{\mathrm{g}_k(A, b) : b \in \mathrm{int}(\mathrm{cone}(A)) \cap \mathbb{Z}^d\}.$$

The number $\mathrm{g}_k(A) = \mathrm{g}_k(A, v)$ was called in [6] the *diagonal k-Frobenius number of A*. A lower bound for $\mathrm{g}_k(A)$ (and thus, by definition of $\mathrm{g}_k(A)$, for $\mathrm{F}_k(A)$) was given in [6, Theorem 1.3]. Next we derive the lower and upper bounds for $\mathrm{F}_k(A)$ presented in Theorem 2 and Theorem 3.

Before we start the proofs it is worth remarking they will be based on the results obtained in [3, 6] and the structural Theorem 1.

4 Proof of Theorem 2

First we show that the k-Frobenius number $\mathrm{F}_k(A)$ is bounded from above by the (suitably normalized) number $\mathrm{g}_k(A)$.

Lemma 3.

$$\mathrm{F}_k(A) \leq \left(\frac{\det(AA^T)}{n-d+1}\right)^{1/2} \mathrm{g}_k(A). \tag{11}$$

Proof. Put for convenience $\gamma = \left(\frac{\det(AA^T)}{n-d+1}\right)^{1/2}$. As it was shown in the proof of Lemma 1.1 in [3], for any $b \in \mathrm{int}(\mathrm{cone}(A)) \cap \mathbb{Z}^d$ the vector γb is contained in $v + \mathrm{cone}(A)$. Therefore $\gamma b + \mathrm{cone}(A) \subset v + \mathrm{cone}(A)$ and, consequently,

$$\mathrm{int}(\mathrm{g}_k(A)\gamma b + \mathrm{cone}(A)) \cap \mathbb{Z}^n \subset \mathrm{int}(\mathrm{g}_k(A)v + \mathrm{cone}(A)) \cap \mathbb{Z}^n \subset \mathrm{Sg}_{\geq k}(A).$$

Hence for any $b \in \mathrm{int}(\mathrm{cone}(A)) \cap \mathbb{Z}^d$ we have $\mathrm{g}_k(A, b) \leq \mathrm{g}_k(A)\gamma$. Therefore $\mathrm{F}_k(A) \leq \mathrm{g}_k(A)\gamma$ and the lemma is proved.

The diagonal k-Frobenius number $g_k(A)$ in its turn is bounded from above in terms of A and k due to the following result.

Theorem 6 (Theorem 1.2 in [6]). *The diagonal k-Frobenius number associated with A satisfies the inequality*

$$g_k(A) \leq \frac{n-d}{2}(\det(AA^T))^{1/2} + \frac{(k-1)^{1/(n-d)}}{2}\left(\det(AA^T)\right)^{1/(2(n-d))}. \quad (12)$$

Combining (11) and (12) we obtain the inequality (5).

5 Proof of Theorem 3

The set of exponents $E^k(A)$ of the monomial ideal $I^k(A) = \langle x^{g_1}, \ldots, x^{g_t} \rangle$ has the form

$$E^k(A) = \bigcup_{i=1}^{t} (g_i + \mathbb{Z}^n_{\geq 0}). \quad (13)$$

By (13), any $g \in E^k(A)$ has $||g||_\infty \geq m(A)$. Therefore the point $(m(A)-1)\mathbf{1} \notin E^k(A)$ and, by Theorem 1 (i), we obtain $A((m(A)-1)\mathbf{1}) = (m(A)-1)v \notin \mathrm{Sg}_{\geq k}(A)$. Therefore, by the definition of $g_k(A)$,

$$m(A) - 1 \leq g_k(A) \leq F_k(A).$$

This proves the lower bound in (6).

To derive the upper bound, we will show first that $g_k(A)$ satisfies the inequality

$$g_k(A) \leq g_1(A) + m(A). \quad (14)$$

Let us choose any $y \in \mathrm{int}((g_1(A) + m(A))v + \mathrm{cone}(A)) \cap \mathbb{Z}^d$. To prove (14), it is enough to show that $y \in \mathrm{Sg}_{\geq k}(A)$. Consider the point $y' = y - m(A)v$. Since $y' \in \mathrm{int}(g_1(A)v + \mathrm{cone}(A)) \cap \mathbb{Z}^d$, we have $y' \in \mathrm{Sg}_{\geq 1}(A)$. Therefore, there exists $z \in \mathbb{Z}^n_{\geq 0}$ such that $Az = y'$. Hence $A(m(A)\mathbf{1} + z) = m(A)v + Az = y$. Finally, observe that $m(A)\mathbf{1} \in E^k(A)$ by (13), and hence $m(A)\mathbf{1}+z \in m(A)\mathbf{1}+\mathbb{Z}^n_{\geq 0} \subset E^k(A)$. Consequently, $y = A(z + m(A)\mathbf{1}) \in \mathrm{Sg}_{\geq k}(A)$ and the inequality (14) is proved. The upper bound in (6) now follows from (11), (14) and (12) with $k = 1$.

6 Proof of Theorem 4: Computing k-holes via Hilbert bases

In this section we combine the results of Hemmecke et al. [39] with our techniques to compute the elements of $\mathrm{Sg}_{<k}(A)$ proving Theorem 4. We present an algorithm to compute an *explicit* representation of $\mathrm{Sg}_{<k}(A)$, even for an infinite case, using semigroups. We remark that this explicit representation need not be of polynomial size in the input size of A.

In view of the proof of Theorem 1 (ii), it is enough to compute all fundamental k-holes and then for each fundamental k-hole f with nonempty set $L^k_{A,f}$ compute the standard monomials of the ideal $I^k_{A,f}$. If the set $L^k_{A,f}$ is empty, the set of all k-holes H_k contains the translated semigroup $f + \mathrm{Sg}(A)$. By Lemma 2, all fundamental k-holes are located in a zonotope $P = \{A\lambda : \lambda \in [0, 1)^n\}$. Thus, with a straightforward generalization of the approach proposed in Hemmecke et al. [39], the fundamental k-holes the can be computed by using a Hilbert basis of the cone cone(A).

Let f be a fundamental k-hole. Recall that for a nonempty set $L^k_{A,f}$ the monomial ideal $I^k_{A,f} \subset \mathbb{Q}[x_1, \dots, x_n]$ is defined as

$$I^k_{A,f} = \langle x^\lambda : \lambda \in L^k_{A,f} \rangle$$

and $f + A\lambda$ is not a k-hole if and only if $x^\lambda \in I^k_{A,f}$.

Thus we need to compute the exponents of standard monomials for the ideal $I^k_{A,f}$. Any such exponent $\lambda \in \mathbb{Z}^n_{\geq 0}$ corresponds to the k-hole $f + A\lambda$. The exponents of standard monomials can be computed explicitly from a set of generators of the ideal. Hence, it is enough to find the generators of $I^k_{A,f}$. Let us fix an ordering $\prec$ in $\mathbb{Z}^n_{\geq 0}$. The minimal generators for the ideal $I^k_{A,f}$ correspond to the $\prec$-minimal elements of the set

$$L^k_{A,f} = \{\lambda \in \mathbb{Z}^n_{\geq 0} : \exists \text{ distinct } \mu_1, \dots, \mu_k \in \mathbb{Z}^n_{\geq 0} \text{ such that } f + A\lambda = A\mu_1 = \dots = A\mu_k\}.$$

For computational purposes it is enough to compute a set of vectors of $L^k_{A,f}$ that contains all the $\prec$-minimal elements. We will proceed as follows. Let K be a complete graph with the vertex set $V = \{1, 2, \dots, k\}$. By a weighted orientation H of K we will understand a weighted directed graph $H = (V, E)$ such that any two vertices of H are connected by a directed edge $e \in E$ with a weight $w(e) \in \{1, \dots, n\}$.

Let $\mathcal{S}$ be set of all weighted orientations of K. For each $H \in \mathcal{S}$ we construct the following two auxiliary sets: the set

$$L_H = \{\lambda \in \mathbb{Z}^n_{\geq 0} : \exists \mu_1, \dots, \mu_k \in \mathbb{Z}^n_{\geq 0} \text{ such that } f + A\lambda = A\mu_1 = \dots = A\mu_k \text{ and } (\mu_i)_{w(e)} \leq (\mu_j)_{w(e)} - 1 \text{ for each } e = (i, j) \in E\}$$

and the set

$$M_H = \{(\lambda, \mu_1, \ldots, \mu_k) \in \mathbb{Z}_{\geq 0}^{(k+1)n} : f + A\lambda = A\mu_1 = \cdots = A\mu_k$$
$$\text{and } (\mu_i)_{w(e)} \leq (\mu_j)_{w(e)} - 1 \text{ for each } e = (i,j) \in E\}.$$

Then, in particular, $L_{A,f}^k = \bigcup_{H \in \mathcal{S}} L_H$, where the union is taken over all orientations in $H \in \mathcal{S}$.

We will need the following result.

Lemma 4. *Let λ_0 be a $\prec$-minimal element of L_H. Then there exists a $\prec$-minimal element of M_H of the form $(\lambda_0, \hat{\mu}_1, \ldots, \hat{\mu}_k)$.*

Proof. Let λ_0 be a $\prec$-minimal element of L_H. Suppose on contrary, for every $(\mu_1, \ldots, \mu_k) \in \mathbb{Z}_{\geq 0}^{kn}$ the vector $(\lambda_0, \mu_1, \ldots, \mu_k)$ is not a $\prec$-minimal element of M_H. Let $(\hat{\mu}_1, \ldots, \hat{\mu}_k)$ be a $\prec$-minimal element of the set

$$M_H|_{\lambda=\lambda_0} = \{(\mu_1, \ldots, \mu_k) \in \mathbb{Z}_{\geq 0}^{kn} : f + A\lambda_0 = A\mu_1 = \cdots = A\mu_k$$
$$\text{and } (\mu_i)_{w(e)} \leq (\mu_j)_{w(e)} - 1 \text{ for each } e = (i,j) \in E\}.$$

By the assumption, there exists a vector $(\lambda', \mu_1', \ldots, \mu_k') \in M_H$ such that $(\lambda', \mu_1', \ldots, \mu_k') \prec (\lambda_0, \hat{\mu}_1, \ldots, \hat{\mu}_k)$ and $(\lambda', \mu_1', \ldots, \mu_k') \neq (\lambda_0, \hat{\mu}_1, \ldots, \hat{\mu}_k)$. If $\lambda' \neq \lambda_0$ we get a contradiction to the $\prec$-minimality of λ_0 in L_H. On the other hand, if $\lambda' = \lambda_0$ we get a contradiction to the $\prec$-minimality of $(\hat{\mu}_1, \ldots, \hat{\mu}_k)$ in $M_H|_{\lambda=\lambda_0}$.

In view of Lemma 4, to compute a generating set for $L_{A,f}^k$ (or to determine that $L_{A,f}^k$ is empty) it is now enough to compute the set of all minimal elements for $M_H, H \in \mathcal{S}$ and remove the last kn components from each of them.

7 Proof of Theorem 5: Generating functions and k-feasibility

We wish to prove a representation theorem of a set of lattice points as a sum $\sum_{b \geq k\text{-feasible } b \in [-M,M]^d} t^b$. First we need a lemma about how to find an objective function that orders all points in a box.

Lemma 5. *Let n be a constant denoting the number of variables. Given a positive integer R and the associated n-dimensional box $B_R = [0, R]^n$, there exists an integer linear objective function $\bar{c}_R^T x$ such that $\bar{c}_R^T(y_i - y_j) \neq 0$ for all pairs of non-zero lattice points y_i, y_j inside the box B_R. One can find one such vector in polynomial time.*

Proof. Let s be a single auxiliary (real) variable and the associated vector $\mathbf{c}(s) = (1, s, s^2, s^3, \ldots, s^{n-1})^T$. Now for each of the $L = R^n(R^n - 1)/2$ pairs of non-zero lattice vectors $y_i - y_j$ we construct one univariate polynomial $f_{i,j}(s) = \mathbf{c}(s)^T(y_i - y_j)$ (since y_i, y_j are distinct the polynomial $f_{i,j}$ is not identically zero). These are polynomials of degree $n-1$ so they can only have at most $n-1$ real roots each. Note also that these are polynomials all of whose coefficients are integer numbers between $-R$ and R, that means, by the famous Cauchy bound on the absolute values of roots of univariate polynomials that any of the real roots of any $f_{i,j}(s)$ must be bounded in above by $1+R$. Thus taking, for example, the value $s_0 = R+2$ gives $\bar{c}_R = \mathbf{c}(s_0)$ as an integer vector that totally orders all lattice points in the box B_R. Note that the bit-size description of $\bar{c}_R$ is polynomial in the input namely n, and $\log(R)$ because the entries are the first $n-1$ powers of R. The lemma is proved.

Below we will use the algorithmic technique of rational generating functions developed by Barvinok and Woods in [13, 17] (see also the book [14]).

A key subroutine introduced by Barvinok and Woods is the following *Projection Theorem.*

Lemma 6 (Theorem 1.7 in [17]). *Assume the dimension n is a fixed constant. Consider a rational polytope $P \subset \mathbb{R}^n$ and a linear map $T : \mathbb{Z}^n \to \mathbb{Z}^d$ such that $T(\mathbb{Z}^n) \subset \mathbb{Z}^d$. There is a polynomial time algorithm which computes a short representation of the generating function $f\big(T(P \cap \mathbb{Z}^n), x\big)$.*

Now all the set up of the proof of Theorem 5 is ready. Recall that A is an integral $d \times n$ matrix and n, k are constants. Let M be a given positive integer as in the statement of the theorem, and let R be another integer positive number (to be set a bit later). We can define the polyhedron (note X_i denotes an n-dimensional vector so this polytope lives in nk-dimensional space):

$$Q(A,k,R,M) = \big\{(X_1, X_2, \ldots, X_k) \in \mathbb{R}^{nk} \ : \ AX_1 = \cdots = AX_k,\ \bar{c}_R^T X_i \geq \bar{c}_R^T X_{i+1} + 1,$$
$$\text{for } i = 1, \ldots, k-1 \text{ and } R \geq X_1 \geq 0,\ M \geq AX_1 \geq -M\big\}.$$

One can use Barvinok's algorithm to compute the generating function of the lattice points of $Q(A,k,R,M)$ [14]. The polytope $Q(A,k,R,M)$ has the following two key properties: First, all its integer points represent *distinct* k-tuples of integer points that are in some parametric polyhedron $P_A(b) = \{x : Ax = b, x \geq 0\}$. When we turn lattice points into monomials, $z_1^{X_1} z_2^{X_2} \ldots z_k^{X_k}$ has only those exponents where $X_i \neq X_j$. Namely, this is precisely the set of all monomials coming from k-tuples of distinct vectors in $\mathbb{Z}^n_{\geq 0}$ that give the same value $AX_1 = AX_2 = \cdots = AX_k$. Second, all vectors X_1 are in the box $B_R = [0, R]^n \subset \mathbb{R}^n$, and third, all vectors AX_1 are in the box $[-M, M]^d \subset \mathbb{R}^d$. It is important to note if R is too small the pre-images X_1, when projected by A, may not hit all points in $\mathrm{Sg}_{\geq k}(A) \cap [-M, M]^d$ which is what we want. We will calculate a sufficiently large R below.

We now apply a very simple linear map $T(X_1, X_2, \ldots, X_k) = AX_1$, by multiplication with A. This map yields of course for each k-tuple (which has $X_i \neq X_j$) the corresponding right-hand-side vector $b = AX_1$ that has at least k-distinct solutions and $b \in [-M, M]^d$. From the generation function of the lattice points of $Q(A,k,R,M)$ we use Lemma 6 to obtain a generating function expression

$$f = \sum_{b \in \text{projection of } Q(A,k,R,M):\ \text{with at least } k\text{-representations and } b \in [-M,M]^d} t^b.$$

Now we must take care that this captures all such b within the box $\{b : -M \leq b_i \leq M\}$. To achieve this we need to find a large enough value of R that suffices to capture all representations of b's within the box B_R when projected. Such R can be calculated as the maximum value among n linear programs, one for each variable x_i, given by $\max x_i$ subject to $-M \leq Ax \leq M$, $x_i \geq 0$. With this choice of R, the generating function f above gives all the desired values of b. Which is the desired short rational function which efficiently represents the sum $\sum_{\geq k\text{-feasible } b\in[-M,M]^d} t^b$. This proves the main result in the body of the paper for $\geq k$-feasibility. Now, because if one knows a description for $\mathrm{Sg}_{\geq k}(A)$ and $\mathrm{Sg}_{\geq k+1}(A)$ one knows $\mathrm{Sg}_{=k}(A) = \mathrm{Sg}_{\geq k}(A)\backslash\mathrm{Sg}_{\geq k+1}(A)$ and $\mathrm{Sg}_{<k}(A) = \mathrm{Sg}(A)\backslash\mathrm{Sg}_{\geq k}(A)$. The similar generating functions of the other cases ($< k$, $= k$) are just obtained as the difference of the generating functions.

Now that we have proved the main statement of Theorem 5, we move to prove Parts (a) to (d) of the theorem.

Part (a) If we have a generating function representation of

$$\sum_{b:\ b \text{ is } \geq k\text{-feasible},\ b\in[-M,M]^d} t^b,$$

it has the form

$$f(t) = \sum_{i\in I} \alpha_i \frac{t^{p_i}}{(1-t^{a_{i1}})\cdots(1-t^{a_{ik}})}.$$

Note that by specializing at $t = (1,\ldots,1)$, we can count how many b's are $\geq k$-feasible (again the set is finite because it fits inside a box). Remark the substitution is not immediate since $t = (1,\ldots,1)$ is a pole of each fraction in the representation of f. This problem is solvable because it has been shown by Barvinok and Woods that this computation can be handled efficiently (see Theorem 2.6 in [17] for details) and this proves Part (a).

Part (b) This item is a direct corollary of the following extraction lemma.

Lemma 7 (Lemma 8 in [48] or Theorem 7.5.2 in [50]). *Assume the dimension n is fixed. Let $S \subset \mathbb{Z}^r_+$ be nonempty and finite set of lattice points. Suppose the polynomial $f(S;z) = \sum_{\beta\in S} z^\beta$ is represented as a short rational function and let c be a cost vector. We can extract the (unique) lexicographic largest leading monomial from the set $\{x^\alpha : \alpha \cdot c = L, \alpha \in S\}$, where $L := \max\{\alpha \cdot c : \alpha \in S\}$, in polynomial time.*

Part (c) Barvinok and Woods developed a way to do monomial substitutions (not just $t_i = 1$ as we used in Part (a)), where the variable t_i in the current series, is replaced by a new monomial $z_1^{a_1} z_2^{a_2} \cdots z_r^{a_r}$. Note that the rational generating function $f = \sum_{b\in Q\cap\mathbb{Z}^d} b^b$ can give the evaluations of the b's for a given objective function $c \in \mathbb{Z}^d$. If we make the substitution $t_i = z^{c_i}$, the above equation yields a *univariate* rational function in z:

$$f(z) = \sum_{i \in I} E_i \frac{z^{c \cdot u_i}}{\prod_{j=1}^{d}(1 - z^{c \cdot v_{ij}})}. \tag{15}$$

Moreover $f(z) = \sum_{b \in Q \cap \mathbb{Z}^d} z^{c \cdot b}$. Thus we just need to find the (lexicographically) largest monomial in the sum in polynomial time. But this follows from Part (b).

To conclude we see how to compute the k-Frobenius number efficiently when n number of variables and k are fixed.

Proof of Corollary 2. We start by observing that there is an upper bound for the k-Frobenius number. Indeed Theorem 1.1 in [5] gives already an upper bound that is certainly smaller than $M = k(n-1)!a_1a_2\cdots a_n$. The k-Frobenius number must be smaller and thus we will use M in the bounding box created in Theorem 5.

Next we claim the same generating function descriptions obtained in Theorem 5 can be also obtained for the sets of those b which are $=k$-feasible, $\geq k$-feasible, or $< k$-feasible with the added condition that $b_i \leq M$. This is because the generating functions of those sets of b's can be obtained from the set of b's encoded by Theorem 5 through Boolean operations (intersection, unions, complements). Indeed we clearly have $\text{Sg}_{\geq k+1}(A) \setminus \text{Sg}_{\geq k}(A) = \text{Sg}_{=k}(A)$ and $\text{Sg}_{<k}(A) = \text{Sg}_{\geq k}(A) \setminus \text{Sg}_{=k}(A)$. The same identities hold under the intersection with the box $[-M, M]^d$, thus the claim follows.

We may see now that Corollary 2 follows directly from what we achieved in Theorem 5 and the Boolean operation Lemma of Barvinok and Woods. Indeed, from Theorem 5 we have a rational function representation of the k-feasible b for the knapsack problem (note that R is calculated by LPs as in the proof of Theorem 5).

$$f(t) = \sum_{i \in I} E_i \frac{t^{c \cdot u_i}}{\prod_{j=1}^{d}(1 - t^{c \cdot v_{ij}})} = \sum_{b \in \text{projection of} Q(A,k,R,M) \cap \mathbb{Z}^d,\ \geq k-\text{feasible} b \in [-M,M]^n} t^{c \cdot b}.$$

Clearly the k-Frobenius number is simply the largest (lexicographic) b, such that t^b is **not** in $f(t)$, it is in its complement. Note the choice of bound M is such that we indeed have the k-Frobenius number inside. Then, for the complement $\overline{S} = \mathbb{Z}_+ \setminus S$, we compute the generating function $f(\overline{S}; x) = (1-t)^{-1} - f(t)$ and then we compute the largest such t^b in the complement using Lemma 7.

8 Integers with exactly k representations

Let $a = (a_1, \ldots, a_n) \in \mathbb{Z}^n_{>0}$ with $\gcd(a_1, \ldots, a_n) = 1$. Recall that $g_k(a)$ denotes the largest positive integer b such that $a_1x_1 + \cdots + a_nx_n = b$ has *exactly* k integral nonnegative solutions if such b exists and zero otherwise. In other words, a positive g_k is the largest integer that has *exactly k representations* by $a_1, \ldots, a_n$. In the case $n = 2$ Beck and Robins [18] obtained the formula

$$g_{k-1}(a_1, a_2) = ka_1a_2 - a_1 - a_2 . \tag{16}$$

Note that $F_k(a_1, a_2) = g_{k-1}(a_1, a_2)$ and, in general, $F_k(a) = \max\{g_i(a) : i \leq k-1\}$.

Given a fixed vector a, the behavior of the sequence $\{g_i(a)\}_{i=0}^{\infty}$ is far from being simple for $n \geq 3$. For instance, it has been observed in Brown et al. [23] that $g_i(a_1, a_2, a_3)$ is not necessarily increasing with i. For example, $g_{14}(3, 5, 8) = 52$ whereas $g_{15}(3, 5, 8) = 51$ and therefore in that particular case $g_{14} > g_{15}$. Furthermore, Shallit and Stankewicz [62] proved that for $i > 0$ and $n = 5$, the quantity $g_0 - g_i$ is arbitrarily large and positive. They also give an example of $g_0 > g_1$ for $n = 4$.

For $n = 3$ Brown et al. [23] posed the question whether or not the inequality $g_0 < g_1$ always holds. In what follows we answer this question in affirmative.

Theorem 7. *Given $a_1 < a_2 < a_3 \in \mathbb{Z}_{>0}$, we have*

$$g_0(a_1, a_2, a_3) < g_1(a_1, a_2, a_3).$$

To prove this theorem, we first consider the case in which all coefficients are pairwise relatively prime.

Lemma 8. *Let $a_1 < a_2 < a_3$ be the pairwise relatively prime positive integers. Then*

$$g_0(a_1, a_2, a_3) < g_1(a_1, a_2, a_3).$$

Proof. Let $f = g_0(a_1, a_2, a_3)$ denote the Frobenius number of the vector (a_1, a_2, a_3). Then

$$a_1x_1 + a_2x_2 + a_3x_3 = f \tag{17}$$

has no nonnegative integral solution. On the other hand

$$a_1x_1 + a_2x_2 + a_3x_3 = f + a_1 \tag{18}$$

has at least one nonnegative integral solution. We will prove that this solution is unique. Let $\bar{x}$ be any nonnegative integral solution to (18). Observe that $\bar{x}_1 = 0$, otherwise we could trivially construct a nonnegative integral solution for (17). Therefore $\bar{x} = (0, \bar{x}_2, \bar{x}_3)$. Consider by contradiction that $\tilde{x} = (0, \tilde{x}_2, \tilde{x}_3)$ is another nonnegative integral solution to (18). Then $a_2(\bar{x}_2 - \tilde{x}_2) + a_3(\bar{x}_3 - \tilde{x}_3) = 0$. In particular, since $\gcd(a_2, a_3) = 1$, this implies $|\bar{x}_2 - \tilde{x}_2| = la_3$ with l a natural number, which in turn implies that either $\bar{x}_2 \geq a_3$ or $\tilde{x}_2 \geq a_3$. Assume w.l.o.g. that $\bar{x}_2 \geq a_3$, and since $(0, \bar{x}_2, \bar{x}_3)$ is a nonnegative integral solution to (18), this implies that $f + a_1 \geq a_2a_3$ and hence $f \geq a_2a_3 - a_1$ which is in contradiction with the fact that $f = g_0(a_1, a_2, a_3) \leq g_0(a_2, a_3) \leq g_0(a_2, a_3) = a_2a_3 - a_2 - a_3$.

The following result was proved in [23].

Lemma 9 (Theorem 1 in Brown et al. [23] with $k = 3$). *Let $d = \gcd(a_2, a_3)$ and $j \in \mathbb{Z}_{\geq 0}$ then either $g_j(a_1, a_2, a_3) = dg_j(a_1, a_2/d, a_3/d) + (d-1)a_1$ or $g_j(a_1, a_2, a_3) = g_j(a_1, a_2/d, a_3/d) = 0$, i.e., no right-hand side achieves exactly j integral solutions.*

To prove Theorem 7 we will reduce the general case to the case of pairwise relatively prime coefficients. Consider a triple (a_1, a_2, a_3). If the coefficients are pairwise relatively prime, the result follows from Lemma 8. Assume that there is some g.c.d. different from 1 for a pair of coefficients. From Lemma 9, we can get rid of the g.c.d. and that does not change the relative order between g_0 and g_1. By applying at most three times Lemma 9, we come back to the case where all coefficients are pairwise relatively prime and the result follows. The theorem is proved.

9 Computing $F_k(a)$ by dynamic programming and the behavior of $F_k(a_1, a_2, a_3)$

The computational aspects of the classical Frobenius problem were studied by many authors (see [20, 33] and the references therein). In this section, we propose a practical dynamic programming approach that allows us to compute the number of integral solutions to the knapsack problems and the k-Frobenius numbers.

9.1 A simple dynamic programming algorithm for $F_k(a)$

Given $a_1, \ldots, a_n \in \mathbb{Z}_{>0}$, we denote by $T_i(b)$ the number of integral solutions of the knapsack problem $\sum_{l=1}^{n} a_l x_l = b, x \in \mathbb{Z}^n_{\geq 0}$, satisfying $x_j = 0$ for $j < i$ and $x_i \geq 1$, i. e., $T_i(b)$ counts the number of integral solutions of the knapsack problem where the smallest non-zero index is i. The idea of the algorithm is to update an array $T_i(b)$ for increasing b and for all i. The following observation allows us to initialize the dynamic programming approach.

Lemma 10. *$T_i(a_i) = 1$ and $T_i(b) = 0$ for all $0 \leq b \leq a_i - 1$.*

The following lemma explains how to update the function T.

Lemma 11. *Given $b \neq a_i$,*

$$T_i(b) = \sum_{j=i}^{n} T_j(b - a_i). \tag{19}$$

Proof. We first prove that (19) holds with $\geq$ instead of $=$. Indeed consider any nonnegative integral solution $\bar{x}$ to $\sum_{l=1}^{n} a_l x_l = b - a_i$ with $\bar{x}_j = 0$ for all $1 \leq j \leq i-1$, it can be transformed into a nonnegative integral solution for $\sum_{l=1}^{n} a_l x_l = b$ by considering $\bar{x} + e_i$. Obviously the first $(i-1)$ components are still zero and the ith component is positive.

We now prove that (19) holds with $\leq$. Consider a solution $\hat{x} \in \mathbb{Z}^n_{\geq 0}$ with $\hat{x}_j = 0$ for all $1 \leq j \leq i-1$ and $\hat{x}_i \geq 1$ to $\sum_{l=1}^{n} a_l x_l = b$. By subtracting 1 from the ith component, we obtain a nonnegative integer solution to $\sum_{l=1}^{n} a_l x_l = b - a_i$ where the first $i-1$ components are zero.

Using Lemma 10 and Lemma 11, we can fill in an array $T_i(b)$ starting from $b = 0$ for increasing values of b. Obtaining the array T allows us to count the number of different solutions.

Lemma 12. *The number of nonnegative integral solutions to $\sum_{l=1}^{n} a_l x_l = b$ is equal to $\sum_{i=1}^{n} T_i(b)$.*

Proof. This follows from the fact that any integral solution to $\sum_{l=1}^{n} a_l x_l = b$ is counted exactly in one set corresponding to the smallest index for which x_i is nonzero.

To determine the k-Frobenius number, we need to know the largest b that is $< k$-feasible. It is therefore important to determine a stopping criterion for the dynamic programming algorithm. A first obvious criterion is to use the upper bound for b given in [5]. It turns out that this upper bound is very often too large compared to the actual k-Frobenius number and leads to longer computation times. The following lemma allows us to interrupt the computation with the guarantee for a k-Frobenius number earlier.

Lemma 13. *Assume that $a_1 < a_2 < \cdots < a_n$ and that $\sum_{i=1}^{n} a_i x_i = b$ has at least k integral solutions for all $\bar{b} \leq b \leq \bar{b} + a_1$. Then $\sum_{i=1}^{n} a_i x_i = b$ has at least k integral solutions for all $b \geq \bar{b}$.*

Proof. This follows from the fact that $T_1(b) = \sum_{i=1}^{n} T_i(b - a_1)$ and that the number of integral solutions of $\sum_{i=1}^{n} a_i x_i = b$ is at least $T_1(b)$.

We now have all the necessary ingredients to describe a dynamic programming-based algorithm.

We have done a number of computational experiments in order to evaluate the empirical complexity of Algorithm 1. Our implementation of the algorithm was in standard C code with CPU Intel Core i7-975, 4 cores, 8 threads, CPU clock 3.33 GHz. In the experiments we checked the average time needed to compute the k-Frobenius number for some common values of n and a_1, as well as the dependency on k. The algorithm demonstrated surprisingly good performance. For instance, to check the dependency on k, we fixed the dimension to $n = 10$ and the range of values for a_i to $[2^{14}, 2^{16}]$. For each value of k, we run 20 experiments. The summary of the results is reported in Table 1. We observe that the average time grows very slowly with k. We utilize this property of Algorithm 1 in the next section.

Algorithm 1 DP algorithm for the k-Frobenius number

Require: $a_1 < a_2 \cdots < a_n, k \geq 1$
  $T_i(a_i) = 1$ for all $i = 1, \dots, n$
  $T_i(b) = 0$ for $i = 1, \dots, n$ and $b < a_i$
  $b := a_1 + 1$
  while $\exists \bar{b} \in [b - a_1, b - 1]$ with $\sum_{i=1}^{n} T_i(\bar{b}) < k$ **do**
    for $i := 1$ to n **do**
      $T_i(b) := \sum_{j=i}^{n} T_j(b - a_i)$
    end for
    $b := b + 1$
  end while
  Return the largest b that has less than k solutions

Table 1 Dependency on k

Dimension $n = 10$									
k	1	4	16	64	256	1024	2^{12}	2^{14}	2^{16}
Average time (s)	0	0	0	0	0	0	1	1	1

9.2 Experiments on the k-Frobenius number $\mathrm{F}_k(a_1, a_2, a_3)$

In this section we report the results of several experiments concerning the average behavior of the k-Frobenius number $\mathrm{F}_k(a)$ and make a conjecture on the basis of these results. The experiments were performed using the dynamic programming algorithm introduced in Section 9.1.

Let $G_T = \{a = (a_1, a_2, a_3) \in \mathbb{Z}^3_{>0} : \gcd(a_1, a_2, a_3) = 1, ||a||_\infty \leq T\}$ and let $A_k(T)$ denote the average value of the (normalized) k-Frobenius number over G_T, that is

$$A_k(T) = \frac{1}{|G_T|} \sum_{(a_1, a_2, a_3) \in G_T} \frac{F_k(a_1, a_2, a_3)}{\sqrt{a_1 a_2 a_3}}. \tag{20}$$

It follows from the results of Ustinov [68] that

$$A_1(T) \to \frac{8}{\pi} \quad \text{as } T \to \infty. \tag{21}$$

For the case $k > 1$ the upper bound in Theorem 1.1 of [5], together with (21) implies that for any $\epsilon > 0$ for large enough T

$$A_k(T) \leq \frac{8}{\pi} + \sqrt{2(k-1)} + \epsilon.$$

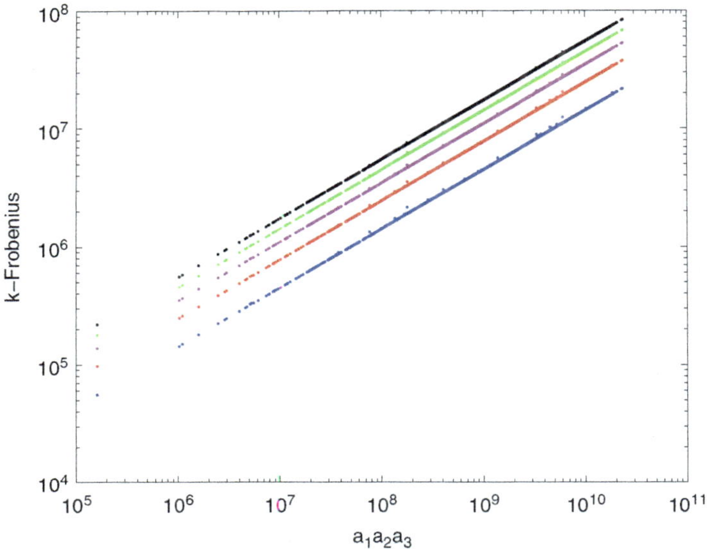

Fig. 1 Value of the k-Frobenius number (over the y-axis) with respect to the product $a_1a_2a_3$ for $a_i \in [2, 3000]$ for $k = 10000, 30000, 60000, 100000, 150000$.

On the other hand, from the lower bound in Theorem 1.1 of [5] and the proof of Proposition 1 in [4] we conclude that for large enough T

$$A_k(T) > \sqrt{2k} - \epsilon \, . \tag{22}$$

In our experiments we empirically obtained the values for $A_k(T_0)$ for fixed T_0 and various values of k. For instance, for $T_0 = 3000$ we have drawn 2000 triples of coefficients in the range $[2, T_0]$. For each triple, we have computed the k-Frobenius number for $k = 10000, 30000, 60000, 100000, 150000$. In Figure 1, we represent, with a logarithmic scale, the product $a_1a_2a_3$ on the x-axis and the k-Frobenius on the y-axis for the respective values of k. In order to assess an approximate value for $A_k(T_0)$ we now perform a linear regression over $\log(a_1a_2a_3)$ and $\log(F_k(a_1, a_2, a_3))$ whose results are shown in the following table.

k	$A_k(T_0)$	$\sqrt{2k}$
10000	141.5535	141.4214
30000	245.0937	244.9490
60000	346.5570	346.4102
100000	447.3675	447.2136
150000	547.8815	547.7226

The obtained computational results allow us to make the following conjecture:

Conjecture. Let $G_T = \{a = (a_1, a_2, a_3) \in \mathbb{Z}^3_{>0} : \gcd(a_1, a_2, a_3) = 1, ||a||_\infty \leq T\}$ and let $A_k(T)$ denote the average value of the (normalized) k-Frobenius number over G_T. Then

$$\limsup_k |A_k(T) - \sqrt{2k}| \to 0 \quad \text{as } T \to \infty,$$

that is $A_k(T)$ asymptotically approaches the term $\sqrt{2k}$ of its lower bound (22).

Acknowledgements We are truly grateful to Lenny Fukshansky, Martin Henk, and Matthias Köppe for their comments, encouragement, and corrections that improve the paper we present today. The second author is grateful for partial support through an NSA grant. Some of our results were first announced in [7].

References

1. K. Aardal, A.K. Lenstra, Hard equality constrained integer knapsacks, in *Integer Programming and Combinatorial Optimization*. Lecture Notes in Computer Science (Springer, Berlin, 2002), pp. 350–366
2. D. Adjiashvili, T. Oertel, R. Weismantel, A polyhedral Frobenius theorem with applications to integer optimization. SIAM J. Discret. Math (to appear)
3. I. Aliev, M. Henk, Feasibility of integer knapsacks. SIAM J. Optim. **20**, 2978–2993 (2010)
4. I. Aliev, M. Henk, A. Hinrichs, Expected Frobenius numbers. J. Comb. Theory A **118**, 525–531 (2011)
5. I. Aliev, L. Fukshansky, M. Henk, Generalized Frobenius numbers: bounds and average behavior. Acta Arith. **155**, 53–62 (2012)
6. I. Aliev, M. Henk, E. Linke, Integer points in knapsack polytopes and s-covering radius. Electron. J. Comb. **20**(2), Paper 42, 17 pp. (2013)
7. I. Aliev, J.A. De Loera, Q. Louveaux, Integer programs with prescribed number of solutions and a weighted version of Doignon-Bell-Scarf's theorem, in *Proceedings of IPCO 17th Bonn*, ed. by J. Lee, J. Vygens (2014), pp. 37–51
8. V.I. Arnol'd, Statistics of integral convex polygons. Funktsional. Anal. i Prilozhen **14**(2), 1–3 (1980)
9. G. Averkov, C. Wagner, R. Weismantel, Maximal lattice-free polyhedra: finiteness and an explicit description in dimension three. Math. Oper. Res. **36**(4), 721–742 (2011)
10. W. Baldoni, M. Beck, C. Cochet, M. Vergne, Volume computation for polytopes and partition functions for classical root systems. Discret. Comput. Geom. **35**(4), 551–595 (2006)
11. I. Bárány, A.M. Vershik, On the number of convex lattice polytopes, Geom. Funct. Anal. **2**(4), 381–393 (1992)
12. V. Barucci, Numerical semigroup algebras, in *Multiplicative ideal theory in commutative algebra*, ed. by J.W. Brewer, S. Glaz, H. Heinzer, B. Olberding (Springer, Berlin, 2006), pp. 39–53
13. A.I. Barvinok, Polynomial time algorithm for counting integral points in polyhedra when the dimension is fixed. Math. Oper. Res. **19**, 769–779 (1994)
14. A.I. Barvinok, *A Course in Convexity*. Graduate studies in Mathematics, vol. 54 (American Mathematical Society, Providence, 2002)
15. A.I. Barvinok, *Integer Points in Polyhedra*. Zürich Lectures in Advanced Mathematics (European Mathematical Society (EMS), Zürich, 2008)

16. A.I. Barvinok, J. Pommersheim, An algorithmic theory of lattice points in polyhedra, in *New Perspectives in Algebraic Combinatorics* (Berkeley, CA, 1996–1997), vol. 38 (Mathematical Sciences Research Institute Publications/Cambridge University Press, Cambridge, 1999), pp. 91–147
17. A.I. Barvinok, K. Woods, Short rational generating functions for lattice point problems. J. Am. Math. Soc. **16**, 957–979 (2003)
18. M. Beck, S. Robins, A formula related to the Frobenius problem in two dimensions. *Number theory (New York, 2003)* (Springer, New York, 2004), pp. 17–23
19. M. Beck, C. Kifer, An extreme family of generalized Frobenius numbers. Integers **11**, 5 p. (2011)
20. D. Beihoffer, J. Hendry, A. Nijenhuis, S. Wagon, Faster algorithms for Frobenius numbers. Electron. J. Comb **12**, Research paper #R27 (2005)
21. B. Braun, *Unimodality Problems in Ehrhart Theory*. article in this volume
22. M. Brion, M. Vergne, Residue formulae, vector partition functions and lattice points in rational polytopes. J. Am. Math. Soc. **10**(4), 797–833 (1997)
23. A. Brown, E. Dannenberg, J. Fox, J. Hanna, K. Keck, A. Moore, Z. Robbins, B. Samples, J. Stankewicz, On a generalization of the Frobenius number. J. Integer Seq. **13**(1), Article 10.1.4, 6 p. (2010)
24. W. Bruns, R. Koch, NORMALIZ, computing normalizations of affine semigroups. Available from ftp://ftp.mathematik.uni-osnabrueck.de/pub/osm/kommalg/software/
25. W. Bruns, J. Gubeladze, N.V. Trung, Problems and algorithms for affine semigroups. Semigroup Forum **64**, 180–212 (2002)
26. C. Bowles, S.T. Chapman, N. Kaplan, D. Reiser, On delta sets of numerical monoids. J. Algebra Appl. **5**, 1–24 (2006)
27. W. Castryck, Moving out the edges of a lattice polygon. Discret. Comput. Geom. **47**(3), 496–518 (2012)
28. S.T. Chapman, M.T. Holden, T.A. Moore, Full elasticity in atomic monoids and integral domains. Rocky Mt. J. Math. 36, 1437–1455 (2006)
29. S.T. Chapman, P.A. García-Sánchez, D. Llena, The catenary and tame degree of numerical semigroups. Forum Math. **21**, 117–129 (2009)
30. C. De Concini, C. Procesi, M. Vergne, Partition function and generalized Dahmen-Micchelli spaces. Transform. Groups **15**(4), 751–773 (2010)
31. D. Cox, J. Little, D. O'Shea, *Ideals, Varieties and Algorithms* (Springer, New York, 1992)
32. A. Dobra, A.F. Karr, P.A. Sanil, Preserving confidentiality of high-dimensional tabulated data: statistical and computational issues. Stat. Comput. **13**, 363–370 (2003)
33. D. Einstein, D. Lichtblau, A. Strzebonski, S. Wagon, Frobenius numbers by lattice point enumeration. Integers **7**, A15, 63 (2007)
34. F. Eisenbrand, N. Hähnle, Minimizing the number of lattice points in a translated polygon, in *Proceedings of SODA* (2013), pp. 1123–1130
35. E. Ehrhart, *Polynômes arithmétiques et méthode des polyèdres en combinatoire*. International Series of Numerical Mathematics, vol. 35 (Birkhäuser, Basel, 1977)
36. L. Fukshansky, A. Schürmann, Bounds on generalized Frobenius numbers. Eur. J. Comb. **3**, 361–368 (2011)
37. P.A. García-Sánchez, J.C. Rosales, *Numerical Semigroups*. Developments in Mathematics, vol. 20 (Springer, New York, 2009)
38. C. Haase, B. Nill, S. Payne, Cayley decompositions of lattice polytopes and upper bounds for h^*-polynomials. J. Reine Angew. Math. **637**, 207–216 (2009)
39. R. Hemmecke, A. Takemura, R. Yoshida, Computing holes in semi-groups and its application to transportation problems. Contrib. Discret. Math. **4**, 81–91 (2009)
40. M. Henk, R. Weismantel, *On Hilbert Bases of Polyhedral Cones*, Preprint SC 96-12 Konrad-Zuse-Zentrum für Informationstechnik, Berlin, April 1996. www.zib.de/bib/pub/pw/index.en.html
41. R. Kannan, Lattice translates of a polytope and the Frobenius problem. Combinatorica **12**, 161–177 (1992)

42. H. Kellerer, U. Pferschy, D. Pisinger, *Knapsack Problems* (Springer, Berlin, 2004)
43. A.G. Khovanskii, The Newton polytope, the Hilbert polynomial and sums of finite sets. Funktsional. Anal. i Prilozhen. **26**, 57–63 (1992)
44. A. Knutson, T. Tao, The honeycomb model of $GL_n(\mathbf{C})$ tensor products. I. Proof of the saturation conjecture. J. Am. Math. Soc. **12**(4), 1055–1090 (1999)
45. J.C. Lagarias, G.M. Ziegler, Bounds for lattice polytopes containing a fixed number of interior points in a sublattice. Can. J. Math. **43**, 1022–1035 (1991)
46. H. Liu, C. Zong, On the classification of convex lattice polytopes. Adv. Geom. **11**(4), 711–729 (2011)
47. J.A. De Loera, T.B. McAllister, On the computation of Clebsch-Gordan coefficients and the dilation effect. Exp. Math. **15**(1), 7–19 (2006)
48. J.A. De Loera, D.C. Haws, R. Hemmecke, P. Huggins, B. Sturmfels, R. Yoshida, Short rational functions for toric algebra and applications. J. Symb. Comput. **38**, 959–973 (2004)
49. J.A. De Loera, J. Rambau, F. Santos, *Triangulations: Structures for Algorithms and Applications*. Algorithms and Computation in Mathematics Series, vol. 25 (Springer, Berlin, 2010), 539 p. ISBN: 978-3-642-12970-4
50. J.A. De Loera, R. Hemmecke, M. Köppe, *Algebraic and Geometric Ideas in the Theory of Discrete Optimization*. MOS-SIAM Series on Optimization, vol. 14 (Society for Industrial and Applied Mathematics (SIAM)/Mathematical Optimization Society, Philadelphia, 2013), xx+322 pp.
51. B. Nill, Lattice polytopes having h^*-polynomials with given degree and linear coefficient. Eur. J. Comb. **29**, 1596–1602 (2008)
52. B. Nill, G.M. Ziegler, Projecting lattice polytopes without interior lattice points. Math. Oper. Res. **36**(3), 462–467 (2011)
53. C. O'Neil, On factorization invariants and Hilbert functions. Preprint available as ArXiv:1503.08351 (2014)
54. I. Pak, E. Vallejo, Combinatorics and geometry of Littlewood-Richardson cones. Eur. J. Comb. **26**, 995–1008 (2005)
55. O. Pikhurko, Lattice points in lattice polytopes. Mathematika **48**(1–2), 15–24 (2001/2003)
56. S. Rabinowitz, A census of convex lattice polygons with at most one interior lattice point. Ars Comb. **28**, 83–96 (1989)
57. J.L. Ramírez Alfonsín, Complexity of the Frobenius problem. Combinatorica **16**, 143–147 (1996)
58. J.L. Ramírez Alfonsín, *The Diophantine Frobenius Problem*. Oxford Lecture Series in Mathematics and Its Applications (Oxford University Press, New York, 2006)
59. J.L. Ramírez Alfonsín, Gaps in semigroups. Discret. Math., **308**, 4177–4184 (2008)
60. J.R. Schmidt, A. Bincer, The Kostant partition function for simple Lie algebras. J. Math. Phys. **25**, 2367–2373 (1984)
61. A. Schrijver, *Theory of Linear and Integer Programming* (Wiley, Chichester, 1998)
62. J. Shallit, J. Stankewicz, Unbounded discrepancy in Frobenius numbers. Integers **11**, 27–34 (2011)
63. R.P. Stanley, *Combinatorics and Commutative Algebra*. Progress in Mathematics, vol. 41, 2nd edn. (Birkhäuser, Boston, 1996)
64. B. Sturmfels, *Gröbner Bases and Convex Polytopes*. University Lecture Series, vol. 8 (American Mathematical Society, Providence, 1995)
65. B. Sturmfels, On vector partition functions. J. Comb. Theory A **72**, 302–309 (1995)
66. A. Szenes, M. Vergne, Residue formulae for vector partitions and Euler-MacLaurin sums. Adv. Appl. Math. **30**, 295–342 (2003)
67. A. Takemura, R. Yoshida, A generalization of the integer linear infeasibility problem. Discret. Optim. **5**, 36–52 (2008)
68. A.V. Ustinov, Solution of the Arnold problem on weak asymptotics for Frobenius numbers with three arguments. Sb. Math. **200**, 597–627 (2009)
69. X. Wei, R. Ding, Lattice polygons with two interior lattice points. Math. Notes **91**(5–6), 868–877 (2012)

CPSIA information can be obtained
at www.ICGtesting.com
Printed in the USA
BVOW11*1228180416
PP6940200001B/1/P

9 783319 24296